Research Methodology in Agricultural Sciences

Jayne Njeri Mugwe • Steven Runo

Research Methodology in Agricultural Sciences

Springer

Jayne Njeri Mugwe
Department of Agricultural Science and Technology
Kenyatta University
Nairobi, Kenya

Steven Runo
Department of Biochemistry, Microbiology and Biotechnology
Kenyatta University
Nairobi, Kenya

ISBN 978-981-95-1891-3 ISBN 978-981-95-1892-0 (eBook)
https://doi.org/10.1007/978-981-95-1892-0

This Springer imprint is published by the registered company Springer Nature Singapore Pte Ltd.
The registered company address is: 152 Beach Road, #21-01/04 Gateway East, Singapore 189721, Singapore

Foreword 1

The success of training in institutions of higher learning, I believe, lies in their ability to nurture and promote a dynamic learning environment, generate knowledge, and support impactful research. Aligned with the United Nations Sustainable Development Goals, Universities must prepare students for successful careers and empower them to contribute meaningfully to social progress. This is particularly vital in agricultural research, which is central to securing our food systems and environmental protection. By embracing innovative training methods and promoting research on real-world issues, we can shape a generation of researchers who are equipped to drive positive change in their communities and beyond.

Postgraduate research plays a crucial role in this agenda; it drives academic and scientific progress by giving students an opportunity to engage deeply in their fields through research projects and thesis writing. Indeed, graduate research is a seedbed for insights and practices as it helps students to develop specialised expertise and prepare for roles in academia, industry and research.

Despite its significance, many students face challenges in conducting their research projects due to inadequate research skills. These gaps limit their ability to design and implement robust studies and produce quality research outputs. Too often, such gaps delay completion rates. For instance, some students undertaking masters degrees exceed the standard two-year timeframe due to difficulties translating classroom knowledge into practical application. Major weaknesses primarily arise in three main areas: conceptualizing the research, data analysis and interpretation, and scholary writing. To address these issues, it is crucial to provide students with targeted training and mentorship in research methodology.

This book on research methodology in agricultural sciences offers a comprehensive and practical guidance. Authored by senior researchers at Kenyatta University, who bring a wealth of experience in research, teaching, and supervising postgraduate students, the book successfully tackles critical barriers such as selecting a research topic, developing research objectives and hypotheses, presenting and synthesizing literature reviews, and effectively navigating through data analysis. Additionally, it thoroughly covers ethical considerations and best practices for conducting research.

A standout feature of this book is its integration of qualitative and mixed methods approaches, farming systems research and on-farm experimentation, and impact evaluation designs; areas often underrepresented in many books. Given that agricultural research frequently necessitates both quantitative and qualitative designs, this book's comprehensive integration of these methodologies is a notable strength. It offers a holistic perspective that is crucial for comprehensive agricultural research. It also equips students to tackle diverse research questions and enhances their analytical research skills.

Another key feature is the inclusion of four dedicated chapters for reporting and publishing. Publishing journal articles poses particular challenges for most students due to limited mentorship, inadequate infrastructure, and resources. These chapters address these issues by providing practical guidance on writing, choosing a journal, submitting and revising manuscripts.

As students engage with this textbook, they will be better prepared to undertake rigorous and impactful research, driving advancements in agricultural science and beyond. I recommend this book to graduate students, scholars, early career researchers, and experienced researchers. It is also a necessary addition to the library of every University.

Former Ag. Vice Chancellor
Kenyatta University,
Nairobi, Kenya

Waceke Wanjohi

Foreword 2

Agriculture is at a critical juncture, facing unprecedented challenges and opportunities in the twenty-first century that will shape the future of global food systems. With the world's population steadily increasing and expected to hit 9.7 billion by 2050, the demand for food is rising, putting additional pressure on the agricultural system to produce more food with fewer resources. Climate change with its unpredictable weather patterns and environmental degradation adds further strain. It is clear that the future of agriculture will depend on innovative and sustainable solutions that can enhance productivity and ensure food security.

At the heart of these solutions is agricultural research, which is a powerful driver of discovery and change. Over the years it has advanced knowledge and shaped farming practices that have continued to sustain communities and ecosystems. From precision agriculture to biotechnology and data-driven decision making, research continues to unlock new possibilities for transforming agriculture into more sustainable and productive systems.

However, despite its critical role, many graduate students and early career researchers entering the field of agricultural research face significant gaps in research skills and practical experience. They often struggle to design and implement robust research studies and transform their findings into actionable recommendations. This challenge is exacerbated by the rapid pace of change in agricultural technologies and methodologies, requiring researchers to continually update their knowledge and adapt to emerging best practices. To effectively support them and address the evolving needs in the field, it is crucial to provide educational resources that cover fundamental research principles and offer up-to-date insights and practical solutions tailored to contemporary agricultural issues.

This book on research methodology in agricultural sciences, packaged in 28 chapters, addresses this critical need in the field. The book uses a comprehensive approach with real-life examples and cases to illustrate key concepts. It also covers both traditional and modern research techniques, ensuring that researchers will be well-equipped to handle the complexities of modern agricultural research. I commend the authors for providing a clear and relevant resource. It is essential for the next generation of agricultural researchers as they confront challenges of food security.

Vice Chancellor
University of Embu,
Embu, Kenya

Daniel Mugendi

Preface

Purpose

This textbook provides a comprehensive guide to the methods and methodologies currently used in agricultural sciences, covering the entire research process from conceptualization to reporting. It presents a step-by-step approach to planning and conducting research. This resource is useful for students who are either learning research methods as part of the coursework, working on their research projects or preparing their theses and dissertations.

Agricultural research is a complex interdisciplinary science requiring whole-oriented research that includes biophysical and social approaches. Essential information is provided for these approaches, including information on experimentation and layout of experiments, on-farm research approaches, survey research, qualitative approaches, mixed approaches, and designs for evaluating the impacts of interventions. Further topics include data analysis, presentation and visualization. Topics in reporting include scientific writing skills, thesis writing, and journal article publishing. In all the chapters, care is taken to explain theories, concepts, and principles using extensive examples, flow diagrams, photos, and case studies to make the chapter easy for readers. This book will assist practitioners, particularly students, negotiate the steep learning curve of research methodologies and gain the skills needed to conduct credible and meaningful research.

Why Book on Agricultural Sciences?

Eradication of poverty, zero hunger, and achieving food security are the top priorities of the United Nations' sustainable development goals. In the era of high population growth and increasing pressure on agricultural systems, agricultural research is central to increasing income levels for the farming sector and meeting the food requirements. However, one of the problems curtailing credible and reliable agricultural research, particularly in Africa and low- and middle-income countries of the Global South, is a shortage of well-trained scientific and technical staff. This book will help build knowledge and skills that are key to improving the relevance and effectiveness of research in eradicating poverty and promoting food security.

Our experience in teaching research methods, implementing research projects, mentoring, and supervising students, have informed the creation of this book. Through years of guiding students and conducting research, we have identified key challenges and areas where additional support is needed. Top on our list is understanding research method concepts, writing research proposals and data analyses, and publishing journal articles. We have thus attempted to address these gaps by providing clear explanations, practical examples, and step-by-step guidance throughout this book. We have also used a comprehensive approach to explaining concepts in detail, ensuring that complex concepts are broken down into manageable parts for better understanding. We have also enhanced our discussions by referencing other studies, allowing readers to see how concepts are applied in the world of science. The aim is to give readers the skills and knowledge needed for successful research in agricultural sciences.

Audience

This book is of interest and helpful to students, faculty staff, and researchers seeking knowledge in agricultural and allied sciences research. Professionals, national and international agricultural scientists, and policymakers will also find this to be a helpful read.

Key Features of This Book

1. Provides clarifying examples, recommended further reading, and critical points at the end of each chapter
2. Includes boxes with detailed information to unpack complicated concepts and, in some cases, to summarize key points
3. Flow diagrams, photos, and diverse illustrations are used to enhance understanding
4. Includes an extensive discussion on publishing in three chapters
5. Provides end-of-chapter review questions and exercises.

Organization of the Book

The book is organized into 6 parts, with 28 chapters discussing different phases of the research process.

Part I: Introduction to Research

Part I, "*Introduction to Research,*" consists of four chapters (1–4) and lays the foundation for understanding research methods. The aim is to strengthen the grasp of the principles of research methods. The first chapter, "Foundations of Agricultural Research," examines the value of research in agriculture. It emphasizes that effective scientific research is problem-oriented, systematic, and truthful. Such research increases our understanding of the phenomenon under study. Regardless of the specific nature of research, it must be linked to the research-development curriculum model to benefit society. Key stakeholders in agricultural research, such as governments, research bodies, and farmers, play a crucial role in shaping research and advancing the agricultural research agenda.

The second chapter covers the "*The Research Process, Key Concepts, and Terminologies.*" It outlines an eight-step research process, emphasizing the need for integrating literature at the various stages of research. Key concepts covered include the family of variables and strategies for controlling extraneous variables. Additionally, the chapter addresses the role of academic supervisors and provides practical tips for students to improve the completion rate of their research projects. It ends with a discussion on measurements and the integral role of theory in framing research.

Chapter 3 expands on the research process and guides readers through "*From Topic Selection to Research Objectives and Hypotheses.*" It begins by presenting sources of research topics and offers guidance on identifying and formulating research problems. The chapter explores the critical role of objectives and provides instructions for crafting clear and effective research objectives. Different types of hypotheses and methods for their formulation are also examined.

The last chapter in this section is Chapter 4, on "*Literature Review*". It starts by outlining the different types of literature reviews and explaining their importance in the research process. The chapter then explores the various sources of literature and highlights key databases that are particularly useful for agricultural research. It concludes with a concise discussion of the theoretical and conceptual frameworks, emphasizing their close interdependence with the literature review.

Part II: Research Proposal and Ethics

Part II covers two topics: *Research proposal writing and Ethics in research* (Chaps. 5 and 6). Chapter 5 outlines the significance of a research proposal, including its novelty and clarity. It emphasizes that a well-crafted proposal is essential for a successful thesis and publishing scientific articles. The chapter details the format and components of an academic research proposal, divided into three sections: the front matter, mid matter, and final matter.

Chapter 6 begins by highlighting the role of ethical, legal, and administrative aspects of research. It provides a brief history of research ethics and discusses the fundamental principles and sources for acquiring the relevant information. The chapter explores the role of institutional review boards and the process for obtaining ethical clearance. It outlines ethical guidelines for various stages of research, including conceptualization, proposal writing, and implementation. Specific ethical issues are addressed; on-farm experimentation, animal research, participatory research with farmers, data analysis, and publishing. The chapter also covers contemporary topics, such as artificial intelligence, student supervision, and issues in collaborative research projects.

Part III: Research Designs

Part III covers research designs and is the longest, comprising of ten chapters. We start with Chap. 7, which provides an extensive *overview of various research designs*, drawing from selected books to help students become familiar with variants. The key message is that there is significant overlap in the classification of research designs. After evaluating these overlaps, we propose a new classification combining broad paradigms and structural criteria that best fit agricultural research. This gives three main categories: 1. Experimental, which includes true and quasi-experiments; 2. Non-experimental designs (surveys, qualitative and participatory approaches, which do not include manipulation of independent variables); and 3. The mixed approach, which combines experimental and non-experimental designs, including qualitative and quantitative ones. The chapter ends by discussing research philosophical paradigms and their role in selecting the designs and research methodologies.

After the overview chapter, we get to Chap. 8, which presents *fundamentals of experimentation*. It begins with a brief history of experimentation and then explores the common types of agricultural experiments. The chapter emphasizes the concept and process of experimentation, highlighting the importance of controlling experimental error to enhance validity. The chapter discusses the general principles of experimentation, focusing on the 3 Rs: randomization, replication, and reduce noise (local control). It concludes by outlining the key elements for designing an effective experiment, including selecting the appropriate design, managing the experiments, and determining the need for experimental protocols.

Chapter 9 focuses on *common types of experimental designs and field layouts*. It discusses three basic experimental designs used in agriculture: the completely randomized design, the randomized complete block design, and the Latin square design. The chapter also examines factorial and split-plot experiments, highlighting each design's layout, advantages, and disadvantages. The chapter concludes with practical guidance on planning and designing agricultural experiments, outlining key steps, such as meticulous planning, accurate layout with proper right angles, and using net plots and guard rows. Overall, the chapter gives practical insights into effectively applying the designs in agricultural research.

Chapter 10 explores *farmer participatory research and on-farm experimentation*. It begins with an overview of the origins of farmer-participatory research and its closely associated concept of on-farm experiments. It provides a detailed discussion on the evolution of farmer-participatory research from 1960 to 2020. Next, it covers the concepts and principles of on-farm experimentation, including the various types and relevant case studies. The chapter also outlines the key steps in conducting an on-farm experiment, emphasizing its data-intensive nature and the additional analytical techniques to manage variability within and across farms. The chapter concludes by addressing challenges in running on-farm experiments and proposing possible solutions.

Before shifting to the non-experimental designs, a mid-chapter (Chap. 11) on *research designs for impact assessment* is presented. The chapter begins by highlighting the significance of impact evaluation studies and the need to apply a valid methodological approach. The chapter stresses the importance of counterfactuals and the theory of change. The chapter then explores basic designs within both experimental and non-experimental categories. This encompasses randomized controlled trials (RCTS), quasi-experiments, and correlational and qualitative mixed research approaches. A detailed discussion on quasi-experiments covers methods of creating comparison groups, such as propensity matching and difference-in-difference techniques. The chapter concludes with an overview of pre-experimental, qualitative, and mixed methods approaches, providing a comprehensive perspective of impact evaluation research designs.

After these five chapters on experimental designs, we shift to the non-experimental designs, starting with *surveys* in Chap. 12. It starts by defining surveys, outlining their purpose, and detailing the types of information commonly collected from farmer surveys. The chapter covers various survey designs, including cross-sectional and longitudinal, along with their variants. It also explores longitudinal surveys and animal epidemiological surveys, including case-control studies. Finally, the chapter discusses ways of administering surveys and provides practical guidance on effectively conducting a survey.

Chapter 13 focuses on *qualitative and participatory research approaches*. It begins with a brief overview of qualitative approaches and their philosophical assumptions. The chapter highlights the current state of qualitative research in agriculture, noting its limited application despite its benefits for studying complexity in farming systems. Qualitative research is primarily applied to gender, value chain analysis, and policy analysis studies within agriculture. The chapter then explores the basic characteristics of qualitative approaches and concludes by providing general guidelines and steps for conducting a qualitative study and provides a comparison of qualitative and quantitative research approaches.

Chapter 14 builds on Chap. 13 by focusing on *qualitative research approaches*. It explores five approaches: case studies, ethnography, grounded theory, narrative, and phenomenology. The chapter begins with an overview of case studies, their types, and general characteristics. It then covers ethnography, highlighting its role in studying cultures and indigenous knowledge. Grounded theory is examined, highlighting the unique characteristics, such as simultaneous data collection and analysis, constant comparison, theoretical sampling, and theory development. The chapter discusses narrative inquiry and its methodological aspects, including triangulation. It concludes with an exploration of phenomenology and its core characteristics of bracketing during data collection and synthesizing the core essence of the experience. The chapter wraps up by stressing the importance of trustworthiness in qualitative research and provides criteria for its assessment.

Chapter 15 concludes this extensive section by focusing on *mixed-method research designs*. The chapter introduces the concept of mixed methods and justifies the importance of mixed designs for a comprehensive understanding of the agricultural landscape. The chapter briefly explores philosophical assumptions and the objectives underlying mixed research. It provides a detailed discussion of typologies of mixed research designs, categorized into four basic designs: Sequential, concurrent, multiphase, and composite. This framework results in eight specific designs, further illustrated through relevant case studies. The chapter concludes by underscoring the challenges of mixed methods research and offers potential solutions.

Chapter 16 is about *sampling methods*. It starts with a brief overview of sampling, differentiating between a sample and a census, and highlighting the importance of obtaining a representative sample. The chapter highlights key terms used to describe populations, such as accessible population, before discussing the various types of sampling. It covers probability and nonprobability sampling techniques and discusses sampling strategies in qualitative research approaches. Additionally, the chapter addresses errors in sampling and considerations in determining sample size. It concludes by emphasizing the relevance of these sampling methods for surveys and biophysical studies, including experiments.

Part IV: Methods and Techniques of Data Collection

Part IV consists of three chapters and discusses data collection methods and tools (17, 18, and 19).

Chapter 17 is on *data collection in the biophysical sciences*. This chapter begins by addressing the concept of validity and reliability in data collection and ways of enhancing validity. The chapter then explores data collection methods for specific biophysical contexts, including legumes, maize, soil, pest, and diseases. The chapter presents methods of soil analysis, as well as techniques for quantifying biological nitrogen fixation. It also explores strategies for carrying out a screening experiment emphasizing the use of lattice design to evaluate a large number of varieties. The chapter concludes with an overview of molecular and metabolomic analysis.

Chapter 18 discusses *data collection in surveys*, emphasizing its connection with qualitative approaches. The chapter begins by distinguishing between interview survey (investigator-completed) and questionnaire survey (interview-completed) instruments. The chapter outlines six key steps in planning and designing a survey instrument, focusing on aligning the instrument with the research objectives and formatting it for ease of use by the respondents. It also discusses the processes involved in conducting popular surveys in agriculture, covering both the interview and questionnaire survey. The chapter concludes with a detailed guide on question formulation and includes an illustration of the format, arrangement and types of questions using a case study.

Chapter 19 focuses on *data collection in qualitative research approaches*. It starts by highlighting data collection methods, such as interviews and observation, and the types of data collected in each of the methods. The chapter explores various types of interviews, including the survey administration and strategies to enhance validity. It provides guidelines for conducting in-depth interviews, focus group discussions, and key informants' interviews. Additionally, the chapter covers participant and non-participant observation. Further, it discusses content analysis and audio-visual data collection techniques. The chapter concludes with an overview of selected participatory research tools.

Part V: Data Processing, Analysis, and Presentation

Part V is on data organization and analysis. It has five chapters (20, 21, 22, 23, and 24)

Chapter 20 is on quantitative data analysis, focusing on descriptive analysis. The chapter starts by laying a foundation for data analysis, guiding the process, and introducing the available data types. It briefly reviews the computer programs commonly used for data analysis. It then covers descriptive statistics, including frequencies, measures of central tendencies, measures of variability. It concludes by demonstrating how to create and utilize a pivot table, a powerful tool for summarizing and visualizing data.

Chapter 21 focuses on inferential analysis, which is key to drawing conclusions about a population based on sample data. We begin by differentiating between parametric and non-parametric tests, providing several examples to illustrate these concepts. The chapter then transitions to the key steps of hypothesis testing, and examines techniques for testing differences in means, including a detailed look at various types of t-tests and ANOVA. Following this, we shift our focus to the analysis of relationships between variables and the chapter concludes with an overview of regression and prediction techniques, including a brief on binary regression methods.

Chapter 22 presents an overview of *analyzing survey data using SPSS*. It begins by highlighting the common data types from farmer surveys and outlining the steps in the analysis. The chapter introduces SPSS, highlighting the significance of its two main windows: Data view and variable view. It discusses how to perform descriptive statistics, including calculating frequencies, measures of central tendency, and variability using illustrative examples. Finally, the chapter covers simple inferential statistics, such as t-test, ANOVA, correlations, simple regression, and logistic regression.

Chapter 23 focuses on *qualitative data analysis*, beginning with an overview of the most common data encountered in qualitative research, including interview transcripts, photos, and drawings. The chapter outlines the purpose of qualitative data analysis and its significance in understanding complex phenomena. We then outline the five general steps of qualitative data analysis, from data organization to synthesis and reporting. The chapter then provides a comprehensive look at the various types of analysis, including content analysis, thematic analysis, discourse analysis, narrative analysis, grounded theory, and phenomenological analysis. It also covers analyzing specific data sources, such as data from focus group discussions, observations, and visual data. It also provides a brief overview of computer-assisted data analysis tools, e.g., NVIVO and MaxQDA, that aid the analytical process. The chapter concludes with data synthesis methods, including thematic and framework synthesis, to summarize and interpret findings effectively.

Chapter 24 addresses *data presentation and visualization*, and covers three primary methods of presentation: graphs, tables, and narratives. The chapter explores a range of graphical presentations, from simple charts to complex graphics, including maps, photos, and flowcharts, and discusses the types of data suitable for each technique. Additionally, it provides a brief overview of R Studio's versatility in data visualisation, particularly for molecular studies.

Part VI: Dissemination of Findings and Scientific Publishing

Chapter 25 addresses *disseminating the research findings*, emphasizing the critical role of effectively sharing research findings with the clientele. It stresses the need for research bodies to have a dedicated section to coordinate the dissemination process. The chapter highlights the need to disseminate for impact. In addition, the chapter explores the innovation diffusion theory and identifies, which explains how technologies or innovations among farmers are adopted. It further explores elements of multiple dissemination pathways, including oral, written, practically oriented, and information and communication technology (ICT) based approaches.

Chapter 26 focuses on the *structure of a scientific paper*, beginning with an introduction to the IMRAD format (Introduction, Methods, Results, and Discussion). The chapter details how to write and structure the core components of a scientific manuscript, including the beginning matter, mid-matter and end-matter. The beginning matter contains abstract, authors, and keywords. The mid-matter is the main part of the paper with IMRAD framework and the end matter. The chapter comprehensively discusses the components of introduction, materials and methods, results, and discussions, giving examples and tips on how to write these sections. The chapter concludes with the end matter consisting of references, conflict of interest, data availability statement, and acknowledgments. It also guides how to structure qualitative and mixed methods research papers.

Chapter 27 addresses *journal selection, submission, and production*. It begins with an overview of the significance of publishing for both the institution and the individual researcher. It covers various types of journals, including international, regional, paid-for, fee-to-publish, credible, and predatory journals. It outlines key factors to consider when selecting a target journal to publish. They include seven key factors. The chapter details the submitting process and the subsequent steps leading to publication. The chapter concludes by highlighting challenges to publishing, including inadequate research skills and infrastructure, among others.

Chapter 28 focuses on enhancing *presentations and scientific writing skills*. It begins by emphasizing the significance of presenting scientific work and developing scientific writing skills. The chapter outlines 14 key principles in scientific writing, emphasizing clarity, conciseness, coherent writing, logical flow, objectivity, and consistency. It guides on writing a thesis, drafting a journal manuscript from it, and preparing for thesis examination. The chapter concludes with providing advice on making presentations at seminars and conferences, including oral and poster formats, with examples of PowerPoint slides and poster designs to illustrate effective presentation techniques

Nairobi, Kenya

Jayne Njeri Mugwe
Steven Runo

About the Authors

Jayne Mugwe

Work life

Prof. Jayne Mugwe is an Associate Professor of Integrated Soil Fertility Management at the Department of Agricultural Science and Technology at Kenyatta University. She joined Kenyatta University after a 20-year tenure at Kenya Forestry Research Institute. Prof. Mugwe has taught research methods courses at the University level for about 17 years and brings immense experience from over 30 years in active research. She has held several leadership roles at Kenyatta University including as a Departmental Chair and Acting Registrar of Research, Innovation, and Outreach. Prof. Mugwe has implemented several multidisciplinary research projects and supervised more than 40 masters and doctoral students. She has published widely with more than 140 publications. Google Scholar ORCID ID: https://orcid.org/0000-0003-1405-7166 Scopus ID: 6508194565.

https://agriculture-environment.ku.ac.ke/faculty-profiles-department-of-agricultural-science-and-technology/prof-jayne-mugwe-njeri

A Dream Come True

This book is a dream come true. I dreamt of writing a book on research methods in 2009 while working for the Kenya Forestry Research Institute. Back then, I felt the need to write a book that could help early career researchers carry out credible research and write journal papers.

The dream was still alive after I joined Kenyatta University in the same year (2009). As I taught research methods courses, I always wished I could write a reader-friendly book for agricultural science postgraduate students. I am intrigued by the richness of agriculture as a discipline, particularly how it integrates with fields of biology, economics, environmental science, and sociology. The interdisciplinary nature makes agricultural research both challenging and fascinating.

The dream started bearing fruit in 2017. During my sabbatical leave at the University of Embu, I consolidated all the materials and ideas I had organized over time and started writing the book. This process culminated in a book proposal that was submitted to Springer Nature in 2022, was accepted, and ultimately led to the publication of this book.

I give all the glory to God for providing the insight, strength, and good health to pursue the dream.

Dedication from Jayne Mugwe

I dedicate this book to the farmers of this world.

During the Africa Green Revolution Forum of 2018, the theme was Lead, Measure, and Grow, and the question posed to all agricultural practitioners was: How will you lead to transform agriculture in Africa? In this book, we provide leadership in agricultural transformation by providing easy-to-follow guidelines for navigating the research journey to generate the required knowledge and technologies for raising agricultural productivity.

Wangari Maathai quote: The little things that citizens do. That is what causes impact.

This book is our "little thing" to help graduate students navigate the research journey smoothly.

Steven Runo

Steven Runo is an Associate Professor in the Department of Biochemistry, Microbiology and Biotechnology at Kenyatta University. He specializes in Molecular Biology and Biotechnology with research interests in leveraging modern tools in molecular biology to overcome biotic and abiotic constraints of food production. He has coordinated several research projects, particularly in parasitic plant biology and published in the same field. He is currently running a laboratory that uses molecular genetic tools to understand interactions between parasitic plants that limit crop production in Africa and their hosts. In recognition of his contributions to science, he was awarded the Royal Society Africa Prize in 2020.

ORCID: https://orcid.org/0000-0002-9043-2338

SCOPUS ID: 36909262500

https://spas.ku.ac.ke/department-of-biochemistry-faculty/prof-steven-runo

Acknowledgment of Co-authors of Various Chapters

Dr. Kezia Mbuthia contributed to Chapter 23 on "Qualitative data analysis". Dr Mbuthia is a sociologist from South Eastern Kenya University (SEKU) and holds expertise in analysis of qualitative data, particularly use of Nvivo. Currently, she is teaching in the Department of Sociology and Anthropology at SEKU. She holds a PhD in Sociology from Kenyatta University, obtained in 2023, and a master's from SEKU. Dr. Mbuthia's area of specialization is sociological theories and methodologies, particularly social gerontology.

Dr. Jackson Kago contributed to Chapter 23. Dr Kago is an urban researcher and teaches in the Department of Spatial and Environmental Planning at Kenyatta University. Dr. Kago earned his PhD in Human Geography from the University of Bordeaux Montaigne in 2022. His doctoral research focused on mobility patterns and their impact on urban-rural linkages in small and intermediate towns in Kenya, specifically examining the milk value chain. His thesis employed qualitative techniques to assess both the usability and operational aspects of mobility by interviewing various actors and stakeholders along the dairy value chain.

Dr. Eric Oduor contributed to Chapter 20 and 21 on data analysis in experiments. Dr Oduor is a soil scientist with a bias toward soil chemistry, focusing on soil carbon, phosphorus and nitrogen dynamics, climate change, agroecology, circular and green technologies, and general agriculture. He is enthusiastic about data analytics, research designs, and survey studies. Eric holds a PhD in Land Resource Management (and Planning (2024) and a Masters in Soil Science. https://agriculture-environment.ku.ac.ke/faculty-profiles-department-of-agricultural-science-and-technology/mr-erick-oduor-otieno

Dr. Abed Kiwia contributed to Chapter 17 on biophysical data collection. Dr Kiwia is an experienced natural resources specialist and a distinguished scholar who has authored over 70 scholarly articles and reports on soil fertility and fertilizer systems, improved agricultural production, program management, and climate change resilience. He holds a PhD in Public Policy and Administration from Walden University, USA, earned between 2012 and 2017, and a master's degree in Agroforestry. He currently serves as Coordinator of the Soil Health Program and Systems Agronomist in Sustainable and Regenerative Agriculture at Alliance for a Green

Revolution in Africa (AGRA). He has also worked as a Researcher at the World Agroforestry Centre—ICRAF (https://www.cifor-icraf.org/).

Chris Kamau contributed to Chapter 11 on "Research designs for impact assessment" Chris is a Tutorial Fellow at Kenyatta University in the Department of Agricultural Economics. He has extensive experience in agricultural research and development, focusing on the adoption and impact evaluation of agricultural technologies. His expertise includes designing various data collection instruments and managing data collection processes using data collection tool kits, such as ODK and Kobo Collect. Additionally, he is highly experienced in data modelling, particularly with econometrics models, in the context of impact evaluation of agricultural programs. He is also proficient in various data analysis software applications such as MS Excel, SPSS, and STATA. https://agriculture-environment.ku.ac.ke/department-of-agricultural-economics/mr-kamau-christopher-njuguna

Reviewers of this Book

We are grateful to the following reviewers

Name	Chapter reviewed	Affiliation and specialization
Prof. Monicah-Muna, PhD	Reviewed most of the chapters.	Kenyatta University Department of Environmental Science and Education Specialization: Applied soil fertility
Dr. Ivan Adolwa, PhD	Chapter 10: Farming system research and on-farm Experimentation	African Plant Nutrition Institute (APNI) Specialization: Farming systems
Mr. Chris Kamau	Chapters 11 and 22: Designs for evaluating impact and data analysis using SPSS.	Kenyatta University Department of Agricultural Economics Specialization: Agribusiness management and impact evaluation
Dr Eric Oduor, PhD	Reviewed most of the chapters.	Kenyatta University Department of Agricultural Science and Technology
Dr. Anderson N. Maina, PhD	Reviewed most of the chapters.	Department of Plant and Animal Sciences Murang'a University of Technology
Dr. Jackson Kago, PhD	Chapters 13 and 14 on qualitative research approaches.	Kenyatta University (KU).
Dr. Keziah Mbuthia, PhD	Chapters 13 and 14 on qualitative research approaches.	South Eastern University (SEKU)

Acknowledgments

Acknowledgment by Jayne Mugwe

To My Family

To my husband Simon Mugwe Weru: I say thank you very much for the support and comfort you provided to me and our family all the years.

To my children, and lovely grandchildren: Your support and love are immense, and I always treasure you.

To My Support Team

I thank the Alliance for a Green Revolution in Africa (AGRA). This book's many examples emanate from research conducted under two AGRA-funded projects. The first project implemented from 2010 to 2013 was "Enhancing Productivity and Market Development of Soybeans and Climbing Beans in Central Highlands of Kenya" (SHP 024). The second one, from 2014 to 2017, was titled "Scaling up Soybean and Climbing Beans Using a Value Chain Approach in Maize-based Systems of Central Highlands of Kenya" (SHP 014). These two projects carried out research under controlled and on-farm conditions and were supported directly and indirectly by 35 postgraduate students.

I thank Dr. Bashir Jama, Dr. Rebbie Harawa, and Dr. Abednego Kiwia for their support in implementing these two projects.

Many other examples in this book are borrowed from research projects I coordinated or partnered with. These include

1. "Balanced nutrition for management of soil acidity and nutrient deficiency for sustained agricultural productivity" between 2020 and 2022. Research Grant by Kenya Climate Smart Agriculture Project (KSCAP)
2. "Climate-smart options allowing agricultural intensification for smallholders in the dry zones of the Central Highlands of Kenya" implemented from 2019 to 2022. Funded by The Flemish Interuniversity Council Own Initiative (VLIR - UOS)
3. "Enhancing farmers' benefits from integrated soil fertility management through improved water use efficiency and better marketing strategies in the highlands of central Kenya". Funded by The Flemish Interuniversity Council Own Initiative (VLIR-UOS)
4. "Scaling up integrated soil fertility management for improved livelihoods." Funded by The Association for Strengthening Agricultural Research in Eastern and Central Africa (ASARECA) and implemented from 2012 to 2015
5. "Accelerated uptake and utilization of soil fertility management best-bets practices in ECA region". Funded by the Association for Strengthening Agricultural Research in Eastern and Central Africa (ASARECA). Grant implemented between 2008 and 2012.

To Our Partners

Sincere thanks to key partners who have collaborated with us on these projects. With no particular order, I thank Prof. Monicah Muna, a friend and colleague I have worked with for over 20 years. Prof. Felix Ngetich, whom I supervised during his PhD studies, later became a valuable member of the projects. I thank Prof. Roel Merkx and Dan Diels of KU-Leuven for their partnership in implementing the projects funded by VLIR-OUS and supervising students. I also thank the many students who worked in these research projects and the staff members we collaborated with from the Ministry of Agriculture.

To My Students

I appreciate the many students I have interacted with in my research methods classes. They contributed significantly by providing examples used in this book and ideas on the issues to emphasize. I sincerely appreciate the contribution of George Gacheru Mwaura, currently living in the USA, and Christopher Njuguna, who was instrumental in guiding part V of this book, especially the impact assessment section.

To My Mentors

I extend my heartfelt gratitude to my Chief Mentor, Prof Daniel Mugendi, Vice-chancellor of the University of Embu. I have worked with Prof Mugendi since 1992, during which he has guided me through Masters and Doctorate training with unwavering encouragement and supervision. While we were both working for Kenya Forestry Research Institute, his support was instrumental in giving me a foundation as a young research officer. Over the years, our collaboration in numerous research projects, detailed in this acknowledgment section, has been immensely valuable. In addition, I am grateful for the opportunity provided by the University of Embu during my sabbatical leave in 2017, which allowed me to engage with both undergraduate and postgraduate students through teaching research methods and supervising their research projects.

I greatly appreciate Prof. James Kungu, my PhD supervisor, who contributed immensely to my strong foundation in the field of research.

I sincerely thank Prof. Waceke Wanjohi, former Ag. Vice Chancellor of Kenyatta University and Deputy Vice Chancellor of academics, for inspiring me to take a leadership position. While she was the Dean of the School of Agriculture and Enterprise Development (KU), I worked as a chair in one of the departments. Prof. Wanjohi's mentorship and guidance were immensely beneficial. I will forever be grateful. In addition, I appreciate her confidence in my ability to teach research methods in the School of Agriculture.

I thank Prof. Paul K Wainana, Vice Chancellor of Kenyatta University, for giving me a golden opportunity to serve in the Registrar of Research Innovation and Outreach office. My role as Registrar was to coordinate the implementation of research at the university. This appointment gave me the confidence and motivation to complete this book. It also enhanced my ideas on research, especially the university's role in dissemination, and contributed immensely to the chapter on dissemination of this book.

I appreciate Reverend Wachira of the Presbyterian Church of East Africa (PCEA) Parish Church for the prayers and encouragement to complete the book.

To My Mentees

I thank the over 40 postgraduate students I have successfully supervised and mentored and others I continue supervising. They give me momentum, and their thesis work has enriched my research methods skills.

I appreciate Dr. Kinyanjui Nganga, Pastor of Family Church in the USA, for encouraging me to start this book.

To My Great Institution

I thank Kenyatta University for providing a conducive working environment and the space needed to achieve this dream. I am deeply grateful for the support and resources that facilitated the writing of this book. I have extensively used the special book collection of books on research methods in the library. I have also used the rich information provided by the library through various platforms, e.g., repositories for thesis, My Loft, and databases such as AGORA. I thank the University Library staff, in particular the chief Librarian, Dr. Njoroge Gitau, and Mr JohnThuku, for always being helpful. This book would not have been possible without the nurturing academic atmosphere at Kenyatta University.

To My Heroes and Heroines in Agriculture

I thank the many farmers I worked with in the field, researching and developing technologies for improving agricultural productivity. May God bless you abundantly and continue guiding you to produce food for your families and to feed the nation.

Acknowledgment of Microsoft Excel

In Chaps. 20 and 21 on data processing and analysis, we have utilized several screenshots generated through Microsoft Excel. In Chap. 20, these screenshots illustrate the creation of a data analysis matrix, scatter plots, pivot tables, and box plots. Chapter 21 further employs screenshots to demonstrate how to organize various data types from different experimental designs for running ANOVA and conducting correlation and regression analysis. Microsoft Excel was instrumental in enhancing the content in data handling and analysis by offering practical demonstrations.

In Chap. 24 on data presentation and visualization, we extensively used Microsoft Excel to plot various graphs, including bar graphs and line graphs. Overall, Microsoft Excel played a key role in enhancing the practical value and clarity of this book, which is primarily aimed at postgraduate students undertaking master's and doctoral research.

Acknowledgment of SPSS

Chapter 22 is on survey data analysis using Statistical Package for Social Scientists (SPSS). Throughout this chapter, several screenshots have been used to demonstrate key procedures and steps in data analysis. These include creating frequency tables and cross-tabulations and performing essential analyses such as measures of central tendency, measures of variability, ANOVA, T-tests, Chi-square, Spearman rank correlation, and binary logistic regression. The use of SPSS has significantly enhanced the value of this book by providing in terms of giving students with a practical application of these analytical techniques.

Acknowledgment by Steven Runo

I am grateful to my family for the support they provided during the writing of this book.

I thank the many postgraduate students I have supervised. The experience gained through working with them helped shape the focus and direction of this book. I dedicate this work in part to all those students who continually strive for excellence.

I also thank the projects that supported some of the work used in this book. In no particular order, I acknowledge the following projects and funding sources:

1. Mawazo Fellows Fund
2. Baseline funding from King Abdullah University of Science and Technology and the Bill and Melinda Gates Foundation (grant number OPP1136424)
3. The National Research Fund (NRF) of Kenya (contract number KU/DVCR/NRF/VOL1/27)
4. Kenyatta University through the Vice Chancellor's Research grant number KU/DVCR/VRG/VOL.11/216
5. Regional Universities Forum for Capacity Building in Agriculture (RUFORUM) through a capacity building competitive grant (RU/2016/DRRG/01/010)
6. NRF and an In-Country/In-Region scholarship from the German Academic Exchange Service—Deutscher Akademischer Austauschdienst (DAAD)

I acknowledge the invaluable contribution of RStudio whose software facilitated the preparation of many graphics especially those in Chap. 24.

Contents

Part I Introduction to Research

1 Foundations of Agricultural Research 3

2 The Research Process, Key Concepts, and Terminologies 27

3 From Topic Selection to Research Objectives and Hypothesis 55

4 Literature Review 75

Part II Research Proposal and Ethics

5 Research Proposal 105

6 Ethical, Legal, and Administrative Issues in Research 127

Part III Research Designs

7 Overview of Research Designs 153

8 Fundamentals of Experimentation 177

9 Common Experimental Designs and Layout of Experiments 195

10 Farming Systems Research and On-Farm Experiments 223

11 Research Designs for Impact Evaluation Studies 259

12 Nonexperimental Designs 1: Surveys 293

13 Nonexperimental Designs 2: Qualitative and Participatory Approaches ... 313

14 Common Approaches for Qualitative Research 335

15 Mixed Methods Research Designs 353

16 Sampling 375

Part IV Methods and Techniques of Data Collection

17 Methods of Data Collection from Experiments 399

18 Methods and Tools of Data Collection in Non-experimental Designs (Surveys) 421

19 Methods and Tools of Data Collection in Nonexperimental Designs II (Qualitative Approaches) 447

Part V Data Processing, Analysis and Presentation

20 Quantitative Data Analysis I: Key Steps and Descriptive Data Analysis 473

21 Quantitative Data Analysis II (Inferential Analysis) 499

22 Survey Data Analysis Using Statistical Packages for Social Sciences (SPSS) 527

23 Qualitative Data Analysis 561

24 Data Presentation and Visualization 585

Part VI Dissemination of Findings and Scientific Publications

25 Disseminating Research Findings 603

26 Structure of a Scientific Paper 621

27 Journal Selection, Submission, and Review of Manuscript 643

28 Scientific Writing Skills and Presentations 667

Part I

Introduction to Research

Part I of this book consists of four chapters. The chapters cover the fundamentals of agricultural research.

In Chap. 1, we examine the foundations and importance of agricultural research. We also cover the evolving role of agricultural research in addressing global challenges.

In Chap. 2, we cover the research process and key terminologies in research. Key steps involved in conducting research are provided. Key concepts and terminologies are also covered to ensure a clear understanding of the language and concepts used in agricultural research.

In Chap. 3, we explore the journey from topic selection to research objectives and hypotheses. Here, we discuss ways to choose relevant research topics and formulate research objectives and testable hypotheses. We also discuss strategies technical and non-technical strategies for enhancing success in a research project.

Finally, in Chap. 4, we cover the literature review, an important component of the research that is relevant at all stages of the research process. This chapter guides you on how to identify sources of literature, conduct a thorough review of existing research, identify knowledge gaps and build a strong foundation for new investigations.

1 Foundations of Agricultural Research

Abstract

The term "research" carries different meanings for different people. Many nonprofessionals equate it with gathering information on a specific topic of interest, often overlooking the importance of reliability and reproducibility. On the other hand, scientific research is characterized by a systematic approach of seeking knowledge, that is replicable, evidence-based, and trustworthy. Research contributes to the cumulative building of universal knowledge over time. Nonscientific sources of knowledge, such as authority figures or personal experiences, should be approached with caution due to their subjective nature and lack of empirical support. The primary goal of agricultural research is to improve food security. It employs a research-development continuum framework that translates scientific findings into practical applications, thereby promoting societal well-being and sustainable development. Current trend in agricultural research focuses on interdisciplinarity, technological innovation, sustainability, and inclusivity. Key stakeholders in agricultural research include governments, industry leaders, research institutions, and farmers. Farmers are instrumental in this process; in addition to being beneficiaries of research, they help identify challenges, manage experiments, provide valuable feedback, and flag issues that require further investigation. The Consultative Group on International Agricultural Research (CGIAR), which encompasses 15 global centers, is dedicated to advancing agricultural research, promoting sustainable practices, and boosting food security. Their initiatives have notably improved productivity through innovations in crop varieties and farming methods. Overall, research can be categorized according to its specific purposes, with each type playing a distinct and vital role in advancing knowledge.

Keywords

Adaptive research · Basic research · Cumulative knowledge building · Descriptive research · Evidence-based · Exploratory research · Explanatory research · Research-development continuum · Systematic; scientific research method · Stakeholders in agricultural research: wholeness-oriented agricultural research

Learning Outcomes

After reading this chapter, you are expected to be able to:

1. Appreciate the difference between casual use of the term "research" and formal systematic research.
2. Recognize the "scientific research method," as the true and reliable source of knowledge.
3. Identify characteristics of scientific research.
4. Appreciate other sources of knowledge (nonscience knowledge sources).
5. Explain the importance of agricultural research.
6. Apply the concept of the research-development continuum in agricultural research.
7. Appreciate the wholeness-oriented agricultural research, which encompasses both biophysical and social aspects.
8. Identify the role of key stakeholders in agricultural research.
9. Appreciate the role of National Agricultural Research Institutions (NARI) and the Consultative Group in International Agricultural Research (CGIAR) in promoting agricultural research.
10. Appreciate the different types of research.

J. Njeri Mugwe, S. Runo, *Research Methodology in Agricultural Sciences*, https://doi.org/10.1007/978-981-95-1892-0_1

1.1 What Is the Nature of Research?

1.1.1 Defining Research: From Everyday Usage to Academic Rigor

Research is a term commonly used by both technical professionals and those outside of technical fields. In everyday language, it often refers to the process of gathering information on a specific topic of interest. In today's digital age, "Googling" has become synonymous with research, as many people rely on Google and other search/browsing engines to search for information. For example, businesses and companies may claim to stay informed by continually researching online to gather information on various issues that affect their operations. Similarly, individuals often state that they entered a particular market after conducting thorough research.

Many other nontechnical individuals often equate research with simply searching for information on the internet. For example, farmers frequently share knowledge of new farming methods that they discover through their experiences and online searches. In contrast, professionals may use the term "research" to refer to more formal processes, such as journalists conducting investigative work and creating documentaries based on their findings.

In academic settings, the term "research" can have different meanings depending on the context. For instance, graduate and undergraduate students often describe their process of searching the internet for information to complete assigned projects or term papers as "research." Similarly, those writing their proposals and theses will refer to searches conducted in libraries to gather information for their literature review chapters as part of their research. Additionally, some students claim to have written review papers based on their library "research." This variability in understanding shows the *ambiguity* of the term in common usage.

As scholars, we question, and we ought to question, the reliability of the information generated by these nonformal methods. We can pose a series of questions. Is the information they generate and present accurate and true? Do they all convey consistent messages on their respective topics? Is the information reproducible? Can the individuals present evidence? The answer is yes and no. While some information on the internet is accurate and true, such as from credible journals and research reports, other information may be subjective, reflecting personal opinions, views, beliefs, and convictions, or even fake information. The reliability of the information online varies, and not all sources are factual and dependable. Such information helps in making decisions but it is neither credible neither reliable.

So What Exactly Is Research?
According to the Oxford Advanced Learner's Dictionary, research is a *careful study of a subject*, especially in order to *discover new facts or information* about it. Other definitions of research include the following:

1. Research is a systematic, controlled, empirical, and critical investigation of hypothetical propositions about the presumed relations among natural phenomena (Kerlinger 1973).
2. Research is the process of looking for a specific answer to a specific question in an organized, objective, reliable way (Payton 1979).
3. Research is a systematic, formal, rigorous, and precise process employed to gain solutions to problems and/or to discover and interpret new facts and relationships (Waltz and Bausell 1981).
4. Research is the pursuit of truth with the help of study, observation, comparison, and experiment; the search for knowledge through an objective and systematic method of finding solutions to a problem (Kothari and Garg 2019).
5. Research is an activity undertaken to discover new insights in a structured manner, thereby expanding their understanding (Saunders et al. 2019).

According to these definitions, we see that *Research is not a haphazard aggregation of facts or information. Rather Research is organized, structured, and evidence-based.* The main purpose is to generate credible knowledge. It bears the following characteristics:

- Research is a systematic and controlled investigation.
- Research is a formal and rigorous process.
- Research involves the process of seeking a specific answer.
- Research is the pursuit of truth through study and observation.
- Research is the systematic process of collecting and analyzing information.

Research Is Organized and Evidence-Based
All fields, from natural sciences to social sciences, are involved in research to improve performance, generate new knowledge, or solve problems. The way research is conducted varies across disciplines. For instance, while medical research focuses on laboratory experiments and strict protocols to test treatments, historical research will use more descriptive approaches to study past events. However, a com-

mon aspect is the emphasis on evidence-based inquiry. Evidence-based inquiry is the process of seeking knowledge through systematically collected empirical data. It also means that conclusions are drawn from systematically gathered and analyzed data rather than assumptions or opinions.

This shared commitment to evidence forms the foundation of the *scientific method*, a structured and logical process used to investigate questions, test ideas, and build knowledge. In the scientific method, the processes can be *scrutinized* and *replicated*. This is because the research procedures are described in a way that enables other people to understand and make judgements about their quality. This differs from the other informal methods of building knowledge described at the start of the chapter, which are haphazard and difficult to repeat the procedures.

Based on this, we can define research as an organized investigation, designed to produce reliable and valid information.

Thus, this is our definition: "*Research is the systematic and reliable process of seeking specific answers to questions, aiming to gain knowledge or discover new facts or relationships*."

This definition emphasizes the purposeful and systematic nature of research as a structured and reliable quest for truth and knowledge. As noted by Saunders et al. (2019), meaningful research is guided by logic, built on coherent relationships, and directed toward generating new understanding. To ensure its validity and relevance, research must be carried out by a *skilled researcher* using appropriate methods with precision and integrity.

We define Agricultural Research as "*a controlled, systematic, and evidence-based inquiry to generate new information to solve problems in agriculture*."

Key Attributes of Scientific Research

Scientific Research Is Truthful Think about this. People believe different things. People talk about diverse experiences and explanations regarding phenomena. How do we know what we know and what we are told is the truth? This philosophical question touches on a branch of philosophy known as "*Epistemology*." Epistemology looks at how knowledge is (nature of knowledge), where it comes from (origin of knowledge) and how we can acquire this knowledge (what methods do we use to acquire the knowledge?). This is discussed in Chap. 7. Sometimes, what we know and believe may be incorrect, i.e., it does not agree with the way the world is, and, at times, it is wrong. We call this a myth.

If we believe something to be true, while it is not, it can be very risky to us and to the people around us, as we will make wrong decisions. For instance, if farmers rely on "myths" they hear about using certain plant extracts to control pests and diseases and which may not be proven to work, they risk losing the crop. If such extracts are tested and proven to work, the results are science-based; therefore, farmers will be confident they work. Other information, knowledge, and explanations that cannot be proven to be true or wrong are belief-based information and explanations. Such information remains an assumption.

The testing and proving that the plant extracts work is a process of scientific research that seeks to establish the truth (Bhattacherjee 2012). For instance, physical scientists strive to comprehend the realities of the physical world, while biologists investigate various biological phenomena, and chemists delve into the realm of chemical interactions. They are all seeking to uncover the truth to contribute to improved performance in their respective fields. In agriculture, we seek the truth to enhance agricultural performance.

The pursuit of truth is guided by theory. Theory is an important term in research because it provides an explanation and prediction of a phenomenon (subject of study) (Brazil et al. 2005; Giles and Harrison 2023; Díaz Andrade et al. 2023). It explains how and why things work the way they do. For instance, in testing the plant extracts, we could ask ourselves, why test plant extracts? What determines which plants to be tested? The answer to this is that the researcher is guided by some theory. For instance, the "theory of plant defense mechanism" could be the guiding principle. If, for example, it were soil fertility, the theory of soil organic matter dynamics could guide the research. Theory is the highest form of knowledge in research. It turns specific findings into generalizable knowledge that explains and predicts (see Chap. 2 for more discussion).

Scientific Research Is a Public Good

One of the guiding principles in research is the sharing of findings for positive social benefit. It is thus a public good. The concept of a public good assumes that citizens are in a relationship with one another, which encourages the creation and maintenance of facilities or arrangements that serve common interests and produce public value from which many individuals benefit (Boulton 2022). Indeed, most research is publicly funded, either by donors or national governments, so that findings should serve the public.

Naturally, scientists publish their findings in scientific journals. These journals are important sources of research findings, theories, and methods (the topic of journals is covered in Chaps. 25 and 26). Most journals are available in the public domain, especially online, allowing as many people as possible to access them. The accessibility of the journals enhances the availability of answers to questions that researchers may have and, therefore, avoids possible duplication of the work.

More importantly, the open sharing of research on public domain platforms is a way of ensuring that research is credible and verifiable. As a researcher, you can read the work, and if you think that there is something wrong with the conclusions, you can do a similar study to find out for yourself. Indeed, Boulton (2022) asserts that the knowledge claims and evidence of science on which they may be based must be made openly available to be tested against reality and logic through scrutiny of peers.

A recent debate related to the concept of a public good is centered on the advocacy of "Open Science." This debate is promoting the practice of making science accessible to all. It encompasses practices such as publishing open research and open-book science, including the sharing of data. These activities have the potential to make the scientific process more transparent, inclusive, and democratic, and therefore act as an accelerator for achieving the UN Sustainable Development Goals (UNESCO 2021). Other advantages of open access include faster sharing of innovations and publications, increased research collaborations, and more accountability (European Commission 2015). Overall, it will enhance the concept of public good and help bridge the gap among science, technology, and innovation.

The science-based research has several other attributes that enhance our understanding of the scientific method (Creswell and Creswell 2022). They include the following:

1. *Purposive:* It is guided by a need to solve a research problem. In other words, the research problem is the backbone that supports any scientific research. Therefore, the aim of the research will be to search for a solution. Solving a problem is what research is all about.
2. *Requires expertise:* Research requires a theoretical understanding of the subject matter and proficiency in research skills. Therefore, training and practical skills are required to ensure that research is rigorous. Researchers must be well-versed and adept in various research methodologies to conduct their studies effectively. They must also be well versed with theoretical underpinning in their areas of specialization. These skills are mainly acquired during graduate training and this book is a resource for enhancing skills in agricultural research.
3. *Replicable:* Scientific research is repeatable, enabling researchers to arrive at conclusions consistently. The results of hypothesis testing should be supported repeatedly, either by the same research or by others. This repetition ensures the findings are the truth and not due to chance. Other scientists who wish to replicate or confirm the results can use (replicate) the same methods and procedures, further validating the original research.
4. *Control*: control implies that, in exploring causality concerning two variables (factors), studies are conducted in a way that minimizes the effects of other factors affecting the relationship. Specific procedures are implemented to ensure that any effects not related to the variables being measured are eliminated from the study.
5. *Parsimony:* is a fundamental concept that states that researchers should favor explanations that are straightforward and rely on minimal assumptions. Often summarized by the phrase "Occam's Razor," this principle suggests that among competing hypotheses, the one that makes the fewest assumptions should be adopted. A parsimonious explanation should have no extraneous causes or forces should be invoked in explanation of a phenomenon than are minimally required to account for the data (McFadden 2023; Dubova et al. 2025).

Steps in the Scientific Method

The scientific method is also known as the hypothetico-deductive method because it relies on both the generation of hypotheses and the process of deriving logical conclusions from those hypotheses, followed by testing (Haig 2000; Campos and Pfister 2023). Testing is done by collecting data to determine whether a certain hypothesis is true. If the results align with the deduced predictions, the hypothesis is provisionally accepted. If they do not, the hypothesis is rejected or revised.

Different books on research methods give various variants of steps in the scientific methodology. Here we will demonstrate the steps using two scenarios: (1) We have observed low crop yields and want to determine the optimum amount of nitrogen that will give farmers 5 t ha^{-1} of maize; (2) We have observed low adoption of organic manure, and we want to assess if the level of knowledge of the farmer is related to the decision to use organic manure. The scientific method will follow these steps:

Step 1: *Make observations*: We start with observations driven by areas of concern, i.e., the problem, which is low crop yields. We gather information from several sources, including field and literature. We then define concepts and variables of interest.

Step 2: Stating hypothesis: ask questions about the phenomenon, make predictions and set a testable hypothesis that is falsifiable, with clear variables.

Step 3: *Determine research type*: The nature of the study will determine whether it is an experiment or nonexperimental. The research should ensure the design yields the data necessary to test the hypothesis.

Step 4: *Ethical clearance:* Any research involving human beings or animals must be conducted in accordance with the ethical clearance guidelines (University and National guidelines to avoid violation of human and animal rights). This is discussed in Chap. 5.

Step 5: *Data collection*: Researchers collect data using various form of inquiry such as surveys and experiments.

Step 6: *Data analysis and interpretation:* Test the hypothesis and approve or disapprove. Need some knowledge of biostatistics; a statistician may be consulted.
Step 7: Make a conclusion and communicate the findings.
Step 8: Apply information to address concerns and observations.

A distinctive feature of this method is the focus on falsification, a concept introduced by Karl Popper (Voit 2019). Rather than seeking to prove a hypothesis is true, researchers aim to disprove it. The rigorous testing ensures that the most robust and reliable theories withstand scrutiny. By adhering to the principle of falsification, the scientific method helps maintain the integrity of scientific knowledge, preventing premature conclusions and ensuring continuous improvement.

This cycle of hypothesis formation, testing, and revision is essential to science's self-correcting nature. It is iterative in nature (Fig. 1.1—see arrow from results back to testing). It allows for the constant refinement of theories, ensuring that our understanding of the world becomes increasingly precise and reliable over time. The scientific method is thus a fundamental tool of discovery and innovation, guiding research across all disciplines and advancing knowledge.

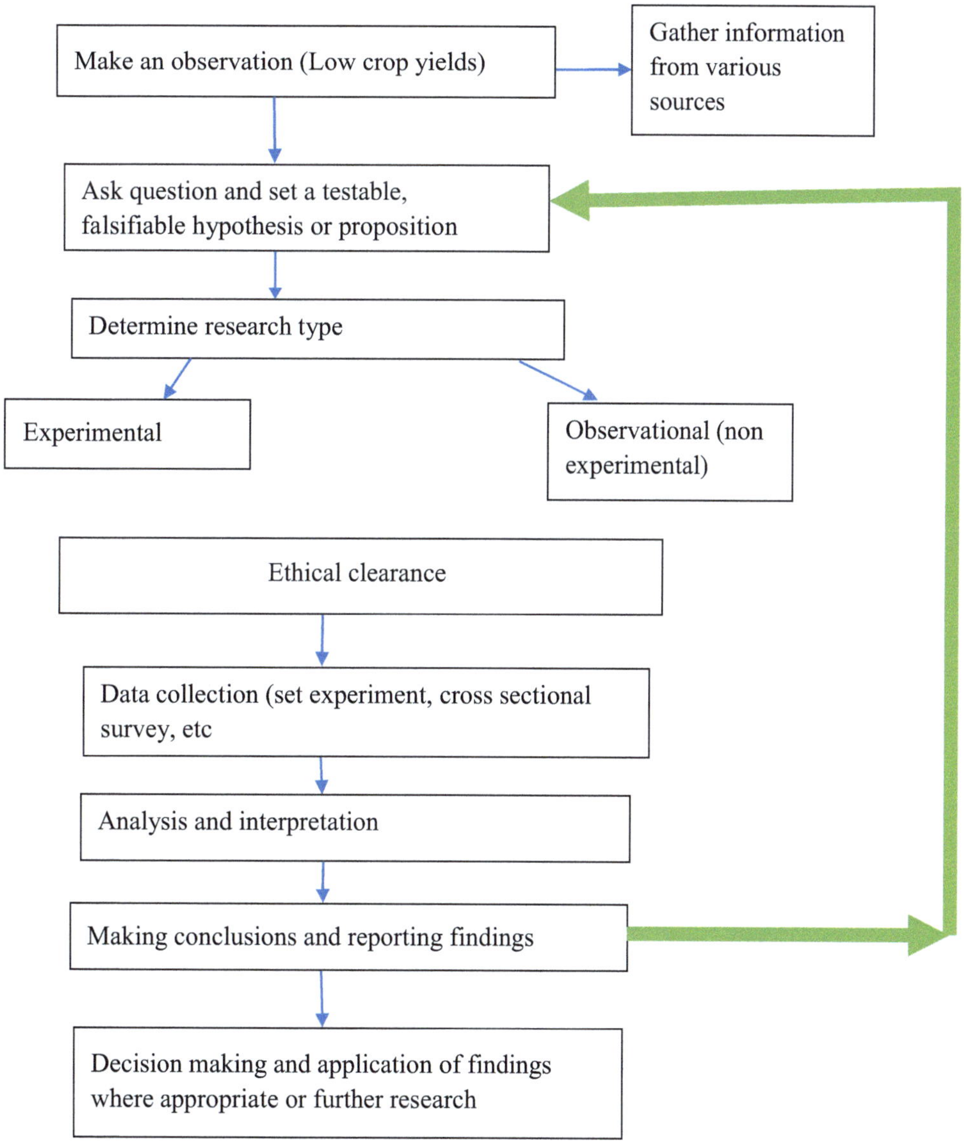

Fig. 1.1 The scientific research process starts with observations (aspects that are not working), which are investigated to provide information that can aid in decision-making

1.1.2 Nonscientific Sources of Knowledge

Let us look at other sources of knowledge besides scientific research to enable us to appreciate the important role of knowledge generated using the scientific method:

1. *Authority*

People in authority are a source of knowledge. They are regarded as sources of authoritative knowledge and may influence those seeking knowledge and information. For example, parents and teachers during the early human development stage are regarded as the sole authority by children (Raviv et al. 1990). Growing up, we rely on the knowledge we gain from our seniors. In new employment, we see employees who have been working there for a long as authorities and rely on them for orientation. Students view their teachers as authorities and, at times, look up to them as role models. Farmers view and regard agricultural extension workers as authority in the agricultural landscape. They will also look up to successful farmers as a source of knowledge. Researchers sometimes look up to those who have published peer reviewed journal articles and books in certain areas as authorities and believe everything they write. Likewise, they will treat information from government administrators as trustworthy. The feeling is that they believe it is true because a person they treat as an expert says so.

Using authority is good when information is true because it saves us from doing the evaluation ourselves, but when the authority is not good or credible, we are not able to know whether the information is 100% true (Lee et al. 2023). Experts may not know everything. At times they give their opinions which might be irrelevant.

2. *Traditional Beliefs and Cultures*

Sometimes people use traditionally accepted knowledge and understanding that has been passed through generations as the basis of sourcing for knowledge. For example, *In some drylands of Kenya, some farmers believe that "fertilizers destroy soils."* Most farmers hold these beliefs passed from their foreparents. This is, however, not evidence-based. Accepting something as true because it has been traditionally accepted as such is a faulty way of acquiring knowledge.

In a study carried out to identify traditional ways of forecasting the coming of the rains, some of the ways mentioned by the farmers were found to be contradictory (Mugi-Ngenga et al. 2016). This means that some traditional beliefs are not true and cannot be relied upon.

3. *Personal Experience*

Personal experience is a common source of knowledge. It is also called experiential knowledge gained through using the five senses: seeing, hearing, smelling, feeling, and tasting (Wilken and Cabiati 2024). Personal experience as a source of knowledge has limitations because one's personal experience is too narrow to generalize and sometimes personal experience may be expressed with bias. For example, in mentorship programs, successful people in business or professional careers are asked to mentor and train the youth. Depending on the agenda the mentors want to sell, there could be some biases, where mentors may hide some of their failures, called "scars," and paint an exaggerated picture.

An example of biased personal experience is a project that was promoting dairy goats in the central highlands of Kenya. The strategy was to use pioneer farmers to train and motivate other farmers on the benefits of keeping dairy goats. During training, the two major benefits were emphasized: (1) The ability of the dairy goats to survive on low feeds. This was to solve the problem of the shortage of animal feed. (2) Increased incomes from selling goat milk, which will fetch a higher price than that of cow milk. During field days, the pioneer farmers would train the other farmers on how to keep the dairy goats. These farmers would entice the other farmers by telling them that their incomes had increased from the selling of milk. However, after investigations, most of the pioneer farmers were found to have benefitted more from selling the young ones rather than from selling the milk. The information passed to other farmers needed to be revised. This disadvantaged the farmers who were out to learn and keep the superior dairy goats.

4. *Intuition*

Intuition is knowledge of something without being consciously aware of where the knowledge came from. Usually called the "sixth sense" and cannot be explained on the basis of reasoning. In his classical book Think and Grow Rich (Hill 1937), Napoleon Hill recognizes intuition and describes it as that portion of the subconscious or creative mind. According to this author, this creative mind has ideas, plans, and thoughts that flash; sometimes called hunches or inspirations. They are important for warning of impending dangers in time to avoid them, and notifying of opportunities to embrace them (Hill 1937). However, there is no evidence that it exists and it is not proven that people used intuitions successfully. People believing in this will easily say, "I believe because I just feel it," but cannot tell the source of knowledge. The problem with this kind of knowledge is that it cannot be explained on the basis of reasoning. It is individual-based. It is, therefore, difficult to generalize or develop policies and practices based on intuitive knowledge.

5. *Rationalization*

Rationalization refers to knowledge acquired through logical reasoning (Rindermann et al. 2014; Wilkinson 2025). Using this method, premises are stated, and logical rules or arguments are followed to arrive at a conclusion. For example, all humans are mortal; because I am human, I am mortal. The problem with this method is that if the premises are wrong or there is an error in logic, then the conclusions will not be valid.

6. *Divine Powers*

Religion and faith-based beliefs are based on supernatural and divine powers that cannot be measured or verified (Barrett 2007; Bluemke et al. 2016; Van Eyghen 2019, 2022). They are based on religious and spiritual beliefs that argue about infinite wisdom and omniscience. Take the case of Christians, Muslims, Buddhists, and Hindus. They carry beliefs that guide their behavior. They all believe in a divine power providing knowledge on how to live. For example, Christians will read the Bible, while Muslims will read the Quran. Based on these beliefs, you may hear a statement like "The prophet told me to do this activity," or I had a dream that I believe is from the divine powers. The beliefs vary depending on the religion and are, therefore, not universal, nor are they reliable. Table 1.1 shows the summary of sources of knowledge so far discussed.

Table 1.1 Summary table of sources of knowledge

Source	Characteristics	Common expressions
Scientific knowledge	Empirical procedures, systematic, verifiable, accurate	I measured, and data revealed the results (empirical - evidence)
Authority	Gained from people believed to be experts; as the experts say it	"I believe it because an expert says it"
Traditional beliefs and cultures	Practiced and passed through generations	"I believe it because it has always worked like this"
Personal experience	Personal testing and experience	"I have experienced what I am saying "
Intuition	Gut feeling: cannot tell the source of knowledge	"I believe it because it is what I feel"
Rationalization	Conclusions based on logical reasoning	"It is true because it is derived from logic"
Divine powers	Revelation of God and supernatural powers (cannot be investigated) based on personal conviction (faith)	"This is my belief based on my faith"

1.2 Core Aspects of Agricultural Research

1.2.1 Importance of Agricultural Research

The main goal of agricultural research is to improve productivity by addressing various challenges in agriculture. Specifically, agricultural research is used to:

(a) *Solve problems facing agriculture:* Broad problems in agriculture include food in security, environmental degradation, and rural poverty. Research is important because it suggests improvements to deal with these problems. For instance, research will suggest better methods and innovations for increasing food production. If food production is increased then more food will be available for the growing population. A powerful historical example is the Green Revolution of the mid-twentieth century, which directly provided a solution to increase global food production. During that time, there was a need to raise production, but the existing varieties were low-yielding. Through research, improved high-yielding crop varieties, particularly of wheat and rice, were developed. These improved varieties produced up to four times the yield of the traditional varieties (Evenson and Gollin 2003; Pingali 2012). As a result, global cereal production nearly tripled, enabling many countries to achieve food self-sufficiency.

(b) *Add to scientific knowledge base:* As researchers seek answers to problems, they develop more knowledge and a deep understanding of the problems. The knowledge gained is captured in reports, academic journals, and other research publications. Over time, these results are synthesized and compiled into more comprehensive resources such as books, contributing to the long-term advancement of scientific knowledge. Indeed research is never conducted in a vacuum; every study builds on prior work, recognizing that earlier investigations have already addressed certain aspects of the question being explored. WE call this 'Cumulative Knowledge building". It is like individual trees contributing to making a forest.

(c) *Research informs policy decisions.* Research provides information to policymakers. Policymakers range from government officials, legislators, and regulatory bodies to international agencies and donor partners involved in making agricultural policies and regulations. Researchers offer findings, trends, and recommendations to policymakers. Researchers often present the findings in the form of short reports (policy briefs). The policymakers use the findings to improve on future policies. In addition, researchers may evaluate the effectiveness of existing policies and provide recommendations for improving them.

(d) *Describe phenomena:* Studies with this objective are termed exploratory or formative research studies. This may include describing climate change phenomena, exploring and describing traditional methods of forecasting rainfall events, and describing rainfall change over time, etc.

(e) *Enable prediction:* Prediction is the ability to estimate phenomenon A, given phenomenon B. Through research, we establish relationships that enable us to make predictions. For example, given an area's average yields and rainfall patterns, we can predict the yield in the next 10 years. Predictive studies fulfil this role.

(f) *Enable control of variables:* In scientific research, control is concerned with the ability to regulate the phenomena under study. Usually, a phenomenon is manipulated in order to exert control over another. This ensures that the response measured is real and not influenced by variables that are not measured. Without the ability to control, findings may not be credible.

(g) *Develop theory:* Theory development involves formulating concepts, laws, and generalizations about a given phenomenon, such as the innovation-diffusion theory (Moseley 2004; Dearing 2009).

(h) *Confirm or validate existing theories:* Previously established concepts or hypotheses are investigated to test whether they hold true under different conditions or over time.

(i) *Search for meanings:* These involve the study of activities, practices, beliefs, attitudes, and experiences. For example, in addressing the challenges of climate change, we may choose to study Indigenous knowledge as a way of adding knowledge of managing climate change

(j) *Develop new research methodologies and tools*: In addition to building knowledge, research also creates new methodologies, instruments, models, and tools for better research. These include methodologies of data collection, analysis, and interpretation. The methodologies help improve the accuracy, reliability, and efficiency of scientific investigations. For instance, in soil science, dry methods of soil analysis, such as near-infrared spectroscopy (NIRS) was developed as an alternative to the quantitative wet chemistry (Ng et al. 2022). This method is better than wet chemistry because it offers rapid, nondestructive analytical techniques. In data analysis, new software tools such as R and Python visualize data more effectively. Emerging research evaluating how to integrate machine learning and artificial intelligence will play a key role in transforming how data is analyzed and interpreted.

1.2.2 Concept of Research-Development Continuum

The research-development continuum in agricultural research is a cyclic process to enhance society's benefits (Patel et al. 2004). It represents the big picture of why research is being conducted. This cycle translates scientific knowledge and innovation into real-world solutions to enhance productivity, sustainability, and food security. For many years, agricultural projects labelled as "research and development projects" have emphasized the generation of innovations during the research phase and the transfer of these to society through the development phase. This integrated approach enhances both scientific understanding and improvement in agricultural practices. In other words it connects research knowledge with practice.

Figure 1.2 illustrates a research-development continuum framework for an agricultural project, showing activities within the various phases of a research and development programme. The starting point is research to enhance understanding (Basic research); the aim is to understand the mechanisms and scientific discoveries. The basic research may be carried out in a controlled environment, such as a glasshouse or in controlled field experiments, e.g., screening soybean varieties for tolerance to soil acidity. Following this initial phase is a second phase whose aim is to understand ranges and options. This phase mainly involves applied research. Applied research tests whether outcomes observed in controlled settings hold true under real-world agricultural conditions, such as different agroecological zones, seasons, or farming systems. The results guide 'action research' refined with farmers input leading to wider dissemination and scaling up.

Let us take an example of crop improvement research, that is developing and screening new varieties. The "best bet" varieties identified in phase one will be tested in different agroecological zones to understand under which range they are adopted. The output of this phase will lead to the adaptive research phase, whose aim is to pilot and test the varieties in farmers' fields. Later, the testing progresses to action research with farmers participating and adopting the best-performing varieties. The researcher, at this stage, will assess the impact of adopting the varieties. Results of the impact evaluation will inform decisions regarding the varieties for wider dissemination and scaling up, extending the technology to farmers beyond the testing area. This way, what started as *basic research feeds into the development*

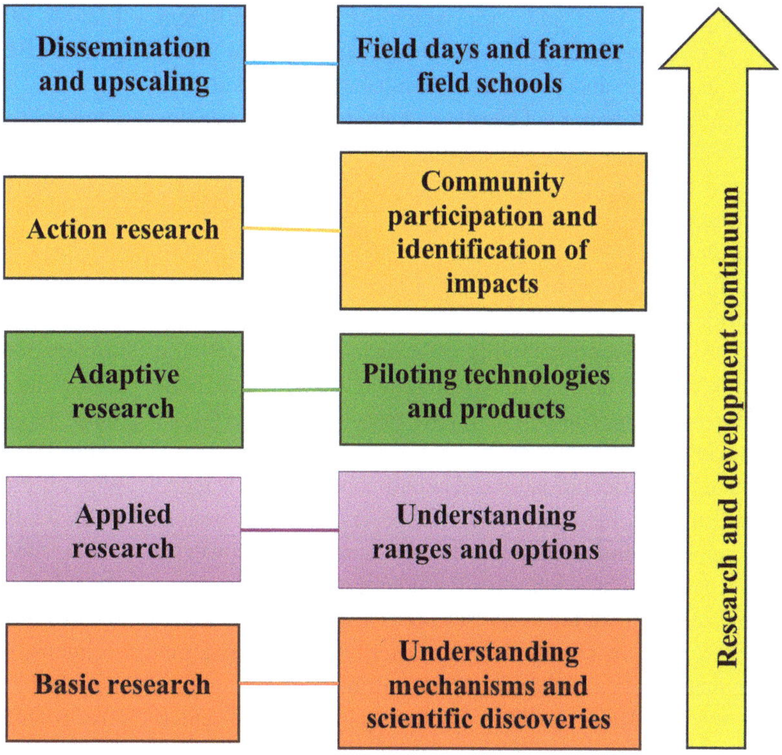

Fig. 1.2 Research-development continuum in an agricultural research and development project

Table 1.2 Examples of major research developments in agriculture

Research conducted (basic research)	Outputs feeding into development
Fertilizer trials	Fertilizer recommendations for different regions (e.g., amount of nitrogen, phosphorus, etc.)
Breeding studies	Release of new crop varieties or breeds of animals.
Variety testing	Adapted varieties for different regions (e.g., maize, bean, sorghum, etc.)
Studies in crop protection; testing efficacy of chemicals to control diseases and pests	New chemicals for control of crop pests and diseases
Agronomic trials	Recommendation on spacing, and weeding frequency
Animal feeding trials	New diets and recommendations on their use
Agricultural economics	Policy recommendations, economically viable decisions. cost-benefit analysis, market trends

agenda and is translated into benefits for the society. Certain research phases may take longer than others. The topic on dissemination is covered in Chap. 25.

We have many examples of research outputs feeding into development. In reality, major agricultural development and advancement have occurred through this process (Table 1.2).

1.2.3 Biophysical and Social Aspects in Agricultural Research

Agricultural research combines both biophysical (natural) and social sciences (Galt et al. 2024). Biophysical sciences focus on fields such as agronomy, crop protection, soil science, animal science and crop science and typically generate numbers. The researchers use quantitative data and borrow heavily from the basic sciences, such as biology and chemistry, to study crop and animal production techniques. They also include planting techniques, crop rotation, animal feeding regimes and fertilizer studies.

The social sciences encompass fields like agribusiness, agricultural economics, agricultural policy and extension, and related studies. These fields examine human behaviors and interactions within the agricultural context. The studies focus on understanding how human beings, including farmers, interact with one another and organizations. Given that farmers form the main social clientele in agriculture, understanding their behavior is key. Consequently, agricultural researchers must engage in interdisciplinary research that integrates both biophysical and social aspects. This integrated approach is common in agricultural research institutions where researchers work together with communities to address complex agricultural challenges.

The biophysical and social scientists employ different research approaches. Biophysical researchers aim for precision, accuracy, and predictable results through controlled experiments (Chap. 8). This allows them to isolate variables and replicate findings consistently. In contrast, social sciences often produce less precise and predictable results, using strategies like surveys and qualitative approaches. This is because human behavior and social systems are complex and influenced by many factors. This makes them hard to control or predict. While biophysical scientists prioritize objectivity and reproducibility, social scientists deal with the complexities of human interactions and thus adopt more subjective methods. These methods are discussed in PART II of this book.

Due to its orientation toward experiments, natural science often employs the term "scientific observations" to denote the rigorous and systematic collection of data. However, there can be confusion among students when it comes to social research. While social scientists also engage in scientific research, their methodologies differ. Social research predominantly relies on observations and interviews to gather information. This includes using surveys and closed-ended and open-ended interviews to explore various aspects of human behavior and societal phenomena. Despite the differences in methodology, both natural and social sciences aim to advance knowledge and understanding through systematic inquiry and analysis. They all rely on scientific observations, though such observations are guided by different paradigms. The concept of paradigms is elaborated further in Chap. 7.

The subject of social experiments is not as common in agricultural research, yet it is equally important, especially when we want to evaluate how interventions affect farmers' behavior or variables. For instance, a social experiment might assess the impact of providing incentives on farmers' adoption of sustainable farming practices or the effectiveness of extension services in improving crop management practices. We could also test the effect of training on the adoption of improved crop varieties. We call these social experiments randomized controlled experiments (RCTs) (These are covered in Chaps. 8 and 11).

1.2.4 Wholeness-Oriented Agricultural Research

The goal of agriculture in the twenty-first century is to produce food efficiently and and in a manner that is environmentally and socially friendly. Over the past decades, the world has witnessed rapid population growth and an increase in environmental degradation. These factors have created challenges as the population tries to meet the food demand within the context of climate change. Effective agricultural research thus requires putting efforts into understanding natural resource management, the role of changing farming systems to accelerate agricultural production and their connection with populations that depend on agriculture. With an estimated population projected to of 8.6 billion in 2030, 9.8 billion in 2050 and 11.2 billion by 2100 (UNDES 2017), more food production is needed from the already shrinking arable land. This calls for the intensification of agroecosystems while maintaining ecological balance.

Intensification of agroecosystems needs a rethinking of the general methodology of agricultural research (Alrøe and Kristensen 2002). These authors argue that agriculture is a complex science due to the following two main reasons:

1. Agriculture involves both *social and ecological systems*. Research into these social-ecological systems faces several challenges in understanding complex agro-ecosystem interactions.
2. The concept of *sustainable agriculture*: The need to produce more food for the growing population while ensuring the sustainability of the environment.

Agricultural research thus requires wholeness-oriented research (Fig. 1.3). Wholeness-oriented research emphasizes agriculture's interconnectedness, integration, and holistic nature.

Increased Productivity Increased productivity encompasses not just growing more crops with higher production but also using resources efficiently, being economically viable, and ensuring fairness. Whole-centeredness, therefore,

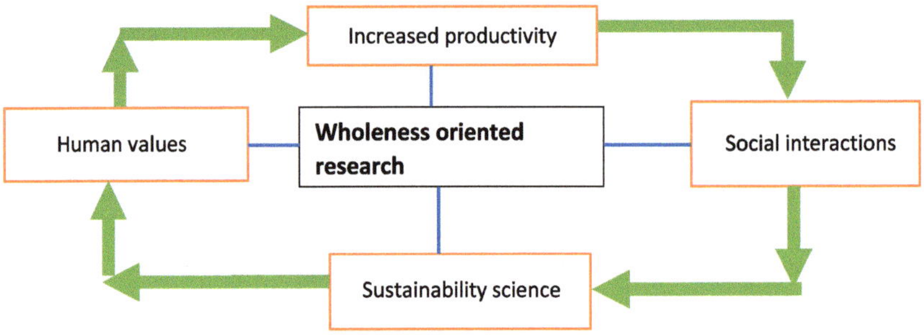

Fig. 1.3 Concept of wholeness-oriented research in agriculture

focuses on achieving productivity gains while also supporting community well-being and environmental care.

Social Interactions Whole-centeredness in agricultural research highlights the need to include social interactions for sustainable results. It involves working together with farmers, local communities, and other stakeholders. Instead of researchers working alone, the focus is on collaboration, sharing knowledge, and learning from each other (Chap. 10).

Human Values Human values, shaped by individuals' interests, desires, and perceptions, influence their behavior and interactions within society (Schwartz 2012). Wholeness-centered agricultural research should incorporate these human values to ensure outcomes align with the needs and interests of the communities concerned. This approach makes research more impactful and beneficial to the people it serves. Consequently, there is a need to understand human behavior, particularly by employing qualitative research approaches. (covered in Chaps. 13 and 14).

Sustainability The sustainability of agricultural systems is important to ensure the survival and well-being of humans throughout the world. Research methodologies, therefore, need to include aspects of sustainability science involving interactions of natural and social systems. A strong connection exists between agriculture and many Sustainable Development Goals (SDGs), including zero hunger, ending poverty, sustainable consumption and production, and combating climate change.

1.2.5 Emerging Trends in Food and Agriculture

Trends in the Twenty-First Century

Agricultural research in the twenty-first century is marked by interdisciplinarity, technological innovation, sustainability, and inclusivity. We highlight the following key trends shaping agriculture:

(a) The first trend is that agricultural research is currently focusing on *global food security, climate change, and environmental degradation* (UNFAO 2021). Unlike the twentieth century's emphasis on yield increases, modern research takes a holistic approach, integrating technology, data science, environmental studies, and social sciences to create productive and resilient farming systems (Lacoste et al. 2021). Sustainable agriculture aims to produce enough food while conserving resources, protecting ecosystems, and supporting rural livelihoods. Key innovations such as precision farming, agroecological practices, integrated soil fertility management, conservation tillage, and techniques that enhance soil health and reduce greenhouse gas emissions are key.

(b) The second area is the integration of *digital technologies* and *smart technologies* in agricultural research (Trendov et al. 2019; Gemtou et al. 2024). Tools such as remote sensing, geographic information systems (GIS), artificial intelligence (AI), and big data analytics are being utilized. These technologies aim to improve decision-making and monitor environmental variables. They also help optimize resource use. Additionally, they support real-time data collection and analysis. This enables more adaptive and efficient farming systems.

(c) Third is that agricultural research is now more *inclusive, participatory, and farmer-centric* (Adolwa et al. 2022). There is a growing recognition of the importance of involving farmers, especially smallholders and local communities, throughout the research process. Participatory research approaches place farmers at the center. The researchers value farmers' knowledge, experiences, and priorities. This collaboration ensures that innovations are technically sound and contextually relevant. As a result, they are more practical and more likely to be adopted by end users. Farmer-centric research enhances ownership and strengthens local capacity. Ultimately, it leads to more impactful outcomes.

(d) Finally, the twenty-first century has seen a growing focus on *policy-oriented and impact-driven* research. Agricultural research now plays a key role in shaping national and international policies on food systems, trade, rural development, and environmental management. Institutions and funding bodies are promoting research that produces actionable knowledge. This research aims to address urgent development goals, such as those in the UN Sustainable Development Goals (SDGs).

Trends in Food Production

Agricultural production has increased significantly, especially in developed countries (Fuglie et al. 2012). Overall, food supply has increased faster than population growth in these countries. These gains have largely been achieved through improved agricultural technologies. For instance, adoption of high-yielding varieties, which has a had a significant impact on agricultural productivity. Further, irrigation and the use of fertilizers, together with better resource management and more appropriate agricultural policy, have also been major contributors.

Some key developments globally have been:

- Increased cereal output per capita.
- Consistent annual growth in production of major crops (paddy rice, wheat, maize).
- Improvements in agricultural productivity.
- Improvement in global food security.
- Advancements in crop yields and sustainable agricultural practices.

However, in Sub-Saharan Africa and some nations in Latin America and South East Asia, increasing production alongside a rising population is a big challenge (Mohapatra et al. 2019). Increased production to respond to population has been due to a rise in cultivated land rather than an improvement in yields or intensification of agricultural practices. The cultivated fields are subjected to intensive cultivation without adequate replenishment of nutrients. This, coupled with erosion on steep slopes, has led to nutrient mining, land degradation, and poor soil fertility in most farms. Consequently, agricultural yields are low, posing a threat to food security in the region. The food systems are highly threatened.

This scenario has consistently been highlighted in the annual status of agriculture reports published by the Alliance for a Green Revolution Africa (AGRA). For instance, the latest report (AGRA 2022) mentioned that approximately 650 million people, nearly 50%, lack economic and physical access to sufficient food. The report identified a complex web of challenges affecting every stage of the food system, from production to consumption. Addressing these multifaceted crises requires concerted efforts from governments, international organizations, and local communities to enhance food security and ensure that food systems are resilient. Therefore, all stakeholders in the agriculture sector must collaborate to provide innovative solutions to unlock the potential of agriculture (AGRA 2022). Realizing this vision will require greater investment in research across all sectors ranging from developing climate smart resistant varieties to strengthening markets and value chain systems.

In the next section we explore who these stakeholders are and their roles in driving research and innovation, and how their collaboration drives progress in agriculture sector.

1.3 Stakeholders and Collaboration in Agricultural Research

1.3.1 Who Are the Stakeholders in Agricultural Research?

Agricultural research is important at the local and international levels and involves several stakeholders. Important stakeholders are those who conduct research, beneficiaries, and those that provide funds. They include:

1. *Government:* Governments play a multifaceted role in supporting research. First, they provide funding through research grants to institutions and academia. The funds mainly support the development of research infrastructure, such as laboratories and research innovation hubs. In academia, the grants are instrumental in supporting faculty and students' research through the university research grant framework. Young faculty members are beneficiaries of such funds, and it helps them to develop research credentials. Second, governments also manage policies and regulations to guide research, including those related to food safety, intellectual property rights, and ethical standards for research conduct. Governments are also crucial in facilitating collaboration both locally and internationally to enhance research outcomes. They establish research networks, consortia, and partnerships to promote knowledge exchange. Finally, governments also play a key role in the dissemination of research findings and in promoting public engagement with science through science communication initiatives. It supports extension operations to offer practical advice to farmers and support the adoption of improved agricultural practices.
2. *International Agricultural Research Centres (IARCs):* These organizations have an international mandate and generate technologies that serve wider applications (Byerlee and Lynam 2020). Examples of these are the International Maize and Wheat Improvement Centre (CIMMYT), the International Institute for Tropical Agriculture (IITA), the International Centre for Potato Research (CIP), the International Food Policy Research Institute (IFPRI), the International Livestock Research Institute (ILRI), and International Crops Research Institute in the semi-arid Tropics (ICRISAT) (see full list in Table 1.3)
3. *National Agricultural Research Institutions*: These institutions serve the mandate of respective countries, and their priority research areas are those that directly affect the country. Their primary goal is to increase production to meet the food and nutritional security needs of their nations. Key roles include developing new technologies, disseminating and exchanging knowledge, building capacity, and building partnerships among the stakeholders.
4. *Universities*: One of the main mandates of universities is research (Abramo et al. 2011). Universities offering agriculture and agricultural-related courses carry out research either solely or with other partners. African universities have a regional network called "Regional Universities Forum for Capacity Building in Agriculture (RUFORUM) that plays a key role in spearheading research and development initiatives in agriculture (Okalebo et al. 2012). Currently, with 65 member countries from Africa, RUFORUM engages in innovative and responsive

Table 1.3 Consultative group on International Agricultural Research centres

SNo	Center	Location	Date	Mandate, commodities
1	Africa Rice Center (AfricaRice	Abidjan, Cotedvoire	1975	Rice
2	International Food Policy Research Institute (IFPRI)	Washington, D.C., United States	1980	Policy solutions, efficient markets and trade solutions
3	International Institute of Tropical Agriculture (IITA)	Ibadan, Nigeria	1971	Banana and plantains, cassava, cowpea, maize, soybean, and yam
4	International Livestock Research Institute (ILRI)	Co-hosted by Kenya and Ethiopia	1995	Livestock: Feeds and diseases
5	International Maize and Wheat Improvement Centre (CIMMYT)	Texcoco, Mexico	1971	Major: Maize and wheat Others: Tricale, sorghum, millets
6	International Potato Centre (CIP)	Lima, Peru	1971	Potato, sweet potato, Andean roots, and tubers
7	International Rice Research Institute (IRRI)	Los Banos, Philippines	1971	Rice-based agrifood systems
8	Center for International Forestry Research (CIFOR)[a]	Bongor, Indonesia	1973	Forests, trees, and agroforestry
9	International Center for Agricultural Research in the Dry Areas (ICARDA)	Beirut, Lebanon	1975	Crop, forages, Genebank rangeland, small-ruminant production barley, lentil, and faba bean
10	International Crops Research Institute for the Semi-arid Tropics (ICRISAT)	Patancheru, India	1972	Dryland agri-food systems: chickpea, pigeon pea, millet, sorghum, and groundnut
11	World Agroforestry (ICRAF)	Nairobi, Kenya	1991	Trees, forestry, and agroforestry
12	WorldFish	Penang Malaysia	1992	Fisheries and aquaculture
13	Alliance of Biodiversity International and the International Center for Tropical Agriculture *(CIAT)*[b]	Cali, Columbia	2019	Climate change, biodiversity, environmental degradation, and malnutrition
14	Alliance of Bioversity International and the International Center for Tropical Agriculture *(CIAT)*[b]	Rome Italy	2019	Climate change, Biodiversity, environmental degradation, and malnutrition
15	International Water management Institute	Colombo, Srilanka	1991	Water, food, and ecosystems

Source: CGIAR Website https://www.cgiar.org/; Renkow and Berylee (2010), Byerley and Lynam (2020)
[a]CIFOR and WAC (ICRAF) merged in 2019
[b]In 2019, Biodiversity International and CIAT joined forces to create the Alliance of Biodiversity International and the International Center for Tropical Agriculture (CIAT); located in Rome and Columbia with an expanded mandate

research as well as enhancing networking with the university's education.

5. *Donor Agencies*: These support agricultural research immensely by providing funds for research activities, infrastructure, and capacity-building initiatives. Research is an expensive agenda that cannot go on without funding. Key funding agencies in agriculture include: (1) The United States Agency for International Development (USAID) (1961–2025), which supported global agricultural programs. Feed the Future programs supporting research to address food security challenges in developing nations are a vital program that was spearheaded by USAID. (2) The World Bank is particularly key in funding large-scale projects in diverse areas of research and development and capacity building. (3) The European Union funds research through Horizon Europe and Erasmus programs.
6. *Non-governmental Organizations*: These are organizations and initiatives for research and development. Examples include the Bill and Melinda Gates Foundation which has been supporting the Alliance for a Green Revolution in Africa (AGRA) to transform smallholder farming systems in Africa, the One Acre Fund, Food for the Poor, Eco Agriculture, Heifer International, and Feeding the 5000. They are more concerned with agricultural initiatives that directly reach farmers, but they are also involved in research in varying degrees.
7. *Industry and Private Sector:* The industry funds research initiatives either independently or in collaboration with research institutions, academia, or government. Examples of industry players include private companies in business: Syngenta, Monsanto, Yala, Cargill, etc. They are central to agricultural research and technology innovation, driving the development and commercialization of seeds, fertilizers, and agrochemicals. They invest heavily in agricultural research and development to create new and improved seeds, crop varieties, pesticides and other agricultural inputs. For instance, over the years, they have invested heavily in breeding programs. Despite their contributions, challenges such as market concentration, intel-

lectual property concerns, and potential impacts on small-scale farmers need to be carefully managed to strike a balance between innovation, ethics, and sustainability in the agricultural.

8. *Farmers:* Farmers consume research findings and contribute to agricultural research by providing valuable insights and practical knowledge and participating in field experiments. Their involvement in the research process ensures that innovations are relevant, sustainable and aligned with the challenges faced in agricultural practices.
9. *Food and Agricultural Organization (FAO):* FAO is a body of the United Nations and plays a crucial role in supporting global research and offering technical expertise. It also facilitates international collaboration and develops standards and policies that influence agricultural research. It is well-known as a source of agricultural information.

1.3.2 The Role of Farmers in Agricultural Research and Development

Farmers play a crucial role as stakeholders, significantly contributing to the overall success of research in agriculture. Initially, they help to identify problems that the research addresses. As the research is being implemented, they are active participants. After the technologies are developed, they are the major targets for researchers to disseminate information.

The major roles of farmers in agricultural research include the following:

1. *Problem identification:* farmers identify problems and challenges they are facing, such as low productivity of food crops, low milk yields, low soil fertility, etc. This forms the basis of the research problem for the researchers to work on.
2. *Providing research materials:* Farmers provide experimental land, experimental animals such as sheep, goats, cattle, and forages. In this case, the farmers offset part of the costs involved in establishing the experiment. Farmers in Kansas, USA, were willing to cooperate with researchers and contribute inputs for on-farm experiments (Norman et al. 1998) (more details in the Chapter on on-farm experiments).
3. *Managing and running the experiments*: Farmers are involved in managing the experiments, such as weeding, watering, milking, managing treatments, collecting, and recording data.
4. *Adopting and sustaining new ideas and models* brought to them by researchers to enhance productivity.
5. *Giving feedback about the new technologies and models:* Farmers provide feedback based on their experiences, such as yields per hectare and milk productivity (Norman et al. 1998).
6. *Help in identifying issues for further research:* issues that may affect the adaptability of new ideas in the long run are important for the researchers (Franzel et al. 2001).
7. *Make research more applicable:* When farmers provide farms for experimental activities, they allow assessment of diversity in soil types and other variables, making the experimentation more vibrant.
8. *Farmers provide a social aspect in research:* Farmers' involvement provides the human element, which is lacking in labs and stations, therefore making the research more acceptable to common farmers.
9. *Providing oversight:* Some companies and researchers release chemicals and products that may harm the ecological balance. Farmers, therefore, act as whistleblowers when they notice anomalies, especially if they are engaged in ecological indicator farming such as bee farming.

1.3.3 The Consultative Group on International Agricultural Research

Established in 1971 (Stapleton 2022), the Consultative Group on International Agricultural Research (CGIAR) comprises 15 international agricultural research centers dedicated to advancing agricultural productivity and reducing poverty. The CGIAR member centers pursue a variety of research agendas while operating semi-autonomously (Gedil et al. 2014; Rünzel et al. 2021). They focus on specific crops, livestock, fish commodities, natural resource management, and policies.

Over its five-decade existence, CGIAR has achieved remarkable success, particularly during the Green Revolution era, where it significantly enhanced the productivity of rice and wheat farmers in Asia (Alston et al. 2022). However, in recent decades, its impact has expanded to benefit Sub-Saharan Africa, particularly through crops like cassava, potato, and bean. Notably, between 2016 and 2020, CGIAR's agricultural technologies were adopted across vast areas in Asia, Africa, and Latin America, covering approximately 221 million hectares (Fuglie and Echeverria 2024). As a global leader in agricultural science and technology, CGIAR has been instrumental in lifting millions of people out of hunger and poverty, as evidenced by numerous studies highlighting its significant positive impacts on poverty reduction and agricultural development.

In 2021, CGIAR celebrated its momentous 50th anniversary, commemorating decades of pioneering agricultural research and development achievements (CGIAR

Organization System 2021). Operating across 89 countries, CGIAR collaborates with approximately 3000 partners from various sectors, including governments, academia, policy institutes, private companies, and NGOs. CGIAR has released over 770,000 accessions and 60 innovations throughout its history, showcasing its commitment to innovation and impact. A cornerstone of CGIAR's success lies in its work on crop improvement technologies, which has consistently yielded significant (Renkow and Byerlee 2010). However, CGIAR's contributions extend beyond crop improvement technologies. Integrated pest management and natural resource management have also played major roles in achieving impactful outcomes (Renkow and Byerlee 2010).

Despite these successes, we still have pressing global challenges particularly the need to build vibrant food systems amidst climate change. In response to this, CGIAR has set a strategic course for continued impact in the decades ahead in its new strategic plan, CGIAR 2030. The 2030 Research and Innovation Strategy for CGIAR (CGIAR Organization System 2021) emphasizes the importance of a holistic approach to transforming food, land, and water systems. It covers all research areas across CGIAR and explains how CGIAR plans to use its capabilities, resources, and expertise to address major global and regional challenges in collaboration with partners over the next decade. The strategy particularly focuses on supporting the livelihoods of millions of low-income producers and consumers.

1.3.4 The Interconnectedness Between Government, Research, Industry, and Farmers

Research (research institutions and universities), industry, government, and farmers in the agricultural sector are interconnected and have a symbiotic relationship (Fig. 1.4) as listed below:

- Research institutions generate scientific knowledge and technological innovations.
- Industry generates knowledge and develops practical solutions, such as improved seeds, fertilizers, and farming equipment.
- Government provides funding, policy support and regulatory frameworks.
- Farmers implement innovations on the ground, provide feedback to researchers and industry stakeholders, and drive the adoption of new technologies.

This interconnectedness operates as a "triple helix" with the potential to yield better outcomes than any single group working in isolation (Paswan et al. 2022). Research institutions, industry, and government have a distinct focus and mission. Through collaborative efforts, they can effectively supplement and complement one another, filling gaps within each system. This collaboration not only enhances the overall impact of their work but also creates opportunities to harmonize data, streamline processes, and manage resources more efficiently. Such coordinated efforts support *impactful research* and address pressing societal challenges. Individual researchers and students benefit from this collaborative approach, as their projects do not exist in isolation but contribute to and draw from the collaborative strength of this interconnected system.

1.4 Types of Research

Reading through books on research methods, one realizes that there is no universal way of classifying types of research. Types of agricultural research are best described by purpose, and they include the following:

(a) *Basic research*

Basic research is also called fundamental research because it increases understanding of basic principles. The primary purpose is to generate new knowledge, refine theories, and formulate new theories and generalizations. Universities mainly carry out this kind of research. The findings of basic research are not directly applicable in the field. They include studies, such as molecular studies, plant physiology, and crop and animal breeding. Other research that falls within the basic research includes verification and refinement of research methods, developing models, and procedures and tools for use in research.

(b) *Applied research*

Applied research focuses on determining solutions to practical problems facing society or occurring in the field. It uses the basic scientific principles to create solutions. It can easily be called "research beyond the lab" because it aims at advancing results from basic research to address real problems. For example, new knowledge acquired from applied research is used directly to increase agricultural production. This may include research to develop water harvesting techniques to address the ravages of climate change. Other examples include:

- A study to determine optimal rates of nitrogen for the production of maize in Kenya, results to be applied by farmers growing maize.
- A study to determine optimum spacing for newly introduced tissue culture bananas in Region A.

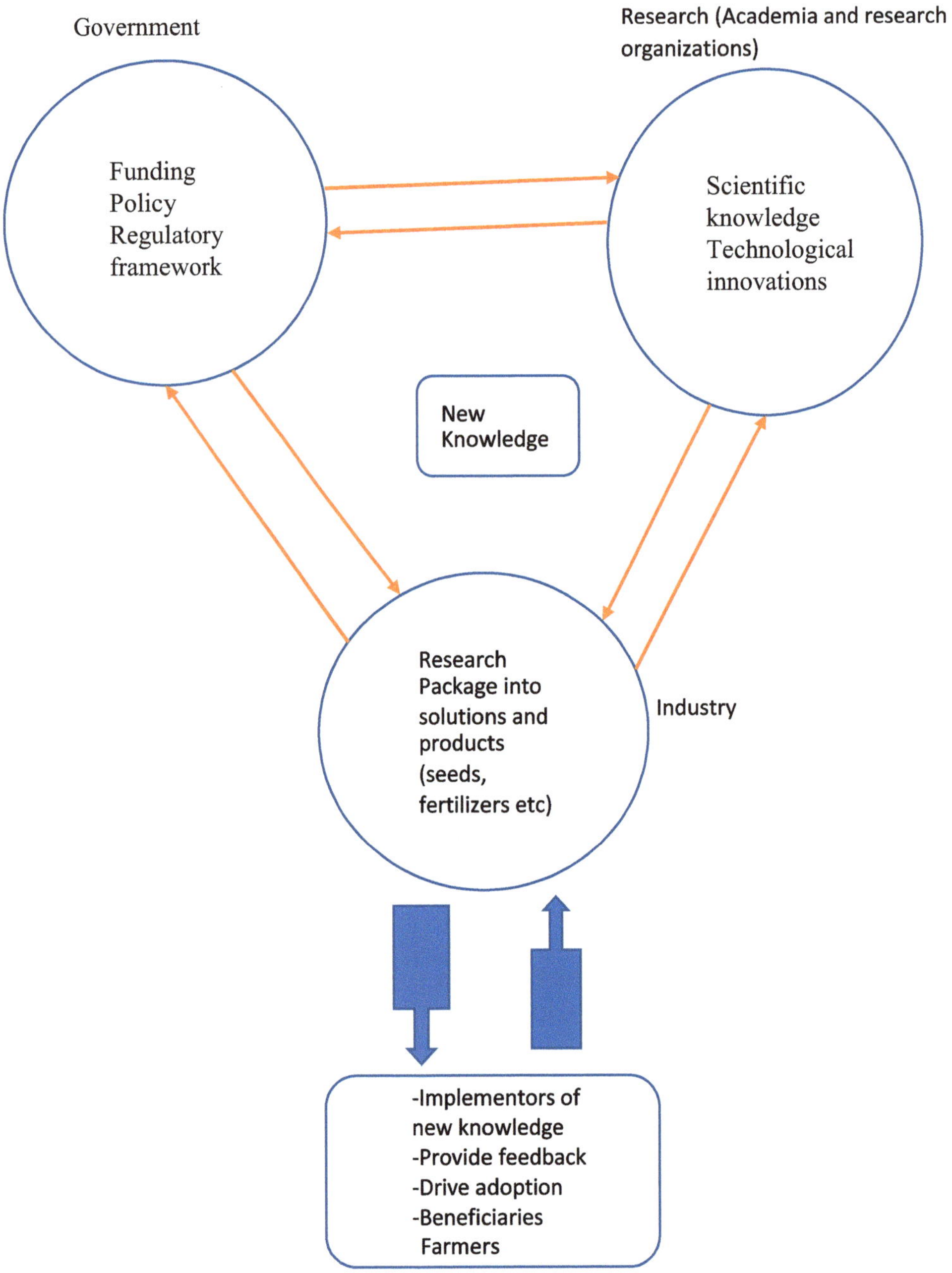

Fig. 1.4 The interconnectedness of governments, research bodies, and farmers

- A study to assess the optimum cutting frequency of fodder species introduced to increase milk production.

(c) *Adaptive research*

Adaptive research is designed to modify or refine existing knowledge or technologies to solve a specific localized problem. It aims to validate existing knowledge to ensure it meets the practical needs of the end-users. For instance, Kenya's climate-smart agricultural project (GOK 2017) aimed to validate existing technologies through testing and adoption in collaboration with farmers. The project evaluated various technologies and practices designed to enhance agricultural resilience in the face of climate change. The research involved a series of field trials with farmers to test which technologies were most effective under farm conditions.

(d) *Participatory action research*

Participatory action research (PAR) is conducted primarily to solve a *specific, immediate and concrete problem in a local setting*. The aim is to involve farmers and stakeholders using an adaptive, flexible, and *participatory approach*. The principle behind action research is that it combines action with research (investigation). The logic is that the process will involve learning (research) that will, in turn, enable action to be carried out more effectively. In other words, the investigation informs action and the researchers and stakeholders learn from critical reflection.

In a typical participatory action research project, the researchers will work with communities, farmers, and other stakeholders from the conception of the research through implementation to dissemination. In this case, it may be limited in its contribution to theory and scientific progress, but useful in bringing change to the community. It may also not produce sound theories, but it provides real answers to farmers' problems. These concepts are covered in Chap. 10 on on-farm experimentation. Here are examples of PAR:

- Farmers work alongside agricultural researchers to identify pest problems and various pest control methods. Through field trials and discussions, they develop integrated pest management strategies that combine biological control, resistant varieties, and cultural practices.
- Farmers and researchers collaborate to introduce and evaluate different legume varieties that can improve soil fertility through nitrogen fixation. They conduct on-farm trials to assess the performance in terms of growth, yields, and soil health improvement. Farmers provide feedback on their experiences, which helps researchers fine-tune recommendations for integrating legumes into the farming systems.

(e) *Research and development research*

Research and development research is two-pronged and mainly refers to a research project. We talk about a research and development project. The purpose of this investigation is both to generate new knowledge (research agenda) and to develop products, processes, or services for use by the consumers of research (development agenda). The main consumers of agricultural products are farmers. Such products include new varieties, technologies, agrochemicals for controlling pests and diseases, and extension methods. For example, a project in the central highlands of Kenya (Mugendi et al. 2011; Mugwe and Otieno 2021) carried out research to develop integrated soil fertility management for growing soybeans but also worked with farmers to promote the best bet practices. The farmers in this project were provided with seeds of improved soybean varieties. The farmers who participated in the project benefited from increased yields.

Research and development activities are generally expensive but beneficial because they result in products that consumers of research and development can use. It is expensive in terms of personnel, funds, and time spent working on the project. Adequate time allocation is therefore necessary to enable benefits to be accrued. For example, products, e.g., varieties or farm implements, are developed according to detailed specifications. Once completed, these products are field tested, mostly with the participation of the farmers, until a specified level of effectiveness is achieved. This process is long and, therefore, makes this type of research expensive but of value.

(f) *Impact evaluation research*

Evaluation research aims to determine the worthiness of resources used in a program. It is defined as a systematic inquiry that is carried out to arrive at an assessment or appraisal of a program, practice, or activity to provide information that will be used in decision-making. In other words, it assesses the overall merit or value of a program or policy at the end of the research-development continuum. Suppose a program to promote dairy breeds to farmers is carried out for three years. At the end of three years, we would like to assess the worthiness of the program by assessing how farmers benefited and if it was cost-effective. Research designs for evaluating impact are discussed in Chap. 11.

(g) *Descriptive research*

Descriptive research reports the way things are (status quo). This type of research attempts to describe such things as possible behavior, attitudes, values, and characteristics. The output of this is a description of characteristics of a particular situation, individual, or group. For example: Describe the behavior of insects before it rains; Describe a specific pest; Describe a particular plant; Describe how farmers are marketing their beans (market chains).

Other descriptive studies include (1) What clay minerals are contained in the soils of County A? (2) What are the characteristics of farmers growing Mangoes in Region Z?; (3) What is the genetic makeup of the thrips attacking French beans in County C, and how do they reproduce? How are the soil profile descriptions of County B? Is the soil profile in a pine forest different from that in a natural forest? How many farmers are growing improved banana varieties in region B?

Other examples of descriptive studies include agricultural value chains and descriptions of plants and animals. The primary purpose of value chain analysis is to describe the value chain and how commodities move through the various segments (producer to consumer). The results are described in terms of maps, trends, and channels showing how the commodities transform.

(h) *Exploratory research*

The purpose of exploratory research is to discover ideas and new insights. The answers and analytics may not offer a final conclusion to the perceived problem. It is conducted to handle new problem areas that have not been explored before. This exploratory process lays the foundation for more conclusive research and data collection. There are several approaches to conducting exploratory research, including reviewing existing literature, interviewing subject matter experts, and conducting in-depth individual or focus group interviews. These methods are typically unstructured, relying heavily on the quality of insights participants provide to inform and shape the next stages of the research.

The key advantage of exploratory research is its flexibility and adaptability, allowing researchers to adjust their focus and methods as new information and perspectives emerge. This adaptability ensures that the research can evolve in response to findings, leading to a deeper understanding of the topic. Example: exploratory work to assess soil and water conservation practices to help design a study on communication, as illustrated in previous studies on climate change and bio-products (Mugi-Ngenga et al. 2016; Sijtsema et al. 2016). The main methods used for collecting data include focus group discussions and key informant interviews.

(i) *Explanatory research*

Explanatory research aims to explore and clarify the relationships and causal connections between variables. Its research aims to establish cause-and-effect relationships and seeks to provide a deeper understanding of the underlying factors that influence a particular phenomenon or behavior. Explanatory research therefore goes beyond describing and identifying associations; it aims to uncover the mechanisms, processes, or influences that explain why certain outcomes occur. It often involves experimentation, testing hypotheses and examining the reasons behind observed correlations. The ultimate goal is to enhance comprehension and contribute to the development of more comprehensive theories or explanations.

(j) *Analytical research*

Analytical research aims to go beyond description to the identification of patterns and trends and relationships. Statistical analysis of data and other analytical tools are used to examine the data.

(k) *Correlational research*

Correlational research is also called associational research. It aims to establish the relationship between two or more variables without any attempt to influence them. This involves collecting data to determine whether and to what extent a relationship exists between two or more quantifiable variables. It will also give, in quantitative terms, the degree to which variables are related by providing a correlation coefficient between one quantitative and another quantitative.

Examples:

1. Is there a relationship between the age (yrs) of the household head and farm size (ha)?
2. Is there a relationship between the number of trees and farm size in hectares?
3. Is there a relationship between income levels (USD) and household expenditure (USD)?

(l) *Comparative research*

The purpose of comparative research is to analyze similarities and differences between two or more entities to gain insights into their characteristics and behaviors. It is closely related to causal-comparative, where two entities are selected. In comparative research, researchers systematically examine various aspects of these entities, seeking to identify patterns, trends, and relationships. This approach provides a comprehensive understanding of how different variables may influence outcomes, helping researchers draw meaningful comparisons. For instance, researchers could compare the adoption rates and outcomes of certain farming practices, revealing patterns and factors influencing the success or challenges of these practices.

(m) *Predictive research*

The primary purpose of predictive research is to make predictions or forecasts about future events, trends, or behaviors. Predictive research leverages statistical and computational techniques to anticipate future events. For instance, we can use historical data, such as weather patterns, soil characteristics, and crop yields, to develop models or algorithms that predict future agricultural outcomes, such as yields.

(n) *Operations research*

The goal of operations research is to improve efficiency, maximize performance, and make informed decisions within organizations. It uses mathematical and analytical methods to optimize decision-making and problem-solving in complex organizational systems. It is sometimes called management science. Operations research encompasses mathematical modeling, statistical analysis, optimization algorithms, simulations, and computer programming. This topic is crucial to students taking agricultural economics.

(o) *Experimental research and nonexperimental research*

Experimental and nonexperimental research is a broad categorization consisting of many studies (Patten and Newhart 2017, 2018). One of the most important distinguishing characteristics is that experimental research involves the manipulation of variables, while nonexperimental research observes variables naturally.

An experiment is designed to test a cause-and-effect relationship and is characterized by intentional alteration of the independent variables in a controlled environment. On the other hand, nonexperimental research involves observing and analyzing variables as they naturally occur without manipulating them. It is particularly useful for studying variables that cannot be manipulated due to ethical or practical reasons. It involves observational studies, descriptive studies, exploratory studies, cross-sectional surveys, and many others. As you will see in later chapters of this book, all research types are broadly categorized into either experimental or nonexperimental. Table 1.4 summarizes the common types of research classified according to purpose.

Broad Paradigms in Research

Broad paradigms of inquiry in research are quantitative, qualitative, and mixed methods approaches. They have also been referred to as strategies or approaches of inquiry (Creswell 2014; Creswell and Guetterman 2019; Creswell and Creswell 2022) and research methodologies (Kothari and Garg 2019):

Table 1.4 Summary of common types of research according to purpose

Type of research	Purpose	Examples	Research question
Basic	To generate new knowledge, formulate new theories and generalizations	Molecular breeding studies	What is the genetic makeup of thorn melon?
Applied	To determine solutions to practical problems facing a society	Experiments to develop drought-resistant crop varieties through breeding programs	Which crop varieties exhibit the highest levels of drought resistance under controlled experimental conditions?
Adaptive	Modify existing knowledge or technologies to solve a specific local problem	Evaluate the performance of 12 varieties of newly released common bean varieties under farm conditions (On-farm trials with farmers)	Are there differences in performance of the newly released common bean varieties under on-farm conditions
Participatory Action Research	Solve a specific, immediate and concrete problem in a local setting involving stakeholders	A project evaluating and promoting best-bet soybean varieties	What are farmers' views on the benefits and disadvantages of the introduced best-bet soybean varieties?
Impact evaluation	To assess the effect, result or outcomes of technology, program, on a certain aspect expected to change; assess worthiness and value to enable decision-making	To assess the effect of promoting dairy breeds to farmers on their adoption and milk yields	How does promoting dairy breeds affect adoption levels and milk yields among households in Region A?
Descriptive	To describe the nature, characteristics, components of a situation, individual, or phenomenon	Study on perceptions Study on characterizing households Study on value chain analysis	What is the perception of farmers about the control of mango fruit flies using chemicals? What are the characteristics of households who engage in collective marketing of bananas? How does banana move through the various market segments? (producers to consumers)
Exploratory	To explore new problem areas which have not been studied before	Exploratory work to assess soil and water conservation practices that exist; to help design a study on communication	What are some of the indicators used by farmers to forecast start of rains?
Explanatory	To establish cause-and-effect relationships; explain relationship between two or more variables	Experiment conducted to assess the response of maize to different levels of fertilizer	What is the effect of different nitrogen fertilizer levels on maize yields
Correlational	To determine the relationship between two or more quantitative variables without necessarily probing into the causal reasons underlying them	Relationship between: 1. farm size and number of trees;	What is the relationship between farm size and number of trees?
Predictive	To determine the relationship between independent and dependent variables, predict the dependent variable from the independent variables	Prediction of maize grain yields from given levels of nitrogen Prediction of factors that influence adoption of certain technologies	What is the response of maize yields to the application of different levels of fertilizer? What socio-economic variables/factors influence adoption of certain technologies
Comparative (true experiment)	To compare performance groups in a natural way of two or three variables or interventions on the dependent variable	Effect of four nitrogen levels on the yield of maize	Is there a difference among the fertilizer N levels

(i) Quantitative research paradigm

The quantitative research paradigm is focused on quantifying variables and analyzing numerical data to uncover patterns. It collects data to describe, explain, and predict control phenomena of interest. It is a means for testing theories by examining relationships between variables. These variables, in turn, can be measured so that numerical data can be analyzed using statistical procedures. Random sampling is done to ensure the sample is representative.

This paradigm is particularly advantageous when seeking to establish causal relationships, identify patterns, and quantify trends. Some of the research designs which may be categorized as quantitative research are experimental designs, quasi-designs, survey research, and correlational research (Chaps. 7, 8, and 9).

(ii) Qualitative research paradigm

The qualitative paradigm focuses on understanding complex phenomena through detailed, non-numerical data, such as statements and words. It emphasizes exploring individuals' experiences, perceptions, and social contexts. Unlike quantitative research, which relies on statistical analysis, qualitative research aims for deeper insights into human behavior, including attitudes and emotions that are difficult to observe directly. Qualitative research is advantageous because it permits research to go beyond the statistical results usually reported in quantitative research (Discussed in detail in Chaps. 13 and 14).

(iii) Mixed methods research

Mixed methods research is an inquiry approach that combines quantitative and qualitative forms. Both approaches are used in tandem so that a study's overall strength is greater than qualitative or quantitative research. An example is the use of questionnaires to collect data and complement such data with focus group discussions or Participatory Rural Appraisal (PRA) approaches.

Additional Reading Materials

1. Byerlee D, Lynam JK (2020) The development of the international center model for agricultural research: A prehistory of the CGIAR. World Dev 135:105080. https://doi.org/10.1016/J.WORLDDEV.2020.105080
2. Creswell WJ, Creswell DJ (2022) Research Design, 6th edn. SAGE Publications Ltd.
3. Patel BK, Muir-Leresche K, Coe R, Hainsworth SD (2004) Green Book A Guide to Effective Graduate Research in African Agriculture, Environment and Rural Development.
4. Patten ML, Newhart M (2018) Experimental and Nonexperimental Studies. In: Understanding Research Methods, 10th edn. Routledge, pp 12–15.
5. Patten ML, Newhart M (2017) Understanding research methods: An overview of the essentials, tenth edition. Taylor and Francis.

Key Highlights of This Chapter

This introductory chapter provides a general introduction and basic principles of the scientific method. It also discusses the nature of agricultural research. Most of these will be revisited in later chapters with detailed discussions.

What Is Research?

Nonprofessionals use the term research to represent information aggregation and internet searches, but such information is neither reproducible nor entirely true. Research is a systematic investigation, evidence-based and motivated by the need to provide a solution. "Agricultural Research" is a controlled, "*systematic inquiry to generate new information to solve problems in agriculture*."

The Scientific Method of Research

Scientific research aims to generate reliable information through the scientific method, which involves a systematic approach to problem-solving. Researchers design and conduct studies to address issues and obtain answers.

Characteristics of the Scientific Method of Research

Scientific research is about carrying out a diligent inquiry or a critical examination with the aim of generating new information that is authentic and reliable. Attributes of scientific-based research include the following:

1. Evidence-based
2. Truthful
3. Public good

4. Generates new information
5. Purposive
6. Systematic
7. Empirical
8. Requires expertise
9. Replicable
10. Control

Core Aspects of Agricultural Research

Importance of agricultural research: The main goal of agricultural research is to improve productivity and address challenges in agriculture. The primary purpose is to:

1. Discover new knowledge by asking questions and answering them.
2. Create new methods and practices.
3. Describe phenomena or achieve new insights into them.
4. Enable prediction by establishing relationships that predict future outcomes.
5. Enable control by manipulating variables.
6. Contribute to theory development.
7. Confirm or validate existing theories.
8. Understand the meanings behind practices, beliefs, and experiences.
9. Develop new research methodologies and tools.
10. Determine the frequency of occurrence.

Concept of research-development continuum: The research-development cycle translates scientific knowledge into practical solutions that benefit society. The framework starts from scientific discovery through a series of activities that end with dissemination and upscaling to create impact.

Biophysical and social aspects: Agricultural research involves biophysical and social sciences. Though they use different research approaches, they aim to advance knowledge and understanding within their respective fields through systematic inquiry; therefore, both involve scientific observations.

Wholeness-oriented research: Emphasizes the interconnectedness, integration, and holistic nature of agricultural research and includes aspects of sustainability science. The aim is to produce food efficiently while ensuring environmental sustainability.

Emerging trends in agriculture: Aagricultural research in the twenty-first century is marked by technological innovation, sustainability, and inclusivity. While developed nations produce sufficient food due to technological advancements, some developing nations still struggle to produce sufficient food due to low productivity and inadequate technologies.

Who Are the Stakeholders in Agricultural Research

Important stakeholders are those who conduct research, beneficiaries, and those providing funds. They include:

1. Governments: support research by providing funding, establishing policies and regulations, and promoting collaboration.
2. International Agricultural Research Centres (IARCs): develop technologies with wide applications.
3. National Agricultural Research Institutions focus on research areas that directly affect their countries.
4. Universities: conduct research either individually or in collaboration.
5. Donor Agencies: provide funds for agricultural research.
6. Nongovernmental Organizations: have initiatives for research and development.
7. Industry players: like Syngenta, Monsanto, and Cargill, fund agricultural research, driving innovation, and investing in new seeds and pesticides.
8. Farmers play a crucial role in agricultural research by using research findings and providing practical knowledge.
9. The Food and Agricultural Organization (FAO), a United Nations body, supports global research, offers technical expertise, and develops standards and policies influencing agricultural research.

Role of Farmers

Farmers are an important stakeholder in research. Farmers play key roles in identifying problems, managing experiments, providing research materials, implementing new ideas, offering feedback, identifying issues for further research, making research more applicable, providing a social aspect in research and providing oversight to prevent ecological harm.

The Consultative Group on International Agricultural Research

The Consultative Group on International Agricultural Research (CGIAR) comprises 15 international agricultural research centers. Over its five-decade existence, CGIAR has achieved remarkable success in raising agricultural productivity. In 2021, CGIAR celebrated its momentous 50th anniversary, commemorating decades of pioneering agricultural research and development achievements (CGIAR@50). A cornerstone of CGIAR's success lies in its work on crop improvement technologies, integrated pest management, and natural resource management.

Types of Research

Types of agricultural research are best categorized by purpose. They include the following:

1. Basic research.
2. Applied research.
3. Adaptive research.
4. Participatory action research.
5. Research and development.
6. Impact evaluation research.
7. Descriptive research.
8. Exploratory research.
9. Analytical research.
10. Comparative research.
11. Operations research.
12. Experimental and nonexperimental research.
13. Qualitative research.
14. Quantitative research.

Review Questions

1. Compare and contrast scientific and nonscientific sources of knowledge. Why is scientific research considered more reliable?
2. Discuss the role of farmers as key stakeholders in agricultural research. Why is their involvement crucial?
3. Identify the main barriers to agricultural productivity in developing countries and explain how research can help address these challenges.
4. What is the significance of the research-development continuum in agricultural innovation?
5. Describe the impact of CGIAR on global agricultural development. What makes its approach effective?
6. Differentiate among basic, applied, and participatory agricultural research, providing an example of each.

Chapter Exercise

1. Choose one of the Sustainable Development Goals (SDGs) and explain how agricultural research supports its achievement, providing real-world examples.
2. Write a reflection on your personal understanding of what "research" meant before reading this chapter, compared to now. How has your perspective changed?

Answers to Review Questions

- Question 1. References 1.1.1 and 1.1.2.
- Question 2: Tip; farmers are contributors and beneficiaries. Reference 1.3.1 and 1.3.2.
- Question 3: Common barriers include poor soil fertility, lack of access to improved seeds and technologies, inadequate extension services, climate variability, and limited financial resources (Reference: 1.2.5).
- Question 4: Reference: 1.2.2.
- Question 5: CGIAR has significantly contributed to increased agricultural productivity, food security, and poverty reduction globally. Its impact includes the development of high-yielding crop varieties, improved farming practices, and sustainable natural resource management. (Reference: 1.3.3 and Table 1.3).
- CGIAR's effectiveness lies in its global network of research centers and interdisciplinary approach (Reference: 1.3.3)
- Question 6: Reference: 1.4.

References

Abramo G, D'Angelo CA, Di CF (2011) University-industry research collaboration: a model to assess university capability. High Educ (Dordr) 62:163–181. https://doi.org/10.1007/S10734-010-9372-0/METRICS

Adolwa I, Oberthür T, Cook S (2022) Towards a farmer-centric framework for scaling productive and sustainable cereal cropping systems. Grow Africa 1. https://doi.org/10.55693/GA11.GKDE6695

AGRA (2022) Empowering Africa's food systems for the future

Alrøe HF, Kristensen ES (2002) Towards a systemic research methodology in agriculture: rethinking the role of values in science. Agric Human Values 19:3–23. https://doi.org/10.1023/A:1015040009300/METRICS

Alston JM, Pardey GP, Rao X (2022) Payoffs to a half century of CGIAR research. Am J Agric Econ 104:502–529. https://doi.org/10.1111/AJAE.12255

Barrett JL (2007) Is the spell really broken? Biopsychological explanations of religion and theistic belief. Theol Sci 5:57–72. https://doi.org/10.1080/14746700601159564

Bhattacherjee A (2012) Social science research principles, methods and practices, 2nd edn. University of Southern Queensland, Toowoomba

Bluemke M, Jong J, Grevenstein D et al (2016) Measuring cross-cultural supernatural beliefs with self- and Peer-reports. PLoS One 11:e0164291. https://doi.org/10.1371/JOURNAL.PONE.0164291

Boulton G (2022) Science as a global public good: the roles of the representative bodies of science—a perspective. Proc Indian Natl Sci Acad 88:832–841. https://doi.org/10.1007/S43538-022-00125-X/METRICS

Brazil K, Ozer E, Cloutier MM et al (2005) From theory to practice: improving the impact of health services research. BMC Health Serv Res 5:1–5. https://doi.org/10.1186/1472-6963-5-1/PEER-REVIEW

Byerlee D, Lynam JK (2020) The development of the international center model for agricultural research: a prehistory of the CGIAR. World Dev 135:105080. https://doi.org/10.1016/J.WORLDDEV.2020.105080

Campos JJ, Pfister NT (2023) Scientific method. Transl Radiat Oncol:11–14. https://doi.org/10.1016/B978-0-323-88423-5.00016-9

CGIAR Organization System (2021) CGIAR 2030 research and innovation strategy: transforming food, land, and water systems in a climate crisis

Creswell JW (2014) Research and design qualitative, quantitative, and mixed methods approaches. SAGE Publication Inc

Creswell WJ, Creswell DJ (2022) Research design, 6th edn. SAGE Publications Ltd

Creswell JW, Guetterman TC (2019) Educational research_ planning, conducting, and evaluating quantitative and qualitative research. Pearson. pp 1–681, Upper Saddle River, New Jersey, U.S.A.

Dearing JW (2009) Applying diffusion of innovation theory to intervention development. Res Soc Work Pract 19:503–518. https://doi.org/10.1177/1049731509335569

Díaz Andrade A, Tarafdar M, Davison RM et al (2023) The importance of theory at the Information Systems Journal. Inf Syst J 33:693–702. https://doi.org/10.1111/ISJ.12437

Dubova M, Chandramouli S, Gigerenzer GI et al (2025) Is Ockham's razor losing its edge? New perspectives on the principle of model parsimony. Proc Natl Acad Sci 122:e2401230121. https://doi.org/10.1073/PNAS.2401230121

European Commission (2015) Open innovation, open science, open to the world - A vision for Europe

Evenson RE, Gollin D (2003) Assessing the impact of the Green Revolution, 1960 to 2000. Science (1979) 300:758–762. https://doi.org/10.1126/SCIENCE.1078710

Franzel S, Coe R, Cooper P et al (2001) Assessing the adoption potential of agroforestry practices in sub-Saharan Africa. Agric Syst 69:37–62. https://doi.org/10.1016/S0308-521X(01)00017-8

Fuglie KO, Echeverria RG (2024) The economic impact of CGIAR-related crop technologies on agricultural productivity in developing countries, 1961–2020. World Dev 176:106523. https://doi.org/10.1016/J.WORLDDEV.2023.106523

Fuglie KO, Wang SL, Ball VE (2012) Productivity growth in agriculture: an international perspective. CABI Publishing

Galt RE, Pinzón N, Robinson NI, Baukloh Coronil MB (2024) Agroecology and the social sciences: a half-century systematic review. Agric Syst 216:103881. https://doi.org/10.1016/J.AGSY.2024.103881

Gedil M, Tripathi L, Ghislain M et al (2014) Biotechnology success stories by the Consultative Group on International Agriculture Research (CGIAR). System:95–114. https://doi.org/10.1007/978-3-319-04001-1_6

Gemtou M, Guillén BC, Anastasiou E, et al (2024) Smart farming technologies and sustainability. pp 99–120. https://doi.org/10.1007/978-3-031-61749-2_6

Giles EL, Harrison SL (2023) Reflecting on the importance of theory-informed qualitative research in people with chronic respiratory disease and their carers. Chron Respir Dis 20. https://doi.org/10.1177/14799731231185128

GOK (2017) Kenya climate smart agriculture strategy-2017-2026, Nairobi

Haig BD (2000) Statistical significance testing, hypothetico-deductive method, and theory evaluation. Behav Brain Sci 23:292–293. https://doi.org/10.1017/S0140525X00262444

Hill N (1937) Think and grow rich. Hyperink, San Francisco

Kerlinger NF (1973) Foundations of behavioral research, 3rd edn. Holt, Rinehart and Winston

Kothari CR, Garg G (2019) Research methodology: methods and techniques. New Age International, New Delhi

Lacoste M, Cook S, McNee M et al (2021) On-Farm Experimentation to transform global agriculture. Nat Food 3:11–18. https://doi.org/10.1038/s43016-021-00424-4

Lee J, Kang D, Lee HY, Kim JW (2023) The effects of authoritative source cue and argument strength of correction tweets on MMR vaccine-related misinformation credibility. Health Educ J 82:935–951. https://doi.org/10.1177/00178969231210215/SUPPL_FILE/SJ-DOCX-1-HEJ-10.1177_00178969231210215.DOCX

McFadden J (2023) Razor sharp: the role of Occam's razor in science. Ann N Y Acad Sci 1530:8–17. https://doi.org/10.1111/NYAS.15086

Mohapatra U, Wali VS, Swetalina M, Naik VS (2019) Agricultural research and development in the developing world : too little, too late? Agric Update 14:90–98. https://doi.org/10.15740/HAS/AU/14.1/90-98

Moseley SF (2004) Everett rogers' diffusion of innovations theory: its utility and value in public health. J Health Commun 9:149–151. https://doi.org/10.1080/10810730490271601/ASSET//CMS/ASSET/CDE1A7BF-8EBC-42F9-8A36-BB8BB0FFAB87/10810730490271601.FP.PNG

Mugendi DN, Mugwe J, Mucheru-Muna M et al (2011) Dissemination of integrated soil fertility management technologies using participatory approaches in the central highlands of Kenya. In: Innovations as key to the Green Revolution in Africa, p 1279–1296. https://doi.org/10.1007/978-90-481-2543-2_130

Mugi-Ngenga EW, Mucheru-Muna MW, Mugwe JN et al (2016) Household's socio-economic factors influencing the level of adaptation to climate variability in the dry zones of Eastern Kenya. J Rural Stud 43:49–60. https://doi.org/10.1016/J.JRURSTUD.2015.11.004

Mugwe J, Otieno EO (2021) Integrated soil fertility management approaches for climate change adaptation, mitigation, and enhanced crop productivity. In: Handbook of climate change management, pp 1–22. https://doi.org/10.1007/978-3-030-22759-3_325-1

Ng W, Minasny B, Jeon SH, McBratney A (2022) Mid-infrared spectroscopy for accurate measurement of an extensive set of soil properties for assessing soil functions. Soil Secur 6:100043. https://doi.org/10.1016/J.SOISEC.2022.100043

Norman DW, Bloomquist LE, Freyenberger SG et al (1998) Farmers attitudes concerning on-farm research: Kansas survey results. J Nat Resour Life Sci Educ 27:35–41. https://doi.org/10.2134/JNRLSE.1998.0035

Okalebo JR, Othieno CO, Ochuodho JO, et al (2012) University Outreach Support to Farmer Associations in Western Kenya: the case of The RUFORUM's Community Action Research Project (CARP) at Moi University. Ruforum

Paswan J, Singh VK, Karmakar M, Singh P (2022) Does university–industry–government collaboration in research gets higher citation and altmetric impact? A case study from India. Scientometrics 127:6063–6082. https://doi.org/10.1007/S11192-022-04508-1/METRICS

Patel BK, Muir-Leresche K, Coe R, Hainsworth SD (2004) Green book a guide to effective graduate research in African agriculture, environment and rural development. African Crop Science Society, Kampala

Patten ML, Newhart M (2017) Understanding research methods: an overview of the essentials, 10th edn. Taylor and Francis

Patten ML, Newhart M (2018) Experimental and nonexperimental studies. In: Understanding research methods, 10th edn. Routledge, pp 12–15

Payton OD (1979) Research, the validation of clinical practice. F.A. Davis Co

Pingali PL (2012) Green revolution: impacts, limits, and the path ahead. Proc Natl Acad Sci 109:12302–12308. https://doi.org/10.1073/PNAS.0912953109
Raviv A, Bar-Tal D, Raviv A, Peleg D (1990) Perception of epistemic authorities by children and adolescents. J Youth Adolesc 19:495–510. https://doi.org/10.1007/BF01537477/METRICS
Renkow M, Byerlee D (2010) The impacts of CGIAR research: a review of recent evidence. Food Policy 35:391–402. https://doi.org/10.1016/J.FOODPOL.2010.04.006
Rindermann H, Falkenhayn L, Baumeister AEE (2014) Cognitive ability and epistemic rationality: a study in Nigeria and Germany. Intelligence 47:23–33. https://doi.org/10.1016/J.INTELL.2014.08.006
Rünzel M, Sarfatti P, Negroustoueva S (2021) Evaluating quality of science in CGIAR research programs: use of bibliometrics. Outlook Agric 50:130–140. https://doi.org/10.1177/00307270211024271/ASSET/E2A1AC73-A6F5-4BC1-8627-C74A99421300/ASSETS/IMAGES/LARGE/10.1177_00307270211024271-FIG6.JPG
Saunders MNK, Philip L, Adrian T (2019) Research methods for business students. p 833
Schwartz SH (2012) An overview of the Schwartz theory of basic values. Theoretical and Methodological Issues Issue 1 Conceptual Issues in Psychology and Culture Article 2:12–2012. https://doi.org/10.9707/2307-0919.1116
Sijtsema SJ, Onwezen MC, Reinders MJ et al (2016) Consumer perception of bio-based products—an exploratory study in 5 European countries. NJAS - Wageningen J Life Sci 77:61–69. https://doi.org/10.1016/J.NJAS.2016.03.007
Stapleton SC (2022) More than the sum of its parts: the Consultative Group on International Agricultural Research, CGIAR (https://www.cgiar.org/). J Agric Food Inf 23:50–58. https://doi.org/10.1080/10496505.2023.2225247/ASSET//CMS/ASSET/641D2BCE-E8EE-49E8-AE94-DF54DB30FE41/10496505.2023.2225247.FP.PNG
Trendov NM, Varas S, Zeng M (2019) Digital technologies in agriculture and rural areas. FAO
UNDES (2017) World population prospects: the 2017 revision. United Nations
UNESCO (2021) UNESCO recommendation on open science. https://doi.org/10.54677/MNMH8546
UNFAO (2021) Tracking progress on food and agriculture-related SDG indicators 2021: a report on the indicators under FAO custodianship. https://doi.org/10.4060/CB6872EN
Van Eyghen H (2019) Is supernatural belief unreliably formed? Int J Philos Relig 85:125–148. https://doi.org/10.1007/S11153-018-9671-4/METRICS
Van Eyghen H (2022) From Theism to spirit beliefs. Religion 13:460. https://doi.org/10.3390/REL13050460
Voit EO (2019) Perspective: dimensions of the scientific method. PLoS Comput Biol 15:e1007279. https://doi.org/10.1371/JOURNAL.PCBI.1007279
Waltz CF, Bausell RB (1981) Nursing research: design, statistics, and computer analysis. F.A. Davis Company
Wilken JP, Cabiati E (2024) Foundations of experiential knowledge: recovering a forgotten resource for social work. Eur Soc Work Res 2:224–242. https://doi.org/10.1332/27551768Y2024D000000022
Wilkinson E (2025) Logical Rationalism. J Philos Log 54:1–22. https://doi.org/10.1007/S10992-025-09796-5/METRICS

The Research Process, Key Concepts, and Terminologies

2

Abstract

This chapter serves as a cornerstone for understanding the research process and introduces some key terminologies employed in research. The research process is a journey comprising a series of interconnected steps that lead to knowledge generation. It includes an eight-stage process consisting of a preparatory and implementation phase, with a literature review being carried out throughout the research journey. The literature review comprises three main phases: initial phase in developing a research idea, a mid-phase during planning and implementation, and a final phase during reporting. Your department will allocate academic supervisors to guide, coach, and mentor you through the research process, but students carry almost 80% responsibility for the success of their research project. Maintaining a good relationship with your supervisors is extremely beneficial. It helps to make your research journey smooth and productive. Many students fail to complete their research projects on time, and others halt the process. Reasons for the failure include inadequate research skills, time constraints, inadequate institutional support, and limited infrastructure and funding. Students can address these constraints by investing in building research skills, networking, and time management to maintain a work-life balance. Total research setting or visioning the research journey from the start boosts motivation. Further, a clear plan, reflective diary, consistent documentation, and visualizing success help maintain focus and progress. A solid understanding of research concepts and terminologies is essential for developing research skills. Key concepts include constructs, variables (independent, dependent, extraneous, categorical) and their control to ensure validity (truthfulness) and reliability (consistency). Theories, built through cumulative research, explain, predict, and deepen understanding of phenomena. A good theory is testable, relevant, predictive, parsimonious, accurate, and comprehensive. While related, laws, propositions, and models differ from theories in scope and function.

Keywords

Academic supervisor · Academic research project · Concept · Construct · Deductive · Extraneous variables · Independent variables · Dependent variables · Implementation phase · Inductive · Planning phase · Theory · Total research setting · Reliability · Research problem · Validity · Variables

Learning Outcomes

By the end of this chapter, you are expected to be able to:

1. Summarize the “preparatory” and “implementation” phases in the research process.
2. Describe the role of your supervisor in the research journey.
3. Outline strategies for enhancing the completion of your research project.
4. Appreciate the concept of total research setting.
5. Explain the difference between concept, construct, and variables.
6. Describe the various types of variables.
7. Summarize the concept of extraneous variables and how to control them.
8. Explain the meaning of validity and reliability.
9. Appreciate the importance of theory in research.
10. Determine the characteristics of a good theory.
11. Apply the concept of theory testing.
12. Contrast theory, law, and model.

J. Njeri Mugwe, S. Runo, *Research Methodology in Agricultural Sciences*, https://doi.org/10.1007/978-981-95-1892-0_2

2.1 The Research Process

2.1.1 Stages of the Research Journey

In Chap. 1, we emphasized that research is a systematic and evidence-based process that generates knowledge through a process of data collection. Practically, the research process comprises a series of interconnected steps. The process has a beginning and an end. It is like a *journey* with a starting point and a destination. Between the start and the destination is a series of stages that allow the researcher to carry out the research study systematically. Various authors have described different numbers of steps in this research journey. For instance, Creswell presents six steps (Creswell 2014), Kothari describes seven steps (Kothari and Garg 2019) and Kumar presents eight steps (Kumar 2018). In this book, we use an eight-step research process.

In this eight-stage research process, the first five stages involve a *preparatory phase*, while the last three involve an *implementation phase*. The preparatory phase mainly involves planning the research, while the implementation phase involves conducting the study (Table 2.1).

In the *preparatory phase* of a research study, two critical questions guide researchers in their work. The first question is, *"What is the goal of your research?"* This involves identifying specific phenomena or topic to investigate, clarifying the research problem, and formulating clear research objectives/questions or hypotheses (Discussed in the next Chapter). To answer this question, researchers review existing literature to *identify gaps, controversies or areas needing further investigation*. The second question is, *"How will you conduct your research?"* This requires careful planning and methodological considerations. Researchers evaluate various data collection methods and select those best suited to address their research questions while ensuring methodological rigor.

During the *implementation phase* of the research process, researchers put their plans into action. A key question you must answer is, *"How will you find out?"* This requires careful consideration of the methods and techniques you will use to collect data, such as experiments, surveys, interviews, observations, etc. (Patten and Newhart 2017). Researchers must also handle logistical aspects like scheduling data collection sessions, acquiring resources, and coordinating with participants or collaborators. Ethical considerations are crucial during this phase as researchers must prioritize research participants' well-being, rights, and dignity. The topic of ethical considerations is discussed in Chap. 6. The implementation phase will also answer the question, *What did you find out?*". Here, an answer is provided to the question raised at the start of the research journey during the preparatory phase. The research will be considered incomplete and unworthy if this answer is not provided.

Like many other books on research methods, this book presents the research process as a series of linearly linked steps. In reality, however, the process is not straightforward or linear. It involves an *iterative process* that requires you to return to each stage more than once (cyclic). The back-and-

Table 2.1 The eight stages in the research process

No	Steps	Questions you ask/activities	Phase
1	Identifying the broad problem area (Topic)	What is the broad topic area?	Preparatory phase
2	Selection and formulation of the research problem	What is the research problem? Why should it be studied? (Justification) Who will benefit from the research?	
3	Formulation of research questions, objectives, and hypothesis	What do you hope to achieve? Why are you carrying out the study? What are the outcomes?	
4	Designing research	Where will the study be carried out (research site) What research design? (experimental, nonexperimental, mixed methods) How will data be collected? What instruments will be used to collect the data?	
5	Developing a research proposal	What is your intended plan of research?	
6	Collection of data	How will you translate the plan into action? How will data be collected, recorded, stored, and managed Setting up experiments and taking measurements Collecting data from sample participants	Implementation phase
7	Data management, analysis, and interpretation	Manage the data to ensure quality? Analyze data. What does the output of analysis mean? What did you find out?	
8	Reporting	How will you report your findings? Report the findings in different fora (thesis, scientific papers, conferences, etc.)	

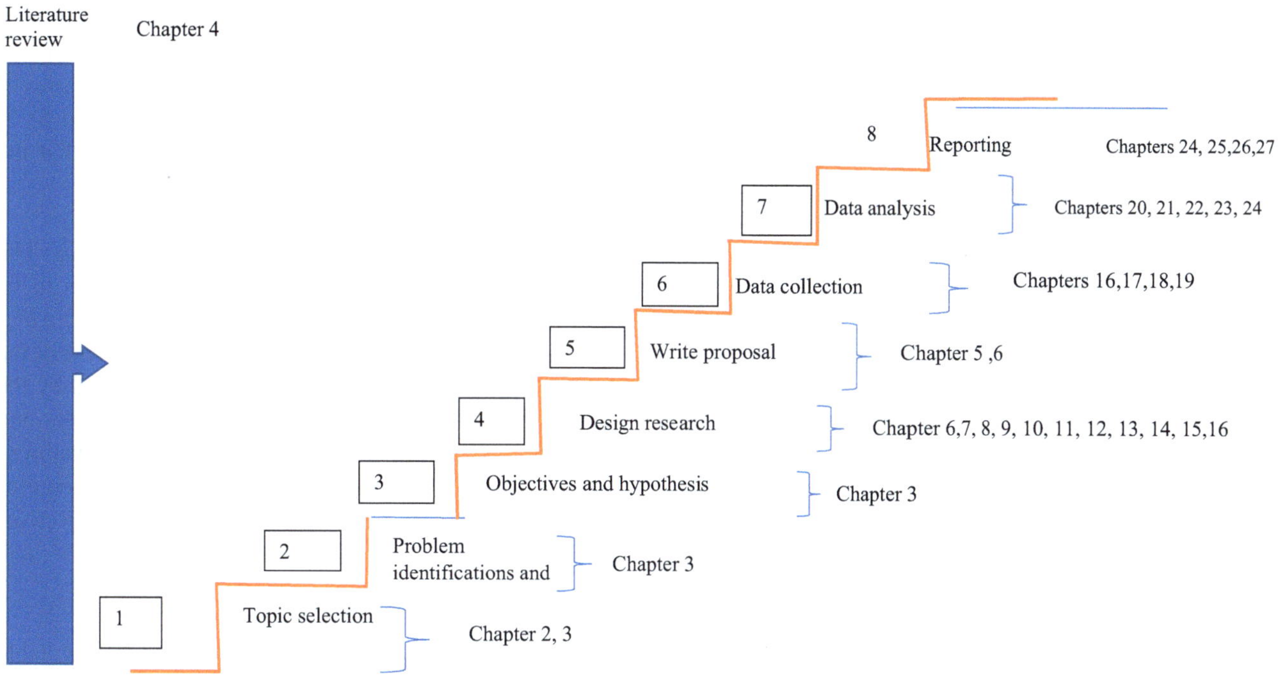

Fig. 2.1 The eight-step research process and the flow of the chapters in the book

forth exercise demands reflection and critical thinking. It also involves substantial rewriting. This continual reflection and rewriting enhances your research outcomes' overall robustness and quality (revisited later in this chapter).

This book simplifies the research process by arranging the chapters in stepwise sequential order from topic selection to reporting (Fig. 2.1). The figure displays how the chapters fit into the research process and can be used to *guide your reading of this book*. Additionally, the chapters are organized into six parts, further enhancing targeted reading.

Parts I and II, consisting of Chaps. 1, 2, 3, 4, 5, and 6, are recommended when you are laying the foundation for understanding concepts, principles and the nature of agricultural research. Parts III and IV, containing most chapters in this book (Chaps. 7, 8, 9, 10, 11, 12, 13, 14, 15, 16, 17, 18, and 19), are extremely important when designing the study, preparing data collection tools and collecting the data. Part V, with four Chaps. 20, 21, 22, and 23, is suitable when at the stage of analyzing data. Finally, Part VI, consisting of 4 chapters, is highly relevant when you are reporting and disseminating your research findings using written and oral fora. This structured approach promotes a cohesive and comprehensive understanding of the research process.

The Preparatory Phase

An academic research undertaking requires careful planning. This is because the stages involved in the research process are not mutually distinct. The steps tend to be cyclic, involving back-and-forth processes.

Table 2.2 Examples of topics in various disciplines

Discipline	Examples of topic areas
Animal sciences	Animal nutrition, animal health, forage production, drugs, and vaccines
Soil Science	Mineral fertilizers, soil micro-organisms, soil conservation, organic fertilizers
Agricultural economics	Marketing of legumes, adoption, policy issues, regulatory frameworks, value chain analysis, impact evaluation
Agronomy	Intercropping patterns, spacing, tillage practices and fertilizer rates
Breeding	Genomic approaches, disease resistance, drought-tolerant varieties
Horticulture	Avocado production, tomato production, marketing of French beans, etc.
Biotechnology	Biofortification, bioremediation, nanobiotechnology
Crop protection	Fall armyworm, push and pull, integrated pest management

Step 1: Selecting the Topic Area The topic is the subject matter of the study you want to conduct. This falls at the planning stage and is the first step and primary decision in starting a research study. It is important to spend time thinking about the topic area so that you choose a clear and relevant area. Table 2.2 shows examples of some topic areas in selected disciplines.

There are three ways of arriving at your research topic. One is where your supervisor provides you with a list of possible research topics, and you choose one from that list.

Second, your supervisor provides you with a research project for which you study a particular topic. This is the case when faculty staff running a research project support your research. Number three is when your supervisor gives you a free hand to choose your topic. In this case, you freely choose your topic and share it with your supervisor for approval.

The following factors guide the choice of your topic (Saini et al. 2023):

1. Your course/specialization: choose a topic in your area of specialization or closely related.
2. Your interest: pick a topic that interests you. You will live with this topic for a long time, from the conceptualization of the research problem to data collection to reporting. Choose something that is exciting to you and will power your interest throughout the research journey.
3. Available resources: consider financial resources, human resources, time constraints, infrastructure, and laboratory facilities as these collectively influence the choice of the research topic.
4. Ethical and logistical considerations: Choose a topic that aligns with ethical and legal standards.
5. Significant: The topic must be meaningful, with potential impact and relevance.
6. Literature review: Conduct a thorough literature review to shape your research.
7. Scope of the topic: Avoid overly broad topics or those associated with controversy.

Other ways of identifying a research topic involve consulting sources of research ideas. The sources of these research ideas are discussed in Chap. 3. After selecting a broad topic and reading through the literature, you will likely come up with a *working title*. A working title helps you navigate more targeted literature that guides you to the next step in formulating a problem.

Step 2: Formulating the Research Problem This stage answers the question of why this research? The research problem is the "*central element*" of any research. It defines the focus of your research and sets the direction for your next steps. Research problems are either (1) Problematic phenomena, (2) Observed events that are puzzling regarding our currently accepted ideas, or (3) Problematic theories and (4) Current ideas that are challenged by new hypotheses. They are *specific issues or gaps in knowledge* that have not been explored and form a compelling reason for further exploration. They help define a real need for research. This sounds easy, but it is challenging to articulate a clear, concise, & manageable research problem (Mugenda and Mugenda 2019). This is mainly because of the need for precision in identifying the research problem while ensuring relevance and significance. Details of how to formulate a problem are given in Chap. 3.

Formulation of a research problem involves a back-and-forth exercise guided by literature review and reflection. An important part of this exercise is defining and clarifying "a statement of the problem" (covered in Chaps. 3 and 5). A statement of the problem is *the specific problem* that the study will address.

Step 3: Formulation of Objectives and Hypothesis The objectives of a research project summarize what is to be *achieved by the study, while the hypothesis states a prediction based on theory*. Objectives are a statement form of the research question and depict what is to be achieved by the study. Here, the dependent and independent variables and their relationships are defined. Objectives map the pathway for the investigation being designed, while the hypotheses keep the research on track. The topic of setting objectives and hypotheses is covered in Chap. 3.

Step 4: Designing Research Selecting the most suitable research design is crucial for any research, as it enables you to arrive at acceptable and meaningful findings. The question is: How will the research be conducted to provide valid answers? In this step, you are required to explain the choice of the design and provide details of how you will collect data systematically. There is a wide variety of designs, but usually, the choice is between broad structures, whether experimental, nonexperimental, or mixed, which will consequently determine the mode of data collection. You will also choose from various methods, techniques, and procedures. The nature of research will determine choices from observation (exploratory), survey (descriptive), or experimental study (causal). Research designs are covered in Part II of this book, from Chaps. 7, 8, 9, 10, 11, 12, 13, 14, and 15.

Step 5: Developing a Research Proposal Once all the preparatory work is complete, the next step is to consolidate everything together in the form of a research proposal. The research proposal's main function is to provide your *research's operational plan*. It shows how answers will be obtained, reassuring the readers of the problem's justification and the methodology's validity to obtain answers accurately and objectively. Universities and other research institutions may have varying requirements for the style and content of a research proposal. A generic guideline for developing a research proposal for a thesis or dissertation is provided in Chap. 5.

The Implementation Phase

Step 6: Data Collection Having completed the first five steps, including writing a good research proposal, you then proceed to collect data from which you will make inferences and conclusions. In this step, you collect data according to the methods you developed in the research design stage and detailed in the proposal. Standard procedures for data collection will be followed (Chaps. 17, 18, and 19) and ethical issues adhered to (discussed in Chap. 6.)

Step 7: Data Management, Analysis, and Interpretation After the data collection phase, you will return from the field with raw data that must be handled well (managed) to ensure quality and later analyzed to answer the research questions of your study. This will be followed by an analysis. The aim of the analysis is to translate the data into actionable knowledge. The analysis reveals patterns and trends that help the researcher answer the research questions. Interpretation follows the analysis and tells us what the data means. The analysis, as well as the interpretation, is dependent on the objectives and hypotheses. Chapters 20, 21, 22, 23, and 24 deal with data management and analysis.

Step 8: Reporting Writing the thesis or report is the last step and usually the most difficult research process (Murray 2011; Kember and Corbett 2018; Parija and Kate 2018). The report or thesis should follow the guidelines provided by the university. In addition, you may need to write other reports in the form of research papers for conference presentations, journal papers, and brochures that appeal to other stakeholders in agriculture. The way you write highly depends on what you want to communicate to your clientele. However, the scientific reports must detail what you did and report on findings, conclusions, and recommendations (Chaps. 25, 26, 27, and 28).

2.1.2 The Place and Time of the Literature Review

The literature review is a crucial component of the research process and contributes significantly to every stage. It can generally be divided into three main phases: the initial phase, the mid-phase, and the final phase (Cronin et al. 2013). These phases reflect varying levels of detail in the process, although they are not mutually exclusive.

Initial phase: This is the first phase when you start to search for your sources of literature after identifying the broad topic. The broad topic helps you to define the keywords that you use to search for sources. In the formulation of the problem and formulation of hypothesis, a literature review helps bring clarity and focus and builds the study's theoretical background. The main purpose of this initial phase is threefold: (1) Survey of the current state of knowledge in the area of inquiry, thus helping you to broaden the knowledge base in your area of study (2) Identification of key authors, articles, themes, and findings in that area and (3) Identification of gaps in knowledge in the area. At this initial stage, you start developing a database for your literature.

Mid phase: This comes after the research problem is clear and you are preparing to start writing the proposal. The literature review at this stage helps you in designing the research, mainly providing information on data collection procedures. Here, you borrow the best methodologies documented in the literature. Depending on your university's format, the literature review chapter will be a major section of your proposal. You will write this chapter during the mid-phase.

This mid-phase of the literature review is extensive and requires dedication and allocation of quality time to read, understand and integrate the various studies reviewed. During this stage, a researcher synthesizes the literature, identifying key themes, patterns, and gaps in knowledge. The aim is to eventually present a cohesive literature review chapter that provides an overview of the current state of knowledge in the chosen topic of study.

Final Phase: After you have analyzed your data (stage 7), the next stage is interpretation and reporting. Here again, you will need to go back and rely on the literature to help you in the interpretation of the results. Finally, during the writing of the results, the literature review is a must, especially in the discussion section, as you contextualize your results and compare them with other studies. You will find that you will refer back to the literature already read, but you will still need to find new literature.

At the stage of disseminating your work to stakeholders, you will still need to consult the literature. For instance, if you are preparing to present your findings at a conference, the literature review will help you to compare with others and, therefore, improve confidence during the presentation. If, on the other hand, you are publishing the work in a journal, the literature review will help you write a strong background and a convincing discussion. Furthermore, new knowledge is coming on stage every day, and therefore, a literature review should be a continuous activity in research.

2.2 Your Academic Research Project

2.2.1 Navigating Your Research Journey

As you start your research journey, there are key things you need to know that could help you fast-track the completion of your project. First, your department will allocate supervisors to guide and mentor you through the research journey. The supervisors provide support, ensuring you stay on track and maintain the quality of your work (Jensen 2015; Gedamu and Gezahegn 2021; Nangimah and Walldén 2023). Second, your research work will be rigorously examined, and passing is a prerequisite to graduation. 3. Third, as a student, you carry core responsibilities, including committing substantial time, effort and discipline to your research journey.

While your supervisors play a crucial role, your own efforts contribute more than 80% to the completion of your project. This includes being proactive and continually seeking ways to improve your research. Maintaining consistency will play a big role in the success of your project.

Your key responsibilities include the following:

1. *You will choose a research topic*: you have responsibility for this. You will choose a topic that you justify with clear implications; whether provided by the faculty or chosen freely, personal justification, and ownership are important.
2. *You will work through your project with the assistance of a supervisor*: at least two supervisors who are experts in your area of study.
3. *You will have to use appropriate design and analytical techniques* for collecting and processing your data.
4. *You will write and present the research project proposal* and receive approval before data collection.
5. *You will have to generate data that is valid and reliable.*
6. *You will write and present your research project report and present for examination.* Part of the examination process involves presenting your work orally in a board panel.
7. *You will need to publish your work in a refereed journal.*
8. *Your final report should be rigorous and reflect quality.* Experts will examine it.
9. *You will need to invest in learning theoretical foundations pertinent to your research study.*
10. *You will need to invest in learning research skills*: Though your supervisor will assist you, you will be the main driver.

Note: Students do fail research projects. The reasons for failure are many, but the key reason is poor-quality research work emanating from poor planning and execution.

2.2.2 Delay in Completing Research Project by Graduate Studies

The majority of postgraduate students delay completing their programs, and some halt the process (Jiranek 2010; Groenvynck et al. 2013; Matheka et al. 2020; Mkhai 2023). This challenge is global, affecting both developed and developing nations. In most universities and scholarship programs, for example, Erasmus Mundus (European Union Programme), the master's program structure is 2 years, while the doctoral program is 3 years. However, many students go beyond this time. The delay is a pressing concern in higher education training worldwide. What is the cause of this delay? Lets consider the structure of the postgraduate programme. A postgraduate program with coursework can be structured into four stages: (1) Coursework, (2) Writing a research proposal, (3) Carrying out the research, (4) Writing the dissertation and publishing journal papers. Experiences in most universities is that coursework is completed smooth, but the research project delays the completion of the program for most students for several reasons (Massyn 2018). Inadequate research skills, time constraints, institutional support, inadequate funds and infrastructural issues and poor student-supervisor relationships, low research interest, and insufficient access to literature are some of the main causes of delays (Watson 2008; Agné and Mörkenstam 2018; van Rooij et al. 2021).

A research project is a skill-intensive engagement that demands substantial proficiency from the student (Kaur et al. 2023). Success hinges not only on acquiring knowledge but also on the ability to apply that knowledge throughout the research process. This includes formulating the research questions, designing sound methodologies, collecting, and analyzing data, interpreting results and effectively communicating the findings. The challenge is intensified by the individual nature of research, placing the responsibility, almost 100%, on the student to develop this diverse skill and navigate the complexities of the research project independently. As a graduate student, investing time in learning research skills is essential to ensure the timely completion of the research project and also as a strategic investment in one's academic and professional future.

The challenge of time constraints affects all of us, but it intensifies for students, especially those juggling studies with full-time or intermittent work. Balancing academic pursuits with work obligations and personal commitments is a delicate act. Work demands adherence to employer timelines, while personal and family responsibilities require dedicated time. Consequently, students often need more time for their research projects. Delivering a research project is complex and time-consuming, yet it is frequently underestimated (Trochim 2009; González-Varona et al. 2023; Meadmore et al. 2023). Recognizing research's demanding nature is crucial for students aiming to maintain a healthy work-life balance amidst academic responsibilities. This challenge can be eased through effective time management and supportive supervision.

The availability of funding and research facilities also poses a significant hurdle for students, particularly self-sponsored students (Omoya et al. 2024). Conducting research often entails expenses for data collection, equipment, and other resources, which many students need help to afford. Limited access to funding can constrain the scale and scope of the research endeavors, affecting the quality and depth of fieldwork (Koohpaei et al. 2024). While universities offer some facilities, students often may supplement these resources by seeking support from faculty staff engaged in funded research projects.

Students who acquire support from funded projects receive multiple benefits that contribute to timely completion. They receive not only financial support for their studies but also opportunities for professional development (Koohpaei et al. 2024). For instance, funded projects often conduct strategic training, e.g., on data analysis and scientific writing, to enhance the skills of the partners and collaborators. In addition, working alongside experienced researchers enables students to learn best practices for research excellence and funding acquisition. Moreover, such relationships facilitate the building of collaborative networks and the development of collaborative skills. We strongly encourage you to reach out to faculty staff for support. You could also reach out to other researchers in the research institutions. Doing so not only enhances your chances of completing your project on time but also produces a manuscript suitable for publication in high-impact journals.

Other sources of funding include national research funds and competitive research grant fellowships. You are encouraged to aggressively seek these grant opportunities. Remember, success in such opportunities will highly depend on the strength and justification of your research proposal.

2.2.3 Tips on How to Succeed in Your Research Project

Take full responsibility: If you want to succeed in your research project, you must take 100% responsibility for the whole process. Jack Canfield, in the book "Success Principles," recommends this as the number one principle for success in life. In other words, you avoid excuses and blaming others when things do not seem to work. Approach your research project with this mindset. You are the driver of the research project, and as you work with your supervisor, resist the habit of consistently blaming them for delays. Most students are fond of blaming their supervisors, and in most cases, this behavior creates a poor relationship, which escalates the problem. Poor relationships between a student and supervisor should be avoided at all costs.

Invest in research methods skills: Research is an academic affair, and you must learn and grasp the principles and concepts of research methods. Learn by reading research methods books available in the library, online or even buy personal copies from bookshops. You can choose to invest and buy at least two books for your library. You may also learn through listening to numerous videos and podcasts on research methods available on the internet.

Dedicate to reading widely: The secret to conducting good research is reading widely. Successful researchers are avid readers. Reading widely has many benefits, foremost among them being the expansion of one's knowledge base. Exposure to a diverse range of literature encompassing various theories, methodologies, and perspectives equips researchers with a comprehensive understanding of their field of study (more is discussed in the chapter on literature review). Reading widely will also enhance your analytical thinking skills and broaden analytical capacity.

Be patient and persistent: Research is a journey, not an event. This journey is not always smooth; it can sometimes be tough and demanding. As a student, you may face difficulties such as complexities in methodologies, data collection hurdles, or the need to reassess and refine research questions. For example, if your research project requires that you have four seasons, this needs patience and hard work. In some instances, students face personal challenges. This then requires one to adapt and develop resilience to endure the challenges and allow one to navigate the complexities of a research project. This is so-called "Grit," a powerful combination of passion and perseverance that empowers you to stay committed to your goals (Duckworth 2016; Akos and Kretchmar 2017; Hernández et al. 2020; Xu et al. 2020). In agriculture, we have to give time for a seed to germinate and grow. There is planting time and harvest time. You are sowing a seed when you initiate the research project. The harvest will come after some activities, and you have to give time and effort for the process to take place.

Keep focused on your goal: In your academic research project, your goal is to produce two products: a good thesis and high-impact journal articles. A good thesis passes the examination process with high grades, while the ability to publish high-impact journal articles is a sure way of judging credible research. As you initiate your research, from the selection of the topic to writing the proposal and implementing the research, keep focused on these two products. A good thesis and journal publications always hinge on a good proposal (Chap. 5, 6).

Cultivate a good relationship with your supervisor: Make your supervisor a friend and a mentor. Three key things that can enhance a good relationship with your supervisor include (1) Being able to create trust, (2) Being cooperative, and (3) Being able to work in a team. You build trust by ensuring you deliver what you have promised and addressing all comments. Sometimes, supervisors organize students, in small cohorts to enhance sharing and learning. Be willing to be a team player in such cohorts. Cooperation between members of a team is important for success. If working on a collaborative research project involving many partners, ensure you are cooperative and a reliable member of the team. You should aim to contribute to the success of the entire project. All the partners should aim to deliver the goals of the project.

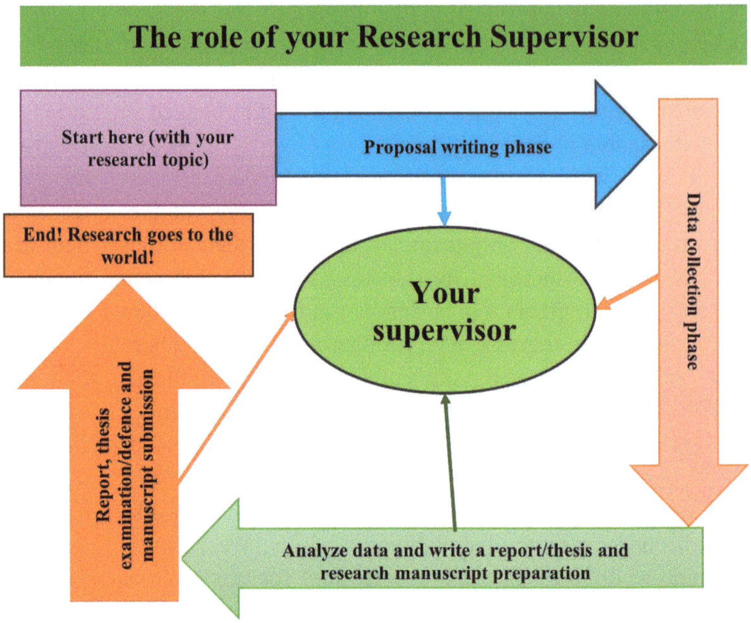

Fig. 2.2 Four key stages in an academic research journey and role of the supervisor

Commitment to maintaining healthy relationships and teamwork is crucial in such projects.

Developing strong data analysis skills: Many students often struggle to analyze data efficiently after collection. Investing time and effort in learning data analysis early can significantly improve research capabilities, as it can help you to swiftly identify and apply appropriate analysis methods. This saves time and enhances the accuracy and depth of your research findings. You can learn these skills using a multifaceted approach. Aim to supplement theoretical foundation in statistics courses with online courses, podcasts, and workshops. Additionally, consider learning at least one data analysis software. If in the biophysical sciences, choose from among GENSTAT, SAS, and R. If in social sciences, choose from among SPSS, STATA, and R A more detailed discussion of these software is given in Chap. 20.

Invest in scientific writing skills: One of the main topics discussed in this book (Chap. 27) is the principles of scientific writing. Scientific writing is a form of academic writing meant to communicate science. Possession of good writing skills is not a "walk in the park, "meaning it takes time to develop writing skills. Shortcuts do not work. You need writing skills to write the scientific publications required to fulfill the degree requirements, such as research proposals, journal articles, conference proceedings papers, posters, theses, etc. Jerry Jenkins, a best-selling author who has published 200 books, including "21 New York Best Sellers" states that he still takes daily steps to hone his writing skills. He advises that there is no shortcut to being a good writer and emphasizes that a "Good reader is a good writer." This means that if you want to be a good writer, you have to read books by other authors in your area of specialization. As you read, you learn the conventions and expectations in writing. We advise that as a student, you read materials to hone your research skills. These include books, journal articles, and theses. These can be obtained in the university repositories, while journal articles are available from many databases (Chap. 4).

2.2.4 Role of Supervisor

As stated earlier, the faculty will appoint an academic supervisory team consisting of two or more supervisors. Depending on your topic and the location of your research, you may be allocated more than two supervisors. For instance, if you are carrying out research outside your university (Sandwich program), at times you need a field supervisor.

Supervisors play a critical role in guiding your research journey (Nangimah and Walldén 2023). We can subdivide your journey with the supervisors into four stages (Fig. 2.2). The first stage of writing and presenting the proposal is critical because it determines the success of the rest of the stages. At this stage, the supervisors will advise on topic selection and help you through problem identification and proposal writing. They are particularly key at the proposal development stage as they guide you on the content and format. They help you prepare for the proposal defense.

During the data collection stage, they will guide you on how to collect valid and reliable data, check the validity of your data collection tools and, most importantly, guide you in collecting quality data suitable for your research objectives. Quality data is a key ingredient in any research and, thus, crucial for the overall success of your research project.

The third stage comprises data analyses, interpretation, and writing. The supervisors will guide you in choosing suitable data analysis software and offer advice on suitable alter-

Box 2.1: Summary Roles of Your Supervisor (Andrew 2007; Bégin and Gérard 2013; Sharma 2017; González-Ocampo and Badia 2019; Chugh et al. 2021; Xypaki et al. 2021; Gravett et al. 2022; Guarimata-Salinas et al. 2024)

- Coach, mentor, trainer. Your supervisor will train and mentor you through regular meetings to discuss progress. The major role of the supervisor is to give constructive feedback on your research work.
- Will supervise your writing as well as fieldwork.
- Guide you on suitable literature.
- Advice you on administrative processes and procedures.
- Guide and advice on opportunities for publishing your work.
- Help refine your proposal by asking you critical questions.
- Help develop appropriate plans, designs, strategies, and research techniques.
- Guide you on research ethics.
- Guide on data analysis and any need to consult a statistician.
- Track and give feedback on your overall progress.
- Connect you to other researchers and help in acquiring permission to work in other laboratories outside the university.
- Provide information about upcoming scientific conferences.
- Read and comment on the final thesis and advise you on when to submit.
- Coach and advice on how to effectively defend your research work.

natives. They will also connect you with collaborators and networks where you can get assistance in case of complicated analysis.

Finally, at the last stage, they hold your hand through the examination process. They prepare you to present your work to a panel of examiners by asking you critical questions. They basically coach you through the process (preparation for the examination process is discussed in Chap. 27). Box 2.1 and 2.2 give summary roles of the supervisors and tips for enhancing the completion of your research project, respectively.

Box 2.2: Tips to Enhance the Completion of Your Research Project

Always remember you are the driver of your research project. The ultimate responsibility of the research project rests on you. Take charge and drive the process to finish within the stipulated time. You will also experience less stress by taking charge and working consistently according to the timelines. Use the following tips to help you enjoy the research journey:

- Build a good relationship with the supervisor.
- Prepare to work independently and be ready to learn new skills where deficient.
- Ensure you follow instructions and guidelines from your supervisor.
- Take responsibility for finding out what is expected.
- Be proactive, meet deadlines and arrange formal meetings according to schedule.
- Enhance your research skills.
- Undertake recommended reading and produce written work as agreed.
- Learn organizational skill.
- Enhance your time management skills. Research work is involving, demanding, and daunting, and time management is of the essence.
- Train on writing skills.
- Remain positive even when you receive criticism in your work.
- Ask when you do not understand. Never pretend that you know.
- Learn communication and presentation skills.
- Take the initiative to inform your supervisors in case of challenges.
- When you fail to deliver, communicate in good time and apologise, respectively.

2.3 The Total Research Setting

Academic research gives you a practical experience and tests your ability as a future academician. It is also a long journey that requires resilience. As you start your research journey, we advise you to situate your research in what we call "total research setting" or visioning. This is where you can see your study from beginning to end. It is like setting goals for your research. You see the *beginning* and the *end or from inception to completion*. Most importantly, you see clearly the link between the topic and the problem you identified, through the other steps, to offering results and conclusions that should feed to offering a solution to the problem (Fig. 2.3).

This concept of a total research setting enhances your motivation. Visioning helps to keep pace even when things become challenging. This proactive visioning approach helps researchers in the following ways:

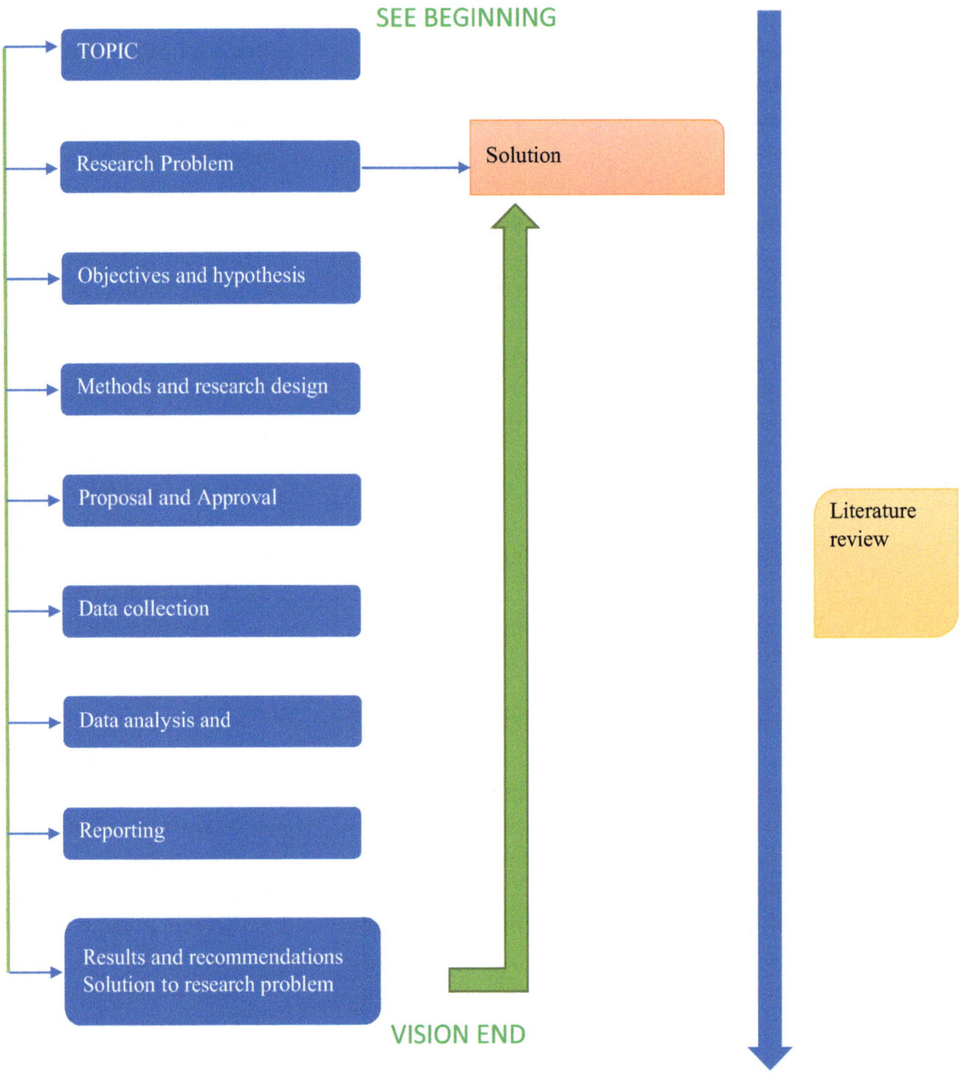

Fig. 2.3 Total research setting in the research process

- Identify challenges, plan contingencies, and align efforts with overarching objectives.
- Contributes to a more organized and purposeful execution of the study.
- Facilitates resource allocation, time management, and prompt issue resolution.
- Contributes to the credibility and rigor of the study.

As part of your work planning, include the following to enhance success:

1. Comprehensive planning and road map

Develop a comprehensive research plan that outlines each stage of the research process in detail. This is personal and additional to the plan you present to your supervisor. It is for guiding your personal journey of research. In this plan, set clear, achievable milestones and deadlines to ensure steady progress. By mapping out the entire research journey from inception to completion, you can visualize the various steps and prepare accordingly. This personal plan allows you to evaluate the risks and challenges and lay out contingency plans. Specific goals are necessary; for instance, in variety testing, you could have a selection of varieties in month one, conduct preliminary tests in month two and plant in month three, etc.

2. Reflective Practice

As a researcher, you are not just collecting data, you are part of the process. Reflective practice means being aware of this fact and making sure your actions stay aligned with the research objectives. Maintaining continuous reflection on your research project helps you to spot potential errors or biases and make corrections as needed. It also promotes continuous improvement. One of the ways of actionilizing continuous reflection is keeping a "Reflective diary" or "Research notebook," where you record events and lessons learned (Saunders et al. 2019).

3. Documentation and Record-keeping

We encourage documentation and record-keeping throughout the research process. The notes taken will be useful when you are writing the research methodology section of your thesis or journal article. Journal articles require detailed methods and procedures. Moreover, a comprehensive audit trail (line-up of activities) is extremely important in qualitative research approaches. The audit trail helps you in the validation of your findings. In addition, should questions arise in your research process, well-maintained notes will be a critical resource for resolving them.

4. Visualization of Success

Visualize success at each stage of the research process. We encourage students to imagine the successful completion of their project and the impact it will have. The concept of positive visualization is widely discussed in motivational literature. For instance, *The Power of Positive Thinking* by Norman Vincent Peale emphasizes how maintaining a positive outlook fuels increased motivation and perseverance. *The Secret* by Rhoda Bryne highlights the law of attraction, suggesting that focusing on positive outcomes attracts success . Finally, Jack Canfield's book in *The Success Principles: How to Get from Where You Are to Where You Want to Be, the author outlines strategies such as* creating a vision board, setting clear goals and actively managing them to achieve these goals. This is similar to seeing your research project visually from beginning to end, where the end will be celebrated.

2.4 Key Terminologies in Research

2.4.1 Concept

According to Webster's Dictionary, the word concept refers to something conceived in the mind. A concept is, therefore, a generalized idea about persons, objects, behaviors, and events and is usually agreed upon by many people. It is a mental deposition that helps the universe understand the objects' meaning and communicate. It is like a "label" or symbol that we use to share experiences. This implies that the meaning of the concept is understood in a similar manner by the people in the interaction. It is based on our experiences. For example, we have concepts of persons such as researcher, farmer, teacher, mother, etc. and concepts of income, age of farmer and concepts of objects such as plant and soil. Concepts of events can be the rainy season, short rainy season, planting season, harvesting, and many other concepts describing objects and their qualities, e.g., plant, leaf length, maize height, and price of a commodity.

Concepts are useful in identifying, distinguishing, naming, labeling, and categorizing persons, objects, or events. Because concepts are mental images or abstractions conceived in our minds, their nature varies from person to person and sometimes from region to region (Reilly et al. 2024). Concepts have been developed over time and have become a shared language. Almost every term we use in our understanding of human and environmental characteristics and behavior is a conceptual entity.

We can think of people of different age groups (children, youth, middle-aged, and elderly). They use certain terms (concepts) that are only understandable to those age groups. For instance, in Kenya, the young people have adopted a dynamic version of the Kiswahili language called "sheng," which they keep changing terms, concepts or labels of items keep changing. Older people sometimes find it difficult to understand these new concepts. Similarly, different disciplines tend to have their key terms and language that they use to communicate. These are also concepts as symbols of common use. Agriculturalists, on the other hand, have developed their concepts to describe unique characteristics not described by others, e.g., for a soil scientist, soil pH is a well-understood concept. In contrast, for an animal scientist, etiology is a well-understood concept.

Regardless of the discipline, scientists master measurements. In your research study, your purpose is to measure things. Some concepts, such as maize yield and plant height, are precise, objective, and easy to measure. Other concepts are highly subjective, and that makes it difficult to use them as they are in a research study because they cannot be measured easily on a statistical scale. In research, we say they are abstract and difficult to measure. Examples include crop production, wealthy households, and price volatility, as used in agronomy and agricultural economics. If you have such concepts in your study, you need to convert them into measurable variables so that they can be measured; although on different scales, the same variable will have different precision. This is called the operationalization of the concept.

2.4.2 Construct

Construct is a concept that is specifically chosen, created, or invented to explain a given phenomenon. Constructs exist at a higher level of abstraction than concepts because they are specific and associated with the behavior of a phenomenon. They are also called "Latent Variables" when they cannot be measured directly (Fox et al. 2016; Uher 2022). A construct may be simple, e.g., milk yield, age of farmer, or a combination of a set of related concepts such as farmer wealth, farm productivity, and farmer perception of soil status, consisting of several underlying concepts. We can use the terms unidimensional and multidimensional to describe such constructs. Milk yield is a direct construct and, therefore, unidimensional. Farmers' wealth and farm productivity are complex constructs and, therefore, multidimensional.

Whether you are in the natural (biophysical) sciences or social sciences, you must ensure that your constructs are clearly defined. For instance, in biophysical sciences, you may conduct research involving maize and one of the parameters to be measured is maize height within a growing season. In this case, the measurement of maize eight must be well defined in terms of how the maize height will be measured, including which plant will be measured and when.

On the other hand, we can have a social study, for example, where a student (real) was carrying out a study to assess the impact of collective marketing of avocado on income. In this study, the student was asked to measure income. This income may have referred to monthly income, annual income, personal income of household head, total family income, total income from farm enterprise and income from avocado sales. Unless the term income is precisely and clearly defined, there is confusion because the student does not know what to measure in the farms during data collection. Moreover, enumerators helping collect data will not know what to measure (ask farmers) if not well defined.

To ensure that scientists communicate what they are measuring, scientific research requires *operational definitions* that define constructs in terms of how they will be empirically measured. For example, in the above example, the student must specify which income he or she is measuring. Is it money obtained from avocado sales? If it is from avocado sales, Is it weekly, monthly or annually? In this example, the student defined the construct (income) for this study as income obtained from the sale of avocados annually (aggregated sales for the whole year) for those farmers who participated in collective marketing and for those farmers who did not participate in collective marketing.

Price was another construct used in the same avocado study described here, which could have had many meanings. For example, regarding the selling point, we have options of farm gate, brokers, and open market that offer different prices. The time of selling is significant because prices vary according to demand. Variety could also be of interest because different varieties of avocado fetch different prices. All these factors need to be considered when defining the construct price so that it is precise and clear what price will be measured. Once concepts and constructs are operationalized, they form measurable variables. The variables, in this case, generate data.

How Well Do Operational Definitions Represent the Constructs?

One complication of measuring constructs and abstract concepts is the potential mismatch between the operational terms or variables used and the rich complexity of the underlying construct. For instance, in education, intelligence might be evaluated by an individual's ability to solve math problems, while anxiety might be gauged by heart rate (Newton Suter 2012). However, these operationalizations may not fully capture the multifaceted nature of the constructs they aim to represent. Similarly, in agriculture, there may be cases where the operational terms used to assess certain constructs, such as soil productivity or livelihood resilience, fail to encompass the entire scope of the underlying phenomena. This discrepancy highlights the need to carefully consider measuring instruments to ensure accurate and meaningful assessments in research and practice. This is an essential topic of the validity of measurements and measurement tools (see Chap. 17).

2.4.3 Variables

A variable is a characteristic or factor that may take one or more values (Hutcheson 2011; Bai and Jin 2016). It is a noun that stands for variation within a class of objects. In the aforementioned examples, i.e., in the study to assess the impact of participation in collective marketing on income, variables include weekly income from avocado sales, monthly income from avocado sales, avocado price at open-air markets on specified dates etc.

Examples of variables measured in agriculture include:

- Yields of crops, e.g., maize, sorghum, green grams.
- Amount of soil, e.g., nitrogen, phosphorus, and potassium.
- Amount of milk produced, etc.
- Age and farm size of farmers.
- Education level of farmers.

Constructs are conceptualized at the theoretical plane (abstract), while variables are operationalized and measured at the empirical plane (Haucke et al. 2021; Auer et al. 2024; Symbaluk and Hall 2024). This simply means that constructs are abstract ideas, while variables are their measurable forms.

Types of Variables

Independent and Dependent Variables

In cause–effect research, where our interest is to study the effect of A on B, we have two types of variables: the *independent variable (IV),* in this case, A, which is the cause, and B, which is the *Dependent variable (DV),* the result or outcome. These two terms are basic in understanding cause-and-effect relationships in both biophysical and social research studies. The independent variable (s) is/are the variable whose effect we would like to determine. They are also called explanatory variables or predictors. In experiments, an independent variable is a factor that is being manipulated, deliberately set by the experimenter. We often refer to them as the *treatments*.

Examples of independent variables in experimental research:

1. Three types of inputs: Manure, fertilizer, fertilizer + manure: They are expected to cause an effect on the crop under investigation
2. Four fertilizer rates: 0, 10, 30, and 60 kg N ha^{-1}; Effect on test crops like soybean
3. Four protein sources in a diet of animals: Soybean, fodder trees, Cottonseed cake, and fish meal. We measure their effect on milk yields

In the above examples, the investigation will involve experimentation, and the effect of the independent variables will be measured.

The dependent variable (DV) is the effect and depends on the independent variable. They are outcomes or results of the influence of the independent variable. In experimentation, this is the factor that the experimenter measures, the factor assumed to be affected by the independent variable. The dependent variable depends on the treatment or independent variable. A research problem will describe the causal relationships between independent and dependent variables and explain how these relationships come to be, e.g., factors influencing the performance of a certain technology.

Examples of dependent variables:

1. *Growth of dairy goats* as a result of feeding on various protein diets.
2. *Maize yields* obtained as a result of applying different levels of nitrogen fertilizer.
3. *Bean yields* obtained from varietal trials testing the performance of five varieties of beans.
4. *Tomato yields* obtained from treatments using different insecticide spraying frequencies.

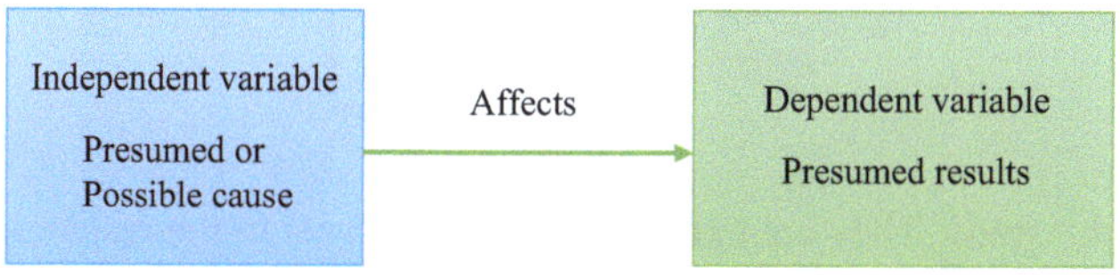

Fig. 2.4 The relationship between an independent variable and dependent variable with a "presumed cause" and "presumed result"

5. *Level of knowledge* of farmers after a training program (those trained and not trained).
6. *Uptake of hybrid maize* among farmers after a training program (uptake by those trained and those not trained).

Differences Between the Independent Variables and Dependent Variables in Experimental and Nonexperimental Studies

In experiments, independent variables are manipulated, while in nonexperimental studies, variables are studied as they are. Suppose a researcher decides to investigate the effect of protein diets on the growth of goats. The independent variable is the protein diet, while the dependent variable is the growth of goats. The researcher will decide on the types of protein diets to use (manipulate the independent variable) and feed them to different groups of goats. The researcher will then measure the effect of the diets on the growth of the goats (dependent variable).

On the other hand, suppose a researcher decides to investigate the effect of age on the uptake of hybrid maize among farmers in a certain region. A researcher, in this case, will ask: Does age affect the uptake of hybrid maize among farmers? The researcher will not be able to manipulate age. The researcher will go to the field and carry out the studies on farmers as they are (a survey study). At the end of the data collection, the relationship will be explored to find out if age actually affects the uptake of hybrid maize. In this case, we talk of a presumed cause (Fig. 2.4).

Intervening or Mediating Variables These are variables that stand between independent and dependent variables and mediate the effects of the independent variable e.g. if farmers adopt improved maize variety (DV) after a training (IV), that result may be due to (a) their farm sizes and (b) education level of the farmer (intervening variables) (Fig. 2.5).

Another example: Increased income (dependent variable) could be due to participation in collective marketing by smallholder farmers mediated by increased production at the farm level.

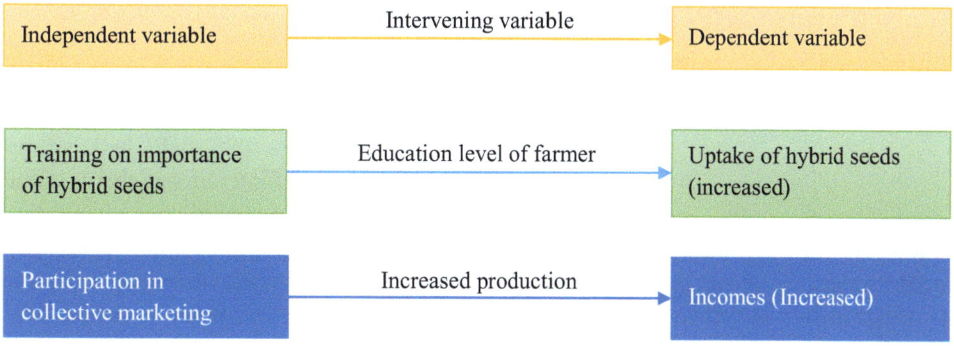

Fig. 2.5 Relationship between an independent variable and an "intervening variable" influencing the dependent variable

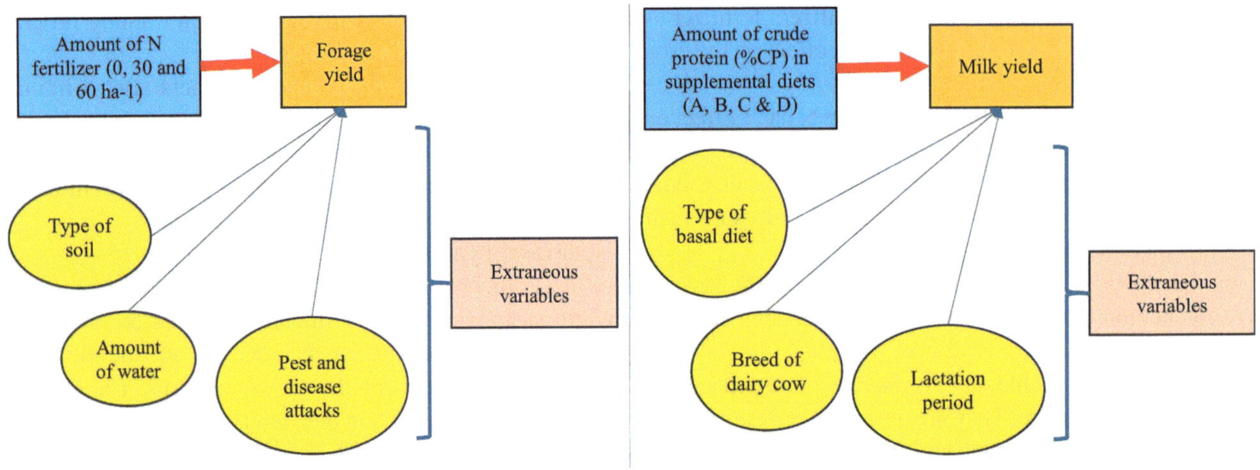

Fig. 2.6 Simple presentation of extraneous variables in livestock studies

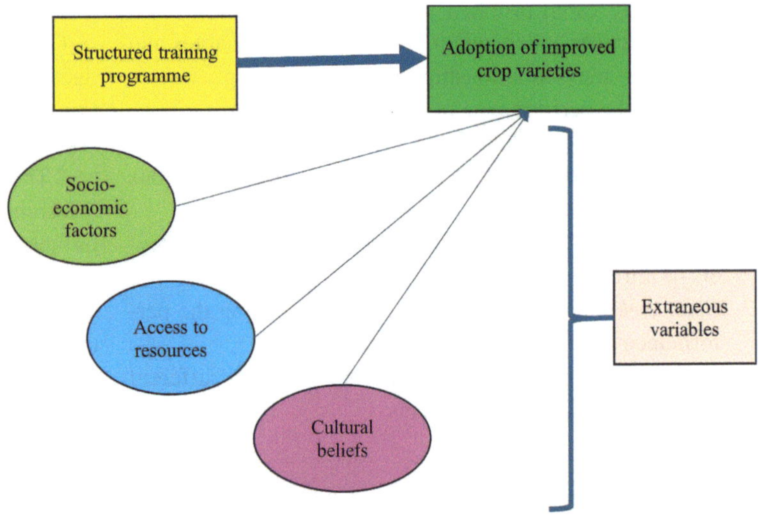

Fig. 2.7 Simple presentation of extraneous variables in randomized controlled trial evaluation effect of training on adoption of improved crop varieties

The above scenario presents actual observations encountered in studies involving social research. Most studies have multiple independent variables, and the researcher will aim to estimate the effects of each on the dependent variable.

Moderating Variables New variables constructed by the researcher that affect the relationship/direction between the IV and DE. For example, in a study to determine the effect of socio-economic variables on fertilizer use, a researcher may construct a joint variable (age * income level = A) and find a significant relationship between A and the use of inorganic fertilizer.

Extraneous (Confounding or Spurious) Variable

An extraneous variable refers to any variable that may influence the outcome of a study but is not the main variable of primary interest. Its effect is "*unintended*" and not needed in the study. They are extraneous (not relevant) to the research question. They are thus, a nuisance and must be controlled to ensure the validity and reliability of the study. Examples of extraneous variables in a biophysical experiment include the physical environment, weather, soil composition, etc. See examples of livestock experiments (Fig. 2.6). In a social study, extraneous variables could be attributes of the participants, i.e., their characteristics we are not aware of or procedural variations and how we plan the study, which might invite the extraneous variable (Fig. 2.7).

To an agricultural researcher, extraneous variables are important because the study will not produce authentic results because of interference that is not accounted for. To produce authentic results, they should be controlled. This means controlling for potentially confounding variables; for instance, in controlled laboratory experiments, researchers have the ability to control almost all the extraneous variables.

Confounding Variable
We stated that extraneous variables must be controlled because they are not part of the research objective. What about when we fail to consider them because they cannot be measured or observed? They can pose significant challenges in research. If these variables are not accounted for, they can easily cloud and confuse (confound) the study's results, making it difficult to interpret the findings (Jager et al. 2008; Howards 2018). These variables can become intertwined with the variables of interest, ultimately distorting the true relationships and leading to unreliable or invalid conclusions. For instance, in the illustration (Fig. 2.6), evaluating the effect of different diets on milk yields, extraneous variables, such as breed, age, health status, type of basal diet, and lactation period, could all potentially confound the relationship between the diets and milk yields.

Therefore, researchers need to recognize the limitations caused by unmeasurable or unobservable extraneous variables. Practical and effective strategies for managing extraneous variables include employing sound research designs, statistical techniques, and control strategies outlined in Chaps. 8, 9 and 11. Indeed, the ability to control extraneous variables is a key criterion by which research designs are structured and evaluated. Moreover, many concepts learned in research methods are geared toward estimating and addressing extraneous variables to safeguard the validity and reliability of research outcomes.

In brief, main ways of managing confounding variables in field studies include:

1. *Experimental designs:* the designs control extraneous variables through randomization, blocking, and controlling the environmental variables (Discussed in Chap. 8).
2. *Randomization:* In both a biophysical and social study, random assignment of treatments, participants, or sample selection without bias is involved, so every member has an equal chance of being selected. In a natural experiment, the treatments are randomized within the experimental units (Chaps. 8 and 11).
3. *Replication:* Random variability is inherent in any experiment. Replication allows the researcher to distinguish between systematic and random fluctuations caused by extraneous variables (Chap. 8).
4. *Elimination:* Involves completely avoiding or removing the extraneous variable by use of experimental designs or equalizing the factors in the environment, e.g., in a glasshouse experiment (Chap. 8).
5. *Holding the extraneous variable constant*, e.g., by limiting the study to one level of the variable or blocking, we can use one level of soil fertility and ensure that management practices are uniform. We can ensure that extraneous variables such as breed and lactation period are constant in livestock studies. In a social study, we could hold gender constant. The randomized complete block design does exactly this (Chap. 9).
6. *Statistical control procedures:* we can use statistics, such as partial correlation, analysis of covariance (ANCOVA), and multiple regression. In this case, we measure the extraneous variables and include them in the statistical analysis. The statistical procedure estimates the effect of the extraneous variables on the dependent variable; can remove this from the analysis. For instance in on-farm trials, household and farm variables that influence the performance of the treatments are measured and included in the analysis as covariates (Chap. 10).
7. *Matching the subjects:* The goal is to ensure that groups are compared and are similar in terms of specified characteristics. The researcher tries to find pairs of participants similar on a variable, such as, age, and education level, physically or using statistical techniques (Chap. 11 on research designs for evaluating the impact).
8. *Using subjects as their own controls:* involves comparing each subject's performance or response under different conditions or treatments. This allows a researcher to account for individual differences that may confound the results. For example, in evaluating the effectiveness of a new agricultural intervention on crop yield, researchers may use a within-subject design where each farmer's yield is measured *both before and after implementing the intervention*. This is covered in Chap. 11.

2.4.4 Other Types of Variables

Variables are crucial components in research as they help to define and measure the concepts or phenomena under investigation. They are the building blocks for research questions, hypotheses, and study objectives. Variables provide researchers with a means to quantify, analyze and understand the complexities of the real world. Apart from the independent, dependent, or extraneous variables, here we describe other types of variables classified according to different criteria:

Demographic or Attribute Variable Variables that describe sample characteristics are called demographic or attribute variables (O'Bryant et al. 2004; Fassett et al. 2022). For instance, in studies interviewing farmers, demographic variables of interest include age, gender of the farmer, ethnic background, education level, religion, and socioeconomic status. In these studies, researchers obtain demographic information from the sample to determine if samples are representative of the populations of interest. These demographic variables cannot, however, be manipulated, but researchers can explain relationships between demographic variables and dependent variables. For example, we can determine the effect of gender on the uptake of improved banana varieties or the relationship between age and land allocated to pasture production.

Categorical Variables Categorical variables are variables that represent groups or classes (Preisser and Koch 1997; Sun and Sutradhar 2015). They may have two or more distinct values (e.g., single, married, divorced, separated, or widowed marital status). Some variables (e.g., gender) are dichotomous (i.e., have only two values). Categorical variables are important as they allow researchers to classify attributes that cannot be measured continuously but belong to discrete categories. Other examples of categorical variables include types of crops (e.g., wheat, maize, rice), levels of pest severity (e.g., low, moderate, high), soil fertility ratings (e.g., poor, fair, good) or farming practices (e.g., organic, conventional).

Continuous Variable Variables may be termed continuous if their values range across a continuum (Smeltzer and Ray 2022). Examples of continuous variables are age in years, weight of tomatoes in kilograms, height of a tree in meters, yield of maize in tons per hectare, amount of nitrogen in g per kg, amount of milk in liters and level of income of households in Kenya.

Polytomous variables are variables that take more than two values (Wilmink and Snijders 1989; Revuelta 2010). They are frequently encountered when examining factors that have multiple levels or options. Examples include; the education level of the household head (Illiterate, Primary, Secondary, Tertiary), types of crop diseases (e.g., fungal, bacterial, viral) and levels of pest severity (e.g., mild, moderate, severe). Levels of knowledge (low, medium, high).

Qualitative Variables Variables that express qualitative characteristics or properties e.g., color of infected fruits (yellow, orange) or education level of a farmer (illiterate or literate).

Quantitative Variables Express amounts or those that bear numerical values, e.g., height, weight.

Discrete variables Continuous variables that can be measured with an arbitrary measure of fitness or exactness, such as maize yields, can be measured using a ruler.

2.4.5 Data

Data are units of information or any statistic, figures, facts, or knowledge collected during a study (Totten et al. 2020; Hackman et al. 2024). They are the information you collect to answer your research question or achieve your objective. We have two types of data: primary data, which the researcher directly collects for a specific purpose, and secondary data, which the researcher picks from existing sources, e.g., census data.

Depending on whether the study is qualitative or quantitative, data can be numbers, images, words, figures, ideas, models, databases, video clips, etc. Research data is planned, collected, and analyzed to answer research questions.

Data can also be categorized according to the scale of measurement (Marateb et al. 2014; Shukla 2023). Four primary measurement scales are recognized:

- Nominal scale: This is the simplest scale used for categorical data where the categories have no inherent ranking. Examples include gender, type of crop varieties, and diseased and not diseased plants. Nominal data is used to label or categorize variables without any quantitative value.
- Ordinal Scale: this scale involves data that can be ordered or ranked, but the interval between the ranks is not necessarily equal. For instance, ranking of food crops into first, second, third, etc. or level of satisfaction with responses like poor, fair, and good.
- Interval Scale: these have ordered categories with equal intervals between them but lack an actual zero point. Examples include soil temperature, measured in degrees Celsius or Fahrenheit or soil pH. These data do not have a true zero point.
- Ratio scale: This scale has all the qualities of an interval scale with the addition of a true zero point, which allows for the measurement of the absolute magnitude of the variable. Examples include crop yields, milk yields, and plant height.

2.4.6 Validity

Quality is one of the most important concerns in research, as it directly influences the credibility, reliability, and usefulness of the findings. One of the key indicators of research quality is *validity*, a term used to assess the soundness and accuracy of the conclusions drawn from a study. Technically, validity is defined as *"the best available approximation to the truth or falsity of a given proposition, inference or conclusion"* (Leung 2015; Noble and Smith 2015; Kool et al. 2020). This definition highlights the fundamental aim of research: to approach the truth as closely as possible by minimizing alternative explanations and enhancing the robustness of inferences.

Building on this technical definition, *validity also has a practical dimension* that relates to the truthfulness of the data collected and the accuracy of the measurements used. It asks a critical question: *Did the research measure what it was intended to measure?* In this context, validity becomes the guiding principle for ensuring that all information collected is accurate and meaningful. Validity reflects the quality of a procedure or instrument used in the study, indicating whether it is accurate, correct, true, and appropriate (Cohen et al. 2017). It encompasses not only the fidelity of the measurement process but also the overall trustworthiness of the research outcomes.

Consider you have a bathroom scale. It is considered valid if the scale accurately measures your actual (true) weight. However, if the scale provides a measure lower or higher than the person's actual weight (taken in a medical center), it lacks validity. The measure taken on your weight using this scale is not meaningful.

Using a similar scenario, consider a soil pH meter that is poorly calibrated or has a contaminated electrode. Using such a meter will give measurements that are not valid. A valid soil pH measurement will accurately represent the true acidity or alkalinity of the soil. This will give meter readings that agree with those considered reference samples. Researchers usually take several corrective measures to address the issue of invalid soil pH measurements caused by a poorly calibrated or contaminated pH meter. These include regular pH meter calibration using standard solutions, proper maintenance, and cleaning of the electrode to prevent contamination, and frequent verification of measurements against known reference samples as laboratory analyses are ongoing.

Basic Types of Validity

(i) *Internal validity* refers to confidence in a study's cause-and-effect relationship. In other words, it is the extent to which we can infer causality. We ask, did the independent variable cause the dependent variable? Experiments conducted in a laboratory where conditions are controlled tend to have higher internal validity than from a field (farm) study.

(ii) *External validity* refers to the generalization or representativeness of our results; samples should represent the population; sampling and data collection procedures are important in determining the sample's representativeness. External validity addresses the ability to generalize your study to other people and broader populations. An experiment conducted under field conditions will better generalize to the population than in a controlled setting, in a laboratory and vice versa. The topic of validity is revisited in Chaps. 7 and 11.

2.4.7 Reliability

Reliability is the extent to which a measure will produce consistent results (Leung 2015; Noble and Smith 2015). It thus relates to the consistency or dependability of a measure. If it is reliable, you can be confident that all the items that make up the measure are consistent with each other and that, if you were to use the measure again with the same individuals, they would be rated similarly to the first time.

In real life, a reliable person is consistent in their behavior. Reliable people keep promises, and you expect them to always remain true to their promises. On the other hand, the unreliable person is inconsistent, does not keep promises, and finds it difficult to work within a team. Reliability is an important tool in assessing individuals' work performance in an organization. This measures their consistency and dependability in carrying out tasks and responsibilities. Reliability is important because it ensures that tasks and responsibilities are completed consistently and effectively, improving the organization's productivity and efficiency.

Using these two analogies, we can say that a reliable measure or instrument will always give similar results. In this case, we are sure that the results will be error-free. Suppose you measure a soil sample's pH at three different times: 9 am, 10 am, and 11 a.m. If, for example, you got different readings, such as pH of 5.2 at 9 am, 5.3 at 10 am, and 5.0 at 11 am, this indicates measurement error, and your results would be unreliable. With a measurement error, it is difficult to know the true value. Reliable soil pH readings are expected to be the same at 9 am, 10 am, and 11 am.

Validity and Reliability Form the Foundation of Robust Research

Validity and reliability are the cornerstones of research. These concepts ensure that the conclusions drawn are accurate, meaningful, and consistent. They are emphasized across disciplines such as education (Creswell 2014; Cohen et al. 2017), business (Schindler 2019) and psychology (Wilson and MacLean 2009), among others. From understanding variables and measurements to complex designs and data collection methods, the main focus remains on achieving validity and reliability to maintain accuracy. Prioritizing these principles throughout the research process is essential to safeguard the integrity and credibility of the findings.

In this book, validity and reliability are repeated in many topics and are prominently featured in Chaps. 8, 9, and 10, which discuss research designs. Moreover, the chapters addressing sampling and data collection emphasize how to maximize the validity and reliability of the data collected. Nevertheless, the meaning of validity and reliability differs in qualitative research compared to quantitative. In qualitative research, the focus shifts toward concepts of trustworthiness.

2.5 Errors in Measurements

Naturally, measurements are rarely perfectly reliable (Nimon et al. 2012). All instruments and humans do not conform to perfection. They will always display inconsistency. We expect to find differences in separate measurements. Theoretically, it is reasonable to look at an observed score X as a function of the "True value" (T) plus "Error component" (E). Measurement Error is the difference between the true value and the observed value. We cannot calculate these error components because we do not know the true score. Therefore, we must devise some way of estimating how much of our measurement is attributable to error and how much represents an accurate reading. That estimate is reliable.

We encounter systematic and random errors. Systematic errors are predictable measurement errors that occur in one direction, overemphasizing or understating the true value (Patterson and Lowe 1970; Kirkup and Frenkel 2006). Such an error is constant and biased. If a systematic error is detected, such as a soil pH meter overestimating the pH value because of a faulty electrode, recalibration is done. Thus, systematic errors are constant and, therefore, do not present a problem for reliability. However, they are a great concern of validity because, although they are consistent, test values are not a true representation of the quantity being measured.

Random errors are due to chance and can affect a subject's score unpredictably. Suppose in measuring soil pH, the technician records wrong measurements in one instance, and in another instance, due to exhaustion, fails to read correctly and also forgets to calibrate the pH meter as required. There are random errors, which are a reliability concern, that focus on the degree of error that is present within a measurement system. As random errors diminish, the observed score moves closer to the true score, and the measurement is more reliable.

2.6 Theory

Theory is an important term in science because it lays the foundation for science. A theory is a general explanation of the world. It consists of organized relationships that explain the world. It is an answer to "why question" about the world (Stewart and Klein 2016). A theory will explain why the world is the way it is. It will tell us what is happening, why and how it is happening. Theories drive all scientific disciplines. For instance, in biology, we have the "evolution theory" that explains the origin of species as proposed by Charles Darwin. There is also the "Big Bang Theory" in astronomy that explains the expansion of the Universe. Social theories explain societies and organizational behavior in social sciences, e.g., the theory of planned behavior. Psychology has many theories that explain human behavior, cognition, and mental processes. In agricultural economics, we have the utility maximization theory and many other theories that explain agricultural systems.

We should be able to distinguish between scientific theory and a layperson's meaning of theory. Laypersons will commonly talk about believing in a theory that explains something, a kind of belief system. For example, theorizing that it will rain in the next few days. "My theory is that it will rain before the end of the month." In this context, theory is another term for hypothesis or prediction. Sometimes, people say they believe in the theory of not lending money to friends because they do not pay. This, again, is a prediction that is not verified. In Kenya, in Lake Baringo, there is an island with no people, and it is called "Devil Island" because theoretically, the people there believe that there are devils who invaded the area and, therefore, rule the island. These are myths and not scientific theories. A scientific research theory is, however, more than a simple hypothesis. It is a hypothesis that has been supported by repeated observations, testing and confirmations.

But how does a theory develop? A theory is created, not discovered. It is created to organize and give meaning to a complex array of individual facts and observations. For example, what can explain why increasing nitrogen levels will increase maize yields to a certain level and then cause a decrease? Researchers will make several observations and come up with theories to explain them. A theory usually develops from a long process of research that uses empirical data to make assertions based on deductive and inductive data analysis (Kivunja 2018). The assumptions, assertions, and predictions of relationships postulated by a theory become an intellectual base upon which research data can be grounded to search for meanings of future studies. These frameworks not only guide the formulation of research questions and hypotheses but also help researchers interpret their studies' findings. Typically, theories enable the scientific community to build upon existing knowledge, refine existing ideas, and explore new avenues of inquiry, ultimately contributing to the advancement of a particular field of study.

Theory, research and practice are interlinked and form part of a continuum that aims to understand the determinants of behaviors, test strategies for change and disseminate effective interventions (S Reeves 2008). For example, farmers will be advised to use biological, physical, cultural, and chemical methods to control pests and minimize environmental impact following the theory of integrated pest management. Similarly, in disseminating interventions to farmers, assessing the impact they cause relies strongly on the change theory (Chap. 11). In assessing the impact of an intervention, the theory of change is critical because it points to how change will happen and what will have caused it. This helps to clarify measurements to be undertaken in assessing impact.

A scientific theory is therefore accepted as being true by the scientific community because it is grounded in actual data and observations that can be tested through research, but sometimes it is not perfect. For this reason, the progress of science is marked by progression over time to make theories better through better observations using more accurate instruments and more informed logical reasoning. This is the essence of science: constantly working toward building scientific knowledge. Universally, we aim to develop theories that will be generalized across time and space. The more the data supporting a theory, the stronger the theory.

A theory contains principles and statements that structure our understanding and explanation of the world (Wilson and MacLean 2009). Several authors argue that a theory is made up of definitions of variables and concepts, a set of relationships among variables, specific predictions and a domain where the theory applies (Corley and Gioia 2011). A theory is the highest level of abstraction, where propositions are investigated. The following definitions help us further understand theories, their key characteristics and their components

- A theory is a statement of concepts and their interrelationships that show how and /or why a phenomenon occurs (Corley and Gioia 2011).
- A theory as an explanation of a phenomenon arrived at thorough examination and contemplation of relevant facts (Stewart et al. 2010).
- A theory is a belief, idea, or set of principles that explain something; the set of principles may be organized in a complex manner or quite simple.
- A theory is a set of interrelated constructs (concepts), definitions, and propositions that present a systematic view of a phenomenon by specifying relationships among variables to explain and predict the phenomenon (S Reeves 2008).
- A scientific theory is a system of constructs (concepts) and propositions (relationships between those concepts) that collectively present a logical system and coherent explanation of a phenomenon of interest (natural or social behavior) of interest with some assumptions and boundary conditions.

These definitions give us the following key components of scientific theories:

- *Constructs*: the "what" of theories. What constructs are important in explaining a phenomenon?
- *Propositions*: The "how" of theories. They explain and show how concepts are related, i.e., relationships and predictions. One of the main aims is to predict phenomena.
- *Reason or logic*: The "why" of theory. Why are these concepts related? How is a scientific theory explained by empirical evidence?
- *Boundary conditions*: The "who," when and where of theory (Under what conditions will these concepts and relationships work?).

We define agricultural theory as a set of interrelated concepts (constructs) and propositions that present a logically structured and coherent explanation of phenomena with some assumptions and boundary conditions.

Summary of the Importance of Theory in Research

Theories serve important roles in increasing our understanding of phenomena. Specific roles include the following:

1. *Organize and explain* various specific facts and descriptions in the area of interest. It helps reduce or summarize multiple facts and observations into a single explanation. This makes the world more comprehensible by providing a few abstract concepts around which we can organize and explain a variety of observations (Wilson and MacLean 2009).
2. *Allow prediction of what should occur given a set of circumstances:* for example, using the balanced plant nutrition theory, we can predict that if a crop is provided with an adequate balance of essential nutrients, it will achieve optimal growth (Leibigs Law of the Minimum).
3. *Provides underlying logic for the phenomenon;* a theoretical foundation always backs up good research. Provides a framework for interpreting research results based on theory.
4. *Provides a guide for systematic relationships* of various research facts. This helps identify gaps, weak points and inconsistencies and, therefore, alerts researchers to future research needs. Theories encourage further research (known as the heuristic value of theory). In other words, they encourage thinking and arguments that may lead to new research ideas.
5. *Adds sense-making through the strategic synthesis of empirical data.* This is one of the main pillars of research. It guides the researcher's understanding not only of the "what" of natural phenomena but also of the "why" of their occurrence It suggests ways to make sense of the research data.
6. *Contributes to cumulative knowledge* by bridging gaps or revealing phenomena. Increases researchers' understanding of interconnections.
7. *Identify constructs and relationships* worth further research. Theories help researchers create theoretical and conceptual frameworks for their studies.

Theories can be applied at many stages of quantitative and qualitative (and mixed) research processes, including providing rationale for the study; defining the aim and research questions; considering the methodological stance; developing data collection and generation tools; providing a framework for data analysis and interpretation (S Reeves 2008; Kelly 2010; Bradbury-Jones et al. 2014; Stewart and Klein 2016). Many theories are best understood when diagrams display their basic premises visually (Newton Suter 2012).

The main advantages of theory compared to common sense are listed below (Nilsen 2015):

1. Theories are more consistent with existing facts than common sense, which typically means that a hypothesis based on an established theory is a more educated guess than one based on common sense.
2. Theories give individual facts a meaningful content.

2.6.1 Characteristics of a Good Theory

A Good Theory Must Be Falsifiable (Testable) a theory is testable if it is capable of failing some empirical test, i.e., a theory specifies outcomes under particular conditions, and if these outcomes do not occur, then the theory is rejected.

A Good Theory Must Have Explanatory Relevance Explanatory relevance is the ability to explain and predict reality. The explanation provided by a theory must offer good grounds for believing that the phenomenon would occur under specified conditions. Good theories must explain the target phenomena better than the alternative theories.

A Good Theory Has Predictive Power within its domain, a good theory should predict the phenomenon beyond which the theory was originally designed. Concepts and principles are substantive enough to enable us to predict future events.

A Good Theory Must Be Parsimonious Parsimony relates to a given theory's degree of freedom or simplicity. This attribute is based on the principle advanced by Father William of Ockham (law of parsimony), popularly known as Occam's razor, literally meaning slicing away unneeded evil pixies to produce a more parsimonious explanation (Dubova et al. 2025). This means that whenever you have two or more competing explanations for the same phenomenon, and they all account for the data, then the simplest one is the one you should favor, as it is most likely to be true. In other words, when you have competing explanations that sufficiently explain the observed evidence, the simplest theory (one with the smallest number of variables and assumptions) is the best.

A good theory can account for data, which is a measure of accuracy. A theory must explain most of the existing data within its domain.

A Good Theory Is Comprehensive it should be broad enough to explain a range of phenomena without special exceptions (the ability to apply across more than one field). Remember, theories are instruments of consolidating facts and simplifying explanations across the world. This means its value should extend far beyond the particular observations, laws, or sub-theories it was originally designed to explain. The more fields a theory is applied, the better.

Theory Testing

As we have seen, theory is integral to research, and research is integral to theory. Theory guides the development of many research questions, and research helps generate new theories and determine whether support for theories exists. What is important to remember is that theory is not fact; it is a belief about how things work; theory is a belief or "best guess" that awaits the support of empirical evidence through testing.

When we talk of testing a theory, we are not referring to testing a theory itself, as it is not testable in isolation. Instead, the validity of a theory is determined through empirically testing hypotheses derived from it and through observations of the phenomenon it describes. A hypothesis predicts the relationship between the variables included in the theory. The research outcome will reveal certain facts that support or refute the hypothesis. If the hypothesis is supported by the data, then the theory from which it was derived also gets support. Conversely, if the hypothesis is not supported, it questions the validity of the underlying theory. The process of testing hypotheses is essential for refining theories. Hypothesis testing is one of the main hallmarks of scientific research. The topics of formulating and hypothesis testing are covered in Chaps. 3 and 20, respectively.

Countless theories exist, and some are more well-developed and well-defined than others (Box 2.3 shows some theories in agriculture). More mature theories have devoted more

Box 2.3: Some Theories Useful in Agriculture

- Theory of planned behavior
- Diffusion of innovations theory
- Framer cooperative theory
- Rural sociology theory
- Social capital theory
- Systems theory
- Feminist theory
- Technology acceptance theory
- Theory of interpersonal behavioral
- Human ecology theory
- Value chain theory
- Production function theory
- Resource allocation theory
- Human capital theory
- Market acquisition theory
- Input-output theory
- Land tenure theory
- Sutainble development theory

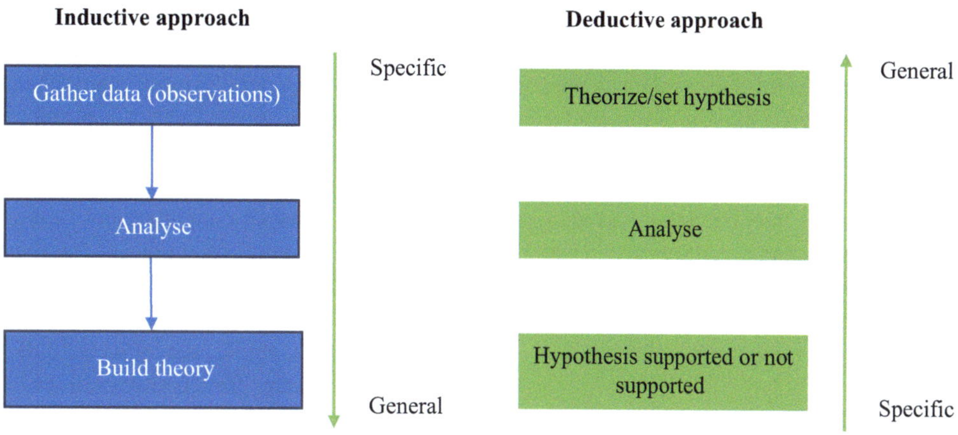

Fig. 2.8 Deductive and inductive approaches

time and effort to supporting research, and newer theories may be more tentative as supporting evidence is being developed. Theories are sometimes disproven and need to be scrapped completely or heavily overhauled as new research evidence emerges. Exploratory research leads to the birth of new theories as new phenomena and questions arise or as practitioners discover ways that existing theories do not fit reality.

2.6.2 Induction and Deduction Approaches

Induction and deduction are key terms in scientific research. The relationship mentioned earlier is that "theory is integral to research, and research is integral to the theory," which is well understood when we consider inductive and deductive approaches in research. These two approaches differ, but they are also complementary.

The inductive approach is called "a theory-building approach." It is the technique for moving from a set of observations to a theory. Once a theory has been formulated, it becomes a framework for enterpreting and explaining phenomena. For example, if a researcher observes a pattern in a specific data set, they can use an existing theory to explain the underlying principles governing that pattern.

Starting with a theory and using it to explain particular observations is known as a deduction. In other words, deduction takes the data from a particular case and applies the general theory to deduce an explanation for the data. Thus, induction is an approach to generating theories, and deduction is the approach to applying them (Fig. 2.8). We call deduction a theory-testing or hypothesis-testing approach, as already explained, that we do not test theory per se. It involves moving from general (theory) to specific.

2.6.3 Law, Proposition and Model

When a theory consistently produces the same outcomes, it may be elevated into law. Laws typically have broader applications than theories and enable precise predictions. For example, Newton's law of motion and the law of gravity began as theoretical observations but have been universally accepted laws due to their consistent and predictable nature. What initially starts as a theoretical observation can ultimately become a universally accepted law through rigorous testing and validation. Examples of laws applied in agriculture include Hardy- Weinberg law applied in agricultural genetics (Ferreira et al. 2017) and law of diminishing returns applied to help optimize levels of production (Liang 2023) and Leibegs Law of minimum, which guides fertilizer management (Ferreira et al. 2017; Rodriguez et al. 2019).

Proposition

A proposition is a tentative relationship between constructs stated in a declarative form. Examples are (1) An increase in soil fertility will cause an increase in crop production (2) Prompting the use of cattle vaccines will increase animal productivity (3) Value addition of soybeans will increase prices at the farm gate. Looking at these examples critically shows that propositions are basically associations between abstract constructs. These cannot be tested directly, but instead, they are tested indirectly by examining relationships between corresponding measures (variables) of these constructs.

The empirical formulation of a proposition, as stated as a relationship between variables, is what we call a hypothesis. Therefore, we can infer that propositions are used in the theoretical phase, while hypotheses are used in the empirical phase. Hypotheses are empirically tested using observed data and may be rejected or supported by empirical observations in the study.

In terms of scope and complexity, the variable (and its associated hypothesis) is the smallest unit, followed by a concept and construct, then propositions and finally, theory. Propositions will connect two or more constructs, while theories are more complex and of larger magnitude and scope compared to propositions or hypotheses.

Model

A model is a symbolic representation of the elements within a system. While a theory explains a phenomenon, a model is a structural representation of the concepts that make up the theory. Many of the concepts we deal with are infinitely complex, and models help simplify and clarify these concepts for better understanding and analysis. By using models, researchers and practitioners can better predict outcomes, plan interventions and optimize processes in various fields, including agriculture.

There are various types of models used across different fields:

(a) *Physical models* are tangible representations of objects or systems. For example, in agriculture, a physical model might include a scaled-down version of an irrigation system used to demonstrate water flow and distribution in a farm. Such a model can be used to demonstrate how the real irrigation system will work
(b) *Schematic representations* that show the relationships between different system components.
(c) *Computer simulations* use computer algorithms to simulate complex processes and predict outcomes. In agriculture, computer simulations can simulate crop growth under different environmental conditions, helping farmers and researchers make data-driven decisions about planting and harvesting. For example, Decision support systems that use computer programs and modules to make specific recommendations for effective land management.
(d) *Process models* represent a sequence of steps or stages within a process. For instance supply chain model that maps the steps from production to markets, identifying critical points where efficiency can be improved.
(e) *Statistical models* involve mathematical equations that describe relationships between variables. In agricultural research, statistical models can be used to analyze data on crop yields, soil fertility, and weather patterns to identify trends and correlations. Statistical models allow us to predict outcomes fairly accurately. In social sciences, statistical models are less precise, containing some degree of error resulting from the variability of human behavior and physical characteristics. Research studies are used to test these models and estimate their degree of accuracy

Theory and Model

While some literature may use the terms "theory" and "model" interchangeably, there is a distinction between them. A theory is an empirically supported description of the relationships among concepts, constructs and hypotheses that are advanced to explain and predict phenomena (Schindler 2019). The theory is a product of research and aims at generalizing studies. A model, on the other hand, aims to make an explanation or concept easy to understand. Generally, models are schematic representations of reality or of one's view of a possible world, constructed to improve one's understanding of the world and/or to make predictions. They involve assumptions, simplifications, and mathematical or computational formulations that allow researchers to make predictions or test hypotheses about a system's behavior under various conditions (Wunsch 1994). Whereas theories are broad, models are narrower and more concrete. For example, the AQUACROP model was utilized to simulate crop growth and yield under different agricultural scenarios (Aziz et al. 2022; Shan et al. 2022).

Recommended Further Reading Materials

1. Akos P, Kretchmar J (2017) Investigating grit at a non-cognitive predictor of college success. Review of Higher Education 40:163–186. https://doi.org/10.1353/RHE.2017.0000
2. Corley K, Gioia D (2011) Building Theory about Theory Building: what constitutes a theoretical contribution? 36:12–32. https://doi.org/10.5465/AMR.2009.0486
3. Creswell JW (2014) Research and design qualitative, quantitative, and mixed methods approaches. SAGE Publication Inc
4. Cronin P, Ryan F, Coughlan M (2013) Undertaking a literature review: a step-by-step approach. 17:38–43. https://doi.org/10.12968/BJON.2008.17.1.28059
5. Murray R (2011) How to write a thesis. Open University Press

Key Highlights of This Chapter

The research process consists of a preparatory phase and an implementation phase consisting of eight steps. The preparatory phase consists of five steps: (1) topic selection, (2) problem identification, (3) formulation of research objectives and hypotheses, (4) designing the research and (5) proposal development. The final product of these five steps is a research proposal. The implementation phase consists of three steps: (1) data collection, (2) data management, analysis, and interpretation, and finally, (3) reporting.

Role of Literature Review

Literature review is integral and serves multiple purposes in the research process. At the beginning of the study, during the writing of the proposal, it is useful for identifying knowledge gaps, refining the research problem and objectives and shaping the research design. At the mid-phase, it supports the development of the research context, guiding data collection and ensuring you align with the existing body of knowledge. Finally, during the reporting phase, literature is essential for framing the discussion, comparing findings with existing research and concluding.

The Role of the Academic Supervisor

The guidance and support of supervisors are essential during the academic research journey. A team of academic supervisors comprising two or more members assists at different research project stages. At the proposal stage, supervisors offer advice on topic selection, problem identification, proposal writing and preparing students for the defence. During the data collection phase, they guide students in collecting reliable data. In the third stage, supervisors guide suitable software and best practices for data analysis, interpretation and writing. In the final stage, supervisors support students through the examination process and coach them for presentations to examiners.

Role of the Student

As the primary driver of your research project, your role is vital, with supervisors guiding and supporting you through the research journey. You should, therefore, build a good relationship with the supervisor, work independently, follow instructions and meet deadlines. You also need to enhance your research and organizational skills. Other important skills needed to achieve successful outcomes include time management, writing skills, a positive attitude, effective communication and proactive problem-solving skills.

Tips on How to Succeed in Your Research Project

As the driver of the research project, avoid consistently blaming your supervisor for delays. Instead, focus on developing strong research methods skills, dedicating time to extensive reading and being patient and persistent.

Cultivate a good relationship with your supervisor by nurturing trust, being cooperative and contributing to collaborative projects.

Consider supplementing statistics courses with online resources and learning at least one data analysis software, such as SAS or R. Invest in scientific writing skills: You need writing skills to write scientific publications. Develop this through wide reading, "a Good reader is a good writer".

Total Research Setting

Starting your research project with a total research setting or visioning approach involves planning from inception to completion, clearly linking your topic to the problem and eventual solution. This proactive vision enhances motivation and helps maintain momentum even through challenges. It aids in identifying potential obstacles, planning contingencies, and aligning effort with your main objectives. Additionally, it ensures a more organized study execution and efficient resource allocation. In your plan, include the following;

- Comprehensive planning and roadmap
- Reflective practice
- Documentation and record keeping
- Visualization of success

Key Terminologies

Data is crucial for research, and it is important to determine what kind of data to collect.

Concepts: generalized ideas about people, objects, behaviors, and events. They assist in communication and vary from person to person and region to region.

Construct: Constructs are carefully chosen concepts used to explain specific phenomena. To become variables, they need to be clearly defined and measurable in research studies, referred to as operationalization.

Variables: Variables are things that change and include dependent, independent, categorical, demographic and extraneous variables, among others.

Confounding variable: Extraneous variables must be controlled because they are not part of the research objective. If not controlled, they confuse (confound) the study's results, making it difficult to interpret the findings. Therefore, researchers need to look for ways of controlling or minimizing them.

Practical and effective strategies for managing extraneous variables include employing sound research designs, statistical techniques and control strategies. Moreover, many concepts learned in research methods are geared toward estimating and addressing extraneous variables to safeguard the validity and reliability of research outcomes.

Validity

Validity refers to the truthfulness of the data or information collected and asks whether the research measured what it intended to. This concept of accuracy in research is the fundamental rule of thumb: every information we collect must be accurate.

Reliability is the extent to which a measure will produce consistent results. It thus relates to the consistency or dependability of a measure. A reliable measure or instrument will always give similar results.

Validity and Reliability Form the Foundation of Robust Research

Validity and reliability are the cornerstones of research. These concepts ensure that the conclusions drawn are accurate, meaningful and consistent. They are emphasized across disciplines. From understanding variables and measurements to complex designs and data collection methods, the main focus remains on achieving validity and reliability to maintain accuracy. Prioritizing these principles throughout the research process is essential to safeguard the integrity and credibility of the findings.

Errors in Measurements

Measurements are rarely perfectly reliable due to inconsistencies in instruments and humans. Systematic errors are predictable and biased, while random errors are unpredictable and impact the accuracy of measurements.

Theory

"In science, a theory is a systematic collection of concepts, ideas and propositions that explain the world. It provides answers to the question "why" and is fundamental to all scientific disciplines.

A theory is created, not discovered. It organizes and gives meaning to a complex array of individual facts and observations. Theory, research and practice are interlinked and form part of a continuum that aims to understand the determinants of behaviors, test strategies for change, and disseminate effective interventions.

The scientific community accepts a theory as true because it is grounded in actual data and observations that can be tested through research, but sometimes, it is not perfect. Science's progress is marked by progression over time to improve theories through better observations using more accurate instruments and more informed logical reasoning. This is the essence of science: constantly working toward building scientific knowledge. Universally, we aim to develop theories that will be generalized across time and space.

Importance of Theory in Research

Theories play a critical role in organizing and explaining specific facts, allowing for predictions, providing a logical foundation for research, guiding systematic relationships, adding sense-making to empirical data and contributing to cumulative knowledge.

Theory Testing

Theory guides the development of many research questions, and research helps generate new theories and determine whether support for them exists. We do not test theory directly; instead, the validity of a theory is determined by empirically testing hypotheses derived from it and through observations of the phenomenon it describes.

Induction and Deduction Approaches

Deduction and induction are key concepts in scientific research. They are complementary approaches used for generating and testing theories. Deductive is theory testing, while induction is theory building. Both approaches are essential for theory building and testing.

Law

When a theory consistently produces the same outcomes, it may be elevated into law. Laws typically have broader applications than theories and enable precise predictions.

Proposition

A proposition is a tentative relationship between constructs. It helps in understanding abstract associations and is tested indirectly through variables, forming a hypothesis.

Review Questions

1. Briefly summarize what you anticipate would be included in the planning and implementation phase of an agricultural research project.
2. Explain the purpose and activities of the literature review in the different phases.

3. Identify two common reasons for research project delays. For each, propose a specific, actionable strategy that you, as a student, could implement to mitigate that risk.
4. Explain the meaning of validity and reliability in research
5. Explain the concept of "theory testing" in research.
6. Clearly contrast theory, law and model in terms of their scope, function and the level of certainty or generality they imply.

Chapter Exercise

1. Imagine that you have just been assigned your research project supervisors. Describe three proactive steps you would take to build and maintain a strong, productive relationship with them throughout your research journey.
2. You are studying the impact of a new fertilizer (Variable A) on maize yield (Variable B) in certain regions.
   (a) Identify Variable A and Variable B, and state whether they are independent or dependent variables. Tip: What fertilizer? What will you measure?
   (b) Propose two potential "extraneous variables" that could affect your results in a farming context and explain how you might control for them to ensure the validity of your findings.
3. Consider a study on farmer adoption of drought-resistant crop varieties.
   (a) Give an example of a "categorical variable" you might include in your survey of farmers.
   (b) Explain how controlling for relevant variables would contribute to the validity of your study's conclusions.
4. Choose a common agricultural problem (e.g., nutrient deficiency in soil, pest outbreaks).
   (a) Briefly describe how a relevant "theory" (even a simple one) could guide research into solving this problem, explaining why the phenomenon occurs rather than just what happens.
   (b) Select three of the characteristics of a good theory and explain, in your own words, why each one is important for a theory to be useful in scientific research.
5. Reflect on the statement, "Theories, built through cumulative research, explain, predict and deepen understanding of phenomena." How does this idea apply to the ongoing development of climate-smart agricultural practices?

Answers to Review Questions

- Q1. Reference 2.1.1
- Q2. Initial phase (developing a research idea); Mid-phase (implementation planning and implementation); Final phase (reporting). Reference 2.2.2
- Q3. Causes of delays. References 2.2.2
- Q4. Validity and reliability. Reference 2.4.6 and 2.4.7
- Q5. Theory testing. Reference 2.6 and 2.6.1
- Q6. Reference 2.6.3

References

Agné H, Mörkenstam U (2018) Should first-year doctoral students be supervised collectively or individually? Effects on thesis completion and time to completion. High Educ Res Dev 37:669–682. https://doi.org/10.1080/07294360.2018.1453785

Akos P, Kretchmar J (2017) Investigating grit at a non-cognitive predictor of college success. Rev High Educ 40:163–186. https://doi.org/10.1353/RHE.2017.0000

Andrew A (2007) Supervisory power and postgraduate supervision. Int J Manag Educ 6:18–29. https://doi.org/10.3794/ijme.62.179

Auer M, Ricijas N, Kranzelic V, Griffiths MD (2024) Chapter 4: research design and measurement. Eval Health Prof 47:81–92. https://doi.org/10.1177/01632787231179460

Aziz M, Rizvi SA, Sultan M et al (2022) Simulating cotton growth and productivity using AquaCrop model under deficit irrigation in a semi-arid climate. Agriculture 12:242. https://doi.org/10.3390/AGRICULTURE12020242

Bai Y, Jin W-L (2016) Random variables and uncertainty analysis. In: Marine structural design, pp 615–625. https://doi.org/10.1016/B978-0-08-099997-5.00033-2

Bégin C, Gérard L (2013) The role of supervisors in light of the experience of doctoral students. Policy Futur Educ 11:267–276. https://doi.org/10.2304/pfie.2013.11.3.267

Bradbury-Jones C, Taylor J, Herber O (2014) How theory is used and articulated in qualitative research: development of a new typology. Soc Sci Med 120:135–141. https://doi.org/10.1016/j.socscimed.2014.09.014

Chugh R, Macht S, Harreveld B (2021) Supervisory feedback to postgraduate research students: a literature review. Assess Eval High Educ 47:683–697. https://doi.org/10.1080/02602938.2021.1955241

Cohen L, Manion L, Morrison K (2017) Research methods in education. https://doi.org/10.4324/9781315456539/RESEARCH-METHODS-EDUCATION-LOUIS-COHEN-LAWRENCE-MANION-KEITH-MORRISON

Corley K, Gioia D (2011) Building Theory about theory building: what constitutes a theoretical contribution? 12–32. https://doi.org/10.5465/AMR.2009.0486

Creswell JW (2014) Research and design qualitative, quantitative, and mixed methods approaches. SAGE Publication Inc

Cronin P, Ryan F, Coughlan M (2013) Undertaking a literature review: a step-by-step approach. Br J Nurs 17:38–43. https://doi.org/10.12968/BJON.2008.17.1.28059

Dubova M, Chandramouli S, Gigerenzer GI et al (2025) Is Ockham's razor losing its edge? New perspectives on the principle of model parsimony. Proc Natl Acad Sci 122:e2401230121. https://doi.org/10.1073/PNAS.2401230121

Fassett KT, Wolcott MD, Harpe SE, McLaughlin JE (2022) Considerations for writing and including demographic variables in education research. Curr Pharm Teach Learn 14:1068–1078. https://doi.org/10.1016/J.CPTL.2022.07.022

Ferreira IEP, Zocchi SS, Baron D (2017) Reconciling the Mitscherlich's law of diminishing returns with Liebig's law of the minimum. Some results on crop modeling. Math Biosci 293:29–37. https://doi.org/10.1016/J.MBS.2017.08.008

Fox JP, Marsman M, Mulder J, Verhagen J (2016) Complex latent variable modeling in educational assessment. Commun Stat Simul Comput 45:1499–1510. https://doi.org/10.1080/03610918.2014.939518

Gedamu AD, Gezahegn TH (2021) EFL supervisors' written feedback focus and language functions: a mixed methods study. Asia-Pac J Second Foreign Lang Educ 6:1–18. https://doi.org/10.1186/S40862-021-00125-2/TABLES/6

González-Ocampo G, Badia MC (2019) Research on doctoral supervision: what we have learnt in the last 10 years. In: Palgrave studies in education research methods, pp 117–141. https://doi.org/10.1007/978-3-030-23731-8_7

González-Varona JM, Martín-Cruz N, Acebes F, Pajares J (2023) How public funding affects complexity in R&D projects. An analysis of team project perceptions. J Bus Res 158:113672. https://doi.org/10.1016/J.JBUSRES.2023.113672

Gravett K, Kinchin I, Winstone N (2022) Evolving identities: a collaborative autoethnography in supervising and being supervised by colleagues. In: Landscapes and narratives of PhD by Publication: Demystifying students' and supervisors' perspectives, pp 119–134. https://doi.org/10.1007/978-3-031-04895-1_7

Groenvynck H, Vandevelde K, Van Rossem R (2013) The PhD track: who succeeds, who drops out? Res Eval 22:199–209. https://doi.org/10.1093/RESEVAL/RVT010

Guarimata-Salinas G, Carvajal JJ, Jiménez López MD (2024) Redefining the role of doctoral supervisors: a multicultural examination of labels and functions in contemporary doctoral education. High Educ (Dordr) 88:1305–1330. https://doi.org/10.1007/S10734-023-01171-0/TABLES/6

Hackman L, Mack P, Ménard H (2024) Behind every good research there are data. What are they and their importance to forensic science. Forensic Sci Int 8:100456. https://doi.org/10.1016/J.FSISYN.2024.100456

Haucke M, Hoekstra R, Van Ravenzwaaij D (2021) When numbers fail: do researchers agree on operationalization of published research? R Soc Open Sci 8. https://doi.org/10.1098/RSOS.191354

Hernández EH, Moreno-Murcia JA, Cid L et al (2020) Passion or perseverance? The effect of perceived autonomy support and grit on academic performance in college students. Int J Environ Res Public Health 17:2143. https://doi.org/10.3390/IJERPH17062143

Howards PP (2018) An overview of confounding. Part 1: the concept and how to address it. Acta Obstet Gynecol Scand 97:394–399. https://doi.org/10.1111/AOGS.13295

Hutcheson G (2011) Categorical explanatory variables. J Model Manag 6. https://doi.org/10.1108/JM2.2011.29706BAA.002/FULL/XML

Jager KJ, Zoccali C, MacLeod A, Dekker FW (2008) Confounding: what it is and how to deal with it. Kidney Int 73:256–260. https://doi.org/10.1038/SJ.KI.5002650

Jensen I (2015) Methodological challenges – from a supervisor's experiences. In: The Roskilde Model: problem-oriented learning and project work, pp 141–153. https://doi.org/10.1007/978-3-319-09716-9_9

Jiranek V (2010) Potential predictors of timely completion among dissertation research students at an Australian faculty of sciences. Int J Dr Stud 5:1–13. https://doi.org/10.28945/709

Kaur R, Hakim J, Jeremy R et al (2023) Students' perceived research skills development and satisfaction after completion of a mandatory research project: results from five cohorts of the Sydney medical program. BMC Med Educ 23:1–9. https://doi.org/10.1186/S12909-023-04475-Y/TABLES/1

Kelly M (2010) The role of theory in qualitative health research. Fam Pract 27:285–290. https://doi.org/10.1093/fampra/cmp077

Kember D, Corbett M (2018) Structuring the thesis: matching method, paradigm, theories and findings, pp 1–427. https://doi.org/10.1007/978-981-13-0511-5/COVER

Kirkup L, Frenkel RB (2006) Systematic errors. In: An introduction to uncertainty in measurement, pp 83–96. https://doi.org/10.1017/CBO9780511755538.008

Kivunja C (2018) Distinguishing between theory, theoretical framework, and conceptual framework: a systematic review of lessons from the field. Int J High Educ 7:44. https://doi.org/10.5430/IJHE.V7N6P44

Koohpaei A, Abardeh MH, Sharifi S et al (2024) Investigating the implementation challenges of the research doctoral program and providing related solutions: a qualitative study. BMC Med Educ 24:1–10. https://doi.org/10.1186/S12909-024-05815-2/TABLES/4

Kool H, Andersson JA, Giller KE (2020) Reproducibility and external validity of on-farm experimental research in Africa. Exp Agric 56:587–607. https://doi.org/10.1017/S0014479720000174

Kothari CR, Garg G (2019) Research methodology: methods and techniques. New Age International Publishers, New Delhi

Kumar R (2018) Research methodology: a step-by-step guide for beginners, 4th edn. S. publishers, New Delhi, p 34–56

Leung L (2015) Validity, reliability, and generalizability in qualitative research. J Family Med Prim Care 4:324. https://doi.org/10.4103/2249-4863.161306

Liang X (2023) Research on the influence of the law of diminishing marginal returns in agriculture. Adv Econ Manag Polit Sci 8:262–267. https://doi.org/10.54254/2754-1169/8/20230323

Marateb HR, Mansourian M, Adibi P, Farina D (2014) Manipulating measurement scales in medical statistical analysis and data mining: a review of methodologies. J Res Med Sci 19:47

Massyn L (2018) Enhancing completion rates of mini-dissertations for a professional master's degree: an integrated approach. Int J Manag Educ 16:114–120. https://doi.org/10.1016/J.IJME.2018.01.002

Matheka HM, Jansen EPWA, Hofman WHA (2020) Kenyan doctoral students' success: roles of motivation and self-efficacy. Perspect Educ 38:115–129. https://doi.org/10.18820/2519593X/PIE.V38I1.9

Meadmore K, Church H, Crane K et al (2023) An in-depth exploration of researcher experiences of time and effort involved in health and social care research funding in the UK: the need for changes. PLoS One 18:e0291663. https://doi.org/10.1371/JOURNAL.PONE.0291663

Mkhai E (2023) Factors contributing to slow completion rate among postgraduate students of the Information Studies Programme at the University of Dar es Salaam, Tanzania. IFLA J 49:554–563. https://doi.org/10.1177/03400352231174468

Mugenda A, Mugenda G (2019) Research methods: qualitative and mixed methods approach. Centre for Innovative Leadership & Governance, pp 100–102

Murray R (2011) How to write a thesis. Open University Press

Nangimah M, Walldén R (2023) How supervisors provide and students react to EAL thesis supervision: voices from Sweden and Indonesia. Front Educ (Lausanne) 8:1118436. https://doi.org/10.3389/FEDUC.2023.1118436/BIBTEX

Newton Suter W (2012) Introduction to educational research: a critical thinking approach. https://doi.org/10.4135/9781483384443

Nilsen P (2015) Making sense of implementation theories, models and frameworks. Implement Sci 10:1–13. https://doi.org/10.1186/S13012-015-0242-0/TABLES/2

Nimon K, Zientek LR, Henson RK (2012) The assumption of a reliable instrument and other pitfalls to avoid when considering the

reliability of data. Front Psychol 3:21611. https://doi.org/10.3389/FPSYG.2012.00102/BIBTEX
Noble H, Smith J (2015) Issues of validity and reliability in qualitative research. Evid Based Nurs 18:34–35. https://doi.org/10.1136/EB-2015-102054
O'Bryant SE, O'Jile JR, McCaffrey RJ (2004) Reporting of demographic variables in neuropsychological research: trends in the current literature. Clin Neuropsychol 18:229–233. https://doi.org/10.1080/13854040490501439
Omoya O, Jacob US, Odeyemi OA, Odeyemi OA (2024) Exploring perspectives: a scoping review of the challenges facing doctoral training in Africa. High Educ (Dordr) 89:1103–1127. https://doi.org/10.1007/S10734-024-01264-4/TABLES/2
Parija SC, Kate V (2018) Thesis writing for master's and Ph.D. program, pp 1–317. https://doi.org/10.1007/978-981-13-0890-1/COVER
Patten ML, Newhart M (2017) Understanding research methods: an overview of the essentials, 10th edn. Taylor and Francis
Patterson HD, Lowe BI (1970) The errors of long-term experiments. J Agric Sci 74:53–60. https://doi.org/10.1017/S0021859600020943
Preisser JS, Koch GG (1997) Categorical data analysis in public health. Annu Rev Public Health 18:51–82. https://doi.org/10.1146/ANNUREV.PUBLHEALTH.18.1.51
Reilly J, Shain C, Borghesani V et al (2024) What we mean when we say semantic: toward a multidisciplinary semantic glossary. Psychon Bull Rev 32:243–280. https://doi.org/10.3758/S13423-024-02556-7
Revuelta J (2010) Estimating difficulty from polytomous categorical data. Psychometrika 75:331–350. https://doi.org/10.1007/S11336-009-9145-9/METRICS
Rodriguez DGP, Bullock DS, Boerngen MA (2019) The origins, implications, and consequences of yield-based nitrogen fertilizer management. Agron J 111:725–735. https://doi.org/10.2134/AGRONJ2018.07.0479
S Reeves MAAKDH (2008) Why use theories in qualitative research? Br Med J 337:631–634. https://doi.org/10.1136/bmj.a631
Saini RK, Issac M, Kumar KJD et al (2023) Selecting a thesis topic: a postgraduate's dilemma. Ind Psychiatry J 32:19–23. https://doi.org/10.4103/IPJ.IPJ_62_22
Saunders MNK, Philip L, Adrian T (2019) Research methods for business students. Pearson, p 833
Schindler PS (2019) Business research methods. McGraw-Hill Higher Education, p 608
Shan Y, Li G, Su L et al (2022) Performance of AquaCrop model for maize growth simulation under different soil conditioners in Shandong Coastal Area, China. Agronomy 12:1541. https://doi.org/10.3390/AGRONOMY12071541
Sharma UN (2017) The role of supervisor and student for completing a thesis. Tribhuvan Univ J 31:223–238. https://doi.org/10.3126/TUJ.V31I1-2.25358
Shukla D (2023) A narrative review on types of data and scales of measurement: an initial step in the statistical analysis of medical data. Cancer Res Stat Treat 6:279–283. https://doi.org/10.4103/CRST.CRST_1_23
Smeltzer MP, Ray MA (2022) Statistical considerations for outcomes in clinical research: a review of common data types and methodology. Exp Biol Med 247:734. https://doi.org/10.1177/15353702221085710
Stewart D, Klein S (2016) The use of theory in research. Int J Clin Pharm 38:615–619. https://doi.org/10.1007/S11096-015-0216-Y/METRICS
Stewart CM, Schifter CC, Markaridian Selverian ME (eds) (2010) Teaching and learning with technology: beyond constructivism. Routledge Taylor & Francis Group
Sun B, Sutradhar B (2015) Bivariate categorical data analysis using normal linear conditional multinomial probability model. Stat Med 34:469–486. https://doi.org/10.1002/sim.6333
Symbaluk D, Hall R (2024) Research methods: exploring the social world in Canadian context. https://doi.org/10.31542/B.GM.6
Totten V, Simon EL, Jalili M, Sawe HR (2020) Acquiring data in medical research: a research primer for low- and middle-income countries. Afr J Emerg Med 10:S135–S139. https://doi.org/10.1016/J.AFJEM.2020.09.009
Trochim WMK (2009) The research methods knowledge base, 2nd edn. Cornell University
Uher J (2022) Rating scales institutionalise a network of logical errors and conceptual problems in research practices: a rigorous analysis showing ways to tackle psychology's crises. Front Psychol 13:1009893. https://doi.org/10.3389/FPSYG.2022.1009893/XML/NLM
van Rooij E, Fokkens-Bruinsma M, Jansen E (2021) Factors that influence PhD candidates' success: the importance of PhD project characteristics. Stud Contin Educ 43:48–67. https://doi.org/10.1080/0158037X.2019.1652158
Watson P (2008) Measuring postgraduate cohort throughput: a case study. S Afr J Higher Educ 22:728–735. https://doi.org/10.4314/SAJHE.V22I3.25813
Wilmink FW, Snijders TAB (1989) Polytomous logistic regression analysis of the general health questionnaire and the present state examination. Psychol Med 19:755–764. https://doi.org/10.1017/S0033291700024351
Wilson S, MacLean R (2009) Research methods and data analysis for psychology, vol 1. McGraw-Hill Higher Education, p 640
Wunsch G (1994) Theories, models, and data. Demografie 36:20–29
Xu KM, Meijs C, Gijselaers HJM et al (2020) Measuring perseverance and passion in distance education students: psychometric properties of the grit questionnaire and associations with academic performance. Front Psychol 11:563585. https://doi.org/10.3389/FPSYG.2020.563585/BIBTEX
Xypaki M, Sinakou E, Mejía-Cáceres MA, Schwambach A (eds) (2021) Storied doctorates. Springer International Publishing, Cham

3 From Topic Selection to Research Objectives and Hypothesis

Abstract

The chapter titled "From Topic Selection to Research Objectives and Hypothesis" outlines a comprehensive roadmap for transitioning from the research topic to the formulation of specific research objectives and hypotheses. Potential sources for research topics include literature reviews, personal field experiences, and current contemporary issues. The process of distilling a research topic into a clear research problem and objectives involves careful conceptualization. The research problem is a concise statement that guides the research. The research problem is supported by the statement of the problem (SOP). The SOP not only identifies the research gap but also concludes with a purpose statement that captures the overall aim of the study. From this research aim, researchers can then derive actionable research objectives, formulate a hypothesis, and generate pertinent research questions to narrow the focus of the study. The research objectives function as a guiding compass, ensuring that the researcher remains on track and aligned with the study's purpose. Additionally, the chapter delves into the formulation of hypotheses and testable assertions, which can emerge from both theoretical frameworks and empirical observations. Through practical insights and illustrative examples, this chapter equips researchers with the necessary tools and strategies to effectively transition from topic selection to clearly defined research objectives and hypotheses, ensuring clarity and coherence throughout the research process.

Keywords

Hypothesis · Conceptualization · Correlational objectives · Descriptive objectives · Causal objectives · Exploratory objectives · Purpose statement · Research objective · Research problem · Statement of the problem · Research topic · Research gap · Sources of research topics

Learning Outcomes

At the end of this chapter, you are expected to be able to:

1. Appreciate the role of a research problem in driving research.
2. Use the various sources of problems to identify a research topic.
3. Formulate a research problem from the topic using three easy steps.
4. Explain what constitutes a "good research problem."
5. Summarize how to write a statement of the problem.
6. Identify statements in the literature that point to a research gap.
7. Demonstrate writing a "statement of the problem."
8. Show the link between the purpose statement and the research objectives.
9. Demonstrate writing of a good research objective and hypothesis.
10. Distinguish between a causal objective and a descriptive objective.

3.1 Refining Ideas into Clear Researchable Components

Chapter 2 introduced an eight-step model for conducting research. The first three steps, (1) topic selection, (2) formulation of the research problem, and (3) development of research objectives, form the "conceptual foundation" of any research project. In this chapter, we delve deeper into the transition from a broad topic of interest to the formulation of clear research objectives and hypotheses. Having a general idea is not sufficient for meaningful academic inquiry. For research to be feasible, focused, and impactful, a broad topic must be systematically narrowed into a well-defined research problem (Ellis and Levy 2008; Willis 2023). This is the process of *research conceptualization* (Fig. 3.1).

J. Njeri Mugwe, S. Runo, *Research Methodology in Agricultural Sciences*, https://doi.org/10.1007/978-981-95-1892-0_3

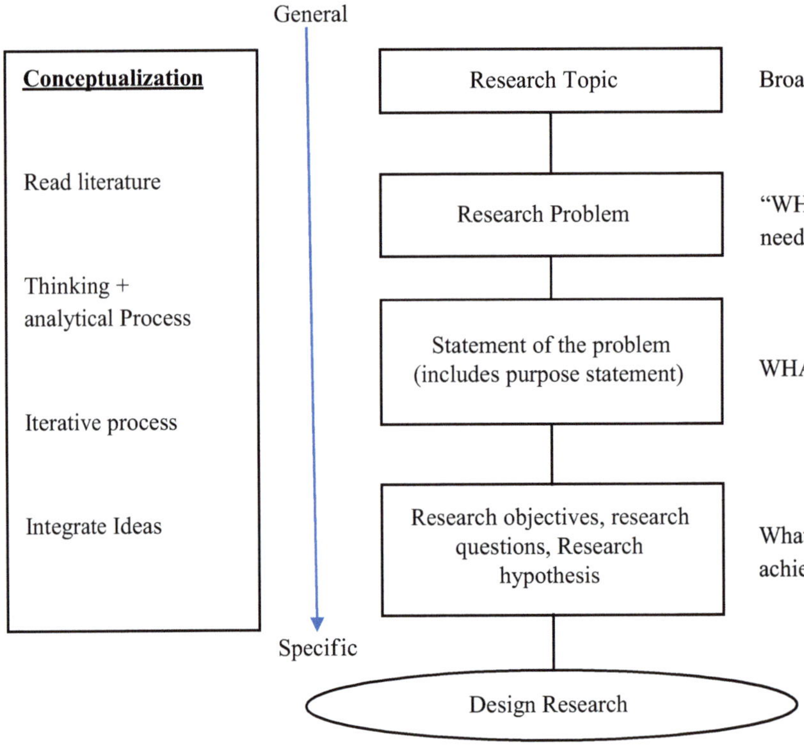

Fig. 3.1 Conceptualization process showing the steps from "Topic area to Research objectives"

Conceptualizing research in this way allows the researcher to:

- Avoid vague or overly ambitious investigations.
- Focus time and resources more effectively.
- Identify measurable variables and clearly defined concepts.
- Align the study with specific research goals or address gaps in existing literature.

This chapter provides a practical, step-by-step guide to help you transform a general idea into specific, researchable components. These steps include:

1. Topic selection: Identifying a general area of interest or concern.
2. Identifying and narrowing the research problem: Pinpointing a specific issue within the broader topic that warrants investigation.
3. Writing the statement of the problem: Clearly articulating the issue, its context, and its significance.
4. Formulating research objectives: Stating what the research seeks to accomplish.
5. Developing research questions: Translating objectives into answerable questions.
6. Formulating hypotheses: Developing testable statements or predictions (where applicable).
7. Linking to research design: Connecting your refined problem and objectives to the appropriate research methods and strategies.

3.2 The Topic Area Selection

Considerations in the selection of research topics

The topic is your broad problem area. It is the broad subject matter that you aim to study. To be able to set up your *research problem*, you must first select a research topic in your field of specialization. We emphasize that you must select a topic in your area of specialization. When a topic does not seem congruent with the degree name, students are often tasked to explain the relevance of their research topics to the degree of specialization during the examination.

Professional and personal interests are key in selecting a research topic. You should ensure that your research topic is related to your future career path and has the potential to contribute to the achievement of your career objectives. For example, most students taking PhD are interested in academic careers in teaching and research. If this is what you want, select a topic that you can teach easily and would like to pursue for future research. In addition, the research journey is long, and you should select a research area that is interesting to you to keep you motivated.

As you select the topic, think about your *scholarly identity*. What do you want to be your focus in the future? What would you want to be known for and make an impact on? If you already have an area of your passion, this will be an added advantage as you choose the topic. Think about an area where you want to make a positive change in the long term and what lasting impact you want to make.

Sources of Research Topics

Important sources of research ideas range from our experiences, theories, and academic writings. In summary, the following are the main sources of research topics:

1. Proposal from faculty or your supervisor.
2. Problems in the field reported by farmers and other stakeholders.
3. Research themes by agricultural organizations and funding agencies.
4. Emerging challenges and issues.
5. Personal experiences.
6. Academic literature: Thesis: Journals, Research reports, grey literature.
7. Agricultural meetings and forums, e.g., conferences, seminars, and symposiums.
8. Researcher's interest in a particular area.
9. Graduate courses taken.
10. Discussion with experts.
11. Existing theories and principles.
12. Mass and social media.

Proposal from Faculty Staff In certain instances, faculty staff and their partners may be actively involved in ongoing research projects. They might invite you to work with them. If you opt to join them and work on a topic aligned with their current work, it offers many benefits. For instance, you will get an opportunity to contribute to and gain from these existing projects, thereby promoting collaboration and synergy within the academic community. Moreover, in many cases, you stand to benefit from full funding. A "word of caution" as you work on an existing project, always establish a clear niche for yourself within the project. This involves demonstrating a comprehensive understanding and originality of the work during the proposal preparation and presentation. Proposal evaluators most often put students working in funded projects on task to explain their individual contributions.

Problems in the Field Problems and challenges from the field are rich sources of research topics. These may include practical experience in the field while working with farmers, traders, institutions, communities, and other agricultural stakeholders. They could also be problems reported by agricultural extension workers or reported in the dailies. Problems emanating from the field form a good source of research topics because, as a researcher, you get first-hand information on the problems, challenges, and issues being experienced in the field. The issues reported are current and urgent. For instance, extension officers working in the field may report attacks on crops caused by new pests or diseases. A researcher in crop protection can use this information to explore methods for controlling these pests or diseases. In this case, the researcher is addressing a practical problem.

Research Themes by Funding Agencies and Research Organizations International and regional research and funding agencies that support research work are guided by strategic plans or research themes. Similarly, national research organizations in your country will have strategic plans that will clearly outline priority research areas. For example, in Kenya, we have the Kenya Livestock and Agricultural Research Organization (KALRO). The priority research areas in the national research organization strategies would be good points to start in identifying a topic. Similarly, funding agencies, such as the Bill and Melinda Gates Foundation and the International Development Research Agency (IDRC), will also be guided by key research themes. These can also form good research topics.

Emerging Challenges and Issues These are modern or current issues affecting the world and societies. It is one of the most evident sources of research topics because it is real: A research problem can arise from problems facing communities, a nation, and society: Such problems include:

- COVID-19 pandemic effects on food security.
- Poor market access by farmers.
- Nutrition and economic status.
- Population dynamics, environmental degradation, and development.
- High rate of inflation affecting agricultural, inputs, and food prices.
- Poverty and food security in semi-arid areas.
- New pests and diseases affecting crops or livestock.
- Water scarcity, climate change, and rainfall unreliability.

Personal Experiences Every day, personal or professional experiences may lead us to identify a problem. This can guide your topic selection. Alternatively, we may encounter a question or questions that we would like to try and answer. This can originate from interesting challenges encountered in life e.g. low prices of agricultural commodities, low crop yield in a certain region due to soil acidity. Each observation can inspire a meaningful research topic.

Researcher's Interest Your interest as a researcher is paramount. The research journey is long and sometimes challenging, and therefore, interest will keep you motivated through the journey. By carefully observing existing practices in one's area of interest, one can formulate a researchable problem. For example, a student interested in the role of grain legumes in soil fertility can observe practices among the producers and read literature on the legumes to come up with a research topic.

Academic Reports and Writings Academic writings are extremely important sources of research topics. Research ideas exist in written records or literature such as books, theses, research projects, research papers, journal articles, and magazines. Reading books and journals in search of ideas is called a literature review (Chap. 4). As a student, you are encouraged to read avidly in your study area. As you read, note information about the ideas and study areas that the researcher did not explain well. These form good research topics.

Current information from researchers can be found in subject journals. Journal articles describe the entire research process. You will find research ideas well stated either in the introduction, conclusion, recommendation, or section on further research of most articles and research papers. At the conclusion, the articles state what else needs to be done, as well as critical gaps and missing information. These gaps are sources of research ideas. Your library will have a good mix of journal articles and other reports (see Chap. 4).

Agricultural Meetings and Forums Stakeholders in agriculture frequently organize forums and meetings in the form of scientific conferences, seminars, and symposia to share and disseminate knowledge. During such meetings, declarations are often made to highlight emerging issues, set new priorities, and call for "urgent action." These declarations are usually formalized and signed by major parties in the form of communiques. Such communiques can for a research topic and help identify relevant research problems for investigation. For example, the UN climate change conference in 2017 held in Bonn, Germany (COP23), declared the need to develop and implement new strategies for mitigation and adaptation in the agriculture sector, including reducing greenhouse gas emissions. Other examples of significant meetings include the "African Green Revolution Forum" (AGRF) held every year, and Conferences organized by National Agricultural Research Institutions. The AGRF of 2023 had the theme of "Recover, Regenerate, and Act, and at the end of the meeting, a communique was developed that emphasized the importance of soil health (https://agrf.org/summit/2023/). The communique underscored the foundational importance of soil health and the need to implement comprehensive soil conservation and restoration practices.

Advanced Graduate Courses As you undertake coursework, units taught form another potential valuable source of research topics. Most of the units will explore current issues and challenges facing different sectors. For example, in animal nutrition units, challenges of livestock production facing smallholder farmers, such as low-quality diets, are discussed. These may easily be developed into a research topic in animal sciences. Further, in graduate seminars, important research areas are often critically reviewed and may provide insight into knowledge gaps.

Consultation with Experts One of the most important characteristics of passionate researchers is a commitment to developing deep expertise in a specific area. As a result, most researchers become subject-matter experts in their area of profession. They will therefore be well-versed with current research challenges and opportunities in their domain. When consulted, they may provide useful guidelines on urgent problems and issues of concern. For instance, if you are interested in legumes, it will help to consult an expert in legume crops. You can get such an expert from a university or a research organization, or from academic research databases such as Google Scholar.

Theories The many theories in agriculture inspire many research problems (extensively discussed in Chap. 2). Researchers will aim to provide information that will confirm or disconfirm the given theory. For example, in agricultural extension, several technology adoption theories have been developed to explain the adoption of innovations and practices. (Box 3.1). One such theory is "The innovation diffusion theory," which focuses on the process through which technologies are tested and adopted. It identified four key factors that influence adoption: nature of innovation, the communication channel used, time, and nature of society. A researcher may, therefore, wish to investigate which of the four aspects is important in explaining the adoption of a given agricultural innovation. Looking at the theories in Box 3.1, we see that theories may be useful in suggesting interesting questions and guiding fieldwork.

Box 3.1: Examples of Common "Theories in Agriculture" and Hypothesis That Can be Developed

Conservation agriculture theory: approach to managing agroecosystems for improved and sustained productivity and food security while preserving and enhancing the resource base and the environment (*Investigate if various practices fulfil these characteristics of conservation agriculture*).

Mellor's theory of agricultural development (Mishra 2019): The supply curve for agricultural production in traditional agriculture is backward-sloping (investigate *if selected traditional agricultural systems are backward-sloping*).

The technology acceptance model (Rahimi et al. 2018): theory argues that when a farmer is presented with new technology, a number of motivating factors influence their decision about how and when they will implement and use the technology (*investigate which motivating factors will be significant in influencing use of a new technology of planting sweet potato).*

Media Both mainstream and social media reports provide information on diverse societal issues. This information helps in ideas that might give rise to research ideas. Just like social media, dailies report current issues affecting the nation and society. They mostly emphasize controversial issues. Therefore, dailies may give you an idea of a research issue.

For example, media reports about farmers facing rising input prices can motivate researchers to examine how these changes affect the profitability of specific agricultural enterprises. Likewise, reports of farmers dealing with low commodity prices at certain times of the year may encourage researchers to explore potential interventions to stabilize or enhance market prices. In this way, media content can act as a starting point for identifying urgent agricultural problems that require further investigation.

3.3 Research Problem Formulation

3.3.1 What Is a Research Problem?

A research problem is an unresolved question, issue, difficulty, gap in knowledge or controversy that calls for investigation (Creswell 2013; Creswell and Creswell 2022). It is an important step in any research study. Every research is problem-driven. The problem guides the research from conceptualization to data collection and analysis to provide the answer. As Fraenkel et al. (2012) state, a research problem is "anything a person finds unsatisfactory or unsettling, a difficulty of some sort, a state of affairs that needs to be changed," or anything that is not working as well as it might. This problem shapes the direction and scope of the study.

The research problem may be presented as a single sentence or as multiple well-structured sentences in a thesis or research report, depending on the nature of the research. It should clearly define the issue to be investigated, highlight its significance, and provide context about where it occurs. It should also guide the framing of the research objectives and hypotheses. Therefore the research problem must be well-defined and clearly outlined to ensure both feasibility and depth of study (Fraenkel et al. 2012). This makes the formulation of a research problem a challenging and time-consuming task that calls for patience.

A "*good research problem*" is focused and points to the need for investigation. Formulation of a good research problem is a "pain point" for most students. This is because it requires clarity, critical thinking and a deep understanding of the subject area. A broad problem, such as "Low agricultural productivity is a major cause of poverty," is not a good research problem. It is confusing regarding what is to be done, and it lacks focus and depth. Furthermore, you may not have the enormous resources required to carry out such a study. A well-defined research problem should narrow the scope by specifying *where*, *when*, and *how* the issue manifests. For example, refining the broad problem to "Low maize yields among smallholder farmers in region A due to limited access to irrigation technologies" makes the issue more manageable and researchable. This allows for clearer objectives, targeted data collection, and meaningful conclusions that can contribute to solving real-world problems within a specific context. This is discussed later in the chapter.

In your reading and analysis, you will be exposed to both practical issues, difficulties, controversies, and theoretical problems. They will arise from among the stakeholders and may include the following (Creswell 2014).

Practitioners and Farmers What are the difficulties and practical issues affecting farmers, consumers, input suppliers, and what knowledge is missing?

Academicians and Researchers What information is lacking to deepen their understanding of a particular issue? Are there controversies and contradictions in theory that need investigation?

Institutions What are the organizational and policy issues facing them? What knowledge is missing that can help improve their management and decision-making?

Problems facing practitioners and farmers are primarily practical in nature. These challenges often affect specific groups, such as farmer organizations, input suppliers, consumers, and traders. They manifest in the field as areas of concern for those directly involved in agricultural production and marketing, including inefficiencies in processes within institutions and barriers to improving overall performance. The focus of the study, therefore, is to identify and address these practical problems with the goal of implementing changes that enhance efficiency, effectiveness, and outcomes for all stakeholders involved. We provide examples of practical problems and the anticipated change (Table 3.1).

On the other hand, controversies and many theoretical problems also exist in agricultural research. These problems aim to expand understanding and knowledge rather than directly drive immediate change. They often arise from situations or relationships that are not well understood, contradictions between competing theories or perspectives, or unresolved questions that challenge existing assumptions. Addressing such theoretical problems helps build a stronger conceptual foundation, guiding future practical interventions and policy decisions. Examples of theoretical problems:

Table 3.1 Practical problems faced by practitioners and stakeholders and the anticipated change

Stakeholder	Practical problem	Anticipated change
Farmers and practitioners	Soil degradation due to low use of mineral fertilizers	Introduction of sustainable soil management practices such as rotation and use of organic resources to restore soil heath
Academics and researchers	Gaps in knowledge regarding the impact of climate change on crop production	Perform studies to understand the effects of climate change on various crops and develop adaptive strategies for farmers
	Need for a deeper understanding of the socioeconomic impacts of new agricultural technologies.	Conduct socioeconomic research to evaluate the impact of new technologies on farmers' outcomes and other livelihood variables
Institutions	Inefficiencies in agricultural extension services	Identify training and capacity building needs; carry out the training for extension workers to improve service delivery

- The relationship between soil microbial diversity and crop yields in intercropping farming systems.
- There is a controversy regarding the effectiveness of organics vs mineral fertilizers in terms of yields, nutrient use efficiencies, and sustainability.
- The economic implication of adopting conservation agriculture techniques is not well understood.

We can summarize a *research problem to include the following practical issues, controversies, and theoretical problems*

- Practical problem in the field, e.g., low soil nitrogen causing declining crop yields or emerging pests and diseases.
- A theory that is controversial and needs to be verified.
- An issue that does not seem right.
- Difficulty due to unknown facts.
- Lack of effective tools and information.
- Controversy that is currently being debated in the field.
- Something that conflicts with the general view or in the literature.
- Something that must be (re) evaluated, changed, transformed, or constructed, or needs new designs.
- A gap in the field's knowledge.
- Something the field has not or should not have finished researching.
- An unexplained observation, an observation that sticks out.
- Any questions that you want to answer.
- Statement to be investigated or examined.
- Any prediction or assumption you want to test.
- A concern raised in the field or in an organization that does not have an answer.

3.3.2 Importance of a Research Problem

A well-defined research problem is central to the success of any research endeavor. Its importance can be considered in three main areas:

A research problem provides direction and focus, grounding the study in a meaningful context (Ellis and Levy 2008). It thus *justifies* and provides relevance to the study. It explains why the research matters, not only to the investigator but also to the broader academic community and society. By situating the problem within current discourse or practice, the researcher underscores its contribution to understanding or addressing an important issue. For example, a study examining the adoption of climate-smart agriculture among smallholder farmers of region A would need to highlight the relevance of this issue for agricultural sustainability, smallholder farmers or food security in the region. Such framing clarifies the practical and scholarly value of the inquiry.

Second, the research problem is essential in *securing approval and institutional support* (McFadyen and Rankin 2016). In academic research, a clearly stated and justified problem is necessary for obtaining authorization from thesis supervisors, department committees, or ethical review boards. For applied or funded research, the problem statement must convince sponsors or donors that the proposed study is worthwhile, feasible, and aligned with their goals. A well-justified research problem conveys urgency and feasibility, and it positions the research as a credible and valuable endeavor that addresses real-world needs or knowledge gaps. In doing so, it increases the likelihood of the study receiving the necessary institutional and financial support.

Third, the research problem provides a *framework for evaluating the success of the research*. It serves as an anchor point from which the research objectives, design, and findings are derived and to which they return. A clearly defined problem allows the researcher to assess, upon completion of the study, whether the research addressed the issues originally proposed. In this way, the research problem contributes not only to the internal coherence of the study but also to its overall utility and impact. By returning to the research problem in the concluding sections of a report or thesis, the researcher demonstrates how the study has contributed to resolving or clarifying the issue under investigation.

3.3.3 Distilling the Research Problem

As a student, identifying and formulating a research problem is your responsibility. A single research topic can encompass many possible problems within it, which makes the process of narrowing down particularly crucial. This process, known as distillation, requires time, effort, and critical thinking. It involves reading widely, asking pertinent questions, and refining your ideas to identify a clear and researchable problem. Even experienced researchers find this challenging, as it necessitates the ability to spot new issues demanding investigation. To navigate this effectively, you need to cultivate curiosity, possess a solid understanding of your subject and understand research methods.

In reality, many unresolved practical problems and unanswered questions exist in the world today. However, not all of these issues qualify as research problems in their current form. Often, they are too broad, vague, or complex to be addressed directly through research. To make such problems researchable, they must undergo a process of clear definition and narrowing down, transforming them into specific, investigable questions—questions that a researcher can effectively study, whether descriptively or experimentally" (Chen et al. 2023).

While practical problems are indeed valuable, they primarily serve as starting points for deeper inquiry. Such problems guide researchers in pinpointing more specific and researchable issues that warrant focused investigation. For example, common challenges in agriculture, such as low yields, market failures, or pest invasions (e.g., locusts), are real-world concerns that impact food security. However these are broad issues that need further refinement. For instance, to effectively transform a broad issue like "low yields" into a research problem, a researcher must delve deeper. This might involve exploring the specific factors contributing to low productivity, examining the extent of the yield gap, or assessing the effectiveness of various strategies employed by farmers to enhance yields. By refining the inquiry in this manner, the problem becomes suitable for systematic study.

We provide a three-step process into this process of identifying a research problem:

(i) The selection of the topic area.
(ii) The selection of a general problem.
(iii) The reduction of the general problem to some specific, precise, and well-delimited. or a focused problem.

Step 1: The Selection of a Topic

The first step in locating a specific problem for the project or a thesis is to identify the topic (broad problem area) that is most closely related to your interests and professional goals. The choice of the topic area might be dictated by various factors: intellectual or academic training. Nowadays, topic choices are being influenced by social conditions, such as issues being perceived as a global endemic, e.g., climate change. Social and financial incentives also play a role in the choice of the subject matter for investigation. Common topic areas include: Maize production in smallholder farms, soil and water conservation, soil fertility, climate change issues, food security, marketing, value chains, communication and production efficiency (Box 3.2).

Box 3.2: Topic and its Dissection into General Problem Areas

1. Selection of the general problem
   - Dissect the broad topic area into subareas after:
     - In-depth literature review (use keyword to search for literature)
     - Consulting/brainstorming with those who have research and training experiences or specialized knowledge of the research area.

Soil and water conservation	Soil fertility	Marketing
Water harvesting structures	Mineral fertilizers	Pricing
Mulching	Organics	Value chains
Conservation agriculture	Grain legumes	Consumer preferences
Tillage practices	Tree legumes	Branding and value addition

Step 2: Selection of the General Problem Area

After identifying the topic write down as much detail as possible about this problem. This stage calls for an in-depth review of the literature. Dissect the broad problem area into a subareas (Box 3.2). Breaking down the broad problem into its subareas helps to clarify the problem so that the researcher can focus on a few questions. For example, if a student chooses soil and water conservation, typical problem areas of interest are water harvesting structures, mulching, conservation, and agriculture tillage practices. This could form a *general problem* that should be further reduced to a specific topic that is researchable.

Step 3: Reducing the General Problem to a Focused Problem

At stage 3, the general problem is distilled or reduced to a more focused issue (research problem). The refined research problem typically stems from the need to address a specific gap. We see that you will distil your ideas from a broad perspective to a specific idea, which forms the research problem (Fig. 3.2). With the research problem at hand, researchers proceed to determine the specific goals they aim to achieve and the hypothesis they seek to test.

3.3.4 Key Characteristics of a Good Academic Research Problem

A good academic research problem must meet the following criteria:

1. *Researchers' interest and competence:* You must align your interests and expertise. Always ensure you have the necessary skills, knowledge, and motivation to tackle the problem effectively. In a master's program, you could have just one year of research, while in a doctoral program, you might have 2.5 years.
2. *Practical feasibility*: be feasible within the practical constraints of time, budget, and resources. Select a problem that you can address and manage within the available financial resources and within a reasonable time frame. Additionally, the scope should be sufficiently narrow to be completed without overwhelming you. In the presentation of your results, avoid the pitfall of stating that you did not complete the study as required or that you omitted some variables due to insufficient funding.
3. *Contribute to knowledge:* should contribute to the existing body of knowledge by addressing gaps and providing new insights. It should have the potential for meaningful findings to advance understanding and offer solutions to relevant sites.
4. *Relevance and timing*. Address issues relevant to the current context. This ensures the findings will be applicable and relevant to contemporary discussions, policies and practices. Timely research resonates more with stakeholders and has a higher chance of making an impact.
5. *Clarity:* should be written in a clear, non-technical language without jargon; in a way, it stimulates a reader's interest. Avoiding jargon ensures that the problem is understandable and engaging. This is important for communicating its significance and garnering interest and support from the academic community.
6. *Theoretical relevance:* You must have a deep theoretical underpinning of the research problem. The research problem should fit into the broader context of current theory and relevant research. Avoid making assumptions or unsupported claims or statements.

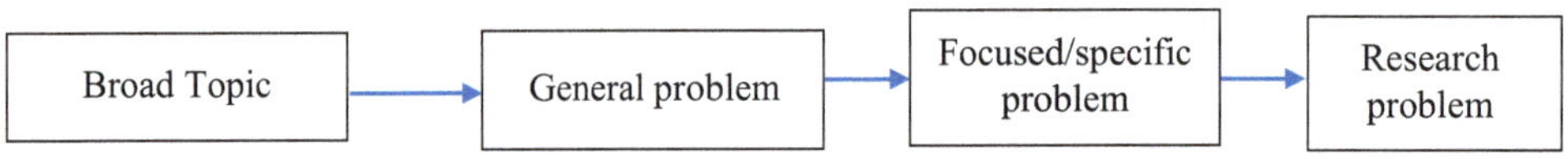

Fig. 3.2 Distillation of research problem from the broad topic involves a series of steps

7. *Ethical considerations:* should adhere to the moral principles of morality and ensure the welfare of participants and integrity throughout the research process. The ethical grounding promotes trust in the research findings.

3.4 The Statement of the Problem

3.4.1 Defining the Research Problem

Defining the research problem involves writing *a statement of the problem* to provide a clear description of the problem that is to be investigated through research (Chen et al. 2023). In a proposal, you will locate the "specific research problem" in the statement of the problem. Here, the problem is defined and is called the "statement of the problem." This is an extremely important section for clarifying your research problem.

Statement of the Problem

In the statement of the problem, the researcher clearly describes the research problem already identified (Chen et al. 2023). It is a very important section of any research because it introduces the reader to the importance of the problem. A problem statement summarizes "the context for the study" and the specific problem the study seeks to address. It should be short, one paragraph and at most three short paragraphs.

The statement of the problem also describes the issue motivating the research and shows the *knowledge gap* that the research intends to close. The gap provides direction and clarifies what is to be investigated. It clarifies the knowledge void. Knowledge void is what is unknown or poorly understood. It could also be an issue that is not completely understood. Some key characteristics of statement of the problem include focus, clear definition, conceptual clarity, and identification of variables (Box 3.3).

Box 3.3: Key Characteristics of Statement of the Problem

- Provides the focus and direction of the study. Recall that a problem can be something to be explained, to be further understood or to be addressed.
- A good statement of the problem clearly defines the problem, states the concepts to be related in the study, and identifies a feasible solution to the problem.
- A well-written problem statement helps to identify the variables to be investigated in the study
- Naturally leads to the formulation of a research question or objective (bridges the gap between the problem and the specific issue the research will address)

A well-formed problem statement of your proposal/thesis contains five parts. This includes the actual research problem identified above and four other parts (Creswell 2014).

- *Topic.* Provide an overview of your topic on how things should be *happening (Ideal).* This principal proposition provides general information on how things should be happening or expected to happen. *For example, population increase requires a doubling of food production to meet the increasing demand*
- *The problem:* This is the area of concern, controversy, and issue identified in the background. This is supported by justification and context.
- *Justification of the problem:* The evidence and relevance found in the past and present show that the problem exists, as well as what others have done. Context of your topic: current conditions, settings. Who is directly or indirectly involved? How are things happening, and what are the reasons for each case? Summarizes what is known from the literature. For instance, we have not doubled the food production for several reasons.
- *Research gap:* Deficiencies in evidence: our current knowledge about the problem: Rationale/justification of your topic: Identify what is not known. Justify the presence of the gap between what is expected and what is the context. This must be the reason for you to go for research. If we know this, we can be able to address the gap in two.
- *Intent:* Intervention to fill the gap and solve the problem

Examples

Topic: Soybean production in Embu County. KARLO-Embu's major development goal is to increase soybean production by providing farmers with high-yielding varieties. Over the years, KARLO has made concerted efforts to facilitate the promotion of improved varieties of soybeans.

Research problem: This is a reality; indicate what is happening. For instance, "However, recent studies have reported low adoption of the varieties in Embu County" (*this is a practical problem*).

Justification and context: Estimates indicate that only about 40% of smallholder farmers use improved varieties. In addition, yields are low, averaging 0.5 tha^{-1} (REF). Studies conducted elsewhere indicate that various socio-economic factors affect the adoption of agricultural technologies (REFS).

Research gap: In Embu, factors responsible for the low adoption are not clearly understood because few studies are available.

Intent: The intervention to fill the gap and solve the problem. "This study aims to determine the factors influencing the adoption of improved soybean varieties in Embu County."

How to Link the Statement of the Problem with the Objectives

Using the above format, we can build up a statement of the problem by first thinking about our topic, which gave rise to the research problem in your three initial steps of formulating the research problem. Following this, the rest of the subsections are shown in Table 3.2, which shows the topic, research problem, gap, intent, and objectives.

Topic

Region A is among the most densely populated counties in the country, with an average of more than 700 inhabitants/km^2. In addition, farm sizes in the region have reduced, ranging between 0.5 ha and 1 ha per household. Consequently, this has prevented further opportunities, resulting in continuous cultivation without replenishing soil nutrients.

Research Problem

The farmlands in the region are experiencing a decline in agricultural productivity due to continuous cropping, limited access to sufficient inorganic fertilizers, and low availability of local organic fertilizers in terms of quality and quantity.

Justification

Nutrient depletion is very high at the extent of 4.4 million tons N, 0.5 million tons P, and 3 million tons K every year from cultivated Africa's soil (REF). Yields from farms average 1.2 t ha^{-1}, with a yield gap of 3.5 t ha^{-1}. In addition, apart from soil nutrient depletion, the majority of smallholder farmers in Tharaka-Nithi depend on rain-fed agriculture (REF); hence, water is also a limiting factor to crop production. The declining soil fertility and poor soil water conservation cause crop failure and reduce soybean yields for farmers in the region.

Deficiencies

Few studies, however, have focused on the potential of soil replenishment inputs combined with soil water conservation techniques to improve soil fertility and agricultural productivity in the study area. Most studies have concentrated on the use of sole soil fertility replenishment inputs without water conservation strategies.

Purpose

The purpose of this study will, therefore, be to determine the effect of soil fertility inputs and tied ridging on soybean yields and soil chemical properties.

Table 3.2 Illustration of how to link the topic with the research problem and research objectives

Topic	Subject area	Broad topic (Crop production, Agricultural Marketing)	Overview of the topic (present the research problem)
Subject area	Issue of concern Problem Controversy Difficult situation	Research problem	
Evidence for the issues	Literature Practical experiences Statistics	Context of the study	Current scenario
Deficiencies	What is missing? What is not known? What do we need to know?	Gap Justification/rationale	Rationale
Relating Discussion to Audience (Beneficiaries)	How will you address what we need to know?	**Intent** of the study	Purpose statement; leading to research questions/ objectives/hypothesis
Intervention or Action points	What specific things will we do? What will the study aim to achieve?	Purpose and objectives hypothesis	

Specific Objectives

1. To assess the effect of soil fertility inputs and tied ridging on growth parameters and soybean yields.
2. To assess the effect of soil fertility inputs plus tied-ridging on soil pH, total N, available P, soil exchangeable bases, and soil organic carbon.

Hypothesis

1. Soil fertility inputs plus tied ridging will have no significant effect on growth parameters and soybean yields.
2. Soil fertility inputs plus tied ridging will have no significant effect on soil pH, total N, available P, soil exchangeable bases, and soil organic carbon.

An example from a master's student thesis is given in Box 3.4.

Box 3.4: Example of Statement of the Problem Linked to Objectives and Hypothesis

Title of the study: Effects of maize (*Zea mays* L.)—Soybean (*Glycine max* (L.) Merrill) intercropping patterns on yields and soil properties in two contrasting sites of Embu and Meru counties, Kenya (Matusso 2015).

In the central highlands of Kenya, maize is an important staple food. It plays a key role in feeding the growing population ***(Topic)****. Low soil fertility is one of the major constraints causing decreased maize production and other staple foods, leading to hunger and poverty.* ***Reality (Research problem)****: Over 90 percent of farmers are resource-poor smallholders. Continued mining of nutrients, poor nutrient conservation practices, increasing population growth and land scarcity worsen the situation (Ref). The maize-soybean intercropping system is being promoted as one of the options for increasing crop production by the farmers. Evidence indicate that improved intercropping systems with integration of legume as part of ISFM technologies have large scope for increasing crop yields and household income.* ***Justification (Context)****: However, small-scale farmers of this region lack information on optimum cropping patterns and the profitability of maize-soybean intercropping systems* ***(Gap)****. The purpose of this study is therefore to evaluate the effect of maize-soybean intercropping pattern on maize and soybean yields and soil properties* ***(Intent)****.*

Specific objectives

1. Determine the effects of different maize–soybean intercropping patterns on yields, N uptake, and soil properties.
2. Determine the effects of different maize–soybean intercropping patterns on light interception and leaf area index in two contrasting sites.
3. Quantify the land equivalent ratio (LER) of different maize–soybean intercropping patterns in two contrasting sites.

Hypotheses

1. The MBILI intercropping pattern gives significantly higher yields than the conventional intercropping pattern.
2. The 2:6 intercropping pattern of maize-soybean is more efficient in light use and gives a higher leaf area index than the conventional (1M:1S) intercropping pattern.
3. The 2M:6S intercropping pattern of maize-soybean significantly improves soil properties more than the conventional intercropping pattern.
4. Nitrogen uptake and land equivalent ratio are significantly higher in 2M:4S intercropping pattern than the MBILI (2M:2S) intercropping pattern.

Source: Master's thesis (Matusso 2015).

3.4.2 Research Gap in the Statement of the Problem

A research gap is an extremely important aspect of any research. Some students often face challenges in finding a research gap but once you know what you are looking for and how to look for it, it is easy (Chen et al. 2023). Many Universities offer guidelines on identification and writing of research gaps in their graduate school websites, and you could benefit from such information. A research gap reveals "*missing information in literature and practice*" about your topic. Filling this gap gives value to your research. It also justifies your study and motivates you because you will be eager to contribute to a worthwhile goal. Your supervisor will be keen and look forward to reading your research gap. Without a research gap, it will be difficult to show what your study contributes and likely to face difficulties when defending your work to the academic board.

Knowing how to identify a research gap will help you locate the gap easily. It all starts from the topic, as demonstrated in this chapter and also in the samples of the statement of the problem. After identifying the topic, you are able to pick keywords to guide your literature review. As you read through the literature, look for issues that are not fully addressed. You will be able to discover areas that need more research attention. Some areas not fully addressed could be policies that have nothing to do with research. Avoid such and concentrate on those that give a research gap. This could be in the form of unanswered questions, contradictory findings, and changing trends that do not have answers yet. Literature review and wide reading, as emphasized in Chaps. 2 and 4, are crucial activities for any researcher. An extensive literature review and good research are related. We emphasize that the *"secret to good research is extensive literature review."* See Chap. 4 on how to review literature.

Specific Guidelines for Finding Research Gaps

As you read through literature in research papers and theses, pay attention to these key sections: (1). Introduction section, and (2). Discussion, conclusions, and recommendations. These sections are likely to highlight research gaps. Look for the following statements that express a research gap;

- Research recommendations include...
- A key question remains....
- It is important to address.......
- It has not been reported ....

Such statements will allow you to identify research gaps resulting from the following deficiencies:

(a) *Knowledge gap:* This expresses the difference between *what is known and what is required to be known.* This can result from limited research in certain aspects, inadequate information or contradictory findings from different authors, or outdated information from recent developments. Look for areas where the current literature is scarce, conflicting findings or recent developments have not been addressed. For example, in agricultural research, changing environmental conditions might necessitate updates in crop resilience. Indeed, one of the key knowledge gaps is the need to develop climate-smart agricultural practices. Similarly, new climatic patterns that cause the emergence of new pests and diseases and could impact crop production, which older studies did not anticipate

To confirm a knowledge gap, you need to extensively review the available literature, including meta-analysis and systematic reviews (Chap. 4). These two types of literature have the benefit of presenting comprehensive information.

(b) *Research methods gap:* Look at the *data collection and analysis methods* and assess any inadequacies. Are the methods available giving valid and reliable data? Are the methods flexible enough to be applied by farmers? For instance, when measuring soil pH, examine if there are reliable methods that farmers can use to determine soil pH on their farms accurately. Are these methods cost-effective and easy to use in agricultural settings? Many current studies on precious agricultural technologies are based on these arguments.

Look and evaluate the *population or sample* used in previous studies. Do they reveal any discrepancies? For example, consider whether the sample size was adequate and if the sampling method was appropriate. If the previous research primarily focused on a specific demographic or geographic area, it may not provide sufficient information to extend findings to other populations. This becomes a good research gap.

Evaluate data analysis techniques employed. Are they robust? Are there more advanced methods that could yield better insights? For example, recent studies have justified the use of the "double hurdle models" and argued that they are more robust than the singular logistic regression or probit regression in studying the adoption of technologies among farmers. The gap in the research method might also involve the lack of interdisciplinary approaches that integrate quantitative and qualitative data (Chap. 15).

Sample statements expressing research gaps are in Box 3.5. You will find that researchers use certain language and statements to express the research gaps. Box 3.5 provides guidelines on the statements that you could borrow from as you write your research gap.

Box 3.5: Sample of Statements that Express Research Gap

1. Despite extensive research on mineral fertilizers, there is still a lack of understanding about reducing greenhouse gas emissions in farming systems.
2. Previous studies have mainly focused on cereal crops, but traditional crops such as arrowroots have been given little attention.
3. While significant progress has been made in a specific area, there is still a gap in knowledge regarding another area.
4. Existing literature has identified an issue, but more investigation is needed.
5. The current research lacks consensus on a particular topic.
6. Past studies have primarily examined collective marketing on cereals, leaving gaps in our under-

standing of how it can be enhanced in perishable fruits.
7. Despite efforts to address low soil nitrogen, there is still a need to explore ways to improve nitrogen use efficiency on smallholder farms.
8. While some research has been conducted on economic factors influencing the adoption of organic inputs, gaps persist in our knowledge concerning the influence of institutional factors.
9. Existing literature has largely overlooked the impact of soil labile P on plant P uptake, which warrants further investigation.
10. Although previous research has explored a specific area, there is still limited understanding of another related area
11. Author A's method underestimates the values, while another author's method overestimates them. We therefore lack……
12. Most studies have focused on certain populations, with scarce information on vulnerable communities living in ……

3.4.3 Purpose Statement

A purpose statement or intent of the study gives the overall direction or focus of the study. It logically follows the research gap. It encompasses the aims or goals the researcher hopes to achieve with the research. We can see that the purpose statement is generated from the research problem and emanates from the gap identified. It supports the problem and clarifies the knowledge to be generated. It "feeds" *the general objective and describes how the knowledge will be generated.* The purpose statements should be clear and indicate the study's concepts, variables, population, settings, and relationships. The variables and the target population should be consistent with the variables and target population operationalized in the methods section of the study. It should be stated objectively without bias and lead the researcher to generate the answers.

Depending on the nature of the study, it is advisable to end your problem statement with a purpose statement. The purpose statement should come at the end of the problem statement and should convey the focus of the study. It is usually expressed in a declarative manner, starting with the phrase: *The purpose of the study is ….,* followed by neutral verbs such as, *to investigate, to examine, to compare, to explore, to find out, to inquire, to determine,* among others depending on whether the study will follow quantitative or qualitative approach.

A quantitative purpose statement is often one sentence that identifies variables, their relationships, participants, and the research location. Three examples are given: (1) Biophysical (2) Survey.

Biophysical purpose statement: The purpose of this study is to relate nitrogen fertilizer levels to maize performance in region A.

Survey purpose statement: "The purpose of this study is to relate socioeconomic and farm factors to the adoption of mulching technology on smallholder farms in Region A." In this example, variables such as gender will be considered independent variables that will influence the dependent variable, e.g. "adoption."

3.5 Research Objectives, Research Questions and Research Hypothesis

The development of the research objective, along with a supportive hypothesis and research question, is a crucial step in producing relevant results for evidence-based practice (Ellis and Levy 2008; Chen et al. 2023). The research objectives, questions, and hypotheses are complementary and play an important role in narrowing the purpose statement. They are almost mirror images of each other. Depending on the University guidelines, you may not need to write all three of them.

3.5.1 Research Objective

The research objective is a statement of the research study's goals. It is the focus of the study. For example, if your objective is to evaluate the effect of five diets on milk yields of dairy goats, it means that at the end of the study, you should be able to demonstrate how the five diets affected the milk yields. Objectives tell us what the study aims to achieve. They operationalization of the study. They are active statements that use active words. Action words such as to determine, To Differentiate, to examine, to evaluate, to assess and to compare are used. Objectives avoid us of words like: To understand, to study to learn.

In essence, we say that "*Objectives drive the study*" because they guide every stage of the study from the design, methodology, data analysis, and conclusion. Specifically, they:

- Influence the choice of research design and methodological approach, whether the study will be a survey, experiment, case study, or qualitative inquiry.
- Shape the formulation of hypotheses (in quantitative research).

- Clarify the key variables under investigation, thus enhancing focus and supporting effective evaluation.
- Help the researcher remain precise and focused on what needs to be accomplished.
- Determine the type of questions to be asked; research questions must align directly with the stated objectives.
- Guide the review and organization of relevant literature by highlighting the key themes and issues.
- Direct decisions regarding data collection and analysis, including the selection of participants, research instruments, and study sites. This helps avoid collecting irrelevant or excessive data.
- Shape the data analysis strategies and structure of the research report.
- Guide the content of the conclusion section by providing a clear benchmark for assessing whether the study achieved its intended goals.

Qualities of Effective Research Objectives
Effective research objectives are "*Purpose aligned*" and "*Contextualized*," meaning they should be tailored to your specific study and its environment. Table 3.3 gives key qualities of effective objectives.

How to Write a Research Objective
Examples of Good Objectives

1. To determine the effect of sorghum-green gram intercrop and phosphorus application on the yield of sorghum and green gram in region A.
2. To examine the effect of phosphorus application on green gram nodulation in sorghum-green gram cropping systems in region A.
3. To evaluate the yield advantage of sorghum and green gram in intercropping systems over the pure stand in region A.

These objectives use active words, are specific, and are complete. They state what will be measured.

Examples of poorly stated objectives:

1. To evaluate bean varieties in Kiambu County. (lacks specificity; not clear what will be measured).
2. To determine the effects of soil fertility decline. (ambiguous and incomplete; what will be measured?).
3. To evaluate the effect of market liberalization (wide and broad; does not specify the outcome variable to be measured).
4. To carry out characterization of smallholder farmers (What specific characteristics, for what purpose).
5. To carry out a survey on pests and diseases (vague, unspecific targets, what crops, which pets and diseases).

Table 3.3 Qualities of effective research objectives

Logical flow (do they flow from the purpose statement):	Must directly follow your statement of the problem and purpose. Consider them as sequential steps that build upon one another, creating a coherent path toward your study's conclusion (see Fig. 3.1)
Conciseness	State your objectives as briefly and clearly as possible. Avoid unnecessary jargon or overly complex sentences
Unambiguity (Clarity?)	Meaning that each objective must be absolutely clear. There should be no room for misinterpretation regarding what you intend to do or achieve. Anyone reading your objective should grasp it immediately
Specificity (define what you achieve and how)	Must be specific and crystal clear, leaving no room for misinterpretation about what you intend to do or achieve. Clearly define *what* you will achieve and *how* you will do it, using precise action verbs
Relevance (are objectives linked to the research problem	Ensure each objective directly links to your research problem and the broader goals of your research project. Ask yourself: "Why am I doing this?" The answer should always connect back to your study's core purpose
Interconnectivity (Are the objectives linked?)	When viewed together, your objectives should illustrate a complete and logical sequence of steps for your entire research project. They should form a coherent set that, when fulfilled, leaves no gaps in addressing your research problem from start to finish
Achievability (Are they realistic?)	Your objectives must be realistic and achievable within your given resources (time, budget, access to data). Consider: "Will I be able to gather the necessary data?" and "Is the intended outcome actually possible?" The type of data required should be clear or strongly implied
Ethical soundness (Is there any ethical misconduct?)	All your objectives must be ethically responsible. Ensure they do not cause harm or place undue burden on any participants, the environment, or any community involved in your research (see Chap. 6)

Types of Research Objectives
We can categories research objectives into four main categories (Chasokela et al. 2025):

1. Exploratory objectives.
2. Descriptive objectives.
3. Causal objectives.
4. Correlational objectives.

Exploratory Objectives They aim to investigate a topic about which little is known or understood. They focus on gaining an understanding of the phenomenon's nature, identifying potential variables or generating hypotheses for further studies. *Example*: To examine indigenous farming practices used by local communities to understand their effectiveness in crop production and soil management.

Descriptive Objectives They aim to describe the characteristics and behavior of a phenomenon. Examples include:

1. To describe the types of horticultural crops grown by farmers in County B.
2. To find out the opinion of farmers about the availability of markets for bananas in County C.
3. To determine nitrogen fertilizer equivalency values of *Tithonia diversifolia*, *Senna spectabilis*, and *Calliandra calothyrsus*.

Causal Objectives Seek to establish causality or predict outcomes based on identified patterns or associations

1. To investigate the effect of organic, inorganic, or combined N sources on maize yield and nitrogen use efficiency
2. To determine the effects of different soil-incorporated organic inputs on selected soil chemical properties and maize yields

Correlational objectives aim to examine, determine, or analyze the relationship or association between two or more variables within a study population.

Comparative Objectives
Examples:

- To examine the influence of transaction costs on smallholder farmer participation in collective marketing
- To determine household and farm characteristics influencing farmers' decision to adopt or not to adopt the introduced integrated soil fertility replenishment technologies

The causal objectives, particularly those that involve experimental designs, are often highly valued in academia, and hence, manuscripts emanating from such studies are likely to appeal to the international audience. This is because establishing a causal relationship is a fundamental goal of scientific inquiry. Correlational objectives are also powerful because they can be conducted in diverse contexts. They also have a high likelihood of getting published in high-impact journals, especially if they uncover novel and unexpected associations. Correlational objectives' ability to predict is highly applicable in theory and practice. Descriptive objectives carry the least likelihood of publication in high-impact journals. They may not directly appeal to an international audience because they focus on description, which is limited in generalization. Most times, such publications will only appeal to local journals.

It is, therefore, advisable to think critically about the potential of publishing your work as you choose the type of objective. Your supervisor will pay close attention to aligning the chosen objective with the research problem. Seek their guidance and insights about the most appropriate research objective to pursue.

3.5.2 Research Hypotheses

A research hypothesis is a statement of expectation or prediction that will be tested by research, also called "Educated guess," "Intelligent guess," or "Proposed explanation." Hypotheses are used in quantitative studies, and they narrow the purpose statement to specific predictions.

Definition by Orodho and Kombo (2002): Hypotheses are educated guesses or predictions about possible differences, relationships, and causes of research problems. These predictions are derived from existing theories, previous research, or concepts.

Fraenkel et al. (2012) simply define a hypothesis as a prediction of the possible outcomes of a study. In formulating a hypothesis, a researcher makes a prediction of the research outcome and writes down a statement of the expectations about the relations among the variables in the research study. This written statement forms part of your research plan and report. The research collects data to support or refute the hypothesis, but not to prove a hypothesis. This is very important to note. Discussion on how to write a hypothesis and testing is in Chaps. 6 (research proposal) and 20 (Quantitative data analysis), respectively.

Types of Hypotheses

Null Hypothesis and Alternative Hypothesis

Null hypothesis (Ho): The null hypothesis assumes and states no relationship between the independent and dependent variables. It is the most valuable hypothesis in the scientific research method and is tested using statistical analysis.

Alternative hypothesis: also called research hypothesis (Ho): is inverse or opposite of the null hypothesis. You will use the alternative hypothesis to state that there will be a difference based on the results from past research or an explanation or theory reported earlier.

Directional and non-directional hypothesis

Directional: It specifies the existence of a relationship and also the expected direction. It will state the nature of the relationship as either positive or negative. Other terms used include decrease or increase, higher or lower, and more or less that a researcher expects to emerge in a relationship. The particular direction expected is based on what the researcher has found in the literature, in theory or from personal experience.

Examples:

1. *There will be a significant positive relationship between nitrogen levels and maize grain yields*
2. *There will be a significant positive relationship between infestation by red spider mites and the yields of French beans*
3. *There is a significant negative relationship between the intensity of use of manure and the amount of fertilizer used in county A*
4. *Application of manure plus fertilizer significantly increases soybean yields than the application of sole*

Non directional: It reflects a relationship between two variables, but does not specify the anticipated direction and nature of the relationship, e.g., whether positive or negative.

Examples:

1. *There will be a significant relationship between nitrogen levels and maize grain yields*
2. *There will be a significant relationship between infestation by red spider mites and yields of French beans*
3. *There will be a significant relationship between the intensity of use of manure and the amount of fertilizer used in County A*
4. *Application of manure plus fertilizer has no effect on soybean yields*

Associative and Causal Research Hypothesis

Associative: It reflects a relationship between two or more variables of interests in natural settings without manipulation. It is mainly used in correlational studies to hypothesize and explain relationships. For example, in farm size is related to the age of farmers.

Example: Farm size is related to the age of farmers in smallholder farms of A county.

Causal: It predicts a cause-and-effect relationship between two or more variables in an experimental setup. An experimental research hypothesis states the expected relationship between the independent and dependent variables Examples of objectives and their Null and Alternative hypotheses are shown in Table 3.4.

How to Write a Hypothesis

In your academic research work, you will most likely get guidelines on whether you need to write the Null and Alternative hypotheses. The easiest way of starting to write a hypothesis is to ask a question and then rephrase that question in a form that assumes no relationship (Null) or has a relationship (alternative).

The hypothesis needs to include variables (Willis 2023). You need to state the variables in an order, starting with the independent variable (position one), followed by the dependent variable (position two). If the intention is to compare groups, you need to explicitly state the groups, e.g., users and non-users of a certain technology. If variables are related, specify the relationship among variables. You then predict the changes you expect in your groups, such as less or more or no changes. You will then test the predictions using statistical procedures.

Hypothesis in Qualitative Research

In qualitative research studies, hypotheses are not tested; rather, qualitative inquirers are more likely to generate new hypotheses as a result of what they find out as they collect and process data (Willis 2023; Okoye and Hosseini 2024). Researchers will observe patterns and relationships in the natural setting rather than hypothesizing what such patterns and relationships might be beforehand. However, some qualitative researchers do state some of their ideas before they begin a study, but these are usually called "propositions" rather than hypotheses. Propositions are not intended to be tested but instead are used to guide the study in data collection and analysis.

Table 3.4 Examples of null and alternative hypotheses in causal relationships

Objective	Null hypothesis	Alternative hypothesis
To determine the effect of different levels of nitrogen and phosphorus on maize yields	There is no significant effect of applying different levels of nitrogen and phosphorus fertilizer on maize yields (comparing groups)	Application of different levels of nitrogen and phosphorus fertilizer significantly affects maize yields (non-directional)
To evaluate the effect of applying manure plus fertilizer on soybean yields	There is no significant effect of applying manure plus fertilizer on soybean yields (comparing groups)	Soybean yields from sole manure will be significantly higher than those with fertilizer (directional)
To assess the influence of the age of the farmers on participation in the collective marketing of tissue culture bananas.	The age of a farmer does not have a significant effect on participation in collective marketing of tissue culture bananas (relationship-based)	Age of farmers has a significant effect on participation in collective marketing of tissue culture bananas (non-directional)
To determine the relationship between the number of cattle and farmers' annual income?	The number of cattle does not significantly increase the annual income of farmers (relationship-based)	The age of farmers significantly increases the annual income of farmers (Directional)
To evaluate the relationship between red spider mites and the yield of French beans	There is no significant relationship between infestation by red spider mites and the yields of French beans	There will be a significant positive relationship between infestation by red spider mites and yields of French beans (Directional)

Qualitative researchers commonly start their studies by stating research questions. They pose a central question that asks for an exploration of the phenomenon. This is because in qualitative research, the intent is to explore the complex set of factors surrounding the phenomenon and, in the end, present the varied perspectives or meanings that the participants hold.

3.5.3 The Logic of Hypothesis Testing

In an experiment, testing of the null hypothesis tells us whether the results are due to the effect of the manipulation of the dependent variable or due to chance (Whitley and Ball 2002; McNulty 2022). In studies where the independent variables are not manipulated, for example, in causal-comparative studies involving cross-sectional surveys, hypothesis testing helps determine whether observed differences or associations between variables are statistically significant or if they could have occurred by chance.

In hypothesis testing, you choose an alpha level (Ruxton et al. 2015). This is the significance level or the probability of rejecting the null hypothesis when it is true, also known as Type 1 error. A common value used for alpha is 5% or 0.05. Analysis of the data gives the *p-value,* which is compared to the pre-chosen alpha value. A result is statistically significant when the p-value is less than the chosen alpha.

- *If p-value > alpha*: Fail to reject the null hypothesis (i.e., not a significant result). Treatments were not significant.
- *If p-value < = alpha*: Reject the null hypothesis (i.e., significant result). Treatments were significant.

The significance level can be inverted by subtracting it from 1 to give a confidence level of the hypothesis given the observed sample data; e.g. if you had chosen 5%, it would be 100% − 5% = 95%. We can express this as the test failing to reject the null hypothesis at a 95% confidence level. Details of hypothesis testing are covered in Chap. 20.

Here is an example of research questions followed by a restatement of each as a possible hypothesis:

Question: what is the effect of manure, lime, and P fertilizer on soybean grain yields?
Hypothesis: Application of manure, lime, and P fertilizer significantly increases soybean grain yields?
Question: Is maize grain yield higher under manure combined with mineral fertilizer than in sole manure application?
Maize grain yield under sole manure is significantly higher than from manure combined with mineral fertilizer

Research hypotheses are important in providing direction to a research study, but it is possible to proceed with research without a hypothesis. A single research problem may give several hypotheses. For example, a research problem on the causes of low yields of sorghum in semi-arid areas may have hypotheses about soil, pests and diseases, soil moisture conservation, weeding frequencies, and so on.

In a quantitative study, there are two main reasons for stating a hypothesis: (1) An indication that you have sufficient knowledge in the area to carry out the study and (2) It gives direction to the data collection, analysis, and interpretation, informing you of the procedures to follow and the type of data to collect (Box 3.6 is on characteristics of good hypothesis).

Box 3.6: Characteristics of a Good Hypothesis

- Clarity: good hypothesis should be simple, precise, and clear
- Specific: should be specific with simple terms
- Predicted outcome: Should state the predicted outcome that is reasonable
- Measurable variables: Defines variables in measurable terms. For a hypothesis to be testable, the variables must be testable. If there are no means for measuring the variables, then it is impossible to collect the data necessary for testing the validity of the hypothesis. Sometimes, researchers make many propositions and hypotheses that are essentially untestable.
- Testable: Testable using statistical procedures in a reasonable amount of time. A testable hypothesis is verifiable. This means deductions, conclusions, and inferences can be drawn from empirical observations. A testable hypothesis allows the researcher to support or not support the hypothesis by observation and data collection.
- Consistent with known facts: Should be consistent with the existing body of knowledge

3.5.4 Research Questions

Like the research objectives and hypothesis, they narrow the purpose statement to specific questions researchers seek to answer. Unlike the purpose statement, researcher questions could be more than one. They are used in both quantitative and qualitative studies, but their elements differ depending on the type of research you are conducting. In quantitative research, the question relates to attributes or variables. In qualitative research, the questions include the central concept being explored. This central concept is called the central phenomenon.

Recommendation for Further Reading

1. Willis LD (2023) Formulating the research question and framing the hypothesis. https://home.liebertpub.com/rcare 68:1180–1185. https://doi.org/10.4187/RESPCARE.10975
2. Chen S, Sharma G, Muñoz P (2023) In pursuit of impact: from research questions to problem formulation in entrepreneurship research. Enterp Theory Pract 47:232–264. https://doi.org/10.1177/10422587221111736
3. Ruxton GD, Wilkinson DM, Neuhäuser M (2015) Advice on testing the null hypothesis that a sample is drawn from a normal distribution. Anim Behav 107:249–252. https://doi.org/10.1016/J.ANBEHAV.2015.07.006
4. Whitley E, Ball J (2002) Statistics review 3: hypothesis testing and P values. Crit Care 6:222–225. https://doi.org/10.1186/CC1493/FIGURES/1

Key Points in This Chapter

Refining ideas into clear, researchable components involves research conceptualization and refinement to develop researchable components.

The Topic Area The topic area encompasses your broad problem space, where your professional and personal interests are essential for selecting a research topic. Ensure it aligns with your career goals and reflects your scholarly identity. Key sources for research ideas include personal experiences, theories, and academic writings.

1. Proposal from faculty or your supervisor
2. Problems in the field reported by farmers and other stakeholders
3. Research themes by agricultural organizations and funding agencies
4. Emerging challenges and issues
5. Personal experiences
6. Academic literature: Thesis: Journals, Research reports
7. Agricultural meetings and forums, e.g., conferences, seminars, and symposiums
8. Researcher's interest in a particular area
9. Graduate courses taken
10. Discussion with experts
11. Existing theory and principles
12. Mass and social media

What Is a Research Problem?

A research problem is an unresolved question, issue, difficulty, gap in knowledge or controversy that calls for investigation. Every research is problem-driven. Once identified the research problem becomes a compass for the study.

Practically, a research problem can be summarized as practical issues, controversies, and theoretical problems.

A "good research problem" is focused and points to the need for investigation. Broad problem areas are not good research problems because they lack focus and are confused about what is to be done.

Importance of a Research Problem

Research problems justify the time and resources invested in research. They also help in securing of approval and support from relevant stakeholders. Finally, they provide a benchmark for evaluation.

Distilling the Research Problem

As a student, identifying and formulating a research problem is your domain. To formulate a research problem successfully, the problem is distilled from the topic through a process of conceptualization. This requires time, effort, and critical thinking.

In the search and identification of a research problem, one can distinguish three steps:

(i) The selection of the topic area.
(ii) The selection of a general problem.
(iii) The reduction of the general problem to some specific, precise, and well-delimited or focused problem.

Characteristics of a Good Research Problem

A good academic research problem should align with the researchers' interests and expertise. The problem should be timely and relevant and have a strong theoretical underpinning. Finally, it must contribute to knowledge and adhere to ethical considerations, promoting trust and integrity throughout the research process.

Defining the Research Problem

Defining the research problem involves writing a statement of the problem to provide a clear description of the issue to be investigated. This "statement of the problem" also reveals the knowledge gap that the research intends to fill.

The statement of the problem is eventually linked with the objectives and hypothesis of the study through the purpose statement.

Research Gap

A research gap is a crucial aspect of any research, representing the missing information for existing literature. Identifying this gap justifies the research and captures your supervisor's interest.

Purpose Statement

A purpose statement defines the overall direction and focuses of the study derived from the research gap. It outlines the aims or goals the researcher hopes to achieve with the research.

Research Objectives, Research Hypothesis, and Research Questions

Research objectives, hypotheses, and questions are complementary, and their main role is narrowing the study to actionable components. They are crucial for defining what the research intends to achieve and shaping the study design.

Research Hypothesis

The hypothesis is a testable statement predicting the outcome of a study, "educated guess," or "proposed explanation. Hypotheses are formulated to be tested through empirical observation and statistical analysis in quantitative research.

Review Questions

1. What is the difference between a research topic and a research problem?
2. List three common sources of research topics.
3. Briefly explain the "role of a research problem in driving research." Why is it considered a "guiding compass" for the study?
4. What is the function of a purpose statement in research?
5. What are the key components of a good research problem?
6. Given a research topic of your choice, outline the steps to convert it into a research problem.
7. Write a brief statement of the problem for a study on low maize yields in acidic soils.
8. Identify three indicators of a research gap when reading academic literature.
9. Using an example, demonstrate how to derive a hypothesis from a research objective.
10. Explain the link between a purpose statement and research objectives using an illustrative case.

Chapter Exercise

1. Research problem distillation exercise:
   - Choose a broad topic related to your field (e.g., climate-smart agriculture).
   - Identify at least three potential research problems within that topic (General problems).
   - Narrow one of these problems down into a specific, researchable problem statement.
2. Based on your chosen problem above:
   1. Write a clear purpose statement.
   2. Develop 2–3 research objectives.
   3. Draft 2 research questions.
   4. Propose one testable hypothesis.
3. Write a short reflection (200–300 words) on the most challenging aspect of transitioning from topic selection to hypothesis development. What made this process difficult?
4. You have read several journal articles on conservation agriculture (CA). You find statements like "Existing methods of CA are not widely adopted" or "Limited data exists on the long-term effectiveness of minimum tillage
   (a) "How would you use such statements to identify a "research gap" for a potential study?
   (b) Based on this draft a "statement of the problem," ensuring it includes a research gap and concludes with a purpose statement.

Answers to Review Questions

- Q1: Reference: 3.1.
- Q2: Reference: 3.2.
- Q3: Reference: 3.3.1 and 3.3.2.
- Q4: Reference: 3.4.3.
- Q5: Reference: 3.3.4.
- Q6: Reference: 3.3.3
- Q7: Reference: 3.4.1.
- Q8: Reference: 3.4.2.
- Q9: Reference: 3.5.2.
- Q10: Reference: 3.4.

References

Chasokela D, Sibanda L, Chasokela D, Sibanda L (2025) Writing a research proposal. https://services.igi-global.com/resolvedoi/resolve.aspx?doi=104018/979-8-3373-0806-7.ch011 261–290. https://doi.org/10.4018/979-8-3373-0806-7.CH011

Chen S, Sharma G, Muñoz P (2023) In pursuit of impact: from research questions to problem formulation in entrepreneurship research. Enterp Theory Pract 47:232–264. https://doi.org/10.1177/10422587221111736

Creswell J (2013) Steps in conducting a scholarly mixed methods study. Discipline-Based Education Research Group Speaker Series

Creswell JW (2014) Research and design qualitative, quantitative, and mixed methods approaches. SAGE Publication Inc

Creswell WJ, Creswell DJ (2022) Research design, 6th edn. SAGE Publications Ltd

Ellis TJ, Levy Y (2008) Framework of problem-based research: a guide for novice researchers on the development of a research-worthy problem. Inf Sci Int J Emerg Transdiscipl 11:017–033. https://doi.org/10.28945/438

Fraenkel JR, Wallen NE, Hyun HH (2012) How to design and evaluate research in education. McGraw-Hill Education, p 642

Matusso JMM (2015) Effects of maize (zea mays l.) – soybean (glycine max (l.) merrill) intercropping patterns on yields and soil properties in two contrasting sites of Embu and Meru counties, Kenya

McFadyen J, Rankin J (2016) The role of gatekeepers in research: learning from reflexivity and reflection. GSTF J Nurs Health Care 4:82–88

McNulty R (2022) A logical analysis of null hypothesis significance testing using popular terminology. BMC Med Res Methodol 22:1–9. https://doi.org/10.1186/S12874-022-01696-5/TABLES/4

Mishra S (2019) Agricultural development and economic transformation: promoting growth with poverty reduction, by John W. Mellor. Published by Palgrave Macmillan, New York, NY, 2017, pp. 259, ISBN: 978-3-319-65259-7. Aust J Agric Resour Econ 63:385–387. https://doi.org/10.1111/1467-8489.12302

Okoye K, Hosseini S (2024) Understanding dependent and independent variables in research experiments and hypothesis testing. R Programming 99–107. https://doi.org/10.1007/978-981-97-3385-9_5

Orodho AJ, Kombo DK (2002) Research methods, Nairobi

Rahimi B, Nadri H, Afshar HL, Timpka T (2018) A systematic review of the technology acceptance model in health informatics. Appl Clin Inform 9:604–634. https://doi.org/10.1055/S-0038-1668091/ID/BR180026R-34/BIB

Ruxton GD, Wilkinson DM, Neuhäuser M (2015) Advice on testing the null hypothesis that a sample is drawn from a normal distribution. Anim Behav 107:249–252. https://doi.org/10.1016/J.ANBEHAV.2015.07.006

Whitley E, Ball J (2002) Statistics review 3: hypothesis testing and P values. Crit Care 6:222–225. https://doi.org/10.1186/CC1493/FIGURES/1

Willis LD (2023) Formulating the Research Question and Framing the Hypothesis. https://home.liebertpub.com/rcare 68:1180–1185. https://doi.org/10.4187/RESPCARE.10975

4 Literature Review

Abstract

A literature review is an analytical examination of scholarly work of existing studies and related work on a specific topic. It is a mandatory part of an academic journey and helps students develop critical thinking skills. There are many types of literature reviews; the most common are: narrative, systematic, and critical. Within the research context, a literature review plays a crucial role in identifying the state-of-the-art, enriching the broader ongoing scholarly dialogue, highlighting gaps in the existing literature, sharpening the research problem, refining research approaches and methodologies, and overall increasing the credibility of the research study. The main sources of literature review include primary, secondary, and tertiary sources available online or in print copies. While internet offers many advantages there are also disadvantages, the main ones being information overload and fake information. Several databases provide information on agriculture, including both general and specialized platforms. Key databases include CABI, AGORA, and PUBMED. Others include SCOPUS and Web of Science. Each of the database serves specific purposes. Steps to searching, analyzing, and writing literature start with identifying a topic, reading, and analyzing and end with synthesizing and writing. Literature reviews of masters and Ph.D. thesis work are organized thematically and guided by research problems and objectives. The chapter ends with a short section on the theoretical and conceptual frameworks to emphasize the close interdependence between the literature review and the theoretical and conceptual frameworks. Differences between the theoretical and conceptual frameworks are also presented.

Keywords

Conceptual framework · Google Scholar · Gray literature · Critical review · Databases Internet · Library Catalogue · Narrative review · PUBMED · Scholarly literature · Scopus · Search engine · Sources of literature · Systematic review · Synthesizing literature · Theoretical framework · Repositories · Web of Science

Expected Learning Outcomes

By the end of this chapter, you will be able to:

1. Describe types of literature review and their suitability.
2. Explain the role of literature in research.
3. Identify sources of literature.
4. Explain the role of databases in agriculture in the available literature.
5. State the main agriculture-based and multidisciplinary databases.
6. Highlight the role of CABI as a source of literature.
7. Describe how to search and evaluate literature.
8. Explain how to synthesize and organize literature.
9. Describe the steps in writing a literature review.
10. State components of a theoretical framework and a conceptual framework.
11. Distinguish between theoretical and conceptual frameworks.

4.1 Overview

4.1.1 What Is a Literature Review?

A literature review is an in-depth analytical examination of scholarly literature on a topic or research question. It demonstrates knowledge and understanding of the topic of study and identifies state-of-the-art research in that particular field. As a researcher, it helps you to summarize key concepts, theories, and findings. An effective literature review should be discursive prose, not a listing and summary of individual pieces of literature (Ellis and Levy 2008). As you carry out the literature review, you should strive to create a cohesive and engaging narrative that synthesizes diverse sources

J. Njeri Mugwe, S. Runo, *Research Methodology in Agricultural Sciences*, https://doi.org/10.1007/978-981-95-1892-0_4

within the research topic. You should also aim to present a compelling argument (Lingard 2018).

When Do You Carry out a Literature Review?

In the course of your studies, you will be required to undertake a literature review for various purposes and at different times. You may carry out a literature review in the following key activities:

1. *Term papers of units*: Tutors teaching various units give assignments on literature review in the form of term papers.
2. *Learning research methodology*: Part of the literature review in a research methodology course.
3. *Generation of research ideas*: A preliminary search to generate research ideas for your research project.
4. *Literature review chapter*: Gathering information for your literature review chapter in your research proposal or thesis report.
5. *Thesis and research paper writing:* Literature reviews to strengthen the discussion in your thesis report and research papers.
6. *Review paper:* Literature review to write a review paper in your area of specialization.

4.1.2 Types of Literature Reviews

Literature reviews can take various forms, each serving a different purpose in the research process (Paré et al. 2015). The main ones include narrative, systematic, meta-analysis, critical, methodological, and scoping reviews.

Narrative literature review A narrative literature review, often a traditional review, presents literature in a story-telling format, providing a comprehensive overview of the topic (Baumeister and Leary 1997; Green et al. 2006). It usually covers historical developments, key concepts and major debates. Unlike other reviews, it does not follow a strict structure, and the author may include personal insights. Narrative reviews are essential in research as they provide updated insights and stimulate research ideas, enabling researchers to formulate hypotheses or determine research questions.

Systematic literature review follows a structured and methodological approach to collecting and analyzing literature. It includes a well-defined research question and search strategy to identify relevant studies, which include multiple databases and information sources, including gray literature, followed by analysis and interpretation of results. Data from selected studies are synthesized using a predetermined methodology, often involving statistical analysis. Many systematic reviews employ statistical techniques to amalgamate findings from independent studies into a unified interpretation known as meta-analysis (Rosenthal 1995; Walsh and Downe 2005; Borenstein et al. 2009; Fisher et al. 2022). Through statistical analysis from multiple studies, meta-analysis studies generate more precise and reliable estimates of intervention effects (Fisher et al. 2022). This means the pooled summary statistic provides a more robust estimate of the true effect. Meta-analysis is often used to draw generalizable conclusions from a body of research.

The term meta-analysis should not be confused with meta-synthesis. Whereas meta-analysis involves quantitative data synthesis, meta-synthesis is the nonstatistical technique used to aggregate, evaluate, and interpret findings from textual information (Cronin et al. 2013). This approach synthesizes various studies to identify common themes and core categories. Unlike quantitative meta-analysis, which reduces data to numerical summaries, a meta-synthesis adopts a qualitative perspective. It offers richer and more comprehensive information on a specific research topic, along with a clearly explained methodology that outlines the number of papers reviewed. For example, this method has been utilized in studies by Low et al. (2023) and Jones-Garcia and Krishna (2021).

Critical review focuses on evaluating the literature's quality, validity, and relevance (de Klerk and Pretorius 2019; Garrod 2023). It involves a critical appraisal of research methods, study designs, and the strength of the evidence. Critical reviews aim to ascertain strengths and weaknesses in the literature and assess the credibility of research findings (Paré et al. 2015; Saunders et al. 2019).

Methodological review A methodological review focuses on the research approaches, strategies, and data collection techniques employed (Gentles et al. 2016). It is meant to provide a structure for understanding and undertaking a particular research approach, method, or methodology. By drawing on a wide body of methodological knowledge, researchers can use methodological reviews to improve their research skills and techniques. Research gaps may also be identified through methodological reviews, but they are not the primary focus.

Scoping review Designed to explore the breadth and depth of a research topic by mapping the existing literature (Peters et al. 2021). They provide a broad overview of the available studies without delving into in-depth analysis. A scoping literature review will aim for thorough coverage and encompass gray literature while applying specific criteria for inclusion and exclusion to ensure relevance to the research focus (Pham et al. 2014; Tricco et al. 2016).

The literature review types depend on the subject matter, specific research objectives, and nature of the project. The following are useful considerations as you carry out your literature review:

- The narrative literature review is most suitable for *introductory research projects*, such as those undertaken by undergraduates. It provides a comprehensive overview of existing literature on the topic, which can help students build their understanding of the subject matter.
- A systematic literature review requires advanced skills and is therefore suitable for *master's or PhD students*. These reviews involve rigorous methodology and can provide high-quality evidence.
- Critical reviews are appropriate for research projects requiring an *in-depth literature evaluation*. Students critically assess the strengths and weaknesses of the studies. This is particularly useful for advanced students (master's and PhD) who are required to critically evaluate methods, quality, and gaps in the literature.
- Methodological reviews are most suitable when evaluating *existing methodologies* used in previous studies. The review helps establish the strengths and weaknesses of the methodologies, followed by the selection of the best methodology for the study
- Scoping reviews are most suitable during the *initial stages* of the research work when the aim is to provide an initial understanding and breadth of a topic. It helps establish the foundational base of the research area.

Review Paper

When conducting a literature review for your research project, you may write a standalone review paper. We encourage graduate students to publish a review paper early in their research journey. Publishing a review article helps the student become an expert in their area of study by enhancing their understanding of the topic and building their publication records, which are essential for academic and research careers. The review paper also teaches students how to review and synthesize existing literature and sharpen their skills in identifying research gaps. It lays the groundwork for the thesis research project. The importance and popularity of review articles have grown over the years and Bahishti 2021 notes that a well-written article tends to attract high citations and is thus likely to increase the visibility and reputation of the authors.

4.2 Importance of Literature Review in the Research Process

As you start your research project, your supervisors will ask you to conduct a literature review as a preliminary search to guide your research ideas. This exercise helps you to explore the existing scholarship related to your research topic and *identify gaps* or areas where further investigation is needed. Moreover, the literature review at this stage equips you with a strong *foundation of knowledge* to build upon, ensuring that your research is informed and contributes meaningfully to the academics or practical within your field of study.

Furthermore, it is a building block to all your research activities and should be a priority for all academicians. It is the key principle for conducting quality research and should be integrated throughout the entire research process (Chap. 2). Most universities offer comprehensive guidelines for their students on conducting and writing effective literature reviews. These guidelines are good reference materials that may contribute to your mastery of the art of literature review and equip you for an exciting and enlightening academic and research journey.

Literature review is important because it establishes connections with studies closely aligned to the undertaken research and contributes to the broader ongoing dialogue in the literature, filling in gaps, and extending prior studies (Marshall et al. 2021). In the context of research, literature review is important for the following reasons:

1. *Theoretical underpinning*: It refines the theoretical foundation of research and makes research more profound and relevant. Literature will help you to gain a deeper understanding of the theories that support your research study.
2. *Summarize current understanding:* The literature review is a comprehensive guide, offering updated insights into prior research within the selected field. It not only acquaints readers with the depth, breadth, and scope of the research but also facilitates the documentation of ideas from diverse perspectives. This ensures that no crucial variable from past research is overlooked, presenting relevant information in a logically organized manner.
3. *Identification of research gap:* It aids in pinpointing additional information (gap) needed in the body of knowledge, thereby preventing duplication. The research gap justifies the study by highlighting unanswered questions about subjects and problems and thus showing a link with the accumulated knowledge in your area of interest. This is emphasized in Chap. 3.
4. *Sharpening of the research problem:* Understanding of previous works helps in formulating a significant

research problem, refining the statement of the problem, and formulating objectives.

5. *Refining research approaches and methodologies:* Extensive reading exposes the researcher to various approaches to dealing with research issues, contributing to knowledge, and understanding. It helps shed light on methodologies used and informs the design for proven useful approaches. A literature review will reveal strategies, procedures, and measuring instruments that have been found useful in investigating the problem in question.
6. *Data handling:* It helps in developing an analytical framework or a basis for analyzing, interpreting, and presenting data
7. *Reporting of research findings*: Reporting the research work in either thesis or a journal helps in proving a benchmark for comparing what you have found out with the work of other authors. It helps the researcher be in a better position to interpret the findings. Reviewing literature is key to writing the discussion section of a thesis or journal paper.
8. *Theoretical and conceptual framework*: Helps in specifying the theoretical (TF) and conceptual framework (CF), forming the foundation within which the findings are interpreted and discussed. Grant and Osanloo (2014) indicate that the literature review and theoretical framework are intrinsically linked. A theoretical framework guides us in logically developing the different, yet interconnected parts of the literature review. CF and TF are handled at the end of this chapter.
9. *Credibility of research:* When researchers cite literature, it strengthens and enhances the credibility of their work
10. *Publishing:* helps in publishing of work in a journal. Students can identify the most suitable journal for publication based on the alignment of their research. The most common journals in your reference list are candidates for your target journal. Secondly, a strong literature review enhances reviewers' and editors' confidence

Box 4.1 Gives a quick list of why a literature review is important in research.

Box 4.1 Importance of the Literature Review Process in a Research Study:

1. Helps you in the formulation of research objectives and hypothesis
2. Set the context of your study
3. Enhances quality and impact of your research
4. Build the knowledge base of your topic
5. Helps you identify existing research gaps
6. Facilitates understanding of the theoretical underpinning
7. Exposes you to various research methodologies and approaches
8. Used by supervisor and examiners to evaluate the relevance of your research
9. Keeps you updated on the latest information
10. Helps you to demonstrate the importance of your topic
11. Enhances the credibility of your research
12. Provides you with the basis for comparing with other studies
13. Helps in the identification of a suitable journal for publication
14. Enhances reviewers' and editors' confidence
15. Aids in identifying key themes, trends, patterns and changing paradigms
16. Helps identify the most authoritative and pertinent sources and authors
17. Helps build a strong theoretical and conceptual framework
18. Helps build a logical structure for the literature review chapter
19. Equips you with research skills
20. Enhances information literacy skills
21. Enhance writing and communication skills
22. Instil academic integrity and research ethics
23. Helps in building critical and analytical skills

Literature review is a complex, time-consuming, and demanding process carried out during the entire research journey (Snyder 2019). A strong literature review is positively related to quality of the research. It requires you to do the following:

Carry out extensive reading, analyses, and synthesis of existing literature.

Cover a broad range of existing literature from many sources.

Keep abreast of the most recent advancements in your area of study (keeps you on your toes).

Develop critical thinking and analytical skills.

4.3 Sources of Literature

Three basic sources of literature are useful for your literature. These are primary, secondary, and general references:

1. *Primary sources:* are publications in which researchers report the results of their studies. These include scholarly scientific materials:
   - Scholarly journal papers.
   - Conference papers.
   - Graduate research thesis.
   - Project reports.

- Institutional reports.
- Working papers and occasional papers.
- Extension and training materials (Briefs, pamphlets, and magazines).
- Pre-prints with a rider as they are not peer-reviewed.

2. *Secondary sources*: refers to publications in which the author describes the work. The most common are textbooks, edited books, and research reviews.
3. *General references*: These are sources researchers often refer to first. They are good when examining a topic for the first time. They tell you where to locate other sources, such as articles, books, and other documents. They also include indexes and abstracts. In agriculture, the CAB abstracts are commonly used.

These sources are found in Libraries (Universities and research institutions), institutional repositories, journals, conference proceedings, databases, government publications, and online platforms. The next sections discuss the sources of literature and where they can be found. A summary of the sources of literature is shown in Table 4.1.

4.3.1 Scholarly Scientific Materials

1. *Journal Articles*

Journal articles provide scholarly literature. They are the gold standard for quality in sources for your research. Your supervisors require you to use them. This is because journal articles undergo rigorous peer review and are usually published by researchers from reputable academic and research institutions (Discussed in Chap. 27). Unlike textbooks, they also tend to offer the most recent developments and advancements. Your supervisors and reviewers will likely evaluate your proposal and thesis based on how many journal articles are used, including how many are recent materials. Journal article reviewers also check

Table 4.1 Sources of literature and their most important use and availability

Source	Use for the literature review	Availability
Journal papers	Most useful for research projects because they are written by experts and checked for quality through peer review.	Majority are available online through subscriptions. Increasingly available through repositories and social networks. Many journals are adopting an open-access format
Thesis	Very useful for master's and PhD if the topic matches because it presents a researcher's original findings, arguments and contributions to a particular field of study. Also, carry a generic format of your thesis	Mostly in hard copies; becoming increasingly available in library repositories, e.g., Kenyatta University has a comprehensive repository of theses
Conference proceedings	Collection of papers and presentations from conferences. Useful source of current literature, but may vary in quality and is mostly not strictly peer-reviewed	It can be difficult to find if not placed online by the organizers
Textbooks and edited books	Useful because they are written by experts. Good for detailed information	Widely available. Can easily be obtained from the library (catalogs) or purchased
Handbooks	Useful as a starting point; contains well-organized, authoritative information on diverse fields of agriculture	Old handbooks are mainly in hand copies, but new ones are in digital form available from publishers
Research reports	Annual or project reports from government and research institutions. Offer in-depth information Practical solutions and recommendations	Least available Mostly in hard copies in respective institutions
Working papers	Preliminary research documents provide insights into ongoing studies, and they offer early access to cutting-edge research findings and ideas before formal publication.	Many are available through online and institutional repositories
Annual reports	Comprehensive documents produced by corporations and organizations that provide a detailed overview of their performance, strategic goals, and operational activities during the previous year, for stakeholders, investors, researchers, and analysts.	Not widely available. Mostly in hard copies in respective institutions. A few institutions are increasingly posting them online on their websites
Pre-prints	Information is not peer-reviewed	Easily accessible
Encyclopedias	Excellent starting point: Information is well organized, comprehensive, and authoritative	The new encyclopedia is available online from the publishers. Specific chapters can be purchased from the indexed information
Extension and training materials	Not very useful because they are designed to teach or enhance specific skills or knowledge related to the source's subject matter. Depending on the clientele, they may lack details	Not easily available online
Gray literature	Unpublished; mainly institutional reports; not peer-reviewed	Not easily available online

the use of relevant and current literature from credible journals.

The library and the internet are the main areas in which to source journal articles. Libraries subscribe to key journals. You can check the information on available journals from the librarian. Nowadays, virtually all journals can be accessed using full-text online databases through your university or Google Alerts.

Many agricultural journals exist. The main ones are those published by Elsevier, Springer, Taylor and Francis, Emerald, and Wiley. Your librarian is an asset in identifying the most relevant journals for your topic.

2. *Graduate Research Thesis*

A thesis is a culmination of a student's original work. It is an important source of literature as it presents details of the research work, especially in the case of a Ph.D. thesis. Theses usually include a detailed literature review, research methods, data analysis, and discussion. This information can help you gain a good understanding of your study. However, the quality and rigor of the research methods used in the thesis can vary. It is advisable to critically evaluate their methodologies and relevance to your work before using them in your study.

3. *ProQuest Dissertation and Thesis Database*

ProQuest Dissertation and Thesis Global is a database comprising a comprehensive, curated collection of multidisciplinary dissertations and theses. ProQuest is often available from institutional libraries and universities. The platform provides full-text access to the dissertations and theses, mostly in PDF. However, the platform requires a subscription. understanding of the topic.

4. *Conference Proceedings*

Similar to journals, conference proceedings result from research conducted mainly by academic and research institutions scholars. They contain research papers, discussions, and posters presented at a conference. Researchers often present their work shortly after it is concluded or even before the final data is collected. Consequently, conference proceedings are an excellent source of up-to-date information as they present the most recent findings. However, unlike journal articles, peer review may not be as rigorous, which could lead to a wide variation in quality. Moreover, while some papers are thoroughly researched, others may contain only preliminary findings. They should, therefore, be used with caution.

5. *Research Reports*

Research reports may include reports from government, nongovernmental organizations, and research institutions. They can be annual reports, project reports, or special reports. The nature and content of these reports vary depending on the institution and the specific research project. These reports are useful as they cover a wide range of topics, which allows you to access multidisciplinary information. In addition, some reports are detailed, making them useful for acquiring in-depth literature and research information. Another advantage is that they usually offer practical solutions and recommendations.

However, they have the disadvantage of varying quality and research methodology soundness. They may also not be easily available online. They also lack peer review and are usually classified as gray literature.

6. *Working and Occasional Papers*

Researchers write working papers and occasional papers to present their early-stage research findings and ongoing work. These papers may contain findings that will never be published in peer-reviewed journals. They have the advantage of presenting early research and current findings and in-depth research methodologies. However, their quality and relevance vary and should always be evaluated based on the specific topic.

7. *Textbooks and Edited Books*

Textbooks are typically authored by experts in their respective fields and provide a thorough and structured introduction to a subject. They are especially beneficial for gaining an initial understanding of a topic or for exploring it in a structured manner. Textbooks are extremely useful for understanding the theoretical basis of a subject. Although textbooks and edited books are crucial literature sources, students and researchers should consider the publication dates as the relevance may vary with time.

8. *Preprints*

Preprints are journal articles that precede peer review and publication. Authors often share their preprint versions of their manuscripts on preprint platforms to disseminate their research findings quickly. Because the work is not peer-reviewed, you should interpret the findings with caution.

9. *Handbooks*

Handbooks contain facts unique to a topic, often including statistics and key terminology. Numerous handbooks in agriculture discuss general agronomy topics, crop protection, agricultural economics, soil science, and other disciplines. They provide researchers with comprehensive, organized and authoritative information on a specific topic that covers main concepts, theories, and trends. Handbooks are an excellent starting point in a literature review, and you may use the list of references for tracing seminal and influential works on your topic. Old "Handbooks" were mainly published by publishing houses such as Elsevier, Springer, CRC, and Wiley. Examples of handbooks in agriculture include the following:

- Handbook Digital Farming edited by Dörr and Nachtmann (2022), published by Springer (Dörr and

Nachtmann 2022). It is the first-ever handbook on digital farming and sheds light on the technological, economic, soil and legal perspectives of the digital transformation in agriculture
- Practical Handbook on Practical Microbiology edited by Amaresan et al. (2022) by Springer (Amaresan et al. 2022). It presents details of techniques involved in the study of beneficial microbes in agricultural microbiology
- The Handbook of Global Agricultural Markets edited by Luc Nijs (2014). It is a good reference for researchers in finance, economics, and agribusiness (Nijs 2014).
- CRC Handbook of Plant Science in Agriculture by Christie (2022) has two volumes (Christie 2022). The first volume presents information on general botany and the growth of crops, while the second volume presents crop production and utilization. An e-book was published in 2019.
- Handbook of Agricultural Economics by Elsevier has six volumes running from 2001 (Volume 1) to 2022 (Volume 6) (Gardner and Rausser 2001). A good reference book for researchers in agricultural economics with a wide range of topics
- The Palgrave handbook of agricultural and rural development in Africa by Osabuohien (2020). Examines agricultural and rural transformation in Africa, both from theoretical and empirical standpoints and emphasizes the close relationship between agriculture and poverty reduction and employment. A good reference for understanding the role of agriculture in SDGs.

10. *Encyclopedias*

Encyclopedias are good places to start your literature review when you know little about the topic. They are an excellent quick reference as they contain comprehensive and well-organized information. For instance, the Encyclopedia of the UN Sustainable Development Goals presents numerous book series across the years covering topics relevant to SDGS presenting diverse topics. A single encyclopedia can also carry many pages, e.g., the Encyclopedia of Agricultural Food and Engineering (Heldman and Moraru 2010) is 2120 pages. The information is thus useful for building and defining keywords and concepts. They can also help you identify other references in your topic of interest, identifying experts in the field. Examples of encyclopedia in agriculture include:
- *Encyclopedia of agriculture and food systems* edited by Neal Van Alfen (2014) by Elsevier (Van Alfen 2014) discusses global agriculture and provides useful information for understanding the challenges of agriculture, food, and the environment. It lays a good foundation for researchers working on the interrelationship between food systems and sustainability issues in environmental conservation.
- *Encyclopedia of smart agriculture technologies* by Qin Zhang (2023) by Springer (Zhang 2023). It presents the latest advancements in agricultural engineering and cutting-edge technologies.
- *Encyclopedia of agricultural, food and biological engineering* by Heldman and Moraru (2010) by CRC Press (Heldman and Moraru 2010). It is a good reference for research in food science and processes used to produce raw agricultural materials and for converting them to consumer products.

11. *Extension and Training Materials*

Agricultural extension and training materials can be important sources of literature they offer mostly practical knowledge and advice. They contain simplified information distilled from complex scientific concepts. They are mainly tailored for farmers and extension workers. Therefore, for research, they are less useful for theoretical understanding of the subject matter. However, they can be good sources of case studies and successful stories.

12. *Gray Literature*

Gray literature refers to information that is not published in mainstream journals or monographs. They are usually hard to find. These may include mainly government documents in hard copies. They vary in quality because they are not peer-reviewed. You should therefore evaluate them properly. To do so, you should check the author's affiliations and qualifications and verify the referee lists to see what other related literature has been considered. You can also examine the credibility of data collection and analysis methods. Lastly, verify the data and the currency of the information. Other examples of gray literature include:
- Government reports.
- Conference proceedings either in hard or e-print.
- Research reports by Government and nongovernmental organizations.
- Maps.
- Fact sheets.
- Newsletters and bulletins.
- Pamphlets and other extension materials.
- Annual reports from the Government and the private sector.
- Informal communication (blogs, webinars, and podcasts).
- Newspapers and magazines.

The status of agriculture reports (SAAR) published annually by the Alliance for a Green Revolution in Africa (AGRA) (https://agra.org/publications/africa-agriculture-status-report-aasr/) is an important report in this category of gray literature. These reports are accessible for download from the AGRA website. They provide a comprehensive summary of the current state of agriculture and food production. The reports cover various topics, including agricultural productivity and food security, agricultural policies and markets, climate change and sustainability, and technological innovations. These reports provide data, analysis, and recommendations that contribute to a deeper understanding of the challenges and opportunities in African Agriculture. As such, they are valuable sources of literature developed by diverse experts and grounded in credible data.

4.4 How to Locate the Literature

1. *Libraries*

Libraries are one of the most important homes of literature. They should be your original search engine and source. Libraries subscribe to journals and e-books and maintain a catalog for books. Most University libraries offer *information literacy programs*, which are essential for your academic and research success. Information literacy involves knowing why you need the information, where to find it, how to manage it, how to evaluate its credibility and relevance, and how to communicate in writing. You are encouraged to use the resources and available guidance at your library so that you benefit from this invaluable resource.

The librarians managing the libraries have a wealth of knowledge on literature related to various disciplines, including agriculture. You can consult them for guidance on the latest resources for your research topic as they are well-trained and always ready to offer help. Your university library's web pages are also great resources to check out. They provide valuable information on services and resources available to students for their academic and research needs. In addition, these web pages offer tutorials and workshops, as well as videos on information literacy, database searches and citation management. You can gain a rich experience in literature review sources by visiting the library and exploring online resources. Librarians should be your friend in your thesis research journey.

2. *Search Engines*

Search engines are important for searching literature on the internet. Users enter specific keywords, phrases and topics related to their research interests into the search engine query box; the search engines can be used to retrieve a wide array of information: journal papers, conference proceedings, books and other scholarly resources from repositories, libraries, publisher websites and academic databases. One of the most common search engines is Google Scholar.

3. *Google Scholar*

Google Scholar is a dedicated search engine that resembles the Google web search engine but is designed specifically for scholarly articles and books. Unlike the broader Google engine, Google Scholar focuses on academic content, making it useful for researchers. However, it is also good to use the Google search engine, which is broad in scope. Consider using the two formats in tandem to get in-depth academic content from Google Scholar and supplementary information from general Google. You may use Google Scholar during the initial *stages of searching for your literature.*

4. *Databases in Agriculture*

Databases are important sources of literature. Several databases provide information on agriculture. We have speciality-based databases offering information on agriculture and other databases that offer information in multiple disciplines. Some of the major ones include CABI, FAO, AGORA, AGRIS, HINARI, etc.

(a) Agriculture-based databases: They include CABI, AGORA, AGRIS, AGROVOC, and FAO (Table 4.2)

Centre for Agriculture and Biosciences International

The Centre for Agriculture and Biosciences International (CABI) offers one of the most useful global databases in agriculture. It is a leading publisher and provide information, and training in the fields of Agriculture, Environment, Life Sciences, and Health. The mission of CABI is to "improve people's lives worldwide by providing information and applying scientific expertise to solve problems in agriculture and the environment." It is a valuable resource for students, researchers, practitioners, and policymakers.

CABI contains *CAB abstracts*, one of its flagship resources and a foundation of its agricultural and biosciences database. Besides journal articles and books, CABI also includes gray literature. CABI database also compiles other information and data from various sources, making it easy for researchers to find and use. You can set alerts to receive notifications and stay updated with new literature.

CABI Direct is its major digital platform, providing access to a vast collection of information, including CABI abstracts and Global Health. There are over 15 million records available in CAB Direct as of October 2023 (CABI Website; https://www.cabi.org/). The CAB abstracts cover a wide array of topics related to agriculture, while the global health subset focuses on health-related research and information. The global health sub-segment offers information on public health, epidemiology, disease prevention, nutrition and more. There is a special emphasis on food safety, zoonotic diseases, one health perspective and methods to ensure food security in the face of climate change. Access to this database requires a subscription. Check if your university has subscribed.

Another useful database within CABI is a compendium offering information on agricultural production, health, and biosecurity. The compendium consolidates data and research across species, pests, and diseases and provides a specialized

Table 4.2 Online databases offering information in agriculture

Subject-specific: agricultural databases	Description
CABI database	Comprehensive database of agriculture and environmental sciences by the Centre for Biosciences International (CABI) containing over 15 million records; includes CABI Direct plus other sub-databases: Global health, CABI compendium, and CABI Cases
AGORA	Established by the Food and Agriculture Organization of the United Nations (FAO) in collaboration with major publishers, offers an exceptional digital library collection encompassing the fields of food, agriculture, environmental science, and related sciences
FAO	Agency of the United Nations and spearheads global initiatives to ensure food security for everyone, aiming to have enough high-quality food for everyone to lead active, healthy lives. Offers millions of records of information
FAOSTAT	Offers unrestricted access to food and agriculture data spanning over 245 countries and territories, encompassing all FAO regional groupings from 1961 to the latest available year
AGRICOLA	Created and maintained to serve as the catalog and index for the collections of the United States National Agricultural Library of the United States Department of Agriculture
AGRIS	Managed by Food and Agricultural Organization offering bibliographic records in areas of agriculture, forestry, and fisheries
AGROVOC	Agriculture vocabulary database of thesaurus (multilingual) containing a structured collection of agricultural concepts, terms, definitions, and relationships which are used to unambiguously identify resources, allowing standardized indexing processes and making searches more efficient
Specialty databases, e.g., AgECon search and VetMed resource	Databases catering for specific needs and disciplines. They provide specialized and up-to-date information on the specific field.
Multidisciplinary databases with agricultural-related literature	
Scopus	A search engine and bibliographic database containing scholarly literature. Frequently paired with databases like PubMed, Web of Science, and Google Scholar, offering literature across diverse fields, including science, technology, medicine, social science, art, and humanities

(continued)

Table 4.2 (continued)

Subject-specific: agricultural databases	Description
Web of Science	A comprehensive reference database and citation index for life science research. One can track research trends, discover related works, and assess the impact of research through citation analysis
Science Direct	The world-leading scientific, technical, and medicine research database that provides access to a vast collection of peer-reviewed academic journals, articles, and research papers. It is operated by the publishing company Elsevier and serves as a valuable resource for researchers and academics
Pubmed	An open-source search engine, primarily the MEDLINE database of references and abstracts on life sciences and biomedical topics
HINARI	A program that was established by the World Health Organization (WHO) to provide free or low-cost access to a vast collection of biomedical and health literature for institutions in low- and middle-income countries
JSTOR	A digital library that provides access to academic journal articles, books, and primary source materials for research and education purposes
ERIC	An online digital repository containing a collection of educational research, including journal articles, reports, and other resources
AJOL	A digital platform that promotes research and knowledge dissemination on the African continent through providing access to a range of African-published scholarly journals
Research for life	Started in 2002, it provides more than 15,000 institutions in over 125 lower and medium-income countries
ProQuest dissertations and theses	Comprehensive curated collection of multidisciplinary dissertations and theses globally; over 15 million citations and over 3 million full texts
Open knowledge maps	Useful for mapping a research topic, getting an overview, and identifying relevant concepts and related papers
Institutional repositories and websites	
Universities	Contains curated repositories containing mainly theses, research papers, and conference proceedings emanating from the faculty and students
Research institutions	Contains annual reports and research papers. May also contain extension and training materials, e.g., training manuals and pamphlets
Donors and projects	Donors maintain websites showing their area of focus, but also contain project reports of research projects they have supported
Social research networks (ResearchGate)	Popular social network covering a wide range of subjects. Allows researchers to create profiles, share publications, and connect with colleagues

in-depth information on crop protection, plant health, and animal health. It also provides data, description and guidance and contains maps, diagnostics and decision support tools for practitioners and policymakers. The following compendiums are available (CABI Website) (https://www.cabidigitallibrary.org/journal/cabicompendium):

- Animal health and production.
- Aquaculture.
- Crop protection.
- Forestry.
- Horticulture.
- Invasive species.

CABI also offers a special section containing a curated collection of real-life examples called *Cases* to serve students, researchers, and researcher-led practitioners. The cases are in One Health, Plant Health, Food Science and Nutrition and Animal Science. This section of the database is useful, particularly for students and researchers in animal health and crop protection who may be looking for information on rare and novel diseases reported as cases or case series reports. Researchers can also utilize this platform to report cases of new diseases. You can easily access this on the CABI Website. There is a possibility of being paid once you publish a case study.

AGORA Database

AGORA, which stands for Access to Global Online Research in Agriculture, provides access to a vast collection of free journals and books (Ochs 2004). This initiative was established by the FAO in collaboration with various publishers and research organizations. The main objective of AGORA is to make it easier for people, especially in low and medium-income countries, to access critical information on agricultural and life science topics. Topics include agriculture, fisheries, food, nutrition, veterinary science and related subjects.

Eligible students and staff must provide their credentials to access AGORA. You can confirm your eligibility by contacting your library and learn how to access and benefit from the available information. However, AGORA may not cover all sources available on published websites, particularly recently published papers. Nonetheless, AGORA provides a substantial array of valuable information.

Food and Agriculture Organization Database (UN-FAO)

Food and Agriculture Organization of the United Nations (UN-FAO) is not a primary database for literature, but it offers valuable information and resources that can be a crucial part of your literature search. FAO publishes various reports, data and publications related to agriculture, food and nutrition. Additionally, FAO maintains a comprehensive statistical database called FAOSTAT that provides access to agricultural and food-related data and statistics. The database is particularly useful in providing data on agriculture-related aspects such as crop and livestock production statistics.

AGRIS

Agricultural Information Systems (AGRIS) is an international bibliographic database by UN-FAO that collects bibliographic references related to agricultural literature. The database indexes journal articles, books, and book chapters, conference proceedings, monographs, and various forms of gray literature, including thesis.

AGROVOC

Agriculture Vocabulary (AGROVOC) is a controlled multilingual vocabulary created to encompass concepts and terminologies (Thesaurus). This activity is coordinated by FAO in collaboration with an international community of experts and institutions. It is the largest open dataset; it was developed in 1980 to standardize indexing open data available for public use in multiple languages. It contains a collection of concepts, terms, definitions, and relationships in food and agriculture-related topics and is used to identify resources unambiguously. It contains more than 38,000 concepts and more than 802,000 terms in up to 40 languages (Subirats-Coll et al. 2022). The AGROVOC system was first initiated in early 1980 to explain documents and other forms of information resources in a controlled language for indexing and searching, but over time, it developed into a semantic web technology. It is available both human and machine-readable formats (Subirats-Coll et al. 2022).

(b) *Specialty Databases*

Many other databases cater for specific needs and disciplines and are crucial in providing specialized and up-to-date information on the specific field. Most of them will be listed on your university website, but you can also get information on important specialty databases from your supervisor. Notable ones include:

- Food and technology abstracts.
- Soil Science Society of America (SSSA) Digital library.
- Animal science database.
- Agricultural economics database (AgEcon) for agricultural and applied economics literature.
- Vetmed resource for veterinary information (CABI).
- Animal health and production compendium in FAO.

(c) Multidisciplinary databases: These include SCOPUS, WEB OF SCIENCE, PUBMED, HINARI, JSTOR, AJOL, Research for life, and Open knowledge maps discussed hereafter.

1. *SCOPUS Database*

SCOPUS is an extensive abstract and citation database by Elsevier and covers various academic fields, including agriculture. It indexes many scholarly publications covering various aspects of agriculture and other related disciplines. What sets SCOPUS apart is its advanced search and discovery tools, which help researchers easily navigate the database and find relevant literature. Additionally, SCOPUS provides author profiles, enabling researchers to identify experts in their respective fields of study.

2. *Web of Science*

Web of Science contains a vast collection of journal articles and other scientific publications in the physical and life sciences. Developed and maintained by Clarivate Analytics, it provides access to multiple databases offering a wide range of academic disciplines.

3. *PUBMED Database*

PubMed focuses on *biomedical sciences*, but it also contains substantial literature on agricultural and environmental health sciences. It covers a wide range of topics in biomedical and life sciences, including animal health, nutrition, genetics, and environmental health. Although not a specialized database like AGRICOLA or CAB Abstract, PUBMED it is a valuable resource for researchers in agriculture and you can access it for free.

4. *HINARI Database*

HINARI, which stands for Health Network Access to Research Initiative, is a program that grants institutions in low and medium-income countries access to a vast range of scientific literature on *biomedical and life sciences.* Although HINARI mainly focuses on providing resources for health-related research, it also serves as a valuable tool for accessing agricultural literature. Researchers can find literature on various topics such as crop production, livestock husbandry and food security.

5. *JSTOR*

JSTOR refers to Journal Storage and is a digital library providing scholarly literature in agriculture and related fields. It includes many older and historically significant agricultural journals and provides valuable information on the historical development of agricultural practices. It operates an open access policy and, hence, a useful resource where you can get the full-text format of journal articles and other reference materials.

6. *ERIC Database*

ERIC (Educational Resources Information Centre) is a widely recognized database in the field of education. It houses a vast collection of educational resources such as reports, research papers, and conference proceedings. The database covers topics related to curriculum development and training in agriculture. It also features literature in agricultural education, including the effectiveness of educational programs related to agriculture, agribusiness, and rural development, thus an invaluable resource for agricultural researchers.

7. *AJOL*

African Journal Online (AJOL) database covers a wide spectrum of agricultural topics. It specializes in African research and academic fields and is thus beneficial to students and researchers studying agriculture in Africa. Many of the articles are open access, enhancing access to the majority of African scholars.

8. *Research4life*

Research4life is a partnership of United Nations agencies and Universities like Cornell and Yale. It provides access to scholarly literature for low and medium-income countries. The focus areas are of health, agriculture, environment, physical, and social sciences. Research4life has four databases: AGORA, Health Network Access to Research Initiative (HINARI), Access to Research for Development and Innovation (ARDI), Online Access to Research in the Environment (OARE), and Research for Global Justice (RGL). The aim is to improve teaching, research, and policymaking. Eligible institutions register for access. You can contact your university or institution for details on how to register and get login credentials.

9. *Open Knowledge Maps*

Open knowledge maps enable you to map a research topic, get an overview, and identify relevant concepts and related papers. It is, therefore, a good source for the early stages of locating literature sources. It offers an immediate overview of a subject, presenting the main areas at a glance, along with pertinent papers and concepts associated with each area. An illustration of a knowledge map is shown in Fig. 4.1, which was generated by feeding the simple concept "integrated soil fertility management." Such a map saves time in the initial stages of the literature review by offering an overview of the research landscape to help you quickly identify relevant papers and connections between topics.

Knowledge maps are easy to use. You start by visiting the platform and accessing it through your library. Enter your research question or topic and click the search button. The platform will generate a visual map with nodes and connections representing academic papers and relationships. The knowledge map can be improved by zooming in and out and clicking on the nodes to access the papers. On the left-hand side (see Fig. 4.3, you will find the filter and sort button and you can filter results by publication year and source.

Open knowledge maps may not cover all academic databases and sources, so it is recommended to complement your research review with other databases and search engines like Google Scholar and library resources.

(d) *Institutional Repositories, Websites, and Networks*

Research and academic institutions such as universities and research centers have specialized repositories that con-

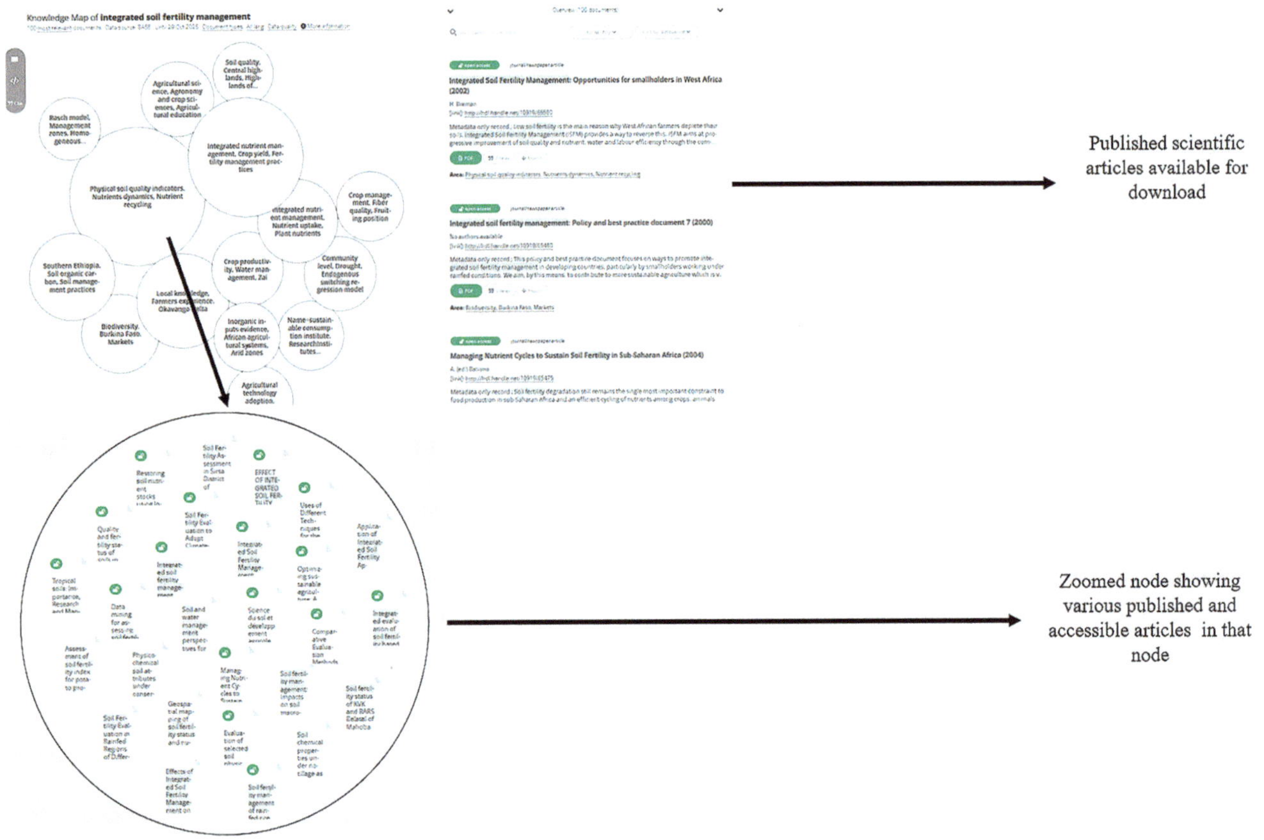

Fig. 4.1 Open knowledge map created on 26.10.2023 for integrated soil fertility management (BASE) showing nodes and relationships

tain a wealth of academic materials, including theses, conference proceedings and various types of reports. The literature spans a wide range of disciplines and topics, meant to serve all academic and research interests of members of the specific institution. These repositories are not only accessible within the confines of the institution but can also be accessed online, making them important sources of literature for researchers and students alike.

Social Research Networks

Social research networks, like ResearchGate, have become increasingly crucial for literature reviews and academic networking. These platforms allow researchers to engage with their peers, exchange their work, and explore relevant studies in their area of interest. ResearchGate, for instance, is a vast repository of research papers and other publications where researchers can discuss important issues in their field.

Websites of institutions Research-based institutions (national and international) and Universities have repositories and resource libraries for theses, research papers, conference proceedings, and a wide range of research reports. They offer a useful source of literature, especially for students

4.5 Searching the Internet

The World Wide Web (www) constitutes a segment of the internet. As an interconnected network of web pages and resources, the WWW facilitates the seamless dissemination and access to diverse information, making it a crucial tool for individuals seeking knowledge across various subjects. It provides an immense wealth of information accessible with just a single click. You can search the internet using keywords through search engines available on the internet, each with its own algorithms and features. Some prominent Leading search engines include *Google, Bing, and Yahoo*.

The advantages of searching the WWW are that the information is current, and you can access various materials. Internet flexible. In addition to using keywords to search for information, you can include PDF or PPT to get PDF files and PowerPoint presentations, respectively.

Advantages of Searching the WWW

Searching for information from the Internet has many advantages and they include:

1. Information is current and updated.
2. A wide variety of materials are available, including manuscripts, books, journal articles, PowerPoint presentations, etc.

3. Information can be viewed on one's own computer.
4. Cheap source of information; open access to articles that are freely available.
5. There are varied formats of materials.

Disadvantages of Searching the Internet
1. Overload: Researchers often struggle to filter through large volumes of data and references.
2. Time commitment: Usually time-consuming before getting the specific information being looked for. The need to search for new information continually may require significant time and occasionally yield less productivity compared to using more traditional databases (Fraenkel et al. 2012).
3. Disorganization: Much of the information is not organized, and good online searching skills are needed. Most students lack search skills.
4. Ethical violations: Copyright violations are more likely to happen than with traditional sources of information.
5. Some information lacks credibility: Anyone can publish on the internet, even information that may not have credibility. A lot of information on the Internet is not peer-reviewed; hence, it lacks quality control, and some online sources may have bias and hidden agendas, for example, the promotion of a certain product.
6. Uncertain reliability: Some information on the internet is ill-conceived. It has not passed through a vetting process and may not be reliable, unlike materials found in the library.
7. Predatory journals: With the high prevalence of open-access predatory journals, novice researchers will likely be misled into using them for the review. These journals are of low quality and do not uphold rigorous academic standards.

4.6 Steps in Literature Review

The first step in reviewing the literature is identifying a research topic (Refer to Chap. 2). You should first identify a topic, "the subject matter of the proposed study," before deciding on the literature to review (Lingard 2018). This is because the central research question will set the parameters for the review. Always remember that the objective of the literature review is to present previous research and its development to specific research questions/objectives.

We present the following six steps for reviewing and writing your literature in Table 4.3:

Table 4.3 Steps in conducting a literature review

Step	Activity	Details
Step 1	Choose a topic	Provides an overarching research question to provide parameters for the search of the literature
Step 2	Collect relevant material	Use keywords to search and find relevant literature
Step 3	Select relevant literature	Read and take notes; evaluate critically for relevance; mainly read the abstract Evaluate the literature
Step 4	Read and organize literature in a literature matrix	Record articles in a matrix; read critically, critiquing and identifying gaps
Step 5	Synthesize	Compare articles by themes; put related literature together
Step 6	Write	Write a literature review in a coherent, critical manner guided by an outline

However, authors present a diverse number of steps. For instance, Levy and Ellis (2006) condensed the steps into a three-phase process involving inputs, processing, and output. The first phase of inputs encompasses the selection of relevant literature, identifying key themes, and understanding the existing body of knowledge in the chosen research area. In the processing phase, researchers critically analyze and synthesize the selected literature, identifying patterns, trends, and gaps. This involves thoroughly examining methodologies, findings, and arguments in the literature. The final phase, output, involves the construction of a cohesive and comprehensive literature review that contributes to the understanding of the research topic, highlights gaps in current knowledge, and sets the stage for the research study at hand. In the next section, we discuss the six steps of reviewing literature.

4.6.1 Step 1: Choose Topic

Begin by choosing the topic or referring to the topic that you have already chosen in step one of the research process.

4.6.2 Step 2. Collect Relevant Information

The second step involves locating the literature by first identifying the keywords, followed by locating the literature. It will involve the following sub-steps:

(a) *Identifying the keywords and search words*

The initial crucial step in literature exploration involves identifying keywords and search terms. These fundamental terms encapsulate your research questions and objectives, serving as essential tools for literature searches. Search terms encompass a range of elements, including authors' names and other relevant identifiers, providing a foundation for effectively navigating and uncovering pertinent literature. There are several strategies for identifying the key terms. They include the following (Creswell 2014; Saunders et al. 2019):

- Early literature reviews involve reviewing articles by key authors and recent review articles identified from class notes and general searching of the internet. Besides helping you to identify key terms, they are also useful for leading you to other work that is pertinent to your research topic. They contain relevant references that you can trace at this point. You can pick your key terms from the words that the authors use in literature.
- A tentative title for your project: Write a tentative title that captures the study's central idea. Pick at least two keywords from the title.
- Databases available in the library, e.g., AGORA, can be used to find words that match your topic.
- Handbooks and encyclopedias, textbooks, and recommended reading by your supervisor. There are several handbooks and encyclopedias on agriculture (see Sect. 4.3. Encyclopedias contain summaries and therefore, you can rapidly read several on your topic of study.
- Library Catalogs: Your library catalogs can be used to generate keywords
- Posing a general research question you want your study to address. You can pick some keywords from the research question.
- Synonyms and related concepts borrowed from authors associated with the ideas.

(b) *Searching the library*

It is a good idea to start by consulting the librarian, who is an expert at retrieving the information. They can help you identify the most suitable databases, library catalogs, and online resources for your field of study. Additionally, they will offer insights into the availability of physical and digital resources and help you access books, journals, and databases. They will also be useful in:

- Developing an effective search strategy, including the use of Boolean operators to enhance the precision of your search.
- Help you access databases that are subscribed to by the University.
- Introduce you to reference management tools.

Collaborate with the librarian to initiate a comprehensive search, utilizing your designated keywords to explore both primary and secondary sources. Cast a wide net by examining various types of literature, encompassing encyclopedias, dictionaries, glossaries of terms, handbooks, journals, abstract series, electronic resources, books, reviews and syntheses, and databases (Creswell 2014). This diversified approach ensures a thorough exploration of available resources, providing a well-rounded foundation for your research.

(c) *Searching library catalogs*

Library catalogs give you a good place to start searching for literature related to your area of focus. To locate either secondary or primary sources, start by typing the title or author's name, and you will get the information about availability and the call number of the document. You can search by subject or keywords if you do not have the author's name or the title.

An effective way of refining or expanding your search is to use Boolean operators. These are terms that instruct the computer systems to include or exclude specific keywords in your search. By using operators like "AND," "OR," "NOT" you can tailor your search to retrieve more relevant results. "AND" narrows the search, while "OR" broadens it by allowing keywords to appear. "NOT" excludes specific keywords to filter unwanted results.

(d) *Searching the databases*

Select relevant databases, including subject-specific ones like CABI Abstracts, AGRICOLA, AGORA, and PUBMED, as well as multidisciplinary databases like SCOPUS and Web of Science. Start with broad keyword searches, then refine using filters and advanced options for publication types and dates. FAO is useful for statistics.

These databases are likely to house high-quality indexed journals. By exploring these resources, you can easily access hundreds of journal articles related to agricultural topics.

(e) *General searching on the internet*

General searching on the internet is the most used by students. It is sometimes called "googling" or using Google as the search engine. It is still relevant in searching for literature. It inserts keywords or phrases related to the information you are looking for, using search engines such as Google and Google Scholar to find information, websites, and related resources on a wide range of topics.

Coverage of the Review

Coverage is an important aspect that defines the search's breadth and depth. It would help if you decided on how many references to have and the publication dates. These depend on research goals, and you must balance comprehension and focus. For an extensive review as for a doctoral dissertation, 100 or more references may be searched, but for a master's detailed review, 30 to 40 references are adequate.

4.6.3 Step 3. Select Priority Ones Through Reading and Skimming:

Once you collect literature, you need to determine if it is a good source. Give priority to peer-reviewed journal articles, and books. Subsequently, proceed to conference papers, thesis, and dissertation posters and other materials available on websites. Begin by skimming (Box 4.2) through this initial set of articles or chapters and duplicate those that are central to the topic. Throughout this process, try to obtain a sense as to whether the article or chapters will make a useful contribution to your understanding of literature. Here, you mainly read the abstract and introduction.

Box 4.2 What is skimming in the review of literature?

Skimming is a strategy for identifying the main ideas and concepts in a document without reading details. It allows you to read more in less time. It is one of the strategies for rapid reading. Another important strategy for rapid reading is scanning.

For more details on how to skim through literature refer to:

https://learningcenter.unc.edu/tips-and-tools/skimming/

As you read, evaluate the literature for credibility and accuracy and include only the good ones in your literature review. Due to the vastness of the publishing world, it is easy to become overwhelmed by the sheer amount of available literature. We recommend starting with journals and publishers known for publishing high-impact articles. Some examples of such publishers include Elsevier, Springer, Taylor and Francis, Wiley and Sons, and Cambridge. Including credible and accurate sources in your literature review ensures that your research is well-supported and reliable. Some journals from these publishers include:

- Plant and Soil (Springer)
- Agricultural Systems Journal (Elsevier)
- Experimental Agriculture (Cambridge)
- Agroforestry Systems Journal (Springer)
- Agricultural Water Management (Elsevier)

All sources should be evaluated to determine if they contribute to the understanding of your research topic. Mills and Gay (2020) recommend the following questions for evaluating literature sources:

- What was the statement of the problem? Check its relevance to your topic
- Who was studied? Is the sample population studied relevant, given your research problem and population under investigation?
- Where was the source published? Check if it is a scholarly or popular magazine; check if a journal article is refereed or non-refereed. Check if predatory or credible.
- Where was the source published? Check if it agrees with the period you have set.
- How was the study conducted? Check the objectivity of the information presented; check if interpretations are supported with evidence; check how data was collected and analyzed; check if appropriate design and methodologies were used.

Dakota State, Karl Mundt Library recommends using at least the following five criteria:

(i) Accuracy: Is the information reliable and error-free? Is there an editor who verifies the information? Is there adequate documentation? Are the conclusions justified by the information presented?
(ii) Authority: Is the source of the information reputable? What type of source is it, scholarly? Is it indexed/peer-reviewed material in a bibliography?
(iii) Objectivity: Does the information show bias? What is the purpose of the information: to inform, persuade, explain, advertise? Does the source show political or cultural bias?
(iv) Currency: Is the information current. Does it cover the time period you need?
(v) Coverage: Does it provide the evidence or information you need?

Review Your Results

Save your search in citation management software such as Zotero, Mendeley, or EndNote. Ensure you have found seminal pieces; they have been cited many times, and their work is considered foundational. An example of seminal work in integrated soil fertility management is publications by Dr. Bernard Vanlauwe (Mucheru-Muna et al. 2010; Vanlauwe et al. 2015, 2019; Adewopo et al. 2020), who has written widely on this topic.

General Guidelines for Searching and Prioritizing Literature

You cannot use every document (journal or book) related to your topic. It is crucial to exercise selectivity. The emphasis should be on a source that assists you in four fundamental aspects: (1) Refining your research question, (2) Contextualizing your study, (3) Identifying appropriate research methods and methodologies, and (4) Writing a good report.

In terms of setting priority for the different sources of literature, Creswell (2014) gives a ranking of the various sources of literature as follows, starting with no 1:

1. Starting with a broad synthesis of the literature, such as an overview found in encyclopedias, especially when examining a topic for the first time. The abstract could also be consulted at this stage.

2. Next are journal articles in credible journals.
3. Turn to books.
4. Conference papers.
5. Dissertation abstracts.

The ranking of these sources should be considered in your literature review. In addition, prioritize recent research within the last five years. However, for older topics that may be receiving less attention, you might extend to within the past ten years. Remember that classic literature may be old, but they are still perceived as important. This needs to be included despite not being recent, as provided in most university proposal guidelines.

Review articles are a priority because they tend to be focused and comprehensive. They consolidate information from multiple sources, giving readers a synthesized overview consisting of theories, trends, and controversies. They enable researchers to build a strong foundation for their investigations. Researchers mainly benefit greatly from the analysis and interpretations embedded in these review papers.

Empirical literature is important, especially in providing guidance in operationalizing variables and collecting data processes. They are valuable references for the planning phase as well as the execution phase, in data analysis and reporting. It is advisable to leverage and adopt methods and techniques that have worked in these studies instead of devising new ones with uncertain outcomes.

4.6.4 Step 4. Read and Organize Literature

Once you have selected quality and relevant literature, the next step is organizing it for your literature review. This involves reading and categorizing the sources by authors, topics, or themes. Begin with a critical examination of each selected source by taking notes. Highlight essential ideas and include citations as needed.

The University of South Columbia offers useful guidelines for effective reading: Start by reading the abstract, which summarizes the background, methods, results, discussion, and conclusions of the study. Consider the following questions: Is the study relevant to my research? What is its focus and purpose? What are the primary findings? Are there terms that could help refine my search parameters? If the abstract indicates the paper is suitable, proceed to the introduction. Here, scrutinize the research problem and the main argument or theoretical framework presented in the initial paragraphs. In the concluding parts of the introduction, identify the authors' goals. Consider these questions: What are the study's claims? What gaps in existing knowledge does it address? Does it offer new insights relevant to my research problem?

Further guidance for reviewing the literature is given in Box 4.3.

Box 4.3 Guidelines for Reviewing and Organizing the Literature

For each of the documents, i.e., research articles, journals, etc. assess and look at for the following:

- Assess the author's research problem, and whether it is well articulated.
- Evaluate the strengths and weaknesses of the way the author has formulated the problem.
- Assesses if the articles discuss major debates in the field.
- Assess the trends, themes, approaches, methodologies, theories and frameworks of the study.
- Evaluate the basic components of the study design, i.e., population, interventions, and outcomes. Assess the accuracy and validity of the measurements.
- Evaluate the analysis of data. Is it accurate and relevant? Are the conclusions valid based on the data and the analysis.
- Assess the indication of areas needing further research, gaps and oversight.
- Compares and contrasts studies.
- Identifies areas of consensus and dissent.
- Examines methodological or theoretical strengths.
- Evaluate whether the book or article contributes to our understanding of the problem under study and whether it is useful for theory and practice. Look out for strengths and weakness.

Organization The organization, on the other hand, may take many forms, such as constructing a literature map or concept map (Creswell and Guetterman 2019), using index cards (Mugenda and Mugenda 2019), and tables (Green et al. 2006). Here we use a literature matrix, that is conceptually simple for students. A literature review matrix is a structured table that helps you to organize the information from each of the sources you have selected. It includes columns for citation, main findings, methodologies critique and gaps. They are useful for drafting summaries of important articles, and templates vary according to the discipline and nature of the study (Klopper et al. 2007). Many universities, such as James Cook University and Toronto Metropolitan University, provide templates for students to use. A simple matrix is shown in Table 4.4 with a summary containing the following:

Table 4.4 A sample of a simple literature review matrix

	Paper citation	Title	Objectives	Methodology	Results and policy implications	Critique/gaps
1						
2						
3						
4						

- Title of study.
- Purpose of study.
- Research design and methodologies used.
- Brief overview of findings.
- Synopsis/critique/gaps of the content.

The draft summaries of important articles captured in the literature review matrix are important for deepening your understanding and building a strong foundation of the topic and also facilitate the writing of the literature. Each summary should be concise and focused on the key aspects of the source.

As a student, creating a literature review matrix can be beneficial for your supervisor to understand the extent of your research. Your supervisor may request to see your matrix to gain insight into the sources you have considered, how you have organized the information, and the breadth of your literature review. This feedback from your supervisor can help identify any gaps in your literature review and suggest additional sources to consider.

4.6.5 Step 5: Critically Analyze and Synthesize

After summarizing the literature, assemble the literature review, structuring it thematically or organizing it by important concepts. Synthesis means combining separate elements to form a whole (Barnett-Page and Thomas 2009). In writing a literature review, your purpose is to convey to your reader what knowledge and ideas have been established on a topic and what their weaknesses and strengths are. Hence, it will contain a summary and a synthesis (Cooper 1988) that will compare, contrast, critique, analyze, synthesize, and evaluate different pieces of literature. Critically synthesizing literature review will often include:

- *Reading and recording*: Once you have identified the literature, the next stage is to read, digest and fully record the literature you have found. While reading each article, you should consider both the content of the article and the implications of your research project.
- *Organizing similar articles*: Once you have read and made notes on several articles, you should consider organizing them into a structured form, with articles that take a similar approach being placed together.
- *Comparison across studies*: The aim is to extract key points by comparing and contrasting across studies, instead of reading one paper after another. Key points for a review may concern areas of similarities and/or differences in: Research aim(s) or hypotheses, Research design and sampling, Instruments and procedures used, how data were analyzed, results or findings, and interpretations.

4.6.6 Step 6: Writing the Literature Review

Writing a literature review involves providing an overview, summary, and evaluation of the current state of knowledge on a specific topic area of research. The purpose of a literature review is to inform the reader about the established knowledge and ideas related to the topic, including their strengths and weaknesses. However, many students often struggle with writing literature reviews because they mistakenly believe that it is only a summary of the gathered knowledge.

In reality, a literature review is not a summary or copy-paste of existing information. Instead, you should analyze and critique the work based on the research problem and objectives. You should be able to add your perspective as a researcher by identifying any inadequacies in the existing literature and providing your position on the topic. A literature review is not a summary of each scholarly resource, an essay, or a report, nor a list of summarized articles (Baumeister and Leary 1997; Green et al. 2006; Lingard 2018).

In writing your literature review, therefore, you need to demonstrate your intellectual ability to synthesize, critically evaluate, and organize relevant sources according to the topic's overarching aim and research question or hypothesis (Cooper 1988; Paré et al. 2015). To achieve this, you need two things (1) Make an outline and (2) Decide on the way to organize the written literature.

Create an outline Creating an outline can be a game-changer in writing. It not only saves you time but also increases the chances of presenting your ideas in an organized manner. You ensure a logical progression of thoughts by breaking down your content into headings and subheadings. This helps streamline your writing and makes it easier for readers to follow. An outline can save time and increase your chances of being understood.

Way of organizing literature: There are many ways of organizing written literature. There are five common ways to organize literature reviews (Klopper et al. 2007; Snyder 2019):

(a) *Chronological*—The research is discussed in chronological order. This may be useful to demonstrate the historical or developmental context of your research. It could be by dates or trends. In chronology, you will group and discuss sources in order of their appearance, highlighting the changes in research on the field. For example, a literature review on how soil fertility paradigms have changed over the years will present changes that have occurred through the years by giving examples of key developments and changes. It will then end with current theories and the direction your research will take.
(b) *Classical approach*—major writings discuss the key research about the topic.
(c) *Topical or thematic*—structure that considers different themes, categories, or concepts: Sections are devoted to various concepts or categories relevant to your study. You will group and discuss your sources regarding the themes or topics. By grouping themes or topics together, you will be able to demonstrate the types of topics that are important to your research.
(d) *General-specific*—broad–increasingly specific begins from a wide perspective and progressively focuses on the specific topic of your research.
(e) A combination of the above approaches.

Compiling and writing Following the outline, determine the subheading under which each of the references in the literature review matrix fits in your subheadings. This process involves careful consideration of the content, themes, and concepts covered in each reference. This should be followed by analyzing the references under each subheading for similarities and differences. If three references say the same thing, you will not need to describe each one. Most novice graduate students starting their research have difficulty describing one study after the other. This is wrong. You should write one sentence and cite the three sources, a hypothetical example is shown below:

> "It has also been reported that minimum tillage systems also increase output, efficiency, and productivity of farm workers, while at the same time improving the timing of planting and harvesting (Badalikova 2016; Horlings and Marsden 2011; Alvarez and Steinbach 2009)."

In cases where you need to write about contradictory studies in a literature review, you should approach this with an analytical perspective. Start by clearly presenting the contradictory studies and summarizing the findings, methodologies, and results. Highlight areas of agreement among these studies. Next, analyze and present reasons for the discrepancies. Conclude by offering your interpretation and implications for your research and how they inform your research questions and objectives.

Use of connector words Use connector words or linking words to improve coherence in your literature review. Connector words act as a bridge between sentences and paragraphs and help establish coherence in your writing. One major weakness in students' written documents is "rambled" writing. This type of writing lacks coherence and makes it difficult for readers to understand the connection between ideas. To improve the quality of writing in literature, it is essential to emphasize the use of proper writing skills (Discussed in Chap. 28). Adopting connector words and *topic sentence* structures for paragraphs (see Chap. 28) are the two main skills required to write coherent literature that flows smoothly. Examples of connector words include:

- For sequencing ideas: we use words like "firstly", then "secondly" or furthermore and "moreover."
- For showing relationship: to show similarity and contrast; we use "similarly" and "likewise" to compare or in "contrast" to contrast different sources.
- Showing disagreement: use "However," "Conversely" and "nevertheless."
- Introducing evidence: phrases like "for example" or "for instance."
- Providing additional information: Phrases like "in addition" or "additionally" help expand a topic or idea.
- Summarizing and concluding: Phrases like "To summarize" or "in conclusion" or in summary."

At the end of your literature review, you must summarize and conclude the key points in your writing. There needs to be a sense of completion to the whole piece; you need to "round off" rather than stop abruptly.

4.7 Writing the Literature Review Chapter of the Proposal/Thesis

The subsections in your literature review chapter are directed by the research problem and objectives. While crafting your review, prioritize and summarize past research on the topic and pinpoint the existing research gap, highlight and show how your study fills the gap and clearly show why it is necessary to conduct the study. The following are examples of sentences that summarize, highlight the research gap, and contribute to your study.

The contributions of A, B, and C have explored this much about my questions, while the investigations conducted by D have further increased our understanding. Building upon the foundation laid by D, my proposal involves advancing the knowledge in the following manner. The following are the components of a good literature review write-up:

- Summarizes key concepts, principles, and findings.
- Brief and focused.
- Show what has been done.
- Demonstrate what others have said.
- Highlights and show similarities and differences among the studies.
- Identify any agreements and disagreements.
- Identify any flaws in the methodologies.
- Show the gap your study will fill.
- Well-structured paragraphs.

End the literature review by providing a summary synthesis of key findings and use this to position your research within the broader scholarly conversation and to emphasize the significance and its novelty. Find agreements and disagreements in literature to showcase awareness of the existing dialogue and the unique perspective your research brings to the table. This establishes a clear rationale for your research.

Typical Structure of Literature Review

The literature review follows a broad-narrow structure. Literature review of master's and PhD thesis work tends to be organized thematically, guided by research problem and objectives (Fig. 4.2).

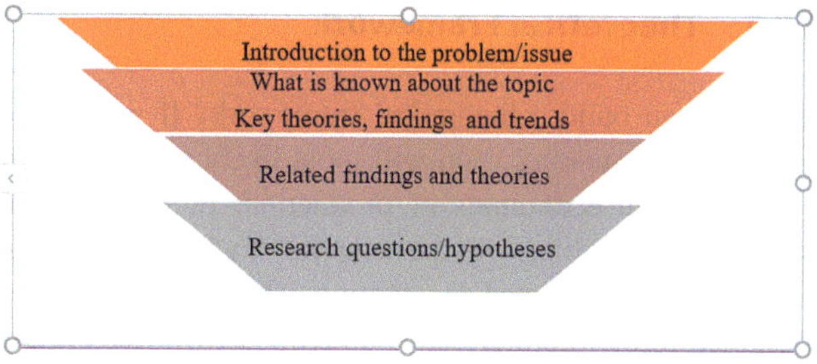

Fig. 4.2 A literature review from broad to narrow

4.8 How Much Is Adequate Literature?

"How much literature is enough?" is a common question for students. This question does not have a direct answer. An adequate amount of literature can vary depending on the specific requirement of the project, the research field, and the depth of the study. Here we advise that it is better to describe the adequacy of literature in terms of quality or effectiveness. An effective literature review should be comprehensive enough to establish a solid theoretical and empirical foundation for the research. An effective literature review achieves the following four aspects (Levy and Ellis 2006):

1. Methodologically analyze and synthesize quality literature.
2. Provide a firm foundation for the research topic.
3. Demonstrate that the proposed research contributes something new to the overall body of knowledge or improves the research field knowledge base.
4. Provide a firm foundation for research methodology.

With the copious amount of literature available, the focus should be more on "quality" rather than "quantity." Levy and Ellis (2006) argue that researchers should prioritize using quality literature to achieve effective literature. They emphasize that the use of poor-quality literature undermines the statement by Sir Isaac Walton that "I have seen a little further by standing on the shoulders of giants" and therefore may not hold water. It will be like "standing on the shoulders of midgets" suggesting that insights gained may be limited and insufficient. Therefore, the emphasis should be on selecting and prioritizing sources that bear influential studies. Additionally, seek advice from your supervisor on the depth and breadth of your literature review.

4.9 Theoretical Framework and Conceptual Framework

We have included a discussion on Theoretical and Conceptual frameworks in this literature review chapter because these three terms are intertwined. Some universities, however, will require students to present conceptual frameworks in the introductory chapter despite their common association with literature review. Students usually struggle to understand the differences between these terms and how to write them in their academic work.

Theoretical and conceptual frameworks are related but different in conceptualization and application. The term theory was first introduced in Chap. 2. We found that theories are developed by researchers to explain phenomena, make predictions, and draw conclusions. As defined, a theoretical framework *details the chosen theory*, while a conceptual framework *entails the concepts in the theory for your particular study*. A theoretical framework will explore the relationship between constructs in a given phenomenon in a broad way. A conceptual framework is more specific and represents a researcher's idea of how the research problem will be addressed and will consist of variables, constructs, and theories. In other words, the theoretical framework is like the world map that shows the general landscape and connections, while the conceptual framework is your personal travel plan that maps out the exact route and stops for your specific journey (research study).

4.9.1 Theoretical Framework

How do you build a theoretical framework? If working in production economics, you will most likely select theories explaining the economics of production. If, on the other hand, you are dealing with the behavior of adoption, you will look for theories related to adoption. This theory (theories) presented in the literature have undergone prior testing and validation by other researchers, rendering them widely accepted within the scholarly literature. A multitude of theories exist, and therefore, careful selection ensures relevance to the study. Picking the most suitable theory takes effort, time, and deep reflection.

The theory (theories) within the theoretical framework provides a foundation or hook for your research study and, therefore, serves as a road map for developing the arguments you will use in your study (Kivunja 2018). We call this "theoretical underpinning," which forms the basis of your study and helps guide you through problem conceptualization to your data analysis and interpretation of the meaning inherent in your data. Because of its role in guiding the study, it has been termed as a lens through which you see the study (Luft et al. 2022). Thus, a theoretical framework is one of the most important components as you begin and end your research.

According to Grant and Osanloo (2014), without a theoretical framework, the structure and vision of your study are unclear, much like attempting to construct a house without a blueprint. These authors describe the Theoretical framework as a "blueprint" for the entire dissertation inquiry, emphasizing its role in delineating the researcher's approach to the research philosophically, epistemologically, methodologically, and analytically. The framework serves as a guiding structure, providing clarity and coherence to the study's design and methodology, ensuring a solid "foundational plan.

Steps to Creating and Writing the Theoretical Framework

The first thing to note in creating a theoretical framework is that there are three basic elements to include. These are:

1. Definition of terms: this helps identify key concepts, define any new terms or phrases to provide the basis of understanding.
2. Explanation of theories: explain each of the theories you plan to utilize and show how they relate to your research.
3. Logical connections/interrelationships; should connect concepts, terms, and theories logically and clearly. This helps when defending your proposal in scientific meetings.

Steps in Creating a Theoretical Framework

1. *Pick your key concepts and variables*: Consult the statement of the problem and the research objectives; here you pick key variables and concepts and clearly define each of the terms. These terms will be your keywords for searching literature.
2. *Explore and review literature*. Comprehensively review all related theories in light of study objectives and key variables. This will help build your understanding of the theories and existing knowledge in your area.
3. *Explain and evaluate relevant theories;* Explain how other researchers have defined these concepts and identify connections between them. Discuss different models and theories, select the one that best fits your research study, and justify it.
4. *Select relevant theories*: Finally, select the most relevant theory or theories for your research study. Choose theories that align with key concepts and variables that you plan to study. As you write the theoretical framework, define each term and maintain consistency in your TF. Determine the relationships between the key concepts and variables and explain the components. Conclude the theoretical framework section by summarizing the importance of guiding your research.

4.10 2 Conceptual Framework

The conceptual framework consists of two words, i.e., concepts and framework. It is, hence, a framework composed of a network of *interlinked concepts*. However, various definitions exist in the literature. According to Miles and Huberman (1984) a conceptual framework (CF) is a written or visual presentation that explains the main elements to be studied. These elements include *key factors, concepts, or variables* and the presumed relationships among them. Mugenda and Mugenda (2019) define a conceptual framework as a hypothesized idea that identifies the concepts under study and their relationships. It outlines the preferred approach in the research and delineates the relationships and desired effects forming the independent and dependent variables. As defined by Jabareen (2009), a conceptual framework is a network or "plane" of interlinked concepts that collectively offer a comprehensive understanding of a phenomenon. It involves arranging concepts logically, forming a less formal structure based on specific concepts and propositions derived from empirical observations and intuition. This framework supports the researcher's position on the problem, providing direction to the study and establishing a framework specific to the idea of researcher.

A conceptual framework serves as a foundational structure in research, offering a systematic arrangement of interlinked concepts that enhance understanding. Specifically, it serves the following focus:

1. Provides a roadmap for the study, aiding the researcher in maintaining focus and direction.

2. Enables quick comprehension of proposed relationships for readers.
3. Establishes clear connections between the literature and the research goals and questions.
4. Serves as a reference point for discussing literature, methodology, and data analysis.
5. Enhances the trustworthiness of the study by offering a structured framework.
6. Assists researchers in organizing thoughts and completing investigations.
7. Functions as the foundation, similar to building a house and for addressing the research problem.
8. Explains or predicts how key concepts/variables interrelate to illuminate the problem/phenomenon.
9. Aids in organizing ideas and clarifying concepts for the researcher.
10. Introduces the research and its contribution to advancing the field of practice, aligning with relevant concepts applicable to the study's scope.

Figure 4.3 shows an example of a conceptual framework that depicts links between gendered preferences for technology attributes and the socioeconomic characteristics driving the choice and application of soil management practices (Ndeke et al. 2021). The framework illustrates how various elements interact to tackle the issue of low agricultural productivity. Farmer-specific needs and their perceptions of technology are crucial factors, serving as intermediaries that influence preferences, choices, and the intensity of soil and water conservation strategies used. Ultimately, the framework emphasizes that enhancing soil fertility and implementing effective soil water conservation measures are essential for improving agricultural productivity.

4.10.1 Key Differences Between Theoretical Framework and Conceptual Framework

Using our experience in the supervision of masters and Doctoral students and teaching of research methods, combined with information borrowed from literature (Grant and Osanloo 2014; Kivunja 2018; Luft et al. 2022), we have come up with the following differences between theoretical framework and conceptual framework. These are also summarized in Table 4.5:

(a) *Purpose*: TF provides a more detailed explanation of "how" and "why" certain variables are related and offers the theoretical basis of the study, explaining underlying mechanisms or principles guiding research. CF primary purpose is to provide a visual representation and clear definition of key concepts, variables, and their relationships. In the study, they help researchers understand and communicate the research structure.

(b) *Source*: Whereas the theoretical framework is derived from existing scholarly literature you review about your

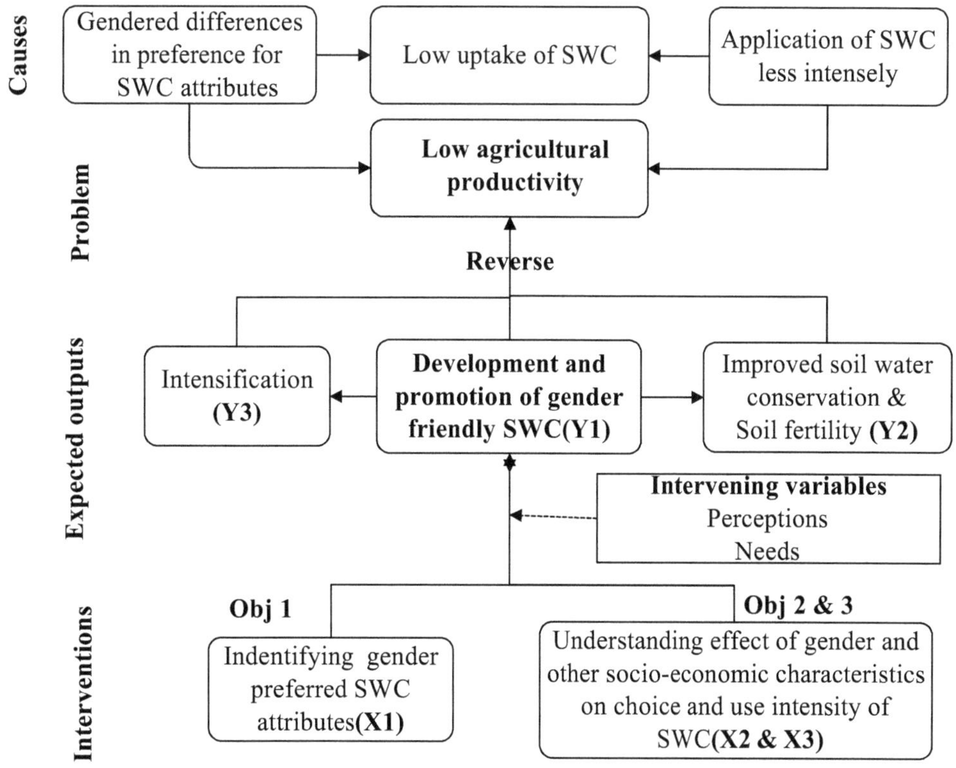

Fig. 4.3 A thesis conceptual framework from a study by Ndeke (2021)

Table 4.5 Key differences between theoretical and conceptual framework

Criteria	Theoretical framework	Conceptual framework
Purpose	Provide a detailed explanation of variable relationships outlining theoretical basis and underlying mechanisms guiding research	Provide a narrative or visual representation and clear definition of key concepts, variables, and their relationships for a specific study
Source	Explicitly grounded in established theories and models	Concepts and constructs drawn from experience, literature, and theories by researcher
Information	More detailed and comprehensive explanation of the relationships, why and how	Less detailed, broad overview of key concepts and relationships
Use	More useful in quantitative studies-theory and hypothesis testing	Often useful in both quantitative and qualitative studies, structures and guide the study
Origin	Based on published works, identifiable theories	Arguments about why the topic matters and why the methods proposed are appropriate and rigorous (Crawford, 2020)
Example	Might use innovation-diffusion theory and explain the intricacies of the theory	Might visually or graphically represent key concepts and variables like IV (institutional and household factors) and show a relationship with the dependent variable of the study

research topic, the conceptual framework is a much broader concept encompassing all aspects of your research.

(c) *Thinking Process*: Theoretical Framework comprises mainly other people's perceptions that you interpret as relevant to your research, while CF could be more of a product of your own thinking. Indeed, CF could be drawn from theory, literature, and own experiences. In written works, you may pick an existing theory, e.g., innovation diffusion theory or clusters of theories (e.g., innovation-diffusion theory and Behavioral theory) to study adoption that you modify and contextualize to fit the CF of your study, or you could pick theories directly from research papers by other authors studying the similar phenomenon.

(d) *Coverage and magnitude*: Theoretical framework is a sub-set of a conceptual framework. In terms of coverage, TF is smaller based on existing concepts, propositions, and explanations in the literature. On the other hand, the conceptual framework of a logical master plan of your entire research and forms is an all-inclusive umbrella relating to all the concepts and ideas that occupy your mind as you contemplate, plan, implement, and conclude your research.

(e) *Process underlying literature review*: TF is mainly deductive, as in the natural sciences and provides a structured way of hypothesis testing to verify a theory's power. While a conceptual framework may draw on existing theories, it is not necessarily focused on hypothesis testing. Instead, it is more useful in helping researchers to organize ideas and inform in the selection of research methods and data collection

(f) *Presentation*: The theoretical framework broadly depicts relationships between things in a given phenomenon. In contrast, the conceptual framework encapsulates the precise approach by which research will be undertaken.

(g) *Scope of application*: Theoretical framework extends its applicability beyond the current research problem and context, while CF is usually confined to the specific problem and context. However, it is possible for other research to adapt the CF in future.

(h) *Methodological approach*: TF is primarily situated within the quantitative paradigm, where hypotheses are tested and conclusions are drawn. In contrast, the conceptual framework can be situated in various paradigms, including qualitative, quantitative, and mixed methods. In the theoretical framework, data are predominantly collected through experimental designs, empirical surveys, and tests. Conversely, within the conceptual framework, data are gathered through a combination of empirical and descriptive methods, including interviews and direct interviews.

Both theoretical and conceptual frameworks could serve the same purpose and three main similarities include:

1. Helps researchers identify the main variables and concepts in a given study.
2. To provide the researcher with a general approach encompassing methodology—research design, target population, treatments, data collection, and analysis.
3. Guide the researcher in data collection, interpretation and explanation, and the formulation of conclusions.

Recommended Further Reading

1. Cronin P, Ryan F, Coughlan M (2013) Undertaking a literature review: a step-by-step approach. 10.12968/bjon200817128059 17:38–43. https://doi.org/10.12968/BJON.2008.17.1.28059
2. Ellis TJ, Levy Y (2008) Framework of Problem-Based Research: A Guide for Novice Researchers on the Development of a Research-Worthy Problem. Informing Science: The International Journal of an Emerging Transdiscipline 11:017–033. https://doi.org/10.28945/438

3. Snyder H (2019) Literature review as a research methodology: An overview and guidelines. J Bus Res 104:333–339. https://doi.org/10.1016/J.JBUSRES.2019.07.039
4. Mills EG, Gay LR (2020) Educational Research: Competencies for Analysis and Applications. Pearson

Key Highlights in This Chapter

What Is a Literature Review?

Literature review is an analytical examination of scholarly work that shows an understanding of past and current information on the topic of study. Students carry out literature reviews for normal class assignments and research projects.

Types of Literature Reviews

Narrative literature review: Traditional review presenting literature in a story-telling format, providing a comprehensive overview. One of the most essential in a research project.

Systematic literature review follows a structured and methodological approach to collecting and analyzing literature. Many employ statistical techniques to amalgamate findings from independent studies into a meta-analysis.

Meta-analysis: is a statistical technique used to combine and analyze the findings from multiple independent sources in a particular topic to make a quantitative conclusion.

Critical review: It involves a critical appraisal of research methods, study designs, and the strength of the evidence. Very useful for the identification of research gaps.

Methodological Review: Focuses on the research approaches, strategies, and data collection techniques employed. Research gaps may also be identified through methodological reviews, but they are not the primary focus.

Scoping review: designed to explore the breadth of a research topic by mapping the existing literature. They provide a broad overview of the available studies without delving into the in-depth analysis.

Critical reviews are appropriate for research projects requiring an in-depth literature evaluation by Masters and PhD students, while systematic literature reviews require advanced skills and are, therefore, suitable for PhD students.

Importance of Literature Review in the Research Process

Within the context of research, the literature is important for the following reasons:

1. Enhances theoretical understanding of the subject matter.
2. Summarizes current understanding.
3. Helps in the identification of the research gap.
4. Aids in sharpening the research problem.
5. Helps to refine research approaches and methodologies.
6. Guides in Data handling and analysis processes.
7. Supports reporting of research findings.
8. Theoretical and conceptual framework.
9. Enhances credibility of research.
10. Helps in publishing especially in journals.

Sources of Literature

There are three main sources of literature: the primary, secondary and tertiary sources. Primary sources of literature include scholarly scientific materials such as journal articles, graduate research thesis, conference proceedings, research papers, books, encyclopedias, handbooks, and gray literature.

Where to Locate Literature

You can locate literature using the following avenues:

- Libraries
- Use of search engines
- Google scholar
- Databases

Databases

Databases that can be used to locate literature in agriculture consist of agricultural-based databases, specialty databases, and multidisciplinary databases.

Agriculture-Based Databases

Several databases providing information on agriculture exist. Some of the major ones include AGORA, CABI, AGRIS, AGROVOC and FAO

- AGORA database: Access to Global Online Research in Agriculture provides access to a vast collection of free journals and books.
- *Centre for Agriculture and Biosciences International*: The Centre for Agriculture and Biosciences International (CABI) offers one of the most useful global databases in agriculture. *CABI Direct* is its major digital platform, providing access to a vast collection of information, including CAB Abstracts and Global Health.

- AGROVOC: Agriculture Vocabulary (AGROVOC) is a controlled multilingual vocabulary created to encompass concepts and terminologies (Thesaurus) coordinated by FAO in collaboration with an international community of experts and institutions.
- FAO: Food and Agriculture Organization publishes various reports, data and publications related to agriculture, food, and nutrition. It also maintains a comprehensive statistical database called FAOSTAT that provides access to agricultural and food-related data and statistics. FAO operates an open data policy providing high-quality data to all countries.

Multidisciplinary Databases Include:

- Scopus Database: an extensive abstract and citation database by Elsevier that covers various academic fields, including agriculture.
- Web of Science contains a huge collection of journal articles and other scientific publications in physical and life sciences.
- HINARI database: HINARI (Health Network Access to Research Initiative), is a program that grants institutions in low and medium-income countries access to a vast range of scientific literature on biomedical and life sciences.
- Pubmed Database focuses on biomedical sciences but also contains a significant amount of literature related to agricultural and environmental health.
- ERIC Database: Educational Resources Information Collection is a widely recognized database in education and is a useful source for information on agricultural education.
- JSTOR is a Journal Storage and is a digital library providing scholarly literature in agriculture and related fields.
- AJOL is an African Journal online. It specializes in African research
- Research for Life is a partnership of various United Nations agencies to allow access to literature to low and medium-income countries. They include databases such as AGORA, HINARI, ARDI, and OARE

Institutional repositories and websites: Research and academic institutions have specialized repositories that act as databases containing a wealth of academic materials, including theses, conference proceedings, and various types of reports.

Steps in Literature Review

The first step in reviewing the literature is identifying a research topic "the subject or subject matter of the proposed study," before deciding on the literature to review.

We present the following six steps for reviewing and writing your literature:

1. Select topic
2. Collect relevant material
3. Select the most important relevant literature for your topic
4. Read and organize in a literature matrix
5. Organize and synthesize
6. Write the literature review

Step 1: Choose Topic

Begin by choosing the topic or referring to the topic that you have already chosen in step one of the research processes.

Step 2. Collect Relevant Information

The second step involves locating the literature by first identifying the keywords followed by locating the literature. It will involve the following sub-steps:

1. Identifying the keywords
2. Searching the library
3. General searching on the internet
4. Searching databases
5. Decide on coverage of the review

Step 3. Select Priority Ones Through Reading and Skimming

Once you locate literature, you need to determine if it is a good source while giving priority to searching journals, articles and books as they are easily accessible and obtainable.

Step 4. Reading and Organizing Literature

Once you have selected relevant, quality literature, the next step is to organize it for a literature review. This involves reading and organizing the literature by sources, authors, or topics.

Step 5: Critically Analyze and Synthesize

After summarizing the literature, assemble the literature review, structuring it thematically or organizing it by important concepts. LR will contain a summary and a synthesis that will compare, contrast, critique, analyze, synthesize, and evaluate different pieces of literature

Step 6: Writing the Literature

Writing a literature review involves providing an overview, summary and evaluation of the current state of knowledge on a specific topic area of research. In reality, a literature review is not a summary or copy-paste of existing information. There are five common ways to organize literature reviews: Chronological, Classical Approach, Topical or thematic, General-Specific

Theoretical and Conceptual Framework

Theoretical framework and conceptual framework are related to the literature review. The two terms differ in conceptualization and application. A theoretical framework relies on established theories and provides a general lens for interpreting research. In contrast, a conceptual framework highlights specific variables and relationships tailored to a particular study. Theoretical frameworks emphasize validating or extending theories, while conceptual frameworks focus on practical clarity and direct applicability to the research problem.

Review Questions

1. In your own words, explain the meaning of a literature review.
2. What are the three main types of literature reviews discussed in this chapter?
3. List at least seven reasons for carrying out a literature review in the research process.
4. Explain the advantages and disadvantages of using the internet for searching literature.
5. Identify five databases useful for agricultural literature and explain what kind of information they provide.

Chapter Exercise

1. Visit your institution's library and approach the Information Desk or the Research Help Desk to connect with a librarian. This will help you identify the library's resources to support your agricultural research. In your visit seek guidance on:
   (a) The most efficient ways to locate specific types of agricultural literature, such as policy documents and technical reports
   (b) How to navigate specialized agricultural databases that may not be immediately visible. Target AGORA, CABI
   (c) Methods for accessing materials not physically available in your library, including using the inter-library loan system or digital repositories.
   (d) Choose an area of your interest and access at least 10 journal articles. Skim through and select at least 4 that are most relevant. Read the four and prepare a literature review matrix
2. Visit your university library, either physically or online and explore its institutional repository or digital archive. Do the following:
   (a) Locate the institutional repository (ask at the Information Desk or search the library's website).
   (b) Search for past theses related to your area of agricultural research, for example, topics on crop improvement, soil health, climate-smart agriculture, or rural development in Kenya.
   (c) Pay attention to how the documents are structured, including how research problems, literature reviews, objectives, and methodologies are presented.
3. There is a "close interdependence between the literature review and the theoretical and conceptual frameworks." Explain this connection in your own words. How does engaging with existing literature help you build or refine these frameworks for your study?
4. Your research aims to understand why some small-scale farmers readily adopt new agricultural technologies while others resist.
   (a) Briefly suggest a relevant theoretical framework (e.g., Diffusion of Innovations Theory, Theory of Planned Behavior) that could guide your study and explain why it fits.
   (b) Based on your chosen theory and understanding of the problem, outline what a conceptual framework for your study might include (e.g., key variables and their presumed relationships).

Answers to Review Questions

- Q1. Reference: 4.1.1.
- Q2. Reference: 4.1.2.
- Q3. Reference: 4.2.
- Q4. Reference: 4.5.
- Q5. Reference: 4.4 and Table 4.2.

References

Adewopo J, Peter H, Mohammed I, et al (2020) can a combination of UAV-derived vegetation indices with biophysical variables improve yield variability assessment in smallholder farms? Agronomy 10:1934. https://doi.org/10.3390/AGRONOMY10121934

Amaresan N, Patel P, Amin D (eds) (2022) Practical handbook on agricultural microbiology. Springer US, New York

Bahishti AA (2021) The importance of review articles & its prospects in scholarly literature. Extens Rev 1(1):1–6. https://doi.org/10.21467/exr.1.1.4293

Barnett-Page E, Thomas J (2009) Methods for the synthesis of qualitative research: a critical review. BMC Med Res Methodol 9:1–11. https://doi.org/10.1186/1471-2288-9-59/TABLES/1

Baumeister RF, Leary MR (1997) Writing narrative literature reviews. Rev Gen Psychol 1:311–320. https://doi.org/10.1037/1089-2680.1.3.311

Borenstein M, Hedges L V., Higgins JPT, Rothstein HR (2009) Introduction to meta-analysis. In: Introduction to meta-analysis, pp 1–421. https://doi.org/10.1002/9780470743386
Christie BR (2022) CRC handbook of plant science in agriculture. CRC Press
Cooper HM (1988) Organizing knowledge syntheses: a taxonomy of literature reviews. Knowl Soc 1:104–126. https://doi.org/10.1007/BF03177550/METRICS
Creswell JW (2014) Research and design qualitative, quantitative, and mixed methods approaches. SAGE Publication Inc
Creswell JW, Guetterman TC (2019) Educational research. In: Planning, conducting, and evaluating quantitative and qualitative research. Pearson, pp 1–681
Cronin P, Ryan F, Coughlan M (2013) Undertaking a literature review: a step-by-step approach. 1012968/bjon200817128059 17:38–43. 10.12968/BJON.2008.17.1.28059
de Klerk W, Pretorius J (2019) Guideline for conducting critical reviews in psychology research. J Psychol Afr 29:645–649. https://doi.org/10.1080/14330237.2019.1691793
Dörr J, Nachtmann M (2022) Handbook digital farming: digital transformation for sustainable agriculture, pp 1–403. https://doi.org/10.1007/978-3-662-64378-5/COVER
Ellis TJ, Levy Y (2008) Framework of problem-based research: a guide for novice researchers on the development of a research-worthy problem. Inform Sci 11:017–033. https://doi.org/10.28945/438
Fisher DJ, Zwahlen M, Egger M, Higgins JPT (2022) Meta-analysis in Stata. In: Systematic reviews in health research: meta-analysis in context: Third edition, pp 483–509. https://doi.org/10.1002/9781119099369.CH25
Fraenkel JR., Wallen NE., Hyun HH (2012) How to design and evaluate research in education, p 642
Gardner BL, Rausser GC (eds) (2001) Handbook of agricultural economics. Elsevier
Garrod B (2023) What makes a good critical literature review paper? Tour Hosp 4:141–147. https://doi.org/10.3390/TOURHOSP4010008
Gentles SJ, Charles C, Nicholas DB et al (2016) Reviewing the research methods literature: principles and strategies illustrated by a systematic overview of sampling in qualitative research. Syst Rev 5:1–11. https://doi.org/10.1186/S13643-016-0343-0/FIGURES/2
Grant C, Osanloo A (2014) Understanding, selecting, and integrating a theoretical framework in dissertation research: Creating the blueprint for your "house". AIJ 4:12. https://doi.org/10.5929/2014.4.2.9
Green BN, Johnson CD, Adams A (2006) Writing narrative literature reviews for peer-reviewed journals: secrets of the trade. J Chiropr Med 5:101–117. https://doi.org/10.1016/S0899-3467(07)60142-6
Heldman DR, Moraru CI (eds) (2010) Encyclopedia of agricultural, food, and biological engineering
Jabareen Y (2009) Building a conceptual framework: philosophy, definitions, and procedure. Int J Qual Methods 8:49–62. https://doi.org/10.1177/160940690900800406/ASSET/B6B6449D-88E1-4ABC-9916-EF26EB904492/ASSETS/IMAGES/LARGE/10.1177_160940690900800406-FIG1.JPG
Jones-Garcia E, Krishna VV (2021) Farmer adoption of sustainable intensification technologies in the maize systems of the global south. A review. Agron Sustain Dev 41:1–20. https://doi.org/10.1007/S13593-020-00658-9
Kivunja C (2018) Distinguishing between theory, theoretical framework, and conceptual framework: a systematic review of lessons from the field. Int J High Educ 7:44. https://doi.org/10.5430/IJHE.V7N6P44
Klopper R, Lubbe S, Rugbeer H (2007) The Matrix method of literature review. https://doi.org/10.10520/AJA10231757_377
Levy Y, Ellis TJ (2006) A systems approach to conduct an effective literature review in support of information systems research. Inform Sci 9:181–212. https://doi.org/10.28945/479
Lingard L (2018) Writing an effective literature review: part II: citation technique. Perspect Med Educ 7:133–135. https://doi.org/10.1007/S40037-018-0407-Z
Low G, Dalhaus T, Meuwissen MPM (2023) Mixed farming and agroforestry systems: a systematic review on value chain implications. Agric Syst 206:103606. https://doi.org/10.1016/J.AGSY.2023.103606
Luft JA, Jeong S, Idsardi R, Gardner G (2022) Literature reviews, theoretical frameworks, and conceptual frameworks: an introduction for new biology education researchers. CBE Life Sci Educ 21:rm33. https://doi.org/10.1187/CBE.21-05-0134/ASSET/IMAGES/LARGE/CBE-21-RM33-G001.JPEG
Marshall C, Rossman GB, Blanco GL (2021) Designing qualitative research, 7th edn. SAGE Publications Ltd
Miles MB, Huberman AM (1984) Drawing valid meaning from qualitative data: toward a shared craft. Educ Res 13:20–30. https://doi.org/10.3102/0013189X013005020/ASSET/AE13716F-D080-4112-BCE8-39AFAE38C1FD/ASSETS/0013189X013005020.FP.PNG
Mills EG, Gay LR (2020) Educational research: competencies for analysis and applications. Pearson
Mucheru-Muna M, Pypers P, Mugendi D et al (2010) A staggered maize–legume intercrop arrangement robustly increases crop yields and economic returns in the highlands of Central Kenya. Field Crop Res 115:132–139. https://doi.org/10.1016/J.FCR.2009.10.013
Mugenda A, Mugenda G (2019) Research methods: qualitative and mixed methods approach, pp 100–102
Ndeke AM (2021) Gender influence on soil fertility and water management technologies uptake among smallholder farmers in Tharaka Nithi County. University of Embu
Ndeke AM, Mugwe JN, Mogaka H et al (2021) Gender-specific determinants of Zai technology use intensity for improved soil water management in the drylands of upper eastern Kenya. Heliyon 7:e07217. https://doi.org/10.1016/J.HELIYON.2021.E07217/ATTACHMENT/42C3C6AC-1B1C-41F7-B346-FE44585D4852/MMC1.DOCX
Nijs L (2014) The handbook of global agricultural markets: the business and finance of land, water, and soft commodities. Palgrave Macmillan
Ochs M (2004) AGORA: Access to Global Online Research in Agriculture [Abstract only]: Libraries and "Out-Reach" -- serving the world around us. Issues Sci Technol Librariansh. https://doi.org/10.29173/ISTL1953
Osabuohien ES (ed) (2020) The Palgrave handbook of agricultural and rural development in Africa. Springer International Publishing
Paré G, Trudel MC, Jaana M, Kitsiou S (2015) Synthesizing information systems knowledge: a typology of literature reviews. Inf Manag 52:183–199. https://doi.org/10.1016/J.IM.2014.08.008
Peters MDJ, Marnie C, Colquhoun H et al (2021) Scoping reviews: reinforcing and advancing the methodology and application. Syst Rev 10:1–6. https://doi.org/10.1186/S13643-021-01821-3/PEER-REVIEW
Pham MT, Rajić A, Greig JD et al (2014) A scoping review of scoping reviews: advancing the approach and enhancing the consistency. Res Synth Methods 5:371–385. https://doi.org/10.1002/jrsm.1123
Rosenthal R (1995) Writing meta-analytic reviews. Psychol Bull 118:183–192. https://doi.org/10.1037/0033-2909.118.2.183
Saunders MNK, Philip L, Adrian T (2019) Research methods for business students. Pearson Education, p 833
Snyder H (2019) Literature review as a research methodology: an overview and guidelines. J Bus Res 104:333–339. https://doi.org/10.1016/J.JBUSRES.2019.07.039
Subirats-Coll I, Kolshus K, Turbati A et al (2022) AGROVOC: the linked data concept hub for food and agriculture. Comput Electron Agric 196:105965. https://doi.org/10.1016/J.COMPAG.2020.105965

Tricco AC, Lillie E, Zarin W et al (2016) A scoping review on the conduct and reporting of scoping reviews. BMC Med Res Methodol 16:15. https://doi.org/10.1186/s12874-016-0116-4

Van Alfen NK (ed) (2014) Encyclopedia of agriculture and food systems, 2nd edn. Elsevier

Vanlauwe B, Descheemaeker K, Giller KE et al (2015) Integrated soil fertility management in sub-Saharan Africa: unravelling local adaptation. Soil 1:491–508. https://doi.org/10.5194/SOIL-1-491-2015

Vanlauwe B, Coe RIC, Giller KE (2019) Beyond averages: new approaches to understand heterogeneity and risk of technology success or failure in smallholder farming. Exp Agric 55:84–106. https://doi.org/10.1017/S0014479716000193

Walsh D, Downe S (2005) Meta-synthesis method for qualitative research: a literature review. J Adv Nurs 50:204–211. https://doi.org/10.1111/j.1365-2648.2005.03380.x

Zhang Q (ed) (2023) Encyclopedia of digital agricultural technologies. Springer International Publishing

Part II

Research Proposal and Ethics

Research Proposal

5

Abstract

A research proposal outlines the proposed research and procedures to be followed in the research process. It tells the reader (supervisor and reviewers) and yourself about "What will be investigated" and "How it will be investigated." The format of a research proposal varies depending on the institution, school, department, and subject matter. However, this chapter discusses the generic format of an academic research proposal. The proposal comprises three parts: (1) preliminary pages (front matter), (2) body, and (3) back matter. The body has three main chapters: Introduction, Literature Review, and Research Methodology. The preliminary pages include the title, certification, abstract, and table of contents pages. It may also include a list of tables and figures. The introduction contains background to the problem, a statement of the problem, research objectives/questions, and the significance of the study. The literature review chapter includes reviewing and presenting findings of other authors about the topic to show how the study fits in. It also presents key sections on theoretical and empirical literature and theoretical and conceptual frameworks. The sections in the third chapter, i.e. research methodology, include the location of the study, the research design, methods and tools for collecting data, data analysis, and ethical considerations. Finally, the components of the back matter are presented, including references and appendices, where the work plan, budget, and data collection tools are placed.

Keywords

Background to the study · Conceptual framework · Data analysis matrix table · Delimitations · Empirical literature · Ethical considerations · Limitations · Significance of the study · Statement of the problem · Hypothesis · Specific objectives · Research Design · Research methodology: theoretical framework · Theoretical literature

Learning Outcomes

After reading this chapter, you will be able to:

1. Appreciate the value of an academic research proposal.
2. Describe the format and the elements of a research proposal.
3. Relate a "strong Research proposal" to a "Good thesis" and journal publication.
4. Differentiate between the Background to the study and the statement of the problem.
5. Write a compelling "statement of the problem."
6. Show the relationship between research objectives and research hypotheses in a proposal.
7. Differentiate between empirical and theoretical literature.
8. Explain how to structure description of the study area.
9. Differentiate between main objective and specific objectives.
10. Show components of a research methodology for experimental and non-experimental designs.
11. Differentiate between limitations and delimitations.
12. Outline the components of the back matter of a proposal.

5.1 Meaning of a Research Proposal

In Chap. 2 (Research process), we find that after formulating the objectives and designing the research, a researcher needs to write a research proposal before commencing the study. The main purpose of the research proposal is to show that the person intending to undertake the research has a clear and systematic conceptualization of the research problem and process. It also demonstrates that the person has done adequate reading and is informed about recent work in the area of interest. The proposal provides an intent and plan of what you want to do in your research project. In other words, the research proposal acts like a "blueprint" that will guide the researcher and ensure the research is conducted rigorosly and ethically.

J. Njeri Mugwe, S. Runo, *Research Methodology in Agricultural Sciences*, https://doi.org/10.1007/978-981-95-1892-0_5

The overall purpose is to present a framework for obtaining answers to the research problem already identified. The proposal aims to demonstrate that the research problem is significant, is well justified, and that the research plan is practical. It also demonstrate that the intended results will contribute to the welfare of society. Preparing a research proposal before engaging in any research, whether for academic purposes or other pursuits, is mandatory. Academic research proposals must be approved before starting any research. This step helps to assess the credibility of the proposed research.

Typically, departments organize departmental seminars where students present their proposals. During the presentations, the faculty staff members evaluate the feasibility, achievability, practicability, and replicability of the proposed study. To prepare for a proposal presentation, it is important to be practical, persuasive, and clear. Following the presentation, students should address all comments given to improve the overall quality of the proposal.

In addition to approval by your institution, you may need approval from other certifying bodies, such as the Institutional Review Board and professional bodies (Discussed in the next chapter). This depends on the nature of your research.

When writing a research proposal for your thesis, you should strictly follow the guidelines of the university. The guidelines help students to present their proposals clearly and logically. During the examination process, conformity with the guidelines is checked. Following the guidelines shows professionalism and enhances the quality of your proposal.

Your supervisor will guide any variations from the guidelines in content and presentation based on the nature of your study, whether quantitative, qualitative, biophysical, or social. Keep your proposal organized, simple, and logically presented to make it easier for readers to understand. Some students make the mistake of mentioning their supervisor's name during seminar presentations in case of an issue. It is best to avoid doing this. Instead, be prepared to defend whatever you write. Remember, it is your proposal, and you should fully own it. It is also not a time to own up and admit that something is wrong.

At times, researchers in research institutions and individuals seeking financial support from funding institutions write a research proposal as a *grant proposal* (Cheng 2014). A grant proposal is a document written to secure funds from a funding organization. The primary purpose is to convince the granting agency that the proposed research is worthy of funding and aligns with the agency's goals and priorities. The grant proposal exhibits some similarities and differences with an academic research proposal.

One of the main similarities is the inclusion of clear research objectives, a justification for the study, and a robust methodology. Both proposals also need to include a thorough literature review to show the research's context and importance. Additionally, they may both show researchers' expertise and qualifications.

However, key differences exist between an academic proposal and a grant proposal. Grant proposals emphasize a research and development agenda, where knowledge will be generated and touch society at the same time. The focus is on practical application that can lead to the real-world impact of the proposed research. Often, the funding agencies are interested in projects that can address real, specific problems and contribute to society's goals, the so-called applied research or research for development (Chap. 1). On the other hand, academic research proposals tend to focus more on advancing theoretical knowledge.

This chapter mainly discusses an academic proposal, though some concepts and principles apply to a grant research proposal.

5.2 Purpose and Characteristics of Research Proposal

As elucidated earlier, the proposal gives three key things that guide the research project: (1) What is motivating the research (2) What will be done, and (3) How it will be done. It should address the following questions:

- Is your chosen topic feasible for research?
- Why is your research important and worth investigating?
- What methods will you use to gather information and answer your research questions?

5.2.1 Novelty in a Research Proposal

Novel research refers to ideas that are original and unique. A novel research proposal introduces a new perspective, addresses a gap in existing knowledge, or explores new research questions that have not been extensively studied. The novelty of your proposal is crucial in getting it approved and publishing the work. Examiners look for new ideas, not duplications, and editors of journals are interested in new information and discoveries.

In Chap. 3, we provided several sources of research topics and problems. An in-depth literature review is one of the most important sources to help bring a novel research idea on board. By reading existing work, you can assess and identify what is lacking or inadequate, such as looking at new ways of addressing an old problem or an improved methodology. For instance, in soil fertility management, organic and inorganic sources are traditional ways of replenishing soil fertility. Recently, new innovative ways have been proposed to look at the use of insect waste, especially black soldier fly

(Beesigamukama et al. 2020). This idea is novel because the old problem persists, but the idea of using insects has not been explored before.

5.2.2 Well-Written Proposal

When writing a research proposal, it is not just the content that matters. The quality of your writing matters and determines whether your proposal is accepted or rejected by the proposal defense committee. Even if your research idea is strong, a poorly written proposal can make it hard for reviewers to understand and assess the value of your work (Badenhorst 2020).

To enhance success in a proposal defense, aim to have clear writing that is cohesive and persuasive. This is a requirement in scientific documents (see Chap. 28). Clear writing involves investing time to structure your proposal, using appropriate language, and ensuring that it is free of grammatical errors. Creative and critical thinking are required as you write your arguments. All arguments should be supported by strong evidence and logical reasoning.

Your department and school expect you to use the specific guidelines for research proposals. Ensure you are familiar with these guidelines and adjust your writing accordingly. Overall, the way you write is of utmost importance. Dedicating time and effort to crafting a quality research proposal can significantly increase your chances of a successful defense and consequently, a worthwhile research project.

5.2.3 Interdependence Between Well-Crafted Proposals and Publications

Due to time limitations, students often rush through research proposal development, leading to poorly crafted proposals. A poorly crafted proposal negatively affects the quality of the emanating publications. Rushing may result in a weak literature review that is deficient, poor problem formulation, and inadequate details in the methodology. This impacts the research project's success. Here, we emphasize the importance of writing a "strong proposal."

A strong proposal should be well justified and follow university guidelines on writing and formatting, among other characteristics (Fig. 5.1).

A strong research proposal is the cornerstone for good publications that make an impact (Fig. 5.2). We put it that "A good thesis or dissertation hinges on a strong proposal." Similarly, "A quality journal publication hinges on a strong proposal." We advise that you take time to understand scholarly discourse as you craft the proposal. As the saying goes, "if you fail to plan, you plan to fail." "Without a strong research proposal, you are planning for failure. Failure could mean many things, including wasted resources, inability to secure funding, and ultimately, producing research outputs that lack rigor and impact. Failure could also mean the inability to publish in high-impact journals.

5.2.4 Well-Formatted Proposal

Take time to format your proposal correctly according to the guidelines provided by your institution and to write it well. Formatting is about organizing layout and appearance. Many students believe that formatting is less important than content, but this is not true. Your supervisor and examiner will review your formatting, which is essential in ensuring that the document meets standards and guidelines and is appealing to readers. It also involves the use of clear headings and subheadings and adherence to the recommended font type and size. Consistency in margins, citation styles, and the use of recommended fonts is also part of formatting and helps

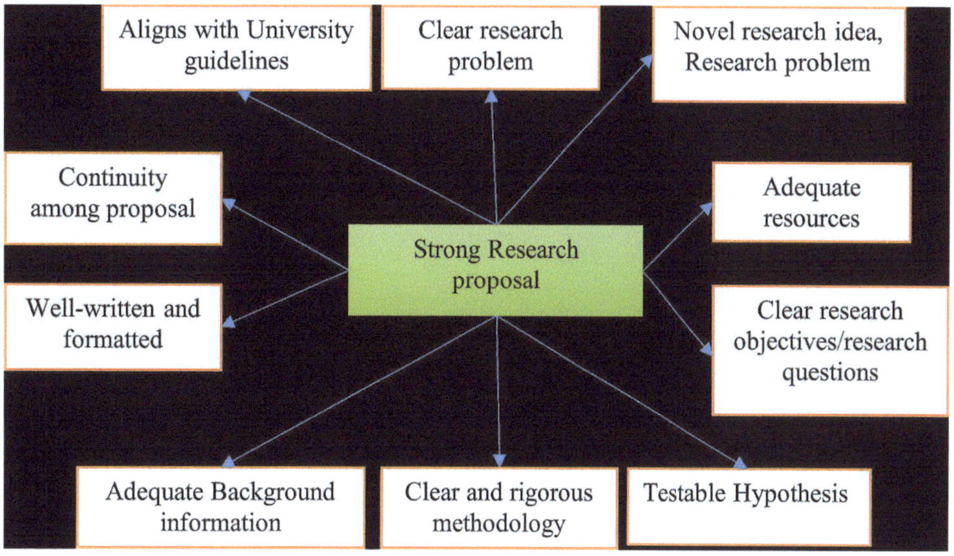

Fig. 5.1 Characteristics of a strong academic proposal

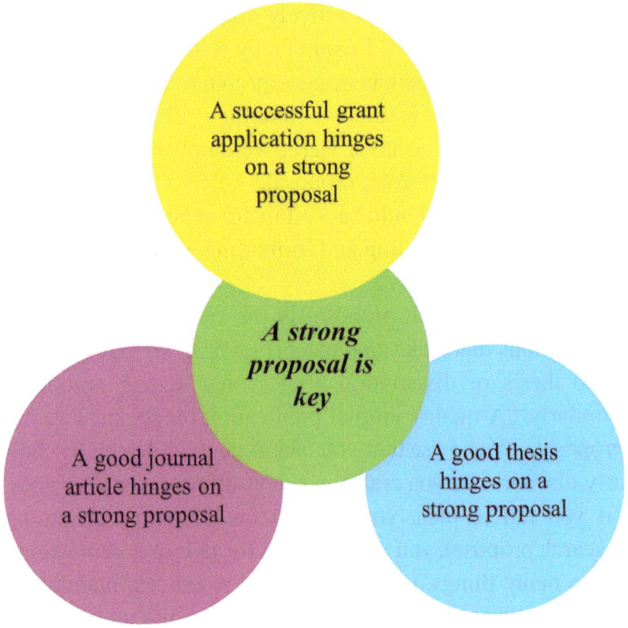

Fig. 5.2 Interdependence between proposal and publications

you attain a professional look. Use the help of the university guidelines on how all these should be done and follow accordingly.

5.3 Components and Structure

An academic proposal's exact components and structure vary according to institutions, schools/colleges, departments, and subject matter. However, there are generic components that are essential for all proposals. The format of an academic research proposal can be subdivided into three parts (Box 5.1):

1. Frontmatter.
2. Middle matter or the Main Body of the proposal.
3. Back matter.

Box 5.1: Components of a Research Proposal

Front matter	
Title on cover page	Reflect the scope and content of the proposal; should be concise and informative; not more than 20 words.
Declaration	Student and supervisor sign: Borrow rubric from university guidelines.
Abstract	Summary of the proposal; written last.
Table of contents	List of sections and chapters.
List of abbreviations and acronyms	Shortened form of terms used.

(continued)

Middle matter	
Introduction	
Background	Introduces the research problem or issue, justifies the study, provides contextual information, highlighting the current state of knowledge and gaps (present 4–6 paragraphs)
Statement of the problem	Focus on the knowledge gap to be filled (1 to 2 paragraphs).
Research questions	End with a question mark (?).
Research objectives	Statements using action verbs.
Research hypothesis	Statements with measurable variables.
Purpose of the study	Overarching goal or objective study aims to achieve.
Significance of the study	Indicate benefits and beneficiaries.
Delimitations of the study	Scope or coverage.
Limitations of the study	Explain constraints, restrictions, or restrictions inherent in the research study.
Assumptions of the study	Show presuppositions that underlie the research study.
Definition of terms	Provide precise explanations of terms and concepts used in the study.
Literature review	
Literature review sections	Theoretical literature. Empirical literature. Written in sections informed by the title and objectives of the study. 2.1…… 2.2…… 2.3…… 2.4……
Theoretical framework	Theoretical literature (mainly narrative but can also include diagram.
Conceptual framework	Mainly narrative but can also include a diagram.
Summary of research gaps	Summarize areas lacking sufficient exploration.
Materials and methods	
Study area	Describe the location of the study.
Study design	Explain design (experimental or nonexperimental).
Sampling methods	Explain sampling techniques.
Methods of data collection	Explain methods to be used in data collection.
Procedures of analyzing data	Show techniques of handling and analyzing data.
Ethical considerations	Show how adherence to moral principles and guidelines will be achieved.

(continued)

Box 5.1 (continued)

Back matter	
References	
Appendices	
Workplan	Dates and activity
Budget	Requirements and costs
Data collection tools	Questionnaires, checklists, etc.
Field and plot layout plans	Drawing
Experimental protocol	Detailed methods

5.4 The Front Matter

The front matter of a research proposal contains the title, captured on the cover page, declaration page, abstract, list of tables, list of figures, list of acronyms and abbreviations.

5.4.1 Title

The title of a proposal is crucial as it creates the first impression and influences the reader's decision.

Many students often ask about how to create a title for their research proposals. The answer is simple: the title should mirror the proposal's scope and content. We recommend that you to start with a "working title" or "tentative title" highlighting the research project's main focus. This title acts as a reference point as you write the proposal and can evolve as the proposal takes shape and the objectives and research design become clearer. The goal is to refine the title until the end to ensure that it accurately represents the content of the proposal.

In addition to accurately representing the proposal's content, the title should be brief (concise), specific, and informative. A well-written title makes it easy for readers and reviewers to grasp the focus of the study. A brief title avoids overly general terms, such as vague descriptors and jargon, while highlighting unique aspects of the study. The aim is to have an informative title that conveys the nature of the study and its unique aspects (Box 5.2).

In summary, a good title has the following characteristics:

- Brief and specific but informative.
- Portray the nature and aspects of the study.
- In line with the set objectives of the proposal.
- It is unambiguous.
- May indicate the geographical area and species under study.

Box 5.2: Examples of Well-Formulated Titles

Examples of Titles *Effect of mulch and hydrogel on soil moisture, growth and yield of yellow passion fruit in Kiambu County, Kenya (19 words).*
Nature of study: Mulch and hydrogel
Variables: Soil moisture, growth and yield
Species under study: Yellow passion fruit
Geographical area: Kenya
Source: Michoma et al. (2023)
Farmers perceptions and adoption of micro catchments for agroforestry trees establishment in East Shewa Zone, Ethiopia (16 words)
Nature of study: perceptions and adoption
Variables: micro-catchments
Species under study: agroforestry trees
Geographical area: Ethiopia
Source: Kinuthia (2020)

5.4.2 Declaration Page

The declaration page follows the cover page in most cases and typically includes a statement declaring the authenticity and integrity of the work. Guidelines or rubrics are provided by the institution. The page has two sections; the first has a statement where the student declares that the work presented in the proposal is their own. A second section has a statement where the supervisors declare they have approved the proposal submission. The student and the supervisor sign on this page to reinforce the principle of academic integrity and accountability. The page is thus also called the certification page.

5.4.3 Abstract

An abstract is a summary that gives the reader a snapshot of the document. A proposal abstract should be a concise and precise summary of the proposal.

A common pitfall with students is missing the inclusion of all components of an abstract while exceeding the word limit. Though concise, the abstract should be comprehensive. It should have the research's background, problem, objectives, methods, and expected outcomes. Keep in mind that the abstract should not exceed 500 words. Some instructions give a one-page limit. It is written last after you are satisfied with the content and format of your proposal.

5.4.4 List of Tables and Figures

The lists of tables and figures in a research proposal are an organized overview of all the tables and figures used in the document. They provide a sequential list of tables and figures, along with their corresponding titles, page numbers, and chapters where they are found. To ensure consistency and easy navigation for the reader, the captions in the list must match exactly with how they appear in the actual document.

5.4.5 List of Acronyms and Abbreviations

The list of acronyms and abbreviations provides definitions or full forms of shortened terms used in a research proposal. It appears after the table of contents and before the main body of the proposal. The list is typically organized alphabetically for easy reference and helps readers understand the technical language used in the proposal. It also serves as a helpful reference for those unfamiliar with the specific terminology used in the proposal, enhancing its overall clarity.

5.5 Middle Matter

5.5.1 Chapter 1: Introduction

Sections of this chapter of the research proposal include:

1. Background to the study.
2. Statement of the problem.
3. Research objectives, questions, and/or hypotheses.
4. Purpose of the study.
5. Significance of the study.
6. Delimitation of the study.
7. Limitations of the study.
8. Assumptions of the study.
9. Definition of key terms and concepts.

5.5.1.1 Background to the Study

The background section of a research proposal provides the introduction and context for the specific area of interest. It introduces the general problem being addressed in the research. This section should provide a brief global, regional, and national overview of the research topic. Typically, the background section consists of three to four paragraphs that focus on the main points of the research topic, as well as the scope of the inquiry.

The background section of your research proposal should include the following elements:

- Introduction to the research topic.
- Brief overview of the problem: Provide a summary of the issue you propose to investigate.
- Historical overview of the subject and the problem: Explain how the problem has evolved over time.
- The magnitude of the problem: Use statistics, numbers, or figures to illustrate the extent of the problem.
- The rationale for choosing this problem: Explain why you decided to focus on this particular issue.
- Review of previous work: Describe past efforts to control or maintain the problem and specify whose efforts they were.
- Authoritative literature: Only use works that are widely accepted as credible sources of information. Avoid literature that is more than 10 years old, unless it contains classic information that has had a lasting impact.

The background information should be organized as follows:

(i) Highlight the major debates and gaps on the subject area globally, about the variables being investigated.
(ii) Highlight the major debates and gaps on the subject area regionally, about the variables being investigated.
(iii) Highlight the major debates and gaps on the subject area nationally regarding the variables being investigated. This is where the research problem will be found.

5.5.1.1.1 How to Write the Background

Begin by conducting a comprehensive review of the literature related to your topic. This step involves searching and assembling the relevant literature using keywords and variables extracted from your topic (Refer to Chap. 4 on literature review). As you read, take thorough notes on findings, theories, methodologies, and gaps.

Next, have a plan and structure for your background, including an outline of the key concepts you intend to address. The structure should follow a funnel shape, starting with broad information and gradually narrowing to the specific research problem (Table 5.1). In the first paragraph, briefly write what is already known, both in practice and literature. As you continue with subsequent paragraphs, narrow down the information, contextualize the discussion of the study, and justify your ideas. Adding context to your study situates the research problem in a broader framework, allowing readers to understand its significance under specific conditions and influences.

For example, if your proposal involves studying farmers, you should mention if they are smallholder farmers and provide information on their socioeconomic profile. This will

Table 5.1 Structure of background to the study as written in a research proposal

Funnel approach to writing background	
Paragraph one	Present what is known in literature, theoretically and practically. Show the importance of the topics; global perspective. Cite government reports and include statistics. Brief overview of the concepts.
Paragraphs two and three	Present relevant theories and key findings in the literature. Include recently published articles in credible journals. Support each statement with several citations. Highlight research gaps and clearly show their importance.
Paragraph 4 (last paragraph)	Explain what makes this context a unique case to study. Showing the distinctive characteristics, circumstances, and factors that set this case apart from the others for research.

Keywords in background: Chronological order, context, missing points (gaps in the study)

help to understand the unique challenges and opportunities they may face. Finally, in the last paragraph focus on the specific research need that your study addresses and clearly explain why the identified gap or challenge is significant.

Two case studies of background are presented: A PhD research proposal involving a biophysical study involving soil erosion and fertility management (Oduor et al. 2022) and 2. A masters research proposal on a social study involving the adoption of soil fertility management (Mwaura 2021) (Box 5.3):

Note pitfalls to avoid in background of the study (Box 5.4).

5.5.1.2 Statement of the Problem

Refer to Chap. 3 Sect. 3.4 for a detailed explanation of how to come up with a statement of the problem (SOP) or "problem statement." Statement of the problem is the specific problem that you will address in your proposal. From the literature, there is no real "best way" of writing a statement of the problem as many variants are presented by different authors (Creswell 2013; Mugenda and Mugenda 2019).

However, most authors agree that the statement of the problem should point out "the research gap" and help define the variables of the study. The statement of the problem" should also be short. While it is important to provide sufficient context, a lengthy SOP can dilute its impact and make it challenging for readers to grasp the core issue you plan to address in your proposal. SOP should be "sharp" and hence written with clarity and brevity. Aim for not more than two paragraphs.

Writing the SOP is challenging, even for an experienced researcher, and often involves multiple rounds of revision and refinement. It also involves rational and creative thinking. Make sure that your "Statement of the problem" fulfils the following two key purposes;

1. Gives a clear issue or gap in knowledge that your research aims to address.
2. Justify the entire research study.

The statement of the problem serves to answer the question, "Why is this research necessary?" However, many students tend to become hyperbolic and focus on global issues that their studies will not be able to address. As a result, the statement of the problem becomes unclear and diffuse. Poor SOP has been spotted even at conferences. To be effective,

Box 5.3: A Sample Case Study Showing Background in a Research Proposal (Mwaura 2021)

In sub-Saharan Africa, low soil fertility is a significant challenge that impedes agricultural productivity (Reference). The primary cause of this problem is continuous cropping without replenishing the soil, which depletes soil organic matter and negatively affects soil quality parameters (Reference). The situation is further worsened by low utilization of soil-replenishing inputs, such as inorganic and organic fertilizers (*References*). Although the use of inorganic fertilizers is promoted, they remain expensive, and timely accessibility is still a challenge for farmers in most sub-Saharan African countries, according to reference. *Organic inputs, on the other hand, have the potential to improve soil fertility and crop productivity* (Reference). Moreover, organic resources can be cost-effective nutrient sources and can address the issue of low soil fertility. Therefore, there is a need to encourage the increased use of organic inputs, which can either replace or complement inorganic fertilizers (Reference).	What is known in regard to soil fertility and the declining situation and potential of using of organics.
Compared to inorganic fertilizers, the use of organic resources is more sustainable and eco-friendly (References). *These resources include manure, compost, agroforestry, crop residue incorporation, intercropping, relay of legumes, biomass transfer, and natural and improved fallows (References).* They are commonly cited in the literature as crucial amendments for *improving soil fertility, structure, water holding capacity, and infiltration (References). Furthermore, organic resources enhance microbial activities, which are essential for soil fertility parameters.* Despite these benefits, farmers may not fully embrace their use due to a lack of knowledge about their potential profitability.	Presents what organic are and gives their advantages from biophysical realm and brings in inadequate information on profitability.

(continued)

Box 5.3 (continued)

In the Central Highlands of Kenya, the predominant farming system is subsistence mixed farming with minimal external input use, a common practice in sub-Saharan Africa (*Reference*). *The primary reason for this low external input use is attributed to high levels of poverty (Reference). As a result, the usage of fertilizers in sub-Saharan Africa is low in comparison to developed nations (Reference).* This has resulted in stagnated agricultural productivity in these countries, with cereal yields in sub-Saharan Africa averaging at 1.5 MT/ha, significantly lower than Latin America's 8 MT/ha (reference). Organic resources can serve as an alternative to inorganic inputs, *but their adoption rates are not well documented, with past studies showing low adoption rates (Reference).* Demographic and socioeconomic factors have been generalized as the primary factors influencing adoption rates, ignoring the influence of cultural factors and the intensity of adopting organic resources (*Reference*). Additionally, past studies have not considered profitability as a key determinant of adoption patterns.	Presents case of inability to use adequate mineral fertilizers but also brings the gap of low adoption of organics due to lack of recommendations on rates Additionally.
The impact of socio-economic factors on the adoption of organic resource management has not been thoroughly examined in this study area. However, research conducted in other regions has revealed that these factors can play a role in determining whether or not technologies are adopted. Some examples of these studies are (*give references showing such studies*). On the other hand, most of these studies have *neglected to consider cultural factors* and their influence on the uptake of technologies. Furthermore, there has been *little exploration of how farmers perceive the benefits and profitability of organic resource management (Reference).* It is therefore necessary to investigate how socio-economic and cultural factors impact the adoption and intensity of organic resource management technologies. This will involve examining the relationship between farmers' perceived profitability and their actual adoption behavior to better understand the complex dynamics that underlie their decision-making processes. *Ultimately, this knowledge will contribute to more effective and targeted interventions that support smallholder farmers in making informed choices about organic resource management.*	Highlights several research gaps (cultural factors, perceptions, intensity etc.) and also highlights the importance of this study in supporting smallholder farmers.

Box 5.4: Pit Falls to Avoid in Background to the Study

- Do not include too much details or information that is not relevant to the research problem
- Do not overload with too many citations; rather, select the most relevant and impactful sources
- Do not make broad, sweeping statements without providing specific evidence or citations to support them
- Do not just provide one-sided discussion; avoid bias in writing

the statement of the problem should be well articulated, concise, clear, and specific. It should also conclude with a purpose statement (See example in Boxes 5.5 and 5.6).

Box 5.5: Example 1 of a Statement of the Problem

Statement of the problem (Mwaura 2021)

Over half of the SSA population are smallholder farmers relying wholly on subsistence agriculture for livelihood. However, agricultural productivity has declined due to reduced soil fertility attributed to poor soil management, soil and nutrient loss by erosion, and nutrient mining by continuous cropping without adequate amendments. Application of external soil inputs (organic and inorganics fertilizers) can enhance soil fertility in rain-fed smallholder farming systems, thus increasing crop production *(Topic)*. However, the use of inorganic fertilizers by the smallholder farmers has been low due to their high price and untimely accessibility *(Research Problem)*. As a result, the farmers in these areas have experienced low yields for a long time (REF). Efforts have been made to promote organic-based inputs in these two counties, but farmers have been slow to adopt them (REF) *(Evidence and justification)*. It is unclear why farmers in the Central Highlands of Kenya, who rely on agriculture for their livelihood, use organic-based inputs sparingly. Studies examining the socioeconomic factors that influence the adoption and intensity of organic-based inputs in this region are limited. Previous studies mainly focused on identifying factors that influence the adoption of organic-based technologies, leaving a gap in understanding the intensity of adoption. Additionally, these studies did not explore how farmers perceive the profitability and benefits of these technologies *(Deficiencies/gap)*. Therefore, this study aims to determine the socioeconomic factors that influence the intensity of adoption of organic-based inputs, the perceived benefits of using them, and their profitability in Gatanga and Meru South sub-counties *(Intent/purpose)*.

Box 5.6: Examples of Statement of a Problem

Example 1: Soybean is an important cash crop recently introduced to smallholder farmers in central Kenya as a source of food, income and for soil fertility improvement. *(Topic)*. However, its production levels are markedly low with yields not going beyond 800 kg ha^{-1} (REF) *(Research problem)*. Acidity and depletion of the major nutrients (N, P) due to continuous cultivation with removal of nutrients by crop, and leaching of exchangeable cations (Ca^{2+}, Mg^{2+}, K^{+}) are the main factors that adversely affect soybean production. Soils in central highlands have low soil pH ranging from 4.5 to 5.3. The low pH limits the availability of phosphorus, while biological nitrogen fixation, and microbial activity are reduced, which in turn affects soybean development and yields *(Evidence)*. Acidic infertile soils maybe corrected through liming or the use of organic fertilizers (manures) but information on how combinations of lime and organic manures perform is limited because past studies investigated sole components, either sole lime or sole manure *(Research gap)*. Therefore, the aim of the study was to determine the effect of applying manure, lime, and inorganic phosphorus fertilizer on soil properties and soybean grain yields in the acidic soil of the central highlands of Kenya *(Purpose)*.

Example 2: The adoption of improved maize varieties, developed through advancements in agricultural research and breeding programs, holds significant potential for enhancing yield, resilience, and overall agricultural sustainability among the smallholder farmers in Region A(REF) *(Topic)*. Despite the evident benefits, the widespread adoption of these improved maize varieties among farmers remains a challenge in various regions due lack of understanding of the factors that influence adoption or non-adoption within specific farming communities *(Research problem)*. The adoption process is often complex, given the dynamic nature of farming systems and varying agroecological conditions. Factors such as access to information, availability of seeds, farmer knowledge, and perceived risks and benefits play a pivotal role in influencing adoption decisions (REF). Therefore, understanding the complex interplay of these factors is vital to developing effective strategies that promote the widespread adoption of improved maize varieties, leading to increased agricultural productivity and improved livelihoods for smallholder farmers *(Evidence)*. Limited research has, however, been conducted to explore the socioeconomic, agronomic, and institutional factors that shape farmers' decisions regarding the adoption of the newly introduced varieties *(Research gap)*. To bridge this knowledge gap, this study aims to investigate the determinants (socioeconomic, agronomic, and institutional) of the adoption of new improved maize varieties released between 2015 and 2018, among smallholder farmers *(Purpose)*.

(continued)

5.5.1.3 Writing Research Questions

Research questions refer to questions that a researcher would like to answer by undertaking the study. While some universities recommend that research questions follow objectives, they are presented first in some cases. Some qualitative researchers recommend that research questions be used instead of objectives, e.g. Creswell (2014). A student should follow the university guidelines.

Examples: A research question that depicts the relationship between two or more variables of interest to the researcher. However, the relationship is stated in the form of a question. Examples include:

- *What are the effects of the different soil-incorporated organics (with and without inorganic fertilizers) on soil chemical properties and maize yields?*
- *Is there a relationship between transaction costs and the participation of smallholder farmers in the collective marketing of climbing beans in the central Uganda region?*
- *How are the N_2 fixation potential and N balances of soybeans, ordinary beans, and cowpeas affected by P fertilizer in different soil fertility status?*
- *What is the magnitude of N leaching following different soil fertility amendment options in different soil fertility status?*
- *What is the effect of market liberalization on prices of milk among smallholder farmers of western Kenya?*
- *To what extent are the soil organic matter-related soil physical, chemical, and biological properties influenced by the quality/quantity of the organic resources applied to the soil?*

Examples of poor research questions:

- Which is the best mango variety?
- What is the effect of food security?
- What is the effect of market liberalization?
- What is the impact of soil fertility decline?

Insufficient because:

- What are the variables?
- What will be measured?
- What relationship will be examined?
- What is the population?
- What is the context?

5.5.1.4 Writing Research Objectives of the Study

The research objectives are a statement form of the research question and depict what is to be achieved by the study (Badenhorst 2020). They should be closely related to the statement of the problem and summarize what you hope to achieve in the study. Objectives make the statement of the problem specific and measurable. They help translate the SOP into measurable components that guide the study. Refer to Chap. 3 for the qualities of a good objective.

When writing a proposal, it is important to clearly list your objectives under two headings: Main objective (aim) and Specific objectives or sub-objectives.

The *main objective, also known as general or overall objective* is a broad statement that outlines the overall focus of your study, including the main associations and relationships you aim to establish or discover. The main objective is further broken down into specific objectives that are logically connected and systematically address various aspects of the problem.

The *specific objectives* are the detailed aspects of the topic that you want to investigate within the main framework of your study. They are listed numerically in your proposal and guided by the variables and relationships being explored. Each relationship between variables is spelt out in a specific objective and aligned to the conceptual framework and variables of the study.

The number of objectives is influenced by the conceptualization and the nature of relationships the researcher may be interested in investigating. As a guide, the objectives may not be less than two and should not exceed four for the master's thesis. PhD proposals could have three to seven objectives. The number of specific objectives will influence the volume of the literature review and data to be collected, so one should limit the number of specific objectives to be investigated.

Some students include an objective of providing recommendations. This should be avoided because it is assumed that the study's findings will lead to recommendations, so there is no need for a specific objective regarding recommendations.

In formulating and writing your quantitative objectives, observe the following:

- Objectives should logically flow from the statement of the problem or core research question.
- Use *action verbs* that are specific enough to be evaluated. *Examples* of action verbs are: to determine, to compare, to verify, to calculate, to describe, to assess, to evaluate, to explore.
- Avoid the use of vague non-action verbs such as to appreciate, to understand, or to study.
- The wording of objectives determines the type of research (descriptive, correlational, and experimental) and the type of research design you need to adopt to achieve them.

Here is an example of the main objective and specific objectives (Box 5.7).

Box 5.7: Sample of General Objective and Specific Objectives

General objective

To enhance soil and water productivity for improved crop yields in the highlands of Kenya using selected soil management technologies.

Specific objectives

(i) To assess the effect of tied ridging and integrated soil fertility management on soil water content at different soil depths in Tharaka-Nithi County.
(ii) To evaluate the effect of tied ridging and integrated soil fertility management on soil bulk density and aggregate stability in Tharaka-Nithi County.
(iii) To determine the effect of tied ridging and integrated soil fertility management on soybean and maize yields in Tharaka-Nithi County.
(iv) To evaluate the economic feasibility of tied ridging and integrated soil fertility management technologies in soybean and maize production in Tharaka-Nithi County.

Source: Ndung'u (2019)

5.5.1.4.1 Choosing the Type of Research Objective

The type of objective you chose (refer to Chap. 3) should be informed by the research problem and the statement of the problem. You may choose a causal, correlational, or descriptive objective. However, always remember that the type of research objective you opt for is key in influencing the likelihood of publication in a high-impact international journal.

5.5.1.5 Writing Research Hypotheses

In this section, you need to write a hypothesis matching your objectives. A hypothesis is a prediction or "educated guess" (Chap. 3). This means the format of writing is somehow similar to that of objectives. The difference is in the framing to either show a prediction or test a difference. Depending on the guidelines given, some schools will recommend a null hypothesis while others will recommend alternative hypothesis.

Examples

Example 1: Suppose you are matching a causal objective; Let's consider an example of a causal objective, "To examine the effects of diets that include protein supplements on milk yields of dairy goats among smallholder farmers in region X." A matching hypothesis will be:

Ho: Protein supplemental diets have no significant effect on milk yields of dairy goats among smallholder farmers in region X."

HA: Protein supplemental diets have a significant effect on milk yields of dairy goats among smallholder farmers in region X."

Example 2:

From a comparative and directional perspective: To determine the impact of participating in the collective marketing of bananas on farmers' income in region A

We can write matching hypotheses as shown below:

H0: Participation in the collective marketing of bananas does not significantly influence income for farmers

HA: Farmers who participate in collective marketing of bananas have significantly higher incomes than those who do not participate"

Example 3: Descriptive objectives. An example of a descriptive objective is "To describe characteristics of farmers keeping improved breeds of dairy goats in Region A." Such will not have a hypothesis because there is no relationship being tested. You cannot be able to formulate a hypothesis with variables that cannot be tested using statistical tools.

5.5.1.5.1 Do Qualitative Objectives Have Research Hypotheses?

Qualitative research objectives will not have a research hypothesis. This is because qualitative objectives focus on understanding and exploring phenomena, experiences, and social processes. As such, it is not easy to formulate a hypothesis. It is also not necessary to have one because you will not test the hypothesis.

Students often ask if the number of hypotheses should be equal to the number of research objectives. The answer is that hypotheses may be the same number as the research questions (recommended), or they can be more or less, depending on whether you have a qualitative research question (which will not have a hypothesis). For example, in mixed methods studies having both quantitative and qualitative objectives, it is only the quantitative objectives that will have corresponding hypotheses.

5.5.1.6 Significance of the Study

The section on *significance* provides an explanation of who will benefit from the outcome and how they will benefit and, therefore, links the research to the end users. Examples are presented in Box 5.8. It can thus be considered "a statement for the target beneficiary(ies). This section should state the different categories of beneficiaries and their respective benefits. In agricultural research, the beneficiaries are usually stakeholders in agriculture, such as farmers and policymakers. This section of the proposal is intended to answer questions such as:

- What are the intended uses of the study results?
- Who will gain from the findings?
- What will the results mean to practitioners, especially (farmers).
- What is the expected outcome of the research?
- Is the study worthwhile?
- How can the research inform policy?

Box 5.8: Samples of Significance of the Study Statements

Sample 1: Significance of a study with the title: "Enhancing communication for effective dissemination of soil fertility management in the central highlands of Kenya"

The information generated from this study will help inform the stakeholders (policy makers, extension, and researchers) on effective communication channels for scaling up soil fertility management on smallholder farms. Improved communication should increase the adoption of soil fertility management technologies that will result in increased crop production and contribute to the reduction of extreme poverty.
Source: Kimaru (2012)

Sample 2: Significance of the study: Title of research proposal: tied ridging and integrated soil fertility management technologies effects on soybean-maize yields and selected soil physical properties in Tharaka Nithi County, Kenya

The findings of this study will contribute to scientific knowledge on integrated soil fertility management, specifically on the effects of tied ridging technologies on soil water content, soil physical state, and maize-soybean performance. The results will help provide farmers with options for dealing with the problems of declining crop yields due to reduced soil fertility and soil moisture that are prevalent in the area. Additionally, the results of this research will provide more insight into the economic feasibility and sustainability of the two technologies to researchers, county and national governments, agricultural experts, and environmental managers. This will be useful in facilitating the adoption of the ISFM options for improved maize-soybean production, soil fertility management, and soil water conservation in the area.
Source: Ndung'u (2019)

- What is the significance of the study to our collective knowledge?
- Why should a donor support the study?

5.5.1.7 Limitations of the Study

Limitations of the study are constraints in some aspects of the study that the researcher recognizes negatively affect the results or generalizability of the results but over which they have no control. Think about threats to internal validity e.g. extraneous variables, that may have been impossible to avoid or minimize. The most common limitations include; changing weather conditions, size of the sample, span of study, or data collection procedures. It is important for researchers to acknowledge and write down the limitations of their study, as no study is perfect and without limitations.

5.5.1.8 Delimitations of the Study

Delimitation of a study refers to the specific boundaries or constraints that a researcher establishes to narrow down the scope of the investigation. Delimitations are crucial for maintaining focus, managing resources, and clearly understanding the study's scope. By defining delimitations, we can focus on a specific region, country, or community, excluding others to manage the scope and maintain relevance. Another common aspect of delimitation is targeting a specific population, age group, or demographic while excluding other groups. In this section, you need to include the things you are not doing and why you have chosen not to do them.

5.5.1.9 Assumptions

Assumptions of the study are a notion of any important fact presumed to be true but not verified. Articulating these assumptions is essential for the researcher to establish the rationale behind the study and, consequently, the findings. In the end, it is imperative to interpret the results within the context of the stated underlying assumptions. This enhances the accuracy of the interpretation of the outcomes.

5.5.2 Chapter 2: Literature Review

The purpose of Chap. 2 which is the literature review chapter in a research proposal is to comprehensively review literature related to the investigated problem. In this chapter, you are required to present a thorough explanation of theories and previous research studies relevant to the research topic.

Depending on your University, it may have the following components:

1. Theory-based literature.
2. Empirical-based literature.
3. Theoretical framework.
4. Conceptual framework.
5. Summary of literature and research gaps.

Refer to Chap. 4 for details on the literature review chapter.

5.5.2.1 Theoretical Literature

Theoretical literature is usually presented in the first sections of the literature review and focus on presenting and discussing theories within a particular.

5.5.2.1.1 Summary of Literature and Research Gaps

In the "Summary of Literature and Research Gaps" section, the researcher summarises the reviewed literature and identifies potential research gaps. See an example of a summary below (Box 5.9). It is preferable to present this information in a table format. This synthesis aims to demonstrate the researcher's expertise and qualifications to express opinions in the field. Furthermore, the synthesis aims to highlight that the proposed study will uniquely address the problem, despite the existing research, making it distinct.

Box 5.9: Example of Summary of Research Gap

Despite the important roles of organic materials, they are limited in provision of nutrients to meet crop requirement due to limited quantities available, low nutrient content of most of the organic materials, high labor demand for processing and application. A more practical and sustainable approach suggested is integrated nutrient management involving the combined use of organic and inorganic nutrient sources. Integrated nutrient management results in more yields and nutrient use efficiency than that expected from mere additives effects of sole applications. However, the success of nutrient management will depend on several factors: types and quantities of organic materials, rates and proportions at which the two nutrient sources are combined, and soil and environmental conditions. Research in central highlands of Kenya on these issues is scanty and as such guidelines for their use and management are lacking.

5.5.2.1.2 Differences Between Biophysical and Social Research Proposals

It is good to recognize that there can be and there are some differences in the organization and content of your Chap. 2 (Literature Review) of your research proposal in a biophysical and social study or between a quantitative and qualitative study. There are usually variations in emphasis and content based on the nature of research and the specific requirements per the field, data collection, results, and scientific findings. For instance, biophysical sciences will often emphasize empirical studies and experiments and focus heavily on research methodologies for testing theories. Similarly, the use of theoretical frameworks and conceptual frameworks (Chap. 2), varies across discipline with biophysical studies tending to apply established theoretical frameworks and models to guide experimentation. On the other hand, the social sciences often develop conceptual frameworks to illustrate relationships among variables.

Social quantitative studies, for example, agricultural economics and policy studies, typically place a stronger emphasis on theoretical frameworks, concepts, and theoretical debates. It seeks to provide a comprehensive understanding of relevant theories and their applications. It may also present historical and contextual perspectives tracing the development of theories and examining social trends and socioeconomic factors that influence the research area. Regardless of the nature of the study tips in Box 5.10 are a useful guide.

Box 5.10: Tips for Writing Your Literature Review Chapter (Chapter 2) in the Research Proposal

- Refer to work by recognized experts in your area of research.
- Be aware of "fake information" from predatory journals and other sources.
- Always support your argument with valid evidence in a logical manner.
- Be concise.
- Distinguish clearly between facts and opinions.
- Isolate emerging research gaps.

5.5.2.2 Research Methodology

In Chap. 3 of the research proposal, the researcher explains how they are going to achieve the objectives. It mainly includes the planned research design, experimental treatments, population and sample, measurement of variables, and procedures for data collection, laboratory analyses (if any), and data analysis.

However, as you write this chapter, note that there is a variance between research methods and research methodology. Research methods are the specific techniques that a researcher uses to conduct research and to answer the research problem. Research methodology is broader; it is a systematic approach encompassing the overall strategy and framework. It is concerned with the underlying logic, assumptions, and rationale behind the chosen methods (Kothari and Garg 2019). As you embark on writing this chapter maintain clarity in distinguishing between these two terms, as this will contribute to enhanced understanding of your research proposal.

5.5.2.3 Format and Components of Chap. 3 of the Research Proposal

The format of the research methodology chapter varies depending on the institution, school, or department guidelines and the specific requirements of the research project. Strictly follow the specific format guidelines provided by your institution or school. Additionally, seek guidance from your supervisor to ensure that your research methodology aligns with their expectations and academic standards. Remember, this is what will deliver results, and hence the methods proposed should be sound.

The chapter consists of several components that provide a detailed overview of how you plan to conduct your research. This helps your supervisors, reviewers, and readers evaluate the soundness and feasibility of the proposed research. The components include the following;

1. Location of the study and its description.
2. Research paradigm and design (if applicable).
3. Population, sample, and sampling procedures.
4. Data collection procedures such as:
   - Experiments, surveys.
   - Piloting the instruments.
   - Validity and reliability of the instruments.
5. Data analysis plan including:
   - Empirical models.
   - Definition of variables.
6. Ethical considerations.
7. Research methodology matrix.

5.5.2.4 Location of the Study

An example of a study site or location description is shown in Box 5.11. The location of the study description is also referred to as the description of the study site or research setting. In Box 5.11, the author first presents the geographical location and provides coordinates of the study area. Next is a description of climatic conditions, including rainfall patterns and temperature ranges.

The key characteristics of the farming systems are discussed in detail. The information includes descriptions of dominant crop and livestock production systems, standard farming practices, and rainfall patterns. It also includes characteristics of the common soils of the site. The characteristics of the population are also given, including population size and distribution. Key socioeconomic features that may influence farming decisions and the adoption of technologies are also included.

Overall, the section on the location of the study should give enough details to enable the reader and evaluators to

Box 5.11: Example of Study Site Description

3.1 Study area	
The study site will be at Nkarini Mixed Day Secondary School in Tharaka South sub-County, *Tharaka-Nithi County (see the area map in appendix 3). It is located at latitude 0°14′43″ S, longitude 37°52′37″ E and altitude of 790 m above sea level (a. s. l).* The site is in the Lower Midland 5, marginal cotton and livestock-millet zones; a dry region with two short and very uncertain cropping seasons but with good yield potential (Jaetzold et al. 2007). *Mean annual temperature is about 27°C with high potential evapotranspiration rate. The annual rainfall ranges between 600 and 900 mm, which is highly variable and* poorly distributed (*References*). Rainfall pattern is bimodal with long rains (LR) falling from around March to June and short rains (SR) from October to February. The site was selected for the study because most of the soil fertility and soil and water management practices have been tried and recommended in the region by various researchers over time (*References*).	Gives site location, map, and coordinates Describes the climatic conditions and justifies site selection.
Livestock keeping is the primary agricultural activity in the region, where they keep goats and cattle. Less than 1% of the cattle kept are of an improved type. Annual crops grown are maize *(Zea mays L), beans (Phaseolus vulgaris), bulrush millet (Pennisetum glaucum L.), cowpeas (Vigna unguiculata L.), Hyacinth bean (Dolichos lablab), finger millet (Eleusine coracana), groundnut (Arachis hypogaea), pigeon peas (Cajanus cajan), sorghum (Sorghum bicolour L.) and sweet potatoes (Ipomoea batatas) (Jaetzold et al. 2007). The dominant soils are categorized as Nitisols, which are well-drained, very deep,* dusky red to dark red, friable clay with moderate to high inherent fertility (*References*). Tharaka South sub-County covers an area of about 766.1 Km^2. The population is 75,250 in 2019 (*References*).	Describes farming systems and soils plus population.

contextualize the study. The information will also be useful in the generalization of the findings after the study is eventually completed.

5.5.2.5 Research Paradigm

Paradigms are a set of beliefs or world views that guide a researcher when conducting research. Many authors (Mertens 2010, 2014; Creswell 2014; Mugenda and Mugenda 2019) emphasize the crucial role of including a discussion on the research paradigm in a research study, as it shapes the research orientation and provides direction to the inquiry process. However, the inclusion of a dedicated section in agriculture is not a widespread practice apart from a few cases with Ph.D. proposals and in the area of social studies particularly in qualitative studies. This could be attributed to most agricultural research leaning more toward experimentation.

Nevertheless, while research paradigms may not receive explicit attention in agricultural research proposals, the underlying philosophical and theoretical assumptions that inform the research should still be evident. These assumptions are key in shaping the research questions, methodology, and interpretations. In this regard, we recommend that a section discussing the research paradigm be included, particularly in the context of Ph.D. research on social issues. This is important as we engage more in qualitative and mixed methods approaches to address the many complex agricultural problems and challenges (see Chaps. 13 and 14).

In discussing your research, your philosophical paradigm should include the key three concepts: *the ontological stance (nature of reality), the epistemological stance (nature of knowledge)*, and *the axiological stance (nature of value)*. These should go hand in hand with explaining your chosen research approaches (quantitative, qualitative, or mixed methods) and justifying why they are the most suitable for addressing your research question (Chap. 7).

5.5.2.6 Research Design

The section on research design is guided by the research objective/questions, the philosophical stance chosen, and the research approach, whether it is experimental, nonexperimental, or mixed methods. The research design will specify the methods, procedures, techniques, and tools to gather and analyze data. It is advisable to involve a biometrician or a statistical expert to confirm that the chosen design is appropriate and sound. If the wrong design is used, you will likely face challenges later during statistical data analysis, affecting the study's robustness. Moreover, examiners keenly check the appropriateness of the design.

In describing the design, there is variation among the various designs, whether experimental or nonexperimental.

5.5.2.6.1 Experimental Designs

In writing the section on experimental research, the researcher should be keen to ensure that the application of the basic principles of experimentation is clearly shown. We have presented the principles of experimental designs in Chaps. 7 and 8. Manipulation of the dependent variable, the randomization and local control must be included and written clearly.

We recommend the following aspects to be included in an agronomy experiment (see sample of an agronomy experiment; Box 5.12):

Box 5.12: Sample of Agronomy Experiment

The plot size will be 6.75 meters (m) by 3 m. The border between the plots will be one meter (1 m) while the one between blocks will be 3 m. The test crop will be sorghum (*Sorghum bicolour L.* of variety gadam) that is commonly grown in the area because of its resistance to drought. Except for *MBILI* intercrop treatment, all the plots will have sorghum mono-crop. Spacing between the sorghum will be 0.75 m by 0.20 m for inter and intra row spacing. For the *MBILI* intercrop, two rows of sorghum will be alternated with two rows of green grams. Inter and intra row spacing for sorghum will be 0.50 m and 0.20 m respectively, while the inter and intra row spacing for green grams (*Vigna radiate*) (N 26) will be 0.33 m and 0.15 m, respectively. Inter-row spacing between the sorghum and green grams will be 0.33 m. Three seeds for sorghum and green grams will be planted per hill and thinned to two plants. Plant population density will be maintained at 320 plants per plot (158,025 plants ha^{-1}). The land will be tilled using hand hoe to a depth of 15 cm under conventional tillage and weeding done using the same on a need basis to ensure weed-free fields. The land will be cleared using a machete at 0 cm depth and weeding done by uprooting under minimum tillage (MT). Under MT, a mixture of cattle and goat manure will be incorporated along the planting lines 2 weeks before planting.

Under conventional tillage (CT), animal manure will be incorporated into the soil to a depth of 15 cm during land preparation and 2 weeks before planting. Manure will be analyzed for N content and equivalence of 15, 30, and 60 N kg ha^{-1} for low, moderate, and high rates applied. Crop residue will be surface applied as mulch at the rate of 5 Mg ha^{-1} after crop emergence. The mineral fertilizer will be applied using NPK 23:23:0 to supply 15, 30, and 60 N kg ha^{-1} for the low, moderate, and high rates, respectively. Phosphorus will be supplied using Triple Super Phosphate (TSP) at a rate of 60 P kg ha^{-1} for all the plots with mineral fertilizer as treatment.
Source: Matusso (2015)

- Specific experimental design chosen, treatments, and their description.
- Blocking procedure and number of blocks.
- Number of replications.
- Randomization procedures.
- Field plan and plot layout, including spacings (Chap. 9).
- Agronomic practices (inputs, planting, weeding, etc.).
- Timelines, source of experimental materials, varieties, etc.

In experimental research, the following key questions must be handled, among others;

- What is the experimental design? (completely randomized design (CRD), randomized complete block design (RCBD), Latin Square, and Split plot, etc.).
- What are the treatments? (You may need a table of treatments for clarity).
- What is the number of replications?
- How will randomization be done?
- What agronomic practices will be implemented?
- What will be the source of experimental material?
- How will extraneous variables be controlled?

The field and plot layouts must be drawn in this section. If they cannot fit in this section, they can be placed in the appendix and referred to here (see Chap. 9 on field and plot layout).

5.5.2.6.2 Non-experimental Designs

In non-experimental research designs (surveys, qualitative, and participatory) data will be collected without manipulation in natural settings. It will include gathering information directly from human respondents, groups, organizations or communities. Common methods will include interviews, focus group discussions, observations, and questioning (Chaps. 18 and 19). Secondary data may also be collected. The following three areas must be explained:

1. Study population and target population.
2. Sample size, sampling frame, etc.
3. Sampling techniques (probability or nonprobability).

Study Population In this section, describe the population from which the sample will be drawn. This could be all farmers growing Hass avocado in a certain region or all farmers keeping Friesian dairy cows. A distinction or comparison could be made between a target population and an accessible population, and you must state the exact or estimated population from which the sample will be drawn.

Sample Size Next, describe the sample size and the sampling strategies you will use to get the sample to get the sample from the population. This includes the actual sample that will be studied and how it will be determined (see Chap. 16 on sampling).

Different approaches are recommended for sample size determination, and students must find the most appropriate approaches without reproducing approaches used by other students. Many students are tempted to rely on what other graduate students state in their proposals. We recommend that standard accepted scientific procedures be followed to determine the sample size. If the population is small, you could choose to include all the members (census). In the case of a sample selection, you could use these methods:

- Use an internet sample calculator.
- Use mathematical formulas found in many statistics books.

As emphasized in Chap. 16, in addition to using the calculator, other factors to be considered during sampling include: previous literature, past studies and pilots studies; level of significance: power, response rate, and variability of the population (Das et al. 2016; Piepho et al. 2022).

Sampling Techniques The sampling technique you will use will depend on the research paradigm and research design. Sampling is critical to both internal and external validity. Representativeness should be the guiding principle in sampling. Quantitative research will use a probability or random sampling design while qualitative research will use non-probability or purposive sampling (Table 5.2). Justify the sampling method for your study. The sampling technique chosen will have an implication on the data analysis and interpretation.

For mixed research, you must include both components outlined above for experimental and non-experimental.

5.5.2.7 Methods and Tools of Data Collection

In the section on methods and tools of data collection, you need to describe details of data collection methods and measurement procedures. It should be detailed in such a way that it allows others to repeat the study, guide your research process and provide clarity to your supervisor and other readers. Just like a recipe specifies where the meal will be cooked, what ingredients will be needed, and the exact measurements, your methodology should ensure that for each of the data/parameters to be collected, you will need to specify how the measurement will be done, explaining when and how. For example, in the case of destructive sampling, you will need to indicate when and how this will be done.

If you are sampling the soil, you must explain when and how the soil will be sampled. You also need to include procedures to ensure that the sample is representative. In case of multiple sampling points, indicate the number and justification. It is important to mention how the samples will be treated after sampling to ensure they are well-labeled and do not mix up. It is also good to indicate where the analysis will be carried out.

In some cases, you will need to describe in detail the methods, but in the case of standard methods, for instance, methods in soil analysis (Kjedahl for total nitrogen and, hydrometer method for determining soil texture), you only need to indicate the reference.

In non-experimental designs (surveys, qualitative, and participatory), describe the tools to be used, the piloting of the tools, how the interview will be conducted (face-to-face, telephone, etc.), and how the participants will be contacted (Chaps. 18 and 19). A survey's most common tool is a questionnaire. In the case of a qualitative study, focus group discussion, in-depth interviews, observations, and document analysis are common. Some key activities include the preparation of data collection tools (questionnaires, checklists, etc.), contacting the selected participants, training of enumerators (in case of face-to-face interviews), and piloting of instruments must be described. In case of self-filled questionnaire, explain the distribution and collection of questionnaires. All the tools referred to in this section must be included in the ANNEX.

Table 5.2 Characteristics of probability and non-probability sampling methods

Probability sampling	Non-probability
Random	Non-random (purposive).
Subjects have an equal chance of being selected	Subjects do not have an equal chance of being included in the sample (purposively chosen).
It has the advantage of generalizing the results	Results cannot be generalized.
Mainly used in quantitative studies	Use in qualitative studies.
Simple random sampling, systematic, stratified, cluster, multiphase	Convenient, snowball, and many other variants.

5.5.2.8 Data Analysis

In the data analysis section, you need to describe the methods and techniques that will be used to analyze and present the data. It is common to use more than one method of analysis, and the choice of method will depend on the type of research, the objectives, and the hypotheses being tested. It is important to highlight the advantages of the selected technique over others. It is recommended that you also indicate the software that you will use for data analysis. These include SPSS, SAS, GENSTAT, R Studio, NVIVO, etc. You can consult a biometrician to guide you on this sub-section.

Experiments and quantitative surveys will generate quantitative (numerical) data. These will be analyzed using statistical tools, comprising descriptive and inferential analysis. Describe the methods, pointing out the specific data and parameters that will be subjected to each procedure. Box 5.13 gives the data type and the Likely method of data analysis. For instance, maize yield data will be subjected to ANOVA to test for significant differences among the treatments, while

Box 5.13: Data Type and Likely Method of Analysis

Design and type of data	Likely method of data analysis
Experimental data (quantitative)	Measures of central tendency (mean, median, and mode). Measures of variability (range and box plots). Comparison of means (t-test for two means). Analysis of variance (ANOVA). Measure of relationship; correlation and regression analysis.
Survey quantitative data	Means, frequencies. Chi-square. One-way ANOVA. Correlation techniques (e.g. Pearson product-moment). Regression techniques (simple and multiple). Binary regression methods (probit and logistic regression). Factor analysis.
Qualitative data	Thematic analysis, content analysis, and narrative analysis, grounded theory etc.

correlational analysis will be used to assess the significant relationship between growth parameters and yields.

Qualitative studies generate qualitative (narrative) data that cannot be analyzed statistically. Data analysis procedures aim to identify themes and patterns as well as interlinkages. The procedures often involve content analysis, thematic analysis, and narrative analysis, among others (See Chap. 21).

5.5.2.9 Methodology and Data Analysis Table

A table known as the "Methodology and Data Analysis matrix Table" is an important part of your research proposal. The table serves as a visual aid that summarizes and outlines key elements in your research methodology. The table typically includes sections on research design, data collection methods, data sources, sampling, and data analysis methods. The table further enhances proposal organization and also provides a well-defined road map for executing the study. Depending on the objective, data analysis may vary. For example, some objectives may require mean comparison through the use of ANOVA, while others may require regression analysis. A sample of the table is shown below (Tables 5.3 and 5.4).

The table helps to ensure that your mind is clear about the data to be collected for each of the objectives and how each will be handled to achieve the objectives of the study.

Table 5.3 A template for a matrix for methodology and data collection and analysis table

Objective	Data to be collected	Sources of data	Method of data collection	Method of data analysis
Objective 1				
Objective 2				

5.5.2.10 Ethical and Administrative Considerations

It is extremely important to consider ethical and administrative aspects in agricultural research (Engels et al. 2011; Harfouche et al. 2021; Diptaningsari and Martono 2021). As we explore the complexities of our food systems, human beings, and animals, we must ensure that their well-being is always our top priority. Following ethical guidelines and administrative requirements is not only a moral duty, but it is also crucial for maintaining the integrity of our research and avoiding potential ethical violations that could have serious consequences. By doing so, we can ensure that our research is not only safe and fair but also has a positive impact on society and the environment. Details of ethical considerations are discussed in Chap. 6.

In this section of the proposal, you should describe how you will manage the various ethical principles and guidelines. These will vary from study to study depending on the subjects and the context, but three key principles are important: beneficence (doing good), autonomy (self-rule), and Justice (fairness). In addition, explain how you will obtain informed consent from the participants and how you will maintain confidentiality.

If you are conducting animal research, you must describe your ethical review approval process and how you will employ humane practices to comply with relevant regulators and ethical guidelines for animal welfare.

Ethical considerations should also extend to the dissemination of research findings to benefit the farmers and other stakeholders who are directly impacted by agricultural research (Lairumbi et al. 2008; Neef and Neubert 2011). Describe how findings will be disseminated transparently in a manner that is accessible and understandable to the beneficiaries (see ways of disseminating information in Sect. 5.1).

In this subsection of the proposal, consider explaining how these principles and guidelines will be managed:

- Obtaining ethical certificates and research permits from relevant authorities.
- Beneficence (doing good).
- Autonomy (self-rule); information disclosure.
- Justice (Fairness).

Table 5.4 An example of a matrix for methodology and data collection and analysis table

Objectives	Type of data to be collected		Method of data collection	Method of data analysis
	Dependent variable	Independent variable		
1. To determine the influence of household socioeconomic factors on farmers' level of knowledge of application of combined organic + inorganic fertilizer	Farmers' level of knowledge (combined organic + inorganic fertilizers, weighted mean <3 = low levels 3 = moderate levels >3 = high levels	Household socioeconomic variables	Household survey Interview schedule Focused group discussion	Multinomial logistic regression Thematic analysis

- Privacy and confidentiality.
- Informed consent
- Veracity (honesty and integrity); avoid fabricating or deceiving participants.
- Harm versus benefits.
- Compensation of participants.
- Dissemination of findings to participants.

Refer to Chap. 6 on ethics for detailed information on ethical reviews ad clearance.

5.6 Back Matter

The back matter includes the references and appendices.

5.6.1 References

Every proposal needs to have a list of cited references. This list indicates the sources that have been reviewed and used by the researcher. Including an updated bibliography shows that the researcher is up to date with the most current information in their field of study. This adds credibility to the proposal by demonstrating that the researcher has not overlooked any important information.

5.6.2 Citations and References

Citations and references are key terms in the scholarly practice of citing and acknowledging sources you have consulted and incorporated into your review. Citations are provided in the form of **in-text citations**, meaning they are included in the main text to show the origin of the borrowed statement or idea. Referencing refers to recording all cited literature sources at the end of the literature review in a section called "references." Both the citations and the reference list use a standard specific style or format.

5.6.3 Types of Referencing Styles

There are many styles and forms of referencing. Each of the referencing styles has its own set of rules and conventions for in-text citations and reference list entries. The commonly used styles are the following:
Some of the common referencing styles used in academic writing across various disciplines include: CSE (Council of Science Editors), APA Style (The American Psychological Association style), Harvard, AMA Style (American Medical Association), MLA (Modern Language Association), and Chicago (Madler and Renwick 2025).

5.6.4 Choosing a Referencing Style

The reference style to use depends on the specific requirements of the institution, school, or department. Consult the guidelines to know which style to use. Whatever style you choose, accuracy, clarity, and consistency are the most important factors when writing your references. Different publications, such as journal articles, books, websites, and others, use distinct formats of citation. You can check this in the various citation style guides provided above. Use a consistent referencing style both in in-text citations and reference lists. This is an important consideration when choosing and applying a reference style, as each style provides guidelines for various types of sources.

5.6.5 Appendices

Appendices, or annexes, are sections in a document where additional information is placed. This information may include the work plan, budget, research tools, questionnaires, survey interview schedules, and experiment protocols. Instruments like checklists for focus group discussions, key informants, and case studies are also placed in the appendices. Other items include samples of informed con-

Table 5.5 A sample workplan showing the schedule of activities of an academic research project

Activities	2022												2023					
	Jan	Feb	Mar	Apr	May	Jun	Jul	Aug	Sept	Oct	Nov	Dec	Jan	Feb	Mar	Apr	May	Jun
Proposal writing and defense	■	■	■															
Fieldwork		■	■	■	■													
Lab analysis			■	■	■	■	■	■										
Lab analysis								■	■	■	■							
Data analysis							■	■	■	■	■	■	■					
Thesis writing and publication		■	■	■	■	■	■	■	■	■	■	■	■	■	■	■	■	
Defense and graduation																	■	■

sent forms and cover letters to be sent to stakeholders if applicable.

5.6.6 Work Plan or Time Schedules

A work plan or time schedule outlines the specific timeframe for completing a research project. This is important because it helps the researcher determine the feasibility of completing the study within the given time constraints. Additionally, it enables the researcher to stay on track and meet project deadlines. The work plan should include a list of major activities, along with an estimated time for completing each activity. This information can be presented in the form of a Gantt chart, an Excel spreadsheet, or a simple table (See Table 5.5). A work plan is an essential tool for the student, supervisor, and other project stakeholders, as it provides clear benchmarks and milestones for measuring progress and ensuring that the research is completed on time.

5.6.7 Budget

Research needs funds. A budget aligned with activities is therefore required. A budget is a list of items that will be required to carry out the research and their approximate cost. The budget should be detailed enough and precise on items needed, prices per unit, and total cost.

The budget helps your supervisor and reviewers decide on the viability of your proposal. In writing this section, consider funds needed, data access, and equipment. Think about all the expenses and ensure that you can meet them. Your supervisor and other reviewers will check on this to confirm that the study is feasible.

5.6.8 Data Collection Tools and Protocols

Data collection tools and protocols include questionnaires, checklists, open-ended interview guides, informed consent forms, data recording sheets, Field layout plans, plot layout plans, and various protocols for taking measurements in the biophysical realm.

Further Reading Materials

1. Badenhorst C (2020) Research conceptualization in doctoral and master's research writing. Writing and Pedagogy 12:423–444. https://doi.org/10.1558/WAP.19542

2. Diptaningsari D, Martono E (2021) Considering Ethics in Agricultural Experimental Research with Special Mention to Agricultural Biotechnology Research. Journal Perlindungan Tanaman Indonesia 25:10–20. https://doi.org/10.22146/JPTI.57790
3. Das S, Mitra K, Mandal M (2016) Sample size calculation: Basic principles. Indian J Anaesth 60:652–656. https://doi.org/10.4103/0019-5049.190621
4. Mugenda, A., Mugenda G (2019) Research methods: Qualitative and mixed methods approach. 100–102

Key Points in This Chapter

Meaning of a Research Proposal

A research proposal is a plan detailing the procedures of intended research. It justifies the research and shows the methods of carrying out the project. Preparing a research proposal before engaging in any research, whether for academic purposes or other pursuits, is mandatory.

Academic research proposals require approval from your institution's examination board before data collection can begin. Depending on the nature of your research, you may also need approval from other certifying bodies, such as the Institutional Review Board and professional bodies.

The proposal should be well-written and follow the guidelines from your institution. It should be compelling, novel, and well-formatted. The quality of your writing can also decide whether your proposal is accepted or rejected by the proposal defense committee. "A good thesis or dissertation hinges on a strong proposal." Similarly, "a quality journal publication hinges on a strong proposal." Take time to craft a strong proposal that will give you a good thesis and a manuscript that you can publish in a reputable high-impact journal.

The Front Matter, Middle Matter, and Back Matter of a Research Proposal

The front matter of a research proposal contains the title captured on the cover page, declaration page, abstract, list of tables, list of figures, list of acronyms, and abbreviations. The title should be informative and brief, while the abstract needs to be complete.

The middle matter is the main part of the proposal, referred to as "Chap. 1: Introduction" and generic components include:

1. Background to the study.
2. Statement of the problem.
3. Research objectives, questions, and/or hypotheses.
4. Purpose of the study.
5. Significance of the study.
6. Delimitation of the study.
7. Limitations of the study.

The background introduces and contextualizes the topic of interest. It highlights the problem being addressed in the research. It should flow from broad to narrow, presenting a global, regional, and national research topic overview.

The statement of the problem is the specific problem that you will address in your proposal. It should point out the research gap and help define the variables of the study. It should also be short but sufficient context and end with a purpose statement.

Research Objectives of the Study

The objectives are a statement of what is to be achieved by the study and are closely aligned to the problem statement.

Chapter 2: Literature Review

The purpose of Chap. 2, which is the literature review in a research proposal is to comprehensively review the literature related to the investigated problem. In this chapter, you are required to present a thorough explanation of theories and previous research studies relevant to the research topic.
It will have the following components:

1. Theory-based literature.
2. Empirical-based literature.
3. Theoretical framework.
4. Conceptual framework.
5. Summary of literature and research gaps.

Research Methodology

Chapter 3 of the research proposal is called Research Methodology or Materials and Methods. In this very critical chapter, you explain how you are going to achieve the objectives or answer the research questions. It mainly includes the planned research design, experimental treatments, population and sample, measurement of variables, and procedures for data collection, laboratory analyses (if any especially for biophysical research), and statistical analysis.

The format of the research methodology chapter varies depending on the institution, school, or department guidelines and the specific requirements of the research project. Ensure you strictly follow the specific format guidelines provided by your institution or school. Additionally, seek guidance from your supervisor to ensure that your research methodology aligns with their expectations and academic standards.

The chapter has several components that provide a detailed overview of how you plan to conduct your research. The components include the following:

1. Location of the study.
2. Research paradigm.
3. Research design.
4. Population, sample, and sampling procedures.
5. Data collection procedures:
   (a) Piloting the instruments.
   (b) Validity and reliability of the instruments.
6. Data analysis:
   (a) Empirical models.
   (b) Definition of variables.
7. Ethical considerations.
8. Research methodology matrix.

Back Matter

The back matter includes the references and appendices. A reference is a list of cited references. A consistent referencing style should be used.

Review Questions

1. What is the purpose of an academic research proposal?
2. Outline characteristics of a "Strong research proposal"
3. Explain the statement that "A good thesis and quality journal publication" hinges on a "strong research proposal"
4. What is the purpose of the "Background to the problem" section?
5. What is the importance of the "Statement of the problem" section?
6. What are the main sections of the literature review chapter in a research proposal?
7. What specific elements are typically found within the appendices of a research proposal?

Chapter Exercise

Visit your university library or access its online research repository. Locate the section containing past theses or research proposals and read at least three different research proposals. Pay attention to their overall structure and how they address the key components. Then select one research proposal that you find well-structured, clear, or relevant to your interests. Based on the selected proposal, answer the following questions:

1. What is the exact title of the research proposal?
2. Does the abstract contain all the components of a research proposal?
3. Analyze the background and briefly describe the general context or history that leads to the research problem in this specific proposal.
4. Based on your understanding from the course material, Does "Statement of the Problem" isolate the knowledge gap? critique
5. What are the main themes or categories under which the literature is reviewed in this proposal?
6. Does the proposal clearly distinguish between "theoretical literature" and "empirical literature"?
7. If it has a Theoretical Framework: What is the main theory or theories guiding this research?
8. If conceptual framework is included, what are the main variables and their proposed relationships as illustrated?
9. What research design does the author propose /use? You can refer to the appropriate chapters in this book.
10. What specific methods will be used to collect data (e.g. surveys, interviews, focus groups, observations)? What tools will be employed (e.g. questionnaires, interview guides, observation checklists)?
11. How does the author propose to analyze the collected data?
12. Were ethical considerations included? Explain
13. Based on your reading of this proposal and the course material, what do you consider to be the *strongest* aspect of this particular proposal? Why?

Answer to Review Questions

- Q1. Reference: 5.2.
- Q2. Reference: 5.2.
- Q3. Reference: 5.2.
- Q4. Reference: 5.5.1.1.
- Q5. Reference: 5.5.1.2.
- Q6. Reference 5.5.2.
- Q7: Reference: 5.6.5.

References

Badenhorst C (2020) Research conceptualization in doctoral and master's research writing. Writ Pedag 12:423–444. https://doi.org/10.1558/WAP.19542

Beesigamukama D, Mochoge B, Korir NK et al (2020) Exploring black soldier Fly Frass as novel fertilizer for improved growth, yield, and nitrogen use efficiency of maize under field conditions. Front Plant Sci 11:574592. https://doi.org/10.3389/FPLS.2020.574592/BIBTEX

Cheng YH (2014) Dissertation grant proposals as "writing games": an exploratory study of two L2 graduate students' experiences. Engl Specif Purp 36:74–84. https://doi.org/10.1016/J.ESP.2014.07.001

Creswell J (2013) Steps in conducting a scholarly mixed methods study, Discipline-Based Education Research Group Speaker Series. UNL DBER Group

Creswell JW (2014) Research and design qualitative, quantitative, and mixed methods approaches. Sage

Das S, Mitra K, Mandal M (2016) Sample size calculation: basic principles. Indian J Anaesth 60:652–656. https://doi.org/10.4103/0019-5049.190621

Diptaningsari D, Martono E (2021) Considering ethics in agricultural experimental research with special mention to agricultural biotechnology research. J Perlindungan Tanaman Indones 25:10–20. https://doi.org/10.22146/JPTI.57790

Engels JMM, Dempewolf H, Henson-Apollonio V (2011) Ethical considerations in agro-biodiversity research, collecting, and use. J Agric Environ Ethics 24:107–126. https://doi.org/10.1007/S10806-010-9251-9/METRICS

Harfouche AL, Petousi V, Meilan R et al (2021) Promoting ethically responsible use of agricultural biotechnology. Trends Plant Sci 26:546–559. https://doi.org/10.1016/J.TPLANTS.2020.12.015

Kimaru WS (2012) Enhancing communication for effective dissemination of soil fertility management in the central highlands of Kenya

Kinuthia R (2020) Farmers' perceptions and adoption of micro catchments for improved establishment of agroforestry trees in East Shewa Zone, Ethiopia

Kothari CR, Garg G (2019) Research methodology: methods and techniques. Oxford University Press, Docklands

Lairumbi GM, Molyneux S, Snow RW et al (2008) Promoting the social value of research in Kenya: examining the practical aspects of collaborative partnerships using an ethical framework. Soc Sci Med 67:734–747. https://doi.org/10.1016/J.SOCSCIMED.2008.02.016

Madler AM, Renwick S (2025) Referencing and citation styles. In: Reference Module in Social Sciences, pp 108–120. https://doi.org/10.1016/B978-0-323-95689-5.00213-3

Matusso JMM (2015) Effects of maize (zea mays l.) – soybean (glycine max (l.) merrill) intercropping patterns on yields and soil properties in two contrasting sites of Embu and Meru counties, Kenya

Mertens DM (2010) Philosophy in mixed methods teaching: the transformative paradigm as illustration. Int J Mult Res Approaches 4:9–18. https://doi.org/10.5172/MRA.2010.4.1.009

Mertens D (2014) Research and Evaluation in Psychology and Education. Libro Electronico 4

Michoma JM, Mugwe JN, Korir N, Mwangi M (2023) Effect of hydrogel and mulching on soil moisture, yield and economics on yellow passion fruit in Embu and Kiambu counties, Kenya. Asian J Agric Hortic Res 10:9–21. https://doi.org/10.9734/AJAHR/2023/v10i4240

Mugenda A, Mugenda G (2019) Research methods: qualitative and mixed methods approach. Centre for Innovative Leadership and Governance, pp 100–102

Mwaura GG (2021) Adoption intensity, perception and profitability of organic based soil fertility Management Technologies in Murang'a and Tharaka-Nithi counties, Kenya

Ndung'u MM (2019) Tied-ridging and integrated soil fertility management technologies' Effects on soybean-maize yields and selected soil physical properties in Tharaka-Nithi County, Kenya

Neef A, Neubert D (2011) Stakeholder participation in agricultural research projects: a conceptual framework for reflection and decision-making. Agric Human Values 28:179–194. https://doi.org/10.1007/S10460-010-9272-Z/TABLES/1

Oduor DO, Mutavi IN, Long'ora AE (2022) Effects of socio-cultural attributes on dominant tree species diversity in Ugenya Sub-County Siaya County, Kenya. Afr J Hist Geogr 1:39–48. https://doi.org/10.37284/AJHG.1.1.831

Piepho HP, Gabriel D, Hartung J et al (2022) One, two, three: portable sample size in agricultural research. J Agric Sci 160:459–482. https://doi.org/10.1017/S0021859622000466

Ethical, Legal, and Administrative Issues in Research

6

Abstract

Ethical, legal, and administrative guidelines ensure integrity and compliance in research. Upholding these interconnected principles protects individual and community rights ultimately enhancing societal impact. Researchers, institutions, and policymakers should collaborate to maintain these standards. Ethics are morals that guide research, while legality involves adherence to laws and regulations. The administrative aspect ensures that our research complies with ethical and legal standards. History records many studies that violated the dignity of participants, especially in the 1940s–1960s. These led to the setting of international standards, for instance, the Belmont report and the Nuremberg code. As research advances, ethical considerations have evolved to include many laws and codes. Research ethics in agricultural research is not as advanced as that of medicine, but the concern has increased within the last two decades. The fundamental principles include informed consent, beneficence, respect for autonomy and confidentiality, justice, integrity, and social responsibility. We can obtain ethical guidelines from several sources, including institutional review boards, professional organizations and research policies of various institutions. Institutional review boards are the "watchdogs" of research and play a key role in reviewing and approving research proposals. To ensure that research is conducted with the highest morals, values, integrity and accountability, ethical principles, and guidelines must be followed during the entire research process, from conceptualization and data collection to dissemination of research findings. During the conception stage the research proposal must show a worthwhile research gap, as well as the benefits of the research project. During the conduct of research, including data collection, the researcher must be sensitive about collecting data that is error-free. With the changing paradigms in agricultural research embracing more farmer involvement, qualitative and mixed methods research, ethical guidelines and principles must be adapted to ensure that standard guidelines are followed. Farmers must be treated with fairness and respect. Similarly, all collaborators in multidisciplinary research projects must enjoy fairness and justice. In publishing, integrity and confidentiality are key considerations for the authors, peer reviewers, and editors. These three parties must uphold fairness in the peer review process and maintain trust in scholary communication. With the advent of Artificial Intelligence (AI) technologies, key ethical considerations include equitability, accountability, transparency, sustainability, privacy, and robustness. Finally, supervisors and their students must maintain professionalism, fairness, and courtesy and cultivate a harmonious relationship.

Keywords

AI technology · Animal research · Anonymity and confidentiality · Beneficence · Belmont report · Conflict of interest · Consent · Data fabrication and falsification · Data integrity and privacy · Ethical review certification · Historical injustices · Honesty and integrity · Human subjects research · Nuremberg code · On-farm research · Privacy · Institutional review board · Intellectual property Rights · Social responsibility

Learning Outcomes

Reading of this chapter should enable you to:

1. Recognize the significance of ethical, legal, and administrative issues in agricultural research
2. Appreciate the history of ethical violations in research
3. Explain the status of ethical and legal considerations in agricultural research
4. Determine how to balance ethics and the pursuit of truth in one's research project

J. Njeri Mugwe, S. Runo, *Research Methodology in Agricultural Sciences*, https://doi.org/10.1007/978-981-95-1892-0_6

5. Determine the priorities of various stakeholders regarding ethical principles in research
6. Describe fundamental guidelines and principles in research
7. Identify where to locate ethical guidelines and how to seek approval from Institutional Review Board
8. Discuss the importance of adhering to ethical guidelines
9. Apply ethical considerations in various stages through the research process
10. Determine ethical considerations in surveys, qualitative research, and on-farm experiments in agriculture
11. Apply ethical guidelines in publishing
12. Determine ethical considerations in various facets of agriculture (collaborative projects, use of genetic resources, biotechnology, etc.)
13. Identify ethical guidelines for the use of Artificial Intelligence (AI) in smart farming
14. Apply ethical considerations in the supervision of students' research projects

6.1 The Interconnectedness of Ethical, Legal, and Administrative Considerations

This chapter covers ethics, legal issues and administrative processes in research. The three are related and support one another in ensuring that researchers carry out fair research. They work together to support agricultural research and ensure responsible and compliant research practices. Researchers need to familiarize themselves with these elements to guide them through research projects. As we will see later in this chapter, research organizations, academic institutions, and donor agencies place emphasis on adherence to ethical and legal standards.

Ethical considerations are about morals and guide us in our research, just as the principles guiding us in our personal lives do. Whether they are under supervision or working independently, researchers should strive to adhere to ethical guidelines. The guidelines outline moral obligations. The obligations, first and foremost, protect the rights and well-being of the research participants. Secondly, they protect the integrity and credibility of research. Overall, they protect all stakeholders involved in research (Yip et al. 2016).

Legal issues are like society's rules and regulations, such as traffic laws and property rights. They protect our research outcomes and involve adherence to laws, regulations, and statutes governing the conduct of research. Legal issues can arise in every phase of a research project (Van Emon 2020). The intellectual property rights and data protection regulations are of particular relevance in agricultural research. Other important aspects include regulations to govern advances in agriculture, such biotechnology and precision agriculture.

Administrative issues refer to organizational and management activities that support implementation of the research project. They include processes like managing timelines, university requirements, supervisor allocation, ethical approval requests, progress reporting, plagiarism checks, etc. Many students find these processes tedious and often complain about delays. However, They are not just tasks to complete, but a guide on how we can effectively implement ethical and legal requirements in our research. Just as we need permission to host and participate in many forums, researchers must obtain approvals to conduct research from their university and regulatory bodies.

These three aspects are closely interwoven (Fig. 6.1). The ethical guidelines and considerations are the moral compass for research decisions. They influence the formation of laws and administrative procedures that aim to protect the well-being of individuals, communities, and the environment (Resnik et al. 2020). The legal framework translates ethical principles into enforceable legal mandates. On the other hand, administrative issues provide practical mechanisms for implementing ethical and legal requirements.

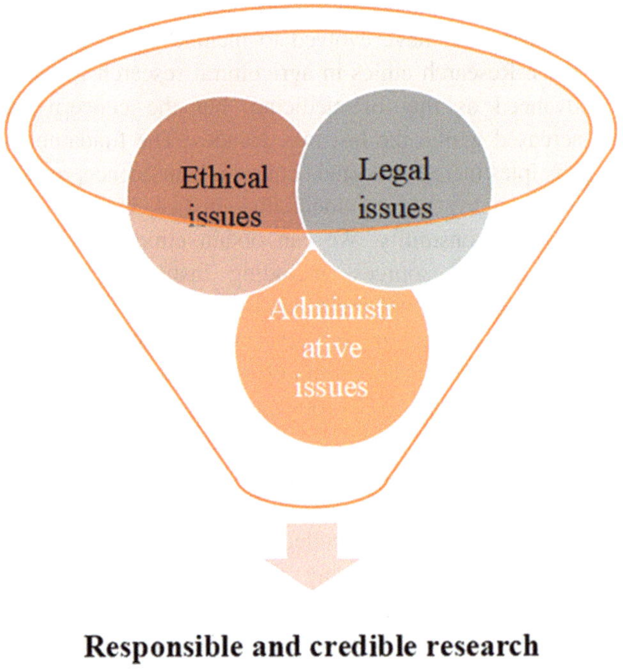

Fig. 6.1 The interconnectedness of ethical, legal, and administrative issues in research

6.2 Ethical Considerations in Agricultural Research

We can think of ethics as morals that guide us into acceptable and unacceptable behavior. In everyday life, there are many areas where we apply ethics without paying explicit attention because we have grown accustomed to them. For example, we respect other people's privacy and avoid snooping on their personal belongings or reading their diaries without permission. Similarly, researchers work with human participants who also deserve their privacy to be guarded. Researchers should therefore safeguard sensitive information from participants. This includes maintaining confidentiality and anonymity.

Additionally, in many cultures and religions, we have the "Golden Rule," a principle that guides the conduct of society. Just as we are taught to treat others as we would like, researchers should conduct their studies with the same level of respect, honesty, and fairness they expect from others. This means considering the well-being and rights of research participants, being truthful in reporting findings, and upholding the highest standards of integrity.

As a student you may not yet be fully accustomed with the rules and regulations of research. You must, therefore, learn and apply them for your benefit and also for the benefit of your research subjects. This chapter is packaged to help you learn and navigate the research process with integrity.

6.3 Brief History and Growth of Research Ethics

History records many studies that violated the rights and dignity of participants (Algahtani et al. 2018; Jamrozik and Selgelid 2021). Notable examples of these studies include the Tuskegee syphilis study (1932–1972), Nazi medical experiments (1930s–1940s), the Thalidomide tragedy in the 1950s (Allen 2001), and the Human Radiation experiment (1940–1970) (Box 6.1). These cases involved uninformed and unaware participants being exposed to diseases or subjected to unproven treatments. The initial response to these unethical breaches was the formulation of the *Nuremberg Code* (Schmidt 2004), developed in the aftermath of the Nazi research atrocities revealed during the post-World War II Nuremberg trials. The Nuremberg Code is an international set of research standards that aims to prevent severe research misconduct and the abuse of vulnerable, unsuspecting research subjects. In the USA, the *Belmont Report* was directly drafted from the Nuremberg Code in 1978 to guide biomedical research (Sims 2010; Fons-Martinez et al. 2024). It is anchored in three fundamental principles: respect for persons, beneficence, and justice. These principles have become integral to the ethical framework guiding research involving human beings.

The Belmont report key concerns include the following;

- Ensure potential participants understand their voluntary participation rights
- Clarify that refusal to participate does not affect access to care
- Capacity for self-understanding and asking questions
- Option for withdrawal without penalties
- Avoidance of coercion through explicit threats of excessive rewards
- Use incentives carefully to prevent undue pressure on potential participants

Over the years and as research advances, ethical considerations have evolved to encompass emerging technologies, such as genetics, biotechnology, and artificial intelligence. These cutting-edge technologies bring forth new challenges requiring continuous reflection and adaptation of ethical frameworks to ensure equitable scientific practices. For example, genetic research raises questions about privacy and potential discrimination in the use of the resources. Artificial intelligence requires careful ethical considerations regarding transparency, accessibility, and potential biases (Lee and Park 2018).

Commitment to adhere to best practices has increased and this is evidenced by the many emerging professional bodies whose role is to specify the rules and regulations that safeguard and govern best practices. One of the most important safeguards for ensuring ethical research conduct is the Institutional Review Boards (IRBs) (Grady 2015). The IRBs review proposals to ensure adherence to ethical principles and guidelines in academic and research institutions (Rothenberg 2004; Halse and Honey 2007; Babyar 2018; AlFattani et al. 2025). There is also a proliferation of ethical codes, rules, and regulations among professional bodies that intend to create a culture of best research practices. There are also many well-known codes globally, including the Medical Association Declaration of Helsinki (Davis 2023; Chodankar et al. 2025), The National Institute of Health (NHS) (Collins and Tabak 2014), The National Foundation of Science (NSF), the Food and Drug Administration and the Environmental Protection Agency (Kretser et al. 2017).

Many governments and funding bodies prioritize the integration of systematic research ethics training as a requirement for grant applications. For instance, the European Union Horizon 2020 program strongly encourages applications to incorporate research ethics within the proposals. Other prominent donors and funding agencies worldwide emphasize the significance of ethical conduct by requiring researchers to undergo comprehensive ethics training. These include the National Institutes of Health (NIH) in the USA, the Wellcome Trust in the United Kingdom, the National Science Foundation, and the World Health Organization (WHO). The training aims to promote responsible research ethics practices. Such initiatives demonstrate a commitment to increasing ethical awareness and conduct across diverse research disciplines and enhancing accountability.

Box 6.1: Examples of Historical Injustices in Medical Research (Puras 2017; Barrow et al. 2022)

Tuskegee syphilis study (1938–1972): Researchers used rural African American men to study the natural course of syphilis. During the study, the men received free examinations and medical care. However, the participants did not receive information about the diseases, were not informed that they were participating in research, and were not told that the research would not benefit them directly. Even after a known treatment was discovered, the participants were not informed about the new treatment, and the research continued for decades, which led to the death of many study participants.

Nazi medical experiments (1930s to 1940): Nazi human experimentation involved a series of medical experiments conducted on prisoners by Nazi Germany in concentration camps, primarily between 1942 and 1945. These experiments affected 15,754 documented victims from various nationalities and age groups, though the actual number of victims is believed to be much higher. Among those documented, about a quarter were killed as a result of the experiments. Survivors often endured severe and permanent injuries, reflecting the brutal and inhumane nature of these experiments. These atrocities highlight the profound suffering inflicted on individuals and the gross violations of human rights during this dark period in history.

Human radiation experiments (1940s to 1970s): Numerous human radiation experiments were conducted in the United States, involving the non-consensual administration of plutonium and other radioactive elements to mostly terminally ill patients. These included: 1. Children were fed food containing radioactive traces 2. Impoverished pregnant women were given radioactive iron, 3. US soldiers and prisoners were exposed to high levels of radiation 4. Prisoners' testicles were irradiated, and 5. Bodies were exhumed from graveyards to test for radiation without the consent of their families.

6.4 Overview of Ethical Issues in Agricultural Research

Compared to the field of medicine, research ethics in agriculture is less developed. In medicine, extensive regulations, established codes of conduct and a strong emphasis on patient welfare have shaped a robust ethical foundation over the years. For instance, researchers in the medical field have a wide range of codes and regulations, such as the Declaration of Helsinki issued by the World Medical Association (Chodankar et al. 2025), Good Clinical Practice, the World Health Organization, and Council for International Organizations of Medical Sciences.

However, in agriculture, the ethical landscape has emerged possibly in the last few decades to address complex issues of genetic modification, sustainable farming practices, rights, and dignity of human subjects, e.g., farmers, and welfare of farm animals (livestock). However, it still lags behind the other disciplines (Korthals 2016; Zimdahl and Holtzer 2018). This may be attributed to the characteristics of agricultural research. In the early stages, agricultural research practices mainly focused on improving crop yields and addressing agricultural challenges. For instance, during the era of the Green Revolution (1950–1984) (Eliazer Nelson et al. 2019; Cabral et al. 2022), the need for ethical considerations was limited. However, as agriculture became increasingly complex and intertwined with broader economic, societal, and environmental issues, ethical considerations have gained prominence. For instance, the adoption of a farming systems approach, the use of Genetically Modified Organisms (GMOS) and biotechnology has sparked ethical debates in the agricultural arena (Zilberman et al. 2018).

We, therefore, find that, within the last two decades, there has been a growing and increased emphasis on ethical guidelines specific to agricultural research. Some of the guidelines in agricultural research exist in the following areas:

- Pesticide regulation to guide development, registration and safe use for human health and the environment.
- Food safety regulations in food processing, labeling, and product safety.
- Seed regulations to govern the production, distribution, and sale of agricultural seed.
- Laboratory practice guidelines to guide laboratory studies.
- Biosafety regulations to govern the release and use of genetically modified engineered crops.
- Guidelines for animal research for the humane treatment of animals.
- Guidelines for research involving humans as research subjects, e.g. in surveys and participatory research.

6.5 Balancing Research Ethics and the Pursuit of Truth

Much like their counterparts in other fields, researchers in agriculture must navigate the challenge of balancing research ethics in a highly competitive environment. Alongside the pursuit of objective truth and knowledge, those involved have a responsibility to the subjects of their research (humans or animals). As an researcher, it is critical to adhere to the principles of scientific research, such as validity, reliability, and reproducibility, while also prioritizing the protection of human rights, fairness, and dignity of research subjects. When uncertainty arises about the best course of action, remember that ethical considerations limit our choices in the pursuit of truth. Ultimately, prioritizing respect for human dignity and doing good is the better path forward.

The research environment in Universities and research institutions is highly competitive and performance-based, which can easily compromise research ethics (Drolet et al. 2023; Berghaeuser et al. 2024). These authors report that competition for grants and scientific publications makes researchers compromise integrity, falsify research results, and plagiarize to achieve their goals. We emphasize that the pursuit of truth should be the guiding principle in our research. Recognizing the responsibility bestowed upon us as researchers, we should always commit to upholding the highest standards of integrity and transparency throughout the research process.

The other balance needed concerns the allocation of time to bureaucratic procedures, paperwork, and institutional accreditation. While ethical oversight systems and reporting requirements are important for ensuring compliance, some students and researchers view this as burdensome because, at times, considerable time and resources are used to maneuver these administrative tasks (Petillion et al. 2017; Raykov 2020). Despite this perception, it is necessary to recognize that the requirement is mandatory. Faculty staff should, therefore, take an active role in supporting students through these processes. Additionally, institutions should strive to streamline the administrative processes and offer support to the students to alleviate the burden.

6.6 Research Ethics Considerations Vary among the Stakeholders in Agricultural Research

In chapter one, we learnt that agricultural research is a complex activity that involves multiple stakeholders. These stakeholders have different interests, objectives, and responsibilities. They include scientists, researchers, farmers, policymakers, funding agencies, the general public, governments, universities, and research institutions. Additionally, agricultural research involves several stages that have an impact on multiple stakeholders beyond the laboratory. Looking at the list of stakeholders in Chap. 1 (Sect. 1.3.1), we realize that their priorities in terms of ethics will vary.

- Scientists and researchers: Their vital role is to generate data and knowledge. Priority in research ethics is to generate accurate, credible, and trustworthy data. To achieve this, scientists and researchers often need support from other stakeholders, especially those who provide the data. They must also operate within the law and be open and honest in dealing with participants. However, they must also guard against exposing themselves or co-workers to unacceptable danger in pursuit of data (Nebeker et al. 2019; Weinhardt 2021).
- Farmers: They are both sources of data and beneficiaries. Their primary priority in guarding research ethics is to provide accurate data. Second, as recipients of information from researchers, they must trust that the information disseminated by researchers is accurate and will have a positive impact on their livelihood. Third, they also need their rights protected during data collection.
- Policymakers: They are consumers as well as supporters of research. Ethics priority is ensuring the application/implementation of the research findings and fulfilling their active role of supporting society and researchers.
- Funding agencies: They mainly provide financial resources to support research. Their priority and responsibility lie in their commitment to respecting grant agreements and adhering to the promises as per the grant agreements established with the researchers. They should also not interfere with the results, whether "Positive" or "negative."
- General public: interested in research that aligns with societal priorities and addresses the needs of society.
- Governments: must provide an effective governance structure for the research community and society. They need to develop effective policies to fund and regulate research and innovation.
- Universities and research institutions: establish adequate policies to support research ethics, legal and administrative issues ensure compliance among their staff and students.

6.7 Fundamental Principles in Research Ethics

1. *The principle of informed consent* is the foundation of ethical research dealing with human subjects (Fons-Martinez et al. 2024). It refers to voluntary permission given by an individual after providing clear and detailed

information about the research, known as full disclosure of the project. Researchers must obtain informed and voluntary consent from individuals who participate in their research. Informed consent ensures that participants do so willingly without being forced or manipulated. Failing to obtain informed consent may have serious ethical and legal implications (Yip et al. 2016).

2. *The principle of beneficence*: emphasizes that actions should be taken for the benefit of others, with a focus on maximizing benefits while minimizing potential harm to research participants (Belmont report). As such, researchers should prioritize the well-being and welfare of participants by designing research protocols that minimize risks and continuously monitor their safety throughout the study.
3. *The principle of respect for autonomy*: emphasizes the significance of acknowledging people's autonomy and their entitlement for confidentiality. This principle requires researchers to prioritize privacy and confidentiality, protecting participants' personal information. It involves not revealing any identity details, such as name, place of residence, village, or any other information that could potentially reveal his or her identity. This means there is no way a person will be identified in stored files. A coding system should be used to replace personal identification.
4. *The principle of justice*: "Justice entails a just and equitable allocation of the advantages and disadvantages of research." It means distributing benefits and burdens fairly and equitably. Researchers must treat all groups and individuals with respect without discrimination or exploitation. Selection and recruitment of participants should be done equitably, especially considering vulnerable groups. Transparency is also key while allocating resources and opportunities related to research participation. Researchers should also treat others with respect and fairness, acknowledge contributions, and conduct research in a manner consistent with the values of their profession.
5. *The principle of honesty and integrity*: Researchers should conduct their research with honesty and integrity. This includes accurately collecting data, reporting research findings, and avoiding data fabrication, falsification, and plagiarism. They should also adhere to ethical guidelines as provided by the clearance body and take responsibility for any ethical lapses that may occur during the research process.
6. *The principle of social responsibility*: researchers are responsible for considering the larger social implications of their work. This should include paying attention to ethical issues related to the physical impact of their research on society, the environment, and future generations. Food systems and environmental protection are thus key issues in agriculture.

6.8 Where Do You Get Ethical and Legal Guidelines for Conducting Agricultural Research?

Many students and budding researchers may need help determining where to access ethical guidelines or how to comply. You may acquire ethical guidelines from the following eight key sources;

1. *Institutional Review Boards*

Universities and research institutions must have an oversight committee to safeguard research ethics. Many universities have, therefore, established institutional review boards (IRB) or ethics review committees to oversee and approve research proposals. These bodies are composed of experts in various fields, including ethics, veterinary sciences, and law, whose role is to evaluate the ethical aspects of research projects. The IRB should act as your first stop in terms of seeking ethical guidance.

Your supervisor and the section coordinating graduate training in your university will advise on whether your proposal needs full review by the IRB. If required, you are advised to wait for approval from IRB prior to engaging in data collection. Steps for obtaining approval are discussed later in this section.

2. *International and National Ethical Codes and Guidelines*

Agricultural researchers can also refer to the international and national ethical guidelines that provide overarching principles and standards. The World Health Organization (WHO) offers guidelines for research involving humans while the World Animal Protection provides recommendations for the ethical treatment of animals. In addition to the global guidelines, many countries have their national ethical guidelines for agricultural research, often tailored to their specific agricultural practices and priorities. Researchers can benefit immensely from these guidelines.

An excellent example of a helpful code of ethics for agricultural researchers and students is the Consultative Group in International Agricultural Research (CGIAR) research ethics code (CGIAR 2020). Its purpose is to reinforce the moral foundations of international agricultural research. The code comprises a set of ethical principles, conduct codes, policies, and guidelines that guide CGIAR research staff, governance officials, individuals involved in the research, and third parties in their activities and operations. It provides both general and specific instructions to researchers to uphold ethical standards. The general standards emphasize conducting high-quality research and promoting an open exchange of ideas. The guidelines also establish standards for publication and authorship, emphasizing respect for intellectual property rights, accountability, transparency and conflicts of interest.

The section on specific standards in the research ethics code covers guidelines for conducting research that involves human subjects. It includes standards for selecting participants, obtaining informed consent, ensuring confidentiality, and protecting personal data. The code also provides guidelines for research involving animals and animal welfare and for research that utilizes modern biotechnology and participatory methods. Other important codes and regulations include a Declaration of Helsinki and the Belmont Report (see Box 6.2).

3. *Professional Organizations*

One can obtain ethical guidelines for agricultural research from professional organizations and bodies related to specific areas of expertise. For instance, the American Society of Agronomy has a set code of ethics that its members must adhere to. We also have "the International Code for the Sustainable Use and Management of Fertilizers," providing a set of voluntary practices and agreed-upon expectations for behavior by various stakeholders in the use and management of plant nutrients (FAO 2019).

4. *Government Regulations and Funding Agencies*

Many agricultural research government agencies have regulations and funding criteria that consider ethical practices. Funding agencies also provide their own guidelines and recommendations for research ethics, which can serve as a valuable source of information.

5. *Collaborators in Other Organizations*

When collaborating with a colleague from diverse disciplines, they can be a source of ethical information and insights. Diversity in the research team brings together individuals with different perspectives and experiences, each of which may have unique ethical considerations. They can thus be a good source of ethical information and play a crucial role in contributing to a more comprehensive understanding of ethical issues in your area of study.

6. *Research Policies*

University and research institutions have policies that often include a section on ethics to guide researchers and their students. The policies cover various topics, such as responsible conduct of research, treatment of human and animal subjects, handling of sensitive data, authorship, and collaboration. For instance, Kenyatta University's research and innovation policy includes sections on financial management, administration of grants, conflicts of interest, intellectual property, environmental health and safety, research involving humans, and biosafety in teaching and research. The policies also outline ways for obtaining relevant approvals.

7. *Supervisors*

Your research project supervisor can provide valuable guidance on ethical guidelines. With their expertise and experience in the field, they can also assist you in navigating the process of obtaining approvals. Your supervisor can also mentor you on designing studies that take into account the rights and dignity of participants and animals, as well as maintaining integrity when handling data. Ultimately, your supervisor will guide you on the importance of transparency, honesty, and responsible conduct of research, thus instilling a strong ethical foundation.

8. *Documented Ethical Issues in Literature*

Lots of literature on ethical considerations exists in various books, especially in social science and education research methods books. This literature can be a source of information on ethical issues that are important in your area of study. The social science research methods books, including those of psychology, are particularly useful if you are working with farmers.

Box 6.2: Codes and Ethical Guidelines Relevant for Agricultural Research

The top three ethical guidelines for research involving human subjects

1. Nuremberg Code: a set of guidelines established in response to unethical human experimentation carried out by the Nazi Physicians during World War II. It was created during the Nuremberg Trials from 1946 to 1947. Key aspects include voluntary informed consent, avoidance of unnecessary harm, freedom to withdraw, and benefit to society and qualified researchers. Contains 10 main ethical principles (Schmidt 2004).
2. Declaration of Helsinki: contains principles guiding medical and biomedical research that was first adopted in 1964 by the World Medical Association. Key components include informed consent, Beneficence and Non-Maleficence, respect for vulnerable populations and research Ethics review (Chodankar et al. 2025).
3. Belmont report: is a publication of the National Commission for the Protection of Human Subjects and Behavioural Research, written in 1979. It contains three key components for the protection of human subjects: 1. Respect for persons, 2. Beneficence and 3. Justice. It is a foundation document in the field of research ethics (Sims 2010).

Other guidelines and laws

4. International Treaty on plant genetic resources for food and agriculture by FAO in 2020. https://www.fao.org/plant-treaty/en/
5. Red list of Threatened species by The International Union for Conservation of nature and Natural Resources in 2020. https://www.iucnredlist.org/
6. International code of conduct for plant germplasm adopted in 1993. https://www.fao.org/3/aj634e/aj634e.pdf
7. Biosecurity regulations by various countries and organisations.

8. Animal Welfare Act adopted in 1966 https://www.govinfo.gov/content/pkg/COMPS-10262/pdf/COMPS-10262.pdf
9. Plant Variety Protection Act. https://www.fao.org/faolex/results/details/en/c/LEX-FAOC064685/
10. National Environment Policy Act. https://www.epa.gov/laws-regulations/summary-national-environmental-policy-act
11. World Trade Organisation (WTO).
12. Guide for the care and use of laboratory animals (2011). Guide for the Care and Use of Laboratory Animals (Eighth Edition) | Research A to Z (umich.edu)
13. Good laboratory practice standards.
14. Convention on Biological Biodiversity (CBD). Nagoya protocol for on access and benefit sharing https://www.cbd.int/abs/

6.9 The Role and Processes of Institutional Review Ethics Boards

Institutional Review Boards (IRBs), also known as Research Ethics committees, are "watchdogs" of ethics in research. They are mainly established in academic and research organizations to oversee all the research that is conducted in the institution. Their main role is to protect the rights and welfare of human research subjects and animals alike. Most Universities and research institutions have at least one IRB, while others have more than one. For example, the International Livestock Research Institute (ILRI) and Kenyatta University have three boards each; 1. Animal Care and Use Committee, 2. Research Ethics Committee and 3. Biosafety Committee (ILRI 2021; Kenyatta University 2023). They are mainly developed to guide their researchers, but they also help other institutions that are not accredited to manage research ethical clearance. There are also many independent or commercial IRBs, especially in developed nations (Emanuel et al. 2006). The primary mandate of IRBS is to protect the rights, well-being and dignity of human research subjects by ensuring that the research proposal meets stringent ethical criteria.

IRBs conduct thorough reviews of research proposals, guided by regional and local ethical guidelines. They assess key components such as informed consent, risk-benefit ratios, and the overall scientific and social value of research. They apply ethical principles as outlined by international guidelines, as well as regional and local frameworks. The composition of these boards often includes diverse members with expertise in various disciplines, as well as individuals from outside the research community, known as laypersons.

The extent of the review process by the Institutional Review Board (IRB) is determined by the level of risk associated with the research study. The IRB assesses the potential risks to human participants and the adequacy of protections in place. There are generally three levels:

1. *Exempt review*: Some studies are considered minimal risk and may qualify for an exempt review. Exempt studies are typically low-risk research involving activities such as surveys, interviews, or the analysis of existing data (secondary data and records), like in document analysis. However, even exempt studies need to be formally reviewed and approved by the IRB. Always check with your institution about the IRB process, even if you think that your study does not require IRB review.
2. *Expedited review*: Designed for studies with minimal risk but that do not qualify for an exempt review, may undergo an expedited review. This process involves one or more experienced reviewers and is quicker than a full board review. Farmer surveys, especially those that have qualitative data collection, commonly conducted are commonly reviewed at this level. Others reviewed at this level may include observational studies of farming practices and programme impact evaluations.
3. *Full board review*: Designed for studies that pose more than minimal risk to participants or involve vulnerable populations, often undergo a full board review. An example is a study on pesticide use on farm workers or studies involving animals. The entire IRB committee evaluates the research proposal during scheduled meetings. Full board reviews are more thorough and may take more time than exempt or expedited. All members often read the research protocol, and discuss the value and weaknesses; sometimes voting takes place to approve, require revisions or disapprove.

Most universities provide these guidelines and much more about the approval process by IRB on their websites to guide staff and students. The guidelines cover various aspects, including the preparation of informed consent forms, measures to protect participant confidentiality, and procedures for obtaining voluntary and informed consent. Additionally, the guidelines may address considerations specific to studies involving vulnerable populations, ensuring that research is conducted ethically and with due diligence. You are encouraged to familiarize yourself with these guidelines.

Typically, IRBs meet monthly at large institutions and quarterly in small ones. Some will meet only when a need arises. If you suspect your study will not be exempt, planning for a full-board review is often best. After approval, the IRB also plays a critical role in ensuring compliance with ethical guidelines and regulatory requirements. They monitor ongoing research to ensure that any modification in the research design is ethically sound.

Steps in Seeking Approval from Institutional Review Boards

To obtain approval from your university's IRB, you must summarise the procedures in your research and provide evidence that your research methods and procedures will protect participants' welfare and rights. IRB approval process varies between institutions, but generally, we have five main steps:

1. Develop a proposal that complies with the requirements: The proposal must provide clear research objectives, a detailed research design and plan, and necessary consent documents. During this preparatory phase of the research, you seek information regarding the requirements for obtaining approval from the board. The specific requirements vary across institutions, but common elements are usually required (see Box 6.3). As you prepare the proposal, use these elements as guidelines for seeking approval from IRB.
2. Completion of IRB form: Institutions typically have standard forms for application, often available online. Be sure to follow the instructions strictly. Some IRB forms are one page, and others could be up to 30–40 pages, depending on the nature of the project and the specific documentation required by the institution.
3. Submission of IRB form: Once you fill out the form, submit it and include the required attachments, such as, full proposal, study tools, questionnaires, and checklists. The review process may take 2–4 weeks.
4. Responding to IRB feedback: Promptly address any comments given and send them back to IRB.
5. Approval and implementation: If satisfied, you will receive official approval to conduct your research. This may include an ethical review certificate (Fig. 6.2).

Box 6.3: Elements for Approval by Institutional Review Board (Resnik 2021)

1. *Clear proposal*: Provide a clear outline of the proposal with clear research questions, objectives, hypotheses, methodology and procedures of data collection and analysis.
2. *Training in ethics*: Undertake and complete the required training on ethics and present evidence of completion if required by the board.
3. *Risks and benefits*: Include risks and benefits: Indicate risks and benefits and provide strategies for mitigating and minimizing risks.
4. *Vulnerable population protection*: Protection of vulnerable participants: If dealing with vulnerable individuals e.g. children and economically marginalized communities, indicate how you will protect their rights and well-being.
5. *Data protection*: Show how you will handle research participants' data, ensuring that participants' personal information remains confidential.
6. *Recruitment of participants*: Recruitment and selection of participants: methods of recruiting participants must be clear. Ensure they are fair, unbiased, and respectful.
7. *Informed consent process*: develop a consent form that is clearly written and ensure information is comprehensively documented according to the guidelines of the IRB.
8. *Information sheet*: providing information to farmers regarding the nature of the study.

Which Research Topics Need Ethical Approval?

Inexperienced researchers and students may raise questions about which projects require ethical approval. The answer to this question is that any research project that involves direct interactions with human participants, experimentation on animals, use of hazardous materials, or the use or analysis of sensitive data requires ethical approval. They include and not limited to the following:

- *Human subject research*: involving all approaches, quantitative, qualitative and mixed approaches. This can be on-farm experiments, surveys, interviews, focus groups, participatory research etc. These are probably the most common in agricultural research.
- *Animal research with farmers*: This is two-pronged; Animals will need protection as well as human participants. This may include breeding studies, nutrition investigation studies or animal health e.g. involving vaccinations.
- *Animal studies in the laboratory*: Need protection to ensure human treatment and welfare. Reviewers also check compliance with the 3R in animal protection (Hubrecht and Carter 2019).
- *Food safety studies*: this includes the responsible conduct and welfare of participants.
- *Data privacy and security of farmers' data*: collecting and sharing farmers' data through digital agriculture is a major ethical concern. Data must be protected and shared responsibly. This is discussed at the end of this chapter.

6.10 Purpose of Research Ethical and Legal Guidelines

Researchers must follow research ethical and legal guidelines. Some key reasons include:

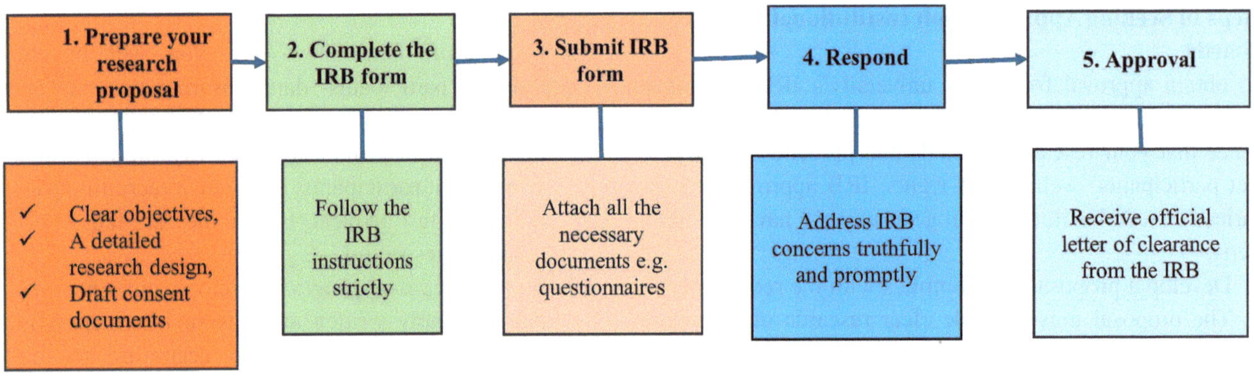

Fig. 6.2 Steps in obtaining ethical review approval

1. *Publication of research findings*: Many reputable journals have strict ethical standards and guidelines. Manuscripts that lack proper ethical considerations may face challenges with peer review amid the publication process. Some journals require evidence that the research underwent ethical review by an IRB or an equivalent body. Manuscripts that do not demonstrate proper informed consent procedures may be rejected. In addition, manuscripts lacking evidence of proper care, ethical treatments, and approval from body-accrediting animal studies may face rejection.
2. *Donor compliance*: Some funders require the strict following of ethical guidelines. This is sometimes included in the contract to ensure commitment.
3. *Promote high-quality research*: The field of ethics plays a crucial role in promoting honesty and advancing knowledge by preventing researchers from manipulating or misrepresenting their findings. When researchers prioritize the ethical treatment of participants and protect their rights and privacy, participants are more likely to provide accurate information, promoting the research's validity and reliability.
4. *Helps protect the rights of the research subjects from research risks and abuse*: Adhering to ethical norms protects the rights of study participants and reduces the risk of abuse, exploitation, and harm. Institutional ethical boards review research proposals to identify and address ethical concerns before the study starts. Obtaining informed consent ensures that participants are fully informed and able to make educated decisions about their involvement, reducing the risk of coercion or exploitation. Researchers also identify and mitigate any physical, psychological, and emotional risks associated with the research.
5. *Promote essential values in collaborative work*: Ethics promote cooperation and coordination among individuals carrying out research in various disciplines and organizations. This leads to accountability, mutual respect, trust, and fairness. For example, ethical norms guiding publishing and authorship policies, patenting policies, and data-sharing policies exist to guide researchers. These norms aim to safeguard intellectual property interests while also promoting collaboration.
6. *Help accountability to the public*: The practice of ethics promotes accountability among researchers and generates public support for their work. When researchers and institutions prioritize integrity and transparency in their work, the credibility of their research within the general public and among donors increases. This builds confidence in the scientific process and is essential for the sustainability and advancement of knowledge.
7. *To ensure impartial sharing of potential benefits*: Ethics promotes fairness, justice, and responsible treatment of all individuals or groups involved or affected by the research process. For example, in an intervention study whose results demonstrate a positive impact on the participants in the experimental group, ethical considerations dictate that the researchers extend the same benefits to the control group upon the study's conclusion.

6.11 Specific Ethical and Legal Considerations in Agricultural Research

Researchers in agricultural research must pay attention to ethical concerns throughout the research process, from conception to dissemination. This will help ensure the smooth running of the project and avoid dilemmas and conflicts in the research project. Here, we consider four stages in the research process;

1. Conceptualization and proposal writing.
2. Implementation of the research:
   - Survey research involving farmers.
   - On-farm experimentation.
   - Qualitative and participatory research.
   - Research involving livestock and other animals.

3. Data analysis.
4. Publishing and dissemination of research findings.

6.11.1 Ethical Guidelines During Conceptualization, Research Design and Proposal Writing

In your research proposal (Chap. 5), you should include a section on ethical considerations, which must cover all aspects of ethics in the research study. To ensure that your proposal aligns with ethical and legal standards, you must obtain ethical clearance from the Institutional Review Board or Ethics Review Committee. The University typically provides guidelines and procedures for obtaining clearance. At the conceptualization stage, two key ethical issues arise. As you select the topic of study, you need to ask yourself, "Is this project worthwhile?" and "Who will benefit from the research?"

During the preparation of the proposal, ensure that you have a clear ethical framework for the study by incorporating the following information:

- Indicate which and how the necessary permits and approvals will be obtained, such as permission from the central government and local authorities. For example, in Kenya, all research must be approved by the National Council for Science and Technology and Innovation (NACOSTI).
- Specify how the participants will be informed about the study, the voluntary nature of their participation and their rights.
- Explain mechanisms for collaboration in the field, for example, with Ministry of Agriculture officials, and clearly explain how the community will be reached through the local authorities.
- Clearly outline the procedures you will use to obtain consent from the participants.
- Describe strategies for engaging with and obtaining consent from groups, e.g. in focus group discussions.
- Clearly describe any potential risks, harm, or discomfort associated with participation.
- Specify how the participants will be informed about the study's progress, findings, and any potential publications.
- Develop a data protection plan that adheres to the relevant to your university, institution, and international laws.
- Develop protocols to prevent participants' identities and responses.
- Indicate how data will be anonymized, stored securely, and used to maintain confidentiality.

6.11.2 Ethical Guidelines in the Collection of Data Using Survey or Qualitative Data Collection Tools Involving Farmers

In collecting data using surveys or qualitative or participatory rural appraisal (PRAS) involving farmers, you are likely to use quantitative questionnaires, focus group discussions, key informant interviews, collect documents, take photos, etc. (Chaps. 18 and 19). The following are some of the key activities needed to comply with ethical guidelines (Husband 2020; Kaur et al. 2022; Becot et al. 2023):

Community approval: Farmers typically live in communities governed by local counties. Ensure community leaders and relevant authorities have information about the nature of your research and that they have provided consent. Community leaders and local authorities are called gatekeepers because they regulate and control entry and are intermediaries in data collection (Creswell and Plano Clark 2011). The approval can be done by either writing officially or visiting to inform them about the research study. In Kenya, the Chiefs and sub-chiefs take charge of the smallest administrative units. During surveys, they can be useful in providing the sampling frame and guiding the survey teams.

Informed consent: Obtain informed consent from farmers who participate in the study. Ensure you explain the purpose of the survey with clarity, informing them about how the data will be handled and used and any potential risks and benefits. Ensure participation is voluntary, and inform farmers that they are free and have the right to withdraw from the survey at any time. Additionally, be transparent about your institution and give contact information to the researcher or administrator in charge of the research study. Ensure farmers have a clear understanding of the processes as well as the activities for the data collection.

Beneficence and justice: consider the potential benefits and burdens of the survey for participating farmers and the broader agricultural community. Ensure the survey findings are used to improve the farming practice of the area without causing environmental harm.

Community engagement and communication: Share the survey's progress, findings, and potential implications with farmers. Establish channels for farmers to ask questions and voice their concerns, and ensure you address any concerns openly and honestly.

Anonymity and confidentiality: Prioritize the protection of the rights, privacy, and dignity of participating farmers. Safeguard their personal information and ensure that data collected is anonymized to prevent identification.

Respect for culture: Respect local customs, traditions and norms when conducting surveys in farming communities. Be culturally sensitive and adapt survey instruments and procedures to the local context. Ensure data technicians are well-trained and made aware of the cultural values of the community.

Selection of participants: select participants based on the objectives of the study rather than non-research interests, always ensuring varied representation from subject groups. This includes ensuring you have included women, men, and minority groups where possible and consistent with the objectives of the study. For instance, in a study involving different genders ensure gender representation is fairly done.

Sensitivity in compensation: Avoid giving unwarranted or disproportionate financial and other incentives to obtain the participation of research participants, particularly when it might coerce participation. However, researchers may offer rewards and benefits according to plan and budget. Such compensation should be disclosed to subjects when signing informed consent and should not be an inducement to participate (Njue et al. 2015).

6.11.3 Implications of Changing Paradigms in Agricultural Research to Research Ethics

In this book, we have discussed different types of research designs and highlighted the evolving landscape in agricultural research and development (Chap. 10 on farming systems research). There is growing emphasis on engaging more in qualitative and mixed methods research methodologies to address the complex challenges in agriculture and food security. As researchers increasingly engage in these domains, they encounter ethical dilemmas that differ from traditional quantitative studies (Sanjari et al. 2014; Pietilä et al. 2020). They, therefore, require a paradigm shift. Unlike positivists, who often prioritize detachment and objectivity, qualitative and participatory researchers take a more subjective and reflective stance. This mean that researchers will maintain intimate relationships with participants in qualitative and participatory research. Ethical considerations, therefore, take on special significance. Researchers in this new paradigm must engage with farmers in depth as they involve them as co-partners in generating knowledge. The most common participatory research activities and studies include:

- Participatory variety screening projects (Ojulong 2017).
- Farmers' demonstration events.
- Value-chain analysis involving various stakeholders.
- Focus group discussions.
- Market studies using both quantitative and qualitative approaches.
- Gender studies using both quantitative and qualitative approaches (Quisumbing et al. 2014).

Most research projects involving communities are typically anchored in research and development agendas involving multidisciplinary teams. They typically comprise faculty staff and students, partners from the Ministry of Agriculture, researchers from national research institutions, nongovernmental organizations, communities, and their leaders. Several relationships arise from these team members, both vertically and horizontally, together with their diverse expertise and interests. The research team must, therefore, ensure ethical principles and guidelines take care of all the parties involved.

6.11.4 Ethical Guidelines in Participatory Research

Several authors have provided guidelines on ethical considerations in conducting qualitative research in various fields (Carey 2016; Lenton et al. 2021). We draw from these authors and our experiences to provide recommendations for key ethical issues in participatory and qualitative research. Specifically, we draw from practical experiences working closely with farmers in conducting on-farm experiments and organizing dissemination events, while working in research and development projects. These projects are acknowledged in the first section of this book.

The stakeholders in these projects included a range of diverse interests and demands. They are listed in Box 6.4. We give the following recommendations to consider in maneuvering ethics in participatory action research:

- First and foremost, you must commit time and effort to build relationships. This is not easy; sometimes, you may fail when dealing with diverse personalities. Sometimes, you may need to engage extra hands and expertise to handle relationships among collaborators.
- All collaborators must be fully informed about the research project's purpose, procedures, risks, and benefits before agreeing to participate. Provide accessible information and ensure all individuals are informed and give consent to participate. In our case (Box 6.4), we signed consent forms and agreements with headteachers of schools that were hosting mother experiments.
- Think about equitable benefit sharing. The benefits should be shared equally among all the stakeholders, especially the participating communities. Results must be shared according to the agreements.

- Think about fairness if you plan to evaluate impact at the end of the project and you are allocating experimental and control farmers. Are you fair in selecting the control group? (Chap. 11 on impact evaluation).
- Think about flexibility in handling ethical issues as they arise. You are likely to encounter ethical issues that were not initially anticipated. As a researcher, you must embrace flexibility and go out of your way to ensure that harm is avoided and justice always prevails. These will help build trust and enhance the sustainability of the research outcomes. For example, conflicts among collaborators usually arise, and they must be handled with maturity.
- Think beyond the methods you plan to use. Think about democratic processes.
- Always ensure that anonymity and confidentiality are maintained in meetings in reporting, whether internal or external reports.

According to CGIAR guidelines (CGIAR 2020), researchers conducting participatory research are supposed to actively involve farmers and the public; groups, communities and other stakeholders in the design and execution of research. Additionally, they should support communities through capacity building and effective information exchange channels to enhance the adoption of improved technologies.

Box 6.4: Stakeholders in Our Multidisciplanary Research Projets*

- *Schools*: Headteachers, teachers and students in schools that hosted the experiments.
- *Churches*: Leaders (Pastors) and members (congregation).
- *Farmer groups*: Officials and members.
- Individual farmers who hosted experiments, field days and demonstration events.
- Individual farmers who participated to learn (attended field days and learning events).
- Ministry of Agriculture personnel.
- Private investors, agro-dealers, and agro-machinery dealers.
- Nongovernmental Organizations (NGOs) in the region.
- *Students*: undergraduates and postgraduates.
- *Traders*: buying produce from farmers.
- *Administration*: Chiefs, sub-chiefs, and village leaders.
- Banks: suppliers of credit.
- Agrodealers: supplied inputs to farmers.

**Research projects*

1. "Scaling up soybean and climbing beans using a value chain approach in maize-based systems of central Highlands of Kenya" (2014–2017) funded by Alliance of a Green Revolution in Africa (AGRA)
2. "Integrated Soil Fertility Management for improved Livelihoods" (2012–2015) funded by Association for Strengthening Agricultural Research in Eastern and Central Africa (ASARECA)
3. "Enhancing Productivity and Market Development of Soybeans and Climbing Beans in Central Highlands of Kenya" (2010–2013) funded by The Alliance for a Green Revolution in Africa ((AGRA)
4. "Enhancing farmers' benefits from integrated soil fertility management through improved water use efficiency and better marketing strategies in the highlands of central Kenya" (2008–2013) funded by, The Flemish Interuniversity Council Own Initiative (VLIR—UOS)

6.12 Guidelines in On-Farm Experimentation

The topic of on-farm experiments is discussed in Chap. 10. The experiments take place in the real setting of farms directly involving practitioners/farmers and other stakeholders, e.g. extension officers. Given this unique context, researchers must adhere strictly to ethical guidelines to ensure trust and good collaboration. Pay attention to the following activities:

- Ensure farmers are adequately informed about the objective of the on-farm experiments. Explain in detail the roles of the farmers. Additionally, explain the role of the researcher and other stakeholders, including details of the scheduled visits and activities.
- Ensure farmers are informed about the methods and expected outcomes and that they provide written and signed informed consent. Inform them about the duration of the experiment.
- Respect for farmers' autonomy and their "fundamental right" to make informed decisions about their land and the confidentiality of their data is important; e.g. allow farmers to select part of the land for the experiments.
- Strive to minimize any negative impact on the farmers' operations, for example, crop protection, where hazardous chemicals are being used.
- Respect local customs and practices.

- Be prepared to share findings and benefits with the participating farmers. Sharing results with farmers is courteous and helps farmers benefit, and research may benefit from farmers' feedback (Coe et al. 2003).
- Agree on how to compensate farmers in case of crop failure.
- Respect the time farmers put into managing the experiments. You may choose to compensate them fairly and give attribution accordingly.
- Respect the time and contribution of the other stakeholders. You may choose to compensate them fairly, and recognize their roles.

6.13 Research Involving Animals

Compliance with Regulatory Bodies Before conducting any research involving animals, it is necessary to adhere to the regulations set by local, national and international authorities. These regulations outline standards for the housing, care and treatment of animals. Many universities have established committees, such as the Institutional Animal Use and Care Committee (IAUCC), to oversee animal use and ensure protocols meet ethical and legal standards for animal welfare (Mohan and Huneke 2019).

The Principle of 3R When using animals in our research studies, the internationally established principles of replacement, reduction and refinement (the three Rs) should be taken considered (de Boo et al. 2005):

- Replacement: this term means that alternatives to animal use in scientific research should be sought when possible.
- Reduction: refers to the utilization of methods and practices to obtain comparable or more information with a minimal number of animals.
- Refinement refers to the application of methods and practices that alleviate, prevent, or minimize pain, suffering, distress, or lasting harm to enhance the welfare of the animals used. This encompasses the use of a range of strategies, for instance, anesthesia and putting animals in good housing conditions.

Animal Welfare Researchers must ensure that animals are treated with care, respect, and dignity throughout the research period (de Boo et al. 2005). This includes providing appropriate housing and veterinary care and minimizing any potential distress or harm. At times, researchers may need to use anesthetics and pain relief management techniques.

6.14 Guidelines on Ethical Norms in Data Analysis

Ethical norms in data analysis are important for;

- Maintaining data integrity
- Ensuring well-being and fairness to participants
- Upholding the trustworthiness of research findings
- Data sharing

Maintaining Data Integrity

Maintaining *data integrity* in data analysis is important for data quality and accuracy. This concerns how well your data reflects the reality on the ground. Inaccurate data leads to misleading results and recommendations and negative consequences to your stakeholders. As a researcher, you should analyze your data without manipulation and report findings truthfully. All results, including null, should be reported and avoid selective reporting.

The front-end (data gathering) and back-end (reporting findings) in data analysis are crucial for maintaining the integrity of the data. The front end involves carefully collecting and managing data to ensure that you are analyzing clean and error-free data. Data collection and entry should be well-designed to minimize or avoid errors. This includes using standardized data collection forms and validation checks during data entry. The back end, on the other hand, encompasses reporting findings with transparency and clarity.

Cohen et al. (2017) highlight an important ethical concern in data analysis: manipulation and misrepresentation. These authors give some key areas of misrepresentation of data (Box 6.5). This involves selectively reporting only the data that supports one's hypothesis, which can introduce bias and a lack of transparency in the results. Such practices can mislead both the academic community and the broader public. Therefore, researchers must always avoid this misconduct and follow the highest level of integrity in their work.

Ensuring Well-being and Protection of Participants

Ensure the privacy and confidentiality of participants. This is done by anonymizing or encrypting the data, using codes or pseudonyms (data masking) to remove or protect any identifying information. Ethical guidelines stipulate that participants should not be easily identifiable from the data (Kaiser 2009).

Another important aspect is informed consent and the use of the data. Researchers must respect the boundaries of consent during the data analysis and ensure that they use the data for purposes disclosed to the participants during the data collection phase.

Ownership of the data must be discussed during the data collection phase. Negotiating data ownership is a serious ethical exercise (Cohen et al. 2017). In addition to giving consent for data collection, participants must also be aware of how the data will be used during the research process. This is because the raw data, even after the study, is still the possession of the participants, but once analyzed and interpreted, they become the property of the researcher.

Upholding the Trustworthiness in Data Analysis and Interpretation

Upholding trustworthiness in data analysis is an absolute necessity. It should be a commitment as we advance knowledge. Research must be guarded like a public trust, as it is a privilege and an honour (Palys and Lowman 2002). Researchers must thoroughly document and disclose their data analysis methods, software, and any data transformation. Transparency is key to helping others understand how the results were arrived at. Table 6.1 and Box 6.5 give more information on these aspects. Additionally, researchers must subject data to the correct statistical analysis during data analysis and refrain from misusing statistics. After the analysis, data must be secured from unauthorized access, loss, or theft.

We should exercise caution when dealing with commissioned studies due to the potential for vested interests (Elliott 2013). These studies may prioritize "fine-tuning" the results and the subsequent publication to align with the interests of the commissioning party, which can compromise the objectivity and integrity of the research findings.

Data sharing: Sharing data with other researchers is encouraged to allow them to check on your data. Publishing has an open data policy and researchers are supposed to provide data.

Box 6.5: Ethical Violations to Be Avoided in Data Handling and Analysis

- Using inappropriate data analysis techniques
- Falsifying data and being selective in using the data
- Making false and unsubstantiated claims
- Failing to give adequate say to participants, e.g. in studies involving qualitative research
- Making untruthful claims of causality
- Ignoring outliers
- Over-summarizing data or collapsing data
- Judging rather than analyzing data
- Breaching ethical requirements of presentation
- Giving exaggerated weight to some data
- Excision or removal of items in visual data
- Failing to exert suitable control in data analysis
- Concealing data that does not fit your hypothesis

Table 6.1 Ethical norms in data analysis and how to avoid violating them

Ethical issue	How to avoid violation
Maintain data integrity in analysis	Ensure error-free data during data collection and entry Analyze data without manipulation, e.g., removing part of the data Report finding truthfully: Avoid fabrication and falsification e.g. if a drug or chemical is not effective, the truth should be reported Avoid selective reporting
Data privacy and confidentiality	Obtain informed consent regarding data analysis and ownership Anonymize or encrypt both quantitative and qualitative data Follow data protection principles
Data interpretation and presentation	Use appropriate and transparent data analysis techniques Report data results with honesty and accuracy including any issues of bias
Data storage and security	Use secure data storage and transfer methods Protect data from unauthorized access
Maintain the trustworthiness of data analysis	In qualitative data use established qualitative methods to maintain trustworthiness Use a transparent and systematic approach in statistical analysis Clearly document how the analysis was done, software used and how themes or patterns were derived

6.15 Guidelines in Ethical Norms in Dissemination and Publishing

Publishing is a building block in the development of a coherent and respected nature of knowledge (Springer 2013). The roots of scholarly publishing can be traced to 1665, when Henry Oldenburg published a paper in the Journal of Philosophical Transactions of the Royal Society (Singhal and Kalra 2021). Since then, publishing has grown in importance and remains a significant factor in the sharing and dissemination of research findings. However, the accuracy and impact of the results disseminated rely on honest reporting, which in turn depends on the integrity of the authors, strong institutional research governance and a reliable publishing process by the editors and publishing houses.

Ethical guidelines are in place for *authors*, *editors* and *peer reviewers* to ensure that published work is accurate and trustworthy. These guidelines cover authorship, copyright and patenting policies, data sharing policies and confidentiality rules during peer review. The guidelines exist in many documents, including author guidelines from journals, university research ethics guidelines and internationally recognized guidelines such as COPE (2019). These norms aim to protect intellectual property interests while encouraging collaboration. Key "Do points" are given in Table 6.2.

Table 6.2 Key "Do points" for authors, reviewers, and editors in the publishing process

Audience	Dos
Authors	Honesty and integrity in the research process (conceptualization, research proposal writing, implementation, and reporting
	Avoid data fabrication and falsification; report data as it is
	Avoid plagiarism; including re-use of one's work
	Avoid gift/ghost authorship
	Avoid sending manuscripts to multiple journals at the same time
	Ensure research work is reported
	Ensure research involving human and animal subjects receives ethical clearance and relevant ethical protocols are followed in the field
	Declare any conflict of interest
	Follow the ethical guidelines provided
	Include the source of funding
	Provide evidence of ethical clearance and informed consent
	Sign copyright agreement
	Avoid predatory journals
Peer reviewers	Maintain confidentiality
	Provide well-informed recommendations
	Have objectivity and impartiality
	Maintain transparency in the documentation of recommendation
	Ensure timeliness of reading and giving feedback
	Disclosure of conflict of interest
	Constructive feedback
Editors	Maintain editorial independence
	Maintain confidentiality in handling of manuscript
	Fairness and objectivity in the selection of reviewer
	Managing the peer review process efficiently and timely
	Enforce ethical guidelines
	Protect the anonymity of the peer review (blind review)
	Avoid unfair rejection of a manuscript
	Maintain a transparent editorial process

The *key ethical guidelines in publishing* include the following;

Honesty and integrity: At the core of ethical norms in publishing research findings lie honesty and integrity. Researchers have a duty to present their research findings accurately and truthfully without manipulation or distortion. This includes reporting both positive and negative results.

Data fabrication and falsification: Fabricating and manipulating data or results is severe ethical misconduct. It is referred to as the gravest form of scientific misconduct (Singhal and Kalra 2021). Researchers must maintain the integrity of data collection, analysis and presentation. Data should not be altered to fit preconceived hypotheses or to produce desired results. Transparency and accurate reporting are paramount. The frequency of data fabrication and falsification by researchers is controversial; nevertheless, it exists (Fanelli and Tregenza 2009).

Anonymity, confidentiality, and non-traceability: these are important aspects of protecting research participants' privacy. The question here is whether an external, internal, or local audience will be able to recognize the participants and the institutions that participated in the research, particularly in case of negative results" (Cohen et al. 2017).

Plagiarism: unauthorized use of someone else's works, ideas or intellectual property should be avoided. Researchers must properly attribute sources, including published research and the work of others, through citations and referencing.

Ethical treatment of research subjects: research involving humans and animals must adhere to strict ethical guidelines and be reported in the publication.

Conflict of interest: researchers must disclose financial, commercial, personal, and professional conflicts of interest that could potentially bias the research findings and conclusions. This is most relevant in crop protection research where an agrochemical company funds research. Such information on the source of funding should be disclosed in the publication.

Dual publication: Researchers should avoid dual publication, which involves submitting the same research findings to multiple journals simultaneously. Self-plagiarism, where researchers reuse their previously published work without proper citation norms should also be avoided. Ethical norms encourage re-use and citation of one's work. Closely related to this are identical or matching publications, and sliced publications known as *"salami scheming*" (Singhal and Kalra 2021). Duplicate or redundant publication occurs when a paper overlaps substantially with the already published work, while salami scheming refers to slicing data from one research project into various pieces with the aim of artificially increasing publication volume (Singhal and Kalra 2021).

Peer review: peer review is the cornerstone of research publications and ethical norms require peer reviewers and editors to maintain confidentiality, objectivity, and fairness in the assessments.

Authorship: authors should accurately reflect the contributions of individuals to a research project. All those who made intellectual or practical contributions should be acknowledged authors, while those with minor roles should be recognized and appreciated.

6.16 Guidelines in Other Different Facets of Agricultural Research

This section discusses significant facets of agriculture and their ethical considerations. The most important research ethics in agriculture fall into six categories as illustrated by CGIAR guidelines (Engels et al. 2011; Sichinga-Ligowe et al. 2024). These are:

1. General research ethics dealing with scientific dishonesty, publication, confidentiality, etc.
2. Guidelines on the use of human and animal subjects in research.
3. Guidelines on utilization of genetic resources.
4. Intellectual property rights.
5. Biotechnology.
6. Environmental conservation and sustainability.

The first two categories, i.e. general guidelines and use of human and animal subjects, have been discussed at the beginning of this chapter. Therefore, this section discusses the other categories and also includes the use of Digital agriculture AI technologies, a crosscutting matter in agricultural research and development. It also adds to the above list by CGIAR the following categories:

- Guidelines for supervision of students.
- Guidelines for researchers working in a collaborative project.

6.16.1 Guidelines in the Use of Genetic Resources

International and national agreements and regulations must be followed by researchers when using genetic resources (Kumar 2018; Friso et al. 2020). Ethical research should consider the long-term use and conservation of these resources, which includes measures to prevent over-exploitation, degradation and unintended consequences. This is crucial for the protection of biodiversity and the promotion of sustainability in agricultural systems for future generations.

6.16.2 Guidelines in Use of Intellectual Property Rights

Intellectual property rights revolve around ownership, use and questions arise on who has the right to own and control these intellectual creations (Prasad et al. 2012; Brandl and Glenna 2017). Main questions include:

- Who has the right to own and control intellectual property like patents, copyrights and trademarks?
- What should researchers do when IPR restricts knowledge access, but they have a duty to disseminate their findings?
- Should research findings and inventions resulting from publicly funded research be openly available to the public?
- How can researchers navigate the tension between local intellectual property laws and global ethical norms?
- How do researchers navigate the growing movement toward open science?

In general, there is no definitive response to these inquiries. It essentially involves finding a middle ground between safeguarding the interests of the public who financed the research and safeguarding intellectual property rights (IPR). Moreover, intellectual property rights (IPR) should be treated with caution, and researchers should be knowledgeable about the applicable laws and regulations.

6.16.3 Guidelines in Environmental Conservation and Sustainable Agriculture

The issue of climate change is a significant global challenge and a crucial area of focus in agricultural research. The main focus of research on climate issues revolves around developing technologies for mitigation and adaptation (Mugwe and Otieno 2021) and climate justice (Resnik 2022). Ethical concerns regarding environmental preservation often overlap with those of climate justice. The concept of climate justice entails the fair distribution of the advantages and difficulties of combating climate change, acknowledging that its effects are not uniform across countries and communities. Researchers must contemplate how conservation endeavors can tackle climate-related injustices and assist communities in adapting to the shifting climate (Resnik 2022).

Producing enough food to feed a growing population can be a complex issue that often intersects with concerns about environmental preservation and the sustainability of ecosystems. To make informed and ethical decisions, it is crucial to carefully evaluate and weigh the potential benefits and drawbacks of different approaches. Ultimately, the goal should be to prioritize actions that promote the overall well-being of communities and ensure that our food systems are both productive and sustainable. To achieve this, researchers and policymakers must take a holistic view and consider the long-term implications of their actions. By doing so, we can work toward a more resilient and equitable food system that meets the needs of people and the planet.

6.16.4 Guidelines in the Use of AI Technologies

Artificial intelligence (AI) technologies are smart information systems that can aid farmers in improving their farm efficiency (Russell et al. 2022). The use of these advanced technologies in agricultural research has the potential to revolutionize the global agricultural landscape, as they offer numerous benefits, such as increased yields and resource use optimization and offer "real value" to farmers (Falola et al. 2023). They increase overall efficiency, productivity, and profitability. However, integrating these systems into farming practices raises complex ethical considerations (Dara et al. 2022).

Artificial intelligence involves gathering extensive data from farms, soil, crops, local climate, etc., and employing integration AI methods like machine learning, deep learning and reinforcement to predict and provide farmer recommendations (Ryan 2023). The data are used to build algorithms to make forecasts. For instance, algorithms can be built using weather and historical data to predict yields and image recognition algorithms in robotics used for crop protection. These recommendations help farmers make informed decisions to improve farm operations, such as determining the ideal timing for seeding or harvesting and controlling pests and diseases (Sykuta 2016).

While these advancements offer significant benefits to agricultural productivity, they also raise important ethical issues that must be addressed. Key ethical concerns regarding the use of AI in agricultural practices have been raised in a detailed discussion by Dara et al. (2022) and Gardezi et al. (2022). These authors raise issues of fairness, accountability, transparency, sustainability, privacy, and robustness in the use of AI-based technologies and provide recommendations to agricultural technology providers and policymakers. For example, farmers question how data on their farming practices will be stored and secured and how benefits will be shared between them and the innovators (Wiseman et al. 2019).

Some of the Key Ethical Concerns and Challenges in AI Use Include:

1. *Transparency and accountability*: As AI algorithms and decision-making processes can be complex, it may be difficult for farmers and stakeholders to understand them fully. This raises important questions, such as whether farmers have been provided with sufficient information regarding the use of AI and who is responsible for losses if the system does not work properly due to errors. Additionally, were farmers made aware of how their data were collected and was the process transparent? The accuracy and availability of data have been identified as a major concern (Mark 2019) and farmers have questioned whether all data collected, especially from third parties, was accurate. Poor explainability and a deficiency of model transparency (Dara et al. 2022), exacerbate existing ethical concerns about data ownership and generate distrust in future AI solutions. Researchers and technology developers must be transparent about their methods, data sources and any potential biases in their algorithms.
2. *Data privacy, confidentiality, and ownership*: AI systems collect and analyze vast amounts of data, including farm and climate data, crop information, and personal data from farmers. Questions arise about who owns and has control of farm data (Mark 2019). Researchers must therefore develop strong data security measures and obtain informed consent from farmers when collecting and using their data.
3. *Access and inclusivity*: AI agriculture can transform farming practices, but there is a risk that some farmers, particularly smallholders in low-income nations or those in resource-constrained regions, may need to be included (Zampati 2023). Strategies should be developed to bridge the digital divide and empower marginalized farming communities.
4. *Robustness and biases in data and algorithms*: Biases are likely to occur due to historical inequities or data collection methods, resulting in unfair treatment of farmers. Researchers should include comprehensive bias assessments and mitigation strategies to ensure fair treatment.
5. *Sustainability*: Farmers and the agriculture industry are increasingly prioritizing sustainability, which means that any AI deployment must not only meet but exceed current achievements in sustainability (Ryan 2023). The use of AI systems such as drones, self-driving cars and farming robots carries the risk of potential errors or mismanagement, which could result in the release of harmful chemical fumes and pollution. Improper use of AI in agriculture can cause over- or under-irrigation of soil, emphasizing responsible management. Algorithms replacing farmers' knowledge raise concerns about social sustainability and upheaval in the agricultural industry (Lakshmi and Corbett 2020).

Gardezi et al. (2022) and Zampati (2023) have proposed four ways to enhance farmers' trust in AI:

1. *AI developers* should improve the transparency and explainability of their models. This will help users understand how AI works and build trust in the technology.
2. *AI fairness* should be addressed to ensure equitable and inclusive outcomes. This will create a better partnership between humans and machines. Developing a code of conduct is one way of enhancing equitability.
3. *Responsibility and accountability* for AI decisions should be clearly defined. This will establish a clear chain of responsibility for the impact of AI smart technologies.
4. *Regulatory and voluntary compliance* with data ownership, privacy and security are essential components to

ensure responsible and ethical practices in handling sensitive information.

Addressing the ethical issue highlighted in the use of AI and digital farming requires the collaboration of all the parties involved (Lakshmi and Corbett 2020; Dara et al. 2022). These include Governments, researchers and innovators, farmers, communities, and stakeholders. Policymakers should create policies for farm data use, fund research and development and support education programs. Researchers must strive to fill knowledge gaps in AI, propose new technologies and work on policies for responsible innovation. On the other hand should stay informed about the AI technologies, review data agreements and demand transparency and accountability.

6.16.5 Guidelines for Supervision of Students

One of the key relationships in academics is between the faculty staff and the students they supervise in a research project, which is both educational and professional. Several ethical principles are necessary to uphold the integrity of the supervision process. A basic ethical issue is that the supervisor should maintain high professionalism in their interaction with the students. This professionalism includes principles of fairness, respect, and courtesy with a commitment to promoting an environment free from discrimination and harassment (Drolet et al. 2023).

Authorship practices are another key ethical issue in this relationship (Chap. 25). The supervisor must maintain ethical authorship practices and respect intellectual property rights. For instance, students should be listed as senior authors in papers that result from their thesis work, except in collaborative research projects, where authorship may vary depending on each member's contribution. Supervisors should also help students understand and implement the ethical requirements.

Finally, students must also maintain ethical and moral principles in their supervisor relationships. Honesty, transparency, respect and clear communication are required in all academic activities.

6.16.6 Guidelines for Collaborative Projects

Most agricultural research is collaborative, aiming to enhance benefits for farmers. Donors today's emphasize multidisciplinary collaborative projects (Makate and Mango 2017; Mishra et al. 2024). The projects involve researchers and stakeholders, multiple institutions, and multidisciplinary teams. Collaboration is beneficial. It enhances the generation of innovative solutions by bringing diverse backgrounds together to tackle the complex challenges encountered in agriculture.

Researchers must prioritize ethics in collaborative projects to create a productive working environment. They must follow the following key guidelines;

- Maintain teamwork: key components include effective communication, mutual respect, and shared goals.
- Practice fair authorship: attribute contributions accurately and transparently. Also, ensure they publish quality work that is significant to society.
- Develop data management and ownership guidelines: establish clear data management, documentation, storage and archiving guidelines.
- Embrace mentorship of young researchers: Many collaborative projects host many young researchers and students. The senior researchers should mentor the novice researchers with fairness and respect.

Recommended Further Reading

1. Drolet MJ, Rose-Derouin E, Leblanc JC, et al. (2023) Ethical Issues in Research: Perceptions of Researchers, Research Ethics Board Members and Research Ethics Experts. J Acad Ethics 21:269–292. https://doi.org/10.1007/S10805-022-09455-3/METRICS
2. Sichinga-Ligowe I, Joy EJM, Manzeke-Kangara MG, et al. (2024) An introduction to conducting responsible and reproducible agricultural research. CAB Reviews: Perspectives in Agriculture, Veterinary Science, Nutrition and Natural Resources 19:. https://doi.org/10.1079/CABIREVIEWS.2024.0058/ASSET/19F08C22-4AA3-424B-99DB-F9DA93A15F69/ASSETS/GRAPHIC/FIGURE
3. Sims JM (2010) A brief review of the Belmont report. Dimensions of Critical Care Nursing 29:173–174. https://doi.org/10.1097/DCC.0B013E3181DE9EC5
4. Schmidt U (2004) The Nuremberg Code. Justice at Nuremberg 199–263. https://doi.org/10.1057/9780230505247_7

Key Points in This Chapter

Interconnectedness of Ethical Considerations, Legal and Administrative Issues

The three aspects guiding research: 1. ethical considerations, 2. legal and 3. administrative issues are related and are key in ensuring integrity in research. They enforce acceptable behavior and also help to protect participants.

Brief History and Growth of Research Ethics

The Tuskegee syphilis study and Nazi medical experiments are some of the major research studies that failed ethical standards. In response, the Nuremberg Code and the Belmont Report were established in the twentieth century, setting international ethical standards that form key pillars in guiding research practices.

As technology advances, ethical considerations have evolved to encompass genetics, biotechnology and artificial intelligence.

Professional bodies and Institutional Review Boards (IRBs) have become key in enforcing and monitoring adherence to ethical guidelines. Governments and funding bodies prioritize research ethics training to promote responsible practices.

Overview of Ethical Issues in Agricultural Research

The ethical landscape in agricultural research has lagged behind that of medicine, but has emerged in recent decades to address complex issues such as genetic modification, sustainable farming practices and the rights of human subjects and farm animals.

Research Ethics Vary Among the Stakeholders Involved in Agricultural Research

Agricultural research stakeholders have different priorities in research ethics:

- Scientists and researchers; accurate and credible data generation.
- Farmers' priority: provide accurate data.
- Policymakers: supporting research and governance.
- Funding agencies: fund and respect research findings and adhere to grant agreements.
- General public: research that aligns with societal priorities.
- Governments, universities and research institutions: Governance and policies to support research ethics and ensure staff compliance.

Fundamental Principles in Research Ethics

The key principles of ethical research involving human subjects include informed consent; beneficence; confidentiality; respect for autonomy, justice, honesty, integrity; and social responsibility.

Where Do You Get Ethical and Legal Guidelines for Conducting Agricultural Research?

The key sources to find information on ethics include:

1. Institutional review boards (IRB) in universities and research institutions.
2. International and national ethical codes and guidelines.
3. Professional organizations.
4. Government regulations and funding agencies.
5. Collaborators in other organizations.
6. Research policies of universities and research institutions.
7. Supervisors and research mentors.
8. Documented ethical issues in literature.

The Role and Processes of Institutional Review Ethics Boards

The Institutional Review Boards (IRB) oversee research to ensure its ethical conduct. IRBs conduct thorough reviews of research proposals. Three levels of review are considered: exempt, expedited and full board. After approval, the IRB ensures compliance with ethical guidelines and regulatory requirements.

Steps in Seeking Approval from Institutional Review Boards

To gain approval from your university's IRB, you need to summarize your research procedures and demonstrate how they will protect participants' welfare and rights. The IRB approval process generally involves five main steps:

1. Develop a proposal that meets the requirements.
2. Complete the IRB form.
3. Submit the IRB form and required attachments.
4. Address any IRB feedback promptly.
5. Receive official approval for your research.

Why Do You Need Research Ethical and Legal Guidelines?

1. Publication: manuscripts lacking ethical approval may be rejected.
2. Donor compliance: some funders require adherence to ethical guidelines.
3. High-quality research: ethics promote honesty and prevent manipulation of findings.
4. Protection of rights: ethical norms protect participants and reduce the risk of harm.

5. Values in collaborative work: ethics promote cooperation, trust and fairness.
6. Accountability to the public: ethical practices build confidence in the scientific process.
7. Impartial sharing of benefits: ethics promote fairness and responsible treatment of all involved.

Ethical Issues When Collecting Data from Farmers

1. Get community approval.
2. Obtain informed consent.
3. Consider benefits and burdens.
4. Communicate findings to farmers.
5. Protect anonymity and confidentiality.
6. Respect local culture.
7. Select participants based on study objectives.
8. Be sensitive in offering compensation.

Implications of Changing Paradigms in Agricultural Research for Research Ethics

Ethical guidelines for participatory research include:
1. Build strong relationships with stakeholders, and seek help when needed.
2. Ensure informed consent and share benefits equally among all stakeholders.
3. Maintain fairness in evaluation and be flexible in addressing ethical issues.
4. Follow through on commitments and maintain anonymity and confidentiality.

Key Ethical Guidelines for On-Farm Research

- Clearly inform farmers about the experiments and their roles
- Obtain informed consent from farmers
- Respect farmers' autonomy and confidentiality
- Minimize negative impacts on farmers' operations
- Respect local customs and practices
- Share findings and benefits with participating farmers
- Agree on compensation for crop failure
- Respect the time and contribution of farmers and other stakeholders

Key Ethical Guidelines for Animal Research

The 3Rs principle, replacement, reduction and refinement, should be considered when using animals in research.

Ethical Norms in Data Analysis

1. Maintaining data integrity. Data should be collected and reported truthfully without manipulation.
2. Respecting participants' privacy and confidentiality, obtaining informed consent and being transparent about data usage are important ethical considerations.
3. Upholding trustworthiness in data analysis involves replicability, transparency in analysis methods and secure data storage.
4. Follow the data sharing policy to allow other researchers to validate findings.

Guidelines for Ethical Norms in Publishing

Publishing is crucial for sharing research findings accurately and reliably. Ethical guidelines for authors, editors and peer reviewers cover honesty, data integrity, participant privacy, avoiding plagiarism, ethical treatment of subjects, conflict of interest disclosure and proper authorship attribution.

Guidelines for Other Facets of Agricultural Research

In agricultural research, ethical considerations include general research ethics, guidelines for human and animal subjects, genetic resource use, intellectual property rights, biotechnology and environmental conservation. Researchers must adhere to international and national regulations when using genetic resources, prioritizing long-term conservation and sustainability to safeguard biodiversity.

Guidelines for the Use of Intellectual Property Rights

Intellectual property rights, including patents, copyrights and trademarks, raise important questions about ownership and control. Researchers must balance public interest and intellectual property rights, especially when their work is publicly funded. It is essential to consider applicable laws and regulations carefully.

Guidelines for Environmental Conservation and Sustainable Agriculture

Research focuses on developing technologies for mitigation, adaptation and climate justice to address the challenges of climate change. It is crucial to prioritize actions that promote overall well-being and ensure productive and sustainable food systems for the planet.

Guidelines for the Use of AI Technologies

AI technologies in agriculture raise ethical concerns regarding fairness, accountability and transparency. Enhancing trust through improved transparency, fairness and responsibility is important. Collaboration is essential to address these concerns.

Guidelines for the Supervision of Students' Projects

Supervisors have a role in guiding their students on maintaining high professionalism, ethical authorship practices. Students must also maintain ethical and moral principles in their supervisor relationships.

Guidelines for Collaborative Projects

Ethics are crucial in collaborative research projects. Key components include effective teamwork, communication, and mutual respect. Individual contributions must be accurately attributed, including fair mentorship for budding researchers.

Review Questions

1. Why are ethical, legal and administrative guidelines essential in research? Discuss how these aspects work together to protect research participants and enhance credibility of research?
2. What is the role of ethics in research in emerging technologies such as biotechnology?
3. Explain the role of Institutional Review Boards (IRBs) in research. Why are they referred to as the "watchdogs" of research?
4. What ethical issues are needed in survey research involving farmers?
5. What ethical issues should be considered in qualitative research and research involving participatory rural approaches?
6. What ethical considerations must be taken into account when publishing research findings, and how do these influence the credibility and impact of the research?

Use examples to discuss issues such as authorship, plagiarism, data fabrication, peer review integrity and transparency in reporting results.

7. Why is professionalism and fairness important in the student-supervisor relationship?
8. What practices promote a healthy academic collaboration?

Chapter Exercise

1. Contact the office of Ethical Review of your institution and obtain consent templates.
2. Read about the historical case of the "Tuskegee Syphilis Study" and identify which ethical principles were violated. Suggest how the research could have been conducted ethically.
3. Locate and review the research ethics policy of the university or research institution. Summarize the key guidelines. Identify how it protects research participants and promotes scientific integrity.
4. Develop a checklist of ethical considerations for conducting agricultural research involving human subjects. Include aspects like informed consent, fairness, confidentiality and respect.
5. Write a short essay discussing the ethical implications of using AI in research. Address issues such as data privacy, algorithm bias and researcher accountability.

Answers to Review Questions

- Q1: Reference 6.1.
- Q2: Reference 6.4.
- Q3: Reference 6.9.
- Q4: Reference 6.11.2.
- Q5: Reference 6.11.3 and 6.11.4.
- Q6: Reference 6.15.
- Q7: Reference 6.16.5.
- Q8: Reference 6.16.6.

References

AlFattani A, AlShahrani A, AlBedah N et al (2025) Exploring researchers' perspectives on institutional review boards functions in Saudi Arabia: a survey utilizing the IRB-RAT tool. BMC Med Ethics 26:26. https://doi.org/10.1186/S12910-025-01179-4/FIGURES/2

Algahtani H, Bajunaid M, Shirah B (2018) Unethical human research in the field of neuroscience: a historical review. Neurol Sci 39:829–834. https://doi.org/10.1007/S10072-018-3245-1/METRICS

Allen BR (2001) Thalidomide. Br J Dermatol 144:227–228. https://doi.org/10.1046/J.1365-2133.2001.04120.X

Babyar J (2018) The IRB structure and medical research reform. Clin Transl Med 7:10. https://doi.org/10.1186/s40169-018-0188-3

Barrow JM, Brannan GD, Khandhar PB (2022) Research ethics. StatPearls

Becot FA, Inwood SM, Buchanan EA (2023) Navigating the ethical and methodological dimensions of a farm safety Photovoice project. J Bioeth Inq 20:249–263. https://doi.org/10.1007/S11673-023-10261-8/METRICS

Berghaeuser H, Prass M, Lindner R (2024) Researchers' practice and perception of research ethics and the role of institutional support: insights from a pan-European Researcher Survey. J Acad Ethics:1–22. https://doi.org/10.1007/S10805-024-09590-Z/TABLES/3

Brandl B, Glenna LL (2017) Intellectual property and agricultural science and innovation in Germany and the United States. Sci Technol Hum Values 42:622–656. https://doi.org/10.1177/0162243916675954

Cabral L, Pandey P, Xu X (2022) Epic narratives of the green revolution in Brazil, China, and India. Agric Hum Values 39:249–267. https://doi.org/10.1007/S10460-021-10241-X/TABLES/1

Carey MA (2016) Focus group research. Focus Group Research, pp 1–118. https://doi.org/10.4324/9781315428376/FOCUS-GROUP-RESEARCH-MARTHA-ANN-CAREY-JO-ELLEN-ASBURY/RIGHTS-AND-PERMISSIONS

CGIAR (2020) CGIAR Research Ethics Code

Chodankar D, Bhatt A, Davis S (2025) Declaration of Helsinki 2024: too much too late? Perspect Clin Res 16:1–2. https://doi.org/10.4103/PICR.PICR_218_24

Coe R, Huwe B, Schroth G, Sinclair F (2003) Designing experiments and analyzing data. World Agroforestry Centre (ICRAF)

Cohen L, Manion L, Morrison K (2017) Research Methods in Education. https://doi.org/10.4324/9781315456539/RESEARCH-METHODS-EDUCATION-LOUIS-COHEN-LAWRENCE-MANION-KEITH-MORRISON

Collins FS, Tabak LA (2014) NIH plans to enhance reproducibility. Nature 505:612–613. https://doi.org/10.1038/505612A

COPE (2019) COPE Ethical guidelines for peer reviewers. https://doi.org/10.24318/COPE.2019.1.9

Creswell JW, Plano Clark VL (2011) Designing and conducting mixed methods research, pp 51–99

Dara R, Hazrati Fard SM, Kaur J (2022) Recommendations for ethical and responsible use of artificial intelligence in digital agriculture. Front Artif Intell 5:884192. https://doi.org/10.3389/FRAI.2022.884192/BIBTEX

Davis S (2023) Declaration of Helsinki: can it still serve as a north star for ethics in regulatory trials? Perspect Clin Res 14:1–2. https://doi.org/10.4103/PICR.PICR_247_22

de Boo MJ, Rennie AE, Buchanan-Smith HM, Hendriksen CFM (2005) The interplay between replacement, reduction and refinement: considerations where the Three Rs interact. Anim Welf 14:327–332. https://doi.org/10.1017/S0962728600029651

Drolet MJ, Rose-Derouin E, Leblanc JC et al (2023) Ethical issues in research: perceptions of researchers, research ethics board members and research ethics experts. J Acad Ethics 21:269–292. https://doi.org/10.1007/S10805-022-09455-3/METRICS

Eliazer Nelson ARL, Ravichandran K, Antony U (2019) The impact of the green revolution on indigenous crops of India. J Ethnic Foods 6:1–10. https://doi.org/10.1186/S42779-019-0011-9/TABLES/2

Elliott DB (2013) Industry-funded research bias and conflicts of interest. Ophthalmic Physiol Opt 33:1–2. https://doi.org/10.1111/OPO.12016

Emanuel EJ, Lemmens T, Elliot C (2006) Should society allow research ethics boards to be run as for-profit enterprises? PLoS Med 3:e309. https://doi.org/10.1371/JOURNAL.PMED.0030309

Engels JMM, Dempewolf H, Henson-Apollonio V (2011) Ethical considerations in agro-biodiversity research, collecting, and use. J Agric Environ Ethics 24:107–126. https://doi.org/10.1007/S10806-010-9251-9/METRICS

Falola PB, Adeniyi AE, Madamidola OA et al (2023) Artificial intelligence in agriculture: the potential for efficiency and sustainability, with ethical considerations. In: Exploring ethical dimensions of environmental sustainability and use of AI, pp 307–329. https://doi.org/10.4018/979-8-3693-0892-9.CH015

Fanelli D, Tregenza T (2009) How many scientists fabricate and falsify research? A systematic review and meta-analysis of survey data. PLoS One 4:e5738. https://doi.org/10.1371/JOURNAL.PONE.0005738

FAO (2019) The international code of conduct for the sustainable use and management of fertilizers. https://doi.org/10.4060/CA5253EN

Fons-Martinez J, Murciano-Gamborino C, Diez-Domingo J (2024) Evolution of informed consent in research: from the Hippocratic Oath to the tailored consent. Open Res Europe 4:72. https://doi.org/10.12688/OPENRESEUROPE.17311.1

Friso F, Mendive F, Soffiato M et al (2020) Implementation of Nagoya Protocol on access and benefit-sharing in Peru: implications for researchers. J Ethnopharmacol 259:112885. https://doi.org/10.1016/J.JEP.2020.112885

Gardezi M, Adereti DT, Stock R, Ogunyiola A (2022) In pursuit of responsible innovation for precision agriculture technologies. J Responsible Innov 9:224–247. https://doi.org/10.1080/23299460.2022.2071668

Grady C (2015) Institutional review boards purpose and challenges. Chest 148:1148–1155. https://doi.org/10.1378/chest.15-0706

Halse C, Honey A (2007) Rethinking ethics review as institutional discourse. Qual Inq 13:336–352. https://doi.org/10.1177/1077800406297651

Hubrecht RC, Carter E (2019) The 3Rs and humane experimental technique: implementing change. Animals 9:754. https://doi.org/10.3390/ANI9100754

Husband G (2020) Ethical data collection and recognizing the impact of semi-structured interviews on research respondents. Educ Sci 10:206. https://doi.org/10.3390/EDUCSCI10080206

ILRI (2021) Institute, International Livestock Research (ILRI) animal care and use operational guidelines

Jamrozik E, Selgelid MJ (2021) History of human challenge studies. SpringerBriefs in Ethics, pp 9–23. https://doi.org/10.1007/978-3-030-41480-1_2

Kaiser K (2009) Protecting respondent confidentiality in qualitative research. Qual Health Res 19:1632–1641. https://doi.org/10.1177/1049732309350879

Kaur J, Hazrati Fard SM, Amiri-Zarandi M, Dara R (2022) Protecting farmers' data privacy and confidentiality: recommendations and considerations. Front Sustain Food Syst 6:903230. https://doi.org/10.3389/FSUFS.2022.903230/BIBTEX

Kenyatta University (2023) Policy and operational guidelines for biomedical. Environmental, Social and Behavioural Research

Korthals M (2016) Agricultural ethics. In: Encyclopedia of global bioethics, pp 80–88. https://doi.org/10.1007/978-3-319-09483-0_14

Kretser A, Murphy D, Dwyer J (2017) Scientific integrity resource guide: efforts by federal agencies, foundations, nonprofit organizations, professional societies, and academia in the United States. Crit Rev Food Sci Nutr 57:163–180. https://doi.org/10.1080/10408398.2016.1221794

Kumar DM (2018) The Nagoya protocol: legal protections for genetic resources and ramifications for aquatic science. Limnol Oceanogr Bull 27:31–35. https://doi.org/10.1002/LOB.10235

Lakshmi V, Corbett J (2020) How artificial intelligence improves agricultural productivity and sustainability: a global thematic analysis. In: Proceedings of the 53rd Hawaii international conference on system sciences

Lee YJ, Park JY (2018) Identification of future signal based on the quantitative and qualitative text mining: a case study on ethical issues in artificial intelligence. Qual Quant 52:653–667. https://doi.org/10.1007/S11135-017-0582-8/TABLES/5

Lenton LA, Smith V, Bacon AM et al (2021) Ethical considerations for committees, supervisors and student researchers con-

ducting qualitative research with young people in the United Kingdom. Methods Psychol 5:100050. https://doi.org/10.1016/J.METIP.2021.100050
Makate C, Mango N (2017) Diversity amongst farm households and achievements from multi-stakeholder innovation platform approach: lessons from Balaka Malawi. Agric Food Secur 6:1–15. https://doi.org/10.1186/S40066-017-0115-7/TABLES/3
Mark R (2019) Ethics of using AI and big data in agriculture: the case of a large agriculture multinational. ORBIT J 2:1–27. https://doi.org/10.29297/ORBIT.V2I2.109
Mishra V, Ishdorj A, Tabares Villarreal E, Norton R (2024) Collaboration in agricultural value chains: a scoping review of the evidence from developing countries. J Agribus Dev Emerg Econ. https://doi.org/10.1108/JADEE-12-2023-0311/FULL/PDF
Mohan S, Huneke R (2019) The role of IACUCs in responsible animal research. ILAR J 60:43–49. https://doi.org/10.1093/ILAR/ILZ016
Mugwe J, Otieno EO (2021) Integrated soil fertility management approaches for climate change adaptation, mitigation, and enhanced crop productivity. In: Handbook of climate change management, pp 1–22. https://doi.org/10.1007/978-3-030-22759-3_325-1
Nebeker C, Torous J, Bartlett Ellis RJ (2019) Building the case for actionable ethics in digital health research supported by artificial intelligence. BMC Med 17:1–7. https://doi.org/10.1186/S12916-019-1377-7/TABLES/1
Njue M, Molyneux S, Kombe F et al (2015) Benefits in cash or in kind? A community consultation on types of benefits in Health Research on the Kenyan coast. PLoS One 10:e0127842. https://doi.org/10.1371/JOURNAL.PONE.0127842
Ojulong H (2017) Participatory variety selection for enhanced promotion and adoption of improved finger millet varieties: a case for Singida and Iramba districts in Central Tanzania. Afr J Rural Dev 2:77–93
Palys T, Lowman J (2002) 1. Anticipating law: research methods, ethics, and the law of privilege. Sociol Methodol 32:1–17. https://doi.org/10.1111/1467-9531.00107
Petillion W, Melrose S, Moore SL, Nuttgens S (2017) Graduate students' experiences with research ethics in conducting health research. Res Ethics 13:139–154. https://doi.org/10.1177/1747016116677635
Pietilä A-M, Nurmi S-M, Halkoaho A, Kyngäs H (2020) Qualitative research: ethical considerations. In: The application of content analysis in nursing science research, pp 49–69. https://doi.org/10.1007/978-3-030-30199-6_6
Prasad R, Bagde US, Varma A (2012) An overview of intellectual property rights in relation to agricultural biotechnology. Afr J Biotechnol 11:13476–13752. https://doi.org/10.5897/AJB12.262
Puras D (2017) Human rights and the practice of medicine. Public Health Rev 38:1–5. https://doi.org/10.1186/S40985-017-0054-7/METRICS
Quisumbing AR, Meinzen-Dick R, Raney TL et al (2014) Gender in agriculture: closing the knowledge gap. In: Gender in agriculture: closing the knowledge gap, pp 1–444. https://doi.org/10.1007/978-94-017-8616-4/COVER
Raykov M (2020) Education researchers' perceptions of and experiences with the research ethics application process in Europe and beyond. Eur Educ Res J 19:10–29. https://doi.org/10.1177/1474904119893461/ASSET/D6AF3B81-2D4E-444E-B47E-8C73CA5C4EAC/ASSETS/IMAGES/LARGE/10.1177_1474904119893461-FIG7.JPG
Resnik DB (2021) Standards of evidence for institutional review board decision-making. Account Res 28:428–455. https://doi.org/10.1080/08989621.2020.1855149
Resnik DB (2022) Environmental justice and climate change policies. Bioethics 36:735–741. https://doi.org/10.1111/BIOE.13042
Resnik DB, Smith E, Master Z, Shi M (2020) Survey of equal contributions in biomedical research publications. Account Res 27:115–137. https://doi.org/10.1080/08989621.2020.1722947
Rothenberg R (2004) The inescapable logic of institutional review boards. Ann Epidemiol 14:517–519. https://doi.org/10.1016/j.annepidem.2004.07.089
Russell SJ, Norvig P, Chang M-W et al (2022) Artificial intelligence: a modern approach, p 1166
Ryan M (2023) The social and ethical impacts of artificial intelligence in agriculture: mapping the agricultural AI literature. AI Soc 38:2473–2485. https://doi.org/10.1007/S00146-021-01377-9/TABLES/3
Sanjari M, Bahramnezhad F, Fomani FK et al (2014) Ethical challenges of researchers in qualitative studies: the necessity to develop a specific guideline. J Med Ethics Hist Med 7:14
Schmidt U (2004) The Nuremberg code. In: Justice at Nuremberg, pp 199–263. https://doi.org/10.1057/9780230505247_7
Sichinga-Ligowe I, Joy EJM, Manzeke-Kangara MG et al (2024) An introduction to conducting responsible and reproducible agricultural research. In: CAB reviews: perspectives in agriculture, veterinary science, nutrition and natural resources, p 19. https://doi.org/10.1079/CABIREVIEWS.2024.0058/ASSET/19F08C22-4AA3-424B-99DB-F9DA93A15F69/ASSETS/GRAPHIC/FIGURE
Sims JM (2010) A brief review of the Belmont report. Dimens Crit Care Nurs 29:173–174. https://doi.org/10.1097/DCC.0B013E3181DE9EC5
Singhal S, Kalra BS (2021) Publication ethics: role and responsibility of authors. Indian J Gastroenterol 40:65–71. https://doi.org/10.1007/S12664-020-01129-5/METRICS
Springer (2013) Publishing Ethics for journals A guide for Editors-in-Chief, Associate Editors, and Managing Editors Guide Publishing Ethics
Sykuta ME (2016) Big data in agriculture: property rights, privacy and competition in ag data services. Int Food Agribusiness Manag Rev 19:57–74. https://doi.org/10.22004/AG.ECON.240696
Van Emon JM (2020) Introduction to legal challenges in agriculture, agrochemicals, and agribusiness. ACS Symp Ser 1362. https://doi.org/10.1021/BK-2020-1362.CH001
Weinhardt M (2021) Big data: some ethical concerns for the social sciences. Soc Sci 10:36. https://doi.org/10.3390/SOCSCI10020036
Wiseman L, Sanderson J, Zhang A, Jakku E (2019) Farmers and their data: an examination of farmers' reluctance to share their data through the lens of the laws impacting smart farming. NJAS: Wageningen J Life Sci 90–91:100301. https://doi.org/10.1016/J.NJAS.2019.04.007
Yip C, Han NLR, Sng BL (2016) Legal and ethical issues in research. Indian J Anaesth 60:684–688. https://doi.org/10.4103/0019-5049.190627
Zampati F (2023) Ethical and legal considerations in smart farming: a farmer's perspective. In: Towards responsible plant data linkage: data challenges for agricultural research and development, pp 257–272. https://doi.org/10.1007/978-3-031-13276-6_13/FIGURES/1
Zilberman D, Holland TG, Trilnick I (2018) Agricultural GMOs—what we know and where scientists disagree. Sustainability 10:1514. https://doi.org/10.3390/SU10051514
Zimdahl RL, Holtzer TO (2018) Ethics in agriculture: where are we and where should we be going? J Agric Environ Ethics 31:751–753. https://doi.org/10.1007/S10806-018-9753-4/METRICS

Part III

Research Designs

This lengthy section, comprising 10 Chapters, is devoted to research designs. To fulfil the objectives of your research study, you need to have a plan and procedures for collecting the data. In this topic, you will learn how to design a research study to accurately fulfil your objectives. Common types of research designs in agriculture are discussed; these include experimental, non-experimental and qualitative strategies, as well as on-farm experimentation. Gaining an understanding of the designs is crucial as it allows you to select the most appropriate and economical design for collecting valid information to answer your research questions.

Chapters in Part III

Chapter 7	Overview of Research Designs
Chapter 8	Fundamentals of Experimentation
Chapter 9	Common Experiments and Field Layout
Chapter 10	Farming Systems Research and On-Farm Experimentation
Chapter 11	Research Designs for Impact Evaluation Studies
Chapter 12	Non-experimental Designs-Surveys
Chapter 13	Non-experimental Designs-Qualitative and Participatory Research Approaches
Chapter 14	Common Designs for Qualitative Research
Chapter 15	Mixed Research Designs
Chapter 16	Sampling

Overview of Research Designs

7

Abstract

A research design is the structure that allows you to collect data to answer the research objective/question. It is a requirement before data collection commences. This chapter starts with a presentation on conceptualizing a research design, followed by a classification of research designs. We present several authors' classifications of research designs and show that classification is not universal. Different authors have used different criteria and devised different terminologies. The key criteria used are broad paradigms and structure. In broad paradigms, we present characteristics of the three categories, i.e. quantitative, qualitative, and mixed, while in structure, we isolate experimental and nonexperimental. Using the broad paradigms and structure, we classify the research designs into three: experimental designs, nonexperimental designs (surveys, qualitative, and participatory), and Mixed (integrated). Experimental designs are categorized into three categories: "pre-experimental," "quasi-experimental," and "true experiments," and their characteristics are presented. Research philosophy refers to the set of beliefs or assumptions that guide the researcher's approach to conducting research and understanding the nature of knowledge. Also called "worldview," it influences decisions about a research study's design, methods and interpretation. Four main research philosophies are discussed in this chapter: 1. Positivism, associated with the belief that knowledge can be discovered through observable and measurable phenomena (objectively determined); 2. Interpretivism asserts that reality is subjective and context-dependent on the meanings and interpretations people give to them; 3. Transformative seeks to create positive change and contribute to social transformation, particularly addressing inequities and injustices; and 4. Pragmatism: a more flexible approach that acknowledges the value of both positivist and interpretive perspectives.

Keywords

Broad paradigms · Epistemology · Experimental designs · Interpretivism · Manipulation · Nonexperimental designs · Ontology · Pre-experimental designs · Positivism · Pragmatism · Qualitative research approach · Quantitative approach · Quasi-experimental design · Randomization · Research methods · Research philosophy · Surveys · True experiments · Validity of data

Learning Outcomes

After reading this chapter, you are expected to be able to:

1. Appreciate the meaning of a research design.
2. Determine the features of a good research design.
3. Appreciate the diverse classification of research designs by selected authors.
4. Explain classification based on structure and paradigms.
5. Differentiate between experimental and nonexperimental designs.
6. Distinguish between a true experiment and a quasi-experimental design.
7. Recognize the role of philosophical worldviews in research.
8. Demonstrate an understanding of the four main philosophical worldviews.
9. Choose an appropriate research design for a research project.

7.1 Conceptualizing a Research Design

What Is a Research Design?

After defining your research problem and formulating the "research questions and objectives," the next daunting activity is preparing a research design. Research design is simply a plan or structure that will guide you in systematically collecting, analyzing, and interpreting data to achieve the objec-

J. Njeri Mugwe, S. Runo, *Research Methodology in Agricultural Sciences*, https://doi.org/10.1007/978-981-95-1892-0_7

tives. The nature of this structure is dictated by the objective of the study or the research question the study intends to answer. Research design is mandatory for every research study before the study commences. It helps provide a comprehensive road map outlining the step-by-step execution of a research study. A researcher must therefore seek to acquire sufficient knowledge of research designs and their variants to ensure that a suitable design is used.

Several authors extensively define research design (Kothari 2004; Creswell 2014; Saunders et al. 2019). The definitions provided by these authors can be summarized as follows concerning what a research design is:

- An arrangement of conditions for the collection and analysis of data in a manner that aims to combine relevance to the research purpose with economy in procedure.
- A plan specifying the structure to address the research problem.
- A blueprint indicating how a research study is to be completed.
- A framework for collecting, analyzing, and reporting adopted in a research study.
- A plan that maintains coherence and integrity of the research study together.
- A master plan containing the methods, processes, and strategies for collecting and analyzing information needed in a research study.
- A structure for organizing data collection and analysis in a research study.

These definitions show that the research design is the *foundational framework* for a study, guiding the systematic approach to data collection, analysis, and reporting. It ensures that research is structured to effectively address the research problem and achieve the objectives, providing a coherent and logical path from conception to conclusion.

7.2 Importance of Research Design

The main purpose of the research design is to guide data collection. It translates the research objectives into a research plan or project. Without a design, collecting reliable and valid data will be impossible. It should answer the following questions:

1. What is the focus of the study?
2. What is the study location?
3. What kind of data is needed?
4. What is the source of the data?
5. Over what periods will the study span?
6. What approach will be taken to select a sample?
7. Which methods of data collection will be employed?
8. How will the collected data be analyzed and interpreted?

First and foremost, a research design helps the study *answer the research question* to give the most valid and credible answers. This involves carefully choosing the approaches, strategies, methods, and measures to gather and analyze data. A clearly defined design will specify the variables to measure and the relationships to assess, ensuring that the objectives are effectively met to yield relevant results.

Second, a research design *controls extraneous variables* and ensures the credibility and accuracy of the findings by reducing biases and errors. For example, in an experiment, sources of variance can be controlled by elimination, randomization, or matching. This ensures that the study collects credible data.

Third, research design assists in *optimizing resources* (time, funds, participants, etc.), improving the data collection efficiency and carrying out the research study. By clearly outlining the methodologies and methods, a researcher can plan and allocate resources efficiently. Efficient allocation helps to avoid the wastage of resources and guarantees that the research is conducted with optimum use of available resources.

Fourth, a research design helps to *replicate and apply* research findings effectively. By clearly outlining the methods and procedures used in a study, it allows other researchers to repeat the research to verify its results.

Finally, a research design determines how the data should be *analyzed*. A well-structured design ensures that the chosen analysis techniques are appropriate for the type and nature of the data collected, thereby enabling meaningful and valid interpretation of the findings.

7.3 Components of a Good Research Design

Data is the most important output from a sound research design: a good design should provide objective, reliable, and valid data. Objective data ensures that the findings are unbiased and not influenced by the researcher's feelings or opinions. They are based on empirical evidence. Valid data accurately reflect the phenomena being studied and measure what it is intended to measure. Together, these data qualities are essential for drawing meaningful and credible conclusions and advancing knowledge in the field.

You can think of a good research design as comprising the following main components to allow you to obtain data:

1. Structure and conceptual framework.
2. Population and sampling technique.
3. Location and timing of data collection.
4. Instruments and methodologies of data collection.
5. Approaches to data analysis.

The above five core components are supported by the researcher's *philosophical assumptions, skills and preferences* (Fig. 7.1). The philosophical assumption forms the framework for the research design (Creswell 2009). It influences how the research is framed, conducted, and interpreted. We discuss the philosophical paradigm later in this chapter.

The structure determines whether to choose experimental or nonexperimental. The choice depends on the nature of the research problem. The conceptual framework closely relates to the structure and outlines the concepts, variables, and relationships (Chap. 4). The specific components may vary based on the research paradigm and methodology chosen for the study. Depending on whether the study is biophysical or social, studies may require additional considerations related to the unique characteristics of biological systems and human interactions.

Researchers usually select research designs that align with their skills and preferences. These preferences and skills are often shaped by their past training, background, and experiences. It is essential to choose a research design that matches your expertise, as this will enhance the rigor and effectiveness of your study.

Put in another way, your research design can be equated to an *architectural plan* for building a house, which serves as the foundational map for a research project's success. Just as architects carefully plan a building's layout, dimensions, and materials, researchers must craft a design to organize various components, addressing research questions and hypotheses. Precision and clarity in both plans are crucial for ensuring methodological rigor and achieving meaningful outcomes (Box 7.1).

Box 7.1 Similarities Between an Architectural Design and Research Design

Aspect	Architectural plan	Research design
Plan	Foundational plan guiding the whole building	Foundational plan guiding the research project
Address objectives	Objectives influence the plan; intended use determines the plan. How much space? What is the desired visual impact?	Research problem to be addressed, objectives to be achieved, hypothesis to be tested determine the research design
Organize components	Organizes materials, elevations, roof plan, the various components, floor plans, electricity and lighting, etc. into a comprehensive guide for construction	Organize various components as shown in Fig. 7.1, specifying how they will come together seamlessly
Precision and clarity	Precision and clarity ensure methodological rigor (accurate execution and dimensions, minimizing likelihood of errors) Consistency across various components	Precision and clarity ensure methodological rigor (accurate and reliable measurements minimizing errors) Increase validity and reliability
Collaboration and communication	The detailed architectural plan, acts as a communication tool between the architect, contractors, builders, and other stakeholders	Well-crafted research design facilitates communication among team members (student, supervisors, and collaborators, technicians)

7.4 Categories of Research Designs

Many research designs are documented in research methods books. As a student, you will be exposed to many of these books on research methods, ranging from disciplines in natural sciences to social and behavioral sciences. Generally, education and psychology appear to have relatively more books on research methods than other disciplines. Because research design is core, all these books cover the topic of research designs and consequently present their classifications and types.

The classification and types of designs vary considerably and often can be confusing because we do not seem to have universal categories. For example, Creswell (2009) and Creswell (2014) adopt the classification of quantitative, qualitative, and mixed research designs, while Kothari (2004) adopts research types and designs. On the other hand, some authors define quantitative and qualitative methods and include varied typologies such as strategies. For example, Saunders et al. (2019) use the analogy of "research onion" and describe six layers (from outside of the onion inwards) that represent decisions a researcher makes in selecting a design. Within the research onion, the authors introduce terms such as approaches, strategies, techniques, and procedures. They refer to quantitative and qualitative methods and introduce a time dimension in their framework of the research onion.

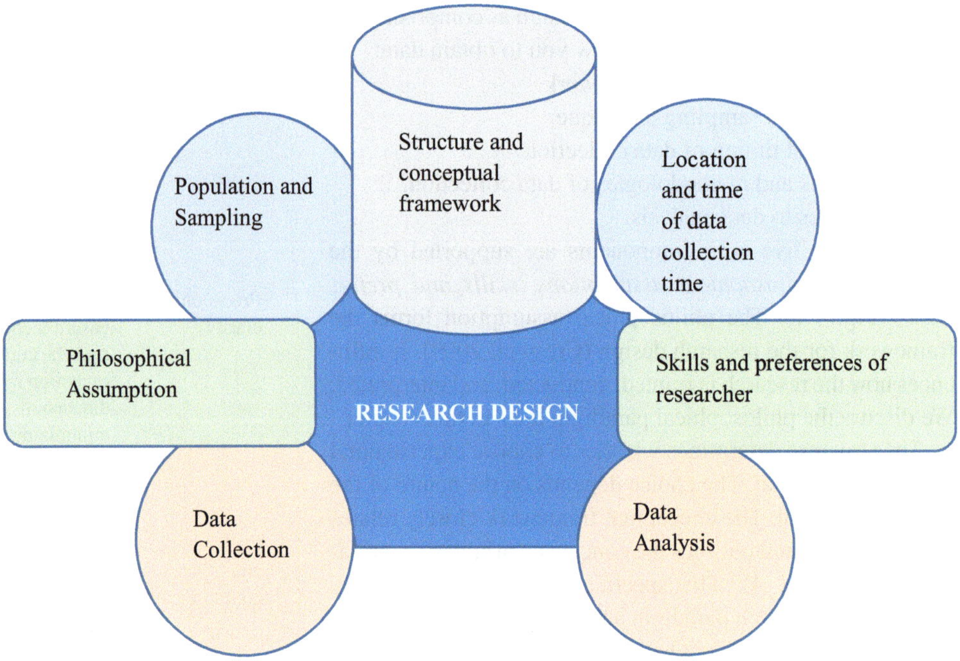

Fig. 7.1 Core components of a research design supported by philosophical assumptions, skills, and preferences of the researcher

In this book, we aim to help you understand these varied classifications by presenting how selected authors have classified and presented research designs (Table 7.1). We then analyze and pick the main types and themes to help craft and suggest a classification most appropriate for agricultural sciences.

As shown in Table 2.1, research designs are described by different authors using diverse ways and classifications, lack a unanimous grouping and are possibly informed by the discipline. The research designs have overlaps and lack clarity in what is defined as a research type and design. Further, the classification and categories therein are not "mutually exclusive." This implies that a research study categorized based on its application can also be categorized based on its objective. From the table, two main domains emerge 1. *Structure and* 2. Broad *paradigms*. The classification by structure gives two categories (Experimental and nonexperimental research designs), while classification by paradigm gives two main categories (Quantitative and Qualitative research designs).

Besides these domains, researchers commonly mix designs, creating an extra category of mixed methods design. In this book, therefore, we classify research designs using the single dimension as well as the mixed dimension.

7.5 Classification Encompassing Structure and Broad Paradigms

A classification encompassing structure and broad paradigms is most suitable in agricultural research. This is because agricultural research has both biophysical and social dimensions. The biophysical realm (mainly crop and animal production) leans more towards true experiments with well-structured experimental designs that are comprehensively explained in biostatistics books (Little and Hills 1978; Gomez and Gomez 1984; Mead et al. 2017). On the other hand, the social dimension involving human aspects (farmers and other stakeholders) requires flexible designs that accommodate experiments with well-defined variables of measurements but also include observational designs.

A suitable classification for agricultural sciences must, therefore, accommodate a biophysical and social dimension. In our classification, we also consider that a research design is an intersection between philosophy, strategies, and the specific methods employed in a research study (Creswell 2014). The final classification of research designs has three broad categories that intersect with philosophical assumptions (Fig. 7.2.).

The three all-encompassing categories are:

- "*Experimental designs*" (true and quasi experiments)-quantitative.
- "*NonExperimental designs*" (both quantitative and qualitative) which are:
  (a) Survey (quantitative).
  (b) Qualitative.
  (c) Participatory action research.
- "*Mixed methods designs*" (experimental, nonexperimental, qualitative, quantitative).

Table 7.1 Classification of research types and designs by selected authors of research methods books

Title and author/s	Criteria	Categories/classification
Research methods quantitative, qualitative, and mixed methods (Mugenda and Mugenda 2019)	Levels of research complexity (seven levels)	Exploration, descriptive, association or relationship, prediction, control, explanation, theory testing
Educational research: competencies for analysis and applications (Gay and Airasian 2012)	Broad paradigms, purpose and method	*Quantitative* Method (survey, correlational, causal-comparative, and experimental) *Purpose* (Basic and applied, evaluation, research and development, action) *Qualitative* Narrative, ethnography, case studies
Research design: qualitative, quantitative, and mixed methods approaches (Creswell 2009)	Broad paradigms	Quantitative Qualitative Mixed
Educational research: planning, conducting, and evaluating quantitative and qualitative research (Creswell and Guetterman 2019)	Broad paradigms	*Broad paradigms* Quantitative Qualitative Mixed *Designs* "Experimental designs" "Correlational designs' "Survey designs" "Grounded theory designs" "Ethnographic designs" "Mixed methods designs" "Actions research designs'
Research methodology: methods and techniques (Kothari 2004)	Research types	*Research types* "Descriptive versus analytical" "Applied versus fundamental' "Quantitative versus qualitative' "Conceptual versus empirical' *Research designs* "Research design in case of exploratory research studies" "Research design in case of descriptive and diagnostic" "Research design in case of hypothesis-testing research studies"
Research methods and data analysis for psychology (Wilson and MacLean 2009)	Research methods	Quantitative and qualitative methods Experimental methods Correlational methods Observational methods Survey methods
Research methodology: a step-by-step guide for beginners (Kumar 2018)	Quantitative and qualitative study designs	*Quantitative study designs* Based on the number of contacts with the study population, "cross-sectional, before-and -after designs, longitudinal" The reference period of the study "retrospective, prospective, retrospective-prospective" The nature of the investigation "experimental, nonexperimental, quasi or semi experimental" Others; "the replicated cross-sectional designs, trend studies, cohort studies, panel studies, blind studies, double blind studies" *Qualitative study designs* Case study Oral study focus group interviews Participants interviews Holistic Action Feminist Participatory

(continued)

Table 7.1 (continued)

Title and author/s	Criteria	Categories/classification
Introduction to research in education (Ary et al. 2009)	Research approaches	*Quantitative research* Experimental Nonexperimental *Qualitative* Basic interpretative studies Case studies Document and content analysis Ethnography Grounded theory Historical Narrative Phenomenology
Research methods (White and McBurney 2013)	Structure	*Nonexperimental* Observational Archival Case-study Survey *Experimental* True experiments: Single factor designs and factorial designs) Single subjects experiments Quasi experiments
How to design and evaluate research in education (Fraenkel et al. 2012)	Broad paradigms	Quantitative research methodologies Experimental True experiments Quasi-experiments Single subject Correlational Causal comparative Survey Qualitative methodologies Content analysis Ethnography Historical Mixed methodologies
Research methods for business students (Saunders et al. 2019)	Broad paradigms and strategies	*Methodological choice*: quantitative, qualitative, and mixed *Nature of research project*: exploratory, descriptive, explanatory, evaluative, and combination *Mixed*: concurrent, sequential exploratory, sequential explanatory, sequential multiple phase *Strategies*: "Experiment, survey, archival and documentary, case study, ethnography, action, grounded research, narrative inquiry"
Introduction to educational research: a critical thinking approach (Newton Suter 2012)	Structure	*Experimental designs*: true experiments, quasi-experimental designs, and single-subject experimental designs *Nonexperimental designs*: causal-comparative designs, correlational research designs, and descriptive research designs *Qualitative designs*: ethnographic designs, case study designs, historical
Business research methods (Schindler 2019)	Structure and paradigm	Qualitative research Observational research Experiments Survey research

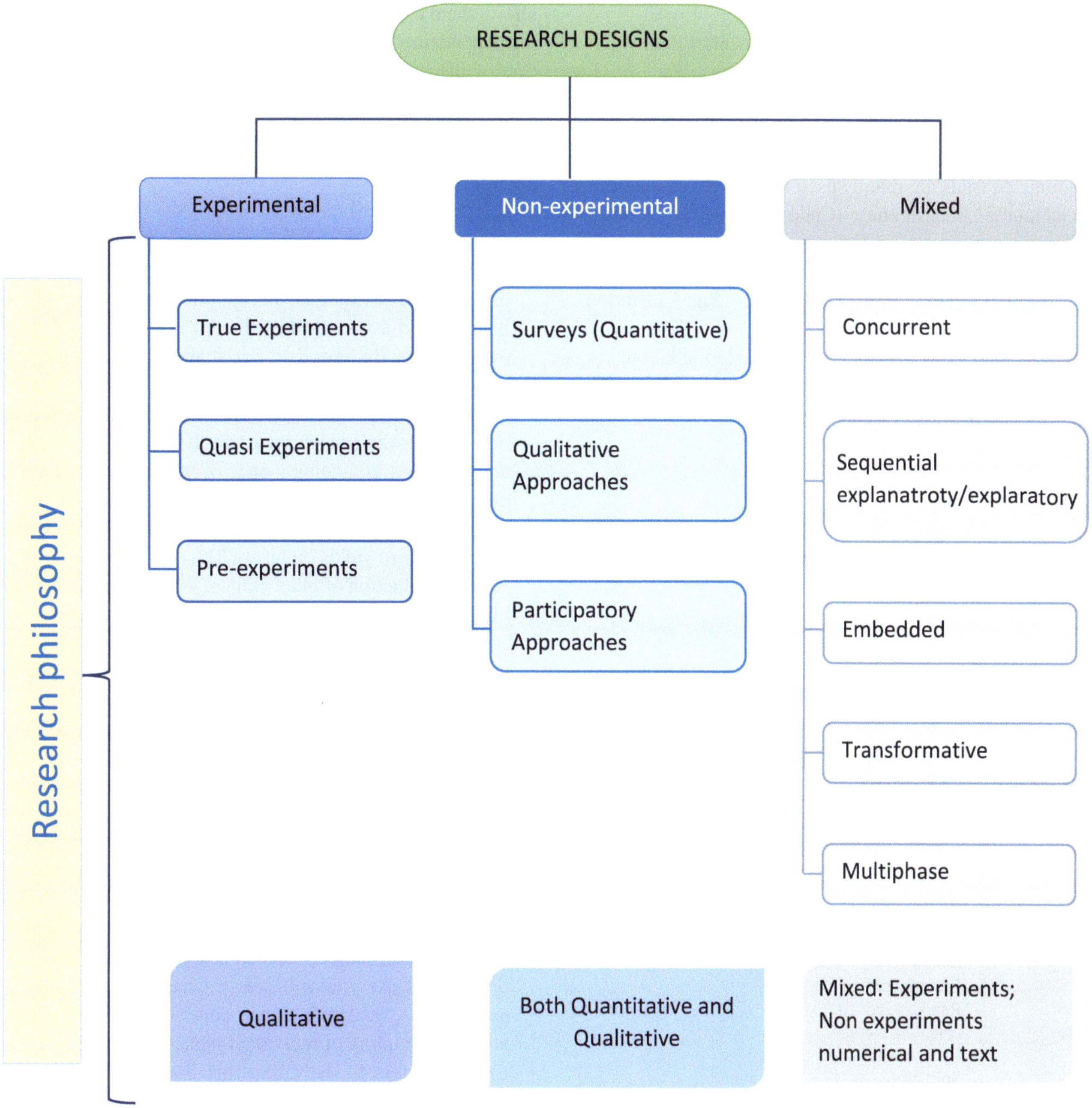

Fig. 7.2 Types of research designs using structure and broad paradigm domains

7.5.1 Experimental Research Designs

All experiments involve the manipulation of the independent variable to assess its effects on a dependent variable (Creswell 2009). They measure cause-and-effect relationships. They are also referred to as "intervention studies" because the researcher intentionally and in a controlled manner manipulates the exposure factor (intervention), to investigate the effects of the changes implemented. This characteristic often makes this type the most rigorous of all research types, known as the "gold standard" of research studies, because of the ability to manipulate the independent variable and control extraneous variables.

Key characteristics of experiments include:

- Measures cause-and-effect relationship,
- Random assignment of treatments or subjects,
- Manipulation of treatment conditions,
- Control of extraneous variables,
- Measures well-defined outcomes,
- Involves group comparisons and
- Possess threats to validity.

Types of Experiments

There are three main types of experiments (Table 7.2): 1. True experiments, 2. Quasi-experiments, and 3. Pre-experiment research designs. True experiments are randomized controlled trials (RCTs), while quasi-experimental and pre-experimental designs are non-randomized. The non-randomized trials are mainly used in program evaluations in social studies in agriculture (Chap. 11). We can summarize the characteristics of the three types of experiments as follows;

1. True experiments = Manipulation + Randomization + Control
2. Quasi experiments = Manipulation + Control-Randomization
3. Pre-experimental = Manipulation – Randomization and Control

True Experiments

In True experimental research designs, researchers have almost complete control over the extraneous variables. This allows them to confidently predict that the observed effect on the dependent variable is solely attributable to the manipulation of the independent variable. Most agricultural experiments in the biophysical sciences are true experiments (Discussed in Chaps. 8 and 9). True experiments have three core characteristics: Manipulation, randomization, and control.

Table 7.2 Summary of types of experiments and their scientific validity

Structure	Types	Characteristics	Scientific validity
True experimental design	Post-test only control design (crop studies) Pretest-post-test control group (soil-measure soil before and after) Completely randomized design Factorial design Randomized complete block Latin square Split plot	Manipulation of the independent variable Presence of control Randomization	High
Quasi-experimental design	Post-test only–randomized Post-test–pretest–non-equivalent group	Manipulation of the independent variable Non randomized Existence of either randomization or control group	Moderate
Pre experimental	One-shot case design One group pre-test and post-test design Static group comparison	Manipulation of the independent variable Limited control over the extraneous variable No randomization No control group	Low

- *Manipulation*: involves the researcher deliberately altering or intervening on one or more variables referred to as treatments. For example, a biophysical researcher may be interested in assessing the effect of using different types of fertilizers (Independent variable) on maize yield per acre. Thus, the independent variable (fertilizers) undergoes specific changes by the researcher and the resulting outcome/results (maize yield) in the dependent variable are measured as a consequence of these changes. On the other hand, a social researcher may be assessing the effect of introducing plant clinics for managing maize necrosis virus on maize yields on farms. The manipulated variable is the introduction of plant clinics, where a group consisting of selected farmers will be exposed to plant clinics. In contrast, another group will not have access to plant clinics (control group). In the end, the researcher measures the effect of the plant clinics on maize yields.
- *Randomization*: In the context of social studies, it pertains to allocating participants (subjects) into various groups, such as experimental and control groups. In a random distribution, each individual has an equal or known probability of being assigned to either group. Random selection ensures that every member of the population has an equal chance of being chosen to be a member of the sample. This procedure helps ensure that there are no major differences between subjects in each group before the intervention begins. As you will see in Chap. 9, under natural experiments, randomization is done within blocks, plots, and also sampling of individual plots.
- *Control*: refers to the group that does not undergo the intervention, meaning that no intervention is applied. The performance of the control group compared to the experimental group is used to assess the outcomes observed in the experimental group. The inclusion of a control group helps prevent alterations in the research results due to external elements.

The main advantages of true experiments are:

1. They are the most useful designs in studies that seek to establish a causal relationship between independent and dependent variables.
2. Provides purity of the observation and, therefore high credibility.
3. Expedite the occurrence of the phenomenon that would otherwise require years to happen naturally.

4. Executed in a laboratory, experimental unit or specialized research setting.

The disadvantages of true experiments are mainly in social research due to the involvement of humans, who have limitations in manipulation and randomization. These include:

1. The controlled settings of true experiments may not reflect real-world conditions, making it difficult to generalize findings beyond the study environment.
2. Ethical constraints in human studies pose challenges to replicating experiments, a task more feasible in the natural sciences.
3. Many aspects of human behavior, thoughts, and emotions lack precise and universally accepted measurements. This is a natural phenomenon due to inherent differences.
4. Studies conducted in communities, e.g., farms, and villages, have difficulties controlling the extraneous variable.
5. Sometimes it is challenging to get an operation for treatment/intervention.

Biophysical researchers employ various types of true experiments, both in laboratory and field settings. These include designs such as "post-test only control design" and "pretest-posttest control group." The specific arrangements can differ; e.g., researchers might use factorial designs or randomized complete block designs depending on the study's objectives and the variables of interest. A significant advantage for biophysical researchers is their ability to easily manipulate, randomize, and control extraneous variables. Throughout the experimentation process, the primary focus is on ensuring high internal validity. For a more detailed discussion, see Chap. 8.

In social sciences, the laboratory is society (people), and therefore, the scenario of using true experiments is a bit different. The common term used is randomized controlled trials (RCTs). Just like the ones in natural sciences, they are most rigorous scientifically for determining the outcome or effectiveness of a certain intervention.

Most studies in nutrition and animal clinical/health studies involving drugs or surgical procedures use RCTS, just like in medicine (clinical trials) (Lim and In 2019). For example, a researcher evaluating the effect of two diets on the weight gain of dairy goats will allocate the diets to two groups whose members are randomly selected. Likewise, to evaluate the effect of a certain drug or diet on animal performance, a researcher will randomly allocate members to two groups and then assign the treatments (drugs). Like the field trials, the random assignment eliminates (or reduces) confounding from known or unknown characteristics i.e. it makes the groups equivalent in terms of their baseline characteristics. Because of this characteristic, animal clinical studies using RCTS are said to be the most credible compared to other types of trials because of high internal validity.

Box 7.2: Case Studies of RCTS in Livestock

Nutrition Trial

- Richards et al. (2016). Randomized controlled trial on impacts of dairy meal feeding interventions on early lactation milk production in smallholder dairy farms of Central Kenya." Preventive veterinary medicine 125 (2016): 46–53.

Health Trial

- Dee et al. (2018). A randomized controlled trial to evaluate the performance of pigs raised in antibiotic-free or conventional production systems following challenge with porcine reproductive and respiratory syndrome virus. PLoS ONE 13(12): e0208430. https://doi.org/10.1371/journal.pone.0208430.

Quasi Experiments

Quasi-experiments, also known as semi-experiments, deviate from the characteristics of a true experiment. They have manipulation of the independent variable but lack either or both random assignment to groups or control groups. The absence of random assignment implies there may be systematic differences between the groups, introducing a degree of uncertainty into the research design. That is why it is known as a "quasi" experiment, as it falls short of the ideal conditions for experimental control and randomization rather than being a true experiment.

A control group, whenever present in quasi-experiments, is commonly called the *comparison* group because it does not meet the randomization requirements and cannot be equivalent to the experimental group. Thus, the results of quasi-experimental research lack the same validity as those obtained in true experiments. This is because the absence of randomization of the subjects in the experimental and control groups cannot guarantee equivalence between the groups at baseline. Control and treatment groups, therefore, differ not only in terms of the experimental treatment they received but also in other factors, including unknown ones. As we see later in Chap. 11, the researcher tries to control for as many of these differences as possible statistically.

Pre-experimental

Pre-experimental experiments are sometimes not considered experiments because the only characteristic they possess is manipulation, and in most cases, there is a limitation as to how much manipulation can be done. There are three main types:

One-group design with post-test only: Involve one group with post-test only. These informal designs are not technically sound. They lack randomization and control. Any researcher using these designs has challenges evaluating the effectiveness of their independent variable. They are not popular in agriculture but can be used in project evaluation studies.

"One group design" (pre-test and post-test): One group is given a pre-test and given a pre-test to find out if the treatments had any effect on the group. Here again, there is no comparison group. Therefore, this design suffers from uncontrolled variation.

Static group comparison: Two groups (one receives the treatment or condition, one does not) are both given post-tests to see if the treatment has any effects. Because no pretest is given, it is indefinite whether there were any group alterations before exposure to the treatment. Chapter 11 discusses research designs for evaluating the impact of interventions from a social perspective.

7.5.2 Nonexperimental Designs

In nonexperimental research, the researcher does not manipulate the independent/predictor variable. There is also no random assignment of treatments or participants to groups. Rather than manipulating the independent variable, the researcher relies on interpretation, observation, or interactions to conclude. This implies that in nonexperimental designs, if you observe a correlation between two variables, it is inappropriate to hastily conclude a cause-and-effect relationship. This is because numerous alternative explanations may exist for the observed relationship, emphasizing the need for careful consideration of the study design and potential confounding factors. In other words, a nonexperimental design cannot demonstrate a true cause-and-effect relationship. In this book, we categorize nonexperimental designs into 1. Surveys, 2. Qualitative approaches and 3. Participatory approaches (Table 7.3).

7.6 Broad Paradigms Relating Quantitative and Qualitative

Quantitative Research

Quantitative research is research that produces numerical and quantifiable data. It is a means for testing theories and assumptions by evaluating relationships between variables. The variables, in turn, can be measured so that numbered data can be analyzed using mathematical and statistical procedures. This is the approach used in experimentation by biophysical researchers and quantitative social studies. Surveys are quantitative methods that involve the collection of data from respondents (discussed in Chap. 12).

Qualitative Research

In contrast, qualitative research is characterized by its descriptive nature, where the data are in the form of statements rather than numerical values. These words are often grouped into categories. Qualitative data aims to gather in-depth insights on topics and is more of an exploratory study. Data can be collected through direct observation, participating in the activities of a particular population (interaction), and through interviews, which are in-person interactions between the researcher and the subjects. For example, focus group discussions to get views of farmers in regard to the adoption of a new technology. Qualitative designs are discussed in Chaps. 13, 14, and 19.

Types of Approaches in Qualitative Designs

Five main qualitative approaches are common in agriculture: These include case studies, Ethnography, Grounded theory, Narrative, and Phenomenology. Detailed discussion is in Chap. 13. Here we give a brief description.

Case study: An in-depth examination of individual units, organizations, groups, or phenomena selected for their distinctive characteristics. The in-depth examination involves the exploration of all peculiarities in its real-life context.

Ethnography: a study involving human society and culture. Involves expensive fieldwork to collect data on culture (beliefs, traditions, experiences, practices, behavior of individuals or communities.

Grounded theory: derives a theory from the gathered data; thus, the theory is "Grounded" in the data. The method includes comparing the gathered sets of data against one another until categories and propositions that state relations between these categories develop. These hypotheses or propositions are tentative and not tested in the study.

Narrative: involves the use of stories and life narratives. The stories are the data with the aim of unearthing meaning from their transcriptions. The most common forms of narratives are "biographic, psychological and discourse analysis" (Creswell 2014).

Phenomenology: The central question explored is the "essence" (nature) of the "lived experience" of the phenomenon of the person (Creswell 2014). A phenomenon can be an event, experience, or situation. The approach comprises temporarily putting aside or "bracketing" personal attitudes and beliefs regarding the phenomenon.

Table 7.3 Classification encompassing structure and broad paradigms

Experimental (quantitative)	Non-experimental			Mixed
True experiments (randomized, manipulated, local control) Completely randomized design Randomized complete block Factorial Split plot Latin square	**Surveys (Quantitative)**	**Qualitative**	**Participatory (Participatory research)**	**Experiments, non-experiments, quantitative, qualitative**
	Cross-sectional	Case study	Focus group discussions	Triangulation
Pre-experimental One-shot case study One group pre-test and post-test	Longitudinal (Trend, cohort and panel	Phenomenology	Venn diagrams	Exploratory
	Descriptive	Ethnography	Seasonal calendars	Embedded
	Correlational	Grounded theory	Resource maps	
	Causal-comparative	Narrative/historical	Transect walks/ maps	
Quasi experiments	Baseline		Timelines	**Explanatory**
	Impact assessment		Pairwise ranking	
	Epidemiological (Case-control studies, cohort)	**Observational (social studies)**	Ranking and scoring	
			Key informant interviews	
			Daily activity calendars	
			Wealth ranking	
			Resource flow maps	
			Constraints analysis	

Participatory Research

Participatory research involves researchers collaborating with farmers, local communities, and other stakeholders in the research process. The process acknowledges the importance of local knowledge and enhances ownership among the participants. Farmers actively collect data, contributing to a more comprehensive understanding of agricultural systems. Effective communication and ongoing dialogue are central to this process, facilitating mutual learning and understanding. Participatory research is widely acknowledged for improving the relevance and impact of agricultural studies by integrating the insights of those directly engaged in farming. Discussed in Chap. 14.

Mixed Methods Designs

Mixed methods research is a method of inquiry that integrates both quantitative and qualitative forms. Both methods are used in tandem so that the overall strength of a study is greater than either qualitative or quantitative research (Creswell 2014). An example is using questionnaires to collect quantitative data to determine relationships of selected variables and complement such data with focus group discussions or Participatory Rural Appraisal (PRA) approaches.

Key qualities of mixed designs include:

- Broad research questions to accommodate both quantitative and qualitative approaches.
- Quantitative and qualitative data are gathered and integrated to gain a thorough insight into the research question.

Classes of Mixed Designs

There are three basic classes:

Concurrent mixed designs: Researchers combine quantitative and qualitative to offer a comprehensive inquiry into the research problem.

Sequential mixed design: The researcher seeks to build on the findings of one method using the other methods. This may involve beginning with a quantitative method for testing the hypothesis, followed by an exploratory qualitative to explain the results of the quantitative or vice-versa.

Transformative mixed design: Researchers seek to transform society. The transformative aspect emphasizes a commitment to social change and seeks to go beyond mere observation and measurements to actively contribute to societal improvement. This design often involves a cyclical process of collecting, analyzing and interpreting data to inform transformative actions. A detailed discussion of mixed research designs is in Chap. 15.

Table 7.3 shows the three categories of designs according to structure and broad paradigms: Experimental, nonexperimental, and mixed and also includes the various paradigms, approaches, and types within these three categories.

7.7 Research Philosophies in the Selection of Research Design

7.7.1 What Is a Philosophy?

Before defining research philosophy, let us first look at the meaning of the term philosophy. According to the Oxford Dictionary, philosophy is a set or system of beliefs resulting from the search for knowledge about life and the universe. In everyday living, we develop personal philosophies containing beliefs, values, and principles that guide our behavior. For example, a personal philosophy is that "Good people are not always moral people," or "the only way to do great work is to love what you do" Steve Jobs. These beliefs and assumptions are usually shaped by culture, family values, religious beliefs, and life experiences. The assumptions shape how we view the world and, consequently, the decisions we make and how we behave.

Philosophical thinking is found in all areas of life as we question the truth. It is found in all areas of study, e.g. arts, humanities, etc. In the Biblical accounts, Jesus is described as a philosopher who used vivid and figurative language, such as parables, to enable individuals to perceive the world differently. In the New Testament, Jesus used many parables to explain the philosophy of growth; for example, "*The Parable of the mustard seed*" (Matthew 13:31–32, Mark 4:30–32, Luke 13:18–19) and "*The Parable of the growing seed*" (Mark 4:26–29). In many instances, too, and especially in the proverbs, the bible contrasts the two ways of the wise person and the fool by showing their varying outcomes of thriving versus destruction. These philosophical dimensions are expressed to guide living by humankind.

In actuality, the search for the truth raises many philosophical questions and arguments. Most times, philosophy does not have a body of consensus. Yet philosophy is important because it involves thinking and reasoning and, therefore, enhances problem-solving capabilities. It enhances our capacity to structure for addressing inquiries and distils crucial information from extensive amounts of data. This is what guides our decision-making processes in our everyday lives. You could think of decisions you make every day; for example, why did you choose to join university X and not Y, or you choose program A and not B? You were guided by some philosophical thinking. Our research, too, is guided by our *beliefs, assumptions* and *understanding* that we have about how the world operates.

7.7.2 What Is a Research Philosophy?

Research philosophy is defined as a framework that directs research by providing guidance based on assumptions about nature, reality, and knowledge (Collis and Hussey 2022). These beliefs and assumptions are also referred to as worldviews or philosophical foundations (Creswell 2014). They are important because they govern the process of investigation and guide the methods by which a researcher will collect and analyze data to explain a phenomenon. We believe that every research has a philosophy behind it. According to Creswell (2014), all research has a philosophical assumption, and researchers should be aware of their assumptions about gaining knowledge that their study seeks to realize.

The methodologies, "methods of data collection," analysis techniques and interpretation of findings, the research adopts to answer the research question, are influenced by the research philosophy.

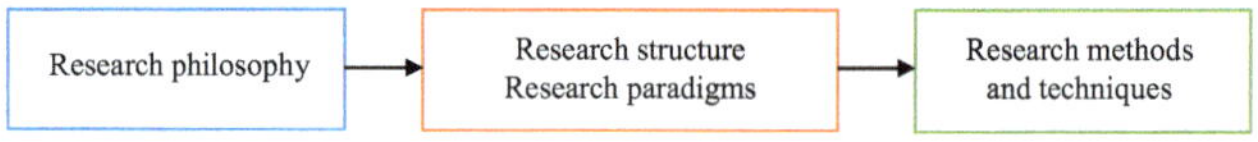

Despite their foundational importance, philosophical assumptions are often overlooked in research writing (Creswell 2014). As a result, these assumptions tend to remain implicit and receive little attention in scholarly papers and books. Researchers frequently conduct studies without explicitly addressing their ontological, epistemological, or axiological positions, allowing these perspectives to remain hidden beneath the surface of the research and the theories it generates. However, articulating these assumptions is essential, as it enhances transparency and allows for meaningful comparison across studies. Furthermore, a clearly stated and credible research philosophy contributes to the development of a coherent and methodologically sound research design (Saunders et al. 2019).

Another reason why including philosophical assumptions in your research is essential relates to the clarity it brings to the research process. Questions about research philosophies frequently arise in seminars and scientific meetings, particularly among PhD students in the social sciences. It is important for students to identify and articulate their research philosophy clearly. This understanding is vital, as it enables students to design cohesive research projects where all elements work together effectively.

7.8 Components of the Philosophical Assumptions

The philosophical assumptions in research are about three components:
1. "Ontology" (the nature of reality): concerned about our assumptions regarding the nature of things.
2. "Epistemology" (How the nature of knowledge is conceived) and
3. "Axiology" (role and the nature of values).

Ontology is concerned with the "nature of reality." Here we are asking, "What is the world?" (Crotty et al. 2020). It is the "study of being" or reality. It pertains to the actual existence of phenomena in the world from which humans can attain knowledge. It concerns assumptions researchers have regarding the way the world operates. Answers the question" What is reality? What is the content or topic to be examined in the study? What exists and "what is the nature of the world"(Tombs and Pugsley 2020). Ontological questions consist of the following (Mertens 2010, 2014; Kivunja and Kuyini 2017):

- Is there a reality out there, or is it a construction?
- What exists?
- What is the nature of reality?
- What does it consist of?
- What are its qualities?
- Is reality objective or is it a product of individual perception?
- What is the nature of the situation under investigation?

Epistemology delves into the examination of knowledge, focusing on what we consider as "valid and legitimate" forms of knowledge (Kivunja and Kuyini 2017). It encompasses inquiries into the validity, scope, and methods employed in acquiring knowledge. Key questions in epistemology include defining "what qualifies as a valid knowledge claim"? And exploring the processes of knowledge acquisition and production; How can knowledge be acquired." Examining the dynamic between the researcher and the researched subject allows us to delve into the concept of epistemology and understand its impact on the design of research. Epistemology questions are as follows (Mertens 2010, 2014; Kivunja and Kuyini 2017):

- How can we know how the world is?
- How can we know it?
- To what extent can any given thing be known?
- How do we know the truth?
- What defines knowledge and the connection between the one seeking knowledge (Knower) and the object of that knowledge (would be known)?
- What counts as true knowledge?

Axiology refers to the nature and role of values (what is worth/good) within the research process (Alele and Malau-Aduli 2023). In other words, axiological assumptions concern themselves with how researcher deal with their values and those of their subjects. Questions of ethics, behavior of the scientist and bias are part of axiology. Methods of data collection adopted by the researcher speak to the value the researcher holds regarding their interactions with the subjects. For example, using a questionnaire may mean the researcher does not value personal interaction, whereas the use of in-depth interviews indicates that the researcher values personal interaction.

Axiological questions are as follows (Mertens 2010, 2014; Kivunja and Kuyini 2017):

- What characterizes ethics and ethical behavior?
- What action should be taken to uphold the rights of all participants rights?
- What values will guide your conduct during research?
- What are the moral considerations and attributes that need to be considered?
- How will I secure the goodwill of participants?

7.9 Philosophical Worldviews

Based on the assumptions regarding ontology, epistemology, and axiology, four worldviews are recognized. These are:
1. Positivism or postpositivism
2. Interpretivism or constructivism
3. Transformative
4. Pragmatism

The researchers' ontological, epistemological, and axiological stances shape decisions regarding the selection, collection, and analysis of data, as well as the interpretation of meanings derived from the data and the degree of generalizability of research findings (Fig. 7.3).

7.9.1 Positivism

The positivist tradition is the domain of natural sciences (e.g. chemistry, biology, physics) that generates scientific knowledge based on *empirical observations*. In natural sciences,

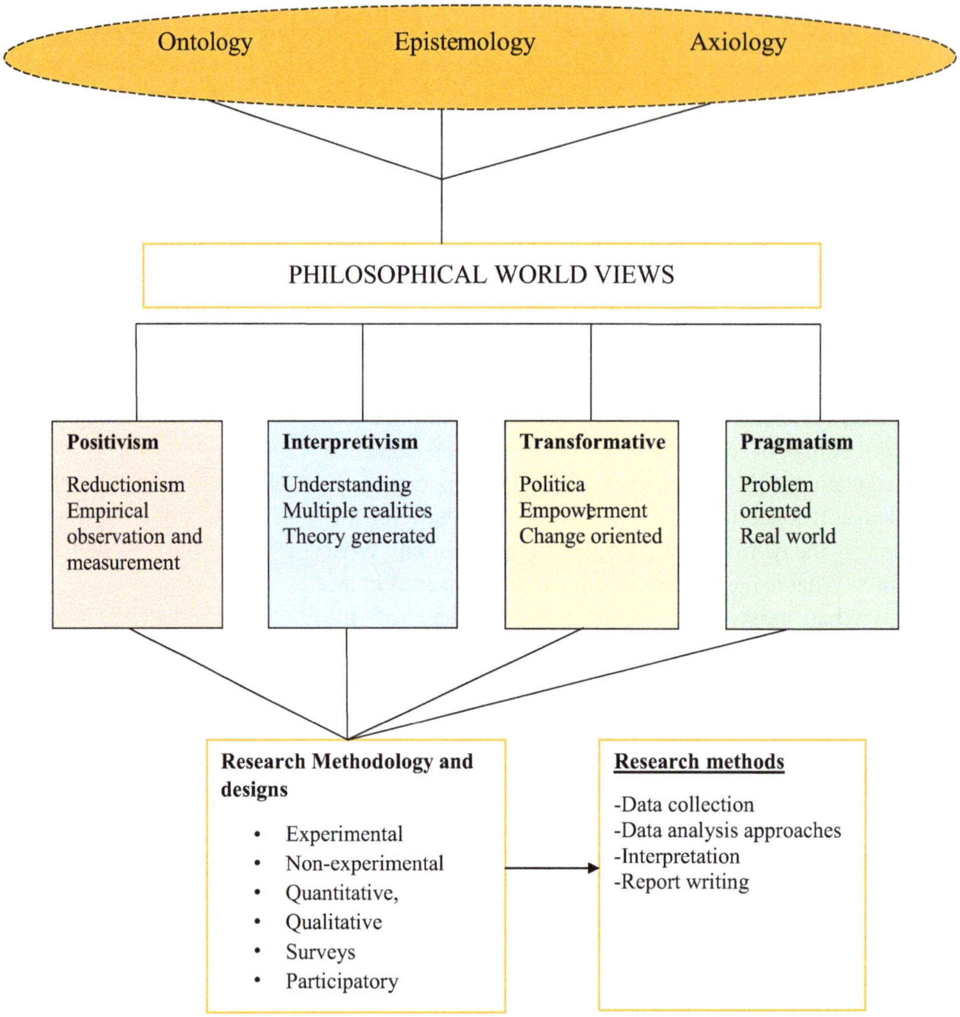

Fig. 7.3 Ontology, epistemology, and axiology shape the world views giving rise to four philosophical views (Positivism, interpretivism, transformative, and pragmatism). These philosophical views influence the research methodologies and methods employed in a research study

data is derived systematically through repeated observations. The natural sciences operate on the premise of a singular, objective reality that can be discovered through experimentation. When applied to social sciences, this paradigm encompasses certain assumptions that characterize the positivist paradigm. The field of agriculture embraces both the natural science approach (e.g. biotechnology research) as well as the social science (e.g. household food security studies).

Ontological Assumption In the positivist paradigm, the fundamental ontological assumption posits a "single objective reality" that the research wishes to depict (Park et al. 2020). This reality is perceived to exist autonomously, irrespective of the observer's perspective. This reality is "out there, and we can go out and find it. This reality can be discovered through scientific research methods. It can be identified and measured. For example, measuring the age or attitude of a farmer. The concepts we use to understand things exist independently of us. This is done by breaking complex phenomena into small elements called *constructs or variables* that are studied through scientific research methods. Positivists perceive the world as being real, like physical objects and natural phenomena. In this view, researchers employ universal laws and theories to explain and forecast behavior and events in the world. This is exemplified by the testing of a hypothesis to determine statistical significance and the level of confidence one can place in the results obtained.

Epistemology Positivists do not acknowledge the influence of the investigator on the research process. They assume that "reality" or truth exists externally or independently of the human mind. This truth is certain and accurate. To accurately develop this truth, knowledge must be developed objectively without interference from the researcher or participant. During data collection, a clear separation must be maintained between research participants and the researcher.

Axiological Assumption Positivists do not permit researchers to introduce their bias into the research process. This framework heavily emphasizes objectivity and, therefore, disregards the significance of the individual's subjective experiences and values of both research participants and the researcher. This requires the researcher to maintain a distance to avoid any interaction with the subjects. This is easily done in physical sciences but a bit challenging in social sciences, where strategies to avoid interference need to be put in place. In the case of data collection using a questionnaire, the researcher maintains a separation from the respondent (Chap. 18).

Methodology
This framework is used by researchers who use quantitative research. Also called scientific realism. Positivism aligns with the *hypothetical deductive method of science*, which follows a circular process. This process begins with the formulation of theory from existing literature, leading to the creation of testable hypotheses. Subsequently, an experiment is designed by identifying variables for manipulation and measurements, incorporating group assignments. The empirical study and experimentation are then carried out, with the results being used to inform theory and contribute to the existing literature. This completes the circular process of the hypothetical deductive method (theory, hypothesis, measuring variables, experimentation, theory).

Positivists adhere to the principle that the only way to examine the world is through a study employing experimentation (Tombs and Pugsley 2020). Positivism incorporates the following principles (Gay and Airasian 2012; Robson and McCartan 2016; Parker et al. 2020):

1. Both social and natural sciences should emphasize the discovery of laws that facilitate explanation and prediction. Believes in a deterministic philosophy in which events or occurrences bring about certain outcomes, thereby suggesting the existence of laws and theories that govern the world.
2. Social and natural sciences should employ methodology based on the hypothetical-deductive model involving the progression from theory and hypothesis to the operationalization of variables, and experimentation).
3. The basic laws of nature are formulated through the replication of the synthesis of scientific discoveries and theories.
4. The primary purpose of science is to develop universal causal laws that can explain and predict phenomena.
5. All scientific propositions are grounded on facts and hypotheses are rigorously tested against empirical evidence.
6. Science predominantly relies on quantitative data, utilizing strict rules and procedures that set it apart from common knowledge.
7. Laws of nature are derived from empirical data, emphasizing the importance of observable and measurable characteristics.
8. Science according to positivism is considered value-free, aiming for objectivity and neutrality in the pursuit of knowledge.
9. Research with positivism operates in a dualistic and objective world where researchers minimize interaction with the study participants to reduce potential bias.
10. Positivism favors larger samples over small samples, as findings can be generalized.

7.9.2 Interpretivism

Interpretivism is also referred to as "subjectivism, social constructivism or constructivism" (Pervin and Mokhtar 2022).

Ontological assumption: Interpretivism is referred to as the *constructivism paradigm* because researchers embracing this belief are convinced that reality is socially constructed. Therefore, the social worlds or social entities cannot be studied as natural phenomena. People construct subjective meanings of their experiences toward certain objects, things, and phenomena. These meanings exhibit diversity and multiplicity, prompting the researcher to explore the complexities of these perspectives rather than simplifying them into a limited set of categories and concepts. Researchers operating within this paradigm, therefore, argue against the use of universal laws applied to everybody because rich insights of individuals are lost as each person develops their views of their world.

Epistemology assumption: Interpretivism is also called *subjectivism epistemology*; it challenges the notion that an objective "truth" exists awaiting discovery and asserts that the truth emerges through engagement with our world. Contends that interaction between the researcher and the researched is essential, enabling the researcher to perceive the world from the view participant's perspective. This approach requires data collection methods that bring them closer such as observation, in-depth interviews, pictures etc.

Axiological assumption: The inquiry in this framework is *value-laden* signifying that the research process is influenced by the researcher and the context under study. The researcher and the subject maintain a close interaction as the researcher dialogues, intermingles, questions, and records the information. However, as they interact, the researcher is expected to adhere to ethical principles.

Methodology
Qualitative methods predominate in the interpretivism paradigm. The researcher and observer become locked into an interactive process as the researcher interviews the participant. The questions captured in the data collection tool are broad and general to allow the participant to construct the meaning of a situation. During data collection, an effort is made to understand the participant's viewpoint

(subject being observed). The research is a product of researchers' values and cannot be independent of them. A researcher background shapes their interpretation as their interpretation is influenced by personal, cultural, and historical experiences. Effort is placed on understanding the subjects observed and their interpretation of the world around them.

Positivism and interpretivism are the two major contrasting research paradigms and are compared in Table 7.4.

7.9.3 Transformative Paradigm

Transformative is the third research paradigm that originated from a group of researchers who are concerned with the *marginalized and oppressed* (Mertens 2014). It emerged during the 1980s and 1990s as a response to the perceived limitations of positivists' assumptions imposing structural laws and theories that did not adequately address the marginalized individuals in our society or issues of social justice (Creswell 2009). This is very much like the paradigm proposed by Chambers (1986), Farmer First, which aimed to increase benefits to marginalized farmers who were not benefiting from production-oriented research (Chap. 10-Farming systems research). This paradigm places emphasis on *participatory or emancipatory*, and therefore, some authors refer to this paradigm as participatory or emancipatory (Noel 2016).

Table 7.4 Differences between positivism and interpretivism research paradigms

Positivism	Interpretivism
Ontological assumption Reality is "objective, ordered and governed" by natural laws that can be realized through experience	*Ontological assumption* Reality is "internally experienced and socially constructed" through interaction and interpretation and is based on the definition people attach to it
Epistemological assumption Knowledge exists to inform us of natural laws; can be discovered and explained and it is possible to control and predict their occurrence	*Epistemological assumption* Knowledge is understood in the subjective meanings people attach to physical and social objects, and the actions they take with them in a social context
Methodologies Experiments Surveys Randomized controlled trials Quasi experiments Structured interviews	*Methodologies* Unstructured interviews Ethnography Observation Focus groups Case study Phenomenology Action research
Data analysis approaches Deductive Exploring relationships (correlations and regression) Comparing groups (T-test, ANOVA) Modelling	*Data analysis approaches* Inductive Thematic analysis Discourse analysis Grounded theory Template analysis
Interpretation and reporting Objective reporting Summary statistical data (descriptive and inferential Statistics) and p-value Generalizable	*Interpretation and reporting* Subjective reporting Contextualization Main analytical findings Illustrative quotes Less generalizable

Source: Tombs and Pugsley (2020), Mertens (2010), and Creswell (2009)

Ontological Assumption Just like interpretivism, beliefs in multiple realities. However, it stresses acceptance of such differences in perceptions as equally legitimate. The transformative is grounded in the acknowledgement that injustice and inequality pervade society, coupled with the belief that research and evaluation serve as crucial tools for addressing these systemic issues. Mertens (2010) articulates this paradigm by emphasizing the explicit role that research and evaluation can and should play in recognizing and mitigating discrimination and marginalization based on various factors such as race, ethnicity, religion, gender, sexual orientation, socioeconomic status, age, and disability. The transformative paradigm serves as a comprehensive framework that encompasses a range of theoretical perspectives, addressing the concerns of specific populations highlighted by feminist, Indigenous, queer, disability rights, and critical race scholars.

Epistemological Assumption

The transformative paradigm epistemological assumption centers on a close relationship between the researcher, the "knower," and the participant, the "would-be-known" or the marginalized. The critics of positivism and interpretivism pointed out that these two stances failed to provide adequate action advocacy to help marginalized peoples and communities. Thus the research within the transformative stance must incorporate an *action agenda* for reform (transformation) that can change the livelihoods of the participants and the researcher's life. In the case study presented (Box 7.3.), all those who participated in the research experienced a change. They all increased their knowledge and understanding of the climate change risks and vulnerabilities and adaptable options to mitigate the effects. As a result of increased knowledge and awareness, farmers changed their farming systems resulting in increased agricultural productivity.

Axiology

The transformative paradigm places great emphasis and priority on human rights and social justice. By its nature, this paradigm involves working with people and communities, and therefore, the rights and welfare of the research participants are paramount. One of the basic tenets of the transformative paradigm is the greater involvement of research participants.

Box 7.3: Case Study: Transformative Participatory Research Study (The Mitigating Role of Climate-Smart Villages to the Impacts of COVID-19 Pandemic in the Myanmar Rural Communities)

Research problem: The effects of climate change on agriculture are largely location-specific. Smallholder farmers in the dry zone of Myanmar are highly vulnerable to the ravages of climate change, with farmers reporting the negative impact of climate change on agriculture. Their production processes are exposed to shocks and stresses of climate change and variability and compounded by degradation of ecosystem services, insecure land tenure, and inadequate health and food security. Men and women are vulnerable as they tend to be affected differently by climate change shocks. In addition, some smallholder farmers are marginalized because they are almost landless.

Research objective: There were two main objectives: 1. To empower vulnerable farmers living within climate-smart villages through testing and identifying adaptable options for creating resilience against climate change 2. To evaluate farmers' experiences and perceptions from three-year participatory action research on a community-based adaptation project of four agro-ecologies of Myanmar.

Research questions included 1. What are the key climate risks and vulnerabilities associated with smallholder farmers in Myanmar? What are the adaptive options that match the identified climate risks and vulnerabilities? How do the farmers test the adaptation options? What are farmers' experiences as they test the adaptation options? What are key learning outcomes among the farmers and other collaborators? What changes occur among the participating farmers in embracing the tested options?

Methodology: The study used a transformative research paradigm which aimed to empower the vulnerable smallholder farmers suffering from climate change, drought, and low-crop yields. Based on the participatory action research process, the research followed a three-step process of participatory assessment and implementation; the first step involved *Participatory climate risk and vulnerability assessment*, which was carried out for 2–3 days per village. The Process utilized tools such as, community mapping, seasonal calendar, timelines, problem tree analysis and focus group discussions. Step 2 was *identification of adaptive options* to the identified climate-induced risk and vulnerabilities. At the end of the exercise, a menu of technological and practice options was developed. Farmers choose what they wanted to test and work with depending on their agro-ecology socio-economic and cultural context. In step 3, *the adaptation options were tested*. The testing was done in collaboration with a local NGO, extension facilitators, and researcher. The collaborators helped provide resources to farmers, e.g. the provision of agroforestry trees and coached and trained farmers throughout the study period.

Data collection: Qualitative and quantitative data using household surveys were collected over the three-year period. Baseline data was also collected at the beginning of the study in 2011 and at endline in 2020. Data collected includes demographics, livelihoods, and impacts of climate change, coping activities and extension services. Matched pairs and one-to one comparison were incorporated in the data collection framework.

Data analysis: Descriptive and inferential analyses were done for quantitative data, while thematic analysis was done for qualitative data. *McNemars* test was used; recommended for pre-test post-test study as well as analyzing matched pairs and case-control studies. Key inferential statistics used were *Fisher's exact test* in the analysis of contingency tables *Measures of association* the *Rank biserial association test*, and the *Spearman rank correlation.*

Results: The participatory approaches provided local village leaders, communities, and participating researchers and facilitators with information on farmer perceptions of climate change, drivers of their risks, and coping strategies. Social learning helped nurture capacities and increase farmers' knowledge to find solutions, test and increase the range of adaptation options in climate-smart villages.

There were increased opportunities for meeting gender equity goals in climate-adaptive economic activities.

Participants described an increase in their ability to test and learn together and develop viable, sustainable solutions to problems that affected the community. There was an aspect of community empowerment. Indeed, one of the aims of the aims of the research approach involving the use of social-technical methodologies and tools (use of participatory tools while allowing men, women and the landless to participate) was to inculcate true resilience among the individual farmers' mind-sets and change their attitudes and farming practices.

At the end of the study, farmers increased the productivity of crops and animals; specifically, the percentage of farmers who planted improved corn varieties and agroforestry trees and kept chickens and pigs increased.

The approach ensured that there were opportunities for men, women, and landless households to participate in the community adaptive process. Discussions were held on varieties of crops they have tested which are suitable and performing well in their farms.

Source: Barbon et al. (2021): https://www.sciencedirect.com/science/article/pii/S2666049022000305?via%3Dihub, https://creativecommons.org/licenses/by/4.0/

Methodology

A diversity of methods within the qualitative and quantitative approaches are used. The methodology originates from participatory action research traditions that aim to involve people throughout the research process. Collaboratively, all partners, including the target participants, are involved in the planning, data collection, analysis, interpretation, and use of research.

Barbon et al. (2021) presented in a case study in Box 7.3, illustrated the transformative methodology by focusing on methods that allowed opportunities for individual, systematic transformation as well as organizational transformation. Moreover, all partners (researchers, facilitators from non-governmental organizations (NGOs) and extension, village leaders, and researchers) were involved at all stages of the research process. There was a step-wise analysis of the climate risks and vulnerabilities, followed by testing and data collection, and finally, discussions about what worked and what did not work. Farmers reaped the rewards of the research by planting new crop varieties and adopting other agricultural practices.

Key Characteristics of Transformative Paradigm

- *Participatory and collaborative*: It uses the principle of participatory inquiry. Researchers engage participants as active collaborators in their inquiry/ research.
- *Focus on bringing change in practice.* After the termination of the study, the researcher proposes an action plan for change for the benefit of the study subjects.
- *Social justice and empowerment*: Focus on promoting social justice and equity. Starts with important issues affecting society, such as disfranchisement and the need for empowerment. During the study, the participants are made aware of the issue and informed about the need to be freed.
- *Multidisciplinary approach*: draws on insights from multiple disciplines, recognizes the complex social issues and interdisciplinary nature of the issues.
- *Power relations*: Focus on understanding power relations within the society and how this affects the marginalized and oppressed.
- *Ethical considerations*: Prioritize ethical principles, ensuring participants' autonomy is respected and their voices are heard.
- Social theories: Employ social theories to me and interpret findings. Common terms include feminist theory, critical theory, queer theory, pos colonial theory, etc.

Action and Participation in a Transformative Worldview

You may wonder how action and participation will be embraced in research that is being driven by a transformational worldview. As we have seen, this worldview emphasizes participatory approaches and action-oriented agendas and seeks to address social injustices and empower marginalized communities. This is the concept behind farming systems research and on-farm experimentation discussed in Chap. 10. Some of the key activities in research that embrace a transformative agenda include the following:

- Empowering farmers.
- Involving farmers in the identification of problems and challenges.
- Engaging farmers in the design and implementation.
- Strengthening local cooperatives and communities.
- Developing farmer-friendly agricultural practices.
- Co-creating knowledge with farmers.
- Having collaborative planning and implementation.
- Training community members to enhance skills (training of trainers concept).
- Protecting social justice.
- Facilitating advocacy agenda.
- Emphasizing change-oriented research.
- Targeting women as key beneficiaries.
- Targeting marginalized communities, indigenous people, and youth.

7.9.4 Pragmatic Paradigm

The pragmatic paradigm emerged from the perspectives of philosophers who contended that accessing the truth about the real world was not achieved solely through a single scientific method (positivists) or constructed (interpretivism) (Biesta 2015; Kelly and Cordeiro 2020).

Ontology

Believe in "non-singular reality ontology." That there is no singular reality and each individual possesses unique interpretations shaped by their experiences and circumstances.

Epistemology assumption: Advocates of this contend for generation through the utilization of research methods deemed most suitable for studying the phenomenon (Paton 1990; Tashakkori and Teddlie 2010). Effectiveness is considered in terms of establishing the results that address a specific problem, diverging from paradigms that emphasize the nature of reality. Knowledge is constructed based on real-world experiences.

Axiology
Pragmatism is value-laden, involving conducting research that benefits people.

Methodology
Pragmatism combines mixed methods involving integrating quantitative and qualitative methods to gain a comprehensive understanding of the research problem. Researchers select data collection methods based on what is most practical and effective for answering the research questions and addressing the research problem.

Creswell (2009) and Mertens (2014) provide the following characteristics of the pragmatism paradigm:

- *Freedom to choose*: researchers possess the freedom to select methods, techniques and procedures that best align with the needs and objectives of the study. Mixed methods researchers explore various methods and approaches for collecting and analyzing data rather than relying on a singular method.
- *Best fit purpose*: allows for the adoption of a worldview that accommodates research design and methodologies most suitable for the study. Researchers in pragmatism consider what and how to research based on desired consequences and outcomes. In mixed methods research, establishing a purpose for the blending of quantitative and qualitative is crucial, requiring a rationale for both the proposal and the research report.
- *Flexibility*: rejecting the need to confine the study to either a positivist or interpretivist paradigm, pragmatism aims to use the best approach for knowledge acquisition, incorporating methodologies that facilitate knowledge discovery. Pragmatism embraces multiple methods, diverse worldviews, varying assumptions and different methods of data collection and analysis.

Agriculture is multifaceted and is a productive enterprise and extractive activity, representing a way of life deeply rooted in one of the most integrated relationships between humans and the environment (Duram 1998). The author asserts that both social and ecological aspects are worldwide and universal in agricultural decision-making, and therefore, this must be studied in an interrelated way. The recommendation is, therefore, to consider behavioral pragmatism as a valid philosophical approach to addressing complex environmental and agricultural issues.

In this classical paper (Duram 1998) on the behavioral pragmatism paradigm and its application to agriculture, the main goals of this paradigm are to:

- Adopt a "systems approach" to study human and environmental relationships. This is a way of dealing with chaos in the relationship between humans and the environment. The current thinking in agroecology, integrated soil fertility management.
- Gaining insight through human experiences, this approach sheds light on individual characteristics, such as attitude, knowledge, and behavior, that significantly impact farmers' decisions.
- Embracing a holistic perspective, the methodology views problems as intricate and interconnected, recognizing the complexity of the problematic situations in the agricultural context.
- Supporting social activism and proper policy formation that encourage ecologically sustainable land use decisions. To this end, it aligns closely with the transformative paradigm.

7.10 How Do You Choose a Research Design?

This chapter has focused on giving an overview of research designs and the role of our set of beliefs (philosophical views). In this last section, we discuss selecting a research design while incorporating the components given at the start of this chapter (Sect. 7.3). The components include the philosophical stance, structure, population and sampling, skills and preferences of the researcher, location and time, data collection, and data analysis.

We consider our starting point to be the research objective that emanates from the research topic and research problem (Chap. 3). The research objective is important because it tells us exactly what we want to achieve. It will tell us where we want to go. It tells us whether we want to 1. Establish a causal relationship, 2. Test interventions but also explore the phenomenon or 3. Describe the characteristics of a phenomenon (Fig. 7.4). The following questions will guide in determining the structure:

1. Will I manipulate the independent variable?
2. Will I use a control group?
3. Is concern a relationship or a difference between variables?
4. Will my subjects be randomly assigned?
5. Which design is feasible and best controls for threats to internal validity?

If you plan to manipulate the independent variable, an experimental design is appropriate. Within the experimental designs, further decisions include selecting the type of

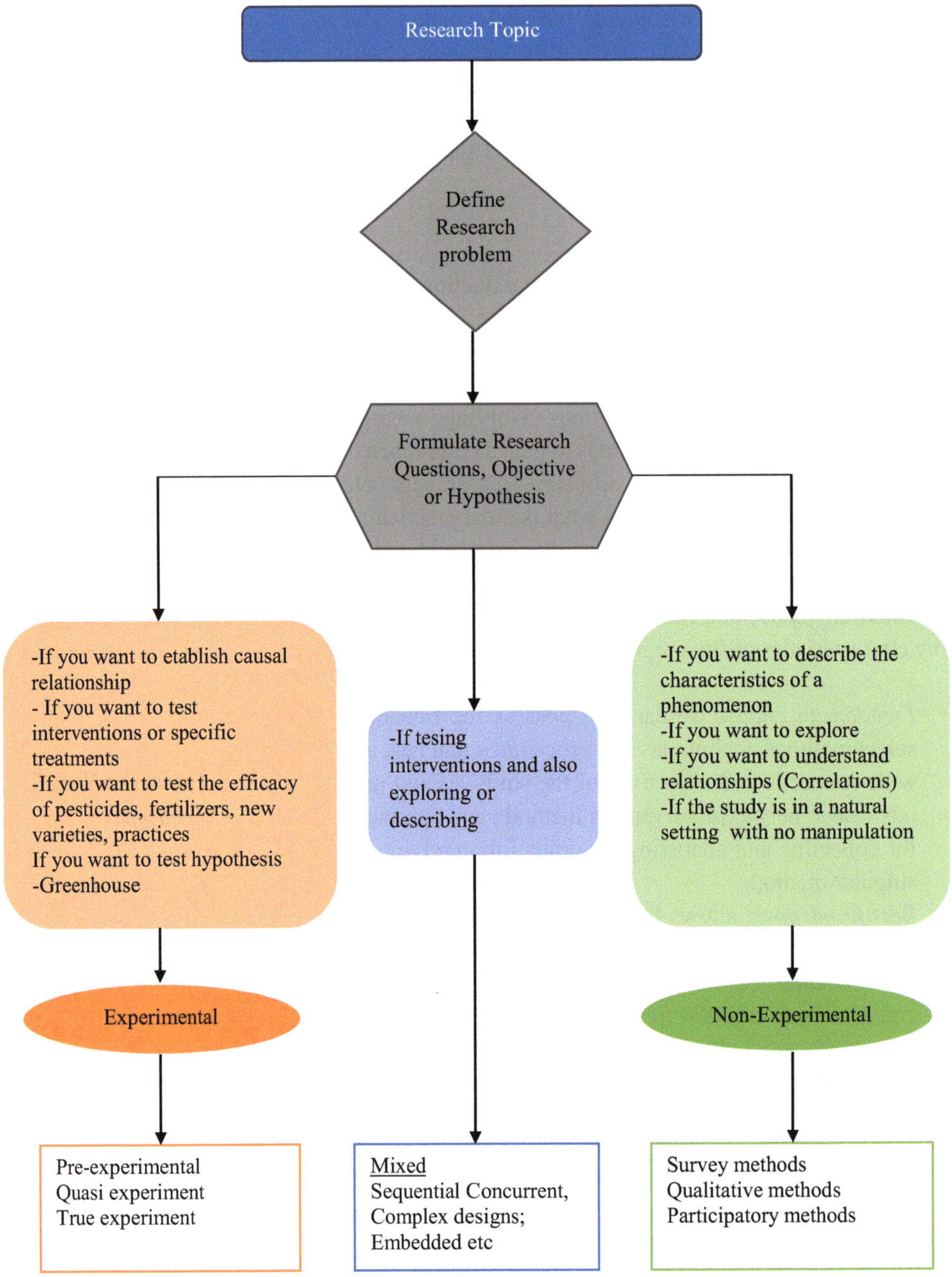

Fig. 7.4 Choosing a research design and strategies of inquiry

experiment based on whether you will randomize the treatments/subjects and include a control. The options include; pre-experimental (minimal control over extraneous variables), quasi-experimental (some control but lacking random assignment) and true experiments (with random assignment and having a high level of control over variables).

If your research does not include manipulating the independent variable, consider nonexperimental designs, which could be a survey (using questionnaires or interview schedules), Qualitative (in-depth exploration using methods like focus group discussions and interviews), or participatory design (involving stakeholders).

If it will use a mixed approach involving experimental and nonexperimental data, choose the most suitable design based on your objectives and the integration of quantitative and qualitative data (this is discussed more in Chap. 15).

While this initial draft of the research design provides a solid foundation, it is essential to incorporate all critical components necessary to justify the research study comprehensively. This is particularly important in the context of a PhD thesis, where the design must demonstrate both academic rigor and methodological integrity. By deliberately aligning the research design with the study's objectives and including all requisite elements, the researcher ensures that the study is conceptually robust and capable of generating valid and reliable findings. Such a thorough approach not only reinforces the credibility of the research but also facilitates a strong defense during examinations and increases its acceptance in academic forums such as seminars, conferences, and peer-reviewed publications.

Key Points in This Chapter

What Is a Research Design?

Research design is the structure that allows you to collect data to answer the research objective/question. The nature of this structure is dictated by the objective of the study.

Importance of Research Design

The most important functions include:

- Ensures systematic inquiry through clearly defined research methods that address specific questions objectively.
- Enhances validity and reliability by minimizing bias and controlling for extraneous variables.
- Promotes resource efficiency through structured procedures that guide data collection and analysis.
- Facilitates the development of meaningful interpretations and well-supported conclusions and.
- Supports replication and generalization, contributing to the broader applicability and practical relevance of the study.

Components of a Good Research Design

A good design should give data that is objective, reliable and valid. There are five core components to allow you to obtain the data. These are:

1. Structure and conceptual framework.
2. Population and sampling technique.
3. Location and timing of data collection.
4. Instruments and methodologies of data collection.
5. Approaches to data analysis.

The philosophical assumption, skills, and preferences of the researcher support these five core components.

A research design is equivalent to an architectural plan for building a house, serving as a foundational map for a research project's success.

Classification of Research Designs

While no universal classification of research designs exists, a suitable framework for agricultural sciences should integrate both biophysical and social dimensions. It must also reflect the alignment among philosophical assumptions, research strategies and specific methods. The three categories are thus:

- "*Experimental designs*" (true and quasi experiments)-quantitative
- "*Nonexperimental designs*" (both quantitative and qualitative)
- (a) Survey
- (b) Qualitative
- (c) Participatory action research
- "*Mixed methods designs*" (experimental, nonexperimental, qualitative, quantitative)

Types of Experiments

There are three main types of experiments; 1. True experiments, 2. Quasi experiments and 3. Pre-experiment research designs. True experiments are randomized controlled trials (RCTs) while quasi-experimental and pre-experimental are non-randomized. The non-randomized trials are mainly used in program evaluations in social studies in the field of agriculture. We can summarize the characteristics of the three types of experiments as follows:

1. True experiments = Manipulation + Randomization + Control.
2. Quasi experiments = Manipulation + Control—Randomization.
3. Pre-experimental = Manipulation − Randomization and Control.

Quasi Experiments

The absence of random assignment implies there may be systematic differences between the groups, introducing a degree of uncertainty into the research design.

Pre-experimental

Pre-experimental studies are sometimes not regarded as true experiments, as there are limitations on the extent of manipulation that can be conducted. There are three main types:

1. *One group design with post-test only*: Involve one group with post-test only. Any researcher using these designs has challenges evaluating the effectiveness of their independent variable. They are not unpopular in agriculture; they can be used in project evaluation studies.
2. *One group design (pre-test and post-test)*: A group receives a pre-test and a post-test to determine if the treat-

ments had any effect. This design suffers from uncontrolled variation.

3. *Static group comparison*: one group receives the treatment, while the other does not. After the treatment, both groups undergo post-tests to assess any effects of the treatment. However, since no pretest was conducted, it is unclear whether there were any differences between the groups before they were exposed to the treatment.

Research Philosophy

The three core philosophical assumptions—ontology, epistemology and axiology, form the foundational lens through which researchers view reality, acquire knowledge and consider the role of values in the research process. These assumptions guide inquiries into the nature and existence of phenomena, the credibility of knowledge claims and the ethical dimensions embedded in the conduct of research.

Nonexperimental Designs

Nonexperimental research lacks manipulation of the independent variable. Researchers rely on interpretation, observation and interactions to conclude, being cautious about inferring cause-and-effect relationships from correlations.

Common forms include surveys and qualitative methods. Surveys collect data from selected participants, offering insights into attitudes and behaviors through various techniques.

Classification According to Broad Paradigms

Quantitative research: Quantitative research involves producing numerical and quantifiable data to test theories and assumptions by evaluating relationships between variables. It is characterized by testing cause-and-effect relationships and correlations. It mainly uses probability sampling techniques.

Qualitative research: In contrast, qualitative research aims for in-depth insights and is exploratory, relying on direct observation, participation and interviews. Qualitative data express meanings, concepts and descriptions, providing idiographic causal explanations.

Participatory research: Participatory research involves collaboration between researchers and stakeholders, including farmers and local communities. It emphasizes local knowledge, ownership and active participation in data collection. Data collected could be quantitative or qualitative.

Mixed methods designs: Mixed methods research integrates quantitative and qualitative approaches to understand the research question comprehensively. Three classes of mixed designs are concurrent (combining both methods), sequential (building on findings of one method with the other) and transformative (seeking to contribute to societal improvement through a cyclical process of data collection and interpretation).

Research Philosophy

Research philosophy refers to the set of beliefs or assumptions that guide the researcher's approach to conducting research. There are three components of philosophical assumptions: ontology, epistemology and axiology. These form the foundation of a researcher's beliefs about the nature of reality, knowledge acquisition and values' role in the research process. These components guide fundamental questions about phenomena' existence, qualities and nature, the validity of knowledge claims and the ethical considerations in research.

Philosophical Worldviews

The philosophical assumptions, when combined, give rise to four main worldviews:

- Positivism: aligns with objective reality and emphasizes empirical observation and measures. World is outside of researchers.
- Interpretivism: acknowledges subjective realities and focuses on understanding meanings.
- Transformative: transformative research seeks social change and prioritizes justice and ethical considerations.
- Pragmatism: takes a practical approach, valuing flexibility, and multiple perspectives in research methodologies.

These philosophical worldviews serve as overarching frameworks that shape how researchers approach their studies, influencing choices in research design, methods of data collection and the interpretation of findings.

How Do You Choose a Research Design?

The aim is to choose a research design that is best suited to answer the specific research objective. It requires you to think and make decisions regarding the following:

- Aims and objectives are driven by the research problem (experiment or nonexperimental).

- Sampling methods to select your subjects.
- Data to be collected and procedures for collecting each of the data points and.
- Data analysis methods to answer your objectives.

The first choice is whether you will use an experiment or nonexperimental. These questions will guide you:

1. Will I manipulate the independent variable?
2. Will I use a control group?
3. Is the concern a relationship or a difference between variables?
4. Will my subjects be randomly assigned?
5. Which design is feasible and best controls threats to internal validity?

Recommended Further Reading

1. Research Design: Qualitative, Quantitative, and Mixed Methods Approaches (Creswell 2009)
2. Research and Evaluation in Education and Psychology: Integrating Diversity with Quantitative, Qualitative, and Mixed Methods (Mertens 2014)

Review Questions

1. Why is it important to understand the classification of research designs before beginning a research study?
2. Using the analogy of an "architectural plan for building a house," explain the meaning of a research design (use examples).
3. How do the skills and preferences of the researcher support the five core components of a good research design?
4. Describe the core beliefs of the four main philosophical worldviews: Positivism, Interpretivism, Transformative, and Pragmatism.
5. How does a clear research objective, derived from the research problem, guide the fundamental choice between an experimental and nonexperimental design?

Chapter Review

1. Clearly define the three main types of experiments: true experiments, quasi-experiments and pre-experimental designs.
2. Why must researchers be "cautious about inferring cause-and-effect relationships from correlations" in such designs? Provide an agricultural example where correlation might exist without causation.
3. Review three different research studies (journal articles) from your field of interest. Classify each study based on:
   - Research paradigm (quantitative, qualitative, or mixed).
   - Research structure (experimental or nonexperimental).
   - Underlying research philosophy (positivism, interpretivism, transformative, or pragmatism).
4. Match the following characteristics to the correct research philosophy: positivism, interpretivism, transformative, and pragmatism:
   - Emphasizes measurable data.
   - Values lived experiences.
   - Focuses on social justice and empowerment.
   - Combines methods based on practical needs.
5. Draft a basic outline for a research study in your field of interest, clearly indicating:
   - The research problem.
   - The research objectives.
   - Your chosen paradigm and research philosophy.
   - Type of design (experimental or non-experimental).
   - Justification for each choice.

Answers to Review Questions

- Q1. Reference: 7.1 and 7.2.
- Q2. Reference: 7.3 (Box 7.1).
- Q3. Reference: 7.3 (Fig. 7.1).
- Q4. Reference: 7.8 (Fig. 7.3).
- Q5. Reference: 7.10 (Fig. 7.4).

References

Alele F, Malau-Aduli B (2023) An introduction to research methods for undergraduate health profession students. An introduction to research methods for undergraduate health profession students. https://doi.org/10.25120/FH2Z-YVA8

Ary D, Jacobs L, Razavieh A, Sorensen C (2009) Introduction to Research in Education, p 696

Barbon WJ, Myae C, Vidallo R et al (2021) Applying participatory action research methods in community-based adaptation with smallholders in Myanmar. Front Clim 3:734053. https://doi.org/10.3389/FCLIM.2021.734053/BIBTEX

Biesta G (2015) Pragmatism and the philosophical foundations of mixed methods Research. In: SAGE handbook of mixed methods in social & behavioral research, pp 95–118. https://doi.org/10.4135/9781506335193.N4

Chambers RL (1986) Outlier robust finite population estimation. J Am Stat Assoc 81:1063–1069. https://doi.org/10.1080/01621459.1986.10478374

Collis J, Hussey R (2022) Business research: a practical guide for students, p 351

Creswell JW (2009) Research design: qualitative, quantitative, and mixed methods approaches. Sage publications, pp 3–22

Creswell J (2013) Steps in conducting a scholarly mixed methods study. Discipline-Based Education Research Group Speaker Series

Creswell JW (2014) Research and design qualitative, quantitative, and mixed methods approaches. SAGE Publication Inc
Creswell JW, Guetterman TC (2019) Educational research: planning, conducting, and evaluating quantitative and qualitative research. Pearson, pp 1–681
Creswell JW, Plano Clark VL (2011) Designing and conducting mixed methods research. In: Designing and conducting mixed methods research, pp 51–99
Crotty M, Shakespeare W, Henry V (2020) The foundations of social research: meaning and perspective in the research process. In: The foundations of social research: meaning and perspective in the research process, pp 1–248. https://doi.org/10.4324/9781003115700/FOUNDATIONS-SOCIAL-RESEARCH-MICHAEL-CROTTY/RIGHTS-AND-PERMISSIONS
Dee S, Guzman JE, Hanson D et al (2018) A randomized controlled trial to evaluate performance of pigs raised in antibiotic-free or conventional production systems following challenge with porcine reproductive and respiratory syndrome virus. PLoS One 13:e0208430. https://doi.org/10.1371/JOURNAL.PONE.0208430
Duram LA (1998) Taking a pragmatic behavioral approach to alternative agriculture research. Am J Altern Agric 13:90–95. https://doi.org/10.1017/S0889189300007724
Fraenkel JR, Wallen NE, Hyun HH (2012) How to design and evaluate research in education, p 642
Gay LR, Airasian PW (2012) Educational research: competencies for analysis and applications plus Myeducationlab with Pearson Etext – Access Card Pac
Gomez KA, Gomez AA (1984) Statistical procedures for agricultural research. In: Statistical procedures for agricultural research with emphasis on rice, pp 7–8
Kelly LM, Cordeiro M (2020) Three principles of pragmatism for research on organizational processes. Method Innov 13. https://doi.org/10.1177/2059799120937242
Kivunja C, Kuyini AB (2017) Understanding and applying research paradigms in educational contexts. Int J Higher Educ 6. https://doi.org/10.5430/ijhe.v6n5p26
Kothari CR (2004) Research methodology: methods and techniques (2nd ed.). New Delhi: New Age International Publishers
Kothari CR (2009) Research methodology methods and techniques. New Age International Publishers
Kothari CR, Garg G (2019) Research methodology: methods and techniques. New Age International Publishers
Kumar R (2018) Research methodology: a step-by-step guide for beginners, 4th edn. Sage Publishers, New Delhi, pp 34–56
Lim CY, In J (2019) Randomization in clinical studies. Korean J Anesthesiol 72:221–232. https://doi.org/10.4097/KJA.19049
Little TM, Hills FJ (1978) Agricultural experimentation: design and analysis, p 350
Mead R, Curnow RN, Hasted AM (2017) Statistical methods in agriculture and experimental biology: third edition. In: Statistical methods in agriculture and experimental biology: third edition, pp 1–472. https://doi.org/10.1201/9780203738559/STATISTICAL-METHODS-AGRICULTURE-EXPERIMENTAL-BIOLOGY-ROGER-MEAD/RIGHTS-AND-PERMISSIONS
Mertens DM (2010) Philosophy in mixed methods teaching: the transformative paradigm as illustration. Int J Mult Res Approaches 4:9–18. https://doi.org/10.5172/MRA.2010.4.1.009
Mertens DM (2014) Research and evaluation in education and psychology: integrating diversity with quantitative, qualitative, and mixed methods (4th ed.). Thousand Oaks, CA: SAGE Publications, Inc
Mugenda A, Mugenda G (2019) Research methods: qualitative and mixed methods approach, pp 100–102
Newton Suter W (2012) Introduction to educational research: a critical thinking approach. In: Introduction to educational research: a critical thinking approach. https://doi.org/10.4135/9781483384443
Noel L-A (2016) Promoting an emancipatory research paradigm in design education and practice. In: DRS biennial conference series 2. https://doi.org/10.21606/DRS.2016.355
Park YS, Konge L, Artino AR (2020) The positivism paradigm of research. Acad Med 95:690–694. https://doi.org/10.1097/ACM.0000000000003093
Parker MJ, Fraser C, Abeler-Dörner L, Bonsall D (2020) Ethics of instantaneous contact tracing using mobile phone apps in the control of the COVID-19 pandemic. J Med Ethics 46:427–431. https://doi.org/10.1136/MEDETHICS-2020-106314
Paton A (1990) A global taxonomic investigation of Scutellaria (Labiatae). Kew Bull 45:399. https://doi.org/10.2307/4110512
Pervin N, Mokhtar M (2022) The Interpretivist research paradigm: a subjective notion of a social context. Int J Acad Res Progress Educ Dev 11. https://doi.org/10.6007/IJARPED/V11-I2/12938
Richards S, Van Leeuwen JA, Shepelo G et al (2016) Randomized controlled trial on impacts of dairy meal feeding interventions on early lactation milk production in smallholder dairy farms of Central Kenya. Prev Vet Med 125:46–53. https://doi.org/10.1016/J.PREVETMED.2016.01.006
Robson C, McCartan K (2016) Real world research: a resource for users of social research methods in applied settings, p 533
Saunders MNK, Lewis P, Thornhill A (2019) Research methods for business students, p 833
Schindler PS (2019) Business research methods, p 608
Tashakkori A, Teddlie C (2010) SAGE handbook of mixed methods in social & behavioral research. https://doi.org/10.4135/9781506335193
Tombs M, Pugsley L (2020) How to: understand research philosophies and paradigms in medical education. Cardiff University
White TL, McBurney D (2013) Research methods, p 458
Wilson S, MacLean R (2009) Research methods and data analysis for psychology, vol 1. McGraw-Hill Higher Education, p 640

Fundamentals of Experimentation

8

Abstract

The design and analysis of experiments were developed by R. A. Fischer, a famous biostatistician, during the 1920s and 1930s. Other key authors, including Campbell and Stanley also contributed to understanding the experiments. Experimental studies have become increasingly complex over the years, involving more complex data sets (big data) due to the advanced statistical methods. In this Chapter, we present the fundamentals of experimentation and explain the concepts of experimentation, the experimental process and the concept of experimental error and controlling "noise" in experiments. We follow this with three main principles of experimentation: "the 3Rs," randomization, replication, and local control." The final section of this Chapter discusses 14 elements of designing effective experiments, including, among others, the selection of experimental design and treatments; treatment of response variables; the role of uniformity trials; management of experimental; and experimental protocol.

Keywords

Agronomy experiment · Experimental error · Experimental site · Experimental process · Field experiment · Laboratory experiments · On-station research · Local control or blocking · Randomization · Response variable · Replication · Three Rs · Treatments · Treatment error · Unexplained variation or "Noise" · Uniformity Trials

Learning Outcomes

By the end of this Chapter, the learner is expected to be able to:

1. Appreciate the origin of agricultural experiments.
2. Recognize the common types of agricultural experiments.
3. Illustrate the basic experimental process.
4. Differentiate between "experimental error" and "treatment error".
5. Demonstrate understanding of "Extraneous variables" to determine how to reduce "Noise" in an experiment.
6. Relate the principle of "local control" to data validity.
7. Apply the principle of 'Three R" in designing an effective experiment.
8. Identify the main causes of experimental error in an agronomy experiment.
9. Determine a suitable number of replicates in an experiment.
10. Design an effective experimental protocol.

8.1 Design of Experiments

8.1.1 Brief History of the Design of Experiments

Design and analysis of experiments were developed by Sir R. A. Fischer, a famous biostatistician while working at "Rothamsted Experimental Station," then the Institute of Arable Crops Research in the 1920s and 1930s (Yates 1964; Fisher 1980). Founded in 1943, the experimental station is one of the oldest global agricultural research institutions. Its primary mission was to evaluate the impact of inorganic and organic fertilizers on crop yields. Fischer dedicated 14 years to this station, conducting extensive crop experiments. Through the analysis of its immense data, Fisher laid the foundation for experimentation. He developed the concept of analysis of variance (ANOVA) and explained it as an important analysis for partitioning the variation present in the data into two components: (1) due to attributable factors, and (2) due to chance factors (Fisher 1980).

Fishers' initial contribution primarily focused on applying statistical methods in agriculture, advancing many of the statistical concepts scientists use today. For example, His books advanced the concept of randomly assigning individu-

J. Njeri Mugwe, S. Runo, *Research Methodology in Agricultural Sciences*, https://doi.org/10.1007/978-981-95-1892-0_8

als to groups and replication in experiments (Hall 2007). The recognition of the need to systematically address variability in experiments led to the formulation of three foundational principles (3Rs): randomization, replication, and local control (Reduce noise). These principles became the cornerstone of robust experimental design. Other statistical procedures advanced were testing the significance of differences, critical value, and chi-square goodness of fit. These methods shaped the field of experimentation and statistical analysis and have been widely applied in "biological, psychological, and agricultural" research.

Several other publications have made significant contributions to the understanding of research experimental designs. Among them are works by Campbell and Stanley (1963), Cook and Campbell (1979) and Shadish et al. (2002). These authors emphasized the importance of experimental designs in establishing causal relationships between variables and delineated the essential characteristics of true experiments and quasi-experimental designs. They also paid particular attention to validity and how validity can be enhanced, outlining the various types of experimental designs and discussing threats to validity. Their books are considered essential reading for anyone interested in experimental research design. This applies to both social and natural sciences.

Campbell and Stanley (1963) outline potential threats to validity in experimental designs. They specified 15 different types and evaluated all designs regarding "potential threats to validity." They also discuss the importance of rigorous research design and analysis techniques, such as matching techniques and statistical controls, to address these threats. Overall, researchers continue to draw upon the information presented in these classical books to ensure the scientific rigor and credibility of their research.

Experimental studies have become increasingly complex over the past few decades, thanks to the emergence of advanced statistical techniques and computer technology. This progress in computing technology and statistical methodology has played a vital role in managing and analyzing large datasets from intricate experimental designs. In recent years, the concept of big data science has gained popularity, and experimental studies now involve analyzing even more massive and complex datasets. With the aid of advanced statistical methods, researchers can extract meaningful patterns and insights from vast amounts of data.

This Chapter and Chap. 9 discuss the nature of experiments.

8.1.2 Classes of Experiments

There are four overarching classes of experiments in agriculture as shown hereby:

1. *Laboratory experiments*: These are where the independent variable is intentionally manipulated, and the dependent variable is measured within a controlled and artificial setting (Brüggemann and Bizer 2015). This method effectively manages and controls extraneous variables that might influence the dependent variable. They can be a study carried out in a science laboratory or a screen house experiment (greenhouse experiment), where conditions are controlled. In some greenhouses, the conditions, mainly temperatures, soil moisture, and humidity, are controlled and monitored.
2. *Field experiments*: These are studies conducted in a natural environment, where the independent variables are manipulated, and measurements of the dependent variable are taken on test units in their authentic, real-world setting, such as a farm. Depending on the specific research site, field experiments can be further distinguished as "*on-station and on-farm*" experiments. The distinction lies in whether the experiments are conducted within controlled research facilities or directly on farms.
3. *On-station research*: It is a type of research carried out in an environment intentionally designed and controlled by the researcher, often in a research station, university farm or a comparable setting where researchers maintain complete control. While farmers are not actively engaged in managing the research, they might be invited to participate and learn from the process, depending on the study's objectives.
4. *On-farm research*: On-farm research (OFR) entails conducting research directly on farmers' fields. This approach can take various forms: it may be researcher-designed and farmer-managed, entirely designed and managed by researchers or involve collaborative efforts between farmers and researchers. Central components of OFR encompass the farmer, their land, active farmer involvement and the specific environmental conditions prevailing on the farmer's land. On-farm research and experimentation are discussed in Chap. 10.

8.2 Common Types of Laboratory and Field Experiments

Agronomy Experiments Broadly, agronomy experiments involve crop production and soil management. The main aim is to develop management practices that optimize crop production. Experiments mainly include field crop management, soil and water management, cropping systems, eco-physiology, and management of biotic and abiotic stresses. The following are common agronomy experiments:

- *Varietal screening experiments*: Aim to test the performance of different varieties or genotypes. It often involves developing and evaluating new varieties with improved traits, such as disease resistance, drought tolerance, and yields. Depending on the purpose of the screening, various types are recognized. For instance, in maize varietal experiments, screening may involve progeny (progeny trials), observation or preliminary trials, national, regional, and international performance trials, on-farm trials or demonstrations (Badu-Apraku et al. 2012).
- *Forage production experiments*: Forage production is essential to livestock production. These experiments are conducted to develop efficient and sustainable forage production systems. They mainly involve evaluating and comparing different forage varieties, management practices, and environmental conditions to determine the best strategies for maximizing yields and quality of forages.
- *Intercropping experiments*: These experiments focus on assessing the effect of different intercropping and rotation systems on soil health, pest management, and overall yields. They help increase understanding of species interactions and design systems that enhance resource utilization (Fig. 8.1). Conducting these experiments over several years to evaluate the system's stability is advisable. Stability considerations account for variable climatic conditions and pressures from pests and diseases, leading to more resilient agricultural systems (Finney 1990).

Crop Rotation Experiments Experiments primarily investigate opportunities to enhance agricultural production in the short and long term by manipulating the design of crop rotation. Treatments are implemented over multiple cropping seasons or years, and their effects are assessed in the final season (Payne 2015).

Soil Fertility Experiments The discipline of soil science explores the diverse role of soils and investigates ways to use the soil more sustainably. Soil fertility experiments will, therefore, aim to evaluate the influence of soil management practices on the availability of soil nutrients and agricultural

Fig. 8.1 Soil fertility and agronomy experiments; (**a**) soil fertility, (**b**) intercropping of sorghum and green grams in a farmers field

productivity. The experiments provide valuable information on the best management practices to optimize soil fertility and improve crop production by providing nutrients. Typical experiments assess effect of fertilizers, nutrient omission trials, combinations of mineral and organic fertilizers, and how they can be managed under various cropping systems. Pot and field experiments are common (Johnston and Poulton 2018).

Animal Nutrition Experiments Mainly involve feeding trials designed to compare production results between two or more feeding schemes in livestock production. In most cases, the feeding experiments will aim to estimate digestibility and feed intake. This may include measuring weight, body length, food intake, palatability, and feeding preference (Meier et al. 2012).

Veterinary Clinical Experiments These experiments entail research that acquires data from animal patients in veterinary settings (Taylor and Meyer 2023). Clinical studies aim to enhance animal healthcare by identifying the most efficient therapies and practices for a particular condition or by advancing a basic understanding of the disease. There are two primary categories of clinical studies: clinical trials, also known as interventional studies and observational studies. In a clinical trial, veterinary patients may undergo specific interventions based on a research plan or protocol developed by the investigators, while in observational trials, the natural course of the disease or condition is observed and analyzed (see Chap. 12).

Agroforestry Experiments Agroforestry integrates agriculture and forestry to address environmental and socio-economic objectives. The experiments conducted in agroforestry are diverse and are geared toward optimizing productivity. They include experiments that test and screen species, such as screening experiments to compare trees' adaptability and growth rates. Other agroforestry-based experiments include system management and examination of components interactions, seed testing, evaluation of germplasm and provenances, and the domestication and selection of superior plus trees for improvement. Methods for quantifying carbon stored in agroforestry trees is another crucial area of research (Atangana et al. 2014).

Randomized Controlled Evaluations These are field experiments used to evaluate the impact of development programmes or policy interventions. The population subjected to the program or policy intervention is randomly selected from the eligible population, and simultaneously, a control group is randomly chosen from the same eligible population. Measurements are done after the intervention period. One of the notable strengths of this approach lies in its robust ability to address questions of causality, offering evaluators and program implementers a clear understanding that the observed outcomes result from the intervention itself and not from any other factors. Suppose you want to assess the effectiveness of two training methods: A and B. You will need to choose two groups of farmers and assign each training method to them, including a control (not trained). You would have three treatments (Training method A, Training method B and control). The allocation of the treatments would be random. This is discussed in detail in Chap. 11.

Crop Breeding Experiments Crop breeding is a method of crop improvement. It involves changing crops' traits to bring on board desired characteristics intending to improve productivity. Most studied agricultural traits are those associated with biotic and abiotic stress tolerance, grain and biomass yield, taste and nutrition, and ease of processing. Usually, the process begins with developing inbred lines, which are then propagated for additional testing and eventual variety release. The predominant method can be categorized into three main groups: 1. Selection based on observed variation involves choosing plants with desirable traits found in the natural variation or within traditional varieties; 2. Controlled mating selects plants with desirable genes from parents through controlled breeding; 3. Monitored recombination selects specific genes or marker profiles using molecular tools to track variation within a genome (Hasan et al. 2021).

Biotechnology Experiments Biotechnology applies science and engineering to utilize living organisms or systems to create products and services (Ranjha et al. 2022). Experiments involve the development of genetically modified crops, medicines, biofuels and environmental tools, as well as using techniques like DNA sequencing, cloning and fermentation. A brief discussion of biotechnology is given in Chap. 17.

8.3 Fundamentals of Experimentation

8.3.1 Role of Experimentation

Two fundamental pillars of science are the meticulous design and execution of scientific research and the capability to replicate research findings, as highlighted in Chap. 1. Implementation rigor guarantees an impartial approach to experimental design, methodology, analysis, interpretation, and reporting of results (Hofseth 2018). Reproducibility validates original results. This aspect holds particular significance when disseminating results to end users.

Experiments play a central role in scientific study by providing a method to test cause-effect hypotheses. The success of modern science owes much to the careful design and precise execution of experiments. As mentioned earlier, R.A Fisher, a geneticist and statistician, significantly contributed to the current theory and experimental methods. Experiments are usually associated with theory-testing research. Theories, instrumental in problem-solving, often outline the expected outcomes when a change is introduced. See the following:

- There is a problem with fruit flies in mangoes, but a theory tells us that a new mango variety is tolerant to fruit flies, so our prediction will be: If we use an improved mango variety, will there be less or no damage from fruit flies?
- Low milk yields on smallholder farms are a problem because of inadequate protein supplementation due to the high costs of dairy meal. However, a theory tells us that Calliandra fodder contains protein and can be used to substitute dairy meal. Will milk yields increase if we use Calliandra fodder to replace dairy meals?
- Farmers use low-quality cattle manure, resulting in low maize yields, but a theory exists that if farmers are trained to manage manure, the yield of maize will increase. If we train farmers on manure management, will maize yield increase?

To test a theory, you introduce the change and observe whether the predicted outcome occurs. In essence, we test the hypothesis (see Chaps. 2 and 20). This is the basis of experimentation, which entails implementing a specific action and methodically observing the resulting effects.

8.3.2 The Concept of Experimentation

Experimental research is characterized by manipulation, differentiating it from other research types. This manipulation involves deliberately altering one or more independent variables to observe their impact on the dependent variable. Researchers determine the nature of the treatment, who receives it (such as individuals), and the extent of its application (Kim et al. 2021). After administering the treatment for a specific duration, researchers observe or measure the groups subjected to different treatments (post-test) to determine whether the treatment has induced a discernible difference. The primary goal is to determine whether the treatment has had a meaningful impact. This differs from an observational study, which involves analyzing without changing existing conditions.

Consider an experimental research scenario in which two groups of farmers are subjected to different soil fertility treatments (treatment A and B). The researcher applies the treatments and conducts a post-test to measure crop yields. Suppose the average scores of groups A & B show significant differences and the researchers thoroughly investigate without finding any sensible alternative explanations (such as variations in weather conditions or pest infestations). In that case, they can reasonably conclude that the treatment affected crop yields.

There are circumstances in which experimentation is unrealistic and unethical. In these circumstances, alternative methods are used. It is not feasible to experiment with a prediction such as: if the average annual rainfall increases by 200 mm, maize production in country A will increase. We could test this prediction using simulation models describing the relationship between rainfall and maize production. However, these models will be based on theories tested by experiments.

Conversely, a correlational study might reveal a strong relationship between farmers' education levels (a socio-economic factor) and adopting hybrid maize varieties. However, the correlational study alone cannot demonstrate that improving farmers' education levels will necessarily lead to an enhancement in adoption of hybrid maize varieties. Only experimental research, with controlled manipulation of variables, can provide a more robust basis for making such a causal inference.

Social studies involving human beings and communities make predictions that cannot be tested by conventional experimentation. As an alternative, researchers often employ observational methods, observing and analyzing social phenomena without intentional interventions. Surveys serve as a valuable tool to collect precise data on attitudes, behaviors, and perceptions, enabling the examination of correlations and patterns in social dynamics (Chap. 12). While experimentation is limited in some social contexts, the use of observational approaches and surveys remains crucial for gaining precise insights into human behavior and societal trends (Babbie 2021).

8.3.3 The Experimental Process

The procedures involved in an experimental study closely mirror those in other types of research types. They encompass selecting a topic and framing a problem, establishing objectives and hypotheses, crafting a research plan (selecting treatments, experimental units and measuring instruments), implementing the study, scrutinizing the data and ultimately formulating conclusions (Chap. 2). An experimental study is guided by at least one hypothesis that states an expected causal relationship between two variables. The researcher experiments to test the hypothesis, carefully selecting the groups and allocating the treatments. In addition, the researcher ensures that the extraneous variables are controlled to be able to attribute the observations (measurements) to the treatments.

In an experiment, the group that receives the new treatment is called the *experimental group*, and the group that receives a different treatment or is treated as usual is called the *control group*. The groups to receive the different treatments should be equated on all variables that may influence the performance of the dependent variable (Box 8.1).

Using the example of farmers' training in Box 8.1, suppose at the end of the experimental period, trained farmers record an average post-harvest loss of 20%, while the other group (control) records an average of 25% post-harvest loss. There is a 5-point difference; is it meaningful, or is it just a chance difference due to measurement errors? Statistical

Box 8.1: Illustration of Basic Experimental Process

1. Research objective: To assess the effect of training farmers on the use of new, improved bags on post-harvest losses.

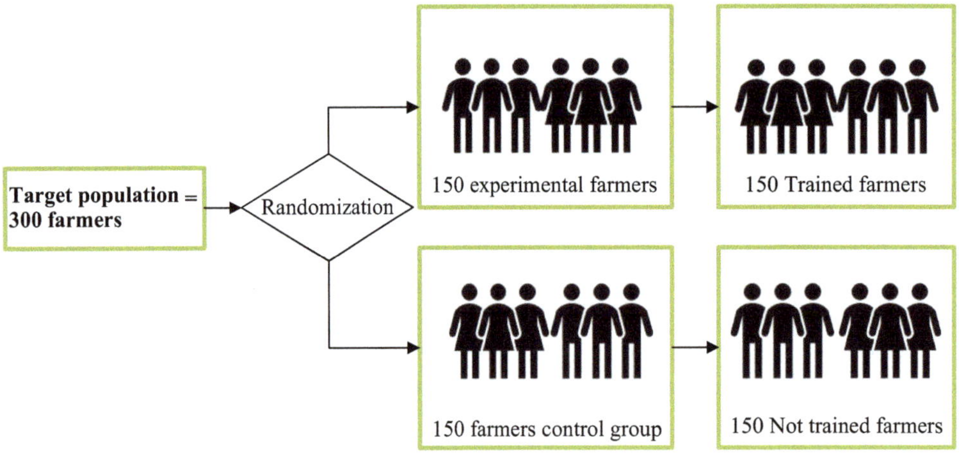

2. To assess the effect of pesticide A on yield of tomato.

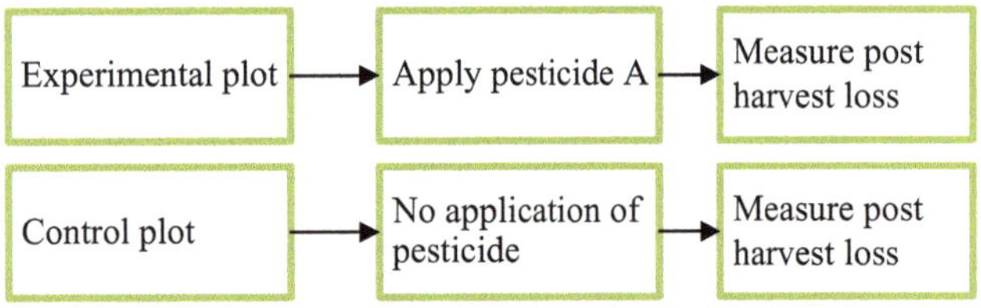

3. To assess the effect of feeding dairy cows three different protein feed supplements (A control diet, Diet A and Diet B) on milk yields.

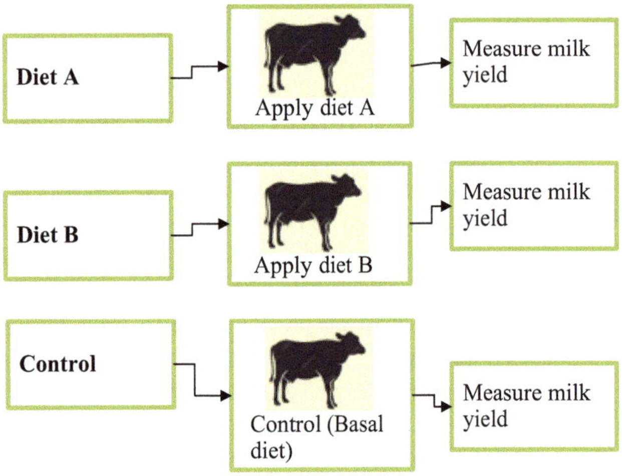

analysis allows the researcher to answer these questions with confidence. This is tackled in Chaps. 19 and 20.

8.3.4 The Concept of Variation and Experimental Error

As mentioned, experiments are powerful tools for researchers because they are objective. They carry the highest level of credibility among the research designs when properly designed and managed. Experiments allow us to study phenomena objectively. However, there is no perfect experiment that will explain the causal relationship. There will always be other things that can influence your dependent variable. We call these *extraneous variables or covariates* if measured (mentioned earlier in Chap. 2). In the real world, holding all the potential variables that could affect the dependent variable constant is impossible. We have *individual variation* (intrinsic) that occurs naturally. This intrinsic variation tends to obscure the effect of different interventions (treatments), making comparisons problematic. This can consequently lead to mistakes in interpreting the results of experiments, and subsequently erroneous judgments and recommendations for commercial purposes. Every researcher must thus understand this fundamental and universal concept of variation in scientific experimentation and their application (Mead et al. 2017).

We could think of the following variations that occur naturally:

- Experimental plots in the field vary greatly, and so do pots in a greenhouse experiment, even if we are using the same soil.
- Plants in an experimental plot vary, though in the same treatment application.
- Participants (human beings) in an experiment differ from each other in many different ways.
- Farmers are different from each other, first as individuals (human beings) and secondly in the way they manage their farms.
- The animals (cattle, goats, etc.) in an experiment vary in many ways, such as genetics and weight.
- A pair of twin heifers, being given the same diet in the same environment, will perform differently in growth as well as general characteristics.
- Similar seed lots of pumpkins planted on the same day, under the same environment; plants will show differences in growth and yields.
- Measurements taken on the same subject may vary naturally.

The causes of variation are numerous (Little and Hills 1978). For example, in the case of the growth of plants from the same seed lot, the list of factors that might be expected to affect the growth of the plant includes differences in seed weight, soil variation, depth of planting, competition from other plants, interaction with soil microbes, etc. At maturity, the weighing of different plants, the use of different machines and the subjective recording of the weights are also expected to result in variations. Similarly, a human or animal experiment will be expected to have numerous causes of variation.

In comparison, however, biological materials tend to exhibit a much greater amount of variation than physical science materials and processes (Brown 1993; Ilan 2020). This is mainly because of the genetic system and sensitivity of the living matter during its growth and development to the effects of the environment. Overall, no matter how many factors we attempt to control, we shall always find that the final weights of two plants will be different, even in a controlled environment.

All these variations and many more cause *random fluctuations* that have nothing to do with your independent variable. All these variations are causes of error in experiments, referred to as *random variation* or *experimental error*. In other words, experimental error is a result of variation caused in our dependent variable by things that we are not able to explain because we cannot measure them directly. On the other hand, we directly measure variation in the dependent variable due to treatments. Experimental error is *unexplainable*. How, then, do we handle this error in our experiments? This is the basis of understanding the fundamentals and principles of experimentation and application of statistical science.

The purpose of applying statistical procedures in experiments is to separate the unexplained part of the variation from the variation due to treatments (i.e. due to the independent variable) (Little and Hills 1978). At the beginning of this Chapter, it was referred to as *partitioning the data variation*. Statistical analysis procedures allow researchers to mathematically identify and isolate this effect that may have been caused by other variables, so that the true treatment effects (true value) are more clearly detectable. The statistical analysis also allows the researcher to conclude whether the observed differences are real and then assigns a probability level that your conclusion is correct. The probability level is incorporated as a measure of accuracy (Chap. 20 on inferential analysis).

In summary, we see that there are two main sources of variation in an experiment:

1. *Variation due to treatments*: Variation in the dependent variable resulting from the application of distinct treatments for which you can account in the statistical analysis; also called variation due to treatments or explained variation.
2. *Variation due to non-treatment factors*: Variation in the dependent variable that cannot be accounted for; also

called un-explained variation. It includes all other sources of variation in the dependent variable that does not directly result from the experimental treatments. It is called "Noise" because it obscures the effect of the treatment (discussed later in this Chapter).

The concepts discussed in the rest of this Chapter and Chap. 9 are important to help a researcher minimize the unexplained variation and maximize the explainable variation.

8.4 General Principles of Experimentation

The three main principles of experimentation, namely randomization, replication and local control (reducing noise), are known as the 3 Rs.

8.4.1 Principle of Randomization

What is Randomization?

Randomization is the random assignment of treatments or conditions of the study to the experimental group without any prior plan, that is, by chance (Hirst et al. 2014). The law of probability is applied. Every experimental unit has an equal chance of being allocated to any of the experimental units. An experimental unit is the entity that carries the treatment.

What Is the Purpose?

Random assignment avoids experimenter bias. It randomly assigns treatments to experimental units and helps distribute unknown extraneous variables evenly. This implies that it prevents systematic and subjective bias from the experimenter. For example, an experimenter may tend to favor an innovation or variety. A researcher working with a new bean variety carries out tests compared to other older varieties, but assigns the new variety to fertile fields. Another essential purpose of randomization is that it forms a basis for valid statistical tests. Independent observations are one of the preconditions for subjecting data to statistical tests e.g. ANOVA and t-tests. The lack of random assignment makes comparisons biased no matter how precise the measurements are.

How Do You Randomize?

Random allocation may be done by drawing lots, such as a ball from a container or numbers from a table of random numbers. Methods of randomization are covered in later sections dealing with individual forms of experimental designs.

8.4.2 Principle of Replication

What Is replication?

Replication means repetition. In this case, the study's treatments or conditions are repeated in experimental units (Demétrio et al. 2013). The end result of replication is several units of each treatment.

What Is the Purpose?

- Enables a valid estimate of experimental error, allowing statistical inferences, such as performing tests of significance. It is impossible to estimate experimental error without replication. This is important for *estimating precision*, that is, the ability to quantify uncertainty. Replicates are needed to distinguish "noise from signal" (unexplained variation from variation due to treatments) or estimate the precision of treatment differences. There is no way of estimating error from the un-replicated experiment. In experiments, we want to determine the effect of treatments (control variables) on the dependent variable, hence detecting this signal.
- *Decrease experimental error*: Reduced error leads to increased precision—a measure of the variability of the experimental error. Calculating an average over more values will increase precision and reliability since the calculated value will be closer to the true mean.
- *Replication* enhances the validity of results by repeating the comparison under various conditions, thereby expanding the range of applicability or generalization. This practice enables a more accurate estimation of the mean effect of a treatment.

Determining the appropriate number of replications in any experiment depends on several factors, including the homogeneity of the experimental material, the number of treatments, and the desired level of precision. Increasing the number of replications reduces experimental error, indicating that a higher number leads to greater precision in the experiment. As a general guideline, it is recommended that the number of replications in a design should offer at least 10–15° of freedom for computing the experimental error variance (Patel et al. 2004).

8.4.3 Principle of Local Control or Reducing Noise

Local control is an additional technique to the randomization and replication to control the experimental error further and increase the experiment's precision (Hirst et al. 2014). It pertains to the strategic implementation of balancing, blocking and grouping the experimental units to account for the variability in response due to the variability caused by inherent differences in experimental units, and the individual items that receive the treatment. Blocking involves physically *grouping treatments* in an experiment. Balancing in experimental design involves assigning treatments to experimental units in a manner that results in a well-distributed or balanced arrangement of treatments. This practice ensures that each treatment is equally represented and distributed across the experimental units, contributing to a more equitable and unbiased assessment of the treatment effects.

What Is the Purpose?

The primary objective is to enhance efficiency by minimizing experimental error. Blocking measures and isolates block variation. By doing so, the goal is to extract this *block-level variation* from the statistical comparison of treatment means. This process enables a more accurate assessment of treatment effects by reducing the interference caused by extraneous sources of variability (experimental error). This ultimately improves the precision and reliability of experimental results.

How Do you Carry Out Local Control?

If you can anticipate sources of variation, you can group the treatments to decrease variation within each group and reduce some variation from the statistical analysis. Blocking should be done in the field so that the "Blocks" are *homogenous*. Blocking is usually done against slopes, soil fertility gradients, or other forms of variation in the field. The treatments are "*nested*" within a homogenous environment. Researchers sometimes carry out uniformity trials to help determine the extent of soil homogeneity.

Relationship Between Blocks and Replications

Experimental designs typically use blocks consisting of one plot for each experimental treatment. For example, a block containing four treatments would include four plots, each receiving a different treatment. When a block has one plot per treatment, it represents one replication of each treatment. Therefore, blocks are often called "replications" informally. It is important to note that "replication" and "blocking" are two distinct concepts. Replication involves repeating the entire experiment to improve statistical reliability, whereas blocking aims to *minimize variability within specific groups* or blocks of experimental units.

8.4.4 Noise Versus Signal

Experimental error is also called "noise" because it makes it difficult to hear the signal (Marchant et al. 2019; Berry et al. 2022). It is caused by unexplained or uncontrolled variation. In a crop experiment, plots treated alike may give different responses due to numerous factors that, if not correctly controlled, tend to increase the experimental error. These include:

1. *Human error*: Human errors in measurements and observations can be caused by various factors, including variations in the skills, interpretations and subjective judgments of individual operators collecting data. Additionally, inaccuracies may arise due to imprecision or limitations of the measuring instruments used in the experiment. Addressing and minimizing these human errors is crucial in ensuring the accuracy and reliability of the results. Therefore, it is essential to provide *proper training, standardize procedures* and carefully select and *calibrate* measuring devices to achieve this goal.
2. *Variability among plots*: Inherent variability among plots, including distinctions in soil characteristics such as drainage, elevation and compaction, constitutes a significant source of variation in an experiment. Researchers often conduct preliminary assessments, such as *uniformity trials*, to address this challenge before initiating the main experiment. By characterizing and understanding the inherent variability within the experimental site, researchers can implement strategies to control or account for these factors during the main experiment.
3. *Lack of uniformity in the physical conduct of the experiment*. For instance, an unequal distribution of fertilizer doses across plots can introduce variability, which may affect the experiment's reliability. This non-uniformity can arise from factors such as carelessness in applying treatments. Similarly, variations in animal characteristics in animal experiments can contribute to inconsistencies in the experimental conditions. Following rigorous *standardization protocols* and paying meticulous attention to detail is essential to avoid such issues. Ensuring uniformity in the application of treatments and maintaining consistent conditions across experimental units are critical steps in reducing the impact of these potential sources of variability on the experiment's outcomes.
4. *Contamination from neighboring plots*: Interference among treatments in adjacent plots causes variation. For example, suppose irrigation is done in one plot. In that case, it might cause fertilizer or rhizobia (in the case of inoculation experiments) treatments to percolate into the neighboring plots, resulting in unintended cross-contamination. To address this issue, it is crucial to carefully consider the spacing, buffer zones and appropriate experimental design (Discussed in Chap. 9).
5. *Non-uniform behavior of all other extraneous factors*, known or unknown, that have not been/could not be con-

trolled. For example, weather changes can affect the experiment's reliability by introducing difficult-to-manage variability. To address this issue, researchers often employ statistical techniques, randomization, and replication techniques to account for the impact of uncontrollable factors. While it may not be possible to control every external variable, these strategies enhance the experiment's robustness by statistically accounting for the influence of uncontrolled factors on the observed outcomes.

Measures to prevent experimental error should be a top priority to ensure that the observed effects can be accurately attributed to the intended treatments without external interference.

Box 8.2: The Concept of "Noise" and Its Control in an Experiment

"Noise" in experiments refers to the unexplained variation in observed outcomes that cannot be directly attributed to the treatments or factors being studied. This interference can arise from differences among experimental units, uncontrolled environmental conditions, measurement errors or other unforeseen factors. Noise obscures the "signal" just like background noise in a classroom corridor can make it difficult to hear what the teacher is saying. When noise is present in an experiment, we cannot hear the signal. The noise introduces random fluctuations or disturbances in the data, making it difficult to distinguish between what is genuinely caused by the *treatments* and what is merely the result of *unforeseen factors*. This interference acts like background "noise" in the data, *masking or clouding* the true signal generated by the treatment.

For example, in experiments that test the effect of different fertilizers on crop yield, the signal is the actual change in crop yield that can be attributed to the fertilizers. However, noise from differences in soil fertility, can distort the signal by introducing variability in the observed yields that is not directly related to the fertilizers. As a result, the true impact of fertilizers on crop yield can become obscured by this unwanted variation, and the real effects of the fertilizer can be hidden or clouded by all these other factors. It will be difficult to detect significant differences in an experiment, with most variation being in the residuals (error) in ANOVA.

Reducing noise in an experiment is crucial for revealing the clear signal from the treatments. The goal is to enhance the *signal-to-noise ratio*, to hear the signal. We employ the strategies such as randomization, replication, control of experimental conditions, etc., standardizing processes. When we do this, we can confidently say, "Yes, the fertilizer is making a positive difference," without being confused by all the other stuff going on in the background.

8.5 Elements of Designing Effective Agricultural Experiments

In addition to the application of the three principles of experimentation (the 3Rs), designing an effective agricultural experiment requires you to consider the following 15 elements:

1. Objectives and hypotheses.
2. Experimental design.
3. Treatments.
4. Experimental units.
5. Number of replications.
6. Control treatment/controlling variables.
7. Experimental site selection.
8. Uniformity trials.
9. Management of the experiment.
10. Response variables.
11. Precise measurements.
12. Experimental protocol.
13. Recording data.
14. Data analysis and interpretation.
15. Conclusion and recommendation.

1. *Objectives and Hypotheses*

   Always recall that "objectives drive the study" and therefore, all aspects of the design depend on objectives; in addition to being SMART (See Chap. 3), they must therefore be:

   (a) *Complete*: indicate both the independent (treatments) and dependent variables (parameters to be measured); if not complete, it is impossible to design the experiment. An objective stating "the objective is to compare treatments" or an objective is to "evaluate the performance of varieties" is incomplete.
   (b) *Relevant*: should be relevant to solving the problem of your study. They should link clearly to the research problem and the results expected to provide the solution.
   (c) *Capable of being met by an experiment*: The nature of the objective and variables should be capable of being met by an experiment.

2. *Design of Experiment*

   The experiment should be designed and conducted so that there will be a high probability of measuring differences between treatments with acceptable precision (Campbell and Stanley 1963). Ensure you apply the basic principles of experimentation. Many designs are available to choose from depending on the number of variables and are discussed in the next Chapter.

3. *Treatments*

   The choice of treatments will depend on objectives. The following should be considered when choosing treatment:

- *Quantitative levels*: Some experiments require varying quantities that could have many different levels, e.g. sowing dates, amount of fertilizer to be applied, e.g. comparing the effect of fertilizer N levels on yields. Choosing the level to use depends on the objective (emanated from the research gap) and what you already know about the response. For example, in an experiment to determine the effect of different fertilizer levels on the yield of maize, where the issue is to determine optimal levels for N for farmers, the levels chosen will depend on the levels the farmers are using, the recommended rate and then the new levels you wish to compare (Box 8.3).
- *Comparison and contrasts*: Experiments involve comparisons; ensure all the treatments needed to make all comparisons implied by the objectives are included.
- *Factorial treatment structure*: Many experimental objectives involve investigating interactions by examining several treatment factors. Ensure all relevant combinations of treatments are included.

Box 8.3: Determining Fertilizer Levels in Experiments

Suppose you are conducting an experiment to identify the optimum level that maximizes crop yields; It is important to check recommended fertilizer application rates and methods. You also need to check local agricultural practices and expert recommendations from extension services or experienced farmers in the region to ensure that their rates and practices are factored into the rates in the experiment. It is recommended to test the levels of fertilizer covering a range from low to high. For instance, you can consider levels such as 0 kg/ha (control, no fertilizer), 50 kg/ha, 100 kg/ha, 150 kg/ha and 200 kg/ha. Ensure incremental increases that are uniform, for example, you may increase fertilizer by 25 kg/ha or 50 kg/ha.

4. *Experimental Units*

   The experimental unit is the fundamental entity to which a specific treatment is applied, whether a *person, object, plant, tree, animal, plot or pot* (Kim et al. 2021). This essential element of experimental design represents a subset chosen from a broader population of interest. For instance, in agricultural research, plots within a field are commonly selected as experimental units. These plots represent the larger farming area, and their treatments provide insights into how the interventions may affect the entire population. Table 8.1 shows different types of experimental units.

5. *Number of Replications*

As we have seen, all experiments must be replicated. In addition to the other three reasons given earlier for replication, replication in agriculture has an additional advantage of *insurance*; more replications provide insurance against things going wrong, e.g. crops being destroyed, farmers

Table 8.1 Various combinations of independent variables showing possible experimental units

Independent variable (factor)	Treatments	Experimental unit	Dependent variable
Treated vs. *treated (effect of treatment IV on DV)*	*Intervention*	*Treated and non-treated plots*	*Results being measured*
Fertilizer (Effect of fertilizer levels on crop yields)	Levels of fertilizer	Pots	Crop yield
Animal diets Effect of protein diets on milk yields	Types of diets	Animals	Milk yields
Training (effect of training on uptake of technology)	Trained or untrained	Farmers	Uptake of technology (numbers and extent)
Pesticides Effect of pesticide on yields of crops	Types of pesticides	Plots	Yields, economics
Training on manure used (effect of training on knowledge levels of manure)	Trained and untrained	Farmers	Knowledge levels on the use of manure
Vaccine (Effect of vaccine on disease incidences of goats	Vaccine types	Goats	Number of diseased vs non-diseased

dropping out of the trial, animals dying, etc. The number of replications will be determined by the site factors (land availability, resources, existing variability, etc.). The more replications, the more land is required. A greenhouse experiment may even have 10 replications. Field experiments mainly have three to six replications. Trials with farmers need more replications due to high farm variability (Njuho 1998; Patel et al. 2004).

There should be enough replicates to satisfy all the reasons for replication. Ten is considered a reasonable minimum. Very many replicates have no advantage. The number required for insurance depends on the risk. A long-term trial in a risky environment may be worth ensuring by adding more replications (Watson et al. 2024). This is the case with on-farm trials discussed in Chap. 10. In on-farm experiments, we deliberately use more replicates.

The number of replications is a common question that students ask. Determining the number of replications is often a decision involving trade-offs between statistical rigor and practical constraints. Most authors (Berndtson 1991; Patel et al. 2004; Aaron and Hays 2004; Hall 2007; Gent et al. 2018) determine the number of replications based on the following:

- Use of power analysis.
- Previous studies and pilot studies.
- Resource constraints.
- Expected biological variability and desired precision.

In most crop, soil, fodder and forage experiments, researchers have primarily utilized 3, 4 or 5 replications as these numbers strike a balance between statistical rigor and practical considerations. The choice is primarily based on a literature review of similar studies, but researchers also consider the number of treatments, and resource constraints, including land, budget, and available labor. The availability of land and the possibility of introducing more variability in many treatments are always major considerations in determining the number of treatments and replication to manage variability in various experimental conditions effectively.

6. *Control of Environment and Control Treatment*

What is control, and why is it necessary? We need to ensure two types of controls in experiments: controlled environment and controlled treatment as discussed below:

1. *Controlled variables*: this requires researchers to keep conditions that are not being investigated constant. This encompasses elements like temperature, humidity, soil conditions and other environmental variables that may vary across experimental units. This allows researchers to isolate the effects of the specific treatments they are studying. For instance, in a fertilizer trial, the same cultivars should be used, and all non-treatment factors must be kept constant.
2. *Control treatment*: This is a treatment incorporated into the experiment and used to establish a *baseline or reference point* against which the effects of other treatments can be compared (Fig. 8.2). The control treatments usually involve maintaining the status quo or applying a standard practice, i.e. the independent variable is kept at a normal level. For example, in an experiment assessing the effect of a diet fortified with vitamin A on the growth of animals, the control recipients could have the normal diet used by farmers. The selection of control treatment is a critical decision that highly depends on the objective of the study. You should always consider suitable controls, which are not necessarily treatment without any intervention (see Box 8.4).

Fig. 8.2 A control plot next to a treated plot for comparison; in a soil fertility experiment in Tharaka-Nithi county, Kenya

Box 8.4: Selecting Control Treatment

Control treatment may not mean "zero input" (Patel et al. 2004). Farmers' practice could be the most appropriate. At times, you could include two controls, which may include a farmer practice and an absolute control (no inputs). See the following examples.

Fertilizer experiment
Research objective: Assess the effectiveness of various fertilizer formulations on rice yields under smallholder farms. Suitable control could have the following:

- Plot with no input (zero inputs)
- Farmer practice: plot with farmers conventional fertilizer use

Crop protection experiment
Research objective: Assess the efficacy of different pesticides in controlling early blight in tomatoes. Suitable control could have the following:

- A group (plot) with no pesticide application
- Placebo; apply water or an inert material
- Farmer practice: Conventional pesticides used by farmers

Animal nutrition experiment
Research objective: Assess the effect of different crude protein diets on milk yields in dairy cows. Suitable control could have the following:

- A group of cows on a well-balanced diet
- A group of cows on a farmer's conventional diet (farmers' practice)

7. *Experimental Site*

The site for the experiment will be determined by the research objectives. It has to represent the problem area on a large scale, e.g. in the same agroecological zone. Representativeness on a small scale considers such things as appropriate soil types and previous management. Accessibility, security and adequacy of land are important parameters to consider. The objectives will also dictate the number of sites for the experiment. For instance, if the experiment requires comparing conditions, such as soil type and rainfall, more than one site will be needed.

In site selection, aim to keep sources of variation at a minimum. Avoid fields with poor drainage, ditches, slopes, etc., and prioritize having a homogenous field. The field's history is important, as it helps us know which areas to avoid. Soil fertility gradients are common, especially in smallholder farms. In some cases, researchers may grow catch crops (e.g. non-fertilized maize) to remove soil nutrients, especially nitrogen.

8. *Uniformity Trials*

Uniformity trials are conducted before setting the experiment to assess a particular factor's uniformity and variability across a plot or field. These trials are crucial for controlling spatial variability, one of the main factors that determines the efficiency of experiments (Casler 2015; Jones et al. 2021). They are common in crop breeding and agronomy experiments. Uniformity trials provide valuable information that guides subsequent experimental design and statistical analysis. This helps to minimize bias and random errors caused by uneven treatment application.

In a soil fertility study, a uniformity trial would be a critical component enabling researchers to determine soil fertility variation of a targeted experimental field. A common practice is to plant maize in a field without applying fertilizer. As the crop is growing, assessments are made to determine areas of low and high soil fertility based on the maize growth. The field is divided at harvest into small units (plots and yields recorded). From these yields, a researcher can draw a "fertility contour map" which gives a graphical picture of the variation in soil fertility (Jones et al. 2021). Eventually, this information is used to subdivide the field into relatively homogenous sub-groups to control experimental errors that are critical in designing the experiment. The homogenous groups often form the blocks.

9. *Management of Experiments*

In a field experiment, management means preparing the land, sowing, weeding and all other agronomic practices needed to raise crops, deworming animals, and spraying animals. It is usually not considered as part of the design but can have a large impact on the data, yet they have a large impact on the success of the experiment. Always aim for uniform management of the animals, crops or people for all variables that are not part of the treatments. This will play a key role in *preventing confounding treatments* with management practices. For example, in the management of agricultural experiments, all management practices should be applied following recommended practices except for the treatment variable, e.g. if testing the performance of 6-series maize varieties in each region, the spacing, fertilizer application or input application should be uniform and according to the recommended practices.

10. *Precise Measurements*

Taking precise measurements is key to obtaining accurate and reliable data. Precise measurements reduce measurement errors and enhance the validity of the experimental results. Errors in experiments are mainly from the measurements and positional conditions of the experimental unit or from randomization of the experimental units (Hartung et al. 2019). During planning (in the protocol) clearly indicate the data that will be collected and how each will be collected. Have a data collection schedule that specifies the timing and frequency of data collection, e.g. data may be collected daily (milk yields), weekly or at specific growth stages.

11. *Response Variables*

Several variables describe the experimental units in relation to some of their characteristics and we usually distinguish (i) response variables, (ii) factor variables, and (iii) covariates. The response variables, also called outcome variables, describe the response of units to the experimental treatments (e.g. the yield, the weight, the height, and so on), while the factor variables describe the experimental stimulus allocated to each unit (e.g. fertilizer rate, genotype, herbicide applied, etc.) (Coe et al. 2003, 2014). In some cases, we also record other covariates, also called accessory variables, which measure potential confounding effects. For example, if we intend to study the yield of maize under different fertilizer levels, the soil parameters can act as a confounder. Other covariates, depending on the objective of the experiment, may include pest and disease incidences. Growth parameters, phenological stages data, e.g. flowering and fruiting, may also be taken.

We also collect other types of data depending on the nature of the experiment. *Characterization data* is used to contextualize the site and determine the recommendation domain. These include the location of the experiment, rainfall, soil characteristics, dates of planting, etc.

In any experiment, data collection typically occurs in three main phases. First, before the experiment begins with the aim of assess the appropriateness of the site for research. This involves evaluating soil quality, climate conditions and other environmental variables to ensure they align with the research objectives. This phase also establishes a baseline for comparison and provides context for interpreting subsequent measurements. The second phase involves collecting measurements during the experiment itself. This includes recording various parameters such as labor use, dry matter, degree of damage, and disease incidences throughout the experiment. Finally, at the conclusion of the experiment, measurements are taken to assess outcomes such as crop yields, milk yields and the impact on farm and household incomes. Collecting data at each of these phases is essential for understanding the progression of the experiment, evaluating its outcomes and drawing meaningful conclusions.

12. *Experimental Protocol*

An experimental protocol is an important tool that guides research. It is a written description of everything that will be done in the experiment, starting with the objectives and analysis, interpretation and use of data (Casler 2015; Giraldo et al. 2018; Sichinga-Ligowe et al. 2024). Protocols act as a roadmap for research. Within research organizations, each study has its protocol, typically stored in a database. Often, it is a requirement that management approves the protocol before implementation. This approval process serves as a guide and minimizes the likelihood of errors during implementation. Furthermore, written protocols ensure continuity, particularly in cases of staff turnover. By adhering to protocols, researchers maintain consistency and transparency in their work, facilitating the sharing of findings with fellow scientists. Ultimately, protocols are crucial in guiding research, ensuring data reliability and enhancing collaboration within the scientific community.

In academic proposals, protocols should be included to capture detailed information about the study (Sichinga-Ligowe et al. 2024). This may be included in the protocols in the appendix of the research proposal, as university guidelines often limit the space available for detailed protocols within the main document. The protocol ensures that essential information is accessible without overwhelming the main proposal. Besides helping you plan your research, the protocol is a valuable tool for supervisors to evaluate the research critically. This ensures correct data collection and implementation of the research work. Moreover, in collaborative projects, it may be necessary for each study to have a protocol included in the project file. If in such a project, you must prepare one for your academic research.

A sound experimental protocol must fulfill the following purposes:

- It allows you to include details; it may not be possible to write everything in the research proposal within the word limit.
- Ensures that the experiment is conducted consistently, thus reducing variability during data collection.
- Allows other researchers to replicate the study.
- Allow other researchers to take over if the principal investigator is not in a position to continue, e.g. in case of transfer or natural attrition.
- Record keeping serving as a permanent record of methods and procedures used.

Key Components of a Crop Experiment Protocol

- Introduction: indicate the research objective and hypothesis to be tested, and include background information on the crop, soil and agricultural systems.
- Materials and methods; list all the materials and equipment required for the experiment. Include specifications and quantities, e.g. types of fertilizer and quantities to be applied, crop varieties and source of seeds.
- Experimental design; indicate the experimental design, treatments and number of replications. Explain how randomization will be done and also have a field plan and plot layout plan. Show the size of the plots and the size of the paths and border areas.
- Methods: outline step-by-step procedures to be followed during the experiment. Include information on soil preparation, application of inputs, planting and crop protection. Explain in detail all the agronomic procedures to be followed.
- Data collection methods: explain details of data collection methods. Indicate the data to be collected and how it will be collected. Indicate the step-by-step process of how each data will be collected and recorded. Be specific and include time frames for the various activities. Include how harvesting will be done with the.
- Data analysis: specify the statistical methods of data analysis techniques that will be used to analyze the data. Include statistical tests, software, models, data standards, etc.
- Safety precautions: Describe safety measures and precautions to ensure the well-being of the researchers, workers, data collection technicians and farmers. Include guidelines for handling chemicals and any hazardous materials.
- Timelines: include a timeline for when each phase of the experiment will occur.
- Appendices: include any additional information, documents, data collection forms, consent forms and standard operating procedures.

A researcher, whether a graduate student or a principal investigator, must ensure the protocol is sufficiently detailed, kept up to date and securely archived so that information about the activity can be found in the future and that a biometrician is involved (Patel et al. 2004).

13. *Recording Data*

In all kinds of experiments, data is collected through direct observation and measurements. During data collection, record all the data in a manner that ensures accuracy and consistency using data sheets or digital tools. In addition, regularly calibrating the measuring instruments and equipment, scales, rulers, soil testing devices, soil pH meters, etc., ensures that the technicians collecting the data are well-trained on proper techniques and procedures for making measurements. Use well-designed data sheets that are complete, clear and easy to use.

At the end of the experiment, data should be organized and stored securely ensuring that it is well-documented and easily accessible for analysis and future reference.

During data analysis, data should be checked for accuracy and consistency to identify anomalies and outliers that may need verification and correction.

14. *Data Analysis and Interpretation*

At the end of the study, data analysis is conducted to interpret the results and draw meaningful conclusions. The analysis often requires the use of statistical techniques to make informed recommendations or determine the appropriate course of action (Gomez and Gomez 1984). The method of statistical analysis employed depends on both the objective of the study and the experimental design used. As stated at the beginning of this chapter, statistical analysis in experiments primarily involves *partitioning the sources of variation* in the data to isolate treatment effects from random error. Chapter. 9 shows how the sources of variation are partitioned depending on the design. The researcher must conduct a thorough and appropriate statistical analysis to ensure that the findings are both valid and reliable. For detailed procedures and examples, refer to Chaps. 20 and 21 on experimental data analysis, which guide conducting and interpreting statistical tests for various experimental designs.

15. *Conclusions and Recommendations*

At the reporting stage, researchers organize their findings systematically, known as the experiment's conclusions. The conclusions summarize the discoveries, patterns and relationships while indicating whether they align with the initial hypotheses. Clear and concise conclusions are vital for conveying the significance of the experiment's outcomes and acknowledging any limitations. Ultimately, the conclusions provide a brief overview of the experiment's impact on the research question.

Recommendations guide future research, decision-making and real-world applications. Clear and well-supported recommendations enhance the experiment's usefulness, guiding subsequent studies and facilitating the implementation of research findings. Researchers may also highlight areas where further research is needed (Covered in Part 6 on Reporting).

Recommended Further Reading

1. Campbell DT, Stanley JC. 1963. Experimental and Quasi-Experimental Designs for Research.
2. The Green Book: A guide to effective graduate research in African Agriculture by Patel et al. (2004).
3. Gomez KA, Gomez AA. 1984. Statistical procedures for agricultural research. Statistical procedures for agricultural research with emphasis on rice. (September): 7–8.
4. Little TM., Hills F Jackson. 1978. Agricultural experimentation: design and analysis https://www.wiley.com/en-us/Agricultural+Experimentation%3A+Design+and+Analysis-p-9780471023524
5. Coe, R.D., et al. (2002). Designing Experiments and Analyzing Data. Lecture Notes. World Agroforestry Centre (ICRAF) publication.

Key Points in This Chapter

Brief History of Experiments

R.A. Fisher, a biostatistician, developed experimental design and analysis techniques while at Rothamsted Experimental Station. These methods contributed to the field of experimentation and have been widely adopted in research.

Classes of Experiments

There are two broad classes of agricultural experiments: (1) laboratory experiments conducted in a controlled, artificial setting to control possible extraneous variables and (2) field experiments carried out in the natural environment. Experiments can either be carried out on an on-station site where the researcher solely manages the experiment or in the on-farm environment on the farmer's fields, and the farmer being involved.

Common Types of Field Experiments

Common types of field experiments include the following:

- *Agronomy experiments*: Aim is to develop management practices that optimize the production of the crops.
- *Varietal screening experiments*: Test the performance of different varieties to compare their performance across some characteristics.
- *Forage production experiments*: Aim to test different forage varieties, management practices and environmental conditions to determine the best strategies for maximizing yields and quality of forages.
- *Intercropping experiments*: designed to help understand interactions between species so that we can design systems that enhance resource utilization.
- *Crop rotation experiments*: investigate how different rotation patterns can be manipulated to enhance agricultural productivity.
- *Soil fertility experiments*: Evaluate the effect of soil management practices on the availability of soil nutrients and agricultural productivity.
- *Animal nutrition experiments*: Feeding trials to compare the effects of different diets on livestock performance.
- *Veterinary clinical experiments*; Aim to advance animal health care by identifying the most effective therapies and practices for a given condition.
- *Agroforestry experiments*: develop integrated systems of agriculture and forestry components that are productive, profitable, healthy and sustainable.
- *Randomized controlled evaluations*: Randomized experiments aimed to evaluate the impact of development programs or policy interventions on specific outcomes.
- *Crop breeding experiments* involve changing crop traits to incorporate desired characteristics to enhance yield, resistance, and adaptability.
- Biotechnology experiments: apply advanced biological techniques to increase agricultural productivity.

General Principles of Experimentation

There are three main principles of experimentation, namely: randomization, replication, and local control (reducing noise). They are also known as the 3 Rs. *Randomization* refers to *the* random assignment of treatments or conditions of the study. *Replication* means repeating treatments or study conditions. *Local control* is a technique that involves blocking and grouping experimental units into homogeneous entities.

Noise Versus Signal

Experimental error is also called "noise" because it makes it difficult to hear the signal (detect response). In a crop experiment, the experimental error is likely to be from (1) human while making observations and measurements, (2) inherent variability among plots, e.g. the soil differences, such as drainage, elevation and compaction, (3) lack of uniformity in the physical conduct of the experiment, (4) interference among treatments in neighboring plots, and (5) non-uniform behavior of all other extraneous factors.

Elements of Designing Effective Agricultural Experiments

In addition to the application of the three principles of experimentation (the 3Rs), designing an effective agricultural experiment requires you to consider the following key elements:

1. Objectives and hypotheses.
2. Experimental design.
3. Treatments.
4. Experimental units.
5. Number of replications.
6. Control (Treatment and controlling the variables).
7. Uniformity trials.
8. Site selection.
9. Management of the experiment: Always aim for uniform management of the animals, crops, or people for all variables that are not part of the treatments. This will play a key role in preventing confounding treatments with management practices.
10. Experimental protocol.
11. Taking precise measurements is key to obtaining accurate and reliable data. Precise measurements reduce measurement errors and enhance the validity of the experimental results.
12. Response variables: distinguish (i) response variables, (ii) factor variables, and (iii) accessory variables.
13. Recording data.
14. Data analysis and interpretation.
15. Conclusions and recommendations.

Review Questions

1. Who was R.A. Fisher, and what was his significant contribution to the field of experimentation?
2. Explain the "noise versus signal" analogy in experimental error. What is the "signal" you are trying to detect in a crop experiment, and what constitutes the "noise" (experimental error)?
3. Identify and explain the three main principles of experimentation (the 3 Rs).
4. Why is uniform management of an experiment an important principle?
5. Explain how to determine the number of replications.
6. Why is developing a detailed experimental protocol essential before commencing any agricultural experiment?

Chapter Exercise

1. Identify and describe three distinct sources of experimental error ("noise") in a crop experiment. For each, suggest a practical measure a researcher could take during the experiment's management to minimize that specific error.
2. Choose a recent peer-reviewed journal article that reports the results of an agricultural experiment in your area of specialization. After reading the article, identify how the authors applied the three core principles of experimental design: randomization, replication, and local control (blocking). Were these principles clearly stated and appropriately implemented? Give a personal critique.

Answers to Review Questions

- Q1: Reference 8.1.1
  - Laid the foundation principles for experimentation
  - Introduced the concept of 3R
  - Developed ANOVA
  - Provided a systematic and objective approach to scientific inquiry
- Q2: Reference 8.3.4
- Q2: Randomization, Replication and educe Noise - Local Control (3R). Reference 8.4.1, 8.4.2, 8.4.3
  - Randomization involves random assignment of treatments to experimental units. This helps to evenly distribute the unknown extraneous variables.
  - Replication: Repeating each treatment. It increases precision and reliability, allows estimation of experimental error, and enhances generalization.
  - Local; control: groups experimental units in more homogenous blocks or entities; this reduces experimental error, increases precision, and improves efficiency.
- Q4: Reference 8.5 (9)
- Q5: Reference 8.5 (5)

References

Aaron DK, Hays VW (2004) How many pigs? Statistical power considerations in swine nutrition experiments. J Anim Sci 82:E245–E254. https://doi.org/10.2527/2004.8213_SUPPLE245X

Atangana A, Khasab D, Chang S, Degrande A (2014) Agroforestry and biodiversity conservation in tropical landscapes. Trop Agroforest:227–232. https://doi.org/10.1007/978-94-007-7723-1_11
Babbie ER (2021) The practice of social research, p 575
Badu-Apraku B, Fakorede MAB, Menkir A, Sanogo D (2012) Conduct and management of maize field trials. IITA
Berndtson WE (1991) A simple, rapid and reliable method for selecting or assessing the number of replicates for animal experiments. J Anim Sci 69:67–76. https://doi.org/10.2527/1991.69167X
Berry JC, Qi M, Sonawane BV et al (2022) Increased signal-to-noise ratios within experimental field trials by regressing spatially distributed soil properties as principal components. elife 11. https://doi.org/10.7554/ELIFE.70056
Brown AC (1993) Variability in biological systems. S Afr J Sci 89:308
Brüggemann J, Bizer K (2015) Laboratory experiments in innovation research: a methodological overview and a review of the current literature. J Innov Entrep 5:1–13. https://doi.org/10.1186/S13731-016-0053-9/TABLES/1
Campbell DT, Stanley JC (1963) Experimental and quasi-experimental designs for research. Mifflin and Company
Casler MD (2015) Fundamentals of experimental design: guidelines for designing successful experiments. Agron J 107:692–705. https://doi.org/10.2134/AGRONJ2013.0114
Coe R, Huwe B, Schroth G, Sinclair F (2003) Designing experiments and analyzing data. World Agroforestry Centre (ICRAF)
Coe R, Sinclair F, Barrios E (2014) Scaling up agroforestry requires research 'in' rather than 'for' development. Curr Opin Environ Sustain 6:73–77. https://doi.org/10.1016/J.COSUST.2013.10.013
Cook TD, Campbell DT (1979) Quasi-experimentation: design and analysis issues for field settings. Houghton Mifflin, Boston
Demétrio CGB, Menten JFM, Leandro RA, Brien C (2013) Experimental power considerations—justifying replication for animal care and use committees. Poult Sci 92:2490–2497. https://doi.org/10.3382/PS.2012-02731
Finney DJ (1990) Intercropping experiments, statistical analysis, and agricultural practice. Exp Agric 26:73–81. https://doi.org/10.1017/S0014479700015428
Fisher J (1980) R.A. Fisher and the Design of Experiments, 1922–1926. Am Stat 34:1–7. https://doi.org/10.1080/00031305.1980.10482701
Gent DH, Esker PD, Kriss AB (2018) Statistical power in plant pathology research. Phytopathology 108:15–22. https://doi.org/10.1094/PHYTO-03-17-0098-LE/ASSET/IMAGES/LARGE/PHYTO-03-17-0098-LE_T3-1512962124331.JPEG
Giraldo O, Garcia A, Corcho O (2018) A guideline for reporting experimental protocols in life sciences. PeerJ 2018:e4795. https://doi.org/10.7717/PEERJ.4795/FIG-8
Gomez KA, Gomez AA (1984) Statistical procedures for agricultural research. In: Statistical procedures for agricultural research with emphasis on rice, pp 7–8
Hall NS (2007) R. A. Fisher and his advocacy of randomization. J Hist Biol 40:295–325. https://doi.org/10.1007/S10739-006-9119-Z/METRICS
Hartung J, Wagener J, Ruser R, Piepho HP (2019) Blocking and re-arrangement of pots in greenhouse experiments: which approach is more effective? Plant Methods 15:1–11. https://doi.org/10.1186/S13007-019-0527-4/FIGURES/1
Hasan N, Choudhary S, Naaz N et al (2021) Recent advancements in molecular marker-assisted selection and applications in plant breeding programmes. J Genet Eng Biotechnol 19:128. https://doi.org/10.1186/S43141-021-00231-1
Hirst JA, Howick J, Aronson JK et al (2014) The need for randomization in animal trials: an overview of systematic reviews. PLoS One 9:e98856. https://doi.org/10.1371/JOURNAL.PONE.0098856
Hofseth LJ (2018) Getting rigorous with scientific rigor. Carcinogenesis 39:21–25. https://doi.org/10.1093/CARCIN/BGX085
Ilan Y (2020) Order through disorder: the characteristic variability of systems. Front Cell Dev Biol 8:495391. https://doi.org/10.3389/FCELL.2020.00186/XML/NLM
Johnston AE, Poulton PR (2018) The importance of long-term experiments in agriculture: their management to ensure continued crop production and soil fertility; the Rothamsted experience. Eur J Soil Sci 69:113–125. https://doi.org/10.1111/EJSS.12521
Jones M, Harbur M, Moore KJ (2021) Automating uniformity trials to optimize precision of agronomic field trials. Agronomy 11:1254. https://doi.org/10.3390/AGRONOMY11061254/S1
Kim SB, Kim DS, Magana-Ramirez C (2021) Applications of statistical experimental designs to improve statistical inference in weed management. PLoS One 16:e0257472. https://doi.org/10.1371/JOURNAL.PONE.0257472
Little TM, Hills FJ (1978) Agricultural experimentation: design and analysis, p 350
Marchant B, Rudolph S, Roques S et al (2019) Establishing the precision and robustness of farmers' crop experiments. Field Crop Res 230:31–45. https://doi.org/10.1016/J.FCR.2018.10.006
Mead R, Curnow RN, Hasted AM (2017) Statistical methods in agriculture and experimental biology: third edition. Statistical methods in agriculture and experimental biology: third edition 1–472. https://doi.org/10.1201/9780203738559/STATISTICAL-METHODS-AGRICULTURE-EXPERIMENTAL-BIOLOGY-ROGER-MEAD/RIGHTS-AND-PERMISSIONS
Meier JS, Kreuzer M, Marquardt S (2012) Design and methodology of choice feeding experiments with ruminant livestock. Appl Anim Behav Sci 140:105–120. https://doi.org/10.1016/J.APPLANIM.2012.04.008
Njuho PM (1998) Some statistical considerations to on-farm trials in Kenya. https://doi.org/10.4148/2475-7772.1291
Patel BK, Muir-Leresche K, Coe R, Hainsworth SD (2004) Green book a guide to effective graduate research in African agriculture. Environment and Rural Development
Payne RW (2015) The design and analysis of long-term rotation experiments. Agron J 107:772–785. https://doi.org/10.2134/AGRONJ2012.0411
Ranjha MMAN, Shafique B, Khalid W et al (2022) Applications of biotechnology in food and agriculture: a mini-review. Proc Natl Acad Sci India Sect B Biol Sci 92:11–15. https://doi.org/10.1007/S40011-021-01320-4/METRICS
Shadish WR, Cook TD, Campbell DT (2002) Experimental and quasi-experimental designs for generalized causal inference. Boston, MA, USA: Houghton Mifflin Company
Sichinga-Ligowe I, Joy EJM, Manzeke-Kangara MG et al (2024) An introduction to conducting responsible and reproducible agricultural research. In: CAB reviews: perspectives in agriculture, veterinary science, nutrition and natural resources, p 19. https://doi.org/10.1079/CABIREVIEWS.2024.0058/ASSET/19F08C22-4AA3-424B-99DB-F9DA93A15F69/ASSETS/GRAPHIC/FIGURE
Taylor P, Meyer RE (2023) Veterinary clinical research or experiments on pets. Vet Anaesth Analg 50:383–385. https://doi.org/10.1016/j.vaa.2023.07.006
Watson CA, Topp CFE, Mead A et al (2024) Future proofing a long-term agricultural experiment for decades to come: relocation and redesign. Eur J Agron 158:127214. https://doi.org/10.1016/J.EJA.2024.127214
Yates F (1964) Sir Ronald fisher and the design of experiments. Biometrics 20:307. https://doi.org/10.2307/2528399

Common Experimental Designs and Layout of Experiments

9

Abstract

After discussing the fundamentals of experimentation in Chap. 8, we present common designs in agricultural experiments. We first cover three basic designs: completely randomized design (CRD), randomized complete block design (RCBD), and Latin square design (LSD), followed by factorial and split-plot experiments. In each of the designs and experiments, characteristics of the design are described, and examples are presented. In addition, a generic analysis of variance (ANOVA) table is shown to illustrate how data is analyzed by partitioning in each design. The advantages and disadvantages of each of the designs are also outlined. The second section of the Chapter tackles the designing and laying out of the experiments, presenting ten steps in the planning phase, plus a detailed presentation of how to implement an agronomy-based experiment.

Keywords

ANOVA table · Data collection · Completely randomized design · Complex factorial experiments · Factorial experiment · Field plan layout · Guard zone · Homogeneity · Latin square · Design · Net plot area · One-factor experiment · Plot layout plan · Randomized complete block design · Split plot experiment · Two factor experiment

Learning Outcomes

By the end of this chapter, the learner is expected to be able to:

1. Apply a completely randomized design (CRD) to design an experiment.
2. Distinguish between a completely randomized design and randomized complete block design.
3. Apply the concept of experimental error to design valid experiments.
4. Differentiate between a "factorial experiment" and a "split-plot experiment."
5. Recognize the unique characteristics of a "Latin Square Design."
6. Apply the Latin square design to animal experiments.
7. Relate each of the designs to variation partitioning and the respective ANOVA table.
8. Plan and lay out a well-designed field experiment.
9. Demonstrate how to determine the number of rows and guard rows in an agronomy experiment.

9.1 Designing and Implementing an Experiment

9.1.1 What Is an Experimental Design?

Experimental design is a framework that guides researchers in a systematic and controlled approach to inquiry (Patel et al. 2004; Diggle and Chetwynd 2011). Through this framework, hypotheses about variable relationships are objectively tested by intentionally manipulating independent variables and observing their effects on dependent variables. The core purpose is to establish causality within a structured and replicable context.

Data generated from the experiments exhibit variability, *known and unknown*; the experimental design enables a researcher to (1) control unwanted variability (due to confounders) while maximizing desired variability (due to treatments), and (2) *partition* variability into various components during statistical data analysis. Indeed, statistical analysis's main purpose is to account for the *variability in data* to ensure the findings' validity. This Chapter demonstrates partitioning and accounting for data variability by including a generic ANOVA table for each design. The ability to account for all variability (desirable and undesirable) is the key to making sound inferences.

J. Njeri Mugwe, S. Runo, *Research Methodology in Agricultural Sciences*,
https://doi.org/10.1007/978-981-95-1892-0_9

9.1.2 Rigor in Conducting an Experiment

The rigor adopted in designing, implementing, and analyzing an experiment is inherently connected (Hofseth 2018; Sichinga-Ligowe et al. 2024). A well-designed experiment, aligned with the objectives, generates valid data, while appropriate data analysis provides valid statistical inferences. Conversely, poorly designed experiments raise doubts about the validity of the statistical inference drawn from the data. Therefore, a researcher must always be meticulous in *design and analysis phases* to ensure the integrity and validity of the research findings.

Many "experimental designs" exist in agriculture (Little and Hills 1978; Gomez and Gomez 1984; Mead et al. 2017). The first section of this chapter covers commonly used designs in agriculture and outlines their structure and form. Each design has advantages and disadvantages, and a researcher must carefully consider these factors when selecting the appropriate design. The choice depends mainly on the study's objective, site conditions, structure of the treatments, variability of the experimental units, and available resources.

9.2 Completely Randomized Design

9.2.1 Characteristics of Completely Randomized Design

The simplest experimental design is the "completely randomized design" (CRD) (Reddy 2019). This design randomly allocates treatments to experimental units, meaning that randomization is done without any restrictions. It implies that each experimental unit has the same chance of being assigned any treatment. In this design, treatments are replicated but with no blocking.

The design is suitable when the entire experimental area exhibits *homogeneity*, meaning it is uniform in all aspects that could affect the results. However, achieving uniformity in field experiments, such as on-farm, is rare, making the use of a completely randomized design uncommon in such settings.

Nevertheless, blocking in CRD is often useful even in the more controlled environment of a greenhouse (Brien et al. 2013; Ledesma 2015). Blocking with a fixed-position arrangement was more efficient in the precision of the greenhouse experiment than the re-arrangement of pots (Hartung et al. 2019). This approach allows for more systematic control over potential sources of variability within an agricultural setting (see Fig. 9.1). For example, in a greenhouse setting, factors such as light intensity, temperature gradients, and humidity levels can vary across the space and affect the outcome. By strategically incorporating blocking, researchers can account for spatial variations in factors and other localized effects that might impact the experiment.

Suitable experiments for CRD include the following:

1. Laboratory experiments involving incubation, pathology, and germination tests.
2. Animal experiments, e.g., feeding experiments in a uniform environment, where animals are confined in cages (Fig. 9.2).
3. Pot experiments: pots arranged in a similar environment and conditions controlled.

Fig. 9.1 Blocking in a glass house experiment might be necessary

Fig. 9.2 Poultry experiment: chicken is confined in cages following a completely randomized design. (Photo credit: Maina Anderson N)

9.2.2 Steps in Laying out a Completely Randomized Design Experiment

Laying out a CRD follows the following steps:

1. Using the research objective, specific treatments to be compared and the response variables to be measured, identify and define the experimental units: pots, plots, etc.
2. Randomly assign the experimental units to one of the treatment groups using your preferred randomization techniques.
3. Replicate each treatment and apply it to multiple experimental units.
4. Apply treatments to the experimental units according to the randomization plan.

To demonstrate laying out a CRD in the field (assuming the field is uniform), we present two examples:

Example 1: *Test the effect of three nitrogen rates on growth and yield of maize yield;* with the following attributes:

- Nitrogen rates include: 0, 30, and 60 kg N ha^{-1} (treatments).
- The treatments are replicated three (3) Times. Thus, the required number of plots is 3 × 3 = 9.
- Site: open field.
- Experimental unit: plot.

In this case, we start by defining the total number of plots, 9 plots, followed by assigning a plot number to each experimental unit, and finally allocating the treatments at random (Fig. 9.3).

Example 2: *To evaluate the effect of inoculating soybean with rhizobium under two soil types on growth and nodulation:*

- Soils with different pH: Soil A with pH 4.8 and Soil B with pH 6.6.
- Treatments: (1) no inoculation (soil pH 4.8), (2) no inoculation (soil pH 6.6), (3) with inoculation (soil pH 4.8), and (4) with inoculation (soil pH 4.8),
- The treatments are replicated six times (6 times). Thus, the required number of pots is 4 × 6 = 24.
- Site: Potted in a greenhouse.
- Experimental unit: pot.

We start by defining the total number of pots (24 pots), then acquire the pots and assign them a pot number, and finally, allocate the treatments at random (Fig. 9.4).

We present more examples from published work (Table 9.1).

9.2.3 Analysis of a Completely Randomized Design

In the analysis of completely randomized designs (CRD), the variation among observations comes from two main sources: *treatment effects and experimental error.* Treatment variation refers to differences in the response variable attributed to the different levels of the independent variable (treatments) being tested. On the other hand, experimental variation, also known as error variance, accounts for random variability within treatment groups that the treatments themselves cannot explain.

Chapter 20 on data analysis provides a detailed analysis of CRD. The ANOVA table associated with CRD is as follows: Basic Model: Observed yield = Treatment effect — Variation unit variation (Table 9.2).

9.2.3.1 Advantages of Completely Randomized Designs

1. The layout is simple and easy to manage.
2. Full flexibility is permitted, allowing for using any number of treatments and replicates.
3. Statistical analysis is relatively straightforward, even when dealing with varying replicates and experimental errors across different treatments.

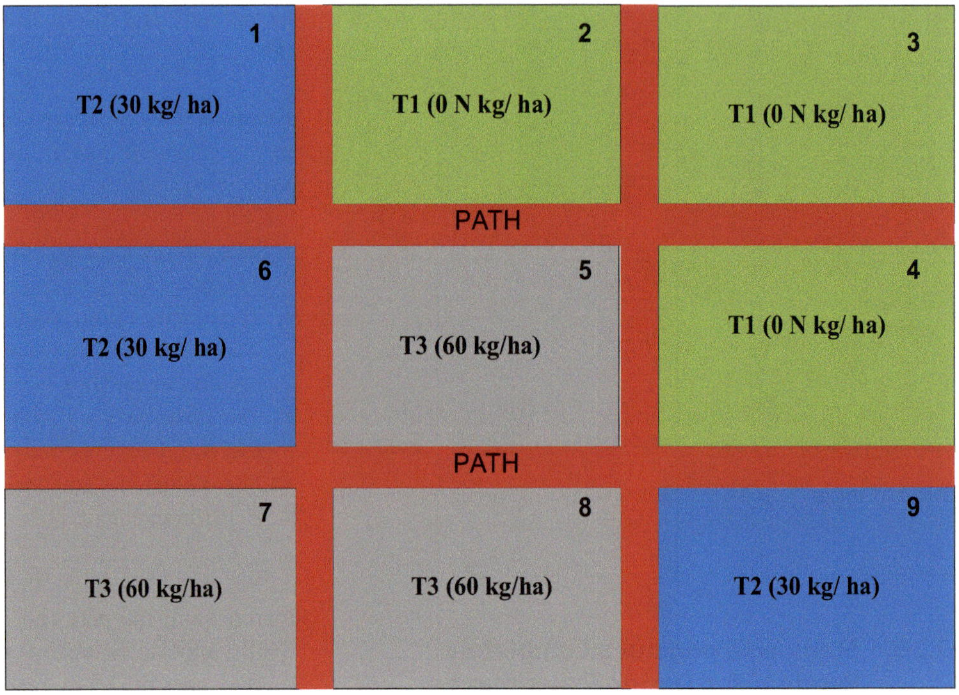

Fig. 9.3 Completely randomized design testing effect of three fertilizer rates on maize yields (treatments = 3, replications = 3)

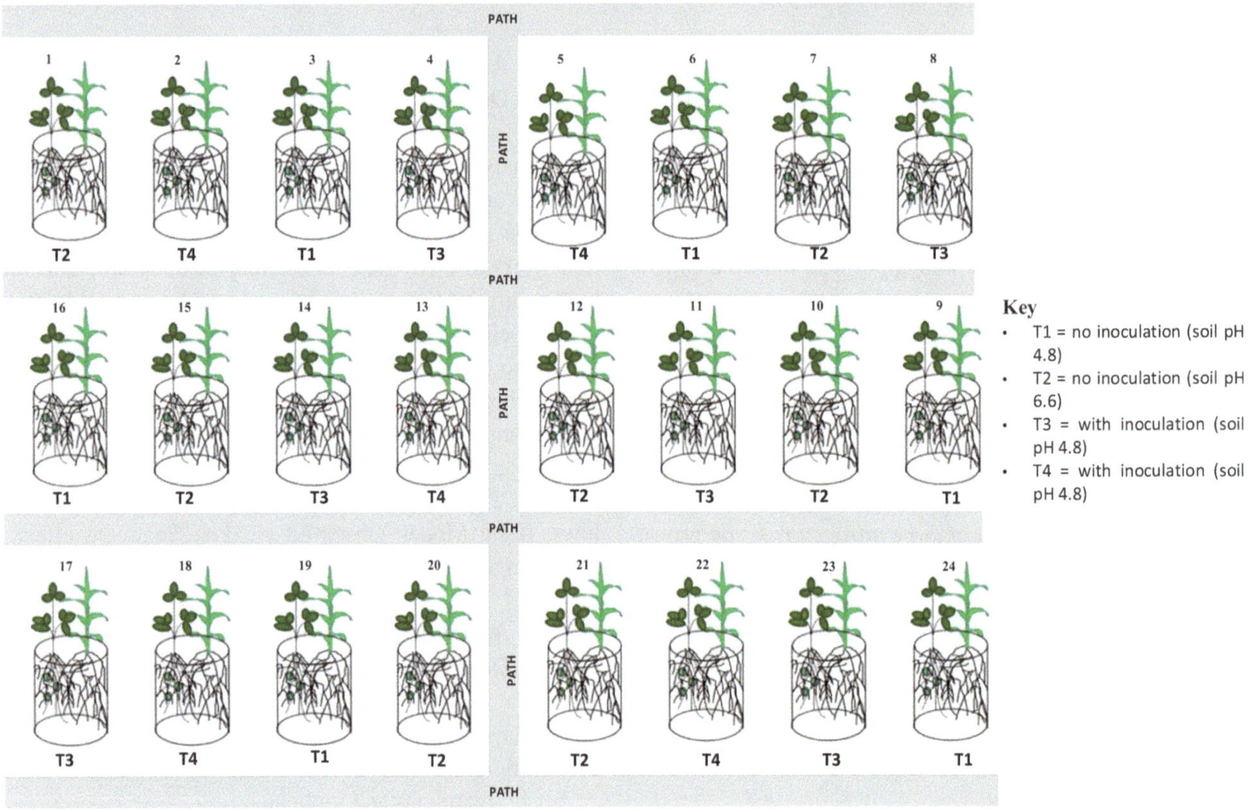

Fig. 9.4 Potted experiment, with 4 treatments replicated 6 times (24 pots) in a glass house

Table 9.1 Examples of published studies that used a completely randomized design

Objective	Treatments	Replications	No of experimental units	Reference
To determine the effects of supplementing a low-quality basal diet (hay) with either Berchemia discolor or *Zizyphus mucronata* on performance of growing goats in Kenya	Five treatments T1. Control; Rhodes grass hay ad libitum plus 60 g maize bran. T2. Control plus 15% B discolor. T3. Control plus 30% B discolor. T4. Control plus 15% Z mucronate. T5. Control plus 30% Z mucronate	Four (4) replications	5 × 4 = 20 goats	Osuga et al. (2012)
To determine the effect of feeding two algal oils on egg production and DHA, enrichment in eggs and selected tissues Algal oils OmegaPro advance algal oil (APOA) and crude algal oil (CAO)	Five treatments T1. Control: Standard diet T2. APOA (0.25). T3. APOA (0.76) T4. CAO (0.23) T5. CAO (0.69)	Eight replications per diet 40 enriched colony cages with 10 birds per cage. All the cages were similar in all aspects.	40 × 10–400 birds	Maina et al. (2023)
To evaluate the effect of compost and inorganic nitrogen fertilizer and their mixtures on maize growth, yield, and nitrogen uptake	Ten treatments Factorial combinations of five compost rates (0, 5, 10, 15, and 20 t·ha^{-1}) and four rates of inorganic nitrogen fertilizer (0, 46, 92, and 138 kg·N·ha^{-1})	Three	10 × 3 = 30 pots Pot experiment	Nigussie Ochieng' et al. (2021)
To assess the effect of various inorganic soil amendments on reducing cadmium (cd) uptake by tobacco plants grown in agricultural soils with low total cd concentrations	Five treatments representing soil amendments T1. Control T2. Sepiolite T3. Zeolite T4. Hapatite T5. Appatite II	Three	5 × 3 = 15 Two soil types (30 pots)	Keller et al. (2005)

Table 9.2 ANOVA table template for a completely randomized design

Source of variation	Degrees of freedom (df)	Sum of squares SS	Mean sum of squares MSS	F	P
Treatments (between treatments)	k-1	SS (Tr)	MS (Tr)	MS (Tr)/ MSE	
Error (within treatments)	n-k	SS (error)	MS (error)		
Total	n-1	SS (total)			

k number of observations for treatment, *n* total number of observations

4. Analysis remains uncomplicated even when data is incomplete or missing.
5. It maximizes the degrees of freedom for error given a specific number of experimental units and treatments.

9.2.3.2 Disadvantages of Completely Randomized Designs

1. Relatively low accuracy due to lack of restrictions, which allows environmental variation to enter experimental error.
2. Not suitable for large numbers of treatments because a relatively large amount of experimental material is needed, increasing the variation.

9.3 Randomized Complete Block Design

9.3.1 Nature of Randomized Complete Block Design

Randomized complete block design (RCBD) is the most commonly used design in agricultural field research (Ochieng' et al. 2021). Researchers consider this as the standard design for agricultural experiments, though it is also extensively used in medicine, biology, and social and business studies. In this design, treatments are replicated and blocked, which means that plots are arranged into blocks, and then treatments are randomly assigned to plots within a block (Fig. 9.5). In other words, the researcher places similar experimental units together into blocks or replicates. This arrangement controls variation in an experiment by accounting for spatial effects in the field.

The RCBD is an expansion of CRD specifically employed when the experimental units exhibit heterogeneity (Ochieng' et al. 2021). The fundamental concept of RCBD involves grouping all experimental units into homogeneous blocks, thereby *minimizing variation within blocks* and maximizing variation between them (Shieh and Jan 2004; Mead et al. 2017). During analysis, differences attributed to blocks are accounted for, effectively reducing random variation and enhancing the precision of treatment comparisons. Removing

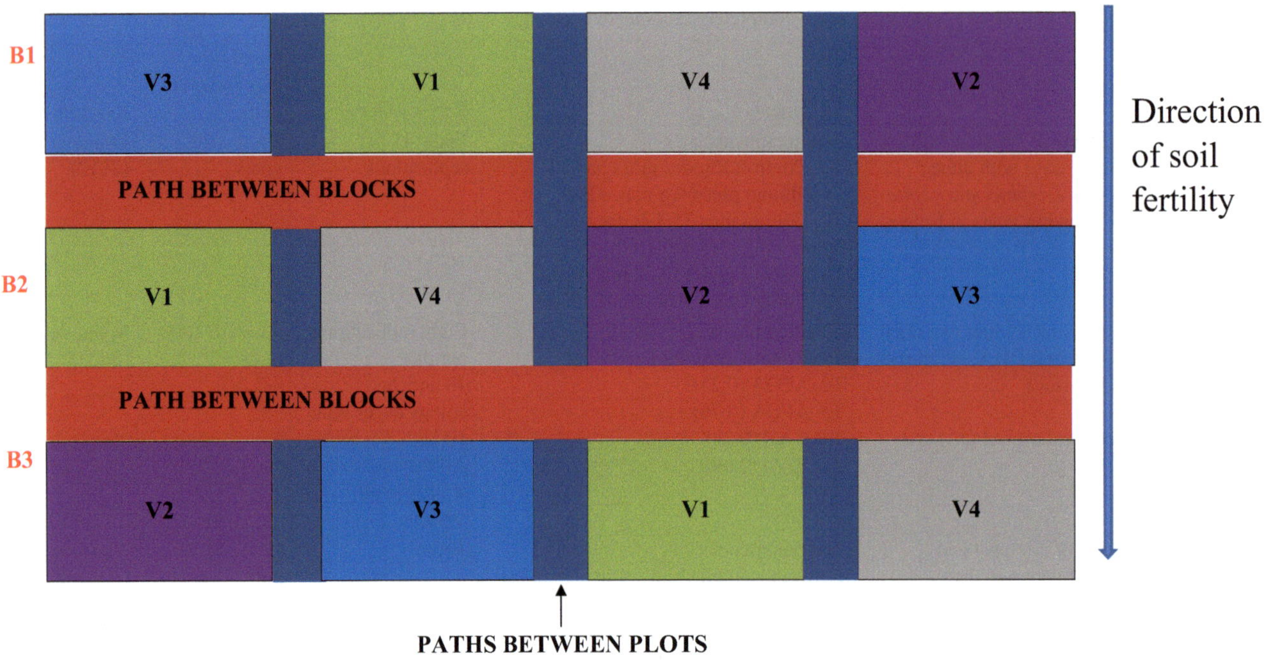

Fig. 9.5 Randomized complete block design showing blocks and plots layout (treatments = 4, replications = 3)

this source of variation within replicate variation reduces experimental error (noise) and improves the ability to detect smaller treatment effects. In other words, a "signal" can easily be detected, unlike when the noise is large.

The effectiveness of the design hinges on the ability to identify the patterns of non-uniformity in a field, such as changing soil types, drainage patterns, fertility gradients, direction of insect migration into a field, etc. Identifying and accounting for these sources of variability allows researchers to create blocks. These factors are used as the blocking criteria. For example, in an experiment involving the testing of varieties in a field known to have different soil fertility, the differences in soil fertility are used as the blocking factors, where experimental units in each block occupy a homogenous soil fertility level. At times, researchers deliberately plant a uniformity trial, the assessment of which identifies areas of similar soil fertility levels. This information is consequently used to lay the blocks.

Similarly, the slope of the land (degree of slope) is another common blocking factor. In this case, blocking is done against the slope; the upper part constitutes the first block, and the rest of the blocks are demarcated such that each of the blocks bears the same slope (Fig. 9.5). In animal experiments, too, blocking to reduce the source of variation from factors such as the animal's weight, age, breed, etc., is necessary. For example, in an experiment involving dairy cows, Rauch Ochieng' et al. (2021) used an RCBD and blocked the cow's parity, days in milk, and milk protein yield.

In on-farm experiments (discussed in Chap. 10), RCBD is preferred because it offers practicality and efficiency in accommodating the spatial and environmental variability present in farm settings. The farms are treated as blocks (replicates), and treatments are applied in the various farms. This means a farmer is a replicate, and the treatment appears only once on each farm. However, in big farms, a standard RCBD experiment can be conducted depending on the objective of the experiment.

Depending on land availability and the number of treatments, a researcher may have a "complete block" or "incomplete block." The complete block will have each treatment applied in blocks, i.e. balanced. The incomplete, on the other hand, will have one or more treatments missing in some blocks. This should not be confused with missing data. An experiment could have missing data and still qualify to be a complete block. If a block is only missing data from a bunch of observational units the experiment will still be referred to as RCBD. In cases of missing data, it is easy to sort out as the analysis can still be performed (Gomez and Gomez 1984).

9.3.2 Field Layout of the Randomized Complete Block Design

Here, we use two examples: testing rice varieties and a feeding experiment:

9.3.2.1 Example 1: Experiment on Testing Rice Varieties

Objective: To evaluate the yield response of four Nerica rice varieties, V1, V2, V3, and V4, in a subhumid zone. The four rice varieties are the treatments; the researcher wishes to test them and pick the best-yielding variety.

Number of replications: The treatments are replicated three times, giving 12 plots (4 varieties × 3 replicates). Blocking is done against soil fertility gradient and therefore, the blocks are assumed to be homogenous. Blocking helps in reducing random errors within the treatments. Each treatment appears only once within a block. In our case, all the four treatments must appear in the blocks (Fig. 9.5).

Also, note that inter-block paths are slightly larger than inter-plot paths.

9.3.2.2 Example 2: Animal Feeding Experiment

Objective: To evaluate the effect of different dietary protein inclusion levels on goat growth performance, feed efficiency, and body composition.

Treatments: Treatments are based on the percentage of protein included in the total diet, resulting in five total groups: one control and four increasing inclusion levels. The diet should be otherwise balanced for energy, vitamins, and minerals:

1. T1 Control (Standard basal diet, which is the baseline/recommended 16% total protein).
2. T2 = Standard diet + 18% total protein.
3. T3 = Standard diet + 20% total protein.
4. T4 = Standard diet + 22% total protein.
5. T5 = Standard diet + 24% total protein.

Experimental Unit: Group of goats. Here, we chose six goats per treatment.

Blocks: Involves grouping animals into sets based on certain characteristics (e.g., age, weight, or breed, sex). Treatments are then randomly assigned to animals within each block. The group of goats receiving similar treatment can be in one pen (representing an experimental unit). During data analysis, the mean is calculated.

Number of Replications: We could have four blocks representing four replications (Fig. 9.6).

In this figure, a picture of a goat represents five goats (five goats per treatment), the experimental unit. In total we need 20 × 5 = 100 goats.

9.3.2.3 What Are the Experimental Units in Animal Experiments?

Students conducting animal experiments often face difficulties determining the appropriate experimental units and blocking variables for their study. These difficulties arise due to the complexity of animal research and the need to account for various sources of variability that can influence experimental outcomes. To determine the appropriate experimental units, students must decide whether they should use *individual animals* or *groups* while selecting blocking variables involves identifying factors to control unwanted variation (Bate and Clark 2014). Understanding the interactions between treatments and blocking variables is essential, and selecting the right experimental design requires careful consideration of objectives and constraints. Boxes 9.1 and 9.2 provide common blocking variables and likely experimental units in animal experiments.

Box 9.1: Common Blocking Variables in Animal Experiments

The following factors should be taken into account and controlled for in the study design (Bailoo et al. 2014; Festing 2014; Laajala et al. 2016):

- *Sex:* Male and female animals may respond differently to treatments due to hormonal differences. Therefore, blocking by sex helps to account for sex-related effects.
- *Age:* Animals of similar age may have similar physiological characteristics and growth rates. Blocks help to control for age-related differences. However, note that it is rare to include animals of different ages within a single scientific experiment because age introduces physiological, metabolic, and behavioral variability that confounds treatment effects, complicates analysis, reduces reproducibility, and compromises the validity and interpretability of results.
- *Weight:* Animals of similar weight may have similar metabolic rates and nutritional requirements. Variation in initial body weight among experimental animals introduces physiological differences that confound treatment effects and reduce experimental precision and reproducibility; this can be mitigated by using animals of uniform weight (highly recommended), blocking by weight during randomisation, or adjusting for initial weight as a covariate in analysis.
- *Genetics:* Animals from different genetic backgrounds and breeds may exhibit variability in traits of interest. Blocking by genetic lineage or strain helps to control for genetic differences.
- *Housing conditions*: Animals housed in different environments may experience differences in temperature, humidity, and social interactions. Blocking by housing conditions helps to standardize the environment.
- *Lactation stage in milking cows*: The stage of lactation can significantly affect experimental outcomes. It is important to consider early, mid, or late lactation stages.
- *Parity:* Older cows (higher parity) may have different milk production and composition characteristics compared to younger cows (lower parity).
- *Health status:* The health status of an animal, including disease resistance and reproductive status, should be considered and controlled for in the study design. As far as possible, no research should be carried out with sick animals.

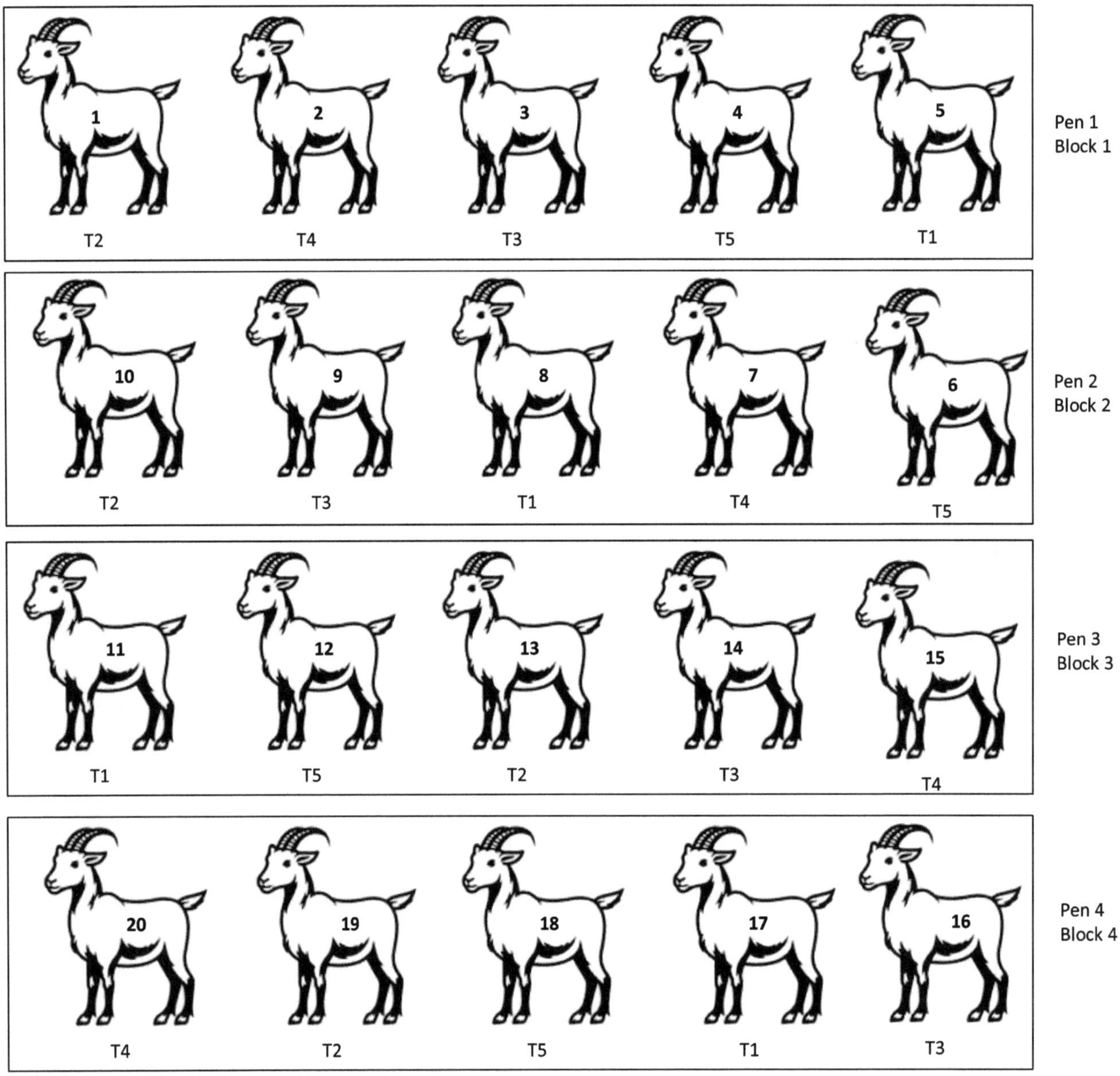

Fig. 9.6 Schematic layout of blocked goats in RCBD set-up with four replications

See recommended further reading at the end of the chapter.

9.3.3 Analysis of Randomized Complete Block Design

Analysis of RCBD is more complex than CRD because of the extra restriction factor.

Sources of variation: Whereas there are two sources of variation in CRD (treatment variation and experimental variation (Error), RCBD will have three sources of variation:

1. Treatment variation.
2. Block variation.
3. Experimental error.

Basic model will take the form of:

$$\text{Response} = \text{General mean} + \text{Treatment effect} + \text{Block effect} + \text{Error}$$

Yield of a particular experimental unit as.

Observed yield = Block mean + Treatment mean + Error, represented by equation as:

Box 9.2: Experimental Units in Animal Experiments

Experimental units in animal experiments are varied and depend on the type of study being conducted. Here are some of the different types:

- *Individual animal:* Each animal is considered an individual unit of observation and analysis. especially for large animals like cattle, horses, and mature swine. They are subjected to different treatments or conditions to measure their responses.
- *Cages or pens:* Cages or pens are often used as experimental units in studies involving small animals or birds. Animals within the same cage or pen are exposed to the same treatment conditions, making the cage or pen the experimental unit.
- *Plots of land:* In grazing studies, plots of land may serve as experimental units.
- *Herd or flock*: In studies involving large livestock, e.g. cattle, goats, or sheep, the entire herd may be considered an experimental unit. Treatments are applied at the herd or flock level, and outcomes such as productivity, health, or behavior are measured collectively for all animals within the group and finally averaged.
- *Part of an animal or biological sample*: In studies focusing on biochemical, genetic, or physiological aspects, parts of an animal or biological sample, such as blood, tissue, or feces, may serve as experimental units. Fistulated rumens of cattle can function as a site for several experiments at a go, especially in studies investigating rumen function. Rumen digesta can also an experimental unit.
- *Animal for some time*: A single animal provides several experimental units in a cross-over experiment. Each animal is used as its control and receives distinct treatments separated by wash-out periods.

$$Y_{ij} = b_i + t_i + e_{ij}$$

where:

Y_{ij} = observed yield for treatment *j* in block *i*.
B_i = mean yield for block *i*.
T_i = treatment effect on yield for treatment *j*.
E_{ij} = represents the random unit variation within a block.

In the analysis of variance, therefore, we divide the total variation into components for the variation between treatments, the variation between blocks, and the error variation remaining after these two sources have been subtracted from the total. The generic model of the ANOVA table of a randomized complete block design will take the following form (Table 9.3).

Table 9.3 ANOVA table template for a randomized complete block design

Source of variation	Degrees of freedom	Sum of squares (SS)	Mean sum of squares (MSE)	*F*	*P*
Treatments	k-1	Treatment SS	SS (TR)/k-1	FTr = MS (Tr)/MSE	
Blocks	n-1	Block SS	SS (Bl)/ n-1	FBl = MS (Bl)/MSE	
Error	(k-1) (n-1)	Error SS	SSE/(k-1) (n-1)		
Total	Kn-1	SS(Total)			

Advantages of Randomized Complete Block Designs

1. Provides more efficient or accurate results than the completely randomized design due to grouping.
2. Complete flexibility. Can have any number of treatments and blocks.
3. Relatively easy statistical analysis even with missing data.
4. Allows calculation of unbiased error for specific treatments.

Disadvantages of Randomized Complete Block Designs

1. Not suitable for large numbers of treatments because if the block size is large, it may be difficult to maintain homogeneity within the blocks, consequently error term will be increased.
2. Not suitable when a complete block contains considerable variability because a large error term may result.
3. Interactions between block and treatment effects increase error.
4. If there is any missing data, RCBD may be less efficient than CRD.

9.4 Latin Square Design

9.4.1 Key Characteristics of a Latin Square Design

Latin square design is a matrix design that controls two sources of variation (Salinas Ruíz et al. 2024a). It differs from randomized complete block designs in that the experimental units are grouped in blocks in two different ways: by rows and columns. Whereas RCBD attempts to control one form of local variation, *LSD controls two forms of variation.* They allow bi-directional blocking and are better for controlling spatial heterogeneity than increasing precision compared to RCBD (Salinas Ruíz et al. 2024b). Other names are *rotation experiments, cross-over designs*, and *switch-over designs*. Examples of situations that have two forms of variation include animal experiments involving diets or field experiments with two forms of variation, such as soil fertility and drainage or soil texture.

The major feature of this design is its capacity to handle two different sources of variation simultaneously: with a *row-blocking* and a *column-blocking* (Salinas Ruíz et al. 2024a). This is accomplished by ensuring that every treatment occurs only once in each row block and once in each column block. The number of treatments, rows, and number of replications, and columns must be equal; therefore, the total number of experimental units must be a perfect square. This makes it possible to estimate the variation among the row blocks as well as the column blocks and to remove them from the experimental error.

For example, if there are 4 treatments, there must be 4 replicates or 4 rows and 4 columns. This is a 4 × 4 Latin square, which gives a total of 16 experimental units. Because of this restriction, Latin square experiments can easily become large and unmanageable. Also, if the number of treatments is too small, there are too few degrees of freedom for error. The most common squares are in the range of 5 × 5 to 8 × 8. Mostly, they will range from 5–10 (Dean et al. 2015).

2 × 2 Latin Square

A	B
B	A

3 × 3 Latin Square

A	B	C
B	C	A
C	A	B

4 × 4 Latin Square

A	B	C	D
B	C	D	A
C	D	A	B
D	A	B	C

Examples of where Latin Square can be used:

1. Field experiments in which the experimental area has two different fertility gradients running perpendicular to each other or a unidirectional fertility gradient with a residual effect from previous experiments.
2. Insect field experiments where the insect migration has a predictable direction that is perpendicular to the dominant fertility gradient.
3. Laboratory experiments with replication over time, where the difference between the experimental units conducted at the same time and among those conducted over time constitutes the two known sources of variability.
4. Animal experiment involving evaluation of the effect of diets on milk production of dairy cows.
5. In a greenhouse, response varies due to light differences across the building and temperature differences along the building.

9.4.2 Field Layout of a Latin Square Design

For illustration, we will use an animal experiment (Fig. 9.7).

Objective: To determine the effect of five feed rations on milk yields of dairy cows

- Treatments: feed rations; four diets (A, B, C, and D).
- Number of replications = 4.
- Number of experimental units = 4 × 4 = 16 animals.
- The source of variation is from two directions: the feed rations and the individual cows.

9.4.3 Examples of Published Studies with Latin Square Designs

Example 1 A study involving ensiling rice straw with whole sugar beet was conducted on dairy cows at the Animal Production Research Institute and Sugar Crops Research Institute of the Agricultural Research Centre, at Sakha Experimental Station (Kafr El-Shaikh Governate, Egypt). The aim was to assess the impact of ensiling rice straw with whole sugar beet on the performance of lactating dairy cows (Abo-Donia et al. 2024). The design was a 3 × 3 Latin square design involving 9 multiparous Friesian crossbred cows.

Treatments consisted of T1, T2, and T3 as follows.

- T1: Control = Ensiled rice straw plus corn grains for 24 h (control treatment).
- T2: Ensiled rice straw plus beetroot substituted at 50% of corn grains.
- T3: Ensiled rice straw plus beetroot substituted at 50% of corn grains.

The experiment was replicated three times, i.e. three periods with three animals per period. To eliminate any carryover effect, all the cows were fed a basal diet of 20 days between the experimental periods. The cows were housed in individual stalls, and body weights were recorded at the start of the study and then weekly after the morning milking to adjust the amount of concentrated feed provided.

	Experimental period		
Treatment	1	2	3
T1	3 cows	3 cows	3 cows
T2	3 cows	3 cows	3 cows
T3	3 cows	3 cows	3 cows

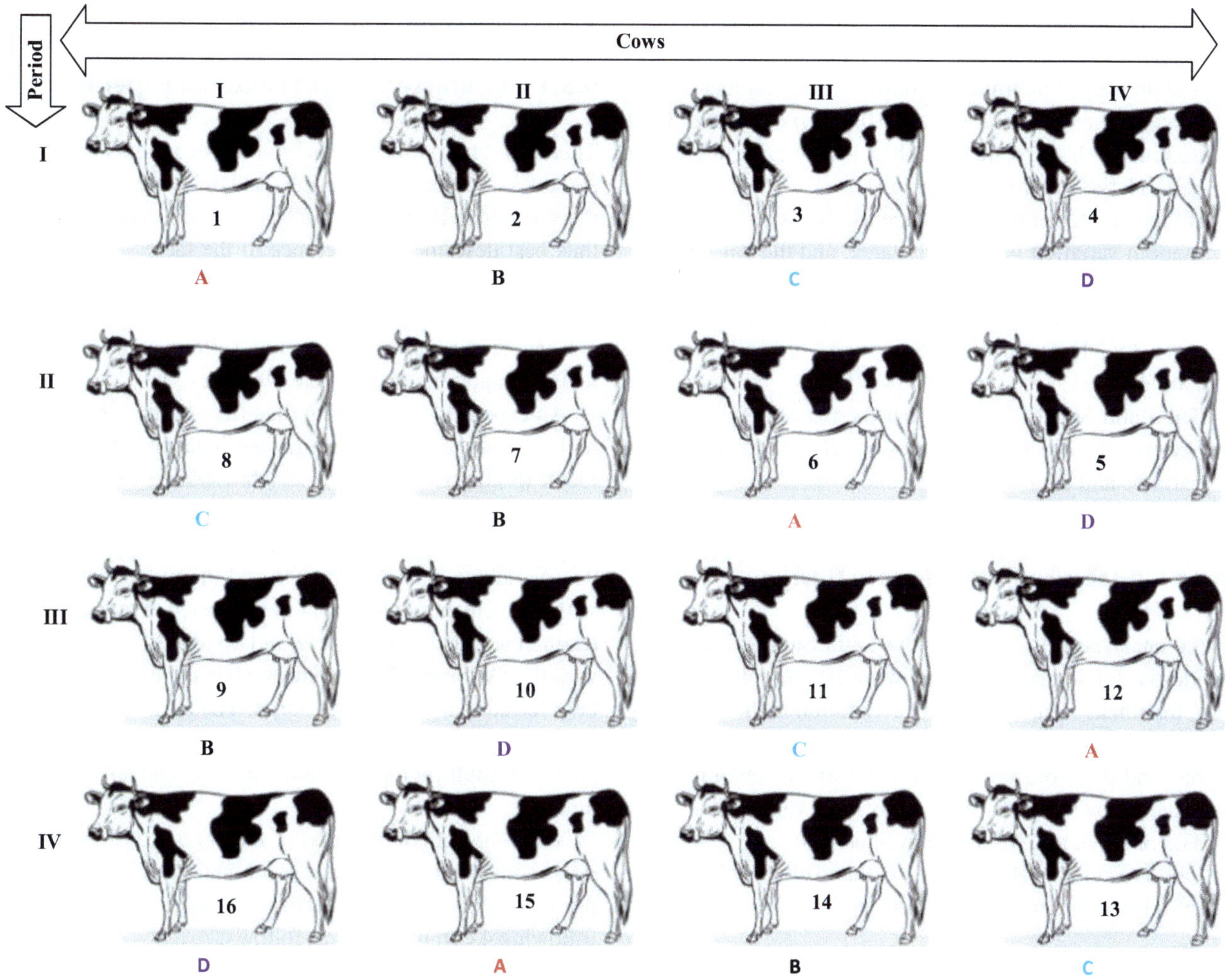

Fig. 9.7 Schematic illustration of Latin Square layout (4 × 4 LSD)

Source: Abo-Donia et al. (2024) Co-ensiled rice straw with whole sugar beet and its effect on the performance of lactating cows. Trop Anim Health Prod 56:1–9. https://doi.org/10.1007/S11250-024-03945-9/TABLES/5 http://creativecommons.org/licenses/by/4.0/

Example 2 Experiment to evaluate the inclusion of coffee pulp silage on the performance and profitability of crossbred cows (Flórez-Delgado et al. 2023). Four lactating multiparous bovines with a body weight of 380 ± 10 kgs, 6 ± 0.03 years (age), and a body weight condition of 3.8 on a scale of **1–5** were distributed in a 4 × 4 Latin Square design.

Treatment Diets	Experimental period (21 days each)			
	1	2	3	4
T1	Cow A	Cow B	Cow C	Cow D
T2	Cow B	Cow C	Cow D	Cow A
T3	Cow C	Cow D	Cow A	Cow B
T4	Cow D	Cow A	Cow B	Cow C

Source: Flórez-Delgado et al. (2023). Effects of the inclusion of coffee pulp silage in the diet on the performance and profitability of crossbred milk cows in the middle tropics. Trop Anim Health Prod 55:1–9. https://doi.org/10.1007/S11250-023-03451-4/TABLES/6. https://creativecommons.org/licenses/by/4.0/

The experiment consisted of four sub-periods of 21 days each. Treatments (diets) consisted of four as follows:

- T1: control = basal diet of chopped Pennisetum.
- T2: 4% coffee pulp silage included in the DM of the diet intake.
- T3: 8% coffee pulp silage included in the DM of the diet intake.
- T4: 12% coffee pulp silage included in the DM of the diet intake.

9.4.3.1 Advantages of Latin Square Designs

- Controls more variation than CRD or RCBD designs because of 2-way stratification. it is therefore more efficient.
- Allows for control of two extraneous sources of variation.
- Results in a smaller mean square for error.
- Simple analysis of data.
- Analysis is simple even with missing plots.

9.4.3.2 Disadvantages of Latin Square Designs

1. The number of treatments is limited to the number of replicates, which seldom exceeds 10. We recommend 5–10 treatments. If there are more than 10, the experiment will be too large and unmanageable.
2. If there are fewer than 5 treatments, the df for controlling random variation is relatively large, and the pdf for error is small. This limitation can be overcome by repeating a small Latin Square and then combining the experiments, e.g. a 3 × 3 LSD repeated 4 times or a 4 × 4 LSD repeated 2 times.
3. The Latin Square assumes that there are no interactions between the blocking variables or between the treatment variable and the blocking variable.

9.4.4 Analysis of Latin Square Design

In the analysis of variance for the Latin Square design, we divide the total variation into row variation, column variation, treatment variation, and error variation in the same way as the randomized block design. For an LSD with t treatments, and therefore t rows, t columns and t^2 total units, we have the following form of the analysis of variance:

The linear model for the Latin Square is:

Observed yield = Row effect + Column effect + Treatment effect + Random unit variation (error).

$$Y_{ijk} = r_i + c_j +_{ij} + \varepsilon_{ijk}$$

where:

Y_{ijk} is the observed yield for the *i-th* row, *j-th* column, and *k-th* treatment.
r_i is the effect of the *i-th* row.
c_j is the effect of the j-*th* row.
tk is the effect of the k-*th* row.

ε_{ijk} *is* the random error or the variation associated with the *i-th* row, *j-th* column and *k-th* treatment.

The analysis of variance for this design aims to separate the variability due to treatments, rows, columns, and errors, i.e. TSS = SST + SSrows + SScolumns + SSE.

Source of variation	Degrees of freedom	Sum of squares (SS)	Mean sum of squares (MS)	F	P
Rows	k-1	SS (rows)	RMS	RMS/EMS	
Columns	k-1	SS (columns)	CMS	CMS/EMS	
Treatments	k-1	SS (Tr)	TrMS	TrMS/EMS	
Error	(k-1)(k-2)	SS(error))	EMS		
Total	(k^2–1)	SS (total)			

9.5 Factorial Experiment

9.5.1 Characteristics of Factorial Experiment

The term factorial describes how the treatments are formed and does not refer to the design used in laying out the experiment (Salinas Ruíz et al. 2024c). Factorial experiments are thus best described in reference to the factors involved and related to the experimental design, either CRD, RCBD, or LSD. For example, the experiment will consist of a factorial arrangement involving two factors (nitrogen and rice varieties) in a randomized complete block design.

An experiment evaluating the effect of nitrogen levels on maize yields tests the effect of one independent variable or factor (N) on the dependent variable (Maize yield). This is a ***one-factor experiment*.** With two or more factors, we use the term "factorial experiment" to denote two or more independent variables or factors being used in an experiment, each with discrete possible levels being tested. For example, in testing the effect of P fertilizer at two levels (P1 and P2) and varieties (V1 and V2) on soybean yields, the experiment will be referred to as a 2 by 2 factorial. The term factorial is borrowed from mathematics, where it is used to refer to the result of multiplying a whole number and all the whole numbers.

The simplest combination of factors is a 2 × 2 (2^2) factorial, giving 4 (four) treatment combinations. Using the example of phosphorus and soybean varieties, we generate the following 4 combinations as follows:

P level	Variety	Combination	Treatment
P1	V1	P1V1	1
P1	V2	P1V2	2
P2	V1	P2V1	3
P2	V2	P2V2	4

The combinations are called the **"treatments" or "treatment combinations."** The number of treatments is the product of the levels of each factor; in a 3 × 2, 3^2 factorials, the number of treatments is 2 × 3 = 6, in a 2 × 3, $2^{3,}$ the number of treatments is 2 × 2 × 2 = 8 treatments. See more examples in Table 9.4.

The number of treatments increases as factors and levels increase (Table 9.3). However, one needs to be cautious in planning large factorial experiments. Extremely large experiments should be avoided because they tend to be complex and costly. Furthermore, it is challenging to obtain adequate experimental land and to control heterogeneity. Practically, suppose you are dealing with many levels, for example, evaluating many varieties: the response of 20 hybrid maize to fertilizer rates. In that case, it is advisable to have a single-factor experiment consisting of evaluating the performance of the varieties (Screening) and later subjecting the best performing to fertilizer trials.

Table 9.4 Example of published studies that used factorial arrangement structure

Factor 1	Factor 2	Factor 3	Factor 4	combinations	No of treatments
Nitrogen levels (0, 20, 40, 80 kg N ha^{-1})	Spacing	None	None	4 × 3	12
Tree species (Calliandra, Albizzia, Tephrosia)	Crop (cereal, legume)	Tree spacing (2 × 2m, 3 × 3m, 4 × 4 m)	None	3 × 2 × 3	18
Animal diets (dairy meal supplementation, fodder tree supplementation)	Basal diet (Hay and Napier grass)	Enzyme (with and without)	None	2 × 2 × 2	8
Nitrogen (0, 60 kg N ha^{-1})	Water harvesting (furrow, tied ridging, conventional)	Manure (with and without)	Cropping pattern (monocrop, intercrop)	2 × 3 × 2 × 2	24

One of the key benefits of a factorial experiment is that it allows the researcher to study the effect of each factor on the dependent variable as well as the effect of *interactions* between factors on the dependent variable. Suppose you want to evaluate the effect of rice varieties and planting dates on yields: rice variety at 5 levels (five rice varieties) and planting dates at two levels (planting before rains and planting after the rains). Such as an experiment will yield three things at once:

1. Comparison of the rice varieties with each other.
2. Evaluation of the effect of planting date.
3. Determination of whether varying the planting date changes the relative performance of the varieties (e.g. one variety may only perform well if planted before the rains), commonly known as interactions.

If one factor influences the effect of the other factor, a significant interaction exists between the two factors. In the example above, if, after analysis, we find that varying the planting data changes the performance of the variety, then a *significant interaction exists*. The factors in this case are not independent. If, on the other hand, the simple effect of variety is the same on both early planting and late planting, there is no interaction (these factors are operating independently).

It is important to know if there is an interaction between the factors because if there is an interaction, you cannot make predictions or recommendations based on the results of single factors. This means you will not be able to talk about the effect of one factor without mentioning the other factors. Information about interaction helps a researcher find the most suitable optimum factor combination.

A factorial arrangement of treatments can be a powerful tool. Still, because the number of treatments can get very large, it is best used when there is a reason to believe that the factors may influence each other and have a significant interaction. If there is no suspicion that the factors may influence each other, testing them in separate experiments is frequently easier and more thorough (Gomez and Gomez 1984).

9.5.2 Field Layout of Factorial Experiments

To demonstrate the field layout of factorial experiments, we will use two examples: agronomy and animal science.

Example 1: Agronomy Research objective: To evaluate the effect of three crop varieties (bean varieties) and three nitrogen rates on grain yields.

We design an experiment in a randomized complete block design and replicate the treatments four times. We get the following structure:

Factor 1: Bean varieties: B1, B2, and B3.
Factor 2: Nitrogen rates: N1, N2, and N3.
Factorial arrangement: 3 x 3 = 9 treatment combinations (treatments).
Design: Randomized complete block design.
Replications: 4 reps.
Number of plots = 9 × 3 = 36 plots (Fig. 9.8).
Response variable: bean grain yields.

	Nitrogen (3)		
Bean variety (3)	N1	N2	N3
B1	B1N1	B1N2	B1N3
B2	B2N1	B2N2	B2N3
B3	B3N1	B3N2	B3N3

Field Layout Plan

Example 2 Animal Science.

Research objective: To determine the effect of protein supplemental diets on the quality and yield of milk of three breeds of dairy goats.
We design a factorial experiment in a randomized complete block design replicated three times; The structure is as follows (Table 9.5) and Fig. 9.9.

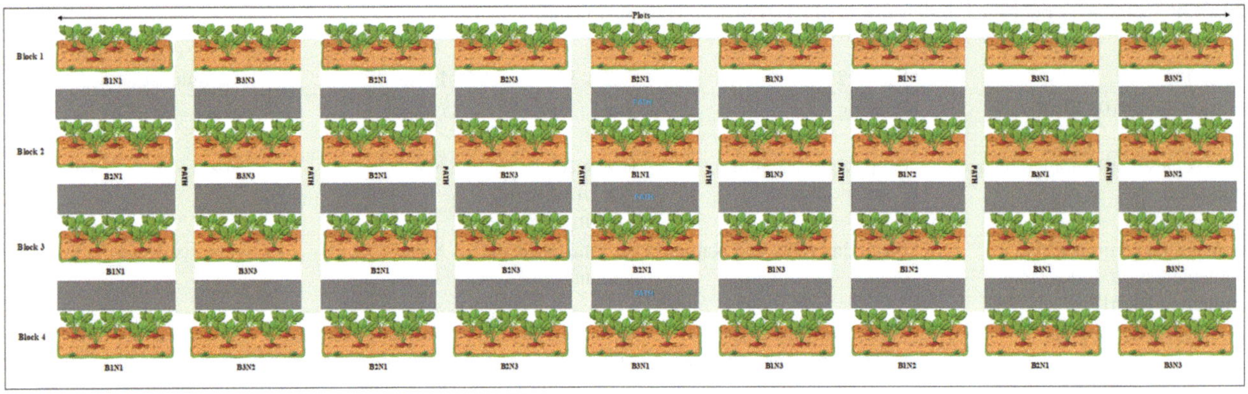

Fig. 9.8 Field layout plan of a 2-factor experiment (3 bean varieties and 3 nitrogen rates = 9 treatments) in RCBD

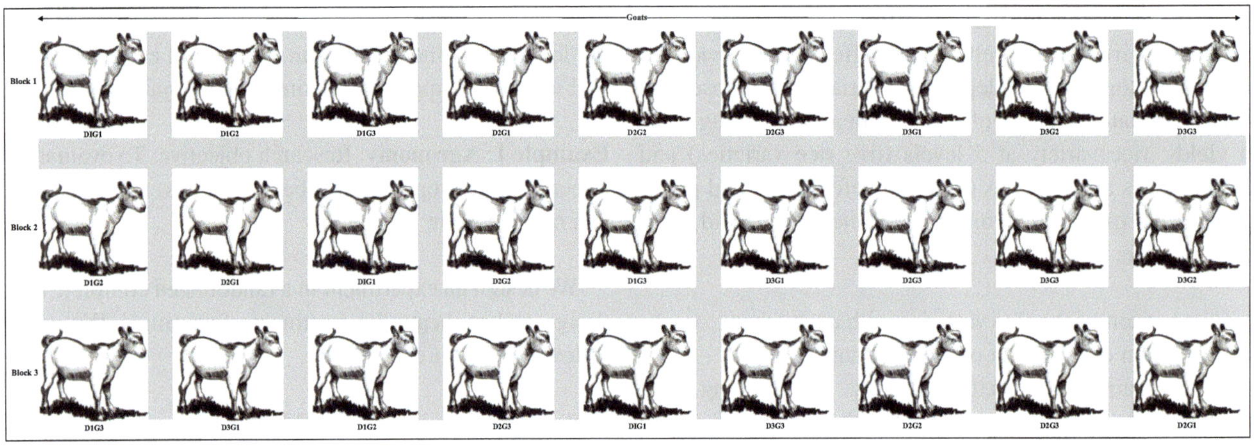

Fig. 9.9 Factorial experiment in RCBD consisting of 3 × 3 factorial (3 protein diets and 3 dairy goat breeds)

Table 9.5 A 2-factor factorial experiment in a randomized block design

Factor A (protein diet)	Factor B (dairy goat breed)	Treatment combination
D1	G1	D1G1
	G2	D1G2
	G3	D1G3
D2	G1	D2G1
	G2	D2G2
	G3	D2G3
D3	G1	D3G1
	G2	D3G2
	G3	D3G3
		9 treatments

Factor 1: Protein diet: D1, D2, and D3.
Factor 2: Dairy goat breeds: G1, G2, and G3.
Factorial arrangement: 3 x 3 = 9 treatment combinations (Treatments).
Design: Randomized complete block design.
Replications: 3.
No of plots: 9 × 3 = 27.
Response variable: quality and quantity of milk.

The field layout will have three replications each carrying 9 plots making 27 plots

REP 1	1	2	3	4	5	6	7	8	9
REP 2	19	18	17	16	15	14	13	12	11
REP 3	20	21	22	23	24	25	26	27	28

Randomized on a per-block basis

We present examples of factorial experiments (animal experiments) in published works and clarify the designs (Table 9.6).

9.5.2.1 Complex Factorial Experiments

The complexity of factorial experiments increases substantially when the number of independent variables increases from two to three. In the 2-factor design, there can only be one interaction effect, but in a 3-factor experiment. A change from a 2-factor to a 3-factor design introduces the possibility of obtaining four different interaction effects. If the three are given symbols A, B, and C, the three-factor design allows a test of the main effects of A, B and C: the two-way interaction effect of A × B, A × C and B × C, and the three-way interaction of A × B × C. If you plan to conduct com-

Table 9.6 Examples of factorial experiments in published studies

Objective	Factors	Replications	Design	Reference
To determine the effects of feed processing and enzyme supplementation on broiler chick performance from 1 to 36 days.	2 factors (2 × 2) Factor 1: Feed (pelleted and mash) Factor 2: Enzyme (with and without enzyme).	Four (4) replications 21 chicks per replicate.	Completely randomized design	Ghobadi and Karimi (2012)
To evaluate the effects of feed form and feed particle size on growth performance, carcass characteristics, and digestive tract development of broilers.	2 factors (2 × 3) Factor 1: Feed form (mash and crumble pellet) Factor 2: Feed particle size (fine, medium, and coarse).	Six (6) Replications 32 birds per replicate	Completely randomized design	Lv et al. (2015)
To determine the effects of two lipotropic factors on growth performance, stability of leg and breast muscles and carcass characteristics in broiler chicken-fed diets different in metabolizable energy levels.	2 factors (3 × 3) Factor 1: Supplemental lipotropics (control, choline and carnitine) Factor 2: Dietary metabolizable diets (control, 0.42 and 0.84 MJ/kg greater).	Five (5) Replications 12 birds per replicate	Completely randomized design	Jahanian and Ashnagar (2018)

plex factorial experiments, consult your supervisor or biometrician for advice on the necessity of such an experiment.

To enhance our understanding of this complexity, let us look at two case studies. The first case involves three factors in a soil fertility experiment, and the second case involves an agroforestry experiment. In these experiments, there is more than one interaction (Boxes 9.3 and 9.4).

9.5.3 Analysis of Factorial Experiment

In the analysis of variance for a factorial experiment, we examine the effect of two or more factors and their interactions. Each factor can have multiple levels, and the factorial design allows us to evaluate all possible combinations of these levels. In this case, we consider factors A and B.

The basic linear model is expressed as follows:

Observed yield = Overall mean + Main effect of Factor A + Main effect of Factor B + Interaction between A and B + Random error.

$$Y_{ijk} = \mu + A_i + B_j + (AB)_{ij} + \varepsilon_{ijk}$$

where

Y_{ijk} is observation associated with replication k of the factor combination ij ($k = 1...r$).
μ is the overall mean,
A_i is the effect of factor A ($i = 1 \ldots a$).
Bj is the effect of factor B, ($j = 1 \ldots b$).
$(AB)_{ij}$ is the interaction of factor A level i with factor B level j
ε_{ijk} is the error associated with replication k of the factor combination ij ($k = 1...r$).

The total sums of squares may be expressed as: SST = SSA + SSB + SS(AB) + SSE.

Box 9.3 Case Study on 3-Factor Factorial Experiment

Title: Maize rotation effect on maize productivity and soil fertility parameters under selected agronomic practices in a sandy loam soil (Uzoh et al. 2019).

Research problem: The inclusion of legumes in cropping systems is essential for the sustainable management of farming systems and reducing N fertilizer requirements for maize production. The magnitude of rotation benefit provided by any legume depends on several factors, and therefore, it is not clear whether maize can sustainably be grown after grain legumes.

Research objective: To determine the effect of growing legumes and maize in the same year in rotation on maize yield and soil fertility indices.

Hypothesis: legume-cereal rotation with or without residue and/or N addition gives higher yield and improves soil fertility than maize monocrop.

Materials and methods: *The experimental layout was a three-factor, 4 × 2 × 2 factorial in randomized complete block design replicated three times, giving 48 plots. The three factors were: (1) Crop rotation at four levels (maize soybean, cowpea and velvet bean), (2) Residue management (residue added and residue removed),* and two N fertilizer rates (0 N ha^{-1} and 60 kg N ha^{-1}).

Statistical analysis: data were analyzed using ANOVA to determine the main and interaction treatment effects. Mean separation was done by Fischer's Least Significant (F-LSD) at a probability level of 5%.

The results present the following primary and interaction effects:

Crop rotation (C)
Residue management (R)
Fertilizer application (F)
C × R
R × F
C × F
C × R × F

Source: Uzoh et al. (2019)

Box 9.4: Case Study of a 4-Factor Factorial Experiment

Title: Soil nutrients and maize yield responses to indigenous agroforestry tree post-fallows management in Tanzania (Vyamana Ochieng' et al. 2021).

Research problem: Most agroforestry studies on the evaluation and selection of agroforestry trees shrubs in Tanzania have focused on exotic multipurpose trees, while Indigenous tree species are more adaptable to local environments.

Research objective: To assess the potential for utilizing agroforestry tree species with contracting crown shading characteristics through management of their fallows and coppice, intercropping for increasing soil nutrients and yield of maize.

Materials and methods: *The experimental layout was a 4-factor, 2 × 2 × 4 × 2 factorial in randomized complete block design replicated three times,* giving 96 plots. The four factors were:

- *Factor one*: Two levels of study species: *Albizzia harvey* and *Albizzia Verscolor.*
- *Factor two*: Two levels of cropping systems: ICO- continuous cropping (control) and ICI—Maize intercropped with coppices.
- *Factor Three*: Four Levels of tree cutting; CO-Trees not cut (Tree fallows), C1 -5com (ground level), C3–30 com above the ground and C3–90 cm above the ground.
- Factor four: Two levels of coppice thinning; THO-All coppice shoots left (no thinning) and TH1- coppices thinned to leave two coppice shoots per stump.

Statistical analysis: Plot means were subjected to ANOVA using General Linear model procedure in SAS at 5% level of statistical significance. Block and Block by treatment interaction were error terms in the model. The ANOVA for maize, biomass and soil data tested the effects of the two study species, two cropping system, four tree cutting heights and two coppicing regimes.

Interactions in this experiment were complex (more than 10); look at the following main effect and interaction effects that were in this experiment:

1. Block (Blk)
2. Tree species (spp)
3. Blk × spp
4. Cutting height (cut)
5. Blk × cut
6. Spp × cur
7. Blk × Spp × cut
8. C × R × F
9. Coppice thinning (Copthin)
10. Blk × Copthin
11. Spp × Copthin
12. Cut × Copthin
13. Blk × Spp × cut
14. Spp × cut × Copthin
15. Blk × cut × Copthin
16. Residual error

Source: Vyamana et al. (2021)
Licence number: 5821590490215 by Elsevier

Source of variation	Degrees of freedom	Sum of squares (SS)	Mean sum of squares (MS)	*F*	*P*
Replications	r-1	RSS	RMS	RMS/ EMS	
Factor A	a-1	ASS	AMS	AMS/ EMS	
Factor B	b-1	BSS	BMS	BMS/ EMS	
AB interaction	(a-1)(b-1)	ABSS	ABMS	AB/ EMS	
Error	(r-1)(ab-1)	ESS	EMS		
Total	(rab-1)	TSS			

9.5.3.1 Advantages and Disadvantages of Factorial Experiments

Factorial experiments are associated with the following advantages:

- Greater precision is achieved with factorial structure which is one of the major advantages (Mead et al. 2017).
- The researcher gets an opportunity to assess the effects of one or more factors at a time.
- Effective because the researcher can determine the interaction effect.
- Saves time and finances.

Disadvantages include the following:

- As the number of factors or levels of the factor or both increases, the number of treatment combinations increases rapidly, requiring large experimental fields and bigger blocks. As the blocks increase, maintaining homogeneity among the plots within a block becomes difficult under field conditions. There is a possibility of increasing the experimental error and decreasing the experiment's precision. Researchers must be sensitive about the number of levels per factor. It is advisable to use a small number of levels with less replication (Mead et al. 2017). These authors recommend three levels of utmost, and four levels should be avoided for any one factor.

- Statistical procedures and calculations are more complicated than in single-factor experiments.
- As the number of factors or levels of the factor increase, in case of interaction effects increase. It is challenging to manage higher-order interaction effects.

9.6 Split-Plot Experiment

9.6.1 Characteristics of Split-Plot Experiments

A split plot is an arrangement that allows researchers to assess the impact of multiple factors simultaneously and consider interactions between these factors (Salinas Ruíz et al. 2024c). It is adopted especially when it is challenging to randomize treatments within the experimental units. This can happen due to various constraints, such as equipment limitations, biological considerations, or plot size requirements. For example, irrigation methods may require a larger plot size than other treatments, while tillage methods may not be effective in a small area. In other instances, some treatment factors like inoculation and fertilizer application are prone to contamination from neighboring plots. In such cases, the experimental field is divided (split) into plots of different sizes (big and small plots), thus forming a split-plot design to enhance boundaries. In the following example, irrigation regimes are in big plots while varieties are in small plots (Fig. 9.10).

In a split-plot, we, thus, have two levels of experimental units: the factor in the large field called the *whole plot* and *the sub-plots* (factor in the small field) nested within the whole plot. The whole plots are also called the main plot, while the subplots are also called split plots or split units. Corresponding to the two levels of experimental units are two levels of randomization. One randomization is conducted to determine the assignment of treatments to whole plots. Then, a randomization of treatments to split-plot experimental units occurs within the whole plot.

Examples of Split-Plot Experiment Factors:

- Irrigation regimes (whole plots) and wheat varieties (sub-plots).
- Nitrogen rates (whole plots) and banana varieties (sub-plots).
- Planting dates (whole plots) and nitrogen fertilizer rates (subplots).
- Tillage practices (whole plots) and phosphorus fertilizer rates (subplots).
- Light regime, temperature, or humidity in a growth chamber or glass house (whole plot) and tomato varieties (subplots).

The following components are important in a split-plot experiment.

1. *Factorial structure:* each combination of levels in the main plots and subplots is represented at least once in the experiment.
2. *Main plot:* assigned to larger plots, and each main plot received one level of the sub-plot treatment factor.
3. *Sub-plot:* This is nested within the main plot and randomized within each main plot.
4. *Replication*: replication to increase the precision and validity of the experiment.
5. *Basic experimental design*: the experiment has to be fitted into a basic experimental design.

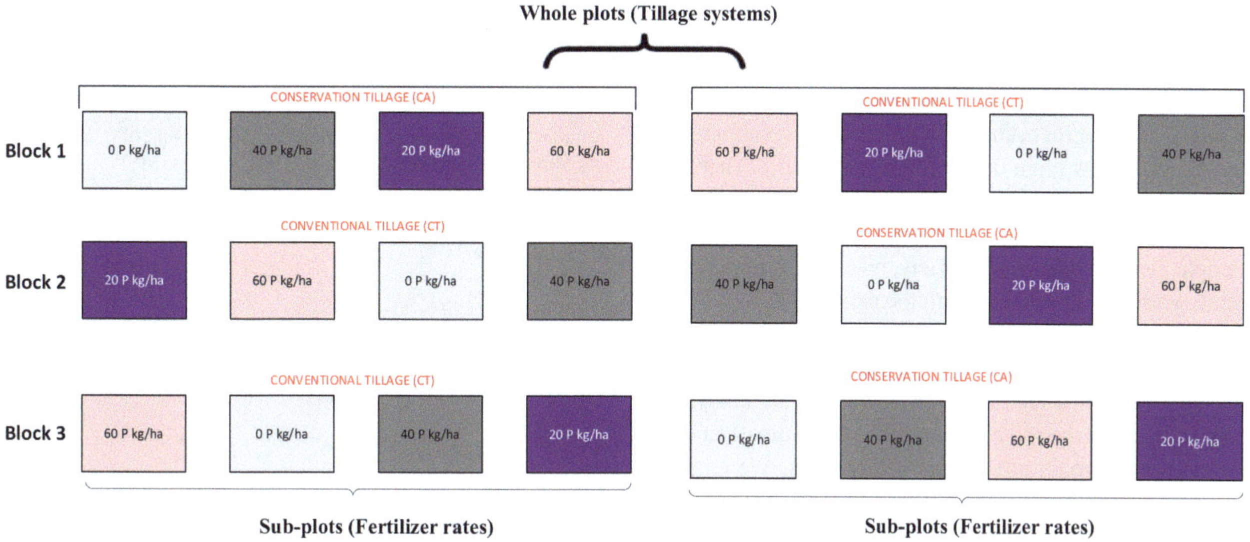

Fig. 9.10 Split plot experiment in a randomized complete block design showing main plots at two levels and subplots at four levels (2 × 4 = 8 treatments)

Just like the factorial experiments, the split-plot arrangement is fitted in one of the basic designs, i.e., CRD, RCDB, or LSD. Most field experiments will have split-plot arrangements fitted in RCBD. We could extend the same principle of splitting the plots to accommodate another factor in a split–split–plot (three-way factorial) with three factors, e.g. irrigation, variety, and fertilizer.

9.6.1.1 Example of a Split-Plot Experiment

An agronomist wants to evaluate the effect of fertilizer rates under conservation agriculture (CA) and conventional tillage (CT) on maize yields. Because of the difficulty in allocating tillage systems to small plots, this is allocated as the whole plot treatments, while fertilizer is allocated to subplot treatments. In this example, we are manipulating fertilizer rates and tillage practices. The experiment has the following parameters:

- Factors: fertilizer rates and tillage practices.
- Main plots: tillage practices at two levels: conservation agriculture and conventional tillage.
- Sub-plots: fertilizer rates at four levels (0, 20, 40, 60 kg N ha^{-1}).
- Number of treatments: 2 × 4 = 8.
- Number of replications: 4.
- Number of plots = 8 × 4 = 36 plots.
- Experimental design (basic): RCBD (Fig. 9.10).
- Independent variables: fertilizer and tillage practices.
- Dependent variable: maize yield and other response variables.

Randomization will be done at the whole plot level (two levels) and the sub-plot level (four levels). Levels of the whole plot factor are randomly assigned to the whole plots using a different randomization for each block. On the other hand, levels of the subplots are randomly assigned within each main plot using a separate randomization for each main plot.

Because there are two sizes of plots, there are *two experimental errors*, one for each size plot. Usually, the subplot error is smaller and has more degrees of freedom (df). Therefore, the main plot factor is estimated with less precision than the subplot and interaction effects. When there is flexibility, for example, using fertilizer and variety, precision is important in deciding which factor to assign to the main plot.

9.6.1.2 Analysis of Split-Plot Experiment

This is a form of a factorial experiment, so the analysis is handled in much the same manner. We will estimate and test the appropriate main effects and interactions. We examine the effect of two or more factors and the interactions. Each factor can have multiple levels, and the factorial design allows us to evaluate all possible combinations of these levels. In this case let's consider factors A and B.

The basic linear model is expressed as follows:

Observed yield = Overall mean + Main effect of whole plot factor A + Main effect of sub-plot factor B + Interaction between main plot and sub plot+ Error for whole plot + Error for sub-plot.

Consider the example given in Fig. 9.10 with tillage and nitrogen as the two factors; we can label them A and B and their effect on the yield of maize.

Let '*r*' equal the number of blocks, '*a*' the number of levels of A or whole plot per block, and '*b*' the number of levels of B or sub-plots per whole plot.

Thus, we have *ab* treatment combination replicated *r* times. We have *abr* total number of experimental units.

Let Y_{ijk} be an observation associated with *i*th block, *j*th factor A effect, and *k*th factor B effect.

The observation Y_{ijk} is expressed in a mathematical form as:

$$Y_{ijk} = \mu + \rho_i + \alpha_j + \gamma_{ij} + \beta_k + (\alpha\beta)_{jk} + \varepsilon_{ijk}; i = 1,2,\ldots,r; j = 1,2,\ldots,a; k = 1,2,\ldots,b$$

where

μ = overall mean
ρ_i = *i*th block effect
α_j = *j*th factor A effect
γ_{ij} = *ij*th random effect associated with whole-plot factor
β_k = *k*th factor B effect
$(\alpha\beta)_{jk}$ = *jk*th interaction effect
ε_{ijk} = random effect associated with sub-plot factor

The total sums of squares may be expressed as: SST = SSA + SSB + SS(AB) + SSE.

The ANOVA table is as follows:

Source of variation	Degrees of freedom	Sum of squares (SS)	Mean sum of squares (MS)	*F*	*P*
Block	r-1	BSS	BMS	BMS/EMS	
Factor A	a-1	ASS	AMS	AMS/EMS	
Error (a)	(r-1) (a-1)	BSS	BMS	BMS/EMS	
Factor B	b-1				
Error (b)	(a-1)(b-1)	ABSS	ABMS	AB/EMS	
AB	(r-1) (ab-1)	ESS	EMS		
Total	(rab-1)	TSS			

9.6.2 Examples of Published Studies with Split-Plot Designs

Example 1 Experiment to evaluate the efficacy of nitrogen (N) forms on maize performance, N uptake, and use efficiency and grain protein content (Ochieng' et al. 2021). The experiment was a split plot with three N forms (ammonium, nitrate and urea) as the main plots, and the subplots consisted of four rates of N (0, 25, 50, 100 kg N ha^{-1}). The experiment adopted a randomized complete block, replicated three times.

- *Factors*: nitrogen forms and nitrogen rates.
- *Main plots*: three levels of nitrogen forms: ammonium, nitrate, and urea.
- *Sub-plots*: fertilizer rates at four levels (0, 25, 50, 100 kg N ha^{-1}).
- Number of treatments: 3 × 4 = 12.
- Number of replications: 3.
- Experimental design = RCBD.
- No of plots: 12 × 3 = 36.
- Independent variables: nitrogen forms and nitrogen rates.
- Dependent variable: maize grain yield, stover yield, Nitrogen uptake, N-use efficiency, protein content etc.

The filed layout of this experiment is shown in Fig. 9.11.

Example 2 Experiment to assess the response of soil bacteria to different fertilizers and cropping arrangements (Li et al. 2020). A three-year study was established using a split-plot arrangement in RCBD replicated 4 times. The main plots consisted of three fertilizers (unfertilized control, organic fertilizer and synthetic fertilizer), while the subplots were 4 crop types: clear fallow, red clover (*Trifolium pratense),* Timothy (*Phleum pratense*), and a mixture of red clover and Timothy. The design can be displayed as follows:

- *Factors*: fertilizer types and cropping types.
- *Main plots*: three levels of fertilizer types (unfertilized control, organic fertilizer, and synthetic fertilizer).
- *Sub-plots*: cropping types: clear fallow, red clover (*Trifolium pratense),* Timothy (*Phleum pratense*) and a mixture of red clover and Timothy.
- Number of treatments: 3 × 4 = 12.
- Number of replications: four replicates.
- No of plots: 12 × 4 = 48.
- Independent variables: fertilizer types and cropping types.
- Dependent variable: soil bacterial community, crop growth, soil properties, etc.

More examples of split-plot design are given in Table 9.7.

9.6.2.1 Advantages of Split Plot Design

- Two or more factors needing relatively large and small units can be combined in the same experiment.
- An estimate of the subplot treatment and its interactions with the whole plot treatment is obtained.
- Provides increased precision of subplot treatments compared to factorial randomized complete block.
- Can permit the introduction of new treatments in an experiment that is in progress.

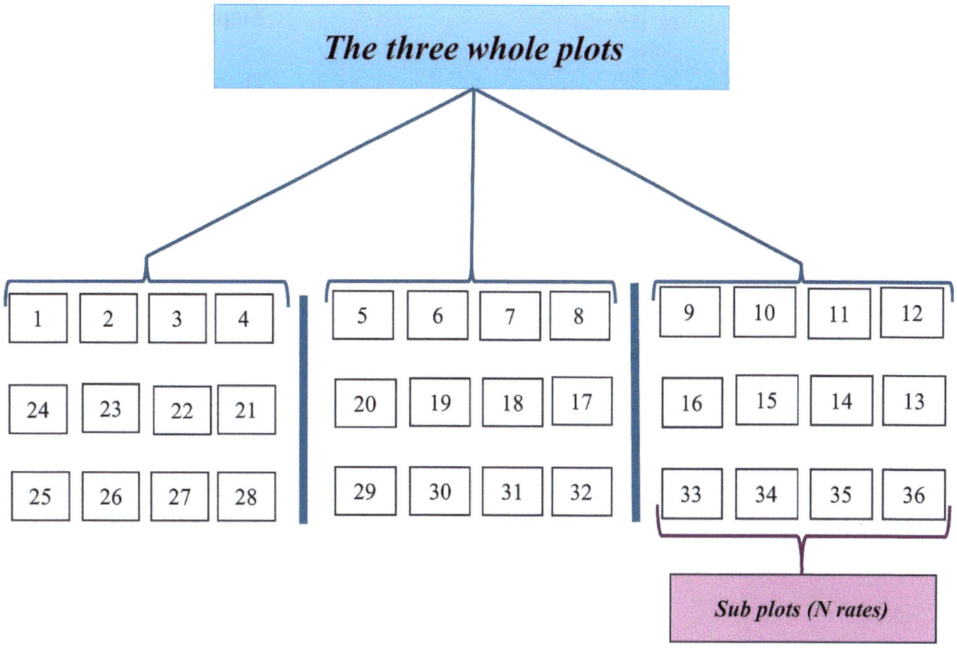

Fig. 9.11 Field plot layout of split-plot experiment with the whole plot (3 nitrogen forms) and sub-plots (4 nitrogen rates). (Source: Ochieng' et al. (2021))

Table 9.7 Examples of published studies involving split-plot experiment

	Main plot	Subplot	Design	No. of plots
Ochieng' Ochieng' et al. (2021)	9 m × 3 m *Treatments: N forms (3)* T1: Ammonium T2: Nitrate T3: Urea	3 m × 3 m *Treatments: N rates (4)* T1: N0 T2: N25 T3: N50 T4: N100	RCBD replicated 3 times	3 × 4 × 3 = 48
Ndengu et al. (2022)	6 m × 8 m *Treatments: Bean genotypes (2)* T1 = SER-45 T2: SER-83	6 m × 2 m *Treatments: Combination of soil fertility and cropping system (8)* T1: Sole beans T2: Beans + manure T3: Bans + manure + fertilizer T4: Beans + maize T5: Beans + maize + manure T6: Beans + maize + manure + fertilizer T7: Beans + maize + fertilizer T8: Maize +manure + fertilizer	RCBD replicated 3 times	2 × 8 × 3 = 48
Li et al. (2020)	6 m × 8 m *Treatments: Fertilizer types (3)* T1: No fertilizer T2: Organic (urine in 2014 and manure slurry in 2015) T3: Synthetic calcium nitrate	6 m × 2 m *Treatments: Crops (4)* T1: Fallow T2: Pure red clover T3: Pure timothy T3: Mixture of pure red clover and timothy	RCBD Replicated 4 times	3 × 4 × 4 = 48

9.6.2.2 Disadvantages of Split-Plot Design

- The main plot factor is estimated with less precision, so larger differences are required for significance—may be difficult to obtain adequate df for the main plot error.
- Statistical analysis is more complex because different standard errors are required for different comparisons.

9.7 Designing and Establishing a Field Experiment

9.7.1 Experimental Design Is the Foundation for Quality Data

This section empowers you to implement the principles learned in Chapters 8 and 9 to design an effective experiment. Chap. 8 covers the fundamentals of experimental design, including the importance of randomization, replication, and control, as well as key principles for designing treatments and data collection. The principles ensure the data is (1) *Unbiased*, (2). *Reliable,* and (3) *Valid*. Integrating these principles allows you to plan and implement a well-designed and structured field experiment.

A well-designed experiment is the foundation for generating quality data (Sahu 2013). Although designing and implementing a good experiment requires effort, the benefits far outweigh the costs. The main benefits include improved *data quality, reproducibility and transparency.*

Think of it like running a successful agricultural enterprise. Just as a farmer spends time planning and preparing the soil, selecting the right seeds, and carefully tending to the crops to ensure a bountiful harvest, you must invest time and effort in planning and structuring your experiment. The meticulous planning and attention to detail pay off in the end with quality, reliable results that stand up to scrutiny, much like a well-managed farm yields abundant produce. With valid experimental data, you will find it easier to identify patterns and trends, making your analysis more straightforward. Moreover, presenting your well-organized work to examiners and other interested parties will be more compelling and credible.

1. **Steps in the Planning Phase of Designing an Experiment**

There are two main phases in designing and establishing an experiment: the *planning phase* and the *implementation phase*. The planning phase is done in the office before any field work, while the implementation phase comprises experimentation. The two phases are covered here.

The output of the planning phase is an experimental protocol. The planning phase consists of 10 steps as follows:

Step 1: Relook at the objectives/hypothesis of the study: and ensure that they clearly define the independent variables that will form the treatments and show measurable variables (dependent variables). Objectives will drive the respective study and determine the choice of variables, treatments, replication, sampling, and entire data collection process.

Step 2: Select the treatments: Brainstorm and carefully consider variables as per the objectives. Think critically about the levels and any combinations needed.

*Step 3: Select experimental material***:** in accordance with the objectives and the population about which inferences are to be made. Consider factors such as the type of breeds, animals, and varieties of crops to be used. You should use materials that are representative of the target population and area. This ensures that the results can be generalized to a broader context.

Step 4: Select an experimental design: considering the objectives and the available resources (e.g. land and funds). Choose a simple and operationally feasible design that is likely to provide good precision. A simple design reduces the risk of errors and data collection inconsistencies, enhancing the results' accuracy and reliability.

*Step 5: Select experimental units and the number of replications***:** that are likely to provide good precision. This will be guided by the design and variability of the site. Select experimental units (individual plots, plants, leaves, etc) that are appropriate. In selecting the number of replications, strike a balance between precision and feasibility.

Step 6: Determine control variables: How will you control the influence of neighboring units and extraneous variables on your measurements? You need to decide on the measures to control the effects of adjacent experimental units on each other and management practices to control extraneous variables. This mainly involves randomization (to balance potential impact on variables) and border rows to act as protective barriers to reduce external influence.

Step 7: Determine the field plan and draw: You must draw a field plan (map) and a plot layout plan for your experiment (Figs. 9.12 and 9.13). The plan should have appropriate plot sizes to accommodate adequate experimental material and data collection requirements. Field layout shows the arrangement of experimental units within the entire experimental area (field) or research area (Fig. 9.12). The plan must show actual randomization and replication. It must also show the plot arrangement and sizes, including the paths between the blocks and the plots. The plots must be large enough to accommodate the growth and development of crops and small enough to minimize variability within the plots. For instance, if destructive sampling is done, you should include enough plants for this.

A buffer zone is also included around the experimental area to help minimize edge effects, where conditions at the edges of the field may differ from those in the interior. Border rows serve as a buffer or protection to reduce the impact of external factors such as grazing animals and other external factors (Fig. 9.13).

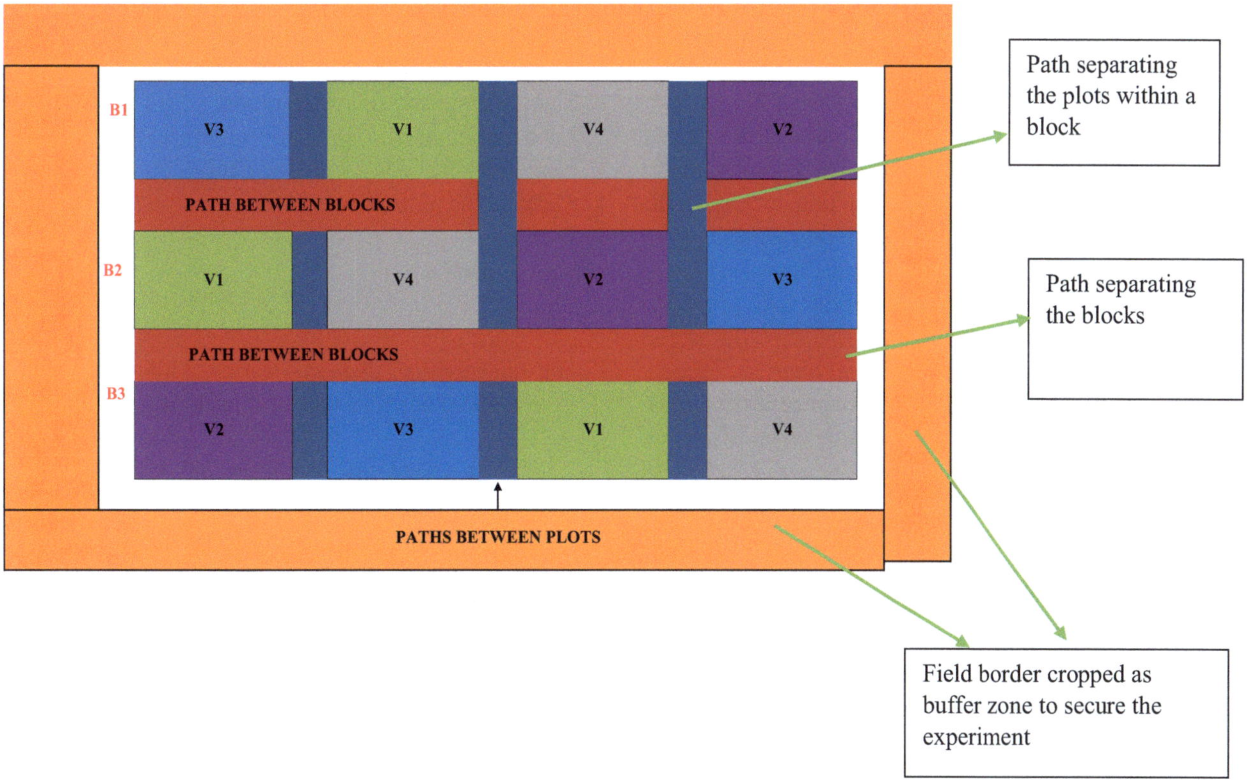

Fig. 9.12 Field layout plan for randomized complete block showing the blocks, plots and paths separating the plots and blocks. The size of paths varies depending on the crop

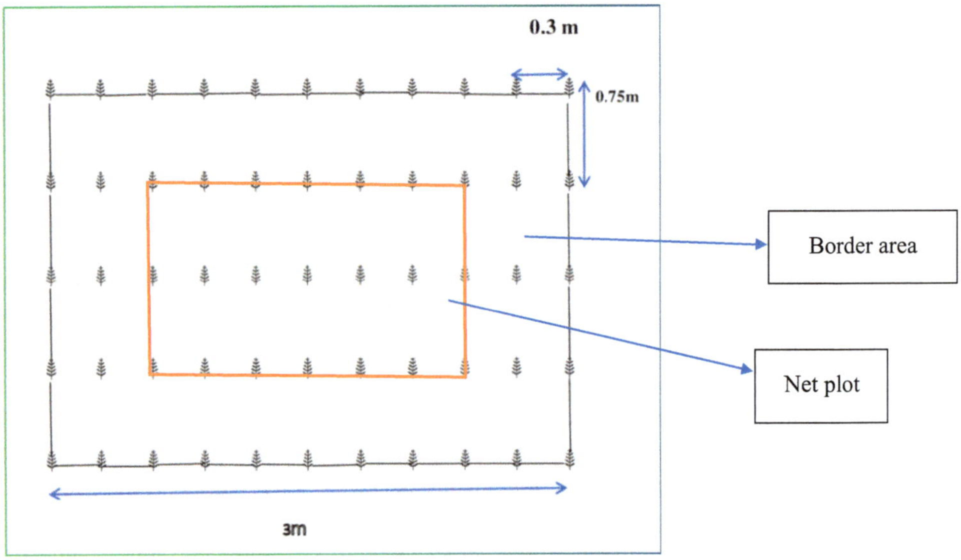

Fig. 9.13 Plot layout plan showing a net plot of 0.3 × 1.35 m and a border area

A plot layout is a comprehensive plan that shows the placement of individual components within a plot. In the case of a crop experiment, a plot layout should demonstrate the *spatial arrangement of plants***,** including the distances between rows and within each row. Additionally, the layout should indicate the *total plot area, net plot area and the guard rows* surrounding the plot (Figs. 9.12 and 9.13). The net plot should have at least three net harvestable rows and 3-meter row length for small cereals (Abera et al. 2020). A plot of 6 m for small grains and about 30 plants per plot for maize and sorghum is a common practice in most experiments (Abera et al. 2020).

Step 8: Determine and document the types of data to be collected: Specify the data to be collected, as well as the methods and procedures of data collection, indicating the manner and order in which they are to be collected. Work out a data recording mechanism to facilitate data entry, validation, and analysis. This will involve preparing appropriate data recording sheets.

Step 9: Prepare an outline for statistical analysis: Prepare dummy results tables to help you visualize the information you will likely present. The dummy tables assist you in anticipating the structure and data you will include in your research output. Furthermore, they are beneficial in promoting conversation with your supervisor and collaborators.

Step 10: Prepare a detailed experimental protocol: Prepare a detailed protocol and share it with your supervisor for advice. You could also share the protocol with a biometrician to confirm your design and the structure of your dummy tables.

2. **Steps in the Implementation Phase of Conducting a Field Experiment**

Once the proposal and protocol are ready and approved by all administrative offices, you are ready to proceed to the next phase of experimentation. The protocol will have a field layout plan, a plot layout plan\ and details of how to collect data. This will guide you through the whole process of data collection (laying out the experiment, management, and data collection procedures). The steps provided here consider an agronomy trial. They consist of six key steps:

1. Site selection and characterization.
2. Land preparation.
3. Demarcation and layout of the field.
4. Marking the plots.
5. Planting.
6. Data collection.

1. *Site selection and characterization:* This will be guided by the field plan. Select a site considering the following:

   - Select a secure site, accessible and with a low risk of animal damage or theft.
   - Assess the adequacy of the size and shape of the land for experimentation purposes.
   - Select sites without trees and hedges within the experimental plot to prevent interference with the study.
   - Ensure the absence of termite molds within the experimental plots to maintain the experiment's integrity.

- Prioritize sites with uniformity in previous land use within the experimental plot for more consistent results.
- Consider the slope of the land when laying out the experiment; ensure homogeneity in the block layout.
- Assess land management practices, considering the nature of the crop for site management and spacing.
- Evaluate the nature of the soil to inform soil-specific considerations.
- Consider the nature of the treatments when selecting and characterizing the site for experimental purposes.
- If on-farm, these three aspects are important: (1) Confirm the farmer's willingness to cooperate throughout the experiment, (2) Verify the site's availability for the experimentation period to avoid disruptions, and (3) Ensure that the farmer's management practices represent those employed by other local farms, contributing to the study's overall relevance and applicability.

Two important activities in site characterization include: (1) collecting data on the site's history, which can impact the experimental outcomes, and (2) soil sampling and analysis.

You should always ensure that the soil characteristics of the experimental site are determined. Some students miss this important aspect in an experiment. Depending on whether it is a purely agronomic experiment or a soil fertility experiment, there are a few important factors to consider when it comes to soil sampling.

- Sampling must be done at the right depth and time of the growing season, using clean tools to avoid contamination.
- Follow a systematic sampling pattern.
- Have composite samples.
- Pay attention to sampling variability, avoid typical areas, and keep detailed records.
- Proper labeling, careful transport and storage of the samples can help in accurate soil analysis.
- Use recommended soil sampling procedures described in manuals. See Chap. 17 for a detailed discussion.

2. *Land Preparation*

Land preparation is a crucial step in ensuring successful agronomic experiments. The process begins with clearing and cleaning the site, followed by soil tillage to create a good seedbed. Leveling the land is essential to achieve uniform water distribution. After leveling, mark the plot layout and guard rows.

3. *Demarcation and Layout of the Field*

Demarcation and laying out the plots in the field should be done carefully to ensure the measurements are accurate and reliable. Before the demarcation day, gather the demarcation materials, such as rope, tape, ranging poles, and pegs.

Key activities: Positioning the blocks and the plots in the field.

Blocking: On sloping land, lay blocks horizontally against the slope and implement soil and water conservation measures like terraces based on slope steepness and orientation.

Plot layout: Requires mainly rectangular plots that must have *well-positioned right angles*. This is done using the Pythagorean theorem to establish the base. The Pythagoras theorem uses the "3–4-5" method to measure specific units along baselines to form accurate right angles. Instead of 3 m, 4 m, and 5 m a multiple can be chosen: e.g. 6 m, 8 m, and 10 m or e.g. 9 m, 12 m, and 15 m (Dasgupta and Oswalt 1992; Badu-Apraku et al. 2012) (Fig. 9.14).

4. *Marking the Plots*

First, ensure you have all the necessary materials for laying out the experiment and arrange them in an orderly manner. Begin by creating planting holes at the optimal plant population using twine marked at the required spacing as per the plot layout plan. Apply basal fertilizer to the planting

Fig. 9.14 Marking right angles using the Pythagoras theorem

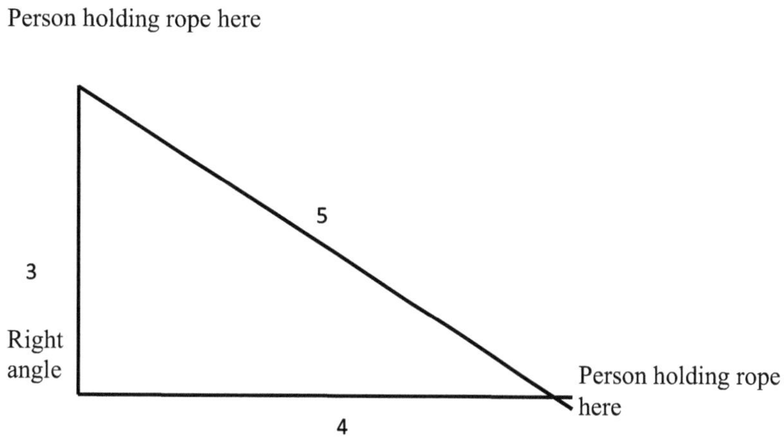

Fig. 9.15 (**a–c**) A well-labeled experiment involving soybean and sorghum in Tharaka-Nithi County, Kenya. The labels include the Block, the plot and the treatment

holes on one side, utilizing a suitable fertilizer cup to achieve the prescribed amount per hole.

Each plot should be clearly labeled with plot markers for easy identification (Fig. 9.15a, b). A comprehensive field plan should be developed to align with the experiment's objectives, clearly labeling treatments, replications, and blocks. The plan should also indicate plot sizes and guard rows on the field plan to ensure a comprehensive layout for the experiment. It should also include any marks and any unusual observations. This carefully managed demarcation process is crucial for a systematic and well-organized agronomic experiment, supporting accurate plot layout and reliable data collection.

5. *Planting.*

Finally, the actual planting follows the experimental design, paying attention to planting depth and spacing. Effective land preparation is vital to creating uniform conditions, minimizing external influences, and ensuring reliable results in agronomic experiments.

6. *Data Collection.*

Data collection includes both the gathering and accurate recording of the data. In agronomy experiments, growth and yield data are usually collected. You should ensure that data is collected from the net plot to maintain accuracy and consistency (Fig. 9.16):

Fig. 9.16 Harvesting soybean from a net plot; data collection from a net plot enhances accuracy and consistency

- It is recommended to record data directly into the field books at the time of collection. You should avoid recording data on rough sheets of paper and then transferring these into the field book later because copying data records increases the chance of errors.

- Be on the lookout for any anomalies or outliers that may indicate inconsistencies.
- Always strive for the highest level of accuracy and precision when collecting data.
- The type of data collected should
- Align with the research objectives, ensuring relevance in addressing the study's goals.

Acknowledgments The authors appreciate the following for allowing the re-use of their resources:

1. Elsevier, publication by Vyamana et al. (2021).
2. Springer Link, publication by Florez-Delgado et al. (Flórez-Delgado et al. 2023) and Abo-Donia et al. (2024).

The photographs used in this chapter were captured in experiments in two projects.

1. Climate-smart options allowing agricultural intensification for smallholders in the dry zones of the Central Highlands of Kenya were implemented from 2019 to 2022. Funded by The Flemish Interuniversity Council Own Initiative (VLIR—UOS).
2. "Scaling up Soybean and Climbing Beans Using a Value Chain Approach in Maize-based Systems of Central Highlands of Kenya" (SHP 014) funded by the Alliance for a Green Revolution in Africa (2014–2017).

Recommended Further Reading

1. Bate and Clark (2014) The Design and Statistical Analysis of Animal Experiments. The Design and Statistical Analysis of Animal Experiments 1–310. https://doi.org/10.1017/CBO9781139344319
2. Mead et al. (2017) Statistical methods in agriculture and experimental biology: Third edition. Statistical Methods in Agriculture and Experimental Biology: Third Edition 1–472. https://doi.org/10.1201/9780203738559/STATISTICAL-METHODS-AGRICULTURE-EXPERIMENTAL-BIOLOGY-ROGER-MEAD/RIGHTS-AND-PERMISSIONS
3. Sahu (2013) Research Methodology: A Guide for Researchers In Agricultural Science, Social Science and Other Related Fields. Research Methodology: A Guide for Researchers In Agricultural Science, Social Science and Other Related Fields. https://doi.org/10.1007/978-81-322-1020-7
4. Patel et al. (2004) Green Book A Guide to Effective Graduate Research in African Agriculture, Environment and Rural Development.
5. Gomez and Gomez (1984) Statistical procedures for agricultural research. Statistical procedures for agricultural research with emphasis on rice 7–8.

Key Points in This Chapter

What Is an Experimental Design?

An "experimental design" is a structure prepared by a researcher to allow the investigation of cause-and-effect relationships between variables. It systematically controls and manipulates independent variables to observe their effect on the dependent variable while minimizing the impact of confounders.

Data generated from the experiments exhibit variability, known and unknown; the experimental design enables a researcher to partition and account for this variability.

Common Experimental Designs Used in Agriculture

Completely Randomized Design

The "completely randomized design" (CRD) is the simplest and most straightforward experimental design. In this design, treatments are randomly allocated to experimental units, and there is no blocking.

In the analysis of completely randomized designs (CRD), the variation among observations comes from two main sources: treatment effects and experimental error.

Randomized Complete Block Design

The most commonly used design in agricultural field research is the randomized complete block design (RCBD). In this design, treatments are replicated and blocked. This arrangement controls variation in an experiment by accounting for spatial effects in the field.

During analysis, differences attributed to blocks are accounted for, effectively reducing random variation and enhancing the precision of treatment comparisons.

Common blocking variances in agronomy experiments include slope, soil fertility, and drainage papers. Blocking is also done in animal experiments using age, cow's parity, days in milk, and milk protein yield.

Analysis of RCBD is more complex than CRD because of the extra restriction factor.

RCBD have three sources of variation:

1. Treatment variation.
2. Block variation.
3. Experimental Error.

Latin Square Design

Latin square designs (LSD) differ from randomized complete block designs in that the experimental units are grouped in blocks in two different ways: by rows and columns. Whereas RCBD attempts to control one form of local variation, LSD controls two forms of variation. Because of this restriction, Latin square experiments can easily become large and unmanageable.

In the analysis of variance for the Latin Square design, the total variation is partitioned into row variation, column variation, treatment variation, and error variation in the same way as the randomized block design.

Factorial Experiment

The term factorial describes how the treatments are formed and does not refer to the design used in laying out the experiment. Factorial experiments are thus best described by the factors involved and related to the experimental design, either CRD, RCBD, or LSD.

Large experiments should be avoided because they are complex and costly. Furthermore, it is challenging to obtain adequate experimental land and to control heterogeneity.

One of the key benefits of a factorial experiment is that it allows the researcher to study the effect of each factor on the dependent variable and the effect of interactions between factors on the dependent variable.

Split-Plot Experiment

A split-plot experiment is a specific arrangement that is used to investigate the effects of multiple factors, especially when it is challenging to randomize treatments within the experimental units. The experimental field is divided (split) into plots of different sizes (big and small plots), thus forming a split-plot design to enhance boundaries.

In the analysis of the split-plot experiment, the analysis is handled similarly to factorial, taking into consideration the main effect and the interaction effects.

Designing and Establishing a Field Experiment

A well-designed experiment is the foundation for generating quality data. It enhances data quality, reproducibility and transparency. Steps in the planning phase of designing an experiment involve two main phases: planning and implementation.

The implementation phase is executed after the proposal and protocol have been approved. It consists of six key steps:

- Site selection and characterization.
- Land preparation.
- Demarcation and layout of the field.
- Marking the plots.
- Planting.
- Data collection.

Review Questions

1. How does experimental design help manage known and unknown variability in data?
2. What is the role of blocking in RCBD, and how does it help increase experimental precision?
3. What is a major limitation of LSD, and why might it become unmanageable in field experiments?
4. In the analysis of a Latin Square design, how is the total variation partitioned?
5. What is the benefit of using a factorial experiment when studying multiple factors?
6. How does a factorial experiment relate to CRD, RCBD, or LSD?
7. Why should "large experiments be avoided"? Discuss the challenges associated with them.
8. How does a well-designed experiment enhance "data quality, reproducibility and transparency"?
9. Once the proposal and protocol are approved, what are the six key steps involved in the implementation phase of establishing an agricultural field experiment?

Chapter Exercise

Choose a recent peer-reviewed journal article from an agricultural science journal that has been published within the last five years. After carefully reading the 'Materials and Methods' section of the article, critique the paper based on the following criteria:

1. Identify the specific experimental design used in the study and explain why the researchers likely chose this particular design.
2. Describe the layout of the experiment as detailed in the article. Comment on the application of the 3Rs. Assess the number of blocks, plot size and how treatments were assigned within the blocks.
3. In your opinion, was the chosen design appropriate and robust for the stated objectives?

Answers to Review questions

- Q1: Use 3R for managing and controlling known and unknown variation. See also the partitioning of variation in the ANOVA tables (unknown variation is accounted for). *Reference: see application of 3Rs in all the designs.*
- Q2: Controls local variability and helps improve precision and more accurate estimation of the treatment effect. It minimizes variation within blocks, accounts for between-blocks variation, reduces "Noise" and enhances "Signal." Reference: Sects. 9.3.1 and 9.3.3 (Table 9.3).
- Q3: With more than 5 treatments, the experiment is likely to be large and unmanageable. e.g. testing 9 varieties; 8 × 8 = 64, while 10 varieties100 plots. Reference:
- Q4: Reference: 9.4.4.
- Q5: Allows studying of the main and interaction effect. Reference: 9.5.1 and 9.5.2.
- Q6. Factorial is not a stand-alone experimental layout. It describes how treatments are formed. A factorial experiment is always laid out using a basic experimental design such as CRD, RCBD or LSD. E.g., a 3 × 3 factorial experiment laid out in RCBD; Reference 9.5.1.
- Q7: Because of logistical complexity. It is costly (funds and time); and a challenge of controlling heterogeneity. Inability to control heterogeneity increases experimental error. Reference…
- Q8: A well-designed experiment enhances data quality by improving accuracy and reliability. It enhances reproducibility by having standardised procedures that include detailed protocols; Transparency is achieved by having clear objectives, well-documented methodology and accountability. Reference 9.7.
- Q9: Check reference 9.7 (2).

References

Abera B, Berhane M, Nebiyu A et al (2020) Diversity, use and production of farmers' varieties of common bean (Phaseolus vulgaris L., Fabaceae) in southwestern and northeastern Ethiopia. Genet Resour Crop Evol 67:339–356. https://doi.org/10.1007/S10722-019-00877-4/METRICS

Abo-Donia FM, Elsheikh HA, Esh AMH et al (2024) Co-ensiled rice straw with whole sugar beet and its effect on the performance of lactating cows. Trop Anim Health Prod 56:1–9. https://doi.org/10.1007/S11250-024-03945-9/TABLES/5

Badu-Apraku B, Fakorede MAB, Menkir A, Sanogo D (2012) Conduct and management of maize field trials

Bailoo JD, Reichlin TS, Würbel H (2014) Refinement of experimental design and conduct in laboratory animal research. ILAR J 55:383–391. https://doi.org/10.1093/ILAR/ILU037

Bate ST, Clark RA (2014) The design and statistical analysis of animal experiments. ILAR J:1–310. https://doi.org/10.1017/CBO9781139344319

Brien CJ, Berger B, Rabie H, Tester M (2013) Accounting for variation in designing greenhouse experiments with special reference to greenhouses containing plants on conveyor systems. Plant Methods 9:1–22. https://doi.org/10.1186/1746-4811-9-5/FIGURES/9

Dasgupta SK, Oswalt DL (1992) Experimentation skill development series no. 2 1992 agronomic practices for human resource development program

Dean A, Morris M, Stufken J, Bingham D (2015) Handbook of design and analysis of experiments. https://doi.org/10.1201/b18619

Diggle PJ, Chetwynd AG (2011) 5 experimental design: agricultural field experiments and clinical trials. Statist Sci Method:57–70. https://doi.org/10.1093/ACPROF:OSO/9780199543182.003.0005

Durner EF (2021) The randomized complete block design. Applied plant science experimental design and statistical analysis using SAS® OnDemand for Academics 177–191. https://doi.org/10.1079/9781789249927.0012

Festing MFW (2014) Randomized block experimental designs can increase the power and reproducibility of laboratory animal experiments. ILAR J 55:472–476. https://doi.org/10.1093/ILAR/ILU045

Flórez-Delgado DF, de Jesús Gil-Herrera R, Maza-Ortega RE (2023) Effects of the inclusion of coffee pulp silage in the diet on the performance and profitability of crossbred milk cows in the middle tropics. Trop Anim Health Prod 55:1–9. https://doi.org/10.1007/S11250-023-03451-4/TABLES/6

Ghobadi Z, Karimi A (2012) Effect of feed processing and enzyme supplementation of wheat-based diets on performance of broiler chicks. J Appl Anim Res 40:260–266. https://doi.org/10.1080/09712119.2012.672307

Gomez KA, Gomez AA (1984) Statistical procedures for agricultural research. Statistical Procedures for Agricultural Research with Emphasis on Rice, pp 7–8

Hofseth LJ (2018) Getting rigorous with scientific rigor. Carcinogenesis 39:21–25. https://doi.org/10.1093/CARCIN/BGX085

Jahanian R, Ashnagar M (2018) Effects of dietary supplementation of choline and carnitine on growth performance, meat oxidative stability and carcass composition of broiler chickens fed diets with different metabolisable energy levels. Br Poult Sci 59:470–476. https://doi.org/10.1080/00071668.2018.1476677

Keller C, Marchetti M, Rossi L, Lugon-Moulin N (2005) Reduction of cadmium availability to tobacco (Nicotiana tabacum) plants using soil amendments in low cadmium-contaminated agricultural soils: a pot experiment. Plant Soil. https://doi.org/10.1007/s11104-005-3101-y

Laajala TD, Jumppanen M, Huhtaniemi R et al (2016) Optimized design and analysis of preclinical intervention studies in vivo. Scientific Reports 6:1–13. https://doi.org/10.1038/srep30723

Ledesma DR (2015) Statistics training guide: experimental design analysis of variance

Li H, Penttinen P, Mikkonen A et al (2020) Response of soil bacterial community diversity and composition to time, fertilization, and plant species in a sub-boreal climate. Front Microbiol 11:499959. https://doi.org/10.3389/FMICB.2020.01780/BIBTEX

Little TM, FJackson H (1978) Agricultural experimentation: design and analysis. Wiley, p 350

Lv M, Yan L, Wang Z et al (2015) Effects of feed form and feed particle size on growth performance, carcass characteristics and digestive tract development of broilers. Anim Nutr 1:252–256. https://doi.org/10.1016/J.ANINU.2015.06.001

Maina AN, Lewis E, Kiarie EG (2023) Egg production, egg quality, and fatty acids profiles in eggs and tissues in Lohmann LSL lite hens fed algal oils rich in docosahexaenoic acid (DHA). Poult Sci 102:102921. https://doi.org/10.1016/J.PSJ.2023.102921

Mead R, Curnow RN, Hasted AM (2017) Statistical methods in agriculture and experimental biology: third edition. Statistical Methods in Agriculture and Experimental Biology: Third Edition 1–472. https://doi.org/10.1201/9780203738559/STATISTICAL-METHODS-AGRICULTURE-EXPERIMENTAL-BIOLOGY-ROGER-MEAD/RIGHTS-AND-PERMISSIONS

Ndengu G, Mponela P, Chataika B et al (2022) Effect of combining organic manure and inorganic fertilisers on maize–bush bean intercropping. Exp Agric 58:e29. https://doi.org/10.1017/S0014479722000102

Nigussie A, Haile W, Agegnehu G, Kiflu A (2021) Growth, nitrogen uptake of maize (Zea mays L.) and soil chemical properties, and responses to compost and nitrogen rates and their mixture on different textured soils: pot experiment. Appl Environ Soil Sci 2021:9931763. https://doi.org/10.1155/2021/9931763

Ochieng' IO, Gitari HI, Mochoge B et al (2021) Optimizing maize yield, nitrogen efficacy and grain protein content under different N forms and rates. J Soil Sci Plant Nutr 21:1867–1880. https://doi.org/10.1007/S42729-021-00486-0

Osuga IM, Abdulrazak SA, Muleke CI, Fujihara T (2012) Effect of supplementing Rhodes grass hay (Chloris gayana) with Berchemia discolor or Zizyphus mucronata on the performance of growing goats in Kenya. J Anim Physiol Anim Nutr (Berl) 96:634–639. https://doi.org/10.1111/J.1439-0396.2011.01189.X

Patel BK, Muir-Leresche K, Coe R, Hainsworth SD (2004) Green book a guide to effective graduate research in African agriculture, environment and rural development

Rauch R, Martín-Tereso J, Daniel JB, Dijkstra J (2021) Dietary protein oscillation: effects on feed intake, lactation performance, and milk nitrogen efficiency in lactating dairy cows. J Dairy Sci 104:10714–10726. https://doi.org/10.3168/jds.2021-20219

Reddy MV (2019) Statistical methods in psychiatry research and SPSS. https://doi.org/10.1201/9780429023309

Sahu PK (2013) Research methodology: a guide for researchers in agricultural science, social science and other related fields. https://doi.org/10.1007/978-81-322-1020-7

Salinas Ruíz J, Montesinos López OA, Crossa J (2024a) Latin square design. Introduction to Experimental Designs with PROC GLIMMIX of SAS, pp 167–195. https://doi.org/10.1007/978-3-031-65575-3_8

Salinas Ruíz J, Montesinos López OA, Crossa J (2024b) Randomized complete block design. Introduction to Experimental Designs with PROC GLIMMIX of SAS, pp 75–98. https://doi.org/10.1007/978-3-031-65575-3_4

Salinas Ruíz J, Montesinos López OA, Crossa J (2024c) Introduction to Experimental Designs with PROC GLIMMIX of SAS https://doi.org/10.1007/978-3-031-65575-3/COVER

Shieh G, Jan SL (2004) The effectiveness of randomized complete block design. Statistica Neerlandica 58:111–124. https://doi.org/10.1046/J.0039-0402.2003.00109.X

Sichinga-Ligowe I, Joy EJM, Manzeke-Kangara MG et al (2024) An introduction to conducting responsible and reproducible agricultural research. CAB Reviews: Perspectives in Agriculture, Veterinary Science, Nutrition and Natural Resources, p 19. https://doi.org/10.1079/CABIREVIEWS.2024.0058/ASSET/19F08C22-4AA3-424B-99DB-F9DA93A15F69/ASSETS/GRAPHIC/FIGURE

Uzoh IM, Igwe CA, Okebalama CB, Babalola OO (2019) Legume-maize rotation effect on maize productivity and soil fertility parameters under selected agronomic practices in a sandy loam soil. Scientific Report 9:1–9. https://doi.org/10.1038/s41598-019-43679-5. https://creativecommons.org/licenses/by/4.0/

Vyamana VG, Chamshama SAO, Andrew SM (2021) Soil nutrients and maize yields responses to indigenous agroforestry tree post-fallows management in Tanzania. Tree Forest People 6:100164. https://doi.org/10.1016/J.TFP.2021.100164. https://creativecommons.org/licenses/by-nc-nd/4.0/

10 Farming Systems Research and On-Farm Experiments

Abstract

Farming Systems Research (FSR) is an interdisciplinary approach that directly involves farmers in addressing their challenges with a focus on understanding and improving the entire farming system. On-farm research and experimentation are components of FSR and aim to improve crop productivity and profitability by conducting experiments on farmers' fields. Unlike top-down models, , FSR uses a bottom-up approach, that is farmer-centric and addresses smallholder farms' diverse and complex nature. FSR approach has evolved through five phases since the 1960s, from the conceptualization stage to a focus on sustainability science. To implement FSR, there are five stages: diagnosis (characterizing the farm and identifying solutions), designing, implementation, data collection and analysis and finally dissemination and impact evaluation. On-farm experiments (OFE) aim to facilitate farmer-researcher collaboration, evaluate biophysical factors under diverse conditions and provide realistic input-output data for economic analysis. We explore four types based on participation levels: Type 1 (researcher-designed and managed), Type 2 (researcher-designed and farmer-managed), Type 3 (farmer-designed and managed) and Type 4 (farmer groups-designed and managed). OFE use the fundamental principles of experimentation to manage variability. Critical practices for successful implementation of OFE include nurturing partnerships, seeking farmer consent, capacity building and providing technical support. Data varies from individual plot to village level and includes biophysical, social and economic aspects. Key analytical techniques include analysis of variance (ANOVA) but also incorporate more advanced techniques, including tests for treatment-site mean interactions, adaptability analysis, multiple regression and mixed effects models. Challenges such as farmer cooperation and communication barriers highlight the need for strategic solutions. Future efforts should focus on enhancing collaboration, addressing communication issues and implementing innovative practices, including digitalization, to improve the effectiveness and scalability of OFE.

Keywords

Adaptability analysis · Bottom-up model · Covariates · Evolution of farming systems · Farmer centric · Farmer evaluations · Field variability · Genotype*Environment* Management interaction · Holistic · Green Revolution · Multidisciplinary research · Nontreatment factors · On-farm experiments · Participatory approaches · Reductionist appraoch · Researcher-designed and farmer-managed · Top-down approach · Mother-baby model

Learning Outcomes

By the end of this chapter, you are expected to be able to:

1. Explain the meaning of farming systems research (FSR) and on-farm experimentation (OFE).
2. Appreciate the rationale behind farming systems research.
3. Identify characteristics of FSR.
4. Apply the three pillars (farm, farmer and environment) in farming systems research.
5. Differentiate between "Top-down' and "Bottom up" research and extension model.
6. Briefly explain the evolution of farming systems research.
7. Explain paradigms of FSR evolution.
8. Describe the stages in implementing farming systems research.
9. Determine the purpose of conducting on-farm experiments.
10. Identify types of on-farm experiments.

J. Njeri Mugwe, S. Runo, *Research Methodology in Agricultural Sciences*, https://doi.org/10.1007/978-981-95-1892-0_10

11. Explain fundamental principles in on-farm experimentation, including key differences between on-station and on-farm experiments.
12. Explain the concept of G*E*M (Genetics, Environment and Management) interactions.
13. Outline principles of conducting on-farm experiment.
14. Describe steps in conducting on-farm experiments.
15. Design an on-farm experiment.
16. Determine the types of data to be collected in on-farm experiments, including treatment and nontreatment factors.
17. Appreciate key differences in handling and analyzing data from OFE compared to on-station data.
18. Demonstrate challenges in the execution of on-farm experiments.
19. Discuss future directions and opportunities.

This is a long chapter that discusses two related topics: farming systems research (FSR)and on-farm experimentation (OFE). The first part discusses farming systems research and has the following topics:

1. Overview of farming systems research.
2. Evolution and growth of farming systems research.
3. Key stages in the implementation of farming systems research.

The second part discusses on-farm experimentation and discusses the following sub-topics:

1. Principles and practices of OFE.
2. Purpose of OFE.
3. Type on-farm experiments.
4. Designing on-farm experiments.
5. Steps in conducting an on-farm experiment.
6. Analytical techniques on OFE.
7. Stakeholders in OFE and barriers to implementation.
8. Future directions.

10.1 Section One: Farming Systems Research

10.1.1 Overview of Farming Systems Research

10.1.1.1 What Is Farming Systems Research?

Farming systems research (FSR) refers to multidisciplinary and participatory agricultural research that involves farmers and other stakeholders to solve problems of farmers. Rather than focusing on individual components, such as crops, soil, or livestock, FSR aims to understand the overall farming system by studying the interactions between various components of the farming system (Andersen 2017; Chambers 2020; Behera and France 2023). Researchers and their partners study the relationships among the physical, biological and socio-economic environments of the farmer (whole-farm approach). The aim is to understand farmers' perspectives and practices to help design sustainable solutions to enhance the productivity and resilience of the whole farming system. Farming systems research was first developed in 1970, but it has evolved significantly, and in the present world is expected to meet multiple global and local demands and challenges, e.g., supporting livelihood, conserving biodiversity, offsetting emissions and adapting to climate change (Kpienbaareh et al. 2022; Karaca and Ince 2023; Keating 2024). As such, FSR remains highly relevant in achieving sustainable agricultural development.

10.1.1.2 Origin of Farming Systems Research

The primary goal of agricultural research is to create innovative technologies that farmers can adopt to improve productivity and production efficiency, ultimately enhancing food security. But the truth is that this is not all a success story. The adoption rate of the technologies developed is typically not 100%, leading to less-than-desired results. Low adoption is mainly attributed to weaknesses in the research approaches used (Chambers 1994). This realization led researchers to look for new ways of implementing the research agenda.

Historically, before 1970, the dominant approach was the top-down model. This model focused on disciplinary specialization in commodity-oriented research at experimental research stations and laboratories (Keating 2024). They were more *production-oriented*, *single-discipline-based* and used a *reductionist approach*. The researchers would identify priorities, develop technologies in controlled environments and disseminate them directly to farmers or through extension officers using a "top-down model." The main goal was to optimize agriculture by striving for continuous productivity gains, mainly measured through crop yields or return to labor.

While the top-down approach was effective in homogeneous production environments like laboratories or experimental farms, it proved inadequate in more complex and heterogeneous environments, such as smallholder farms (Darnhofer et al. 2012). The smallholder farms are more diverse, with varying soil types and mixed livestock-cropping systems and are influenced by social and cultural factors (Chikowo et al. 2014; Danso-Abbeam et al. 2021; Wawire et al. 2023). Consequently, many farmers failed to adopt the technologies and production methods promoted by researchers and their extension counterparts due to "lack-of-fit."

The limitations of the top-down approach became particularly evident during the *Green Revolution*, exemplified by Norman Borlaug's initiatives in the 1960s and 1970s (Pingali 2012). Although the Green Revolution succeeded in Southeast Asia (Frankema 2014; Cabral et al. 2022), the initiative yielded mixed results in Latin America, Asia and Sub-Saharan Africa (Pingali 2012). While some regions experienced increased crop yields, the benefits were unevenly distributed, with wealthier and large-scale farmers often reaping the rewards. In contrast, the smallholder farmers struggled to adopt the high-input technologies (Pingali 2012). The failure in these regions was attributed to the high cost and the impracticability of the prescribed packages for smallholder farmers, highlighting a significant "lack-of-fit" of these packages in the smallholder production systems.

In response to these challenges, FSR emerged as an innovative approach to address the shortcomings of the traditional *top-down model,* which ignored resource-poor farmers and their ability to contribute to researching and developing appropriate technologies in a *bottom-up model* approach. Therefore, FSR promotes a systemic view, capturing the interactions between various components such as crops, livestock and farmers' subjective perceptions, values, and preferences (Behera and France 2023). This approach emphasizes that agricultural frameworks cannot be studied in isolation but must be understood as part of a larger system (Hermans et al. 2020; Danso-Abbeam et al. 2021). It therefore advocates for collaboration between biophysical and social sciences and the involvement of farmers in the research process to enhance relevance and practicability.

Building on this perspective, developing agricultural technologies suitable for smallholder farmers requires considering both the environmental and social context. Early works by researchers in Latin America, Africa and Asia, such as Stifel (1990), Simmonds (1986) among others, laid the foundation for understanding farm households, family decisions and decision-making processes. By incorporating these perceptions, the goal was to develop technologies and practices that align with the specific needs of the smallholder farmers, addressing the lack of fit that had hindered previous initiatives. When involved in developing the technologies, resource-poor farmers and communities help give input into the research and extension processes, thus promoting the development of technologies that are context-specific and practice-based. This is the so called "Bottom-up model". Table 10.1 shows a comparison of the top-down and bottom-up approaches.

Reasons for initiating FSR in the 1960s to 1970s could therefore be summarized as follows:

1. Limited adoption of research recommendations by smallholder farmers in developing nations.
2. Researchers' priorities were different from the clients' (farmers') priorities.
3. Conditions at the research stations and laboratories differed from those of farmers' fields.
4. Evaluation criteria used by professional researchers contrasted with farmers' criteria and preferences.
5. Researchers focused on commodity-based evaluations to optimize productivity, while farmers often aim at minimizing risk through multiple cropping systems.
6. Though the Green Revolution recorded a rapid increase in crop and animal production systems, it resulted in serious trade-offs related to negative environmental effects.

Table 10.1 Simple comparison of differences between top-down and bottom-up research and extension approach

Bottom-up approach	Top-down approach
Systems approach	Reductionist approach
Linkages between subsystems	Commodity-based
Priorities reflect the whole farm and farmer	Researchers priorities
Farmers are largely involved	Minimal farmer involvement
Multiple variables measured	Few variables measured
Expensive	Modest
Interdisciplinary and multidisciplinary	Specialized disciplines, mainly agronomists and breeders

10.1.1.3 Key Characteristics of Farming Systems Research

The following are key characteristics of farming systems research (Hayman et al. 2012; Martin et al. 2013; Madry et al. 2013; Santos et al. 2021; Behera and France 2023):

1. *Farmer involvement is a core element:* Researchers collaborating with farmers is one of the most important principles in FSR. The researchers learn from the farmer and give opportunity to the farmer to give input into the research process.
2. *Multidisciplinary approach:* farmers have complex farming systems with many variables interacting, such as crops, animals and off-farm enterprises, within a decision-making process involving land allocation, labor, capital, etc. To address these wide ranges of issues, FSR teams consist of several disciplines working together: agronomists, agricultural economists, animal scientists and sometimes sociologists.
3. *Holistic:* inclusion of farmers' physical, economic and social-cultural dimensions. Considers whole farm perspective and incorporates all components, including crops, soils, water resources, and socioeconomic factors (see next section).
4. *Work with representative farmers:* representative farms are chosen because it is impossible to work with all farmers. To ensure the representation of variability between farms, recommendation domains are used as platforms for categorizing farmers. The farms should be well-characterized to allow extrapolation to the recommendation domains.
5. *Complements on-station research*; research on-station and on the farms are sometimes conducted concurrently and serve different purposes. The on-station will evaluate new technologies and innovations that are not well-refined, while on-farm research will focus more on adaptive research and validating "best bet" technologies.

Additionally, the farmers and their advisors are key in providing feedback for on-station work.

6. *Dynamic and iterative:* FSR is characterized by constant change and adaptation throughout the research process. Researchers and practitioners actively seek to enhance their understanding of farming systems by gathering information. Throughout the process, farmers and other stakeholders may adjust research objectives, methodologies, and interventions.
7. *Participatory approaches:* participatory approaches are key to guaranteeing the participation of all parties in a way that none feels intimidated. FSR actively involves farmers and local communities in the research process, and farmers act as partners and co-researchers.
8. *Expensive and time-consuming:* FSR requires a significant investment to cater for community engagement and data collection. Additionally, it requires extensive data collection to gain a deep understanding of the farming system. The process can involve surveys, field observations, and socioeconomic assessments. Collecting and processing such data demands financial resources, skilled personnel, and time.
9. *Systems thinking:* FSR uses systems thinking to analyze the farm as a dynamic and interconnected system with feedback loops and dependencies. This allows researchers to identify leverage points for improvement by analyzing cause-and-effect relationships within the system.
10. *Highly context-specific:* Though holistic, FSR is context-specific. What works well in one region or farming system may not work in another due to differences in physical, biological, and socio-economic components. Consequently, FSR is tailored to meet the distinct needs of various farming communities.

10.1.1.4 The Multi-faceted Nature of Farming Systems Research

Farming systems research (FSR) is characterized by its adaptability, diversity, and broad scope, taking on various forms and labels as described by both early and recent scholars (Simmonds 1986; Stifel 1990; Darnhofer et al. 2012; Hayman et al. 2012). This variation reflects the different ways practitioners interpret and engage with the complex interactions within farming systems, making FSR inherently resistant to rigid definitions or standardization. It encompasses a wide range of methodologies and frameworks, which are often adapted to suit specific local conditions, research objectives and contextual challenges (van der Lee et al. 2022; Behera and France 2023). Consequently, multiple models and practices have emerged, each with distinct focuses and methods. Despite this diversity, FSR is unified by core characteristics, chiefly its emphasis on understanding and integrating physical, biological and socio-economic components of farming systems (Fig. 10.1).

So far, we can infer that FSR emphasizes the interdependence of three core entities: the farmer, the farm and the environment, collectively referred to as the total environment (Patanothai 1997; Rudnick et al. 2021; Behera and France 2023). The farmer is central to FSR, contributing knowledge, experience and decision-making that drive innovation and adaptive responses to risks such as climate variability and market shifts (van Zonneveld et al. 2020; Yeleliere et al. 2023; Ewulo et al. 2025). The farm is the biological and physical foundation, comprising soil, crops, livestock and biodiversity, whose integration with technology and infrastructure enhances efficiency and ecological resilience. The environment, on the other hand, encompasses the broader climatic, ecological and socio-economic context, including markets and policies, that subsequently shapes farming operations and influences decisions at both individual and community levels (Fig. 10.2).

10.1.2 Evolution and Growth of Farming Systems Research

It has been nearly six decades since the inception of the FSR movement, and its relevance remains as strong today as in the 1960s. Initially, much of the twentieth-century focus was on developing methodologies for its implementation, with FSR widely supported for strengthening linkages between researchers and extension systems to improve the efficiency of agricultural research (Simmonds 1986; Shaner 2019). By the mid-1980s, it was widely adopted by donor agencies and agricultural organizations, particularly in developing nations. Today, in the twenty-first century, FSR continues to serve as a powerful tool for advancing sustainable agriculture and contributing to the achievement of sustainable development goals (Lacoste et al. 2021; Behera and France 2023; Dixon et al. 2023). So far, numerous projects have demonstrated FSR's impact and versatility across diverse farming systems, with the model continuously adapting to shifting priorities in agricultural development.

The evolution of FSR can be described in five distinct phases (Fig. 10.3):

- *Phase 1 (Pre-1960):* Top-down, linear model with minimal farmer involvement.
- *Phase 2 (1960–1980):* Conceptual refinement through multidisciplinary integration.
- *Phase 3 (1980s—2000):* Farmer participatory phase with widespread methodological development.
- *Phase 4 (2000–2015):* Benefits realization and measurable impact on farmers and researchers.
- *Phase 5 (Post-2015):* Shift towards sustainability science and systems thinking in innovation.

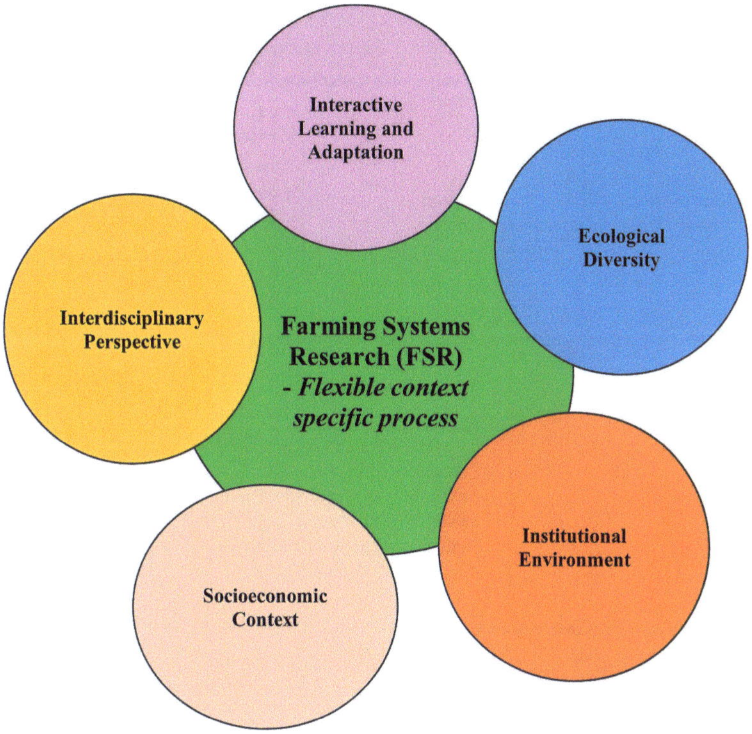

Fig. 10.1 The multifaceted nature of FSR; it is interdisciplinary, with different disciplines customizing into different contexts, and has evolved to embrace diverse methodologies, making it difficult to restrict and standardize

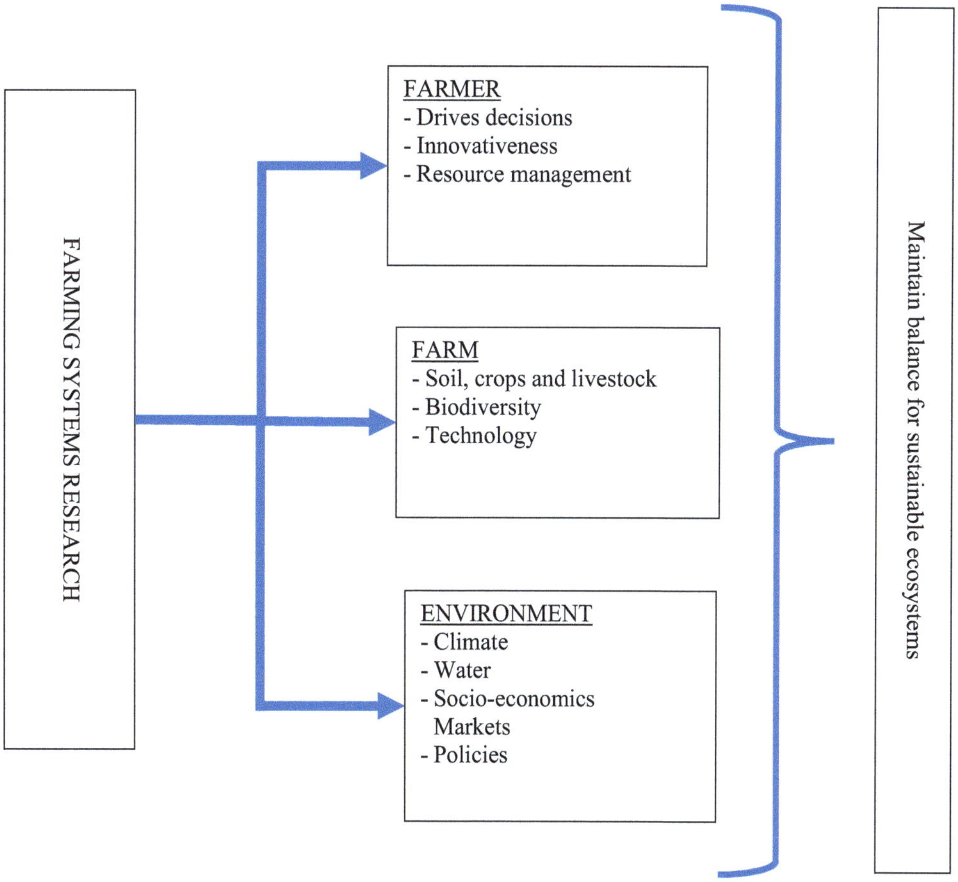

Fig. 10.2 The FSR aims to maintain a balance between the farmer, the farm, and the environment. The three components operate in a dynamic and interdependent relationship

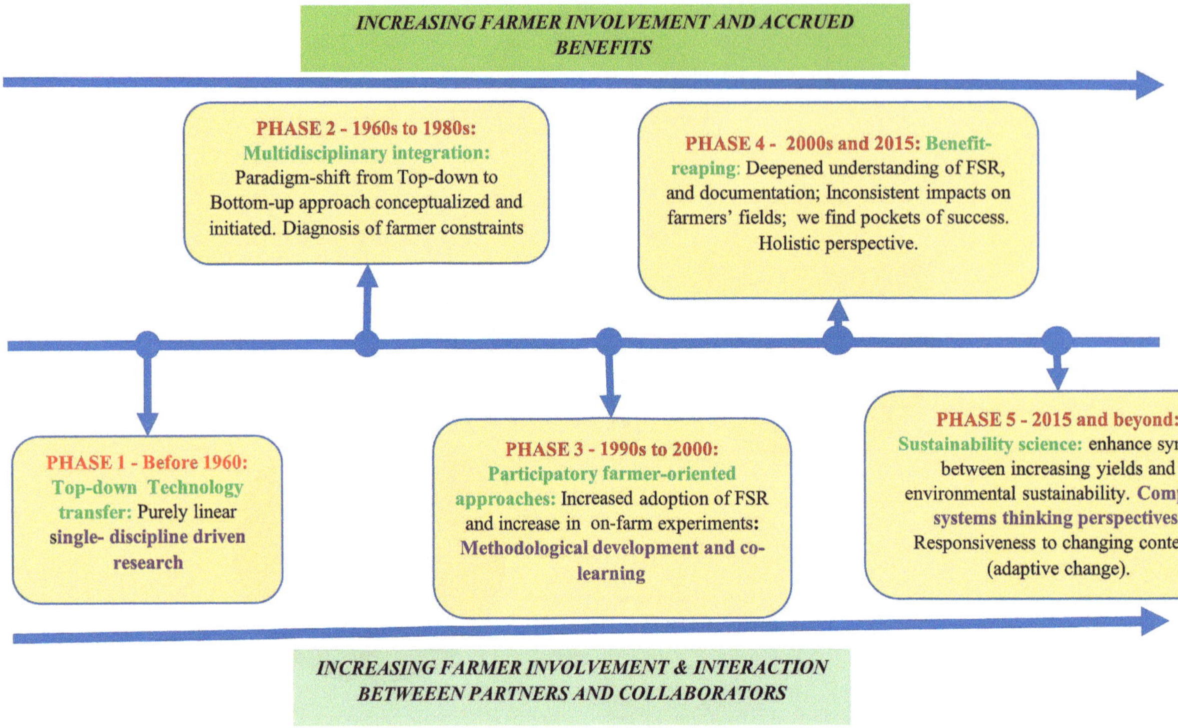

Fig. 10.3 Five phases in the evolution of farming systems research from 1960 to the twenty-first century

Phase One: Pre-1960

During this period, agricultural research primarily focused on crop breeding and improvement conducted through on-station experiments. Research facilities were centralized and operated by government agencies, universities and national agricultural institutions. The process followed a top-down model, with researchers developing and testing innovations independently of farmers. Research efforts were often discipline-specific, isolated from related fields like agronomy, soil science and plant pathology, resulting in fragmented outcomes. This phase is best described as a "linear, single-discipline-driven approach." Most research was concentrated on staple crops such as wheat, rice and cotton.

Phase Two: 1960s to 1980s—The Multidisciplinary Phase

This phase marked a significant shift toward more farmer-oriented approaches, although top-down methodologies still dominated. One widely adopted method was the Training and Visit (T&V) system (Feder et al. 1986), which involved structured training for extension staff and regular farm visits to disseminate production techniques aimed at increasing yields and incomes. While T&V expanded the area under cultivation, it was criticized for being rigid, top-down and financially unsustainable (Feder et al. 1986; Dejene 1989). A study by Hussain et al. (1994) found that T&V increased the quantity but not the quality of extension-farmer interactions. Despite these limitations, this phase saw increasing farmer engagement, with research activities increasingly taking place in farmers' fields, improving relevance and participation. Multidisciplinary collaboration became more common, integrating agronomy, soil science, entomology, plant pathology and agricultural economics, enabling more comprehensive solutions to complex challenges.

Key developments during this phase included:

- Establishment of international research centers focused on global crop improvement and technology dissemination.
- Expansion of research into a broader range of crops, including fruits and vegetables.
- Increased attention to environmental concerns and sustainable farming practices.

This period also coincided with the success of Green Revolution technologies in boosting yields across parts of Europe, North America, Asia and Latin America (Pingali 2012; Cabral et al. 2022). However, these technologies favored homogeneous, resource-rich environments with fertile soils, access to fertilizers and efficient markets. They were less effective and often unsuitable for the resource-poor farmers of Sub-Saharan Africa and parts of Latin America (Khan et al. 2014; Larson et al. 2020).

Phase Three: 1990s to the 2000s

This phase was about 40 years since the conceptualization of FSR, marked a period of exponential growth and widespread adoption of FSR. It can be called the "Participatory Farmer-oriented phase" due to emphasis on on-farm experimentation and collaboration between researchers, extension agents, and farming communities. This is evidenced by the many symposiums, workshops and publications developed during this period (Norman et al. 1995). The main activities involved developing and refining methodologies for conducting FSR and developing guidelines for carrying out the various operations within FSR. Key areas of operation involved the following areas:

1. Methodologies for involving partnerships: Researchers focused on developing methodologies, modalities, and ways of working with farmers and other collaborators.
2. Methodologies and practices of designing on-farm experiments and statistical data analysis. On-farm experimentation was a major component, and many researchers, including biophysical scientists, agricultural economists, as well as sociologists, worked together to conduct field trials on the farmers' fields. Key questions addressed "what" and "how" and "to do" questions, such as: how do you plan an on-farm experiment? What designs are most appropriate for on-farm experiments? What data needs to be collected from the research studies (field and social data)? How do we minimize unexplained errors? What are the appropriate statistical techniques for on-farm work? How do we effectively involve the farmer in the research process (Farrington 1988; Tripp and Woolley 1990; Minjauw et al. 2004; Snapp et al. 2019)? Techniques such as adaptability analysis were developed to tailor strategic practices to specific needs.
3. *Institutionalization of FSR:* Principles of FSR were incorporated into the policies of international and national research organizations to institutionalize FSR. Organizations like the International Maize and Wheat Improvement Centre (CIMMYT) and the International Rice Research Institute (IRRI) were instrumental in developing and promoting FSR methodologies (Clayton 1983; Byerlee and Tripp 1988; Bingen and Gibbon 2012). Other key activities to promote institutionalization included streamlining of FSR in research projects and intensive capacity building of stakeholders on how best to engage with FSR.
4. *Ways of eliciting farmers' participation.* Farmer participation was a priority, and formal approaches to involving farmers in the research process were developed. The participatory approach became a dominant term. Two common ones were rapid Rural appraisal (RRA) and Participatory rural appraisal (PRA) techniques, which played a key role in cementing relationships between farmers and researchers (Chambers 1994).
5. *Categorization and classification of on-farm trials:* In 1995, Peter Biggs, an agricultural researcher, proposed a categorization and classification framework for on-farm experiments. Biggs's framework aimed to systematically organize and understand the diverse types of on-farm experiments. The proposed framework has since been valuable for researchers and policymakers in designing, conducting, and evaluating on-farm experiments.
6. *Techniques of empowering households:* Gender disparities in agricultural activities were recognized, and modalities of balancing gender were developed, e.g., gender analysis techniques and the use of farmer groups (including women groups) to improve the framework of interaction and participation.
7. *Methodologies for collecting data and evaluation by the farmers:* Farmer-friendly techniques for farmers to evaluate the experiments were developed. Examples include Scoring and matrix ranking, Bao games, Wealth Ranking etc. (Franzel and Scherr 2002; Coe 2007).
8. *Ways of reporting and disseminating the research work:* Improved reporting and dissemination methods were developed, which considered effective communication to embrace participatory farmer-centric and context-specific approaches to research. It required methods that could convey the complexity of the farming systems and promote holistic solutions while considering the local context. Some of the most common methods developed and emphasized included participatory workshops that emphasized two-way communication and field days that allowed farmers to see research in action and learn directly from researchers.

Phase Four - 2000 to 2015: Benefit Reaping Phase

During this phase we think that the FSR had gained widespread recognition, with various methodologies in place and researchers benefiting from experience. Many donor-supported projects integrated both research and development objectives, requiring clear demonstrations of farmer involvement and tangible benefits to the target communities. We call this phase a "benefit-reaping phase" where tangible outcomes and impacts were expected. However, the nature of these benefits varied among the stakeholders. While researchers often gained in knowledge, publications and methodological advancements, the outcomes for farmers were less consistent. Not all farmers involved in these projects experienced equal benefits, largely due to uneven levels of engagement, support, and contextual factors such as access to resources, local infrastructure and institutional backing.

For the researchers, this phase was marked by the production of numerous practical guides and resource materials to the processes. Examples include:

- CIMMYT participatory and extension, and field guides (Lilja and Bellon 2008).
- Participatory monitoring and evaluation guides.
- FAO farmer field schools guides (FAO 2016).
- World Bank and CGIAR publications on integrated agricultural research for developmen.t
- Participatory learning and actions guide (PLA) notes.
- Participatory market chain analysis for smallholder producers (Devaux et al. 2013).
- Enabling rural innovation in Africa: A market facilitator guide to agro enterprise development (Ferris et al. 2006).

Cumulatively, during this phase, FSR underwent a systematic expansion and complexity, embracing a diverse array of domains. It evolved by adopting multidisciplinary perspectives, addressing not just crop production but also livestock management, soil health, water resource management and the socioeconomic dimensions within farming systems. As FSR continued to adapt, it responded to emerging challenges such as climate change adaptation, sustainable agricultural practices, gender inclusivity and market access (Altieri et al. 2015).

Phase five: This is the final phase, i.e., from 2015 onwards

We see a new paradigm shift toward a more inclusive approach emphasizing synergy between increased yields and environmental sustainability. This is likely due to the accumulated knowledge and understanding of the farmer's environment and the development of better and appropriate research designs and data analysis for on-farm research. The paradigm shift is also associated with the changing climate amidst increasing population and the pressure to reduce poverty and attain zero hunger as stipulated in the UN SDGs goals I and 2.

Understanding the heterogeneity of farming systems helps in better targeting agricultural innovations; past efforts are considerable on this front. Over the last six decades, numerous authors have consistently reported that farmers' environments are highly complex and heterogeneous. In addition to climate variation across geographical regions, farms and farmers have been described as "notoriously heterogeneous" (Vanlauwe et al. 2019) and cannot be supported by "one-size-fits-all" solutions (Nelson et al. 2019). This is especially true for the smallholder farmers of the developing nations. Three key sources of variation and variability emerge:

1. Within farm soil fertility variation: referred to as soil fertility gradients (Zingore et al. 2007), caused mainly by inherent differences (soil forming processes) and past preferential management practices by farmers.
2. Variation due to resource access: variable access to production resources (Giller et al. 2011). Access to resources such as land, labor, cash, inputs, etc., is different for different farmers.
3. Management practices: These vary depending on farmers' production objectives. For example, a farmer with good access to infrastructure and output markets is likely to choose cash crops over other crops. Management practices may also vary depending on the farmer's experience, knowledge and skill level. Farmers have varying competency levels in managing resources such as inputs, land, water, etc.

With this deepened understanding of farmers' environment and conditions, plus methodologies of handling farm data, including the use of ICT, researchers are arguing for new ways of conducting agronomic research and scaling up the adoption of innovations. A compelling argument is being made for incorporating "Options and context interaction (O x C") as a novel agronomic research paradigm (Nelson et al. 2019; Vanlauwe et al. 2019). According to these authors, options include crop species and varieties, innovative irrigation options, inputs, and labor. On the other hand, contexts are ecological, economic, and social factors. Performance is a result of the interaction between options and context. This is similar to the *Genotype*Environment*Management* interaction traditionally considered by agronomic breeders.

By recognizing the inherent complexities and variations in agricultural environments, the "O * C*M paradigm" aims to enhance the precision, adaptability, and effectiveness of agronomic research. The evolution phases are summarized in Table 10.2.

Table 10.2 Evolution and development of the farmer participatory research approach

Phase	Paradigm	Key components/activities
Before 1960	Traditional top-down	Research by agronomists in controlled laboratories and field sites Farmers receive recommendations following linear framework; researcher to extension to farmer
1960s to 1980s	Bottom-up approach conceptualized	Realization that the Green revolution worked selectively and was not sustainable New thinking and conceptualization of farming systems research. The FSR concept is born, and attempts are made to institutionalize it National Agricultural Research Institutions adopts major components and philosophies of FSR Diagnosis and mindset change among researchers were mainstreamed
1990s to 2000	Farmer participatory-oriented approach	Heightened activities using participatory approaches Methodologies/steps in FSR developed FSR is institutionalized mainly with donor-supported projects incorporating integrated Agricultural Research for Development (IAR4D) Capacity building of researchers and other agricultural stakeholders Numerous workshops and documentation of both processes and research outputs

(continued)

Table 10.2 (continued)

Phase	Paradigm	Key components/activities
2000 to 2015	Benefit-reaping stage	Consistent methodologies used in research and development projects Farmers and researchers increase knowledge Adoption of multidisciplinary perspectives However, uneven impact due to inconsistent institutional support and farmer participation
2015 onwards	Sustainability science and options context interaction	Population increases, climate change and reducing poverty and zero hunger are major drivers of the agricultural process; Sustainable Development Goals 1 and 2 Emphasis on increased synergy between increased yields and environmental concerns (adaptive sustainable systems such as agroecology and regenerative agriculture (see next section) The Green Revolution model has negative environmental impacts New ways of capturing farmer variabilities beyond the recommendation domain range (G*E*M interaction) Need to consider fine-scale variations among smallholder farmers Agricultural innovation systems that incorporate digitalization

Other Terms Related to Farming Systems Research

Here, we discuss three terms related to FSR but applied in the context of soil management, sustainable agriculture, and Agroforestry. These terms are (1) Integrated soil fertility management, (2) Agro-ecology, and (3) Diagnosis and design in agroforestry.

a. Integrated Soil Fertility Management (ISFM) is a comprehensive set of principles designed to increase agricultural production using integrated nutrient management. The approach involves strategically combining organic and inorganic fertilizers to maintain soil health and fertility, which is critical for crop productivity (Mugwe and Otieno 2021). The principles emphasize three aspects: (1) Maximizing agronomic efficiency, (2) Sustainable intensification, and (3) adaptation to local conditions (Vanlauwe et al. 2015). Maximizing agronomic efficiency ensures that the inputs and management practices are optimized to achieve the best possible yields. Sustainable intensification aims to increase agricultural productivity per unit of land area in an environmentally sustainable and economically viable way over time.

On the other hand, local adaptation acknowledges the diverse nature of smallholder farming systems, each with its unique combination of challenges and opportunities. Further, it acknowledges the spatio-temporal heterogeneity in smallholder farming landscapes' climate, soil, and socioeconomic conditions. Therefore, ISFM seeks a specific decision-making process for agronomic input and management practices at the farm and plot level and advocates for promoting and adopting context-specific soil fertility management practices that optimize yields (Dunjana et al. 2023).

Like FSR, ISFM empowers farmers to prioritize and sustain their farming systems by involving them in decision-making and leveraging their expertise to optimize nutrient cycling, enhance soil structure, and promote sustainable practices (Mugendi et al. 2011; Jahangir et al. 2021). This approach fosters ownership and responsibility for soil health, addressing food security challenges while promoting environmental sustainability and resilience. Collaboration between scientific expertise and farming communities is key to ISFM's success.

b. The Concept of Agroecology is an interdisciplinary and holistic approach to agriculture and food systems (Francis et al. 2003; Wezel et al. 2020) and it is also related to the new paradigm of the concept of "Regenerative agriculture" (Giller et al. 2021). It is a response to the challenges posed by conventional industrial agriculture, which has caused depletion of natural resources and degradation of life support systems. According to Andrée (2021)), agriculture faces a major current and future challenge, and the concept of agroecology is poised as a solution and emphasizes integrating ecological principles into farming practices to create sustainable and resilient farming systems (Altieri et al. 2015; Franzluebbers et al. 2020). It considers the environmental, social and economic dimensions of agriculture. The approach promotes agricultural systems that are productive, environmentally friendly, socially just and economically viable. Long-term sustainability and resilience are emphasized.

Agroecology is now regarded as an ideal for farmer participation, with a core tenet being the co-creation and sharing of knowledge (Gliessman 2018; Wezel et al. 2020; Utter et al. 2021). In this approach, farmers use a deep understanding of local ecosystems and farming systems to generate knowledge. Collaboration is key, with farmers and researchers working together to develop sustainable, context-specific solutions. By co-creating knowledge, the resulting innovations are scientifically sound and practically relevant. This partnership strengthens the resilience of agricultural systems, promotes biodiversity and supports sustainable and equitable food production practices.

Another critical component of agroecology is restoring and reintegrating agricultural systems, emphasizing their interdependence (Palomo-Campesino et al. 2018). This holistic approach creates farming systems that mimic natural ecosystems, promoting synergy between crops, livestock, trees and other components (Hawes et al. 2021). It involves recycling resources and nutrients, reducing waste and dependency on external inputs and enhancing natural processes like nutrient cycling, pest regulation and water management. This knowledge-intensive approach prioritizes understanding ecological relationships and reintegrating components that promote healthier ecosystems over input-intensive practices (Thiessen Martens 2022).

c. *Diagnosis and design (D and D)* are terms used in agroforestry to plan, referring to the two essential phases used to plan sustainable and productive agroecosystems (Scherr 1990; Cardoso et al. 2001). The term represents two phases of implementation. These phases are crucial for evaluating the current situation, identifying opportunities and challenges, and developing a tailored agroforestry system that aligns with the specific needs and goals of a given area or farming community. The process begins with a diagnostic phase involving a comprehensive assessment of the landscape's ecological, social and economic aspects and farming practices (Scherr 1990). During this phase, researchers and practitioners collect data on soil quality, climate, biodiversity, land use and existing farming systems. This thorough evaluation helps to understand the current situation and identify key issues.

Following the completion of the diagnosis phase, the design phase begins (Scherr 1990). In this phase, practitioners work closely with the local community to create a tailored agroforestry system that addresses the identified challenges and capitalizes on existing opportunities (Scherr 1990). It focuses on integrating local knowledge, community input and ecological considerations to create an effective and sustainable system. Overall, D and D promotes the development of agroforest systems that are productive, environmentally friendly, socially inclusive, and economically viable agroforestry.

10.1.3 Key Stages in Implementation of Farming Systems Research

The methodologies, procedures and steps of implementing farming systems research (FSR) are comprehensively detailed by seminal works and practical guides that have shaped the field over decades (Darnhofer et al. 2012). In this section, building upon the foundational principles of farmer participation and contextual relevance, we will briefly discuss the key stages involved in implementing FSR.

As the FSR evolved, researchers developed guidelines for implementing farming systems research with farmers (Clayton 1983; Collinson 1987; Jackson-Smith and Veisi 2023; Keating 2024). The stages can be simplified into five stages as follows and summarized further in Fig. 10.4:

1. Diagnosis stage.
2. Design and protocol preparation stage.
3. Experimentation stage.
4. Assessment and data analysis stage.
5. Dissemination and impact evaluation stage.

1. Diagnosis Stage

The diagnostic stage aims to gain an understanding of farming constraints and opportunities within the farming system. Farmers are engaged in identifying challenges and problems they face in their farming systems (Collinson 1987). This stage typically consists of two main steps:

(a) *The first step* is to assemble background information about the farming system and the use of secondary data. It involves reviewing secondary data from existing reports, from government ministries and other documents. Information on production systems, research data, production data and any other information related to the area is collated and documented.

(b) *The second step* involves identifying the needs and production constraints of the households, along with a deeper understanding of their farming systems and practices. This includes analyzing biological, physical, and socioeconomic factors and their interrelationships. A range of data collection methods is used, including formal and informal surveys, quantitative and qualitative methods, participatory research methods, etc. The output of this stage is a *comprehensive description of farming systems and the production constraints* faced by farmers.

2. Design and Protocol Preparation

In this stage of designing and writing the protocol, the information gathered during the diagnostic stage is analyzed to prioritize problems and concerns for the next stage (Collinson 1987; Jackson-Smith and Veisi 2023). Five key activities achieved include:

1. *Data analysis and creating groups of farmers:* The data collected is analyzed to identify patterns and key issues. Farmers are grouped based on shared characteristics like location, crop types, farming practices and socioeconomic characteristics, referred to as *recommendation domains.*
2. *Prioritizing problems:* Farming households face many problems, some of which are researchable by agricultural researchers and many beyond their reach. Those beyond the reach of researchers are documented and passed on to other stakeholders in agriculture. The ones that are researchable are isolated for the next stages.
3. *Formulating the research problem and objectives* involves specifying key issues identified during the problem prioritization stage and translating them into specific research problems and objectives. The research problem and objectives must be clear.
4. *Designing the experiments and choosing the location:* Experiments are designed considering the objectives of the study. The research objectives drive the study and preparation of the protocol.

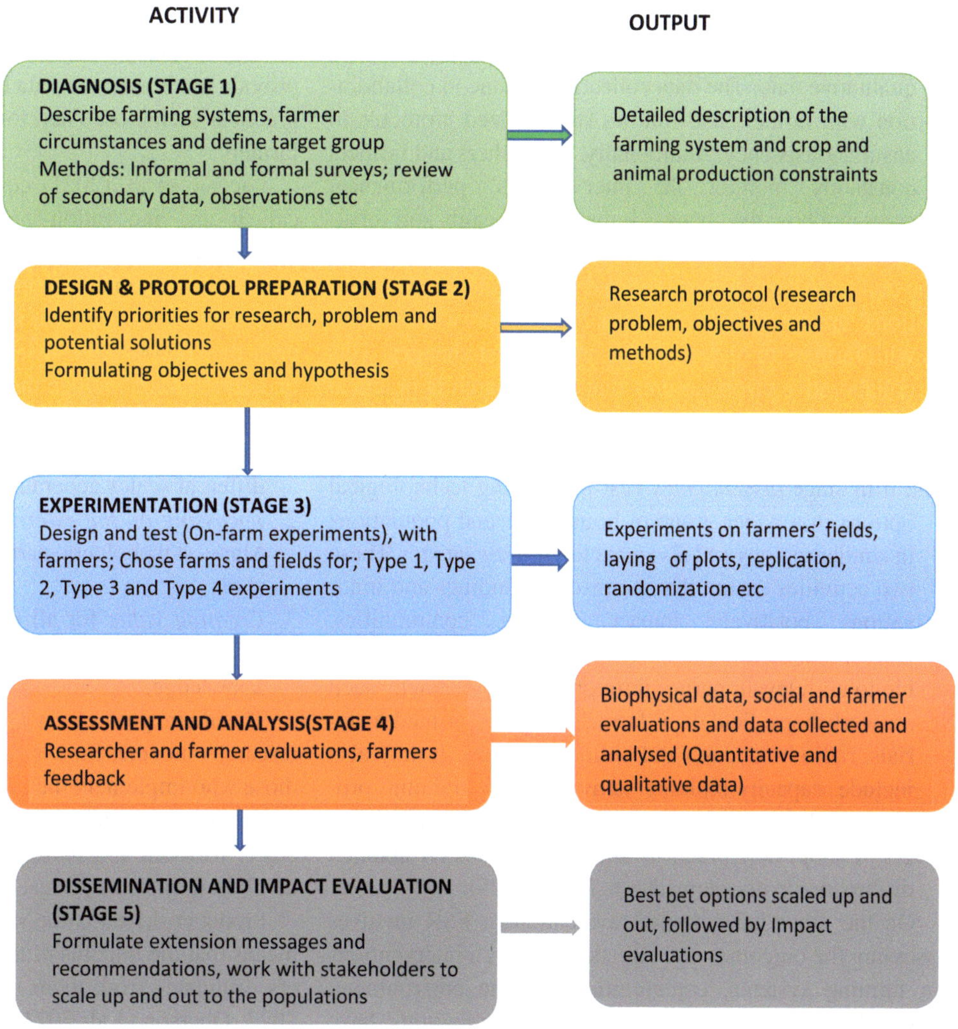

Fig. 10.4 Stages in implementation of FSR

5. *Preparation of the protocol* involves preparing a detailed protocol that will guide the research team, the farmers and other collaborators. The protocol will guide the logistics of how researchers and farmers work together, the structure of the experiments and how data will be captured.

3. Experimentation Stage

The experimentation stage in FSR is a critical phase where researchers and farmers collaboratively test and evaluate various interventions, innovations and practices within the farming system (Kyveryga 2019; Snapp et al. 2019; Toffolini and Jeuffroy 2022). The implementation adheres closely to collaboratively developed protocols. This experimentation stage involves:

- *Farmers' selection:* Researchers select farmers who will participate in the experimentation. This is often done collaboratively with local communities and extension agents, considering criteria such as willingness to participate, representativeness of the target farming system, diverse farm characteristics (e.g., farm size, resource availability.
- *Selecting interventions:* Researchers, in collaboration with farmers, select a set of interventions or innovations to test during the experimentation phase; this can include new crop varieties, cropping patterns, improved soil management practices, pest control strategies, water management techniques or other innovations relevant to the farming system.
- *Experimental design:* Researchers and farmers collaborate to design experiments on the farms that systematically test the selected interventions. This involves determining appropriate plot sizes, treatments, replications and control groups, while adapting the design to specific farm conditions and available resources.
- *Implementation:* The chosen interventions are implemented on selected farms or plots within the farming system. Researchers work closely with farmers providing necessary guidance, technical backstopping and training on the proper techniques and practices.

4. Assessment and Data Analysis Stage

At the assessment and data analysis stage, continuous monitoring and systematic data collection are conducted. This includes *agronomic, socio-economic and farmer preferences* data, and thus comprises both quantitative and qualitative data. The data collection is done in collaboration with farmers and follows standardized protocols to ensure consistency. Additionally, researchers and farmers constantly monitor the interventions' performance, assessing how the changes impact crop health, soil fertility, resource use efficiency and overall farm productivity.

The collected data is then analyzed to evaluate the interventions' impact and effectiveness. Researchers work closely with farmers to review results and suggest changes. This collaborative effort ensures that the research is practical and beneficial to the local community.

5. Disseminating and Evaluating Impact

The fifth stage *disseminates* best-performing technological options to a wider group of households and populations in similar regions and also aims to *evaluate impact*. These two activities ensure that the research findings and innovations positively impact farming communities. Dissemination aims to make research outcomes accessible to all relevant parties through various channels, such as, workshops, field days, farmer meetings, printed materials, radio broadcasts and online platforms. It may also include capacity-building activities like training programs and workshops to empower farmers and stakeholders to adopt new practices and technologies (A detailed discussion on dissemination is in Chap. 25).

On the other hand, impact evaluation in FSR involves assessing the outcomes and effects of research interventions on farming systems, communities and the environment. Researchers aim to determine if the interventions have achieved their desired objectives. Impact evaluation requires systematic data collection to measure productivity, income, resource use efficiency, environmental sustainability and social well-being indicators. This data is compared to baseline information collected before the interventions were introduced. During the assessment, control groups, attribution and contribution should be taken into account (see Chap. 11 on impact evaluation).

10.2 Section Two: On-Farm Experimentation

10.2.1 Principles and Practices of On-Farm Experiments 10.1.2

10.2.1.1 What Is On-Farm Experimentation?

On-farm experimentation (OFE) is a subset of farming systems research. It involves conducting experiments directly on farmers' fields, thus allowing researchers to observe and evaluate agricultural interventions in real farming conditions. The farmers actively participate and contribute their knowledge and experiences. This is in contrast to on-station experiments, which are conducted at research centers and experimental farms in universities without involving farmers directly (Kyveryga 2019). The results of the experiments provide more accurate data on how interventions perform in practical scenarios. Therefore, OFE maximizes benefits for farmers.

Lacoste et al. (2021) comprehensively define OFE and call it an "innovation process that brings stakeholders together around mutually beneficial experimentation to support farmers in management decisions". These authors note that researchers and other specialists work together and harness their knowledge base. Three key benefits of on-farm experimentation include the following:

1. *Harnessing farmers' knowledge* by allowing farmers to share their knowledge with the experts. Social learning at different scales generates knowledge.
2. *Focussing on the external perspectives* of other experts. Many stakeholders participate and provide specialized skills.
3. *Creating value* for all by stimulating the production of new insights through co-learning and hybridization of knowledge.

The main goal of on-farm experimentation is to make research outcomes more relevant, practical and beneficial for those who implement them. It emphasizes a partnership that is "producer-driven" and "researcher-guided" (Lacoste et al. 2021; Toffolini and Jeuffroy 2022). It is farmer-centric and involves farmers as co-researchers.

Producer-driven is the same concept guiding FSR, and it means that the impetus of the research conducted with farmers originates from them (farmers) (Toffolini and Jeuffroy 2022; Oberson et al. 2024). The role of the researcher is to guide research, especially bringing in the expertise of scientific rigor and systematic methodologies. Farmers, on the other hand, mainly contribute land and labor and are central to defining the interventions to be tested. The research process is collaborative, with researchers avoiding working on predefined interventions.

10.2.1.2 Purpose of On-Farm Experimentation

One of the objectives of on-farm experiments is to ensure that testing is done in a "real-world setting" in the fields of "would-be clients and users" of technologies (Kyveryga 2019; Tanaka et al. 2023). The following are the specific objectives of conducting on-farm experiments:

1. *Permits farmers and researchers to work together as partners.* This enhances the potential for adoption because it increases the chances of generating viable technologies and management practices.
2. *Facilitates adaptive research.* Enables the evaluation of whether technologies or practices that have proven successful in other regions are also effective in new local contexts. National agricultural institutions often collaborate with international and regional partners, allowing them to test and refine these externally developed innovations through OFE.

3. *Important in getting farmers' assessment of a practice-ideas.* Farmer have their criteria for assessing a new technology. For example, researchers working on orange flesh sweet potatoes selected varieties based on yields. Farmers, on the other hand, may include criteria on the taste and texture of the root tubers. This helps researchers develop additional criteria in further evaluation of the varieties.
4. *Permits biophysical evaluation under a wide range of conditions:* Allows evaluation of biophysical performance of a practice under a wide range of socio-economic and environmental conditions, for example, farmer management practices and soil types. At the on-station, the range of conditions is narrow.
5. *Allows acquisition of realistic input–output data for calculation of economic returns.* Economic returns or profitability is one of the most important aspects for evaluating the suitability of technology for farmers. Economic returns calculated using on-station data are biased because researchers work on small, highly controlled plots. Data are, therefore, not representative of the variability observed in farms and households. The yield response tends to be biased upwards, whereas labor is unrepresentative of the farming community as operations differ; for example, farmers might use oxen while at the on-station, human labor is applied. Research involving farmers, on the other hand, is realistic. For example, farmers record their activities and actual yield and labor input under their management. The average values of many farmers will be computed and, therefore, will be representative.
6. *To study how farmers adopt and adapt technology.* Adopt is the taking up of technology (using it) while adapting involves modifying the technology. Farmers will adopt techniques and technologies they find suitable, whereas they may also modify technologies to fit their circumstances. The modified practice is a more sustainable technology and should be encouraged. For example, some farmers in the central highlands of Kenya decided to feed *Calliandra* biomass to animals instead of applying it directly to the soil for nutrient provision (Mugwe et al. 2009a, b). They would then use manure from the animals to improve the soil fertility.
7. *Provides diagnostic information about farmers' experiences, problems, and preferences.* When technologies are tested with farmers, exposure allows farmers more exposure. Thus, they give feedback on their challenges and preferences, thus helping researchers to refine the technology.
8. *To study what farm and farmer characteristics affect the performance of technologies.* Identification of farm and farmer characteristics that affect the performance of new technology helps to target farmers in promotion activities. Many adoption studies were possible because farmers experimenting with new technologies adopted their choices.
9. *Provides feedback to researchers and extension staff:* On-farm experiments are wholesome and give feedback information on the potential adoption of technologies (Franzel and Scherr 2002; Mugwe et al. 2009a, b). Feedback is important in setting priorities for research; researchers get feedback on the effectiveness of the technologies and practices tested and consequently use the feedback to adjust as necessary to ensure they are appropriate and sustainable.
10. *Facilitate attainment of gender parity:* Provide an opportunity for men, women and children to participate in technology development, thus enhancing their future participation in agricultural activities.
11. *Provision of multiple options:* Farmers receive multiple management options ("best bet options") that they choose from according to their circumstances and perspectives.

10.2.1.3 Types of On-Farm Experiments

Several ways of classifying on-farm experiments exist. According to the state of knowledge, classification recognizes three types of experiments: Exploratory, Determinate and Verification. While using criteria of level of control and management, Biggs (1995) recognizes four types of trials: (1) *Contractual*, where researchers contract farmers to provide land. (2) *Consultative*: The researcher consults farmers about their problems and develops solutions; (3) *Collaborative* researchers and farmers collaborate as partners, and (4) *Collegiate*-researchers strengthen farmers.

In this book, we categorize on-farm experiments into four types based on the level of participation of farmers, researchers, and farmer groups. This is modified from existing frameworks (Ashby et al. 1987; Ashby 1990; Coe 2007) to emphasize the roles and interactions between researchers and farmers. These are:

1. Researcher designed and managed (Type 1).
2. Researcher-designed and farmer-managed (Type 2).
3. Farmer designed and managed (Type 3).
4. Farmer group designed and managed (Type 4).

1. Researcher Designed and Managed On-Farm Experiments

"Researcher designed and managed" on-farm experiments, also branded "Type 1 Experiments (Coe 2007), aims to evaluate biophysical performance, i.e., determine if a technological option will perform in the physical environment of the farmers. Researchers design and manage all species, including implementation and data collection. They are typically on-station experiments relocated to farmers' fields. They require a large piece of land, and researchers exhibit the same rigor as those in a station-controlled environment. This means they will strictly apply the three key principles of experimentation (3Rs) and maintain control of the extraneous variables as much as possible. They are often planted on rented farmers' land or located at a school where ample space is available. These trials tend to be more expensive and difficult to manage than on-station because of the need to travel to supervise the operations and data collection.

An example of "researcher-designed and managed" on-farm experiments is a replicated "mother-baby" demonstration experiment, which is often set within a community e.g., in a school property. The experiment usually carries multiple objectives: (1) Collecting good biophysical data, (2) Creating a learning site for farmers, and (3) Soliciting farmer evaluations of the tested options. The mother-baby approach is one of the models widely used in conducting on-farm experiments and provides a framework for both researchers and farmers (Snapp et al. 2019; Jackson-Smith and Veisi 2023).

The mother-baby model has two types of experiments, the "Mother" and "Baby." The mother experiment is large, and tests many technologies, while the baby trials test a few technologies selected by farmers from the mother (Figs. 10.5 and 10.6). This model was initially used in Malawi to evaluate soil fertility management options. Over time, it expanded to include other critical components, such as variety screening and soil and water management (Snapp et al. 2019). In case study 1, presented in Box 10.1 and Fig. 10.5, the mother experiment had 16 treatments that included solely applied organics and combinations of organics and mineral fertilizer. The aim of this mother experiment was twofold: (1) Collect good biophysical data and (2) Invite farmers to learn during

Box 10.1: Case Study 1 of a Mother Baby Approach to Promote Soybean and Climbing Beans

Objective of the project: to improve agricultural productivity and incomes of smallholder farmers in Central Kenya through adoption of yield-enhancing integrated soil fertility management practices (ISFM). In one of the project sites two mother experiments were established with aim of testing new technologies, innovations, and varieties and also as a learning site for farmers. One of the mother experiments had 16 treatment combinations comprising of 8 soil and water conservation technologies being tested on 2 crops: soybean and maize (8 × 2 = 16). The second mother experiment tested 11 soybean varieties.

Treatments in mother experiments testing tied ridging and soil fertility inputs in Kigogo site, Tharaka Nithi County, Kenya

Treatments	Crop 1	Crop 2
Tied ridging + Tithonia + inorganic fertilizer	Soybean	Maize
Tied ridging + manure + inorganic fertilizer	Soybean	Maize
Tied ridging + inorganic fertilizer	Soybean	Maize
Tied ridging + no inputs	Soybean	Maize
No tied ridging + Tithonia + inorganic fertilizer	Soybean	Maize
No tied ridging + manure + inorganic fertilizer	Soybean	Maize
No tied ridging + inorganic fertilizer	Soybean	Maize
No tied ridging + no inputs	Soybean	Maize

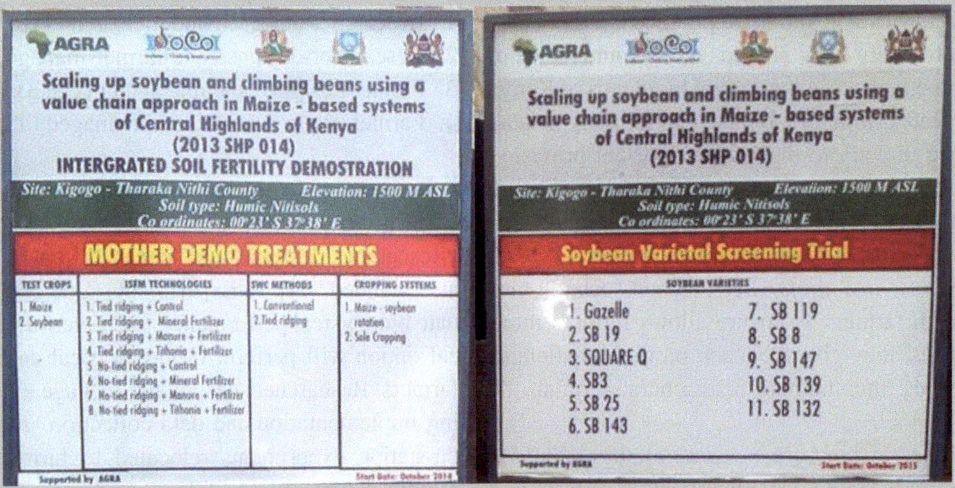

Source: Unpublished project report (Alliance for a Green Revolution in Africa)

During the touring of the plots, farmers assessed the treatments and scored using farmer-friendly criteria in a scale of three on the vigor of the crop (1 = poor, 2 = good, 3-very good), leaf color (1 = yellow, 2 = green 3 = deep green), plant height (1 = short, 2 = medium, 3 = tall), cob size/number of pods (1 = small/few, 2 = medium, 3 = tall/many)

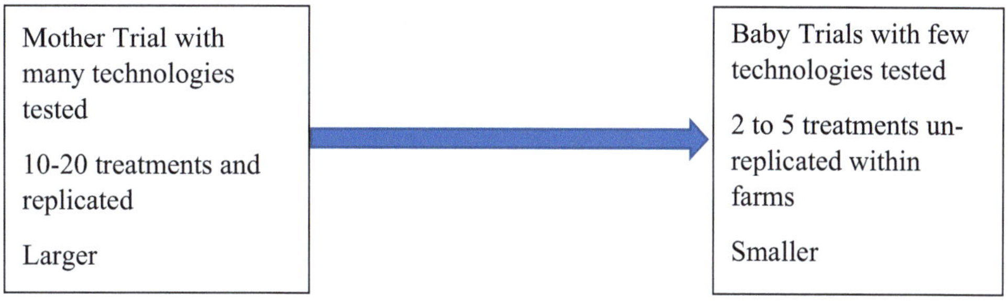

Fig. 10.5 Mother-baby model

Fig. 10.6 A mother experiment involving the screening of soybean varieties in the central highlands of Kenya (Tharaka-Nithi County, Kenya). This was a researcher-designed and researcher-managed experiment. Farmers would visit the site and select a few varieties for planting on their farms (babies)

field days, with an aim to promote soil fertility replenishment options among the farmers.

Typically, farmers learn from the mother trial during field days organized by the researchers and their partners. During such learning days, farmers tour the plots and evaluate the performance of each treatment. As they tour and assess the performance of the various treatments, they are encouraged to select technologies they prefer to test/implement on their individual farms (Figs. 10.6 and 10.7).

Signage at the Kigogo demonstration site, Tharaka-Nithi County, Kenya showing treatments in the mother demo and soybean varietal screening trials.

2. Researcher-Designed and Farmer-Managed on-Farm Experiments

The researcher designed and farmer-managed on-farm experiments are called "Type 2" experiments (Coe 2007). In this one, the researchers design, while the farmers manage the experiments. The primary intent is to assess the performance of practice on representative farms. An important aspect of these experiments is the assessment of benefits to the farmers, such as profits and return to labor. This model ensures that researchers and farmers are actively involved, with almost equal levels of participation in planning and execution. Many farms are involved with at least ten farms per specific technology. For farmers to be able to participate, they are briefed and trained in carrying out experiments as well as in data collection.

In type 2 experiments, some variables are fixed while others are flexible. For example, researchers may standardize variables, such as crop varieties and plot sizes, to ensure consistency and control over biophysical data quality. However, management practices are often left at the discretion of individual farmers. The approach allows for a balance between maintaining control over critical factors and accommodating local variations in farming practices. Researchers usually consult farmers on the design, and each farmer agrees to follow the same procedures to enable comparisons. Researcher-designed and farmer-managed on-farm experiments are most useful for evaluating input and output data, as they allow analysis of costs and returns because farmers conduct all operations (Bellon 2001). An example of a type 2 experiment is presented in Boxes 10.2 and 10.3.

Box 10.2: Example of Researcher Designed and Farmer Managed Experiment

Research objective: investigate the effect of two leguminous plants and two locally available organic resources applied solely or combined with inorganic fertilizer on economics

- Number of farmers: 30 farmers selected during field days.
- Fixed or controlled variables: Maize variety, amount of organic biomass, fertilizer type and amount, plot size. All farmers planted Maize H513. Amount of biomass was measured to supply either 30 or 60 kg N ha^{-1} depending on the treatment. Fertilizer was NPK and plot size was 6 × 4.5 m.
- Non-fixed variables: all agronomic procedures. Farmers carried agronomic procedures independently.
- Farmers selected the plots, managed the experiments, and collected most of the data. Notebooks were provided to farmers who were also trained in data recording.

- Data collected: Maize yields, Labor data, Soil samples. The results are shown in the table below.

Treatment	Net benefit (USD)	Benefit cost ratio	Return to labor
Manure	542	2.9	3.6
Manure +30 kg N ha^{-1}	938.8	2.5	3.1
Tithonia	304.3	1.8	2.1
Tithonia +30 kg N ha^{-1}	795	2.2	2.8
Calliandra +30 kg N ha^{-1}	337.4	1.2	1.3
Leucaena +30 kg N ha^{-1}	462	1.8	2.1
Fertilizer @ 60 kg N ha^{-1}	360	1.3	2.7
Control	63.3	0.6	−0.2
LSD	96	0.4	0.5

Source: Mugwe (2007)
PhD Thesis (Kenyatta University)

Box 10.3: Case Study 2 on Researcher-Designed Farmer Managed Experiment (Type 2)

This is a case study of an experiment carried out with farmers to assess the economic returns of integrated soil fertility management options Muriuki (2011).

Research objective:

The objective was to assess the economic performance of organic resources and mineral fertilizer under farmer conditions and to compare farmers' selection of these resources with the results of scientific economic assessment.

Hypothesis

1. Treatments using a combination of an organic resource and mineral fertilizer are more profitable than those using sole organic resources.
2. Ranking based on calculated economic returns is correlated to farmers ranking of the technologies (based on farmers assessments and observations).

The study was conducted in Mukuuni in Magumoni division and Murugi in Mwimbi division of Meru South district, Kenya. The study collected data from an existing project that was promoting integrated soil fertility management using mother baby approach. The mother experiment consisted of organic materials and mineral fertilizer that was set up in Meru South district in LR 2004. The organic materials consisted of *Tithonia diversifolia* (Tithonia), *Calliandra calothyrsus*, (Calliandra), *Leucaena trichandra* (Leucaena) and farmyard manure. The inputs were either solely applied or combined with mineral fertilizer. During field days farmers were trained on use of the inputs. They also evaluated the performance of the test crops under the different treatments. Afterwards farmers were advised to select inputs of their choice and test them on their farms.

Researcher-designed and farmer-managed experiments were established on 132 farms selected from a sampling frame that comprised 826 farmers who had visited and tried the inputs during 2004. With the help of the researcher, the farmers selected suitable sites for setting up the experimental plots. Plots were measured by the farmers using local methods of measurements, such as, steps and later verified by the researcher.

Plot size varied from farm to farm, ranging from 25 to 100 m^2, while the number of treatments varied from two (2) to four (4). Farmers were trained in record-keeping using simple tables. Inputs applied and outputs obtained were measured by the farmers using local units of measurement such as tins and converted to standard units by the researcher. During the crop growing, season, farmers evaluated the benefits of the technologies based on physical observations of the crop performance, such as vigor, cob size and grain yield. Direct cash requirements for seed and labor were also evaluated. Farmers independently managed the plots. They also recorded all the farming activities, inputs, outputs, and time taken for each activity.

Data collection was done through direct observations, record keeping by the farmers, individual interviews, and focused group discussions. Farmers were visited by the research team during major farm operations such as cutting and incorporation of organic inputs, land preparation, planting, weeding, harvesting, and threshing of the maize crop.

Source: Muriuki (2011).
Masters thesis; Kenyatta University.

3. Farmer Designed and Managed Experiments

"Farmer-designed and managed" experiments, also called "Type 3," are planned and managed by farmers. In these experiments, farmers' participation is higher than that of the researcher. The objective is to study farmers' adaptation (modification) and adoption (acceptance) of technologies.

The process starts with farmers being briefed about new practices, mainly through visits to experiments in field stations or on farms. For example, farmers can visit a type 1 experiment and once briefed, they are given a free hand in experimentation. They are given the flexibility to apply and

Fig. 10.7 Photos taken from a mother-baby experimentation model in Tharaka Nithi County, Kenya

experiment with new technology in a way they deem fit, choosing their own plot sizes, inputs and management practices according to individual preferences and frame conditions.

The individual farmers usually select about 1–3 treatments. This selection process allows farmers to test practices that are most relevant to their specific conditions. For example, Mugwe et al. 2009a, b) used type 3 experiments to assess the determinants of adoption of integrated soil fertility management. By using these experiments, we were able to assess not only the effectiveness of the practices but also the key determinants driving farmer adoption and implementation of the practices.

4. Farmer Group Designed and Managed Experiments

"Farmer group designed and managed" is when groups of farmers design the experiments and implement them collaboratively. The aim is to test the performance of selected technologies on farms managed by farmer groups to enhance farmers' confidence in implementing the technologies. This kind of testing is also called the *"Farmer group training ground model"* or Type 4 experimentation. After visiting and introducing new technologies to farmers, some of the farmers feel ready to fully implement the innovations. However, others prefer further testing and evaluation with fellow farmers to gain more confidence. Type 4 is instrumental in building confidence, leading to wider adoption.

In the Type 4 experiments, farmers manage group farms with support from technicians and extension staff who collect data and provide guidance. The farmers are actively involved in decision-making, planning and day-to-day operations. Evaluation is also done collaboratively using participatory evaluation, such as farmer-led assessments and structured discussions. Researchers may also engage in facilitating participatory monitoring and evaluation (Njuki 2004).

This collaborative effort leads to a sense of ownership among farmers, empowering them to implement best practices and innovate based on collective experiences. For instance, in the central highlands of Kenya, farmers adapted and modified soybean-maize intercropping patterns (Fig. 10.8). Instead of using the conventional maize:legume intercrop of 1:1 pattern, they increased the number of soybean rows and reduced the maize population.

Fig. 10.8 Farmer modification of intercropping of soybeans and maize. Instead of using the conventional intercropping pattern of maize; legumes of 1:2 or 2:1, some famers used 1:4 ratio

10.2.1.4 Designing On-Farm Experiments

In Chaps. 7 and 8, we discussed the various types of agricultural experiments and principles of experimentation, typically in reference to on-station controlled trials. The same principles apply to experiments conducted on farms, but they are modified due to conditions and multiple variables in farmers' fields. In this section, we discuss these key principles of experimentation as applied in on-farm experiments.

Before discussing the principles, look at Fig. 10.9, which shows, on one hand, the appearance and arrangement of treatments in a station experiment and, on the other hand, in on-farm experiment. Whereas the treatments at the on-station are well (neatly) arranged with clear blocks, properly randomized, and replicated, the treatments on the farms are scattered across farms; *farms are treated as replicates and test treatments in just one plot on a farm.* In addition, farms vary greatly in size, orientation, slope, other surrounding crops, etc. All these factors affect the treatment's performance. In practice, researchers have to be keen while selecting the fields to try as much as possible to keep sources of variation to a minimum.

In the figure, farmers are testing the four treatments, but depending on the availability of land and what farmers choose, they may test incomplete sets of treatments.

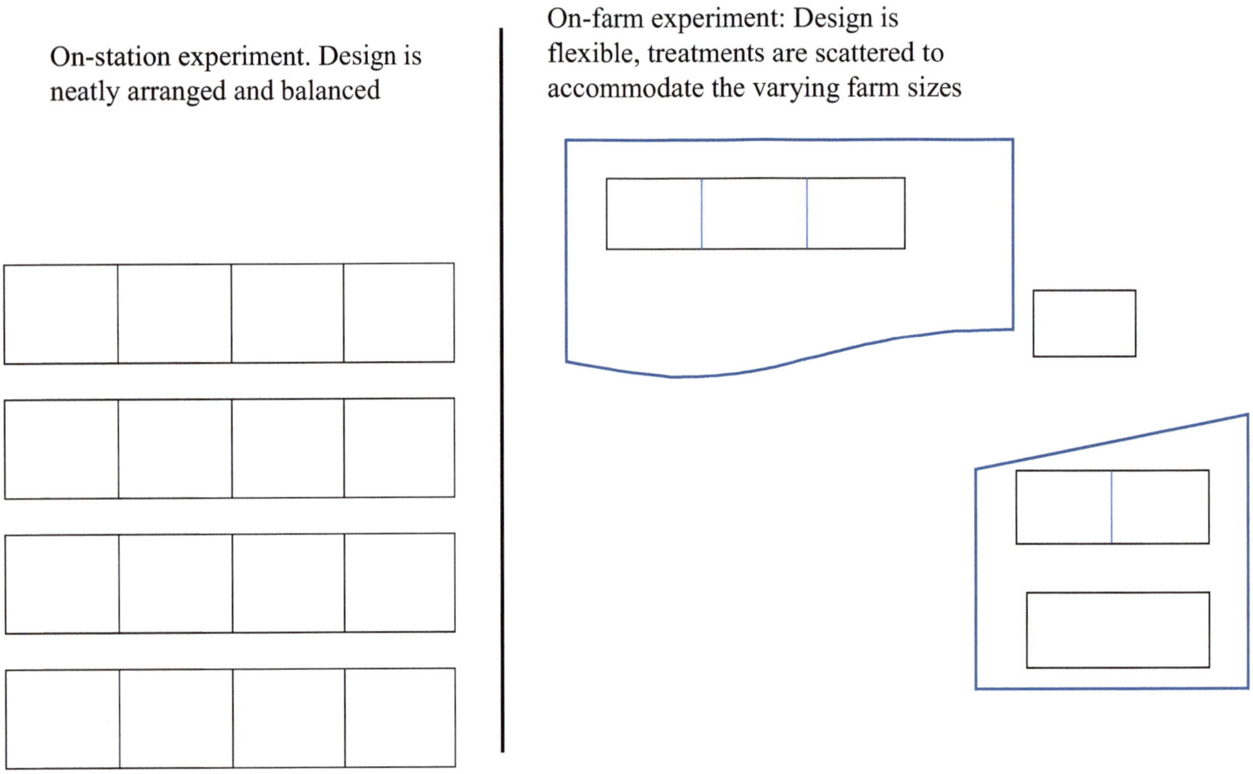

Fig. 10.9 A representation of an experiment having four treatments under an on-station and on-farm scenario

Replications

Many authors have used *farms as replicates* (Bellon 2001; Coe 2007; Mugwe et al. 2009b; Kiboi et al. 2017) as there are practical arguments regarding the use of one farmer-one replicate approach (Mutsaers and Walker 1990; Coe et al. 2019). In this book, we support using a farmer as a replicate (blocks), but always remember that the farms vary, unlike the blocks in a controlled experiment. In controlled experiments under homogenous blocks, each treatment occurs once in each block (Chap. 8). However, on farms, farmers and individual farms vary in farm size, social and ecological characteristics. These inherent variations will influence the performance and preference of the treatments within a farm. Thus, a question arises: "Are the farms true replicates?' "Are the farms homogenous". Farmers may not be replicates because "replicate" typically refers to the multiple instances or plots where the same treatment is applied. The solution lies in first gaining a better understanding of variability to allow selection of farms and farmers who are more comparable (Sinclair and Coe 2019).

After establishing variability, farms can be replicated to ensure that each farm type is represented. This means similar farms are grouped together, to make a replicate. For instance, if there are different soil types such as clay, loamy and silty clay, researchers can select representative farms from each category. Similarly, if there are varying soil organic matter levels, such as high, medium and low, researchers can select farms that represent each level. For example, in the central highlands of Kenya, researchers conducted an on-farm experiment on surface runoff and soil erosion, where they predetermined soil organic matter levels before selecting farms to ensure the representation of each level (Rugendo et al. 2023) (Box 10.4).

Another question that arises is whether experiments can be replicated within each farmer's field. The answer to this is that it depends on the objectives of the experiment. If it is a "Type 1 experiment", replication within the farm may be necessary. If it is an experiment, replication can also be done within the farm, but there will be a trade-off. Replicating within farms means fewer resources for including many farms, yet the *variation between farms* is more important than *variation within an individual farm*, particularly for technology evaluation and for ensuring the results benefit of other farmers not included in the study. Therefore, in sites where there is established variability, it is recommended to replicate different farms of the same type (Sinclair and Coe 2019). The variability among farms needs to be considered during data analysis and in the final dissemination of the results.

Mutsaers et al. (1997) provide an additional consideration for the treatments and choice of design. They argue that because the aim is for farmers to understand exactly what is being tested and be able to participate, the following are recommended:

- The number of treatments and their combinations should not exceed six (6).
- All treatments to occur in every field.
- Lay out the experiment in such a way that the differences between the treatments are obvious to farmers. This may include having big plots where farmers are able to observe differences. In one of our experiments on soybeans, we used big plot sizes of 10 × 10 m with a few treatments' combinations.
- Always include farmers' practices as control. Suppose you are testing improved varieties; there should be no

Box 10.4: Farms Are Replicated According to Soil Organic Carbon Levels

2.3. Selection of experimental fields

A total of 57 farmers' fields (32 in Meru South and 28 in Tharaka South) were initially sampled purposively based on their existing agricultural activities. The main factors considered were; the existing soil and water management practices, type of fertilization used, type of crops grown in a span of 2 years, land preparation practices, use of irrigation, weeding practices and the crop yield in short and long rain seasons. A soil auger was used to collect soil samples from the sampled fields at depths that ranged from 0 to 15 cm using a soil auger and transported to the laboratory for SOC analysis. Subsequently, nine (9) fields were selected in each Sub-County (for each soil type) representing the three SOC levels (low, moderate and adequate) and replicated 3 times.

Source: Rugendo et al. (2023). https://doi.org/10.1016/j.heliyon.2023.e17684 (https://creativecommons.org/licenses/by/4.0/)

more than five improved varieties because of the need to include the local variety. In the case of testing fertilizer, we could test 3 to 4 levels and include the level of fertilizer farmers use as one of the control treatments. It will also be good to include absolute control (no inputs) in fertilizer trials. On the other hand, a cropping patterns experiment could include 3–5 new cropping patterns so as to include the traditional cropping pattern as control, for example, Matusso et al. (2014) used 5 cropping patterns (Sole maize; Sole soybean, Maize-Soybean (1 M:1S) (Traditional pattern), Maize-Soybean (2 M:2S), Maize-Soybean (2 M:4S), Maize-Soybean (2 M:6S).

Randomization

Randomization is a necessary operation in on-farm experiments. Looking at Fig. 10.9, we realize that we could have patterns running through the fields. Randomization within the farm (i.e., allocation of the treatment to the plots randomly) would, therefore, be necessary to ensure that treatment effects are not confounded by any systematic variations in the field conditions.

Randomization in on-farm experiments is important for the following reasons:

- *Reducing bias*: Without randomization, there is a risk of nonrandom factors, i.e., systematic influences that are not due to chance, affecting the results. For example, if the treatments are assigned systematically based on on-farm characteristics such as soil types, the results might be skewed.
- *Statistical validity*: Randomization allows for the application of statistical tests to analyze the data; relies on an assumption of randomization to make inferences about the treatment effect.
- *Controlling confounding variables*: Randomization helps in controlling confounding variables that could affect the outcomes.
- *Generalizability:* By randomly assigning treatments to the plots within the farms results become more generalizable.

Nontreatment Variables

Nontreatment variables or covariates refer to additional farm and farmer variables (besides the treatment) that are likely to affect the outcomes. The most important ones include those associated with *field variability***,** such as soil, environmental conditions, e.g., rainfall, and management practices influenced by *resource availability, farmer skills, and experiences*. The nontreatment variables affect crop growth, yields and other measurements. They raise the "experimental error or noise" (Refer to Chap. 8). Essentially, the covariates should be accounted for by ensuring that they are identified and that data are collected. Additionally, depending on the type of the experiment (Sect. 10.2.3), appropriate decisions will be made on what variables to control.

Due to the presence of nontreatment variables, it is impossible to ensure uniform physical conditions within a replicate. Such covariates and farmer-related variations tend to inflate the error term during data analysis as they are partitioned as unexplained variations. To minimize such effect, data should be collected on these nontreatment variables and factored in during data analysis. Statistical analytical procedures for capturing the nontreatment variables and isolating treatment effects exist.

Control Variables

In on-farm experiments, controlling variables involves standardizing certain factors so that any observed effects can be attributed to the tested treatments. Key control variables can be land preparation, planting density, fertilizer rates and weeding practices. These factors should be uniformly applied across all replications and farms to minimize their influence on the outcomes. This is the case with Type 1 and 2 experiments, where the interest is also to collect biophysical data besides socio-economic variables. Additionally, efforts are made to keep nontreatment variables as consistent as possible both within and between replications. The most important principle here is that there is maximum expression of differences due to treatments.

Control Treatment

This could vary depending on the farmer. For example, if control is farmers' practice, this varies among the farmers due to differences in their traditional practices. The choice of a control treatments should be guided by logical reasoning.

Treatment Effects

Treatment effects are measured under many different environments. Each farmer represents a different environment.

Interactions

The concept of Genotype x Environment x Management interaction is significant (Zingore et al. 2022) in on-farm experiments. It refers to how the performance of a specific genotype, like a crop variety or livestock breed, differs across varying environmental conditions such as climate, soil types, management practices and local farming systems. The same genotype may produce different outcomes in different environments, meaning that a genotype that performs well in one environment may not perform as well in another. Management in terms of farmer management practices or input use, as well, plays a critical role in determining crop response.

In on-farm experiments, researchers must consider GxExM interaction as they aim to understand how different genotypes. Understanding GxExM interaction helps researchers identify which genotypes are best suited for specific farming environments and management practices. This knowledge allows researchers to make informed recommendations to farmers, suggesting crop varieties or livestock breeds that are likely to perform well in a particular region or under specific farming conditions (recommendation domain). This helps farmers make choices that are more likely to lead to successful outcomes rather than following a one-size-fits-all approach. Table 10.3 summarizes differences between on-station and on-farm experiments in terms of design and key principles.

Table 10.3 Key difference between design of on–station and on-farm experiment

Variable	On-station	On-farm
Location	Controlled environment	Farmers' fields
Experimental design	Precise	Flexible
Replications	Replications are done within the experiment	Farmers are "replicates," and the design is evaluated within and between farm variations.
Control treatment	Well-defined and does not vary with replications	Could vary depending on the farmer. For example, if control is farmers' practice this varies among the farmers
Effect of nontreatment variables	Minimal and controlled	Several nontreatment factors are in operation.
Treatments choice	Chosen by researcher	Chosen by both researcher and farmer
Number of treatments:	Uniform across experimental units Same environment on homogenous blocks	Treatment effects are measured under many different environments. Each farmer represents a different environment
Genotype * environment * management interaction	Nonexistent	G*E*M interactions are important and of interest
Controlled variables	Land preparation, planting density, fertilizer rates, weeding, etc., are all standardized and uniformly applied to all replications.	Replicates: Inter-replicate differences are not particularly important, except when checking the homogeneity of the replicates
Experimental error (noise)	Low	Large
External validity	Low; limited representatives of farmer environment	High; simulate the real farming environment

10.2.1.5 Steps in Conducting On-Farm Experiments

In Chap. 8, we looked at detailed steps for designing and implementing a field experiment in relation to a controlled experiment. The steps in conducting on-farm experiments are basically similar; the difference is realized in the contextual application and practical considerations.

Here we present eight steps to conducting a successful on-farm experiment:

1. Identify the research question, objective, and hypothesis.
2. Decide what data you will collect.
3. Develop an experimental plan and write a complete protocol.
4. Choose the location and farmers.
5. Map out the field plots and plant the experiment.
6. Collect data.
7. Analyze and interpret.
8. Draw conclusions and report.

Research Objective and Hypothesis

Guided by the problems farmers are facing their objectives or priorities as well and backed up by background information, clearly formulate and state your research objective. The objective must be clearly stated, showing what will be achieved. This is the first and extremely important step because everything else depends on your goal. The design, treatments, design, and data to be collected will be determined by the variables in the objective (Casler 2015). For an on-farm experiment, you should involve all researchers, farmers, and parties in the formulation of the objective. The best way to do this is by writing it down and ensuring all members agree. Typically, the objectives in agronomy will be related to evaluating a new variety of crops, new cropping patterns, soil fertility management and crop protection.

Decide What Data to Collect

After formulating the objective and the hypothesis, you will need to design the experiment. The variables to be tested will guide the data to be collected. It is advisable to keep the experiment simple so farmers can learn easily.

Design the Experiment Plan and Write the Protocol

The same principles for designing an on-station trial discussed in the chapter apply, but on-farm designs tend to be complex and flexible. Various experimental designs in many statistical books (Gomez and Gomez 1984) can be applied to on-farm experiments. The randomized complete block design is the most common (Cho et al. 2021). There are a number of key considerations in experiment design. These include treatments (what and how many will be?), nontreatment variables, experimental design, the number of replicates and what the replicates mean. Size of the plots (are they adequate for meeting the objectives of the study?)

Treatments: The selection of treatments is guided by the objectives and hypothesis of the trial. For example, in a recommendation domain where soil phosphorus is deficient for soybean production, we can test the hypothesis that:

Application of P fertilizer to gazelle variety of soybeans will increase yield. This requires two plots in each of the farms (+P and –P).

Suppose we add the number of varieties to include Gazelle and SB 19. We test the hypothesis that:

Application of P fertilizer to gazelle and SB 19 varieties of soybean will increase yield. In this case, we need four plots as follows:

1. Gazelle with P
2. Gazelle without P
3. SB 19 with P
4. SB 19 without P

The experiments will need to be conducted on farms and managed by farmers using their management practices (weeding, watering etc).

Another example could involve introducing a newly developed soybean variety that is rust resistant.

Our hypothesis then becomes:

Growing resistant varieties instead of traditional varieties will increase yield in farmers' fields.

In this case we will need to introduce a new resistant variety; compare against one or more traditional varieties that farmers are planting.

Experimental Design

In the selection of the design, key considerations include treatment structure, treatment layout plan, and number of replicates. The key question in regard to treatment structure is: Do the treatments have single or multiple factors? How will the treatments be assigned to the experimental units in the field? How many replicates (replicates within the farm or one replicate per farm?)

Just like in Chap. 9, the simplest is single-factor experiments; where treatments are variations or levels of the same factor, such as:

- Nitrogen: with different levels of nitrogen, for example, 4 levels of N fertilizer (0, 20, 40 and 60 Kg N ha-1).
- Improved Variety testing: Five (5) varieties of soybean being tested (Gazelle-conventionally grown, SB 19, SB3, SB5 and Nyala).
- Cropping patterns: Different cropping patterns, e.g., maize monocrop, bean monocrop, and maize bean intercropping (3 treatments).
- Pest control practices: Testing different chemicals against the fall armyworm in maize.
- Fodder species supplement: different species of fodder evaluated as protein supplements in dairy cows (Calliandra, Leucaena, Desmodium and Dairy meal).

If, on the other hand, we are testing two or more factors, we have a factorial arrangement. For example, two varieties of maize and three nitrogen levels. During data analysis, there will be a need to consider the main effects (Variety, Fertilizer) and the interactions (Variety * fertilizer).

Number of Replicates

The number of replicates is sensitive, as no clear guidelines exist. Rather, the authors argue that the number of replications in an on-farm experiment depends on several factors, including the variability of the system, desired precision, available resources, farm heterogeneity, experiment objectives, farmers' willingness to participate, previous experience and the need to balance trade-offs. Researchers must carefully allocate resources to determine the appropriate replication level for precise treatment comparisons. It is generally preferable to prioritize having more farms with fewer repetitions of the same treatment per farm rather than fewer farms with more internal replication (Njuho 1998). This author argues that increasing the number of farmers maximizes the degrees of freedom for the farm by treatment interaction. This approach minimizes bias introduced by differences in management practices and aligns with the goal of research for generalization.

In many on-farm experiments, each farmer often has only one replicate of each treatment, presenting challenges in capturing within-farm variation (Njuho 1998). To address this, a suggested alternative is to design experiments where each farmer repeats a single treatment, allowing for valid subsequent data analysis. This design can facilitate a better understanding of farmer-treatment interactions and may support the identification of treatments most effective for specific types of farmers. The choice of which treatment to repeat is considered noncritical in this approach.

Some researchers, such as (Cho et al. 2021), have used field-length strips to serve as experimental units, with multiple replications of treatment and control strips strategically placed within a field to minimize variability (Fig. 10.10). Traditional analysis of variance (ANOVA) is then applied to assess treatment significance. However, a challenge arises when fields only allow for three to four blocks, resulting in limited statistical power if each strip is treated as an individual experimental unit (Piepho et al. 2011). The validity of inferences derived from single strips is highly debated (Kyveryga et al. 2018). However, some authors argue that, given the limited resources of farmers, single-strip treatment experiments still provide useful information for decision-making. Moreover, it is argued that in farmer-centric on-farm experimentation, a measure of statistical significance may not be important to farmers since it gives more weight to questions revolving around translational uncertainty (Bramley et al. 2022). Translational uncertainty refers to the degree to which issues around the decision are understood (Kool et al. 2020; Sileshi 2023). For example, a farmer's

Fig. 10.10 On-farm experiment using RCBD with 5 treatments. It is using one control and one treatment strip per block (replicate) (Cho et al. 2021) https://creativecommons.org/licenses/by/4.0/ (modified)

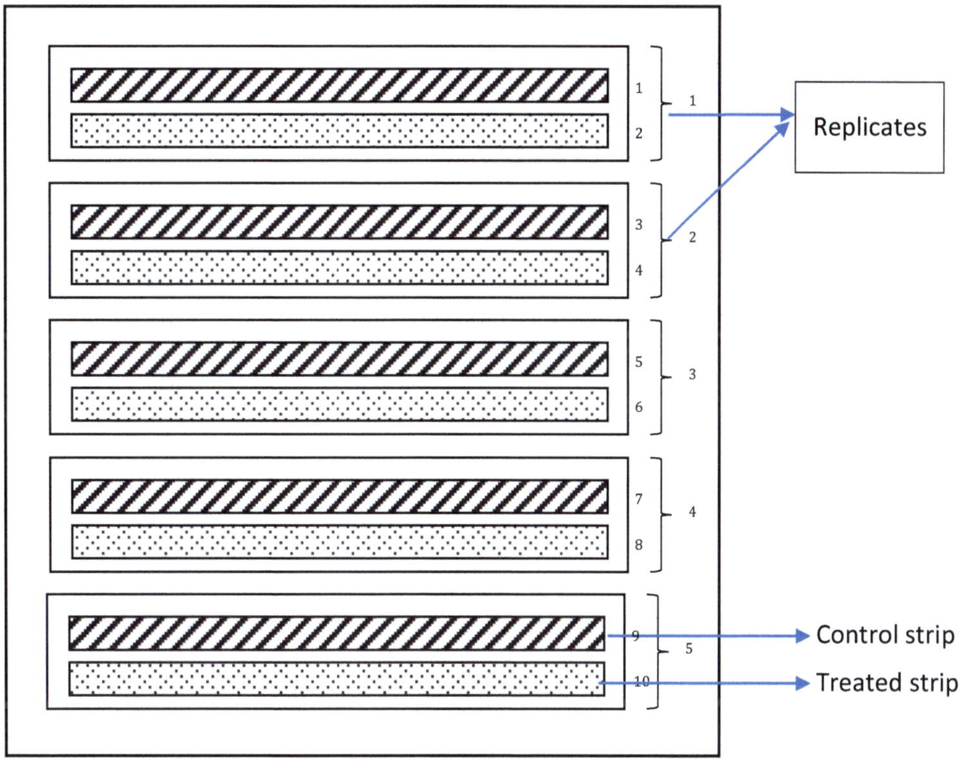

decision to use a certain fertilizer will depend on the expected yield as well as the price and availability of the fertilizer. Scientists, on their part, are more interested in tackling metric uncertainties, which relate to the precision of various attributes, parameters, or outcomes.

To close this discussion on the number of replications, we recommend adopting the best practices that will give valid and reliable recommendations in on-farm experimentation. Whether you want to replicate within the farm to capture within-field variability or to replicate across farms to capture variability across the farmers, the research objective should guide the choice of the design. The strip design seems to be a strong compromise as it is both common and convenient, effectively addressing variability within the field as well as across-farm variability (Laurent et al. 2020). This makes them a versatile option for ensuring robust and applicable results in diverse agricultural contexts.

Writing the Protocol After all the design aspects are concluded and agreed upon by all the partners, a protocol detailing the implementation of the experiment is prepared to guide all the operations during the process. The protocol should be detailed and understood by all the parties involved.

Choice of Location and Farmers

After engaging with farmers and learning about their challenges, farmers are selected based on their biophysical and socio-economic characteristics. Farmers with similar profiles are grouped together. Once these categories are defined, the next stage is matching technology with the appropriate farmer category or recommendation domain. The recommendation domain refers to a group of farmers or population for whom the technology developed will be relevant (Byerlee et al. 1982; Winkelmann et al. 1988; Muthomi 2017; Omulo and Kumeh 2020; Oyetunde-Usman 2022). Here, we want to define which group of farmers the developed technology will work with. In other words, who are the target beneficiaries of the technology being developed?

Map Out the Field Plots and Field Plan

Operations, Data Collection and Evaluation

Operations Operations in the experiments must be well coordinated to ensure the smooth running of the research work. The lead researcher should ensure that the experimental design and statistical aspects are sound and that coordination among the stakeholders is smooth. Box 10.5 shows some of the good practices that a researcher can adopt to ensure a smooth process and statistical rigor in the research process while carrying out on-farm experiments.

Box 10.5: Good Practices in On-farm Experimentation

Experimental design and data collection:

1. Ensure adequate sample size as small sample size hit statistical power thus affecting robustness of and representativeness of the generated results.
2. Ensure proper randomization; as this affects validity of the results; improper randomization

can result in bias due to treatment effects being confounded by other factors.

3. Ensure sufficient replication; insufficient replication may affect reliability and generalizability of the results.
4. Have proper and representative controls.
5. Always ensure you account for site-specific factors that can limit applicability and transferability of the findings remember that socio-economic factors labor requirements as well as environment of the farmers are important for successful dissemination later on.

Division of roles:

1. Clearly determine who will collect the data. Researcher?, farmer? technician?, extension officer?
2. Indicate clearly when data will be collected: Daily, weekly, monthly etc.
3. Indicate if data is quantitative or qualitative; social or biophysical and provide appropriate guidance and tools for collection.

Implementation and collaboration with partners:

1. Always establish a healthy working relationship among the stakeholders and ensure there is adequate communication.
2. Involve participating groups and farmers in early stages of research planning, always ensuring that the exact site of the trial is agreed upon.
3. Pay attention to ethical issues and ensure that your team complies as guided by the ethical clearance board. Maintain the confidentiality and privacy of farmers.
4. Avoid dominating the research process. The reason for going on the farm is to get farmers and other partners involved; encourage farmers to be free and express their opinion. In some cases (Type 3 and 4 or farmer-centric OFE), you should be content to step back and let the farmer take a leading role.
5. Always seek the consent of the farmers.
6. Consider risks and benefits by participating groups and farmers and implement appropriate measures to protect their interests, avoid performing unexplained on-farm operations without the knowledge and consent of the farmers.
7. Consider capacity building for farmers; to train farmers on new techniques and best practices in farm operations and management.
8. Facilitate and encourage farmer-to-farmer knowledge sharing, e.g., organizing farmer-to-farmer visits, to enhance adoption of research outcomes.
9. Never take away harvests without the farmers knowledge and consent, even if it is for lab analysis.
10. Establish a clear schedule of operations and honor the time allocated for each operation.
11. Consider few treatments that are clearly explained to the farmers.
12. Have trained technical persons to support the farmers.
13. Ensure clear division of responsibilities for farm operations and data collection and recording.
14. Adequately involve farmers in all operations to ensure that there is a disconnect between research objectives and farmers' needs.
15. Ensure successful dissemination beyond the experimental stage.
16. Consider compensating your collaborators and your stakeholders for the time and effort.
17. Leverage on the resources and expertise of local organizations to enhance the research process and to escalate impact.

Collect Data

Variability Every farmer field is a different environment and differences exist in soil, weeds, shade, etc., which affect crop production and the performance of new technologies being tested (Kolady et al. 2021; Ruzzante et al. 2021). There is also variation between plots within each field, which amplifies the *within-field variation (affects precision during data analysis).* For instance, trees, shading of crops and farmer management practices cause field variation (Fig. 10.11a, b). Collecting detailed information from on-farm experiments is thus key to quality results (Tripp 1982). Indeed, on-farm experimentation is a data-intensive research activity that ensures variability is accounted for during data analysis and allows valid statistical inferences. However, Mutsaers et al. (1997) provide three key guidelines for data collection:

1. Researchers should use minimal effort to collect plant growth measurements such as height, number of panicles, etc. These biophysical data are better collected from on-station sites where there is more control.
2. Researchers should concentrate more on the collection of data that characterizes farmer practices and variations, such as weeding and shade intensities.
3. Some data sets will be collected at the individual plot level, while other data will be collected at the field and village levels. Some of these data will later be used as

Fig. 10.11 (**a**, **b**) On-farm experiments experience many sources of within-field variation, such as trees that cause shading

covariates (regressors) in statistical analysis to reduce residual variance (Noise).

Soil Analysis Soil analysis forms an essential component in data collection (Webster and Lark 2018). The extent and intensity will, however, depend on the objective of the experiment. An experiment that aims to evaluate the effects of soil inputs may require detailed sampling and analysis at the individual level. In other instances, complete soil analysis may be unnecessary, as it is tedious and expensive and may not add much value. We recommend soil analysis to establish general soil characteristics using simple field kits and observations. Key characteristics include soil texture, which can easily be determined in the field using the feel method, and soil pH, which can be determined using a portable, simple testing kit.

Labor Data Labor data is necessary for estimating a technology's profitability. This can be achieved through two procedures: (1) Recall data from the farmers or (2) Data collected and recorded in real-time. If the researcher needs data collected in real-time, training the farmers and assigning a technical person to support them are necessary.

Yield Data Yield data (grains and stover) are fundamental, and researchers should be keen and innovative in collecting yield data and be aware that losses can be experienced through theft or even farmers harvesting for various reasons, e.g., for home consumption. From our experience working with farmers, we recommend the following:

1. Mark plots or strips where harvesting will take place and agree with the farmers on when it will be done and by who.
2. Train and entrust trained technical personnel working with the farmers to monitor and take the necessary yield data at harvest.
3. Train extension personnel working in the area to carry out the harvesting as required.
4. Be keen to implement continuous monitoring when the crop nears physiological maturity to ensure that farmers are not delayed.

Farmers' Evaluation Farmer assessment is a crucial part of technology evaluation. It involves assessing and evaluating the tested technologies from the farmers' perspective. Farmers will give opinions and insights based on their observations, which is extremely important as the farmers are the end-users (see Fig. 10.12) of farmers assessing soybean performance) and discussing with the researcher. Farmers will adopt a technology if feasible and acceptable (Franzel and Scherr 2002). Feasibility and acceptability have to do with what farmers are doing, how they manage the technological constraints they are experiencing and the existing opportunities (Mugwe et al. 2009a). Several tools can be used to facilitate farmers' evaluation and assessment (Chambers 1994). This data must, therefore, be planned for and collected. The other important aspect of farmer assessment is actual adoption behavior, focusing on what they do on their own, including any modifications they may make to technology.

Data Analysis and Interpretation

As we have already seen, data from on-farm experiments will be varied from individual plot level to village level and from biophysical data, such as crop yields, to farmer evaluation data, such as scores of pests, scores and ranks of farmer observations. The analysis will mainly depend on the objectives of the research design and the type of measurement (qualitative or quantitative). The analysis will take a similar

Fig. 10.12 Farmers assessing the performance of soybean in an on-farm experiment in Kenya

procedure, much like for on-station data, and include the following steps:

1. Preparing data for analysis.
2. Define the analysis objectives: what do you want to discover?
3. Selecting variables for use in the analysis.
4. Conduct exploratory and descriptive analysis to summarize the main patterns.
5. Conduct formal (inferential) analysis to add measures of precision.
6. Interpret and report findings.

One of the most important analyses will be to compare treatments. Analysis of variance (ANOVA) is the standard technique for comparing treatments and analyzing agricultural experiments (Chap. 21). It will also apply to on-farm data. ANOVA is a collection of statistical models used to assess the difference between means of independent groups by partitioning the variability into systematic and random factors. It helps to determine if the effect of the independent variable on the dependent variable is significant. When covariates are included, ANOVA is termed an analysis of COVARIANCE, and such an analysis is typical in the analysis of on-farm experiments. We refer you to (Mutsaers et al. 1997) pgs. 147–187, for a detailed data analysis presentation from on-farm experiments.

10.2.1.6 Analytical Techniques for Data from On-farm Experiments

Analysis of data from on-farm experiments must incorporate analytical techniques that help to account for variability (Bellon et al. 2002), mainly from two sources:

- Variation at the field level (field-level variation) due to differences between fields in soils, management practices, planting dates, pest incidences, etc., which affect all plots in more or less the same way.
- Variation at the plot level, which causes differences between plots within a field due to farmers failing to treat plots uniformly and local differences in soil, shade and pest incidences.

These variations are said to be unavoidable and desirable because one of the aims of on-farm experimentation is to understand these variations. Researchers will use statistical analytical techniques to partition the sources of these variations. In addition to ANOVA, the following analytical techniques are commonly used to analyze on-farm data; researchers usually combine them (Hildebrand and Russell 1996; Coe 2007; Worku et al. 2020):

1. *Descriptive statistics* for each treatment group with each plot or field; provides an overview of the data and helps identify any obvious trends or differences.
2. *Tests for treatment and site mean interaction* to capture variation of treatment effects among farms.
3. *Adaptability analysis:* evaluating and identifying the performance of different treatments across multiple environments or locations. The key objective is to understand the interaction between the treatments and environmental factors and to determine which treatment is better suited for specific conditions. This is usually referred to as "genotype-by-environment" interaction (GE) analysis, common in crop breeding programs and means differences in the response of genotypes to different environments. Adaptability indices are calculated to aid in the selection of treatments in diverse environments.
4. *Multiple regression techniques,* including regressor (and covariance) analysis.
5. *Mixed-effects models or linear mixed models* with fixed and random effects. Fixed effects are treated as constants and include parameters representing the average relationship between a dependent variable and an independent variable across groupings. Random effects are the parameters that capture the variability in the dependent variable specific to each level of grouping.
6. *Data visualization,* e.g., box plots and line graphs to visualize the results and variations between treatments and plots and help understand trends and patterns in the data.

Box 10.6 gives key references for further reading on experimental designs and data analysis in on-farm experimentation. However, we encourage students to consult a biometrician to support them in data analysis.

Box 10.6 Some key references for experimental designs and data analysis in on-farm experimentation

Reference	Focus/topics
1. Tripp (1982). Data collection, site selection and farmer participation in on-farm experimentation. CIMMYT working paper 82/1	Planning and data collection from on –farm experiments
2. Hildebrand Russell (1996) adaptability analysis: a method for the design, analysis, and interpretation of on-farm research-extension. Iowa State University press	Methodology for designing, analyzing, and interpreting on-farm research emphasizing the importance of adaptability analysis and practical application of findings
3. Weber et al. (1997) A field guide for on-farm experimentation. International Institute of Tropical Agriculture	Planning, designing, implementing, data collection and statistical methods (adaptability analysis)
4. Hildebrand PE, Poey F (1985) On-farm agronomic trials in farming systems research and extension. Boulder, CO: Lynne Rienner Publishers.	Looking at data and statistical methods for analyzing data from on-farm experiments
5. Bellon et al. (2002) Quantitative analysis of data from participatory methods in plant breeding. CIMMYT.	Statistical methods for analyzing data from on-farm experiments (yields, rank, and rating data)
6. Coe RD (2007) Analyzing data from participatory on-farm trials. World Agroforestry Centre (ICRAF)	The paper explores how to analyze data from on-farm trials using a simple extension of variance and general linear models
7. Lacoste et al. (2021) On farm experimentation to transform global agriculture	Role of digitalization in on-farm experiments

10.2.2 Reporting and Dissemination

In on-farm experimentation, effective reporting and dissemination are critical to translating experimental findings into actionable insights for farmers, researchers and policymakers (Kyveryga 2019; Tanaka et al. 2023; Sileshi 2023; Lacoste et al. 2025). Reporting should present clear descriptions of the experimental design, treatments, data collection methods and statistical analyses, ensuring transparency and reproducibility. Dissemination should prioritize practical relevance, using farmer-friendly formats such as field days, policy briefs, visual summaries and participatory workshops. Additionally, integrating digital platforms and open-access repositories enhances accessibility and promotes wider adoption of successful innovations across farming communities.

10.2.3 Stakeholders and Implementation Barriers of On-farm Experiments

10.2.3.1 Stakeholders Roles

Stakeholders play vital roles in on-farm experiments, contributing to the success and relevance of the research (Mganga et al. 2023; An et al. 2024; Kuria et al. 2024). Here are the key stakeholders and their roles in on-farm experiments:

- *Scientist/researcher:* The scientist is an investigator, colleague and advisor. Researchers are responsible for designing on-farm experiments in consultation and partnership with farmers.
- *Farmers:* Farmers are also investigators or experimenters and are the primary participants in on-farm experiments. They provide access to their land, labor, and research resources. They may also play an active role in collecting data and observing the treatments' performance.
- *Extension staff:* They provide training and technical assistance to farmers, helping them understand and implement experimental practices effectively. They also disseminate findings and recommendations to the wider farming community.
- *Agricultural organizations*: May fund and facilitate on-farm experiments while also overseeing regulatory compliance.
- *NGOs and development organizations:* often provide support, technical expertise, or resources to facilitate on-farm experiments in rural communities. They may also help build the capacity of farmers and local organizations.

Challenges of On-farm Experimentation

Effective implementation of on-farm experiments is not a walk in the park. Many barriers may affect the effective implementation of the studies, hence affecting the validity and reliability of the research outputs. Key barriers include the following:

a. Cooperation of Farmers Effective implementation requires farmers to cooperate from the start to the end of the experiment. First, farmers may hesitate to dedicate their land, time, or resources to participate, especially if they perceive risks or uncertainties associated with the technology being tested (Bandewar et al. 2017; Simutowe et al. 2024; Adesina et al. 2025). Researchers must engage with farmers before the experimentation and ensure they comply with ethical issues (Chap. 6). Secondly, whilst the research team may agree on how a study should proceed, host farmers could change the planned methodology and attitude or even abandon the study due to farming pressures and changes in attitude. Naturally, maintaining farmer engagement and cooperation over an extended period of time can be challenging, especially if there is no immediate benefit to the farmer. The value created out of an on-farm experimentation process should be captured to make it self-sustaining (Micheni et al. 2016; Uyar et al. 2024).

b. Generalization of Results On-farm experiments occur across multiple locations, thus introducing natural variability in soil types, weather and management practices, such as agronomic procedures. The variability often makes data analysis challenging and makes it difficult to draw statistically robust conclusions or generalize results to a broader population (Doole et al. 2023).

c. Unaccustomed and Illiterate Farmers The majority are illiterate (especially in developing nations) and farmers have little scientific knowledge base, which could impede research. For example, some farmers often fail to adhere to experimental protocols and plans, leading to inconsistencies in data collection.

d. Communication Barriers Cultural aspects such as local customs, habits and taboos often pose significant obstacles in on-farm research activities (Doole et al. 2023). For example, in some instances, women are not allowed to plant trees or open livestock pens; such customs may have a negative impact on experiments involving agroforestry and livestock, respectively, especially when both genders are required to participate. Researchers are, however, encouraged to respect the culture of the people while avoiding sensitive and political issues that may derail the implementation of the experiments. Language is another major issue in communication. Whereas local language should be the medium of communication, when a researcher cannot communicate in the local language, interpretation is made. Interpretation has the risk of reducing the effectiveness of communication.

e. Working in Remote Areas On-farm experimentation activities take researchers out of researchers workstations to sometimes far-off remote areas (Doole et al. 2023). Implementation of the research activities becomes very costly in terms of resources (e.g., personnel, fuel, accommodation expenses) and is also time-consuming (travel time).

f. Complications in Data Analysis Participatory approaches used in on-farm experimentation generate varied data types, including quantitative, descriptive, and broad-based scientific information, which is normally difficult to analyze. Additionally, farmers vary in how they manage the experiments, which often confounds the results, making it difficult to replicate research processes (Hansson 2019; Rommel et al. 2023; Sileshi 2023). Treatment effects are also masked by covariates such as slope and variability in soil types, climate patterns, and vegetation types. The large data sets from multiple locations need efficient and robust management and analysis systems.

10.2.4 Future Directions

While Farming Systems Research (FSR) has evolved, matured and broadened its methodologies to adopt a more holistic approach to agricultural development, it now faces new challenges. Rapid population growth continues to increase food demand, yet food production in many developing countries has stagnated, leaving approximately 800 million people affected by hunger, about 9.2% of the global population (Alexandratos 1999; Johnson 1999; Rask and Rask 2011). Indeed, progress toward achieving the Sustainable Development Goals (SDGs) has been slow and uneven (Dixon et al. 2023). The impact of climate change has further intensified these challenges. There is, therefore, an urgent need to address both immediate food security concerns and long-term sustainability objectives.

We argue that FSR warrants renewed investment to tackle contemporary issues related to climate change, poverty, and food security.

The specific suggestions for the future include:

- Investment in multilevel research: support research integrating regional zones, landscapes and individual farmers to improve technology adoption and resource use efficiency.
- Focus on climate change resilience: Prioritize research interventions that enhance the adaptability and climate change resilience of farming systems.
- Embrace digitalization: Leverage digital tools and technologies to enhance data collection, analysis, and dissemination, improving the efficiency and impact of farming system research.
- Strengthen research methodologies; FSR requires both quantitative and qualitative approaches to address the biophysical and social dimensions. More investments in the ability to conduct mixed research by agricultural researchers are needed.

Acknowledgments Most of the photographs in this chapter were derived from projects funded by the Alliance for a Green Revolution in Africa (SHP 022 and SHP 014).

Recommended Further Reading

1. Behera and France (2023) Farming systems research: Concepts, design, and methodology. Advances in Agronomy 177:1–49. https://doi.org/10.1016/BS.AGRON.2022.08.001

2. Snapp SS, Dedecker J, Davis AS (2019) Farmer Participatory Research Advances Sustainable Agriculture: Lessons from Michigan and Malawi. Agron J 111:2681–2691. https://doi.org/10.2134/AGRONJ2018.12.0769
3. Darnhofer I, Gibbon D, Dedieu B (2012) Farming Systems Research: an approach to inquiry. Farming Systems Research into the twenty-first Century: The New Dynamic 3–31. https://doi.org/10.1007/978-94-007-4503-2_1
4. Jackson-Smith D, Veisi H (2023) A typology to guide design and assessment of participatory farming research projects. Socioeco Prac Res 5:159–174. https://doi.org/10.1007/S42532-023-00149-7/TABLES/1
5. Mutsaers HJW, Weber G, Walker P, Fischer N (1997) A field guide for onfarm experimentation (1st ed.). International Institute of Tropical Agriculture. https://hdl.handle.net/10568/95932
6. Toffolini Q, Jeuffroy MH (2022) On-farm experimentation practices and associated farmer-researcher relationships: a systematic literature review. Agron Sustain Dev 42:1–16. https://doi.org/10.1007/S13593-022-00845-W/TABLES/1

Key Points in This Chapter

Definition of Farming System and On-farm Experimentation

Farming systems research (FSR) is multidisciplinary and interdisciplinary agricultural research that involves farmers in the research process and focuses on the overall farming system.

Origin of Farming Systems Research

The "Farming Systems Research" (FSR) paradigm was introduced in the 1960s as a more effective bottom-up approach to address the shortcomings of the top-down model in agricultural research. Unlike the top-down model, which often fails to consider the diverse and complex nature of smallholder farms, FSR involves farmers directly in the research process.

Development and Evolution of FSR

We categorize the evolution of FSR development into five phases:

- *First phase:* before 1960 when there was little farmer involvement. The interaction between researchers and practitioners (farmers) was more or less a "Top–down approach" following a linear framework where technologies developed by researchers were disseminated by extension staff to farmers.
- *Second phase* of the 1960s to 1980 involved rethinking and conceptualizing FSR.
- *The third phase:* intensive and extensive activities, during which the concept of FSR rapidly gained widespread adoption. This phase is the so-called "Farmer participatory oriented phase," which concentrates on developing methodologies for conducting FSR, with most of the work conducted in developing countries.
- *Fourth phase* from 2000 to 2015, where actual benefits of FSR to farmers and researchers were expected to be felt (increased knowledge).
- *The fifth phase:* Paradigm shift toward sustainability science.

Stages in Farming Systems Research

We simplify stages in FSR into five as follows:
1. Diagnosis stage.
2. Design and protocol preparation stage.
3. Experimentation stage.
4. Assessment and data analysis stage.
5. Dissemination and impact evaluation stage.

Purpose of Conducting On-farm Experiments

We provide eleven key reasons for conducting on-farm experimentation. This includes:
1. Permits farmers and researchers to work together as partners.
2. Facilitates adaptive research.
3. Helps in getting farmers assessment of a practice as well as new ideas.
4. Permits biophysical evaluation under a wide range of conditions (soils, rainfall, farmers management practices etc).
5. Allows acquisition of realistic input-output data for calculation of economic returns.
6. Allows researchers to study how farmers adopt and adapt technology.
7. Provides diagnostic information about farmers' experiences, problems, and preferences.
8. Facilitates researchers to study what farm and farmer characteristics affect the performance of technologies.
9. Provides feedback to researchers and extension staff for improvement.
10. Facilitate gender parity allowing freedom to participate.
11. Provision of multiple options:

Types of On-farm Experiments

In this book, we categorize farm experiments into four types based on the level of participation of farmers, researchers, and farmer groups. These include:

1. Researcher designed and managed experiments (Type 1).
2. Researcher-designed and farmer-managed experiment (Type 2).
3. Farmer-designed and managed experiment (Type 3).
4. Farmer groups designed and managed experiments (Type 4).

Principles of On-farm Experimentation

On-farm experimentation uses the same standard principles of experimental design as on-station research. They include the following:

1. *Participating farmers*: Chosen to represent a recommendation domain.
2. *Design:* Farmers are replicates, and the design evaluates within and between farm variations. Design is simple yet complex: because of many interaction factors (natural factors and agro-economic factors), and data needs to be collected for many variables.
3. *Control variables:* The treatment factors and nontreatment factors are in operation.
4. *Control treatment:* This could vary depending on the farmer. For example, if control is farmers practice this varies among the farmers.
5. *Treatment effects:* Treatment effects are measured under many different environments. Each farmer represents a different environment.
6. *Interactions:* Genotype and Environment and Management interaction are important and of interest. Researchers assess treatment performance in terms of differences among farmers conditions. This helps in providing recommendation domains e.g., a certain variety suitability in a given zone.
7. *Randomization:* Done within the farm.

Steps in Conducting On-farm Experiments

Eight steps to conducting a successful on-farm experiment are presented:

1. Identify the research question, objective, and hypothesis.
2. Decide what data you will collect.
3. Develop an experimental plan and write a complete protocol.
4. Choose the location and farmers.
5. Map out the field plots and plant the experiment.
6. Collect data.
7. Analyze and interpret.
8. Draw conclusions and report.

Implementation of On-farm Experiments

Researchers must ensure that operations in the experiments are well coordinated to ensure the smooth running of the research work. Some of the good practices during the implementation phase include

- Ensure healthy working relationships among the partners and farmers.
- Seeking consent from farmers and following all other ethical principles.
- Capacity building of farmers and ensuring involvement of farmers in all operations.
- Provide technical and moral support (technicians) to enhance data collection and farmer assessment.

Data Collection

On-farm experiments are data-intensive. Data includes both biophysical and agro-economic data to characterize farmers. However, more effort should be spent on the collection of data that characterize the farmer rather than on growth parameters data such as plant height. Key data sets include:

- Individual-plot data, field-level data, natural environment data and socio-economic variables to ensure that variability both within and across farms is accounted for.
- Soil analysis data (minimal data set) and labor data to allow calculation of profitability.
- Farmer evaluations are important because they directly relate to uptake and future impact. Both farmer assessments and perceptions, as well as farmers' adoption behavior, should be monitored.

Data Analysis

Data from on-farm experiments vary based on collection points and type. The analysis will take a similar procedure, much like for on-station data and include the following steps:

1. Define the analysis objectives: what do you want to find out?
2. Prepare the data to fit objectives and to achieve the necessary standardizations.
3. Conduct exploratory and descriptive analysis to summarize the main patterns.
4. Conduct formal (inferential) analysis and add measures of precision.

5. Interpret and report findings.

ANOVA remains one of the most important analyses for comparing treatments, but other analytical techniques for analyzing data from on-farm experiments must incorporate analytical techniques to help account for variability. In addition to ANOVA, the following critical analytical techniques are used:

1. *Tests for treatment and site mean interaction* to capture variation of treatment effects among farms.
2. *Adaptability analysis:* process of evaluating and identifying the performance of different treatments across multiple environments or locations, similar to "genotype-by-environment" interaction (GE) analysis traditionally used by crop breeders.
3. *Multiple regression techniques,* including regressor (and covariance) analysis.
4. *Mixed effects models or linear mixed models* with fixed and random effects.
5. *Data visualization* to visualize the results and variations between treatments and plots and help us understand trends and patterns in the data.

Challenges of On-farm Experimentation

Effective implementation of on-farm experiments is not a walk in the park. Key barriers include the following: *Cooperation of farmers*, *Generalization of results*: *Unaccustomed and illiterate farmers*, *Communication barriers*: *Working in remote areas* and *Complications in data analysis*.

Review Questions

1. What is Farming Systems Research (FSR), and how does it differ from traditional top-down agricultural research approaches?
2. What are the key characteristics that distinguish "Bottom-up Model" and "Top-down Model"?
3. Describe the five stages of implementing farming systems research.
4. What is the purpose of conducting on-farm experiments?
5. Explain the four types of on-farm experiments based on the level of farmer participation.
6. What is the significance of managing nontreatment variables in on-farm experiments?
7. Explain data collection in on-farm experiments.
8. How does the G × E × M (Genotype × Environment × Management) interaction framework help in understanding the outcomes of on-farm trials?
9. What strategies can researchers use to build strong partnerships with farmers during the planning and execution of on-farm trials?
10. Which analytical techniques would you use to assess treatment-site interactions in a multi-location on-farm experiment?

Chapter Exercise

1. Suppose a group of smallholder farmers is struggling with prolonged dry spells impacting their bean yields. You plan to introduce drought-resistant bean varieties.
   (a) Based on the stages of FSR implementation, briefly outline the diagnosis stage you would conduct. What kind of information would you gather from these farmers?
   (b) You decide to propose an on-farm experiment (Type 2: researcher-designed, farmer-managed) to test 2–3 drought-resistant bean varieties against the farmers' local variety. Outline the fundamental principles you would need to incorporate in your experiment design (referencing replication, randomization and management of nontreatment variables).
   (c) What specific biophysical and socio-economic data would you plan to collect during this experiment? Give at least two examples of each.
   (d) How would you handle the statistical analysis?

Answers to Review Questions

- Q1: FSR is interdisciplinary, focusing on the entire farming system; it improves productivity for diverse and complex smallholder farming systems
- Differences:

  Top-down is Researcher-driven, Technology-centered, Dissemination-focused, production-oriented, etc.

  Bottom up is Farmer-centric, problem-oriented interdisciplinary, participatory etc. References: 10.1.1.1 and 10.1.1.2
- Q2: Reference Table 10.1.
- Q3: Stages include diagnosis stage, design and protocol preparation stage, experimentation stage, assessment and data analysis stage, dissemination, and impact evaluation stage. Reference: 10.1.3.
- Q4: Reference: 10.2.1.2.
- Q5. Types of on-farm experiments: Reference 10.2.1.3.
- Q6: Maximize differences due to treatments and minimize differences due to nontreatment factors that cause confounding (see also Chap. 8). Reference: see section on interactions.

- Q7: Data collection: Reference See section on "Collect Data".
- Q8: The G*E*M interactions help understand on-farm results by revealing interactions. The performance of a variety is not consistent across different levels of another factor. Allows researchers to recommend context-specific solutions (Best-fit), which is the essence of FSR and OFE;
- Q9: Clear communication, respect, consent, compensation etc., Reference 10.2.2 (Box 10.5) and 10.2.3.1.
- Q10: Statistical models that account for the multi-location nature of data. Include ANOVA with interaction term, mixed-effects models, adaptability analysis, and multiple regression. Reference: 10.2.1.6.

References

Adesina OS, Whitfield S, Sallu SM et al (2025) Farmers' decision processes about biocontrol innovation adoption: dynamics surrounding push-pull technology in Western Kenya. Agroecol Sustain Food Syst. https://doi.org/10.1080/21683565.2024.2440473

Alexandratos N (1999) World food and agriculture: outlook for the medium and longer term. Proc Natl Acad Sci 96:5908–5914. https://doi.org/10.1073/PNAS.96.11.5908

Altieri MA, Nicholls CI, Henao A, Lana MA (2015) Agroecology and the design of climate change-resilient farming systems. Agron Sustain Dev 35:869–890. https://doi.org/10.1007/S13593-015-0285-2/TABLES/2

An Z, Yang Y, Yang X et al (2024) Promoting sustainable smallholder farming via multistakeholder collaboration. Proc Natl Acad Sci USA 121:e2319519121. https://doi.org/10.1073/PNAS.2319519121/SUPPL_FILE/PNAS.2319519121.SD05.XLSX

Andersen E (2017) The farming system component of European agricultural landscapes. Eur J Agron 82:282–291. https://doi.org/10.1016/j.eja.2016.09.011

Andrée P (2021) Hugh Campbell: farming inside invisible worlds: modernist agriculture and its consequences. Agric Human Values 38:1223–1224. https://doi.org/10.1007/S10460-021-10254-6

Ashby JA (1990) Evaluating technology with farmers: a handbook. CIAT

Ashby JA, Quirós Torres CA, Rivera YM (1987) Farmer participation in on-farm varietal trials: a report of work in progress

Bandewar SVS, Wambugu F, Richardson E, Lavery JV (2017) The role of community engagement in the adoption of new agricultural biotechnologies by farmers: the case of the Africa harvest tissue-culture banana in Kenya. BMC Biotechnol 17:1–11. https://doi.org/10.1186/S12896-017-0347-4/TABLES/1

Behera UK, France J (2023) Farming systems research: concepts, design and methodology. Adv Agron 177:1–49. https://doi.org/10.1016/BS.AGRON.2022.08.001

Bellon MR (2001) Participatory research methods for technology evaluation: a manual for scientists working with farmers. CIMMYT

Bellon MR, Reeves J, Research P, Analysis G (2002) Quantitative analysis of data from participatory methods in plant breeding. CIMMYT

Biggs SD (1995) Farming systems research and rural poverty: relationships between context and content. Agric Syst 47:161–174. https://doi.org/10.1016/0308-521X(94)P4409-U

Bingen JM, Gibbon D (2012) Early farming systems research and extension experience in Africa and possible relevance for FSR in Europe. Farming Systems Research into the 21st Century: The New Dynamic, pp 49–71. https://doi.org/10.1007/978-94-007-4503-2_3

Bramley RGV, Song X, Colaço AF et al (2022) Did someone say "farmer-centric"? Digital tools for spatially distributed on-farm experimentation. Agron Sustain Dev 42:1–11. https://doi.org/10.1007/S13593-022-00836-X/FIGURES/2

Byerlee D, Tripp R (1988) Strengthening linkages in agricultural research through a farming systems perspective: the role of social scientists. Exp Agric 24:137–151. https://doi.org/10.1017/S001447970001588X

Byerlee D, Harrington L, Winkelmann DL (1982) Farming systems research: issues in research strategy and technology design. Am J Agric Econ 64:897–904. https://doi.org/10.2307/1240753

Cabral L, Pandey P, Xu X (2022) Epic narratives of the green revolution in Brazil, China, and India. Agric Human Values 39:249–267. https://doi.org/10.1007/S10460-021-10241-X/TABLES/1

Cardoso IM, Guijt I, Franco FS et al (2001) Continual learning for agroforestry system design: university, NGO and farmer partnership in Minas Gerais, Brazil. Agric Syst 69:235–257. https://doi.org/10.1016/S0308-521X(01)00028-2

Casler MD (2015) Fundamentals of experimental design: guidelines for designing successful experiments. Agron J 107:692–705. https://doi.org/10.2134/AGRONJ2013.0114

Chambers R (1994) Participatory rural appraisal (PRA): analysis of experience. World Dev 22:1253–1268. https://doi.org/10.1016/0305-750X(94)90003-5

Chambers R (2020) Knowledge systems for inclusively responsible food and agriculture. Rethinking Food and Agriculture: New Ways Forward, pp 353–369. https://doi.org/10.1016/B978-0-12-816410-5.00016-5

Chikowo R, Zingore S, Snapp S, Johnston A (2014) Farm typologies, soil fertility variability and nutrient management in smallholder farming in sub-Saharan Africa. Nutr Cycl Agroecosyst 100:1–18. https://doi.org/10.1007/S10705-014-9632-Y/METRICS

Cho JB, Guinness J, Kharel T et al (2021) Proposed method for statistical analysis of on-farm single strip treatment trials. Vol 11. Agronomy 2042(11):2042. https://doi.org/10.3390/AGRONOMY11102042

Clayton E (1983) The farming systems approach to adaptive research. Agriculture, Poverty and Freedom in Developing Countries, pp 137–160. https://doi.org/10.1007/978-1-349-17297-9_7

Coe RD (2007) Analysing data from participatory on-farm trials. World Agroforestry Centre (ICRAF)

Coe R, Njoloma J, Sinclair F (2019) To control or not to control: how do we learn more about how agronomic innovations perform on farms? Exp Agric 55:303–309. https://doi.org/10.1017/S0014479717000102

Collinson MP (1987) Farming systems research: procedures for technology development. Exp Agric 23:365–386. https://doi.org/10.1017/S0014479700017336

Danso-Abbeam G, Dagunga G, Ehiakpor DS et al (2021) Crop–livestock diversification in the mixed farming systems: implication on food security in northern Ghana. Agric Food Secur 10:1–14. https://doi.org/10.1186/S40066-021-00319-4/TABLES/4

Darnhofer I, Gibbon D, Dedieu B (2012) Farming systems research: an approach to inquiry. Farming systems research into the 21st century: the new dynamic 3–31. https://doi.org/10.1007/978-94-007-4503-2_1

Dejene A (1989) The training and visit agricultural extension in rainfed agriculture: lessons from Ethiopia. World Dev 17:1647–1659. https://doi.org/10.1016/0305-750X(89)90034-X

Devaux A, Ordinola M, Mayanja S et al (2013) The Participatory Market Chain Approach (PMCA). From the Andes to Africa and Asia

Dixon J, Li L, Amede T (2023) A century of farming systems. Part 1: concepts and evolution. Farming System 1:100055. https://doi.org/10.1016/J.FARSYS.2023.100055

Doole GJ, Tozer KN, Sauermann C et al (2023) Challenges and opportunities for conducting on-farm research. J NZ Grassland 85:85–93. https://doi.org/10.33584/JNZG.2023.85.3659

Dunjana N, Pisa C, Fanadzo M et al (2023) Integrated soil fertility management for soil fertility restoration in sub-Saharan Africa, pp 21–38. https://doi.org/10.1007/978-981-99-2427-1_2

Ewulo TA, Akinseye FM, Teme N et al (2025) Factors driving climate-smart agriculture adoption: a study of smallholder farmers in Koumpentum, Senegal. Front Agronomy 7:1552720. https://doi.org/10.3389/FAGRO.2025.1552720/BIBTEX

FAO (2016) Farmer field school guidance document: planning for quality programmes, 1st edn. FAO, Rome. https://openknowledge.fao.org/handle/20.500.14283/i5296e

Farrington J (1988) Farmer participatory research: editorial introduction. Exp Agric 24:269–279. https://doi.org/10.1017/S0014479700016124

Feder G, Slade RH, Sundaram AK (1986) The training and visit extension system: An analysis of operations and effects. Agric Adm 21:33–59. https://doi.org/10.1016/0309-586X(86)90056-7

Ferris S, Kaganzi E, Best R et al (2006) A market facilitator's guide to participatory agroenterprise development. International Center for Tropical Agriculture

Francis C, Lieblein G, Gliessman S et al (2003) Agroecology: the ecology of food systems. J Sustain Agric 22:99–118. https://doi.org/10.1300/J064V22N03_10

Frankema E (2014) Africa and the green revolution a global historical perspective. NJAS: Wageningen journal of. Life Sci 70:17–24. https://doi.org/10.1016/J.NJAS.2014.01.003

Franzel S, Scherr SJ (2002) Trees on the farm: assessing the adoption potential of agroforestry practices in Africa. CABI Publishing

Franzluebbers AJ, Wendroth O, Creamer NG, Feng GG (2020) Focusing the future of farming on agroecology. Agr Env Lett 5:e20034. https://doi.org/10.1002/AEL2.20034

Giller KE, Tittonell P, Rufino MC et al (2011) Communicating complexity: integrated assessment of trade-offs concerning soil fertility management within African farming systems to support innovation and development. Agric Syst 104:191–203. https://doi.org/10.1016/J.AGSY.2010.07.002

Giller KE, Hijbeek R, Andersson JA, Sumberg J (2021) Regenerative agriculture: an agronomic perspective. Outlook Agric 50:13–25. https://doi.org/10.1177/0030727021998063/ASSET/F6AC7573-E8A6-41A0-91BC-6CFE006416EA/ASSETS/IMAGES/LARGE/10.1177_0030727021998063-FIG4.JPG

Gliessman S (2018) The co-creation of agroecological knowledge. Agroecol Sustain Food Syst 42:1. https://doi.org/10.1080/21683565.2017.1289727

Gomez KA, Gomez AA (1984) Statistical procedures for agricultural research. Statistical procedures for agricultural research with emphasis on rice:7–8

Hansson SO (2019) Farmers' experiments and scientific methodology. Eur J Philos Sci 9:1–23. https://doi.org/10.1007/S13194-019-0255-7/METRICS

Hawes C, Iannetta PPM, Squire GR (2021) Agroecological practices for whole-system sustainability. CAB Reviews: Perspectives in Agriculture, Veterinary Science, Nutrition and Natural Resources 16. https://doi.org/10.1079/PAVSNNR202116005/ASSET/6736BB51-EEC3-4D7F-9894-59401598F753/ASSETS/GRAPHIC/FIGURE

Hayman P, Rickards L, Eckard R, Lemerle D (2012) Climate change through the farming systems lens: challenges and opportunities for farming in Australia. Crop Pasture Sci 63:203–214. https://doi.org/10.1071/CP11196

Hermans TDG, Whitfield S, Dougill AJ, Thierfelder C (2020) Bridging the disciplinary gap in conservation agriculture research, in Malawi. A review. Agron Sustain Dev 40:1–15. https://doi.org/10.1007/S13593-020-0608-9/FIGURES/2

Hildebrand PE, Poey F (1985) On-farm agronomic trials in farming systems research and extension. Boulder, CO: Lynne Rienner Publishers

Hildebrand PE, Russell JT (1996) Adaptability analysis: a method for the design, analysis, and interpretation of on-farm research-extension. Iowa State University Press

Hussain SS, Byerlee D, Heisey PW (1994) Impacts of the training and visit extension system on farmers' knowledge and adoption of technology: evidence from Pakistan. Agric Econ 10:39–47. https://doi.org/10.1016/0169-5150(94)90038-8

Jackson-Smith D, Veisi H (2023) A typology to guide design and assessment of participatory farming research projects. Socioeco Prac Res 5:159–174. https://doi.org/10.1007/S42532-023-00149-7/TABLES/1

Jahangir MMR, Islam S, Nitu TT et al (2021) Bio-compost-based integrated soil fertility management improves post-harvest soil structural and elemental quality in a two-year conservation agriculture practice. Agronomy 11:2101. https://doi.org/10.3390/AGRONOMY11112101/S1

Johnson DG (1999) The growth of demand will limit output growth for food over the next quarter century. Proc Natl Acad Sci 96:5915–5920. https://doi.org/10.1073/PNAS.96.11.5915

Karaca M, Ince AG (2023) Revisiting sustainable systems and methods in agriculture. Sustain Agric Environ:195–246. https://doi.org/10.1016/B978-0-323-90500-8.00004-X

Keating BA (2024) APSIM'S origins and the forces shaping its first 30 years of evolution: a review and reflections. Agron Sustain Dev 44:1–15. https://doi.org/10.1007/S13593-024-00959-3/FIGURES/7

Khan ZR, Midega CAO, Pittchar JO et al (2014) Achieving food security for one million sub-Saharan African poor through push–pull innovation by 2020. Philos Trans R Soc B: Biol Sci 369. https://doi.org/10.1098/RSTB.2012.0284

Kiboi MN, Ngetich KF, Diels J et al (2017) Minimum tillage, tied ridging and mulching for better maize yield and yield stability in the central highlands of Kenya. Soil Tillage Res 170:157–166. https://doi.org/10.1016/J.STILL.2017.04.001

Kolady D, Zhang W, Wang T, Ulrich-Schad J (2021) Spatially mediated peer effects in the adoption of conservation agriculture practices. J Agric Appl Econ 53:1–20. https://doi.org/10.1017/AAE.2020.24

Kool H, Andersson JA, Giller KE (2020) Reproducibility and external validity of on-farm experimental research in Africa. Exp Agric 56:587–607. https://doi.org/10.1017/S0014479720000174

Kpienbaareh D, Bezner Kerr R, Nyantakyi-Frimpong H et al (2022) Transdisciplinary agroecological research on biodiversity and ecosystem services for sustainable and climate resilient farming systems in Malawi. Adv Ecol Res 66:3–35. https://doi.org/10.1016/BS.AECR.2022.04.001

Kuria AW, Bolo P, Adoyo B et al (2024) Understanding farmer options, context and preferences leads to the co-design of locally relevant agroecological practices for soil, water and integrated pest management: a case from Kiambu and Makueni agroecology living landscapes, Kenya. Front Sustain Food Syst 8:1456620. https://doi.org/10.3389/FSUFS.2024.1456620/XML/NLM

Kyveryga PM (2019) On-farm research: experimental approaches, analytical frameworks, case studies, and impact. Agron J 111:2633–2635. https://doi.org/10.2134/AGRONJ2019.11.0001

Kyveryga PM, Mueller TA, Mueller DS (2018) On-farm replicated strip trials. Precision Agriculture Basics 189–207. https://doi.org/10.2134/PRECISIONAGBASICS.2016.0096

Lacoste M, Cook S, McNee M et al (2021) On-farm experimentation to transform global agriculture. Nature food 2021 3:1 3:11–18. https://doi.org/10.1038/s43016-021-00424-4

Lacoste M, Bellon-Maurel V, Piot-Lepetit I et al (2025) Farmer-centric on-farm experimentation: digital tools for a scalable transformative pathway. Agron Sustain Dev 45:1–7. https://doi.org/10.1007/S13593-025-01011-8/FIGURES/3

Larson DF, Muraoka R, Otsuka K (2020) Rural development strategies and Africa's small farms. The Role of Smallholder Farms in Food and Nutrition Security 45–77. https://doi.org/10.1007/978-3-030-42148-9_4/TABLES/3

Laurent A, Miguez F, Kyveryga P, Makowski D (2020) Going beyond mean effect size: presenting prediction intervals for on-farm network trial analyses. Eur J Agron 120:126127. https://doi.org/10.1016/J.EJA.2020.126127

Lilja N, Bellon M (2008) Participatory research practice at the international maize and wheat improvement center (CIMMYT). Dev Pract 18:590–598. https://doi.org/10.1080/09614520802181657

Madry W, Mena Y, Roszkowska-Madra B et al (2013) An overview of farming system typology methodologies and its use in the study of pasture-based farming system: a review. Span J Agric Res 11:316–326. https://doi.org/10.5424/SJAR/2013112-3295

Martin G, Martin-Clouaire R, Duru M (2013) Farming system design to feed the changing world. A review. Agron Sustain Dev 33:131–149. https://doi.org/10.1007/S13593-011-0075-4/FIGURES/5

Matusso JMM, Mugwe JN, Mucheru-Muna M (2014) Effects of different maize (Zea mays L.)—soybean (Glycine max (L.) Merrill) intercropping patterns on yields and its economics. Acad J Agric Res 9:42–55. https://doi.org/10.5897/AJAR2013.7178

Mganga KZ, Kaindi E, Bosma L et al (2023) Multi-stakeholder participation for successful implementation of applied research projects in Africa. Ecol Solut Evid 4:e12252. https://doi.org/10.1002/2688-8319.12252

Micheni AN, Kanampiu F, Kitonyo O et al (2016) On-farm experimentation on conservation agriculture in maize-legume based cropping systems in Kenya: water use efficiency and economic impacts. Exp Agric 52:51–68. https://doi.org/10.1017/S0014479714000556

Minjauw B, Muriuki HG, Romney D (2004) Development of farmer field school methodology for smallholder dairy farmers in Kenya. BSAP Occasional Publication 33:299–313. https://doi.org/10.1017/S1463981500041820

Mugendi DN, Mugwe J, Mucheru-Muna M et al (2011) Dissemination of integrated soil fertility management technologies using participatory approaches in the central highlands of Kenya. Innovations as Key to the Green Revolution in Africa 1279–1296. https://doi.org/10.1007/978-90-481-2543-2_130

Mugwe JN (2007) An evaluation of integrated soil fertility management practices in Meru South District. Kenyatta University, Kenya

Mugwe J, Otieno EO (2021) Integrated soil fertility management approaches for climate change adaptation, mitigation, and enhanced crop productivity. Handbook of Climate Change Management 1–22. https://doi.org/10.1007/978-3-030-22759-3_325-1

Mugwe J, Monicah AE, Ae M-M et al (2009a) Adoption potential of selected organic resources for improving soil fertility in the central highlands of Kenya. Agrofor Syst 76:467–485. https://doi.org/10.1007/s10457-009-9217-y

Mugwe J, Mucheru-Muna M, Mugendi D et al (2009b) Adoption potential of selected organic resources for improving soil fertility in the central highlands of Kenya. Agrofor Syst 76:467–485. https://doi.org/10.1007/S10457-009-9217-Y/METRICS

Muriuki JP (2011) An evaluation of organic and inorganic inputs for soil nutrient replenishment in Mukuuni and Murugi. Central Kenya

Muthomi E (2017) Challenges and opportunities for youth engaged in agribusiness in Kenya

Mutsaers HJW, Walker P (1990) Farmers' maize yields in S.W. Nigeria and the effect of variety and fertilizer: an analysis of variability in on-farm trials. Field Crop Res 23:265–278. https://doi.org/10.1016/0378-4290(90)90059-K

Mutsaers HJW, Weber G, Walker P, Fischer N (1997) A field guide for onfarm experimentation (1st ed.). International Institute of Tropical Agriculture. https://hdl.handle.net/10568/95932

Nelson R, Coe R, Haussmann BIG (2019) Farmer research networks as a strategy for matching diverse options and contexts in smallholder agriculture. Exp Agric 55:125–144. https://doi.org/10.1017/S0014479716000454

Njuho PM (1998) Some statistical considerations to on-farm trials in Kenya. https://doi.org/10.4148/2475-7772.1291

Njuki J (2004) Participatory monitoring and evaluation for institutional learning and change

Norman DW, Worman FD, Siebert JD, Modiakgotla E (1995) The farming systems approach to development and appropriate technology generation. Food and Agriculture Organization of the United Nations

Oberson N, Moussa HO, Aminou AM et al (2024) Participatory research at scale: a comparative analysis of four approaches to large-scale agricultural technology testing with farmers. Outlook Agric. https://doi.org/10.1177/00307270241295763/ASSET/36745DDC-446F-4810-95C7-DF552F538DD7/ASSETS/IMAGES/LARGE/10.1177_00307270241295763-FIG4.JPG

Omulo G, Kumeh EM (2020) Farmer-to-farmer digital network as a strategy to strengthen agricultural performance in Kenya: a research note on 'Wefarm' platform. Technol Forecast Soc Change 158:120120. https://doi.org/10.1016/J.TECHFORE.2020.120120

Oyetunde-Usman Z (2022) Heterogenous factors of adoption of agricultural Technologies in West and East Africa countries: a review. Front Sustain Food Syst 6:761498. https://doi.org/10.3389/FSUFS.2022.761498/BIBTEX

Palomo-Campesino S, González JA, García-Llorente M (2018) Exploring the connections between agroecological practices and ecosystem services: a systematic literature review. Vol 10. Sustainability 4339(10):4339. https://doi.org/10.3390/SU10124339

Patanothai A (1997) Systems approaches to farm management in variable environments. In: Systems approaches for sustainable agricultural development. Springer, Dordrecht, pp 19–29

Piepho HP, Richter C, Spilke J et al (2011) Statistical aspects of on-farm experimentation. Crop Pasture Sci 62:721–735. https://doi.org/10.1071/CP11175

Pingali PL (2012) Green revolution: impacts, limits, and the path ahead. Proc Natl Acad Sci 109:12302–12308. https://doi.org/10.1073/PNAS.0912953109

Rask KJ, Rask N (2011) Economic development and food production–consumption balance: a growing global challenge. Food Policy 36:186–196. https://doi.org/10.1016/J.FOODPOL.2010.11.015

Rommel J, Sagebiel J, Baaken MC et al (2023) Farmers' risk preferences in 11 European farming systems: a multi-country replication of Bocquého et al. (2014). Appl Econ Perspect Policy 45:1374–1399. https://doi.org/10.1002/AEPP.13330

Rudnick J, Lubell M, Khalsa SDS et al (2021) A farm systems approach to the adoption of sustainable nitrogen management practices in California. Agric Human Values 38:783–801. https://doi.org/10.1007/S10460-021-10190-5/FIGURES/5

Rugendo MK, Gichimu BM, Mugwe JN et al (2023) Surface runoff and soil erosion from Nitisols and Ferralsols as influenced by different soil organic carbon levels under simulated rainfall conditions. Heliyon 9:e17684. https://doi.org/10.1016/J.HELIYON.2023.E17684/ASSET/AAF35068-DF6D-478D-98BA-C3AD0EC24241/MAIN.ASSETS/GR5.JPG

Ruzzante S, Labarta R, Bilton A (2021) Adoption of agricultural technology in the developing world: a meta-analysis of the empirical literature. World Dev 146:105599. https://doi.org/10.1016/J.WORLDDEV.2021.105599

Santos JL, Moreira F, Ribeiro PF et al (2021) A farming systems approach to linking agricultural policies with biodiversity and ecosystem services. Front Ecol Environ 19:168–175. https://doi.org/10.1002/FEE.2292

Scherr SJ (1990) The diagnosis and design approach to agroforestry project planning and implementation: examples from western Kenya. https://www.cifor-icraf.org

Shaner WW (2019) Farming systems research and development: guidelines for developing countries, 1st edn. Routledge

Sileshi GW (2023) Analytic transparency is key for reproducibility of agricultural research. CABI Agric Biosci 4:1–8. https://doi.org/10.1186/S43170-023-00144-8/METRICS

Simmonds NW (1986) A short review of farming systems research in the tropics†. Exp Agric 22:1–13. https://doi.org/10.1017/S0014479700013995

Simutowe E, Ngoma H, Manyanga M et al (2024) Risk aversion, impatience, and adoption of conservation agriculture practices among smallholders in Zambia. Heliyon 10. https://doi.org/10.1016/J.HELIYON.2024.E26460/ATTACHMENT/B440FF10-DBD8-43F0-8976-EA767A96EBEA/MMC1.DOCX

Sinclair F, Coe RIC (2019) The options by context approach: a paradigm shift in agronomy. Exp Agric 55:1–13. https://doi.org/10.1017/S0014479719000139

Snapp SS, Dedecker J, Davis AS (2019) Farmer participatory research advances sustainable agriculture: lessons from Michigan and Malawi. Agron J 111:2681–2691. https://doi.org/10.2134/AGRONJ2018.12.0769

Stifel LD (1990) Farmer first: farmer innovation and agricultural research edited by Robert Chambers, Arnold Pacy and Lori Ann Thrupp intermediate technology publications, 1989, 218 pp. Public Adm Dev 10:474–475. https://doi.org/10.1002/PAD.4230100412

Tanaka TST, Mieno T, Tanabe R et al (2023) Toward an effective approach for on-farm experimentation: lessons learned from a case study of fertilizer application optimization in Japan. Precis Agric 24:2044–2060. https://doi.org/10.1007/S11119-023-10029-5/TABLES/3

Thiessen Martens J (2022) Restoring social and ecological relationships in the agroecosystems of Canada's prairie region. Outlook Agric 51:55–66. https://doi.org/10.1177/00307270221077356

Toffolini Q, Jeuffroy MH (2022) On-farm experimentation practices and associated farmer-researcher relationships: a systematic literature review. Agron Sustain Dev 42:1–16. https://doi.org/10.1007/S13593-022-00845-W/TABLES/1

Tripp R (1982) Data collection, site selection and farmer participation in on-farm experimentation *

Tripp R, Woolley J (1990) Training in farming systems research: review and prospects. Exp Agric 26:247–262. https://doi.org/10.1017/S0014479700018408

Utter A, White A, Méndez VE, Morris K (2021) Co-creation of knowledge in agroecology. Elementa 9. https://doi.org/10.1525/ELEMENTA.2021.00026/118930

Uyar H, Karvelas I, Rizou S, Fountas S (2024) Data value creation in agriculture: a review. Comput Electron Agric 227:109602. https://doi.org/10.1016/J.COMPAG.2024.109602

van der Lee J, Kangogo D, Gülzari ŞÖ et al (2022) Theoretical positions and approaches to resilience assessment in farming systems. A review. Agron Sustain Dev 42:1–20. https://doi.org/10.1007/S13593-022-00755-X/TABLES/3

van Zonneveld M, Turmel MS, Hellin J (2020) Decision-making to diversify farm systems for climate change adaptation. Front Sustain Food Syst 4:496852. https://doi.org/10.3389/FSUFS.2020.00032/XML/NLM

Vanlauwe B, Descheemaeker K, Giller KE et al (2015) Integrated soil fertility management in sub-Saharan Africa: unravelling local adaptation. Soil 1:491–508. https://doi.org/10.5194/SOIL-1-491-2015

Vanlauwe B, Coe RIC, Giller KE (2019) Beyond averages: new approaches to understand heterogeneity and risk of technology success or failure in smallholder farming. Exp Agric 55:84–106. https://doi.org/10.1017/S0014479716000193

Wawire A, Csorba Á, Zein M et al (2023) Farm household typology based on soil quality and influenced by socio-economic characteristics and fertility management practices in Eastern Kenya. Vol 13. Agronomy 1101(13):1101. https://doi.org/10.3390/AGRONOMY13041101

Webster R, Lark RM (2018) Analysis of variance in soil research: let the analysis fit the design. Eur J Soil Sci 69:126–139. https://doi.org/10.1111/EJSS.12511

Wezel A, Herren BG, Kerr RB et al (2020) Agroecological principles and elements and their implications for transitioning to sustainable food systems. A review. Agron Sustain Dev 40:1–13. https://doi.org/10.1007/S13593-020-00646-Z/FIGURES/5

Winkelmann DL, Biggs S, Moscardi ER, et al (1988) Planning technologies appropriate to farmers: concepts and procedures

Worku M, De Groote H, Munyua B et al (2020) On-farm performance and farmers' participatory assessment of new stress-tolerant maize hybrids in eastern Africa. Field Crop Res 246:107693. https://doi.org/10.1016/J.FCR.2019.107693

Yeleliere E, Antwi-Agyei P, Guodaar L (2023) Farmers response to climate variability and change in rainfed farming systems: insight from lived experiences of farmers. Heliyon 9:e19656. https://doi.org/10.1016/J.HELIYON.2023.E19656

Zingore S, Murwira HK, Delve RJ, Giller KE (2007) Soil type, management history and current resource allocation: three dimensions regulating variability in crop productivity on African smallholder farms. Field Crop Res 101:296–305. https://doi.org/10.1016/J.FCR.2006.12.006

Zingore S, Adolwa IS, Njoroge S et al (2022) Novel insights into factors associated with yield response and nutrient use efficiency of maize and rice in sub-Saharan Africa. A Review Agron Sustain Dev 42:1–20. https://doi.org/10.1007/S13593-022-00821-4/TABLES/5

11 Research Designs for Impact Evaluation Studies

Abstract

Certain students put forth research questions that dictate a research framework capable of evaluating the impact of an agricultural intervention or an outcome. Knowledge of impact evaluation methodologies and methods is key to effectively addressing such questions. This chapter first discusses impact evaluation and argues that it is a research activity involving establishing a causal relationship. The theory of change is an important concept in understanding and conducting impact evaluation. Basic types of research designs for evaluating impact are categorized into three: True experiments, quasi-experiments, and preexperiments. The main types of true experiments include "the randomized controlled evaluations," normally labeled (RCT), and "the randomized posttest group" and the "randomized pretest–posttest control" group. General guidelines for RCT implementation, randomization, and steps in conducting RCT are discussed. Quasi-experiments are mid-experiments that do not involve randomization of participants. They are, therefore, weak; propensity score matching and difference-in-difference are common methods employed to create a counterfactual and strengthen the study. The preexperimental designs involve simple before-and-after comparisons. Qualitative approaches are indispensable for revealing "how" and "why" changes occurred and understanding participants' experiences and perceptions. Finally, mixed methods in measuring impact are increasingly gaining recognition as they allow more holistic impact evaluations.

Keywords

Attribution · Average treatment effect · Confounding variables · Counterfactual · Covariates · Difference-in-difference · Nonequivalent control group · Nonexperimental designs · Matching algorithms · Matching methods · One group post-test only · One group pre-test post test design · Pretest · posttest designs · Preexperimental designs · Propensity scores · Propensity score distribution · Propensity score matching · Qualitative approaches · Quasi-experimental designs · Randomization · Randomized controlled trials · Static group design · Theory of change · Threats to validity · Time series design

Learning Outcomes

By the end of this chapter, you are expected to be able to:

1. Demonstrate understanding of the meaning of impact evaluation.
2. Discuss the role of counterfactuals in impact evaluation.
3. Relate the theory of change to impact evaluation.
4. Identify the basic research designs for evaluating impact.
5. Distinguish between "randomized controlled trials" and "quasi-experiments".
6. Determine types of "true experiments" in evaluating impact, together with their variants.
7. Identify the key quasi-experimental designs used in agriculture to evaluate impact.
8. Apply propensity score matching and difference-in-difference methods in forming comparison groups.
9. Demonstrate understanding of the preexperimental designs in impact evaluation and how they can be harnessed to give credible results.
10. Appreciate the use of qualitative and participatory research approaches in impact evaluation.
11. Recognize the role of mixed methods designs in impact evaluation.

J. Njeri Mugwe, S. Runo, *Research Methodology in Agricultural Sciences*, https://doi.org/10.1007/978-981-95-1892-0_11

11.1 Overview and Concepts in Impact Evaluation

11.1.1 Research Objectives Measuring Impact

It is common for students to propose research projects with objectives that aim to evaluate the impacts of an intervention or an outcome. Box 11.1 presents examples of such objectives. From these objectives, we can easily identify independent variables and dependent variables, meaning that we are evaluating a *causal relationship* (cause-effect) and recognizing this fact is the first stage in establishing the research design that a researcher needs.

Box 11.1: Examples of Research Objectives Proposed by Students that Require Impact Evaluation Research Methodology

1. To assess the effect of the adoption of a certain technology or practice (e.g., improved maize varieties or improved farming practices) on yields and incomes among farmers.
2. To examine the impact of households' participation in the collective marketing of bananas on household incomes.
3. To determine the effect of users of plant clinics on the control of pests and the productivity of potato.
4. To evaluate the effect of adopting soil fertility inputs on land area dedicated to maize production in Region A.
5. To assess the effect of using dairy animals' digital technology on milk yields in Region B.
6. To assess the impact of training on technology adoption (technology could be mulching, pest control, and improving maize variety) on household incomes and productivity of a certain crop.
7. To determine the effect of promoting climbing beans on the income and nutrition of households.
8. To assess the effect of COVID-19 restrictions on household incomes and consumption among rural households.
9. To determine the impact of improved indigenous chicken breeds on productivity among smallholder farmers.
10. To evaluate the impact of plant clinics on disease and pest management on tomato productivity and profitability.

Many students proposing such research objectives often face difficulties and struggle to choose a suitable research design and the associated methodology. However, it is easy once you recognize that the objective intends to measure the effect of an independent variable on a dependent variable, or simply *evaluate the impact* of interventions or technologies on a specified output. This chapter aims to discuss the topic of impact evaluation in the context of agricultural projects and interventions. The impact of the interventions is measured on the beneficiaries, in most cases, farmers.

The topic of impact evaluation is unique because it involves *manipulating difficult-to-control* human beings, unlike studies with crops. Human beings vary greatly in all dimensions and possess many preexisting differences. Even when we conduct experiments and use random assignment to minimize differences between the experimental and control groups, we still need to consider control in the study. Researchers working with human beings must understand the potential controls and confounds due to human variability to maximize the study's internal validity. The methodologies discussed in this chapter for evaluating the impact of interventions on human subjects are designed to maximize internal validity. This scenario is similar to biophysical evaluations involving experiments discussed in Chaps. 8 and 9.

11.1.2 What Is Evaluation of Impact?

Evaluation of impact, or "impact assessment," is a common term in donor-funded projects. In the context of a project, it refers to a method to examine whether detected changes in outcomes among project target groups or beneficiaries can be attributed to the particular development program. In other words, donors, e.g., the World Bank and the United States Agency for International Development (USAID), would be interested to know whether the results of particular funded projects or interventions were achieved at the end of the project. This involves measuring the *causal effect* between the intervention and an outcome of interest, e.g., in a project promoting soybean cultivation for increased incomes, the donor would want to know if farmers planting soybeans actually increased incomes at the end of the project.

The hallmark of impact evaluation is causality, attribution, and extent of change (Gertler et al. 2016). Depending on what a researcher intends to find out, there are several key questions in an impact evaluation study in the context of a project. They include the following:

1. Did the project achieve its goals? Were the intended impacts realized in the short, medium, and long term?

2. In what manner and under what conditions did the intervention work (demonstrate efficiency)?
3. What positive and negative impacts were intended, and how were they realized by the intervention?
4. What happened to someone after they participated in the program against a specified outcome?
5. Is there a difference between people who participated in the program and those who did not?
6. In what ways did the intervention or project contribute to intended impacts?
7. What specific features of the intervention/project played a crucial role in its effectiveness?
8. How was the quality of implementation assessed in different sites?
9. To what degree can the difference in impact be attributed to variations in the implementation process?
10. Did these impacts reach all the intended beneficiaries?
11. Is the program cost-effective?
12. Does the project impact people differently (females, youths, minorities, poor, disadvantaged)?
13. Did the project achieve implementation fidelity?
14. What are some of the realized spill-over effects from the implementation process?

Interest in impact evaluation has increased over the years. This has been driven by research funders who increasingly seek evidence that their investments yield positive public outcomes and benefits (Reed et al. 2021). Donors want to ensure that the funds they invest in projects and programs are being utilized effectively and generate positive impacts. Additionally, donors have limited resources and, therefore, need to make strategic decisions regarding where to allocate their funds. In other words, they want their funds to count. By understanding the impact, they can make informed decisions regarding future investments to ensure that resources are used effectively and efficiently. Finally, the impact evaluation results provide valuable insights into what works and what does not work, and help project implementers and donors identify areas of improvement.

The "Feed the Future" program supported by USAID (https://www.feedthefuture.gov/) is a case in point. This program makes comprehensive efforts to evaluate its impact on the ground, ensuring that the goals of reducing hunger, poverty, and malnutrition are met effectively. They use rigorous data collection and advanced analytical procedures. For instance, they had developed a sampling guide for benefits based surveys (Stukel and Friedman 2016). These are used to systematically measure progress and outcomes. The guides provide standardized data collection and analysis methodologies, ensuring consistency in assessing progress and impact. With these standardized procedures, the Feed the Future program tracks key metrics such as agricultural productivity, income levels, and nutrition status.

11.1.3 Concept of Counterfactual

Causality and attribution are key terms used by researchers for understanding impact. These terms refer to establishing a "*causal link between impacts (impact) and intervention (the outcome)*," which could be due to a program, policy change, or activity (Belcher and Palenberg 2018; te Lintelo et al. 2020; Busetti 2023; Michel and Schneider 2025). As already stated, impact evaluation attempts to answer a cause–effect inference question. This requires creating a "*counterfactual*," which is a control or comparison. This allows us to measure the change in both the treated and the nontreated (control) group (see the concept of experimentation in Chap. 8). A counterfactual allows us to answer the question "*What could have happened without the intervention*"? The evaluation aims to eliminate alternative possibilities or explanations for the observed effect.

Counterfactuals can be explained further by using the following example of a region where farmers are growing traditional cassava varieties. Here, we consider a program that introduces a new cassava variety to a group of farmers. In this case, the impact aims to determine the effectiveness of the improved cassava variety on yields. The counterfactual situation or set-up would represent what would have happened to the farmers' cassava yields if they had not been introduced to the improved variety. For our results to be valid, we would like to consider a *credible counterfactual* that perfectly matches the group that received improved cassava varieties.

The primary focus in program evaluation studies is to employ methodologies and models that establish a credible counterfactual to control for factors beyond the intervention that could contribute to the observed change. These models isolate the welfare effect of a specific project by comparing the actual observed outcomes of project participants with counterfactual outcomes, that is, the hypothetical outcomes that would have occurred in the absence of the project (Fig. 11.1). Through this counterfactual analysis, the true impact of the program is established because the external influences are accounted for.

Practically, whether it is a *development project or an academic research project*, the researcher must clearly pinpoint a comparison group so that the change is measured in both the comparison and experimental group (intervention). By doing this, researchers assume that all factors are held constant apart from the intervention. In this case, the change in an outcome/result for the group under treatment and nonparticipant/control groups is connected to the intervention.

We also create a comparison group through randomization. (see concept of randomization in Chap. 8). The randomization prevents selection bias and therefore create an equivalent group. For example, randomly allocating individuals to participate (Provided with improved cassava variet-

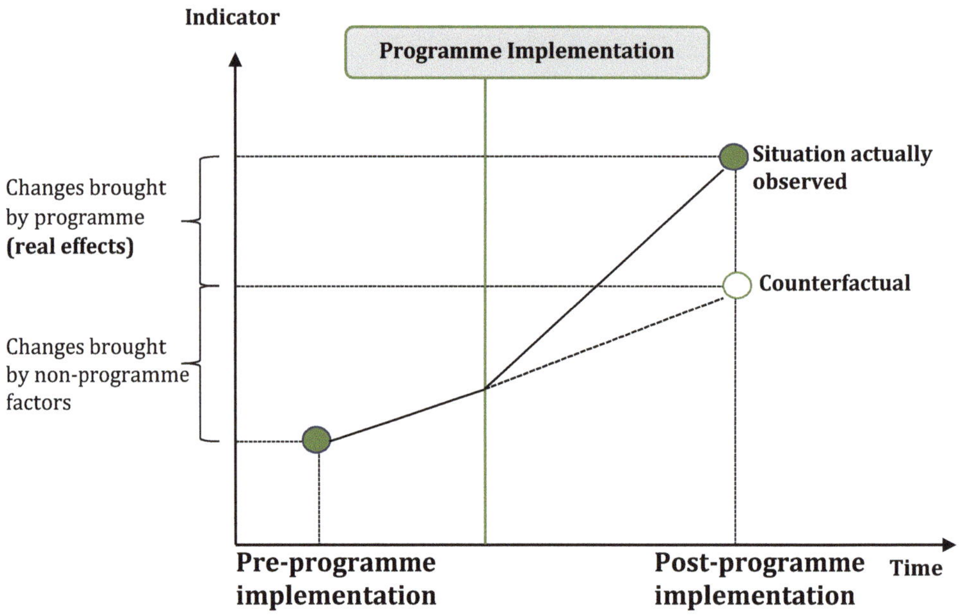

Fig. 11.1 Concept of counterfactual (control) in impact evaluation studies

ies) and non-participating group (not provided with improved cassava varieties) we can then go ahead and measure relevant indicators at baseline, midline and endline to measure the impact of specified outcomes e.g. yields.

However, due to ethical reasons, most projects do not use randomization. Rather, a project will use a *nonequivalent group*. The project will use other criteria, such as demographics and socioeconomic characteristics, rather than randomization, to identify similar and comparable groups. Because of a lack of randomization, the two groups (control and treated) are nonequivalent, known as a *quasi-experimental design* as opposed to a true experiment. The nonequivalent groups suffer from selection bias because, in reality, there are preexisting differences, and therefore, any variation between the groups can be associated with both the impact of the project and preexisting conditions. Estimation of the net impact of the project, in this case, is done by using *balancing models* such as *difference-in-difference* and *propensity score matching* (discussed later in this chapter).

11.1.4 Impact Evaluation is a Research Activity

Every impact evaluation of impact, at its core, is essentially a research activity. It is about generating evidence and contributing to knowledge about what works. It involves a systematic assessment of an intervention (independent variable) to determine its effectiveness and the nature and magnitude of impacts (dependent variable) guided by specific objectives and hypotheses (Gertler et al. 2016; Fabregas et al. 2017). For example, for the first five objectives listed in the first part of this chapter, we could have the following five hypotheses:

1. Adoption of a certain technology or practice (e.g., improved maize varieties or improved farming practices) significantly increases yields and incomes among farmers.
2. Households participating in the collective marketing of bananas have significantly higher incomes compared to those that do not participate.
3. Users of plant clinics have significantly lower pest incidences and higher productivity of potatoes compared to nonusers.
4. Adopters of soil fertility inputs dedicate significantly more land area to maize production in the region than nonadopters of soil fertility inputs.
5. Users of dairy animals' digital technology have significantly more milk yields and incomes than nonusers.

The above hypotheses are tested in regard to the effect of the intervention on the desired outcome. Using the above hypothesis, we could have the following interventions and expected outcomes:

Intervention	Outcome
Adoption of improved maize varieties	Increased yields and incomes
Participation in collective marketing	Increased incomes
Adoption of soil fertility inputs	Increased land under maize
The use of dairy technology	Increased milk yields
Farmers adopting integrated soil fertility inputs	Increased yield

If the results are to be dependable, we must give the evaluation a framework that will tell us what we want to know. In this case, the hypothesis will guide the selection of the appropriate research design and data collection methods. While diverse methods for evaluating impact exist, choosing the most suitable design for a given intervention and context remains a significant challenge (Reed et al. 2021). Ultimately, the chosen method should be capable of producing accurate and credible evidence of impact.

Key challenges in evaluating and evidencing impact that you are likely to face include the following:

- *Counterfactual challenge*: In complex interventions or where randomization is not possible, for example, finding a "suitable comparison group" that is similar to the intervention group can be difficult (Ferraro 2009). The lack of a suitable comparison group introduces a bias and limits the ability to estimate impacts accurately.
- *Limited resources and time:* As a student, you will most likely use ex-post methodology, fitting into an existing project after the project has started and implemented the intervention. Sometimes, projects lack baseline data or have incomplete datasets, which are sometimes inaccurate and unreliable. The researcher might also fail to trace back the same respondent who provided data during the baseline period. These factors will affect the validity of the evaluation estimates. Additionally, rigorous impact evaluation requires a large sample size. Limited resources may constrain your sample size, logistics of data collection, and the ability to employ complex research designs.
- *Skills constraints:* Impact evaluation involves rigorous study designs, data collection, analysis, and reporting. These methods may not have been taught in class, and therefore, you may be required to read unfamiliar methodologies and statistical techniques. We have given resources for further reading at the end of this chapter.

Overall, establishing a clear causal link between the intervention and measured impact is the central challenge in impact evaluation studies. External and contextual factors make it difficult to attribute changes solely to the intervention (Belcher and Palenberg 2018; Flynn et al. 2021; Sharma Waddington et al. 2025). For a student choosing an impact evaluation project, you should choose a suitable evaluation design and a rigorous methodology to ensure you carry out a credible research project. If you fail to employ a suitable methodology, you may be unable to handle examination questions from thesis examiners.

Second, failing to select a suitable design will likely lead to your journal paper being rejected because of *invalid methodological approaches*. The discussion on evaluation designs in the next sections will guide you to various designs and considerations to help you choose the correct design. It is envisaged that the information provided and the recommended reference materials will be useful in carrying out research studies based on impact evaluation research questions.

11.1.5 Categories of Impact Evaluations

Impact evaluation can be categorized into two groups: Prospective and Retrospective (Blome and Augustin 2015).

Prospective Evaluation Prospective refers to looking forward. This type involves developing an evaluation at the beginning of the program, as shown in Fig. 11.2. Baseline data are collected prior to implementing the program for both groups: those receiving the treatment and the comparison/control. The researcher has the advantage of selecting a good comparison group, collecting data during project implementation, and carrying out follow-ups. It is also easy to monitor progress by generating performance trends of the given intervention. This kind of evaluation is most common in development projects that incorporate and plan for impact evaluation at the project design stage. A student wishing to do this kind of impact evaluation will need to spend a considerable amount of time in an existing program. Prospective evaluations are publishable (Planitz et al. 2012; Mears et al. 2014; Salm et al. 2022).

Retrospective Evaluation Retrospective evaluates the impact after implementing the program. It is backward-looking and mainly involves a "recall" approach during data collection. It looks at the treated and comparison group ex-post, i.e., after the effect has already happened. This usually happens when impact evaluation is not part of the program design, for example, students' research projects attempting to fulfil objectives at the beginning of this chapter. A major challenge with this approach is the likelihood of lacking a good comparison group. However, using the methodologies discussed in this chapter, it is possible to conduct credible, publishable research (Pratt et al. 2000; Draper et al. 2009; Sud et al. 2021) (Fig. 11.3).

11.1.6 Theory of Change in Impact Evaluation

What Is the Meaning of the Theory of Change?
A "theory of change" refers to a conceptual map showing how activities implemented in a project or program will lead

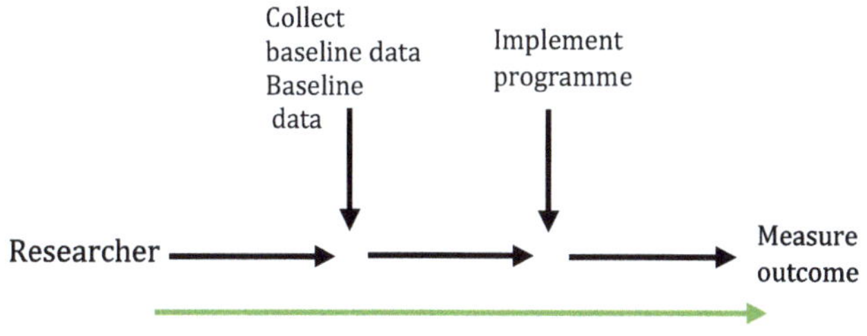

Fig. 11.2 Prospective impact evaluation (forward-looking)

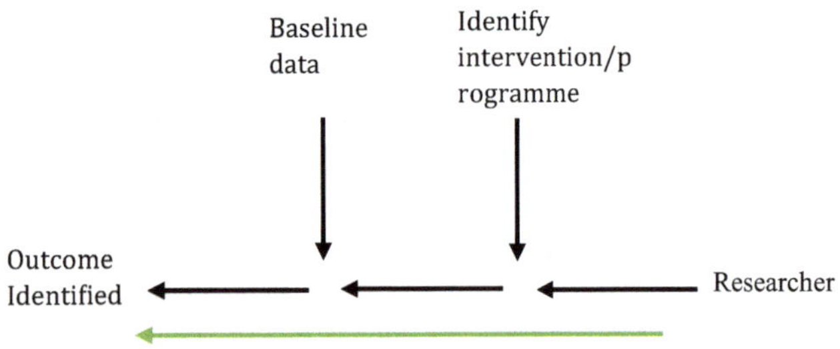

Fig. 11.3 Retrospective impact evaluation (backward-looking)

to change. It simply shows how an intervention will deliver results (Gertler et al. 2016). It is an important concept because it explains how activities are understood to produce a series of results that contribute to achieving the final intended impacts. Typically, it explains the process of change by outlining causal linkages in a project, showing each outcome in a *logical relationship* to all the others, as well as chronological flow and feedback loops (Romão et al. 2023; Zohar et al. 2023).

The concept of the theory of change plays a key role in impact evaluation. First, the theory of change helps the evaluators and stakeholders by providing a clear understanding of how an intervention is expected to bring about change. Understanding the links is particularly important in agricultural programs where goals and outcomes can be complex and multilayered (Mayne and Johnson 2015; Thornton et al. 2017; Omore et al. 2019). Additionally, it helps identify assumptions, pathways, and processes through which change is supposed to occur. This information is central to the identification of appropriate evaluation methods, data sources, and indicators by the evaluators and stakeholders. According to Gertler et al. (2016), selecting the methodology to evaluate impact should, therefore, be pragmatic and chosen to fit the operational context guided by the processes and the research questions.

In summary, TOC is important for the following reasons (Mayne and Johnson 2015; Thornton et al. 2017; Omore et al. 2019):

1. For planning and designing "Research and Development projects."
2. Monitoring and evaluation, as well as ex-post evaluation.
3. Helps in priority setting.
4. Clarify and discuss problems that need to be addressed.
5. Helps to factor in the socio-political context that is important in influencing research uptake.
6. Enhances communication among the stakeholders, hence improving the translation of research into use.
7. Setting the hypothesis to be tested during impact evaluation.
8. Assessing the contribution of scientific outputs.
9. Mapping the trajectory of impact.
10. Seeking and justifying findings.

Results Chain

A typical TOC is often presented using a results chain. Other approaches to illustrate TOC include logic models, theoretical models, logical frameworks, and outcome models (Gertler et al. 2016). A results chain creates causal rationality from the start of the program to the long-term goal. A basic results chain consists of five main components: 1. Inputs or resources, such as funding and personnel, 2. Activities or processes and interventions undertaken to generate change, 3. Outputs in the form of authentic results or products of the activities, 4. Outcomes consist of short-term, intermediate, and long-term changes, and finally, 5. Long-term impact comprises the change that is desired.

Though the starting point of TOC is the inputs or resources that go into delivering the activities and ends with the long-term goal/impact, i.e., the overall impact of the intervention, the construction of TOC is reversed. It starts with project impacts and outcomes and then maps backward, works through the intermediate activities and outputs, and finishes with the project inputs and funding. TOC explains long-term goals and then maps to ascertain necessary preconditions (Breuer et al. 2016; Maini et al. 2018; Reinholz and Andrews 2020; McLellan 2021) (Fig. 11.4).

The TOC can be created to suit various levels of intervention, such as an event, a project, a program, a policy, a strategy, or an organization, based on its intended purpose (Breuer et al. 2016; Aromatario et al. 2019). An initial theory of change is most effective in guiding the planning of an initiative. As the project unfolds and monitoring and evaluation data become available, stakeholders can regularly adjust based on the evidence gathered.

11.2 Basic Designs for Impact Evaluation

The designs for evaluating impact fall into three categories: "true experiments," "quasi-experiments" (mid-experiments), and "nonexperimental methods" (Fig. 11.5).

True experiments, such as randomized controlled trials (RCTs), use random assignment to treatment and control groups to ensure high internal validity and minimize bias. Quasi-experiments, or mid-experiments, lack random assignment but use strategies like matching or statistical controls (e.g., propensity score matching, difference-in-differences) to approximate experimental conditions. Nonexperimental methods rely entirely on observational data without manipulation or control over the assignment process, often using statistical models to estimate causal effects (Olier et al.

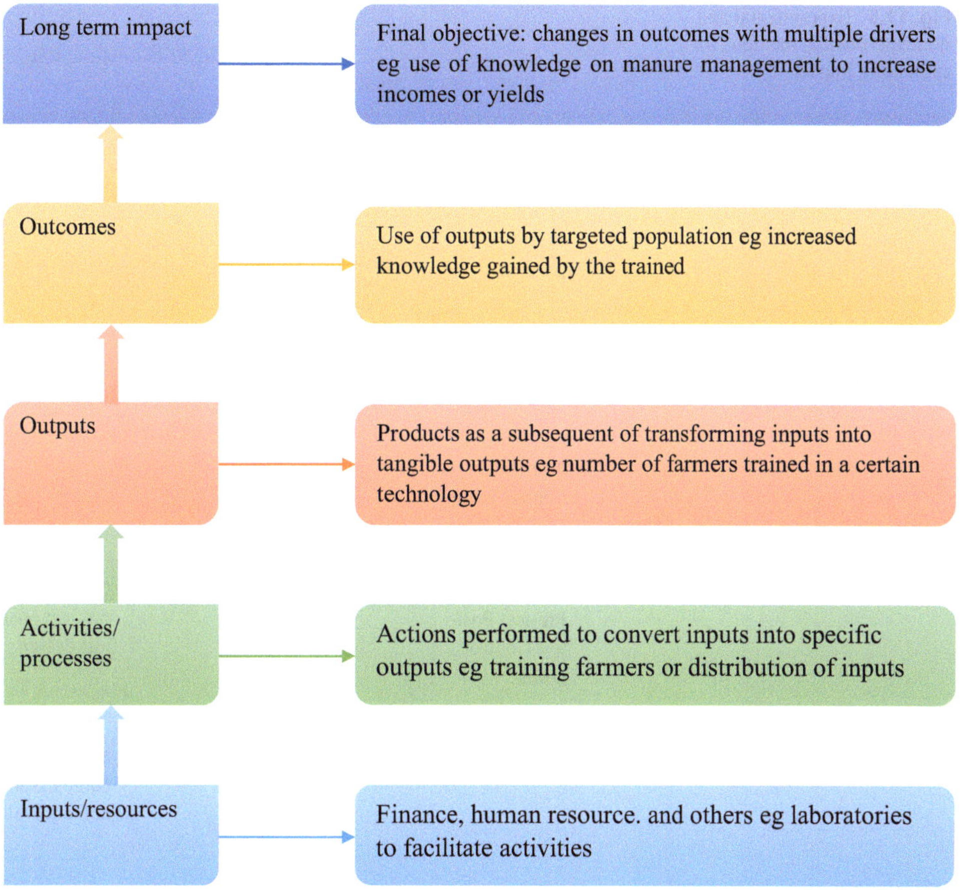

Fig. 11.4 Components of the results chain showing the logical process from inputs to impact

2023). Each design type varies in its rigor, feasibility, and applicability depending on the research context and ethical or logistical constraints.

In the next section, we first discuss the experimental designs, followed by nonexperimental designs.

11.3 Experimental Designs

11.3.1 True Experiments for Evaluating Impact

11.3.1.1 Basic Principles of True Experiments

True experiments for evaluating impact, also called randomized controlled evaluations, randomized controlled trials (RCTS), or randomized experiments, bear structure and principles similar to biophysical agricultural experiments described in this book (refer to Chaps. 8 and 9), as well as in medical experiments and veterinary clinical trials (testing efficacy of a drug in humans and animals, respectively). Randomized controlled evaluations are key in social agricultural programs, as well as in agricultural education and extension to evaluate the effectiveness of the projects, programs, and policy interventions (Stringer et al. 2011; Bezner Kerr et al. 2011; Wegmüller et al. 2022). Just like the experiments in Chaps. 8 and 9, individuals or communities are randomly assigned to an experimental group (treated), i.e., benefits from the program or policy intervention, and a control group (Fig. 11.6). The researcher then measures the results of the intervention by measuring specific outcomes as per the project aims.

Randomized controlled evaluations have become popular over the last 15–20 years in policy studies and development initiatives because of their rigorous evidence (Kumar 2016). The "Abdul Latif Jameel Poverty Lab" (J PAL) is one of the main advocates of RCTS (Gibson and Sautmann 2020). Other organizations, e.g., the International Association for Impact Evaluation (3ie), World Bank, and the Centre for Global Development, are using RCTS to solve diverse societal problems in agriculture and other fields.

Randomized Controlled Trials (RCTs) yield a more precise estimate of impact compared to observational or nonexperimental approaches. In observational studies, respondents from existing groups are randomly selected from both treated and nontreated populations for comparison. However, this method often produces impact estimates that deviate from the "*true value*" due to systematic bias, confounding variables, sampling variability, measurement errors, spill-over, and monitoring effects (Gertler et al. 2016). Conversely, properly conducted RCTs with appropriate randomization and treatment allocation incur only sampling variability as

Fig. 11.5 Experimental and nonexperimental methods for evaluating impact

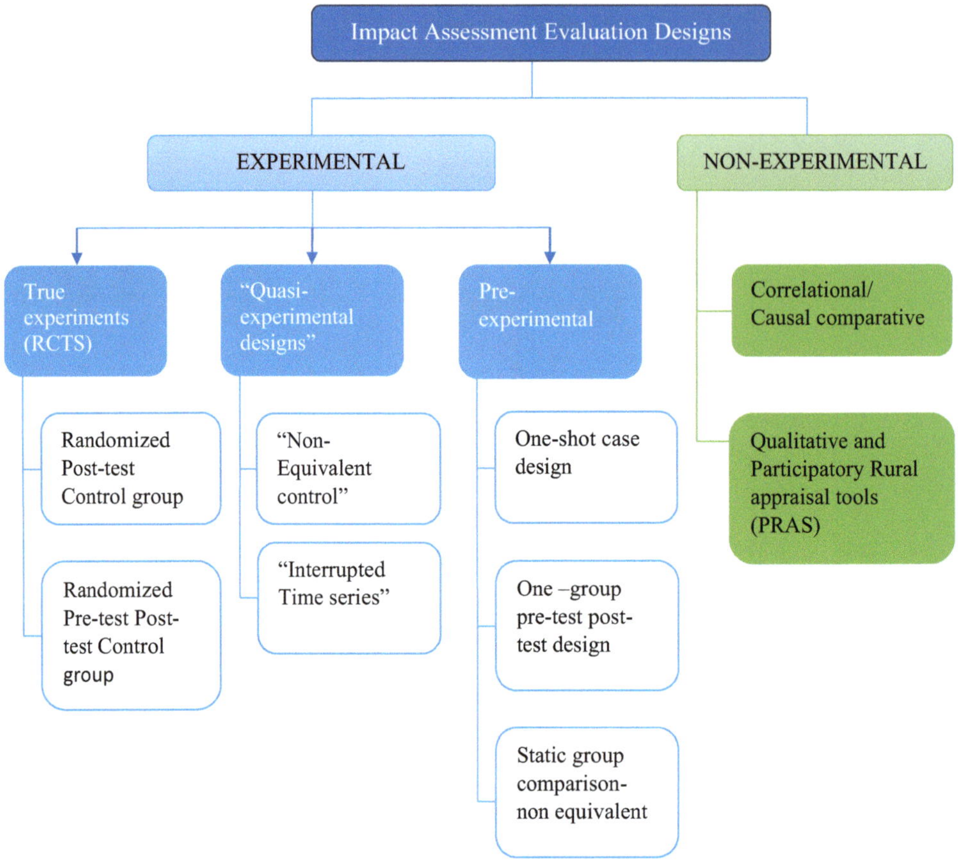

Fig. 11.6 Random allocation of households to experimental (T) and control groups (C)

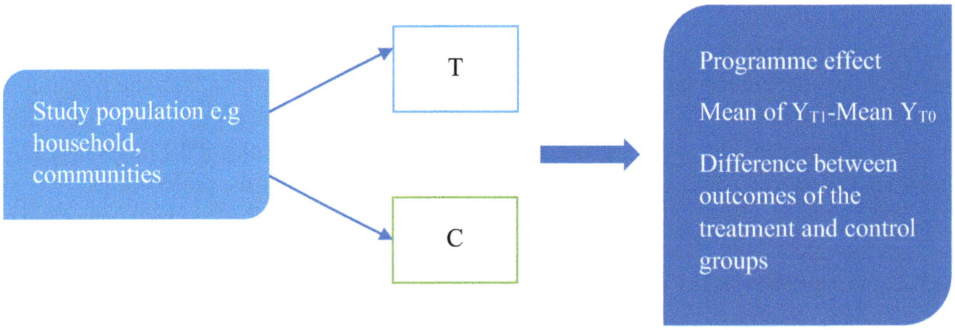

an error. As the sample size increases, the sampling error approaches zero, bringing the trial's estimate of impact closer to the true value.

The robustness of an RCT is in its ability to offer a highly compelling response to questions of causality. This capability aids evaluators and program implementers in concluding that the observed outcomes are genuinely attributable to the intervention and not influenced by external factors (Rogers and Peersman 2014). In social sciences, some researchers have considered the RCTs as the most scientifically rigorous method for testing hypotheses (Hariton and Locascio 2018). It is often hailed as the "gold standard trial" for evaluating the effectiveness of interventions, particularly in the fields of medicine and health, where its usage is widespread. Conversely, RCTs are not without criticism, with some researchers advocating for getting off the "Gold Standard" high horse (Sullivan 2011; Smith 2024). The limitations and threats that are associated with RCTs are discussed later in the chapter.

The fundamental structure of an RCT is illustrated in Fig. 11.6. In this setup, a study population comprising communities, individuals, farmers, or households is randomly assigned to different interventions, and these two groups are then monitored over a specified period. Besides the interventions under comparison, both groups are treated and observed uniformly. When the study concludes, the groups are analyzed based on predefined outcomes. Due to the randomization process, the two groups are assumed to be *equivalent*. Consequently, any disparities in outcomes between, for instance, the treatment A group and the treatment B group are ascribed to the impact of the intervention or program.

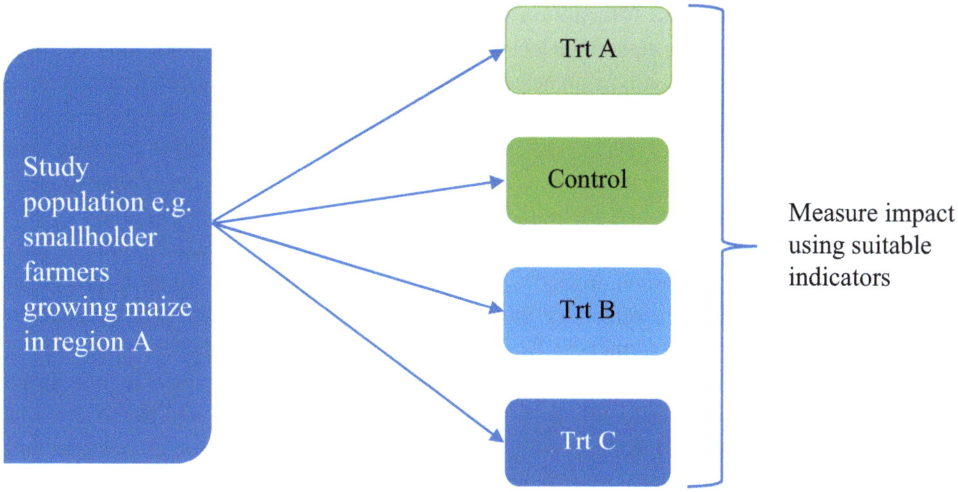

Fig. 11.7 Random allocation of treatments to smallholder farmers

11.3.1.2 Types of True Experiments

Two main types are discussed: the randomized posttest control Group design and the randomized pretest–posttest control group design.

The following symbols will be used:

R = randomization
X = Treated
C = control
O = outcome

1. *Randomized Posttest Control Group Design*

The randomized posttest-only control group design involves two groups: the treatment group, which receives the intervention or treatment, and the control group, which does not receive the intervention. In this design, participants are randomly assigned to either the treatment or control group, ensuring that any differences in characteristics among individuals are spread evenly across both groups.

Treatment group	R	X	O
Control group	R	C	O

After the intervention is administered, data are collected from both groups through posttest measurements. The comparison of outcomes between the treatment and control groups allows researchers to assess the causal impact of the intervention, as any observed differences are attributed to the treatment rather than preexisting differences among participants.

2. *Randomized Pretest–Posttest Control Group Design*

Similar to the randomized posttest design, the *randomized pretest–posttest control group* design necessitates the creation of at least two groups through random assignment (Alessandri et al. 2017; Morgan et al. 2021). In this experimental setup, both groups undergo a pretest before being subjected to different treatments, and after the study, both groups are again assessed through a posttest. By comparing the posttest scores, researchers can evaluate the efficacy of the administered treatments. The pretest is also used to check the groups' equivalence or to provide covariates (Alessandri et al. 2017). Furthermore, the flexibility of the pretest–posttest control group design allows for its expansion to include multiple treatment groups, accommodating the exploration of the effectiveness of various interventions within. For the four groups, we have the following:

Treatment group	R	O	X1	O
Control group	R	O	C	O
Treatment group	R	O	X2	O
Treatment group	R	O	X3	O

We can use the example of "evaluating the effect of farmers adopting different methods for controlling fall armyworm on maize yields." If we are evaluating three (3) methods, we will randomly allocate three groups of farmers to treatments A, B, and C (Trt A, Trt B, and Trt C). We will also randomly allocate one other group to the control treatment, i.e., farmers who will not use any method to control fall armyworm (Fig. 11.7).

Concept of Anticipatory Effects

The combination of random assignment and the control group effectively mitigates various threats to internal validity. However, the incorporation of a pretest introduces the potential for a pretest treatment interaction threat referred to as "anticipatory effect" (Ariel et al. 2021). The premise is that participants react to manipulation as a function of what they perceive to be an expected response and behave as if the stimulus has already happened, thus biasing the pretreatment population mean. During implementation, they may also exhibit variation in skills or behavior manifestation, which can again bias the posttest measurement.

Example of Pretest–Posttest Interaction

Let us consider a new crop management technique to enhance yields. Farmers in two randomly assigned groups (Group A and Group B) undergo a pretest to assess their existing knowledge and practices related to traditional crop management.

After the pretest, both groups are introduced to the new crop management technique. The two groups are trained on the new crop management techniques.

However, the anticipatory effect, or the pretest–posttest interaction, may occur if the farmers in Group A, perceiving the new technique as more advanced and effective, modify their farming practices more diligently than Group B. Farmers in Group B, on the other hand, not expecting the same level of improvement, might not modify their practices to the same extent. As a result, the observed increase in crop yields in Group A may not solely be attributed to the effectiveness of the new crop management technique but could also be influenced by the anticipatory effect, where participants react based on their perceived expectations rather than their practices.

Researchers should thoughtfully address anticipatory effects in agriculture by carefully designing their studies. This may involve employing posttest-only control group designs and limiting the information shared with participants about expected outcomes during the pretest phase. For additional guidance on using randomized controlled evaluations, refer to Box 11.2.

Box 11.2: Guidelines for RCTS

1. Initiated at the outset of the programme, careful control of participation in the program is essential, aligning with the experimental design.
2. Ideal for programs that aim to attribute clear impacts to specific interventions; Unsuitable for hard-to-measure outcomes.
3. Requires a large sample size to detect the effects of the intervention with sufficient precision.
4. RCTS must align with the programme's nature and objectives.
5. Postprogram implementation, an end-line survey is conducted, timing dictated by the theory of change to avoid being too early or too late.
6. Analyzing the programme theory of change is necessary for identifying evaluation questions related to the causal chain and determining which impacts merit evaluation.
7. Population and unit of assignment must be well defined; randomization may involve individual farmer groups or villages.
8. Plan data collection to occur at baseline, midway, and end-line, encompassing household characteristics, socio-economic status, education, health, and other relevant variables.

11.3.1.3 How to Randomize in Social RCT

Three methods may be used to randomize:

1. *Simple randomization:* also called unrestricted randomization. The individuals or sites are enumerated and then allocated to treatment and control groups using randomly generated numbers, such as those provided by a random number generator. Tossing a coin or rolling a dice can also be used (Guest et al. 2015; Arora et al. 2021). It is the most common type of randomization.
2. *Matched pair randomization:* Individuals or clusters are paired based on similar observable characteristics. Within each pair, one unit is randomly assigned to the treatment group and the other to the control group. This initial pairing ensures balance and reduces the necessary sample size. *An example of this is a study conducted to assess the effect of a new agricultural training program on crop yields in smallholder farms. Instead of randomly selecting individual farmers to be part of either the treatment (receiving the training program) or control group (not receiving the training), researchers decided to use matched pair randomization. To begin with, they will identify pairs of farmers who have similar characteristics such as farm size, soil type, and previous agricultural knowledge. Within each pair, one farmer will be randomly selected to undergo the training (treatment), while the other will serve as the control without the training. This method makes sure that the treatment and control groups have similar observable characteristics, which reduces the likelihood of confounding variables influencing the results.*
3. *Stratified random assignment:* This divides the population into small subgroups (strata) based on shared characteristics or attributes such as age, income, and gender. Randomization is then carried out independently within each stratum (e.g., male and females for gender or low, medium, and high income levels) to ensure an equitable distribution of key variables across both treated and control groups.

11.3.1.4 Steps in Conducting RCT

Steps in conducting RCTs are very close to the steps given in Chap. 9 on common experimental designs. A few authors have discussed various steps (Bhide et al. 2018; Leroy et al. 2022). Here, we provide ten steps to follow in conducting an RCT:

Step 1. Define objectives, and specify interventions and outcomes: This is a preparatory stage where the researcher must clearly define the objectives of the impact evaluation, as well as the interventions, outcomes, and impacts to be assessed. Effective coordination between researchers and implementers is essential. This involves creating a detailed protocol that outlines the interventions, their implementation, the target population, and the sampling procedures. The researcher should specify the outcomes, align them with the theory of change, and define the data collection tools. Additionally, it is important to conduct a comprehensive literature review related to the program and identify methodologies and key findings that will inform the design of the assessment. Key items for consideration include:

- Clear objectives and treatments
- Control groups must be well defined
- A TOC outlining expected causal relationships
- Key assumptions
- Obtaining ethical approvals
- Developing a clear informed consent process

Step 2. Select a research area and study population, and design the study carefully: Identify the unit of assignment for randomization. Consistency in research design is paramount.

Step 3. Design treatments and randomly assign people to treatment or control: Randomly assign treatments to the study population involving two steps: random sampling of the population and random assignment of treatments (see step 5).

Step 4. Conduct the baseline survey: Gather baseline data on key variables from both the treatment and control groups. Use randomization to ensure that any observed differences at baseline are due to chance.

Step 5. Random assignment of treatments and control: Ensure random assignment of treatments and control. This process will ensure the similarity between groups when treatment begins.

Step 6. Implement treatments and monitor processes: Ensure a rigorous monitoring system to maintain the integrity of the experiment. Consistent implementation across the treatment group is essential. A robust monitoring and follow-up system using collected data will ensure clarity on treatment assignment and approach.

Step 7. Collect data about implementation (and possibly mid-term outcome data) and prepare mid-term reports: Data that provide information on implementation should be collected, possibly through a mid-term survey, which will usually focus on process aspects. Such a survey may also be used to provide initial estimates of program impacts (through this monitoring and evaluation process) if it is not premature to do so.

Step 8. Collect data on impacts: After program implementation, conduct a follow-up survey considering TOC. The timing of the follow-up survey will depend on the theory of change, as to how long it will take for the expected impacts to occur. Ensure that data collection procedures are consistent with the baseline phase.

Step 9. Data analysis: Analyze collected data, reporting the intervention's effects on adoption and impact. Consider statistical tests, such as t-test or regression analysis, to compare differences between treatment and control groups. Emphasize the statistical significance and the magnitude of each effect.

Step 10. Feedback process: Develop a strategy for implementing RCT results among relevant stakeholders and decision-makers, sharing with them for effective utilization.

Students carrying out impact-based studies using RCTS will need support from the project implementation team if the project has already started. They will require project documents (showing the theory of change) and data. Due to time limitations and a lack of adequate information regarding program implementation, students commonly use "quasi-experimental designs or preexperimental designs." However, it is still possible to carry out RCTS, especially for PhD students. For one can evaluate the impact of using digital dairy technologies on milk production among smallholder farmers (Box 11.3).

Box 11.3: Limitations of RCTS

The main limitations of RCTs in agriculture are in the following areas: participation, complexity, internal validity, external validity, statistical conclusion validity, design, ethical considerations, and field-based challenges (Sargeant et al. 2010; Sullivan 2011; Brainard et al. 2016; Kool et al. 2020; Pynegar et al. 2021). They are also known as "threats to validity." They include the following:

- *Randomization is not perfect:* Randomization does not balance all variables (Deaton and Cartwright 2018). It does not guarantee that all variables will be perfectly balanced across the groups, thus affecting the estimated average treatment effect.
- *Ethical issues and political constraints*: It may be difficult to withhold potentially beneficial interventions or programmes from certain farmers.
- *Criticized as "Black box"*: It often focuses solely on the outcome without providing sufficient evidence of the underlying mechanism that led to the outcome (Quisumbing et al. 2020). It may miss on implementation fidelity (programme adherence).
- *Limited external validity*: Its applicability to a larger population is limited because of focusing on a specific context or groups.
- *Farmer heterogeneity*: The impact of interventions may differ depending on the attributes of the individual farmers, making it a challenge to draw general conclusions about the overall impact. Farmers may vary in education level, access to resources, farm size, etc.

- *Sample size*: If the sample size is too small, it can be challenging to identify significant effects, increase the probability of making Type II errors, and limit the applicability of the study results.
- *Contamination of nonparticipants*: Potential spillover effects may result from benefits or unintended consequences extending to nonparticipating farmers (Bett et al. 2018). This blurs the distinction between the treated and the control groups, compromising the integrity of the study design. Placebo is usually recommended, but this is not always possible.
- *Biasness:* RCTS are not double-blind. The participants know if they are treated or not and alter their behavior. For instance, farmers who did not receive improved seeds performed poorly due to changed behavior (Quisumbing et al. 2020).
- *Cost and time-consuming:* Conducting RCTS requires implementing interventions and tracking outcomes.
- *Potential risk of noncompliance*: Some participants may not fully adhere to the instructions and assignments, e.g., farmers altering the rate of fertilizer due to several reasons, e.g., to apply to other crops/areas besides what is recommended.
- *Interactions between interventions:* Farmers often adopt multiple practices (varieties, inputs), presenting a challenge in evaluating the effect of the target interventions. It becomes difficult to isolate and evaluate the specific effects of the target intervention.

See Culbertson et al. (2014) on how to address the limitations in RCTs.

11.3.2 Quasi-Experimental Designs for Measuring Impact

Characteristics of Quasi-Experiments

Quasi-experiments are also referred to as mid-experiments due to the absence of randomization (Chap. 7). The absence of randomization means that these experiments do not address the issue of confounding variables. Consider a study that evaluates the effects of a government subsidy program on the adoption of organic farming practices. In regions where the subsidy is implemented, farmers who choose to participate could be compared with those in nearby regions where the subsidy is not available. In this case, naturally existing groups are used.

While they may not be as rigorous as RCTs for establishing causality, quasi-experiments have become widely accepted and frequently used in the agricultural social sciences (Shadish et al. 2002; Cook 2015). In fact, true experiments in impact evaluation are quite rare. Despite their limitations, quasi-experiments are popular because they offer several advantages (Schweizer et al. 2016; Bärnighausen et al. 2017). Some key advantages include:

1. *Practicability in agricultural projects*: Quasi-experiments are well-suited for agricultural projects because they allow researchers to work with naturally existing groups, such as villages or farmer cooperatives, making them more practical and feasible than randomized experiments in real-world settings.
2. *Real-world conditions dimension:* Agriculture interventions interact with natural variations, such as weather and soil types. Quasi-experiments are able to capture these interventions in real-world agricultural contexts.
3. *Ethical considerations*: Some agricultural interventions involve practices that are essential for farmers, making it ethically challenging to withhold these practices for the sake of random assignment. For example, evaluating the impact of a new and potentially more efficient irrigation system on crop yields. Withholding this beneficial irrigation from some farmers is unfair and unethical. Most hypotheses involving farmers are practically answered through a quasi-experimental design only due to ethical considerations.
4. *Long-term projects:* Long-term agricultural projects with long-term goals are best studied using quasi-designs, as they allow for the assessment of the sustained impact of interventions over an extended period. For instance, a project that aims to transition farmers from conventional farming practices to integrated soil fertility management practices requires a longer time frame to assess the long-term ecological, economic, and social impacts.

Box 11.4 gives quick points on the characteristics of quasi-experiments.

Box 11.4: Quick Points about Quasi-Experiments

- No random assignment
- Conducted in real-world settings, taking advantage of naturally occurring conditions
- High external validity
- Are often observational in nature with no manipulation
- Has less control over extraneous variables compared to true experiments
- Researchers may use pre-existing groups
- Use a comparison group generated through matching

- Researchers use statistical methods to control confounding variables
- Establishing causal relationships is often challenging due to the lack of random assignment
- They are cost-effective and quick

There are two main types of quasi-experiments; "Nonequivalent control group design and "time-series designs." They are discussed in the next sections.

Nonequivalent Control Group Design

In the "nonequivalent control group design," both experimental and control groups are chosen without the use of randomization, and dependent variables are observed in both groups (Mohr 1982; Krishnan 2019). Researchers label them as nonequivalent groups due to the absence of randomization in participant selection. The allocation process entails the random assignment of intact groups to treatments, rather than the random assignment of individual participants. As an illustration, consider a study involving eight wards within a similar agroecological zone. In this scenario, four wards are randomly designated to the experimental group (X1), while the remaining four are allocated to the control group (C).

We have two variations of the "nonequivalent control group design": 1. Pre- and posttest design and 2. Posttest only design. In the pre- and posttest design, the researcher allocates intact groups to experimental and control treatments and administers a pretest to both groups. The experimental group undergoes the intervention (Treatment), while the control group does not receive any treatment. Subsequently, after the experimental period, a posttest is conducted on both groups to assess the differences between them.

The posttest-only design, on the other hand, uses only a posttest in the design (see the following illustration)

Pretest and posttest design			
Experimental group	Pretest	X1	Posttest
Control group	Pretest	0	Posttest
Posttest-only design			
Experimental group	X1	Posttest	
Control group	0	Posttest	

In an example of a posttest-only design, the experimental group is exposed to the intervention, such as the promotion of soybean cultivation. Following a three-year period, household income is measured for both the experimental and control groups. Any disparity observed in the income levels before and after the intervention is attributed to the soybean project. A similar project was implemented in Malawi, where plant clinic practices were promoted among farmers growing tomatoes, and where impact was evaluated after 3 years (Bett et al. 2018).

Time-Series Design

The time-series design proves valuable when the experimenter aims to assess the impacts of treatment over an extended duration. This design is particularly advantageous in single-subject studies where the researcher has only one group. In this approach, the group is examined over time, incorporating various pretest and posttest measures or observations by the experimenter. The usual practice involves measuring the outcome of interest multiple times, both before and after the intervention (treatment).

Single-group interrupted time-series design							
Group A	O	O	O	X	O	O	O

The design allows the researcher to observe the trend in the outcome variable and capture the overall trajectory of the outcome over time. It is, therefore, suitable for assessing change in entire systems, e.g., a farming system where it would be challenging to find a control group or system. However, this design is labor-intensive and costly because the researcher is required to collect data on several parameters and possibly document other processes to support the findings.

Dealing with Quasi-Experiments

To reduce some of the bias caused by nonrandomization and strengthen the study, researchers try to include equivalent groups as much as possible (Rubin and Thomas 1996; Higgins et al. 2024). They use statistical techniques to construct counterfactual or comparison groups. These statistical techniques are our next topic.

Methods for Creating Comparison Groups in Quasi-Experiments

The predominant method for selecting a comparison group is through *matching*. The researcher chooses a group of comparable observable characteristics believed to impact the outcome. For example, when evaluating an agricultural intervention aimed at improving crop yield, the researcher may compare participating communities with other similar communities regarding demographics, weather patterns, soil types, and traditional crops.. This selection is based on the premise that communities sharing these characteristics are more likely to exhibit behavior similar to that of the treatment group in the absence of the intervention. The matching method is common in agriculture and many other disciplines, such as sociology, clinical studies, public health, and nutrition.

Achieving perfect matching requires pairing each individual in the treatment group with someone in the comparison group who shares all relevant characteristics (such as age, education, occupation, wealth, religion, and risk attitude). This level of precision is often impractical. In reality, researchers aim to estimate the key factors that influenced a person's decision to enroll in the program and use those to find the closest possible match.

To address this challenge, researchers turn to matching methods that rely on observable characteristics and statistical techniques to construct comparison groups. These methods aim to create comparability, facilitating an "apples-to-apples" comparison, while minimizing bias from confounding factors. The primary methods employed for creating matching comparison groups include 1. Propensity Score Matching (PSM), 2. Difference-in-Difference (DID), 3. Regression Discontinuity, and 4. Instrumental Variables. Other advanced methodologies and models are used, e.g., the Endogenous switching model (Tesfay 2020; Fitawek and Hendriks 2021), and their discussion is beyond the scope of this book. The following sections explain the two most commonly used methods in agricultural programs, namely (PSM) and DID.

1. *Propensity Score Matching*

What Is the Propensity Score Matching Technique?

The propensity score matching technique, originally introduced by Rosenbaum and Rubin 1984), involves the calculation of the propensity, which is the conditional (predicted) probability of receiving treatment based on pretreatment characteristics like age, income, and education level. PSM simplifies the matching process by summarizing all observed characteristics into a single score, *the propensity score*, which represents the probability of an individual participating in the program based on their observable traits.

Instead of matching on multiple variables directly, researchers match individuals in the treatment and comparison groups who have similar propensity scores. This indicates the predicted likelihood of participating in the intervention given their observable characteristics. For instance, in the context of households participating in a specific program, propensity score matching aligns treatment households with similar comparison households. Propensity score matching thus ensures that the average characteristics of the treatment and comparison groups are similar. This similarity is considered sufficient to yield an unbiased impact estimate (Sainani 2012).

How Are Propensity Scores Estimated?

The propensity scores are estimated through a binary model such as logistic regression or probit. These are available in most statistical programs, e.g., R and STATA (Kamau et al. 2018). The two regression models are preferred because they are easy to perform using standard software and are understood by most researchers (Ali et al. 2019). It is crucial to emphasize that the method used, such as logistic regression, is not applied for inferential purposes but specifically to generate the balancing score known as a propensity score. The propensity scores are derived or predicted from data related to observable predictors, representing characteristics of individuals within the sample. These scores, expressed as probabilities ranging from 0 to 1, are calculated for each member of the sample. The primary objective is to capture the likelihood of an individual participating in the intervention based on their observable characteristics rather than for making inferential conclusions (Ali et al. 2019).

In choosing the covariate to include in the regression model, the researcher should be guided by economic theory and past research as well as specific goals and objectives of the project. Knowledge of the empirical and theoretical literature is a prerequisite to designing a survey intending to collect data to use in PSM.

Logic of Propensity Score Matching

The logic of "propensity score matching" is that instead of matching pretreatment characteristics (age, education, income, etc.) individually, we lump them together into a propensity score. This is the score that is used to match participants to nonparticipants. With this, we are trying to solve the problem of selection bias experienced in observational studies and quasi-experiments. In observational studies, as previously explained, factors other than chance dictate the allocation of participants to different treatments, leading to variable chances or probabilities associated with each treatment group. The propensity score balances this out by creating the best match for each participant.

While propensity score matching (PSM) helps achieve balance between treatment and control groups based on observed covariates, it differs from randomization in a key way. Unlike randomized controlled trials (RCTs), which balance both observed and unobserved variables, PSM can only account for measured characteristics. As a result, if important unmeasured confounders exist, PSM may not fully eliminate bias, whereas randomization offers stronger control over potential confounding.

Assumptions in Propensity Score Matching

There are two main assumptions in propensity score matching.

1. *Conditional independence or confoundedness.* It states that the assignment of treatment is substantially randomized when considering an extensive array of covariates. It stipulates that observable characteristics must be unrelated to potential outcomes, signifying that the decision to participate is solely based on the observable characteristics of households. Consequently, any discernible difference in outcomes between the treatment and control groups with equivalent values for characteristic X can be attributed to the treatment itself.

2. *Common/overlap support assumption*: The assumption requires that the distribution of observable characteristics between participants and nonparticipants exhibits overlap. Balancing tests are conducted to examine this overlap in the distribution of characteristics, ensuring that the selected covariates sufficiently span both groups. This verification is crucial for establishing the validity of causal inferences.

Steps in Propensity Score Matching

The process of conducting propensity score matching entails a sequence of steps (Austin 2011; Garrido et al. 2014; Harris and Horst 2016; Wordofa et al. 2021; Ouya et al. 2023). At each step, decisions must be made regarding the choice of covariates, models for generating propensity scores, matching distances and algorithms, the estimation of treatment effects, and diagnosing the assessment of the quality of matches. Here we describe five steps for estimating PSM scores:

- *Step 1: Select confounding variables:* Select covariates (matching variables) that are representative of participants and nonparticipants. These are variables that influence both the treatment assignment (e.g., participation in a program or intervention) and the outcome of interest. Proper selection helps ensure balance and comparability between treated (participants) and untreated (nonparticipants) groups; the covariates include demographic and farm variables, e.g., gender, household size, education, farm size, number of animals, etc. A researcher must ensure the presence of a representative sample survey encompassing both eligible participants and nonparticipants in the intervention.
- *Step 2: Select model and estimate propensity scores:* The scores are determined using the "participation equation," typically implemented through either logit or probit regression, with program participation serving as the dependent variable (coded as 1 for participants and 0 for nonparticipants). It is essential to carefully select characteristics that are believed to influence participation, ensuring that they are comprehensive while excluding variables that may be affected by the intervention (Kim 2022). Therefore, it is advisable to use baseline data, when available, for estimating propensity scores (Austin 2011). This approach minimizes the risk of including characteristics that may be altered by the intervention during the matching process.
- *Step 3: Select a matching method:* At this stage, each member of the treatment group is paired with one or more members of the comparison group. Numerous approaches, also known as matching algorithms, exist for constructing a comparison group. These include exact matching or one-to-one matching, nearest neighbor (NN) matching with or without calliper adjustment, and optimal matching. Nearest neighbor matching and optimal matching are commonly chosen methods (Austin 2011). After the matching process, the differences between the two groups are substantially minimized. For a detailed discussion of matching algorithms, refer to the recommended literature (Caliendo and Kopeinig 2008; Austin 2011; Harris and Horst 2016). These authors offer comprehensive guidelines on performing propensity score analysis and interpretation, including how a researcher should choose the appropriate algorithm, as there is no one-size-fits-all solution.
- *Step 4: Compare the balance*: Once the matches are created, it is crucial to evaluate the quality of the matches to ensure that the comparison group exhibits a distribution of propensity scores similar to the intervention group. Assessment of matches typically involves a numerical and visual comparison of balance. The underlying logic of the propensity score is to act as a balancing score, necessitating balanced covariates across different groups. A demonstration of balanced covariates is exemplified in Table 11.1 from a study assessing the impact of plant clinics (Bett et al. 2018). If covariates are not balanced, the model may be misspecified, leading to potentially biased inferences. Procedures for checking balance include standardized bias, t-tests, stratification, joint significance, and Pseudo R (Caliendo and Kopeinig 2008). Here, we give two examples: 1. A study carried out in Malawi to evaluate the impact of plant clinics on tomato productivity (Bett et al. 2018) and a study from Kenya to evaluate the impact of participating in collective marketing by avocado farmers on incomes (Kwizerimana et al. 2023). Balancing tests used included checking if covariates are balanced before and after matching (Table 11.1) and plotting the distribution of propensity scores, plotting a graph of propensity score distribution after matching, and mean comparison before and after matching (Figs. 11.8 and 11.9a, b). Figures 11.10, 11.11, 11.12, and 11.13 demonstrate this concept with different marching algorithms: Nearest Neighbor 1 and 5 (NNM 1 and 5), Kernel-based matching (KNM), and caliper-based matching (CBM).
- *Step 5:* Estimate the impact of the intervention through a comparison of the groups. Calculate the average treatment effect of the treated (ATE) by summing the differences between the outcomes for the exposed and unexposed individuals and then dividing by the number of matched pairs. The estimation of the treatment effect should be confined to the region of common support. Various strategies, such as comparing minima and maxima, can be employed to enforce the common support condition (Garrido et al. 2014).

Table 11.1 Balancing tests for the propensity scores before and after matching[a]

Variable	Before matching			After matching		
	Treated	Control	$p > \|t\|$	Treated	Control	$p > \|t\|$
Age of household head	43.27	40.80	**0.007****	43.266	44.443	0.260
Gender of respondent	0.935	0.924	0.550	0.93727	0.94009	0.892
Participation in on-farm by head	0.978	0.970	0.466	0.97786	0.97115	0.621
Education of household head	1.384	1.220	**0.000*****	1.3727	1.4121	0.449
Area cultivated for tomatoes	0.224	0.150	**0.000*****	0.20776	0.18219	0.242
Use of improved seeds	0.864	0.763	**0.001*****	0.85978	0.87124	0.696
Use of chemical fertilizer	0.781	0.887	**0.000*****	0.79705	0.78486	0.728
Household size	5.936	5.390	**0.000*****	5.9262	6.0024	0.670
Total knowledge index	19.74	16.83	**0.006****	19.598	20.126	0.676
Tropical livestock unit	1.538	2.939	**0.038****	1.5727	1.7018	0.652
No. of credit sources	1.118	1.118	0.983	1.1218	1.1455	0.516
Dependency ratio	97.38	114.26	**0.008****	98.015	97.435	0.932

[a]Source: Bett et al. 2018. Balancing test for covariates before and after matching. Note: Most covariates were significantly different between treated (users of plant clinics) and control (nonusers of plant clinics) before matching. After the matching covariates showed no significant difference, meaning the covariates were balanced after the matching *** significant at 1%, ** significant at 5%; N=558

In Fig. 11.8, though there are a few farmers with no match, generally, the results show that the common support condition is satisfied, as there is substantial overlap in the distribution of the propensity scores of the treated and non-treated groups. Therefore, the researchers concluded that the assumption of common support was satisfied.

This is further supported by Fig. 11.9, which shows a kernel density plot of propensity scores after matching. The figure indicates improved balance between treated and untreated groups. The smooth, overlapping distribution suggests good common support, meaning the groups are comparable in terms of treatment likelihood.

Advantages of Propensity Score Matching

- *Utilizes observational data:* PSM utilizes observational data and is therefore applied in various study designs, including cross-sectional surveys and longitudinal and quasi-experimental. Unlike other methods like DID, it does not always require ex ante measurements. Researchers will most often choose these methods when baseline data are unavailable.
- *Conceptually easy*: Compared to other methods, PSM is conceptually easier to use, especially for a nontechnical audience. It compares with and without users and nonusers and is, therefore, easy to conceptualize and access. Most students in both agricultural economics and other fields often use PSM because of this attribute.
- *Reduces selection bias*: PSM creates a more balanced comparison between the treated and the comparison group (control). It is like it enables a researcher to create a "quasi-randomization" of treatment groups to reduce confounding.
- *Flexibility in matching*: PSM offers flexibility in the choice of matching. Researchers can choose exact matching (1:1) like the one used by Bett et al. (2018), nearest neighbor matching, or kernel matching, depending on data characteristics. Using a combination of the different matching methods allows for customization and optimization of the matching process.
- *Transparency in interpretation*: Provide a transparent outline and structure for comparing the treated and control groups. For example, it allows the researchers to examine the balance of the covariates between the groups before and after the matching. It also incorporates other quality checks, such as balancing tests and sensitivity analyses. All these enhance the understanding and interpretation of the data.

Limitations of Propensity Score Matching

1. *Inability to account for unobserved confounders:* PSM, while adept at adjusting for observed covariates, falls short when it comes to controlling for unmeasured or unknown factors, such as latent characteristics or individual attitudes. If these unobservables influence outcomes, treatment effect estimates may remain biased. This limitation underscores the irreplaceable role of randomized controlled trials (RCTs), which ensure balance across both observed and unobserved covariates, in overcoming such challenges.
2. *Reduction in sample size:* Matching often results in a loss of data, as not all treated participants find suitable matches in the control group. This can reduce statistical power and limit the precision of treatment effect estimates. PSM is considered "data hungry", requiring both a large number of covariates and sufficient sample sizes. To mitigate this,

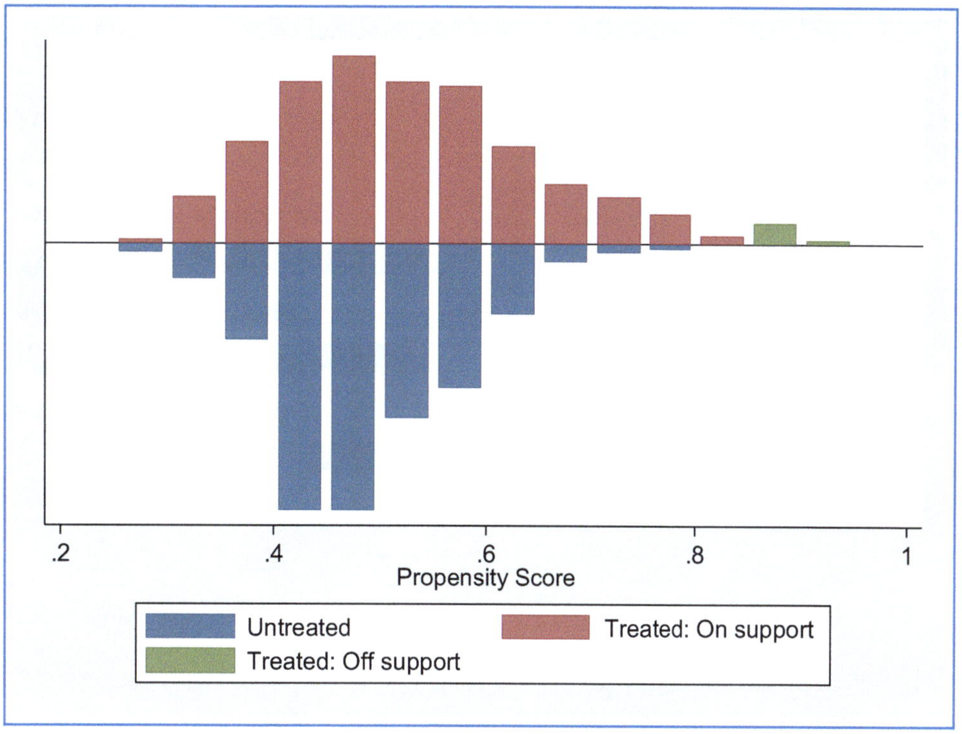

Fig. 11.8 Propensity score distribution and common support for propensity score estimation, produced by Stata software. Note: "Treated and non-treated: on support" indicates the observations in the treated group (users of plant clinics) have a suitable comparison (nonusers of plant clinics). "Treated: off support" in green, indicates the observations in the treated group that do not have a suitable comparison, thus excluded from the final matching. (Source: Bett et al. 2018))

researchers are advised to use more controls than treated participants, as demonstrated in some studies (Bett et al. 2018).

3. *Group imbalance and representativeness*: The challenge of matching may render it impossible for many participants, especially when treatment and control groups differ significantly in their covariate distributions. This can lead to matched samples that are not representative of the full population, potentially compromising the generalizability of findings and necessitating caution in interpreting results.
4. *Sensitivity to model specification:* The accuracy of PSM is heavily contingent on the correct selection and functional form of covariates used to estimate the propensity scores. Poor model specification can lead to biased results. To ensure the robustness of findings, conducting sensitivity analyses is not just recommended, but crucial in the face of alternative model assumptions (Box 11.5).

Box 11.5: Recommended Further Reading on Studies That Have Used Propensity Score Matching to Evaluate Impact

1. Kwizerimana S, Mugwe J, Nigat B (2023) Impact of collective marketing participation on farmers' income: Evidence from smallholder avocado farmers of Murang'a County, Kenya. Social Sciences & Humanities Open 8:100614. https://doi.org/10.1016/J.SSAHO.2023.100614
2. Bett, E., Mugwe, J., Nyalugwe, N. Haraman, E., Williams, F., Tambo, J., Wood, A. and Bundi, M. (2018). Impact of plant clinics on disease and pest management, tomato productivity and profitability in Malawi. *CABI Working Paper* 11, 30pp.
3. Wordofa MG, Hassen JY, Endris GS, et al. (2021) Adoption of improved agricultural technology and its impact on household income: a propensity score matching estimation in eastern Ethiopia. Agric Food Secur 10:1–12. https://doi.org/10.1186/S40066-020-00278-2/TABLES/6
4. Ouya FO, Murage AW, Pittchar JO, et al. (2023) Impacts of climate-resilient push–pull technology on farmers' income in selected counties in Kenya and Tanzania: propensity score matching approach. Agric Food Secur 12:1–14. https://doi.org/10.1186/S40066-023-00418-4/TABLES/6

Research Questions

- What is the effect of training on the uptake of technology A?
- What is the effect of training on the use of fertilizer on the land size planted with fertilizer in season B?

The main advantage of quasi-experimental design is that it can be used when randomization of the group is impossible

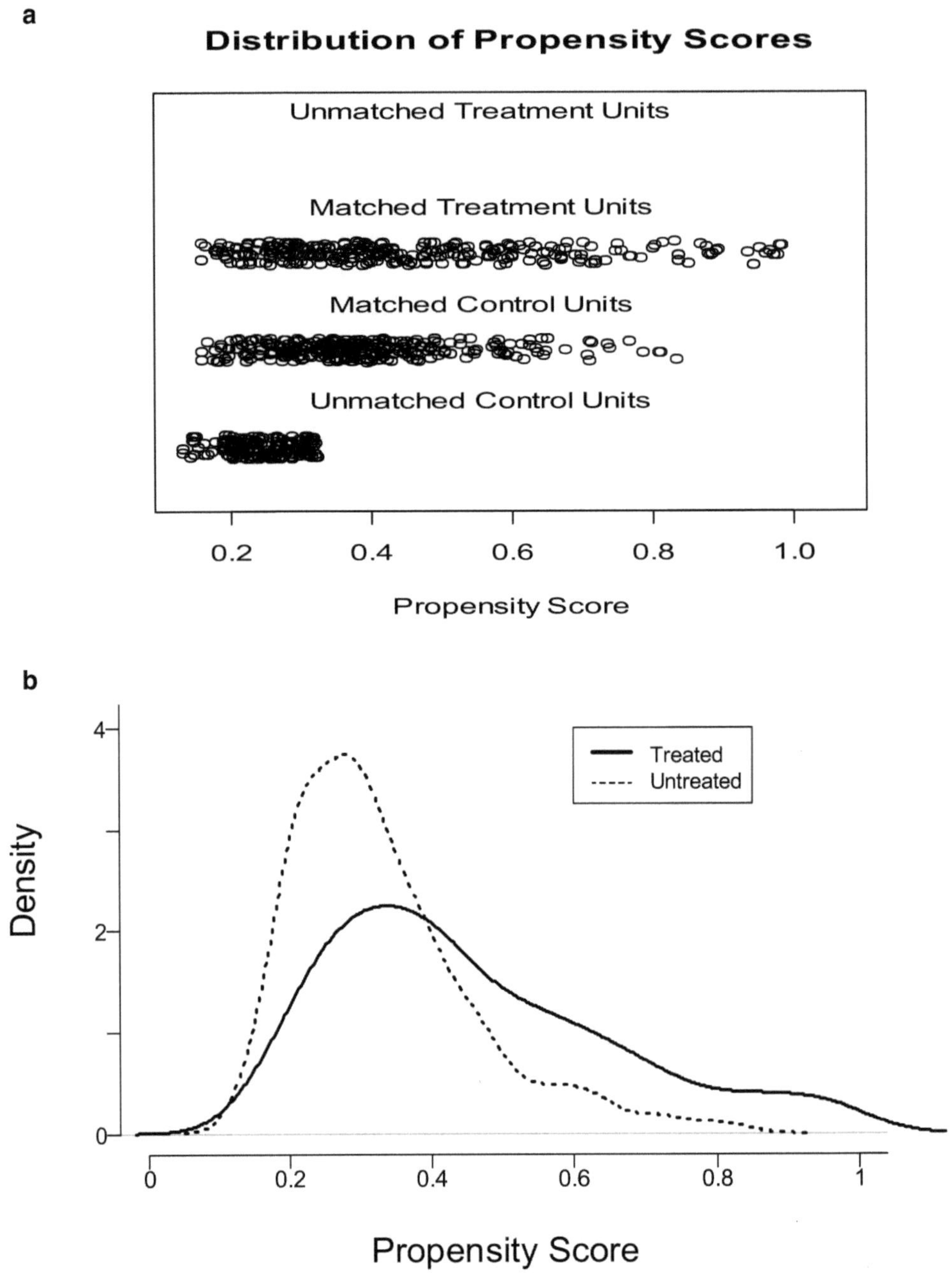

Fig. 11.9 Propensity score distribution after matching. (**a**) Distribution of propensity scores for users and nonusers of plant clinics (Bett et al. 2018). (**b**) Distribution of propensity scores for users and nonusers of plant clinics after matching (Bett et al. 2018). https://doi.org/10.1079/CABICOMM-25-8089. Creative Commons Attribution-Non-Commercial Licence (CC BY-NC). http://creativecommons.org/license

and or impractical. Because of the lack of random assignment, the internal validity of quasi-experimental design is very low. In these designs, there is the possibility of selection bias because the participants are not randomly assigned.

Difference-in-Difference Method

The difference-in-difference (or double difference) method entails comparing observed changes in outcome before and after the project for a sample of participants and nonparticipants (Donald and Lang 2007; O'Neill et al. 2016; Wing et al. 2018). Typically, one collects outcome data of both participants and nonparticipants using a baseline survey before the program. After the program, another survey is carried out. Difference-in-differences takes the before-after difference in the treatment group's outcomes. This is the first difference. In comparing the same group to itself, the first difference controls for factors that are constant over time in that group. Then, to capture time-varying factors, difference-in-differences takes the before-after difference in the control group, which was exposed to the same set of environmental

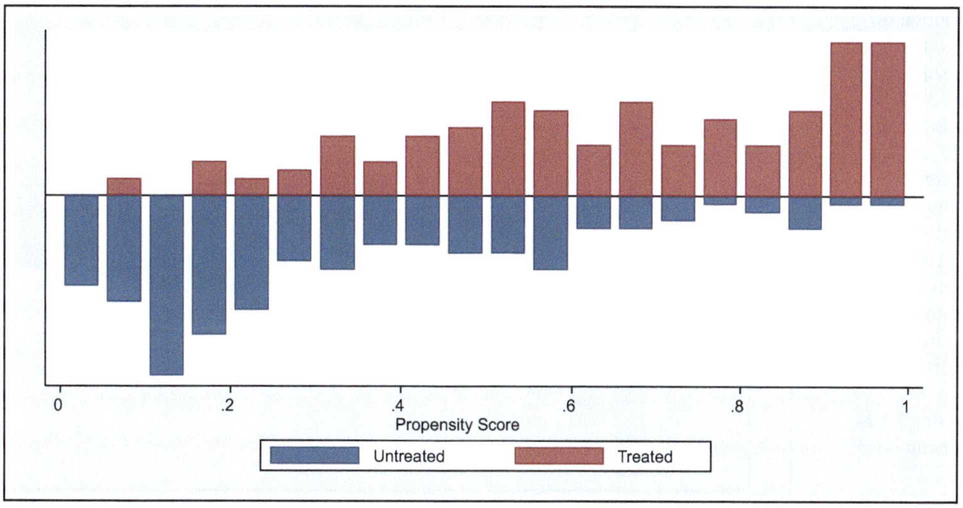

Fig. 11.10 Distribution of propensity scores on the region of common support using Nearest Neighbor (NNM1) with avocado income of years 2018, 2019, and 2020. Propensity score distribution and common support for propensity score estimation. Note: "Treated: on support" indicates that the observations in the treated group (participants in collective marketing) have a suitable comparison (nonparticipants) using NNM1. (Sources: Master's thesis by Kwizerimana (2023) and Kwizerimana et al. (2023) published: https://doi.org/10.1016/j.ssaho.2023.100614 Get rights and content. Under a Creative Commons license. https://creativecommons.org/licenses/by-nc-nd/4.0/: with permission from CABI Africa)

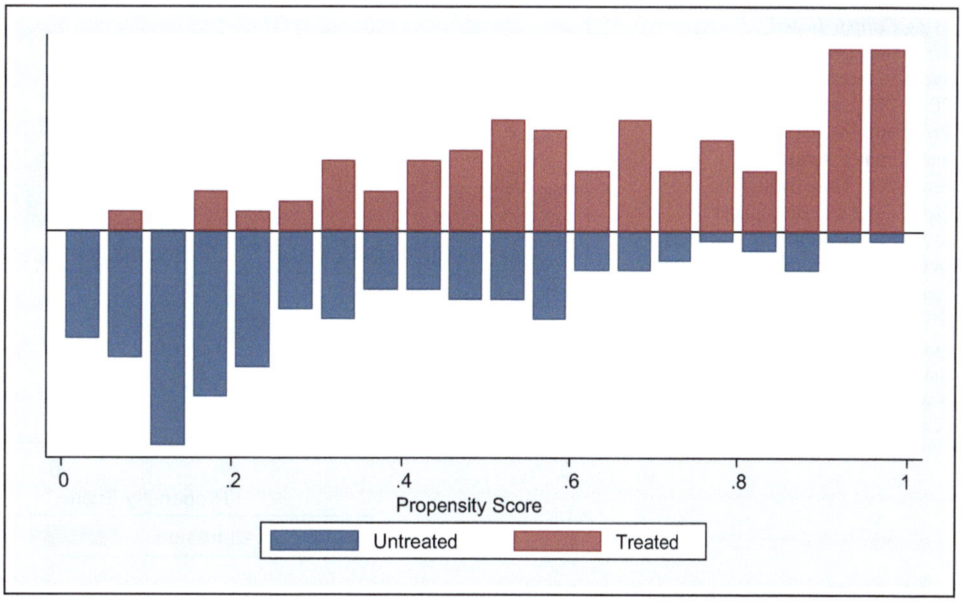

Fig. 11.11 Distribution of propensity scores on the region of common support using *Nearest Neighbor (NNM5)* with avocado income of years 2018, 2019, and 2020. Propensity score distribution and common support for propensity score estimation. Note: "Treated: on support" indicates that the observations in the treated group (participants in collective marketing) have a suitable comparison (nonparticipants) using NNM5. (Sources: Kwizerimana 2023; Kwizerimana et al. 2023)

conditions as the treatment group. This is the second difference. Finally, difference-in-differences "cleans" all time-varying factors from the first difference by subtracting the second difference from it (Zeldow and Hatfield 2021). This leaves us with the impact estimation—or the difference-in-differences.

Difference-in-differences requires data on outcomes in the group that receives the program and the group that did not—both before and after the program. Compute the difference-in-differences as follows:

1. Calculate the before-after difference in the outcome (Y) for the treatment group (B-A).
2. Calculate the before-after difference in the outcome (Y) for the comparison group (D-C).
3. Calculate the difference between the difference in outcomes for the treatment group (B-A) and the difference for the comparison group (D-C). This is the difference-in-differences: (DD) = (B-A)-(D-C).
4. The technique is known as difference-in-difference because of the double subtraction. For example. "To eval-

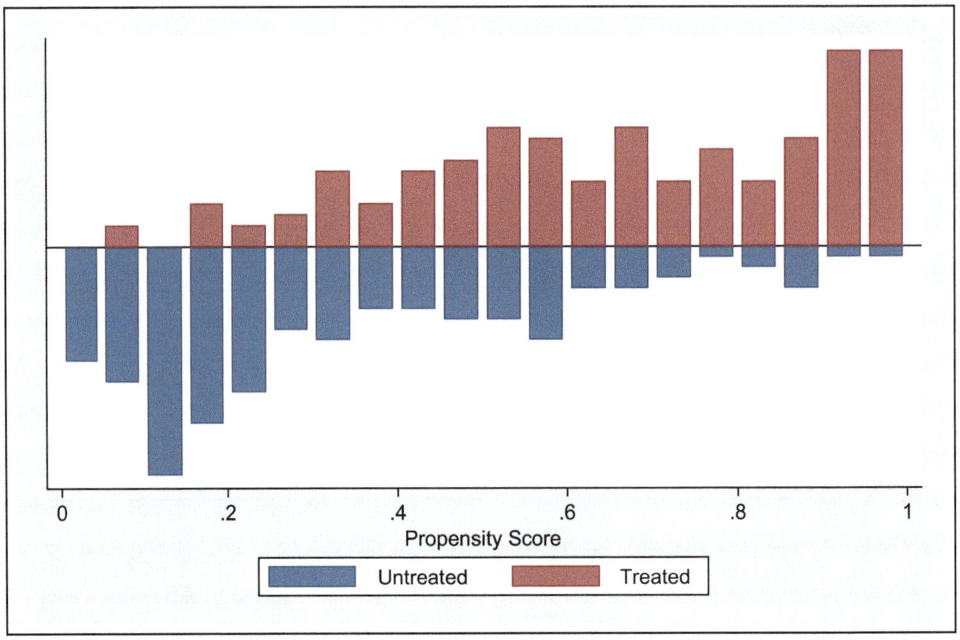

Fig. 11.12 Distribution of propensity scores on the region of common support using *Kernel-Based Matching (KBM)* with avocado income of years 2018, 2019, and 2020. Propensity score distribution and common support for propensity score estimation. Note: "Treated: on support" indicates that the observations in the treated group (participants in collective marketing) have a suitable comparison (nonparticipants) using KBM. (Sources: Kwizerimana 2023; Kwizerimana et al. 2023)

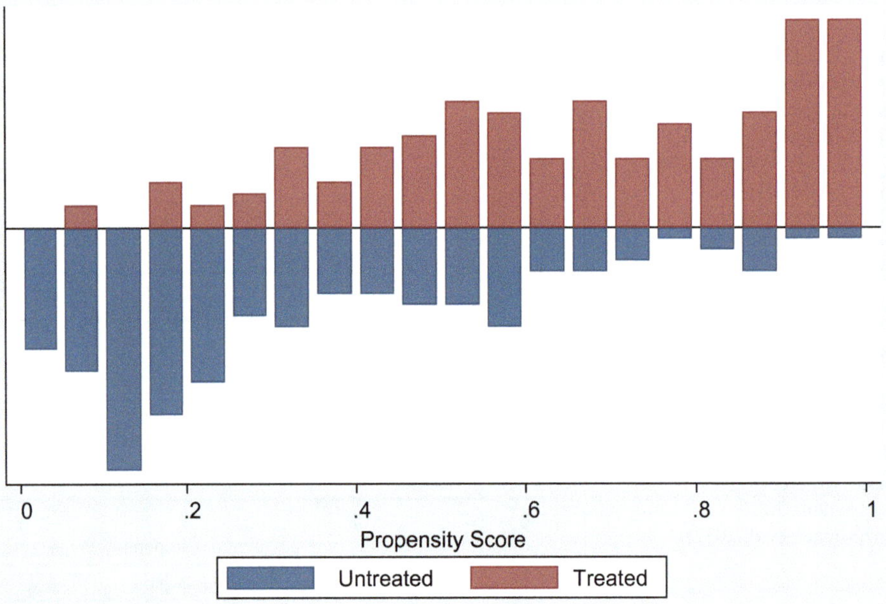

Fig. 11.13 Distribution of propensity scores on the region of common support using *Caliper-Based Matching (CBM)* with avocado income of years 2018, 2019, and 2020. Propensity score distribution and common support for propensity score estimation. Note: "Treated: on support" indicates that the observations in the treated group (participants in collective marketing) have a suitable comparison (nonparticipants) using CBM. (Sources: Kwizerimana 2023; Kwizerimana et al. 2023)

uate the effect of hybrid maize varieties on yields among smallholder farmers," see Table 11.2.

Supposing a hypothetical case where we would like to determine the impact on the percentage change in incomes after farmers start participating in collective marketing. Similarly, we will have the treated as those participating and the control (nonparticipants), and measure their incomes before and after. Figure 11.14 represents the change.

Question: What was the total amount of change in percent incomes for farmers in the project?	
Difference 1:	Project participants (50–20) = 30%
	Nonparticipants (30–20) = 10%
Difference 2:	(Net effect) = 30–10 = 20%
Answer: The change in percent income (impact estimate) from baseline to endline was 20%	

Table 11.2 Example of difference-in-difference in the evaluation of the effect of hybrid maize on yield

	Treatment group	Control group
Before	Average yields of farmers using hybrid maize varieties at t = 0 (A) 1.3 tha^{-1}	Average yields of farmers not using hybrid maize varieties at t = 0 (C) 1.2 tha^{-1}
After	Yields of farmers using hybrid maize varieties at t = 1 (B) 1.6 t ha^{-1}	Yields of farmers not using hybrid maize varieties at t = 1 (D) 1.3 t ha^{-1}
Change	0.3 t ha^{-1}	0.1 t ha^{-1}
Net change	0.2 t ha^{-1}	

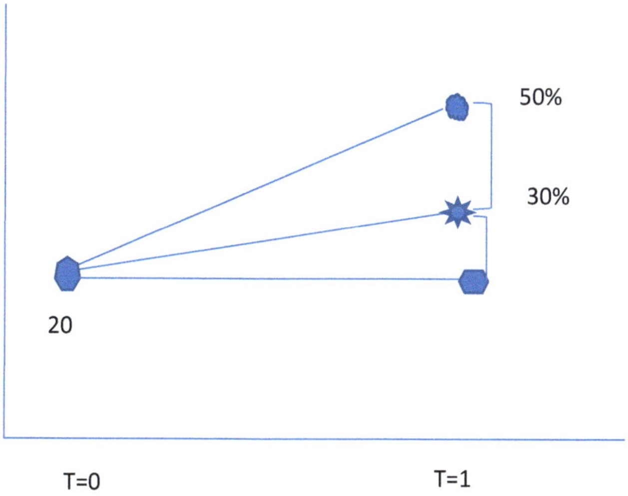

Fig. 11.14 Hypothetical example of the difference-in-difference technique

Steps in Conducting Difference-in-Difference

1. Identify treated (participants) and control (nonparticipants) groups, for example, those farmers growing tissue culture bananas and those that are not
2. Collect baseline data on both the treated and the control groups
3. Define the time period for analysis (After how long do you want to evaluate the impact?)
4. Implement the intervention (program/project)
5. Measure the outcome variable of interest (dependent variable) that captures the change; could be incomes, prices, food availability, etc.
6. Compare data before and after the program
7. Calculate the difference-in-difference estimator by taking the difference between the average change in the outcome variable for the treatment group and the average change in the outcome variable for the control group
8. Apply statistical methods to assess the statistical significance of the D-I-D estimated value

For further reading on the use of D-I-D, the following references are useful:

- O'Neill S, Kreif N, Grieve R, et al. (2016) Estimating causal effects: considering three alternatives to difference-in-differences estimation. Health Serv Outcomes Res Methodol 16:1–21. https://doi.org/10.1007/S10742-016-0146-8/TABLES/3
- Donald SG, Lang K (2007) Inference with Difference-in-Differences and Other Panel Data. Rev. Econ Stat 89:221–233. https://doi.org/10.1162/REST.89.2.221

Advantages and Disadvantages of Difference-in-Difference

Advantages:

- Easy to compute: D-I-D estimation does not require high computational effort to compute the basic estimates. It therefore appeals to many researchers who may not be familiar with complicated econometrics.
- Efficient use of data; D-I-D uses a kind of panel and exploits variation within and between groups over time. This allows for more precise estimates of treatment effects compared to simple before and after or cross-sectional.
- Controls for time-invariant confounders by estimating and subtracting the average difference.

Disadvantages

- *Parallel trends assumption*: D-I-D relies on the assumption that the indicators of interest will follow the same trend over time in the treatment and comparison groups. Where this assumption is attained, a program impact assessment using this method will be correct. However, if there are differences between groups that change over time, the use of the method will give a biased estimate (O'Neill et al. 2016; Zeldow and Hatfield 2021). If, for example, in the case of the above example, some changes occurred in the comparison wards, then the use of it would not provide an accurate estimate. Such changes may occur in cases where there is a development project that interferes with the availability of inputs to farmers, e.g., a development program that gives inputs directly to farmers or increases their incomes. Increased income may mean that these farmers may invest more in their farms.
- *Data requirement:* D-I-D data are collected over multiple periods for both the treatment and the comparison groups; a kind of panel data. Acquiring such data can be challenging and may limit the application of the did method in some studies. In addition, if data are not collected correctly and on schedule, accuracy of the estimate is likely to be affected.

- *Limited generalizability:* D-I-D estimates are specific to the population and context where the study was conducted, and therefore generalizability of the results to other populations and settings may be restricted (Box 11.6).

Looking at the above Box 11.7, we did not discuss the ESR method. Do we really need the case study? (Box 11.8)

Box 11.6: Case Study 1 on Quasi-Experimental Design. Impact of Irrigation Technology Use on Crop Yield, Crop Income, and Household Food Security in Nigeria: A Treatment Effect Approach

A study to determine the impact of irrigation technology use on crop yields, incomes, and household food security (Adebayo et al. 2018). The study used data from 2305 randomly sampled households. The study used propensity score matching and an endogenous treatment regression model. The justification for using this methodology was to enable the generation of a control group and tackle the problem of bias due to the selection of observables. Nearest neighbor and kernel matching algorithms were used. Results showed significant and positive effects of irrigation technology use. Average treatment ATT of irrigation technology use on crop yields for users was between 1954.66 and 2354.66 kg.

Source: Adebayo et al. (2018)

https://www.aimspress.com/article/10.3934/agrfood.2018.2.154

Box 11.7: Case Study 2 on Quasi-Experimental Designs. The Impact of Adaptation Practices on Crop Productivity in Northwest Ethiopia: An Endogenous Switching Estimation

An impact evaluation study was conducted in Ethiopia to analyze the impact of adaptation practices to climate variability on the productivity of major staple crops in the Rib watershed (Adego et al. 2019). Data were collected using a household survey questionnaire, focus group discussion, and key informant interviews. Time-series climate data were also collected to assess how crop yield responds to climate variability. The study used endogenous switching regression, in which climate information and distance to market were validated as instrumental variables. The results revealed that farmers who adopted adaptation strategies would have gained lower yields if they had not adopted them, and those who did not adopt a strategy would have gained higher yields than if they had. Improved seed, contact with development agents, urea, compost, and rainfall are significantly associated with the likelihood of increasing yield.

Source: Adego et al. (2019)

https://www.tandfonline.com/doi/full/10.1080/21665095.2019.1678186

Box 11.8: Case Study 4 on Quasi-Experimental Design (Effects of Sustainable Agricultural Practices on Farm Income and Food Security in Northern Ghana)

This study sought to investigate the determinants of adoption of multiple sustainable agricultural practices (SAPS) and the effects of adoption on farm income and food security in Ghana (Setsoafia et al. 2022). This study specifically investigated the factors affecting smallholder farmers' decisions to adopt SAPs (improved seed, fertilizer, and soil and water conservation) and the impacts of the adoption on farm income and food security. The reduced coping strategy index and household dietary diversity capture food security. A multinomial endogenous switching regression model was utilized to address selection bias issues. Results show that farmers' decisions to adopt SAPs are influenced by the social demographics of the households, plot-level characteristics, extension services, and locations. Adopting all three SAPs positively impacts farm income and food security more than adopting a single or two SAPs.

The justification for using ESR is that adoption has more than two options. In particular, this model helps address the selection bias issues arising from observed factors (e.g., age, gender, and education) and unobserved factors (farmers' innate ability in innovation adoption and motivations to address external shocks). The argument is that both observed factors (e.g., age, gender, education and farm size) and unobserved factors (e.g., farmers' innate abilities and motivations) may affect their decision to adopt a single SAP or a package. Due to the self-selection nature of technology adoption, farmers without adopting any SAPs and those adopting a single SAP or package may be systematically different.

Source: Setsoafia et al. (2022)

https://agrifoodecon.springeropen.com/articles/10.1186/s40100-022-00216-9

11.3.3 Preexperimental Designs for Measuring Impact

One-Group Posttest Only

The one-group posttest is also known as the after-only experimental design. The study group (single group) is first exposed to a treatment, event, or condition. The group is afterward measured, using specific indicators, to see if there are any effects on a predetermined independent variable. An example is a researcher who trains farmers in farmer field schools on the use of manure. The group is assessed after the training to find out if the training using the farmer field school affected the group's knowledge levels.

Illustration	
X	O
Treatment	Observation (dependent variable)
(train farmers using farmer field school methodology)	(knowledge level on use of manure)

The symbol X represents exposure of the group to the treatment, while O refers to observation (measurement of the dependent variable). The change in the dependent variable is measured by the difference between the "before" (baseline) and "after" datasets. This is a major weakness because this design lacks control. The researcher has no way of knowing if the results obtained at O are due to treatments because there is no comparison; i.e., if the researcher finds an increase in the knowledge levels on use of manure, it is wrong to conclude that training increased knowledge levels of manure use; the researcher knows nothing about what the group was like before being trained. In addition, this design measures total change, including change attributable to extraneous variables; hence, it cannot identify the net effect of an intervention.

Because of the weaknesses of this design, the effectiveness of the intervention is realistically assessed using evaluation indicators designed according to the program at hand. It relies on assessment indicators such as:

- Change in the area of land under a certain technology
- Perception of the farmers/ clients with the program
- Qualitative data, e.g., experiences of changes as a result of the program (training)
- Number of farmers using the technology after the program (training)
- Yield increase reported by the farmers
- Number of dropouts after the training

Looking at the variables we are measuring, we see that this design is inappropriate to study the impact of an intervention. It suffers from several threats to internal validity, including maturation, history, regression effects, and test effects. It is sometimes used in programs that were implemented without a preplanned evaluation at the program planning stage, and therefore lack a baseline study.

However, the design is commonly used because it is cheap and provides the current picture in relation to the outcome indicators. It may sometimes be used as an exploratory to help decide on continuation, termination, or changing an intervention or program.

One-Group Pretest and Posttest Experimental Design

The one-group pretest–posttest design is also called the before-and-after design. It consists of a single group, which is measured before the intervention and after the intervention. It is

O	X	O
Pretest	Treatment	Posttest

The change is measured by comparing the difference in the phenomenon or variable(s) before and after the intervention. It can be described as two sets of cross-sectional data collection points on the same population to find out the change in the phenomenon or variable(s) between two points in time. It is thus useful in studies that aim to measure change in a situation, phenomenon, issue, problem, or attitude. It is widely used for measuring the impact or effectiveness of a program.

Consider an example of this design using a three-year project that is promoting soybeans among farmers. In the project design, the justification is that the promotion of soybeans will increase household incomes among farmers. To measure this output, a researcher measures household incomes at the initiation of the project and then implements the soybean project for the end of the 3 years.

O X O

Pretest (dependent variable) implement three-year soybean
Posttest promotion project household
incomes (dependent variable)

At the end of 3 years, after the program has been completely implemented or is assumed to have had its effect on the population, the "after" observation is carried out to ascertain the impact attributable to the project. i.e.. measures the household income among farmers who participated in the project again.

This design is an improvement of the one-shot case study because the before-and-after comparison tells the researcher whether any change has occurred. However, it still does not enable one to conclude that any change, in whole or in part, can be attributed to the program intervention. This is because of the inability of the design to control threats to validity, especially those associated with history, attitudes of subjects, data collectors' bias, etc.

To overcome this, a comparison (control) group is used. Then, if a change in household incomes occurs, the researcher has reason to believe that it was caused by the treatment (soybean project).

Static Control Group Design

The static control group design, also called a nonequivalent control group design, is a popular design for assessing impact (Handley et al. 2018; Krishnan 2019). In this design, two already existing groups are used and measured ex-post. Here, the researcher selects two population groups instead of one: a *control group* and an *experimental group*, usually referred to as static groups. These groups are expected to be comparable as far as possible in every respect except for the intervention (that is assumed to be the cause responsible for bringing about the change). The experimental group either receives or is exposed to the intervention, whereas the control group is not.

If we use the example of the soybean project, the researcher will select one group of farmers where project was implemented (usually one region) and another group that will not have the soybean promotion project. Note that the farmers are not randomly assigned to the two groups. First, the household incomes are measured for both groups at the same time.

A second example of a project focused on collective marketing strategies for bananas. Similar to the soybean example, one group of banana farmers is selected from a region where the collective marketing initiative is implemented, while another group, without exposure to the project, serves as the control. Again, the assignment of farmers to these groups is nonrandom, reflecting the practical constraints often encountered in the real-world.

While the nonequivalent group design offers improved control over threats to validity compared to certain other quasi-experimental designs, it introduces the potential for differential subject characteristics. Because farmers are not randomly assigned, there may be inherent differences between the two groups. To address this challenge, statistical procedures are employed to account for and mitigate the impact of these differences, enhancing the reliability and validity of the study findings.

11.4 Nonexperimental Designs for Evaluating Impact

What Does Nonexperimental Designs Involve?

Nonexperimental methods rely on answering descriptive questions. These questions ask how things are, what happened, and how it happened. This may include describing the initial situation and how it has changed, the activities of the intervention and other related programs or policies, the context in terms of participant characteristics, and the implementation environment. Data collection methods for answering descriptive questions may include the following:

- Collecting data from individuals or groups: through interviews, open-ended questionnaires, and focus group discussions.
- Interviewing project beneficiaries and opinion leaders to get their personal opinions.
- Participatory methods such as ranking, seasonal calendars, and stories.
- Observation: structured or nonstructured; participant or nonparticipant; recorded through notes, photos, or videos; participatory or nonparticipatory.
- Physical measurement: biophysical measurements or geographical information.
- Existing documents and data: including existing datasets, official statistics, project records, and social media data.

As we can see, these methods include both the quantitative and qualitative approaches. The weakness of the qualitative approach is the lack of counterfactual analysis.

Causal comparative designs are some of the most common quantitative methods used in agriculture. They are nonexperimental (observational) and work by comparing equal groups. The estimated impact is the average difference in outcomes between matched individuals. This method is based on the assumption that there is no difference in the treatment and the comparison group, and it is also known as Causal Comparative.

Students are likely to use causal comparative designs, where the design will compare the difference between participants of an intervention being evaluated with a nonparticipant after the intervention is completed (see box for examples of such studies).

Examples of Studies that Used Causal Comparative Designs to Assess Impact

1. Kwezirimana S, Mugwe JN and Bekele N (2023). Impact of collective marketing participation on farmers' income: Evidence from smallholder avocado farmers of Murang'a County, Kenya. Social Sciences & Humanities Open: 100614.

Qualitative Methods for Evaluating Impact

Suppose you have the following research questions:

1. How do farmers perceive the impact of climate change on their agricultural practices and decision-making processes?
2. What cultural factors influence farmers' choice of crop varieties and agricultural practices in a particular ethnic group?
3. What role do gender dynamics play in shaping decision-making in coffee production within farming households?
4. What are the perceived benefits and challenges of integrating precision agriculture among smallholder farmers?
5. How do agricultural extension service services contribute to knowledge transfer and technology adoption among smallholder farmers?

The aforementioned questions indicate that the data gathered should provide comprehensive insights into farmers' perceptions, experiences, and contextual factors influencing their agricultural practices. By utilizing qualitative research methods such as in-depth interviews, focus group discussions, and participant observation, we can better understand how farmers perceive and respond to various factors within their agricultural surroundings. For instance, for questions 1, 2, and 3, we can employ the following methods:

Question 1 on climate change impact	Qualitative methods will allow farmers to express their subjective experiences, perspectives, concerns, and adaptations to the changing climate. These methods will help show farmers' strategies in response to the evolving environmental conditions, providing a comprehensive view of how farmers perceive climate change
Question 2 on cultural factors influencing crop choices	Qualitative methods will provide an exploration of the deeper socio-cultural context. Ethnographic methods employing both participant and nonparticipant observations can be used to unveil the cultural norms, traditions, and community dynamics that shape farmers' decisions regarding crop varieties and agricultural practices
Question 3 on gender dynamics in coffee production	Qualitative methods will capture gender roles, power relations, and decision-making processes within farming households in coffee production. Interviews and participant observations will help shed light on how gender influences various aspects of coffee production

Qualitative impact evaluation is crucial to comprehensively understanding the circumstances surrounding an intervention. It considers social, cultural, and environmental factors that influence the implementation and outcomes of an intervention. This provides a depth of insight that cannot be achieved through quantitative measures alone. It is essential to unravel the "how" and "why" behind observed impacts, which enriches the overall understanding of the causal pathways.

Second, qualitative evaluation aims to understand the *underlying processes and mechanisms* that lead to observed impacts. It goes beyond measuring outcomes and investigates the how and why of specific changes. This detailed exploration of causal pathways is essential for program managers, policymakers, and researchers who want to comprehend what works and the dynamics that drive success or failure. Additionally, qualitative methods are excellent at uncovering unintended consequences, revealing unexpected outcomes that may not be evident through quantitative measures alone.

Third, qualitative impact evaluation makes a significant contribution to *theory-building* by providing valuable insights into the underlying theories of change. This facilitates a better understanding of how interventions produce specific impacts, thereby aiding in developing or modifying program theories. Qualitative methods are particularly useful when interventions are complex and multifaceted. They are adaptable and enable evaluators to capture the intricacies of social processes, relationships, and contextual factors, thereby providing a more complete picture of the intervention's impact.

Fourth, qualitative evaluation emphasizes the *human experience and incorporates participants' perspectives*. It ensures that the voices of those directly affected by the intervention are heard by capturing their narratives, experiences, and perceptions. Qualitative methods provide a more human-centered understanding of impact, enriching the evaluation process and promoting a sense of inclusivity and empowerment among those involved in the intervention.

Lastly, qualitative assessment offers *practical policy and program improvement implications* beyond its analytical benefits. Understanding the complexities of program implementation and participant experiences enables organizations to make informed adjustments to enhance overall effectiveness.

The University of Bath's Centre for Development Studies (CDS) has developed an innovative approach to evaluating project impact, known as the Qualitative Impact Assessment Protocol (QUIP). Unlike traditional methods that rely solely on quantitative measures and control groups, QUIP focuses on capturing the qualitative aspects of impact via narratives and experiences of beneficiaries see details in Box 11.9.

Box 11.9: Quip Qualitative Approach

QuIP is a method that aims to systematically collect evidence of a project's impact by directly asking the intended project beneficiaries about the notable changes in their lives within a specific recall period. Respondents are asked to reflect on these changes and explain why they happened and what factors contributed to them. This approach enables a comprehensive understanding of the perceived impact, allowing beneficiaries to express their experiences and attributions in a nuanced and contextually rich manner.

Unlike statistical inference, which relies on varying exposure to X, QuIP seeks evidence of attribution through respondents' firsthand accounts of the causal mechanisms linking X to Y alongside Z. This method is known as key informant attribution, and it does not require the inclusion of a control group. The narrative data obtained through QuIP aims to complement quantitative evidence on changes in X, Y, and Z, which are often acquired through routine project monitoring or alternative methods.

A typical QuIP study involves 24 semistructured interviews and four focus groups, facilitated in the native language by proficient local researchers. It's worth noting that the specific number of interviews and groups may vary based on the chosen sampling approach. The research team conducting the interviews remains independent and, where appropriate, blindfolded to ensure impartiality and integrity in the evaluation process (Remnant et al. 2024)..

11.5 Mixed-Method Design in Impact Evaluation

We find that both quantitative and qualitative approaches used to evaluate impact have their respective strengths and weaknesses (Bamberger et al. 2010). The RCTs, quasi-experiments, preexperimental designs, and qualitative designs have advantages and disadvantages, especially with the validity of the data collected. Therefore, when used singly, the designs fail to fully capture all of the complexities of how programs operate in the real world.

Indeed, RCTs have been criticized due to strict control over variables that can lead to a lack of generalization in the real world. They have also been found to exaggerate the claims of causality by oversimplifying complex interactions and ignoring contextual factors that may influence outcomes outside of the controlled environment. A study by Bédécarrats et al. (2018) reported that RCTs exaggerated the results of causal relationships, leading to potentially misleading conclusions about the effectiveness of interventions. In another study, RCTs failed to meet the criteria for establishing causality (Schweizer et al. 2016). These authors suggest that the explanatory power of the RCTs could be significantly enhanced by incorporating qualitative information, which would provide in-depth insights into interventions being trialed to actually work in particular contexts.

Furthermore, researchers evaluating development projects come up with multiple questions: questions that require a deductive approach (hypothesis testing) and questions that require an inductive approach (hypothesis generating). For example, we could have a question requiring deductive approaches: "Was there a significant difference between male and female-headed households in the adoption of water harvesting techniques?" and 2. "What are farmers' perceptions toward the use of water harvesting techniques?" Answering these questions requires a mixed-methods approach. To address this weakness, there is growing interest in mixed-method evaluation (Bamberger et al. 2010). Box 11.10 shows the benefits of mixed-method designs in impact evaluation studies.

Box 11.10: Benefits of Mixed Methods Designs in Impact Evaluation

The benefits of using mixed methods are similar to those presented in the section on mixed methods design (Chap. 15 of this book), but specifically, mixed methods designs help to strengthen an evaluation:

- Mixed methods are useful when different evaluation questions require different methods, or when a single evaluation question requires more than one method to answer all components.
- Mixed methods are useful when different methods are used to answer the same elements of a single question, increasing confidence in the validity and reliability of the evaluation results.
- Mixed methods are used when the results from one method are used to help design future evaluation phases using other methods.
- Mixed methods are more likely to reveal unanticipated results.
- Mixed methods can provide a deeper understanding of why change is or is not occurring as planned.
- Mixed methods can often capture a wider range of perspectives than might be captured by a single method.

According to Bamberger et al. (2010), quantitative evaluation risks becoming decontextualized, ignoring how programs are affected by the economic, political, institutional, and socio-cultural characteristics of the populations studied. Another frequent criticism of many quantitative evaluations is that they assume that programs operate as planned and that everyone receives the same package of services (both in terms of quantity and quality). This is often referred to as the "black box" approach, as the evaluation does not look inside the project's "black box."

On the other hand, qualitative methods are also powerful tools for data collection and analysis. However, when used alone, qualitative evaluation designs have a number of potential weaknesses. They often focus on individual subjects and situations, and it is more difficult to generalize from the findings.

A mixed-method evaluation systematically integrates two or more evaluation methods, potentially at every stage of the evaluation process, usually drawing on quantitative and qualitative data (Biesta 2015; Acquah et al. 2021). Mixed-method evaluations may use multiple designs, for example, incorporating both randomized control trial experiments and case studies. They also may include different data collection techniques such as structured observations, key informant interviews, household surveys, and reviews of existing secondary data. In short, a mixed-method evaluation involves systematically integrating different kinds of data, usually drawn from different designs.

The research questions will inform the selection of the mixed-method design (discussed in the next section).

11.6 Choosing an Appropriate Design for the Evaluation of Impact

This chapter discussed numerous methods and designs for evaluating impact. The challenge lies in choosing the most appropriate method suited to a given impact and context. We would want to conduct rigorous impact evaluations that demonstrate evidence. Evidence is about demonstrating impact based on a robust evaluation (emphasized at the beginning of this chapter). For example, donors such as the World Food Programme would want to use rigorous impact evaluation evidence to inform policy and program decisions to optimize interventions and provide thought leadership to global efforts to end hunger and achieve SDGs (WFP 2024).

Similarly, as a researcher, you want to answer an impact evaluation with validity and evidence. Choosing a suitable design depends on the research question. According to Rodgers et al., there are three types of evaluation questions: Descriptive, Causal, and Evaluative questions as briefly discussed here:

1. *Descriptive questions:* ask about how things are and what has happened. They describe the initial situation and how it has changed, the intervention and other related programs or policies, the context (participant characteristics), and the implementation environment.
2. *Causal questions:* ask whether or not, and to what extent, observed changes are due to the intervention in question. Increasingly, causal analysis is about understanding how an intervention contributes to impacts, other factors, and other programs.
3. *Evaluative questions:* ask about the overall value of a program or policies, taking into account intended and unintended impacts, criteria and standards and how performance across different domains should be weighted and synthesized. In addition, impact evaluations that include recommendations need to answer action questions by identifying and assessing possible options for responding to findings (Table 11.3).

Apart from the research question/objective, there are other factors to consider when selecting a research design (summarized in Box 11.11). They include the following:

1. *Available resources:* Assess the resources available. Some designs, such as RCTs, can be resource-intensive, requiring significant financial, human, and logistical investments. In some cases, nonexperimental designs might be more feasible, as they may use existing data or a smaller sample size, and hence are an option when resources are limited.
2. *Ethical considerations:* Consider ethical concerns that may limit your ability to randomly assign individuals or groups to different treatment conditions. Ethical constraints can affect the choice of the design. For instance, in cases where withholding a potentially beneficial treatment from a control group could cause harm or discomfort, it may be ethically challenging to implement an RCT.
3. *Time frame:* Different research designs may have varying requirements regarding data collection and analysis. Some designs, such as longitudinal studies, have the capacity to capture the impact of an intervention for an extended period, but they require a longer time. If a more immediate assessment is required, other designs, such as cross-sectional or pre–post evaluation, may be considered.

11.7 Threats to Internal Validity

Several threats to internal validity exist and may affect researchers' ability to conclude that the outcome measured is a result of the intervention and not some other factor. There are those that involve participants, those related to the use of

Table 11.3 Different types of evaluation questions and relevant designs and methods

Intended use	Typical evaluation question	Conditions	Relevant methods and design
Attribution	Did the intervention cause the impact(s)?	Requires a single cause and a small number of effects. Needs either a homogeneous effect (it works the same for everyone) or knowledge about the contextual factors that influence impacts	RCTs, regression discontinuity, propensity scores
Apportioning	To what extent can a specific impact be attributed to the intervention?	Requires a single effect, large datasets on relevant contributing factors	Regression, econometrics, structural equation modeling
Contribution	Did the intervention make a difference?	Requires an understanding of the different configurations that could produce the results (which can include contextual factors, variations of the program, and other programs)	Contribution analysis, comparative case studies, process tracing, Bradford Hill criteria
Explanation	How has the intervention made a difference?	Requires the development of a clear program theory which sets out a change theory (how change is understood to come about) and an action theory (what activities will be undertaken to trigger this). This can be informed by exploring how actors in the intervention attribute cause and investigate these for plausibility, as well as drawing on research literature and theoretical frameworks	Actor attribution, theory-based evaluation, realist evaluation, process tracing

Adapted from Rogers (2014) and Rogers (2012)

Box 11.11: Key Considerations in Choice of Research Design for Evaluating Impact

1. Research objectives, questions and hypothesis
2. Available resources
3. Ethical considerations
4. Time frame

Tip: Balance is key; ensure you strike a balance between the four criteria ensuring that the research design selected will give a robust and meaningful impact evaluation that is aligned with the research objectives. During the iterative process of designing, you may be required to rephrase your research objectives.

an experimental treatment that the researcher manipulates, and those involving procedures used in the experiment (Creswell 2014). We will use the example of evaluating the effect of training farmers on knowledge levels.

Threats related to participants include the following:

1. *History:* Events outside of the treatment but occurring during the experiment may affect the dependent variable. History becomes a greater threat as the study's duration increases.
2. *Maturation*: This refers to biological or psychological processes within participants that may change due to the passing of time, e.g., ageing, or more experience in a research study. These changes may affect participants' performance on a measure of a dependent variable. This can be addressed by selecting individuals who are at the same level for both the control and the experimental groups.
3. *Testing:* The effects of one test upon subsequent administrations of the same test. In intervention studies, where data are collected over a period of time, it is common to test subjects at the beginning of the intervention. This pretest is likely to affect participants' responses.
4. *Instrumentation:* Threat due to instrumentation refers to unreliability or lack of consistency in measuring instruments that may result in invalid assessment of performance. For example, changes in testing instruments, using two different tests during pretesting and posttesting. Another example is interviewers, including a lack of agreement within and between observers.
5. *Statistical regression:* The fact that groups selected on the basis of extreme scores are not as extreme on subsequent testing.
6. *Selection bias:* Arising from the identification of comparison groups in other than a random manner, for example, participants are assigned based on preexisting characteristics or preferences, it can create group disparities that existed before any intervention took place.
7. *Experimental mortality:* Loss of participants from comparison groups due to nonrandom reasons, for example, farmers dropping out of the study. This leads to an uneven distribution of characteristics among the remaining participants, leading to an increased risk of bias in study outcomes.

8. *Interaction among factors*: Factors operating together with the intervention may make it difficult to determine the true cause of changes in the independent variable.
9. *Diffusion of treatment:* Communication between participants in control and experimental groups may influence perception and behavior, blurring the distinction between the two groups. Rural farmers often meet in social places, and the risk of contamination is quite common.
10. *Resentful demoralization*. If subjects learn that their group receives fewer desirable goods or services, they may experience feelings of resentment and demoralization. This perception of inequality or disadvantage within the group can have psychological implications, potentially leading to decreased morale to participate in the study.

Researchers should identify these threats to internal validity so that these threats are minimized to ensure that observed effects can be confidently attributed to the intervention rather than the confounding variables.

Recommended Further Reading on Impact Evaluation and the Use of Propensity Score Matching:

1. Handley MA, Lyles CR, McCulloch C, Cattamanchi A (2018) Selecting and Improving Quasi-Experimental Designs in Effectiveness and Implementation Research. Annu Rev. Public Health 39:5–25. https://doi.org/10.1146/ANNUREV-PUBLHEALTH-040617-014128/CITE/REFWORKS
2. Leroy JL, Frongillo EA, Kase BE, et al. (2022) Strengthening causal inference from randomised controlled trials of complex interventions. BMJ Glob Health 7:8597. https://doi.org/10.1136/BMJGH-2022-008597
3. Gertler PJ, Martinez S, Premand P, et al. (2016) Impact Evaluation in Practice, Second Edition. Impact Evaluation in Practice, Second Edition. https://doi.org/10.1596/978-1-4648-0779-4
4. Austin PC (2011) An Introduction to Propensity Score Methods for Reducing the Effects of Confounding in Observational Studies. Multivariate Behav Res 46:399–424. https://doi.org/10.1080/00273171.2011.568786
5. Bamberger M, Rao V, Woolcock M (2010) Using Mixed Methods in Monitoring and Evaluation: Experiences from International Development. https://doi.org/10.1596/1813-9450-5245

Acknowledgments
The authors appreciate the following for allowing the reuse of their resources through the Creative Commons model:
1. AIMS Press, publication by Adebayo et al. 2018.
2. Taylor and Francis, publication by Adego et al. 2019.
3. Springer open, Publication by Setsoafia et al. (2022).
4. Kenyatta University, Thesis by Samuel Kwezirimana, supervised by Nigati Bekele and Jayne Mugwe.
5. CABI Africa, publication by Bett et al. (2018).

Acknowledgment of Stata Software
STATA software was used to generate the figures in the section discussing propensity score matching. This includes common support and distribution graphs.

Key Points in This Chapter

Research Objectives: Measuring Impact

Students proposing research objectives that evaluate impact must understand that such studies require appropriate design and methods to attribute observed outcomes (dependent variable) to the intervention (Independent variable).

What Is the Evaluation of Impact?

Evaluation of impact is a method to examine whether detected changes in outcomes among project target groups can be attributed to the particular development program. The hallmark of impact evaluation is causality, attribution, and the extent of change. Interest in impact evaluation has increased, driven by research funders who increasingly seek evidence that their investments yield positive public outcomes and benefits.

The Concept of Counterfactual

To establish a causal link between impacts (outcome) and intervention, which could be due to a program, policy change or activity, requires a counterfactual. A counterfactual is a comparison or control group to measure the change in both the treated and the nontreated (control). A counterfactual allows us to answer the question "What could have happened without the intervention"? The evaluation aims to eliminate alternative possibilities or explanations for the observed effect.

Every assessment of impact is essentially a research activity. Researchers must, therefore, systematically carry out the evaluation using proper methodological approaches to generate evidence-based results.

Theory of Change in Impact Evaluation

A "theory of change" refers to a conceptual map showing how activities implemented in a project or program will lead to change (the impact). It explains the process of change by outlining causal linkages in a project, showing each outcome in a logical relationship. A results chain is used to show this relationship. It includes inputs, processes, outputs, outcomes, and long-term impact.

Basic Designs for Impact Evaluation

The designs for evaluating impact fall into three categories: true experiments, quasi-experiments (mid-experiments), and nonexperimental methods.

True experiments, also known as Randomized Controlled Trials (RCTs), follow a structured approach similar to biophysical experiments in agriculture. RCTS have manipulation, randomization, and control, and therefore are considered the gold standard for establishing cause-and-effect relationships in impact evaluation because they systematically control for extraneous variables and minimize bias. Two main types include the randomized posttest control group design and the randomized pretest–posttest control group design.

Quasi-Experimental Designs also referred to as mid-experiments, do not involve the random assignment of participants to treatment and no control groups. Although not as rigorous as RCTs in establishing causality, quasi-experiments have gained widespread acceptance and have been extensively utilized. Indeed, the use of true experiments in impact evaluation is rare with most studies adopting quasi-experiments because they offer many advantages.

There are two main types of quasi-experiments; "Nonequivalent control group design and "time-series designs." We have two variations of the "nonequivalent control group design": 1. Pre- and posttest design, and 2. posttest only design. In the pre- and posttest design, the researcher allocates intact groups to experimental and control treatments and administers a pretest to both groups.

Methods for Creating Comparison Groups in Quasi-Experiments

The predominant method for selecting a comparison group is through matching. The primary methods employed for creating matching comparison groups include: 1. Propensity Score Matching (PSM), 2. Difference-in-Difference, 3. Regression Discontinuity, and 4. Instrumental Variables.

Propensity Score Matching

A technique that involves calculating the propensity scores. The individuals are not matched on every observable characteristic but on their propensity score, indicating the predicted likelihood of participating in the intervention given their observable characteristics.

Advantages of PSM: Propensity Score Matching (PSM) is conceptually simple and accessible, making it popular among students and nontechnical users. PSM reduces selection bias by creating balanced treatment and control groups, simulating randomization. It offers flexible matching options, such as 1:1, nearest neighbor, or kernel matching, allowing researchers to adapt the method to their data. Additionally, PSM enhances transparency by enabling clear comparisons between groups and incorporating balance checks and sensitivity analyses.

Limitations of PSM: Matching often reduces sample size, lowering statistical power, and may lead to unrepresentative groups when suitable matches are limited. The method is also sensitive to how covariates are selected and specified in the model, requiring careful sensitivity analysis. Additionally, PSM is data-intensive, and alternative methods like stratification, inverse probability weighting, or covariate adjustment may better utilize the full dataset.

Difference-in-difference Method The difference-in-difference (DiD) method estimates impact by comparing outcome changes over time between participants and nonparticipants. It first calculates the before-after change for the treatment group, controlling for fixed factors. Then, it subtracts the same change observed in the control group to account for time-varying influences. This double differencing isolates the program's effect from other factors, providing a clearer estimate of impact.

Preexperimental Designs for Measuring Impact

One-group posttest only: This is also known as the after-only experimental design. The study group (single group) is first exposed to a treatment, event, or condition. Then, using specific indicators, the group is measured to see if there are any effects on a predetermined independent variable.

One-group pretest–posttest design: Compares an experimental group that receives treatment with a control group that does not, with measurements taken both before (baseline) and after the intervention. Nonrandom assignment may introduce bias.

Static control group: A single, unchanging control group is used for comparison with an experimental group across different points over time. Limit the ability to account for external factors.

Nonexperimental Designs They rely on answering descriptive questions. These questions ask how things are, what happened, and how it happened. This may include describing the initial situation and how it has changed, the activities of the intervention and other related programs or policies, the context in terms of participant characteristics, and the implementation environment.

Qualitative Approaches in Evaluating Impact

Qualitative impact evaluation is crucial for a comprehensive assessment of an intervention, as it explores the social, cultural, and environmental factors that influence its implementation and outcomes. Unlike the quantitative approaches, it evaluates the "how" and "why" behind observed impacts, offering insights into underlying processes and mechanisms that drive success or failure. It goes beyond measuring outcomes and investigates the how and why of specific changes.

Mixed-Method Design in Impact Evaluation

The RCTs, quasi-experiments, preexperimental designs, and qualitative designs all have limitations. Therefore, used singly, the designs fail to fully capture the complexities of how programs operate in the real world. To address this weakness, there is growing interest in mixed-method evaluation approaches.

Review Questions

1. Provide a research objective that requires the application of impact evaluation methodologies.
2. Explain the concept of counterfactual in your own words.
3. Suppose your budget is minimal, and you can only conduct a fundamental evaluation of a new irrigation technology. You decide to use a "one-group posttest only" design;
   (a) Why is this considered a preexperimental design?
   (b) What are the significant limitations of this design in demonstrating actual impact?
   (c) Can you think of any scenarios where such a basic design might still provide some helpful information, even if it does not have a definitive impact?
4. What is the role of qualitative designs in impact evaluation?
5. Explain the role of mixed methods in impact evaluation. Provide an example of how combining quantitative data (e.g., yield increases) with qualitative data (e.g., farmer perceptions, challenges faced) could provide a more comprehensive understanding of the training program's impact.

Chapter Exercise

1. Imagine a maize seed distribution project with farmers has the following "Theory of Change steps":
   (i) Inputs: Funds, drought-resistant seeds, training materials, and extension workers.
   (ii) Activities: Procure seeds, train farmers on new varieties, distribute seeds.
   (iii) Outputs: X farmers trained, Y kg of seeds distributed.
   (iv) Outcomes: Increased yield per acre, reduced crop failure in dry spells, and improved household food security.
   (v) Impact: Reduced poverty improved livelihoods for farmers.
      (a) Based on this, formulate two research questions that an impact evaluation would seek to answer at the "outcome" level, using specific action verbs.
      (b) For one of these research questions, propose a hypothesis.
2. Suppose you intend to carry out a study to evaluate the effect of a community-based milk cooler on milk production and farmers' income.
   (a) Describe how you would implement a "randomized posttest-only control group" design for this evaluation. What are the advantages and disadvantages of this specific design compared to a "randomized pretest–posttest control group" design in this context?
   (b) List three practical steps you would take to ensure proper randomization when selecting farmers for your treatment and control groups.
3. Suppose you have identified two farmer groups for your quasi-experimental study. To strengthen your comparison group, consider using Propensity Score Matching (PSM) or Difference-in-Difference (DiD).
   (a) Briefly explain the core idea behind Propensity Score Matching. What kind of data would you need to collect from your farmer groups to perform PSM effectively?
   (b) Briefly explain the core idea behind the Difference-in-Difference technique. What kind of data is *essential* for applying DiD, and why is it particularly useful for showing impact over time?

Answers to Review Questions

- Q1. Reference: 11.1.1
- Q2. Reference: 11.1.3
- Q3. Reference: 11.3.3
- Q4. Reference: 11.4
- Q5. Reference: 11.5

References

Acquah ISK, Naude MJ, Soni S (2021) How the dimensions of culture influence supply chain collaboration: an explanatory sequential mixed-methods investigation. Revista de Gestao 28:241–262. https://doi.org/10.1108/REGE-11-2020-0105/FULL/PDF

Adebayo O, Bolarin O, Oyewale A et al (2018) Impact of irrigation technology use on crop yield, crop income and household food security in Nigeria: a treatment effect approach. AIMS Agric Food 3:154–171. https://doi.org/10.3934/AGRFOOD.2018.2.154

Adego T, Simane B, Woldie GA (2019) The impact of adaptation practices on crop productivity in Northwest Ethiopia: an endogenous switching estimation. Dev Stud Res 6:129–141. https://doi.org/10.1080/21665095.2019.1678186

Alessandri G, Zuffianò A, Perinelli E (2017) Evaluating intervention programs with a pretest-posttest design: a structural equation modeling approach. Front Psychol 8:243862. https://doi.org/10.3389/FPSYG.2017.00223/BIBTEX

Ali MS, Prieto-Alhambra D, Lopes LC et al (2019) Propensity score methods in health technology assessment: principles, extended applications, and recent advances. Front Pharmacol 10:466792. https://doi.org/10.3389/FPHAR.2019.00973/XML/NLM

Ariel B, Sutherland A, Bland M (2021) The trick does not work if you have already seen the gorilla: how anticipatory effects contaminate pre-treatment measures in field experiments. J Exp Criminol 17:55–66. https://doi.org/10.1007/S11292-019-09399-6/METRICS

Aromatario O, Van Hoye A, Vuillemin A et al (2019) Using theory of change to develop an intervention theory for designing and evaluating behavior change SDApps for healthy eating and physical exercise: the OCAPREV theory. BMC Public Health 19:1–12. https://doi.org/10.1186/S12889-019-7828-4/FIGURES/4

Arora C, Malhotra A, Ranjan P et al (2021) Designing and conducting randomized controlled trials: basic concepts for educating early researchers in the field of clinical nutrition. Cureus 13. https://doi.org/10.7759/CUREUS.17036

Austin PC (2011) An introduction to propensity score methods for reducing the effects of confounding in observational studies. Multivariate Behav Res 46:399–424. https://doi.org/10.1080/00273171.2011.568786

Bamberger M, Rao V, Woolcock M (2010) Using mixed methods in monitoring and evaluation: experiences from international development. https://doi.org/10.1596/1813-9450-5245

Bärnighausen T, Tugwell P, Røttingen JA et al (2017) Quasi-experimental study designs series—paper 4: uses and value. J Clin Epidemiol 89:21–29. https://doi.org/10.1016/J.JCLINEPI.2017.03.012

Bédécarrats F, Guérin I, Roubaud F (2018) The gold standard for randomised evaluations: from discussion of method to political economy. HAL Open Science

Belcher B, Palenberg M (2018) Outcomes and impacts of development interventions: toward conceptual clarity. Am J Eval 39:478–495. https://doi.org/10.1177/1098214018765698/SUPPL_FILE/AJE765698_SUPPLEMENTAL_MATERIAL.PDF

Bett E, Mugwe J, Nyalugwe N et al (2018) Impact of plant clinics on disease and pest management. In: Tomato productivity and profitability in Malawi

Bezner Kerr R, Berti PR, Shumba L (2011) Effects of a participatory agriculture and nutrition education project on child growth in northern Malawi. Public Health Nutr 14:1466–1472. https://doi.org/10.1017/S1368980010002545

Bhide A, Shah PS, Acharya G (2018) A simplified guide to randomized controlled trials. Acta Obstet Gynecol Scand 97:380–387. https://doi.org/10.1111/AOGS.13309

Biesta G (2015) Pragmatism and the philosophical foundations of mixed methods research1. In: SAGE handbook of mixed methods in social & behavioral research, pp 95–118. https://doi.org/10.4135/9781506335193.N4

Blome C, Augustin M (2015) Measuring change in quality of life: bias in prospective and retrospective evaluation. Value Health 18:110–115. https://doi.org/10.1016/j.jval.2014.10.007

Brainard J, Wilsher SH, Salter C, Loke YK (2016) Methodological review: quality of randomized controlled trials in health literacy. BMC Health Serv Res 16:1–10. https://doi.org/10.1186/S12913-016-1479-2/TABLES/4

Breuer E, Lee L, De Silva M, Lund C (2016) Using theory of change to design and evaluate public health interventions: a systematic review. Implement Sci 11:1–17. https://doi.org/10.1186/S13012-016-0422-6/TABLES/5

Busetti S (2023) Causality is good for practice: policy design and reverse engineering. Policy Sci 56:419–438. https://doi.org/10.1007/S11077-023-09493-7/FIGURES/2

Caliendo M, Kopeinig S (2008) Some practical guidance for the implementation of propensity score matching. J Econ Surv 22:31–72. https://doi.org/10.1111/J.1467-6419.2007.00527.X

Cook TD (2015) Quasi-experimental design. In: Wiley encyclopedia of management, pp 1–2. https://doi.org/10.1002/9781118785317.WEOM110227

Creswell JW (2014) Research and design qualitative, quantitative, and mixed methods approaches. SAGE Publication Inc

Culbertson MJ, McCole D, McNamara PE (2014) Practical challenges and strategies for randomised control trials in agricultural extension and other development programmes. J Dev Eff 6:284–299. https://doi.org/10.1080/19439342.2014.919339

Deaton A, Cartwright N (2018) Understanding and misunderstanding randomized controlled trials. Soc Sci Med 210:2–21. https://doi.org/10.1016/J.SOCSCIMED.2017.12.005

Donald SG, Lang K (2007) Inference with difference-in-differences and other panel data. Rev Econ Stat 89:221–233. https://doi.org/10.1162/REST.89.2.221

Draper CE, Kolbe-Alexander TL, Lambert EV (2009) A retrospective evaluation of a community-based physical activity health promotion program. J Phys Act Health 6:578–588. https://doi.org/10.1123/JPAH.6.5.578

Fabregas R, Kremer M, Robinson J, Schilbach F (2017) Impact evaluation report 67 evaluating agricultural information dissemination in western Kenya

Ferraro PJ (2009) Counterfactual thinking and impact evaluation in environmental policy. New Dir Eval 2009:75–84. https://doi.org/10.1002/EV.297

Fitawek W, Hendriks SL (2021) Evaluating the Impact of large-scale agricultural investments on household food security using an endogenous switching regression model. Land 10:323. https://doi.org/10.3390/LAND10030323

Flynn R, Mrklas K, Campbell A et al (2021) Contextual factors and mechanisms that influence sustainability: a realist evaluation of two scaled, multi-component interventions. BMC Health Serv Res 21:1–17. https://doi.org/10.1186/S12913-021-07214-5/TABLES/7

Garrido MM, Kelley AS, Paris J et al (2014) Methods for constructing and assessing propensity scores. Health Serv Res 49:1701–1720. https://doi.org/10.1111/1475-6773.12182

Gertler PJ, Martinez S, Premand P et al (2016) Impact evaluation in practice, 2nd edn. World Bank. https://doi.org/10.1596/978-1-4648-0779-4

Gibson M, Sautmann A (2020) Introduction to randomized evaluations | The Abdul Latif Jameel Poverty Action Lab. https://www.povertyactionlab.org/resource/introduction-randomized-evaluations. Accessed 25 June 2025

Guest G, Namey EE, Gamble T et al (2015) Designing randomized controlled trials (RCTs). In: Public health research methods, pp 223–250. https://doi.org/10.4135/9781483398839

Handley MA, Lyles CR, McCulloch C, Cattamanchi A (2018) Selecting and improving quasi-experimental designs in effectiveness and implementation research. Annu Rev Public Health 39:5–25. https://doi.org/10.1146/ANNUREV-PUBLHEALTH-040617-014128/CITE/REFWORKS
Hariton E, Locascio JJ (2018) Randomised controlled trials – the gold standard for effectiveness research. BJOG 125:1716–1716. https://doi.org/10.1111/1471-0528.15199
Harris H, Horst SJ (2016) A brief guide to decisions at each step of the propensity score matching process. Pract Assess Res Eval 21. https://doi.org/10.7275/YQ7R-4820
Higgins JPT, Morgan RL, Rooney AA et al (2024) A tool to assess risk of bias in non-randomized follow-up studies of exposure effects (ROBINS-E). Environ Int 186:108602. https://doi.org/10.1016/J.ENVINT.2024.108602
Kamau CN, Kabuage LW, Bett EK (2018) Impact of improved indigenous chicken breeds on productivity. The case of smallholder farmers in Makueni and Kakamega counties, Kenya. Cogent Food Agric 4. https://doi.org/10.1080/23311932.2018.1477231
Kim HJ (2022) Applications of propensity score matching: a case series of articles published in annals of coloproctology. Ann Coloproctol 38:398–402. https://doi.org/10.3393/AC.2022.01060.0151
Kool H, Andersson JA, Giller KE (2020) Reproducibility and external validity of on-farm experimental research in Africa. Exp Agric 56:587–607. https://doi.org/10.1017/S0014479720000174
Krishnan P (2019) A review of the non-equivalent control group post-test-only design. Nurse Res 26:37–40. https://doi.org/10.7748/NR.2018.E1582
Kumar SM (2016) RCTs for better policy? The case of public systems in developing countries. Econ Politica 33:83–98. https://doi.org/10.1007/S40888-016-0027-1/METRICS
Kwizerimana S (2023) Impact of collective marketing participation on smallholder avocado farmers' Income in Murang'a County, Kenya
Kwizerimana S, Mugwe J, Nigat B (2023) Impact of collective marketing participation on farmers' income: evidence from smallholder avocado farmers of Murang'a County, Kenya. Soc Sci Hum Open 8:100614. https://doi.org/10.1016/J.SSAHO.2023.100614
Leroy JL, Frongillo EA, Kase BE et al (2022) Strengthening causal inference from randomised controlled trials of complex interventions. BMJ Glob Health 7:8597. https://doi.org/10.1136/BMJGH-2022-008597
Maini R, Mounier-Jack S, Borghi J (2018) How to and how not to develop a theory of change to evaluate a complex intervention: reflections on an experience in the Democratic Republic of Congo. BMJ Glob Health 3. https://doi.org/10.1136/BMJGH-2017-000617
Mayne J, Johnson N (2015) Using theories of change in the CGIAR research program on agriculture for nutrition and health. Evaluation 21:407–428. https://doi.org/10.1177/1356389015605198/ASSET/09106DAF-53E2-4775-B03D-8F30404CE8B0/ASSETS/IMAGES/LARGE/10.1177_1356389015605198-FIG6.JPG
McLellan T (2021) Impact, theory of change, and the horizons of scientific practice. Soc Stud Sci 51:100–120. https://doi.org/10.1177/0306312720950830/ASSET/923C10FA-EDC3-4FEB-8F91-894AFF577A8B/ASSETS/IMAGES/LARGE/10.1177_0306312720950830-FIG2.JPG
Mears J, Abubakar I, Crisp D et al (2014) Prospective evaluation of a complex public health intervention: lessons from an initial and follow-up cross-sectional survey of the tuberculosis strain typing service in England. BMC Public Health 14:1–13. https://doi.org/10.1186/1471-2458-14-1023/TABLES/5
Michel J, Schneider K (2025) Demystifying impact evaluation: an impact evaluation framework. Front Epidemiol 5:1460997. https://doi.org/10.3389/FEPID.2025.1460997/BIBTEX
Mohr LB (1982) On rescuing the nonequivalent-control-group design. Sociol Methods Res 11:53–80. https://doi.org/10.1177/004912418201100100З
Morgan GA, Gliner JA, Harmon RJ (2021) Selection of inferential statistics: an overview. Psychology Press
O'Neill S, Kreif N, Grieve R et al (2016) Estimating causal effects: considering three alternatives to difference-in-differences estimation. Health Serv Outcome Res Methodol 16:1–21. https://doi.org/10.1007/S10742-016-0146-8/TABLES/3
Olier I, Zhan Y, Liang X, Volovici V (2023) Causal inference and observational data. BMC Med Res Methodol 23:1–3. https://doi.org/10.1186/S12874-023-02058-5/PEER-REVIEW
Omore A, Kidoido M, Twine E et al (2019) Using "theory of change" to improve agricultural research: recent experience from Tanzania. Dev Pract 29:898–911. https://doi.org/10.1080/09614524.2019.1641182
Ouya FO, Murage AW, Pittchar JO et al (2023) Impacts of climate-resilient push–pull technology on farmers' income in selected counties in Kenya and Tanzania: propensity score matching approach. Agric Food Secur 12:1–14. https://doi.org/10.1186/S40066-023-00418-4/TABLES/6
Planitz B, Sanderson P, Freeman C et al (2012) Testing the prospective evaluation of a new healthcare system. AMIA Ann Symp Proc 2012:734
Pratt CC, Mcguigan WM, Katzev AR (2000) Measuring program outcomes: using retrospective pretest methodology. Am J Eval 21:341–349. https://doi.org/10.1177/109821400002100305
Pynegar EL, Gibbons JM, Asquith NM, Jones JPG (2021) What role should randomized control trials play in providing the evidence base for conservation? Oryx 55:235–244. https://doi.org/10.1017/S0030605319000188
Quisumbing AR, Ahmed A, Gilligan D et al (2020) Randomized controlled trials of multi-sectoral programs: lessons from development research. World Dev 127:104822. https://doi.org/10.1016/J.WORLDDEV.2019.104822
Reed MS, Ferré M, Martin-Ortega J et al (2021) Evaluating impact from research: a methodological framework. Res Policy 50:104147. https://doi.org/10.1016/J.RESPOL.2020.104147
Reinholz DL, Andrews TC (2020) Change theory and theory of change: what's the difference anyway? Int J STEM Educ 7:1–12. https://doi.org/10.1186/S40594-020-0202-3/TABLES/2
Remnant F, Copestake J, Avard R, et al (2024) Qualitative impact protocol | Better Evaluation
Rogers PJ (2012) Introduction to impact evaluation. InterAction 1
Rogers P (2014) Theory of change. UNICEF Office of Research
Rogers PJ, Peersman G (2014) Developing a research agenda for impact evaluation in development. IDS Bull 45:85–99. https://doi.org/10.1111/1759-5436.12115
Rogers P, Hawkins A, McDonald B, Macfarlan A, Milne C (2015) Choosing appropriate designs and methods for impact evaluation. Canberra: Office of the Chief Economist, Department of Industry and Science, Australian Government
Romão DMM, Setti C, Arruda LHM et al (2023) Integration of evidence into theory of change frameworks in the healthcare sector: a rapid systematic review. PLoS One 18:e0282808. https://doi.org/10.1371/JOURNAL.PONE.0282808
Rosenbaum PR, Rubin DB (1984) Reducing bias in observational studies using subclassification on the propensity score. J Am Stat Assoc 79:516–524. https://doi.org/10.1080/01621459.1984.10478078
Rubin DB, Thomas N (1996) Matching using estimated propensity scores: relating theory to practice. Biometrics 52:249. https://doi.org/10.2307/2533160
Sainani KL (2012) Dealing with non-normal data. PM&R 4:1001–1005. https://doi.org/10.1016/J.PMRJ.2012.10.013
Salm S, Cecon N, Jenniches I et al (2022) Conducting a prospective evaluation of the development of a complex psycho-oncological care programme (isPO) in Germany. BMC Health Serv Res 22:531. https://doi.org/10.1186/S12913-022-07951-1

Sargeant JM, O'Connor AM, Gardner IA et al (2010) The REFLECT statement: reporting guidelines for randomized controlled trials in livestock and food safety: explanation and elaboration†. Zoonoses Public Health 57:105–136. https://doi.org/10.1111/J.1863-2378.2009.01312.X

Schweizer ML, Braun BI, Milstone AM (2016) Research methods in healthcare epidemiology and antimicrobial stewardship—quasi-experimental designs. Infect Control Hosp Epidemiol 37:1135–1140. https://doi.org/10.1017/ICE.2016.117

Setsoafia ED, Ma W, Renwick A (2022) Effects of sustainable agricultural practices on farm income and food security in northern Ghana. Agric Food Econ 10:1–15. https://doi.org/10.1186/S40100-022-00216-9/TABLES/4

Shadish WR, Cook TD, Campbell DT (2002) Experimental and quasi-experimental designs for generalized causal inference. Houghton-Mifflin

Sharma Waddington H, Umezawa H, White H (2025) What can we learn from qualitative impact evaluations about the effectiveness of lobby and advocacy? A meta-evaluation of dutch aid programmes and assessment tool. Eval Rev. https://doi.org/10.1177/0193841X251314731/SUPPL_FILE/SJ-PDF-1-ERX-10.1177_0193841X251314731.PDF

Smith JR (2024) All that glitters is not gold: why randomized controlled trials are bronze standard (at best) in determining best practices for the treatment of posttraumatic stress disorder. Method Innov 17:3–5. https://doi.org/10.1177/20597991241227844

Stringer AP, Bell CE, Christley RM et al (2011) A cluster-randomised controlled trial to compare the effectiveness of different knowledge-transfer interventions for rural working equid users in Ethiopia. Prev Vet Med 100:90–99. https://doi.org/10.1016/J.PREVETMED.2011.02.001

Stukel DM, Friedman G (2016) Sampling guide for beneficiary-based surveys for select feed the future agricultural annual monitoring indicators. Food and Nutrition Technical Assistance Project

Sud A, Doukas K, Hodgson K et al (2021) A retrospective quantitative implementation evaluation of Safer Opioid Prescribing, a Canadian continuing education program. BMC Med Educ 21:1–12. https://doi.org/10.1186/S12909-021-02529-7/TABLES/6

Sullivan GM (2011) Getting off the "gold standard": randomized controlled trials and education research. J Grad Med Educ 3:285–289. https://doi.org/10.4300/JGME-D-11-00147.1

te Lintelo DJH, Munslow T, Pittore K, Lakshman R (2020) Process tracing the policy impact of 'Indicators'. Eur J Dev Res 32:1312–1337. https://doi.org/10.1057/S41287-019-00244-0/FIGURES/5

Tesfay MG (2020) Does fertilizer adoption enhance smallholders' commercialization? An endogenous switching regression model from northern Ethiopia. Agric Food Secur 9:1–18. https://doi.org/10.1186/S40066-020-0256-Y/TABLES/8

Thornton PK, Schuetz T, Förch W et al (2017) Responding to global change: a theory of change approach to making agricultural research for development outcome-based. Agric Syst 152:145–153. https://doi.org/10.1016/J.AGSY.2017.01.005

Wegmüller R, Musau K, Vergari L et al (2022) Effectiveness of an integrated agriculture, nutrition-specific, and nutrition-sensitive program on child growth in Western Kenya: a cluster-randomized controlled trial. Am J Clin Nutr 116:446–459. https://doi.org/10.1093/AJCN/NQAC098

WFP (2024) Annual report for the strategic advisory panel on impact evaluation at WFP

Wing C, Simon K, Bello-Gomez RA (2018) Designing difference in difference studies: best practices for public health policy research. Annu Rev Public Health 39:453–469. https://doi.org/10.1146/ANNUREV-PUBLHEALTH-040617-013507/CITE/REFWORKS

Wordofa MG, Hassen JY, Endris GS et al (2021) Adoption of improved agricultural technology and its impact on household income: a propensity score matching estimation in eastern Ethiopia. Agric Food Secur 10:1–12. https://doi.org/10.1186/S40066-020-00278-2/TABLES/6

Zeldow B, Hatfield LA (2021) Confounding and regression adjustment in difference-in-differences studies. Health Serv Res 56:932–941. https://doi.org/10.1111/1475-6773.13666

Zohar H, Simeone L, Morelli N et al (2023) Using theory of change to support participatory visual mapping in urban transformation projects. Des Manag J 18:32–47. https://doi.org/10.1111/DMJ.12086

12 Nonexperimental Designs 1: Surveys

Abstract

Survey design is one of the most commonly used in social studies. Researchers seek information from individuals, including their attitudes, opinions, and behaviors. Surveys are popular with students studying agricultural economics. Classified according to time, two main types of surveys are recognized: cross-sectional and longitudinal. The cross-sectional surveys have four variants: attitude, practice and challenge surveys, baseline and end-line surveys, comparative surveys, and community surveys. On the other hand, the longitudinal category has three variants, namely trend, cohort, and panel. Animal epidemiological research uses many variants of cross-sectional and longitudinal surveys. These include case–control studies, ecological surveys, and other descriptive designs in livestock health studies. Case–control, retrospective, and cohort-prospective studies are the most widely used in livestock health research. Surveys are administered through self-completed questionnaires or interview-completed questionnaires. Interviewer-completed surveys use face-to-face interviews, paper, telephonic, and internet surveys, while self-completed surveys use postal and online platforms. The surveys involve three main phases: planning, mid, and final. The planning phase draws the tools for data collection and logistics organization. The second phase involves piloting, while the third phase is data collection and reporting. This final phase will have participants recruited, data collected, analyzed, and reported.

Keywords

Animal epidemiological surveys · Baseline surveys · Causal comparative survey · Case reports · Case series reports · Cohort surveys · Comparative survey · Cross-sectional surveys · Ecological surveys · Face-to-face interviews · Internet or online surveys · Interviewer-completed survey · Longitudinal surveys · Panel study · Paper surveys · Piloting · Postal surveys · Prospective cohort study · Self-completed survey · Telephonic surveys

Learning Outcomes

By the end of this chapter, learner is expected to be able to:

1. Explain the meaning of survey design.
2. Recognize the purpose of surveys.
3. State the most commonly collected data from farms using surveys.
4. Determine when to use a cross-sectional survey.
5. Identify the variants in cross-sectional surveys.
6. Describe the components of a baseline survey.
7. Differentiate between a panel survey and a cohort survey.
8. Distinguish between experimental and nonexperimental designs in animal epidemiological research.
9. Appreciate the difference between retrospective surveys and prospective surveys in livestock.
10. Appreciate the advantages and disadvantages of surveys.
11. Demonstrate ways of administering a survey (self-completed and interviewer-completed survey).
12. Describe the key steps in conducting a survey.

12.1 Description of Survey Designs

12.1.1 What Is a Survey Research Design?

Survey research involves collecting data systematically about people and their characteristics using a set of predetermined questions or interviews given to them (Aday and Cornelius 2006; Fowler 2009). It is one of the most commonly used techniques for gathering data in social studies. The set of questions is captured in an instrument called a

J. Njeri Mugwe, S. Runo, *Research Methodology in Agricultural Sciences*, https://doi.org/10.1007/978-981-95-1892-0_12

questionnaire or interview guide (Chap. 17). The researcher administers a survey to a sample or the entire population, and it is useful in studies that involve description, explanation, and exploration (Babbie 2021).

Surveys differ from experimental designs in that they do not involve the researcher providing participants with a treatment. It is nonexperimental. Due to the absence of manipulation in surveys, researchers cannot explicitly establish cause-and-effect relationships, even though survey data are often used to suggest such connections (Creswell and Guetterman 2019). This is because although surveys can reveal associations between variables, they lack the control and randomization necessary to determine whether one variable causes a change in another. In almost every statistical course, the statement "correlation does not imply causation" is often repeated to emphasize this point (Zanga et al. 2022). The causal relationship described in surveys is presumed to be the cause (Van der Stede 2014).

In a survey, data collection encompasses any measurement that involves the following:

- Questioning people/respondents
- Observations at a given point in time
- Opinions, perceptions, opinions, attitudes, etc., by receiving answers from the respondents
- Use of a set of a predetermined set of questions that is given to the sample
- No deliberate changes to observe the effect
- No intervention on the part of the researcher
- An observation of what is going on at a particular point in time
- Data are collected from people who are thought to have the desired information

Both quantitative and qualitative research approaches may use a survey design or methodology for collecting data. This chapter focuses on a quantitative survey.

12.1.2 Purpose of Surveys

The main purposes of survey research in agriculture are to:

1. Determine existing conditions and describe conditions as they are.
2. Monitor changes over time (measure change), e.g., in impact evaluation.
3. Explain the situation (analytical surveys).
4. Explain the presumed cause-and-effect relationship between variables.
5. To study the characteristics of a phenomenon.
6. Evaluate programs and new projects (assess the effect of a project on key targets/variables).
7. To formulate and test a hypothesis.
8. To study individual opinions and feelings about policy issues.
9. Identify beliefs and attitudes of individuals about certain phenomena, e.g., attitudes toward introducing a new maize variety.
10. To assess attitudes, behaviors, opinions, and habits of stakeholders in agriculture.
11. To examine practices, i.e., actual behaviors.
12. Forecast future needs (predict).
13. To provide data for planning and decision-making.
14. To identify a research problem.

If you are studying agricultural economics, you are likely to use a survey design in your research project to gather data on various aspects of agricultural practices, economic behaviors, and market trends. Surveys in agricultural economics often aim to understand farmers' decision-making processes, assess economic and agricultural policies, analyze market behaviors, and evaluate the adoption of new technologies. You could check your university repository and find out the most common studies carried out by past students and the methodologies they used. Students in the biophysical field, such as those assessing agronomy practices with farmers, can also conduct surveys when their research objectives call for economic and social data from farmers.

12.1.3 Information Commonly Collected from Farmers Using Surveys

Most farmer surveys will collect the following information on social issues, demographics, practices, and behavior (Boxes 12.1 and 12.2):

1. *Facts (Biography)*: This is usually the background information about your respondents, mainly demographic characteristics, such as gender, education, marital status, and size of household.
2. *Current actions and practices*: Measure the action component of the respondent by asking the respondent how they perform tasks of interest. For example, which maize variety do you plant during the long rainy season? Sometimes, observation is a more reliable method, depending on what you are looking for. For instance, if you are interested in finding out the breed of cattle kept, observation will be a sure way of getting the answer.
3. *Knowledge*: Farmers possess unique experiences and insights that can contribute to shaping agricultural interventions. Additionally, information about knowledge possessed by farmers helps in identifying barriers and gaps that are useful in designing technologies and practices, as well as capacity-building programs.

Box 12.1 Set of demographic and socio-economic factors commonly collected from farmers

- Gender of the household head.
- Age of household head in years.
- Education of household head: Usually years of schooling by household head.
- Level of education (illiterate, primary, secondary, tertiary).
- Household size (number of family members).
- Total land size in hectares.
- Household income using local currency.
- Labor availability.
- Farming experience in years.
- Number of livestock (cattle, shoats, poultry).
- Land tenure; ownership status of land holdings, including information on rented or leased land.
- Family structure and composition; marital status.

Behavior and practices
- Crops cultivation.
- Livestock and animal husbandry.
- Food security and nutrition.
- Preferences of practices.

Farm management practices
- Farm labor: number of family members and hired labor involved in farming activities.
- Main agricultural practices (e.g., agroforestry, conservation agriculture).
- Marketing and market access.
- Farm incomes.
- Climate change adaptation and mitigation practices.

Farmer's knowledge and training needs
Data on knowledge and training needs to help design capacity building and extension messages for farmers.

Box 12.2 Common data collected from farmers related to crop and livestock production

- Types of crops grown, e.g., maize and tomato.
- Varieties of crops.
- Crop management practices.
- Preferences for specific varieties and reasons for selection.
- Yields of various enterprises.
- Factors influencing planting decisions.
- Practices to adapt to climate change.
- Use of fertilizers and pesticides.
- Post-harvest practices.
- Total land under cultivation.
- Challenges faced in crop production.
- Participation in training programs.
- Interaction with extension services.
- Marketing channels for agricultural products.
- Challenges faced in marketing agricultural products.

Specific information on livestock production
- Types and breeds of animals.
- Number of animals.
- Forage, feed, and supplements used for animals.
- Reproductive rates, including conception and calving.
- Common livestock diseases.
- Livestock management practices.
- Marketing of livestock products.
- Use of improved technologies in livestock farming.
- Participation in livestock training programs.
- Key challenges faced in livestock production.
- The farmer's intentions and plans for future livestock production.

4. *Opinion/beliefs*: Refers to collecting views about farmers' convictions regarding agriculture, farming practices, and policies. This information is essential for agricultural extension services, researchers, and policymakers to help them tailor their interventions and programs for farmers.
5. *Attitudes and motives*: Questions measuring attitudes among farmers are common and solicit information about respondents' liking or not liking. They measure how a respondent feels toward a particular subject. This information is best captured in a Likert scale that asks respondents to indicate the degree of agreement or disagreement with a certain statement. Questions on motives solicit information about why they do what they do. E.g., why farmers in the dry areas will use manure for cooking instead of using it for crop production.
6. *Past or future behavior*: Past and future behavior help the researcher assess what farmers have been engaged in and what they plan to do, respectively. Past behavior involves recall questions that demand the respondent recall the phenomenon of interest. This could be activities they have been involved in, e.g., the question of when did you plant the current crop of maize? Or how many beans did you harvest last season?

Box 12.3 Examples of research objectives where cross-sectional surveys are appropriate

1. To assess the socio-economic and demographic characteristics of farmers participating in the collective marketing of bananas *(Characterization, descriptive).*
2. To determine the relationship between socio-economic factors of farmers and adoption of improved maize varieties *(Analytical).*
3. To assess determinants (farm and socio-economic) of the use of inorganic fertilizers by farmers growing maize in Western Kenya *(Analytical).*
4. To determine the relationship between income levels and participation in group marketing of milk by dairy farmers in Kiambu County *(Analytical, correlational).*
5. To assess the attitudes of farmers using animal manure toward soil fertility levels on their farms *(Descriptive).*
6. To assess the knowledge levels of farmers toward the use of biofertilizer on farms *(Descriptive).*
7. To assess preferences toward a specific technology, service, or practice *(Descriptive).*
8. To determine marketing channels and prices for milk and how they affect marketing channels *(Analytical and descriptive).*
9. To assess the perception of farmers on constraints and problems associated with a certain enterprise *(Descriptive).*
10. To determine factors determining access to improved research technologies and services, e.g., credit and extension services *(Analytical).*

12.2 Types of Surveys

According to the time factor, two types of surveys are identified: *cross-sectional* and *longitudinal* (Figs. 12.1 and 12.2). The type of survey you choose greatly depends on the questions to be answered and the variables. The suitable type optimizes the achievement of the study's objective.

12.2.1 Cross-Sectional Survey or Snapshot Survey

A cross-sectional survey, also known as a "snapshot" study, involves collecting data at a specific point in time. It is the most common type used in agricultural research. You are most likely to use cross-sectional surveys because they can take a short time to collect data. It is particularly effective for providing a snapshot of a population's characteristics and information about a specific moment. For example, a study might collect data from 130 farmers on their preferences for types of fertilizer for growing maize, all within three weeks. As Babbie (2021) notes, these surveys are cross-sectional in terms of both the study population and the time of investigation, ensuring the reliability of the data collected.

One of the key advantages of a cross-sectional survey is its practicality, allowing you to compare numerous variables simultaneously. For instance, you could assess how education, age, and gender are related to farmers' knowledge levels, all without incurring significant additional costs.

12.2.1.1 When to Select a Cross-Sectional Survey

You will choose a cross-sectional survey when your objective is to describe (descriptive) or to analyze a relationship

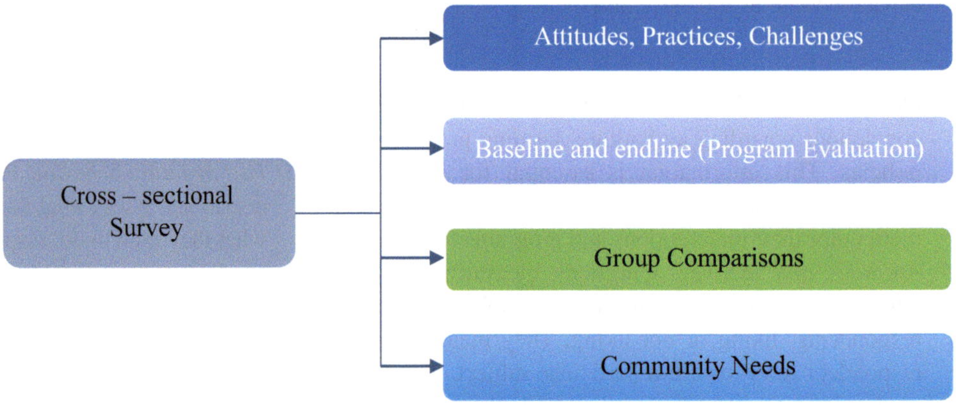

Fig. 12.1 Types of cross-sectional surveys

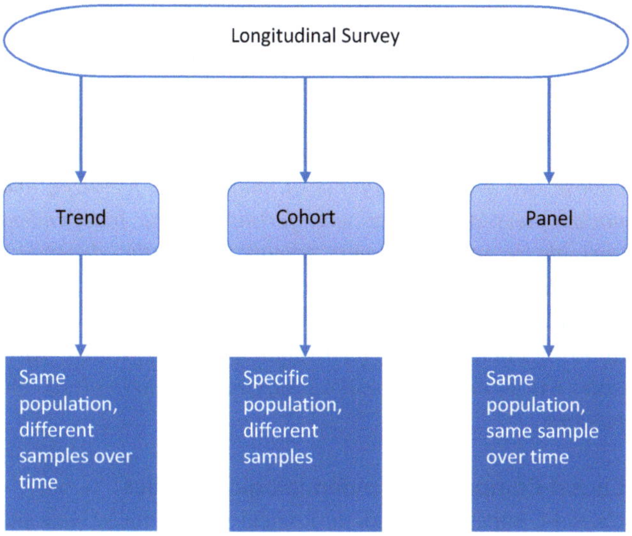

Fig. 12.2 The three types of longitudinal surveys

(analytical). The description is when we are answering questions such as: (1) How many individuals (farmers, consumers) are practising or using a certain technology? (2) What is the frequency of occurrence or prevalence of a certain parameter? What are the attitudes, constraints, knowledge levels, and perceptions of individuals to a certain parameter?

Analysts are concerned about relationships and will ask questions such as: (1) What is the relationship between demographic characteristics and the occurrence of a certain parameter, e.g., adoption? In other words, do these characteristics vary across groups or different populations (see Box 12.3)?

Depending on your objectives, you could include these two scenarios in the same study. Indeed, most studies with cross-sectional surveys aim to describe their samples and establish relationships between variables of interest.

In summary, you will choose a cross-sectional survey when your study aims to assess quantities, characteristics, perceptions, attitudes, constraints, beliefs, relationships, extent of variables, and access to services.

12.2.1.2 Examples

A study by Njenga et al. (2020) compared knowledge, attitudes, and practices (KAP) related to animal and human brucellosis among Kenya's nomadic and non-nomadic pastoralist communities. The study compared knowledge and practices related to brucellosis between participants from the predominantly nomadic pastoralists of Marsabit County and mixed farmers of Kiambu County, Kenya. Box 12.3 shows more examples.

12.2.1.3 Types of Cross-Sectional Surveys

Cross-sectional surveys can further be classified into different types based on the information being collected (Fig. 12.1):

Attitudes, Practices, and Challenges Survey

Cross-sectional surveys on attitudes, practices, and challenges help researchers understand farmers' views on agricultural practices, their farming methods, and the obstacles they encounter (Gesesew et al. 2016; Id and Afsheen 2021). The survey assesses whether farmers follow recommended practices and identifies challenges such as financial issues and lack of input. Most data collected includes those listed in Boxes 12.1 and 12.2.

Baseline and End-Line Surveys

Baseline surveys are carried out at the start of a program or before an intervention. They aim to establish the status quo before program implementation. A baseline survey is a descriptive cross-sectional survey that mostly provides quantitative information on the current status of a particular situation in whatever study topic in a given population (Aggarwal and Ranganathan 2019; Msoma et al. 2020). It aims to describe a given variable's distribution in a study population at one point in time.

Suppose a team of researchers is implementing a program to promote integrated pest management practices. A baseline survey will provide data that act as a reference point to compare program outcomes and impacts. Baseline surveys are useful whenever you want to evaluate impact because a "before" snapshot is required to be compared with an "after" picture.

The "after picture" is the end-line survey conducted after the program to assess the outcomes and impacts. They measure the same variables as the baseline survey to determine the changes over the program duration. The researcher can evaluate the program's effectiveness by comparing the baseline and end-line data.

For example, in a project promoting integrated pest management practices (IPM), a researcher can measure the following variables at the baseline and end line: (1) Current management practices, including the number of times chemical pesticides are used. (2) Level of awareness of IPM methods. (3) Average crop loss due to pests. Suppose at the baseline, the average number of times chemical pesticides were used per season was 8, and at the end of the project, it was two (2) times. This represents a change of −70%. The percentage of farmers aware of IPM practices changed from 20% at the baseline to 80% at the end line, showing a positive change of +60%. These data help to better understand and improve comprehension.

This example is fairly simple and looks straightforward, but we have the challenges of extraneous variables that could have affected the changes measured. These include factors such as weather, market, and policy changes. To accurately evaluate impact, we need to make sure that we account for changes over time that are not due to the project. Researchers use population control and statistical procedures to isolate the project's effect from other influencing factors. Statistical procedures include regression analysis and propensity score matching techniques that can adjust for extraneous variables to accurately reflect the project's true impact. These concepts are presented in Chap. 11 on designs for evaluating impact.

Comparative Survey

A comparative survey is carried out to compare certain groups. The aim of a comparative study is to identify differences between two or more groups in regard to the variable of interest. Examples of these include comparison between subgroups:

- Attitudes of adopters of improved maize varieties and nonadopters or trained and untrained farmers.
- Trained versus nontrained farmers.
- Male-headed households versus female-headed households.
- Livestock production practices in large farms versus small farms.
- Disease prevalence in households with vaccinated animals versus those with unvaccinated animals.

Comparative surveys are utilized in causal-comparative studies to determine causal relationships between two or more groups.

Causal-comparative studies are frequently conducted by students in the social sciences to determine the effect of an independent variable on a dependent variable involving human subjects. Since the independent variable cannot be manipulated, as in an experiment, a comparison approach is taken, hence the term causal-comparative study.

The researcher's goal in a causal-comparative study is to determine whether the independent variable affects the outcome or the dependent variable by comparing two or more groups of individuals. Causal comparative is also known as "ex post facto" research. (Latin for "after the fact") since both the effect and the alleged cause have already occurred and must be studied in retrospect. It is important to note that the relationship between the independent variable and dependent variable is usually a suggested relationship (not proven) because you (the researcher) do not have complete control over the independent variable.

The causal-comparative method seeks to establish causal relationships between events and circumstances. In other words, it finds out the causes of certain occurrences or nonoccurrences. This is achieved by comparing the circumstances associated with observed effects and by noting the factors present in the instances where a given effect occurs and where it does not occur.

12.2.1.4 Example of an Objective that Will Require Causal-Comparative Studies

To assess socio-economic factors influencing the adoption of manure plus fertilizer by smallholder farmers of Region A.

This is a comparative study requiring the researcher to choose two groups (adopters of manure and fertilizer, and nonadopters of manure with fertilizer). The independent variable is socio-economic factors, while the dependent variable is the adoption of manure plus fertilizer.

The survey is then conducted, collecting the data on variables of interest. Data on hypothesized independent variables will be collected, and a comparative data analysis will be done

Causal-Comparative Epidemiological Studies

Causal-comparative epidemiological studies investigate the relationship between specific factors, such as management practices or exposure to a pathogen, and health outcomes in a population. These studies aim to identify potential causes or risk factors for diseases or health-related outcomes, such as disease prevalence or reproductive performance.

Consider a study investigating the potential causal relationship between the presence of a specific tick species (variable x) and the prevalence of a certain disease (health outcome Y) in a cattle population. The study will involve two groups of cattle: one group exposed to the tick species and another unexposed. The survey collects data on tick investigation levels and disease prevalence, but also on potential confounding variables, e.g., management practices, age, sex, and education levels. Statistical analysis is performed to compare disease prevalence between the exposed and the unexposed groups, controlling for potential confounding

Community Needs Survey

A community needs survey aims to gather data from farmers and other stakeholders to guide decision-making and development efforts. The survey often gathers data about their resources and what they require to improve their agricultural productivity and livelihoods. For example, in assessing the community needs of farmers practising mixed farming, the survey might investigate issues such as access to quality seeds and fertilizers, market opportunities, and the impact of climate change on crop yields. Farmers often express a desire for training on better farming techniques, provision of better infrastructure like roads and sustainable farming practices to manage soil and livestock. For example, a study by Mugwe et al. (2019) identified several critical challenges facing farmers engaged in the collective marketing of maize, including poor road infrastructure, poor storage and bulking facilities, lack of credit, and lack of adequate knowledge on collective marketing as some of the most changing challenges affecting collective marketing.

12.2.2 Longitudinal Surveys

Longitudinal surveys gather data repeatedly over a period of time, sometimes lasting many years (Farrington 1991; Caruana et al. 2015). The aim is to evaluate changes over an extended period of time. In practice, longitudinal studies repeatedly draw sample units of a population over time. The researcher may then analyze changes in the population and attempt to describe and/or explain them, e.g., a survey carried out on the same population to monitor the effect of introducing a certain diet on animal growth or to monitor the effect on income levels of farmers following the adoption of improved varieties of maize. Depending on the objectives of the assessment, both qualitative and quantitative data may be collected (Caruana et al. 2015).

The main benefit of a longitudinal survey is that the researcher is able to detect developments or changes in the characteristics of the target population at both group and individual levels (Caruana et al. 2015). Repeated measures over time are useful in establishing a sequence of events. The researcher must decide how frequently to collect data. This is informed by the surprise as well as the practicability. For studies that will take a long time, consider giving large samples to allow for natural attrition without affecting the study's validity. The three main types of longitudinal surveys are: trend studies, cohort studies, and panel studies.

12.2.2.1 *Trend* Studies

Trend studies focus on a particular population, which is sampled and scrutinized repeatedly. While samples are of the same population, they are typically not composed of the same people. Since trend studies may be conducted over a long period of time, they do not have to be conducted by just one researcher or research project. A researcher may combine data from several studies of the same population to show a trend.

Examples

1. Farmers from the same general population who have adopted integrated soil fertility management (ISFM) strategies are surveyed at different times, perhaps in 2001, 2001, or 2005, to understand the trend of this particular population.
2. The yearly survey of farmers keeping improved dairy cows to assess changes in the number of cattle and the impact on their income levels.

Trend studies are useful in forecasting by extrapolating present and past trends, thus making a valuable contribution to planning. Trends regarding the phenomenon under study can be correlated with other characteristics of the study population. For example, you may want to examine the changes in political preference of a study population in relation to age, gender, income, or ethnicity. This design can also be classified as a retrospective–prospective study on the basis of the reference period classification system developed earlier in this chapter.

12.2.2.2 Panel Studies

In panel studies, the same subjects or samples are surveyed at different points in time. Panel studies thus allow the researcher to find out why population changes occur, because they use the same sample of people every time (Lipps 2021). That sample is called a panel. Presume you want to study the changes in the expenditure pattern on fertilizers and pesticides in a farming community. To do this, you would select a few households to find out the amount they spend every season on buying fertilizers and pesticides. You would keep collecting the same information from the same families over a period of time to ascertain the changes in the expenditure pattern. Similarly, a panel study design could be used to study the disease pattern in certain animals, e.g., beef cattle.

Panel surveys can be useful in revealing both net and gross change within the panel members (Lipps 2021). While they can yield extremely specific and useful explanations, panel studies can be difficult to conduct. They tend to be expensive, take a lot of time, and suffer from high attrition rates. Attrition occurs when people drop out of the study, possibly because of the long-term nature of the study.

12.2.2.3 Cohort *Studies*

In cohort studies, a specific population is followed over a length of time. It can also be referred to as a follow-up study. A cohort is any group of individuals who have experienced the same significant event within a given period (Pottegård 2022; Jan et al. 2023). For example, a group of farmers who participate in a certain program, for instance, farmers trained in the safe use of pesticides on crops, might be followed for 4 years to determine the percentage that continues using the safe method. Over the years, from the list of trained farmers, a random sample was drawn for years 1, 2, 3, and 4, and data were collected from that sample. Thus, the population remains the same during the study, but the individuals surveyed are different each time.

You could also follow two cohorts to compare performance. A good example would be following the cohort of farmers who participated in training on ISFM and another cohort that was not trained. The aim would be to assess changes in several aspects over time. Note that the required information does not have to be collected from the same respondents; however, all the respondents must belong to the cohort being studied

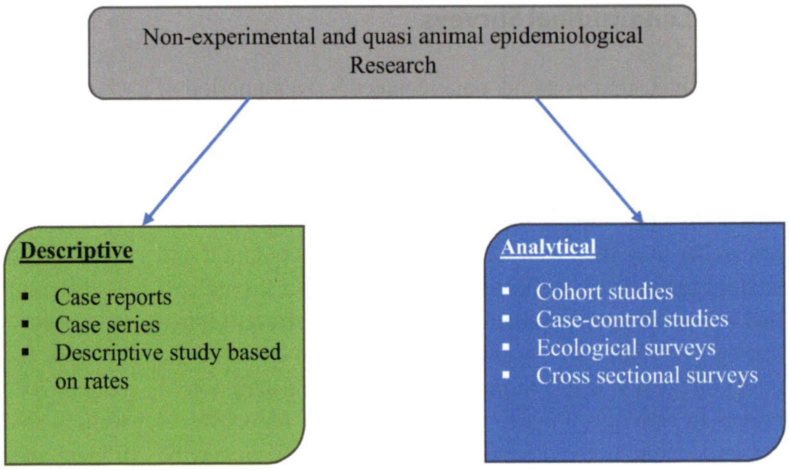

Fig. 12.3 Types of nonexperimental studies in animal epidemiological research

12.3 Experimental and Nonexperimental Research in Epidemiology

Animal epidemiology research focus on the distribution, frequency and determinants of animal diseases in populations. It is either experimental (randomized clinical trials) or nonexperimental, also called an observational study. The nonexperimental is broadly subdivided into descriptive and analytical designs (Fig. 12.3). Whereas the descriptive designs are exploratory, the analytical designs are mainly cross-sectional surveys or longitudinal surveys that aim to test hypotheses and establish causal relationships among variables. This can be done by counting the number of affected animals. For example, to compare disease levels among cattle groups, time frames, and locations, we need to consider counts in the context of the population size from which those cases arose. This is referred to as prevalence, which is defined as the number of cases of a given disease or attribute at a specified point.

12.3.1 Longitudinal Studies in Animal Epidemiological Research

Epidemiological research is important because it will give information about which animals are affected, when and where a disease occurs, which can often suggest the cause of the disease. Subsequently, it will be necessary to identify the determinant (cause) of the disease and explain why the disease occurs. This will help decide on how to reduce the severity or the frequency of occurrence. Veterinary epidemiology has unique terms used to describe the designs, techniques of data collection, and data analysis (Bhopal 2016), but they are generally variants of survey studies. The aim of studies in veterinary epidemiology is to identify the natural history of the disease and its outcome, identify the risk factors involved and identify the incidence rate.

For example, veterinary epidemiologists commonly use longitudinal surveys to monitor the number of cases of a disease, for example, foot and mouth disease in livestock, or they may monitor changes in birth rates and mortality rates. Specifically, they are used to answer the following research questions:

1. What is the occurrence of the disease (number of cases), and are there changes in incidence and prevalence?
2. What are the etiological factors responsible for the disease conditions?
3. What is the source of infection with respect to the disease of interest?
4. What are the new diseases or new strains of existing diseases?
5. Are there changes in the management systems? How are these changes affecting the current disease situation?
6. What is the efficacy of control measures employed during an epidemic?
7. Can the cause of the epidemic be predicted using models?

The results obtained are used to minimize or eradicate diseases or health problems and their consequences. They also provide necessary data for planning, implementation, and evaluation of programs aimed at preventing, controlling, and treating diseases and, therefore, minimizing the chances of disease occurrence in the future.

12.3.2 Descriptive Designs

Descriptive designs are a first step in understanding disease patterns and potential influences within livestock populations. These designs are primarily qualitative, focusing on observing and recording the occurrence of a disease (Christensen 2001). They do not test hypotheses. The observations, while subjective, offer valuable clues about the disease's etiology, paving the way for the formulation of testable etiological hypotheses. This groundwork is also useful for developing effective therapeutic, preventive, and control strategies for livestock diseases. Case reports and case series are common descriptive designs. They are described below.

Case reports These are detailed descriptive accounts of a single animal's experience with a disease or an unusual clinical presentation (Thrusfield et al. 2017). They are often the first documented evidence of a new disease in a livestock species, an unusual manifestation of a known disease, or an unexpected adverse reaction to a feed additive or veterinary drug. Because they focus on an individual animal, case reports are inherently qualitative and provide *in-depth information* that might otherwise be missed in larger studies. They are invaluable for generating new hypotheses and alerting the veterinary community to rare or novel occurrences.

For example, a case report might document neurological symptoms in a dairy cow following the introduction of a new type of silage, suggesting a possible association worthy of further exploration. Another example could involve the first identification of a novel viral infection in poultry, including detailed clinical signs, progression, and post-mortem findings. Although not generalizable, case reports contribute significantly to hypothesis generation and early detection of health threats.

Case series Case series extend the concept of a case report by describing the characteristics of a group of animals with a similar diagnosis or exposure, such as in Paulo et al. (2021). While still descriptive and qualitative, a case series provides a broader picture than a single case report by aggregating data from multiple individuals within a herd, flock, or region. This allows for the identification of common features, patterns, or trends among affected animals. Like case reports, case series do not involve a control group, so they cannot establish causality, but are excellent for generating hypotheses. For instance, a case series detailing the clinical presentation and treatment response of several cattle with a newly identified parasitic infection could help characterize the disease and guide the development of diagnostic criteria and herd health management protocols.

12.3.3 Analytical Designs

12.3.3.1 Case–Control Studies

Case–control studies are also called case comparison or retrospective studies, meaning looking backwards (Horwitz and Feinstein 1982). It is similar to the causal-comparative analytical approach used in cross-sectional surveys to establish relationships. The case–control study compares a group of diseased animals with a group of healthy animals with respect to exposure (Hayden et al. 1982; Lewallen and Courtright 1998) (Fig. 12.4). Cases and controls are selected on the basis of the presence of the outcome condition, and exposure is assessed by looking back over time. The interest is to compare the occurrence of disease A (dependent variable) with respect to exposure B (independent variable), similar to the case of adopters and nonadopters in an adoption study.

In summary, four important attributes of a case–control study are (Horwitz and Feinstein 1982; Lewallen and Courtright 1998):

- The exposure to a factor or factors and outcome (disease) has occurred before the start of the study.
- The study proceeds backwards from effect to cause (causal comparative).

Fig. 12.4 Case–control study in animal health research

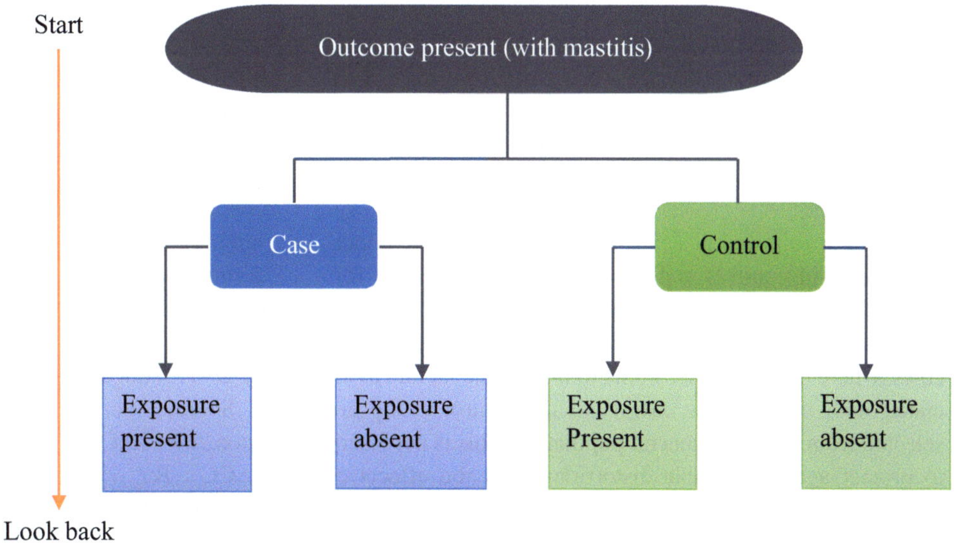

Box 12.4 Case study of a case–control

Tulu et al. (2020) published in the Asian Journal of Animal Sciences, studied risk factors and risk associated with Brucellosis in aborted breeding goats in Jimma Zone, Ethiopia. This was a case–control study.

Research problem: Brucellosis is recognized as a leading cause of abortion in goats worldwide. In Ethiopia, Brucellosis is endemic, and studies on risk factors for brucella infections lack consistency across sites and species, and the studies provide insufficient information about epidemiological information about the disease in the country.

Research objective: The objective of the case–control study was to determine the risk factors associated with brucellosis in aborted goats.

Methodology: Data were collected from 134 cases (representing goats with at least one abortion history within the year) to 268 controls. The data collected include blood samples, agroecology, management systems, flock size, species composition, including new animals, management systems, age, parity, reproductive status, etc. The participants were picked randomly. The data analysis involved using multiple logistic regression models.

Results showed that goats from mixed species, goats with cattle/sheep, were more likely to be Brucella seropositivity than those with no contact. Similarly, breeding goats from large flocks and lowlands were also found to be at a higher risk of harboring brucella infection.

Source: Tulu et al. (2020) https://scialert.net/abstract/?doi=ajas.2020.137.144

- It involves selecting cases (diseased) and controls (nondiseased).
- Involves making comparisons of the differences between groups to elicit the average risk at the level of the individual animal.

The steps followed in a case–control study involve selecting cases and controls and ensuring that they match, followed by the measurement of exposure, and, finally, analysis and interpretation of data. For example, in the case of dairy cows, where the researcher wants to investigate factors associated with mastitis, the cases and the control group are selected from the same general population. This is necessary to protect against possible distortions from the effects of extraneous variables (confounders). The matching criteria that can be used are sex, age, breed, ecological area, etc. The assumption here is that the only difference is that the case group contains the individuals affected by the disease, and the control group contains animals not affected (affected versus nonaffected).

Measurement involves getting information from interviews about the animals and past records on, e.g., management practices. The analysis and interpretation of data use the odds ratio to determine disease factors associated with exposure (Lewallen and Courtright 1998; Sauter-Louis et al. 2012).

12.3.3.2 Cohort Studies in Epidemiology

The cohort studies are most common in epidemiology (Chan 2023). Unlike the case–control studies, cohorts are determined by the level of the explanatory variable (independent variable). They involve repeated information collection through time (see Box 12.5: case study done for 2 years). Researchers can assess if exposure to a particular task (environmental, diet, genetic, etc.) is associated with clinical outcomes in individuals (Pugh et al. 2014). The cohorts (groups of animals) with varying levels of exposure are followed over time to evaluate the occurrence of an outcome. These animals have to be free of the outcome at baseline. Each subject's presence or absence of the risk factor (exposure) is recorded (Fig. 12.5). The subjects are then followed up over time (longitudinally) to determine the occurrence of the outcome. Thus, cohort studies are *forward-direction studies* (moving from exposure to outcome) and are typically *prospective studies* (the outcome has not occurred at the start of the study). At the end of the study, the two groups are compared with regard to the risk factors and disease.

Cohorts can be formed in several ways. In the classical cohort study, two or more groups of animals, such as cattle, goats, and sheep, are selected. A critical aspect is that the individuals in the cohorts should be similar in all aspects, although this is seldom the case. In addition, all the animals

Box 12.5 Case study of a cohort study in animal epidemiology

To investigate the infection dynamics of *Cryptosporidium bovis* and *C. ryanae*, a one-year prospective cohort study was performed on a dairy farm. Fecal samples were collected every 4 weeks from 20 randomly selected calves younger than 65 days. Samples (n = 238) were examined using fluorescence microscopy, and when positive, the species were determined by DNA sequencing.

Source: Åberg et al. (2019) https://doi.org/10.1016/j.vpoa.2019.100010

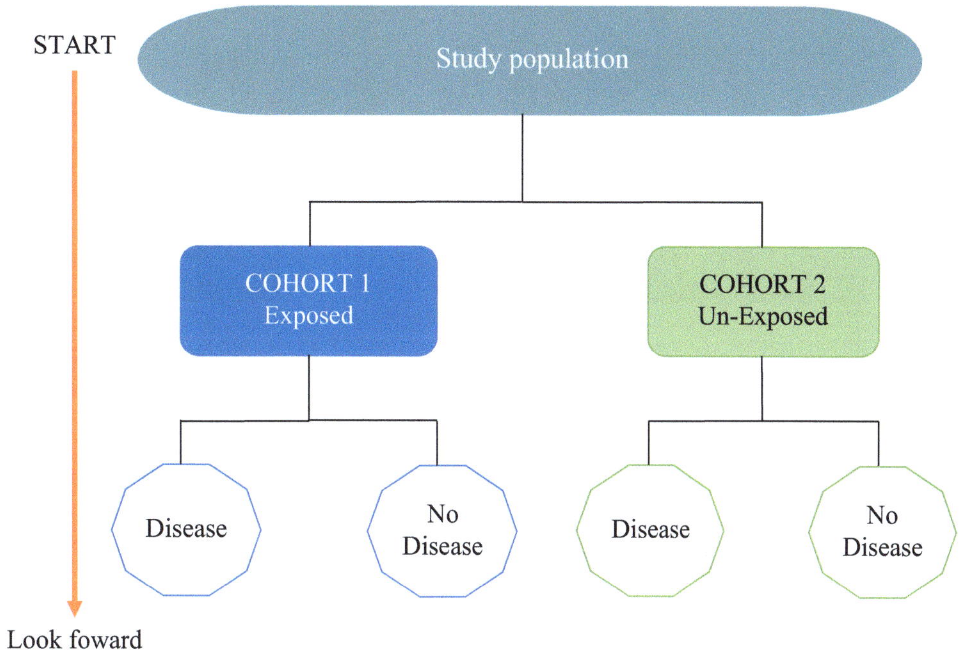

Fig. 12.5 Prospective cohort study in animal health research

are at risk of developing the disease. The separate independent cohort, defined by different levels of risk factors, is then followed over time. Key characteristics of epidemiological cohort studies include:

1. Cohorts are identified before the appearance of the disease
2. The study groups are observed over time to determine the frequency of diseases among them
3. Study proceeds from cause to effect (prospective)
4. There is good evidence of an association between exposure and disease

12.3.3.3 Ecological Surveys

Ecological surveys play a crucial role in studying the epidemiology of livestock. They aim to find a correlation between the occurrence of diseases and exposure in different ecological regions. The main principle behind these surveys is to identify changes in the frequency of health events over time and/or space and relate them to environmental factors. Ecological surveys are also known as multiple comparison studies because they focus on comparing different geographic areas. The objective of an ecological survey is to look for spatial patterns that might suggest an environmental or a special etiological hypothesis. For instance, researchers can use ecological surveys to monitor animal population health and conduct large-scale comparisons, such as comparisons between countries. They can also identify the relationship between animal population-level exposure to risk factors and diseases in different geographical regions, which can help determine risk factors in the population. The unit of observation (unit of analysis) is the population instead of an individual animal. Examples of such studies include Silatsa et al. (2019) and Muturi et al. (2023), among others.

12.3.3.4 Advantages and Disadvantages of Surveys

Advantages of Surveys

1. Surveys allow researchers to collect a large amount of data in a relatively short period of time.
2. Surveys are less expensive than many other data collection techniques.
3. Surveys can be created quickly and administered easily.
4. Surveys can be used to collect information on a wide range of things, including personal facts, attitudes, past behaviors, and opinions.

Disadvantages of Surveys

1. Poor survey construction and administration can undermine otherwise well-designed studies.
2. The answer choices provided in a survey may not be an accurate reflection of how the participants truly feel.
3. While random sampling is generally used to select participants, response rates can bias the results of a survey.

12.4 Ways of Administering Surveys

Surveys are administered in two broad categories: interview-completed surveys and self-administered surveys (Puhan et al. 2011; Christensen et al. 2014) (Fig. 12.6). In the interviewer-completed surveys, the researchers ask ques-

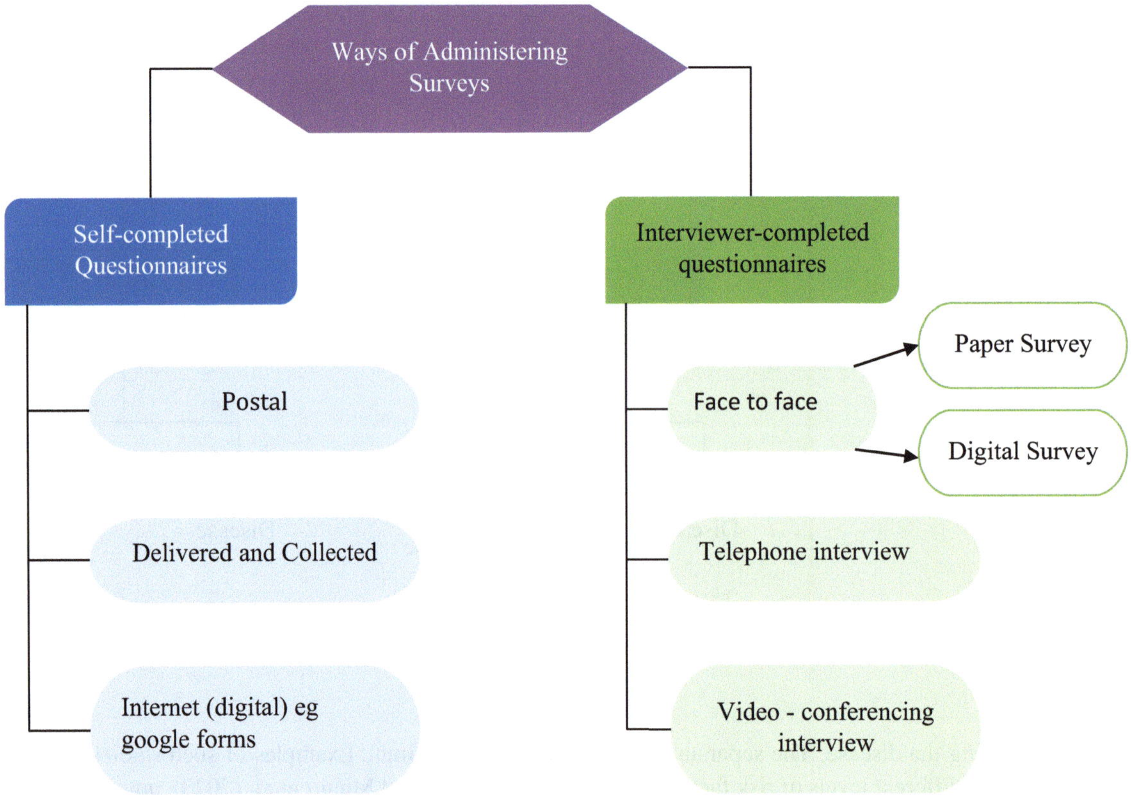

Fig. 12.6 Ways of administering surveys in self-completed questionnaires and interviewer-completed questionnaires

tions directly to the respondents, guided by a predetermined series of questions. Researchers record the answers from each respondent. Conversely, self-administered surveys rely on the respondents reading the questions on their own and recording their answers. Some authors, such as Mugenda and Mugenda (2019), prefer to call the former an interview schedule where the interviewer asks questions and records the responses. However, because these two methods use a similar questionnaire, this differentiation may not always be critically crucial.

12.4.1 Interviewer Completed Survey

Interviewer-completed survey in which the researcher fills the tool can be administered using five ways: Face-to-face, telephone, paper, video conferencing, and the use of digital devices (Fig. 12.3):

1. *Face-to-face interviews*: This involves in-person one-to-one interviews (verbal questioning) between the researcher and the respondent (Puhan et al. 2011). The researcher gathers information or data directly from a respondent by physically meeting the respondent and interviewing them face to face. This method has the advantage of a high response rate and flexibility in questioning. It allows the interviewer to clarify doubts. The researcher also benefits from an opportunity to probe deeper into responses and observe nonverbal cues.

A significant disadvantage is the potential for interview bias (Hoebel et al. 2014). Interview bias can arise when you have a sensitive issue, as respondents may be reluctant to tell the truth due to embarrassment and fear of judgment. The respondent may also seek to please the interviewer, thus distorting the data.

Additionally, face-to-face interviews can be costly and time consuming, especially if the respondents are widely dispersed (Puhan et al. 2011). Researchers may need to travel long distances, which can add substantial costs to the research process. Selecting the respondents and training the enumerators are also additional costs. In farm household surveys, where the interviews are conducted in homes, interviewees may be unwilling to admit strangers, making it challenging to collect the data.

Despite its higher cost implications, the face-to-face interview method remains valuable in situations where ensuring data accuracy and validation on the spot is paramount for the success of the research study (Denscombe 2017). Researchers should weigh the benefits of immediate validation against the associated costs to determine the appropriateness of this method for their specific data collection. It also remains the method of choice in surveys involving farmers where literacy levels might be low.

Face-to-face interviews allow readers to clarify questions and collect more reliable data. Additionally, this approach provides further opportunities to observe the actual framing practices and farm conditions (Denscombe 2017). Details on how to collect the data in interview surveys are revisited in Chap. 18.

2. *Telephonic surveys*: These surveys are usually conducted over the telephone. They have modest costs, require less time, and may have higher response rates. However, just like face-to-face surveys, they hesitate to discuss sensitive issues. They are also time consuming and sometimes nonconclusive due to breakoffs. Breakoff happens when the respondents start losing interest and lack motivation (Edlund and Nichols 2019). The other limitation is that the success of these surveys depends on how many people answer the phone and who want to invest their time in answering questions over the telephone. Targeted respondents often easily hang up or ignore the calls (Edlund and Nichols 2019). However, for those who are positive respondents, there is an advantage of data control.

Telephone interviews also offer significant time-saving benefits for researchers as they eliminate the need to travel to various parts of the country to conduct surveys. With phone surveys, researchers can efficiently reach respondents from different geographical locations without the logistical challenges associated with on-site visits. However, it is crucial to consider potential limitations, such as excluding individuals without phone access or those who prefer face-to-face interactions. Balancing the advantages and limitations, telephone interviews remain a valuable tool in survey research, offering convenience and efficiency in gathering data across diverse geographical regions.

3. *Paper surveys*: As the name suggests, this survey uses the traditional paper-and-pencil approach (Edlund and Nichols 2019). Many would believe that paper surveys are a thing of the past, used before the digital era. However, they are quite handy for field research and data collection. These surveys can go where computers, laptops, or other handheld devices cannot. There is a flip side to it; this survey type is the most expensive data collection method. It includes deploying many human resources along with time and money. After data collection, the data have to be entered into the computer, requiring further human resource input.
4. *Use of digital devices*: Digital devices are increasingly used in surveys, mainly interviews with farmers, due to technological advancements. Digital devices such as smartphones or tablets equipped with survey applications, such as SurveyMonkey, Qualtrics, or ODK, offer several advantages in surveying. The questions are captured on a smartphone or tablet, and data are collected and saved in real time. These applications provide user-friendly interfaces for designing, administering, and managing surveys. Using such digital tools not only enhances the efficiency of data collection but also allows for real-time monitoring of survey progress and responses. They also have the advantage of facilitating GPS tagging of survey locations, providing valuable spatial information for agriculture-related studies. Most surveys nowadays use digital devices. Once data are captured, the entry of the data experienced in paper surveys is avoided. With digital devices, researchers are able to monitor the collection process.

12.4.1.1 Self-Administered Survey

The self-administered survey, where the respondents record their answers, can be administered in three ways: internet or online surveys, postal or mail, and delivered and collected questionnaires (Puhan et al. 2011) (Fig. 12.7).

12.4.1.2 Self-Administered Questionnaires Can Be Conducted in Three Ways

1. *Internet or Online surveys*: With technology advancing many folds, internet surveys are becoming more popular. An internet survey consists of survey questions that can be easily sent out to the respondents online via email, or they can access them through their web browser using a hyperlink (web-based questionnaire) or directly using their mobile phone (phone questionnaire). The success of these requires the availability of a stable internet. These surveys are easy to design and simple to deploy. Respondents are given ample time and space to respond to these surveys so researchers can expect unbiased responses. It also avoids interviewer bias and distortion. They are less expensive than traditional surveys and data can be collected and analyzed quickly. It must be used with caution because respondents may not be representative of the intended population. Often, follow-up is difficult.
2. *Postal surveys or mail questionnaires*: The questionnaires are mailed to respondents who return them after completion. They should be accompanied by a cover letter. They have the advantage of high response rates if a thorough follow-up is done. However, without strict follow-up, they suffer from low response rates. They are also good for reaching wide geographical coverage and are more manageable in handling large samples. It is also relatively easy to obtain a listed population to select sample form, easy to administer, and relatively low-cost compared to face-to-face interviews. Because of the absence of direct contact with the researcher, as in face-to-face, it is easy to avoid interviewer bias and distortion, and answers are unlikely to be socially influenced. A disadvantage, however, is that it is hard to interpret open-ended questions, and it is also time consuming. Additionally, the researcher lacks control over who fills out the questionnaire.

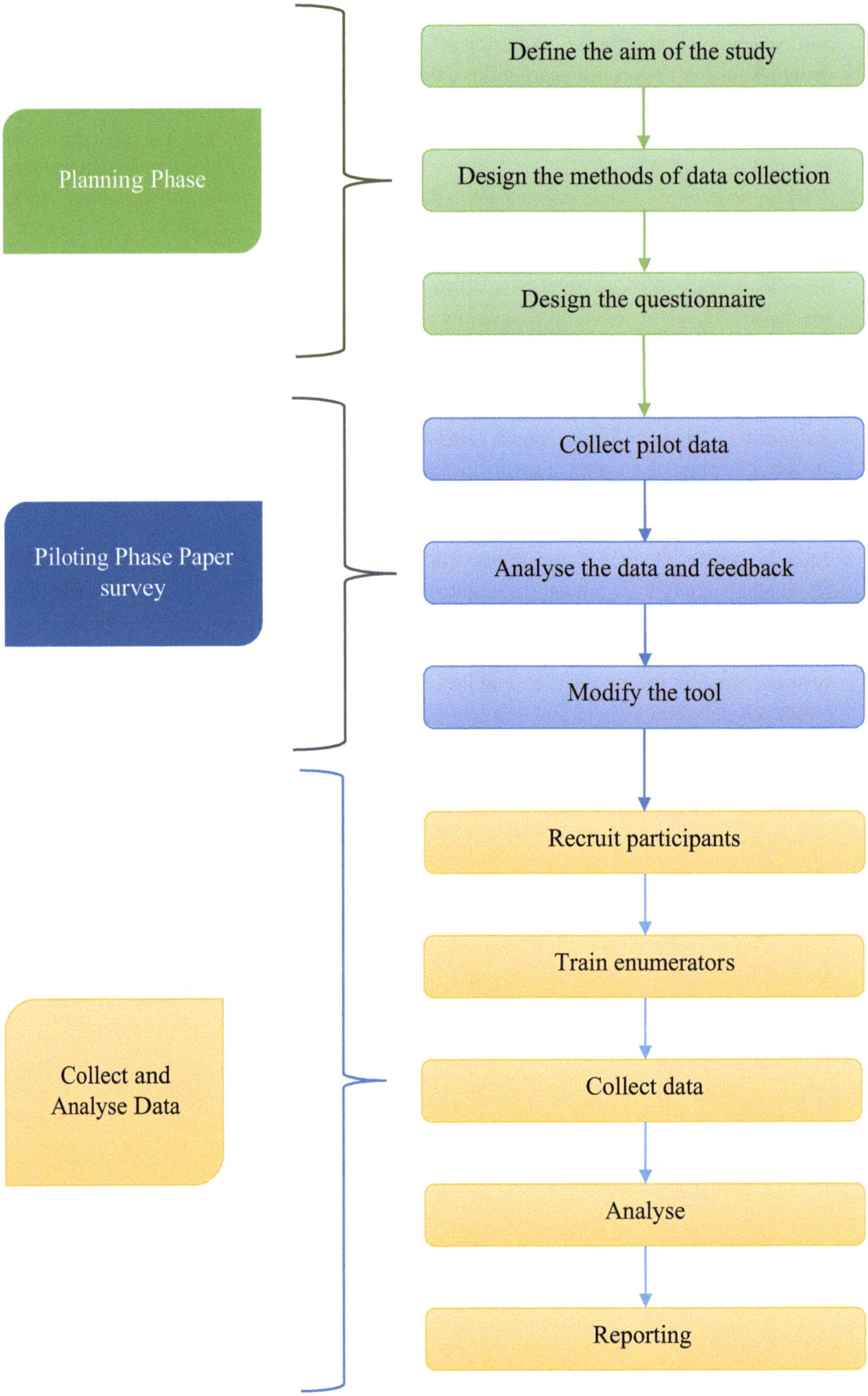

Fig. 12.7 Three phases in planning and implementing a survey

3. *Delivered and collected questionnaires*: Questionnaires are delivered by hand to each of the respondents and collected later.

Several factors will influence the choice of the method to use:

1. Characteristics of the respondents from whom you wish to collect the data, e.g., you can only use internet surveys if your respondents have access to the internet. Further, you can only use the telephone for individuals who can be telephoned, and you are able to select them and get their contacts.

2. Size of the sample you require for your analysis, considering the response rate; the size of the sample will have implications on the confidence you will have in your data and the extent to which you can generalize the results. Response rates in face-to-face and telephone interviews, in general, are higher than in self-administered surveys (Christensen et al. 2014; Meyer et al. 2022).
3. Resources available; these include finances, time, and availability of interviewers and Internet. For paper surveys, you will need to consider the cost of reproducing the questionnaire (printing and clerical work) plus entering the data into the computer for analysis.
4. Number of questions needed to ask to collect your data.
5. Type of questions you need to ask to collect your data.

12.5 Steps in Conducting a Field Survey

This discussion concerns conducting a field survey, which tends to be most common in agriculture, especially when interviewing farmers. The steps occur in three phases (1) Planning, (2) Piloting, and (3) Data collection (Fig. 12.7).

12.5.1 Phase 1 (Three Key Steps)

Step 1. Designing the survey

Designing the survey is the first step in conducting a survey and is crucial as it determines the overall success of the survey study (Baker et al. 2010). It takes place at the proposal writing stage, which involves conceptualizing and refining the research question. Remember that the choice to conduct survey research was informed by the research question. Ensure the research question clearly shows the study's variables and that they are properly operationalized. Operationalized terms are the ones to be used in designing the tool (Chap. 18).

Regardless of which type of survey, cross-sectional or longitudinal survey types, the starting point should always be a good research question that guides in the designing of the survey. Research questions should guide the survey questions. In designing the survey, you should avoid the pitfalls of posing irrelevant research questions (Kelley et al. 2003; Baker et al. 2010). You should avoid directly retrieving and using tools and questionnaires developed and used in other studies. They may not exactly match the objectives of your study. Retrieving of questionnaires from documents is a common pitfall among the students and should be avoided.

Other decisions at the planning stage include the target population, the sample size, how sampling will be done, and the design and layout of the questionnaire (Baker et al. 2010). Students should take time and think deeply about the sample size (how many people to survey), including how much data can be effectively collected and analyzed, considering time, energy, and access to resources. See Chap. 12 for guidance on how to pick a sample.

At the planning stage, the following is accomplished:

- *Information to be collected*: Information to be collected is identified. A researcher asks: What information is the survey trying to collect? What do the survey questions need to determine?
- *Survey participants*: Decide on the unit of analysis and the individuals or organizations you will study. The unit of analysis refers to the specific group or element that is the focus of the investigation. It is the primary subject of analysis for which data are collected, analyzed, and reported. This varies depending on the research objective, whether it is an individual farmer, a group of farmers, a market chain, or a farming system.
- *Population*: A population is a group of individuals who bear similar characteristics. For example, all farmers growing maize make up a population of maize farmers, while all agricultural extension officers make up a population of extension officers in agriculture.

Step 2: Select the survey type

The survey type determines how a survey is conducted and what is collected and recorded. The type of survey used depends on the information you want, how much information can be analyzed, and the time and resources available. A combination of survey types can also be used.

Step 3: Preparation of the data collection tool

Tools should be designed to be attractive, easily understood, easily answered, and give you the required information. Therefore, a lot of time is usually spent on getting the questions right. This step looks at the types of questions to ask, how to design questions, and the sequencing and presentation of questionnaires covering letters and introductions. The types of questions to ask: there are no hard and fast rules to decide which particular type of question to use, because the questions will depend greatly on the type of information being sought (Chap. 18).

12.5.2 Phase 2 (Piloting)

Before proceeding with data collection, the tool should be piloted (Kelley et al. 2003; Baker et al. 2010). This phase is emphasized in this section because many students often fail to take this phase seriously. It should be planned for the time and resources allocated. Piloting or pretesting means trying out the questionnaire on a few respondents. Pretesting the questionnaire provides information about deficiencies and suggestions for improvement. The respondents chosen for the pretesting should be as similar as possible to the actual participants. For example, if the participants are farmers keeping improved breeds of chicken, then individuals critiquing the tool should be poultry farmers keeping improved breeds of chicken.

Key steps to piloting include:
1. Collecting pilot data.
2. Analyzing the data and feedback.
3. Modification of the tool.

During the pretesting, ask the pilot team to comment and state suggestions concerning the survey directions, recording procedures, and specific items. You could also have reviewers examine the completeness of the tool. This is one way of determining its content validity. At the end of the pretesting, all feedback provided should be studied, considered, and used to revise the tool. The end product of an effective pretesting exercise is a revised tool ready for use in gathering data from the participants.

In summary, pretesting is important for the following reasons:

- Learn how informants respond to the survey.
- Identify any signs of fatigue among respondents, boredom, restlessness, and address.
- Allows determine how long the survey takes, understanding of the questions.
- Improve clarity of questions.
- Ensure enumerators are asking the same questions.
- Eliminate unnecessary questions.
- Improve the wording of the questions.
- Correct and improve translation of technical terms.
- Check the accuracy and adequacy of the instrument.
- Estimate the time needed to conduct the interview.

12.5.3 Phase 3 (Data Collection and Analysis)

After pretesting, the actual data collection phase starts. The researcher or survey coordinator should take charge of all aspects of the fieldwork, including recruiting participants and deploying enumerators.

This phase includes four steps:
1. Participant recruitment.
2. Training of enumerators.
3. Data collection.
4. Data analysis.

12.5.3.1 Participant Recruitment

The participants are the source of data (Baker et al. 2010). Suppose you get subjects who will not give you the information you require. This will adversely affect your study. You must ensure that the subjects to be surveyed possess the desired information and are willing to answer the questions. The individual who possesses information but is not interested in the topic is not likely to respond meaningfully. This affects the response rate and the quality of data collected, as disinterested participants may provide incomplete or inaccurate responses. To address this issue, you can conduct a preliminary survey to assess participants' willingness to respond and their level of interest in the topic. For instance, in a pest management practices study, you might first ask farmers about their experiences and interest in discussing pest management strategies. Those who show a keen interest can then be approached for detailed interviews or focus groups. Similarly, in an adoption and impact assessment study, you might conduct a preliminary survey of farmers who have recently adopted new technologies. The interest in discussing the impact and challenges of these technologies is discussed with the farmers. This helps in engaging respondents who are eager to provide valuable insights. It also helps build a rapport with potential respondents, making them feel valued and understood. This can further enhance their willingness to participate actively in the research.

You must select your sample using the right sampling method to draw valid conclusions from your results (Kelley et al. 2003; Moser and Kalton 2017). The choice of the sampling method is crucial because it directly affects the representativeness of your sample and the validity of your findings. Different research objectives and types of studies require different sampling techniques to ensure that the sample accurately reflects the population being studied. For example, if you are conducting a preliminary survey on pest management practices, you might start with farmers who are easily accessible to gather initial insights.

The other key reason why you should choose the right sampling method is in relation to sampling error (Weisberg 2005). Sampling error is the probability that any one sample is not completely representative of the population from which it has been drawn. A researcher should strive to ensure that the sample is representative of the population from which it is drawn. If our sample does not represent the population, that means we cannot trust any of the results. Although the sampling error cannot be eliminated completely, the sampling method chosen will influence the extent of the error (Kelley et al. 2003). One major sampling error occurs when a sampling frame error takes place during sampling. This may happen when the wrong subpopulation is used to select the sample, for example, selecting the wrong gender or socio-economic characteristics. Another major sampling error is when sampling bias occurs, resulting in a sample that does not reflect the characteristics of the population. The sampling frame should, therefore, be carefully selected.

12.5.3.2 Collect Data

Conducting the survey requires a high level of organization, especially if it is a field survey (Moser and Kalton 2017). A researcher will be required to do the following activities:

- Obtain permission from several individuals and groups: village administrators, farmer groups, and individuals.
- Training of enumerators: enumerators have to be trained to ensure they clearly understand what to ask.
- Organizing of budget and logistics; get a coordinator to organize budget and logistics, including travel, to make sure the survey runs smoothly.
- Pretest the tool.

  The supervisor should monitor the data collection process to ensure accuracy during data collection. This involves supervising the enumerators, conducting spot checks, and addressing any issues or questions from the respondents (Kislaya et al. 2021). If *open data kits* are being used, the researcher should be checking the data collected during the evenings to identify any errors, inconsistencies, and missing data (Kislaya et al. 2021). Throughout the data collection phase, the supervisor should ensure consistency, accuracy in data collection, transparency, smooth logistics, and ethical conduct are maintained. This is because the success of data collection will depend on the rigor and thoroughness with which the actual data collection phase is executed. The supervisor assists the coordinator in spot-checking and monitoring that data collections.

12.5.3.3 Process and Analyze Data

Analyze the information. An analysis and discussion are necessary to make sense of the data collected. It is important to consider how the data will be analyzed when designing the survey. Gathering information is useless if it cannot be analyzed. Analysis is also likely to take up a significant amount of time in the survey process. The method of analysis used depends on the type of data gathered.

12.5.3.4 Report the Findings

The information gathered must be given back to the community from which the information was obtained. The survey results should also be given to and used by relevant decision-makers. The survey is only as good as the sorts of planning decisions and interventions that grow from it, and this depends on how well the results can be communicated to others. In the report, it is important to recognize and discuss any difficulties or problems that might affect the interpretation and generalization of the findings. The people you are communicating the results to should be aware of how representative of the community the results are.

12.6 Future of Surveys

Surveys in agricultural research have been and will continue to be important in generating knowledge for policymakers and stakeholders. The design and conduct will, however, be shared by the complexity and sensitivity of the topics addressed, the innovative techniques being used, and the new ethical dilemmas. This is likely to include the increased use of digital data collection tools and AI technologies to address the complexities of climate change, food security, and suitable farming practices. See some of the questions discussed in Chap. 6.

Recommended Further Reading

1. Caruana EJ, Roman M, Hernández-Sánchez J, Solli P (2015) Longitudinal studies. J Thorac Dis 7:E537–E540. https://doi.org/10.3978/J.ISSN.2072-1439.2015.10.63
2. Denscombe M (2017) EBOOK: The good research guide: for small-scale social research projects Martyn Denscombe. Open University Press
3. Babbie E (2021) The practice of social research 15th Cengage Learning
4. Fowler FJ (2009) Survey Research Methods, 4th edn. SAGE Publications, Inc
5. *Aday LA, Cornelius LJ (2006) Designing and conducting health surveys: a comprehensive guide, 3rd edn. Wiley

Though this book is based on human health studies, it contains useful information that can be applied in agriculture. Many of the concepts and principles presented are applicable to animal health. The book provides a thorough understanding of survey design and implementation, addressing issue sampling, data collection, and analysis, all of which are crucial for effective surveys in agricultural studies.

Key Highlights of This Chapter

What Is a Survey Research Design?

Survey research involves collecting data systematically about people and their characteristics using a set of predetermined questions or interviews given to them. Surveys differ from experimental designs in that they do not involve the researcher giving participants a treatment.

Because of the lack of manipulation, researchers cannot explicitly explain cause-and-effect relationships, despite data from the surveys being used to explain cause-and-effect relationships.

Purpose of Surveys

The primary objectives of survey research in agriculture are to comprehend and describe current conditions, track changes over time, and assess the impact of programs and projects.

Surveys are also useful for understanding cause-and-effect relationships, studying phenomena' characteristics, and evaluating stakeholders' opinions, attitudes, and behaviors.

They also assist in forming and testing hypotheses, predicting future needs, and providing data for planning and decision-making. Finally, surveys help in identifying research problems and exploring individual beliefs regarding agricultural policies and practices.

Information commonly collected from farmers using surveys includes facts (biography); current actions and practices, knowledge, opinions/beliefs, attitudes and motives, and past or future behavior.

Types of Surveys

Based on time, two main types of surveys are recognized: cross-sectional and longitudinal. A cross-sectional survey involves collecting data at a point in time, also called a "snapshot" study, while longitudinal surveys gather data repeatedly over a period of time. Three main types of longitudinal surveys are trend studies, cohort studies, and panel studies.

Longitudinal Studies in Animal Epidemiological Research

Researchers in veterinary epidemiology generally use many variants of surveys. Their main aim is usually to identify the natural history of the disease and its outcome, identify the risk factors involved, and identify the incidence rate. They are also useful for monitoring the number of cases of a disease or monitoring changes in birth and mortality rates.

Case–control studies are one of the variants, called case comparison or retrospective studies, meaning looking backwards. Typically, the case–control study compares a group of diseased animals with a group of healthy animals after exposure.

Cohort studies are most common in epidemiology. Unlike the case–control studies, the cohorts (groups of animals) with varying levels of exposure are followed over time to evaluate the occurrence of an outcome. They are forward-direction studies (moving from exposure to outcome) and are typically prospective studies (the outcome has not occurred at the start of the study).

Ways of Administering Surveys

Surveys are administered in two broad categories: interview-completed questionnaires and self-administered questionnaires. In the interviewer-completed questionnaire, the researchers ask questions directly to the respondents, guided by a predetermined set of questions, and record the answers from each respondent. On the other hand, self-administered questionnaires rely on the respondents reading the questions on their own and recording their answers.

Interviewer-completed questionnaires can be administered in three ways: face-to-face interviews, paper surveys, and telephone surveys, usually over the telephone. Self-administered questionnaires can also be conducted in three ways: internet or online surveys, postal surveys, or mail questionnaires.

Steps in Conducting a Survey

Three phases are useful in a survey: planning, mid, and final phases. The planning phase is the first step in designing the study, which mainly includes deciding on the objectives of the study and selecting the necessary parameters. This is followed by selecting the type of survey and designing the questionnaire. The sample is then selected, and a survey is conducted (data collected). Finally, data are analyzed and reporting is done.

Review Questions

1. In your own words, what is survey design, and why is it a popular choice for researchers in social studies, especially within agricultural economics?
2. You are a researcher interested in understanding the current attitudes of farmers toward a new drought-resistant maize variety. Would you use a cross-sectional or longitudinal survey design? Justify your choice.
3. A team of agricultural economists is planning to assess the impact of a new government subsidy program on smallholder farmers' adoption of irrigation technologies. They plan to collect data before the program starts and again after 2 years. Which two variants of cross-sectional surveys would be most relevant for this study? Explain why.
4. Imagine you are tasked with designing a baseline survey for a livestock health intervention program in a rural community. What are the essential components or aspects you would focus on capturing in this survey?
5. A researcher (A) wants to study the long-term health outcomes of dairy cows born in a specific year, tracking them annually until they leave the herd. Another researcher (B) wants to track the same group of farmers over 5 years to understand their evolving adoption of sustainable farming practices. Given the above:

- Which research scenario would best utilize a panel survey?
- Which would best utilize a cohort survey?
- Clearly differentiate between the two, using these scenarios as examples.

6. A veterinary epidemiologist wants to investigate the factors that led to a recent outbreak of a specific cattle disease by looking back at the health records of affected and unaffected animals. Another epidemiologist wants to study the progression of a newly identified poultry disease by monitoring a group of healthy chickens over time. From the above:
   - Identify which scenario describes a retrospective survey and which describes a prospective survey in livestock health research.
   - Explain the fundamental difference between these two approaches.

Chapter Exercise

1. You need to conduct a survey to understand farmers' willingness to adopt new mobile banking services for agricultural payments. Consider the following exercises:
   - Exercise A: Describe a situation where using an interviewer-completed survey would be more advantageous. Specify the administration method (face-to-face, telephonic, internet) you would choose and explain why.
   - Exercise B: Describe a situation where a self-completed survey would be more appropriate. Specify the administration method (postal, online) you would choose and explain why.
   - Exercise C: Based on your chosen methods, discuss one advantage and one disadvantage of each approach.
2. Suppose you are part of a team planning a survey to assess the impact of a new irrigation technology on crop yields and household income among smallholder farmers in a specific sub-county. Carry out the following activities:
   - Phase 1: Planning: List three specific tasks you would undertake in the planning phase.
   - Phase 2: Mid-phase: What crucial activity would you perform in the mid-phase? Explain its importance.
   - Phase 3: Final Phase: Outline the sequence of activities that would take place in the final phase, ensuring you mention participant recruitment, data collection, analysis, and reporting.

Answers to Review Questions

- Q1. Reference: 12.1.1.
- Q2. Reference: 12.2.
- Q3. Reference: 12.2.2.
- Q4. Reference: 12.2.1.3.
- Q5. Reference: 12.3.3 and 12.3.3.2.
- Q6. Reference: 12.3.3.2.

References

Åberg M, Emanuelson U, Troell K, Björkman C (2019) Infection dynamics of Cryptosporidium bovis and cryptosporidium ryanae in a Swedish dairy herd. Vet Parasitol 276:100010. https://doi.org/10.1016/J.VPOA.2019.100010

Aday LA, Cornelius LJ (2006) Designing and conducting health surveys: a comprehensive guide, 3rd edn. Wiley

Aggarwal R, Ranganathan P (2019) Study designs: part 2 - descriptive studies. Perspect Clin Res 10:34–36. https://doi.org/10.4103/PICR.PICR_154_18

Babbie ER (2021) The practice of social research, 15th edn. Cengage Learning, p 575

Baker HK, Singleton JC, Veit ET (2010) Survey research in corporate finance: bridging the gap between theory and practice. Oxford University Press

Bhopal RS (2016) Epidemiological study designs and principles of data analysis: a conceptually integrated suite of methods and techniques. In: Concepts of epidemiology. Oxford Academic Books, pp 314–388. https://doi.org/10.1093/MED/9780198739685.003.0009

Caruana EJ, Roman M, Hernández-Sánchez J, Solli P (2015) Longitudinal studies. J Thorac Dis 7:E537–E540. https://doi.org/10.3978/J.ISSN.2072-1439.2015.10.63

Chan BKC (2023) Biostatistics for epidemiology and public health using R. Springer. https://doi.org/10.1891/9780826110268

Christensen J (2001) Epidemiological concepts regarding disease monitoring and surveillance. Acta Vet Scand 42:11–16. https://doi.org/10.1186/1751-0147-42-S1-S11/FIGURES/1

Christensen AI, Ekholm O, Glümer C, Juel K (2014) Effect of survey mode on response patterns: comparison of face-to-face and self-administered modes in health surveys. Eur J Pub Health 24:327–332. https://doi.org/10.1093/EURPUB/CKT067

Creswell JW, Guetterman TC (2019) Educational research_ planning, conducting, and evaluating quantitative and qualitative research. Pearson, pp 1–681

Denscombe M (2017) EBOOK: the good research guide: for small-scale social research projects Martyn Denscombe. Open University Press

Edlund JE, Nichols AL (2019) Advanced research methods for the social and behavioral sciences. Cambridge University Press

Farrington DP (1991) Longitudinal research strategies: advantages, problems, and prospects. J Am Acad Child Adolesc Psychiatry 30:369–374. https://doi.org/10.1097/00004583-199105000-00003

Fowler FJ (2009) Survey research methods, 4th edn. SAGE Publications, Inc.

Gesesew HA, Woldemichael K, Massa D, Mwanri L (2016) Farmers knowledge, attitudes, practices and health problems associated with pesticide use in rural irrigation villages, Southwest Ethiopia. PLoS One 11:e0162527. https://doi.org/10.1371/JOURNAL.PONE.0162527

Hayden GF, Kramer MS, Horwitz RI (1982) The case-control study: a practical review for the clinician. JAMA 247:326–331. https://doi.org/10.1001/JAMA.1982.03320280046028

Hoebel J, von der Lippe JE, Lange C, Ziese T (2014) Mode differences in a mixed-mode health interview survey among adults. Arch Public Health 72:1–12. https://doi.org/10.1186/2049-3258-72-46/TABLES/4

Horwitz RI, Feinstein AR (1982) Methodologic standards and contradictory results in casecontrol research. Am J Med 66:556–564

Id B, Afsheen S (2021) A cross sectional survey of knowledge, attitude and practices related to the use of insecticides among farmers in industrial triangle of Punjab, Pakistan. PLoS One 16:e0255454. https://doi.org/10.1371/JOURNAL.PONE.0255454

Jan K, Ziauddin L, Nho SJ (2023) Cohort studies. Transl Sports Med:187–190. https://doi.org/10.1016/B978-0-323-91259-4.00078-3

Kelley K, Clark B, Brown V, Sitzia J (2003) Good practice in the conduct and reporting of survey research. Int J Qual Health Care 15:261–266. https://doi.org/10.1093/INTQHC/MZG031

Kislaya I, Santos AJ, Lyshol H et al (2021) Collecting valid and reliable data: fieldwork monitoring strategies in a health examination survey. Port J Public Health 38:81–90. https://doi.org/10.1159/000511576

Lewallen S, Courtright P (1998) Epidemiology in practice: case-control studies. Commun Eye Health 11:57

Lipps O (2021) Panel surveys. Forscenter. https://doi.org/10.24449/FG-2021-00014

Meyer VM, Benjamens S, El Moumni M et al (2022) Global overview of response rates in patient and health care professional surveys in surgery a systematic review. Ann Surg 275:E75–E81. https://doi.org/10.1097/SLA.0000000000004078

Moser CA, Kalton G (2017) Survey methods in social investigation. Taylor & Francis. https://doi.org/10.4324/9781315241999

Msoma A, Lenzi R, Godwin C et al (2020) Baseline, cross-sectional survey of knowledge, attitudes, and practices related to efficient use of government resources in the health sector in Malawi. Malawi Med J 32:146–152. https://doi.org/10.4314/MMJ.V32I3.7

Mugenda A, Mugenda G (2019) Research methods: qualitative and mixed methods approach. Centre for Innovative Leadership & Governance, pp 100–102

Mugwe J, Ayieko D, Bett E, Mogaka H (2019) Determinants of smallholders farmers' participation in collective marketing of maize in the central highlands of Kenya. Afr J Rural Dev 4:225–241

Muturi M, Mwatondo A, Nijhof AM et al (2023) Ecological and subject-level drivers of interepidemic Rift Valley fever virus exposure in humans and livestock in Northern Kenya. Sci Rep 13:1–11. https://doi.org/10.1038/s41598-023-42596-y

Njenga MK, Ogolla E, Thumbi SM et al (2020) Comparison of knowledge, attitude, and practices of animal and human brucellosis between nomadic pastoralists and non-pastoralists in Kenya. BMC Public Health 20:1–10. https://doi.org/10.1186/S12889-020-8362-0/TABLES/4

Paulo C, Machado I, Carvalho H et al (2021) A 5-year retrospective study of canine and feline patients referred to an isolation unit for infectious diseases. Vet Rec Open 8:e5. https://doi.org/10.1002/VRO2.5

Pottegård A (2022) Core concepts in pharmacoepidemiology: fundamentals of the cohort and case–control study designs. Pharmacoepidemiol Drug Saf 31:817–826. https://doi.org/10.1002/PDS.5482

Pugh CA, De BM, Bronsvoort C et al (2014) What can cohort studies in the dog tell us? Canine Genet Epidemiol 1:1–9. https://doi.org/10.1186/2052-6687-1-5

Puhan MA, Ahuja A, Van Natta ML et al (2011) Interviewer versus self-administered health-related quality of life questionnaires - does it matter? Health Qual Life Outcomes 9:1–11. https://doi.org/10.1186/1477-7525-9-30/FIGURES/2

Sauter-Louis C, Carlin A, Friedrich A et al (2012) Case control study to investigate risk factors for bovine neonatal pancytopenia (BNP) in young calves in southern Germany. Prev Vet Med 105:49–58. https://doi.org/10.1016/J.PREVETMED.2012.02.008

Silatsa BA, Simo G, Githaka N et al (2019) A comprehensive survey of the prevalence and spatial distribution of ticks infesting cattle in different agro-ecological zones of Cameroon. Parasit Vectors 12:1–14. https://doi.org/10.1186/S13071-019-3738-7/FIGURES/5

Thrusfield M, Christley R, Brown H et al (2017) Veterinary epidemiology: fourth edition, pp 1–861. https://doi.org/10.1002/9781118280249

Tulu D, Gojam A, Deresa B (2020) Case-control study on risk factors associated with brucellosis in aborted breeding goats in Jimma zone, Ethiopia. Asian J Anim Sci 14:137–144

Van der Stede WA (2014) A manipulationist view of causality in cross-sectional survey research. Acc Organ Soc 39:567–574. https://doi.org/10.1016/J.AOS.2013.12.001

Weisberg HF (2005) The Total survey error approach: a guide to the new science of survey research. University of Chicago Press

Zanga A, Ozkirimli E, Stella F (2022) A survey on causal discovery: theory and practice. Int J Approx Reason 151:101–129. https://doi.org/10.1016/J.IJAR.2022.09.004

Nonexperimental Designs 2: Qualitative and Participatory Approaches

13

Abstract

The experimental and survey designs described in Chaps. 7, 8, 9, and 12 are grounded on the positivist paradigm and involve the collection and presentation of data in numerical form. While these methods are commonly used in agriculture, they are not suited to every research question, as not all aspects of agriculture can be measured quantitatively. This necessitates the consideration of qualitative and participatory approaches, which focus on elements that cannot be expressed numerically. Qualitative research emphasizes narratives, meanings, and in-depth exploration of phenomena and context. It is particularly useful for understanding the complex interactions between human behavior and the farm environment. Rooted in the disciplines of sociology, psychology, and anthropology, qualitative research is based on the philosophical assumptions that multiple realities exist, meanings are complex, and research is value laden. Though the use of qualitative research in agriculture is not as developed as in education and sociology, it is increasingly being used to address agricultural social and development challenges. We mostly find qualitative approaches used in gender, policy analysis, and market research. We argue that the limited use in agricultural research potentially loses more embedded knowledge, particularly useful in understanding the complex farming systems. Qualitative approaches offer several benefits suited to understanding complex issues, such as the multifaceted social processes of farmers. In gender studies, qualitative approaches are suited for illuminating issues of males and females. In agricultural markets, they are useful for mapping the value chains, as well as visualizing roles and relationships, and understanding the formal and informal systems that govern interactions among the actors. Qualitative methods are particularly useful in capturing the contextual perspectives of those affected by policies, providing a more comprehensive understanding of the issues at hand. Basic characteristics of qualitative research approaches include a lack of a hypothesis and the nonuse of statistics in data analysis. It also operates on a central phenomenon that aims to extract details of specific settings. It is context-bound. Data collection is more of a depth than a breadth of analysis, and aims to present information-rich cases, where "thick description" is emphasized. It uses multiple sources of data collected in a natural setting. Steps to follow in qualitative research closely resemble those of quantitative research, but qualitative research is complex, with backward and forward movements and steps overlapping. It is process-oriented, as opposed to quantitative research, which is outcome-oriented. Overall, key differences exist between the quantitative and qualitative approaches in conception, data collection, analysis, and reporting. The main limitations of qualitative approaches include subjectivity, time-consuming, inadequate rigor, and limited generalization.

Keywords

Context bound · Central phenomenon · Emergent design · Gender analysis · Holistic approach · Interpretative · Multiple sources of data · Natural setting · Open-ended questions · Policy analysis · Purposive sampling · Subjectivity and objectivity · Thick description · Value chain analysis

Learning Outcomes
Reading of this chapter should enable you to:

1. Appreciate the meaning and the application of qualitative research
2. Determine when to use qualitative research
3. Articulate the state of the art of the use of qualitative research in agriculture

J. Njeri Mugwe, S. Runo, *Research Methodology in Agricultural Sciences*, https://doi.org/10.1007/978-981-95-1892-0_13

4. Recognize the benefits of using qualitative research in agriculture
5. Appreciate qualitative research in gender studies, agricultural policy studies, and agricultural value chains
6. Differentiate between quantitative and qualitative research approaches at the various stages of the research process
7. Distinguish between the terms the central phenomenon and research objective
8. Appreciate the difference between data collection in quantitative and qualitative research
9. Distinguish between process-oriented qualitative research and outcome-oriented quantitative research
10. Differentiate between qualitative and quantitative research in terms of data analysis and report writing
11. Apply the eight steps in carrying out qualitative research
12. Highlight the limitations of using qualitative research approaches

13.1 Overview of Qualitative Research Approaches

13.1.1 What Is Qualitative Research?

Qualitative research is a methodological approach that explores and provides a deep understanding of phenomena using non-numerical data (Creswell 2014). Researchers use this approach to collect, analyze, and interpret non-numeric data. They use methods such as interviews, observations, and textual or visual materials to find underlying meanings, patterns, and implications (Chap. 19). Qualitative research approaches are utilized to answer research questions about establishing the phenomenon's "whys" and "hows," unlike the quantitative that answers "how many' and "how much." The answers to these questions (why and how) are expressed in statements and words.

In contrast to quantitative experimental and survey methods, which focus on numerical measurement and statistical analysis, qualitative research in agriculture adopts an interpretive approach aimed at understanding the complex interactions and perceptions underlying agricultural practices. It focuses on *narratives, meanings,* and *in-depth exploration of phenomena* and *context.* This aspect is particularly significant in agriculture because it allows researchers to understand the complex nature of farming systems. By employing qualitative approaches, researchers can explore the factors influencing farmers' decision-making, community dynamics, and the socio-cultural context that shapes agricultural practices (Dooley 2007).

For example, a researcher can choose to use qualitative approaches to understand what motivates farmers to practice conservation agriculture. The researcher would then select a few participants and collect data using in-depth interviews, farm visits, and observing farmers' daily activities. This would help gather data on farmers' experiences, challenges, and decision-making processes. It would reveal important factors like community dynamics, cultural beliefs, individual farmers' beliefs, and access to resources. The findings would lead to understanding why some chose to adopt conservation agricultural practices while others did not.

13.1.2 When to Use Qualitative Research

Typically, a qualitative research design is used if your research question seeks to *explore the meaning* of a phenomenon, *generate new constructs and theories, develop a story, explain a pattern, observe a practice* or *process* in depth, or *understand participants' perspectives* (Creswell 2014; Creswell and Guetterman 2019).

The following is a guide of concepts that qualify to engage in a qualitative research approach:

- Meaning of a certain behavior or attitude. (How do farmers practice the new technology, and what are their attitudes?)
- Understand a process, e.g., how changes in the land use system have happened over time.
- Explore how a practice is performed within a certain context. For example, how is rice grown and marketed in a certain region?
- Understand people's beliefs toward a certain practice or behavior. What are farmers' beliefs in dryland areas concerning using mineral fertilizers?
- People's perspectives on a phenomenon (how do farmers perceive zai-pits as a water harvesting system?
- What experiences do people have in doing a certain activity? (What is the experience of farmers participating in the collective marketing of passion fruits? What is the meaning of such experiences?)
- Barriers and challenges associated with certain agricultural practices.
- People's attitudes toward a certain aspect.
- How perceptions and attitudes affect behavior of farmers and other stakeholders in agriculture?

Examples of research questions that fit qualitative research designs:

1. How do farmers experience and perceive the effectiveness of newly introduced climate-smart agricultural technologies in mitigating the impacts of climate variability on their crops?
2. Why do farmers in Region A prefer different climate change adaptation strategies compared to those of Region B?
3. What are the roles, functions, and responsibilities of the various actors in the avocado value chain?

4. How is the process of disseminating agricultural knowledge through various social networks among the farmers, and what are farmers' interpretations regarding the effectiveness of the process?
5. How do farmers perceive government policies on promoting new varieties of iron-enriched beans, and how do these perceptions influence their adoption?

13.1.3 Origin of Qualitative Approaches

Qualitative research has its roots in various disciplines within the domain of social sciences (Huberman and Miles 1983; Miles and Huberman 1984; Dooley 2007). This includes sociology, psychology, and anthropology. It extends back to the nineteenth and twentieth centuries when social scientists began to question quantitative research, which uses a *reductionist approach* to research (Minati 2023). The reductionist approach, dominant in the nineteenth century, held that knowledge could only be acquired through *objective observations* and *hypothesis testing*. It involves breaking down a phenomenon into its most basic parts to understand. However, as social scientists began to study more complex and distinct social phenomena, they realized that the reductionist approach was insufficient for understanding human behavior (Dooley 2007). While justifying qualitative research, Dooley (2007) explained that "life does not come to us as a math problem but more like a story with setting and context."

Wilhelm Wundt, Sigmund Freud, and William James were among the earliest sociologists who pioneered qualitative research in 1900 (Wertz 2014). These sociologists argued that human experience cannot be reduced to a single, observable, and measurable reality. Instead, they argued that subjective meanings that reveal inner experiences were more important. In the 1970s and 1980s, qualitative research continued to evolve with the development of new methods (grounded theory and phenomenology) developed by Glaser and Strauss (Dreher and Santos 2017; Burns et al. 2022). Throughout the twentieth and twenty-first centuries, qualitative research methodologies have grown, evolved, and adapted to new contexts. Today they are widely used across social sciences in many fields, including education, business, and health care. This evolution shows the increasing recognition of the value of qualitative knowledge in understanding complex social phenomena.

13.1.4 Philosophical Assumptions

To remind ourselves, the philosophical assumption relates to "ontology, nature of reality," "epistemology, "axiology, the role of value in a study," and methodology (the research process) (Mertens 2010; Kerry 2013; Pretorius 2024). Researchers employing qualitative approaches and methods embrace an ontological perspective that acknowledges the existence of multiple realities. Qualitative researchers conduct their studies to uncover and report multiple realities as advanced by the individuals (participants). The authenticity of these varied experiences in a qualitative study is captured through the actual words and narratives presented in the research report.

Epistemologically, qualitative researchers develop multiple subjective meanings of their experiences, leading to the complexity of meanings rather than narrowing them into a few categories. The subjective meanings are negotiated socially and historically (Creswell 2014). In practice, qualitative researchers look for their subjects and prefer to collect data in the field where the participants live and work. For example, for an ethnographic study (culture of people), a researcher may decide to go and live with the community concerned. This helps gain deeper insights regarding social interactions as the researcher records the observations.

The axiological assumption guiding qualitative research is based on the principle of being value laden. This means that the researcher admits the value-laden nature of the qualitative study and keenly reports their values and biases, leading to the nature of the information gathered from the participants.

13.1.5 State of the Art of Qualitative Research in Agriculture

Qualitative research approaches are increasingly used to address agriculture's social and development challenges. However, compared to other fields, like education, psychology, history, and sociology, qualitative designs are less developed in agriculture and are mainly used in gender studies, policy, and agricultural markets (discussed in the next section). Traditionally, quantitative approaches have been the main focus of research and publications in agriculture. Most books in agricultural research focus on quantitative methods, e.g., experiments, surveys, and correlational studies. A search through master's and PhD theses in Kenyatta University's and University of Embu library repositories revealed a low level of application of qualitative methodologies. None had used purely qualitative methods, and only a few have attempted to use a mixed methods approach, mainly incorporating key informant interviews, focus group discussions, and case studies.

In Kenya's national agriculture research systems, the situation is similar to that of universities. Researchers mainly use quantitative methodologies, with qualitative research being employed in the context of participatory appraisal tools, particularly when working with extension workers. The agricultural researchers, mainly those in the agricultural economics department, often receive training on participatory research approaches and tools, but not much on other qualitative methods like grounded theory. Consequently, there is less exploration of qualitative approaches in research studies. Quantitative research methods have dominated agricultural research for a long time, overshadowing the contribution of qualitative research methods (Rubin 2016). This overreliance on quantitative methods has limited the depth and breadth of understanding that qualitative methods provide. It limits the exploration of the complex socio-cultural factors and farmer behavior that quantitative data alone cannot capture.

This scenario presents a potential loss of more embedded knowledge regarding the social perspective in agriculture. For example, the farmers, who are major stakeholders in agricultural research, are heterogeneous and, therefore, hold a rich diversity of information acquired through the years. We believe their meanings and interpretations of their experiences would help address many of today's agricultural problems. We, therefore, argue that we need increased use of qualitative research to ensure that this knowledge is captured, understood and used to improve agricultural interventions. There are, however, several barriers limiting the increased and sustained use of qualitative approaches.

One of the barriers to the increased use of qualitative approaches is the *mindset and beliefs* of researchers, lecturers, and students in agriculture. As earlier alluded to, the majority have been trained purely in quantitative approaches, and there is a high level of ignorance concerning the applicability of qualitative methodologies. Some (especially in biophysical disciplines) will tell you that qualitative research is the one in which you collect data from farmers and other stakeholders using a questionnaire. Furthermore, faculty staff will discourage students from engaging in qualitative research because of this ignorance and, more so, due to a lack of professional familiarity.

Past and present training have contributed greatly to the low application of qualitative research in agriculture. Agriculture is considered a life science; therefore, traditional research methods curricula will concentrate on quantitative methods at the expense of qualitative methods. For example, agronomists are well versed in experimental methods, and most topics in curricula center around quantitative techniques. Statistics, for instance, will be taught intensively in both undergraduate and postgraduate programs. In contrast, qualitative methods are often mostly taught as a topic in the research methods curriculum and not by experts in the field. This results in a biased orientation, subsequently creating a big gap in knowledge of qualitative approaches (see Box 13.1). Additionally, the fact that professionals may not be willing or have time to engage in additional training to fill this gap raises concerns.

A third reason for less qualitative orientation than quantitative is related to *publications*, which prefer objective and experimental studies, especially in sciences (Serpa et al. 2020). Faculty staff, as well as students, are required to publish in credible journals. Many credible journals are traditionally mainstream, which tends to be biased toward quantitative approaches, as evidenced by instructions to authors. Most of the journals in the agricultural domain will recommend the IMRAD (Introduction, Materials and Methods, Results and Discussion) template, which may not be strictly applicable to a qualitative study (Chap. 25) (Wu 2011). In addition, reviewers of these journals may not have the evaluation criteria or experience for qualitative research. Thus, they often expect large sample sizes and may perceive qualitative findings to be too subjective, which can lead to rejection.

Despite these drawbacks, we believe that qualitative approaches have a big role to play in tackling the diverse social and developmental problems facing agriculture today. With the emergence of mixed methods research, there is now more emphasis on qualitative research as a paradigm. In this book, we attempt to discuss key issues, concepts, and principles of qualitative approaches, albeit briefly, with the hope that this will contribute to increasing competencies among agricultural researchers. More so, we hope to create interest in qualitative research among agricultural researchers. Key references are provided to help readers who wish to learn more. To be an effective qualitative researcher, you need more knowledge and skills, possibly more than what we provide here; thus, the reference materials at the end of this chapter will be useful.

Box 13.1 Experience by the first author of the book

I have witnessed a student in agricultural economics struggling to formulate a hypothesis for a research study on the dairy value chain. The student was facing difficulty in creating a testable hypothesis with measurable variables. The fact of the matter is that it is impossible to have a testable hypothesis in this kind of study because it involves describing processes rather than explaining a scenario—which many agricultural students may not be familiar with. Agricultural "Research Methods" courses usually emphasize experimental research, and only a few students have knowledge of qualitative approaches. Proficiency in qualitative research is essential to ensure that students and researchers can explore and identify patterns, providing valuable insights for policy-making and interventions as they navigate the complex and diverse landscapes of agricultural research. This chapter is a crucial resource for readers and students, offering valuable insights regarding qualitative research approaches.

13.1.6 Benefits of Qualitative Research in Agriculture

We derive several benefits from qualitative research. As already mentioned, it is suited for exploring complex human issues, such as the *multifaceted social processes of farmers*. The data we collect provide rich and detailed insights into participants' experiences, attitudes, motivations, and perspectives regarding their decision-making processes. These include:

- *Agricultural economics*: Provide information on market dynamics, price information, and the impact of market structures among smallholder farms, e.g., it can reveal how farmers negotiate prices, the challenges they face in market access, and the role of middlemen in the supply chain (see details later in this chapter).
- *Agronomy:* Explores farmers' knowledge, attitudes, and practices related to crop management, soil management, and farming practices.
- *Agricultural extension*: Are used to engage directly with farmers, identify knowledge gaps, and customize education and training programs.
- *Agribusiness management:* Qualitative inquiry offers insights into consumer preferences, the functioning of supply chains, and the internal dynamics of agricultural organizations.
- *Gender studies:* Helps to understand the roles, experiences, and challenges faced by different genders. It explores how gender influences access to resources, decision-making, labor division, and participation in agricultural activities.
- *Policy studies:* Analyses agricultural policies by examining documents and conducting interviews with stakeholders to identify strengths and weaknesses of existing policies, uncover unintended consequences, and suggest improvements.
- *Rural sociology*: Employs qualitative methods to explore community dynamics and social structures.

The following section gives more detailed information on how qualitative approaches are applied in gender, agricultural markets, and policy analysis.

13.2 Qualitative Research in Agriculture

13.2.1 Gender Studies

Gender inequalities and disparities in agriculture have long been recognized, making it important to study gender in agriculture. In developing nations, women comprise more than 60–80% of farmers and play a key role in agricultural development (Yan et al. 2022; Li et al. 2023; Islam et al. 2024). They contribute significantly to agricultural and rural economies by producing crops, preparing food, tending to animals, working in agricultural enterprises for wages, engaging in trade marketing, and caring for their family members and homes. They struggle to manage extremely complex households. However, despite their significant contributions, they are often neglected and subjected to inequalities compared to men (Njuki et al. 2014). They frequently face challenges regarding resource access, decision-making, and recognition of their contributions. This is a big concern because we cannot optimize agricultural growth and development without recognizing the central contributions of both male and female genders in agriculture (Haug et al. 2021; Asante et al. 2023; Msofi Mgalamadzi et al. 2024).

Interest in gender mainstreaming has increased over the past decades due to the growing recognition of the negative effects of gender disparities on agricultural performance (Njuki et al. 2022). Gender inequalities affect the distribution of resources and make it difficult for men and women to contribute to agricultural development optimally. The interest in gender mainstreaming is both among development workers and researchers. In research, gender mainstreaming means considering gender issues throughout the research cycle. This calls for integrating *gender analysis* in the conceptualization, problem analysis, design, and execution of any research project (de Waal 2006; Mangheni et al. 2019).

Gender analysis aims to explore the roles and responsibilities of males and females, and highlight their societal relationship and the inequalities in those (Bala-Miller et al. 2022). Typical questions that gender analysis seeks to answer include who does what and who has what?. Who decides? and how?, Who gains?, and who loses? To answer these questions, three closely related paradigms guide gender-analysis studies (Nelson and Huyer 2016): (1) Gender-responsive research, (2) Gender-focused research, and (3) Gender-informed research. Gender-responsive research seeks to collect data useful in designing agricultural interventions that can meet the needs of both men and women and reduce any existing gender inequalities. Gender-focused research aims to expand knowledge about men's and women's behaviors, constraints, values, and opportunities. For example, it seeks men's and women's areas of knowledge, patterns of time allocation, control and ownership of diverse production assets, and the level of participation in agricultural-related activities. Finally, gender-focused research draws on findings from prior gender studies to deepen understanding and address emerging questions about gender dynamics.

Common research questions posed in agricultural-based gender analysis studies include:

1. Who owns the land and controls the resources therein?
2. How are the resources distributed, and who has them (what decisions do men and women make in their family/community)?
3. What are the reasons behind the observed differences in gender?
4. What gender-related differences exist in social network participation, group membership, and leadership roles?
5. What are the gender disparities in information access and service from extension services?
6. How do climate-change-induced challenges affect men and women? What are farming communities' gender-adaptive strategies for dealing with these challenges?
7. What opportunities exist to enable women and men to have equality of opportunity to participate in market-led development activities?
8. What are the gendered patterns of labor, including the division of tasks and time allocation of women compared to men?

Looking at these questions, we find that the qualitative inquiry methods are the most suitable for illuminating the issues of men and women on how they view their lives and exploring context and cultural differences. Indeed, methodologies for studying gender use tools and frameworks that seek to understand contexts. These frameworks include the Havard analytical framework and gender analysis matrix, Moser Framework, and Women Empowerment Framework (Morgan et al. 2016, 2022a, b). These frameworks mainly use qualitative data collection tools and methodologies, such as observations, participatory approaches, and in-depth interviews, to collect data on the meaning of behavior, beliefs, experiences, and attitudes.

Examples

Study 1: A series of qualitative studies conducted in four countries facing the impacts of climate change. The goal was to determine men's and women's perceptions of climate change, adaptive challenges, and the degree to which assets and group participation play a role in adaptation strategies (Mohapatra and George 2021; Lazoroska et al. 2021; Brisebois et al. 2022).

Study 2: A study to promote gender-inclusive governance of mixed-sex self-help groups in the African context. The study analyzed 20 mixed-sex focus group discussions with 190 group members in Western Kenya. The study undertook an empirical assessment of institutional factors that explain and facilitate the effective participation of female members. Specifically, it examined group membership perceptions and then applied an analytical framework to guide the empirical analysis of gender-inclusive groups in decision-making.

13.2.2 Qualitative Research in Agricultural Markets

Within the agricultural marketing domain, agricultural value chains form one of the main research topics by students in agribusiness and agricultural economics. The goal is often to explore and analyze various value chains in agriculture to increase understanding of how they operate and the relationships therein. One of the most important purposes of value chain analysis is to propose a strategy for "upgrading a value chain," which means increasing efficiency, productivity, and competitiveness (Gehl Sampath and Vallejo 2018; Steffens et al. 2020; Oh et al. 2025). An agricultural value chain shows the movement of an agricultural commodity from production to consumption (Fig. 13.1).

As we can see from Fig. 13.1, the value chain is a map consisting of value chain actors and their relationship with the product at hand. According to the literature, an agricultural value chain encompasses a comprehensive array of value-adding activities that move a product or service through diverse phases of production (Donovan et al. 2015; Borsellino 2020; Akyüz et al. 2023). This includes the procurement of raw materials and other necessary inputs at the production stage by farmers to the final phase of selling and consumption. The Food and Agricultural Organization (FAO) provides a similar definition, characterizing the agricultural value chain as a complete range of activities needed to guide a product or service from its conceptualization through various production phases, involving physical transformation and incorporating diverse producer services. The process extends to delivering the final product to customers and culminates in its eventual disposal after use (Borsellino 2020).

These definitions highlight key aspects and elements of a value chain that are important to the research methodologies for analyzing a value chain. They emphasize the need to include *descriptions, processes, and understanding of relationships* among the actors within a value chain (Hastings et al. 2016; Clay and Feeney 2019). The research methodology should, therefore, be capable of describing how products and services move from the production domain to the consumer. This involves mapping out the sequence of activities from production to distribution and describing the roles played by each actor within the chain. A qualitative research methodology is the best for mapping and describing such a scenario.

Research methodologies for analyzing agricultural value chains vary and highly depend on the particular study's objectives. Donors and other organizations supporting agricultural development, e.g., World Bank and Food and Agricultural Organization, have developed and shared guidelines to guide their researchers and other stakeholders on how to analyze value chains (Stamm and von Drachenfels 2011; Donovan et al. 2015). The main goal of value chain

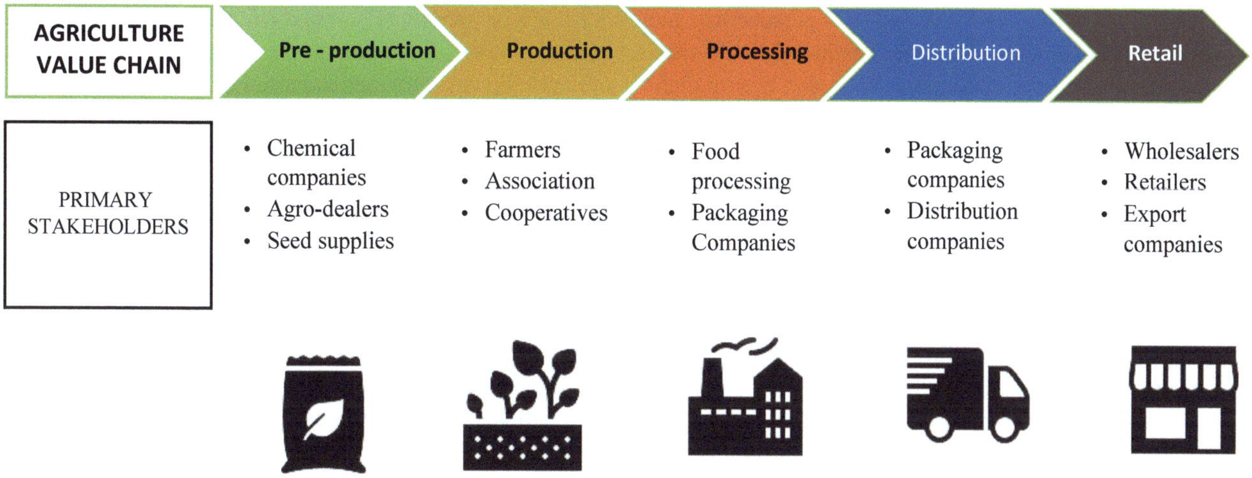

Fig. 13.1 An agricultural value chain showing the movement of an agricultural commodity from production to consumption and the actors

analysis, as explained in these organizations, is typically twofold: (1) To enable the selection of the most appropriate value chain to promote and (2) To select the most suitable intervention for upgrading or improving the selected value chain. In this case, steps for analyzing a particular value chain are given. Typical steps of an agricultural value chain analysis targeting farmers include four areas (Kaplinsky and Morris 2000):

1. Mapping and characterization of the actors in the chain.
2. Identifying governance structure within the chain.
3. Exploring areas of upgrading within.
4. Assessing benefits in the chain participation.

Mapping and characterization of the actors in the chain This step involves identifying all actors involved in the production, distribution, marketing, and sales. The aim is to understand how the products flow through the chain and how value is added at each stage. Researchers, therefore, use methods such as value chain mapping, focus group discussions, in-depth descriptions, and social network analysis to help identify each actor's role challenges, strengths, and weaknesses. The information generated helps provide a clear view of their interactions that researchers can use to design new income-generating opportunities for farmers (Lugusa et al. 2016; Laibuni et al. 2018).

Identifying governance structure within the chain Researchers aim to achieve three key objectives: (1) identify the nature of relationships and coordination mechanisms that exist between different actors within the chain; (2) understand the business environment surrounding the value chain, including market conditions, infrastructure, and institutional support; and (3) identify power dynamics and relationships that influence how the chain operates (Mishra and Dey 2018). Researchers explore how actors interact, make decisions, and share power. This helps reveal both formal and informal systems that influence relationships and coordination within the chain.

Exploring areas of upgrading within the value chain involves evaluating opportunities for improving the efficiency, quality, and value of products and services at each stage of the value chain. The process includes analyzing the current practices, benchmarking against best practices, and developing strategies for improvement. It uses tools like flow charts, gap analysis, and process diagrams to understand the current state of the value chain. Additionally, the Strengths, Weaknesses, Opportunities, and Threats (SWOT) framework helps to identify potential improvements (Kolbina 2015). Qualitative methods such as interviews, workshops, and expert consultations are employed to explore various perspectives and gather in-depth information about the value chain operations.

Assessing benefits in chain participation Involves evaluating the advantages and values that different actors gain from participating in the chain. This helps to understand the economic, social, and other benefits that participants experience. Benefits are identified through interviews and surveys, comparing economic gains with the cost before analysis. For example, for a coffee value chain, a researcher will identify benefits for farmers, quantify profits, evaluate community impact, and assess market expansion.

As we can see, suitable methodologies for accomplishing the above steps incorporate a blend of quantitative and qualitative research methods. Indeed, most of the research questions by students working on value chains mostly point to the need to use both quantitative and qualitative research approaches. Examples include:

- What are key actors' and stakeholders' perspectives and experiences within the rice value chain? How do these perspectives influence decision-making and overall chain performance?
- How do social network formation and communication technology influence the flow of information exchange, coordination among the actors, and adoption of improved tissue culture bananas?
- Why do actors in the coffee value chain adopt diverse strategies to address the impact of climate change, and how does resource availability influence the choice and implementation of these strategies?
- What are the costs and benefits associated with the adoption of the cassava value chain among smallholder farmers?

We illustrate this using an example of a case study of the soybean value chain carried out in Kenya's central highlands, which used quantitative and qualitative methods but was more biased toward the qualitative methods (Murage et al. 2019). The soybean value chain analysis was carried out within the auspices of the Alliance for a Green Revolution in Agriculture (AGRA) supported project. The project team consulted with an expert from Kenya Agricultural and Livestock Research Organisation and analyzed the soybean value chain using various methods. Methods used included, first, focus group interviews (FGD) (consisting of 8 to 12 members), which were carried out with farmer groups who were growing soybeans and were registered members of the specific groups. The FGDs were followed by a series of in-depth interviews with key informants (pioneer soybean growers), traders (both retailers and wholesalers), transporters, supermarkets, agro-input dealers, opinion leaders (influencers), Ministry of Agriculture personnel, financial institutions, and local administrators. Finally, a survey of 87 farmers was carried out: the survey tool had quantitative (closed-ended) and qualitative questions (open-ended).

Data collection tools comprised checklists and questionnaires with closed-ended and open-ended questions. Nonparticipant observation was also used to gather information on activities and processes at various value chain stages to seek clarity and verification. Data analysis was carried out by the value chain expert using various methods. Quantitative data were captured in SPSS and analyzed to give mainly descriptive statistics. On the other hand, the other data from the qualitative tools were organized and thematically analyzed, where themes, trends, and categories were generated. One of the main outputs was a value chain map (Fig. 13.2), and a series of marketing channels (Box 13.2).

Many authors have used qualitative and quantitative methods to comprehensively map the value chains of various commodities (Donovan et al. 2015; Akyüz et al. 2023). The integration enhances the depth and breadth of value in mapping efforts, offering a stronger foundation for understanding the complexities and dynamics of value chains. It is thus necessary for students to be familiar with both quantitative and qualitative methodologies, considering analysis of value chains is an important topic in agriculture.

13.2.3 Qualitative Approaches in Policy Studies

Policy analysis involves a structured approach to assessing and interpreting data to guide policymaking. In the agricultural sector, it enables stakeholders to evaluate how different policies influence key development goals like improving

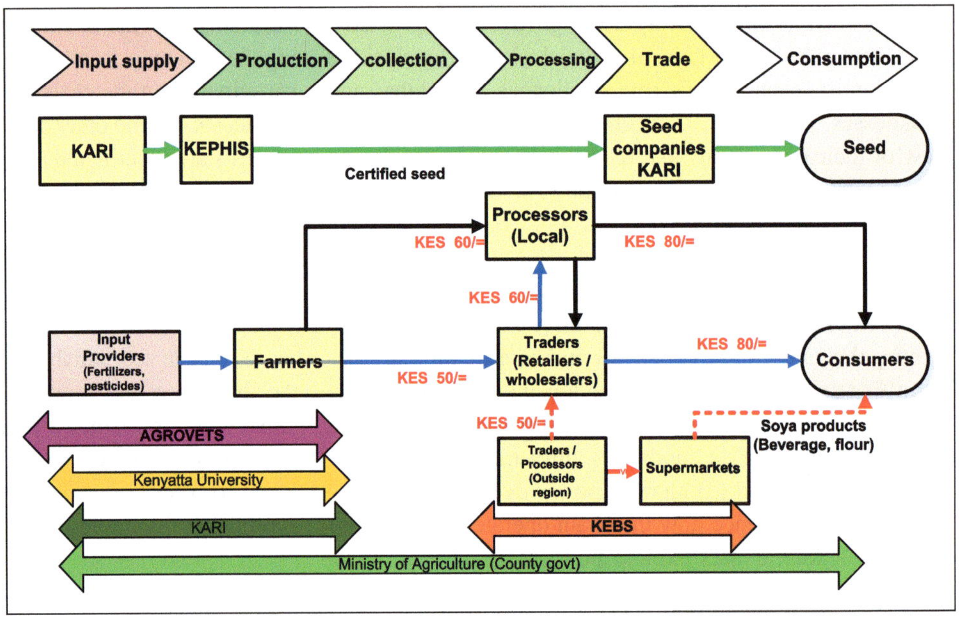

Fig. 13.2 Soybean value chain map with both quantitative data (profits and margin) and qualitative data in central Kenya (data collected in 2014). Dollar 88 KES, KARI Kenya Agricultural Research Institute, KEPHIS Kenya Plant Health Inspectorate Service, AGROVET Agrodealers and Veterinary Services

Box 13.2 Marketing channels identified in the soybean value chain analysis (as depicted in Fig. 13.2)

Channel one: Farmer → Processor → Consumer

In this channel, farmers sold raw soybeans directly to local processors in major towns in the region, Embu, Meru, Nkubu, Chuka, and Kiritiri towns (situated <50 km) at an average of KES 60 (0.68 USD). The processors would either sell raw soybeans directly to consumers through their wholesale outlets at KES 80 (0.90) or make various value-added products such as soybean beverages, flour, and roasted soya. The study found that products from local processors cannot access supermarkets due to a lack of certification by the Kenya Bureau of Standards.

Channel two: Farmer → Traders/Retailer/wholesaler → Processor → Consumer

In channel two, farmers sold soybeans to traveling traders, retailers, and wholesalers at KES 50 (0.56 USD), who in turn sold to consumers at KES 80 (0.90 USD).

Channel three (value-added products): Processor → Supermarkets → Consumer

In this channel, processors from outside the region, mainly from Thika and Nairobi (situated >50 km), sell value-added products to supermarkets in the area. The soybean products sold in the local supermarkets had all been acquired from outside the county, meaning local producers have an opportunity to improve soybean products to sell to supermarkets. This is a strategy for upgrading.

USD = KES 88 (2014)

food security, alleviating poverty, and promoting environmental sustainability. This involves understanding complex interactions between policies, stakeholders, and the broader development context.

Most research in agricultural policy analysis employs a range of methodologies that combine quantitative and qualitative methods to provide a well-rounded understanding of policy impacts (Munyua and Stilwell 2010; Akimowicz et al. 2018). Quantitative methods primarily encompass econometric modeling and cost-benefit analysis to evaluate economic outcomes and policy effectiveness. The qualitative methods involve policy reviews, stakeholder analysis, and in-depth case studies. This help captures the perspectives and experiences of individuals and communities affected by agricultural policies.

13.2.4 Key Methods Used to Collect Data in Qualitative Research

If you choose to engage in qualitative research, you may choose from several techniques to collect data, including interactive and observational methods. Briefly, qualitative researchers commonly use five main approaches: case study, ethnography, narrative, grounded theory, and phenomenology. These are discussed in Chap. 14. The key methods for collecting qualitative data are four (discussed in detail in Chap. 19): They include interviews, observations, and document analysis as discussed, hereafter.

1. Interviews

Interviews are interactive sessions and can be of different types:

- In-depth interviews.
- Focus group discussions: The researcher interviews a defined group of people to get an in-depth perspective of a well-defined topic.
- Group interviews: The researcher interviews several people in small groups or large groups, usually to create a general picture of a practice or phenomenon.
- Key informants interviews: The researcher conducts interviews with influential people who are experts in the topic of interest.

2. Observations

Participant observation: The researcher participates and "Learn by doing." Researchers reside with the participants, where they experience, observe, and record information about the people. They may stay with the participants for months to even years (Rubin 2016). As a student, it may not be feasible to reside in the study area for many months because of time limitations, but depending on your research question, the time could be short.

Nonparticipant observation: The researcher stays away from the participants, and does not engage in what they are doing.

3. Document Analysis

The researcher examines and analyzes documents such as policy reports, historical records, and official reports, such as annual reports, to uncover trends, patterns, and implications.

4. Participatory Research Tools

Participatory approaches are used, with the participants actively interacting to give information and learn. At times, researchers also want to empower the farmer. Several tools are used, ranging from participatory workshops to farm mapping, detailed value chain maps, and transect walks.

13.3 Basic Characteristics of Qualitative Designs

Several key characteristics of qualitative research designs differentiate it from quantitative research. We approach this topic by presenting characteristics at (1) The *conceptualization* stage is followed by (2) *Data collection*, and, finally, (3) *Analysis and reporting as discussed, hereafter.*

13.3.1 Conceptualization of Research Study

Research problem formulation The conceptualization of the problem is less precise in qualitative than quantitative research. In qualitative research, the research problem lacks a fixed definition of concepts and relies more on general ideas. It is more of exploring and identifying concepts.

Theory is developed Qualitative researchers do not usually formulate a pre-established *hypothesis or proposition* to be tested, as is customary in a quantitative study. In the conceptualization of the research study, the role and presence of theory are not always certain or even readily visible. This is because in qualitative research, sometimes the study aims to build a theory from the data collected (Fleming and Zegwaard 2018). Indeed, theoretical perspectives usually emerge naturally from the course of the study.

Research questions As is the norm, research questions in qualitative research will be derived from the general problem and aim to answer a contextual issue or an overarching subject called *"the central question."* The central research question is a wide-ranging question that examines *"the central phenomenon"* or concept in a study (Creswell and Guetterman 2019). This question encompasses a range of themes but mostly focuses on understanding the meaning and social interactions in a particular context, as opposed to quantitative research that has narrow questions based on a few variables. The central question will be stated in such a way that it will aim to extract details of specific settings surrounding the central phenomenon. Sometimes labeled as *"context-bound"* (Creswell 2014). Qualitative research aims not to generalize the findings, as is the case in quantitative research, but to understand specific contexts.

Guidelines for writing qualitative research questions include the following (Creswell 2014):

- Ask one question or two "central questions" plus at least five to seven sub-questions; these are specific questions that become useful in guiding the construction of data collection instruments and subsequent data collection.
- Start the "research questions" with the words *why, what,* and *how* to express an open and evolving design.
- Lay emphasis on a "single phenomenon or concept" and explore details during the study.
- Relate the "central question" to the specific qualitative approach and express language of developing design; will it be Ethnography, Case study, Grounded theory, Narrative, or phenomenology?
- Use probing verbs that express the language of developing design. Avoid words like *determine, effect, influence, relate, and impact* as these tend to measure specific variables. Use nondirectional terms rather than directional words that are common in quantitative research. e.g., (1) *Seek to understand* … or ethnography; (2) *Examine* a … for case study; (3) *Discover……* for grounded theory; (4) *Report the stories* of … for Narrative, and (5) *Describe experiences* of …for phenomenology.
- Anticipate the research questions to change and develop during the study, distinctive to the nature of the emergent and flexible approach.
- Be specific about the participant and the site where the study will be conducted.

Research hypothesis The research question or propositions are an alternative to the research hypothesis. This is because qualitative research is concerned with a more exploratory approach rather than testing hypothesis (proposition) and statistical tests. The research questions are not answered with a simple "yes" or "no" but rather through an in-depth exploration of the particular phenomenon. *Propositions* in qualitative research may be used; these statements suggest potential relationships or connections between concepts. Unlike hypotheses, they are not strictly tested but used to *guide the study in the exploration of patterns*. They are also called "working assumptions" (Creswell and Guetterman 2019). For example, a proposition in a study of the value chain might be: "Information and communication technology plays a crucial role in enhancing information exchange and decision-making within the dairy value chain." Such a proposition is not intended for statistical testing but is used to guide the study's exploration and analysis.

Purpose statement A purpose statement is important because it conveys the primary goal and intention behind the investigation. It could be written in either one sentence or many sentences. A good qualitative purpose statement clearly communicates "the central aim" and outlines the study's specific focus, context, methods, and significance (Creswell and Guetterman 2019). It provides a roadmap for the qualitative study. In contrast to the research questions, actions, verbs, and phrases may be used because the aim is to keep the inquiry open and, at the same time, communicate what the study aims to achieve (in the case of a proposal) (see example in Box 13.3).

Box 13.3 Purpose statement of a qualitative study in gender

The purpose of this qualitative statement is to explore the perceptions and beliefs of female smallholder farmers in rural communities regarding the growing of iron-fortified bean varieties (central aim). By employing in-depth interviews and participant observations (methods), this study aims to discover the complex ways the women contribute to decision-making regarding iron-fortified bean varieties within their communities' unique socio-cultural and economic contexts (Context). This research is motivated by the recognition that women play pivotal roles in agricultural production and sustainability, but are often under-represented in agricultural research. Through an exploration of their narratives, challenges, and strategies, this study seeks to amplify the voices of these women, providing valuable insights that can inform gender-responsive agricultural policies and interventions regarding the adoption of iron-fortified bean varieties for improved household nutrition (Significance).

13.3.1.1 Methodology and Data Collection

13.3.2 Data Collection

Selection of participants Samples in qualitative design are purposively selected (Chap. 16). This means the participants are deliberately selected based on their specific characteristics and experiences relevant to the research question or phenomenon. The researcher's aim is not to have a representative sample, as in quantitative studies, but rather to have a purposively selected sample that will allow depth and richness of information to be collected from the sample. It is more about *"depth" rather than "breadth,"* We aim at identifying and selecting *"information-rich"* cases (Palinkas et al. 2015). Whereas quantitative samples are well defined based on probability theory, qualitative samples are less defined and lack stringent rules on sampling (Luborsky and Rubinstein 1995; Vasileiou et al. 2018).

Sample sizes are typically modest, and several authors give guidelines on how to get a suitable sample size (Pye et al. 2016; Vasileiou et al. 2018). A method commonly employed to ascertain the appropriate sample size is to continue collecting data until "a point of saturation" is reached (Cohen et al. 2017). Saturation is when successive data collection and analysis reveal no new or different insights. However, saturation should be done in such a way that it is consistent with the research question and the analytical framework to avoid losing coherence and potency (Saunders et al. 2018).

Typical sample sizes range from 9 to 17 (Vasileiou et al. 2018; Smith and Little 2018). However, you should always aim to meet the minimum requirement needed to address the research question. Avoid large sample sizes at all costs. They easily lead to redundant data, increase the complexity of analysis, and waste resources without yielding additional value (Francis et al. 2010). This may be the case when open-ended questions are included in a quantitative survey. For example, if you have a sample of 400 farmers in a quantitative survey, you will have qualitative data points from 400 farmers. This is complex, and it is not advisable to include open-ended questions in a quantitative survey tool.

Multiple sources of data Qualitative research leans more toward subjectivity compared to quantitative research and uses distinct methods of collecting information. The predominant methods are individual or group in-depth interviews, observations, and focus groups. In the course of the study, researchers routinely gather and accumulate multiple forms of data captured in the form of field notes, diaries, and individual verbal narratives. In the end, the researchers undertake a comprehensive evaluation of all the data, interpreting their meaning, and structuring it into *main categories or themes* that cut across all the data sources. The information is characterized by rich details and is often referred to as *"thick descriptions"* (Staller 2015; Sankofa 2023).

Observation and interviews are predominant The data collection methods involving observations and unstructured interviews dominate the process. The researcher will dedicate a substantial amount of time in the field, watching people and their activities while concurrently conducting interviews with the participants. The researcher wants to collect comprehensive information through seeing and hearing. Most researchers will record the information in notebooks and supplement it with audio or videotaping of the proceedings.

Natural setting The primary nature of data is the natural setting. By visiting the site, a qualitative researcher gathers data where participants directly experience the issue or problem under study (Austin and Sutton 2014). This is because they believe that activities can best be understood in the axial setting where they occur. Human behavior is also influenced by particular settings. Therefore, whenever possible, it is good to visit such settings rather than ask the participants to come to the researcher's location. For example, a researcher investigating perceptions of conservation agriculture technology will visit the farmers in their location, having confirmed first that the subjects (farmers) are practising the technology. The researcher will observe and interview participants as they carry out their everyday routines.

Open-ended data collection Qualitative research is investigative and open-ended (Renjith et al. 2021). The methodology presents general questions, catalyzing the exploration of participants' perspectives. The interviewer, on the other hand, or group moderator, uses probing techniques to gather data on people's beliefs, opinions, and feelings about the topic or idea being discussed and to determine the degree of agreement that exists in the group.

Data are descriptive and process-based Qualitative data will include field notes, photographs, personal comments, video recordings, etc. The researcher aims to collect any materials that can express the actions of the people/community of interest. They are highly interested in how things happen. They will often observe how people interact with each other, note nonverbal communication, and how people's attitudes are translated into people's actions. For example, in an investigation on why farmers do not use fertilizer on green grams, a researcher might ask; What do you feel about the soil fertility of your farm? A follow-up question would be: What kind of farming methods do you feel are good for your farm? These questions will try to gauge the farmers' perceptions about managing soil health in their farms.

The researcher is a key instrument The responsibility for data collection lies directly with the researcher, who observes behavior, interviews participants, or examines documents (Denzin and Lincoln 2017). While a protocol (nonstandardized) serving as a data-gathering instrument may be employed, the researchers acquire the information. A qualitative researcher avoids standardized questionnaires or instruments developed by other researchers, emphasizing a more personalized and context-specific data-gathering process. The effectiveness of the data-gathering process and the richness of findings from qualitative research are closely linked to the skills, experience, and understanding of the interviewer or group moderator (Sullivan and Sargeant 2011).

13.3.3 Data Processing and Analysis

Emergent design The research process in qualitative research is inherently "emergent," characterized by flexibility that allows the initial plan to evolve and change dynamically. This means every phase of the research process is subject to change once the researcher commences data gathering, guided by participant insights. As researchers accumulate data, they engage in a process of *sequential analysis* that can start generating themes. Sometimes, the initial data analysis may result in new research questions emphasizing the dynamic nature of qualitative research; "the design is not cast in stone." It may also involve the change of participants and sites studied. We can contrast this with the causal-comparative design aiming to "determine factors influencing adoption of improved poultry." Such a quantitative researcher will maintain this design and research question throughout the study. However, a qualitative researcher investigating the perceptions of adoption and benefits of improved poultry may start with a narrative approach but incorporate a phenomenological technique to fill a gap realized during data collection. One question asked is, "Does this mean that different people will be asked different questions?" (Creswell and Guetterman 2019). The answer is that this may be the case because the focus of qualitative research is to understand and explore the central phenomenon contextually, not to develop consensus or agreements of opinion from the study participants.

Interpretive and people-focused Constitutes a mode of interpretive inquiry where researchers derive interpretations from their observations and understanding of the subject matter. The researcher's objective is to elucidate and report how people interpret and give meaning to their perspectives. Therefore, the perspectives of the individuals are important to the researcher who endeavors to capture the *thought patterns* of the participants. The emphasis of the researcher lies in presenting an authentic picture of how participants interpret their attitudes. Here, the researcher acts as an interpreter, and the research report will carry the researcher's interpretation of human behavior, attitudes, and experiences. Additionally, upon issuance of a research report, both readers and participants engage in the process of interpretation, providing additional perspectives in the study.

Holistic account Researchers' aim in data analysis is to develop an understanding of a whole phenomenon rather than reducing information to a linear cause–effect relationship or simplifying it to a few discrete variables. This entails presenting various perspectives and recognizing multiple factors in a given situation. Because the process of collection and analysis is often concurrent, it gives continuous analysis and perspectives, allowing the researcher to refine their understanding of the topic. This results in a holistic, contextual understanding of the research problem.

Detailed reporting The reporting and writing of publications from qualitative research tend to be longer than those of quantitative research to allow for detailed descriptions (Chaps. 25 and 27). Transparency in describing the methods section is key. The methods should be described in such a way that the procedures focus on giving detailed information regarding "how, when and who questions in the data collection" (Mbuagbaw et al. 2020). This helps in the evaluation of the credibility of the study by checking whether the data collection methods influenced data collection, analysis, and interpretation. Some authors prefer to state that the report contains "findings" rather than "results." This is to convey a sense of exploration, discovery, detailed reporting, and interpretation of insights rather than objective, measurable results.

13.3.3.1 Qualitative Research Is Not for the "Faint Hearted"

Overall, qualitative research is time consuming and labor intensive. Qualitative inquiry's iterative and emergent nature demands patience as the researcher navigates through evolving research questions and adapts their methods accordingly (Stevenson et al. 2015; Morgan and Nica 2020). The process also involves prolonged engagement in the field, where researchers immerse themselves in participants' lives, conduct extensive interviews, and analyze diverse data sources to discover the underlying complexities of the studied phenomenon. It is a challenging endeavor demanding resilience. Figure 13.3 shows key attributes of qualitative research.

13.4 General Guidelines and Steps to Follow in Qualitative Research

Like any other research, qualitative research is organized with steps that closely resemble those of quantitative research. Many similarities exist between these two stances. They follow similar sequences and methods. For example, in both types of inquiry, the researcher decides on the topic, states the problem, formulates the research question, designs the research, and collects and analyzes data (Chap. 2). However, the steps in qualitative research are not as direct as those in quantitative research. In some instances, the process is complex, interwoven, with backward and forward movements between several steps for planning and conducting the study. The steps often overlap, and at times, they are conducted simultaneously. Here, we describe steps in qualitative research as a linear process, but it is a highly iterative process. We have briefly described eight steps that guide a qualitative study.

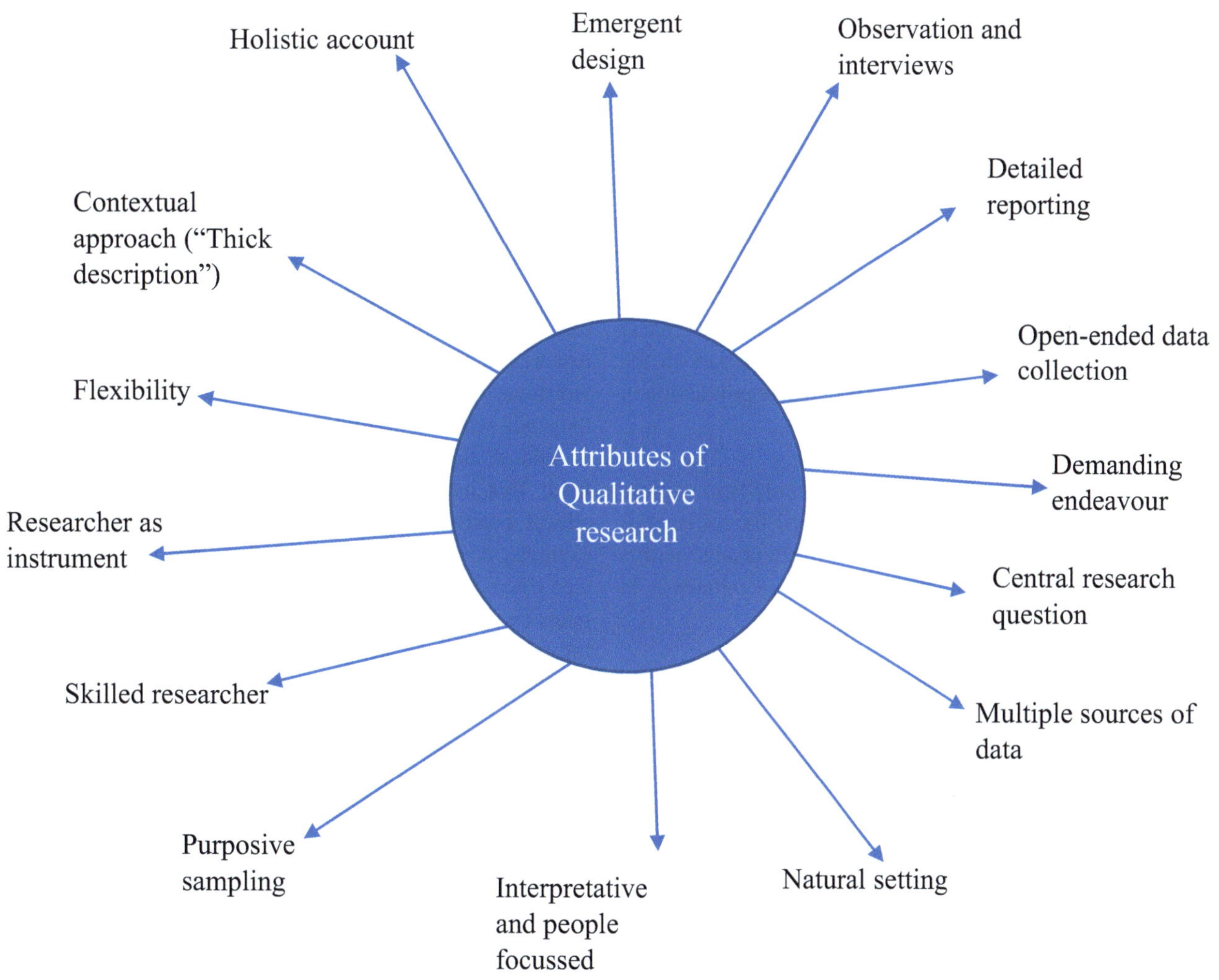

Fig. 13.3 Fifteen key attributes of qualitative research

Step 1: Establish the general problem or phenomenon to be studied

In this first step, the researcher identifies the specific phenomenon to be investigated. It is usually in the form of a *broad topic* serving as a foundation point from which the research develops, for example, farmers' *adaptive strategies in climate change management.*

Step 2: State the purpose of the study and formulate research questions

The purpose of the study and research questions are crucial aspects that drive the research. The research purpose is important for indicating the *intent of the central phenomenon* (broad), while the research question helps narrow the purpose statement. The purpose statement points out the specific issue to be investigated. The research purpose can be an open statement, but a research question should be stated explicitly. The research question must be able to be answered correctly and explicitly, with information that is verifiable. This means that after the study, the report should show that the research question was addressed and answered. This is one of the main aspects that examiners of thesis and dissertations critically assess.

Two examples in the area of climate change are hereby given, with a "purpose statement" and two research questions.

Example

Purpose: The purpose of this qualitative study is to understand climate change scenarios and mitigation strategies among smallholder farmers growing sorghum in Region A.

- Research QuestionsWhat understanding do farmers have in regard to climate change scenarios for the last 10 years?
- What is the role of internal and external factors in the choice of climate change strategies in maize production in Region A?

Purpose: The purpose of this qualitative study is to explore the diverse adaptive strategies employed by coffee producers in managing the impact of climate change with a focus on resilience and community-driven initiatives in Region B.

- Research QuestionsHow do farmers prioritize and choose effective adaptive measures in response to climate change in coffee production?
- What role does community collaboration play in the development and implementation of farmers' adaptive strategies for climate change management in coffee production?

The aim is to have research questions that are focused, answerable, and process driven. This means suitable research questions could be those evaluating processes and displaying aspects of probing to get answers, such as how does milk move from a smallholder farmer to a wholesale trader? Another suitable format is the influence of a setting or perspective (physical, social, temporal, and interpersonal) on participants' views, actions, and behaviors. For example, "How do socioeconomic and marketing factors shape the adoption behavior toward the use of drip irrigation technology for tomato production? A third suitable research question could revolve around evaluating "meaning attributed by participants to situations, events, behaviors, and activities." An example could be "How do farmers keeping improved dairy cattle manage their feed resources during the dry season to ensure sustainability?"

Step 3: Select the qualitative inquiry to use

Guided by the "central phenomenon and research question," the research selects a matching qualitative research approach that gives maximum quality to the research study. Suppose you have a research question that has to do with understanding the culture of a certain farming community, you will need to choose *ethnography* as your main qualitative approach. If, on the other hand, you have a research question seeking to develop a broad theory or generalization to explain a process, such as how farmers determine suitable markets and how they engage with those markets, you choose grounded theory (discussed in the next section).

Step 4: Write a research proposal and process the ethical clearance

An extremely important stage is writing a project proposal, taking into consideration the provided guidelines and format (in the case of thesis and dissertation work). A crucial point to note as far as qualitative proposals are concerned is the need to include detailed methodological procedures, paying close attention to issues of trustworthiness of your research study. Researchers include an assessment of trustworthiness in the proposal to aid in the assessment of whether the study will be worthwhile.

Ethical clearance is processed before proceeding to the field. This includes seeking clearance with the institution's ethical boards and all other administrative authorities, for example, county governments and science regulatory bodies (see Chap. 6). A key ethical issue is the protection of participants by granting confidentiality and privacy while withholding participants' personal details and other identifying characteristics. Another important activity is designing of informed consent form that participants will sign during data collection. A researcher needs to provide a detailed description of the procedures to be used by the institutional review board. As a qualitative researcher, below is some of the information that will be useful to the board as they evaluate the proposal:

- Elaborate presentation of the procedures to facilitate the assessment by reviewers on the institutional review board, enabling evaluation of the potential risks to people and sites in the area of study.

- Specifics regarding the duration of time allocated to spend at the research site; "include number of days, length of the visits, and other relevant details."
- Clearly, indicate how the anonymity of participants will be protected, e.g., masking the identity of the participants.
- Description of how you intend to respect the participants and their way of doing things.
- Details of how you will provide opportunities to give back to the community, e.g., any compensation to the participants or support to the community.
- Include a detailed data collection tool: a checklist of the questions so that reviewers are able to assess the sensitivity of the questions.

Step 5: Managing the gatekeepers

A gatekeeper is a person who facilitates and grants entry and access to the group or participants being studied (McFadyen and Rankin 2016). In other words, the gatekeeper provides entrance to the site (they keep watch of the gates). In some universities, e.g., the University of Waterloo, guidelines are given on how to handle gatekeepers. Sometimes you may need to seek and obtain permission from individuals and sites at "many levels." Suppose the study is about the growing and marketing of avocado by smallholder farmers to a marketing cooperative scheme. Relevant stakeholders for this study will include the farmers themselves and other stakeholders in the value chain, including the administrators of the cooperative scheme. In this case, gatekeepers may include the agricultural extension officer, the manager-in-charge, or the chief executive officer of the cooperative. The extension officer will link the researcher with the local administration as well as organize introduction and logistics to reach the farmers, while the manager of the cooperative will need to grant permission to carry out the study.

Effectively managing gatekeepers is crucial for smooth data collection in research, as these individuals significantly impact the process. Gatekeeper–researcher relationships, dynamic in nature, must be strategically handled throughout the research. This strategic management is essential for ensuring access to information, cooperation, and overall research success.

Think about working with a farming community. The researcher, in the first place, is an outsider who lacks information on the culture and background of the people, despite the fact that data collection will require close interaction. He or she must therefore be accepted by the community/farmers and secure cooperation to answer questions. This demands a very close partnership between the gatekeepers, research participants, and the researcher. Every qualitative researcher should strive to achieve this by ensuring that they remain open and trustworthy throughout the study period. It is advisable to submit written information to some of the gatekeepers as you seek permission (Creswell and Guetterman 2019).

Step 6: Locating the participants and maintaining rapport in the field

Once in the field, the community, the gatekeepers will play a big role in facilitating the recruitment exercise. The qualitative research team collaborates closely with the community and gatekeepers to develop a plan to locate and recruit the participants (McFadyen and Rankin 2016).

Building rapport is crucial in interviewing as it forms a basis for quality data. Rapport is the ability to connect with the participants in a way that creates trust and understanding. Good rapport creates a feeling of ease among the two parties, removes scepticism and suspicion, and hence enhances communication.

Plan to build rapport before the interview and also while conducting the interview. To do this, you should draft a short brief of your research project, written in layman's language, and send it to your participants before the interview (McGrath et al. 2019). This way, the participants will be informed about what to expect, what will be discussed during the interviews, and why it is important to discuss the topic. During the interview, too, rapport is crucial to enable participants to provide a comprehensive account of the experiences that are of interest to them.

One way of building rapport is to approach the participants with openness, sincere interest, and curiosity (Cohen et al. 2017). It is advisable to develop introductory guidelines (for introducing the interviews), paying special attention to anything that can be interpreted as coercive and insensitive to social and cultural contexts (DeJonckheere and Vaughn 2019; Diab and Al-Azzeh 2024). Throughout the interview, carefully select the language and the framing of the question. For example, a question like "Please tell me about your interest" in certain aspects is less offensive than "What rules and regulations do you follow in managing your group activities? (McGrath et al. 2019). Other pleasant questions in interviewing farmers include: Please tell me: For how long have you been involved in farming?

Step 7: Collect data, conduct analysis, and interpretation

Both participant observation and nonparticipant observation are mainly used as powerful tools in data collection because they allow the researcher to collect a wide variety of data. These include social actions, nonverbal communication, environmental factors, and others. In addition, the researcher has the advantage of getting first-hand information about what is happening in regard to information regarding the phenomenon of interest. For example, the researcher can observe activities of interest, such as manure and fertilizer application, and ask further questions regarding the ongoing activity.

Many types of interviews are used, e.g., in-depth interviews, focus groups, and unstructured interviews, and a researcher will need to choose the most suitable type tailored to the research question, characteristics of participants, and skills preferred by the researcher. Traditionally, interviews are mainly carried out using face-to-face interviews, though telephone and online interviews can also be used. During interviews, notes and recordings are taken. Before the analysis, the recorded messages have to be transcribed, and this is time consuming.

Data analysis in qualitative research mainly involves organizing, analyzing, and synthesizing the data (Discussed in Chap. 23). The massive data from the various sources are first assembled, followed by systematic reduction to yield a coherent description of what was observed accordingly. Questions to ask throughout the qualitative analysis may include the following:

- What patterns or common themes arise from the data, and what insights do they provide regarding the overarching study question?
- Are there any deviations from these patterns, and what factors could be associated with these responses?
- What thought-provoking stories surface from the data, and how can these narratives contribute to a deep understanding of the study question?
- Do any of the emerging patterns and categories suggest a necessity for additional data collection, and do any of the research questions warrant revision?
- Do the emerging patterns and categories align with the findings of other authors?

Both in the field and after completion of data collection, qualitative analysis takes different forms depending on the methods used. The process of qualitative analysis, just like quantitative analysis, requires highly skilled researchers to collect, organize, and analyze data. Data analysis in qualitative data is highly technical, requiring the researcher to possess skills in using data analysis software and interpreting the output. Similarly, qualitative data analysis will require a researcher with good analytical skills. As mentioned at the beginning of this chapter, Sect. 13.2, qualitative research skills are inadequate in most agricultural researchers. There is, therefore, a need for intentional and formalized training in qualitative methods in universities. This calls for two actions: incorporate qualitative methods in the curriculum and also build the capacity of faculty staff in qualitative approaches. Though researchers in national agricultural research systems continue to use the PRA tools, there is also a need for further retraining on the full range of qualitative approaches and how to integrate them in mixed methods research studies.

Step 8: Write the report and disseminate the work

This final step of reporting is important because it communicates to your peers and other researchers about your findings. Typically, it is in the form of a report, research paper, or journal manuscript. Regardless of the type, the write-up of a qualitative research study is constructed around themes and patterns and supports the use of direct quotes. The report will normally be detailed and comprehensive, providing a rich understanding of the research topic. The format will be dependent on the journal guidelines or the publication platform (see detailed explanation in part six of this book, Chap. 28) (Fig. 13.4).

13.5 Differences Between Quantitative and Qualitative Approaches

Part 2 of this book has discussed research designs and major differences that emerge in designs within the quantitative and qualitative stances. There are differences in the philosophies, research methodology, and methods (Table 13.1). Box 13.4 gives some limitations of qualitative research approaches.

Key References for Further Reading

1. Sullivan GM, Sargeant J (2011) Qualities of Qualitative Research: Part I. J Grad Med Educ 3:449–452. https://doi.org/10.4300/JGME-D-11-00221.1
2. Cohen L, Manion L, Morrison K (2017) Research Methods in Education. https://doi.org/10.4324/9781315456539/RESEARCH-METHODS-EDUCATION-LOUIS-COHEN-LAWRENCE-MANION-KEITH-MORRISON
3. Mbuagbaw L, Lawson DO, Puljak L, et al. (2020) A tutorial on methodological studies: The what, when, how and why. BMC Med Res Methodol 20:1–12. https://doi.org/10.1186/S12874-020-01107-7/FIGURES/2
4. Sullivan GM, Sargeant J (2011) Qualities of Qualitative Research: Part I. J Grad Med Educ 3:449–452. https://doi.org/10.4300/JGME-D-11-00221.1
5. Denzin NK, Lincoln YS (Eds.) (2018) The SAGE handbook of qualitative research (5th ed.). Los Angeles: Sage.
6. Creswell JW, Guetterman TC (2019) Educational research_ planning, conducting, and evaluating quantitative and qualitative research. Pearson. pp 1–681

Key Points in This Chapter

What Is Qualitative Research?

Qualitative research refers to a methodological approach that explores and provides deep insights into phenomena using non-numerical data. Typically, a qualitative research design will be used if your research question seeks to explore the meaning of a phenomenon, explain a pattern, observe a practice or process in depth, or understand participants' perspectives.

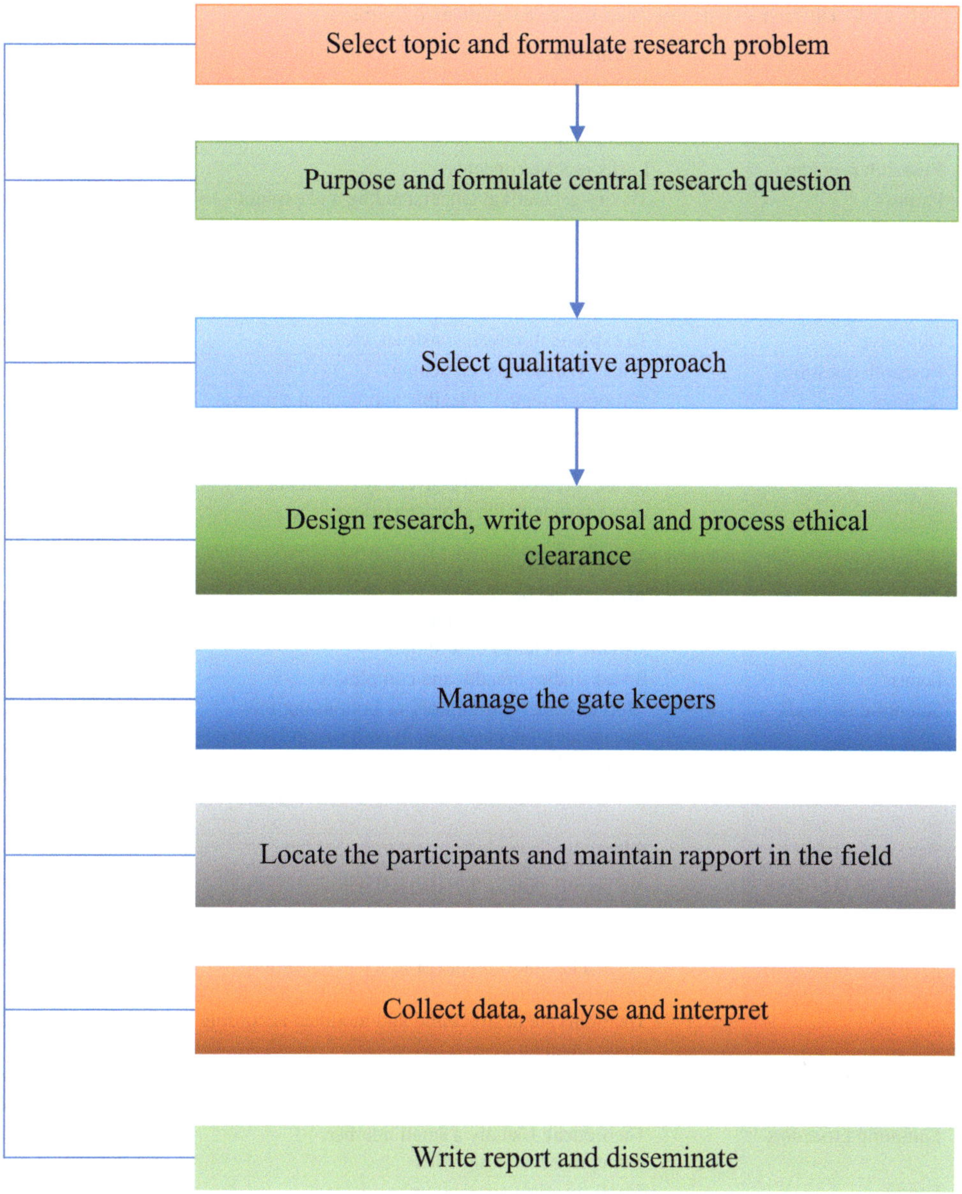

Figs. 13.4 Steps to follow in conducting qualitative inquiry

Brief Overview of the Origin of Qualitative Approaches

Qualitative research has its roots in various disciplines within the domain of social sciences, including sociology, psychology, and anthropology. Recently, it has gained increased interest in agricultural research, especially in understanding human behavior regarding farmers.

Philosophical Assumptions

Researchers employing qualitative approaches and methods embrace an ontological perspective that acknowledges the existence of multiple realities. Epistemologically, qualitative researchers develop multiple subjective meanings of their experiences. The axiological assumption guiding qualitative research is based on the principle of being value laden.

State of the Art of Qualitative Research in Agriculture

The use of qualitative research approaches in addressing social and development issues facing agriculture today is growing. However, compared to other fields, like education, psychology, history, and sociology, qualitative designs are less developed in agriculture. Traditionally, quantitative approaches have been the main focus of research and publications in agriculture.

Table 13.1 Differences between quantitative and qualitative approaches

Criteria	Qualitative	Quantitative
Philosophical assumptions	Multiple realities Reality changes with people perceptions, dynamic, and holistic	Single reality Stable and objective Reductionist approach
Research problem	Lacks precise concepts	Has precise concepts
Purpose	Theory generating: Understanding and explain social relations and interactions	"Hypothesis testing"; determine the cause–effect relationship and predict scenarios
Hypothesis	No hypothesis: lacks variables and explores a central concept	Has hypothesis with measurable variables
Objective	To explore, discover, construct, etc.	To describe, explain, and predict
Research question	What? Why is this happening?	How many? Strength and relationships
Settings	"Process-oriented" Flexible approach in a natural setting	"Outcome-oriented" Highly rigid; experimental setting
Researcher bias	Participants in the study may be aware of the biases by the researcher and conversely researcher may be acquainted with the characteristics of the participants	Participants are unaware of the biases held by the researcher, reciprocally, the researcher remains unaware about the characteristics of the participants "Double blind studies"
Forms of data	Data are in from of "words, pictures, photos, artefacts and objects" Data is rich, time consuming to collect and analyze Provide a detailed description	Data involve numerical and statistics information Classify features, quantify relationships, and construct statistical models Provide specific conclusions
Focus	Broad studies breadth and complexity	Restricted, assess a precise hypothesis
Scientific methods	"Soft Science" Exploratory or bottom-up; the researcher produces a new hypothesis and theory from the data gathered	"Hard Science" Confirmatory or top-down; the researcher tests hypothesis and theory using data collected
Research design	Design develops as the study evolves (emerges)	All variables of the study are identified before data are collected (predefined))
Researcher role	Researcher is the instrument to gather data	Questionnaire or equipment is used collect data
Objective vs Objectivity	Subjective; relies on individual interpretation of events	Objective: measures and analyzes target concepts
	Qualitative data provide rich explanations	Numerical data can test a hypothesis
Data collection	Researcher knows more or less in advance what data are to be collected The items are unstructured and modified according to the situation	Researcher knows clearly what data will be collected Items are fixed and tools pretested to enhance validity and reliability
Approaches	Case studies, phenomenology, ethnography, etc.	Descriptive surveys, experiments, interventional studies, case–control studies, etc.
Sampling procedure	Theoretical: Usually a small number; nonrepresentative samples	Statistical: Random samples; proportionally representative of the population
Samples	Small samples: not randomly selected	Large samples: Randomly selected
Data analysis	Inductive process, codes, themes; illustrative explanation and individual responses Data collection and analysis are done concurrently	Deductive process, numerical aggregation in summaries, responses are clustered Data analysis takes place after data collection
Statistics	Themes, concepts, and theories	Descriptive and inferential statistics
Results	Definite findings useful in specific settings	Generalized findings that can be used in other settings
Study rigor	Trustworthiness focuses on credibility and dependability Emphasis is on procedures to get real and rich data	Validity and reliability Emphasis is on procedures to get hard and replicable data
Nature of report	Descriptive report containing contextual and "direct quotations" from research participants Has detailed, thick descriptions	Report has statistics with comparison of means, correlations, and statistical significance of findings Report with specific key findings

The barriers to increased use of qualitative approaches include:

- Mindset and beliefs among the researchers, lecturers, and students in agriculture.
- Limited training in educational courses and practice.
- Challenges related to publications, which tend to be biased toward qualitative research.

Qualitative methods have the potential to provide a better understanding of social phenomena by providing details that cannot be measured or captured by quantitative methods.

Box 13.4 Limitations of qualitative approaches

- *Subjectivity:* inherently subjective, relying on the interpretation of the researcher and participants.
- *Time-consuming:* Data collection methods and analysis are time consuming and resource intensive, and findings are difficult to present in a visual way.
- *Lack of precision*: Qualitative research is sometimes not accepted and understood, especially within scientific communities.
- *Limited generalizability*: A small sample size focusing on specific contexts or individuals limits the generalizability to larger populations and different agricultural settings
- *Requires highly skilled researcher:* The quality of research is heavily dependent on the skills of the researcher and can be easily influenced by personality and biases of researchers.
- *Biased subjects' response:* Researcher's presence during data gathering may affect the subjects' response, resulting in bias.
- *Rigor:* It is more difficult to maintain, assess, and explain rigor; ensuring data validation and reliability are more challenging.
- *Difficult to compare across cases*: Farming systems, agricultural practices, as well as the context, vary widely, and difficult to compare data across different agricultural settings.
- *Limited ability to establish causality:* The researcher can identify patterns and associations, but cannot control variables to the same extent as in an experiment.
- *Ethical considerations:* Because of the closeness of the researcher to the participants, there are complex ethical issues about confidentiality, informed consent, and potential harm to participants.

Additionally, with the emergence of mixed methods research, there should now be more emphasis on qualitative research as a paradigm.

Gender Studies

Qualitative methodologies are usually the go-to method for capturing the meaning behind behavior, beliefs, experiences, and attitudes of men and women. Indeed, most standardized qualitative approaches used in gender analysis, "Harvard analytical framework" and "Gender analysis matrix," to illuminate the unique perspectives of men and women and explore contextual and cultural differences, mainly use qualitative tools.

Qualitative Research in Agricultural Markets

Key elements emphasized in agricultural markets include descriptions, processes, and understanding of relationships among the actors within a value chain. The typical steps of an agricultural value chain analysis targeting farmers include four areas: mapping the chain, identifying governance structure within the chain, exploring areas of upgrading within, and assessing benefits in the chain participation. Suitable methodologies for accomplishing the above steps incorporate quantitative and qualitative research methods.

Qualitative Approaches in Policy Studies

The qualitative approaches are used to explore various aspects to help policymakers. It allows researchers to carry out policy reviews and in-depth analyses to explore experiences and perspectives regarding existing policies.

Basic Characteristics of Qualitative Designs

Conceptualization of Research Study

1. The conceptualization of the problem is less precise than quantitative research.
2. Qualitative researchers aim to build a theory but not to formulate a pre-established hypothesis.
3. Address a central issue by extracting details of specific settings surrounding the central phenomenon.
4. Instead of a hypothesis, qualitative use propositions suggesting relationships that are not strictly tested.
5. The purpose statement communicates "the central aim" and outlines the specific focus, context, methods, and significance of the study.

Data Collection

1. Samples are purposively selected based on their specific characteristics and experiences.
2. Researchers gather multiple forms in the form of field notes, diaries, and individual verbal narratives.
3. Data collection primarily involves observations and unstructured interviews.
4. Data are gathered in the natural setting where participants experience the issue.
5. The methodology is investigative.
6. Data include field notes, photographs, personal comments, and video recordings, to express the actions.
7. The responsibility for data collection lies directly with the researcher.

Data Processing and Analysis

1. Inherently "emergent," allowing the initial plan to evolve and change dynamically.
2. Researchers derive interpretations from their observations and understanding of the subject matter.
3. Data analysis aims to develop an understanding of a whole phenomenon rather than reducing information to a linear cause-and-effect relationship.
4. Publications tend to be longer and more detailed than those of quantitative research to allow for detailed descriptions.

Overall, qualitative research is time consuming and labor intensive. The iterative and emergent nature of qualitative inquiry demands patience as the researcher navigates through evolving research questions and adapts their methods accordingly. The process also involves prolonged engagement in the field.

General Guidelines and Steps to Follow in Qualitative Research

Like any other research, qualitative research is organized with steps that closely resemble those of quantitative research. They include

Step 1: Establish the general problem or phenomenon to be studied.
Step 2: State the purpose of the study and formulate research questions.
Step 3: Select the qualitative inquiry to use.
Step 4: Write a research proposal and process the ethical clearance.
Step 5: Managing the gatekeepers.
Step 6: Locating the participants and maintaining rapport in the field.
Step 7: Collect data, conduct analysis, and interpretation.
Step 8: Write the report and disseminate the work.

Review Questions

1. Give examples of research questions that would dictate the use of qualitative inquiry.
2. How does the philosophical assumption of multiple realities in qualitative research challenge or complement the positivist view in the context of agricultural research?
3. Why is the use of qualitative research low in agriculture compared to other disciplines like sociology and education?
4. Explain the difference between process-oriented qualitative research and outcome-oriented quantitative research, using an example from agricultural development or policy.
5. What does "thick description" entail in qualitative agricultural research, and why is it emphasized?
6. Why is a qualitative approach more suitable than a quantitative survey for a gender-based study involving smallholder farmers and decision-making in resource allocation at the household level?

Chapter Exercise

1. For each of the following broad agricultural topics, formulate one qualitative research question:
   (a) Farmer decision-making regarding fertilizer use.
   (b) Community perceptions of a new irrigation project.
   (c) The socio-cultural factors influencing livestock health practices.
2. Pick one of the research questions in Q1 (in this subsection) and design a qualitative study showing clearly the various steps you would follow.
3. Suppose you have been asked to map a mango value chain to understand the formal and informal systems governing interactions among actors. Propose how a qualitative approach would be used to map this value chain and visualize the roles and relationships of different actors (e.g., farmers, transporters, traders).
4. Given that qualitative research involves in-depth interactions and is often "value-laden," then, therefore:
   (a) Identify and discuss at least three ethical considerations you would need to address when conducting qualitative research with farmers or other agricultural stakeholders. You could also refer to Chap. 6.
   (b) Create a table that clearly differentiates between quantitative and qualitative research approaches at each of the following stages of the research process:
      (i) Conception (e.g., hypothesis vs. central phenomenon).
      (ii) Data collection (e.g., breadth vs. depth, natural setting).
      (iii) Data analysis (e.g., statistics vs. "thick description").
      (iv) Report writing (e.g., outcome-oriented vs. process-oriented).

Answers to Review Questions

- Q1: Research questions; Reference 13.2.1
- Q2: Multiple realities; interpretivism (reality is not a single, objective, fixed identity), while positivism (reality is single, objective, and measurable). Qualitative is used to provide contextually rich understanding. Mixed methods research provides the most comprehensive and actionable findings: Reference 13.1.4.
- Q3: Reference 13.1.5.

- Q4: Process-oriented qualitative research: understand how and why things happen; uncover underlying processes, context, and meaning; data are non-numerical, involving in-depth analysis; References 13.3.1 Table 13.1.
- Outcome-oriented quantitative approach seeks to measure "what," how much, how many; Quantify relationships, test hypothesis, and generalize to a larger population; Data numerical, statistically analyzed, experiments, correlations, etc. Reference 13.3.
- Q5: Reference 13.3.
- Q6: Reference: 13.2.2.

References

Akimowicz M, Vyn RJ, Cummings H, Landman K (2018) An introduction to mixed methods research in agricultural economics: the example of farm investment in Ontario's greenbelt, Canada. J Rural Stud 61:162–174. https://doi.org/10.1016/J.JRURSTUD.2018.04.012

Akyüz Y, Salali HE, Atakan P et al (2023) Case study analysis on agri-food value chain: a guideline-based approach. Sustainability 15:6209. https://doi.org/10.3390/SU15076209

Asante BO, Puskur R, Garner E et al (2023) Access and control of resources and participation in rice-breeding activities among men and women farmers in Southern Ghana. Sustainability 15:7069. https://doi.org/10.3390/SU15097069

Austin Z, Sutton J (2014) Qualitative research: getting started. Can J Hosp Pharm 67:436–440. https://doi.org/10.4212/CJHP.V67I6.1406

Bala-Miller P, Peletz N, Hanna K (2022) Gender analysis and environmental impact assessment : challenges and opportunities for transformative approaches. Routledge Handbook of Environmental Impact Assessment. Routledge, In, pp 181–200

Borsellino V (2020) Agricultural value chain for food security: challenges and opportunities from SDG2. In: Zero hunger, pp 88–100. https://doi.org/10.1007/978-3-319-95675-6_7

Brisebois A, Eriksen SH, Crane TA (2022) The politics of governing resilience: gendered dimensions of climate-smart agriculture in Kenya. Front Clim 4:864292. https://doi.org/10.3389/FCLIM.2022.864292/BIBTEX

Burns M, Bally J, Burles M et al (2022) Constructivist grounded theory or interpretive phenomenology? Methodological choices within specific study contexts. Int J Qual Methods 21. https://doi.org/10.1177/16094069221077758

Clay PM, Feeney R (2019) Analyzing agribusiness value chains: a literature review. Int Food Agribus Manage Rev 22:31–46. https://doi.org/10.22434/IFAMR2018.0089

Cohen L, Manion L, Morrison K (2017) Research methods in education. Taylor & Francis. https://doi.org/10.4324/9781315456539/RESEARCH-METHODS-EDUCATION-LOUIS-COHEN-LAWRENCE-MANION-KEITH-MORRISON

Creswell JW (2014) Research and design qualitative, quantitative, and mixed methods approaches. SAGE Publication Inc

Creswell JW, Guetterman TC (2019) Educational research_ planning, conducting, and evaluating quantitative and qualitative research. Pearson, pp 1–681

DeJonckheere M, Vaughn LM (2019) Semistructured interviewing in primary care research: a balance of relationship and rigour. Fam Med Commun Health 7:57. https://doi.org/10.1136/FMCH-2018-000057

Denzin NK, Lincoln YS (2017) The SAGE Handbook of Qualitative Research , 5th edn. SAGE Publications, Inc

Diab JL, Al-Azzeh D (2024) Inclusive inquiry: a compassionate journey in trauma-informed qualitative research with GBV survivors from displaced communities. Front Psychol 15:1399115. https://doi.org/10.3389/FPSYG.2024.1399115/BIBTEX

Donovan J, Franzel S, Cunha M et al (2015) Guides for value chain development: a comparative review. J Agribus Dev Emerg Econ 5:2–23. https://doi.org/10.1108/JADEE-07-2013-0025

Dooley KE (2007) Viewing agricultural education research through a qualitative lens. J Agric Educ 32:32

Dreher J, Santos H (2017) Sociology and phenomenology. Civitas - Revista de Ciências Sociais 17:385. https://doi.org/10.15448/1984-7289.2017.3.29429

Fleming J, Zegwaard KE (2018) Methodologies, methods and ethical considerations for conducting research in work-integrated learning. Int J Work-Integrat Learn 19:205–213

Francis JJ, Johnston M, Robertson C et al (2010) What is an adequate sample size? Operationalising data saturation for theory-based interview studies. Psychol Health 25:1229–1245. https://doi.org/10.1080/08870440903194015

Gehl Sampath P, Vallejo B (2018) Trade, global value chains and upgrading: what, when and how? Eur J Dev Res 30:481–504. https://doi.org/10.1057/S41287-018-0148-1/TABLES/4

Hastings K, Howieson J, Lawley M (2016) Creating value chains: the role of relationship development. Br Food J 118:1384–1406. https://doi.org/10.1108/BFJ-10-2015-0389/FULL/XML

Haug R, Mwaseba DL, Njarui D et al (2021) Feminization of African agriculture and the meaning of decision-making for empowerment and sustainability. Sustainability 13:8993. https://doi.org/10.3390/SU13168993

Huberman AM, Miles MB (1983) Drawing valid meaning from qualitative data: some techniques of data reduction and display. Qual Quant 17:281–339. https://doi.org/10.1007/BF00167541/METRICS

Islam MM, Jannat A, Al Rafi DA (2024) Women participation in South Asian agriculture: a comprehensive systematic review. Discov Sustain 5:1–20. https://doi.org/10.1007/S43621-024-00649-W

Kaplinsky R, Morris M (2000) A handbook for value chain research. IDRC

Kerry EH (2013) An introduction to the philosophy of methodology. SAGE Publication. https://doi.org/10.4135/9781473957633

Kolbina O (2015) SWOT analysis as a strategic planning tool for companies in the food industry. Probl Econ Transit 57:74–83. https://doi.org/10.1080/10611991.2014.1088367

Laibuni N, Neubert S, Turoop L, Bokelmann W (2018) An exploratory study on organisational linkages along the African indigenous vegetable value chains in Kenya. Cogent Food Agric 4. https://doi.org/10.1080/23311932.2018.1519972

Lazoroska D, Palm J, Bergek A (2021) Perceptions of participation and the role of gender for the engagement in solar energy communities in Sweden. Energy Sustain Soc 11:1–12. https://doi.org/10.1186/S13705-021-00312-6/METRICS

Li M, Zheng ZJ, Zheng J et al (2023) The effect and mechanism of farmland transfer on the multidimensional relative poverty of rural women. Front Environ Sci 11:1276195. https://doi.org/10.3389/FENVS.2023.1276195/BIBTEX

Luborsky MR, Rubinstein RL (1995) Sampling in qualitative research. Res Aging 17:128–130. https://doi.org/10.1177/0164027595171005

Lugusa KO, Wasonga OV, Elhadi YA, Crane TA (2016) Value chain analysis of grass seeds in the drylands of Baringo County, Kenya: a producers' perspective. Pastoralism 6:1–15. https://doi.org/10.1186/S13570-016-0053-1/FIGURES/8

Mangheni MN, Tufan HA, Boonabana B et al (2019) Building gender research capacity for non-specialists: lessons and best practices from gender short courses for agricultural researchers in sub-saharan africa. Emerald Group Publishing Ltd.

Mbuagbaw L, Lawson DO, Puljak L et al (2020) A tutorial on methodological studies: the what, when, how and why. BMC Med Res Methodol 20:1–12. https://doi.org/10.1186/S12874-020-01107-7/FIGURES/2

McFadyen J, Rankin J (2016) The role of gatekeepers in research: learning from reflexivity and reflection. GSTF J Nurs Health Care 4:82–88

McGrath C, Palmgren PJ, Liljedahl M (2019) Twelve tips for conducting qualitative research interviews. Med Teach 41:1002–1006. https://doi.org/10.1080/0142159X.2018.1497149

Mertens DM (2010) Philosophy in mixed methods teaching: the transformative paradigm as illustration. Int J Mult Res Approaches 4:9–18. https://doi.org/10.5172/MRA.2010.4.1.009

Miles MB, Huberman AM (1984) Drawing valid meaning from qualitative data: toward a shared craft. Educ Res 13:20–30. https://doi.org/10.3102/0013189X013005020/ASSET/AE13716F-D080-4112-BCE8-39AFAE38C1FD/ASSETS/0013189X013005020.FP.PNG

Minati G (2023) Theoretical reflections on reductionism and systemic research issues: dark systems and systemic domains. Systems 12:2. https://doi.org/10.3390/SYSTEMS12010002

Mishra PK, Dey K (2018) Governance of agricultural value chains: coordination, control and safeguarding. J Rural Stud 64:135–147. https://doi.org/10.1016/J.JRURSTUD.2018.09.020

Mohapatra G, George M (2021) Perception and adaptation of agricultural households to climate change in the semi-arid regions of Rajasthan—a gender perspective study. Ecofeminism Clim Change 2:146–155. https://doi.org/10.1108/EFCC-04-2021-0007

Morgan DL, Nica A (2020) Iterative thematic inquiry: a new method for analyzing qualitative data. Int J Qual Methods 19. https://doi.org/10.1177/1609406920955118/ASSET/3570F309-8216-4A1E-9D15-D9812C5CC44E/ASSETS/IMAGES/LARGE/10.1177_1609406920955118-FIG1.JPG

Morgan R, George A, Ssali S et al (2016) How to do (or not to do)… gender analysis in health systems research. Health Policy Plan 31:1069–1078. https://doi.org/10.1093/HEAPOL/CZW037

Morgan R, Davies SE, Feng H et al (2022a) Using gender analysis matrixes to integrate a gender lens into infectious diseases outbreaks research. Health Policy Plan 37:935–941. https://doi.org/10.1093/HEAPOL/CZAB149

Morgan R, Garrison-Desany H, Hobbs AJ, Wilson E (2022b) Strengthening effectiveness evaluations through gender integration to improve programs for women, newborn, child, and adolescent health. Glob Health Action 15. https://doi.org/10.1080/16549716.2021.2006420

Msofi Mgalamadzi L, Matita M, Chimombo M (2024) The gendered nature of household decision making and expenditure choices in the context of smallholder agricultural commercialization in Malawi. CABI Agric Biosci 5:1–15. https://doi.org/10.1186/S43170-024-00270-X/TABLES/7

Munyua H, Stilwell C (2010) A mixed qualitative-quantitative-participatory methodology: a study of the agricultural knowledge and information system (AKIS) of small-scale farmers in Kirinyaga district, Kenya. Libr Manag 31:5–18. https://doi.org/10.1108/01435121011013359/FULL/XML

Murage FM, Mugwe JN, Ngetich KF et al (2019) Adoption of soybean by smallholder farmers in the central highlands of Kenya. Afr J Agric Econ Rural Dev 7:001–012

Nelson S, Huyer S (2016) A gender-responsive approach to climate-smart agriculture: evidence and guidance for practitioners. Climate-Smart Agriculture Practice Brief. Copenhagen: CGIAR Research Program on Climate Change, Agriculture and Food Security (CCAFS)

Njuki J, Waithanji E, Sakwa B et al (2014) A qualitative assessment of gender and irrigation technology in Kenya and Tanzania. Gend Technol Dev 18:303–340. https://doi.org/10.1177/0971852414544010

Njuki J, Eissler S, Malapit H et al (2022) A review of evidence on gender equality, women's empowerment, and food systems. Glob Food Sec 33:100622. https://doi.org/10.1016/J.GFS.2022.100622

Oh S, Lee S, Hwang S (2025) Climbing the value chain: the critical role of innovation routines in Korean SMEs' global competitiveness. Systems 13:274. https://doi.org/10.3390/SYSTEMS13040274

Palinkas LA, Horwitz SM, Green CA et al (2015) Purposeful sampling for qualitative data collection and analysis in mixed method implementation research. Adm Policy Ment Health Ment Health Serv Res 42:533–544. https://doi.org/10.1007/S10488-013-0528-Y/METRICS

Pretorius L (2024) Demystifying research paradigms: navigating ontology, epistemology, and axiology in research. Qualitative Rep. https://doi.org/10.26180/27013843.V1

Pye V, Taylor N, Clay-Williams R, Braithwaite J (2016) When is enough, enough? Understanding and solving your sample size problems in health services research. BMC Res Notes 9:1–7. https://doi.org/10.1186/S13104-016-1893-X/FIGURES/4

Renjith V, Yesodharan R, Noronha J et al (2021) Qualitative methods in health care research. Int J Prev Med 12. https://doi.org/10.4103/IJPVM.IJPVM_321_19

Rubin D (2016) Qualitative methods for gender research in agricultural development. International Food Policy Research Institute

Sankofa N (2023) Participatory thick descriptions: a collaborative and reflective approach. Qual Res Psychol 20:100–120. https://doi.org/10.1080/14780887.2022.2090469

Saunders B, Sim J, Kingstone T et al (2018) Saturation in qualitative research: exploring its conceptualization and operationalization. Qual Quant 52:1893–1907. https://doi.org/10.1007/S11135-017-0574-8/TABLES/1

Serpa S, Caldeira SN, Damião Serpa MS et al (2020) Mobility in the internationalisation of higher education institutions. Int J Higher Educ 9:46. https://doi.org/10.5430/IJHE.V9N4P46

Smith PL, Little DR (2018) Small is beautiful: in defense of the small-N design. Psychon Bull Rev 25:2083–2101. https://doi.org/10.3758/S13423-018-1451-8/FIGURES/3

Staller KM (2015) Moving beyond description in qualitative analysis: finding applied advice. Qual Soc Work 14:731–740. https://doi.org/10.1177/1473325015612859

Stamm A, von Drachenfels C (2011) Value chain development: approaches and activities by seven UN agencies and opportunities for inter-agency cooperation. International Labour Office (ILO) on behalf of the United Nations Inter-agency Cluster on Trade and Productive Capacity, Geneva

Steffens J, Brüssow K, Grote U (2020) A strategic approach to value chain upgrading—adopting innovations and their impacts on farm households in Tanzania. Horticulturae 6:32. https://doi.org/10.3390/HORTICULTURAE6020032

Stevenson FA, Gibson W, Pelletier C et al (2015) Reconsidering "ethics" and "quality" in healthcare research: the case for an iterative ethical paradigm. BMC Med Ethics 16:1–9. https://doi.org/10.1186/S12910-015-0004-1/TABLES/1

Sullivan GM, Sargeant J (2011) Qualities of qualitative research: part I. J Grad Med Educ 3:449–452. https://doi.org/10.4300/JGME-D-11-00221.1

Vasileiou K, Barnett J, Thorpe S, Young T (2018) Characterising and justifying sample size sufficiency in interview-based studies: systematic analysis of qualitative health research over a 15-year period. BMC Med Res Methodol 18:1–18. https://doi.org/10.1186/S12874-018-0594-7/TABLES/3

de Waal M (2006) Evaluating gender mainstreaming in development projects. Dev Pract 16:209–214. https://doi.org/10.1080/09614520600562454/ASSET//CMS/ASSET/650B7206-D23E-41FD-942C-4DA4622E5973/09614520600562454.FP.PNG

Wertz FJ (2014) Qualitative inquiry in the history of psychology. Qual Psychol 1:4–16. https://doi.org/10.1037/qup0000007

Wu J (2011) Improving the writing of research papers: IMRAD and beyond. Landsc Ecol 26:1345–1349. https://doi.org/10.1007/S10980-011-9674-3/FIGURES/1

Yan Z, Wei F, Deng X et al (2022) Feminization of agriculture: do female farmers have higher expectations for the value of their farmland?—empirical evidence from China. Agriculture 12:60. https://doi.org/10.3390/AGRICULTURE12010060

Common Approaches for Qualitative Research

14

Abstract

This chapter explores the five major qualitative research approaches: case study, ethnography, phenomenology, narrative, and grounded theory. The case study research examines a specific phenomenon to understand its complexities and impacts. The case can be an individual, community, organization, or program. Ethnography focuses on the cultural practices of a community from an insider's perspective. It uses multiple methods and a few samples to collect data. Researchers spend extensive time in the field but are fulfilled because of the rich data gathered. It mainly uses participant observation and nonparticipant observation. Grounded theory involves generating new theories through systematic collection and analysis, focusing on building theories grounded in the data rather than existing theories. It uses specific procedures and techniques that require well-developed skills and knowledge. As such, its application, especially by students, is limited. Narratives collect and analyze personal stories or accounts to understand how people perceive and make sense of them. The most common types include biography, autobiography, oral history, re-storying, and letter writing. Triangulation is emphasized in the narrative to thoroughly understand the phenomenon at hand. Phenomenology aims to uncover and describe the scene of individuals' lived experiences regarding a particular phenomenon. A core premise is to "bracket preconceived assumptions," meaning researchers set aside their own beliefs and perceptions to understand the participant's experiences fully. Through this process, the phenomenon aims to condense individual experiences into a "universal essence." In qualitative research, quality is equated with trustworthiness, which includes credibility, transferability, dependability, and conformability. Credibility ensures an accurate representation of participants' views, transferability provides context for applying findings elsewhere, and dependability and confirmability focus on consistency and data-driven results, respectively. These criteria collectively inform the validity and reliability of the data.

Keywords

Audit trail · Case study approach · Case series · Credibility · Grounded theory · Ethnography · Narrative approach · Phenomenology · Trustworthiness · "Universal essence"

Learning Outcomes

After reading this chapter, you are expected to be able to:

1. Apply the case study approach to conduct a qualitative research study.
2. Identify the characteristics of a case study.
3. Determine when to use a case study.
4. Differentiate between ethnography and phenomenology research approaches.
5. Recognize the uniqueness of the grounded theory approach and its limitations for use by the students.
6. Identify the phenomenology approach and the key characteristics.
7. Appreciate quality checks in qualitative research approaches.
8. Describe the narrative approach, the types, and the need to triangulate.
9. Compare the concept of validity and reliability in quantitative research versus qualitative research.

J. Njeri Mugwe, S. Runo, *Research Methodology in Agricultural Sciences*, https://doi.org/10.1007/978-981-95-1892-0_14

14.1 Types of Qualitative Designs

Creswell and others, in several works (Creswell and Plano Clark 2011; Creswell 2013, 2014; Creswell and Guetterman 2019) as well as other researchers (Miles et al. 1984; Denzin and Lincoln 2017), provide detailed explanations of approaches in qualitative methods. Creswell mainly focuses on five strategies (case study, ethnography, grounded theory, narrative, and phenomenology). Other authors include many more methods, including historical and documentary approaches (Cohen et al. 2017).

In this chapter, we discuss five approaches of enquiry that are most relevant in agriculture and natural sciences which includes case studies, ethnography, grounded theory, phenomenology, and narrative as shown in Table 14.1.

Table 14.1 Summary of the five key qualitative research approaches

Research approach	Aim	Research questions
Case study	To provide a detailed, in-depth exploration of a phenomenon, situation, event, individual, etc.	What are the characteristics of this individual or group? What are the characteristics of farmer groups involved in the collective marketing of tissue culture bananas?
Narrative enquiry	To study the lives of individuals by asking one or more individuals to provide stories about their lives.	What insights and understandings emerge from examining life stories? What insights or conclusions can be reached about this past event?
Phenomenology	To explore the meaning of human experiences from the perspectives of the individuals who have lived them	What does this experience mean for those involved? What is the lived experience of farmers in implementing adaptive strategies to climate change?
Grounded theory	Development of new theory through the collection and analysis of data about a phenomenon; hence, theory is grounded on the data collected	What theory can be derived inductively about a phenomenon from the data collected in a particular setting? (rare in agriculture). How do farmers develop and modify their crop protection strategies in response to evolving pests and diseases?
Ethnography	To conduct a detailed study of a particular group or community to gain an understanding of their culture within a natural setting	What are the culture and perspective of this group of farmers in its natural setting? How do cultural norms and traditions influence decision-making processes in production and marketing of passion fruits?

14.2 Case Study

14.2.1 What Is a Case Study?

A case study involves a detailed and thorough analysis of a single unit called "the case." The case could be an individual, a community of participants, an organization, a program, a practice or an event. This technique focuses on understanding the complexities and dynamics of the case to discover processes, relationships, and behaviors. The researcher aims to provide a rich and comprehensive description of the phenomenon under investigation.

Case studies are a widely used strategy in the social sciences, such as education and sociology (Lewallen and Courtright 1998; Kyveryga 2019). They are used for various purposes, including describing a case, generating a hypothesis or developing a theory, testing a hypothesis or theory and developing normative theory (practical wisdom) (Baker 2011; Crowe et al. 2011; Taylor 2024). This approach is equally relevant in agricultural research, where case studies are used to describe phenomena in-depth, mainly focusing on individual farmers, practices, organizations, and events. Typical cases of individual farmers, practices, organizations, and events include the following:

1. *Individual farmer:*
   (a) A farmer using an innovative way of growing apples.
   (b) A farmer trying a new method of irrigation.
   (c) A successful farmer using innovative adaptive strategies to climate change.
   (d) A farmer facing unique challenges of marketing vegetables in a local market.
   (e) A farmer facing challenges of a pest attacking a certain crop.
2. *Practices:*
   (a) Climate-smart agricultural practices.
   (b) Practices involved in the management of farmer groups.
   (c) Inorganic fertilizer use practices.
   (d) Conservation agriculture practices.
   (e) Methods being used in agricultural extension.
3. *Organization:*
   (a) Operations of a dairy cooperative society.
   (b) Operations of a commercial village assisting farmers to sell their commodities.

(c) Operations of agrodealers offering services to farmers.
(d) Operations of an organization targeting to upgrade the mango value chain.

4. *Event:*
(a) Operations and organization of a farmer's field day or training.
(b) Adoption patterns and outcomes of soybean among smallholder farmers.
(c) A field demonstration on silo making for preserving livestock feeds.

The main advantage of case study enquiry is that it provides more comprehensive information than other methods, such as surveys (Priya 2021). In multidisciplinary studies, case studies offer information that would be challenging to acquire through alternative methods (Crowe et al. 2011). Further, they present an opportunity for intimate collaboration between social scientists, biophysical scientists, and farmers. Therefore, case studies are a useful and cost-effective research tool in farming systems research (Chap. 10).

Researchers may use a diversity of methods to collect data; observation (both participant and nonparticipant, interviews and surveys, analytical approaches such as narratives, thick descriptions, discourse tracing, and grounded theory to shed light on the case being studied (Austin and Sutton 2014).

14.2.2 Types of Case Studies

Yin (2003) identified three types of case studies based on outputs: exploratory, descriptive, and explanatory. Leveraging the classification provided by other authors (Luborsky and Rubinstein 1995; Sullivan and Sargeant 2011; Creswell 2014; Austin and Sutton 2014; Denzin and Lincoln 2017) and our field experience working with farmers, we propose the following case studies:

1. *Intrinsic case study*: The researcher's interest is understanding a particular individual or situation. A researcher conducting an intrinsic case study will describe in detail the nitty-gritty of the case to present what is going on. For example, a researcher might investigate a particular farmer to know why he or she is practicing conservation farming. Another researcher might want to understand how extension services operate in a certain county or even determine the effectiveness of a certain extension method. All three examples involve a single case to comprehend the entire case.
2. *Instrumental case study*: Here, the researcher only studies a case to achieve a broader objective or answer a larger research question. A researcher might study or document how an impact evaluation was done in a project promoting integrated soil fertility management (ISFM). Such a study will document details of processes, procedures, activities, and challenges and draw conclusions that apply beyond a particular case. An example would be a researcher interested in understanding the factors influencing the successful implementation of integrated soil fertility management techniques across different regions. Instead of conducting a case study solely focused on a specific farm or region, the researcher might use an instrumental case study approach. In this case, a particular farm that has successfully implemented ISFM becomes the instrument for exploring the broader research question. The researcher investigates this farm's practices, challenges, and outcomes to draw insights and conclusions that can be applied more broadly to adopting ISFM in diverse agricultural contexts. The specific farm studied is instrumental in providing valuable information to address the larger research goal.
3. *Multiple case study*: This is also called "collective case study" because the researcher studies multiple cases simultaneously as part of a single overall study (Sibbald et al. 2021). For example, a researcher might choose several cases of farmer groups to study to streamline and improve the effectiveness of working with groups in an extension program. The same study may be used to advise farmers on how to make their groups more effective by, for example, having mechanisms for resolving conflicts. In one of our projects, where we used farmer groups to promote tree nurseries, some farmers stopped participating in watering the seedlings due to conflicts among the members. To address this, we conducted multiple case studies of farmer groups to understand their nature and operations. We were advantaged to have a sociologist on the team who used the data to organize conflict resolution meetings with the farmer groups. The conflicts were resolved, and farmers agreed to continue participating in seedlings' watering.
4. *Whole farm case study*: The researcher is interested in describing a whole farm by systematically examining the biological, social, and economic factors of an entire farming system. The case study examines the interaction among production practices, economic status and entrepreneurship activities. To be effective, one should use interdisciplinary teams representing a diversity of different fields needed in both the biological and social

domains. A whole farm case study is useful in providing useful insights into how systems work on specific farms and would help chart the way forward. It might be useful for formulating hypotheses and research topics when used with other research data. Another use is that the interaction between researchers and farmers provides an avenue for increasing farmer involvement in research and extension activities.

5. *Model-based case study*: The researcher is interested in studying cases that will yield information to decide the practices that will be chosen to be modeled by other farmers or scaled up to other regions. For example, a project implementing climate-smart agricultural practices (CSA) carried out case studies to help researchers learn and select the best CSA practices for upscaling.

14.2.3 Characteristics of a Case Study

We summarize the characteristics as follows (Yin 2003):

- Multiple sources of data are used to give a comprehensive picture of the context of the phenomenon being studied. A qualitative researcher collects oral and visual data and uses triangulation.
- They have a major limitation of a lack of generalization.
- Research conducted can be descriptive, exploratory, or explanatory.
- Involves an in-depth investigation. Describes the qualitative case in-depth, providing details holistically and presenting the context.
- Suitable for complex phenomena encountered in agricultural systems.
- Each case has a clear focus.
- Suitable for impact evaluation studies and value chain analysis.
- A screening process is recommended in the selection of the case to ensure that the case study participant has the necessary knowledge and experience.

14.2.4 When to Use a Case Study

Case studies can be used if your research question is to:

1. Evaluate the *impact of an intervention*, policy or innovation on specific outcomes often complementing quantitative data in a project evaluation study. This approach is beneficial because it allows a researcher to assess the complexities of the project, including the objectives, project implementation and stakeholders involved. The in-depth analysis allows evaluators to uncover the mechanisms and processes that contribute to the success or failure of the project. Additionally, they facilitate the explanation of unexpected challenges. For instance, in a project promoting integrated pest management, a researcher can select one farmer (one case) or more farmers (cases) who participated in the project and farmers' experiences.
2. A case study is also appropriate when you want to *study a process*, such as the dynamics of a value chain, as explained in Chap. 13. Value chain analysis examines the activities involved in the production and delivery of a product or service, from raw materials to the user. For instance, in the coffee production process, one of the research projects questions a researcher might ask is, "How do smallholder farmers access inputs such as seeds and fertilizers for coffee production?". By interviewing specific farmers, the researcher is able to answer this research question.
3. A case study is also useful in the study of *unique or rare cases*. In studies where the phenomenon under study is rare or unique, or difficult to replicate, case study research is the most feasible. For example, case studies scan be used to explore farmers who have propagated unique fruit types through grafting that researchers would wish to study techniques. Such a case study can generate hypotheses or research questions for further investigation.
4. Finally, a case study is also appropriate whenever you wish to study a *complex phenomenon*. For example, one of the best methodologies for documenting home gardens is carrying out case studies of selected farmers with well-developed home gardens.

14.3 Ethnography

14.3.1 What Is Ethnography?

Ethnography is the study of people's culture in their environment. The goal is to observe and study people in their cultural context. Culture can be characterized not only as an ethnic group but also as a society, a community, an organization, a geographical location, and a social world (Hammersley and Atkinson 2019). Ethnography describes a culture and enhances understanding of another way of life from the native point of view (Gooberman-Hill 2015; Qamar 2021).

The ethnographic researcher resides with an intact cultural group in a natural setting over an extended duration and collects primary observational data (Morgan-Trimmer and Wood

2016; Webster and Rice 2019). Adequate time is needed to document people's culture, which is done using an empirical data collection framework. Human behavior is best observed and recorded within the social and cultural systems.

The roots of ethnography are in anthropological studies, "the scientific study of humans and human behavior and societies in the past and present." Early anthropologists sometimes accompanied scientific expeditions. It is the earliest qualitative approach used by colonialists to study the local people, who were regarded as primitive (Jackson 2012; Zhou et al. 2019). Initially, the anthropologist researchers would collect data being assisted by local representatives like district administrators. The aim was to document indigenous peoples' lives. Eventually, they engaged the district commissioners as field workers who would collect the data and send it to colonialists for analysis. Data collection involved a standard format to ensure that questions asked to native informants in any colonial society were similar, commonly known as the "notes and queries" approach. Sometimes, it took many years for them to get into the culture of the community they were studying. To do so, often they had to learn the language necessary for socializing with the inhabitants and understanding their daily habits, rituals, norms, and actions.

14.3.2 Characteristics of Ethnography

Ethnographic studies have three key characteristics: 1. Require widespread fieldwork by the investigator within the community he/she is studying for extended periods, and 2. Use of multiple methods and techniques of collecting data, and 3. Use of a few samples. The key method for data collection is observation, but it may also include interviews and studying documents and artefacts. Because ethnography emphasizes depth, context and intensity, few samples are adequate and do not limit data collection. The case study described in Box 14.1 (Fontefrancesco 2021) used several methods: interviews with small and big farm owners using both formal and informal interviews, observation and archival data. Two ethnographic narrative accounts were presented to capture depth.

Toward the end of the fieldwork and data collection, the researcher draws up an outline comprising principal themes that emerge from the data and any analytical ideas accrued during fieldwork. The data, including field notes, transcriptions, and documents, are then indexed into the outline to make the first draft of the report. Subsequently, all data pertinent to analytical themes is systematically fitted into form a coherent line of reasoning.

Ethnography is a demanding research activity but fulfilling. You need to immerse yourself in the community being studied to gather in-depth contextual data. The method requires considerable time and effort to build trust with the participants and understand them. However, the richness of the data gathered makes the efforts worthwhile.

The main characteristics of ethnography summarized from our experience and other published works (LeCompte and Schensul 1999; Schensul and LeCompte 2010; Jackson 2012; Hammersley and Atkinson 2019) include the following:

1. Conducted in the field using multiple data sources.
2. Close interaction with participants' culture is required to capture the depth of information.
3. The researcher chooses a suitable role and decides whether to be an overt observer (reveal identity as a researcher) or conceal.
4. Involve a longitudinal perspective with researchers immersing themselves into the daily lives of people being studied to get an insider perspective.
5. Effective ethnography requires detailed field notes that involve mental notes, jotted notes, and full notes.
6. Reflects an accurate view of participants' perspectives and behaviors; sometimes captured as narrative ethnographic information.
7. Contextualizes all human behavior and belief within a sociopolitical and historical framework.
8. Explores various aspects, including social structures, rituals, language, and daily practices.
9. Utilizes inductive, interactive, and recursive data collection and analytical strategies to construct local cultural theories.
10. Interprets findings using the perception of culture (cultural relativism). It has a commitment to cultural interpretation (see case study in Box 14.1).
11. Mechanical analysis of data involves compiling field notes and transcripts and searching for categories and themes, followed by cutting up the data and finally constructing an outline.

14.4 Grounded Theory

14.4.1 What Is Grounded Theory?

Grounded theory is an approach or method of enquiry that aims to develop theories inductively grounded in systematically gathered and analyzed data (Burns et al. 2022). Unlike traditional research approaches, where a researcher may

Box 14.1: Case Study on Ethnography

Title: Modes and forms of knowledge of farming entrepreneurship. an ethnographic analysis among smallholder farms in NW Italy.

Research problem: Rural countryside results from relentless anthropogenic action carried out by men and machinery and shared by the agricultural choices of farmers with small and big farms. Agriculture, therefore, presents a rich area for understanding the interconnection between entrepreneurial choices and sociocultural processes that explain everyday entrepreneurship. However, past studies have concentrated on technical and economic knowledge and ignored other aspects of farmers' knowledge, for example, identifying family and institutions.

Objective: To explore the sociocultural foundations of farming entrepreneurship.

Research questions: What are the modes and forms of knowledge that underpin agricultural entrepreneurial choices? What grounds do farmers use to make entrepreneurial farming choices?

Methodology: Ethnographic research was carried out between 2015 and 2020 in an ongoing project in Monferrato. Interviews were conducted in the fields using a sample of 27 producers growing hazelnuts and 24 producers practicing cereal production. The sample included small and medium-sized farms, and specific interviews were conducted with 36 male farmers and 15 female farmers with an average age of 55 years. Interviews and observation were made using a standard semi-structured tool, which had questions aimed at investigating different aspects of respondents' professional life, the business history of their farms, the producers' specialization, the reasons for the business configuration and the respondents' relationships with their families and surroundings. In addition, archival research data was gathered to enhance the exploration of the socioeconomic dynamics of the sector.

Data analysis involved integrating data from the interviews with the ethnographic and archival information. The ethnographic information was used to situate and better understand the complexity of the farming activities. Data presentation included an elaboration of two parallel ethnographic accounts from farmers chosen based on age, gender, education, family composition, and enterprise characteristics. The ethnographic accounts present farmers envisioning and commenting on their motivation and expectations linked with their business and cultural choices.

Source: Fontefrancesco (2021). Creative Commons Attribution (CC-BY) license (https://creativecommons.org/licenses/by/4.0/). https://doi.org/10.3390/knowledge1010002

enter a study with predetermined theories or hypotheses, grounded theory begins with an open and exploratory mindset. At the study's outset, the researcher deliberately refrains from imposing preconceived notions, allowing theories to emerge organically from the collected data. This approach aligns with the foundational principle of grounded theory, emphasizing the importance of grounding theoretical frameworks in the empirical realities observed during the research process.

The grounded theory approach was advanced in the 1960s by Barney Glaser and Alumen Strauss, two sociologists working in health and nursing (Andersen and Owen 2014; Quinn 2016; Burns et al. 2022). These authors defined "grounded theory" as a method that derives theoretical constructs from qualitative data analysis. Since then, the grounded theory has been refined and applied to other disciplines such as education, medicine, law, agriculture, and business. Additionally, philosophical perspectives have advanced and affected methodological developments over time.

At the same time, new definitions have emerged, adding to better understanding. For instance, grounded theory is characterized as an investigative strategy that deduces a general, abstract theory of a process, action or interaction that is firmly rooted in the participants' perspectives under the study (Chun Tie et al. 2019; McCall and Edwards 2021). Charmaz (2015) gives an almost similar definition, which states that grounded theory is a method using qualitative research enquiry to create conceptual structures or theories through inductive data analysis.

14.4.2 Strategies for Grounded Theory

Grounded theory is difficult for dissertation and thesis students because of the stringent justification needed to convince the examiners (Dunne and Üstűndağ 2020). Though it is intrinsically flexible, it involves intricate and complex procedures. Students and novice researchers face challenges in understanding the discourse and practical application of its concepts and processes (Chun Tie et al. 2019). Indeed, Cohen et al. (2017) state that it is an experiential methodology requiring a set of skills, among them the ability to pay close attention to data.

Essentially, the approach demands a specific set of procedures and strategies for success. As a student proposing this approach, you will need details and keen reflection to defend it successfully to your examiners. Your expertise, skill, and knowledge are key ingredients for a quality grounded theory study. In particular, procedural accuracy and meticulousness are needed. You have to convince the examiners that you have the necessary skills to unearth what truthfully participants tell you. The following six key strategies are crucial:

1. The first key strategy to consider is sampling. The approach usually begins with purposive sampling and later uses theoretical sampling to select participants who can best contribute to the developing theory. This way of sampling makes the categories more robust and precise as they build analytical complexity while staying grounded in data (Charmaz 2015). As the theory building unfolds simultaneously with data collection and analysis, the application of progressive theoretical sampling ensures that participant selection aligns with the research purpose (see case studies in Box 14.2).
2. The second critical strategy is the simultaneous collection and analysis of data. The data examination and analysis process may begin as soon as the first data set (transcripts, field notes, etc.) is available. The early outcomes of this foundational analysis will subsequently inform and guide further data collection, specifically aimed at exploring particular themes and premises in greater depth. As the process continues, the process of data collection changes from initial exploration to a more focused approach toward investigating targeted theoretical concepts or patterns.
3. The third key strategy pertains to the method of data analysis with a primary emphasis on coding (Mohajan et al. 2022). Coding is its central and primary approach to analyzing data within grounded theory. In this context, coding involves systematically assigning conceptual labels to different data segments to identify themes, patterns, processes, and relationships (Cohen et al. 2017). Here, the data coding and analysis phase continually develops concepts and categories from the raw data. Using an interactive process, the original codes are combined into larger codes, more generalized categories whose properties and relationships to one another will provide the beginning of a theoretical explanation of the data (Mertens 2014).
4. Constant comparison is another core value in the data analysis (Chun Tie et al. 2019; Charmaz 2025). It resonates well with the idea of triangulation and allows comparison across data, categories, groups, patterns, and key issues. Progressively, the researcher compares the new data with previously generated existing categories. If there is a poor fit between data and categories, then the categories and theories must be modified until all data are accounted for. Here, we see an application of the concept of modeling data similar to quantitative data analysis, where the amount of data accounted for in the model is an important consideration. If the new data fits well with the existing categories, the researcher will be happy and progress with the simultaneous data collection and analysis. Ultimately, constant comparison serves to merge and explain patterns and variations.

Box 14.2: Case Studies That Have Used Grounded Theory Approach

Coral et al. (2020). Agency and structure: a grounded theory approach to explain land-use change in the Mindo and western foothills of Pichincha, Ecuador. *Journal of Land Use Science*, *15*(4), 547–569. https://doi.org/10.1080/1747423X.2020.1811791

Amin et al. (2021). Exploring Migration Hold Factors in Climate Change Hazard-Prone Area Using Grounded Theory Study: Evidence from Coastal Semarang, Indonesia. Sustainability 2021, Vol 13, Page 4335 13:4335. https://doi.org/10.3390/SU13084335

5. The final strategy involves theoretical categorization and coming up with the core category. The core category is the final product or outcome and can be regarded as the hallmark of this highly systematic process in grounded theory. The researcher must find a theoretical code which best systematizes and arranges a central code category. Usually, other related categories emerge at this final stage. These, too, are important form part of this final product. Once the core category and related categories are determined, the researcher then reviews the literature and integrates it to come up with the final theory.

From this discussion, we come up with the following key characteristics of grounded theory (Fig. 14.1):

- *General to generating theory*: Starts with a phenomenon that is little-known, collects data qualitatively and systematically, followed by an iterative data analysis approach that yields a theory grounded in the data (emerges from data collected in a social setting).
- Require meticulous application of specific and explicit methods and processes.
- Might start with purposive sampling, followed by simultaneous data collection and generation.
- *Coding and memoing*: Coding is done concurrently as data is being collected, coupled with comparative analysis, theoretical sampling, and memoing (Chun Tie et al. 2019). Memoing is documenting your ideas about the data in methodological and theoretical memos.
- Concurrent data generating and analysis gives way to theoretical sampling, guided by the codes and categories that emerge.
- Theoretical sampling is used to identify and follow clues to fill gaps identified in the analyzed sets of data (Charmaz 2015).

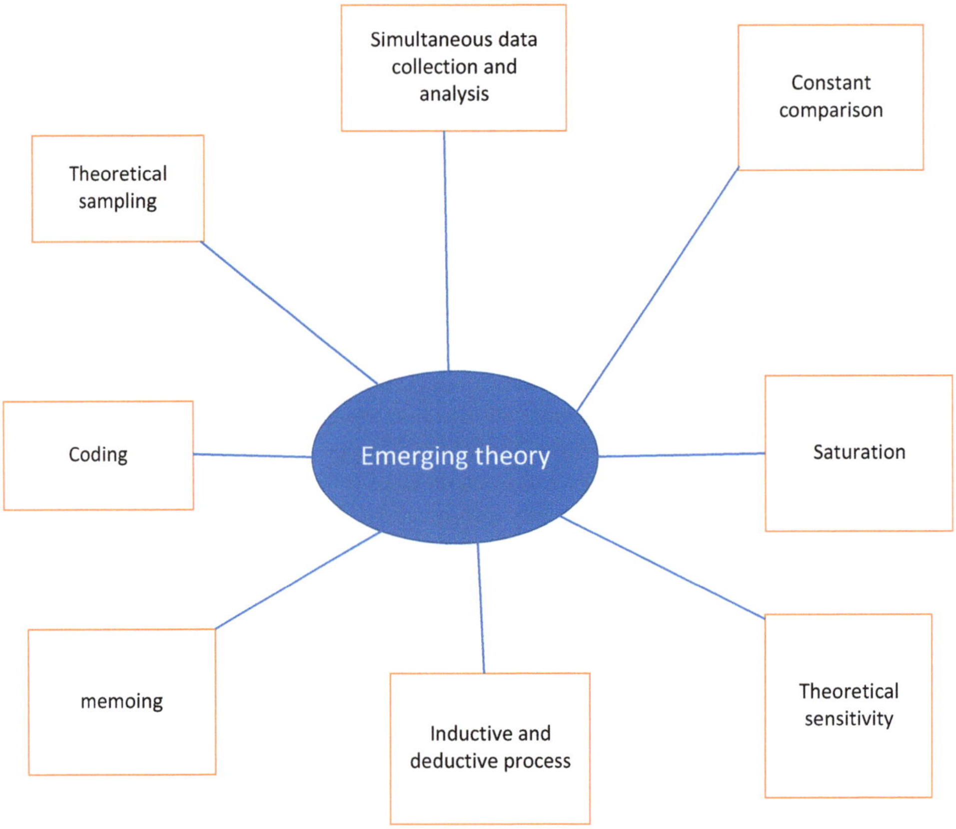

Fig. 14.1 Key strategies used in grounded theory approach

- *Inductive and deductive iterative process*: Thorough iterative process of data collection, coding, comparison, and deriving concepts and categories often involves both inductive and deductive judgment and insight.
- *Experiential features of the researcher*: Quality grounded theory research study needs a researcher to have the experience, skills, and knowledge to maneuver a rigorous and stringent methodology. It also requires experience, knowledge, and skills and a novice researcher wishing to use the approach will need to engage in deep reading reflection and provide a detailed methodology of how they will carry out the study (Cohen et al. 2017).

14.5 Narrative Approach

14.5.1 Nature of Narrative Research

Narrative enquiry investigates human experiences through the lens of storytelling and narratives (Haydon and van der Riet 2014; Foxall et al. 2021). This method recognizes the power of stories in shaping individuals' understanding of themselves and the world around them. Narrative enquiry is rooted in the belief that people make sense of their lives by constructing and sharing stories, and researchers utilizing this approach seek to explore the richness and complexity of these narratives. Creswell (2014) delineates narrative research as a distinctive form of enquiry centered on the exploration of individuals' lives where the researcher invites and asks one or more individuals to share their stories. The essence lies in the participant's narratives, which are subsequently "retold or restoryed" by the researcher to construct a narrative chronology. The process involves capturing the explicit details of the stories and interpreting the underlying meanings, emotions and cultural contexts embedded in them. The narrative is best used when the study has a specific contextual focus, such as a researcher wanting to study the behavior and experiences of farmers during a drought that occurred 20 years ago.

Narrative research utilizes the ability of individuals to tell stories. For instance, farmers will want to share stories to share their experiences, including daily challenges, successes, and aspirations. This method captures their journeys and provides perspectives into their practices, beliefs, and the contextual factors influencing their work. By focusing on these narratives, researchers can better understand the social, economic, and environmental dynamics that shape farmers' lives.

Narrative research uses many methods, including gathering data through the collection of individual stories, reporting their experiences, and chronologically ordering those experiences. The most common approaches/methods include the following:

1. *Biographical* narrative consisting of a person writing the life of another person, while the.
2. *Autography* written and chronicled by individuals who are participants in the study.

3. *Oral history* that gathers personal reflection and/or their causes and effects from one individual or several individuals (Hernandez et al. 2017).
4. *Restorying*: The researcher retells the story.
5. *Letter writing*: Participants are asked to write about their lived experiences on the topic of study.

The procedure focuses on studying one or two individuals, gathering data on their stories, reporting individual experiences, and chronologically ordering the meaning of those experiences. Here we look at the various steps:

1. Assess if the research question aligns with the narrative approach, which is suited for capturing the rich stories or live experiences of either single individuals or lives of a community.
2. Choose one or more participants who possess relevant stories, well documented in a diary, journal or other forms of documentation. Alternatively, the researcher may gather various materials, including correspondence, photographs, and artefacts.
3. Develop some narrative research questions to start with.
4. Gather information about the context surrounding the stories by delving into the environmental, geographical, and cultural factors.
5. Analyze the stories provided by the participants and engage in the process of restorying, wherein the narratives are reorganized within a meaningful framework that makes sense. During the analysis, check for key elements, e.g., places, events, and themes and then rewrite the stories chronologically. Be keen on these three aspects: 1. The chronological sequence mainly consists of the beginning, middle and end but can also consist of past, present and future, e.g., if you are investigating drought incidences, such information will be important to hear about the present and future. 2. Provide causal links among ideas, and 3. Detail themes that arise from the story and highlight their underlying themes and connections are called "deconstructing stories." Some of these issues are captured in the two case studies presented in Box 14.3.

14.5.2 Key Characteristics of Narrative Research Include:

- Involves constructing stories based on data collected from interviews.
- It focuses on the experiences of the individual.
- Chronology of the events as narrated by an individual is important.
- Restorying is a powerful technique involving constructing the narrative account.
- A collaborative approach that involves the researcher and participants in the negotiation of the final report.

14.5.2.1 Restorying

A narrative researcher needs expertise in restorying. Usually, during the interviews, participants will share their stories sometimes without paying attention to the real-time order of events (Haydon and van der Riet 2014). They will, at times, mix dates and locations and forget some events that they share as you are finishing the interviews. The restoring process involves first conducting the interview and transcribing to provide a written to provide a written note. From the written notes, the researcher condenses and identifies key elements in the story. Finally, the researcher organizes the story into a chronological account, paying attention to the characters and setting. After the restorying is complete, the participants are invited to validate the story as compiled by the researcher. This is an important way of enhancing the trustworthiness of data collected in narrative research.

14.5.3 Triangulation in Narrative Research

Narrative research has been criticized as fictitious, and researchers can address this concern through triangulation (Moen 2006). This will involve using multiple data sources to achieve a thorough understanding of the phenomenon. You can use four types of triangulations: (1) method triangulation, which combines several methods for a more comprehensive analysis (2) Investigator triangulation, using multiple researchers to gather and interpret data (3) data source triangulation, which means use of diverse data sources and (4) theory triangulation which apply different theoretical perspectives to interpret the data (Moon 2019). In the two case studies in Box 14.3, multiple sources of data were used.

14.6 Phenomenology

14.6.1 Nature of Phenomenology

Phenomenological studies examine human experiences through the experiences described by the people involved and the meanings participants attribute to those experiences (Hooker 2015; Neubauer et al. 2019). Rooted in philosophy, phenomenology seeks to understand and accurately describe the essence of human consciousness as it engages with the world. In practice, the researcher tries to understand and explain how one or more individuals experience a phenomenon. Because the primary data source is the participants' life-world, in-depth interviews are the most common and effective method of data gathering.

Box 14.3: Case Studies That Used Narrative Research Approach

Case study 1: *What's the story on agriculture? Using narratives to understand farming households' aspirations in Meru, Kenya* (Dilley et al. 2021)

Research problem: Research on farming aspirations often overlooks the narratives that shape them and how they connect with the goals of extension services.

Objective: The study aimed to explore the farming aspirations of individuals in Meru County, Kenya, and how personal circumstances, social networks, and available resources shape these aspirations. It also sought to examine how farmers' aspirations align with the goals of extension services, aiming to improve agricultural development by better-addressing farmers' needs and desires.

Methodology: The study involved surveys in 10 randomly selected villages in Meru County, with 10 households from each village. Respondents, including household heads and a family member, were asked to envision their future in 10 years and share how they planned to reach that point. The survey also explored their future farming expectations and perceptions of crop and livestock risks and rewards. Additionally, two rounds of semi-structured interviews were conducted with 15 selected households to deepen understanding of their current situations and aspirations. Interviews covered livelihood activities, hopes, and support systems like extension services. Private and public extension agents were also interviewed, focusing on their roles in supporting farmers. All interviews were audio recorded, transcribed, and analyzed.

https://doi.org/10.1057/s41287-021-00361-9;

Case study 2: *Rethinking farmer knowledge from soil to plate through narrative enquiry: An agroecological food systems perspective* (Bendfeldt et al. 2021)

Research problem: We need to better understand how smallholder farmers' knowledge and experiences are constructed within an agroecological context and how this knowledge contributes to addressing the complexities of the food system.

Objective: The study sought to uncover how farmers' knowledge, often hidden or underutilized, serves as a valuable asset for sustaining and improving food system possibilities. It also aimed to explore the interconnectedness of farmers' practices with broader food system domains, reflecting agroecological values.

Methodology: The study used narrative enquiry to explore how smallholder farmers' knowledge and experiences are interconnected, like a network, in creating healthy soil and a strong sense of place. Twelve farmers participated in semi-structured interviews on their farms, each committing to two visits lasting 2–2.5 h. The interviews were recorded, transcribed, and co-edited with the farmers into final narratives of 9–12 pages each. After the conversations, researchers toured the farms to observe soil and water conservation practices. The narratives were analyzed using the Whole Measures for Community Food Systems framework, with two rounds of coding in Atlas.ti to identify key themes and sub-themes based on the frequency of references in the transcripts.

https://foodsystemsjournal.org/index.php/fsj/article/view/1035/1006.

The core premise of phenomenology is to bracket preconceived assumptions and engage in the process of "epoche," or suspension of judgment, to approach the phenomenon with "fresh eyes" (Neubauer et al. 2019; Tavakol and Sandars 2025). By consciously withholding preconceived notions, researchers aim to explore participants' lived experiences more openly and unbiasedly. This commitment to bracketing contributes to the authenticity of phenomenological enquiry.

Narrative and phenomenological research are two distinct qualitative methodologies that share some common features (Tavakol and Sandars 2025). Both approaches focus on exploring subjective experiences, valuing individual perspectives, and the depth of lived realities. However, they differ in their underlying philosophical foundations and methodological orientations.

Narrative research is concerned with constructing "*coherent and sequential stories*" to convey the complexities of personal experiences. Its main objective is to understand how individuals make sense of their lives by creating stories, highlighting the role of language and narrative structure in shaping meaning. On the other hand, phenomenological research aims to uncover the "*essence of lived experiences*" and their inherent structures and meanings (Tavakol and Sandars 2025; Byrne 2025). While both approaches use in-depth interviews as a common method for data collection, narrative research focuses on the chronological unfolding of events within a storytelling framework, while phenomenological research prioritizes the exploration of the essence of experiences, emphasizing the unique perspectives and interpretations of participants.

14.6.2 Characteristics of Phenomenology

The basic tenacity of phenomenology is to condense individual experiences with a phenomenon into a description of "*universal essence*" (A grasp of the very nature of a thing) (Byrne 2025). The goal is to move beyond the particulars of individual experiences and uncover the core elements that universally define and shape the phenomenon's essence.

Procedure of conducting a phenomenological study:

1. Assess whether the research question is best suited for examination through the phenomenological approach. For instance, it is beneficial to use this approach when it involves exploring how individuals experience a significant life event, e.g., drought and famine or entrepreneurship involving export markets.
2. Clearly articulate the broad philosophical assumptions of phenomenology, detailing how bracketing will be achieved. As you do this, recall that it is a temporary suspension of prejudgments or assumptions regarding the examined phenomenon, acknowledging the impossibility of complete freedom from presuppositions.
3. Select participants who have experienced the phenomenon and mainly use in-depth interviews to gather data.
4. During the interviews, participants posed two key questions: "What have you experienced in terms of the phenomenon"? and 2. What contexts or situations have typically influenced your experience of the phenomenon?
5. Analyze the data by focusing on the two key questions. Go through the interview transcriptions, explaining significant statements and quotes. Develop clusters of meanings from these significant statements to identify recurring themes, a process called horizonization.
6. Utilize the significant statements to compose a description of what the participants experienced. "Textural description." The statements are also used to write a description of the context or setting that influenced how the participants experienced the phenomenon, which is called imaginative variation or structural description.
7. From the structural and textural description, write a composite description that presents the "Essence" of the phenomenon called essential or essence. This focuses on the common experience of the participants; it means that all the experiences have an underlying structure. It is a descriptive passage, a long paragraph or two paragraphs that make the reader understand what it is like to experience the phenomenon (Box 14.4).

Box 14.4: Case Studies on Phenomenological Research Approach

(Rose et al. 2023). Phenomenological Research on the Youth's Experiences on Agriculture in the Philippines. New Trends in Qualitative Research, 16, e767. https://doi.org/10.36367/ntqr.16.2023.e767

O'Reilly and Scott (2023). "Harvest work, migration, and the structured phenomenology of time". Journal of Ethnic and Migration Studies, 49(15), 4033–4051. https://doi.org/10.1080/1369183X.2022.2116565

14.7 Trustworthiness of Qualitative Research

As alluded to in this book, validity and reliability of data and findings are key terms mentioned, particularly in sections describing quantitative research. However, the accuracy of representation (validity) and reproducibility (reliability) of the phenomenon are big questions in qualitative research. As we have seen in all five qualitative approaches discussed, qualitative research is about presenting multiple realities and that there is subjectivity in the interpretation of the data. This means replicability is not one of the aims of a qualitative researcher. As a quantitative researcher, you may then ask. What does this mean regarding the information's validity (accuracy) and reliability (consistency)?. This is one of the main debates or "bones of contention" between positivism and interpretivism researchers. Quantitative researchers often denounce qualitative research as biased, small-scale, weak in rigor or firmness and unreplicable (Anderson 2010). The narrative of lack of replicability in qualitative research means that a different researcher using the same methodology in a different place at a different time will expect different results.

This can mean that even with a given set of data collected and shared, different researchers will generate unique and dissimilar outcomes/results. This is because reality is constructed (Constructivism/ interpretivism). Because of this, the concept of validity applied in quantitative research is not a goal in qualitative research. Qualitative researchers use the term "trustworthiness" instead to represent confidence in the work's rigor (Moser and Korstjens 2017). This term answers the question: how can we be sure that the data published in referred publications or presented at a conference is of good quality? This aspect is usually witnessed in conferences where scientists question the truth and accuracy of a paper originating from a qualitative enquiry.

In examining trustworthiness, we are asking such questions as: Is your study rigorous?, Does your study report the truth (true picture) of what participants said and meant? Is your study authentic?. When you present the report, can your results be approved by participants? These are in fact questions that are pertinent when we think about ethics and the need to fulfil three core principles of ethics in research (beneficence, non-maleficence, and justice). Incredible data may harm the participants when wrong decisions are made. Trustworthiness is an extremely important concept in thesis work. Some universities have templates of reporting showing where and how to include an assessment of trustworthiness. Indeed, in reporting qualitative studies in technical reports and journal articles, templates and guidelines are given to ensure that researchers can gauge the authenticity of the research. It is mostly within the methodology section, but it is good to include aspects of trustworthiness in your results and conclusions. Moreover, in a thesis, your supervisor and examiners will be interested in how you assess the trustworthiness of your study.

Several authors (Noble and Smith 2015; Korstjens and Moser 2017; Moser and Korstjens 2018) have advanced four main criteria for assessing trustworthiness. These are: 1. Credibility, 2. Dependability, 3. Confirmability, 4. Transferability and 5. Flexibility.

Credibility

Credibility is a measure of the truth value of the study and assesses whether the study findings are true and accurate. The equivalent of this in the quantitative realm is internal validity. Somehow, credibility is highly dependent on the researchers themselves and the research methods used in the study. Credibility refers to the degree to which the results of an enquiry can be confirmed or corroborated by other researchers.

Six techniques are presented that can be used to promote credibility (Noble and Smith 2015; Korstjens and Moser 2017; Moser and Korstjens 2017). These are prolonged engagement, triangulation, persistence observation, negative case analysis, subjective checks, peer debriefing, and member checking.

Dependability

Dependability is the term used to measure or show the consistency of your results. It is thus equivalent to reliability (quantitative study). Three main ways to address dependability, i.e., clear presentation of methodology, writing and preserving field notes and maintaining *an audit trail*. Audit trail refers to documentation of the steps of how the research was carried out including a record of notes and recordings captured during data collection. Researchers should track the whole research inquiry, taking note of all decisions made regarding the methods used for data collection, analysis, and interpretation. The aim is to be able to report adequate contextual information in detail about each part of the process that can allow another researcher to replicate the study and generate similar results. An inquiry audit performed by an external person is one way to assess the dependability of a qualitative study. In an academic study (Thesis/Dissertation), the audit trail is very useful in the defense of your work.

Confirmability refers to neutrality. On the other hand, quantitative research is equivalent to *objectivity*. It measures the degree to which another researcher could corroborate or verify results. Trustworthy research should produce findings that objectively reflect information collected from participants, meaning they are free from bias, assumptions, and preconceived ideas. Confirmability is usually demonstrated using four aspects: audit trail, reflective journal, reference to literature and triangulation (Noble and Smith 2015).

Transferability refers to the potential application of findings and demonstrates to what extent findings can be transferred to other settings or groups. This is parallel to the analogy of *generalizability or external validity*. Transferability concerns whether results apply to other contexts, settings, backgrounds, and environments. Because of the small sample size and nature of the sampling in qualitative research, we avoid the term generalization because this is not the goal in the first place, as explained at the beginning of this chapter. The research should assist the readers and consumers in easily assessing whether results apply to other settings by providing a thick description (Noble and Smith 2015) that provides a rich enough depiction of the situation for application to other circumstances. This includes detailed contextual information about the fieldwork location, participants, methods, timelines, as well as the entire duration of the study (Stahl and King 2020). The rich, detailed descriptions allow readers to draw conclusions and interpret the data, which helps in making decisions of transferability and replicability.

In this chapter and Chap. 13, we discussed five main designs. The key aspects guiding decision-making in qualitative research are the *research question and the philosophical assumption*. The research question is structured around the interpretivism paradigm, which seeks to answer "why" and how" questions. The research question will dictate the type of understanding, the specific phenomenon to explore, or whether it is to uncover meaning and social processes (Fig. 14.2). This will guide the researcher on choosing from among the five main designs presented.

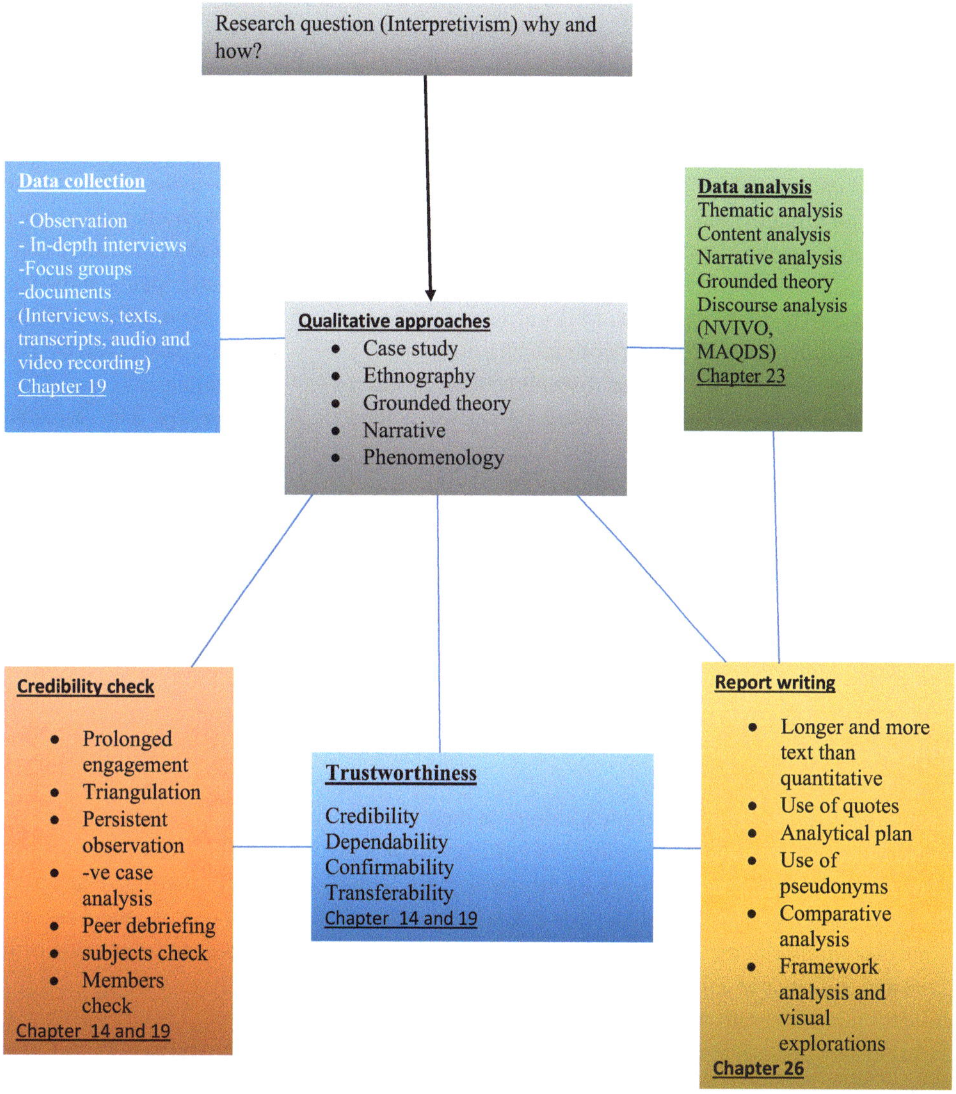

Fig. 14.2 Qualitative enquiry pathways summarizing the five key aspects for decision-making; qualitative approach, data collection, data analysis, credibility check, trustworthiness, and report writing

Apart from the research question, a researcher must also decide on *data collection, data analysis, credibility checks, ensuring trustworthiness, and report writing*. Each stage is guided by specific methods and considerations to ensure depth, accuracy, and integrity. The figure illustrates how qualitative research is not just about gathering data, but about making informed decisions at each step to ensure rich, meaningful, and trustworthy findings.

Recommended Further Reading

1. Chun Tie Y, Birks M, Francis K (2019) Grounded theory research: A design framework for novice researchers. SAGE Open Med 7: https://doi.org/10.1177/2050312118822927/ASSET/6A38AA2A-7480-4C45-8872-213AE874E349/ASSETS/IMAGES/LARGE/10.1177_2050312118822927-FIG1.JPG
2. Cohen L, Manion L, Morrison K (2017) Research Methods in Education. Research Methods in Education. https://doi.org/10.4324/9781315456539/RESEARCH-METHODS-EDUCATION-LOUIS-COHEN-LAWRENCE-MANION-KEITH-MORRISON
3. Korstjens I, Moser A (2017) Series: Practical guidance to qualitative research. part 2: Context, research questions and designs. European Journal of General Practice 23:274–279. https://doi.org/10.1080/13814788.2017.1375090
4. Hammersley M, Atkinson P (2019) Ethnography: Principles in Practice: Fourth Edition. Ethnography: Principles in Practice: Fourth Edition 1–280. https://doi.org/10.4324/9781315146027/ETHNOGRAPHY-MARTYN-HAMMERSLEY-PAUL-ATKINSON/RIGHTS-AND-PERMISSIONS

Key Points in This Chapter

Case Studies

A case study is a detailed analysis of a single unit called "the case." The case could be an individual, a community of participants, an organization, a program, a practice, or an event. Five types of case studies were identified:

1. *Intrinsic case study*: describes in detail what is going on with the aim of gathering in-depth information.
2. *Instrumental case study*: The researcher studies only a case to achieve a broader objective or answer a larger research question.
3. *Multiple case study*: The researcher studies multiple cases simultaneously.
4. *Whole farm case study*: Researcher describes a whole farm or entire farming system.
5. *Model-based case study*: Researcher studies cases that will yield information to decide the practices.

Ethnography

Ethnography is the study of people's culture in their environment. Ethnographic studies have three key characteristics: 1. They require widespread fieldwork by the investigator within the community he/she is studying for extended periods; 2. They use multiple methods and techniques of collecting data; and 3. They use a few samples. The key method for data collection is observation, but it may also include interviews and studying documents and artefacts.

Grounded Theory

Grounded theory enquiry aims to develop theories inductively grounded in systematically gathered and analyzed data. Unlike traditional research approaches, where a researcher may enter a study with predetermined theories or hypotheses, grounded theory begins with an open and exploratory mindset. Key terms in grounded theory include: *theoretical sampling, simultaneous data collection and analysis, coding, memoing, constant comparison, saturation, and theoretical sensitivity.*

Narrative Approach

Narrative enquiry investigates human experiences through the lens of storytelling and narratives. The narrative is particularly useful for studies with a contextual focus, such as a researcher studying the behavior and experiences of farmers during a historical event like a drought.

The most common approaches/methods include the following;

1. *Biographical* narrative consisting of a person writing the life of another person, while the.
2. *Autography* written and chronicled by individuals who are participants in the study.
3. *Oral history* that gathers personal reflection and/or their causes and effects from one individual or several individuals.
4. *Restorying*: The researcher retells the story.
5. *Letter writing*: Participants are asked to write about their lived experiences on the topic of study.

Phenomenology

The phenomenological approach examines human experiences through the experiences described by the people involved and the meanings participants attribute to those experiences. It aims to uncover the "essence of lived experiences" and their inherent structures and meanings. Because the primary data source is the participants' life-world, in-depth interviews are the most common and effective method of data gathering. The core premise of phenomenology is to bracket preconceived assumptions and engage in a process of "epoche," or suspension of judgment.

Trustworthiness in Qualitative Research

The concept of trustworthiness in qualitative research addresses the validity and reliability typically emphasized in quantitative studies. Trustworthiness involves questions about the study's rigor, truthfulness, and authenticity, aligning with ethical principles in research.

The following four criteria are used to assess trustworthiness:

1. Credibility: Equivalent to internal validity in quantitative research; can be promoted through prolonged engagement and triangulation.
2. Dependability: equivalent to reliability, emphasizes result consistency, achieved through clear methodology presentation, meticulous field note writing, and maintaining an audit trail.
3. Confirmability: equivalent to objectivity in quantitative research; demonstrated through aspects like an audit trail, reflective journal, reference to literature, and triangulation.

4. Transferability: concerns the applicability of findings to other settings, emphasizing rich, detailed descriptions for readers to assess and interpret data, even though generalization is not the primary goal in qualitative.

Review Questions

1. Why does ethnography require "widespread fieldwork by the investigator within the community for extended periods"? How does this contribute to understanding "people's culture in their environment," particularly in an agricultural setting like traditional farming practices?
2. Ethnography uses "few samples." How does this characteristic align with the goal of producing an in-depth, rich understanding rather than generalizable findings?
3. Briefly define and explain the significance of theoretical sampling, constant comparison, and saturation in conducting a grounded theory study. How do these processes ensure that the theory is genuinely "grounded"?
4. The example given is "a researcher studying the behavior and experiences of farmers during a historical event like a drought.". How would the five common approaches (biographical, autography, oral history, restorying, letter writing) be applied in such a study? Choose two approaches and explain their application.
5. What does it mean to uncover the "essence of lived experiences" in phenomenology? Provide an agricultural example, such as the experience of losing a crop to pests, and explain how phenomenology would explore this situation.
6. Explain the concept of "epoche" or "suspension of judgment" in phenomenology. Why is it critical for the researcher to bracket preconceived assumptions when trying to understand participants' "life-world"?
7. You are conducting an ethnographic study on traditional farming practices in a remote community. How would prolonged engagement and triangulation enhance the credibility of your findings? Provide concrete examples.
8. Explain the core difference between transferability in qualitative research and generalization in quantitative research. Why are "rich, detailed descriptions" crucial for transferability, even though generalization is not the primary goal?
9. How does the concept of trustworthiness align with broader ethical principles in research, particularly in ensuring the "truthfulness and authenticity" of findings when dealing with human subjects in agricultural contexts?

Chapter Exercise

1. Provide examples of a "community of participants," an "organization," and a "program" that could serve as a "case" in an agricultural study.
2. Given the diverse qualitative methodologies, such as case study, ethnography, grounded theory, narrative, and phenomenology, how should a researcher determine which approach is most suitable for a specific agricultural research question?
3. Your supervisor has informed you that your upcoming graduate research project will utilize grounded theory (GT) methodology. What preparations do you need to make? Tip: Read Charmaz (2014), Strauss and Corbin (2015), Glaser and Strauss (1967), Birks and Mills (2015).

Answers to Review Questions

- Q1 and Q2 Ethnography: Reference 14.3
- Q3. Grounded theory: Reference 14.4
- Q4. Narrative: Reference 14.5
- Q5. Phenomenology: Reference 14.6
- Q6. Phenomenology: Reference 14.6
- Q7, Q8 and Q9: Reference 14.7

References

Amin C, Sukamdi S, Rijanta R (2021) Exploring migration hold factors in climate change hazard-prone area using grounded theory study: evidence from Coastal Semarang, Indonesia. Sustainability 13:4335. https://doi.org/10.3390/SU13084335

Andersen JS, Owen DC (2014) Helping relationships for smoking cessation: grounded theory development of the process of finding help to quit. Nurs Res 63:252–259. https://doi.org/10.1097/NNR.0000000000000040

Anderson C (2010) Presenting and evaluating qualitative research. Am J Pharm Educ 74:141. https://doi.org/10.5688/aj7408141

Austin Z, Sutton J (2014) Qualitative research: getting started. Can J Hosp Pharm 67:436–440. https://doi.org/10.4212/CJHP.V67I6.1406

Baker GR (2011) The contribution of case study research to knowledge of how to improve quality of care. BMJ Qual Saf 20:i30–i35. https://doi.org/10.1136/BMJQS.2010.046490

Bendfeldt ES, McGonagle M, Niewolny K (2021) Rethinking farmer knowledge from soil to plate through narrative inquiry: an agroecological food systems perspective. J Agric Food Syst Commun Dev 11:137–151. https://doi.org/10.5304/jafscd.2021.111.012

Burns M, Bally J, Burles M et al (2022) Constructivist grounded theory or interpretive phenomenology? Methodological choices within specific study contexts. Int J Qual Methods 21. https://doi.org/10.1177/16094069221077758

Byrne T (2025) An introduction to phenomenology in qualitative education research: a critical reading of 'Understanding things from within.' Phenomenol Cogn Sci 1–26. https://doi.org/10.1007/S11097-025-10086-3/FIGURES/4

Charmaz K (2015) Teaching theory construction with initial grounded theory tools. Qual Health Res 25:1610–1622. https://doi.org/10.1177/1049732315613982

Charmaz K (2025) Constructing grounded theory 382

Chun Tie Y, Birks M, Francis K (2019) Grounded theory research: a design framework for novice researchers. SAGE Open Med 7. https://doi.org/10.1177/2050312118822927/ASSET/6A38AA2A-7480-4C45-8872-213AE874E349/ASSETS/IMAGES/LARGE/10.1177_2050312118822927-FIG1.JPG

Cohen L, Manion L, Morrison K (2017) Research methods in education. https://doi.org/10.4324/9781315456539/RESEARCH-METHODS-EDUCATION-LOUIS-COHEN-LAWRENCE-MANION-KEITH-MORRISON

Coral C, Bokelmann W, Bonatti M et al (2020) Agency and structure: a grounded theory approach to explain land-use change in the Mindo and western foothills of Pichincha, Ecuador. J Land Use Sci 15:547–569. https://doi.org/10.1080/1747423X.2020.1811791

Creswell J (2013) Steps in conducting a scholarly mixed methods study. Discipline-based education research group speaker series

Creswell JW (2014) Research and design qualitative, quantitative, and mixed methods approaches. SAGE

Creswell JW, Guetterman TC (2019) Educational research planning, conducting, and evaluating quantitative and qualitative research. Pearson, pp 1–681

Creswell JW, Plano Clark VL (2011) Designing and conducting mixed methods research, pp 51–99

Crowe S, Cresswell K, Robertson A et al (2011) The case study approach. BMC Med Res Methodol 11:1–9. https://doi.org/10.1186/1471-2288-11-100/TABLES/9

Denzin NK, Lincoln YS (2017) The SAGE handbook of qualitative research, 5th edn. SAGE

Dilley L, Mausch K, Crossland M, Harris D (2021) What's the story on agriculture? Using narratives to understand farming households' aspirations in Meru, Kenya. Eur J Dev Res 33:1091–1114. https://doi.org/10.1057/S41287-021-00361-9/FIGURES/3

Dunne C, Üstündağ BG (2020) Successfully managing the literature review and write-up process when using grounded theory methodology—a dialogue in exploration. Forum Qualitative Sozialforschung/Forum: Qualitative Social Research 21:10.17169/FQS-21.1.3338

Fontefrancesco MF (2021) Modes and forms of knowledge of farming entrepreneurship: an ethnographic analysis among small farmers in NW Italy. Knowledge 1:2–11. https://doi.org/10.3390/KNOWLEDGE1010002

Foxall F, Sundin D, Towell-Barnard A et al (2021) Revealing meaning from story: the application of narrative inquiry to explore the factors that influence decision making in relation to the withdrawal of life-sustaining treatment in the intensive care unit. Int J Qual Methods 20. https://doi.org/10.1177/16094069211028345/ASSET/EF273A50-518D-4471-AF2C-402971F5140C/ASSETS/IMAGES/LARGE/10.1177_16094069211028345-FIG4.JPG

Gooberman-Hill R (2015) Ethnographies of pain: culture, context and complexity. Br J Pain 9:32–35. https://doi.org/10.1177/2049463714555439

Hammersley M, Atkinson P (2019) Ethnography: principles in practice, 4th edn, pp 1–280. https://doi.org/10.4324/9781315146027/ETHNOGRAPHY-MARTYN-HAMMERSLEY-PAUL-ATKINSON/RIGHTS-AND-PERMISSIONS

Haydon G, van der Riet P (2014) A narrative inquiry: how do nurses respond to patients' use of humour? Contemp Nurse 46:197–205. https://doi.org/10.5172/CONU.2014.46.2.197

Hernandez SG, Genkova A, Castañeda Y et al (2017) Oral histories as critical qualitative inquiry in community health assessment. Health Educ Behav 44:705–715. https://doi.org/10.1177/1090198117728546

Hooker C (2015) Understanding empathy: why phenomenology and hermeneutics can help medical education and practice. Med Health Care Philos 18:541–552. https://doi.org/10.1007/S11019-015-9631-Z

Jackson JL (2012) Ethnography. Oxford Bibliographies Online Datasets https://doi.org/10.1093/OBO/9780199766567-0001

Korstjens I, Moser A (2017) Series: practical guidance to qualitative research. Part 2: context, research questions and designs. Eur J Gen Pract 23:274–279. https://doi.org/10.1080/13814788.2017.1375090

Kyveryga PM (2019) On-farm research: experimental approaches, analytical frameworks, case studies, and impact. Agron J 111:2633–2635. https://doi.org/10.2134/AGRONJ2019.11.0001

LeCompte MD, Schensul JJ (1999) Choosing and designing an ethnographic research project. In: Designing and conducting ethnographic research, vol 97–126

Lewallen S, Courtright P (1998) Epidemiology in practice: case-control studies. Commun Eye Health 11:57

Luborsky MR, Rubinstein RL (1995) Sampling in qualitative research. Res Aging 17:128–130. https://doi.org/10.1177/0164027595171005

McCall C, Edwards C (2021) New perspectives for implementing grounded theory. Studies in engineering. Education 1:93. https://doi.org/10.21061/SEE.49

Mertens D (2014) Research and evaluation in psychology and education libro electronico 4

Miles MB, Huberman AM, Saldana J (1984) Qualitative data analysis: a methods sourcebook. SAGE, p 408

Moen T (2006) Reflections on the narrative research approach. Int J Qual Methods 5:56–69. https://doi.org/10.1177/160940690600500405

Mohajan H, Mohajan D, Mohajan HK (2022) Munich personal RePEc archive exploration of coding in qualitative data analysis: grounded theory perspective, 1

Moon MD (2019) Triangulation: a method to increase validity, reliability, and legitimation in clinical research. J Emerg Nurs 45:103–105. https://doi.org/10.1016/j.jen.2018.11.004

Morgan-Trimmer S, Wood F (2016) Ethnographic methods for process evaluations of complex health behaviour interventions. Trials 17:1–11. https://doi.org/10.1186/S13063-016-1340-2/TABLES/2

Moser A, Korstjens I (2017) Series: practical guidance to qualitative research. Part 1: introduction. Eur J Gen Pract 23:271–273. https://doi.org/10.1080/13814788.2017.1375093

Moser A, Korstjens I (2018) Series: practical guidance to qualitative research. Part 3: sampling, data collection and analysis. Eur J Gen Pract 24:9–18. https://doi.org/10.1080/13814788.2017.1375091

Neubauer BE, Witkop CT, Varpio L (2019) How phenomenology can help us learn from the experiences of others. Perspect Med Educ 8:90–97. https://doi.org/10.1007/S40037-019-0509-2

Noble H, Smith J (2015) Issues of validity and reliability in qualitative research. Evid Based Nurs 18:34–35. https://doi.org/10.1136/EB-2015-102054

O'Reilly K, Scott S (2023) Harvest work, migration, and the structured phenomenology of time. J Ethn Migr Stud 49:4033–4051. https://doi.org/10.1080/1369183X.2022.2116565

Priya A (2021) Case study methodology of qualitative research: key attributes and navigating the conundrums in its application. Sociol Bull 70:94–110. https://doi.org/10.1177/0038022920970318

Qamar AH (2021) At-home ethnography: a native researcher's fieldwork reflections. Qual Res J 21:51–64. https://doi.org/10.1108/QRJ-03-2020-0019/FULL/XML

Quinn P (2016) A grounded theory study of how nurses integrate pregnancy and full-time employment: becoming someone different. Nurs Res 65:170–178. https://doi.org/10.1097/NNR.0000000000000157

Rose C, Mendoza C, Garcia RQ (2023) Phenomenological research on the youth's experiences on agriculture in the Philippines. New Trends Qualit Res 16:e767–e767. https://doi.org/10.36367/NTQR.16.2023.E767

Schensul JJ, LeCompte MD (2010) Ethnographer's toolkit

Sibbald SL, Paciocco S, Fournie M et al (2021) Continuing to enhance the quality of case study methodology in health services research. Healthc Manage Forum 34:291–296. https://doi.org/10.1177/08404704211028857

Stahl NA, King JR (2020) Expanding approaches for research: understanding and using trustworthiness in qualitative research. J Dev Educ 44:26–28. https://doi.org/10.4135/9781483329574

Sullivan GM, Sargeant J (2011) Qualities of qualitative research: part I. J Grad Med Educ 3:449–452. https://doi.org/10.4300/JGME-D-11-00221.1

Tavakol M, Sandars J (2025) Twelve tips for using phenomenology as a qualitative research approach in health professions education. Med Teach. https://doi.org/10.1080/0142159X.2025.2478871

Taylor RM (2024) Methodological reflections on normative case studies: what they are and why we need better quality criteria to inform their use. Educ Theory 74:301–311. https://doi.org/10.1111/EDTH.12555

Webster F, Rice K (2019) Conducting ethnography in primary care. Fam Pract:523–525. https://doi.org/10.1093/fampra/cmz007

Yin RK (2003) Case study methodology, 3rd edn. SAGE

Zhou D, Zhan X, Cao Y, Chen L (2019) An overview of anthropological regional studies. Int J Anthropol Ethnol 3:1–18. https://doi.org/10.1186/S41257-019-0018-6/METRICS

15 Mixed Methods Research Designs

Abstract

Mixed methods research designs integrate and synergize quantitative and qualitative methods of inquiry in a single study. The aim is to provide a complete understanding of research problems compared to the use of a single research approach. This approach has been utilized since the 1970s and has gained increased interest over the last 40 years. The underlying idea is that neither a purely quantitative nor qualitative approach is sufficient to address the complexity of a research problem. Mixed method designs are particularly important for testing theories and improving validity, reliability, and trustworthiness of a research study. Typologies (systematic classification) play a crucial role in mixed designs as they inform and guide researchers on which designs to use. There are four basic types of designs: sequential, concurrent, multiphase, and compound, yielding eight categories. These categories include sequential explanatory design, sequential exploratory design, sequential transformative design, concurrent triangulation design, concurrent embedded design, concurrent transformative design, multiphase mixed design, and composite design. Key factors to consider in choosing the mixed methods design are how the quantitative and the qualitative strands will be mixed, including time and strategy. Mixed methods research follows the traditional scientific systematic approach but places particular emphasis on justifying the use of mixed methods. Researchers should also include details of the theoretical and conceptual framework and provide aids to help readers understand the integration of quantitative and qualitative data. Despite the clear benefits of the mixed methods approach in agricultural research, several obstacles exist, such as inadequate skills and training, significant resource demand, and complexity in implementation and publishing. Additionally, there is the challenge of ongoing paradigm wars between proponents of quantitative and qualitative methodologies who may hold different views regarding validity and application.

Keywords

Complementarity · Concurrent design · Contextualization · Embedded design · Enhanced Validity and reliability · Flexibility · Multiphase design · Pragmatism · Quantitative and qualitative approach · Sequential explanatory · Transformative design · Triangulation · Typologies

Learning Outcomes

After reading this chapter, you are expected to be able to:

1. Get familiar with “mixed methods research” designs
2. Appreciate the relevance of mixed research design in agriculture
3. Describe the philosophical assumptions of mixed methods
4. Illustrate the purpose of using mixed-method research
5. Determine when to use mixed methods design
6. Identify the various types of mixed methods research designs
7. Distinguish between “sequential” and “concurrent” designs
8. Differentiate between” sequential exploratory” and “Sequential explanatory” designs
9. Recognize the unique qualities of “transformative design”
10. Explain factors to consider in conducting mixed-method research
11. Apply appropriate steps in conducting mixed methods research
12. Appreciate challenges associated with mixed methods research

J. Njeri Mugwe, S. Runo, *Research Methodology in Agricultural Sciences*, https://doi.org/10.1007/978-981-95-1892-0_15

15.1 Introduction of Mixed Methods Design

15.1.1 Concept of Mixed Research Designs

At the start of part II in Chap. 7, we categorized research designs into the following:

1. Experimental designs (True, Quasi, and pre-experiments)-Quantitative.
2. Nonexperimental designs (Both quantitative and qualitative):
   (a) Survey research.
   (b) Qualitative research.
   (c) Participatory research.
3. Mixed (Experimental, Nonexperimental, Qualitative, Quantitative).

We have discussed 1 and 2 within the confines of a single approach or method (experimental, nonexperimental, either quantitative, or qualitative). However, methods that mix these approaches, their paradigms, methods, and techniques are popular in agriculture, just like in many other disciplines. Agricultural researchers may blend methods to different extents and at various stages of their research, depending on the specific research question. They might incorporate diverse mixed methods approaches at different junctures of the research process, encompassing overall design, sampling, data collection, analysis, and interpretation. Philosophical worldviews, discussed in Chap. 7, also have the possibility of being mixed. This chapter presents designs and procedures of mixed methods research to help an agricultural researcher conduct a valid and reliable research study.

The concept and thinking behind mixed methods are not new, but there has been an increased interest in mixed methods over the past 40 years as a distinct approach in social sciences disciplines, such as education. Notably scholars such as John Creswell, have championed mixed methods and authored many works on this subject (Creswell and Plano Clark 2011; Creswell 2013, 2014; Creswell and Guetterman 2019). Other classical books on mixed methods exist too and they have contributed to the interest in mixed research methods (Greene et al. 1989; Newman and Benz 1998; Greene 2008a,b; Clark 2010; Tashakkori and Teddlie 2010). In 2007, an important initiative was undertaken to promote the dissemination of mixed methods research studies by introducing a dedicated journal (Johnson and Onwuegbuzie 2007). The articles published in this journal have played a key role in demystifying and simplifying the complexities associated with mixed methods research. Indeed, these publications have facilitated a greater understanding among researchers of the theories and designs that prove most beneficial and applicable in their respective disciplines.

One of the prominent features of these publications on mixed research methods is the definitions that generally give the same meaning as research methods, but also deviate in terms of emphasis. Here we give definitions from key publications:

- (1980s) "Mixed methods designs are characterized by the incorporation of at least one quantitative and one qualitative method. Neither method is inherently linked to any particular inquiry" (Greene et al. 1989).
- 1990's "Mixed methods research refers to a research approach that combines both qualitative and quantitative research methods, such as data collection, analysis, and inference techniques" (Tashakkori and Teddlie 2010).
- (2000's) Journal of mixed methods: "Mixed methods refer to research in which an investigator collects and analyzes data, integrates the findings and draws inferences using both qualitative and quantitative approaches or methods in the same study" (Johnson and Onwuegbuzie 2007).
- "Mixed methods research in which the investigator collects and analyzes data, integrates the findings, and draws inferences using both qualitative and quantitative approaches or methods in a single study or a program of inquiry (Creswell and Tashakkori 2007).
- 2010's "Mixed-methods research is a research methodology that incorporates multiple methods to address research questions in an appropriate and principled manner" (Clark 2010; Creswell and Plano Clark 2011; Creswell 2014).
- 2020's; Mixed methods integrate and synergize quantitative and qualitative methods in a single study or sustained program of inquiry to generate a more complete understanding than is achievable with a single method (Fetters 2020).

What are mixed methods designs?
The preceding information highlights the increasing acceptance and growth of mixed methods as a distinctive research approach. "Mixed methods designs" denote research methodologies integrating quantitative and qualitative components in a single study. The crucial aspect is the *"intentional interweaving of both the quantitative and qualitative"* within the research process to give a comprehensive account of the research topic.

Looking through the definitions from the 1980s, we recognize that definitions have metamorphosed as more publications are produced and knowledge increases. In their second edition of the book, Creswell and Plano Clark (2011) advocate a definition that includes the varied viewpoints, including research design, philosophy, methods definition, and also key components that go into designing and conducting a mixed research study. They emphasize that the researcher of mixed methods should accomplish the following:

- Collect and analyze "both qualitative and quantitative data" with rigor to answer the research question.
- Mix, integrate or link the two forms of data. Three scenarios are possible: combining data *"concurrently"* (merging them), *"sequentially"* by having one build on the other, or *"embedding"* one within the other.
- Prioritize one or both forms of data based on the research emphasizes.
- Apply these procedures either as a stand-alone study or across multiple phases of a program.
- Situate these procedures within philosophical worldviews and theoretical lenses.
- Integrate the procedures into specific research designs that guide the overall plan for conducting the study.

Mixed methods can indeed be complex, with a high level of inclusivity and complementarity (Dawadi et al. 2021). Besides using quantitative and qualitative methods, researchers can also approach the study from different perspectives and conduct various studies. Broadly, researchers using mixed methods will use qualitative and quantitative *viewpoints, data collection, analysis*, and *inference techniques* in the same study to help achieve breadth and depth in understanding the research problem.

15.1.2 Relevance of Mixed Methods in Agriculture

As discussed in Chap. 1 of this book, and further highlighted in Chap. 10, the agriculture sector is a complex system that faces triple challenges: providing food for a burgeoning population, ensuring sustainable livelihoods for millions of farmers and safeguarding the environment. These challenges are further complicated by the influence of climate change, a global issue that poses a substantial threat to food security in both developed and developing nations. Moreover, the wide diversity of farming systems, particularly within the agriculture–environment interface, calls for a comprehensive understanding of their complex interactions. Furthermore, the varied beliefs and experiences of farmers combined with other factors, such as social, political, cultural, and economic forces, add to the layers of this complexity. All these must be addressed to provide effective and sustainable solutions.

This complexity requires research that generates answers beyond the data sets comprising numbers that agricultural researchers (agronomists, animal scientists, and crop protection experts) are used to. The *human aspects of agriculture* call for additional data forms that can provide a more complete analysis of psychological elements. This justifies the need for both quantitative and qualitative data to provide the most comprehensive analysis of the prevailing problems. For instance, methodological complexities inherent in researching poverty were effectively addressed by the use of mixed methods (Place et al. 2007).

Though not explicitly stated, the use of mixed methods in agriculture has been in use since the 1970s (refer to Chap. 10 on farmer participatory research). The central premise of the proponents of participatory approaches in the 1970s was to use designs and approaches capable of addressing farmers' multiple problems. This was after the realization that recommendations developed by researchers (mainly biophysical) were inappropriate for farmers' environment, dimensions, and operations because they were developed without adequate information about the farmers' environment. Therefore, the farmer participatory research approaches supported the use of several forms of data collection, including both qualitative and quantitative, in single studies and, at times, in a series of studies. For example, various forms of data are collected in the diagnosis and on-farm research stages.

Other common studies using mixed methods in agriculture include adoption studies, which primarily evaluate factors that influence their adoption possibilities. Many studies often combine designs, such as experiments and quantitative surveys, with varied data collection methods (including surveys and interviews, data collection instruments, and data analysis techniques. For example, Tufa et al. (2023) investigated the adoption of conservation agriculture in South Africa using a combination of household surveys (quantitative) and focus groups (qualitative), while Kondo et al. (2020) applied a sequential mixed-method approach, beginning with a questionnaire-based survey followed by exploratory interviews and focus group discussions.

Besides using mixed method models to enhance the adoption of research recommendations, other studies in agriculture have traditionally used mixed methods. In fields such as livestock production, veterinary studies, and epidemiology, it is common to describe a specific behavior and then calculate and analyze its frequency in various situations. Similarly, molecular studies and soil survey studies commonly use quantitative and qualitative data combinations to characterize contexts. For instance, soil profiles are described using both morphological (qualitative variables) generated by a researcher or a farmer (e.g., red soils indicate infertility) and quantitative variables comprising values measured in the laboratory. Additionally, in crop protection, data are often captured numerically to show the extent of pest or disease attacks (prevalence). To enhance complete understanding, the numerical data on pests are combined with qualitative data on the presence or absence and farmers' perceptions and experiences regarding their control measures. Furthermore, samples and photos are also usually captured as forms of data.

The mixed-method approaches that have continued to be used in agriculture may be described as "informal" or "fragmented." A study of Science Direct databases showed that the use of mixed methods by agricultural economists is uncommon (Akimowicz et al. 2018). Moreover, classical publications on mixed methods within agriculture are scanty. Most books and literature on mixed research are on education and psychology. This is an indication that knowledge of the foundations, theories, and philosophical assumptions of mixed methods in the field of agriculture could be lacking or inadequate among many scientists, yet it is crucial for enhancing understanding of the complex issue we face in agriculture today. It is hoped that this chapter will enhance agricultural researchers' understanding and application of mixed methods to generate knowledge for more sustainable agriculture practices.

15.1.3 Philosophical Assumptions

Philosophical assumptions or worldview paradigms were introduced in Chap. 7 of this book and premised as a set of beliefs that guide investigations. Four worldviews were introduced comprising of positivism, associated with quantitative researchers who objectively test hypotheses; constructivism or interpretivism related to quantitative researchers who believe in multiple realities constructed by the respondents; transformative linked to the need to cause a change in society; and finally, pragmatism, associated with mixed methods research that focusses on what works best.

The researcher in mixed methods is guided by a *pragmatism paradigm* that contrasts the singleness of the "qualitative and quantitative" approach (Fig. 15.1). This means that mixed methods are not committed to any philosophical belief per se. Rather, actions are guided by the pragmatism paradigm that is pluralistic and practical, oriented toward "what works" and "practices" (Creswell and Plano Clark 2011; Cohen et al. 2017). "What works" means that if the methods of research and data collected, be it numerical or qualitative, address the research purpose, problem, or question, then they are acceptable. It places importance on both objective and subjective knowledge in achieving its goals (Fig. 15.1).

15.1.4 Purpose of Using Mixed Methods?

The main reason for using multiple research approaches and techniques is to gain a more comprehensive understanding of a research problem than is possible with a single approach. Several authors (Greene et al. 1989; Creswell and Plano Clark 2011; Mugenda and Mugenda 2019) have documented several reasons why mixed methods research is used. All these authors list varied reasons for using mixed methods.

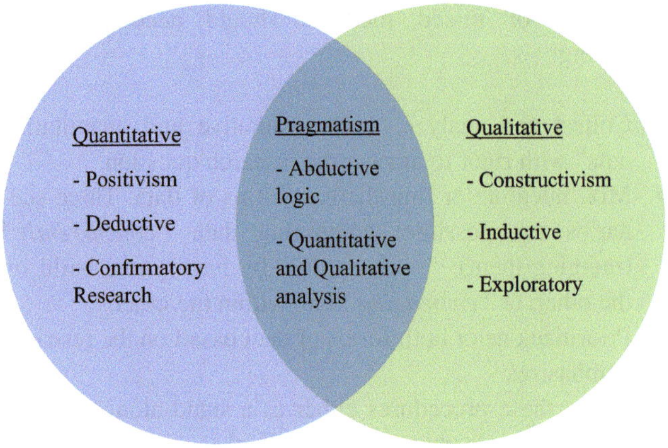

Fig. 15.1 Pragmatism paradigm guides the mixed method designs

For example, Greene et al. 1989) provided five reasons (triangulation, Complementarity, development, initiation, and expansion, while Schoonenboom and Johnson (2017) stated that the purpose of conducting mixing research is numerous, and none have provided an exhaustive list. We hereby provide the most important reasons for conducting mixed methods research in agriculture. These include:

1. *Complementarity*: Complementarity refers to the synergistic integration of both qualitative and quantitative approaches to provide a comprehensive understanding of a research problem. The qualitative and quantitative data collected complement each other by offering different perspectives and insights. While quantitative methods may provide statistical trends and patterns, e.g., crop yield data, qualitative methods probe into the underlying meanings and contexts, enriching the overall interpretation of the research findings. For example, in the assessment of a new in situ water harvesting technology on crop yield and farmers' perceptions, statistical analysis will quantify the differences in crop yields. This will provide evidence of the effectiveness of various technologies tested regarding increased productivity. The qualitative data, on the other hand, will help determine farmers' perceptions, such as concerns and social dynamics surrounding the adoption of the new technologies, thus giving a richer understanding of the results. In a mixed methods study by Kiptot and Franzel (2015), the qualitative data collected at the start of the study were used to formulate hypothesis for the quantitative survey.
2. *Contextualization*: Contextualization places information or events within a specific context to give them meaning and relevance. In agriculture, it helps us to understand farming practices, crop management, and environmental factors. In mixed methods, the qualitative data add details about the context of the study and helps interpret the

quantitative findings. Without proper contextualization, agricultural technologies may not function as intended because they might not align with the local environment and the socioeconomic and cultural conditions of the target community. This is the central argument in Farming Systems Research (FSR) and the heterogeneity of farmers discussed in Chap. 10. For example, gender relations are complex and vary by context, necessitating data that allow understanding of local context (Doss 2014).

3. *Practicability:* Mixed methods research combines qualitative and quantitative data to create practices that are appropriate for specific populations and situations. In donor-funded projects focusing on dissemination and real impact, incorporating qualitative insights is more impactful than just numbers. For example, if a study shows that a certain farming system fits well with local traditions, policies can be created to support these practices, making them more likely to be adopted.
4. *Flexibility:* Mixed methods research allows the researcher to customize their methods to fit the needs of their specific research question. This flexibility helps to boost the robustness and validity of their study. For instance, unexpected findings may surface in relation to farmers' social dynamics and community interactions. In response to these unforeseen insights, researchers may choose to modify their qualitative data collection methods by incorporating participatory observations and focus group discussions. This adjustment enables a thorough exploration of various facets related to the unexpected observations.
5. *Enhanced Validity and Reliability:* Enhanced validity comes from using multiple sources and methods to gather data, commonly known as triangulation. When evidence from different perspectives agrees, it strengthens the study's validity and credibility. For instance, in a study to investigate the effectiveness of a new pest, researchers collect both quantitative and qualitative data. If quantitative results show increased crop yields with the new pest control method, researchers can use qualitative data to explore farmers' experiences and perceptions of its effectiveness. If the results converge on the positive impact of the new method, it enhances the overall validity and reliability of the findings.
6. *Collaboration:* Mixed methods encourage teamwork from different fields. By combining diverse perspectives, methodologies, and analytical approaches, these methods become more thorough and reliable. For instance, a climate change study might involve climatologists who measure weather patterns, agronomists who look at how weather affects crop yields and agricultural economists who evaluate how climate change impacts incomes and farmers' economic decisions and adaptation strategies. This integration enhances the rigor and credibility of the research. The farming systems research advocates for integration and collaboration.

When to Use Mixed Methods Design in Agriculture
Mixed methods in agriculture are most effective for use when aiming for a holistic understanding of complex agricultural systems, exploring new phenomena, validating findings, and addressing multifaceted research questions that cannot be fully captured through a single methodology. Here are some specific scenarios in agriculture where mixed methods can be used to provide a holistic view:

If conducting multidisciplinary research, e.g., involving agronomy and socioeconomic aspects of farmers, mixed methods are well-suited for this kind of research that requires collaboration between different disciplines.

If aiming for validation and triangulation, researchers can use mixed methods to triangulate data for more robust validation of findings. Quantitative results can be corroborated or complemented by qualitative insights, providing a more reliable interpretation of the outcomes.

If conducting exploratory studies: In the early stages of investigating a new agricultural practice or technology, using qualitative methods to explore farmers' perspectives, attitudes, and contextual factors can help inform the design of subsequent quantitative studies. This iterative approach allows researchers to refine their hypotheses based on qualitative insights.

If conducting impact evaluation studies, the mixed methods can capture the broader implications of agricultural practices on communities, livelihoods, and sustainability (Bamberger et al. 2010).

If longitudinal studies are being conducted, mixed methods allow researchers to track quantitative indicators (such as crop yields or environmental parameters) alongside qualitative data that capture evolving sociocultural aspects, enabling a deeper understanding of farmers' trends, patterns, and perceptions.

If studying the adoption of innovations: Understanding the factors influencing the adoption of new agricultural technologies or practices requires quantitative data on adoption rates and qualitative insights into farmers' motivations, barriers, and perceptions.

15.2 Typologies of Mixed Methods Research Design

A typology means classifying things according to their specific types, which is useful in creating an order. Similarities and differences are considered when coming up with typologies of mixed methods. Huge literature exists on the classification of mixed research under various disciplines (Creswell and Plano Clark 2011). The classification differs according to disciplines and uses varied terminologies. According to Creswell and Plano Clark (2011), there are more than 40 different types of mixed research designs, a reflection of the progressive nature of mixed methods research.

Important features used in the classification of mixed methods include sequencing, the status of each method relative to the other, independence, and interdependence (Palinkas et al. 2011). The sequence is about the arrangement, and depending on the sequence, the approaches may be sequentially mixed or proceed in a parallel manner. Status, on the other hand, refers to the role of the method, whether to support or collect new information altogether. Finally, dependence and interdependence are concerned with whether the method is used to prepare for another. All these different scenarios give multiple combinations and a multitude of designs.

15.2.1 The Main Typologies

To make sense of the multitude of mixed-method research designs and simplify their classification in agriculture, we subdivide them into four types: *Sequential, Concurrent, Multiphase, and Composite*, giving eight categories. Creswell describes the first three categories in many of his books, while the composite is one that describes the complex mix of two or more typologies commonly encountered in agriculture.

Sequential design implies a linear phase in data collection and analysis, in which the collection of the first data set (e.g., quantitative) is followed by the collection and analysis of the other data set (qualitative). It has three sub-designs as follows:

1. Sequential explanatory design.
2. Sequential exploratory design.
3. Sequential transformative design.

Concurrent design involves simultaneous collection and analysis of quantitative and qualitative with three sub-designs:

4. Concurrent triangulation design.
5. Concurrent nested design.
6. Concurrent transformative design.

Numbers 7 and 8 are fundamentally separated from the six. They include:

7. Multiphase mixed design.
8. Composite design.

Multiphase is a complex mixed research design that may incorporate all the above six, plus a theoretical framework of transformation. It has been referred to as a sandwich design because a researcher may incorporate the quantitative and qualitative across phases.

Composite includes complex mixed designs that do not necessarily have a transformative agenda, such as methodologies for soil surveys and species selection involving both quantitative and qualitative data. The timing of data collection can sometimes be haphazard. This classification was chosen because it closely aligns with the methods and methodologies in agricultural sciences (Box 15.1).

Box 15.1 Types of mixed research designs in agriculture

Mixed methods design	Sub-design
Sequential	Sequential explanatory design Sequential exploratory design Sequential transformative design
Concurrent	Concurrent triangulation design Concurrent nested design Concurrent transformative design
Multiphase and composite	Multiphase mixed design Composite design

Definition of Key Terms in Mixed Research Designs

Concurrent or convergent: quantitative and qualitative are performed independently at the same time (concurrently), with results brought together during interpretation.

Embed: Entails fixing or adding either quantitative or qualitative to enhance the overall design.

Sequential: Quantitative and qualitative are arranged in an order (do not happen at the same time).

Transformation: Focus is on a transformative theoretical framework meant to empower marginalized or disadvantaged societies.

Nested: is synonymous with embed, indicating that one approach is fitted inside the other.

Triangulation: Use of multiple data, methods, and perspectives to study the same phenomenon to increase the validity of the results.

15.2.2 Sequential Explanatory Design

Sequential explanatory designs are characterized as a two-stage methodology, where the initial stage relies on quantitative data to establish a foundation for and elucidate qualitative data. This dual-phase mixed methods design emphasizes the explanatory component, as illustrated in Fig. 15.2. The process commences with the gathering and analyzing of quantitative data, which are followed by the subsequent collection and analysis of qualitative data. The design of the qualitative phase is purposefully structured to align with or connect to the outcomes of the initial quantitative phase. As this design initiates with a quantitative approach, researchers commonly emphasize quantitative methods over qualitative methods.

The primary objective of this design is to leverage qualitative data to elucidate or expand upon the findings derived from initial quantitative results. This approach is particularly well-suited for studies in which researchers require qualitative insights to provide a deeper understanding and explanation of the outcomes obtained in the quantitative phase of the study. For example, in adoption studies. The sequential explanatory design also applies when a researcher categorizes participants into groups based on quantitative results. Subsequently, in-depth qualitative research is conducted with these identified groups.

Example of a Sequential Exploratory Study

A study by Bagambilana and Rugumamu (2023) used a sequential explanatory design to assess determinants of small-scale farmers' adaptation intent and adoption of adaptation strategies in the semi-arid lowlands of Mwanga District. The authors based the study on a pragmatic philosophy and used a two-step process to collect data from farmers and other stakeholders. A cross-sectional sequential explanatory mixed methods research design was deployed. During the first phase, categorical and quantitative data were collected by administering a closed-ended survey questionnaire to 328 household heads. Binary and proportional odds logistic regressions were run and used to analyze the categorical data for statistically significant results. During the second phase of the process, qualitative data were generated through focus group discussions with 30 participants, in-depth interviews with 16 key informants, and participant observations and subjected to interactive thematic content analysis. The study concluded that the adoption of adaptation strategies is mainly explained by objective adaptive.

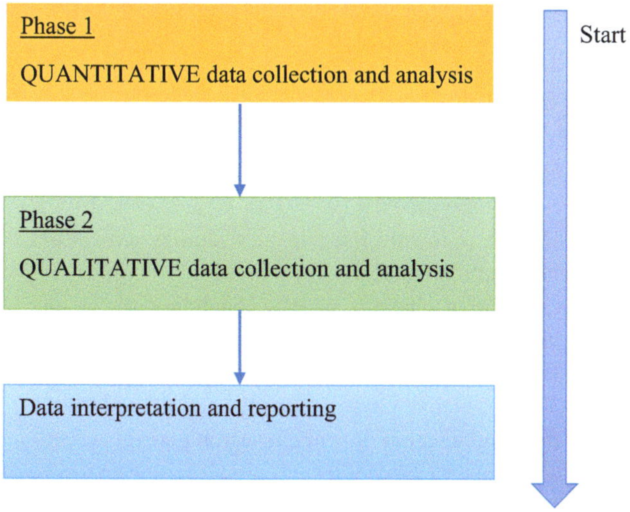

Fig. 15.2 Sequential explanatory design starting with a quantitative phase followed by a qualitative phase

Implementing the sequential exploratory design is straightforward, making it the most accessible among mixed methods designs and particularly appealing to students. This approach keeps the research simple while staying focused, as one type of data build on the other. It also has a clear two-step process, making it easier to carry out. The researcher collects and analyzes one type of data first, then follows up with the other in a separate phase.

This design's simplicity allows individual researchers to undertake it without needing a research team. Additionally, it caters to the preferences of quantitative researchers, which most agriculturalists prefer, as it often commences with a robust quantitative orientation. The final report is also easily manageable, as the results seamlessly merge, providing simplicity, especially for researchers with a strong quantitative inclination.

Naturally, researchers using this design have multiple philosophical assumptions that change and shift across the study (Tashakkori and Teddlie 2010). In phase one, with greater emphasis placed on quantitative research, the researcher typically uses a positivist perspective. When the researcher moves to the qualitative phase, which values multiple realities and in-depth descriptions, the research will shift to the assumption of constructivism. See Box 15.2 for a summary of the attributes of sequential exploratory design.

Box 15.2 Key points on sequential explanatory design

1. Quantitative phase followed by qualitative phase (two-phase model).
2. Utilizes qualitative methodologies to help elucidate quantitative results requiring additional exploration or a more in-depth explanation of quantitative results.
3. Employ quantitative results purposively to guide choice of the tools and participants for the subsequent qualitative study.
4. Useful in answering new questions, trends, or unexpected results from a quantitative study.
5. Most useful when the researcher wants not only to assess trends and relationships, for example in an adoption study but also explain the reasons behind the trends obtained.
6. Philosophical stance shifts from positivism (what? and how many?) to constructivism (how and why?).
7. Simple to use and thus can easily be adopted in research projects by students.

15.2.3 Sequential Exploratory Design

The sequential exploratory design is the reverse of the explanatory model, with the qualitative data informing the quantitative information-gathering process. Instead of starting with quantitative data, as in the explanatory design, the sequential exploratory design begins with qualitative data to explore a phenomenon. Subsequently, it progresses to a second phase involving quantitative data collection. The two phases follow each other linearly, with a final interpretation phase (Fig. 15.3). You can use the qualitative phase for developing an instrument, identifying variables, or stating propositions for testing based on an emergent theory or framework (Tashakkori and Teddlie 2010). In this case, the analysis of the qualitative phase will inform and guide the design of the subsequent quantitative phase. In other words, the qualitative phase results shape the quantitative phase, which could be in refining research questions and variables, developing research instruments, or creating typologies or categories. For instance, we could collect quantitative data to create farmer wealth categories.

The "sequential exploratory approach" is often most appropriate to consider when the aim is to develop data collection instruments or generalize findings from a qualitative study (Creswell 2014).

Examples of Sequential Exploratory Design

A gender study used sequential exploratory mixed methods (Khoza et al. 2021), where the researchers dedicated the qualitative phase to developing the survey tool. The research aimed to explore gender-differentiated determinants of climate-smart agriculture adoption among smallholder farmers. The first qualitative phase collected data through 16 key informant interviews and 3 focus group discussions (with an average of 9 people) held per district, 1 for men only, and 1 for women only. Respondents of this qualitative phase were purposively sampled. The data were subsequently analyzed thematically, and the resultant themes were used to design the survey questionnaire, after which it was pretested and used to collect data from 102 randomly sampled households through a cross-sectional survey. The resultant data were subjected to descriptive statistics, yielding frequency distribution data, subsequently integrated with qualitative findings.

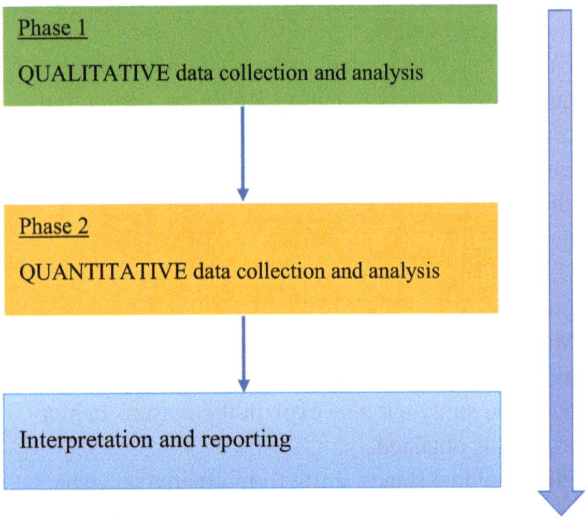

Fig. 15.3 Sequential exploratory design starting with a qualitative phase, followed by a quantitative phase

A study on communication factors (Njenga et al. 2021) to determine the influence of communication factors on adopting selected integrated soil fertility management among farmers (Njenga et al. 2021) is another example that used "sequential exploratory design." At first, the exploratory phase collected and analyzed data to identify soil fertility technologies that interested the farmers. The results obtained were used to guide soil fertility practices selection for further evaluation and to develop data collection tools for the quantitative phase. The qualitative phase collected data using five key informants and three focus group discussions, while the quantitative phase involved a quantitative survey using a standard questionnaire administered to 400 farmers.

Similar to the sequential explanatory design, researchers using a sequential exploratory design undergo a paradigm shift as they transition from the qualitative to the quantitative phase of the study. In the beginning, the priority design is qualitative, and this calls for the constructivist paradigm that values multiple perspectives. As the researcher changes to the quantitative phase, the underlying assumption could shift to those of positivism to guide the methods of measuring the variables and statistical analysis. Overall, multiple philosophical assumptions are used in this design.

Box 15.3 Key points on sequential exploratory design

1. Qualitative phase followed by quantitative phase (two-phase model)
2. Design is mostly used in new studies with scant information about the key variables
3. Qualitative phase is usually the minor phase and useful in situations where instruments are nonexistent
4. The qualitative phase is mostly used to explore the field and develop or refine research questions, variables, and/or typologies
5. Philosophical stance shifts from interpretivism (how and why?) positivism (what? and how many?)

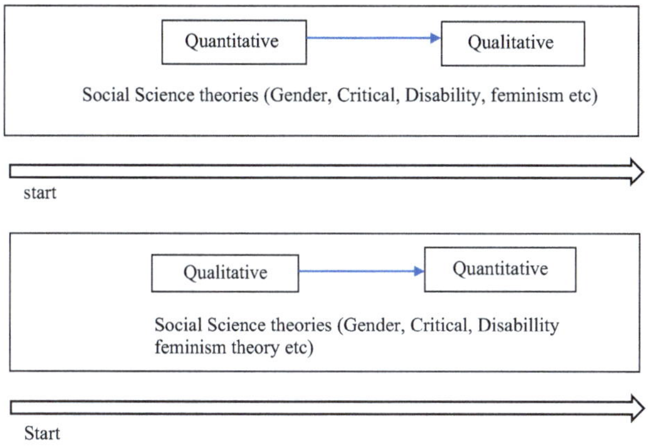

Fig. 15.4 Sequential transformative designs

15.2.4 Sequential Transformative Design

In sequential transformative design, both quantitative and qualitative phases are combined, but the research is also explicitly driven by a transformative theoretical perspective (Fig. 15.4). Because the transformative framework is a priority, either type of data can be used first as long as they meet the transformative agenda.

Similar to the previous two designs, the sequential transformative design comprises two separate data collection phases. This approach emphasizes prioritizing methods that align with the researcher's theoretical perspective, which often centers around the pursuit of social justice and advocating for change for underrepresented and marginalized populations (Mertens 2014). Unlike other designs, the significance of this typology lies more in its thematic focus than the specific methodology employed. The primary goal is not merely to gather empirical data but to adhere to a theoretical framework to address marginalized populations' needs by incorporating transformative theories such as racial and ethnic theory.

Box 15.4 Key characteristics of transformative mixed designs

1. Aims for a transformative agenda by describing problems and proposing solutions; employs transformative lens (emancipatory, interdisciplinary, and participatory)
2. Focus on social justice, including marginalized voices and advocating for social justice
3. Seeks to understand power relations to link results to action and braider social justice questions
4. Uses transformative social theories such as critical theory, advocacy, Queer theory, disability theory
5. Most agricultural research projects with a research and development agenda can easily fit in the transformative mixed designs

15.2.5 Concurrent Triangulation Design

Concurrent triangulation design, also called convergent parallel design, is the most common and well-recognized design in mixed methods. It is one of the easy mixed method designs. It involves concurrent quantitative and qualitative data followed by separate analyses, after which the results are merged and used to interpret the research study (Fig. 15.5). This study triangulates the methods by comparing quantitative and qualitative results. The rationale behind this is that each data component contributes strengths to compensate for the weaknesses of the other, resulting in a more complete understanding of the research results. For example, the differing strengths and nonoverlapping weaknesses of quantitative methods (large sample size, trends, generalization) with those of qualitative methods (small *N*, details, in-depth) are brought together, therefore improving the overall design (Patton 2015).

The main aim of merging the data is to *corroborate, confirm, or support findings* from the two phases. In interpreting the merged data, the researcher scrutinizes for any data convergence, divergence, contradictions, or similarities between the two data forms (Creswell and Plano Clark 2011). An example is a researcher who, during the growing season, carries out a study to assess the attitude of farmers toward the control of a certain pest of maize. He or she uses a survey to collect quantitative data and key informants to collect qualitative data and then combines the two sets of results during interpretation. Merging aims to determine how results about farmers' attitudes converge and diverge.

Examples of Concurrent Triangulation Design

A study by Landaverde et al. (2022) employed a concurrent triangulation design to assess climate change perceptions in Peru. This study aimed to evaluate subsistence farmers' perceptions and strategies for coping with and adapting to climate variability. The researchers collected quantitative data from 103 farmers using standardized questionnaires and qualitative data using semi-structured interviews from 20 farmers. In justifying why they used the mixed convergent parallel design (as they named it), the authors stated that this design provides a deeper understanding of the topic and also the recommendation of other authors on the suitability of the triangulation approach for studying climate change issues. The quantitative and qualitative data were separately analyzed using descriptive statistics and thematic analysis. The SPSS software was used for the quantitative data, while NVIVO was used for the qualitative data. Finally, separate results from the data sets were merged and compared to reveal any convergence or divergence (Landaverde et al. 2022).

Another example of concurrent triangulation design is borrowed from a study by Etana et al. (2020) on the dynamics of smallholder farmers' adaptation decision-making processes. The researchers present a step-by-step flow of the methods used. First, they simultaneously collected both quan-

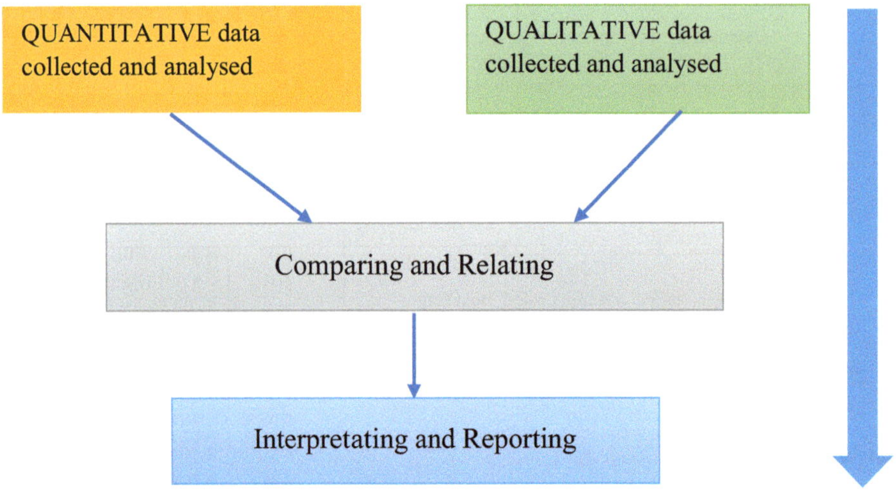

Fig. 15.5 Concurrent Triangulation design

titative and qualitative data using a survey questionnaire (200 farmers), and multiple qualitative methods (in-depth interviews, FGDS, and KII), respectively. Second, they analyzed data separately and independently using thematic analysis for qualitative data and descriptive and dynamic random effect probit model for the quantitative data. Third, the results were merged to make comparisons and to identify any complementarities, convergence, or divergence. Fourth, integrated interpretation was done. The researchers in this study justify this design in that the design was appropriate to explore and understand the multifaceted interaction between the dynamics of adaptation and decision-making (Etana et al. 2020).

A similar philosophical assumption of this design is best portrayed as pragmatism. This is because a researcher using this design will collect, analyze, and merge quantitative and qualitative data and results simultaneously. Pragmatism is an umbrella paradigm to research, and therefore, well suited to guide the work of merging the two approaches to enhance understanding of the research problem (Creswell and Plano Clark 2011). Box 15.5 gives key points about the concurrent triangulation design.

Box 15.5 Key points on Concurrent Triangulation design

1. Both quantitative and qualitative data are collected concurrently or simultaneously
2. Equal priority is given to both qualitative and quantitative phases
3. Results from both qualitative and quantitative phases are analyzed separately and compared, looking for convergence and divergence
4. Uses pragmatism philosophical stance

15.2.6 Concurrent Embedded Design

Concurrent embedded design involves one data collection phase during which quantitative and qualitative data are collected, with one collection method (secondary) guided by the primary one (Creswell 2014). The secondary method is "Embedded" or "Nested" within the predominant method and addresses a different question (Fig. 15.6). This means that the secondary data source is typically embedded within the primary data, in which the two data types are collected and analyzed concurrently rather than sequentially. The framework differs from the concurrent triangulation design in terms of the aim of collecting the two data sets, which is not to validate or check data correspondence but rather to have one data form support the other. The rationale behind this framework is the researcher's feeling that one type of data set may not give enough conclusive information to draw results.

In this type of study, it does not matter whether the primary data are qualitative while the nested data are quantitative or vice versa. The nested method seeks information about different levels of the research problem. The rationale is that a need arises to address different questions within the ongoing research study requiring different methods (Creswell 2014).

The most common practice is to design a study with qualitative data embedded within a large quantitative study. For example, a researcher may plan to evaluate a number of treatments in a randomized controlled trial and then embed qualitative data to explain the outcomes. Embedded studies could also be used in correlational studies to explain relationship outcomes.

An example of a concurrent embedded design could be a quasi-experimental study aimed at introducing an interven-

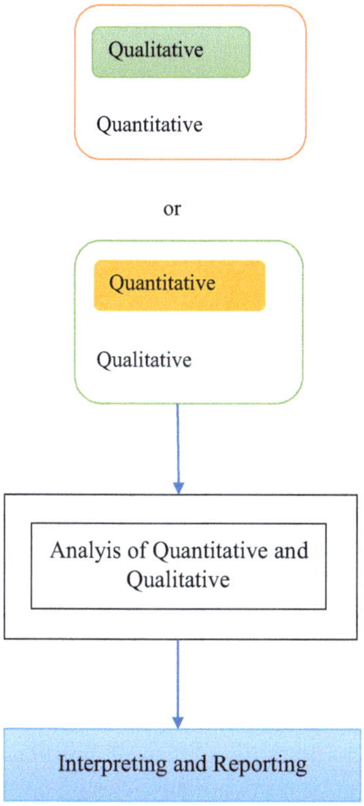

Fig. 15.6 Concurrent embedded design

tion, such as the use of plant clinics, for farmers growing French beans. The primary objective is to assess the impact of the plant clinics on French bean yields. Although the study is initially designed to rely on a quantitative approach to measure yield changes, unanticipated incidents arise during implementation that require deeper exploration. To address these, the researchers embed a qualitative component within the study design. In addition to collecting quantitative data, they also conduct interviews with farmers to gain insights into the observed issues. By the end of the study, both quantitative and qualitative data are analyzed concurrently, and the findings are integrated to provide a more comprehensive understanding of the intervention's effects.

Another example is borrowed from work done in the central highlands of Kenya in a project that promoted soybean varieties among farmers. One of the activities involved screening soybean varieties collected from the East African region to select suitable varieties adaptable to the different agro-ecological zones of the central highlands of Kenya. Screening experiments were conducted to evaluate the varieties within the different agro-ecological zones. Experiments were the primary methods of data collection to identify promising varieties. During the growing season, the researchers embedded a qualitative strand of data collection in which beneficiaries (farmers) were invited to rank and select the varieties according to specified criteria and their preferences. The experimental data and farmers' evaluation results were used to select the "best bet" soybean varieties.

The key framework operating in embedded design is the supplemental role of the secondary approach, regardless of whether it is quantitative or qualitative. Because of this, the worldview assumption of this design takes priority over the primary strand, while the supplemental method takes a service role (Mertens 2014). For example, if the primary strand is an experimental or quantitative survey, the researcher will most probably be working from a positivist worldview as the principal paradigm. If, on the other hand, the primary study involves a case study or ethnography, the researcher will work from an assumption that favors constructivism.

Box 15.6 Key points on Concurrent Embedded Design

1. Both quantitative and qualitative data are collected concurrently or simultaneously
2. Supplemental strand (either quantitative or qualitative) incorporated to enhance the overall design and study
3. Multiple research questions are answered using both primary and secondary data sets
4. Used to explore processes in an experimental design; using qualitative method as the secondary method
5. Philosophical world view considers the primary strand

15.2.7 Concurrent Transformative Design

"Concurrent transformative design" involves the simultaneous collection of both "quantitative and qualitative data" guided by a theoretical perspective of change (Fig. 15.7). It is comparable to sequential transformative design in terms of the central agenda, which is to give voice to the disadvantaged and marginalized. It aims to transform society by providing solutions to its constraints.

The concurrent transformative design and other mixed designs prioritizing transforming marginalized, discriminated, and underrepresented populations carry transformative worldviews (Mertens 2014). The transformative worldview framework emphasizes collaboration and change and is used to correct socioeconomic and environmental changes.

Many research programs working with farmers fit this framework lens. The proponents of farmer participatory research, with the related concept of on-farm research (Chap. 10) are farmer-transformation driven. On-farm research advocates for farmer-motivated research and promotes the idea of conducting research relevant to farmers' needs with the sole

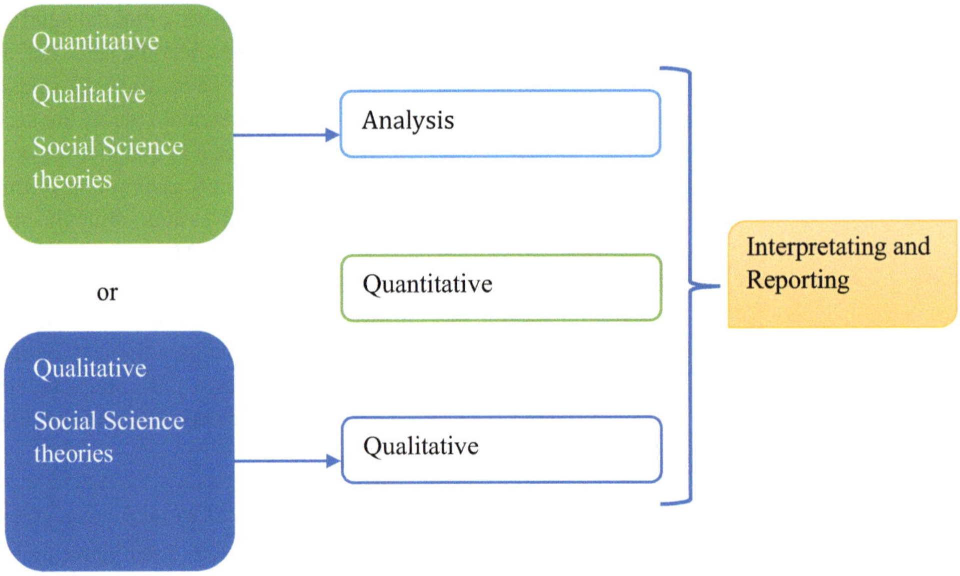

Fig. 15.7 Concurrent transformative Design

purpose of empowering farmers and causing a change in their livelihoods. Because of this, working with farmers is one of the key elements emphasized in farmer-participatory research guided by the farmers' needs. The farmers participate in the whole research process, including problem identification, research design, implementation through on-farm trials, evaluation, and benefit through adoption of the recommendations. Most donor-funded programs targeting farmers emphasize a transformative agenda. These programs stress that clear benefits to participating farmers must be demonstrated. In most cases, a demonstration that the women and marginalized communities (windows, youth) also benefit is a must. To achieve this agenda, researchers must work with farmers to collect quantitative and qualitative data.

A conventional example of concurrent transformative design is a research program aiming to promote water harvesting techniques to marginalized communities in arid areas, to address the problem of rainfall variability due to climate change. The researcher designs and evaluates several water harvesting techniques in a randomized experiment where not only biophysical data are recorded but also incorporates activities for empowering the farmers. This can be in the form of the mother-baby model (Chap. 10). For example, the researcher will organize field days to enable farmers to select suitable water harvesting methods using qualitative methods of data collection. In addition to this activity, the researcher also encourages farmers to implement methods of their choice on their farms to reap the benefits. In this case, the researcher aims to transform farming practices and consequently cause a positive change. The ultimate transformation of farmers is usually reaping the benefits of increased yields and incomes. Researchers will document the new knowledge in scientific publications and share their findings in a mixed methods research paper.

According to literature (Creswell and Plano Clark 2011; Mertens 2014), the transformative design operates under the following considerations:

- Researcher identifies a social injustice in terms of marginalized and disadvantaged communities in various forms.
- Researcher seeks to address social injustice and cause a positive change.
- Researcher is cautious of the need to conduct the study without further marginalizing the community.
- Researcher has a good knowledge of the theoretical and conceptual framework guiding the work in the marginalized populations.

Box 15.7 Key points on Concurrent transformative design

1. Main aim is to use research to transform society.
2. Both quantitative and qualitative data are collected concurrently or simultaneously.
3. Uses value laden framework.
4. Involves a call to action to benefit those being studied.
5. The primary approach predominates the philosophical assumptions.
6. In agriculture, farmers are often involved in the research process.

15.2.8 Multiphase Design

Multiphase mixed method design, also called multi-stage designs, involves combining several designs and phases in one study to address multiple research questions and purposes. Typically, they are single mixed studies, separate from each other and combining both concurrent and sequential strands. At times, they can be extremely complex running over some time (Mertens 2014). These designs are mostly used in programs that aim to meet an overall program objective involving development, adaptation, understanding of processes and program evaluation aspects (Creswell and Plano Clark 2011). For example, a project may be conceptualized to increase yields of a certain crop. The researcher might first carry out a needs assessment to validate the causes of the low yields and gain a deeper understanding from the farmers' perspective. Using these results the researchers might initiate an intervention based on the findings and then plan to assess both the processes and outcomes of the intervention. Assessment of these aspects will be done over time using several studies and various combinations of the quantitative and qualitative approaches (Fig. 15.8).

The soybean project presented in Chap. 10 represents a multiphase mixed design. The overall program objective was to increase yields and incomes among farmers through the adoption of soybean and climbing beans in 3 years. At the start of the project, a needs assessment was done using both quantitative and qualitative approaches. Based on the results, and particularly the causes of low yields, interventions to increase production were initiated. Key interventions included promoting the use of integrated soil fertility management practices and improved soybean varieties among farmers. During the 3 years, multiple studies were carried out to assess the program's processes, outputs, and outcomes. Of particular interest were the adoption and adaptation processes and benefits farmers received as they implemented the interventions. The multiple studies were done in different years and involved combining sequential and concurrent strands depending on the research question (Fig. 15.9).

The purpose of multiphase mixed research in program evaluation is to assess a set of incremental questions that advance one overall objective. Such programs are usually managed and implemented by a team of researchers and practitioners and therefore, they all focus on achieving one overall objective. The multiple questions arising in such programs have long-term objectives that cannot be filled with a single mixed methods study. The team will leverage diverse skills required for the data collection process in multiphase mixed methods studies. This approach offers the advantage of readily available resources, often due to project funding. Furthermore, researchers engaged in multiphase mixed methods work on distinct studies, each contributing to a shared overarching goal.

15.2.9 Composite Mixed Methods Design

Composite mixed methods involve the concurrent or sequentially using quantitative and qualitative approaches. Unlike the multiphase transformative design with an agenda to drive societal change, the composite design acknowledges the diversity of research goals. For example, this typology becomes instrumental in soil surveys, where the aim is to amalgamate statistical data with qualitative insights. Soil surveys will involve soil profile descriptions encompassing laboratory analysis to characterize the profile quantitatively, but they will also consider qualitative information by observing morphological characteristics, e.g., soil color. Researchers can also concurrently integrate qualitative methods to gather

Fig. 15.8 Multiphase mixed design

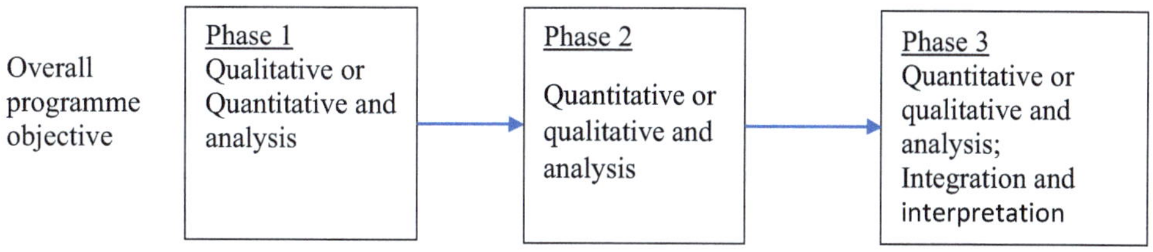

Fig. 15.9 Representation of multiphase mixed design using soybean project to evaluate the contribution to incomes and nutrition of rural farmers

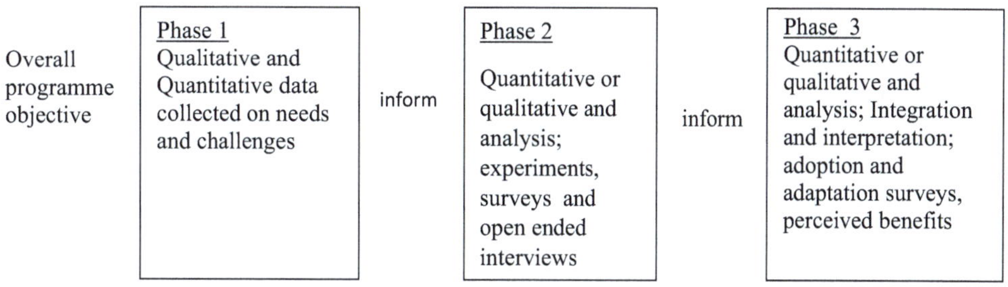

insights from farmers over time, capturing their perspectives on soil health and local agricultural practices.

Other mixed composites could be studies that initiate with singular experiments as the primary method of data collection. However, as the need arises to address additional research questions, researchers may integrate secondary research at various experimental stages. A good example is an experiment designed to evaluate various inputs on legume crop varieties' performance. During the experiment, farmers and traders are engaged to assess the varieties' performance, considering criteria such as taste that inform nutritional aspects. The secondary phase, introduced to address diverse research questions, can be integrated sequentially or concurrently. In the case of legume varieties, farmers may be asked about the taste across different varieties, while traders could provide insights into their marketing experiences and consumer preferences.

15.3 Conducting a Mixed Methods Research

15.3.1 What Are the Key Factors to Consider When Choosing a Mixed Methods Design?

In selecting the mixed methods study design, the researcher makes four key decisions (Creswell and Plano Clark 2011): (1) Whether the quantitative and qualitative strands will remain independent or interactive, (2) Whether the strands will be implemented concurrently, sequentially, or across multiple phases and periods, (3) Whether the two strands have equal or unequal priority, and (4) How the strands will be mixed, including the stage and the strategy (e.g., embedding, merging, connecting, or using a framework). In addition, the following factors also determine the design types chosen:

1. *Research objective/question:* Researchers should consider the research problem that they want to study. A primary consideration is that the design should match the research problem, as discussed in Chap. 2.
2. *Design dimension:* Consideration of both primary and secondary mixed methods design dimensions is recommended to achieve a credible design (Schoonenboom and Johnson 2017). These authors comprehensively describe seven primary dimensions and ten secondary dimensions. They also emphasize the inclusion of the purpose dimension in the title of the study (as this guides the selection of the typology), plus an accurate and comprehensive description of the procedures and methods of the study.
3. *Skills:* Researchers should evaluate their own expertise and consider the quantitative and qualitative skills that they possess. If they lack expertise with certain methods (e.g., quantitative survey skills or collection of qualitative field notes), they should consider working in a team or selecting a design that does not emphasize that method. Indeed, in multiphase designs, incorporating multiple strategies, teams working in the project are usually multidisciplinary, sometimes comprising researchers (who are subject matter specialists) and practitioners.
4. *Available resources:* Consideration must also be given to the available resources, such as the length of time available to complete the study and funding resources for work in a team or the hiring of research assistants. Students' projects have stipulated time for farmers and this is extremely important to consider. For donor-funded projects, time and resources need to be aligned with the project deliverables and approved budget.
5. *Expectations of audiences:* The research can also influence the design choice, particularly if the audience values one type of evidence over the other type.

Questions to ask in deciding to conduct a mixed methods research include:

1. What is the topic of interest that I wish to study?
2. What are my research questions?
3. How could multiple methods help me answer my research questions?
4. What could each of the methods that I might choose tell me about this topic?
5. What can I learn about my topic if I employ more than one method?

A suitable mixed research design is paramount as it determines the credibility of the research study.

15.3.2 Steps in Conducting a Mixed Methods Research

The process researchers follow in conducting mixed methods research is similar to the one described in Chap. 3 and follows a systematic approach. First, the researcher must select a topic and formulate a research problem. The research problem must demonstrate two key aspects: (1) The ability to add to existing knowledge in the particular discipline of study, and (2) the ability to demonstrate practical benefits to society.

Second, after specifying the research problem, a clear statement of the problem needs to be included to show the study context and the knowledge gap.

The researcher must isolate the statement of the problem and rationale of the study from the purpose statement. This interconnects the study's intent, showing how the knowledge gap will be closed. The statement of the problem, together with the purpose of the study, sets the ground for the research questions. To clarify the interconnection between the statement of the problem, purpose, and research questions, the researcher should provide the underlying theoretical framework and a guiding conceptual framework. The research question and the conceptual framework consequently guide the selection and justification of the design. Major criteria determining which design to choose are driven by data collection sequencing, method priority, and the planned data integration steps (Schoonenboom and Johnson 2017).

Third, all other components of the research process must be considered. For example, the principles, procedures, and steps of conducting a literature review are similar to those in Chap. 4. In the methodology section, the researcher must describe clearly the type of design, including sampling techniques for the different types, followed by methods of data collection aligned to each of the approaches chosen.

Finally, given the varied nature and purpose of the mixed methods design, the report should be written clearly to show what was done and the findings. Because of the complexity of the mixed methods, researchers should be conscious of their readers and aim at incorporating aids that help the readers understand the study (Creswell and Plano Clark 2011). Such aids include both the content and the structure. The researcher should aim to have well-defined purpose statement, and well-explained procedures and results. In addition, the report should have clear headings that separate the quantitative and the qualitative elements of data collection and analysis. However, (Bryman 2006) argues that researchers should write the report in such a way that the quantitative and qualitative are "mutually illuminating."

A simplified eight-step process is as follows:

1. *Topic selection:* Guided by the various sources of research problem and criteria for assessing a good topic, select a broad area of the study.
2. *Delineate the research problem:* Use the three-step approach to formulate a research topic (Sect. 15.1.3). The statement of the problem should be clear that the researcher is using mixed methods by using neutral terms.
3. *State the purpose, objectives, and research question:* Outline the overarching goals and specific aims of the study. Articulate clearly the purpose of the study by giving the main intention of conducting the mixed methods study. Formulate research questions and objectives aligned with the purpose of the study, e.g., assess crop yields (quantitative) and explore farmers' attitudes (qualitative). Study aim and research question should call for a mixed methods research.
4. *Design the study:* Choose mixed methods design according to purpose and justify the rationale for mixing as well as for choosing the method. Determine your quantitative and qualitative methods. At this point, researcher should be aware of the key characteristics of the quantitative and qualitative research, including strengths and weaknesses. These characteristics must be considered to ensure that the validity of data is protected. For example, in the quantitative approach, key elements include experimental design principles, deduction rules, explanation, prediction, standardized data collection using closed-ended data, statistical analysis and generalization. Conversely, key characteristics of the qualitative techniques consist of open-ended data, observation, inductive approach, discovery, exploration, in-depth interviews, naturalistic settings, and rich detailed descriptions, as explained in Chap. 13. These aspects should be integrated during the design of the study.
5. *Write the mixed research design protocol:* In writing the protocol, include the key elements of a protocol and justify choosing mixed methods. Include views of reality paradigms including epistemologies, axiology, theories, and theoretical frameworks. In addition to giving information regarding the whole study, the protocol, in particular, should provide details of the methodology and answer the how questions, including methodologies and approaches, data types, data collection instruments, etc. For example, how will sampling be done? how will the mixing be done? How will data be collected? The proposed structure of main components of research proposal using mixed methods design is shown in Box 15.8.
6. Collect data: Data should be collected according to the methods stipulated in the protocol. The techniques of data collection are discussed at length in Sect. 15.2.8. Key considerations include: what data will be collected; how will the data be collected; who will collect the data? for example, should the same research assistant collect both data sets (the Quant. and Qual.). Another consideration is sample size, e.g., in collecting data from farmers, should equal numbers of individuals be samples for both quantitative and qualitative data? In embedded design, it is expected that the embedded data will play a supportive role, the sample for this element should be small.
7. *Analyze and interpret the results:* One of the key issues in data analysis in mixed methods is that the researcher must be conversant with statistical techniques of analyzing quantitative data as well as the various techniques for analyzing qualitative data as explained in chapters. If not conversant with the methods, skills can be outsourced from the experts. The aim is to ensure that proper analysis and interpretation is done. In a nutshell, quantitative data must be subjected both descriptive and inferential statistics, while qualitative data involves coding, themes identification, patterns isolation, and developing theoretical

propositions. A key consideration when planning for data analysis is the type of design used. For example, in concurrent designs data management, final analysis should be done independently but in parallel. If on the other hand, it is sequential, the researcher must consider the chronological order as per the design.

8. *Reporting:* The report should follow the strategy chosen. Creswell and Plano Clark (2011) provides guidance for writing a report from mixed methods research. In a sequential design, the report should be organized to present the two phases distinctly with separate headings and show how each data form contributed to a comprehensive understanding. For concurrent designs, qualitative and quantitative components may be presented in distinct sections, but ensure you demonstrate how the combined data resulted in convergence or identified similarities. In transformative designs, organize the writing to reflect the strategy you used, but ensure you have a separate section to advance an agenda for change or reform emanating from the study. See more on how to report in Chap. 26.

Box 15.8 Proposed structure of main components of research proposal

Chapter 2: Literature Review

- *Empirical Literature:* Review relevant literature on the research topic, covering both quantitative and qualitative studies. Review empirical literature and highlight gaps in the existing literature that the mixed methods study intends to address. Justify the use of mixed methods based on the strengths of combining both approaches.

Theoretical Framework: Present the theoretical framework that informs the study, indicating how it guides the research design and analysis.

Conceptual Framework: Present the conceptual framework that informs the study, indicating how it guides the research design and analysis.

Chapter 3: Materials and Methods

- *Study Site and Description of the Area*

Research Design: Clearly describe the overall research design (concurrent, sequential, or transformative). Outline the specific quantitative and qualitative components of the study. Provide a rationale for the sequence and timing of data collection.

Participants and Sampling: Specify the target population, and justify its relevance to the research questions. Describe the sampling strategies for both quantitative and qualitative components. Provide details on sample size considerations and the criteria for participant inclusion.

Data Collection: Detail the data collection methods for the quantitative component (e.g., surveys, experiments). Outline the data collection methods for the qualitative component (e.g., interviews, focus groups). Explain how the two sets of data will be triangulated to enhance validity.

Data Analysis: Describe the planned analysis for the quantitative data (e.g., statistical tests, regression). Outline the analysis procedures for the qualitative data (e.g., thematic analysis, content analysis).

Integration: Explain how the results from both components will be integrated or compared. Integration of Results: Clearly specify how the quantitative and qualitative results will be integrated. Discuss the criteria for determining convergence, divergence, or complementarity between findings.

Ethical Considerations: Address ethical issues related to participant consent, confidentiality, and potential risks. Explain the steps taken to ensure the ethical conduct of both quantitative and qualitative components.

15.4 Challenges in Mixed Methods Research

Despite the clear benefits of the mixed methods approach in agricultural research, the use of mixed methods research presents several obstacles which could severely affect the credibility of the research. Several authors (Wilkinson and Staley 2019; Dawadi et al. 2021) have attempted to explain the varied challenges researchers face in studies using mixed methods. Here we give six key challenges (Fig. 15.10):

1. *Need for enhanced skills*: One of the major obstacles is that it is demanding in terms of skill sets. Mixed method research requires a researcher who possesses both quantitative and qualitative research skills or a team of researchers who are experienced in both quantitative and qualitative research. In line with this, Creswell and Plano Clark (2011) strongly recommend that researchers without these skills should not engage in mixed methods research. A mixed methods researcher must have knowledge and experience in both quantitative and qualitative research before undertaking a mixed methods study. For instance, the researcher needs to be familiar with and understand elements of rigor in quantitative research, including validity, reliability, experimental control, and generalization. They must also be well acquainted with

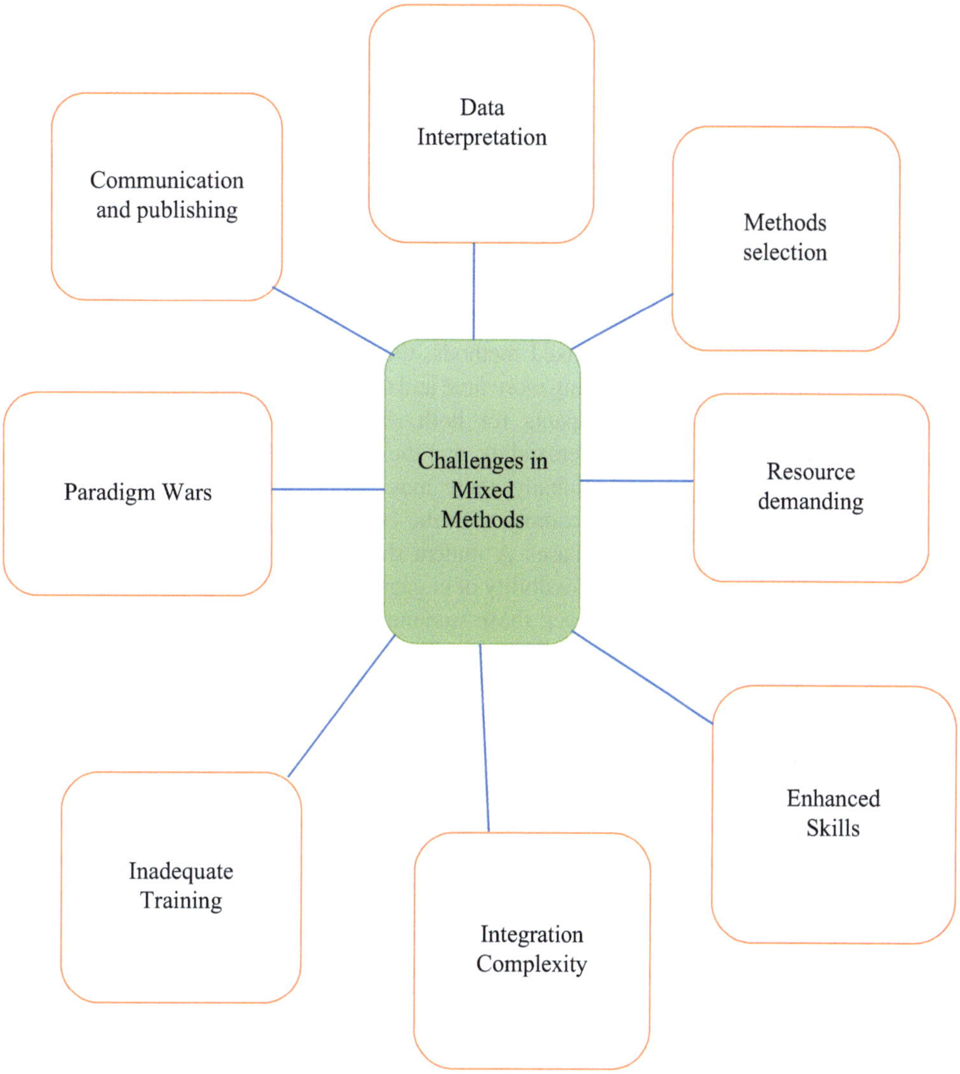

Fig. 15.10 Eight key challenges in implementing mixed methods

the logic of hypothesis testing and have knowledge of how to analyze and interpret data using common statistical procedures. On the other hand, basic qualitative skills are essential. These include, at the minimum, common methods of collecting qualitative data, analysis of qualitative data including coding and ability to develop themes and patterns, as well as ways of controlling validity and trustworthiness. Lack of adequate skills can be addressed through involvement in interdisciplinary research and further training on both approaches.

2. *Integration complexity*: Integration of quantitative and qualitative data presents a major challenge for many researchers (Creswell and Plano Clark 2011). Integration is a technical skill, and most researchers lack expertise on how to optimally and effectively integrate the two data strands to answer the research question, but little guidance exists in the literature on how to undertake the process. In data integration, pertinent questions arise: at what point do you integrate data? When do you stop analyzing, comparing, and contrasting the data? Are you merging, connecting, or embedding the data from the two approaches. These questions matter greatly and researchers struggling to answer these questions find scarce literature guiding on how to merge data from different sources. Dawadi et al. (2021) caution students and novice researchers who wish to engage in mixed methods on the need to develop skills in handling and merging data sets from the different strands to avoid the risk of mixing them wrongly. The development of clear guidelines and frameworks for data integration can help address the complexity of combining data.
3. *Inadequate Training*: Most students and faculty staff could lack adequate skills in mixed methods research due to poor and inadequate training received in their course work. The poor training is associated with the fact that the majority of faculty lack the training and also lack mixed methods topics in the research methods curriculum. A sound foundation in mixed methods can easily be

acquired through reading the enormous literature on mixed methods, especially by Creswell and Co., and enrolling in courses that teach this subject. Universities are also encouraged to incorporate mixed methods topics in the curriculum.

4. *Resource demanding:* High resources (time, funds, personnel) demand is another key obstacle for researchers seeking to engage in mixed methods designs. The integration of both quantitative and qualitative phases may require extensive time and additional resources. Because of the multiple procedures in mixed methods, the data collection may be lengthy, requiring more time and funds. For example, recruiting participants for both strands (qualitative and quantitative) is demanding, and the labor requirement is extensive. Additionally, there may be a need for extra arrangements to comply with the ethical guidelines during the different phases. A student should, therefore, evaluate critically the feasibility of engaging in a mixed method research. However, those students supported by multidisciplinary donor-funded projects have the advantage of being supported with skills and resources. They could also benefit from past data already collected by the researchers' team.
5. *Methods selection*: The most important challenge for a mixed-method researcher is deciding which MMR design is appropriate for a particular study (Dawadi et al. 2021). The suitability of a design will largely depend upon the purpose of the study and the perceived priority given to the qualitative and quantitative strands (i.e., whether equal priority is given to both data sets or one is dominated by another). As a consequence, early-career researchers may not have the confidence to choose one from many designs, especially when each one has its own drawbacks and potential challenges. To address this challenge, there is a need to develop decision-making tools and frameworks that researchers can use to enhance methodological coherence.
6. *Data interpretation*: Combining different data types necessitates researchers to navigate diverse analytical approaches and methodologies, which can complicate the interpretation process. Unlike studies that use only one method, mixed methods research involves combining findings from different sources. This can be challenging, as there may be discrepancies or contradictions between quantitative and qualitative findings. Researchers must work to reconcile these differences and create a clear and comprehensive interpretation of the data.
7. *Communication and Publishing*: Researchers using mixed methods research face difficulties communicating their findings (Glogowska 2011). First, they often face scenarios of communicating to diverse audiences who are well-versed in only one methodological approach. Second, finding an appropriate journal to publish in with expert reviewers who are not biased toward one paradigm is not easy. Third, papers from mixed methods are comprehensive and long, and conforming to the word count limitations imposed by journals may be a hurdle (Wisdom et al. 2012). Some guidelines for writing mixed methods research manuscript are given in Chap. 25 (Structure of journal article). Journals should develop clear guidelines and a standard structure for reporting mixed methods research.
8. *Paradigm wars*: Finally, paradigm wars still reign because of different views held by quantitative and qualitative researchers (Alise and Teddlie 2010). Some researchers are still not yet convinced about the value and benefits of mixed methods (Gage 1989). There are also differences stemming from divergences regarding the underlying ontological and epistemological views of positivists versus constructivists. Some question the newness of the mixed methods paradigm (Cohen et al. 2017), while others feel they do not have the time to invest in learning a new approach to research (Creswell and Plano Clark 2011; Creswell and Creswell 2022). Still, others object to mixed methods because of their personal stand and conviction, e.g., having statements like, does mixing water and wine result in a fortified mixture or diluted wine that is strengthened or weakened? (Cohen et al. 2017). Of course, this is a humorous analogy, but it shows that the paradigm wars could still be at play in the twenty-first century. Those opposed to mixed methods research are probably a small proportion as we still find growing interest and increased literature sources on this subject matter. We encourage those dissenting voices and disputing the mixed methods paradigm to read publications from reputable international journals. It is clear from the literature that acceptance of mixed methods research is increasing in agriculture, especially with promotion by researchers working with communities. We believe that wholly embracing "mixed methods approaches" as a "Third Paradigm" will revitalize and add value to agricultural research. The strength in the quantitative orientation of agricultural researchers will greatly benefit from integrating the qualitative stance that will provide unique social data from one of the most important clients (the farmers).

Acknowledgments The authors appreciate the following publishers for permission granted to re-use their resources through the Creative Commons model:

- Springer Link, publication by Khoza et al. (2021) and Bagambilana and Rugumamu (2023).
- Heliyon, publication by Njenga et al. (2021).
- MDPI, publication by Landaverde et al. (2022) and Etana et al. (2020).

Recommended Further Reading

1. Creswell, J. W., & Creswell, J. D. (2022). *Research design: Qualitative, quantitative, and mixed methods approaches* (6th ed.). Sage Publications.
2. Creswell, J. W., & Plano Clark, V. L. (2018). *Designing and Conducting Mixed Methods Research* (3rd ed.). SAGE Publications.
3. Kiptot E, Franzel S (2015) Farmer-to-farmer extension: opportunities for enhancing performance of volunteer farmer trainers in Kenya. Dev Pract 25:503–517. https://doi.org/10.1080/09614524.2015.1029438
4. Tashakkori A, Teddlie C (2010) SAGE Handbook of Mixed Methods in Social & Behavioral Research. SAGE Handbook of Mixed Methods in Social & Behavioral Research. https://doi.org/10.4135/9781506335193
5. Dawadi S, Shrestha S, Giri RA (2021) Mixed-Methods Research: A Discussion on its Types, Challenges, and Criticisms. Journal of Practical Studies in Education 2:25–36. https://doi.org/10.46809/jpse.v2i2.20

Key Points of the Chapter

Overview

The concept of mixed-method research is not new, but there has been increased interest, mainly spearheaded by John Creswell and other prominent authors. Many books and publications offer varied definitions, but all agree that it is about borrowing from the strength of both the qualitative and quantitative approaches, experimental and nonexperimental designs, to expand understanding of the phenomenon under investigation.

Relevance of Mixed Methods Research in Agriculture

Though not explicitly stated, the relevance of mixed methods in agriculture gained recognition as early as the 1970s, with the advent and advocacy of farming systems research and development agenda.

The multifaceted and complex nature of challenges encountered in agricultural sciences renders either method, quantitative or qualitative, individually inadequate. Moreover, agricultural research inherently involves the collaboration of biophysical and social scientists, reflecting its multidisciplinary nature. The integration of mixed methods provides an avenue for a more expansive and profound comprehension of the diverse problems encountered in agriculture.

Most common studies involving mixed methods research tend to fall within the adoption and adaptation of agricultural technologies. There is also increased use of mixed methods within the climate change research agenda. However, there is limited classical literature and information on mixed methods in agriculture, pointing to inadequate capacity among researchers to conduct credible mixed methods research.

Philosophical Worldview in Mixed Methods Research

The mixed methods researcher operates within a pragmatism paradigm, which diverges from adhering strictly to either a qualitative or quantitative approach. The pragmatism paradigm is characterized by its pluralistic and practical orientation, emphasizing effectiveness and practical application over rigid philosophical beliefs.

Purpose of Using Mixed Methods

The main reason for using multiple research approaches and techniques is to gain a more comprehensive understanding of a research problem than is possible with a single approach. Specifically, six key reasons of using mixed methods are:

1. Comprehensiveness.
2. Contextualization.
3. Enhanced validity and reliability.
4. Practicability.
5. Flexibility.
6. Collaboration.

Types of Mixed Methods Research Design

We subdivide the design into four types; *Sequential, Concurrent, Multiphase, and Composite,* giving eight categories as follows:

1. Sequential Explanatory design.
2. Sequential Exploratory design.
3. Sequential transformative design.
4. Concurrent Triangulation design.
5. Concurrent Embedded Nested design.
6. Concurrent Transformative design.
7. Multiphase mixed design.
8. Composite design.

Sequential exploratory design: The qualitative data are collected and analyzed first followed by quantitative data collection.

Sequential explanatory design: The quantitative data are collected and analyzed first, followed by qualitative data collection. Qualitative data explains quantitative data.

Sequential Transformative design: Both quantitative and qualitative are combined in two phases, but the research is also explicitly driven by a transformative theoretical perspective to reach the marginalized society.

Concurrent triangulation design: Also called "convergent parallel design" involves the simultaneous collection of quantitative and qualitative data, followed by separate analyses and mixing later.

Concurrent Embedded design: Involves one data collection phase during which both quantitative and qualitative data are collected, one collection method (secondary) being "Embedded" or "Nested" within the predominant method (primary).

Concurrent transformative design: Involves a simultaneous collection of both "quantitative and qualitative data" guided by a theoretical perspective of change. It is comparable to sequential transformative design regarding the central agenda, which is to give voice to the disadvantaged and marginalized.

Multiphase is a complex mixed-research design that may incorporate all six above plus a theoretical framework of transformation. It has been referred to as a sandwich design because a researcher may incorporate quantitative and qualitative data across phases.

Composite includes complex multi-methods, mixed designs that do not necessarily have a transformative agenda, such as methodologies for soil surveys and species selection involving both quantitative and qualitative data.

Conducting a Mixed Methods Research

The process researchers follow in conducting mixed methods research is similar to the one described in Chap. 3 and follows a systematic approach. First, the researcher must select a topic and delineate the research problem. The key steps are as follows:

1. Select your topic.
2. Delineate the research problem.
3. State the purpose, objectives, and research question of the study.
4. Write the mixed methods research protocol.
5. Collect data.
6. Analyze and interpret.
7. Report.

Challenges in Implementing Mixed Methods Research in Agriculture

Several challenges limit the use of mixed methods. This includes inadequate skills mix of quantitative and qualitative approaches, complexity in managing the data, inadequate training in mixed methods, high demand for resources, difficulty in selecting the most suitable mixed methods, complexities in resolving discrepancies between quantitative and qualitative findings, methodological biases, and space constraints during publishing in journals, and differing philosophical stances and its value is a stand-alone paradigm.

Review Questions

1. In your own words, explain what mixed methods research is and why it is increasingly used in agricultural research
2. List the eight mixed methods design categories and classify each under the four major typology groups.
3. How did the "advent and advocacy of farming systems research and development agenda" in the 1970s implicitly lay the groundwork for the relevance of mixed methods in agriculture, even if not explicitly stated as such at the time?
4. Explain what "time" (e.g., sequential, concurrent) and "strategy" (e.g., explanatory, exploratory, embedded) refer to in mixed methods design selection.
5. What is the primary "central agenda" of Sequential Transformative and Concurrent Transformative designs? How does this agenda differentiate them from other mixed methods designs, especially in the context of addressing agricultural social and development challenges?
6. What skills or competencies do you think are necessary to successfully carry out a mixed methods study? Which of these do you already possess, and which do you need to develop?

Chapter Exercise

1. A student is designing a mixed methods study to evaluate the adoption and impact of new drought-resistant maize varieties among smallholder farmers. Their research question aims to understand both the quantifiable changes in yield and income, as well as the qualitative factors influencing farmers' decisions and experiences with the new varieties. Based on this information:

   (a) Identify and explain the primary philosophical stance (or worldview) that this student would most appropriately adopt for their overall mixed methods research design. Describe the core tenets of this philosophical stance
   (b) Discuss how the inherent nature of mixed methods research might necessitate or encourage a flexible, "shifting," or adaptive philosophical approach across different phases of this specific study. Provide concrete examples from the evaluation of drought-resistant maize varieties to support your discussion.
2. You are studying the factors influencing the adoption of a new irrigation technology in a specific agricultural community. Then therefore:
   (a) If you wanted to first establish the *rate* of adoption and then understand the *reasons* behind the observed rates, which sequential design would you use and why? Provide a brief outline of the steps.
   (b) If you wanted to first explore farmers' *perceptions* and *needs* regarding irrigation to inform the development of a quantitative survey on technology preferences, which sequential design would be more appropriate? Briefly explain your rationale
3. Identify a current agricultural issue (e.g., resistance to livestock vaccines, soil degradation) and design a *basic mixed methods research outline* that includes:
   (a) Research question.
   (b) Type of design (sequential/concurrent, specific category).
   (c) Data collection tools.
   (d) Strategy for integration.
4. Imagine you are presenting your mixed methods proposal to a panel of reviewers who are skeptical about using both qualitative and quantitative data. Write a short justification (150–200 words) for why mixed methods are necessary for your chosen study.

Answers to Review Questions

- Q1. Reference: 15.1.1 and 15.2.2
- Q2. Reference: 15.2
- Q3. Reference: 15.1.2 (Tip: You may also refer to Chap. 10)
- Q4. Reference: 15.2.1
- Q5. Reference: 15.2.4 and 15.2.7
- Q6. Reference: 15.4

References

Akimowicz M, Vyn RJ, Cummings H, Landman K (2018) An introduction to mixed methods research in agricultural economics: the example of farm investment in Ontario's Greenbelt, Canada. J Rural Stud 61:162–174. https://doi.org/10.1016/J.JRURSTUD.2018.04.012

Alise MA, Teddlie C (2010) A continuation of the paradigm wars? Prevalence rates of methodological approaches across the social/behavioral sciences. J Mix Methods Res 4:103–126. https://doi.org/10.1177/1558689809360805

Bagambilana FR, Rugumamu WM (2023) Determinants of farmers' adaptation intent and adoption of adaptation strategies to climate change and variability in Mwanga District, Tanzania. Environ Manag 72:785–804. https://doi.org/10.1007/S00267-023-01792-2/METRICS

Bamberger M, Rao V, Woolcock M (2010) Using mixed methods in monitoring and evaluation: experiences from international development. https://doi.org/10.1596/1813-9450-5245

Bryman A (2006) Integrating quantitative and qualitative research: how is it done? Qual Res 6:97–113. https://doi.org/10.1177/1468794106058877

Clark VLP (2010) The adoption and practice of mixed methods: U.S. trends in federally funded health-related research. Qual Inq 16:428–440. https://doi.org/10.1177/1077800410364609

Cohen L, Manion L, Morrison K (2017) Research methods in education research methods in education. https://doi.org/10.4324/9781315456539/RESEARCH-METHODS-EDUCATION-LOUIS-COHEN-LAWRENCE-MANION-KEITH-MORRISON

Creswell J (2013) Steps in conducting a scholarly mixed methods study. Discipline-Based Education Research Group Speaker Series

Creswell JW (2014) Research and design qualitative, quantitative, and mixed methods approaches. SAGE Publication Inc

Creswell JW, Guetterman TC (2019) Educational research_planning, conducting, and evaluating quantitative and qualitative research. Pearson, pp 1–681

Creswell JW, Plano Clark VL (2011) Designing and conducting mixed methods research, pp. 51–99

Creswell JW, Tashakkori A (2007) Editorial: developing publishable mixed methods manuscripts. J Mix Methods Res 1:107–111. https://doi.org/10.1177/1558689806298644/ASSET/0C46FEB1-859E-44DE-B987-4AD09F0FBD53/ASSETS/1558689806298644.FP.PNG

Creswell WJ, Creswell DJ (2022) Research design, 6th edn. SAGE Publications Ltd

Dawadi S, Shrestha S, Giri RA (2021) Mixed-methods research: a discussion on its types, challenges, and criticisms. J Pract Stud Educ 2:25–36. https://doi.org/10.46809/jpse.v2i2.20

Doss C (2014) Data needs for gender analysis in agriculture. In: Gender in agriculture: closing the knowledge gap, pp 55–68. https://doi.org/10.1007/978-94-017-8616-4_3

Etana D, Snelder DJRM, van Wesenbeeck CFA, de Cock BT (2020) Dynamics of smallholder farmers' livelihood adaptation decision-making in Central Ethiopia. Sustainability 12:4526. https://doi.org/10.3390/SU12114526

Fetters MD (2020) The mixed methods research workbook: activities for designing, implementing, and publishing projects. SAGE

Gage NL (1989) The paradigm wars and their aftermath a "historical" sketch of research on teaching since 1989. Educ Res 18:4–10. https://doi.org/10.3102/0013189X018007004

Glogowska M (2011) Paradigms, pragmatism and possibilities: mixed-methods research in speech and language therapy. Int J Lang Commun Disord 46:251–260. https://doi.org/10.3109/13682822.2010.507614

Greene JC (2008a) Is mixed methods social inquiry a distinctive methodology? J Mix Methods Res 2:7–22. https://doi.org/10.1177/1558689807309969

Greene JC (2008b) Mixing methods in social inquiry

Greene JC, Caracelli VJ, Graham WF (1989) Toward a conceptual framework for mixed-method evaluation designs. Educ Eval Policy Anal 11:255–274. https://doi.org/10.3102/01623737011003255

Johnson RB, Onwuegbuzie AJ (2007) Toward a definition of mixed methods research. J Mix Methods Res 1:112–133. https://doi.org/10.1177/1558689806298224

Khoza S, van Niekerk D, Nemakonde L (2021) Rethinking climate-smart agriculture adoption for resilience-building among smallholder farmers: gender-sensitive adoption framework. In: African handbook of climate change adaptation: with 610 figures and 361 tables, pp 677–698. https://doi.org/10.1007/978-3-030-45106-6_130/FIGURES/1

Kiptot E, Franzel S (2015) Farmer-to-farmer extension: opportunities for enhancing performance of volunteer farmer trainers in Kenya. Dev Pract 25:503–517. https://doi.org/10.1080/09614524.2015.1029438

Kondo K, Cacho O, Fleming E et al (2020) Dissemination strategies and the adoption of improved agricultural technologies: the case of improved cassava varieties in Ghana. Technol Soc 63:101408. https://doi.org/10.1016/J.TECHSOC.2020.101408

Landaverde R, Rodriguez MT, Niewoehner-Green J et al (2022) Climate change perceptions and adaptation strategies: a mixed methods study with subsistence farmers in rural Peru. Sustainability 14:16015. https://doi.org/10.3390/SU142316015

Mertens D (2014) Research and evaluation in psychology and education. Libro Electronico, p 4

Mugenda A, Mugenda G (2019) Research methods: qualitative and mixed methods approach, pp. 100–102

Newman I, Benz CR (1998) Qualitative-quantitative research methodology: exploring the interactive continuum. Southern Illinois University

Njenga MW, Mugwe JN, Mogaka H et al (2021) Communication factors influencing adoption of soil and water conservation technologies in the dry zones of Tharaka-Nithi County, Kenya. Heliyon 7:e08236. https://doi.org/10.1016/J.HELIYON.2021.E08236

Palinkas LA, Aarons GA, Horwitz S et al (2011) Mixed method designs in implementation research. Adm Policy Ment Health Ment Health Serv Res 38:44–53. https://doi.org/10.1007/S10488-010-0314-Z/TABLES/2

Patton MQ (2015) Qualitative research & evaluation methods: integrating theory and practice, 4th edn. SAGE Publications, Inc

Place F, Adato M, Hebinck P (2007) Understanding rural poverty and investment in agriculture: an assessment of integrated quantitative and qualitative research in Western Kenya. World Dev 35:312–325. https://doi.org/10.1016/J.WORLDDEV.2006.10.005

Schoonenboom J, Johnson RB (2017) How to construct a mixed methods research design. Kolner Z Soz Sozpsychol 69:107–131. https://doi.org/10.1007/S11577-017-0454-1/TABLES/1

Tashakkori A, Teddlie C (2010) SAGE handbook of mixed methods in social & behavioral research. https://doi.org/10.4135/9781506335193

Tufa AH, Kanyamuka JS, Alene A et al (2023) Analysis of adoption of conservation agriculture practices in southern Africa: mixed-methods approach. Front Sustain Food Syst 7:1151876. https://doi.org/10.3389/FSUFS.2023.1151876/BIBTEX

Wilkinson IAG, Staley B (2019) On the pitfalls and promises of using mixed methods in literacy research: perceptions of reviewers. Res Pap Educ 34:61–83. https://doi.org/10.1080/02671522.2017.1402081

Wisdom JP, Cavaleri MA, Onwuegbuzie AJ, Green CA (2012) Methodological reporting in qualitative, quantitative, and mixed methods health services research articles. Health Serv Res 47:721–745. https://doi.org/10.1111/J.1475-6773.2011.01344.X

Sampling

16

Abstract

Sampling is the process of selecting units (e.g., people, organizations, plants) from a population of interest so that by studying the sample, we may fairly generalize our results to the population from which they were chosen. Samples are preferred over a census in scientific research because of the high costs and logistical issues associated with a census. A cornerstone of sampling is a representative sample, which mirrors the population's characteristics. This chapter explains the various types of populations: target population (could be unknown) and accessible population (sampling unit or the study population). We get our sample from the accessible population, which gives the sampling frame. We have two types of sampling methods: probability (mathematical and random) and non-probability (non-mathematical and non-random). Within the probability sampling: simple random sampling, systematic sampling, stratified sampling, cluster sampling, and multiphase sampling. On the other hand, we have purposive, convenience, theoretical, snowballing, and quota sampling in non-probability sampling. Each of these has benefits and limitations. Purposive sampling has variants such as maximal, critical, typical sampling, and others. Sample size is a critical factor in both quantitative and qualitative research approaches. In quantitative studies, the easiest way to determine sample size is to use an online calculator or formulas and standard tables found in statistics and research method books. As you use these methods, it is recommended that you always select the sample to fit the purpose. Other factors, such as heterogeneity of the population, confidence level and number of aggregations, should be considered. In qualitative studies, the most important is "fit-for-purpose," which gives rich quality information to fulfil your objectives and answer your research problem. Seven guidelines are provided for selecting a qualitative research sample. A brief final section highlights that the probability and non-probability methods discussed apply to agronomic and livestock studies, specifically biophysical measurements. Sampling in mixed methods research uses probability and non-probability sampling methods.

Keywords

Accessible population · Census · Cluster sampling · Multiphase sampling · Non-sampling errors · Probability and non-probability sampling · Purposive sampling · Representative sample · Simple random sampling · Sample size · Sampling errors · Sampling frame · Selection bias · Stratified sampling · Systematic sampling · Target population

Learning Outcomes

After reading this chapter, you should be able to:

1. Apply the concept of sampling and the relationship between samples and population.
2. Describe the process involved in sampling.
3. Differentiate between probability and non-probability sampling techniques.
4. Identify the various probability sampling techniques.
5. Determine the non-probability sampling techniques.
6. Distinguish between "simple random sampling" and "stratified random sampling."
7. Distinguish between "cluster sampling" and "multiphase sampling."
8. Identify the variants of sampling methods used in qualitative research.
9. Apply guidelines for determining sample size; to calculate a sample in a research project.
10. Outline the main advantages and disadvantages of each type of sampling technique.
11. Demonstrate understanding of the meaning of sampling and non-sampling errors and how to avoid them.

J. Njeri Mugwe, S. Runo, *Research Methodology in Agricultural Sciences*, https://doi.org/10.1007/978-981-95-1892-0_16

12. Appreciate the universal application of probability and non-probability sampling methods in biophysical studies.

16.1 Introduction to Sampling

16.1.1 What to Measure?

In data collection, we measure variables empirically to obtain objective, quantifiable information about the subjects of our study. Imagine a maize field: you might measure variables, such as, yield per hectare and soil moisture content. In a herd of cattle, you could measure weight, milk production, feed intake, and growth rate. For plants in an experimental plot, variables could include plant height, number of nodules, chlorophyll content, pest incidence, and biomass. If you are working with farmers, you might measure demographic characteristics and farm factors such as farm size.

How do you decide which case and how many to measure? For instance, which animal should be measured in a herd? For which plant the height or nodule numbers should be measured in an experimental plot? We use *sampling* to give us a representative sample that accurately reflects the characteristics of the entire herd. In agriculture, this target population can be quite diverse and includes populations of *people, animals, crops, trees, events, organizations*, etc. (See Box 16.1). Each of these entities represents a different aspect of agricultural research, and the choice of target population depends on the study's specific objectives.

16.1.2 The Concept of Sampling

Sampling refers to the case selection process. It selects a few elements from the entire population. We study these few individuals to derive conclusions about the population (Fig. 16.1). We infer about the population from the results we obtain from the sample. *A population is all the elements that share a common set of characteristics.* Populations can be individuals, units, events or objects, as shown in Box 16.1. Sometimes, it may be possible to study all the members of the population, which we call *census*. However, it is challenging to do this. Most populations are too large, diverse, and distributed over a large geographic area. In most cases, contacting the entire population would be time-consuming and costly. We, therefore, confidently use a sample to tell us about the population as long as the sample is *representative* of the entire population (Fig. 16.1). This is a concept scientist in many disciplines have used for a long time.

Box 16.1 Common entities where measurements are taken in agricultural research

Entity	Target population	Case
Experiment	Experimental plots where different varieties of maize are planted to compare growth rates under various fertilizers	Individual plants to measure height, leaf area index, sample for nutrient analysis, etc
Fields	A large field of 10 ha of rice to study and evaluate the impact of irrigation techniques on yield	Individual plots demarcated to measure yields
Soil	Collect soil samples from various experimental plots or depths and locations within a field to assess nutrient levels	Individual soil samples
People: Extension workers and farmers	Conduct interviews with agricultural workers to understand their interactions with farmers Surveying farmers to gather information on their agricultural management practices	Individual agricultural workers Individual farmers
Transects	A transect through a farm of pasture land to monitor changes in soil and plant species composition over time.	Individual transects to measure soil and plant species
Animals	Studying a herd of dairy cows to evaluate the effects of different feed supplements on milk production	Individual animals
Crops	Analyzing the health and yield of tomato plants under organic versus conventional farming practices	Individual tomato plants to measure yields and pest and diseases incidences
Agroforestry trees	Assessing the growth rates of different mango tree varieties in an orchard	Individual trees to measure height and diameter at breast height (dbh)
Fruit orchard	Monitoring pest infestations and fruit yield in a citrus orchard	Individual fruit trees on which to count number of fruit flies
Households	Surveying rural households to determine the impact of new agricultural technologies on food security	Individual households to interview
Organizations	Agricultural cooperatives to study the effects of collective marketing strategies on smallholder farmers' income	Individual cooperatives and individual farmers
Events	Assessing the effect of using various training methods of adoption of agricultural practices	Individual field days Individual demonstrations

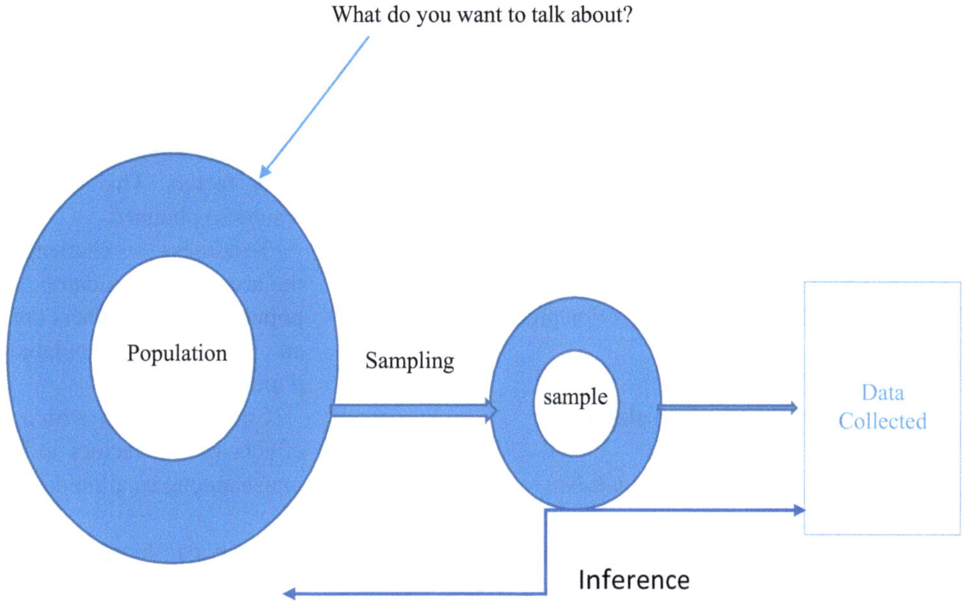

Fig. 16.1 Illustration of the concept of sampling

Census versus a sample

A sample is preferred to a census due to its effectiveness, time efficiency, and potential for quality data. Martínez-Mesa et al. 2016 gave five theoretical and practical reasons that prevent us from carrying out census-based surveys in research:

1. Ethical issues: It is unethical to involve more participants than are genuinely necessary for the study.
2. Budgetary limitations: The significant cost of conducting a census restricts its feasibility as a participant selection strategy.
3. Logistics: Census-based studies often present major obstacles in terms of required personnel, equipment, and other resources, etc.,
4. Time restrictions: Planning and executing a census may demand too much time.
5. Unknown target population: A lack of information about the population hinders the ability to conduct a census. For instance, if a study aims to evaluate control measures used by farmers against fall armyworms on maize, incomplete knowledge about all farmers using such measures makes a census-based study unworkable.

16.1.3 Representative Sample

Selecting a representative sample is one of the most important concepts in sampling. Representativeness is about the *resemblance* of the sample with the population. The question is, How closely does your sample resemble the population? A representative sample closely mirrors the characteristics of the population, thereby minimizing the error associated with sampling. If the sample is not representative, the results are inaccurate. For instance, if you are to pick a sample of farmers keeping dairy cows (whose past reports indicate that farmers keep four breeds of animals), A, B, C, and D in almost equal proportions (25%). Suppose you pick three breeds in your sample, or you pick all breeds at A (40%), B (20%), C (10%), and D (30%). Such a sample will be unrepresentative and biased. The estimate obtained in the study sample will not be accurate to generalize to the target population. Representative samples will give results that are sufficiently accurate for decision-making. This is achieved by ensuring that you use the appropriate research design.

A representative sample may offer greater overall accuracy compared to conducting a census (Kemp 2003). This is because a sample has the advantage of containing fewer cases. This gives an opportunity to carry out more extensive planning and testing of data collection methods. With fewer information-gathering cases, researchers can commit more time and engage in more detailed data collection processes. Moreover, when employing individuals to collect data, such as interviewers, there is room to hire quality staff and allocate more time to reaching difficult-to-access cases. After data collection, more time can be allocated proportionally to verify and validate the accuracy of the data before analysis.

Selection Bias and Response Bias

Selection bias and response bias are two main threats to representative samples in surveys. Selection bias occurs when certain groups are systematically over-represented or under-represented in the sample. For instance, a sample of 35 men and 15 women, picked from 200 people comprising 100 men and 100 women, would be biased. In this sample, men are over-represented, while women are under-represented. This leads to a skewed perception of gender-based differences and minimal or inaccurate responses to survey questions. The final results will be flawed. A response bias may occur if female farmers feel marginalized or overlooked. This will undermine any efforts to develop inclusive gender-responsive agricultural policies. The under-representation of women farmers in agricultural initiatives has been reported widely (Doss 2014; Quisumbing et al. 2014; Njuki et al. 2022). Researchers must carefully consider the selection bias and response base and mitigate it in surveys by applying suitable sampling methods and data collection procedures.

16.1.4 Defining the Population

A researcher must define the *population* to which to generalize results. In an ideal situation, researchers aim to generalize to the entire population, referred to as the target population. However, it is often unrealistic to draw representative sample directly from this target population due to challenge in identifying individual members. Instead, researchers typically work with elements or objects available for selection during the sampling process, known as the *"sampling unit"* or the *"accessible population."*

Consider the following populations:

- All farmers growing maize in Africa.
- All farmers growing rice in the Americas.
- All farmers using mineral fertilizers in East and Central Africa.
- All farmers growing tea in the world.
- All farmers implementing integrated pest management to control cotton ball disease.
- All farmers who purchased nitrogenous fertilizer in the long rainy season of 2020 in country A.
- All agricultural extension officers in Africa.
- All farmers' groups in a country in China.
- All farmer groups that have agricultural-based activities in India.
- All farmers growing Hass avocado in Kenya.
- All farmers growing hybrid 511 maize variety in County B.
- All dairy farmers with zero grazing units in Argentina.
- All farmer groups participating in the collective marketing of bananas in County D.
- All agricultural-based training institutions in a country.
- All field days or training events held in county A to promote the growing of hybrid maize.
- All improved dairy animals.

In the above examples, we note diversity and the wide geographic distribution of the populations. Thus, reaching all the members of the population will be a tall order. Our definition tells us that all members of a population possess a similar set of characteristics. It is also a fact that the above populations will exhibit diverse characteristics, making it difficult to define the population accurately. For instance, the case of all farmer groups in Africa. The farmers' groups will be diverse in terms of composition and activities and many other factors. This will hinder the generalization of the results so obtained.

To address this challenge, researchers draw samples from the accessible population, a manageable subset of the target population. Researchers are able to generalize the results to an accessible population. Examples of accessible populations:

Consider a study with a research objective "to assess the effects of the factors influencing the adoption of hybrid maize among smallholders in Country A."

Target population: All farmers who have adopted growing hybrid maize in country A.
Accessible population: All farmers growing hybrid maize in one of the districts.
Sample: 10% of the farmers growing hybrid maize in the selected district.

While this approach allows for practical data collection and analysis, it also introduces the risk of *losing generalizability*. Because the accessible population may not fully represent the characteristics of the broader target population, there is a potential for bias in the study findings. To address this concern, researchers must strive to demonstrate that the accessible population is similar to the target population in terms of relevant characteristics. This may involve conducting pilot studies, performing statistical tests, or providing detailed descriptions of the accessible population to justify its suitability as a representative sample. This is known as establishing *"population validity.'* By ensuring similarity between the accessible population and the target population, researchers can enhance confidence in the generalizability of their results.

How Do You Identify the Accessible Population?
Identifying the accessible population requires a clear rationale based on theory, previous research, or professional experience. For instance, suppose a researcher aims to understand how socio-economic and cultural factors affect the adoption of integrated pest management (IPM) practices for managing potato blight in "Shangi potato variety" in smallholder farming communities. The researcher may target all potato farmers in the country as the target population. However, for practical reasons, the target population could be defined as all farmers growing potatoes in specific regions. Using theories from sociology, economics, and anthropology, along with existing research and personal expertise, the researcher can justify their choice of accessible population.

Box 16.2 Sampling from the target population to the household

Definition	Type of population/sample
What do you want to generalize to? All farmers growing "shangi" variety in Kenya	Target population (we do not know them individually)
What population can you access to? All sub-counties in Nyandarua County growing shangi potato variety	Accessible population; also known as sampling unit or study population
How can you access them? List of all farmers growing shangi potato variety in three sub-counties of Nyandrua	Get a sampling frame of 8000 farmers
What is in your study? 800 farmers growing shangi potato variety in three sub-counties of Nyandarua	A sample of 800 farmers picked from the sampling frame

We demonstrate this in Box 16.2, which represents the connection between the population and the sample. We use an example of a target population comprising all farmers growing the "Shangi" potato variety in Kenya. We want a sample of 800 farmers. All the farmers growing Shangi variety in Kenya are not known individually, and it will be impossible to reach all of them. Kenya is subdivided into 47 counties, and we picked Nyandarua, one of the counties, as our study area and assumed that there are three sub-counties in Nyandarua County. The farmers in these areas would constitute the accessible population. We use all registered farmers in the three districts (sub-regions) who grow the Shangi potato variety. From this list, we pick a sample of 800 farmers growing the Shangi variety.

16.2 Sampling Process

16.2.1 Designing an Effective Sampling Strategy

As stated above, to select a representative sample, a researcher needs a sampling frame; a list of cases from the accessible population. Ensuring the accuracy of the sampling frame is critical. However, creating an accurate sampling frame in agriculture can be challenging and time-consuming. You can get sampling frame sources like agricultural databases, farmer registries, and government records. By carefully compiling and validating this data, researchers can create a reliable sampling frame that accurately represents the accessible population.

Questions a researcher should ask themselves to achieve a successful sampling strategy:

- What are the research objectives/questions?
- What is the target population?
- Who should be included in the sample?
- Who should be excluded from the sample?
- What budget is available for sampling?
- What is the desired sample size?
- What sampling techniques will be used?
- What should be used as the sampling frame?
- How will the potential participants be recruited?
- What collection methods and techniques will be employed?
- How many researchers/enumerators will work on the project?
- How long will each interview or data collection take?
- How will data be analyzed?

The steps involved in sampling include the following:

1. Defining the target population.
2. Establishing the population parameters of interest by determining the summary descriptors to be used such as frequencies, mean, variance of study variables in the population.
3. Identifying and assessing the sampling frame, which is a list of cases from the target population.
4. Define the number of cases needed; choose the size of the sample.
5. Determine the appropriate sampling methods (the type to be used, probability or non-probability).
6. Establishing standard procedures for sample selection and recruitment protocols.

16.2.2 Types of Sampling Methods

There are two main methods of sampling: Probability sampling and non-probability sampling.

Probability sampling uses some form of random selection. Procedures used ensure that the different units (subjects) in the population have equal probabilities of being chosen. For instance, if the population has 2000 farmers, each farmer has one chance out of 2000 of being selected. The advantage of probability sampling is that sampling error can be calculated (Fowler 2009). The other advantage is that we can fulfil objectives that involve statistically estimating the target population's characteristics. Consequently, probability sampling is commonly linked with survey and experimental research methodologies. A brief section on sampling in experiments is included at the end of the chapter.

In non-probability samples, the probability of each case being chosen from the target population remains unknown. Consequently, addressing objectives necessitating statistical inferences about the population's characteristics becomes impossible. While generalizations about the target population can still be made from non-probability samples, they are limited and not based on statistical grounds. Nevertheless, both probability and non-probability samples enable researchers to address various research objectives as demonstrated.

In non-probability sampling, the researcher decides whether to include or exclude elements in a sample. Not every element of the population has a chance of being selected. This method of sampling is most useful in qualitative research.

Box 16.3 Key differences between probability and non-probability sampling methods

Probability Sampling:

- Mathematical approach.
- Random selection, using probability theory.
- Units have equal probabilities of being chosen.
- Reduces selection bias.
- Easy to generalize.
- Advantage—sampling error can be calculated.
- Used in surveys and experiments.

Non-probability Sampling:

- Does not involve *random* selection.
- May or may not represent the population well.
- Sampling error remains unknown.
- Used when the researcher lacks a sampling frame for the population.
- Studying a few subjects.
- Exploratory survey.
- Judgemental.

16.2.3 Probability Sampling Methods

Probability sampling methods include simple random sampling, systematic sampling, stratified sampling, cluster sampling, and multiphase sampling.

16.2.3.1 Simple Random Sampling

Simple random sampling is a procedure where all the members of a given population have an equal chance (probability) of being included in the sample, and every possible sample has an equal chance of being drawn. The sample selection is completely out of the researcher's control; instead, a random or chance procedure selects the sample. It is the best way yet devised when the sample is large (Fraenkel et al. 2012). Random sampling is favored because it is required for many statistical analyses. If samples are not randomly selected, then a major assumption of many statistical analyses (randomness and independence) is violated, and inferences made from the research are questionable.

Whenever we want to assign members to teams, in class, workshops, and farmer meetings, we ask members to count 1, 2, 3, … 0.1, 2, 3, and so on, and finally ask them to form three groups consisting of 1 s, 2 s, and 3 s. With this method, selection is not completely random because the team assignment is determined by that person's place in the line. Random selection ensures that the individual's position does not affect the team they end up on, promoting fairness and equal opportunity for all participants. It encourages inclusivity and minimizes the potential for biases or favoritism in team formation.

Random sampling, in most cases, involves using a table of random numbers or drawing slips of paper appropriately numbered. You can also use a random number calculator or table of random numbers to identify the appropriate elements to be selected for the sample. Tables of random numbers are usually included in statistics books and some books on research methods like Fraenkel et al. (2012). In addition, software like MS Excel has a function for generating random numbers. The most important thing is to ensure that no element is selected more than once.

Here we demonstrate the use of random numbers. We use an example of a *researcher who wants to select 100 farmers growing tissue culture banana varieties in County A. Supposing there are 4000 farmers growing tissue culture banana varieties in this county:*

- *The researcher compiles the list of all 4000 farmers and each farmer is assigned a number from 1 to 4000. This is the list of the population frame for the study.*
- *The researcher decides on the desired sample size,* e.g., *400,* i.e., *10% of 4000.*
- *A table of random numbers is entered, such as 23,459, 56,243, 91,624, 20,931, 87,451, 64,502, 10,674, and 83,740,* etc.
- *The population has 4000 members, so we are concerned only with the last four digits. We pick four digits because the final number 4000 consists of four digits, and we must use the same number of digits for each farmer.*
- *The farmers assigned numbers 3459, 1624, 0931, 0674, and 3740 are included in the sample.*
- *As for numbers 56,243, 87,451, they are not used because no farmer is assigned 6243 or 7451.*
- *When the researcher finishes with one column, they should move on to the top of the next column on the right, if a*

number is already selected and then it appears it should be ignored.

- *The procedure is continued until 400 farmers are selected.*

Random sampling has a great advantage, especially when the sample size is large. It is likely to produce a representative population sample. We can say it is the gold standard for achieving representativeness. It allows researchers to make more confident inferences about the population based on the sample data. Conversely, one of the biggest disadvantages of random sampling is the logistical and resource-intensive nature of implementation. If dealing with a large or geographically dispersed population, like the ones listed in Sect. 16.4, imagine that each and every member of the population must be identified by name. This means the researcher must be able to access the comprehensive lists or databases of the population from which to draw the sample. In some cases, obtaining accurate or up-to-date information about the population can be difficult. This can lead to potential errors or biases in the sampling process.

Another disadvantage is that random sampling may not always effectively capture certain subgroups or rare characteristics within the population. Because the selection of individuals or items is entirely random, certain groups or outliers may be underrepresented or entirely excluded from the sample.

Furthermore, random sampling may not be feasible or ethical in certain situations, such as when studying sensitive or hard-to-reach populations, such as indigenous farming communities with a language barrier or women farmers who face gender-based barriers. In such cases, alternative sampling methods, such as stratified sampling or convenience sampling, may be more appropriate but could introduce their own biases or limitations. Each of these methods comes with a set of advantages and limitations. Overall, sampling is a give-and-take exercise, and researchers must weigh tradeoffs between different sampling methods. The goal is to select the most suitable method that balances practicability, ethics, and validity of data.

16.2.3.2 Systematic Random Sampling

In systematic sampling, every kth case from the population frame is chosen for the sample (Barnett 1991). To ensure a truly random sample using this method, the list of all the members in the sampling frame must be randomized, as alphabetically or numerically does not provide randomness. Once the population is randomized, the researcher then decides on the sampling interval, which is the distance between the selected cases. The interval is typically calculated by dividing the total population by the sample size. For instance, if the desired sample size is 300 and the total population is 3000, $k = 10$. If k is 10, selection involves taking every tenth name. Unlike other sampling methods, systematic sampling does not give each member an independent chance of selection; once the initial case is chosen, the rest of the sample is automatically determined. This can lead to biased results if there is any underlying pattern. This is one of the limitations of systematic sampling.

To illustrate this, consider a farmer cooperative that has a register that lists all the farmers alphabetically or by the person who joined the cooperative. We could consider this to be randomly ordered, but the arrangement of the list has a pattern called "periodicity" that may cause bias. For instance, the person who recently joined the cooperative might be younger. Researchers should, therefore, carefully examine the list presented to them to ensure there is no cyclic pattern. Periodic reassessment of the sampling approach and adjustments based on the characteristics of the population can also help enhance the representativeness of the sample and minimize potential biases.

Steps in systematic sampling:

1. Identifying and defining the population.
2. Deciding on the desired sample size.
3. Acquiring a list of the population.
4. Calculating K by dividing the size of the population by the desired sample size.
5. Selecting a random starting point within the population list.
6. From that point, choose every Kth name on the list until the desired sample size is reached.
7. If the end of the list is reached before completing the sample, return to the beginning of the list and continue.

16.2.3.3 Stratified Random Sampling

In stratified sampling, sample is selected from a population by first dividing it into subgroups or strata based on certain characteristics (DeYoreo 2018; Frey 2018). This technique aims to ensure that the sample represents the desired proportion of each subgroup. For instance, consider a case where the researcher wants to assess the use of fertilizer in sorghum production in county A. Characteristics such as the social class categories (e.g., rich, poor), farmer's age (old, young), gender, and geographical location (poor and good access to markets), can be used to stratify the farmers. Then, a given number of cases are randomly selected from each subgroup. This approach allows researchers to obtain a sample that reflects the diversity of the population. This enhances more precise estimations and comparisons within each subgroup.

Researchers can use "proportional stratified sampling" to ensure that each stratum in the population is represented in the sample in proportion to its size or importance within the population (DeYoreo 2018). This allows researchers to obtain a sample that accurately reflects the distribution of characteristics present in the population. In proportional stratified sampling, no group is overrepresented or underrepresented.

To carry out stratified sampling, you must first decide on the criteria for stratification. The next step is to decide on the size of each stratum in the sample. Then, using random numbers, select an appropriate number of subjects.

Example *Suppose we want to get a sample of 300 farmers growing sorghum from three agro-ecological zones A, B, and C*

- *Population of size of N = 8000*
- *Distribution of the population is AEZ A = 4000, AEZ B = 2400, and AEZ C = 1600.*
- *Adopting proportional allocation, the sample sizes from each stratum are obtained as follows:*
  - *For AEZ A (strata A) with N1 = 4000: P1 = (4000/8000) = 0.5 and n1 = n. P1 = 300 (0.5) =* ***150.***
  - *For AEZ B (strata B) with N2 = 2400: P2 = (2400/8000) = 0.3 and n2 = n. P2 = 300 (0.3) =* ***90.***
  - *For AEZ C (strata C) with N3 = 1600: P3 = (1600/8000) = 0.2 and n3 = n. P3 = 300 (0.2) =* ***60.***

Stratified sampling offers flexibility beyond proportional distributions, allowing for "non-proportional samples." This is where, the size of the sample in each subgroup reflects the relative variability of the characteristic under study. The final samples are not proportional. Imagine in a study on adoption involving farmers, and you have 5000 adopters and 3000 non-adopters of a certain technology. In a proportional sample of 600 members, we would have 375 adopters and 225 adopters, representing 0.625% and 0.375%, respectively. If experience suggests that for the variable under study, there is more variability among non-adopters, we will need to select more than 225 non-adopters, let us say 390. This approach acknowledges that a more diverse population requires a larger sample to capture its variation, while a homogenous population requires a smaller one for the same purpose.

Non-proportional sampling is commonly used in impact evaluation studies to ensure that certain subgroups are adequately represented (Kalton 1983; Burmeister and Aitken 2012). Suppose a study aims to assess the effectiveness of a program in different socio-economic areas, a researcher might prioritize selecting more participants from areas with lower incomes. This is to ensure that the lower-income participants are not disadvantaged. In a study by Damba et al. (2019), the researchers sampled 75% of beneficiaries and 25% of non-beneficiaries to account for spillovers. Similarly, in a study evaluating the effects of plant clinics on users and non-users, non-proportional sampling was applied to over-represent non-users, thereby enabling more precise matching of the two groups across key socio-economic variables (Bett et al. 2018).

16.2.3.4 Cluster Sampling

In the three methods discussed so far, simple random, systematic, and stratified, the population has a list of its members. Sometimes, we have populations that lack a list of members either because the population is very large or scattered over a large geographical area. This presents a challenge. In such a situation, a researcher can identify "clusters" of individuals and then pick these clusters for the study (Stringer et al. 2011; Moser and Korstjens 2018). The researcher will then include all members in these "intact groups" in the sample. In this case, each member becomes a unit of observation. For instance, if the study aims to investigate farmers across different agricultural sectors but a comprehensive list of registered farmers is unavailable, cluster sampling can be an appropriate approach. In this case, existing farmer groups, such as dairy farmer associations, tree nursery groups, farmer field schools, and cooperatives, can serve as the sampling clusters. Each group within a sector represents a distinct cluster; for example, all dairy farmer groups would constitute clusters within the dairy sector. Suppose there are 200 such farmer groups, which together form the sampling frame of 200 clusters. A random sample of clusters, for example, 25 groups, can then be selected. All individual members within the selected 25 clusters would subsequently be included as observational units in the study as shown in Box 16.4.

Interviewing all the farmers within each of the 25 clusters is more convenient compared to reaching all the 200 farmer groups. It also saves time and expenses. You do not have to travel to all the farmer groups. If these farmers in the 25 clusters represent the remaining farmers, you can confidently draw your conclusion about the aspects of your study for the entire population of dairy farmers. However, it is possible that the 25 clusters are not representative. It is possible that the farmers differ in some ways from farmers in the other clusters. Unrepresentativeness is a limitation associated with cluster sampling. We can address this limitation by maximizing the number of clusters that you select. The more clusters are selected, the more likely the findings will be applicable to the population. But, on the other hand, if you select many clusters, the advantage between saving on time and costs may be compromised. Thus, it is a trade-off between saving time and costs versus losing precision by using fewer clusters. As a researcher, you need to weigh this vis-à-vis gains from using a large sample from a few intact groups and a small sample distributed over all the groups. However, you must always justify the choice of your sampling procedures.

Box 16.4 Example of cluster sampling using dairy farmers groups

Box 16.4: Example of cluster sampling using dairy farmers groups

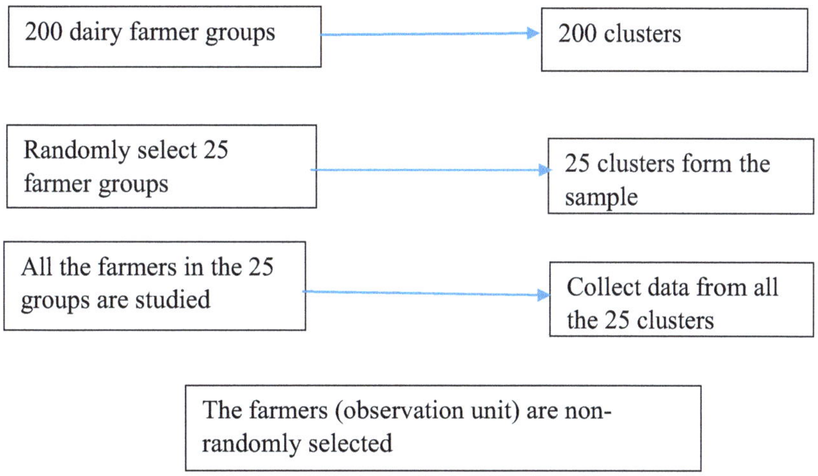

Steps in cluster sampling:

1. Identifying and defining the population.
2. Determining the desired sample size.
3. Identifying and defining a logical cluster (e.g., agro-ecological zone, region).
4. List all clusters; obtain a list that makes the population of clusters (e.g., 60 dairy farmer groups).
5. Estimate the average number of members per cluster (In Africa, size of farmer groups range from 10 to 30 members).
6. Decide on the number of clusters needed by dividing the sample size by the estimated size of the cluster.
7. Randomly select the needed number of clusters using a table of random numbers. Include in your study all population members in each selected cluster. The more samples there are, the better in terms of the sample's representativeness to the population.

Cluster sampling is a commonly used method in research, but it must be done carefully to avoid errors (Luborsky and Rubinstein 1995; Creswell 2014; Austin and Sutton 2014). Always remember that it is the cluster that is randomly selected, not the individuals within it. Noting this distinction is critical for maintaining the sampling process's integrity. You should ensure you strictly adhere to the principles of cluster sampling.

It is possible to carry out cluster sampling in stages, involving the selection of clusters within clusters. This process is called multi-stage sampling, and it is discussed in the next section.

16.2.3.5 Multistage Sampling

Multistage sampling, also called hierarchical sampling, may be used in large-scale investigations (T. Henry 1990; Burmeister and Aitken 2012). It involves sampling within several layers or stages of the population. The aim is to gather a representative sample. In the approach, the population is initially divided into large-sized sampling units at the first stage. From these units, a certain number are randomly selected to proceed to the next stage.

The list is prepared on the selected units from the previous stage, constituting the sampling for the next stage. Finally, sub-samples are drawn by employing random sampling methods from this sampling frame.

This hierarchical process allows researchers to narrow down the population at every stage systematically. This ultimately leads to selecting specific individuals or elements for inclusion in the study. A study to evaluate the effectiveness of technology dissemination approaches used four-stage multistage sampling and narrowed the population from the district to communities to households (Patton 2015; Mugenda and Mugenda 2019).

To illustrate this, let us consider a hypothetical 5-year project implemented in Africa and working in ten countries. The ten countries are (1) Kenya, (2). Tanzania, (3) Uganda, (4) Rwanda, (5) Ethiopia, (6) Zambia, (7) Zimbabwe, (8) Democratic Republic of Congo, 9) Ghana and (10), Nigeria. The project aims to promote legumes to increase nitrogen among smallholder farms. At the end of 5 years, the researchers want to evaluate the effect of this project on food security and income. We design a multistage sampling plan using the following steps (Box 16.5):

Box 16.5 Example of multistage sampling from 10 countries implementing a legume project

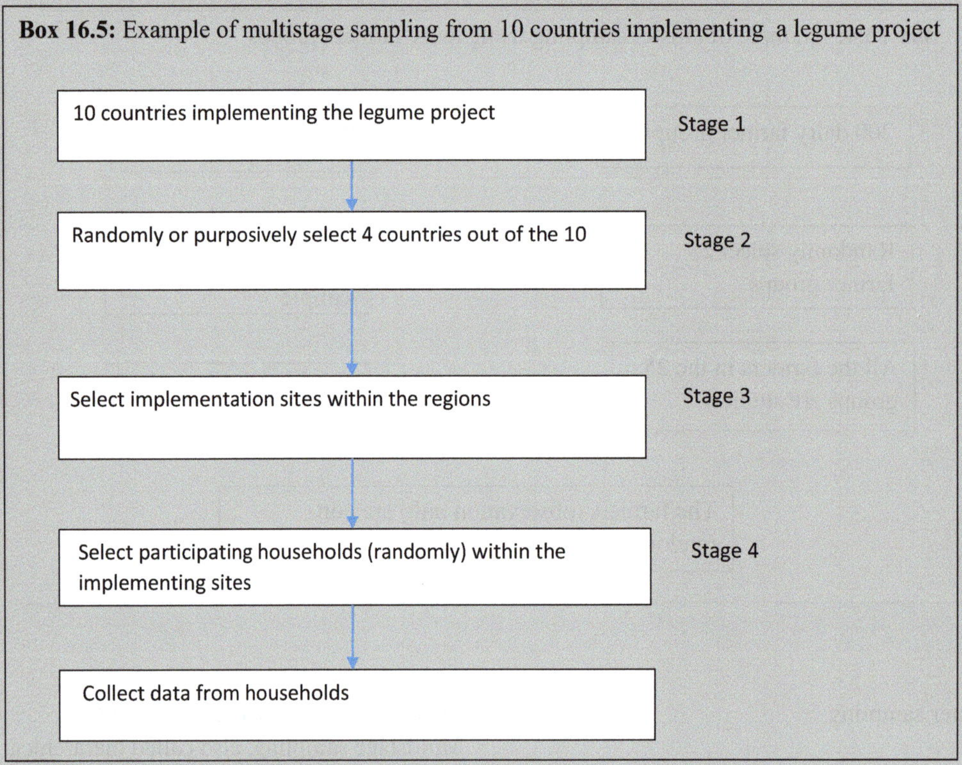

Stage 1: *Selection of countries*: We can randomly or purposively select a subset of countries from the list of ten countries implementing the project. This ensures representation across the region. Let us say we choose four countries for this stage: Kenya, Zambia, Uganda, and Nigeria.

Stage 2: *Selection of regions within countries*: Identify the regions or provinces where the interventions were implemented within each selected country. Randomly select regions from each country to ensure representation. For instance, if Kenya has eight regions where the project was implemented, you might randomly select 2 or 3 regions from Kenya. Similarly, repeat this process for the other selected countries.

Stage 3: *Selection of implementation sites within regions*: Within each selected region, identify specific implementation sites such as districts, or communities where interventions took place. Randomly select implementation sites within each region. Ensure that the number of implementation sites selected is proportional to the size and importance of the intervention in each region.

Stage 4: *Selection of participants within implementation sites*: Finally, within each selected implementation site, participants for the study will be randomly selected. The number of participants selected from each site should be based on statistical considerations such as sample size requirements and representativeness.

One potential drawback of multistage sampling is its complexity, which can pose a challenge for achieving a truly representative sample. The several stages of selection increase the likelihood of introducing bias and sampling errors. Furthermore, the design may not accurately capture the population's characteristics, resulting in skewed estimates. Successful implementation of multistage sampling requires careful planning and allocation of resources, which can make it a time-intensive process. Additionally, interpreting results from this approach may require advanced statistical techniques due to its intricate design.

16.2.4 Types of Non-probability Sampling

Non-probability sampling methods are often chosen for their convenience, cost-effectiveness, or suitability for specific research objectives (Fink 2003; Ahmed 2024; Eniego et al. 2025). Here we describe four non-probability sampling methods:

1. Purposive sampling method.
2. Snowballing sampling method.
3. Convenience sampling method.
4. Quota sampling method.

16.2.4.1 Purposive Sampling

Purposive sampling, also known as judgmental sampling, is a strategic approach where the researcher uses their judgment or discretion to select participants who are most likely to provide the necessary information for the study objectives. This approach, often referred to as "cherry-picking," involves careful selection of subjects based on specific criteria. In qualitative studies, information rich cases are selected for indepth analysis. For instance, if the researcher's aim is to study farmers who received training in silage-making in 2020, they would choose participants who attended relevant workshops or training programs during that period. By intentionally selecting farmers with this experience, the researcher can gather targeted insights into silage-making practices and their impact on agricultural productivity and sustainability.

Other examples of purposive sampling include the following:

- Suppose your interest is to sample farmers for a survey on climate change mitigation. You will be judgmental in the selection of farmers. You may consider purposefully targeting farmers who are practicing conservation agriculture or diversified cropping systems. This way, you will be able to assess the adaptability and effectiveness of climate adaptation measures
- Suppose a new pest of Napier grass has emerged and aim to conduct focus group interviews to gain insights into farmers' management of the new pest. You will most likely select farmers who have their Napier grass attacked by the new pest. You will even indicate in your proposal that you will be selecting study participants who have experience with the new pest
- In crop variety experiments, the researchers may select varieties based on specific traits such as yield potential, disease resistance, or drought tolerance based on field observations.

16.2.4.2 Snowball Sampling

In snowballing, the researcher identifies a few initial subjects with the desired characteristics using purposive sampling (Sedgwick 2013). Once identified the individuals are asked to name others that they know have the required characteristics until the researcher gets the number of cases they want. It is also called network sampling because individuals keep recommending new ones. This sampling method is particularly useful for hard-to-reach subgroups or populations where the characteristics of interest may not be well-known or easily identifiable. It is also useful where the topic is sensitive.

Imagine you are conducting a follow-up survey on farmers who attended a training workshop within a specific time frame. You aim to assess the long-term impact of a soil conservation workshop that was conducted 2 years ago. If you plan to use snowball sampling to select your participants, you will start by identifying a group of farmers who participated in the workshop through available records or by contacting the organizing institution. You will then approach these trained farmers and explain the objective of the follow-up study. You will ask them to recommend other attendees from the same workshop or share contact information of fellow participants who they believe would be willing to participate in the follow-up study.

Snowballing is also useful in sampling individuals who may be considered hard-to-reach groups, including pastoralists who move seasonally and are difficult to engage consistently. Others include ethnic farming minorities who may be marginalized or discriminated against. Can be suspicious of outsiders and wary of participating in research.

Snowballing is particularly useful in qualitative research as it leverages participants' social networks to gain access to these populations (Cohen et al. 2017). It will use contacts and their acquaintances to identify further potential participants. In some cases, the respondents not only provide names of additional contacts but also actively recruit them to participate in the study. This approach is beneficial when dealing with uncooperative individuals, as the recruiters can play a key role in encouraging their peers to become involved. By using trusted community members to facilitate introductions and vouch for the research, snowballing helps overcome barriers of suspicion and reluctance, making it an effective strategy for engaging hard-to-reach groups in agricultural research.

Consider a case where a researcher uses the snowballing method to pick responses for a comparative study between trained farmers and non-trained farmers. The researcher wants to pick 100 trained farmers in a specific ward and 100 un-trained farmers in different wards (Fig. 16.2).

However, snowballing can be susceptible to biases resulting from the influence of the initial contact, as well as the issue of relying solely on volunteer samples. This method may result in a sample not fully representative of the broader population, as the initial participants and their social networks might share similar characteristics or viewpoints. If the population is variable, part of the population will not be captured. Consequently, the findings could be skewed, limiting the generalizability of the research. To mitigate these biases, researchers should be aware of these limitations and consider combining snowballing with other sampling methods to enhance the diversity and representativeness of the sample.

16.2.4.3 Convenience Sampling

In convenience sampling, subjects are chosen because they are convenient or readily accessible to a researcher (Fadele and Rocha 2025). Suppose you want to conduct interviews

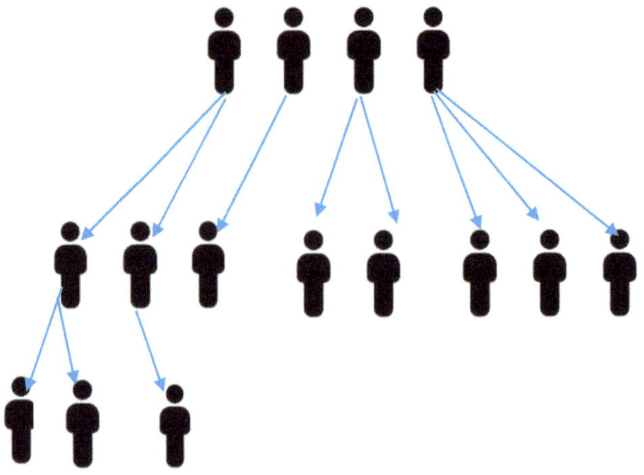

Fig. 16.2 Snowball sampling method

with horticulture farmers. During this time, you realize there will be an agricultural show or fair in the region. Hence, you attend the fair, set up a booth at the fairgrounds, and invite farmers to participate in a survey or interview about their farming practices. You have selected these farmers out of convenience. You do not have a pre-determined list of farmers to sample from; rather, you opportunistically approach individuals as they encounter them at the fair. The farmers willing to participate are conveniently sampled, leading to an accidental sample of fair attendees.

Such a procedure is easy and cheap, but it will likely introduce serious bias. For example, farmers who attend agricultural fairs may be more actively engaged in their farming community or have specific interests or concerns than those who do not attend. For example, younger farmers are likely to attend more than older farmers. We run the danger of missing the older elements in the study. As a result, the findings from this accidental sample may not represent all farmers in the region. Nonetheless, this method allows researchers to gather valuable insights into the perspectives and practices of farmers participating in agricultural events. This can be complemented with another method that makes targeted reach to farmers. But most importantly, researchers should interpret findings of convenience sampling with caution.

16.2.4.4 Quota Sampling

In quota sampling, a researcher selects a sample based on predefined "quotas" or characteristics (Fadele and Rocha 2025). To implement quota sampling, the population is first divided into mutually exclusive segments or sub-groups based on certain criteria, such as location and demographic characteristics. Then, interviewers are assigned quotas that dictate how many participants they should select from each sub-group. For example, an interviewer may be instructed to select 40 farmers from location A and 30 from location B, without further guidelines on making their selections.

While quota sampling allows researchers to control the sample's composition and ensure the representation of key characteristics, it may introduce bias if the selection process is not carefully implemented. Additionally, quota sampling does not guarantee that the sample will be representative of the population, as participants are not randomly selected. Therefore, findings derived from quota samples should be interpreted with caution, and the limitations of this sampling method should be considered when generalizing results to the broader population.

16.2.5 Sampling in Qualitative Research

16.2.5.1 How do You Select a Sample in Qualitative Research?

Compared to quantitative research, the sampling procedures in qualitative research are less rigidly defined and lack concrete rules (Patton 2015). The sampling procedure is dynamic, adhoc, and phasic rather than static. The aim is to pick a sample that develops a detailed understanding. The judgement and experience of the researcher are key elements in qualitative sampling (Onwuegbuzie and Leech 2007; Mey 2023). Key questions to be answered in qualitative sampling include: how large is the sample? How will I select the sample? Will the sample be adequate to provide the required information? The key determinant of sampling is the research question. The sampling should be fit to answer the research question.

Determining the appropriate sample size in qualitative research varies widely depending on the research approach. A case might be one animal (case study report), while in grounded theory, the researcher will continue adding cases until saturation is reached. For phenomenological studies, sample sizes ranging from 6 to 10 are recommended (Patton 2015). Creswell (2014) recommends 20–30 interviews and focus group discussion can have small sample sizes of 4–8

(Hennink and Kaiser 2022). Qualitative samples can range from 1 to 40 or more (Carlsen and Glenton 2011; Marshall et al. 2013).

Authors have argued that even a small sample size is adequate because the aim is not to generalize findings to a larger population but rather to present unique cases with their intrinsic value (Whitley and Ball 2002; Das et al. 2016). Carlsen and Glenton (2011) support this and state that large sample sizes waste resources and overburden the participants. On the other hand, Onwuegbuzie and Leech (2007) contend that the sample should be large enough to generate a thick description, which refers to a rich, detailed account of the study context and participants' experiences. They further argue that while qualitative research primarily focuses on depth rather than breadth, having a sufficiently large sample can enhance the robustness and credibility of the findings, making it possible to draw more meaningful and transferable conclusions.

So, should we go for a small or large sample? The key is to aim for "Suitability of purposes." It should be related to the purpose, the research problem, the major data collection strategy, and the availability of information-rich cases. Have a large enough sample to generate a "thick description," which is the goal of qualitative research, but not too large to cause data overload and not too small to prevent theoretical saturation. Balancing the sample size ensures a comprehensive understanding of the phenomenon without excess resource expenditure.

Many authors fail to agree on the exact sample sizes and mostly rely on the concept of saturation to justify their samples. However, Patton (2015) argues that this concept is filled with ambiguity, as it heavily leans on purposive strategies and justification rather than methodological rules. This makes this author assert that *"qualitative inquiry seems to work best for those researchers who can tolerate ambiguity."* To avoid this ambiguity, sample size determination should be based on practical reasoning. Researchers should start with a minimum sample size and progressively build a rational and criteria as the study progresses. This can be done through citing recommendations from previous studies, using precedent by citing sample size used in studies with similar circumstances and statistical demonstration of saturation with a data set (Marshall et al. 2013).

We provide the following seven guidelines for determining and picking a sample of a qualitative study (Francis et al. 2010; Marshall et al. 2013; Lantz 2013; Das et al. 2016; Sim et al. 2018; Piepho et al. 2022):

1. *Answer the research question:* The sampling strategy should align with the study's research questions. It should be designed to gather rich, detailed information about the studied phenomena.
2. *Analytical generalizability:* The sample should also support the generalizability of the findings. In qualitative research, this means focusing on analytic generalizability rather than statistical generalizability. This involves making meaningful inferences based on the depth and richness of the data rather than on the ability to generalize to a larger population.
3. *Data saturation:* Consider the point in data collection when no new insights from the data are identified (achieve data redundancy). It means the data starts to bring repetitive themes and information that is redundant.
4. *Trustworthiness:* Additionally, the sample should help produce believable and convincing descriptions and explanations that reflect real life. One measure of validity in qualitative research is whether it offers a compelling and accurate account of what is observed. This also involves considering the reliability of the information sources, ensuring they are complete and not biased in ways that could distort the findings.
5. *Ethical considerations:* The sampling strategy must also be ethical. This includes obtaining informed consent when required and considering the subjects' benefits and risks of participation. The ethical relationship between the researcher and the participants is important, as is considering the ethical implications for those who are excluded from the study.
6. *Feasibility:* The sampling strategy needs to be feasible. This involves considering practical financial costs, time constraints, and accessibility. The sampling strategy should be manageable and compatible with the researcher's work style and resources.
7. *Justification:* ensure you justify your sample size and how you will conduct sampling at the proposal stage. In reporting findings, write a full justification and thoroughly report on sample size to allow readers to evaluate your work. This also helps your readers, either examiners or journal paper peer reviewers.

Overall, while selecting a sample for a qualitative study, it is crucial to consider the above guidelines. Use your experience, judgment, and the purpose of the study to select a suitable sample. Always remember that the most important goal is to collect enough quality data to answer your research question and address the research problem at hand. It is important to note that the qualitative data collection and analysis process is iterative, meaning that the sampling strategy may need to be adjusted continually to fit the study's purpose.

16.2.5.2 Purposeful Sampling

When sampling purposefully, it means you will select respondents, groups, or sites that will provide the required information with respect to the purpose of the study. The key consideration is that an information-rich sample is selected to provide an in-depth view of the phenomenon (Benoot et al. 2016; Ames et al. 2019). Cases or subjects are therefore handpicked because they are informative or possess the required characteristics.

Maximal Variation Sampling: The researcher selects cases or subjects that differ on some identified characteristic or trait, e.g., different farming systems or geographical zones. In other words, sampling aims to have samples that exhibit maximal differences in the variable of interest. The procedure requires a researcher to identify the characteristic and then find individuals or sites that display the different dimensions of that characteristic. For example, a researcher conducting a marketing study might identify marketing outlets of soybeans used by farmers as a characteristic of interest. In this case, the researcher then purposefully selects farmers that differ in their marketing outlets, such as supermarkets, wholesalers, and retailers.

Homogenous Sampling: The researcher consciously selects samples that exhibit similar characteristics to enable an in-depth understanding of the phenomenon. To use this procedure, the researcher must first identify the characteristics and find individuals or sites that possess them. For example, in a study on the collective marketing of avocados, a researcher might select the sample from members registered in a marketing cooperative (because they all sell collectively). This is the method mostly used to pick members of a focus group discussion because individuals are generally more comfortable sharing their thoughts and ideas with those with whom they share common interests.

Extreme or Deviant Case Sampling involves selecting a case that is illuminating and noticeable to its success or failure (Patton 2015). The cases are usually outliers, peculiar and unusual cases with extreme characteristics. They can be located by researchers through other individuals or organizations that have cited those characteristics. In a household study, you might pick an extreme of poor and rich households or an extreme case of animal diseases affecting livestock in a specified agroecological zone.

Typical Case Sampling: a Researcher Considers Attributes or Characteristics That Are Possessed by an Average, typical person or case. The sample is selected from these typical or conventional individuals. This is done to avoid rejecting information on the grounds that it has been gained from special or deviant cases. For example, in a farming system where the average farm size is 1.5 ha, a researcher may see farmers possessing land sizes between 1.2 and 1.7 as typical farmers in the sample and consider less than 0.8 ha and > 5 ha as extreme cases.

Critical Sampling: Sampling strategy here is to study a critical sample because it is an exceptional case and the researcher can learn much about the phenomenon. The case is important in a particular context. Data are collected after understanding the events of a critical case.

Intensity Sampling: Allows the researcher to select a small number of rich cases that provide in-depth information and knowledge of the phenomenon. The cases or respondents selected must typically possess the characteristics of interest. The researcher avoids extreme cases which tend to skew the results. The researcher's aim is to obtain rich but not extreme information from typical cases. To be able to identify intense but typical cases, a researcher needs prior information and exploratory work (Patton 2015). There is often a thin line between typical and intense sampling, but intense sampling of samples exhibits information-rich cases that exhibit intense but not extreme inputs (Palinkas et al. 2015).

Snowball or Chain Sampling is the most commonly used sampling technique to locate information-rich key informants (Palinkas et al. 2015). It starts with the researcher identifying one case or subject who displays the attributes or traits of interest. After obtaining the information from the subject, the researcher requests the subject to assist by recommending another subject who would provide similar information.

Confirming and Disconfirming Case Sampling: During the initial qualitative exploratory fieldwork, a researcher gathers data and allows patterns to emerge. After some time, a confirmatory phase of the fieldwork takes over the exploratory process. The confirmatory phase tests the patterns and ideas that have evolved and checks the reliability of findings using new information.

Convenience Purposive Sampling: The researcher collects data conveniently, doing it fast and being motivated by the convenience in selecting the samples. It is commonly used but convenience sampling is neither purposive nor strategic and should not be the first factor to be considered (Patton 2015).

Stratified Purposeful Sampling: This involves selecting participants based on a specific characteristic or attribute and then subdividing the sample into subgroups (strata). This type of sampling ensures a diverse sample is selected.

Opportunistic Sampling: This is also called on-the-spot sampling, or decision sampling. The researcher makes an ad-hoc decision to select a sample to enable data to be collected from new opportunities that arise during the process of data collection. These decisions are not made initially at the start of the study, but evolve in the course of implementing the study.

Theoretical Sampling: The researcher uses the collected information to decide what additional data are required to fill gaps (Maggino 2023). As the qualitative researcher collects and analyses data, they use the identified themes and patterns to make decisions about who to interview and what to collect, based on this gap. Conversely, if there are no gaps, i.e., the same themes

Table 16.1 Sampling methods in qualitative research

Sampling method		Description
Purposive	Maximal	Cases that provide a wide range of perspectives on the phenomenon of interest
	Homogenous	Sample that shares similar characteristics to focus on a specific subgroup
	Extreme or deviant	Unusual or typical cases to explore unique variations of the phenomenon.
	Intensity	Cases that strongly represent the phenomenon (information-rich) but are not extreme.
	Typical case	Cases that are average to illustrate the phenomenon
	Critical	Cases that are essential for understanding the phenomenon and have significant implications. Maximize application of information to other cases
	Confirming	Additional cases that verify or expand upon initial findings
	Stratified purposive sampling	Dividing the population into subgroups and sampling from each to ensure representation across key characteristics
	Opportunistic sampling	Taking advantage of unexpected opportunities during research to gather data from new or surprising sources
Theoretical		Participants based on emerging concepts and theories during the research process
Snowball		Initial participants' networks recruit additional participants.
Convenience		Participants who are easily accessible and available to the researcher. It is cheap but compromises information and credibility
Quota		Sample includes specific proportions of subgroups to reflect the population structure

continue being repeated consistently, the researcher concludes the data collection because they have reached a "Theoretical saturation."

Table 16.1 shows a summary of the main sampling methods in qualitative research.

16.2.6 Errors in Sampling

There are two types of errors: Sampling and non-sampling errors and they include:

1. Sampling errors: These errors arise because, by chance, the sample is not fully representative of the population. Can be estimated and are a random result of sampling procedures
2. Non-sampling errors: These are more serious and are due to mistakes made in the acquisition of data or due to the sample observations being selected improperly. Sampling errors are difficult to measure. They can arise from:

- Errors from other sources.
- Variation between data collection personnel.
- Inadequate tools.
- Inadequate sampling frame.
- Data entry errors.
- Coding errors.
- Non-response.
- Errors in response.
- How questions are worded.

Errors can be avoided by following the correct method of data collection as well as techniques of data collection.

16.3 Sample Size

Appropriate sample size is critical for credible research findings. Generally speaking, the larger the sample size, the better. A large sample has greater reliability and allows researchers to advance statistics. If the sample size is too small, the research outcome might not be reproducible. The study may also lack sufficient statistical power to detect meaningful effects or relationships within the data. This may compromise the study's external validity.

Determining the appropriate sample size is not straightforward (Mills and Gay 2020). It is a balancing act and involves considering various factors like the *level of precision* needed, *population variability, desired confidence level, and available resources*. For instance, having an extremely large sample will be a waste of resources. Factors such as expected effect size and potential for non-response also play a role.

While numerous authors acknowledge the absence of clear guidelines for calculating sample size (Mills and Gay 2020), we emphasize the importance of explicitly expressing the sample size in your research study. It is a must that the actual sample size is indicated in your thesis and journal papers. During examinations at the proposal and thesis stages, sample size is assessed. Examiners want to see the sample size explicitly presented. Moreover, the Institutional Review Board (IRB) scrutinizes the determined sample size for ethical and methodological appropriateness (Rothenberg 2004; Grady 2015; Resnik 2021). Additionally, journals emphasize the importance of clarity regarding how the sample size was calculated. They want to establish transparency and reproducibility in research methodology. Reviewers will consider the paper flawed when the sample size is not mentioned or not clearly justified.

Many students and researchers determine sample size using statistical formulas or online calculators. This is adequate, but we recommend you also consider the other factors given later. The calculators consider factors such as the desired confidence level, the margin of error, population variability, and expected effect size. Among the widely used is Cochran's formula, which offers adaptations for various sampling scenarios. You should carefully select and apply the formula most appropriate for your study (Ahmed 2024; Barlett et al. 2001), taking into account the specific factors in the next section.

16.3.1 Main Factors Determining Sample Size

1. *Research Objectives*

Research objectives influence the precision, scope and design of the study. If the research aims to make a generalization claim about a large population or determine small effects, a larger sample may be required to ensure statistical power and accuracy. Exploratory studies usually require small samples, while probability samples require larger samples than non-probability sampling methods.

2. *Heterogeneity of the Population*

Heterogeneity refers to the degree of dissimilarity among classes or how variable the population is with respect to the characteristic of interest. The more variable the characteristic of interest, the larger the sample size needed to achieve the required level of precision. The more heterogeneous, the more cases required.

3. *The Desired Precision*

Precision or margin of error—a possible difference between the sample estimate and the actual population value. How precise do we need to be in estimating that characteristic? The larger the sample, the smaller the standard error. The 2000–3000 samples have a minimal error, but the cost and energy involved are large. The limit is 30 cases for statistical analysis, but society recommends 100 cases (Comrey and Lee 2013; Kwak and Kim 2017).

4. *Confidence Level*

How certain do you want to be that the population figure is within the sample estimate and its associated precision? *Confidence level* is the likelihood that the results obtained from the sample lie within the associated precision. A 95% confidence implies that if the study was repeated 100 times, 95 out of the 100 samples be expected to represent the population well. The higher the confidence level, the higher the certainty. We normally use 95% confidence to provide forceful conclusions; however, if you are only seeking an indication of likely population value, a lower level, such as 90%, is acceptable, e.g., studies involving household interviews conducted on farms. Researchers often refer to standard statistical tables that calculate sample sizes based on population and margin of error or confidence interval, Saunders et al. (2019), Gill et al. (2010) and Barlett et al. (2001) provide precalculated sample size tables giving minimum sample size for specific scenarios. Such tables can reliably be used.

5. *Sampling Design*

Stratified random sampling typically requires fewer cases than simple random sampling. This targeted approach allows for more efficient use of resources by ensuring representation from each subgroup, thus requiring a smaller overall sample size to achieve the desired precision. On the other hand, cluster sampling, where the population is divided into clusters (groups), often requires a larger sample size for precision. This is because within-cluster variability can lead to increased uncertainty compared to simple random or stratified sampling methods. While cluster sampling may offer cost advantages, as data collection can be more logistically streamlined, its efficiency in capturing variation within the population is lower than other sampling methods.

6. *Available Resources*

Available resources determine the sample size's practicalities and time constraints. The resources include personnel, funding, and facilities, which help to assess the feasibility of data collection within practicalities and time constraints. Limited resources may impact the possibility of obtaining a large sample size. This becomes more critical when the amount of money and/or time is fixed for a project. The rule of thumb is to allocate on half to half basis for data collection and analysis, in case an inadequate sample size is used, this should be reported as a limitation of the study

7. *Level of Aggregation at Which Results Are to Be Presented*—Focuses on what level to present results and who is going to use them. Is it at national, regional, counties or districts or villages? This determines the number of breakdowns:

- The number of variable categories to be analyzed determines the sample size.

- The more breakdowns or strata in the planned analysis, the larger the sample.

Mills and Gay (2020) provide the general guidelines for determining sample size:

- As the population size increases, a smaller proportion of the population is needed to obtain a representative sample.
- For populations of 100 or fewer, it is more practical to survey the entire population rather than sampling.
- For populations around 500, sampling 50% of the population should suffice.
- With populations around 1500, sampling 20% of the population is sufficient.
- Once the population size exceeds a certain threshold, such as 5000, the exact population size becomes less critical, and a sample size of 400 is generally adequate.

16.4 Sampling in Biophysical Studies

All the sampling methods discussed in this paper are applicable to biophysical studies. However, to broaden students' minds in terms of the applicability of sampling methods, we present additional details on sampling in biophysical measurements. The measurements in the biophysical include various parameters in soil, crops, water, climate, and many other biophysical aspects. However, biophysical sampling must consider additional considerations specific to biophysical aspects. For example, soil variability across different landscapes and management practices should be considered when sampling soil properties. Similarly, variations in growth stages, crop varieties, and planting densities must be considered when sampling crops to ensure representative sampling.

Here is the application of the various sampling techniques in taking biophysical data:

- Random sampling can be applied at various stages, including selecting field plots, collecting plant or animal specimens, and sampling soil or water (see Table 16.1).
- Stratified sampling in soil: This method is used when the field exhibits heterogeneity. It involves dividing the field into homogeneous strata based on relevant variables (e.g., soil types, crop varieties) and then randomly sampling within each stratum.
- Systematic sampling may be used when the population exhibits a regular spatial or temporal pattern, such as sampling plants along transects or monitoring livestock at fixed intervals.
- Cluster sampling is practical and cost-effective, especially when the population is large or widely dispersed.

To illustrate this, we present some common measurements.

16.4.1 Sampling in Agronomy Experiments

In agronomy experiments conducted in field settings, estimates of crop yield or other metrics (Chap. 17) are typically derived from measurements taken for all plants within a designated experimental plot, rather than focusing on individual plant assessments. If you are taking grain yield in a plot of maize, you will harvest the net plot and get the total weight. In doing this, researchers traditionally exclude plants located in guard rows and instead harvest only those from the middle row or rows, depending on the number of sowing lines (Chap. 9). This practice eliminates the "margin effect," which refers to potential differences in yield or characteristics observed in plants positioned at the edges of experimental plots. Such exclusion enhances the reliability of yield estimates by ensuring that measurements are based on representative plants. These margin plants have been noted to exhibit yield discrepancies of up to 12% or even 15% compared to plants from inner rows, primarily reflected in their fresh weight (Gomez and Gomez 1984). While the specific causes of these differences are still under investigation, it is suggested that increased sunlight exposure and improved ventilation could be contributing factors. The issue may also be attributed to intra- and interspecific plant competition.

However, there are many other instances where measurements will require sampling. For example, in cases where *destructive sampling* is being done, researchers need to determine which plants will be destroyed and support with criteria for making that decision. Additionally, if you are taking height measurements of a certain crop or monitoring pest or disease infestation across time, you must *tag plants* for monitoring purposes. The plants to be tagged must be selected using sampling methods to avoid bias.

Sampling protocols guide researchers to aid in the collection of representative samples. These protocols outline systematic procedures for selecting sampling locations, determining sample size, collecting samples, and ensuring consistency and accuracy throughout the sampling process. By following established protocols, researchers can minimize bias, variability, and errors in sampling. Sampling protocols often include detailed instructions on equipment use, sample handling, documentation, and quality control measures to ensure the samples' integrity and the results' reproducibility.

16.4.1.1 Plant Samples

Collecting measurements for each plot may not always be feasible or practical for collecting data on growth parameters such as plant height and leaf area. Therefore, sampling within each plot is employed to gather data from specific sampling units, which are then subjected to statistical analysis. For instance, plant height measurements may be taken from only 10 out of 100 plants and leaf area measurements from just 5 out of approximately 50 within the plot.

In this case, the plot sampling technique is used. Individual experimental plots are considered populations from which estimates are derived based on measurements taken from a select number of plants within each plot. Heret we carefully select plants to represent the plot's characteristics to estimate the specific parameters. To effectively employ plot sampling, researchers must define the sampling unit, determine the sample size, and establish the sampling design. This ensures that the collected data accurately represents the plot's attributes and enables robust statistical analysis.

16.4.1.2 Soil Sampling

Soil sampling is crucial in soil science for providing recommendations on input use, soil surveys, and experiments. Researchers often use a combination of probability and non-probability sampling methods. The choice between these methods depends on various factors, such as the research objectives, available resources, and the desired level of precision. To illustrate this, we present a few examples:

(a) *Random soil sampling*: Researchers often use random sampling techniques, such as grid sampling, to collect soil samples for fertility assessment in large agricultural fields or landscapes. Grid sampling involves dividing the area into a grid of equal-sized cells and randomly selecting sampling points within each cell (Mallarino and Wittry 2004; Wollenhaupt et al. 2015). This approach ensures spatial coverage and reduces the potential for bias.
(b) *Random transect sampling for soil moisture*: In studies focused on soil moisture dynamics, researchers may employ random transect sampling to assess moisture levels across different topographic and land use gradients. Transects are established randomly within the study area, and soil samples are collected regularly along each transect.
(c) *Non-probability sampling*: Suppose you want to assess the extent of heavy metal contamination in a particular industrial area. You will use purposive/judgmental sampling to select locations based on prior knowledge of industrial activities, proximity to potential pollution sources, and historical data on contamination levels.
(d) *Non-random: Selection of representative soil profiles*: In soil characterization studies, researchers may use judgmental sampling to select representative soil profiles based on visual inspection, soil color, texture, and other observable characteristics.

Common soil sampling terms are all aimed at getting representative samples (Austin and Sutton 2014). They include the following:

- Grid sampling.
- Soil series zone sampling.
- Soil survey maps.
- Topographic units.
- Management zone sampling approach.
- Zone sampling.
- Whole field sampling.
- Zig-zag and traverse sampling method.
- Composite soil sample sampling.
- Point sampling.
- Cell sampling.
  Geo-referenced soil sapling.

16.4.2 Sampling in Mixed Methods

In mixed methods studies involving quantitative and qualitative approaches, sampling methods must be chosen from both probability and non-probability sampling dimensions. Consequently, sampling is much more complex in mixed methods than mono-methods studies. All the methods presented in this chapter are eligible to be used. This means a researcher must choose from the 17 sampling methods presented: 5 probability and 12 non-probability sampling methods. This selection requires careful planning and execution to ensure that both components are effectively integrated, and the research objectives are met. In either the qualitative or qualitative approaches, decisions have to be made regarding the sample size and the sampling scheme, i.e., how to select the sample members.

Suppose you are using an exploratory sequential design, where the data are first collected in the qualitative phase to develop quantitative instruments. In the qualitative phase, you will need a small sample guided by the research question to gain an in-depth understanding. The aim of this phase will be to collect rich, detailed data using methods such as interviews and observations. This will help uncover themes and patterns. In the quantitative phase, a larger, probability-based sample is selected to test the hypothesis and examine the phenomenon on a broader scale. This will involve administering the instrument to a sample that is representative of the population. Consequently, in mixed methods research, you must be prepared to use both probability and non-probability sampling methods.

Recommended Further Reading

1. Carlsen B, Glenton C (2011) What about N? A methodological study of sample-size reporting in focus group studies. BMC Med Res Methodol 11:1–10. https://doi.org/10.1186/1471-2288-11-26/TABLES/2
2. Frey BB (2018) Stratified Random Sampling. The SAGE Encyclopedia of Educational Research, Measurement, and Evaluation. https://doi.org/10.4135/9781506326139.N671
3. Gomez KA, Gomez AA (1984) Statistical procedures for agricultural research. Statistical procedures for agricultural research with emphasis on rice 7–8
4. Kemp F (2003) Sample Survey Principles and Methods. Journal of the Royal Statistical Society Series D: The Statistician 52:404–404. https://doi.org/10.1111/1467-9884.00369_2
5. Lantz B (2013) The large sample size fallacy. Scand J Caring Sci 27:487–492. https://doi.org/10.1111/J.1471-6712.2012.01052.X

Key Points in This Chapter

The Concept of Sampling

Sampling refers to the case selection process involving selecting a few elements from the entire population. We infer about the population from the results we obtain from the sample.

A population is all the elements that share a common set of characteristics. Populations can be individuals, units, events, or objects. Sometimes, it may be possible to study all the population members, which we call a census. However, it is challenging to do this because most populations are too large and diverse.

A representative sample is the cornerstone of sampling in quantitative research. It is designed to mirror the key characteristics and proportions of the larger population from which it is drawn.

Selection bias and response bias are two main threats to representative samples in surveys. Selection bias occurs when certain groups are systematically over-represented or underrepresented in the sample, while response bias occurs when the respondents do not respond as expected due to external factors.

Defining the Population

A researcher must define the population to which to generalize results. In an ideal situation, researchers would like to generalize to the entire population, referred to as the target population. However, it is often impractical to select a representative sample directly from the target population due to difficulty in identifying individual members. Instead, researchers typically work with elements or objects available for selection during the sampling process, known as the "sampling unit" or the "accessible population."

Sampling Process

The sampling process in agricultural research involves creating a representative sample using a reliable sampling frame. The sampling frame can be sourced from agricultural databases, farmer registries, and government records. Researchers must consider research objectives, target population, inclusion and exclusion criteria, budget, desired sample size, sampling techniques and recruitment methods.

The steps involved in sampling include the following:

1. Defining the target population.
2. Establishing the population parameters of interest by determining the summary descriptors to be used such as frequencies, mean, variance of study variables in the population.
3. Identifying and assessing the sampling frame, which is a list of cases from the target population.
4. Define the number of cases needed; choose the size of the sample.
5. Determine the appropriate sampling methods (the type to be used, probability or non-probability).
6. Establishing standard procedures for sample selection and recruitment protocols.

Types of Sampling Methods

There are two main methods of sampling: Probability sampling and non-probability sampling. Probability sampling uses some form of random selection. Procedures used ensure that the different units (subjects) in the population have equal probabilities of being chosen. In non-probability samples, the probability of each case being chosen from the target population remains unknown. While generalizations about the target population can still be made from non-probability samples, they are not based on statistical grounds. Nevertheless, both probability and non-probability samples enable researchers to address various research objectives.

Probability sampling methods are often chosen mathematically. Probability sampling methods include:

1. Simple random sampling.
2. Systematic sampling.

3. Stratified sampling.
4. Clusters sampling.
5. Multistage sampling.

Non-probability sampling methods are often chosen for their convenience, cost-effectiveness, or suitability for specific research objectives. Four main non-probability sampling methods include:

1. Purposive sampling method.
2. Snowballing sampling method.
3. Convenience sampling method.
4. Quota sampling method.

Sampling in Qualitative Research

In qualitative research, the goal is to select a sample that provides a comprehensive understanding. Various qualitative strategies are employed to sample sites, groups, or individuals, and the sample size may vary based on research issues, intent, and timing. The key is to achieve "suitability of purposes." That means having a sample large enough to generate a "thick description," which is essential for qualitative research but not so large that it leads to data overload.

A sample should not be so small that it prevents theoretical saturation: the point at which no new information or insights emerge from the data.

Sampling methods in qualitative research include the above non-sampling methods plus the following:

- Maximal sampling.
- Homogenous sampling.
- Extreme or deviant sampling.
- Typical sampling.
- Intensity sampling.
- Confirming sampling.
- Critical sampling.
- Opportunistic sampling.
- Theoretical sampling.

Main Factors Determining Sample Size Include:

1. Research objectives.
2. Heterogeneity of the population.
3. The desired precision.
4. Confidence level.
5. Sampling design.
6. Available resources.
7. Level of aggregation at which results are to be presented.

Review Questions

1. Define sampling and explain why it is preferred over a census in scientific research.
2. Why is achieving a representative sample considered a cornerstone of quantitative research? Provide an example of how a non-representative sample could lead to misleading conclusions in an agricultural study (e.g., on farm income).
3. What is the difference between a target population and an accessible population?
4. List and briefly describe any four probability sampling techniques.
5. Name any four non-probability sampling techniques and state one advantage of each.
6. What is multiphase sampling, and how is it different from cluster sampling?
7. Design a sampling plan for a qualitative study aimed at understanding indigenous knowledge of seed selection in a rural community.
8. Evaluate how sampling error and non-sampling error can affect the validity of a research study. Provide examples.
9. Reflect on the phrase "fit-for-purpose" in qualitative sampling. Why might a smaller sample be more effective than a larger one?

Chapter Review Questions

1. Suppose you want to evaluate the impact of a new government subsidy program on maize yield across different farm sizes (small: <2 acres, medium: 2–5 acres, large: >5 acres), based on the assumptions:
   (a) Explain how you would implement a *stratified random sampling* approach for this study. Why might this be more appropriate than simple random sampling here?
   (b) Alternatively, explain how you would implement a *two-stage cluster sampling* approach if your primary sampling units were villages. What are the practical advantages and disadvantages of cluster sampling compared to stratified sampling in this scenario?
2. Suppose you are conducting a qualitative study to understand the lived experiences of farmers who have successfully transitioned from traditional to organic farming practices. Based on these assumption:
   (a) Which non-probability sampling technique would be most suitable for selecting your participants and justify your choice?
   (b) Identify and briefly describe two specific variants of purposive sampling (e.g., maximal variation, critical case, typical case) that you could use to select your

organic farmers, explaining why each variant would be valuable in this context.

3. A student is planning a quantitative survey to estimate the average fertilizer expenditure per acre among smallholders. The student for a 95% confidence level and a 4% margin of error. As a senior researcher who is also supervising the student:
   (a) Briefly explain the general approach you would take to determine the required sample size using "formulas and standard tables."
   (b) Beyond confidence level and margin of error, what two other critical factors might significantly the final sample size calculation for this study? Explain why each is important.
4. You are conducting a biophysical study to compare the effectiveness of three varying fertilizer rates on maize yield.
   (a) Describe how you would use a *probability sampling method* (specify which one and why) to select the individual plants for height measurement within each experimental plot to ensure unbiased data collection.

Answers to Review Questions

- Q1. Reference: 16.1.1 and 16.1.2
- Q2. Reference: 16.1.3
- Q3. Reference: 16.1.4
- Q4. Reference: 16.2.2
- Q5. Reference: 16.2.4
- Q6. Reference: 16.2.3
- Q7. Reference: 16.2.3.5
- Q8. Reference: 16.2.6
- Q9. Reference: 16.2.5

References

Ahmed SK (2024) How to choose a sampling technique and determine sample size for research: a simplified guide for researchers. Oral Oncol Rep 12:100662. https://doi.org/10.1016/J.OOR.2024.100662

Ames H, Glenton C, Lewin S (2019) Purposive sampling in a qualitative evidence synthesis: a worked example from a synthesis on parental perceptions of vaccination communication. BMC Med Res Methodol 19:1–9. https://doi.org/10.1186/S12874-019-0665-4/TABLES/6

Austin Z, Sutton J (2014) Qualitative research: getting started. Can J Hosp Pharm 67:436–440. https://doi.org/10.4212/CJHP.V67I6.1406

Barnett V (1991) Sample survey principles and methods, 2nd edn. Oxford University Press, New York

Bartlett JE, Kotrlik JW, Higgins CC (2001) Organizational research: determining appropriate sample size in survey research. Information Technology, Learning, and Performance Journal 19(1):43–50

Benoot C, Hannes K, Bilsen J (2016) The use of purposeful sampling in a qualitative evidence synthesis: a worked example on sexual adjustment to a cancer trajectory. BMC Med Res Methodol 16:1–12. https://doi.org/10.1186/S12874-016-0114-6/TABLES/5

Bett E, Mugwe J, Nyalugwe N et al (2018) Impact of plant clinics on disease and pest management, tomato productivity and profitability in Malawi

Burmeister E, Aitken LM (2012) How many is enough? In: Research centre for clinical and community practice innovation, pp 1–6

Carlsen B, Glenton C (2011) What about N? A methodological study of sample-size reporting in focus group studies. BMC Med Res Methodol 11:1–10. https://doi.org/10.1186/1471-2288-11-26/TABLES/2

Cohen L, Manion L, Morrison K (2017) Research methods in education. https://doi.org/10.4324/9781315456539/RESEARCH-METHODS-EDUCATION-LOUIS-COHEN-LAWRENCE-MANION-KEITH-MORRISON

Comrey AL, Lee HB (2013) A first course in factor analysis (2nd ed.). New York, NY: Psychology Press

Creswell JW (2014) Research and design qualitative, quantitative, and mixed methods approaches. SAGE Publication Inc

Damba OT, Abarike MA, Nabilse CK, Akudugu MA (2019) Urban poverty analysis in Tamale, Ghana. UDS Int J Dev 6:79–96. https://doi.org/10.47740/377.UDSIJD6I

Das S, Mitra K, Mandal M (2016) Sample size calculation: basic principles. Indian J Anaesth 60:652–656. https://doi.org/10.4103/0019-5049.190621

DeYoreo M (2018) Stratified random sampling. In: The SAGE encyclopedia of educational research, measurement, and evaluation, pp 1624–1624. https://doi.org/10.4135/9781506326139

Doss C (2014) Data needs for gender analysis in agriculture. In: Gender in agriculture: closing the knowledge gap, pp 55–68. https://doi.org/10.1007/978-94-017-8616-4_3

Eniego AA, De Jesus AB, Magat MER (2025) Sampling techniques. In: Introduction to public health and research, pp 77–94. https://doi.org/10.1007/978-981-96-5154-2_4

Fadele AA, Rocha A (2025) The art and science of quantitative research, p. 599. 10.1007/978-3-031-91431-7

Fink A (2003) How to sample in surveys. https://doi.org/10.4135/9781412984478

Fowler FJ (2009) Survey research methods, 4th edn. SAGE Publications, Inc.

Fraenkel JR, Wallen NE, Hyun HH (2012) How to design and evaluate research in education, p. 642

Francis JJ, Johnston M, Robertson C et al (2010) What is an adequate sample size? Operationalising data saturation for theory-based interview studies. Psychol Health 25:1229–1245. https://doi.org/10.1080/08870440903194015

Frey BB (2018) Stratified random sampling. In: The SAGE encyclopedia of educational research, measurement, and evaluation. https://doi.org/10.4135/9781506326139.N671

Gill J, Johnson P, Clark M (2010) Research methods for managers (4th edn). SAGE Publications Ltd, Los Angeles. ISBN 978-1-84787-094-0

Gomez KA, Gomez AA (1984) Statistical procedures for agricultural research. In: Statistical procedures for agricultural research with emphasis on rice, pp 7–8

Grady C (2015) Institutional review boards purpose and challenges. Chest 148:1148–1155. https://doi.org/10.1378/chest.15-0706

Hennink M, Kaiser BN (2022) Sample sizes for saturation in qualitative research: a systematic review of empirical tests. Soc Sci Med 292:114523. https://doi.org/10.1016/J.SOCSCIMED.2021.114523

Henry GT (1990) Practical sampling. https://doi.org/10.4135/9781412985451

Kalton G (1983) Introduction to survey sampling. https://doi.org/10.4135/9781412984683

Kemp F (2003) Sample survey principles and methods. J R Stat Soc Ser D 52:404–404. https://doi.org/10.1111/1467-9884.00369_2

Kwak SG, Kim JH (2017) Central limit theorem: the cornerstone of modern statistics. Korean J Anesthesiol 70:144–156. https://doi.org/10.4097/KJAE.2017.70.2.144

Lantz B (2013) The large sample size fallacy. Scand J Caring Sci 27:487–492. https://doi.org/10.1111/J.1471-6712.2012.01052.X

Luborsky MR, Rubinstein RL (1995) Sampling in qualitative research. Res Aging 17:128–130. https://doi.org/10.1177/0164027595171005

Maggino F (ed) (2023) Theoretical sampling. Springer, Cham

Mallarino AP, Wittry DJ (2004) Efficacy of grid and zone soil sampling approaches for site-specific assessment of phosphorus, potassium, pH, and organic matter. Precis Agric 5:131–144. https://doi.org/10.1023/B:PRAG.0000022358.24102.1B/METRICS

Marshall B, Cardon P, Poddar A, Fontenot R (2013) Does sample size matter in qualitative research?: a review of qualitative interviews in is research. J Comput Inf Syst 54:11–22. https://doi.org/10.1080/08874417.2013.11645667

Martínez-Mesa J, González-Chica DA, Duquia RP et al (2016) Sampling: how to select participants in my research study? A Bras Dermatol 91:326–330. https://doi.org/10.1590/ABD1806-4841.20165254

Mey G (2023) Qualitative methodology, pp. 453–478. https://doi.org/10.1007/978-3-030-28745-0_22

Mills EG, Gay LR (2020) Educational research: competencies for analysis and applications. Pearson

Moser A, Korstjens I (2018) Series: practical guidance to qualitative research. Part 3: sampling, data collection and analysis. Eur J General Pract 24:9–18. https://doi.org/10.1080/13814788.2017.1375091

Mugenda A, Mugenda G (2019) Research methods: qualitative and mixed methods approach, pp. 100–102

Njuki J, Eissler S, Malapit H et al (2022) A review of evidence on gender equality, women's empowerment, and food systems. Glob Food Sec 33:100622. https://doi.org/10.1016/J.GFS.2022.100622

Onwuegbuzie AJ, Leech NL (2007) A call for qualitative power analyses. Qual Quant 41:105–121. https://doi.org/10.1007/S11135-005-1098-1/METRICS

Palinkas LA, Horwitz SM, Green CA et al (2015) Purposeful sampling for qualitative data collection and analysis in mixed method implementation research. Adm Policy Ment Health Ment Health Serv Res 42:533–544. https://doi.org/10.1007/S10488-013-0528-Y/METRICS

Patton MQ (2015) Qualitative research & evaluation methods: integrating theory and practice, 4th edn. SAGE Publications, Inc

Piepho HP, Gabriel D, Hartung J et al (2022) One, two, three: portable sample size in agricultural research. J Agric Sci 160:459–482. https://doi.org/10.1017/S0021859622000466

Quisumbing AR, Meinzen-Dick R, Raney TL et al (2014) Gender in agriculture: closing the knowledge gap. In: Gender in agriculture: closing the knowledge gap, pp 1–444. https://doi.org/10.1007/978-94-017-8616-4/COVER

Resnik DB (2021) Standards of evidence for institutional review board decision-making. Account Res 28:428–455. https://doi.org/10.1080/08989621.2020.1855149

Rothenberg R (2004) The inescapable logic of institutional review boards. Ann Epidemiol 14:517–519. https://doi.org/10.1016/j.annepidem.2004.07.089

Saunders MNK, Lewis P, Thornhill A (2019) Research methods for business students (8th ed.). Harlow, England: Pearson Education Limited

Sedgwick P (2013) Snowball sampling. BMJ 347. https://doi.org/10.1136/BMJ.F7511

Sim J, Saunders B, Waterfield J, Kingstone T (2018) Can sample size in qualitative research be determined a priori? Int J Soc Res Methodol 21:619–634. https://doi.org/10.1080/13645579.2018.1454643

Stringer AP, Bell CE, Christley RM et al (2011) A cluster-randomised controlled trial to compare the effectiveness of different knowledge-transfer interventions for rural working equid users in Ethiopia. Prev Vet Med 100:90–99. https://doi.org/10.1016/J.PREVETMED.2011.02.001

Whitley E, Ball J (2002) Statistics review 2: samples and populations. Crit Care 6:143–148. https://doi.org/10.1186/CC1473/COMMENTS

Wollenhaupt NC, Mulla DJ, Gotway Crawford CA (2015) Soil sampling and interpolation techniques for mapping spatial variability of soil properties. In: The state of site-specific management for agriculture, pp 19–53. https://doi.org/10.2134/1997.STATEOFSITESPECIFIC.C2

Part IV

Methods and Techniques of Data Collection

In Part III, we explored research designs, dividing them broadly as experimental and non-experimental. Within the experimental domain, we focused on randomized controlled experiments commonly used in agriculture, including on-farm experiments. We also covered various designs used to evaluate the impact, assessing how different variables affect outcomes. In the non-experimental domain, we discussed the surveys and qualitative research approaches, including case studies, ethnography, grounded theory, narrative, and phenomenology.

In Part IV, we first present data collection methods in the biophysical sciences focusing on experiments. We then explore data collection in surveys, qualitative approaches, and participatory research methods. The chapters are structured as follows:

Chapter 17	Data Collection in Biophysical Experiments
Chapter 18	Methods and Tools of Data in Non-experimental Designs I (Surveys)
Chapter 19	Methods and Tools of Data in Non-experimental Designs II (Qualitative Research and Participatory Methods)

We have presented data collection in experiments, surveys, qualitative research, and participatory research in separate chapters. However, you will find that there are commonalities, especially in surveys and qualitative research. For instance, the three approaches use interviews and questioning. They have therefore been presented separately to enhance clarity and help students follow through easily when searching for data collection methods for different research approaches. This structured presentation allows a better understanding of each method's specific applications. It also aligns with the research designs presented in Part III, ensuring an organized exploration of various data collection methodologies.

Methods of Data Collection from Experiments

17

Abstract

Data collection methods are specific techniques for acquiring data, while the instrument is a device. Validity in data collection from experiments is mainly achieved through applying principles of experimentation, including choosing a suitable design, randomization, and blocking. It also includes a selection of the most suitable tools and methods, standardizing data collection protocols and procedures, calibrating instruments, and ensuring that laboratories are adequately equipped and managed. Data collection in agronomic experiments is organized into three phases: site characterization, the implementation process, and the harvesting phase. During the site characterization phase, soil sampling and analysis are crucial. This involves collecting soil samples from various locations within the experimental site to assess soil properties, such as nutrient status and texture. Soil analysis is then conducted to evaluate the fertility status of the soil. In the implementation phase, sampling is done at various stages to monitor growth stages and plant health. Alongside this, ongoing soil and plant analyses may be conducted to evaluate nutrient levels and assess responses to treatments. State-of-the-art and efficiently run laboratories are key to maintaining the accuracy and reliability of laboratory results. The harvesting phase measures total yields to assess the effectiveness of the treatments. For legumes, estimating nitrogen fixation is particularly important. Techniques such as isotopic techniques and acetylene reduction assay are used. In screening experiments, lattice designs are commonly used to screen a large number of varieties effectively. Overall, data collection from specific crops depends on the objectives and nature of the crop but mainly includes growth parameters and yield components.

Keywords

Agronomic data, Data collection device · Growth parameters and yield components data · Legumes data · Nitrogen fixation estimation · Molecular analysis · Site characterization · Soil and plant sampling and analysis · Validity and reliability · Variety Screening experiment

Learning Outcomes

After reading this chapter, you are expected to be able to:

1. Differentiate between the method and the instrument/tool of data collection.
2. Identify ways of enhancing validity in the collection of biophysical data from experiments.
3. List the most common data collected during site characterization and the experimentation period.
4. Explain how to sample and analyze soil.
5. Appreciate questions raised by supervisors about soil sampling.
6. Describe key steps in plant nutrient analysis.
7. Appreciate the different methods of data collection in the quantification of nitrogen fixation in legumes.
8. Describe how to collect grain and stover yields in maize and legumes.
9. Appreciate data collection in a variety of screening experiments using a lattice square design.
10. Recognize the importance of data collection in genomic and metabolomic analysis.
11. List common data that a researcher can collect in cereals and legumes.

J. Njeri Mugwe, S. Runo, *Research Methodology in Agricultural Sciences*, https://doi.org/10.1007/978-981-95-1892-0_17

17.1 Overview of Data Collection Methods

17.1.1 Methods and Instruments for Collecting Data

Whether qualitative or quantitative, data collection is the key to gathering, measuring, and analyzing accurate data from all the relevant sources. It is the key to unlocking answers to the research problem, research questions, and testing hypotheses. Data collection empowers us to obtain crucial information about study objects/subjects/participants. It is not just an important part of research; *it is the backbone*. The final results of a research study hinge on the data collected. Without data, the researcher is left without the means to answer the research question, test hypotheses, and draw conclusions and recommendations for the study.

Students sometimes confuse the terms data collection method and tool of data collection, but they are distinct. A data collection method is the *specific technique* for acquiring data from a subject or an event, while a tool/instrument is a *device* (Table 17.1). Understanding this distinction is not just important, but it is crucial to mastering the art of data collection. In agricultural experiments, specific tools/equipment and methods are used depending on the measurement, e.g., soil pH is measured using a pH meter, and soil moisture is determined using soil moisture sensors.

Table 17.1 Examples showing the difference between methods and instruments in selected biophysical studies

Measurement	What to measure?	Method	Instrument
Soil pH levels	Acidity and alkalinity of soil	Soil sampling, preparation, and testing	pH meter or soil testing kit
Pest infestation	Pest density in crops	Counting pests in sampled areas	Hand lens, sweep net, and counting frame
Body temperature in animals	Body temperature of calves	Measure body temperature	Digital thermometer or infrared thermometer
Soil moisture content	Amount of moisture in the soil	Soil sampling and gravimetric analysis	Soil moisture sensors or probes Oven to dry samples and a balance
Feed intake	Daily feed intake of calves	Recording the amount of food given and leftovers	Weighing scale and feed troughs

17.1.2 Validity and Reliability of Data Collection Revisited

Validity and reliability are key concepts in measurements and research. In Chaps. 1 and 9, we introduced and explained validity and reliability as crucial in research. In this chapter and the next chapter (Chap. 18), we revisit the concept of validity to highlight its critical role in the data collection process. In assessing validity, we ask ourselves, "What is the worth of our measured outcomes?" (Suter 2009). To be worthy, data must be valid and reliable. Valid data is accurate, while reliable data is dependable and consistent. It is useless if our data does not satisfy these two qualities, regardless of how it is obtained. Data must also be appropriate and sufficient. Indeed researchers should view validity as a continuous process running through all the stages of research; from conceptualization to instrument design, data collection and interpretation.

17.1.2.1 Validity

Validity is like a stamp of approval for our research. If our data is valid, our findings are real and meaningful, not just random or flawed. This means it will truly represent the real-world scenarios being investigated, and we can confidently apply the results to solve the problem at hand. Reliability is secondary because if the measurement is not valid, reliability becomes irrelevant; consistently flawed data does not lead to meaningful conclusions. Thus, while it is important to have reliable data, it is important to ensure we have valid data in the first place. This is because a measure may be consistently (reliable) but still fail to measure what it is intended to measure (valid). Therefore, while reliability is essential, it is not sufficient to ensure our data accurately reflects the real-world scenarios being investigated (Repke et al. 2024: Gail and Sullivan 2011).

17.1.2.2 Validity in Experiments

Recall that in experiments, the key principles and elements of experimentation are meant to enhance validity and reliability (Kristensen et al. 2008; Nimon et al. 2012; Leung 2015; Kool et al. 2020; Rio et al. 2022). In the context of experiments, consider the following to enhance the data you collect:

- *Internal validity* refers to the degree to which the experiment accurately establishes a causal relationship between the treatment and the observed effect. Blocking and randomization help control for potential confounding variables within each block, reducing systematic biases and enhancing internal validity (Refer to Chaps. 8 and 9). Most of the elements for designing effective experiments are meant to enhance internal validity.

- *External validity* is the extent to which an experiment's results can be generalized to other settings, populations, and times. It can be enhanced by comparing treatments across different settings, populations, geographic locations, and periods, for example, in on-farm experiments (Chap. 10). On-farm experiments, conducted in multiple locations, lead to improved understanding and quantification of G8E*M interactions, thereby significantly enhancing external validity and practical application of the recommendations (Vanlauwe et al. 2019; Kool et al. 2020).
- *Construct validity* refers to the degree to which the experiment measures what it claims to be measuring. The researcher must accurately define and measure the variables of interest, particularly the dependent variable. This involves clearly defining what constitutes "yield" (e.g., weight of harvested crops, number of fruits) and ensuring that all measurements accurately reflect these definitions. This might involve using standardized measurement tools and procedures.
- *Content validity* refers to the extent to which the experiment covers the entire range of the studied concept. In an experiment, you ensure that all relevant aspects of the treatment and outcome are considered. For example, suppose the goal is to assess the impact of fertilizers on crop health. In that case, the experiment should include a comprehensive range of health indicators (e.g., growth rate, disease resistance, nutrient content), measured correctly and not just yield alone.
- *Statistical conclusion validity:* involves the extent to which the conclusions drawn from the statistical analysis are accurate and reliable. It is enhanced by appropriate statistical analysis techniques that account for the block structure and treatment effects (see concept of noise and measurement of covariates in on-farm experiments in Chap. 10). If you do not take into account the variability, you may not be able to detect the differences between treatments when, in fact, the differences exist. Proper analysis ensures that the differences observed are statistically significant and not due to random variation. For instance, using Analysis of Variance (ANOVA) to analyze the results helps in accurately determining if the variations in crop yield are attributable to different fertilizers.

17.1.3 Other Ways of Enhancing Validity in Biophysical Experiments

17.1.3.1 Protocols and Procedures of Data Collection

The protocol is one of the main elements in experimentation (Chap. 9). A protocol should provide detailed descriptions of the methods used in data collection. We emphasized that with a clear protocol, your supervisor can assess and advise if the methods are correct and whether you will be able to get valid data. Additionally, the guidelines will ensure that data is taken using similar procedures. In the protocol, include a step-by-step account of the procedures, the rationale for choosing specific methods, and any modifications made to standard protocols. For instance, if a study involves measuring crop yields, the methodology should describe the sampling process, the specific time of harvest, and how the weight of the crops is measured and recorded. This detailed description ensures that other researchers can replicate the study under similar conditions.

17.1.3.2 Calibration of Instruments

Data accuracy in biophysical studies heavily depends on the instrument calibration (Zhao et al. 2015; Jenkins et al. 2025). You should always ensure you calibrate the instrument. For example, a pH meter used to measure soil acidity must be calibrated using standard buffer solutions before and during the measurement. Similarly, weighing scales should be regularly checked and calibrated. Proper calibration helps reduce systematic errors that could compromise the study's validity.

17.1.4 Laboratories

Laboratories are indispensable in agricultural research. A soil and plant laboratory, one of the most important facilities, is crucial in analyzing the physical and chemical properties of soil and plant samples. The data provided by these laboratories is important for understanding plant growth and soil health.

An adequately equipped and efficiently managed laboratory is essential for ensuring the validity and reproducibility of the analytical results. The measurements' accuracy depends on the equipment's quality and the laboratory personnel's proficiency. Regular maintenance and calibration of the instrument are recommended to avoid errors (Egea González et al. 2004; Taverniers et al. 2004; Betz et al. 2011). Additionally, trained technicians and scientists should be responsible for running the laboratories to ensure the precision of the results.

An efficiently managed laboratory adheres to strict protocols and standards. Laboratory standard operating procedures provide detailed procedures for the analysis, ensuring that the personnel follow similar methods always. Laboratories often follow procedures set by national and international bodies to maintain high standards of quality and accuracy. When using the laboratory, make sure you adhere to the following (Motsara and Roy 2008):

- Follow general safety rules and guidelines for your safety.
- Follow standard operating procedures provided.

- Follow the required quality control procedures. This may include the use of independent or certified references, blanks, blinds, etc.
- Follow validation procedures, as many labs have in-house validation procedures.
- Check when you need to use analytical-grade chemicals to maintain the integrity of your analysis.
- Always check with the technician in charge when unsure what to do.

17.2 Data Collection in Agronomy-Based Experiments

Novice researchers and students often struggle with identifying the appropriate data to collect in their agronomy experiments. This section provides general guidelines on the most common parameters to collect in agronomy experiments involving cereals (such as maize, wheat, and sorghum) and legumes (such as soybeans and common beans). It covers essential parameters and measurements, such as soil characteristics, growth data, and yield data. Additionally, references are provided for more detailed information on various soil and crop measurements, ensuring that researchers and students can access comprehensive resources tailored to their particular needs. These references will help you understand the data collection and analysis requirements for various crops, ensuring that your experiments are thorough and your findings are robust.

In Chap. 8, we mentioned that data collection is done in phases, usually *before the experiment, during the experimentation* period, and at the *end of the experiment.* In this section, we cover data collection for site characterization before setting up the experiment and during the experiment.

17.2.1 Site Characterization Data

Supervisors expect students to indicate, in their protocols, what data they will collect to characterize the site. Additionally, in the publications, site characterization data are presented. The main data should include the site's location, soil characteristics, and field history (Table 17.2).

Table 17.2 Common site characteristics data

Type of data	What to measure	Method of collection
Location	Record GPS coordinates and obtain the experimental site's climate data (temperature, rainfall).	GPS device, climate and weather records from the weather station
Soil characterization	Conduct initial soil tests to determine pH, organic matter content, and nutrient levels (both macro and micronutrients). Assess soil texture by determining the sand, silt, and clay composition.	Soil sampling and analysis in the laboratory
Field history	Note previous crops grown, and document any prior fertilizer and pesticide use.	Farm records interview with farmers
Mineral fertilizer	Type of fertilizer, rate of application (kg ha^{-1}), and method of application	Included in protocol Take records during planting and keep them safely
Organic fertilizer	Type of organic fertilizer (cattle manure, compost, etc.), rate of application (kg ha^{-1}), and method of application	Included in protocol Take records during planting and keep them safely

17.2.2 Data Collection During the Experimentation and at Harvest

Data collection during experimentation often involves gathering information to evaluate the effect of treatments on soil properties and crop performance. Your supervisor expects you to follow the protocol during the actual data collection. Generic data collected in agronomy experiments include data to characterize the crop, e.g., date of first flowering and germination, and response data (Table 17.3).

17.2.2.1 Crop Growth and Performance Data

The data on crop growth stages is important for tracking the progress of the crop throughout its life cycle. For instance, plant height is an indicator of crop growth and vigor, while leaf area can help researchers assess photosynthetic capacity and overall plant biomass (refer to Conduct and management of maize field trials by IITA by Badu-Apraku and Oyekunle (2012) for a more detailed explanation).

17.2.2.2 Pest and Disease Data

Students in the crop protection discipline will be guided in data collection by their research objectives. However, in all agronomy work, researchers record data on the presence and severity of pests, diseases, and weeds in crops. This data helps manage and mitigate crop damage. Regular data collection of pests and diseases allows for the timely identification of outbreaks and the implementation of appropriate interventions.

Table 17.3 Data collected during the experimentation period

Type of data	What to measure	How to measure
Planting date	Ensure random assignment of treatments to plots to avoid bias	Document the randomization process Use specialized software designed for random assignment (e.g., in excel and R)
	Date of planting	Record planting date
Planting method	Tillage practices Seeding rate Row spacing	Document seeding rate (amount of seed per unit area), planting depth, tillage practices, and row spacing.
	Record inter- and intra-row spacing	
Weed management	Weeding frequencies (dates of weeding)	Record dates when the weeding takes place
Growth data		
1. Germination rate	Germination data	Count the number of seedlings in a defined area within 7–14 days after planting.
2. Plant height	Measure plant height at regular intervals (e.g., bi-weekly)	Use a tape measure to measure the height of the base to the top of the main stem.
3. Leaf area index (LAI)	Measure LAI at different growth stages	A leaf area meter is used to measure LAI at different growth stages
4. Number of leaves	Measure at specified intervals	Counting and recording
4. Biomass weight sampling	Collect and dry above-ground biomass from a defined area at different growth stages, then weigh it.	Sampling, drying, and weighing
5. Tillering index	No of tillers per plant	Count and record the number of tillers
6. Pest and disease monitoring	Pest and disease incidence and prevalence	Regularly inspect plants and record the presence of pests, using a standardized scale to rate damage. Identify and document disease symptoms, using a severity scale to assess impact.
7. Soil moisture	Measure soil moisture content at different depths (e.g., 0–20 cm, 20–40 cm)	Method 1: Tensiometers or soil moisture probes at regular intervals. Method 2: Gravimetric
8. Soil nutrient levels and soil pH	Determine the amount of nutrients in soil (macro and micronutrients)	Periodically collect soil samples (e.g., before planting, mid-season, after harvest) to analyze nutrient levels.
9. Labor data	Labor for land preparation, weeding, applying inputs, pest and disease control, and harvesting	Record the time spent and the number of workers using timesheets
At harvest		
1. Date of harvesting	Document the actual date/dates of harvesting	Noting and recording
2. Yield components	Record the number of ears per plant, number of grains per ear, thousand-grain weight, and number of nodules	Counting and recording
3. Grain yield	Harvest grains from a defined area and weigh them after drying. Adjust the weight to the standard moisture content.	Mark the harvest area and take the field weight Determine the moisture content Adjust the weight to the standard moisture content.

(continued)

Table 17.3 (continued)

17.2.2.3 Crop Phenology Data

In Table 17.3, several crop phenology data are indicated. Crop phenology data refers to recording the timing of key events in a crop's life cycle, such as planting date, flowering, fruit set, and maturity, which can influence management decisions. These phenological stages provide information about the development of the crop. Understanding the timing of these events is essential for crop management practices like irrigation scheduling, pest and disease control, and harvest planning.

17.2.2.4 Crop Yield Data

Yield data is a fundamental measure in agronomy experiments. It refers to the amount of harvested crop per unit area, e.g., tonne per hectare, and is useful for evaluating the effectiveness of different agronomic practices. Yield data provides insights into the productivity of different varieties and cultivation techniques, e.g., mono-crop versus intercrop and many agricultural interventions. It also plays a role in the assessment of the overall success of crop management and soil fertility management programs. Researchers and farmers rely on yield data to optimize production and maximize resource use efficiency. Reference [Participatory Research Methods for Technology Evaluation: A Manual for Scientists Working with Farmers (cimmyt.org)]

17.2.2.5 Soil Data

Soil data is crucial for understanding soil fertility and health. As mentioned earlier, these parameters are usually collected before (baseline) and at the end of the experiment to assess changes in nutrient content. The baseline serves as a reference point for assessing changes during the experiment. The most common measurements in soil include the following:

- Soil nutrient levels of essential nutrients, mainly NPK, Ca, Mg, and S, and micronutrients such as Zn, Cu, and Bo.
- Soil pH.
- Soil organic matter.
- Cation exchange capacity.
- Bulk density.
- Texture and structure.
- Soil pH.
- Soil texture and structure.
- Soil moisture.
- Earthworms and other biological measurements.

17.2.3 Soil Sampling Considerations and Sampling

Sampling depth for most soils is typically tillage depth (which can vary depending on the soil type). The top 15–20 cm of soil has the most root activity, and this is the depth at which inputs, such as fertilizer or lime applications, are made. Therefore, the following depths can be considered: For most annual crops such as cereals, pulses, and vegetables, composite soil samples should be collected from 0–20, 20–40, and 40–60 cm depths.

For deep-rooted perennial crops (fruit trees, sugarcane, coffee, plantation crops, and others), composite soil samples should be collected from 0–20, 20–50, 50–80, and 80–110 cm depths using four to five pits. For nutrient calibration studies, soil samples should be collected from 0–15 cm depth at 21 days after planting. Sampling beyond 30 cm depth in a single sample could seriously underestimate a soil acidity problem in the critical root zone.

17.2.3.1 Sampling Tools and Accessories

Proper soil sampling should be done using appropriate methods and equipment. Different tools are used in soil sampling. Depending on field conditions, it is commonly recommended to use a soil auger to collect samples for agronomy, soil fertility, and fertilizer recommendation experiments, as indicated, but this may vary depending on the nature of the soil, e.g.,

- A tube auger and spade are appropriate for sampling soft and moist soils.
- A screw-type auger is more convenient for hard or dry soils.
- A post-hole auger is preferable for sampling excessively wet (waterlogged) soils.
- Core samplers collect undisturbed soil samples for bulk density and moisture content analysis.

17.2.3.2 Sample Labeling

Proper labeling of soil samples is as important as the care needed to obtain the samples. Samples need to be labeled properly for identification. Label sample bags/containers immediately, clearly, and indelibly with unique identifiers (e.g., Plot ID, Sample ID, date, depth, GPS coordinates). A label of thick paper with an identification mark and other details should be put inside the sample bag, and another one carrying the same details tied/pasted outside the bag. If the soil sample is wet (for calibration studies), the label should be written with a lead pencil or a permanent ink marker or put inside a small separate plastic bag before air-drying. With the advance of digital technology, it is advisable to use barcoded or digitized sample labeling and tracking procedures in all soil and agronomic research activities.

17.2.3.3 Precautions During Collection and Storage of Soil Samples

To prevent contamination, special care in collecting and handling the soil samples is required.

The following precautions should be taken to minimize error:

- Avoid contact of the samples with chemicals, fertilizers, manure, or other contaminants.
- Use stainless-steel augers instead of rusted-iron spades to obtain soil samples for micronutrient analysis.
- Do not use containers (bags or boxes) previously used for storing fertilizer, salt, or other chemicals.
- Store soil samples in clean, preferably new, cloth or polythene bags.
- Do not store wet samples for long in restricted/closed conditions. Instead, keep soil samples in plastic bags to air-dry by opening them and storing them on a shelf.
- Use a glass, porcelain, or polythene jar for long-duration storage.

Several guidelines and resources for soil sampling and analysis exist. These are usually prepared by Government agencies, Scientific societies, e.g., the Soil Science of America Society, Universities, and international and national agricultural institutions. These guidelines cover various aspects of soil sampling and analysis, including sampling techniques, sample preparation, and specific tests. A few of these resources include the following:

1. Okalebo, J.R., Gathua, K.W. and Woomer, P.L. (2002) Laboratory Methods for Soil and Plant Analysis: A Working Manual. TSBF, Nairobi.
2. Soil fertility management manual by Kenya Agricultural and Livestock Research Organization (KARLO, Kenya) (Esilaba et al. 2021)
3. IITA guide on soil sampling and sample preparation (IRG2)
4. Field and laboratory research manual for integrated Soil fertility management by KARLO, Kenya (Mangale et al. 2016)
5. Soil analysis Manual by Ministry of Agriculture, Forestry and Food Security, Sierra Leone
6. Soil testing manual by FAO (2020)
7. Soil sampling guidelines by Purdue University (Ackerson 2018)
8. Sampling methods by South Dakota Department of Agriculture (Clark 2019)
9. Manual for soil, plant and water analysis (Singh et al. 2005)
10. Soil sampling and method of analysis (Carter and Gregorich 2007) Soil Sampling and Methods of Analysis – 2nd Edition – M.R. Carter – An (routledge.com)
11. Soil Survey Field and Laboratory Methods Manual by USDA Soil Survey Staff. (2009).
12. Plant Sampling for Agriculture by Dowling and Blaesing (2022)
13. Crop Nutrition Laboratory. https://cropnuts.com/download/plant-testing/
14. Sampling for plant tissue analysis by Flynn et al. (1999)

17.2.4 Common Concerns Raised by Supervisors

Here is a quick list of questions your supervisors will likely ask you regarding soil sampling. Some answers are given.

1. How did you carry out soil sampling?
   You are expected to give a step-by-step process of how you will do the soil sampling (if in the proposal); if you are presenting a progress report or your thesis, you should describe what you did. It should include 1. How did you prepare for sampling, e.g., reviewing the protocols and gathering the necessary equipment and tools? 2. Site selection: How did you ensure that representative samples were taken? 3. Actual sampling and taking a composite sample, 4. Labeling and recording, and 5. Transportation to the laboratory, e.g., give details on how unique samples were handled. If it is the determination of mineral N, mention the special handling of the samples.
2. When did you take your samples?
   - 60 to 90 days before planting
   - For nitrogen, obtain samples as close to the sampling as possible
3. How much composite sample did you take?
   - You need about one kilogram to take to the laboratory
4. What was the sampling depth?
   - Depends on the crop under study. For cereals and legumes, standards depend on the root zone, which is 0–15 or 0–20. Consistency is important.
5. How many times did you sample?
   - This question is specific because it highly depends on the objective of the sampling and the parameter measured.
   - For instance, if monitoring soil pH after application of lime, you may need to sample weekly.
   - If monitoring Mineral N (Nitrate N and Ammonium N), you may need to follow pre-determined periods according to the growth curve of the crop, e.g., for maize, you may consider 2, 4, 6, 8, 14, and 16 weeks after planting.
   - For soil texture and bulk density, you need to sample once before planting. These parameters take a long time to change. Rarely will you measure texture as part of the parameters affected by the treatments.
6. What tools do I need for sampling?
   - Soil auger is recommended for soil and fertilizer recommendation studies.
   - Use the proper type of auger (tube, screw, etc.).
7. What is the recommended practice of labeling soil samples?
   (a) Recommended practice is to ensure your soil samples are well labeled for identification.
   (b) One can use a triple labeling technique where you have one label inside the paper bag, another label on the bag, and another on the string used to tie the bag containing the sample.

17.3 Soil and Plant Analysis

Critical stages in plant analysis include several key phases. The first stage is sample collection, which must be performed at the appropriate growth stage of the plant to ensure relevance to the study's objectives. Proper techniques should be used to avoid contamination and ensure representative sampling (Flynn et al. 1999). For example, collecting leaves from the same position on multiple plants helps achieve consistency. To ensure traceability, samples should be clearly labeled with relevant information such as date, location, and plant part.

The next stage is sample preparation. This involves cleaning the samples to remove any soil, dust, or external contaminants from the plant material. After cleaning, samples should be dried to a constant weight to prevent decomposition and to standardize the moisture content. Grinding the samples to a fine, uniform powder ensures homogeneity and accuracy in subsequent analyses. Once the samples are prepared, the next stage is extraction. Chemical extraction methods are used to extract nutrients or compounds of interest from the plant material, ensuring consistency across all samples to maintain comparability. The accuracy of this process is crucial for reliable results. Extraction is done using solvents and reagents specific to nutrients; mostly acid or base solutions (Okalebo et al. 2002).

Following extraction, the analysis stage involves using precise instruments such as spectrometers, chromatographs, or other analytical tools to measure nutrient concentrations or other relevant parameters. This is followed by data interpretation. The results should be compared with established benchmarks, standards, or historical data to identify trends or anomalies. Based on the analysis, recommendations for nutrient management, corrective actions, or further research should be provided.

Validation and verification are essential to ensure the accuracy of the analysis. This involves cross-checking results by comparing them with those obtained from other methods or laboratories and repeating critical analyses to verify consistency and accuracy. Finally, the application of findings is crucial. The results should inform practical agricultural decisions, such as adjusting fertilization practices or managing crop health. Continuous monitoring of the effectiveness of any changes made based on the analysis results is necessary to ensure long-term benefits. In conclusion, each stage in plant analysis, from sample collection to the application of findings, is critical for ensuring the accuracy and reliability of the results.

Further Reading

- Plant Sampling for Agriculture – A Guide, Fertilizer Australia, Canberra, Australia. Dowling and Blaesing (2022)
- Crop Nutrition Laboratory. https://cropnuts.com/download/plant-testing/
- Sampling for plant tissue analysis. Flynn et al. (1999).

17.4 Data Collection in Legumes

Legumes are some of the most important crops globally. Most agronomists will carry out experiments on these crops with different objectives. Even though data collected in an experiment depends on the objectives, the most common data collected in the legume area is captured in Table 17.3, which includes growth, physiological data, e.g., flowering and yield data.

Legumes are unique because of their ability to fix nitrogen through the symbiotic relationship with rhizobia; thus, nitrogen fixation is one of the most important measurements that is determined in experiments involving legumes. There is no single accurate method of measuring nitrogen fixation (Giller 2001a, b), and the choice of method depends on the research objectives. Some of the common methods of measuring and estimating nitrogen fixation include:

1. *N-Difference method*: The total nitrogen content in the plant, particularly in the nodulated roots and shoots, is measured before and after the plant has been grown in nitrogen-deficient soil. The N content of the non-fixing (derived solely from the soil) is subtracted from the total N content in the N-fixing legume (Giller 2001a, b). It is simple, cheap, and most commonly used in students' projects. Three methods can be used to (1) Compare legumes with non-legumes, e.g., maize, (2) Compare a legume with a non-nodulating legume, and (3) Compare a modulated legume with a non-inoculated legume.
2. *Nodule color and fresh weight*: The appearance of the color of nodules in legumes can provide qualitative information about nitrogen fixation. Researchers measure the effectiveness through the assessment of the color of nodules (red/pink) and the numbers and fresh weight of nodules. This method is also easy to use.
3. *Nodule count:* Counting the number of nodules on legume roots can provide a rough estimate of nitrogen fixation. More nodules often indicate high nitrogen-fixing activity.
4. *Acetylene reduction assay (ARA):* The technique uses gas chromatography to measure the reduction of acetylene to ethylene by nitrogenase in a simple, quick, and inexpensive manner. Its disadvantage is that it cannot be used to measure N fixation integrated over time (Giller 2001a, b).
5. *15N isotope dilution method*: Uses stable isotope labeled nitrogen and is a highly accurate method, but also the most expensive (Danso 1995).
6. *Natural 15 N abundance method*: based on the principle that atmospheric Nitrogen has a different isotopic signature (15 N/14 N ratio) compared to the nitrogen in the soil (Peoples et al. 2009).

The following are resources for information on measuring nitrogen fixation:

- Methods for evaluating nodulated legumes in the field nodules by Peoples et al. (1989).
- Hardarson and Danso (1993) Methods for measuring biological nitrogen fixation in grain legumes | SpringerLink.
- Giller (2001) Nitrogen fixation in tropical cropping systems | CABI Books (cabidigitallibrary.org).
- Pausch RC, Mulchi CL, Lee EH, Meisinger JJ (1996). Use of 13C and 15 N isotopes to investigate O3 effects on

C and N metabolism in soybeans. Part II. Nitrogen uptake, fixation, and partitioning. Agriculture, Ecosystems and Environment 60(1):61–69. https://www.sciencedirect.com/science/article/pii/S0167880996010626

- Beyan, S.M., Wolde-meskel, E. & Dakora, F.D. (2018). An assessment of plant growth and N_2 fixation in soybean genotypes grown in uninoculated soils collected from different locations in Ethiopia. *Symbiosis* 75, 189–203. https://doi.org/10.1007/s13199-018-0540-9
- Unkovich M., Herridge D., Peoples M., Cadisch G., Boddey R., Giller K., Alves B. and Chalk P. (2008). Measuring plant-associated nitrogen fixation in agricultural systems. ACIAR Monograph No. 136, 258pp.

17.5 Data Collection in a Maize (*Zea mays*) Experiment

Maize is a crucial crop globally, and accurate measurement of both grain and biomass yield is essential for enhancing productivity and assessing agricultural practices. Data collected include both morphological and agronomic characteristics (Badu-Apraku and Oyekunle 2012).

The most common data collected in maize experiments include:

(a) Grain yield.
(b) Stover yields.
(c) Plant and ear heights.
(d) Leaf area index.
(e) Chlorophyll content.
(f) Days to silking and anthesis.
(g) Root and stalk lodging disease.
(h) Pest damage scores.
(i) Days to physiological maturity.
(j) Weight of 1000 seeds or 100 seeds.

17.5.1 Detailed Procedure for Measuring Maize Grain and Stover Yield

In this section, we discuss 12 steps in the measurement of maize grain and biomass of maize from experiments situated on-station or in farmers' fields, focusing on the crop-cut method (Kosmowski et al. 2021). The crop-cut method is widely used in experiments and farmer surveys because it is considered more accurate than the farmer estimation method (Murphy et al. 1991). The steps include:

1. Assembling the required materials.
2. Determining the net harvesting plot.
3. Determine the number of rows and plants in the net plot.
4. Harvest the maize cobs.
5. Determine the total maize cobs field weight.
6. Count the number of cobs.
7. Sort the harvested cobs into categories based on size.
8. Take cob sub-sample field weight.
9. Determine the fresh weight of stover.
10. Select stover sub-sample.
11. Take a sample of about 200 g for analysis.
12. Take the sample to the oven for drying.

Step 1 Assemble the required materials and equipment and cross-check against the harvesting tools list in order to ensure that no necessary tool is missing. These materials include the following:

(i) Weighing balances (suspension balance to weigh bulky materials in kg and an electronic one to weigh the subsample fresh weights in grams). The suspension balance was able to weigh up to 100 kg with 0.1 kg precision, while the electronic balance was able to weigh up to 2000 grams with 0.1 gram precision.
(ii) Panga/machetes/cuttlers.
(iii) Polythene sheet/canvas.
(iv) Marker pens.
(v) Data sheets.
(vi) Wooden plunk for chopping plant biomass.
(vii) Clipboards.
(viii) Sisal twine.
(ix) Buckets or containers to collect the samples before weighing.
(x) Gunny bags.
(xi) Manila paper bags with suitable sizes to put the subsamples.
(xii) Pegs to mark intended harvest areas.
(xiii) Tape measure.

Step 2 Once you get to the field, determine the net harvesting plot by removing two or one guard row(s) from each side (Fig. 17.1). For instance, if your demo plot sizes are 10 m ×

Fig. 17.1 Net plot where maize was harvested from to determine grain yield per unit area. (Courtesy, Abed Kiwia)

Fig. 17.2 Counting the total number of rows and plants in the net plot area. (Courtesy, Abed Kiwia)

Fig. 17.3 A researcher demonstrating harvesting of maize cobs in on-farm mother experiments. (Courtesy, Abed Kiwia)

Fig. 17.4 Weighing total cob fresh weight, from net plots. (Courtesy, Abed Kiwia)

Fig. 17.5 Determining number of cobs from harvested net plot. (Courtesy, Abed Kiwia)

10 m and your maize spacing is 0.8 m × 0.4 m (two plants/hole), and you settle on leaving one outer row/column, then your net plot size would be 9.2 m × 8.4 m. (Note: It is advisable to go for *net plots* and not *quadrants* of 3 m × 3 m because quadrants are too small for precise estimation of socio-economic data or for a true reflection of what farmers would actually get from their bigger fields. In addition, quadrants are recommended for rapid assessment to get a rough estimate of the farmer's yields. However, we want to be as precise as possible in estimating yields).

Step 3 Determine the number of rows and plants in the net plot area by counting (see Fig. 17.2) and record.

Step 4 Harvest the maize cobs by removing them from the stover (Fig. 17.3).

Step 5 Put the maize cobs in gunny bags and weigh them using a suspension balance to determine the total cob fresh weight (kg) (Fig. 17.4).

Step 6 Spread the harvested cobs on a clean and dry canvas and determine their total number by counting (Fig. 17.5).

Step 7 Sort the harvested cobs into three categories—big, medium, and small and then randomly take a sub-sample of three cobs, one from each category (Fig. 17.6).

Step 8 Put the sub-sample cobs into a Manila paper bag whose weight has been tarred. Then weigh the sub-sample to determine its fresh weight (in grams) (Fig. 17.7).

Step 9 Determine the fresh weight (kg) of the stover biomass from the harvested net plot. This is done by bundling the stovers into small quantities that can easily be suspended from a weighing balance. The total weight of all the bundles is determined through summation (Fig. 17.8).

Fig. 17.6 Sorting and grouping harvested cobs into big, medium, and small size to ensure a representative sub-sample is picked. (Courtesy, Abed Kiwia)

Fig. 17.7 Determining subsample cob-fresh weight. (Courtesy, Abed Kiwia)

Fig. 17.8 Determining total stover-fresh weight. (Courtesy, Abed Kiwia)

Fig. 17.9 Chopping sample stovers in order to draw subsamples. (Courtesy, Abed Kiwia)

Fig. 17.10 (**a**) Fresh stover sub-sample taken randomly to help in determining stover biomass yield (Courtesy, Abed Kiwia). (**b**) Randomly taking stover sub-samples for fresh weight determination (Courtesy, Abed Kiwia)

Step 10 Randomly select about 1–2 stovers from each bundle and then chop them into small pieces in order to take stover sub-samples (Fig. 17.9).

Step 11 Spread the chopped biomass thinly on a canvas and then randomly take sub-samples of about 200 g/treatment from 6–9 points across two diagonals (Fig. 17.10a, b).

Step 12 Weigh the stover sub-samples and then take them to the laboratory (lab) for oven drying at 70 °C, to a constant weight. The constant dry weight can be achieved in about four days (Fig. 17.11).

17.5.1.1 Entering Data in Excel During Data Collection

Entering data in a spreadsheet, e.g., Excel as it is being collected, is a good practice for efficient data management (Fig. 17.12a, b, c). This allows for immediate organization and analysis, reducing the risk of data loss and errors. As data is entered, you can perform computations and conversions in real time, such as calculating dry weight from wet weight or converting units. Excel's built-in functions and formulas, such as SUM, AVERAGE, and standard deviation, can be applied to perform these calculations.

The maize grain yield can be adjusted to any desired moisture content. For instance, if we want to adjust the maize grain yield to 12% Moisture content, we shall take the grain oven-dry yield multiplied by 1.12 (see the extract of the MS Excel spreadsheet below).

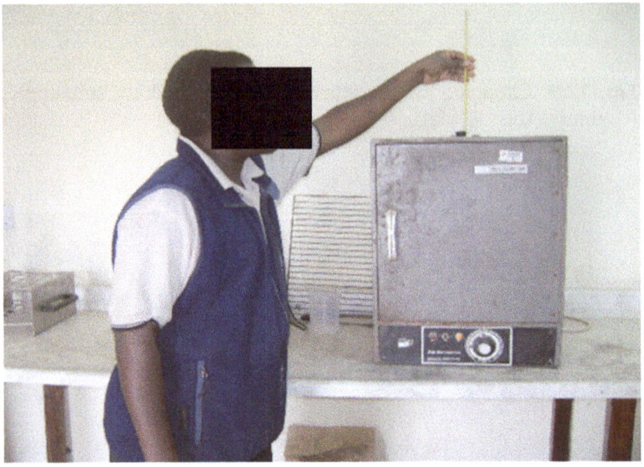

Fig. 17.11 Oven-drying maize cob sub-samples at 70° C to a constant weight

17.6 Economic Data

Cost and benefits data are collected to assess the economic viability of different agronomic practices, indicating economic returns. This includes detailed information on the costs and benefits associated with agronomic practices, including expenses related to seeds, fertilizers, pesticides, labor (CIMMYT 1988). By systematically analyzing cost and benefits data, researchers, farmers, and policymakers can make informed decisions about the profitability and sustainability of various agricultural practices. This data plays a role in optimizing resource allocation, budget planning, and overall financial management. If you are measuring economic data, the following are important resources:

Farm No.
Sampling date...
Name of farmer...
Sampling district ....

										LAB		
Treat No.	Treatment description	Harvest Area (M^2)	Number of Rows	Number of plants	Number of cobs	Total cob Fresh weight (in Kgs)	Cob Sub-sample fresh weight (in G)	Total stover fresh WT.(KG)	Stover sub-sample fresh WT.(G)	Stover Sub-sample dry WT.(G)	Core subsample dry WT.(G)	Grain sub-sample oven dry WT.(G)
1												
2												
3												
4												
5												

Fig. 17.12 (**a**) Data collection tool for use in crop-cut method for determination of maize grain yield (Excel screenshot).

Y2 fx =((Q2*(W2/R2))/O2)*10

	N	O	P	Q	R	S	T	U	V	W	X	Y	Z	AA	AB
1	ROW NC	HARVEST AREA (M	NO OF COBS	TOTAL COB FRESH WT.(K(	COB SUBSAMPLE FRESH WT.(G)	NO OF PLANTS	TOTAL STOVER FRESH WT.(KG)	STOVER SUBSAMPLE FRESH WT.(G)	STOVER SUBSAMPLE DRY WT.(G	CORE SUBSAMPLE DRY WT.(G)	GRAIN SUBSAMPLE OVEN DRY WT.(G)	Core yield (t/ha)	Stover yield (t/ha)	Grain yield (t/ha) oven dry basis	12% adjusted grain yield
2	10	141.18	591	85.0	581.0	600	80.0	334.0	120.0	150.0	310.0	1.6	2.0	3.2	3.6
3	10	141.18	530	75.0	503.0	579	118.0	256.0	115.0	145.0	280.0	1.5	3.8	3.0	3.3
4	9	141.18	479	74.0	483.5	524	97.0	462.0	150.0	135.0	270.0	1.5	2.2	2.9	3.3

Fig. 17.12 (**b**) Feeding formula in Excel to calculate oven-dry maize core yield (Excel screenshot).

`=((Q2*(X2/R2))/O2)*10`

N	O	P	Q	R	S	T	U	V	W	X	Y	Z	AA	AB
ROW NC	HARVEST AREA (M	NO OF COBS	TOTAL COB FRESH WT.(KG	COB SUBSAMPLE FRESH WT.(G)	NO OF PLANTS	TOTAL STOVER FRESH WT.(KG)	STOVER SUBSAMPLE FRESH WT.(G)	STOVER SUBSAMPLE DRY WT.(G	CORE SUBSAMPLE DRY WT.(G)	GRAIN SUBSAMPLE OVEN DRY WT.(G)	Core yield (t/ha)	Stover yield (t/ha)	Grain yield (t/ha) oven dry basis	12% adjusted grain yield
10	141.18	591	85.0	581.0	600	80.0	334.0	120.0	150.0	310.0	1.6	2.0	3.2	3.6
10	141.18	530	75.0	503.0	579	118.0	256.0	115.0	145.0	280.0	1.5	3.8	3.0	3.3
	141.18	479	74.0	483.5	524	97.0	462.0	150.0	135.0	270.0	1.5	2.2	2.9	3.3

Fig. 17.12 (c) Feeding formula in Excel to calculate oven-dry maize grain yield (Excel screenshot).

`=((T2*(V2/U2))/O2)*10`

N	O	P	Q	R	S	T	U	V	W	X	Y	Z	AA	AB
OW NC	HARVEST AREA (M	NO OF COBS	TOTAL COB FRESH WT.(KG	COB SUBSAMPLE FRESH WT.(G)	NO OF PLANTS	TOTAL STOVER FRESH WT.(KG)	STOVER SUBSAMPLE FRESH WT.(G)	STOVER SUBSAMPLE DRY WT.(G	CORE SUBSAMPLE DRY WT.(G)	GRAIN SUBSAMPLE OVEN DRY WT.(G)	Core yield (t/ha)	Stover yield (t/ha)	Grain yield (t/ha) oven dry basis	12% adjusted grain yield
	141.18	591	85.0	581.0	600	80.0	334.0	120.0	150.0	310.0	1.6	2.0	3.2	3.6
	141.18	530	75.0	503.0	579	118.0	256.0	115.0	145.0	280.0	1.5	3.8	3.0	3.3
	141.18	479	74.0	483.5	524	97.0	462.0	150.0	135.0	270.0	1.5	2.2	2.9	3.3

Fig. 17.12 (d) Feeding formula in Excel to calculate oven-dry maize stover (biomass) yield (Excel screenshot)

`=AA2*1.12`

N	O	P	Q	R	S	T	U	V	W	X	Y	Z	AA	AB
ROW NC	HARVEST AREA (M	NO OF COBS	TOTAL COB FRESH WT.(KG	COB SUBSAMPLE FRESH WT.(G)	NO OF PLANTS	TOTAL STOVER FRESH WT.(KG)	STOVER SUBSAMPLE FRESH WT.(G)	STOVER SUBSAMPLE DRY WT.(G	CORE SUBSAMPLE DRY WT.(G)	GRAIN SUBSAMPLE OVEN DRY WT.(G)	Core yield (t/ha)	Stover yield (t/ha)	Grain yield (t/ha) oven dry basis	12% adjusted grain yield
10	141.18	591	85.0	581.0	600	80.0	334.0	120.0	150.0	310.0	1.6	2.0	3.2	3.6
10	141.18	530	75.0	503.0	579	118.0	256.0	115.0	145.0	280.0	1.5	3.8	3.0	3.3
9	141.18	479	74.0	483.5	524	97.0	462.0	150.0	135.0	270.0	1.5	2.2	2.9	3.3

Fig. 17.12 (e) Feeding formula in Excel to calculate maize grain yield at 12% moisture content (Excel screenshot)

- From Agronomic Data to Farmer Recommendations: An Economics Training Manual – CIMMYT Economics Program – Google Books (CIMMYT 1988)
- FAO (2017) ca6428en.pdf (fao.org)
- Guide to measuring input-output data by World Agroforestry Centre (Coe et al. 2019)

17.7 Data Collection in a Variety of Screening Experiments

Varietal screening experiments aim to evaluate new crop varieties or breeding lines to select the most promising. The evaluation assesses a range of traits, such as high yield, drought resistance, and tolerance are evaluated. The data collected mainly includes morphological, agronomic, physiological, and quality traits. Refer to Table 17.3 for the collection of this data.

Proper experimental design is essential for collecting valid and reliable data. The topic of experiments is covered in Chaps. 8 and 9. Here, we discuss a case study of lattice design, a method that effectively controls variability within the experimental field by arranging plots in a systematic grid pattern. The design is commonly used by breeders to screen breeding lines across different environments (Piepho et al. 2006).

17.7.1 Case Study Lattice Design

The lattice design is a special case of the incomplete block design used when the number of treatments is large, such as in progeny or preliminary trials (Akinwale et al. 2021). Lattice designs involve grouping the blocks containing the treatments so that each treatment appears only once in each group (Box 17.1).

Box 17.1

Example of a field layout and randomization for a Striga screening experiment. There are 16 entries each labeled SSS_K1 to SSS_K16. These are randomized in three blocks with treatments for Striga-infested (shaded) and non-infested. Details for each plot are given. Each plot has four rows with a spacing of 0.8 m. Each row has eight plants on a spacing of 0.4 m. The green area represents data collection rows (net plot).

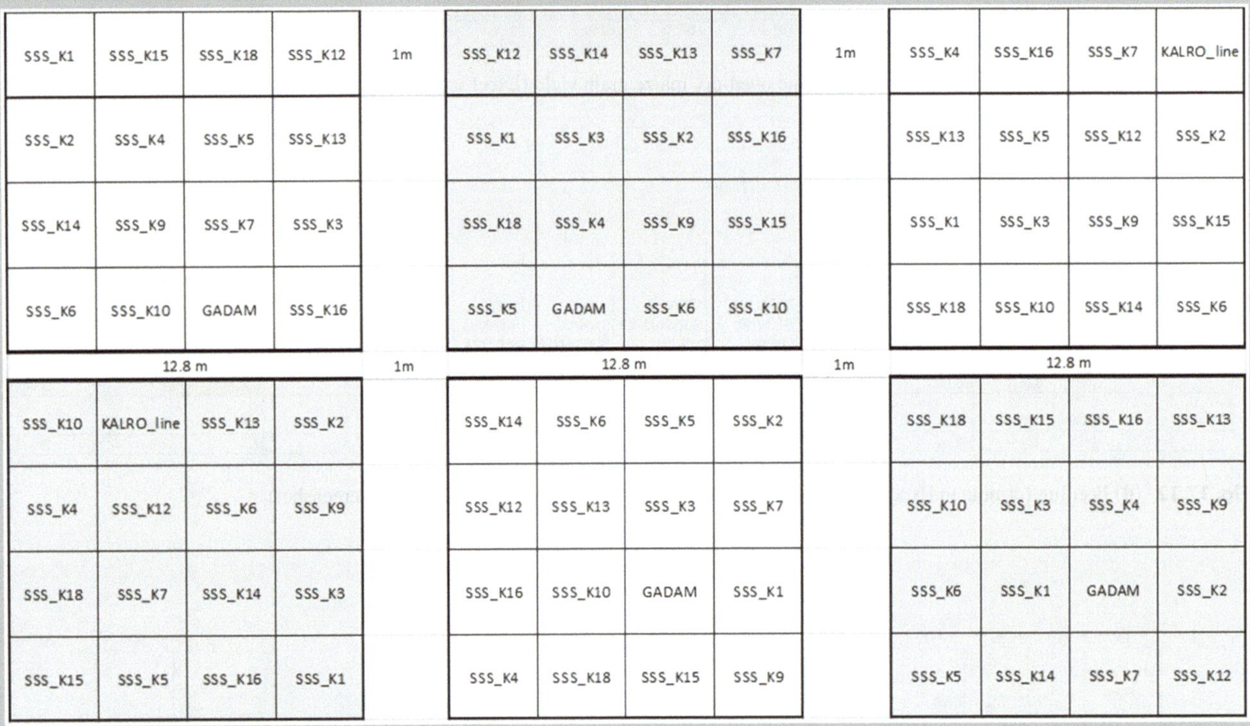

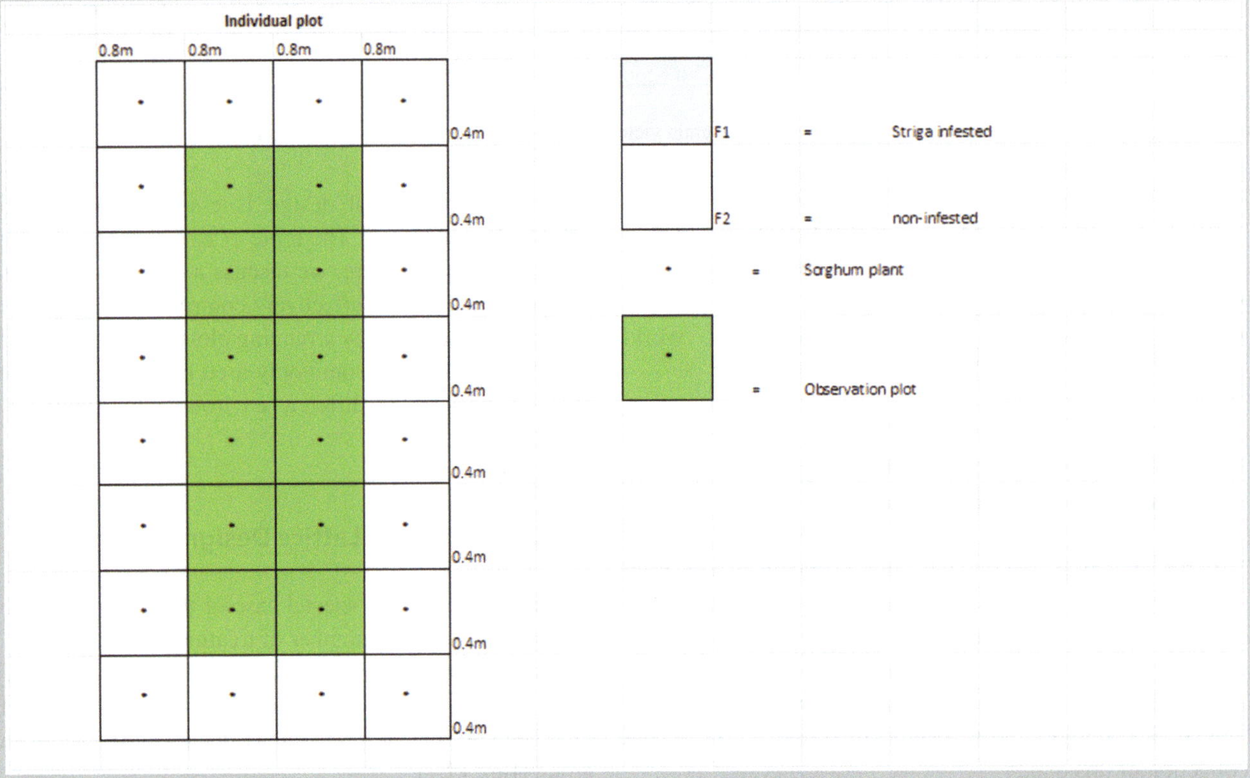

17.7.2 Scheduling of the Trial

Trials are designed and conducted in a particular way to meet specific research objectives. Here, we provide an example of a maize variety trial. The trial will require the knowledge of:

- Trial design and plot sequence (Serpentine or Cartesian)—see Figures.
- Plot size (number of rows, number of planting holes/row, spacings).
- Fertilizer application required.
- Laying the trial basic steps:
  - Choose the most uniform area in the field.
  - Establish a baseline (using the Pythagoras theorem), and from this, mark the corner points of the trial in a square or rectangular fashion, considering provision for border rows and the optimum layout of plots and blocks.
  - Make the planting holes at the optimum at the required spacing.
  - Apply the fertilizer in the planting holes.
  - Label the plots with plot markers. At least mark the first plot.
- Choosing appropriate checks:
  - Each trial should have at least one appropriate local check.
  - The local check is the best available variety nominated by each trial collaborator.
- Border rows:
  - Border rows ensure that the varieties on the edge of the experiment do not have any advantage over the varieties in the center of the experiment and also help to reduce attack by pests on the test varieties.
  - Two rows are ideal as a border.

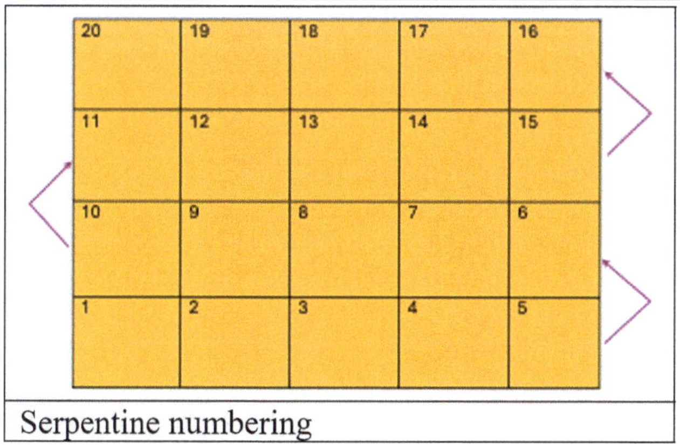

Serpentine numbering

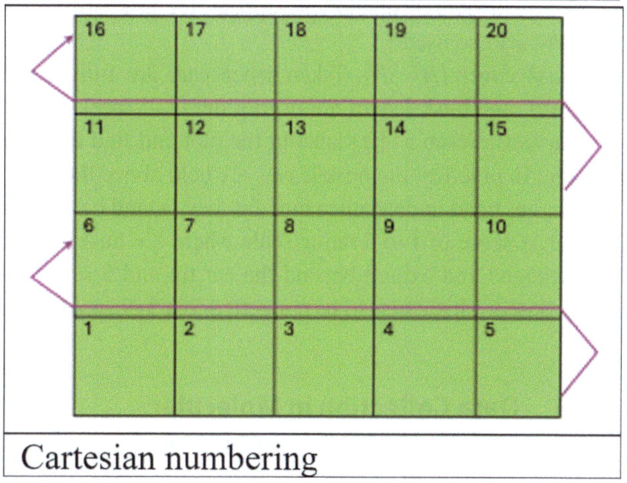

Cartesian numbering

17.7.3 Case Study for Maize Varietal Screening

Good maize experimental data depend on the accuracy and precision of data collection. When to do the collection depends on the trait being analyzed. For instance, flowering data and observations commence during tasseling and are recorded daily to determine when each plot reaches 50% tasseling, anthesis, or silking. Observations must continue until the latest flowering varieties have been recorded. For foliar disease data, data collection is determined by the epidemiology of the disease. If observations are made too early, records may give a false indication of disease resistance. If too late, it may be difficult to differentiate between normal leaf senescence and disease incidence. In most cases, leaf diseases, scoring at the middle of the grain-filling period, is appropriate.

Data should be collected only on competitive plants and, in a four-row plot, for example, from the middle of the two central rows only, leaving the two outside rows as borders as illustrated above (Box 17.1). Maize researchers generally record most of the following traits:

1. *Plant stand (PLST)*. Total number of plants/plot obtained soon after thinning.
2. *Days to tasseling (DYTS)*. Number of days from planting to when 50% of the plants in a plot have emerged tassels.

3. *Days to anthesis (DYANTH).* Number of days from planting to the date when 50% of the plants in a plot have tassels shedding pollen.
4. *Days to silking (DYSK).* Number of days from planting to the date when 50% of the plants in a plot have emerged silks.
5. *Plant height (PLHT).* Average height of plants in centimeters (cm) from the base of the plant to where tassel branching begins. Alternatively, you can estimate plant height by placing a meter stick in the center of the row and determine the average visually.
6. *Ear height (EHT).* Average height in cm from the base of the plant to the node bearing the upper ear (or estimate the distance visually for the whole plot using a meter stick).
7. *Plant aspect (PASP).* A general score for the appearance of the plants in the plot, considering factors such as relative plant and ear heights, uniformity, reaction to diseases and insects, lodging, etc. PASP is rated on a scale of 1 to 5, where 1 = excellent overall phenotypic appeal and 5 = poor overall phenotypic appeal. (A 1–9 rating scale may also be used.)
8. *Husk cover (HUSK).* Taken when ears are fully developed, and husk leaves are drying out (1–3 weeks before harvest). Score 5–10 plants in the plot and find the average. In practice, the husk leaves are held above the ear tip in your hand to determine how far they extend beyond the ear. A score of 1 to 5 rating scale where 1 = husks tightly arranged and extend beyond the ear tip and 5 = ear tips exposed. This can also be converted to a 1–9 scale.

17.8 Data Collection in Molecular and Metabolomic Analyses

In most cases, researchers requiring molecular and metabolomic data collect samples for shipment overseas or for laboratory analysis. Countries have different regulations for shipping plant products, such as plant samples. In general, for the import of an unprocessed plant product, a phytosanitary certificate issued by the National Plant Protection Organization of the exporting country is required. DNA/RNA samples generally do not require a phytosanitary certificate—as pathogens have already been activated during processing. The same would apply to processed plant extracts. However, for samples for measuring labile metabolites or proteins where pathogen-inactivating treatments have not been done, a phytosanitary certificate is required.

This section describes approaches and protocols for collecting and shipping samples for genomic and metabolomic analyses for genome-wide association study (GWAS)/QTL analyses on large collections. The basic protocol allows the application of genomic approaches.

17.8.1 General Protocol for DNA Extraction

The protocols in Box 17.2 are adapted for freeze-dried leaf disk samples for DArT-Seq genotyping and multiplexed PCR analysis. The best DNA results are obtained with young, healthy leaves. Usually, freeze-drying the tissue before homogenization and extraction results in higher-molecular-weight DNA. The protocols describe sample preparation for multiplex PCR or DArT-Seq genotyping (the Basic Protocol is described in Karakas et al. 2023).

Box 17.2 Basic Protocol 1: Preparation of Freeze-Dried Leaf Disks for PCR or DArT-Seq Genotyping (Karakas et al. 2023)

Leaf disks are collected in screw-cap microcentrifuge tubes, freeze-dried, and shipped to a centralized laboratory/service provider for DNA extraction/genotyping. The freeze-drying (lyophilization) method offers a robust and stable format that facilitates easier shipment and storage. Lyophilization removes moisture from various types of samples while preserving their structure (Maisl et al. 2023).

Materials

- Young leaves from plants of interest
- Liquid nitrogen
- 2-ml screw-cap polypropylene microcentrifuge tubes with O-ring and frosted writing surface (e.g., Brand BR780758)
- Parafilm
- Refrigerated microcentrifuge, 4 °C
- Breathe-Easier Sealing Film (e.g., VWR, 10141-844)
- Tube rack, 4 °C
- Freeze-dryer

1. Label 2-ml screw-cap polypropylene microcentrifuge tube on the frosted surface.
2. Collect 1 to 2 leaf disks per plant from young leaves of plants of interest by punching the leaves directly on the plant between the tube rim and a clean Parafilm surface held with the other hand. Twist the tube to make a clean cut. Avoid veins. Keep on ice until Step 3.

The number of leaf disks varies according to the amount of tissue requested by the genotyping provider and the plant species, so it is recommended to perform preliminary tests to determine the optimal amount of starting material or to consult the genotyping company.

(continued)

Box 17.2 (continued)

3. Centrifuge briefly in a refrigerated microcentrifuge to move the disks to the tube bottom.
4. Freeze the tubes containing the leaf disks in liquid nitrogen and store at −80 °C until freeze-drying is performed.
5. For freeze-drying, open the tube and cover the tube opening with a piece of Breathe-Easier Sealing Film.

If Breathe-Easier Sealing Film is not available, you can use Parafilm perforated with a needle.

6. Place the sample tube in a cold tube rack quickly to avoid defrosting and freeze-dry overnight in a freeze-dryer.

Times may vary according to the number of samples and the efficiency of the freeze-dryer.

7. Remove the sealing film and cap the tube tightly.
8. Store at room temperature.

Ship at room temperature as well.

17.8.2 DNA for Microbiome Studies

Soil microbiome analysis is a good way to measure the abundance of various bacterial and fungal species. It is carried out using Next-Generation Sequencing. The protocol (Simmons et al. 2018) provides a good guide for collecting soil samples for microbiome analysis and subsequent DNA extraction.

17.9 Data Collection from Specific Crops

Data collected in research depends on the crop type, reflecting the unique characteristics and growth patterns, and agricultural practices associated with each crop. For instance, researchers focused on staple crops like rice, maize, and rice often gather data on yields, water and nutrient requirements, susceptibility to pests and diseases, and response to various agronomic procedures. Researchers may emphasize market trends and quality parameters in cash crops like tea and coffee. Horticultural crops, including fruits and vegetables, may involve data collection on post-harvest handling, shelf life, and consumer preferences. Table 17.4 shows common data collected from key crops.

Data collected from other common crops

Table 17.4 Morphological and agronomic data collected from common cereals

Crop	Common data collected	References
Maize	Grain yield Stover yield Plant height Leaf area index Cobs/plant Grains/cob Grain weight/cob Weight of 100 grains Harvest index (ratio of seed yield to total above-ground biomass).	Layek et al. (2023)
Rice	Grain weight Straw weight Plant height Leaf area index No of panicles/m2 No of spikelets per panicle No of tillers	De Datta (1981). Luh (1980). Balasubramanian et al. (2007)
Barley	Plant height Grain weight Chlorophyll content Stem girth Panicle length Neck length No of nodes and internodes No of leaves No of tillers Panicle weight	Ullrich (2010). Horsley et al. (2009)
Wheat	Plant height No of tillers Leaf area index No of leaves Leaf length and width Spike length Spikelet count No of grains per spike Grain weight Stem diameter	Bushuk and Rasper (1994). Olgun et al. (2019).
Sorghum	Plant height Leaf area index No of leaves No of tillers No of nodes and internodes No of leaves Leaf size and shape Seed characteristics (size, shape, color) Stem diameter Chlorophyll content	Arkin et al. (1976) Reddy et al. (2004) Smith et al. (2000)

Crop	Data collected	Reference
Kidney beans	Plant height Leaf area index No of leaves No of flowers No of branches No of pods Pod length and width No of seeds per pod Seed size and weight Internodes length Flowers' color and size Iron content	Wang and Cichy (2024)
Soybean	Plant height Leaf area index No of leaves Stem diameter No of pods Pod length and width No. of seeds per pod Length and no of nodes and internodes	Heatherly and Hodges (1998). Coulter et al. (2010). McWilliams et al. (1999).
Green grams	Plant height Leaf area index No of leaves No of branches Stem diameter No of pods Pod length and width No of seeds per pod Seed size and weight	Mugo et al. (2023).
Potato	Plant height Stem diameter No of leaves Leaf size Internode length Node count No of tubers Tuber size Tuber shape Tuber skin color	Stark et al. (2020) Mumia et al. (2018) NPCK (2018)
Tomato	Plant height Leaf area index No of leaves Stem diameter No of fruit set Fruit size and weight Fruit color Fruit shape Branching patterns	Arim et al. (2009). KARLO (2016) UGA Cooperative Extension Bulletin (2017)
Cassava	Plant height Stem diameter No of branches Leaf size and shape Petiole length Root color Root length, diameter, and weight Tuberous root count and size No and length of stolons	Awerije and Rahman (2016) Hillocks and Thresh (2002). Unkovich et al. (2018). Howeler (1988).
Sugarcane	Plant height Leaf length Leaf width Stalk diameter Leaf area index No of internode Internode length No of tillers No of ratoons No of buds No of nodes per stalk	Hunsigi (2012). Bakker (2012).
Avocado	Leaf area index Stem diameter Fruit (size, weight, shape, and color)	Ghosh (2000). Hurtado-Fernández et al. (2018) Davenport (2011).
Coffee	Plant height Leaf area index Internode length Bean color and ripeness Bean size and shape	Amamo (2014). Muhie (2022).

Recommended Further Reading

Crop Genebank Knowledge base
Crop descriptors (cgiar.org)

Key Points in This Chapter

Methods and Instruments for Collecting Data

Data collection is the key to gathering, measuring, and analyzing accurate data from all the relevant sources. A data collection method is the specific technique for acquiring data from a subject or an event, while a tool/instrument is a device.

Validity and Reliability in Data Collection

Validity and reliability are crucial concepts in data collection and measurements. Data must be valid and reliable to be worthy. For data to be valuable, it must possess validity and reliability. Valid data reflects what it is intended to measure, while reliable data is dependable and consistent across different instances. Without these qualities, data loses significance and utility.

Key aspects of validity include: Internal validity, External validity, Construct validity, Content validity, and Statistical conclusion validity.

Other ways of enhancing validity in biophysical experiments may include the preparation of detailed procedures of data collection (protocols), and the Calibration of instruments.

Laboratories

Laboratories are indispensable in agricultural research. A soil and plant laboratory is one of the most important facilities, plays a crucial role in analyzing the physical and chemical properties of soil and plant samples. A well-equipped and well-managed laboratory is essential for ensuring the validity and reproducibility of the analytical results.

Data Collection in Agronomy-Based Experiments

- Data collection is done in phases, usually before the experiment, during the experimentation period, and at the end of the experiment.
- Site characterization data is one of the most important data taken before setting up the experiment. This includes mainly history and soil sampling, and analysis.
- During the experimentation period, various data are taken, including plant sampling and analyses, and monitoring for crop pests and disease.
- Regular data collection of pests and diseases allows timely identification of outbreaks and implementation of appropriate interventions.
- Crop phenology data is also recorded and is important for informing management decisions.
- *Crop yield data* is a fundamental measure in agronomy experiments.

Soil Sampling Considerations and Sampling

Note the following in soil sampling:

- Sampling depth for most soils is typically tillage depth (which can vary depending on the soil type) at an interval of 20 cm. For most annual crops such as cereals, pulses, and vegetables, composite soil samples should be collected from 0–20, 20–40, and 40–60 cm depths. For deep-rooted perennial crops (fruit trees, sugarcane, coffee, plantation crops, and others), composite soil samples should be collected from 0–20, 20–50, 50–80, and 80–110 cm depths using four to five pits.
- *Sampling tools and accessories:* Proper soil sampling should be done using appropriate kits. Different tools are used in soil sampling.
- *Sample labeling:* Proper labeling of soil samples is as important as the care needed to obtain the samples. Samples need to be labeled properly for identification.
- *Precautions during collection and storage of soil samples:* Special care in collecting and handling the soil samples is required to prevent contamination.

Soil and Plant Analysis

Critical stages in plant analysis include several key phases. The first stage is sample collection, which must be performed at the appropriate growth stage of the plant to ensure relevance to the study's objectives. Proper techniques should be used to avoid contamination and ensure representative sampling.

The next stage is sample preparation. This involves cleaning the samples to remove any soil, dust, or external contaminants from the plant material. After cleaning, samples should be dried to a constant weight to prevent decomposition and to standardize the moisture content.

Grinding the samples, followed by extraction, is done. The accuracy of this process is crucial for reliable results.

Following extraction, the analysis stage involves using precise instruments such as spectrometers, chromatographs, or other analytical tools to measure nutrient concentrations or other relevant parameters.

Data Collection in Legumes

Legumes are some of the most important crops globally. Most agronomists will carry out experiments on these crops with different objectives. Although the data collected in an experiment depends on the objectives, growth and yield data are the most common data sets collected. Nitrogen fixation is also an important measurement that is done in experiments that have legumes. There is no single accurate method of measuring nitrogen fixation, and the choice of method depends on the research objectives.

Detailed Procedure for Measuring Maize Grain and Stover Yield

Maize is a crucial crop globally, and accurate measurement of both grain and biomass yield is essential for enhancing productivity and assessing agricultural practices. The crop-cut method is widely used in experiments and farmer surveys because it is considered more accurate than the farmer estimation method. The steps include:

1. Assembling the required materials.
2. Determining the net harvesting plot.
3. Determine the number of rows and plants in the net plot.
4. Harvest the maize cobs.
5. Determine the total maize cobs field weight.
6. Count the number of cobs.
7. Sort the harvested cobs into categories based on size.
8. Take cob sub-sample field weight.
9. Determine the fresh weight of stover.

10. Select stover sub-sample.
11. Take a sample of about 200 g for analysis.
12. Take the sample to the oven for drying.

Data Collection in a Variety of Screening Experiments

Varietal screening experiments select the most promising crop varieties by evaluating their performance across desirable traits such as high yield, drought resistance, and tolerance. A proper experimental design is essential for the collection of valid and reliable data. The lattice design is a special case of the incomplete block design used when the number of treatments is large.

Review Questions

1. What is the difference between a data collection method and an instrument/tool in experimental research?
2. Why is validity important in experimental data collection, and how can it be ensured?
3. What factors must be considered soil sampling and analysis for experimental trials to ensure reliable results?
4. a) Explain the fundamental rationale behind defining and using a "net plot" or "harvest area" in agronomic field experiments. Why are the outer rows and the ends of the inner rows (in this case, 1 meter from each end) typically excluded when collecting yield data? Tip Chap. 9)
5. In what situations would a lattice design be used?
6. Explain why data collection protocols between cereal and legume crops vary. Provide examples.

Chapter Exercises

1. Before establishing a long-term conservation agriculture experiment, a thorough site characterization is important. Based on this:
   (a) Beyond soil nutrient status and texture, list three other essential soil properties that should be assessed during the site characterization phase and explain why each is important for an agronomic experiment.
   (b) Describe a practical strategy for collecting representative soil samples from various locations within a 1-hectare experimental site to assess initial soil fertility. Justify your approach, considering potential spatial variability.
   (c) During your proposal defense, a supervisor asks: "Given the inherent heterogeneity of soils, how do you ensure that your initial soil sampling strategy provides a truly reliable baseline for evaluating treatment effects later?"
2. During the active growth phase of a pigeon pea variety experiment, consistent monitoring is key to understanding crop response to varying conditions.
   (a) Beyond general growth stages, list and briefly describe three specific growth parameters you would collect during the implementation phase to monitor plant development and health, explaining the significance of each.
   (b) Explain the utility of conducting ongoing soil and plant tissue analyses at different critical stages (e.g., flowering, podding) during the implementation phase.
3. You are at the harvesting phase of a variety screening experiment for different upland rice genotypes.
   (a) Describe the key steps involved in accurately collecting grain yield data for upland rice from individual experimental plots, ensuring consistency and minimizing error.
   (b) For a forage sorghum experiment, explain how you would specifically collect data for stover (biomass) yield, and discuss two research objectives where measuring stover yield would be more critical than or equally important to grain yield.
   (c) List at least five common growth parameters and five common yield components that a researcher would typically collect in a general cereal (e.g., wheat, barley) and a general legume (e.g., groundnut, chickpea) variety screening experiment.

Answers to Review Questions

- Q1. Reference: 17.1.1.
- Q2. Reference: 17.1.2 and 17.1.3.
- Q3. Reference: 17.2.3.
- Q4. Reference: 17.5.
- Q5. Reference:17.7.
- Q6. Reference: 17.4 and 17.9.

References

Ackerson JP (2018) Soil sampling guidelines. Purdue University, Purdue Extension
Akinwale RO, Odunlami LK, Eze CE, Oladejo AS (2021) Effectiveness of different alpha lattice designs in the evaluation of maize (Zea mays L.) genotypes in a rainforest agro-ecology. Heliyon 7(7):e07414. https://doi.org/10.1016/j.heliyon.2021.e07414
Amamo AA (2014) Coffee production and marketing in Ethiopia. Eur J Bus Manage 6(37):109–122
Arim KS, Otieno S, Efukho C, Mbuthia G, Mangoli F, Oirere Z, Mbuthia E, Malenge F, Aikawa J, Kita K, Kitajima H, Takashina

Y, Seo T (2009) Guidelines in tomato production. Smallholder Horticulture Empowerment & Promotion Project for Local and Up-Scaling (SHEP PLUS)
Arkin GF, Vanderlip RL, Ritchie JT (1976) A dynamic grain sorghum growth model. Trans ASAE 19(4):622–0626
Awerije BO, Rahman S (2016) Cassava: production, consumption and potential uses
Badu-Apraku B, Oyekunle M (2012) Genetic analysis of grain yield and other traits of extra-early yellow maize inbreds and hybrid performance under contrasting environments. Field Crop Res 129:99–110
Bakker H (2012) Sugar cane cultivation and management. Springer Science & Business Media
Balasubramanian V, Sie M, Hijmans RJ, Otsuka K (2007) Increasing rice production in sub-Saharan Africa: challenges and opportunities. Adv Agron 94:55–133
Betz JM, Brown PN, Roman MC (2011) Accuracy, precision, and reliability of chemical measurements in natural products research. Fitoterapia 82:44–52. https://doi.org/10.1016/J.FITOTE.2010.09.011
Beyan SM, Wolde-meskel E, Dakora FD (2018) An assessment of plant growth and N_2 fixation in soybean genotypes grown in uninoculated soils collected from different locations in Ethiopia. Symbiosis 75:189–203. https://doi.org/10.1007/s13199-018-0540-9
Bushuk W, Rasper VF (eds) (1994) Wheat: production, properties and quality. Springer Science & Business Media
Carter MR, Gregorich EG (2007) Soil sampling and methods of analysis, 2nd edn. CRC Press, Boca Raton
Chalk P (2008) Measuring plant-associated nitrogen fixation in agricultural systems. ACIAR Monograph No. 136, 258 pp
CIMMYT (1988) From agronomic data to farmer recommendations: an Economics Training Manual – CIMMYT Economics Program – Google Books
Clark J (ed) (2019) Recommended soil sampling methods for South Dakota
Coe R, Fitzpatrick I, Gassner A (2019) Guide to measuring farm input and output data. World Agroforestry (ICRAF) Conservation Service, Nairobi
Coulter J, Moncada KRISTINE, Sheaffer CRAIG (2010) Soybean production. Risk management guide for organic producers. University of Minnesota, pp 10–11
Danso SKA (1995) Assessment of biological nitrogen fixation. Fert Res 42:33–41. https://doi.org/10.1007/BF00750498
Davenport TL (2011) Avocado flowering. Horticult Rev 8:257–289
De Datta SK (1981) Principles and practices of rice production. Int Rice Res Inst
Dowling CW, Blaesing DT (2022) Plant sampling for agriculture—a guide. Fertilizer Australia, Canberra
Egea González FJ, Hernández Torres ME, Frenich AG, et al (2004) Internal quality-control and laboratory-management tools for enhancing the stability of results in pesticide multi-residue analytical methods. TrAC Trends Anal Chem 23:361–369. https://doi.org/10.1016/S0165-9936(04)00520-5
Esilaba AO, Miriti J, Gitari J, Mangale N, Kathuku-Gitonga AN, Njiru E, Mwangi H (2021) KCEP-CRAL integrated soil fertility and water management extension manual. Kenya Agricultural and Livestock Research Organization, Nairobi
FAO (2017) Productivity and efficiency measure in agriculture
FAO (2020) Soil testing methods—global soil doctors Programme – a farmer-to-farmer training programme. Rome. https://doi.org/10.4060/ca2796en
Flynn R, Ball ST, Baker RD (1999) Sampling for plant tissue analysis. Cooperative extension service. College of Agriculture and Home Economics. Guide A-123. GNew Mexico University https://pubs.nmsu.edu/_a/A123/
Gail M, Sullivan (2011) A primer on the validity of assessment instruments. J Grad Med Educ 3(2):119–120. https://doi.org/10.4300/JGME-D-11-00075.1
Ghosh SP (2000) Avocado production in India. Avocado production in Asia and the Pacific, pp 24–30
Giller KE (2001a) Nitrogen fixation in tropical cropping systems. Cabi
Giller KE (2001b) Nitrogen fixation in tropical cropping systems. CABI Books (cabidigitallibrary.org)
Hardarson G, Danso SKA (1993) Methods for measuring biological nitrogen fixation in grain legumes. Plant Soil 152:19–23. https://doi.org/10.1007/BF00016330
Heatherly LG, Hodges HF (1998) Soybean production in the midsouth. CRC Press
Hillocks RJ, Thresh JM (eds) (2002) Cassava: biology, production and utilization. CABI Publishing
Horsley RD, Franckowiak JD, Schwarz PB (2009) Barley. In: Cereals. Springer US, pp 227–250
Howeler, R. H. (1988). Agronomic practices for cassava production in Asia. In: Cassava breeding and agronomy research in Asia: Proceedings of a regional workshop held in Rayong, Thailand
Hunsigi G (2012) Production of sugarcane: theory and practice, vol 21. Springer Science & Business Media
Hurtado-Fernández E, Fernández-Gutiérrez A, Carrasco-Pancorbo A (2018) Avocado fruit—*Persea americana*. In: Exotic fruits. Academic Press, pp 37–48
Jenkins EM, Galbraith J, Paltseva AA (2025) Portable X-ray fluorescence as a tool for urban soil contamination analysis: accuracy, precision, and practicality. SOIL 11:565–582. https://doi.org/10.5194/soil-11-565-2025
Karakas E, Ferrante P, Schafleitner R, Giuliano G, Fernie AR, Alseekh S (2023) Plant sample collection and shipment for multi-omic analyses and phytosanitary evaluation. Curr Protoc 3(12):e952. https://doi.org/10.1002/cpz1.952
KARLO (2016) Tomato production manual. Kenya Agricultural and Livestock Agricultural Organization, Kenya
Kool H, Andersson JA, Giller KE (2020) Reproducibility and external validity of on-farm experimental research in Africa. Exp Agric 56:587–607. https://doi.org/10.1017/S0014479720000174
Kosmowski F, Chamberlin J, Ayalew H, Sida T, Abay K, Craufurd P (2021) How accurate are yield estimates from crop cuts? Evidence from smallholder maize farms in Ethiopia. Food Policy 102(2021). https://doi.org/10.1016/j.foodpol.2021.102122
Kristensen E, Nielsen DB, Jensen LN, et al (2008) A mixed methods inquiry into the validity of data. Acta Vet Scand 50:1–8. https://doi.org/10.1186/1751-0147-50-30/FIGURES/1
Layek J, Das A, Ansari MA, Mishra VK, Rangappa K, Ravisankar N et al (2023) An integrated organic farming system: innovations for farm diversification, sustainability, and livelihood improvement of hill farmers. Front Sustain Food Syst 7:1151113
Leung L (2015) Validity, reliability, and generalizability in qualitative research. J Family Med Prim Care 4:324. https://doi.org/10.4103/2249-4863.161306
Luh BS (1980) Rice: production and utilization. AVI Publishing Co.
Maisl C, Doppler M, Seidl B, Bueschl C, Schuhmacher R (2023) Untargeted plant metabolomics: evaluation of lyophilization as a sample preparation technique. Meta 13(6):686
Mangale N, Muriuki A, Kathuku-Gitonga AN, Kibunja CN, Mutegi JK, Esilaba AO et al (2016) Field and laboratory research manual for integrated soil fertility management in Kenya. Kenya Soil Health Consortium 77(25):256–363
McWilliams DA, Berglund DR, Endres GJ (1999) Soybean growth and management quick guide.
Motsara MR, Roy RN (2008) Guide to laboratory establishment for plant nutrient analysis. FAO fertilizer and plant nutrition Bulletin 19. FAO
Mugo J, Opijah F, Ngaina J, Karanja F, Mburu M (2023) Simulated effects of climate change on green gram production in Kitui County. Kenya Front Sustain Food Syst 7. https://doi.org/10.3389/fsufs.2023.1144663

Muhie SH (2022) Strategies to improve the quantity and quality of export coffee in Ethiopia, a look at multiple opportunities. J Agric Food Res 10:100372

Mumia BI, Muthomi JW, Narla RD, Nyongesa MW, Olubayo FM (2018) Seed potato production practices and quality of farm-saved seed potato in Kiambu and Nyandarua counties in Kenya

Murphy J, Casley DJ, Curry JJ (1991) Farmers' estimations as a source of production data: methodological guidelines for cereals in Africa. World Bank, Washington, DC

NCPK (2018) Potato production handbook. A guide for farmers and trainers. National Potato Council of Kenya

Nimon K, Zientek LR, Henson RK (2012) The assumption of a reliable instrument and other pitfalls to avoid when considering the reliability of data. Front Psychol 3:21611. https://doi.org/10.3389/FPSYG.2012.00102/BIBTEX

Okalebo JR, Gathua KW, Woomer PL (2002) Laboratory methods for soil and plant analysis: a working manual. TSBF, Nairobi

Olgun M, Sezer O, Turan M, Basciftci ZB, Ayter NG, AydinIN D, Ardic ÇM, Koyuncu O (2019) Determination of relationship between minerals and yield components by different statistic methods in bread wheat (Triticum aestivum L.) under irrigated conditions. J Appl Biol Sci 11(1):21–28. Retrieved from https://www.jabsonline.org/index.php/jabs/article/view/522

Pausch RC, Mulchi CL, Lee EH, Meisinger JJ (1996) Use of 13C and 15N isotopes to investigate O3 effects on C and N metabolism in soybeans. Part II. Nitrogen uptake, fixation, and partitioning. Agric Ecosyst Environ 60(1):61–69. https://www.sciencedirect.com/science/article/pii/S0167880996010626

Peoples MB, Faizah AW, Rerkasem B, Herridge DF (1989) Methods for evaluating nitrogen fixation by nodulated legumes in the field. ACIAR Monograph No. 11, vii + 76 p

Peoples MB, Unkovich MJ, Herridge DF (2009) Chapter 6: measuring symbiotic nitrogen fixation by legumes. In: Emerich DW, Krishnan HB (eds) Agronomy Monographs: nitrogen fixation in agriculture, forestry, ecology, and the environment. American Society of Agronomy. https://doi.org/10.2134/agronmonogr52.c6

Piepho HP, Büchse A, Truberg B (2006) On the use of multiple lattice designs and α-designs in plant breeding trials. Plant Breed 125(5):523–528

Reddy BVS, Ramesh S, Reddy PS (2004) Sorghum breeding research at ICRISAT-goals, strategies, methods and accomplishments. Int Sorghum Millets Newslett 45:5–12

Repke, Lydia; Birkenmaier, Lukas; Lechner, Clemens M (2024) Validity in survey research - from research design to measurement instruments. Mannheim, GESIS – Leibniz-Institute for the Social Sciences (GESIS – Survey Guidelines). https://doi.org/10.15465/gesis-sg_en_048

Rio S, Akdemir D, Carvalho T, Sánchez JI y. (2022) Assessment of genomic prediction reliability and optimization of experimental designs in multi-environment trials. Theor Appl Genet 135:405–419. https://doi.org/10.1007/S00122-021-03972-2/FIGURES/5

Simmons T, Caddell DF, Deng S, Coleman-Derr D (2018) Exploring the root microbiome: extracting bacterial community data from the soil, rhizosphere, and root endosphere. J Vis Exp. https://doi.org/10.3791/57561

Singh D, Chhonkar PK, Dwivedi BS (2005) Manual on soil, plant and water analysis, 200p. Westville Publishing House, New Delhi

Smith JSC, Kresovich S, Hopkins MS, Mitchell SE, Dean RE, Woodman WL, Porter K (2000) Genetic diversity among elite sorghum inbred lines assessed with simple sequence repeats. Crop Sci 40(1):226–232

Soil Survey Staff. (2014). Soil survey field and laboratory methods manual. Soil Survey Investigations Report No. 51, Version 2.0. R. Burt and Soil Survey Staff (ed.) U.S. Department of Agriculture, Natural Resources

Stark JC, Thornton M, Nolte P (eds) (2020) Potato production systems. Springer Nature

Suter B (2009) Validity in qualitative research on personal relationships. Kentucky J Commun 28(2)

Taverniers I, De Loose M, Van Bockstaele E (2004) Trends in quality in the analytical laboratory. II. Analytical method validation and quality assurance. TrAC Trends Anal Chem 23:535–552. https://doi.org/10.1016/J.TRAC.2004.04.001

UGA extension bulleting Commercial Tomato Production Handbook (2017). https://secure.caes.uga.edu/extension/publications/files/pdf/B%201312_7.PDFhttps://www.infonet-biovision.org/PlantHealth/Crops/potato

Ullrich SE (2010) Barley: production, improvement, and uses, vol 12. John Wiley & Sons

Unkovich M, Herridge D, Peoples M, Cadisch G, Boddey R, Giller K, Alves B, Waisundara V (eds) (2018) Cassava. BoD–Books on Demand

Vanlauwe B, Coe RIC, Giller KE (2019) Beyond averages: new approaches to understand heterogeneity and risk of technology success or failure in smallholder farming. Exp Agric 55:84–106. https://doi.org/10.1017/S0014479716000193

Wang W, Cichy KA (2024) Genetic variability for susceptibility to seed coat mechanical damage and relationship to end-use quality in kidney beans. Crop Sci 64(1):200–210. https://doi.org/10.1002/csc2.21122

Zhao H, Ghirlando R, Alfonso C, Arisaka F, Attali I, Bain DL, et al (2015) A multilaboratory comparison of calibration accuracy and the performance of external references in analytical ultracentrifugation. PLoS ONE 10(5):e0126420. https://doi.org/10.1371/journal.pone.0126420

18 Methods and Tools of Data Collection in Non-experimental Designs (Surveys)

Abstract

Surveys, qualitative, and participatory research use closely related methods and tools to gather data. The main methods are questioning, interviewing, observing, and collecting documents, and audio-visual materials. Six steps are followed in designing a survey tool and collecting the data. These include deciding on the questions to include, writing the questions, formatting the questionnaire, presenting the instrument, analyzing pretesting data and final editing, and lastly, data collection. In writing, the questions use simplicity and avoid leading questions. A checklist of 16 items is given for reference as you construct a questionnaire. The instrument's formatting helps enhance the appeal, which is important for attracting respondents. In conducting an interview survey, consider the following key activities: 1. Planning the interview, 2. Recruitment and training of enumerators, 3. Recruitment of respondents and gaining access; and 4. Finally, collecting the data. Three key activities are involved in the questionnaire survey: pre-survey communication, sending out the questionnaire, and following up. Questions should be organized logically and naturally. The main types of questions in a survey instrument include quantity, category-based, dichotomous, and rank order, Rating scales such as Likert scale, ranking and scoring, semantic differential, and matrix questions. By following these guidelines, researchers can ensure their data collection instruments are well-designed and effectively gather the necessary data. A sample questionnaire is also presented.

Keywords

Double-barreled questions · Biased questions · Discipline-specific questions · Structured and non-structured instruments · Interview survey · Leading questions · Methods and tools · Likert Scale Data · Training enumerators · Questionnaire survey · Standardized survey instruments · Questioning · Quantity question · Interviewing · Interview survey guide · Pretesting · Ranking and scoring questions · Rank order questions · Matrix questions · Semantic differential questions · Questionnaire survey guide

Learning Outcomes

After reading this chapter, you should be able to

1. Explain the meaning of questioning and the various methods and instruments used in surveys.
2. Differentiate between carrying out a survey using a questionnaire and an interview survey guide.
3. Differentiate the types of survey instruments based on the type of questions.
4. Identify the steps involved in preparing a survey research instrument.
5. Describe characteristics of a well-designed and formatted survey instrument.
6. Describe how to write effective survey questions.
7. Outline the key steps in conducting an interview survey.
8. Apply principles and rules of writing questions in a survey tool.

J. Njeri Mugwe, S. Runo, *Research Methodology in Agricultural Sciences*, https://doi.org/10.1007/978-981-95-1892-0_18

9. Recognize and avoid double-barreled questions, leading questions and biased questions in a survey tool.
10. Differentiate between Likert scale, Semantic differential and Matrix-based questions in a survey tool.
11. Apply the key steps in conducting a questionnaire survey.
12. Appreciate the importance of pre-testing a survey tool.

18.1 Overview of Data Collection Methods and Tools in Surveys and Qualitative Approaches

In the categorization of research designs (Chap. 7), the following broad designs were given:

- Experimental: comprising true experiments, quasi-experiments, and pre-experimental designs.
- Non-experimental: comprising surveys, qualitative approaches, and participatory research.

This chapter and the next chapter (19) in this book discuss data collection in non-experimental research designs: surveys, qualitative, and participatory approaches. This chapter focuses on the survey, while Chap. 19 focuses on qualitative approaches.

Methods and Tools

Surveys, qualitative, and participatory research use various methods and tools to gather data (Table 18.1). The main methods are questioning, interviewing, observing, and collecting documents and audio-visual materials. The tools include interview guides/schedules, questionnaires, observation protocols, checklists, cameras, and sampling materials. Surveys focus on quantifiable data despite their different emphasis, while qualitative and participatory surveys focus on in-depth understanding.

Table 18.1 Methods of data collection and tools used in collecting data in surveys and qualitative approaches

Method	Tool/instrument used
Questioning	Questionnaire
Interviewing: Face-to-face, telephone, video conferencing	Questionnaire/interview schedule
Recall	Recall questions
In-depth interviews	Interview schedule with open-ended questions
Observation	Observation schedule
Focus group discussion	Checklists or interview guide
Key informant interviews	Checklist or interview guide
Participatory research methods	Participatory research tools
Content analysis	Document analysis guide records, audiovisual materials, images, scanning devices, and remote sensing devices
Collating documents, photos, artifacts, samples, and audiovisual material	Cameras, sampling bags, recording devices, e.g., tapes

Data Collection in Questionnaire Survey and Interview Survey

As you read this chapter, refer to Chap. 12 on surveys. Chapter 12 explained that researchers conduct a survey using two basic procedures: an interview survey (researcher-completed instruments) or a questionnaire survey (respondent-completed instruments). When conducting an interview survey, we use an *interview survey guide*, while in the questionnaire survey, we use *questionnaires* that we send to respondents to fill. The commonality in these two instruments is that they guide respondents in answering the questions.

In our case, respondents can be farmers, agricultural extension officers, agriculture professionals, and other stakeholders in the value chain. Researchers prefer to use interview-completed interviews in a study involving farmers because of farmers' low literacy levels, especially in developing nations. The preference is also because researchers would like to visit the farms to deepen their understanding of the phenomenon being studied. However, researchers conducting studies involving extension officers or students may prefer questionnaires, where respondents do self-reporting. For instance, tracer studies involving international students. Self-reporting is an efficient way to administer a questionnaire to international students because they can be sent via email or use online platforms. Online platforms are discussed in Chap. 12.

The survey instruments contain a set of standardized questions, often called *items*. These items are used to pose the questions to the respondents. Each item/question in the instrument is developed to address the study's specific objective, research question, or hypothesis. As a researcher, you must also know how information obtained from each item will be analyzed. This information will give you the answers you are looking for.

Interview-Based Surveys Interview-based surveys offer several advantages over questionnaire surveys (Table 18.2). The direct interaction between the interviewer and respondent allows for further probing and clarification of information. Further interview surveys often yield higher response rates compared to questionnaire surveys. Overall, they provide richer data. However, interview surveys have some disadvantages (Saunders et al. 2019). They are resource-intensive and involve traveling and training interviewers. The presence

Table 18.2 Differences between interview and questionnaire surveys

Aspect	Interview survey	Questionnaire survey
Definition	The researcher questions the respondents directly *Investigator-completed*	Respondents answer written or digital questions on their own *Interviewee-completed*
Purpose	Use personal interaction to gather in-depth and detailed data	Collect standardized data
Resources	Higher cost and time investment	Lower cost and time investment
Administration	Administered by the interviewer	Self-administered by the respondent
Suitability	Suitable for individuals with low literacy, e.g., farmers and when observation adds value	Suitable for persons with good literacy, e.g., extension workers and students
Response rates	Typically, higher due to personal interaction	Often lower, especially for emailed and online questionnaires
Depth of information	Opportunity to probe further	Limited to questions listed
Clarification	Interviewers can clarify questions	Respondents may misunderstand the questions with no opportunity for clarification
Flexibility and verification	Interviewers can verify the information and also explore further	Limited ability to explore further the topics
Bias	Personal interaction increases the potential for bias	Less potential for bias as respondents fill out the questionnaire on their own

of an interviewer can also introduce bias. Questionnaire-based surveys offer several advantages. They are cost-effective, especially when distributed online. However, they have a lower response rate, and misinterpretations are possible.

What Is Questioning and Interviewing?

Questioning or interviewing is asking questions to gather information from the target audience, including individuals and groups, using pre-determined questions. Questioning requires careful planning, considering the research objective, target audience, and data type needed. The questions guiding the questioning must be carefully prepared to extract specific information from the respondents.

Types of instruments in surveys

Survey instruments can be classified into three types depending on the nature of the questions. Three data collection tools depend on the type of question: *structured*, *unstructured*, and *semi-structured*.

The *structured instruments* have predetermined questions that are definite and concrete. The questions are presented with exact wording and are in the same order for all the respondents. They have fixed alternatives that respondents answer to, making it easier to analyze quantitatively. They are said to have *closed-ended questions*.

The *unstructured instrument* does not have definite and concrete questions. They lack predefined categories and rely on guiding questions. The *questions are open-ended*. The researcher uses the guiding questions to probe the interview to gain a deeper understanding of the topic.

Semi-structured instruments have a mix of *closed- and open-ended questions*, enabling quantitative and qualitative data collection.

Standardized survey instruments

Depending on your study, you may find a standardized survey instrument. Standardized instruments are *pre-developed instruments* tested for validity and reliability. The benefits of using a standardized questionnaire include consistency, reliability, validity, comparability, and efficiency. An example is the "Household Food Insecurity Access Scale (HFIAS) tool," designed to measure food security at the household level. It assesses the availability and accessibility of food and the household's ability to secure adequate food. Within gender, The Women's Empowerment in Agriculture Index (WEAI) is a standardized tool measuring women's empowerment, agency, and inclusion in the agricultural sector. It includes questions about decision-making, access to resources, and community leadership.

Despite the benefits of standardized questionnaires, students are often required to develop their instruments or questionnaires tailored to their specific research objectives and context. Developing customized tools allows students to address unique research questions catering to their study population's needs and characteristics. It also allows students to learn and better understand the research process.

18.2 Steps in Designing a Survey Instrument and Collecting Data

18.2.1 Key Steps and Activities

Designing a good survey instrument is challenging. It should, therefore, be prepared with great care to ensure you have an effective data collection tool. Before starting to design the instrument, ensure you do the following three activities:

- Articulate the research question/objective that the research intends to address.
- Determine the hypothesis around which the questionnaire is to be designed.

- Prepare a data collection and analysis matrix table to determine the variables to be included. Some variables will be *independent variables* to predict *dependent variables*, while others will be important for *describing the sample*.

An effective survey instrument is one that uses the least number of *well-worded and organized questions* to solicit accurate information from the respondents (Slattery et al. 2011; Rolstad et al. 2011; Mes et al. 2019; Sharma and Ruikar 2025). It means that it should not contain unnecessary questions and should include all questions relevant to achieving the research objectives. Typically, its design should consider the researcher as well as the respondent. Questions should be worded and organized to allow the researcher to obtain the most complete and accurate information possible to enable a sound analysis and interpretation. In the same vein, it should allow respondents to give the necessary information easily. The questions should also be brief and to the point to keep respondents interested throughout the interview.

Six steps in designing a survey and collecting data (Artino et al. 2014; Oosterveld et al. 2019):

1. Decide on the information needed by making a decision on critical elements.
2. Write the items (questions).
3. Format the instrument.
4. Pretest the instrument.
5. Analyze pretesting data, evaluate, and finally edit the instrument.
6. Collect the data.

The steps are captured in Table 18.3.

Table 18.3 The six steps involved in preparing and designing of a survey instrument

Step	What to do?	Activities needed
ONE	Decide on the information needed	Depend on the nature of the research problem, research questions, and objectives of the study Define target population and sampling frame Determine sampling design, sample size, and expected response rate Make a preliminary decision about the method of administering the instrument Ensure concepts and constructs are defined and operationalized Determine the level of measurement and units Prepare data requirements and analysis table
TWO	Writing the items	Refresh on the rules of preparing a good questionnaire Determine the type of question Check wording for clarity Decide on the grouping into various modules Decide on the overall length of the questionnaire
THREE	Formatting the instrument	Decide on the ordering and flow of questions (sequencing) Define format, structure, and visual presentation Explain the purpose of the questionnaire Provide an introduction and middle sections Arrange the closing section
FOUR	Pretesting the instrument and final editing	Collect data from a sample similar to the final respondents; Assess the validity and reliability of the questions
FIVE	Analysis of pretesting data and final editing of the instrument	Analyze the data to determine if the questions answer your research questions and achieve your research objectives Revise questions, wording, and sequencing, if necessary Rework the data requirement and analysis table and make it final
SIX	Administering the instrument and data collection	Recruit the participants Use the best practice for administering the instrument Determine the deadline and follow-up methods Ensure you have a plan for handling the completed questionnaire Arrange logistics for face-to-face interviews, online surveys, etc.

18.2.2 Deciding on the Information Required

Before preparing the survey instrument, you must be clear about what is being studied and what is expected from the study. The reason for this is obvious: the questionnaire is to collect data to fulfil the study's objectives. At the heart of data collection, the instrument should allow us to collect the most complete and accurate data logically. This is done to reach reliable conclusions from what we plan to observe. A well-designed instrument should meet the research goal and objectives and minimize unanswered questions, a common problem in many surveys.

Consider the following factors and make critical decisions:

1. *Define the target population and sampling frame:* This should ensure that the sampling frame and the sampling

methods selected optimize the reliability and generalizability of the findings (refer to Chap. 16) for differences between the target population and sampling population).

2. *Determine sampling design and sample size:* Considering factors such as population, desired precision, and available resources, make a decision on the sampling design from among the probability sampling designs (Chap. 16, Sect.16.2.1). Guided by the same factors, also decide and calculate the suitable sample size (Chap. 16, Sect. 16.3) and work out the suitable sample size.
3. *Make a preliminary decision about the method of data collection:* Decision on the data collection method (face-to-face, telephone, or mailed questionnaires) influences the writing of the questions. The method should suit the target population and the nature of the information.
4. *Define and operationalize all concepts and constructs:* This guarantees that all variables will be measurable and that the items/questions accurately capture the intended variables. In other words the inferences obtained will be valid (correct). This process involves specifying exactly how each concept will be measured and the indicators used. For instance, if you are collecting data on the wealth of households, how will you operationalize this construct? In other words, how will you capture this in the instrument?
5. *Decide on the level of measurement and units:* Decide on the level of measurement (Nominal, ordinal, interval, or ratio) and the units for each question. The decision affects how questions are phrased and how the responses are recorded. Consequently, the level of measurement will determine the method of analysis. For example, if you intend to carry out Pearson product–moment correlations between age and yields, collecting age in categories will be a mistake.
6. *Determine data requirement and analysis table:* This table will show your objectives, the type of data required, and how data will be collected and analyzed (Table 18.4). The table enhances clarity of the data collection procedures and provides a structured approach to managing and analyzing data. At this early stage, you are able to see how the objectives are linked with the data collected and data analysis.

18.2.3 Writing the Questions

Writing effective questions that obtain the information you require is a daunting task requiring care and skill. It involves drafting and rewriting items until they are clear and succinct. A common problem with students is collecting unnecessary data, which wastes time and resources. This happens when the questions do not align with the research objectives. Conversely, there are instances where a student fails to collect data that answers the objectives (Box 18.1 on pitfalls to avoid). This results in gaps in the analysis, which affects the overall study. Students and researchers must carefully design instruments to avoid these issues, ensuring each question is linked to the research objectives.

Box 18.1: Key Pitfalls to Check When Writing Questions

1. Have you included all questions and variables that will answer your research objectives?
2. Have you included questions that are not required?
3. Have you included all variables as per your conceptual framework?
4. Do you know how you will analyze each of the questions?
5. Can you mark and match each of the questions to the objectives of the studies and vice-versa?

Table 18.4 Example of data requirement and analysis table

Objective	Type of data to be collected: Dependent variable	Type of data to be collected: Independent variable	Method of data collection	Method of data analysis
To evaluate the effect of socioeconomic characteristics of farmers on the adoption of tied-ridging for soil moisture conservation	Tied ridges	Total farm size (acres) (continuous) Availability of off-farm income (0 = no, 1 = yes) Livestock ownership (0 = no, 1 = yes) Access to extension(trainings) (0 = no, 1 = yes) Household size (continuous), access to credit (0 = no, 1 = yes), etc.	Face-to-face interviews using an interview schedule	Descriptive statistics Multivariate probit regression

The following questions can be asked in preparation of the instrument:

1. What type of information do you need from these questions?
2. How will the answers serve the aims of the instrument?
3. How will you present the findings from answers to this question?
4. Do you need numerical scores or nominal data from the answers?

The questions will be guided by the research questions and mainly include facts, current practices, opinions and beliefs, and past and future behaviors (Chap. 12).

You are advised to refresh your knowledge of the rules for preparing a good survey instrument (Box 18.2). Following these rules and regulations will help you prepare good questions that will help participants understand the question to provide merciful answers. The key ones include:

Leading question: A leading question is written to lead people to respond in a certain way, which is biased as it is influenced by the way the question is asked. For example, *Do you believe that improved maize varieties yield better than traditional maize varieties? or do you believe that selling bananas collectively will increase prices?* These are leading questions. They include emotionally charged words, such as better yields, which may influence the way people respond and thus lead to biased conclusions.

Double-barreled questions: avoid double-barreled questions that ask two things in one question. They are difficult to understand and answer. For instance, *should farmers be given more credit facilities for buying livestock and investing in fertilizer?* This question carries three items: credit facilities and buying livestock and fertilizer. Such a question is confusing to the respondent. The best way of phrasing this is to have two questions, e.g., *Should farmers be given credit facilities for buying livestock?* 2. *Should farmers be given credit facilities for buying fertilizer?*

Avoid complex and technical language. Many students fall into the trap of using *discipline-specific* or highly technical terms. Technical language is complex and make questions difficult for the respondents to understand. Instead, use straightforward language with no jargon. For example, a soil scientist understands integrated soil fertility management (ISFM) well, but to a farmer, this is jargon. Similarly, integrated pest management (IPM) is well understood for a crop protection specialist, but for a farmer, it is jargon. This is a common problem with students as they construct their survey instruments. It would be wrong to have the following questions in the survey instrument:

Box 18.2: Principles and Rules of Questions Construction in a Survey Instrument

1. Prepare questions that answer the study objectives.
2. Clarity of language: must meet the level of understanding of the respondents to generate needed responses.
3. The response items should be mutually exclusive, meaning the responses should differ (overlap should not exist).
4. Questions should be collectively exhaustive meaning that they should cover all possible answers or options.
5. There should be an underlying rationale for ordering alternatives.
6. Questions (stem) should be worded clearly.
7. Try to state the questions in the affirmative rather than negatively, especially avoid double negatives. State the stem in a positive rather than a negative manner. If it is written in a negative manner, underline or bold the negative words.
8. Avoid using "all the above" or "none of the above" as an opinion. Participants may guess the answer with partial knowledge.
9. Avoid long sentences or phrases or technical terms.
10. Avoid double-barreled questions that contain two distinct ideas or concepts.
11. Questions on complex or specialized issues can be worded so that respondents feel comfortable and provide natural responses.
12. Avoid leading questions that suggest a particular kind of answer.
13. Do not keep the items that everyone will answer or agree to or that no one will answer.
14. All items and responses should be arranged in logical sequence (alphabetic or numerical order), general to specific, and least sensitive to more sensitive.
15. The researcher must ensure that the answer to one question will not influence other questions.
16. Include enough information in each item to answer the question.
17. Keep all options grammatically consistent with the stem to avoid clues to the correct option.
18. Keep all the options the same length or have two short and two long responses.
19. Make all options reasonable and homogenous.
20. It is not necessary that all items should have the same number of options.
21. Avoid general questions such as: *how satisfied are you with the prices of agricultural commodities?*
22. Avoid terms such as often and regularly.

- *Have you adopted integrated soil fertility management?*
- *Do you use integrated pest management to control maize stalk borer?*
- *Have your cattle suffered from bovine respiratory disease in the last year?*

Instead, you must split the complex discipline-specific terms and replace them with commonly used ones. For instance, for integrated soil fertility management, you can ask several questions to check the components of ISFM that the farmer has adopted. Similarly, for IPM, you can split the questions. For instance, in the use of ISFM, we could have the following questions:

- Do you use farmyard manure on your farm?
- Do you mix manure with mineral fertilizer during planting?
- Do you use improved maize seeds?
- Do you use compost manure?

Bovine respiratory disease should be replaced with "cattle lung disease" to read:

- *Have your cattle suffered from "cattle lung disease" in the last year?*

Look at the following questions posed to farmers:

Which of the following is the best way of controlling aphids in French beans?

1. *Spray at the first appearance of the pest.*
2. *Spray on a fixed calendar schedule.*
3. *Follow the economic theory practices.*

Do you use the following integrated soil fertility management strategies?

1. *Organics alone.*
2. *Organic plus organic fertilizer.*
3. *Conservation agriculture with minimum tillage.*

These questions and responses are unclear and contain technical terms unfamiliar to farmers, which will affect their validity. They need to be simplified.

Avoid biased questions: Some questions posed could assume that respondents have some knowledge and experience when, in fact, this is not the case. Such questions are biased. For example, "*How satisfied are you with the new agricultural credit by the Government?*". Instead, ask screening questions, such as: "*Are you aware of the new credit programme by the government?*" After this, you will follow up with the question of satisfaction.

18.2.4 Formatting the Instrument

Your instrument should be well-structured and formatted. A well-structured and formatted instrument gives appeal and is likely to attract prospective respondents. A good structure considers how questions are sequenced and how the sections are arranged. When ordering and deciding on the flow of questions, arrange questions in three sections: an introductory, middle, and conclusion.

18.2.4.1 Introductory Section

The introductory section should invite the respondent's willingness and cooperation to participate in the questionnaire survey. This is to say, the introductory section, located on the front pages, should appeal to persuade respondents either to fill out the questionnaire and return it on time or to participate in an interview.

Other very important pieces of information to be included in the introductory section include (1) the study's purpose, (2) the significance of the study, (3) the sponsoring body, (4) the owner of the data, and (5) ethical issues. Provide information on the purpose of the study (You may phrase: The purpose of this study is to ......) and also describe the significance, ensuring that you mention how respondents will benefit. In addition, provide the name of the research organization or university collecting the data, plus the sponsoring body. Finally, give assurance of anonymity and confidentiality. Attach an ethical clearance certificate and any other document licensing the study at this stage.

The introductory section should also include clear instructions on handling the questionnaire. Indicate the actual completion time and how to send back the questionnaire, including the necessary contact details (email, phone number). Show friendliness and concern by indicating the importance of respondents' participation and thanking them for agreeing to participate.

18.2.4.2 Middle Section

In the middle section, too, the logical flow of questions is extremely important. Questions on one subject or one particular aspect of a subject should be grouped together in modules, for example, you can have sections such as;

1. General interview information (Consent, date, time, etc.)
2. Geographical location (GPS coordinates, County, District, etc.)
3. Demographics.

4. Farm Characteristics.
5. Livestock production.
6. Crop production.

18.2.4.3 Ending Section

The ending section of an instrument should be treated with caution to help build a relationship as the response to questions ends. It is best to place sensitive questions like demographics (age, wealth, and number of children) in the last section of the questionnaire. However, most researchers place such questions at the beginning of the instrument. In most cases, it works fine. Remember to include appreciation remarks to the respondent once more in the closing section, and repeat the sending instructions and contact details in case of sending the questionnaire to the respondent.

Visual Presentation of the Instrument

The instrument's visual presentation or physical appearance can significantly affect the quality and quantity of data collected. A poorly presented format can create an impression of complexity and a commitment that is too long. Data quality can also be affected by confusing layoffs, making it more difficult for interviewers (in the case of interviews) or respondents to complete the tasks accurately.

Visual presentation can be enhanced by arranging the questions into modules (See sample questionnaire in Appendix). For example, you can have an introduction module that provides the purpose of the study, general instructions, anonymity and confidentiality statement, and consent acquisition. You could then have other follow-up modules arranged in a logical flow and showing progression of the study with the topics. Finally, include a closing module to finish the interview, where you make some general observations and thank the respondent.

Length of questionnaire

Another extremely important consideration is the length of the questionnaire. There is no rule regarding the number of questions or the questionnaire length; it depends on the study's objective. Some questionnaires can be as short as two pages, while others can be as long as 100 pages or more. Long questionnaires are mostly used to characterize households and farming systems. The rule of thumb should be to have a brief and concise questionnaire to address the study's variables.

18.2.5 Pretesting the Instrument

Pretesting or piloting means testing the instrument before doing the actual study (Collins 2003). In other words, the process of measuring the effectiveness of the instrument to meet the objectives of the study. Overall, pretesting helps in reducing measurement errors. A pretest involves conducting a dry run of the survey on a small, representative set of respondents in order to reveal questionnaire errors before the survey is launched. It is important to pretest on respondents who are representative of the target population to be studied. Suggestion: It is also a good idea to test out your sample design, data collection methods, data processing, and data analysis, if possible.

Pretesting should be conducted in circumstances that are as similar as possible to actual data collection and on population members as similar as possible to those that will be sampled.

Careful notes should be taken on the problems encountered and possible solutions should be identified.

Reasons for Pretesting

- To ensure that the questions answer your research questions: should meet the research objectives.
- To check if the questions are clear and that the researcher and respondent interpret the question the same way; should make it easy for the respondents to give the information or the interviewer to record the information.
- To evaluate the quality of the answers provided by the questionnaire.
- To ensure that all questions are relevant and that there are no excess questions.
- Check the analyzability of the data collected: data should have sound analysis and be interpreted.
- To assess and revise the phrasing and content of the questions.
- To explore issues of questionnaire structure and question sequence; help to keep the respondent interested.
- To assess the time taken to answer the questions and adjust accordingly; a very long tool will discourage the respondents and affect the quality of data collected.

18.2.6 Analysis of Pretesting Data and Final Editing Questionnaire

After pretesting, analyze the data to determine if the questions answer your research questions and achieve your research objectives. As you do this, consult your data requirement and analysis table to ensure you understand the data type collected and their analysis. You should check if subjecting your data to the descriptive and statistical analysis as per your table gives you the expected results. Depending on the findings, revise questions, wording, and sequencing if necessary. Rework on the data requirement and analysis table and make it final.

Administering the Instrument and Data Collection

- Use the best practice for administering the instrument (see Chap. 12, Sect. 12.6 on various ways of administering survey instruments).
- Determine the deadline and follow-up methods.
- Ensure you have a plan for handling the completed questionnaire.

18.3 Conducting an Interview Survey

As stated earlier, interview surveys are the most popular in agriculture, especially when interviewing rural households. They are preferred in rural areas where literacy levels are likely low, making it difficult for respondents to complete a written questionnaire. Additionally, the variability of farming systems and conditions can be better understood through personal interactions.

Key activities include:

1. Planning the interview survey.
2. Recruitment and training of enumerators.
3. Recruitment of respondents and gaining access.
4. Collecting the data.

1. *Planning the interview survey:*

As mentioned in Chap. 12 and this chapter, interview surveys are resource-intensive and thus require proper planning and logistics. The main activities involve organizing resources and scheduling the activities. You should devise a budget showing all the activities and funds allocated to each activity. The schedule of activities should show the timelines for the various activities covering the whole survey plan. If many villages are covered, these should be shown, considering travel time.

At this stage, you need to create logistics for vehicle acquisitions to ensure the survey runs smoothly. You also recruit the enumerators and make the necessary arrangements to engage them. If there is a signing of contracts, it should be done during this time

2. *Recruitment and training of enumerators*

In most cases, you cannot collect data from all your respondents independently. You will need to recruit research assistants or data collectors to help you. We call them enumerators. To be effective, they will need to be adequately trained. Considering a farm households survey, we provide guidelines to recruit and train your enumerators in Box 18.3.

3. *Recruitment of respondents and gaining access to respondents*

Access to respondents involves securing permission from the local authorities, community, and individual farmers. You can start the process by meeting the local leaders to explain the purpose and benefits of the survey. For example,

Box 18.3: Guidelines for Recruiting and Training Enumerators

Recruitment:

During recruitment, define the requirements and skills required of the enumerators. For instance, for a household survey, enumerators should have a background in agriculture or related fields. Interns, students, and recently graduated persons can be good enumerators. Next, advertise the position widely using community notice boards, e.g., those in churches, schools, markets, and online platforms. After receiving the applications, carefully review them, shortlist them, and conduct interviews as necessary to select suitable candidates.

Training:

- Begin the training with an introduction and explanation of the survey's purpose. Enumerators must understand the objectives clearly to appreciate the importance of their role. For example, suppose the survey focuses on evaluating how farmers have taken up and managing new improved soybean varieties. In that case, it should be explained how the data collected will help improve farming techniques, enhance productivity, and support local communities. It would also be good to reflect on background information on importance of soybeans for nutrition and natural resource management.
- Next, have a detailed walkthrough of the questionnaire. Each question must be reviewed, rationale explained and how it links to the overall objectives. Role-playing and simulations can be conducted where enumerators practice administering the questionnaire with their peers acting as respondents.
- A critical aspect that needs to be covered is handling difficult situations. Strategies and tips for dealing with common challenges like reluctant respondents, language barriers, or cultural sensitivities are discussed. For example, if a respondent hesitates to discuss income, enumerators should be trained to reassure them about confidentiality and the importance of honest responses for the study's success.
- During the training, emphasize the accuracy of recording responses. Teach them standard procedures for recording answers, including handling open-ended questions and using digital tools if applicable. Demonstrate recording techniques for qualitative answers about farm management practices to ensure responses are captured verbatim and ambiguities clarified with follow-up questions.

(continued)

Box 18.3 (continued)

- Train them about ethical considerations, including obtaining informed consent from all participants and ensuring they understand the survey's purpose and their rights. A script for enumerators to read can help explain the survey, assure confidentiality, and seek consent before proceeding. This is usually on the first page of the tool (see appendix at the end of this chapter).
- Emphasize cultural sensitivity and respect. Train them on the cultural context and norms of the communities they will survey. You may even allow enumerators to practice the greetings and small talk to build rapport.
- If digital data collection tools like tablets or smartphones are used, enumerators must be comfortable with the technology and understand how to input data correctly. Hands-on sessions where enumerators practice entering data into the survey software and troubleshoot common issues are beneficial.
- Field practice by the enumerators is crucial. You can organize to have them practice in a nearby village with similar agricultural practices, then review their experiences and receive feedback on their performance, including areas for improvement.

meeting with the village chief in a village setting will help gain community support. At this stage, assure the chief that you have obtained the necessary permission and present evidence, e.g., ethical clearance certificate or any other documentation.

At the individual level, ensure respondents consent before participating in the survey. It is at this point that you gain trust and build rapport.

To ensure high participation rates, avoid scheduling interviews during peak farming hours when respondents are likely busy. If you visit a respondent and find they are very busy, you can reschedule the interview and ask to be given an appointment. Timing during the day is important. Visit mid-morning and late afternoon, and avoid lunchtime when the family could have their meals. Also, late evening should be avoided, as families will likely be engaged in hushed chores, such as feeding and milking the cows.

4. *Carrying out the actual interview*

Enumerators visit the selected households and interview the respondents. Continuous support during the survey period is crucial. As the researcher in charge, you should have regular check-ins and a communication channel, such as a WhatsApp group, to allow enumerators to quickly reach out with questions or concerns and receive immediate guidance. By following these steps, researchers ensure that data collectors are well-prepared to gather reliable, accurate, and valuable data for agricultural studies.

Good Data Collection Practice in Surveys

- Communicate clearly to the respondents about the purpose of the survey.
- Use a targeted and concise survey instrument.
- Establish rapport to encourage participation.
- Emphasize confidentiality in your introduction.
- Offering incentives and rewards, but this should be handled with care.

18.4 Conducting a Survey Using a Self-Completed Questionnaire

Conducting a survey using a mailed questionnaire, hand-delivered, or sent via post can be effective with respondents who can read and write (Cehovin et al. 2018; Edwards and Perkins 2024). This includes farm managers, students, professionals, researchers, and extension workers. Once the respondents receive the questionnaire, they are expected to fill and return it to the researcher.

Three key activities are involved:

1. Pre-survey communication.
2. Mailing the questionnaire.
3. Follow up.

Pre-survey communication

Notify potential respondents in advance about the upcoming survey. This can be done by contacting individual persons through telephone, mail, or email. You can visit the office to meet the officers and organize community meetings, or meet local extension officers or cooperatives. In case you use email or telephone SMS, ensure you provide contact information for respondents to ask or seek clarification.

Mailing the questionnaire

Include all the necessary materials in the package. This includes: 1. The questionnaire, 2. Cover letter, 3. Prepaid return envelope if through post, 4. Information on Completion time.

The cover letter is a major component that accompanies the mailed questionnaire. The cover letter invites the respondent to participate in the study. It contains the following major elements:

- Title of research.
- Information about why the study is important to participate in, its purpose, and its benefits. Use simple language which respondents will understand easily. You will not be there to explain.

- Explain how the respondent was selected.
- Brief on information/ data to be collected.
- Information on participation; approximate time of filling the questionnaire, e.g., 10–15 min.
- Information on how you will handle confidentiality and anonymity.
- Indicate that the researcher will share results with participants only if you know you can do this.
- It is good practice to include your institution and also receive endorsement. This motivates the respondents and creates trust.
- Instructions on how to return.

Follow up

Send follow-up reminders to non-respondents to encourage them to complete the questionnaire. This can be done through phone calls, emails, etc. In addition, keep on monitoring the return rate and plan for additional follow-up to ensure you achieve the desired response rate.

Response Rates

Achieving a complete (100%) response rate in a questionnaire survey is often challenging. If you achieve about 60% or more, this is considered adequate for analysis and reporting. Researchers should employ various strategies to maximize response rates, such as using clear and concise survey instruments, sending reminders to non-respondents, and offering incentives if applicable. It is also important to make the survey process convenient and accessible for participants.

18.5 Types of Questions

You will include various types of questions in the survey instrument. The main ones include:

1. Quantity questions.
2. Category-based questions.
3. Dichotomous question.
4. Rank order question.
5. Rating questions—Likert scale question.
6. Ranking and scoring questions.
7. Semantic differential questions.
8. Matrix questions.

1. *Quantity Question*

Quantity questions are common; they give a factual amount of a variable. They are used to collect demographic and behavioral data in households. Examples include age in years, amount of fertilizer used, crop yields, farm sales, etc. They can be framed as follows:

1. What is your age? _______ years.
2. How much maize yield did you harvest during the long rainy season last year? _________kgs.
3. How much nitrogen fertilizer did you apply to the maize crop last season? ______kgs.
4. What quantity of beans did you sell last year? _______kgs.

The data collected in numerical (ratio data) can be used to calculate averages and assess correlations. Ratio data is a powerful data type when used in collecting data from households.

2. *Category-Based Questions*

Category-based questions are among the most commonly used. Examples include demographic variables such as age in categories, gender, ethnic group, level of education, and also farming practices.

(a) What is the age of the household head in this farm? circle:
   1. 20–29 years
   2. 30–39 years
   3. 40–49 years
   4. 50–59 years
   5. 59–60 years
   6. More than 60 years.

(b) What is your education level?
   1. Illiterate,
   2. Primary level,
   3. Secondary level,
   4. Tertiary level.

(c) Do you use fertilizer:
   1. Yes.
   2. No.

3. *Dichotomous questions*

Dichotomous questions are straightforward, typically asking respondents to choose between Yes / No; Agree or Disagree, True/False

- Local remedies for improving soil pH are available among the farmers (True/False).
- Have you experienced crop damage due to fall armyworms in the past year? (Yes/No).
- Do you believe climate change is affecting your agricultural productivity? (Yes/No).
- Are you currently using genetically modified seeds for your crops? (Yes/No).
- Weather forecasting services by the government help farmers in agricultural planning (True/False).
- Diversifying crops helps to mitigate market risks (Agree/Disagree).

4. *Rank-Order Question*

Rank-order questions prompt respondents to arrange specific items based on a particular characteristic. For instance, respondents might be asked to rank various animal breeds according to their milk production potential. Here is an example:

Please rank the following animal breeds based on their milk production potential, with 1 representing the highest and 5 indicating the lowest:

1. Jersey.
2. Guernsey.
3. Friesian.
4. Ayrshire.
5. Sahiwal.

5. *Numerical Scoring Questions*

Numerical scoring questions assign responses a numerical value. This technique is particularly useful when you want to evaluate diverse responses that need to be systematically evaluated. The question asks the respondents to rate on a numerical continuum, where interviews are assumed equal.

Example

Suppose you want to score farmers' preferences for sweet potato varieties according to different criteria in a screening study. Farmers have been growing the varieties for two seasons, and thus they have experience and also some observations.

Criteria: Taste, yields, marketability, and disease resistance.

Varieties: Variety A, Variety B, and Variety A.

Allocation of score: Farmers use the bao game: farmers assess the variety as per given criteria and allocate the seeds (Kiptot and Franzel 2015). Farmers assess the taste, yield, marketability, and disease resistance. The researchers will then count the number of seeds in each criterion and find the score to identify the best-performing variety based on farmer assessments.

You will use a matrix table in the instrument as follows (Table 18.5).

Table 18.5 Numerical scoring matrix

Variety	Criteria	No of seeds (score)
Variety A	Taste	
	Yield	
	Marketability	
	Disease resistance	
Variety A	Taste	
	Yield	
	Marketability	
	Disease resistance	
Variety A	Taste	
	Yield	
	Marketability	
	Disease resistance	

Data collected using numerical scoring is easy to analyze. You add the scores for each criterion and can calculate the mean score. With means, you can compare them statistically.

6. *Rating Scales Questions*

Rating involves making a measured judgment by the rater about behavior or a product. Rating scales are useful in measuring attitudes, perceptions, preferences, and opinions toward specific items. For instance, we can use rating scales to measure farmers' perceptions of government extension services or to assess agro-dealers' satisfaction with specific products. There are two primary types of rating scales: the Likert scale and the semantic differential.

7. *Likert Scale*

A Likert scale is an ordered scale from which an individual responds to a series of statements (Jebb et al. 2021; Huh and Gim 2025). It allows the respondents to indicate their level of agreement or disagreement with a statement or question by selecting from a range of options, typically from "strongly agree" to "strongly disagree." We can use a rating scale for aspects of farming practices and farmers' opinions. For example, we can rate diseases, pests, livestock health, and crop growth. The rating is flexible and can range from 3 to 7 in most cases, and can be used to measure agreement, satisfaction, preference, and frequency.

Examples of the Likert scale include the following:

Measuring crop performance, soil characteristics, or rating of disease damage using a 3-point scale: if you want farmers to rate these parameters during a participatory field survey or participatory evaluation of farming practices. You can use the following:

Crop performance (growth, yield, fruit size)

Rate the growth of improved maize varieties during the short rainy season 1. Poor 2. Average 3. Good

Soil characteristics (color, texture, stoniness, water-holding capacity)

Rate the water holding capacity of soil in your farm 1. Poor 2. Average 3. Good

Rate response to soil fertilizer on your farm 1. Poor 2. Average 3. Good

Measuring the level of frequency of use of various training methods by extension staff to train farmers in farming practices using a four-point scale. We can assess the use of training methods to train farmers by asking farmers to rate the frequency. We can pose the question:

How frequently do you use the following communication methods in training farmers? (Tick where appropriate) (Table 18.6).

Measuring opinion on satisfaction with a certain aspect. Supposing you want to measure farmers' satisfaction with extension services staff in meeting their needs, you can set Likert scale questions ranging from 1 to 5. Farmers can express their agreement or disagreement with specific statements about the extension services, allowing for a quantitative assessment of their satisfaction levels.

A typical Likert statement might read as follows: directed toward farmers.

Table 18.6 Rating on the frequency of use of various training methods

Method of communication	Very frequent	Frequent	Rare	Never
Demonstration				
Field days				
Leaflet/brochures				
Video documentaries				
Exchange visits				
Lecture method				
Mother baby				
Farmer field school				
TV/radio				
Agricultural shows				
Magazines/books				

Agricultural extension staff provide useful information for managing fall armyworm in maize.

Strongly disagree	Disagree	Neutral	Agree	Strongly agree
1	2	3	4	5

Measuring preference: If you want to determine farmers' preferences for different bean varieties in a screening trial using a 5-point scale. You have to select attributes such as yield, disease resistance, taste, and marketability. Participants may rate each variety on a scale from very poor to excellent for each attribute:

- The pesticide distributed by the extension officer for controlling fall armyworm was effective.
- The training offered during the field day on plant health was relevant.
- What is your level of satisfaction with the price offered by buyers at the farm gate?
- The sweetness of sweet potato varieties supplied by extension staff from the government is pleasing.

	Strongly disagree	Disagree	Neutral	Agree	Strongly agree
Effectiveness	1	2	3	4	5
	Extremely relevant	Relevant	Neutral	Somehow not relevant	Not relevant
Relevance	1	2	3	4	5
	Very satisfied	Dissatisfied	Unsure	Satisfied	Very satisfied
Satisfaction	1	2	3	4	5
	Strongly agree	Agree	Neutral	Disagree	Strongly disagree
Agreement	1	2	3	4	5

Please rate your competence in training farmers using the listed extension delivery methods (tick √ the appropriate competence).

Delivery method	Low ←→ High competence				
	1	2	3	4	5
1 = field day					
2 = farmer field schools					
3 = farmers' training schools (FTCs)					
4 = training workshops					
5 = one-on-one by extension agents					
6 = farmer-to farmer extension					
7 = on-farm demonstration					
8 = leaflets					
9 = participatory extension approaches					

Codes for competency
1 = not competent
2 = a little competent
3 = moderately competent
4 = fairly competent
5 = very competent

Some advantages of Likert-scale questionnaires are that (a) data can be gathered relatively quickly from large numbers of respondents, (b) they can provide highly reliable person ability estimates, (c) the validity of the interpretations made from the data they provide can be established through a variety of means, and (d) the data they provide can be profitably compared, contrasted, and combined with qualitative data-gathering techniques, such as open-ended questions, participant observation, and interviews.

8. *Semantic Differential*

Unlike the Likert scale, which measures the degree of agreement or disagreement of respondents, the semantic differential scale presents respondents with bipolar adjective pairs (e.g., good-bad, efficient-inefficient) and asks them to rate the items along these dimensions (Hawkins et al. 1974).

Consider an example of screening trials for orange-fleshed sweet potato varieties that differ in taste, texture, and other attributes. Understanding farmers' perceptions and attitudes towards the different varieties is crucial for increasing adoption rates. We can use semantic differential scales and ask farmers to rate each sweet potato variety on these four dimensions: "sweet vs. bland," "firm vs. soft," "dry vs. juicy," and "crispy and mushy."

9. *Matrix Questions*

Sweet	1	2	3	4	5	6	7	Bland
Firm	1	2	3	4	5	6	7	Soft
Dry	1	2	3	4	5	6	7	Juicy
Crispy	1	2	3	4	5	6	7	Mushy

Table 18.7 How easily do you access information related to soil fertility management technologies?

Source of information	Animal manure	Compost manure	Green manure	Inorganic fertilizer	Combination of manure and fertilizer	Erosion control measures
Research institutions						
NGOs						
Agro input dealers						
Newsletter/ brochure						
Books						
Workshops/ seminars						
TV& radio						
Scientific conferences						
Internet						
Others(specify)						

Source: Muchai et al. (2014)
Score the sources using 1 for the least accessible and 4 for the most accessible

Matrix questions are a set of items similar to the Likert scale. The set of items defines a dimensional construct that allows researchers to measure attitudes and behavior. Table 18.7 shows an example of matrix questions used to assess the accessibility of information for different soil fertility management practices.

Types of Questions Based on Structure

Two main types of questions are used in a questionnaire: structured and unstructured questions.

Structured or closed-ended questions—these refer to questions that are accompanied by a list of all possible alternatives from which respondents select the answer that best describes their situation.

Examples

Question: What food crops do you grow during the long rainy season?

1. Maize 2. Greengrams 3. Millet 4. Sorghum

What varieties of mango do you grow?

1. Van Dyke 2. Kent 3. Apple 4. Tommy Atkins

Advantages of closed-ended questions

1. Close-ended questions are easier to analyze than open-ended questions. They are coded easily.
2. They are easier to administer because each item is followed by alternative answers.
3. They are economical to use in terms of time and money.

Disadvantages

1. They are more difficult to construct as categories must be well thought out; sometimes it may need an exploratory survey.
2. Responses are limited and the respondent is compelled to answer questions according to the researcher's choices, thus biases.

Unstructured or open-ended questions—These refer to questions that give the respondent complete freedom of response, i.e., they permit the respondent to respond in his/her own words. The amount of space provided, however, is an indicator of whether a brief or lengthy answer is desired.

Example

Question: How do you keep a record of your monthly expenditure in your household? ________________
____
____
____
____

Advantages:

1. They permit a greater depth of response.
2. They are simpler to formulate, mainly because the researcher does not have to labor to come up with appropriate response categories.
3. The respondent's responses may give an insight into their feelings, background, hidden motivations, interests, and decisions.
4. Open-ended questions can stimulate a person to think about his feelings or motives and to express what he considers to be most important.

Disadvantages:

1. There is a tendency to provide information that does not answer the stipulated research questions or objectives.

2. The response given may be difficult to categorize and, hence, difficult to analyze quantitatively.
3. Responding to open-ended questions is time-consuming, and this may put off some respondents.

Measuring Farmers' Knowledge of Agricultural Practices

Studies measuring farmers' knowledge are common among students, but we often observe that students face difficulties in how to collect data on this abstract variable. Knowledge can be referred to as organized or processed information or data and is crucial in any innovation process (Maier 2007; Cantner et al. 2011; Asrar-ul-Haq and Anwar 2016). Knowledge is often created by a combination of education and experience. It is important to determine farmers' knowledge because farmers use it to arrive at decisions that influence agricultural management practices. It is, therefore, useful to understand changes that occur in the landscape at a local level, especially in terms of changes in land use and cultural practices. If agricultural change agents do not understand the knowledge and priorities of producers, improved management will be difficult.

Information about farmers' knowledge and attitudes towards such agricultural practices will provide a better understanding of how to encourage farmers to implement cropping strategies and to improve their delivery approaches. In the fertilizer context, farmers' local knowledge of soil fertility and management strategies plays a significant role in the fertility maintenance of farmlands (Meijer et al. 2015; Šūmane et al. 2018; Nord et al. 2022).

To measure knowledge levels, you construct statements and present them to respondents. It is like administering a test. These statements measuring knowledge usually have "true" or "false" statements; each correct answer is scored 1, and each incorrect answer is scored 0. Each question contains 2 marks. Eventually, knowledge is determined by summing up the weights of their responses to all the questions presented. For example, if the questions are 10, the total marks will be 20. Thus, farmers' knowledge scores could range from 0 to 20, where zero (0) indicates no knowledge, and 20 indicates the highest knowledge. To assess the farmer's knowledge level, Njenga et al. (2021) posed 28 questions to farmers that required an answer of either true or false. Respondents scored (1) for every correct answer and (0) for every wrong answer. Afterwards, marks were awarded to each farmer, and farmers were classified as Low, medium, and high levels of knowledge according to the score obtained.

Measuring Food and Nutritional Security Among Households

Data on food security and nutritional security aims to provide insights into food security status, which enables researchers and policymakers to identify vulnerable populations and design targeted interventions to address food insecurity and to improve the nutritional and well-being of households (Burgaz et al. 2023; Manikas et al. 2023; Kenney et al. 2024). Important data sets collected include:

Food consumption and availability: This includes four key items: 1. Food consumption frequency, 2. Quantity of food consumed (portions of food consumed by households), 3. Sources of food, whether own-produced, bought, or donated, and 4. Dietary diversity is the variety of food consumed by households over a certain period.

Household food security status: The household food insecurity Access Scale (HFIAS) is a standardized tool used to measure food insecurity in households. This is usually combined with information on the strategies employed by households to cope with food shortages (Marques et al. 2015; Ashby et al. 2016).

Food access and affordability: this is concerned with information on household income levels and expenditure, particularly on food as a percentage of the total income. It includes data on food market availability and other food outlets.

Food nutritional intake: involves taking data on daily caloric intake and consumption of key nutrients. It also involves assessment of height and weight measurements of household members and also data on the prevalence of nutrient deficiencies such as micronutrient deficiency and vitamin A.

Seasonality of food: an important variable to assess food availability and access at different periods of the year.

Knowledge, attitude, and practices: involves data on the level of knowledge regarding balanced nutrition and dietary practices. Also involves collecting data on attitudes towards nutrition and practices towards improved nutrition.

Appendix: Sample Questionnaire: Borrowed from Mugwe et al. (2009) Used in the Assessment of Determinants of Adoption of Soil Fertility Management Practices in Central Highlands of Kenya

FARM IDENTIFICATION

Interview schedule number__________ Name of interviewer________________

Location______________ Sub-location_______________________

Village________________ App. Distance from Demonstration site________

Time started_____________________

GREETINGS

Once you get to the farm, exchange greetings with the people you meet and introduce yourself: your name and the institution conducting the research.

PURPOSE OF THE STUDY

The purpose of the study is to assess factors influencing the adoption of integrated soil fertility management. During the survey, simplify ISFM and indicate it is an assessment of soil fertility management practices that farmers use eg fertilizers and manures: In addition, indicate the amount of time likely to take to collect data.

HOW FARMERS WILL BENEFIT?

Inform the farmers that their responses are valuable , as they will contribute to the understanding of the adoption processes of soil fertility practices, which will be used to make decisions regarding future activities in the region.

ANONYMITY AND CONFIDENTIALITY

Your responses are strictly confidential. Inform the respondents that the data will be used for research purposes only.

CONSENT

Before starting the questioning, please acknowledge that you have understood the information provided and that you have consented voluntarily to participate. You may withdraw anytime without penalty

MODULE A. HOUSEHOLD HEAD CHARACTERISTICS

1. Name of household head ____________________________

2. Gender of household head Male ☐ Female ☐

3. Age of household head Actual age (yrs)__________

20-30 ☐ 31-40 yrs ☐ 41-50 yrs ☐ 51-60 yrs ☐ 60 + yrs ☐

4. Education level of household head Illiterate ☐ Primary level ☐ Secondary ☐ Tertiary ☐

5. Marital status of household 1=single 2=married 3=divorced/separated 4=widow 5=widower

6. Name of respondent (if not household head)________________________________

7. Relationship of respondent to household head Husband ☐ Wife ☐ Child ☐

MODULE B. NATURE OF HOUSEHOLD

8. Number of nuclear families ☐

9. Number of adult males (*active on the farm*) ☐

10. Number of adult females (*active on the farm*) ☐

11. Number of adult residents (*active on the farm*) ☐

12 Number of elderly persons (*not active on the farm* ☐

13. Number of children ☐

14. Number of other dependants, sick (not *active on the farm*) ☐

15. House type Iron-roof-mud ☐ Semi-permanent ☐ Permanent ☐

MODULE C: FARM CHARACTERISTICS

16. Farm size: (acres) ☐

17. Land use: Number of tea bushes ☐ Number of coffee trees ☐

18. Occupation of spouses: wife work on farm full time ☐ Husband work on farm full time ☐ Both work on farm full time ☐

19. Subsistence crops grown in order of importance 1. _____________ 2.______________ 3.________________ 4.________________

20. Off-farm employment (tick)

No ☐ no cash income from off farm activities

Not important ☐ little or unreliable income from off farm activities

Important (resident) ☐ important cash income from residents off farm activities

Important (non-resident) ☐ important cash income from non-residents off farm activities

21. Hired labour (tick)

Work	☐	household members work as hired labour on other farms
No	☐	no hired labour
Seasonal	☐	hired labour on a seasonal basis
Permanent	☐	hired labour on a permanent basis

MODULE D: LIVESTOCK PRODUCTION AND MANURE USE

22. Type of livestock owned and numbers

Livestock type	**Number**
Mature cattle	
Young cattle	
Goats/sheep	
Poultry	
Pig	

23.Do you produce enough manure (tick)	yes	☐	No	☐
24. If no, do you buy from outside	yes	☐	No	☐

25. On what crops do you use cattle manure (tick)

	Coffee
	Maize
	Beans
	Potatoes
	Napier
	Other:(*Specify*)

26. Rank the amount used in each of the enterprises (10 for the largest amount and 1 for the smallest amount)

Crop	Score (1 to 10)
Coffee	
Maize	
Beans	
Potatoes	
Napier grass	

MODULE E: INORGANIC FERTILIZER USE

27. Do you use inorganic fertilizer? Yes ☐ No ☐

28. If Yes, Please provide the following information concerning the use of fertilizer on the stated crops

Crop	Fert use (y/n)	Fertilizer type	Area applied per season	Amount applied	Years of fert use on crops
Coffee					
Maize					
Beans					
Potatoes					
Napier grass					

29. What problems do you face with the use of fertilizer on your farm?

....

....

MODULE F: FOOD PRODUCTION

30. Does your farm produce adequate food to meet family needs? Yes ☐ ☐

31. If no, for how many months do buy food to meet the short fall? (tick)

	< month
	2-3 months
	4-5 months
	5-7 months
	8-9 months

32. If yes, do sell and how much? (tick) ☐ Yes ☐ No

33. If yes, how much?

	< bag
	2-3 bags
	4-5 bags
	Other*(specify)*

34. Is soil infertility a current problem on your farm? Yes ☐ No ☐

35. If yes, what indicators are there for soil infertility?

....

....

....

36. What do you use to alleviate soil fertility problems?

……………………………………………………………………………………………….

……………………………………………………………………………………………….

…………………………………………………………………………………………………

MODULE G: SOIL FERTILITY REPLENISHMENT TECHNOLOGIES ADOPTION

35. Ask the farmer what technologies he or she has as a result of attending the field days at Kirege demonstration site

Record the following information; for the various technologies, when did you start and for how many seasons have you had them continuously?

Technology	Start Date (season)	Number seasons practiced

36. Have you modified the technologies? Yes ☐ No ☐

37. If yes, how?

………

……

……………………………………………………………

38. Which ones do you wish to continue practicing and why?

Technology	Reason for continuation

39. Which ones do you wish to discontinue and why?

Technology	Reason for discontinuation

For farmers adopting calliandra, tithonia and leucaena biomass transfer technologies

40. How have you expanded your plantings?

	Calliandra		Leucaena		Tithonia	
	number	length	number	length	number	length
2000 SR						
2001 LR						
2001 SR						
2002 LR						
2002 SR						
2000 LR						
2003 SR						

MODULE H: BENEFITS, CONSTRAINTS AND COPING STRATEGIES

41. Who have been attending field days and how many times?

	1	2	3	4
Husband				
Wife				
Son				
Daughter				
Other, specify				

42. What benefits have you realized since you started using the technologies?

....

....

....

43: Outcomes of the technologies use by farmers

	Before technology	After technology
Approximate quantity of maize		
Months you buy maize for the family		
Amount of maize sold		
Amount of fertilizer bought		

44. What constraints/challenges have you experienced? How have you addressed them?.

Practice	**Constraint**	**Coping strategy**
Farmyard manure		
compost		
Calliandra		
Leucanea		
Tithonia		

45. Interviewer remarks (Record your observation about the household in terms of general farm management, wealth status, state of livestock, presence of manures etc

....

....

....

47. Thank the respondent and provide your contact.

48. Time interview finished ....

This questionnaire was used in a study conducted among farmers who had participated in a project on promoting integrated soil fertility management from 2004 to 2009 (Mugwe, 2009). The survey was cross-sectional and was administered using interviewer-reported

Recommended Further Reading

1. Babbie ER (2016) The practice of social research. 14th edn. Cengage Learning, Boston, MA. ISBN 9781305104945
2. Fink A (2016) How to conduct surveys: a step-by-step guide, 6th edn. SAGE Publications, Los Angeles. ISBN 978-1-4833-7848-0
3. Dillman DA, Smyth JD, Christian LM (2014) Internet, Phone, Mail, and Mixed-Mode Surveys: The Tailored Design Method (4th ed.). Hoboken, NJ: John Wiley & Sons. https://doi.org/10.1002/9781394260645

Key Points in This Chapter

What Is Questioning and Interviewing?

Questioning or interviewing is asking questions to gather information from the target audience, including individuals and groups, using pre-determined questions. Questioning requires careful planning, considering the research objective, target audience, and data type needed.

Data Collection in Questionnaire Survey and Interview Survey

Data collection in a survey falls into two categories: an interview survey (researcher-completed instruments) or a questionnaire survey (respondent-completed instruments) (Gorman 1993; Bowling 2005; Atkeson et al. 2014). When conducting an interview survey, we use an interview survey guide, while in the questionnaire survey, we use questionnaires that we send to respondents to fill.

The survey instruments contain a set of standardized questions, often called items. Interview-based surveys offer several advantages over questionnaire surveys, especially in household surveys.

Designing a Survey Instrument
Six steps to preparing and designing a survey instrument include:
1. Decide on the information needed.
2. Write the items (questions).
3. Format the instrument.
4. Pretest the instrument.
5. Analyze pretesting data analysis and finally edit the instrument.
6. Collect the data.

An effective survey instrument is one that uses the least number of well-worded and organized questions to solicit accurate information from the respondents.

Before preparing the survey instrument, you must be clear about what is being studied and what is expected from the study. The instrument is to collect data to fulfil the study's objectives. A well-designed instrument should meet the research goal and objectives and minimize unanswered questions, a common problem bound to many surveys.

Writing the Questions

Writing practical questions that obtain the required information involves drafting and rewriting items until they are clear and succinct. Researchers must carefully design instruments to avoid these issues, ensuring each question is linked to the research objectives. The following should be avoided.

- Leading question.
- Double-barreled questions: avoid double-barreled questions.
- Avoid complex and technical language.
- Avoid biased questions.

Formatting the Survey Instrument
The instrument should be well-structured and formatted to appeal to respondents. A good structure considers how questions are sequenced and how the sections are arranged. When ordering and deciding on the flow of questions, arrange questions in three sections: an introductory, middle, and conclusion.

Pretesting or piloting means the instrument: before doing the actual A pretest of the instrument is done. It is important to pretest respondents who are representative of the target population to be studied. Suggestion:

- *Interviews:* all approaches use interviews to gather information directly from the respondents. Surveys often use structured interviews with questions that collect quantitative data. Qualitative and participatory research are likely to have semi-structured and unstructured interviews with open-ended questions to allow for in-depth exploration of phenomena.
- *Questionnaires*: structured questionnaires are common in surveys to collect standardized data. Qualitative research might use open-ended questionnaires to gain insights into respondents' experiences and perspectives.
- *Observation:* All methods can utilize direct observation. In farm surveys, researchers gather information through direct observation, which is useful for verifying self-reported data. For example, a researcher might visit a farmer to verify the state of animal housing after the farmer has provided the information through a questionnaire.
- *Focus groups:* This versatile method is widely used in all approaches to collect data from a group of individuals simultaneously. In surveys, focus groups can help pretest quantitative data. In qualitative research, focus groups provide rich and interactive data, comprising qualitative and quantitative forms.

Conducting an Interview Survey

Interview surveys are the most popular in agriculture, especially when interviewing rural households. They are preferred in rural areas where literacy levels are likely low, making it difficult for respondents to complete a written questionnaire (Anderson et al. 2024). Additionally, the variability of farming systems and conditions can be better understood through personal interactions.

Key activities include:
1. Planning the interview survey.
2. Recruitment and training of enumerators.
3. Recruitment of respondents and gaining access.
4. Collecting the data.

Conducting a Survey Using a Self-Completed Questionnaire

Conducting a survey using a mailed questionnaire, hand-delivered or sent via post can be effective with respondents who can read and write. This includes farm managers, students, professionals, researchers, and extension workers. Once the respondents receive the questionnaire, they are expected to fill and return it to the researcher.

Three key activities are involved:
1. Pre-survey communication.
2. Mailing the questionnaire.
3. Follow up.

Types of Questions

You will include various types of questions in the survey instrument. The main ones include:
1. Quantity questions.
2. Category-based questions.
3. Dichotomous question.
4. Rank order question.
5. Rating questions—Likert scale question.
6. Ranking and scoring questions.
7. Semantic differential questions.
8. Matrix questions.

Review Questions

1. What are the main methods used in surveys, qualitative, and participatory research to collect data?
2. A fellow student is debating whether to use "a self-administered questionnaire survey" or "an interview survey" to collect data on farmers' adoption of improved dairy breeds.
   (a) Differentiate between these two survey approaches, highlighting their core operational differences.
   (b) Discuss two distinct advantages and two distinct disadvantages of using an interview survey compared to a self-administered questionnaire survey for this particular research topic.
3. What are the six steps in designing a survey tool and collecting data?
4. Why is question formatting important when designing a questionnaire?
5. If your questionnaire has a low response rate, what follow-up strategies can you apply to improve it?
6. Why is pretesting a questionnaire critical before final data collection? What could go wrong if you skip this step?

Chapter Exercise

1. Suppose you are tasked with writing questions for a survey on farmer satisfaction with agricultural extension services.
   (a) Explain the principle of using "simplicity" in question writing. Rewrite the following question to adhere to this principle: "Do you believe that the newly implemented agricultural extension outreach program has unequivocally and universally improved the overall welfare and productivity of your farming enterprise?"
   (b) Explain what a "leading question" is and why it should be avoided. Provide an example of a leading question related to pesticide use and then rephrase it to be neutral.
2. For each of the following research objectives in an agricultural context, identify the most appropriate type of survey question (e.g., quantity, category-based, Likert scale, rank order, dichotomous, semantic differential, matrix questions) and provide a concrete example of how that question would be phrased:
   (a) To ascertain the farmer's level of agreement with a statement regarding the effectiveness of organic farming practices.
   (b) To identify the primary source of agricultural information for farmers (e.g., extension officer, radio, fellow farmer).
   (c) To determine the average amount of fertilizer (in kg) applied per acre by farmers during the last planting season.
   (d) To understand the farmer's preference order for different drought-resistant crop varieties introduced in the region.
   (e) To ascertain if a farmer adopted a specific new farming technology (Yes/No).
3. How does effective formatting of the questionnaire contribute to attracting respondents and improving data quality? Provide an example of how illogical organization or poor formatting might negatively impact data collection in an agricultural survey.
4. Design a short section (5–7 questions) of a survey instrument aimed at understanding farmers' perceptions and practices regarding climate change adaptation strategies on their farms.

Answers to Review Questions

- Q1. References: 18.1.
- Q2. References: 18.1.
- Q3. References: 18.2.2.
- Q4. References: 18.2.4.
- Q5. References: 18.4.
- Q6. References: 18.2.5.

References

Anderson E, Lybbert TJ, Shenoy A et al (2024) Does survey mode matter? Comparing in-person and phone agricultural surveys in India. J Dev Econ 166:103199. https://doi.org/10.1016/J.JDEVECO.2023.103199

Artino AR, La Rochelle JS, Dezee KJ, Gehlbach H (2014) Developing questionnaires for educational research: AMEE Guide No. 87. Med Teach 36:463–474. https://doi.org/10.3109/0142159X.2014.889814

Ashby S, Kleve S, McKechnie R, Palermo C (2016) Measurement of the dimensions of food insecurity in developed countries: a systematic literature review. Public Health Nutr 19:2887–2896. https://doi.org/10.1017/S1368980016001166

Asrar-ul-Haq M, Anwar S (2016) A systematic review of knowledge management and knowledge sharing: Trends, issues, and challenges. Cogent Bus Manag 3. https://doi.org/10.1080/23311975.2015.1127744

Atkeson LR, Adams AN, Alvarez RM (2014) Nonresponse and mode effects in self- and interviewer-administered surveys. Pol Anal 22:304–320. https://doi.org/10.1093/PAN/MPT049

Babbie, E. R. (2016). The Practice of Social Research. 14th edn. Cengage Learning, Boston, MA. ISBN 9781305104945.

Bowling A (2005) Mode of questionnaire administration can have serious effects on data quality. J Public Health (Bangkok) 27:281–291. https://doi.org/10.1093/PUBMED/FDI031

Burgaz C, Gorasso V, Achten WMJ, et al (2023) The effectiveness of food system policies to improve nutrition, nutrition-related inequalities and environmental sustainability: a scoping review. Food Security 15:1313–1344. https://doi.org/10.1007/S12571-023-01385-1

Cantner U, Joel K, Schmidt T (2011) The effects of knowledge management on innovative success – An empirical analysis of German firms. Res Policy 40:1453–1462. https://doi.org/10.1016/J.RESPOL.2011.06.007

Cehovin G, Bosnjak M, Lozar Manfreda K (2018) Meta-analyses in survey methodology: A systematic review. Public Opin Q 82:641–660. https://doi.org/10.1093/POQ/NFY042

Collins D (2003) Pretesting survey instruments: An overview of cognitive methods. Qual Life Res 12:229–238. https://doi.org/10.1023/A:1023254226592/METRICS

Dillman DA, Smyth JD, Christian LM (2014) Internet, Phone, Mail, and Mixed-Mode Surveys: The Tailored Design Method (4th ed.). Hoboken, NJ: John Wiley & Sons. https://doi.org/10.1002/9781394260645

Edwards P, Perkins C (2024) Response is increased using postal rather than electronic questionnaires – new results from an updated Cochrane Systematic Review. BMC Med Res Methodol 24:1–14. https://doi.org/10.1186/S12874-024-02332-0/TABLES/2

Fink A (2016) How to conduct surveys: a step-by-step guide, 6th edn. SAGE Publications, Los Angeles. ISBN 978-1-4833-7848-0.

Gorman DM (1993) A review of studies comparing checklist and interview methods of data collection in life event research. Behav Med 19:66–73. https://doi.org/10.1080/08964289.1993.9937567

Hawkins DI, Albaum G, Best R (1974) Stapel scale or semantic differential in marketing research? J Market Res 11:318–322. https://doi.org/10.1177/002224377401100311

Huh I, Gim J (2025) Exploration of Likert scale in terms of continuous variable with parametric statistical methods. BMC Med Res Methodol 25:1–10. https://doi.org/10.1186/S12874-025-02668-1/TABLES/4

Jebb AT, Ng V, Tay L (2021) A review of key likert scale development advances: 1995–2019. Front Psychol 12:637547. https://doi.org/10.3389/FPSYG.2021.637547/TEXT

Kenney E, Adebiyi VO, Seligman HK et al (2024) Assessing and monitoring nutrition security in the United States: A narrative review of current measures and instruments. Curr Nutr Rep 13:639–667. https://doi.org/10.1007/S13668-024-00547-7/TABLES/6

Kiptot E, Franzel S (2015) Farmer-to-farmer extension: opportunities for enhancing performance of volunteer farmer trainers in Kenya. Dev Pract 25:503–517. https://doi.org/10.1080/09614524.2015.1029438

Maier R (2007) Knowledge management systems: Information and communication technologies for knowledge management. Springer Berlin, Heidelberg 1–720. https://doi.org/10.1007/978-3-540-71408-8/COVER

Manikas I, Ali BM, Sundarakani B (2023) A systematic literature review of indicators measuring food security. Agric Food Secur 12:1–31. https://doi.org/10.1186/S40066-023-00415-7/TABLES/6

Marques ES, Reichenheim ME, De Moraes CL et al (2015) Household food insecurity: a systematic review of the measuring instruments used in epidemiological studies. Public Health Nutr 18:877–892. https://doi.org/10.1017/S1368980014001050

Meijer SS, Catacutan D, Ajayi OC, et al (2015) The role of knowledge, attitudes and perceptions in the uptake of agricultural and agroforestry innovations among smallholder farmers in sub-Saharan Africa. Int J Agric Sustain 13:40–54. https://doi.org/10.1080/14735903.2014.912493

Mes MA, Chan AHY, Wileman V, et al (2019) Patient involvement in questionnaire design: Tackling response error and burden. J Pharm Policy Pract 12:1–4. https://doi.org/10.1186/S40545-019-0175-0/TABLES/2

Muchai SWK, Muna MWM, Mugwe JN, et al (2014) Client focused extension approach for disseminating soil fertility management in Central Kenya. Int J Agr Ext 02:129–136

Mugwe J, Mugendi D, Mucheru-Muna M, Merckx R, Chianu J, Vanlauwe B (2009) Determinants of the decision to adopt integrated soil fertility management practices by smallholder farmers in the central highlands of Kenya. Exp Agric 45(1):61–75

Njenga MW, Mugwe JN, Mogaka H, Nyabuga G, Kiboi MN, Ngetich F, Mucheru-Muna M, Sijali I, Mugendi D (2021) Communication factors influencing adoption of soil and water conservation technologies in the dry zones of Tharaka-Nithi County, Kenya. Heliyon. https://doi.org/10.1016/j.heliyon.2021.e08236

Nord A, Snapp S, Traore B (2022) Current knowledge on practices targeting soil fertility and agricultural land rehabilitation in the Sahel. A review. Agron Sustain Dev 42:1–16. https://doi.org/10.1007/S13593-022-00808-1/TABLES/3

Oosterveld P, Vorst HCM, Smits N (2019) Methods for questionnaire design: a taxonomy linking procedures to test goals. Qual Life Res 28:2501–2512. https://doi.org/10.1007/S11136-019-02209-6/TABLES/1

Rolstad S, Adler J, Rydén A (2011) Response burden and questionnaire length: Is shorter better? A review and meta-analysis. Value Health 14:1101–1108. https://doi.org/10.1016/j.jval.2011.06.003

Saunders MNK, Lewis P, Thornhill A (2019) Research methods for business students, 8th edn. Pearson Education Limited, Harlow. ISBN 978-1-292-20878-7

Sharma H, Ruikar M (2025) Crafting an effective questionnaire: An essential prerequisite of engaging surveys. Perspect Clin Res 16:118–126. https://doi.org/10.4103/PICR.PICR_157_24

Slattery EL, Voelker CCJ, Nussenbaum B et al (2011) A practical guide to surveys and questionnaires. Otolaryngol. -Head Neck Surg 144:831–837. https://doi.org/10.1177/0194599811399724

Šūmane S, Kunda I, Knickel K, et al (2018) Local and farmers' knowledge matters! How integrating informal and formal knowledge enhances sustainable and resilient agriculture. J Rural Stud 59:232–241. https://doi.org/10.1016/J.JRURSTUD.2017.01.020

Methods and Tools of Data Collection in Nonexperimental Designs II (Qualitative Approaches)

19

Abstract

In qualitative research, several key data collection methods are used, including interviews, observations, document analysis, and participatory approaches. Interviews involve asking questions in person and can take various forms, such as in-depth interviews, focus group discussions (FGDs), and key informants. These can be unstructured, structured, and semi-structured interviews conducted face-to-face, over the phone, online, or one-on-one. In-depth interviews involve intensive probing; focus groups gather a group to discuss a topic; and key informants involve interviews with people knowledgeable about a subject, while group interviews use semi-structured interview schedules with several people simultaneously. Observation is another key method for data collection in qualitative research and can be done through participants or non-participant observation. Document analysis involving examining materials such as photos, letters, films, tapes, diaries, reports, or other documents is another key method. Audiovisual data collection, including photographs, videos, and drone photography adds another dimension to data collection. Accurate data collection and recording are essential in qualitative research, and this can be achieved through systematically recording the data using various tools such as data recording protocols and checklists. Ensuring the trustworthiness of qualitative data collection is crucial, and one powerful approach for this is triangulation, which may involve the use of diverse methods, multiple sources of data, cross-referencing data, and the use of multiple researchers in collecting, analyzing, and interpreting data. Researchers need to consider trustworthiness (validity) throughout the research process, including planning, implementation, and data analysis.

Keywords

Audiovisual data · Document analysis · Face-to-face interviews · Field notes · Focus group discussions · In-depth interviews · Internet based interviews · Interviewer bias · Key informant's interviews · Nonparticipant observation · Open-ended interviews · Observation protocol · Probing · Participant observation · Participatory research tools · Seasonal calenders · Triangulation · Trustworthiness · Venn diagram

Learning Outcomes

After reading this chapter, you should be able to:

1. Identify the various methods and instruments used to collect qualitative data.
2. Link each data collection method to the instrument.
3. Identify the various types of interviews and methods of their administration.
4. Distinguish between focus group discussion and key informants' interview.
5. Apply the concept of initial phase, mid-phase, and closing phase to conduct Focus Group Discussion (FGD)
6. Differentiate between participant and nonparticipant observation.
7. Distinguish between an interview schedule and an observational protocol.
8. Describe steps in conducting group interviews.
9. List some documents that may be acquired from farmers during document analysis.
10. Appreciate the need to tackle validity in qualitative data collection.

J. Njeri Mugwe, S. Runo, *Research Methodology in Agricultural Sciences*, https://doi.org/10.1007/978-981-95-1892-0_19

19.1 Methods, Tools, and Data Collected

The main methods involve exploratory methods to capture the richness and complexity of data. These include:

1. Interviews.
2. Observations.
3. Document analysis.
4. Participatory approaches.

The interviews are important for in-depth exploration of individual experiences and perspectives and are the most common form of collecting data in qualitative research (Jamshed 2014). They are of various types, including focus groups and key informant interviews (Fig. 19.1). Focus group discussions provide a platform for group discussions, enabling researchers to gather various views on a particular topic. Observations are used to collect data on behaviors and interactions in natural settings. Document analysis involves reviewing existing documents and records to extract relevant information. Participatory approaches involve engaging stakeholders in the research process, allowing them to contribute to the study's design, implementation, and analysis. Each of these methods and tools plays a crucial role in collecting comprehensive and reliable data to address the research questions.

19.1.1 Methods of Collection and Types of Data Collected

Your research question determines the method and type of data you collect. Qualitative data will mainly be collected on-site, where people engage with the phenomenon you wish to study. We usually travel to the site to meet our participants. The main data types include field notes, documents, transcriptions, and audiovisual materials.

Documents and audiovisual materials used are varied. They include both personal and public information. Documents may consist of letters, memos, meeting's minutes, annual reports, pictures, and other archived information. Audiovisual materials are things like photos and videotapes, for example, those of farming systems (Table 19.1). These could be obtained from the participants, and their stakeholders, for example, in a study involving farmer groups managing tree nurseries, relevant documents collected could include minutes of meetings, duty roosters, receipts of water payments, cash sale receipts, etc. Other relevant documents, such as annual reports, could be collected from the Ministry of Agriculture offices regarding tree nurseries.

Fig. 19.1 Main methods and tools used to collect data in qualitative approaches

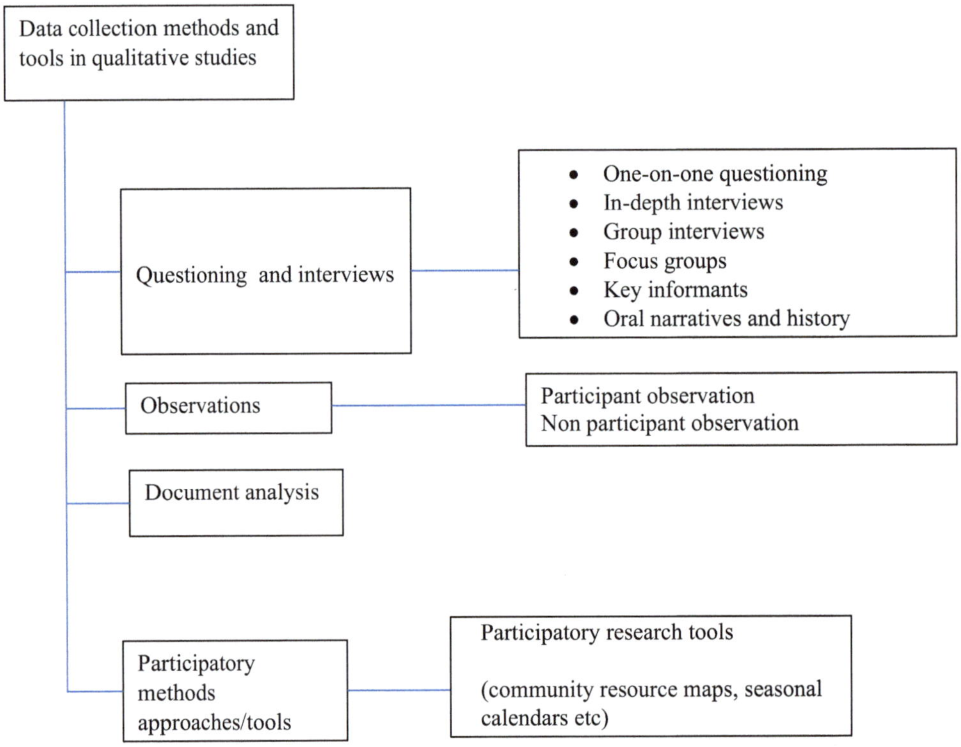

Table 19.1 Methods of data collection and data types in qualitative research

	Methods of collecting data	Types of data
1	*Interviews*	Field notes, Transcriptions
2	*Document analysis*	Field notes, journals, archival documents, maps, and videotapes
3	*Collating documents*	Photographs of people, farms, crops, and events Soil and plant samples Sounds
4	*Observations*	Field notes, drawings of farm transects, and illustrations of activities
5	*Participatory methods*	Qualitative and quantitative (field notes, drawings and maps, action plans, photos, etc.)

19.2 Interviews

19.2.1 Types of Interviews

An interview is an oral administration of questions using a guiding tool as discussed by Creswell and Poth (2016). During the interview, questions are asked to elicit a response. The researcher is the interviewer, while the respondent is considered the student or interviewee in the conversation using open-ended questions (Creswell and Poth 2016). A researcher must obtain maximum cooperation from respondents to obtain accurate information through the interviews. The researcher must, therefore, establish a friendly relationship with the respondent before conducting the interview. The respondents' responses are often recorded through note-making or/or tape recording.

Advantages

1. They provide in-depth data that are impossible to get using a questionnaire. This is because the interview is flexible and allows multisensory channels to be used; verbal, spoken, nonverbal, and heard.
2. Interviews guard against confusing the questions since the interviewer can clarify the questions, thereby helping the respondent give relevant responses.
3. Interviews are more flexible than questionnaires as the interviewer can adapt to the situation and get as much information as possible.
4. Very sensitive and personal information can be obtained from the respondent by honest and personal interaction between the respondent and the interviewer.
5. Interviews yield higher response rates mainly because it is difficult for a subject to completely refuse to answer questions or to ignore the interviewer.

Disadvantages

1. Interviews are more expensive as researchers have to travel to meet participants.
2. Interviewing requires a high level of communication and interpersonal skills.
3. Interviewers have to be trained to avoid bias.
4. Since interviews are adaptable and flexible and show concern for human interaction, they often introduce bias and subjectivity into the study. The subject may be too eager to please the interviewer, or the interviewer may ask questions that tend to support his or her perceived notions.
5. Interviews generally involve a smaller sample because they are time-consuming; if one is interested in using a large sample, interviews become a constraint.

19.2.1.1 How to Enhance the Validity of Data Collected Through Interviews

To enhance the validity of the data collected through interviews, it is crucial to manage potential sources of bias (Kristensen et al. 2008; Leung 2015; Noble and Smith 2015). These biases have been extensively documented in research (Randall and Fernandes 1991; Galdas 2017; Enzenbach et al. 2019; Bergen and Labonté 2020) and they are mainly *interviewer bias, interviewee bias*, and *participation bias*. We briefly discuss the biases here:

1. *Interviewer bias* occurs when the interviewer's comments, tone, or nonverbal behavior influences how interviewees respond. This may happen if the interviewer imposes their beliefs and perspectives through the questions. Bias can also emerge in the interpretation of responses. If interviewees do not trust the interviewer or question their credibility, the data's value may be limited, affecting its validity and reliability.
2. *Interviewee or response bias* can result from interviewees' perceptions of the interviewer or their perceived bias. However, it may also arise from the intrusive nature of interviews, especially in-depth or semi-structured ones, where sensitive topics are explored. Even willing participants might avoid discussing certain aspects because they do not want to delve into sensitive information. Consequently, interviewees might present a partial view that casts them or their organization in a socially undesirable light.
3. *Participation bias* can also stem from the characteristics of those who agree to be interviewed. For instance, the time commitment required for an interview might deter some individuals, potentially skewing the sample. This issue needs careful consideration and mitigation through the sampling approach. Further concerns include cultural differences between the interviewer and interviewees, as noted by Gobo (2011), who suggests that research interviews may be a product of individualistic societies and might not be as effective in other cultural contexts.

19.2.1.2 Structured and Unstructured Interviews

Interviews can vary widely in their level of formality and structure. They can be highly formalized and structured, utilizing standardized questions for each participant or informal and unstructured conversations. There are also intermediate forms that mix elements of both structured and unstructured formats, depending on the interview's objectives. For instance, an interview might include some highly structured sections alongside others that are more open-ended and flexible. We classify these as *structured, semi-structured,* and *nonstructured interviews.*

Structured Interviews The structured interview involves interviewing research participants formally using a specified number of questions that elicit the same information from the respondents. It is completely planned, and all questions relating to the research objectives are prepared before initiating the interview.

Semi-structured Interviews Semi-structured interviews are between structured and unstructured interviews. They will contain both formal and informal questions. The researcher aims to have the participants answer similar questions during the interview but, at the same time, create flexibility in probing. Sometimes, the researcher is guided by themes. The researcher has a list of themes and possibly some key questions to be covered, although their use may vary from interview to interview. This means that you may omit some questions in particular interviews, given a specific context concerning the research topic. The order of questions may also vary depending on the conversation flow. On the other hand, additional questions may be required to explore your research question and objectives, given the nature of events within particular organizations. The nature of the questions and the ensuing discussion mean that data can be captured by audio-recording the conversation or note-taking.

Unstructured Interviews These involve a casual conversation with no formality. They are mainly used to obtain more complex and personal information (Gay et al. 2009). In this kind of interview, you will have no predetermined list of questions to ask, although you need a clear idea about the aspects you want to explore. This will guide you to explore a general area in which you are interested. During the interviews, the interviewee is allowed to talk freely about events, behavior, and beliefs concerning the topic area. A disadvantage, however, is that you cannot be sure that the interviews will produce data for the research (DiCicco-Bloom and Crabtree 2006; Bergelson et al. 2022).

19.2.2 Ways of Administering the Interviews

19.2.2.1 Face-to-Face Interviews

Face-to-face interviews are a traditional and widely used method (DiCicco-Bloom and Crabtree 2006; Bergelson et al. 2022). This approach involves the interviewer and the participant meeting in person, allowing for direct interaction. Suppose you wish to explore how farmers are adapting to the climate change. The best way of collecting such data is to visit the farms and conduct in-person interviews. This allows you to observe the physical environment and the agricultural practices firsthand. The other key advantage of face-to-face interviews is observing nonverbal signals such as body language, facial expressions, and gestures, which can provide additional contexts to the verbal responses (Rubin and Rubin 2005). Also, an emotional connection is created when you meet face to face. This facilitates the establishment of rapport and trust, encouraging participants to share more openly and in greater detail. However, face-to-face interviews can be time-consuming and logistically challenging. This can be tedious particularly if the participants are geographically dispersed. They also require more resources in terms of travel and scheduling, and the presence of the interviewer might sometimes influence the participant's responses due to social desirability bias.

19.2.2.2 Telephone Interview

Sometimes, it may not be possible to gather participants for face-to-face interviews. For instance, in situations where participants are geographically dispersed or have time constraints that make in-person meetings difficult. Consider a case where you wish to study the effectiveness of agricultural extension services. You might consider using telephone interviews to reach farmers in remote areas. Telephone interviews can allow you to reach a wider audience without the need for travel. It thus has the advantage of saving on costs, making it a cost-effective and convenient option. They are also quite flexible. They can be scheduled to accommodate participants when available and at different time zones. However, the lack of visual observation is a significant limitation, as it hinders the interviewer's ability to interpret nonverbal signals and build rapport. Additionally, the quality of the telephone connection and the participant's comfort with speaking over the phone can affect the data quality (Sweet 2002; Novick 2008; Ward et al. 2015).

19.2.2.3 Internet-Based Interviews

Internet-based or digital-based interviews conducted through platforms such as Google Meet, Zoom, Skype, or email have become increasingly popular, especially when face-to-face interactions are impractical. The use of internet interviews has increased since the COVID-19 pandemic (Roberts et al.

2021; Keen et al. 2022). In this type of interview, you invite your interviewee into the platform and interview as if it were face-to-face. A good example is a researcher who wishes to conduct a tracer study of past students, to explore their perception of how the training received has impacted their career. The researcher will organize virtual meetings and interview the participants. This method offers several advantages, including interviewing participants from diverse locations without incurring travel costs. With advanced technology, such as video conferencing, you can observe nonverbal cues and facilitate a more personal connection. Internet-based interviews are flexible and can be scheduled more easily to accommodate participants' schedules. However, this method relies heavily on participants having access to and being comfortable with technology. Technical issues such as poor internet connectivity can disrupt the interview flow. Furthermore, the lack of the full physical presence can still limit the depth of engagement compared to face-to-face interactions.

19.2.2.4 One-on-One Interview

A one-on-one interview involves a personalized interaction with a single participant. The researcher engages the participant in one-on-one conversation to gather detailed information on opinions, experiences, or perspectives (DiCicco-Bloom and Crabtree 2006; Bergelson et al. 2022). These interviews can be structured, semi-structured, or unstructured depending on the research objectives. These interviews are particularly beneficial when a comprehensive understanding of a subject is required, and researchers aim to delve into emotions and personal experiences. These interviews can be conducted in person, over the phone, or via video conferencing based on logistical considerations and the preferences of both participants and researchers. The effectiveness of a one-on-one interview often hinges on the researcher's ability to build rapport and create a comfortable environment for the interviewee to share their thoughts and experiences openly. A disadvantage of one-on-one interviews is that they tend to be expensive and work if the participants are free to speak and share ideas comfortably (Creswell 2014).

19.2.3 General Guidelines for Interviewing

19.2.3.1 Preparation Before the Interview

The key to successful interviewing is preparation. One major step is to ensure you know the research topic and the situational context in which the interview will take place; get as much information as possible regarding the topic from the library and from the site. Other important sources of information are government and nongovernmental organizations working on a similar topic to what you wish to study. For instance, the Ministry of Agriculture will often have comprehensive annual and technical reports about the area (van Eeuwijk and Agerhun 2017).

Familiarity with culture is a key to avoid misinterpretation. People's culture is sensitive, and if they feel you are infringing on their culture, they may not give you the correct answers. You need to be aware of cultural differences and their implications. This enhances trust among the interviewees and increases the likelihood of obtaining valid data.

Build trust and rapport before the actual fieldwork. This may involve supplying relevant materials beforehand. Gatekeepers are important in identifying the participants and allowing you to share your advanced material. During this initial activity, you can initiate relationships that will help you during the interviews.

You also need to prepare the themes carefully to ensure they are relevant. Themes provide focus and direction; without themes, your interview will lack focus and direction. These themes may be developed from the literature, your experiences, discussions with colleagues and others who have worked in the study area, and common sense. Capture these themes in a well-arranged guide that informs you where to start and how to end.

The interview location will likely influence the quality of the data you collect. You should reflect on the site and evaluate its suitability for you and your participants. First of all, consider a site that is comfortable for the participants. Avoid noisy sites that may interfere with the ongoing interviews. You also need to consider accessibility to the site and how the participants will travel there. If the site is far, you must make prior arrangements to facilitate them, ensuring that the facilitation is not misinterpreted as bribing. Ensuring the site is safe for you and the interviews is important.

Ensure you can answer the following questions with clarity:

- Whom will you interview?
- What interview method will you use?
- Where will the interview be conducted?
- What tools will you use to conduct the interview?
- Who will help you in the interviews?
- How will you contact the interviewees?

19.2.3.2 Conducting the Interviews

Here, we provide general guidelines to follow during the interviewing process. Details are, however, given for each of the specific interviews in the respective sections:

- Obtain consent from the interviewee to participate. Ensure that the interviewee fully understands the purpose of the study, what their participation entails, and how the data will be used. Provide the participants with consent forms that outline this information and obtain their written or verbal agreements to participate.

- Be courteous and professional. Treat the interviewee with respect and professionalism. Begin the interview with a friendly introduction, explain your role, and establish a rapport to make the interviewee feel at ease. Maintain a neutral and nonjudgmental attitude throughout the interview.
- Audiotape the interview. A reliable recording device captures the entire conversation. Ensure the device is properly functioning and positioned to pick up your and your interviewee's voices clearly. Always inform the interviewees that the conversation is being recorded and confirm their consent.
- Take brief notes. While the interview is being recorded, jot down key points, observations, and nonverbal cues that may not be captured on the audio. These notes can help you remember important details and provide context when analyzing the data later.
- Use probing to obtain additional information. Encourage the interviewee to elaborate on their answers by asking follow-up questions or prompts like "Can you tell me more about that?" or "Why do you think that is so?" This technique helps to gain deeper insights and more detailed responses.
- Avoid leading questions. Frame your questions neutrally to avoid influencing the interviewee's responses. For example, instead of asking, "Don't you think organic farming is better?" ask, "What are your thoughts on organic farming?" This approach ensures that the interviewees' answers are their own and not swayed by the interviewer's opinions.

19.2.4 The In-Depth Interviews

In-depth interviews are one of the most important methods used to collect qualitative data. It is called an in-depth interview because it involves the researcher interviewing the respondent using open-ended questions that allow for intensive interaction and detailed responses. It is a person-to-person discussion, with the researcher taking the lead by asking questions. In-depth interviews use either an unstructured interview schedule, semi-structured interviews, or both.

Semi-structured interviews place a significant amount of importance on the subject's freedom of expression (Ruslin et al. 2022). They usually have categories with responses and the interviewer checks the respondent's response. The answers are not followed up to obtain a greater depth of information. They do not allow probing, and the interviewer strictly follows the sequences of questions in the instrument.

19.2.4.1 Instruments Used in In-Depth Interviews

The data collection instruments used in in-depth interviews are interview guides, which contain questions to guide the interview (Brinkmann 2022). The interview guide is typically carefully designed and prepared by the researcher and contains well-structured questions that outline the topics, specific questions to be asked, and follow-up inquiries. The guide starts by introducing the overarching topics and themes to be discussed. These topics must align well with the research objectives and helps create a framework for the conversation. Open-ended dominate the interview to encourage participants to respond using their own words and continuous probing. During the interview, probing and follow-ups are dominant activities using a structural flow that is coherent and logical. Usually, the guide will have questions that start from general and gradually move to more specific or sensitive questions.

After each interview, the interviewer usually creates time to reflect on the experiences, tries to identify themes and patterns, and also identifies any unexpected direction that may inform the next interview.

19.2.5 Key Informant Interviews

Key informant interviews are in-depth interviews with people perceived to be (informed about a subject), who provide insights into a topic of interest. Key informants aim to collect detailed information related to a specific, often focused topic of interest to the researcher. However, we should be careful in selecting the key informants to avoid bias. Usually, researchers may be tempted to pick from the available "labeled" key informants and leave out community members, who are equally important in providing the required information.

Interview questions are captured in a semi-structured interview schedule. Telephone or face-to-face interviews may be used depending on your preferred choice, key informants' availability, and overall logistical feasibility. The selection of key informants depends on the research question being investigated. For instance, a researcher will pick key informants representing the various value chain segments in value chain assessment. For instance, to generate Fig. 19.2, key informants from various segments of the soybean value chains were interviewed (Ngum 2019).

19.2.6 Focus Group Discussion

Focus group discussion (FGD) comprises a group of people deliberately gathered to discuss a predetermined topic by the researcher (Kitzinger 1995; Morgan 1997) (Fig. 19.3a, b).

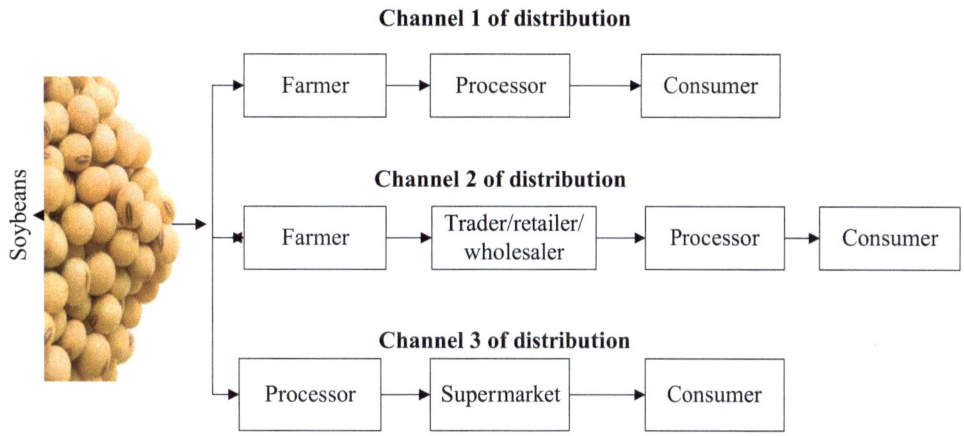

Fig. 19.2 Value chain analysis generated by key informants. (Ngum 2019)

Fig. 19.3 (**a** and **b**) Focus group discussion involving separate groups of males and females. (Njenga et al. 2021)

The members of the group are usually selected to constitute a homogenous group. For example, a dairy marketing group can be constituted to discuss ways of improving milk pricing. FGDs are cheap and can generate large amounts of data quickly. Thus, they are suitable for students who have limited financial resources. The researcher will typically organize FGDs of farmers near their villages, where discussions will be held. Farmers are expected to participate freely during the discussions, and the research team always tries to create a friendly environment. The research team uses an interview guide consisting of preformulated questions that guide the questioning.

Focus group discussion aims to obtain people's perceptions, attitudes, feelings, and ideas on a specific topic in a permissive and nonthreatening environment (Krueger and Casey 2002; Schindler 2022). For example, FGD will be useful in the following scenarios:

- Testing ideas about accepting an innovation or agricultural technology, for example, gathering perceptions about the introduction of genetically modified maize varieties or new bean varieties enhanced with zinc.
- Exploratory studies to help refine research concepts and questions.
- Solve specific problems facing agricultural positioners, for example, poor marketing channels.
- Evaluating agricultural development programs, for example, conducting focus groups to gather experiences on an ongoing agricultural project.
- Diagnosing problems and challenges that researchers need to address.
- Interpreting previously obtained quantitative results.

The discussion is coordinated by a skilled facilitator, who guides the group to engage in a focused and in-depth discussion on the key issues of the research topic. The facilitator remains engaged and vigilant during the discussion to ensure that the predetermined topic is adequately discussed. Characteristics of a skilled facilitator include being professionally trained, a good listener, well-organized, objective, and attentive (Hair et al. 2020). As the discussion and interaction between the facilitator and the respondents continue, the rest of the team members take notes and record the proceedings on audio or video. The team also observes respondents' nonverbal communication, using both communication and observation aspects.

A skilled facilitator is the hallmark of an effective FGD (Baral et al. 2021). The facilitator's role is crucial in guiding the discussion and ensuring that all have an opportunity to voice their thoughts openly. A good facilitator also creates a welcoming atmosphere, encouraging participants to share their thoughts freely. They should be able to manage group dynamics well, balancing dominant voices and encouraging quieter ones to speak. Additionally, a skilled facilitator is able to probe deeper into responses, ask follow-up questions, and clarify points without influencing the answers. Finally, staying neutral and not steering the conversation is crucial. So you can see it as a must to use a facilitator and you may be wondering how you get a good facilitator.

To get a good facilitator, you can invest in training along with your research assistants. Check from among your colleagues who can be a good facilitator. The facilitator must be able to understand it is research work, not just rapport building (Box 19.1 lists characteristics for a facilitator). We recommend you avoid using local extension workers as facilitators, though you should retain them as part of your team. They usually become complacent because of prior interaction with farmers. Instead, choose individuals who can remain neutral and are able to manage group dynamics. The aim should be to collect unbiased and comprehensive data.

Box 19.1: Characteristics of a Skilled Facilitator of a Focus Group Discussion

- Ability to create warm and friendly environment (could incorporate an ice breaker).
- Ability to remain mentally alert and process information fast.
- Ability to make smooth introduction.
- Ability to make pauses and probe.
- Ability to manage group dynamics (be able to control reactions of the participants and have group control).
- Ensure all participates have an opportunity to speak.
- Maintain a balanced and respectful conversation.

Another important consideration is the group size. An ideal focus group should not be too large to be unmanageable and not too small to negate the benefits of group dynamics. For example, two to three persons is too small. The ideal size of FGD should be between six and eight willing persons (Krueger and Casey 2015). Other authors have recommended ten as better but eight as ideal and three to fourteen members (Gill et al. 2008). With more than 12 members the group may become difficult to manage (Nyumba et al. 2018). Another important consideration is the duration of the meetings. Focus groups typically last about 2 h but may run from 1 to 3 h. Researchers should be cautious about participant suffering from fatigue when meetings take too long.

Preparation logistics play a big role in enhancing the success of FGD. Key considerations during preparation include recruitment of participants, identification of the site of the interviews, and development of an FGD guide. Ensure homogenous sets of participants for each subject matter are recruited in regard to socio-economics, age, sex, wealth, and other demographic characteristics. The site of FGD should be neutral and quiet to enhance communication. Finally, the FGD guide should be carefully prepared, consisting of open-ended questions, logically arranged, and asking only the necessary questions. It is advisable to have the research team thoroughly familiarize themselves with the topics in the guide.

19.2.6.1 Benefits and Limitations of Focus Group Discussions

Focus group discussions have the following benefits:

- *Observation*: the researcher can observe the discussion and gain first-hand insights into the respondent's behaviors, attitudes, language (including body language), and feelings.
- *Group interaction*: interaction between respondents will generally stimulate richer responses and allow new and valuable thoughts to emerge.
- *Cost and timing*: FGD can be completed more quickly and generally less expensive than a series of in-depth interviews.

Limitations of focus group discussion include:

- Logistical challenge: expensive and involving to organize.
- Group dynamics: dominant individuals may dominate the discussion and affect the quality of data.
- Complex data analysis: analysis is complex and requires careful coding.
- Facilitator bias: poorly trained facilitator bias can significantly influence quality of data.
- Ethical issues: data lack anonymity and confidentiality.

19.2.6.2 Focus Group Discussion Guide

The facilitator uses a focus group discussion guide to guide the conversation. The FGD guide typically consists of three main sections. The FGD typically starts with an introduction section that introduces the purpose of the FGD, the middle section allows interactions on the main topic of discussion, and the last section is for concluding the discussion. In all these, researchers should carefully take notes (see Box 19.2 for the tips).

Introduction phase: Introduces research objective and roles of the researchers and sets the FGD rolling. It contains:

- *Welcome remarks*: the facilitator begins by welcoming participants to the FGD. The gatekeepers are also recognized.
- *Introductions*: the facilitator guides participants into the introduction and personal sharing to create a friendly environment.
- *Explanation of purpose*: the facilitator explains the purpose of FGD, emphasizing on the research objectives and how participants are expected to participate.
- *Obtaining informed consent*: participants are informed about their right to confidentiality and anonymity. They are also asked to provide informed consent to participate.
- *Ground rules*: the facilitator and participants agree on the ground rules so that participants feel safe to air their views.
- *Overview of the discussion*: facilitator provides an overview of the issues to be discussed, including the main topics and questions.

Mid-Phase This phase is expected to be the longest phase and to stimulate conversation on the main topic. Questions in this phase should be open-ended and arranged in such a way it guides participants in a structured conversation about the research topic. It is the heart of the FGD and important for generating data and deepening understanding of the research topic. It involves the following key activities;

Research topic discussion: the facilitator asks questions and encourages participation.
Probing and clarification: facilitator may follow up on questions or seek clarification helping to increase deeper insights into the subject matter.
Managing group dynamics: the facilitator ensures that dominant members of the group do not monopolize the discussion.

Closing Phase This last phase is for summarizing the discussion, marks the conclusion of the group interaction, and is designed to wrap up the discussion. It provides closure, acknowledges participants' contributions, and ensures that everyone leaves the discussion on a positive note. A well-executed closing phase helps maintain goodwill with participants and contributes to the overall success of the research process.

This phase should include the following key elements;

Summary of the discussion: the facilitator should summarize the key points and insights that emerged from the discussion.

Reflection and clarification: the Facilitator should ask reflective questions, for example, of all the things that were discussed, what to them is the most important (Krueger and Casey 2015), and ask participants for any additional comments or questions.

Participants' feedback: the Facilitator should ask participants for any feedback that can help improve future discussions.

Appreciation: the facilitator thanks the participants for their time and contributions. The facilitator may also inform them about the next steps in the research process and remind them about the confidentiality and anonymity of the process.

Closure: the facilitator officially closes the FGD, signaling the end of the discussion and addresses any logistical matter.

Box 19.2: Tips for Effective Note Taking During Focus Group Discussion

- Be familiar with list of questions in the focus discussion group guide (content and the flow).
- Carry out the recording of the discussion.
- Be mentally alert, listen carefully, and take clear notes.
- Take notes comprehensively.
- Ask participants to repeat questions, after getting permission from the facilitator. Observe, note, but remain impartial.
- Systematically summarize the notes at the end of the conversation.
- Immediately after the FGD, compare across various FGDS and be able to come up with ANT themes, hunches, and ideas (Krueger and Casey 2015).

19.2.6.3 Analyzing Data from Focus Group Discussion

Data from the FGD will mainly consist of notes, video recordings, pictures, and other documents provided by the participants. Analyzing the data is similar to analyzing other qualitative self-reported data. The researcher must gather and compare discussions that address similar themes, while also examining how these themes relate to the variables within the sample population. However, giving percentages in reports of focus group data is inappropriate, and it is important to distinguish between individual opinions expressed by group members versus the actual group consensus. As with all qualitative analysis, it is important to consider deviant cases, which means paying attention to minority opinions and examples that do not fit into the researcher's overall theory.

19.2.7 Oral History

Oral history collects data that are mainly used to study experiences, perceptions, opinions, and accounts of a historical event or happenings in everyday living (Leavy 2011). It is used to gather historical knowledge as how individuals or groups view the phenomenon of interest. Oral history is helpful in collecting data on drought occurrences over a specific period or changes in agricultural practices that have happened. For example, when we hear that Africa is facing the worst drought in the last 40 years, such information could have been collected from an oral history of individuals, who are elderly and is triangulated with other documented information.

As usual, data will be collected by an interviewer (researcher) from the respondent (narrator). The researcher obtains, records, and presents data by interpreting historical and current information as narrated by the respondent. The information is subjective in nature, consisting of experiences and opinions, and could be based on eyewitness evidence or information passed from other sources, such as older people and traditional beliefs (Leavy 2011).

To conduct an oral history interview, you need to decide on the type of experience, perceptions, and historical events or practices you want to find out about (Leavy 2011). Then, the individuals possessing the information will be identified, followed by data collection using interviews. Oral history data are used with other primary and secondary sources to gain understanding and comprehension of the subject at hand.

19.2.8 Group Interviews

Group interviews are conducted using a semi-structured interview schedule with several people simultaneously. They could range from small sessions with a few people (3 to 4 people) to large ones with 10 or more, up to 20 people. Unlike the FGD that researchers widely use to collect specific information group interviews are useful for collecting data on broad-based issues, such as people's culture (Rubin 2016). The background research and the research objective will guide you in selecting the participants. For example, if the group interviews aim to gather information on differences between gender roles, you may need to consider having separate groups of men and women. The data obtained in group interviews give lots of data on general views but for more specific information consider complementing with another method such as survey, key informant interviews, or FGD.

19.2.8.1 Preparing and Conducting the Group Interviews

- *Stakeholder identification*: identify your stakeholders and use one of the purposeful sampling techniques to get your subjects for the study.
- *Information to be collected*: decide the information needed and from whom (What questions will you ask each of the stakeholders?):
  - Depending on the nature of the study, you may have more than one type of stakeholder (farmer, extension officer, trader, etc.).
  - For each of the interview guides, ensure that the researcher, interviewers, and/or moderators are thoroughly familiar with the questions.
  - The guide should include only the necessary questions for the specific interview.
- *Ethical clearance*: process the necessary international and national ethical clearance. Have an informed consent form.
- *Interview guide*: develop an interview guide that lists the questions or issues to be captured.
- *Rapport building at the start of the interview*: begin interview with rapport building:
  - Always start the interview with friendly and familiar greetings using a pleasant voice and body language. Also thank the participants for taking their time to participate in the study.
  - Introduce yourself and your organization and describe the aim of the research including practical implications. Tell the respondents how they are likely to benefit from the research. At this point assure them that the data are for research purposes only.
  - Provide the protocol for informed consent, explain clearly what it requires, and request for their consent and approval to take notes and do any recording.
- *Start with broad questions*: guided by the interview checklist ask questions starting with the broad questions followed by the more specific questions using effective questioning techniques:
  - Start by asking questions on the background and ask one question at a time, verifying unclear responses.
- Ask more emotionally and intelligently demanding questions after you have attained a good rapport and confidence. This is the heart of the in-depth interview because these are the questions that directly correspond to your research question and anchored in theory:

  - Ask open-ended questions, avoiding leading questions and using follow-up and probes.
  - Be ready to listen and give time for the participant to respond completely without interfering; ask the next question or point of clarification after the respondent has responded, which the body language shows.
  - Phrase the question in such a way that it encourages participants to provide elaborate detailed responses.
  - Listen with attention to capture every piece of information from the respondents.
  - Avoid moving too quickly to the next question.
- *Taking notes and audio taping*: during the interview take notes as well as audio tape the conversations. These are complementary in case of failure of the audio taping. Because of the intense questioning and listening at the same time, you may not be able to take all the notes.
- *Have a plan and respect it:* ensure that you have a plan to guide you on conducting the interview. Ensure that you stop the interview at the agreed time.
- *Close interview with courtesy*: conclude the interview and ensure that you
  - Ask the respondents if there is anything they would like to clarify or share with you before you close the interview.
  - It is courteous to summarize what you have gathered and explain what you plan to do with the data (even if you had explained that earlier).
  - Thank the respondents for participating.
- Start with a general question to get people warmed up. The researcher should ensure that the group is homogenous in terms of social class, age, gender, ethnicity, levels of knowledge, etc.
- Make sure the questions are easily understood by the respondents. Use simple language so that all the group members understand the question.
- Assure group members of confidentiality such that they are free to express personal feelings or reveal aspects of their personal life.
- Keep questions short. Do not have several parts to each question as this may confuse the respondents.
- Word questions in such a way that all people will not feel guilty or embarrassed.
- Do not indicate any judgment about what is acceptable and what is not acceptable; keep thanking the participants and make them feel that their contribution is welcome and worthwhile.
- Researcher and facilitator should be fully trained and versed with the questions in the interview guide.

19.2.8.2 Data Collected in Qualitative Farmer Studies

Qualitative farmer studies will focus on understanding farmer experiences, perspectives, attitudes, and behaviors. Data collected will constitute the following:

- Interview transcripts: these are verbatim transcripts of in-depth interviews with farmers on their experiences and attitudes.
- Field notes: detailed observation notes made by the researcher during field visits, observations and interaction with individual farmers and farmer communities.
- Narratives and stories: these include personal narratives and stories shared by the participating farmers that shed light on their experiences.
- Visual data in photographs, videos or visual representation captured during fieldwork. Useful in providing visual context and evidence of agricultural practices and landscape changes over time.
- Focus group transcripts constitute summaries of group discussions with farmers, they reveal shared perspectives, disagreements, and social dynamics.

19.3 Observation

Observation, as the name indicates, is collecting data through observing and recording what you see. A researcher collecting data using observation systematically views, records, describes, analyzes, and interprets people's behaviors or settings (Saunders et al. 2019). The researcher gains valuable information on the phenomenon under study by directly documenting behaviors in their natural context. Most observations on agriculture are done in farm settings and involve collecting firsthand information. This can be done by conducting transect walks on farms to observe individuals and their surroundings systematically. This method is useful as it enables researchers to witness and document farming practices, crop conditions, and interactions within the agricultural environment. As researchers make observations, they can capture varied details and patterns regarding the dynamic processes in the agricultural landscapes. Challenges experienced on the farms can also be captured. This significantly contributes to the depth of the data collected, helping in a holistic understanding of the complexities inherent in agricultural practices.

In qualitative studies, observation is one of the most crucial research techniques, as it helps to understand and interpret the social, cultural, and economic environment.

According to Kumar (2014), observation is a purposeful, systematic, and selective way of watching and listening to an interaction or phenomenon as it takes place. Researchers should maintain a field diary to record day-to-day observations continuously.

In some research studies, particularly in ethnography, the observation method of data collection is considered a participatory study. This is because the researcher has to fully immerse themselves in the environment where the respondents are while taking notes or recording the observations. The observation method may involve using the "five senses" to watch, listen, read, touch, and record the behavior and characteristics of the studied phenomena (Erlandson 1993).

The observation method has its own set of advantages and disadvantages. One of the advantages is that it enables you to collect direct and real-time data in natural settings, allowing you to capture genuine behaviors and interactions. Moreover, observation is a nonintrusive approach, helps minimize the likelihood of altering participants' behavior due to awareness of being studied. However, some disadvantages include the potential for observer bias, and limitations to certain sites where access can be gained. Establishing rapport in such sites may also be challenging, particularly if individuals are unfamiliar with formal research. Additionally, observation provides limited insight into participants' thoughts and feelings and can be time-intensive, especially when prolonged observations are involved (Sutton and Austin 2015; Saunders et al. 2018).

Observation can be structured or unstructured. In structured or systematic observation, data collection is conducted using specific variables and according to a predefined schedule. Unstructured observation, on the other hand, is conducted in an open and free manner in a sense that there would be no predetermined variables or objectives.

Observations are important for the following reasons:

- Firsthand observation of witness behavior including interactions between individuals.
- Observe people in a natural environment, which increases the validity. People act naturally.
- Observation is a good method of reaching hard-to-reach groups such as subcultures, who are less likely to take questionnaires.
- Can develop new knowledge outside the existing knowledge base and lead to further research.

19.3.1 Types of Observation

19.3.1.1 Nonparticipant Observation

Nonparticipant observation involves the researcher collecting data by observing and reading behavior, events, and processes without becoming actively involved. The researcher is an outsider to the whole process of data collection (Creswell 2012). Suppose you plan to conduct a nonparticipant observation on some farmer groups. You will not be directly involved in the situation that will be observed. You will detach yourself and will not participate in the group's activities, intentionally interact, or affect their activities. Instead, you will visit the site where the farmer groups operate, look, and record notes without becoming involved.

Nonparticipant observation is a common method for monitoring trends and practices in agricultural landscapes. Imagine you wish to carry out a market observation on agricultural markets for selected commodities. You would need to visit the agricultural market to observe and document the trading activities and pricing mechanisms. Of interest, too, would be interactions between farmers, traders, and buyers. This kind of observation studies value chains of agricultural commodities. This would involve spending several days at the selected site (local market) to study how prices of various crops fluctuate based on supply and demand. Data collected would include the types of crops sold, pricing trends, and possibly demographics of buyers and sellers.

19.3.1.2 Participant Observation

While in nonparticipant observation, the researcher is detached; participant observation involves a researcher observing ongoing behavior. It involves watching the events or situations or activities from inside by participating in the group to be observed. In this case, you will become part of the community or part of the event, freely interacting with the group members and participating in various activities of the group. Through the interaction, you gather information on the way of life of the observed group or his own and study his behavior or other activities. The data are often qualitative, collecting in-depth data about the behavior and taking photos or videos. It requires a highly trained observer to increase validity/reliability of the data (McCuen 1996).

Summer schools or student internship programs can be viewed as a unique form of participant observation. In these programs, students are embedded within farm families for a specific period. They engage in various farm activities, meticulously documenting their observations in notebooks. These activities span from planting and harvesting to feeding livestock, managing farm equipment, and participating in household chores. Throughout their stay, students gather qualitative data through observations, informal interviews, and participation in daily tasks. This immersive approach enables them to capture rich, detailed information that might be overlooked in more formal research settings, underscoring the value of their role in agricultural research.

If, as a researcher, you plan to undertake participant observation within a farming family setup, you will make similar arrangements to live on the farm. The data to collect will be dictated by your research objective. You will be required to actively participate in farm activities, while you maintain detailed notebooks, wherein you record observations, experiences, and reflections. These records include descriptions of activities, conversations with family members, and any challenges or successes observed during their stay. Informal interviews with farm family members and other local farmers provide additional insights into farming practices, market conditions, and community dynamics. These conversations help you to gather diverse perspectives and contextualize your observations.

A benefit of participant observation is that it allows the researcher to gain information and develop relationships with participants that would not be possible if the researcher did not participate.

19.3.1.3 Research Instrument in Observation Method

An observation schedule is used in the observation method of data collection.

19.3.1.4 Steps in Making Observations

As previously discussed, the observational process is a systematic data collection method. It involves a structured approach of watching, listening, and recording behaviors and conditions in the study setting. This systematic approach ensures the reliability and validity of the data collected. The steps are as follows:

1. Select the observation type and scope: decide whether the observer will be a participant in the activities being observed (actively engaging with the subjects) or a nonparticipant (observing from a distance without interaction). Determine whether the observations will be structured (using predefined schedules or checklists) or unstructured (more open-ended and flexible). You also need to decide on the scope of the observation, including the duration, frequency, and locations of your observations.
2. Select a site and obtain the required information. Initial preparation an visiting the site is necessary to get familiar and also to start building rapport.
3. Develop an observation guide: create a timeline detailing when and where observations will take place. Develop a checklist or rubric that outlines the specific behaviors, events, or conditions to be observed and recorded. Just like in the survey, conduct a pilot test of your observation method in a similar setting to refine your tools and approach. This helps identify any issues with the observation guide and allows for adjustments before the actual data collection begins.
4. Observe and your observations: at the site, identify who or what to observe and for how long. Ensure all necessary materials and tools (e.g., notebooks, audio/video recording devices, checklists) are prepared and functioning. Systematically record your observations according to your predefined guide during each observation session. This might include taking notes, capturing audio or video recordings, and filling out checklists. Strive to remain consistent and as objective as possible. Avoid allowing personal beliefs, emotions, or preconceived notions to influence your observations. Apply the same standards and methods throughout the observation period to ensure data consistency.
5. Decide what information you will collect and come up with a mechanism for recording notes: information to collect might include photos of participants and activities such as vehicles carrying the commodities, specific events and processes. Using the example of the agricultural markets, you might record the activities of the different traders and buyers and their negotiations. Taking of notes is a crucial activity. This forms your data. You should record both descriptive and reflective notes. Descriptive notes capture people, events, and activities. Reflective notes record your thoughts and insights that are emerging from the observations.
6. Analyze the data and report the observations: organize your observational data systematically, categorizing notes, checklists, and recordings. For easy analysis, use coding techniques to identify patterns and themes within the observational data. Clearly document the findings from your observations, providing detailed descriptions and, if applicable, supporting evidence from audio/video recordings.

19.4 Document Analysis

19.4.1 Written Documents

Document analysis, or content analysis, is a systematic procedure of analyzing documentary evidence. Documentary evidence includes textbooks, photos, letters, films, tapes, photos, diaries, themes, reports, or other documents. This may be historical or contemporary (Box 19.3). Documents are the data source; they are analyzed and interpreted by the researcher to give a voice and meaning to an assessment topic (Bowen 2009). The researcher systematically reviews and interprets the various documents to gather information about the research question. For example, documents used in the capacity for building of new agricultural technology will be analyzed to determine the teaching methods used. Also, in studies involving the determination of land degradation, spatial analysis of land via satellite provides information on land use changes as well as the extent of land degradation (see Box 19.3 for list of documents).

Policy research most often involves document analysis. A researcher examines and analyzes documents such as policy

Box 19.3: Some Documents that May Be Used in Document Analysis Involving Farmers

Official Government Reports
- Policies and policy directives
- Food safety standards
- National Agricultural Policy
- Reports on rural development programs
- Regulations of pesticides
- Funding request

Implementation and Training Documents
- Training manuals
- Handbooks: livestock, soil fertility management on various crops
- Midterm and final-term impact evaluation reports
- Financial and operational plans
- Manuals on best practices for crop management
- Guidelines for integrated pest management
- Minutes of meetings
- Committee reports
- Emails
- Newsletter for agricultural extension, etc.

Scholarly Works
- Thesis and dissertations
- Research papers
- Textbooks and other course materials

Media
- Newspapers
- Podcasts, posters, bulletins, newsletters, blogs, etc.

Farmers
- Farm records
- Daily log of farm activities
- Production records
- Market records

briefs, government reports, and regulatory guidelines. For example, analyzing subsidy program documents can reveal their effects on agricultural practices and economic outcomes. Document analysis also allows researchers to trace the historical evolution of policies and compare approaches across different regions. This thorough review helps identify gaps and areas for improvement, providing valuable insights that inform policy-making and advocacy efforts.

Document analysis can be conducted as a stand-alone study or component of a larger qualitative or mixed-methods study. In most policy studies, document analysis is used as the sole method in data collection. In mixed methods, documents are often used to triangulate data and findings gathered through other means, such as surveys (Punch 2012). The aim of including document analysis will be to enhance meaning, understanding, and empirical knowledge of the subject under study. Ethnographers use multiple documents to achieve triangulation. They will collect documents that will help interpret the natural environment encountered and the behavior of participants.

Guidelines for Collecting Documentation

Identify documents that will provide information to achieve your objectives.

Consider collecting documents from both the private and public entries/organizations.

Seek the required permissions to use them, clearly explaining why you need them.

Examine documents for accuracy and usefulness for meeting your objectives.

Record information from the documents. You can take notes, photo copy documents, or even scan for further analysis.

19.4.2 Audiovisual Data Collection

Audiovisual data comprise images or sounds. The data include photographs, videos, audio recordings, and other multimedia formats that capture real-time events, interactions, and environments (Bauer and Gaskell 2000). For example, you may take video recordings of farming practices to provide perspectives on techniques and behaviors that an interview might not fully explain. Similarly, photographs can document changes in crop growth or soil conditions over time. These audiovisual materials can be analyzed to identify patterns, validate findings from other data sources, and create a more comprehensive understanding of the research topic.

Using visual materials makes the information more accessible and engaging because people easily relate to images. In agricultural research, visual materials can be particularly powerful. *Video interviews* with experts, photographs of crop fields at different stages of growth can show the effects of various farming techniques, crop pests, or environmental conditions. *Videos* can demonstrate new farming methods, such as precision agriculture technologies, and more effectively convey than written descriptions. Additionally, *maps* showing soil types, water availability, and land use can help researchers and farmers visualize and better understand the spatial distribution of resources and challenges within a given area. These visual tools enhance the clarity and impact of the research findings but also aid in communicating them to a broader audience, including policymakers, practitioners, and the local farming community.

19.4.3 Collecting Audiovisual Data

Collecting audiovisual data in agricultural research involves several steps and methods to ensure the data are relevant and ethically obtained.

Step 1: Planning and Preparation

The first step involves planning and preparing for the collection of audiovisual data. Begin by clearly outlining the objectives of collecting the data and determine the materials that will help achieve your research aim. Plan for specific aspects of the agricultural study you will capture, such as crop growth stages, farming practices, or interviews with farmers. At this stage, you should also select and acquire appropriate audiovisual equipment that may include digital cameras, video cameras, smartphones, drones for aerial views, and audio recorders. Ensure that the equipment is suitable for the environment, such as being durable and weather-resistant for fieldwork. Additionally, you need to obtain permission from relevant authorities and participants. Consent forms should be secured from individuals who will be filmed or recorded, ensuring they understand the purpose and use of the data.

Step 2: Actual Data Collection Methods which include the following:

(a) Photography: use high-resolution cameras to capture clear and detailed images. Ensure consistency in lighting and framing to help maintain the quality of images over time.
(b) Videography: have a well-thought-out plan for shots or a storyboard to ensure all necessary footage is captured. Using tripods or stabilizers helps avoid shaky footage.
(c) Audio recording: use microphones and minimize background noise. Conduct interviews in quiet locations.
(d) Drones offer a unique perspective for data collection, allowing for aerial surveys of large fields, crop patterns, and landscape changes. This method is particularly useful for recording data on crop health, soil variability, and irrigation patterns from a bird's-eye view.
(e) Field notes and annotations complement audiovisual data, providing detailed context that might not be captured on camera or audio. These notes should include the date, time, location, and any relevant observations.

Step 3: Post-Collection Processing: involves organizing and storing data systematically, including metadata such as date, location, and subject matter. To prevent loss, copies of all data should be stored in multiple locations. Editing and analysis are conducted using software tools to enhance image and audio quality, trim unnecessary footage, and create compilations. Transcribing audio recordings and annotating video footage help to identify key themes and insights relevant to the research.

19.5 Collecting Data Using Participatory Research Tools

Participatory research tools were popularized by authors such as Chambers and others in the 1980s and 1990s as a way to democratize the research process and shifting power from researchers to the communities they were studying (Narayanasamy 2009; Chambers 1994). These tools emerged as a part of a broader movement toward more bottom-up approaches to farming systems research. They emphasize the importance of local knowledge and participation to ensure relevance and sustainability of research outcomes (Chap. 10). They were widely promoted in methodologies such as participatory rural appraisal (PRA) and rapid rural appraisal (RRA).

In the twenty-first century some practitioners have rebranded PRA and its tools under new frameworks. These include community-based participatory research (CBPR) and participatory action research (PAR). Though rebranded, they continue to emphasize collaboration between researchers and communities. The researchers place the tools at the center of engagement and knowledge generation. Additionally, the tools have continued to evolve to integrate the current thinking including digitalization. They remain integral in on-farm experimentation, a major component of agricultural research (Lacoste et al. 2021).

These tools include transect walks, social mapping, wealth ranking, ranking exercises, community mapping, Venn diagrams, and community workshops among others. A few are described in the next section and references on PRA tools given at the end of this chapter.

19.5.1 Community Resource Maps

A community resource map is used to identify available resources and the community's resource needs (Delgado and Humm-Delgado 2013). Various stakeholders, including farmers, government representatives, and community leaders, may be involved in creating this map. However, even a single community can generate a resource map. The map typically includes houses, churches, schools, crops, roads, hospitals, livestock, water holes, rivers, forests, etc. (Fig. 19.4). The map should be one of the first tools used in community participatory research because it opens up the community and ensures participation. The purpose is to provide a clear picture of what resources are available, identify gaps or needs, and explore opportunities for resource sharing.

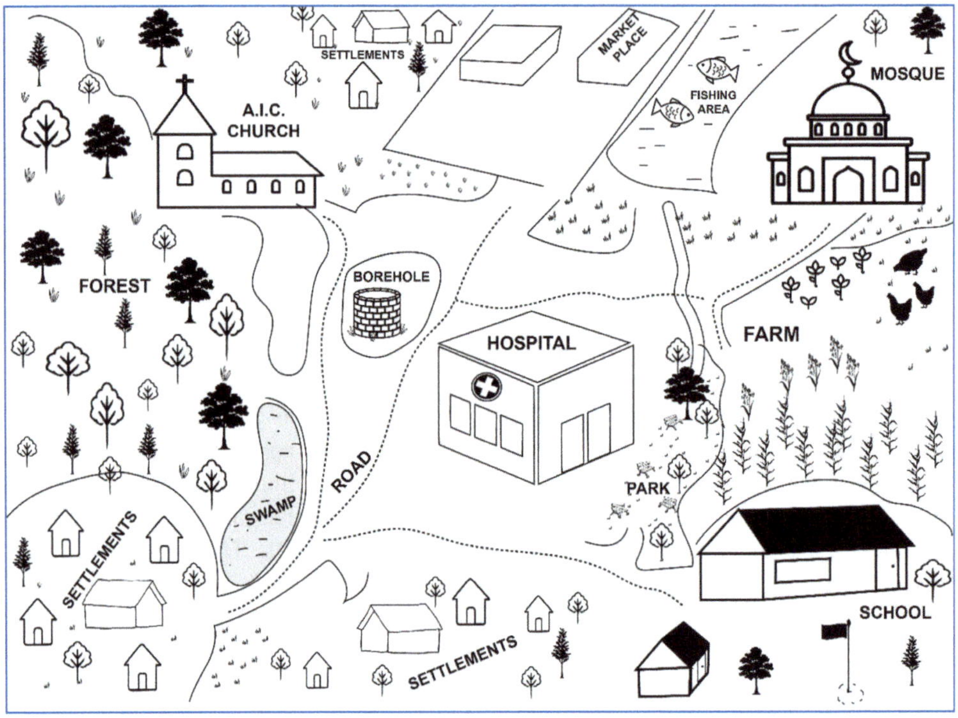

Fig. 19.4 A sketch of a community resources map

19.5.2 Seasonal Calendar

A seasonal calendar (Table 19.2) represents seasonal events in a specific area, such as planting, harvesting, pest infestation, etc. It enables researchers to identify patterns in the agricultural cycles and develop better crop production strategies. During the PRA exercise the farmers and other stakeholders in the community usually create this calendar and use it to observe and record the seasonal changes in their local environment. They share their observations with the researchers, who then analyze the data and create a seasonal calendar as a reference. By understanding the seasonal trends, researchers can adjust their strategies to ensure crops are planted and harvested at the right time, leading to increased productivity and efficiency for the farmers.

Table 19.2 A typical seasonal calendar showing the months in a year and the agricultural activities in the central highlands of Kenya

Month	Season	Season's status	Activity
January		Dry	
February		Dry	Harvest short rains season and land preparation
March		Wet	Planting maize and beans
April		Wet	Weeding
May		Wet	Weeding
June			Harvesting beans
July			
August		Dry	Harvesting maize
September		Dry	Land preparation
October		Wet	Planting
November		Wet	Weeding
December			Weeding

19.5.3 Product Ranking and Scoring

Product ranking and scoring is another tool used to evaluate agricultural products by determining their quality, effectiveness, and potential use in various contexts. Ranking is about ordering, while product scoring is about assigning a quantifiable value. This tool is a valuable resource for researchers, farmers, scientists, and agribusinesses. It helps them assign some value to the items being ranked and scored. The process involves a team of experts and stakeholders who gather input from various sources and use it to score and rank the products. For example, farmers can be asked to list food crops, cash crops, and livestock separately. Community members then score these items based on their importance. After scoring, the products are ranked according to the scores, which helps identify the best option for different users. An example of crop scoring and ranking is shown in Table 19.3. In this example pea is ranked best for cash, while potato is highly ranked for subsistence.

19.5.4 Venn (Chapati) Diagrams

The Venn diagram is a diagram showing the relationship of different institutions within the community (Fig. 19.5). It is

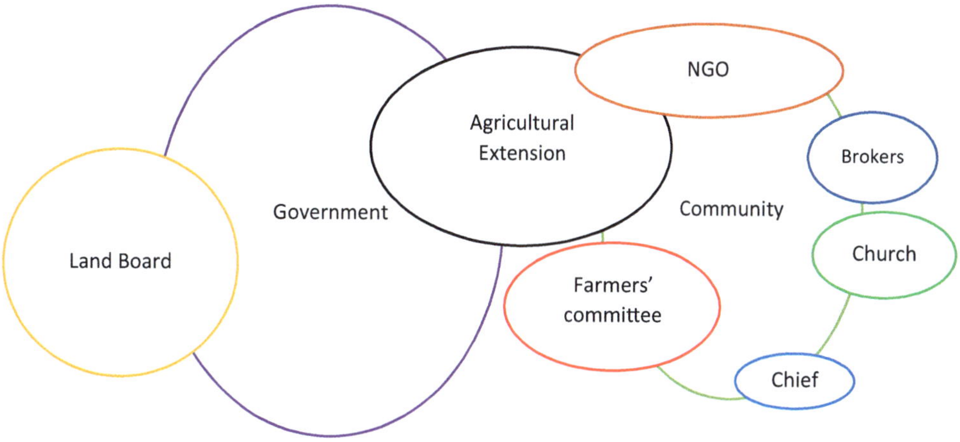

Fig. 19.5 Venn diagram showing the presence and interaction of various stakeholders

Table 19.3 Scoring and ranking of crops

	Cash		Subsistence	
Crop	Ranking	Scoring (%)	Ranking	Scoring (%)
Potato	3	70	1	80
Tomato	5	50	5	54
Maize	2	85	4	66
Bean	4	60	6	40
Pea	1	90	2	77
Yam	6	30	3	72

an effective tool for analyzing institutional relationships and is often referred to as an institutional diagram. In East Africa, this method is widely known as the Chapati diagram (a term derived from the round shape of baked bread). The diagram uses circles of varying sizes to represent institutions and individuals, with the size of each circle indicating its importance or influence. The proximity between circles illustrates the degree of connection or interaction as shown in Fig. 19.5. In this example institutions such as Church, NGO, Agricultural Extension Office, and Chief, Brokers, and Farmers' committee closely interact with the community. Overlapping circles such as those of NGO and Agricultural Extension Office highlight interaction indicating they work together in some programs, for example, organizing joint farmers' field days. This visual approach helps communities and researchers identify key stakeholders, understand their roles, and assess collaborations or gaps among the stakeholders. Such understanding is crucial, especially in research and development projects.

19.5.5 Wealth Ranking

Waelth ranking is tool used to analyze and understand perceptions of wealth differences and inequalities in a community (Table 19.4). It helps identify local indicators of wealth, well-being, and poverty (White and Pettit 2004). The process involves participants ranking each other into wealth categories based on land, livestock, and other assets. The tool is particularly useful in development projects to assess a community's economic status and identify the most vulnerable households. The understanding of the socio-economic differences within communities allows better targeting of resources and interventions before implementing projects, programs, or policies.

Table 19.4 Wealth ranking into poor, intermediate, and rich using various indicators

Criteria	Poor	Intermediate	Rich
Type of house	Mud house	Timber	Permanent (stone and cement)
Homestead land size	20–100 m^2	100–200 m^2	300–700 m^2
Agricultural land size	Small plot, buys food all the time	0.25–3 ha, sometimes sells food	>3 ha, sells food in all seasons
Number of cattle	None, buys milk	3–9, sells milk to neighbors	>10, sells milk to a cooperative
Number of poultry	2	5–10	>11
Transport	None	Bicycle	Motorbike or vehicle

19.6 Recording Data in Qualitative Research

Recording data in qualitative research is critical to ensuring the collected information's integrity and accuracy (Easton et al. 2000; Gorman and Clayton 2004). Specific protocols and tools are utilized to facilitate this process, each tailored to particular research methods. They include the following:

Recording protocols. These are designed to capture data systematically during qualitative research activities. These forms help consistently organize and document responses, observations, and other relevant information. For instance, in an agricultural study exploring farmers' practices, a researcher might use a structured form to note down responses about crop management techniques and pest control strategies.

An interview protocol: guides the data recording process during interviews. It typically includes predefined questions, space for noting responses, and sections for additional comments or observations made by the interviewer (Box 19.4). Developing an interview protocol requires careful consideration of the research objectives, the target population, and the specific information sought. For example, in a study examining the adoption of sustainable farming practices, the interview protocol might include questions about farmers' experiences with different techniques, perceived benefits and challenges, and factors influencing their decision-making processes. This structured approach helps ensure that all relevant topics are covered and data are collected systematically.

An observational protocol records information during observations. This protocol provides a framework for documenting what the researcher sees, hears, and experiences in the field (Box 19.5). It typically includes predefined categories or checklists to note all relevant aspects of the observed situation. For example, in an observational study of livestock

Box 19.4: Example of Interview Protocol

Project: integrated soil fertility management for enhanced crop yields
Objective: to explore farmers' perceptions and experiences in integrated soil fertility management
Date of interview:
Place:
Interviewer:
Interviewee:
Questions

1. Can you describe your experience with integrated soil fertility management?
2. What motivated you to adopt integrated soil fertility management practices?
3. What challenges have you faced in implementing these practices?
4. How do you perceive the benefits of integrated soil fertility management compared to conventional methods?
5. Are there any specific techniques or tools that you find particularly effective?

Make observations about the interviewee's farm environment.
Make notes on nonverbal cues and emotional responses.

Box 19.5: Example of Observational Protocol

Project
Objective: to document livestock management practices in a small-scale farm

A. Animal health

1. Condition of the animals (e.g., signs of illness or injury)

____
____
____
____
____
____
________________________________.

2. Veterinary care and treatments observed

____
____
____
____
____
____
____
____

B. Facility's cleanliness

1. Cleanliness of barns and pens

____
____
____
____
____
____
____
____

C. Waste management practices

1. Feeding practices

____
____
____
____
____
____

2. Types of feeds used

____
____
____
____
____
____
____
____

____
____
____

3. Feeding schedules and routines ____
____
____
____
____
____
____.

Table 19.5 Ways of addressing the various types of validity

Type of validity	How to address the issue
Descriptive validity	Describe the topic thoroughly with detailed data.
Interpretative validity	Stay away from personal bias.
Theoretical validity	Ground research in relevant theories.
Evaluative validity	Define evaluation criteria, use multiple data sources.
Generalizability	Provide rich context and consider combining findings from multiple studies for a broader understanding.

management practices, the protocol might include sections for recording the animals' condition, the facilities' cleanliness, feeding routines, and interactions between the animals and handlers. This structured method helps capture detailed and consistent observational data.

19.6.1 Checklist

A checklist is a simple instrument consisting of a prepared list of expected performance or attributes, which a researcher checks for their presence or absence.

Construction of good checklists

1. The list of items in the checklist may be continuous or divided into groups of related items.
2. All items should be arranged in a logical order.
3. Express each item in a clear, simple language, observable, and measurable concrete concepts. Each item should assess or describe only one behavior or criterion.
4. Items should be discriminative in quality, that is, to secure the required information and details.
5. An intensive literature survey is conducted to determine the type of checklist to be used in a particular research project.
6. Avoid negative statements whenever possible.
7. Ensure that each item has clear responses: yes or no.
8. Review the items independently.
9. The checklist must have quality of completeness and comprehensiveness.
10. A pilot study or trial run will help to develop the most appropriate items.

19.7 Validity and Reliability in Qualitative Data Collection

19.7.1 Key Strategies for Tackling Validity of Qualitative Data

In Chap. 14, we discussed the primary terms used to denote validity in qualitative research, "trustworthiness" and "understanding." We mainly considered criteria by Lincoln and Guba (1985). Qualitative researchers can ensure the trustworthiness of their research by addressing credibility, transferability, dependability, and confirmability. First, a researcher must consider all the complexities of the study and address problems that are not easily explained (credibility). Additionally, the researcher must furnish contextually relevant descriptions to enable listeners or researchers to relate to the study environment (i.e., transferability). Typically, qualitative researchers refrain from seeking generalized conclusions applicable to broader populations. Hence, it is crucial to include as many details as possible so that others can see the setting for themselves. Moreover, researchers should address the consistency of the data collected (i.e., dependability) and the impartiality and objectivity of the data (confirmability).

Maxwell's (1992) criteria for validity in qualitative research include five items (Table 19.5):

- Descriptive validity: accuracy of the descriptions and data provided.
- Interpretative validity: the researcher's ability to make sense of the data and give meaningful interpretations; accurate interpretation of the participant expressions.
- Theoretical validity: assess how the findings align with existing theories and concepts.
- Evaluative validity: whether the research was objective enough to report the data in an unbiased way without judgment.
- Generalizability: the extent to which the findings can be transferred to other contexts.

This section gives practical ways to overcome validity issues in qualitative research (Busetto et al. 2020; Johnson et al. 2020; Noble and Smith 2015; Gay et al. 2018; Patton 1999). We consider data collection involving farmers and come up with 20 strategies.

Before the strategies, we reflect on the research process because all proposal components are linked and inform each other. First, we ask, "Is the research process valid?". Additionally, the following questions are important as we think about validity in qualitative research.

- Are the research problem, objectives, and research questions valid for the desired outcome?
- Is the choice of the methodology appropriate for ensuring that the research problem is answered?
- Is the design valid for the methodology adopted?
- Is the sampling appropriate?
- Are the data analysis methods appropriate?
- Do the results and the conclusions align to the research problem investigated?
- Are the results and conclusions sound?

These questions should form part of the researcher's reflexivity and ensure that the research process is sound.

The 20 strategies include:

1. *Embrace triangulation:* use multiple methods, data collection strategies, and data sources to obtain a complete picture of what is being studied.
2. *Commit to spending sufficient time at the study site: being there* for an extended period allows you to develop a deep understanding of the phenomenon under study. Spending time with the participants helps to build trust, leading to more honest participation.
3. *Establish an audit trail:* describe each process and make available field notes, videotapes, photos, etc. Keep a detailed record of each step in your research process, including notes, photos, and observations of crop growth. For example, take GPS coordinates and farmer contacts. This trail ensures transparency, and others can follow your research journey.
4. *Conduct member checks*: share your research findings with farmers to ensure they accurately represent their experiences and perspectives. Ask for their feedback and comments.
5. *Use peer debriefing*: seek input from fellow researchers or experienced farmers to validate your interpretation. Their insight can provide valuable feedback to allow to improve your research.
6. *Conduct pilot studies*: test your research instruments and methods on a small scale before collecting large-scale data. This allows you to address any issues or challenges before proceeding to data collection.
7. *Use negative case analysis:* look for cases that acknowledge the conduct of your initial hypotheses or findings. Investigate these outliers to gain a deeper understanding of the factors at play.
8. *Document decision-making process:* keep detailed records of decisions made during the research process, including data analysis and interpretation.
9. *Collect detailed data to allow comparison.* Gather comprehensive data on factors such as soil types, crop yields, and framing practices of different farmers and farms. This allows you to compare and contrast results, just like the way you can compare growth patterns and yields from different fields on a farm.
10. *Use thick description*: provide detailed descriptions of farming practices and farmer perspectives in your research report.
11. *Seek diverse perspectives.* Include farmers from different backgrounds and regions and farming system experiences.
12. *Ensure consistency in data collection*: standardize the data collection method and protocols to ensure consistency across different farms.
13. *Maintain reflexivity: reflect on your own experiences*, biases, and assumptions throughout the research process. Consider how they influence your observations and interpretations.
14. *Practice iterative questioning:* use probing and iterative questioning to elicit detailed data. This involves returning to matters previously discussed and extracting related information through paraphrasing. As you do this, participants are encouraged to give more detailed data.
15. *Embrace accurate data recording:* record data promptly and systematically and ensure no crucial details are overlooked or omitted. The use of detailed field notes, video and audio, and all other materials available can enhance accuracy in recording data.
16. *Give sufficient time to participants:* during data collection, remain patient and give participants adequate time to handle the questions without interrupting.
17. *Begin writing as early as possible:* writing early is beneficial because you have time to refine arguments, clarify your thinking, and identify gaps or areas that need further exploration. For example, you can determine the follow-up questions you need to ask the following day.
18. *Write a good scientific report*: write with clarity and follow the principles of scientific writing.
19. *Embrace candid reporting:* in your report, openly acknowledge any biases that may have influenced the design, conduct, and interpretation of data.
20. *Present evidence*: do not shy away from presenting evidence that you think will enhance the credibility of your research. This evidence is vital to your readers, colleagues and journal reviewers. This may include maps, charts, photos, etc.

19.7.2 Triangulation in Qualitative Research

Triangulation stands out as a cornerstone method for both validity and reliability (Rutherford et al. 2010; Moon 2019). This approach is multifaceted and involves various dimensions. It includes the following dimensions:

- *Methodological triangulation* entails employing diverse research methods or approaches, such as interviews, observations, and document analysis, to corroborate findings. This minimizes biases associated with the use of a single method.
- *Multiple source triangulation:* involves gathering data from multiple and varied sources, such as different farmers, individuals, groups, and documents.
- *Data triangulation*: emphasizes cross-referencing information collected through different means or at different farms to validate findings.
- *Environmental triangulation:* collect data from different environmental settings and context, for example, from different geographical locations and climates.
- *Researcher triangulation:* involve multiple researchers in data collection analysis and interpretation.

Recommended for Further Reading

1. *The origins and practice of participatory rural appraisal* by Chambers (1994)
2. *Participatory methods in community practice* by Castelloe and Gamble (2005)
3. *Using participatory rural appraisal* by Pretty and Vodouhê (1997)
4. *How to use and assess qualitative research methods* (Busetto et al. 2020)
5. *Methods of data collection in qualitative research; Interviews and focus group discussion* (Gill et al. 2008)
6. *Practical guidance to qualitative research. SERIES Part 1, 2, 3 and 4 by Korstenjen and Moser* (2017)
7. *Educational research: Planning, conducting, evaluating quantitative and qualitative research* by Creswell (2012)

Key Points in This Chapter

Methods of Data Collection in Qualitative Research

The most commonly used methods in qualitative research include interviews, observations, and document analysis. During the interview, questions are asked to elicit a response. The researcher is the interviewer, while the respondent is considered the student or interviewee. Data collected include field notes, maps, video recordings, transcriptions, drawings, etc.

Interviews are the most commonly used methods. Researchers use open-ended questions and probing to elicit in-depth information. However, they are expensive and can suffer from bias. To enhance the validity of data collected through interviews, 1. manage likely sources of bias, such as 1. interviewer bias, interviewee bias, and participation bias; 2. avoid interviewee or response bias.

Structured and Unstructured Interviews

Interviews can vary widely in their level of formality and structure, but nearly three types are recognized:

1. *Structured interviews* involve interviewing research participants formally using a specified number of questions that elicit the same information from the respondents.
2. *Semi-structured interviews* are in between structured and unstructured interviews. They contain both formal and informal questions.
3. *Unstructured interview* involves a casual conversation with no formality.

Ways of Administering the Interview

- Face-to-face interviews are a traditional and widely used method. This approach involves the interviewer and the participant meeting in person, allowing for direct interaction.
- Telephone interview: Sometimes, gathering participants for face-to-face interviews may not be possible. Telephone interviews can allow you to reach a wider audience without travelling.
- Internet-based interviews conducted through visual platforms and email have become increasingly popular, especially when face-to-face interactions are impractical.
- One-on-one interviews involve personalized interactions with a single participant.

General Guidelines for Interviewing

Preparation before the interview: prepare well by ensuring you know the research topic and the context of the interview. Familiarizing yourself with culture is key to avoiding misinterpretation. Build trust and rapport before the fieldwork by supplying relevant material beforehand, consulting with the gatekeepers, and developing themes carefully to guide the interviews.

Types of Interviews

In-depth interviews are the most important methods used to collect qualitative data. They mainly use unstructured interviews with intensive probing. The questions are open-ended to encourage participants to respond using their own words. During the interview, probing and follow-ups are dominant activities using a structural flow that is coherent and logical. After each interview, the interviewer usually creates time to reflect on the experiences, tries to identify themes and patterns, and also identifies any unexpected direction that may inform the next interview.

Key informant interviews are in-depth interviews with people perceived to be (informed about a subject) providing insights into a topic of interest. Key informants aim to collect detailed information related to a specific, often focused topic of interest to the researcher.

Focus group discussion comprises a group of people gathered to discuss a topic by the researcher. The members of the group are a usually homogenous group. FGDs are cheap and can generate large amounts of data quickly. The discussion is guided by a skilled facilitator. Ideal size of FGD should be between 6 and 8 willing persons. The FGD guide typically consists of three main sections. The FGD typically starts with an introduction section that introduces the purpose of the FGD, the middle section allows interactions on the main topic of discussion, and the last section is for concluding the discussion.

Oral history interviews: a method of collecting data mainly used to study experiences, perceptions, opinions, and accounts of a historical event or happenings in everyday living. It is used to gather historical knowledge as how individuals or groups view the phenomenon of information of interest.

Group interviews are conducted using a semi-structured interview schedule with several people at the same time. They could range from small sessions with a few people (3 to 4 people) to large ones with 10 or more, up to 20 people. Unlike the FGD that researchers widely use to collect specific information group interviews are useful for collecting data on broad-based issues.

Observation

Observation is collecting data by observing and recording what you see. The researcher gains valuable information on the phenomenon under study by directly documenting behavior in its natural context. Nonparticipant observation involves the researcher collecting data by observing and reading behavior, events, and processes without becoming actively involved. In participant observation the researcher become part of the community and participates in daily activities.

Document Analysis

Document analysis is a systematic procedure of analyzing documentary evidence and answering specific questions. Documentary evidence includes textbooks, photos, letters, films, tapes, diaries, themes, reports, or other documents. Most policy research often involves document analysis. Document analysis can be conducted as a stand-alone study or component of a larger qualitative or mixed-methods study.

Audiovisual Data Collection

Audiovisual data comprise images or sounds such as photographs, videos, and audio recordings that capture real-time events, interactions, and environments. In agricultural research, using visual materials are particularly effective for illustrating various aspects of agronomy and making information more accessible and engaging.

Collecting Data Using Participatory Research Tools

Participatory research tools involve engaging stakeholders in the research process emphasizing collaboration and inclusion. These tools enhance the relevance and applicability of the findings by allowing participants to contribute actively to data collection analysis and interpretation. Common tools in agriculture include community resource maps, historical analysis, and seasonal calendars.

Recording Data in Qualitative Research

Accurate data qualitative research is a critical process ensuring the collected information's integrity and accuracy. Various protocols and tools are utilized depending on the particular methods. They include recording protocols that are designed to capture data systematically during qualitative research activities, an interview protocol that guides the data recording process during interviews, an observational protocol that records information during observations, and a check list that is a simple instrument consisting of a prepared list of aspects to be considered.

Validity and Reliability in Qualitative Data Collection

Ensuring the trustworthiness involves addressing credibility, transferability, dependability, and confirmability. Researchers should employ practical strategies to overcome validity issues considering the entire research process and maintaining reflexivity during data collection and analysis.

Review Questions

1. What are the four main data collection methods used in qualitative research?
2. Differentiate clearly between participant observation and nonparticipant observation, including their typical roles of the researcher.

3. What are some common documents collected during document analysis in agricultural research?
4. List examples of audiovisual data that may be used in qualitative research.
5. Identify three distinct participatory research tools that would be most effective in mapping and describing the rice value chain.

Chapter Exercise

1. You are planning a qualitative study to understand the lived experiences of smallholder farmers adopting climate-smart agriculture practices. Critically:
   (a) Identify two distinct qualitative data collection methods you could employ for this study, as mentioned.
   (b) For each method identified in (a) describe a specific instrument (or tool) you would prepare to guide your data collection.
   (c) Briefly explain how your chosen instrument logically links to, and facilitates, your chosen data collection method.
2. You are researching the factors influencing farmers to join dairy cooperatives. You are considering both *focus group discussions (FGDs)* and *key informant interviews (KIIs)* as primary data collection approaches. Therefore:
   (a) Clearly distinguish between an FGD and a KII in terms of their primary purposes, typical participant selection criteria, and the unique types of insights each is best suited to gather.
   (b) For your study on urban dairy farming and cooperatives, explain when you would strategically opt for an FGD and when you would use a KII, providing specific examples of information you would seek from each.
   (c) Create a checklist of items to bring for accurate qualitative data recording in the field.
3. You are conducting a qualitative study to understand how smallholder farmers manage water resources for irrigation during dry seasons, critically:
   (a) Distinguish between an interview schedule and an observational protocol as distinct qualitative data collection instruments.
4. "Ensuring the trustworthiness of qualitative data collection is crucial, and one powerful approach for this is triangulation." Explain the comprehensive concept of triangulation in qualitative research, encompassing its various forms.

Answers to Review Questions

- Q1. Reference 19.1
- Q2. Reference 19.3.1
- Q3. Reference 19.4
- Q4. Reference 19.4.2
- Q5. Reference 19.5

References

Baral S, Uprety S, Bipul L (2021) Focus group discussion. How to guide to focus group discussion. A herd publication

Bauer MW, Gaskell G (eds) (2000) Qualitative researching with text, image and sound: a practical handbook for social research (1st ed.). London/Thousand Oaks, CA: Sage

Bergen N, Labonté R (2020) "Everything is perfect, and we have no problems": detecting and limiting social desirability bias in qualitative research. Qual Health Res 30:783–792. https://doi.org/10.1177/1049732319889354/SUPPL_FILE/SUPPLEMENTAL_FILE.PDF

Bergelson I, Tracy C, Takacs E (2022) Best practices for reducing bias in the interview process. Curr Urol Rep 23:319–325. https://doi.org/10.1007/S11934-022-01116-7/METRICS

Bowen GA (2009) Supporting a grounded theory with an audit trail: an illustration. Int J Soc Res Methodol 12(4):305–316

Brinkmann S. (2022) Qualitative interviewing: conversational knowledge through research interviews (2nd edn). Oxford: Oxford University Press

Busetto L, Wick W, Gumbinger C (2020) How to use and assess qualitative research methods. Neurol Res Pract 2:14. https://doi.org/10.1186/s42466-020-00059-z

Castelloe P, Gamble DN (2005) Participatory methods in community practice. In: Weil M, Reisch M (eds) Handbook of community practice. Sage, Thousand Oaks

Chambers (1994) The origins and practice of participatory rural appraisal. World Dev 22(7):953–969. https://doi.org/10.1016/0305-750X(94)90141-4

Creswell JW (2009) Research design: qualitative, quantitative, and mixed methods approaches, 3rd edn. Sage, Thousand Oaks

Creswell JW (2012) Educational research: planning, conducting, and evaluating quantitative and qualitative research, 4th edn. Pearson, Boston

Creswell JW (2014) Research design: qualitative, quantitative, and mixed methods approaches (4th edn). SAGE Publications, Thousand Oaks, CA

Creswell JW, Poth CN (2016) Qualitative inquiry and research design: choosing among five approaches. Sage, Thousand Oaks

Delgado M, Humm-Delgado D (2013) Community mapping strategies. In: Asset assessments and community social work practice. Oxford University Press, pp 151–180

DiCicco-Bloom B, Crabtree BF (2006) The qualitative research interview. Med Educ 40:314–321. https://doi.org/10.1111/J.1365-2929.2006.02418.X

Easton KL, McComish JF, Greenberg R (2000) Avoiding common pitfalls in qualitative data collection and transcription. Qual Health Res 10:703–707. https://doi.org/10.1177/104973200129118651

Enzenbach C, Wicklein B, Wirkner K, Loeffler M (2019) Evaluating selection bias in a population-based cohort study with low baseline participation: the LIFE-adult-study. BMC Med Res Methodol 19:1–14. https://doi.org/10.1186/S12874-019-0779-8/FIGURES/3

Erlandson DA (1993) Doing naturalistic inquiry: a guide to methods. Sage, Newbury Park

Galdas P (2017) Revisiting bias in qualitative research. Int J Qual Methods 16. https://doi.org/10.1177/1609406917748992

Gay LR, Mills GE, Airasian P (2009) Educational research competencies for analysis and applications. Pearson, Columbus

Gay LR, Mills GE, Airasian PW (2018) Educational research: competencies for analysis and applications (12th ed.). Pearson Education, New York

Gill P, Stewart K, Treasure E, Chadwick B (2008) Methods of data collection in qualitative research: interviews and focus groups. Br Dent J 204:291–295. https://doi.org/10.1038/bdj.2008.192

Gobo G (2011) Ethnography. In: Qualitative research. Sage, Los Angeles, pp 15–36

Gorman GE, Clayton P (2004) Qualitative research for the information professional: a practical handbook. London: Facet Publishing

Hair JF Jr, Howard MC, Nitzl C (2020) Assessing measurement model quality in PLS-SEM using confirmatory composite analysis. J Bus Res 109:101–110

Jamshed S (2014) Qualitative research method-interviewing and observation. J Basic Clin Pharm 4:87–88. https://doi.org/10.4103/0976-0105.141942. PMID: 25316987; PMCID: PMC4194943

Johnson JL, Adkins D, Chauvin S (2020) A review of the quality indicators of rigor in qualitative research. Am J Pharm Educ 84(1):7120. https://doi.org/10.5688/ajpe7120. PMID: 32292186; PMCID: PMC7055404

Keen S, Lomeli-Rodriguez M, Joffe H (2022) From challenge to opportunity: virtual qualitative research during COVID-19 and beyond. Int J Qual Methods 21. https://doi.org/10.1177/16094069221105075/ASSET/1998CD2E-1B21-49A8-879B-3D0FB866D53F/ASSETS/IMAGES/10.1177_16094069221105075-IMG2.PNG

Kitzinger J (1995) Qualitative research: introducing focus groups. BMJ 311:299–302. https://doi.org/10.1136/BMJ.311.7000.299

Korstjens I, Moser A (2017) Practical guidance to qualitative research: Parts 1–4. European Journal of General Practice 23(1):271–275. Taylor & Francis, Abingdon, United Kingdom. https://doi.org/10.1080/13814788.2017.1375091

Kristensen E, Nielsen DB, Jensen LN, et al (2008) A mixed methods inquiry into the validity of data. Acta Vet Scand 50:1–8. https://doi.org/10.1186/1751-0147-50-30/FIGURES/1

Krueger RA, Casey MA (2015) Focus groups: a practical guide for applied research (5th ed.). Thousand Oaks, CA: Sage Publications

Kumar S (2014) A systematic approach to the assessment of impact in a professional doctorate. High Educ Skills Work Based Learn 4(2):171–183

Lacoste M, Cook S, McNee M, et al (2021) On-farm experimentation to transform global agriculture. Nat Food 3:1 3:11–18. https://doi.org/10.1038/s43016-021-00424-4

Leavy P (2011) Oral history: understanding qualitative research. New York/Oxford: Oxford University Press

Leung L (2015) Validity, reliability, and generalizability in qualitative research. J Family Med Prim Care 4:324. https://doi.org/10.4103/2249-4863.161306

Lincoln YS, Guba EG (1985) Naturalistic inquiry. Sage Publications, Newbury Park, CA

Maxwell J (1992) Understanding and validity in qualitative research. Harv Educ Rev 62(3):279–301

McCuen RH (1996) The elements of academic research. ASCE Publications, New York

Moon MD (2019) Triangulation: a method to increase validity, reliability, and legitimation in clinical research. J Emerg Nurs 45:103–105. https://doi.org/10.1016/j.jen.2018.11.004

Morgan DL (1997) Focus groups as qualitative research (2nd ed.). Sage Publications, Thousand Oaks, CA

Narayanasamy N (2009) Participatory rural appraisal: principles, methods and application. Sage, Los Angeles

Ngum AM (2019) Structure, conduct and performance of soybean marketing in Embu, Tharaka Nithi and Meru counties, Kenya. Masters thesis. Kenyatta University

Njenga MW, Mugwe JN, Mogaka HR, Nyabuga G, Oduor N, Kiboi M, Mugendi D (2021) Determinants of farmers' knowledge on soil and water conservation technologies in dry zones of central highlands, Kenya. J Agri Ext 25(4)

Noble H, Smith J (2015) Issues of validity and reliability in qualitative research. Evid Based Nurs 18:34–35. https://doi.org/10.1136/EB-2015-102054

Novick G (2008) Is there a bias against telephone interviews in qualitative research? Res Nurs Health 31:391–398. https://doi.org/10.1002/NUR.20259

Nyumba OT, Wilson K, Derrick CJ, Mukherjee N (2018) The use of focus group discussion methodology: insights from two decades of application in conservation. Methods Ecol Evol 9:20–32. https://doi.org/10.1111/2041-210X.12860

Patton MQ (1999) Enhancing the quality and credibility of qualitative analysis. Health Serv Res 34(5 Pt 2):1189

Pretty JN, Vodouhê SD (1997) Using participatory rural appraisal. In: Swanson BE, Bentz RP, Sofranko AJ (eds) Improving agricultural extension: a reference manual, 2nd edn. FAO, Rome

Punch S (2012) Hidden struggles of fieldwork: exploring the role and use of field diaries. Emot Space Soc 5(2):86–93

Randall DM, Fernandes MF (1991) The social desirability response bias in ethics research. J Bus Ethics 10:805–817. https://doi.org/10.1007/BF00383696/METRICS

Roberts JK, Pavlakis AE, Richards MP (2021) It's more complicated than it seems: virtual qualitative research in the COVID-19 era. Int J Qual Methods 20:. https://doi.org/10.1177/16094069211002959/ASSET/680DEC95-C38B-45BD-B0E9-B430EA7268AE/ASSETS/IMAGES/LARGE/10.1177_16094069211002959-FIG1.JPG

Rubin D (2016) Qualitative methods for gender research in agricultural development, vol 1535. International Food Policy Research Institute

Rubin HJ, Rubin IS (2005) Qualitative interviewing: the art of hearing data (2nd edn). SAGE Publications, Thousand Oaks, CA

Ruslin R, Mashuri S, Rasak MSA, Alhabsyi F, Syam H (2022) Semi-structured interview: a methodological reflection on the development of a qualitative research instrument in educational studies. IOSR-JRME 12(1):22–29

Rutherford GW, McFarland W, Spindler H, et al (2010) Public health triangulation: approach and application to synthesizing data to understand national and local HIV epidemics. BMC Public Health 10:447. https://doi.org/10.1186/1471-2458-10-447

Saunders B, Sim J, Kingstone T, Baker S, Waterfield J, Bartlam B et al (2018) Saturation in qualitative research: exploring its conceptualization and operationalization. Qual Quant 52:1893–1907

Saunders MNK, Lewis P, Thornhill A (2019) Research methods for business students (8th ed.). Pearson Education Limited, Harlow, United Kingdom

Schindler P (2022) Business research methods, 14th edn. McGraw-Hill, New York

Sutton J, Austin Z (2015) Qualitative research: data collection, analysis, and management. Can J Hosp Pharm 68(3):226

Sweet L (2002) Telephone interviewing: is it compatible with interpretive phenomenological research? Contemp Nurse 12:58–63. https://doi.org/10.5172/CONU.12.1.58

van Eeuwijk P, Angehrn Z (2017) How to conduct a focus group discussion (FGD), Methodological Manual. Swiss Tropical and Public Health Institute

Ward K, Gott M, Hoare K (2015) Participants' views of telephone interviews within a grounded theory study. J Adv Nurs 71:2775–2785. https://doi.org/10.1111/JAN.12748

White S, Pettit J (2004) Participatory methods and the measurement of well-being. In: Participatory learning and action 50: critical reflections, future directions. International Institute for Environment and Development (IIED)

Part V

Data Processing, Analysis and Presentation

Quantitative Data Analysis I: Key Steps and Descriptive Data Analysis

20

Abstract

The next step in your research project after data collection is data analysis. Quantitative data analysis involves systematically using statistical methods to examine and interpret numerical data. The aim of statistical analysis is to reveal patterns and trends within a data set. The raw data is reduced during analysis and transformed into actionable conclusions and recommendations. Six simple steps are key to analyzing data: (1) preparing data for analysis, (2) defining the analysis objectives, (3) selecting variables for use in analysis, (4) data exploration and descriptive analysis, (5) confirmatory or inferential data analysis, and (6) interpretation and reporting. Preparing data for analysis involves data entry, variable labeling, cleaning, and organizing data into a matrix to facilitate analysis. Defining analysis objectives is key to determining the type of analysis required. Once the data is well-organized, data exploration and descriptive statistics are done. The objective is to summarize the data using frequencies, measures of central tendency (MoT), and measures of variability (MoV). Measures of central tendency include the mode, median, and mean, while measures of variability include range, percentiles, quartiles and box plots, coefficient of variation, standard deviation, and scatter plots. The normal distribution curve forms the foundation of many statistical tests due to its predictive nature. A pivot table is a useful tool for summarizing, organizing, and analyzing data and can easily be done through a drag-and-drop interface in Excel.

Keywords

Analysis objective · Coding categorical data · Coefficient of variation · Confirmatory analysis · Computer software · Data analysis matrix · Data exploration · Box plots · Descriptive statistics · Interquartile range · Measures of central tendency, Measures of variability, Normal distribution · Percentiles · Pivot table · Scatter plot · Standard deviation

Learning Outcomes

After reading this chapter, you should be able to:

1. Appreciate the meaning of quantitative data analysis.
2. Give highlights of computer software available for quantitative data analysis.
3. Describe the steps in analyzing quantitative data.
4. Apply the general steps to enter data into a computer and organize data into an analysis matrix.
5. Use the main descriptive statistics (frequencies, measures of central tendency, and variability) to analyze data.
6. Demonstrate an understanding of the various measures of variability.
7. Plot scatter plots using Excel and apply them to analyze data variability and relationships.
8. Create a "Pivot table" using an Excel spreadsheet.

20.1 Laying the Foundation in Data Analysis: Post Field Work and Use of Computers

20.1.1 Where Do You Begin After Data Collection?

Upon returning from the field, many students often struggle to start analyzing the data due to a "mental block." They are unsure of where to begin. They have possibly been collecting data for several months and returned with enormous amounts of data. The data could be recorded in various ways, including notebooks, data collection forms, or spreadsheets. The sheer volume of data can lead to information overload, causing a mental block. This chapter guides you through the initial data analysis phase by providing practical advice and simple steps on how to proceed with the analysis. Hopefully,

J. Njeri Mugwe, S. Runo, *Research Methodology in Agricultural Sciences*, https://doi.org/10.1007/978-981-95-1892-0_20

this will help students gain confidence in how to start data analysis.

This chapter also attempts to address many students' common fear and apprehensions toward statistics, and statistics anxiety. Due to its complex mathematical concepts and terminologies, many students find statistics intimidating. Although it is taught at all levels in the university, from a Bachelor's to Doctorate programs, the understanding and application of statistics remain low, and most students are anxious about statistics (Onwuegbuzie et al. 1997; Ralston et al. 2020). This chapter demystifies statistics, empowering students to navigate data analysis independently with little assistance. It is hoped that the simplified approach and relatable examples used will enable students to overcome their fears and develop the necessary skills for seamless data analysis.

In addition, we have avoided the use of detailed statistical procedures. We recommend that students revise basic statistical procedures by reading various books on statistics, such as Dean et al. (2017), Wilson and MacLean (2009), Little and Hills (1978), Mead et al. (2017), and Toutenburg (2009). Another good reference is a handbook of statistics for teaching and research in plant and crop science by Palaniswamy (2005). We further recommend Bate and Clark (2014), Morris (1999), and Hau and Schapiro (2021) for researchers involved in animal experiments including those focussing on nutrition, health, and other research goals. We also avoided complex statistical formulas, algorithms, or manual calculations. Instead, we prioritize practical techniques for processing and interpreting numerical data using computer tools. However, given that data analysis is inherently statistical, we advise readers to consult statistics books or seek guidance from a statistician whenever there is doubt about how a particular statistical technique works.

20.1.2 What Is Quantitative Data Analysis?

Quantitative data analysis involves analyzingraw numerical data using statistics so that researchers make sense of the data (Creswell 2014). The primary goal is to reveal patterns and trends within the data set. A data set is a collection of observations or measurements on a given variable, such as maize height, taken from a field experiment evaluating the effect of different pesticide application rates on growth. Data analysis involves a series of steps to finally translate the raw data into results. The key steps involve organizing the data to facilitate analysis, carrying out the analysis, and finally interpreting (Stoudt et al. 2021). The ultimate goal is to present the data in a form that is understandable and useful to the intended beneficiaries of the research, such as researchers, farmers, and policymakers.

The data analysis is generally categorized into two primary forms. The first is *descriptive statistics,* which provides summaries of variables within a data set to highlight what is typical or representative of the sample (Al-Benna et al. 2010). The second is *inferential statistics*, which enables researchers to examine relationships between variables and draw conclusions that extend beyond the immediate data (Al-Benna et al. 2010). Depending on the research objectives, inferential statistics can help determine the effectiveness of a new treatment, examine whether variables are related, or compare the characteristics of different groups. Inferential statistics also assess whether observed patterns or differences are statistically significant or simply due to chance. This enables researchers to draw conclusions about the findings for the broader population.

This chapter discusses data organization and descriptive statistics. Inferential statistics is discussed in the next chapter (Chap. 21).

20.1.3 What Data Is Available?

Quantitative data emanating from agricultural research activities will mainly comprise the following:

- Quantitative data from experiments, for example:
  - Growth data (animal and crops).
  - Yield data.
  - Soil data.
  - Pest scores (insects, thrips).
- Quantitative data from surveys:
  - Household characteristics (e.g., age and household size).
  - Crop and livestock production data.
  - Categorical and binary data.
- Secondary data (government reports, weather data, market reports).
- Quantitative data from mixed methods and participatory research studies, for example:
  - Scoring and ranking data.
  - Value chain maps, etc.

The above data is what we refer to as *raw data*. Raw data from the field is unprocessed and lacks organization. It cannot be used to make decisions and neither can it be used to say anything regarding the research question or objectives. Raw data is often in field paper workbooks, or in data management software such as MS Excel, Google Sheets among others. Data analysis reduces the raw data into useful information that helps us provide answers to the *questions raised at the start of the study*. Recall that at the start of the research study, you raised research questions and objectives that guided the data collection. At the point of data analysis, the aim is to use the data to answer the questions you raised at the start of the study (Fig. 20.1).

The figure illustrates how raw data from experiments, surveys, and participatory research are subjected to processing and analysis to make them useful for practice. Data undergoes a form of reduction (funnel shape), allowing researchers to extract meaningful patterns.

20.1.4 Use of Computers in Data Analysis

The computer is an invaluable tool in data handling, management, and analysis. Mostly in agricultural statistics/Biometry courses in undergraduate and graduate level studies, basic statistics is taught. Nevertheless, probably the use of computers to analyze data was not integrated into the statistics courses you have undertaken. For your thesis work, supervisors expect you to handle and analyze data with the help of computer software. Therefore, commit to acquiring good computer skills. Different computer programs serve various purposes, and the most commonly used are described here:

- *MS Excel* is a versatile spreadsheet widely used for data entry, cleaning, and manipulation (Hunt and Tyrrell 2002; Bauzon et al. 2021). Due to its flexibility, accessibility, and extensive range of functionalities, it is a friendly and valuable tool for researchers. One particularly useful tool with quantitative experimental data is the pivot table, which allows users to rearrange data and identify patterns and relationships within the data set (demonstrated later in this Chapter).
- *Statistical Package for Social Sciences (SPSS):* It is useful for managing and analyzing social science data gathered through surveys on both numerical and nominal scales (McCormick and Salcedo 2017). In the context of agriculture, this data includes information collected from farmers across various disciplines, primarily focusing on agribusiness, agricultural marketing, and agricultural economics. As such, it is valuable for analyzing survey data obtained via questionnaires or interview schedules. Data can be entered directly into SPSS or Excel and can later be exported to statistical software for further analysis (see Chap. 21 on data analysis using SPSS).
- *GENSTAT* is a general data analysis and statistical software package widely used in different research areas, including agriculture and many other disciplines (Nwaogwugwu and Udoh 2014). Due to its robust functionalities, it is particularly useful for experimental data. The user interface contains comprehensive menus to guide non-technical users in using statistically correct and effective methods.
- *Statistical Analysis System (SAS):* SAS is a statistical software developed and distributed by the SAS Institute and it is suited for analyzing experimental data in agriculture (Fribourg and Carlisle 1981; Yang 2011; Vargas et al. 2013). It was first developed for agricultural research at North Carolina University in 1970 (Herzberg 1990), is a command-driven software, and is one of the most widely used in both academia and industry. It is a flexible and powerful package that has a combination of a

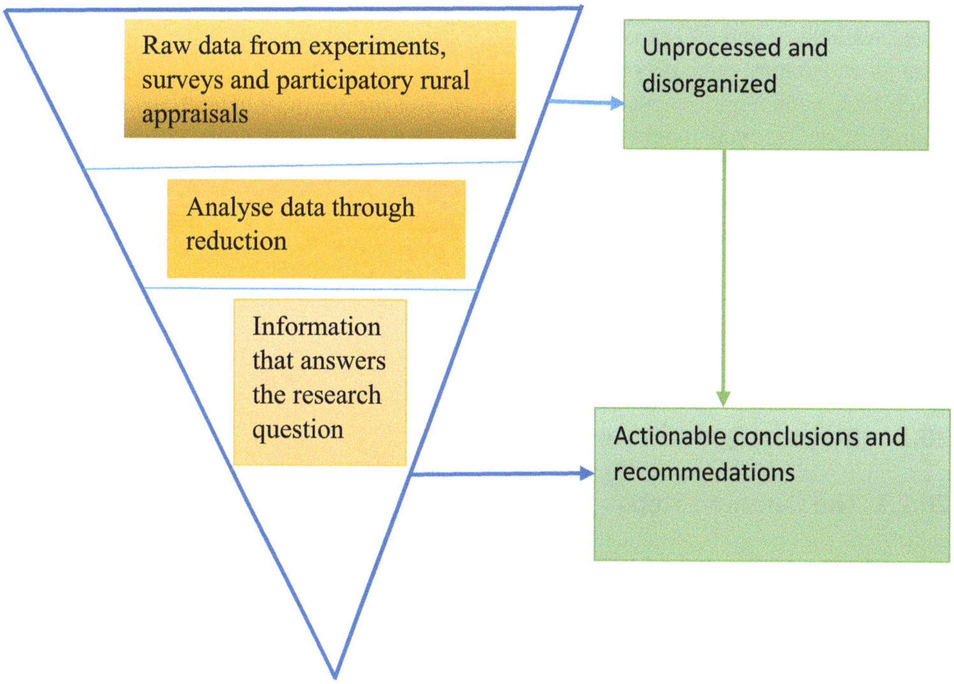

Fig. 20.1 Data analysis involves transforming raw data into meaningful information that helps answer the research question

statistical package, data management, and a high level of programming language (https://support.sas.com/en/technical-support.html). It is widely accepted as the leader in statistical analysis and modeling.

- *STATA* is a general-purpose software for data manipulation, visualization, statistical analysis, and reporting (Gutierrez 2010; Fisher et al. 2022). Due to its powerful functionalities, it is used in diverse fields such as agriculture, economics, and medicine. In agriculture, it is particularly useful for analyzing survey data and handling complex data sets. It can also handle advanced statistical analysis, such as factor analysis and endogenous switching models. Overall, it is a tool for efficient data exploration and interpretation and supports a wide array of data analysis needs.
- *R:* Widely used and accepted in academia and industry, with a large user base (Lai et al. 2019; Giorgi et al. 2022). It was developed by Ross Ihaka and Robert Gentleman in 1993 at the University of Auckland, New Zealand. R is a free and open source, allowing anyone to use and modify it. It is reported to be the most comprehensive statistical package available. It incorporates all the standard statistical tests, models, and analyzes and provides a comprehensive language for managing and manipulating data. It is particularly preferred for data visualization due to its extensive packages and libraries designed to create high-quality visualizations. This enables users to convey complex patterns, trends, and relationships in a compelling manner in journal papers and conference presentations. You can easily plot various types of graphs using R (we have briefly demonstrated this in Chap. 24).

The software you choose depends on the discipline, data types, availability, and user expertise. For instance, if in agricultural economics, you may choose SPSS and STATA, which have econometric models. In contrast, if you are a natural scientist, you might prefer software such as GENSTAT and SAS. Generally, if working with social quantitative data, learning tools like Excel, SPSS, or STATA are advisable. If you are handling experimental data, we encourage gaining proficiency in GENSTAT, SAS, or R. Fortunately, many free online resources, including books and tutorials, make it easy to learn these programs.

20.2 Steps in Quantitative Data Analysis

20.2.1 The Common Steps

Quantitative analysis can be used to answer questions like:

- What is the percentage distribution?
- What trends or anomalies are present in the data?
- What is the average value of a particular variable?
- How do the means of different groups compare to each other.
- What is the extent of variability within the data?
- What is the nature and strength of the relationship between the variables?
- Are the findings statistically significant?
- What are the predictive factors for a particular outcome?
- How do the viable interact with each other?

Every research experiment and/or study is unique because the objectives differ. Several books comprehensively tackle the process of data analysis (Patel et al. 2004; Wilson and MacLean 2009). The following are the *six common steps* to keep in mind to maximize the chances of carrying out good data analysis:

1. Preparing data for analysis.
2. Define the analysis objectives.
3. Selecting variables for use in the analysis.
4. Data exploration and descriptive analysis.
5. Confirmatory or inferential data analysis.
6. Interpretation and reporting.

(i) *Preparing data for analysis*

Data preparation is the first step in conducting an analysis. In this step, familiarize yourself with the appropriate software, enter your data, and clean it to ensure that your data file for analysis is clean and error-free. Remember that this data, as shown in Fig. 20.1, will be analyzed to make critical decisions and should therefore be error-free.

(ii) *Define the analysis objectives*

Conducting a sound analysis without knowing what you wish to achieve is impossible. The analysis objectives are usually determined by the overall research objectives linked to a hypothesis. Defining objectives of analysis in this section deals with *specifics of analysis*, e.g., the specific comparisons that will be made, and the relationships between variables that will be investigated. The analysis is guided by the things we want to know from the data.

Examples:

- Is there a relationship between variables A and B?
- What is the percent distribution? For example, gender from a survey of 300 respondents.
- What is the average? For example, of maize grain yields from various treatments or from farms.

(iii) *Selecting variables for use in the analysis*

At this stage, the variables for analysis are selected. These variables may be in the *data file* or new ones you construct from the data. For example, if you had collected data for the size of the farm by considering all the parcels in a farm, you need to create a new variable called "total farm size" that you calculate by aggregating the sizes of all parcels belonging to the same farm. Other new variables to be created may include the following: average of samples per plot, calculation of plant nutrient concentrations, calculation of harvest index, etc.

(iv) *Data exploration and descriptive analysis*

Data exploration and descriptive analysis summarizes and describes data.. The exploration helps in the identification of the most important characteristics. Data exploration is a way of interacting with data to give researchers an understanding of their data sets' patterns, trends, and unexpected data behavior (Coe et al. 2003). While some students and novice researchers may prioritize hypothesis testing and statistical analyses like analysis of variance (ANOVA), we recommend data exploration as a preliminary step. At this stage, you may produce tables that will answer the detailed analysis objective. Excel is an excellent software for exploring data. For example, in Excel, you can create contingency tables to examine the distribution of counts between two categorical variables through cross-tabulation and quickly identify patterns and trends in the variables. You can also use the filter command to sort data in ascending and descending order to help you spot outliers and anomalies in data. Additionally, by filtering individual columns, e.g., a treatment group, you can assess how values are distributed.

In the descriptive analysis, you get a concise summary of the data and form a foundation for further analysis, including hypothesis testing and modeling. The following are the commonly used descriptive statistics:

1. Distribution: frequency of each value.
2. Central tendency: means, median, and mode values.
3. Variability or dispersion: how spread out the values are, and includes range, interquartile range, variance, and standard deviation.
4. Relationships. include: Karl Pearson's coefficient of correlation, Spearman's rank correlation, factor analysis, discriminant analysis, and scatter diagram.

It is important to note that descriptive statistics do not allow us to make conclusions beyond the data we have analyzed or reach conclusions regarding any hypotheses we might have made. Conclusions will be made once we subject the data to inferential statistics and attach a probability level.

(v) *Confirmatory analysis or inferential statistics*

As we have seen, descriptive statistics help us to describe and summarize data and, therefore, simplify large amounts of data. With inferential statistics, we are trying to reach conclusions that extend beyond the immediate sample data alone. It allows us to make inferences and predictions about a population based on a sample taken from the population in question. It is based on probability (Infer).

Inferential or inferential analysis is a formal analysis aimed at:

- Adding measures of precision (e.g., standard errors and results of significance) to the results found in the descriptive analysis.
- Infer about the population by estimating population parameters.
- Generalization of the results obtained from the sample.
- Tests the hypotheses.

In summary, confirmatory analysis confirms that the patterns observed under descriptives are "real" given the level of uncertainty. Uncertainty arises because of measurement error, sampling variability, and the inherent random fluctuations, often referred to as "noise," inherent in any natural system (topic covered in Chap. 21):

- Which test to use depends on:
- What is the scale of measurement?
- Is it to compare? How many groups do you have?
- Is it to measure relationships?

(vi) *Interpretation and reporting*

Interpretation involves making sense of the analysed data by drawing conclusions from it. You want to ask yourself, what does this output mean? Without interpretation, the analysis is pointless. For instance, what does it mean to have 35% of farmers using tissue culture banana? or 75% of farmers reporting being involved in the collective marketing of mangoes? What do these percentages truly signify? Interpretation goes beyond merely understanding statistical results. It entails integrating the findings with existing knowledge on the subject. The process includes comparing your results with other studies, developing predictive models, and generating new hypotheses.

These steps are typically not executed in a strict sequence but instead interactive, with each step informing and influencing the others. This dynamic process involves the active participation of the researcher rather than relying solely on statisticians or biometricians. We discourage researchers abandoning this stage of data analysis to statistician or biometrician.

Reporting involves drafting a detailed analysis report, including final tables and figures. It should also consider archiving/storing the data and the methods used for future reference and additional analysis.

We emphasize the importance of *proper data storage* after analysis. Your data is extremely important, and you must ensure that it can be accessed later for verification, further analysis, or review. Proper data storage is also essential for accountability, especially for donor-funded projects and facilitates data sharing during publication or responding to queries from peers, institutions, or funders. You can also use the data to publish a data paper, that is, a journal paper about the data (see Kiboi et al. (2022)). Many research institutions have data management policy that guides researchers on their data management plans (Patel et al. 2004) including and not limited to the publication of such data (see examples of published data reporting in poultry research: Maina et al. (2025a) data for published article - Maina et al. (2025b), and Maina et al. (2023a) for published article - Maina et al. (2023b).

Next Steps

After covering the general steps to follow in data analysis, the next section discusses two main aspects of analysis:

1. Preparing data for analysis (survey and experimental data).
2. Descriptive and exploratory analysis.

20.2.2 Preparing Data for Analysis

20.2.2.1 Creating a Data Analysis Matrix

A data analysis matrix is a structured arrangement of data. It is an important first step in processing your data. It facilitates data management by providing a clear and systematic format for storing and organizing data. For example, a data analysis matrix will include treatment assignments, experimental replications, measured parameters, and other relevant variables in the data set. Here, we create a data analysis matrix for experimental and survey data. It is important to note that while the matrix can be directly created in the statistical analysis software, MS Excel provides a powerful tool to do this. We will use MS Excel in the subsequent examples.

20.2.2.2 Experimental Data

Creating a data analysis matrix for experimental data involves several steps. First, you must define the variables measured or observed during the experiment, including independent variables (such as treatment groups or conditions) and dependent variables (outcomes). Once the variables are identified, you need to design a matrix in the form of a table or spreadsheet. Each row in the matrix corresponds to a single observation or data point, while each column represents a different variable. The first row typically contains the variable names, while subsequent rows contain the actual data.

Subsequently, you populate the matrix with the collected data, ensuring accuracy and consistency in data entry to maintain data integrity. Next, you organize the data matrix based on the experimental design and research objectives, grouping similar observations and arranging variables in a meaningful sequence to facilitate analysis. Finally, review for completeness and accuracy, as well as check for missing values, outliers, or errors in data entry. Any discrepancies or inconsistencies should be addressed to ensure the reliability of the data.

Researchers often include *metadata* within their data matrices. This is additional information that provides context about the data. It includes information such as experimental conditions, data collection methods, units of measurement, and any relevant notes or comments (Patel et al. 2004). Including metadata ensures transparency in the analysis process and helps contextualize the data. Once the data matrix is complete and verified, and cleaned, researchers prepare it for statistical analysis by choosing appropriate analysis techniques guided by the research questions, objectives, and hypothesis.

20.2.2.3 Example of Creating a Data Analysis Matrix

Suppose you want to create a data analysis matrix for an experiment that has six treatments, each replicated three times within a randomized complete block design. First, you need to create a table with rows representing individual experimental units and columns representing the variables. Assign a unique identifier to each experimental unit in the data set to distinguish it from others. Allocate columns for treatments, blocks, and response variables, ensuring that the data matrix layout is clear and easily interpretable, with consistent labeling and formatting (Fig. 20.2). The experiment in this figure has $3 \times 6 = 18$ observational units. To populate the matrix, enter the treatment assignments for each experimental unit based on the randomized allocation of treatments within blocks (Fig. 20.2). Record the block assignments for each experimental unit, ensuring that each block contains a complete set of treatments. Enter the measured values of the response variable for each experimental unit.

Example 2: A study assessed the effect of integrated soil fertility management technologies on various response variables. The screenshot (Fig. 20.3) shows that it is possible to organize your data in one MS Excel file and that various aspects/parameters can then be placed in different sheets. It is common to find students having different sheets or files where data was collected in different periods. For instance, when crop yield data was collected for two seasons. Nevertheless, having the seasons on the same Excel sheet, as illustrated in Fig. 20.3, is a good practice. In the illustration, the response variables included germination count, leaf area index (LAI), photosynthetically active radiation (PAR), plant height, and grain yield.

Survey Data Creating a data analysis matrix for survey data is much the same as the experimental data discussed earlier. However, there could be a difference if data were collected using digitized tools that automatically save data to a

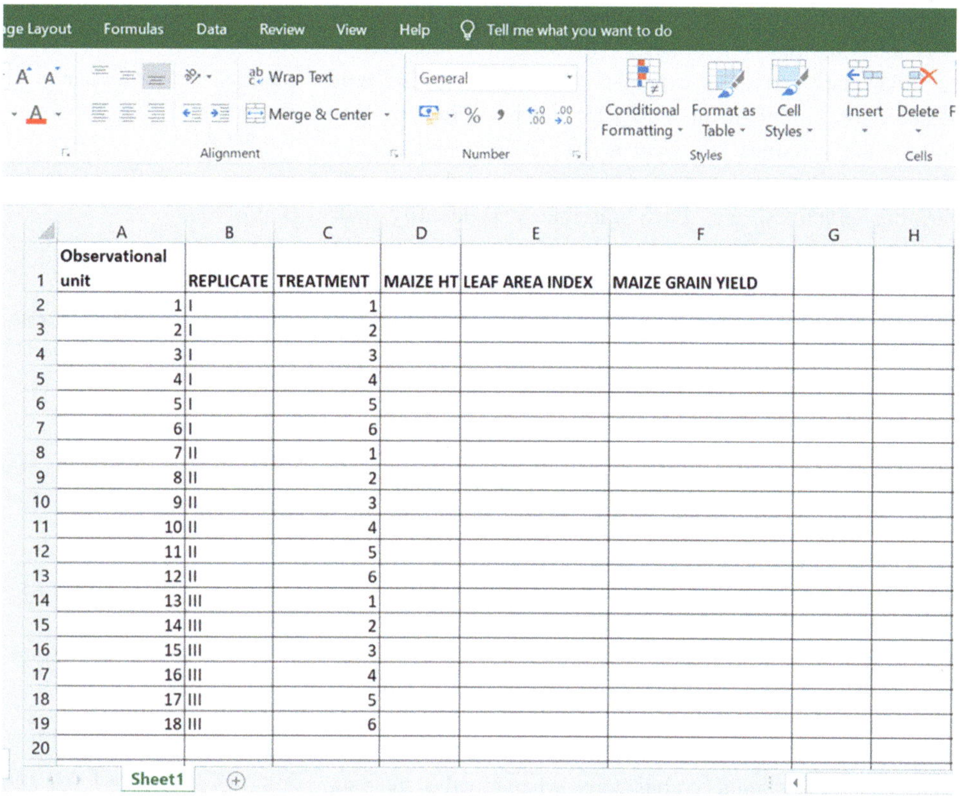

	Observational unit	REPLICATE	TREATMENT	MAIZE HT	LEAF AREA INDEX	MAIZE GRAIN YIELD
2	1	I	1			
3	2	I	2			
4	3	I	3			
5	4	I	4			
6	5	I	5			
7	6	I	6			
8	7	II	1			
9	8	II	2			
10	9	II	3			
11	10	II	4			
12	11	II	5			
13	12	II	6			
14	13	III	1			
15	14	III	2			
16	15	III	3			
17	16	III	4			
18	17	III	5			
19	18	III	6			

Fig. 20.2 Data analysis matrix of experimental data from a randomized complete block with six treatments replicated three times

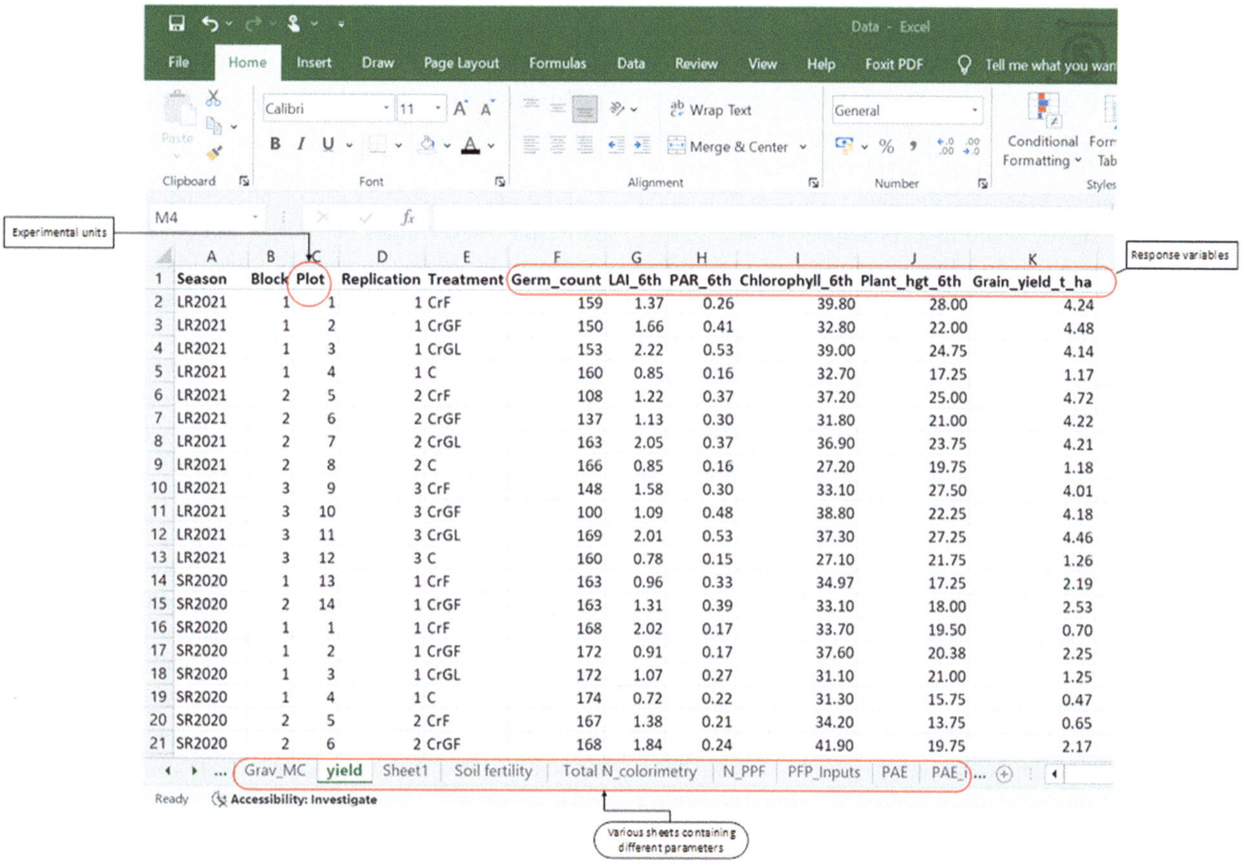

	Season	Block	Plot	Replication	Treatment	Germ_count	LAI_6th	PAR_6th	Chlorophyll_6th	Plant_hgt_6th	Grain_yield_t_ha
2	LR2021	1	1	1	CrF	159	1.37	0.26	39.80	28.00	4.24
3	LR2021	1	2	1	CrGF	150	1.66	0.41	32.80	22.00	4.48
4	LR2021	1	3	1	CrGL	153	2.22	0.53	39.00	24.75	4.14
5	LR2021	1	4	1	C	160	0.85	0.16	32.70	17.25	1.17
6	LR2021	2	5	2	CrF	108	1.22	0.37	37.20	25.00	4.72
7	LR2021	2	6	2	CrGF	137	1.13	0.30	31.80	21.00	4.22
8	LR2021	2	7	2	CrGL	163	2.05	0.37	36.90	23.75	4.21
9	LR2021	2	8	2	C	166	0.85	0.16	27.20	19.75	1.18
10	LR2021	3	9	3	CrF	148	1.58	0.30	33.10	27.50	4.01
11	LR2021	3	10	3	CrGF	100	1.09	0.48	38.80	22.25	4.18
12	LR2021	3	11	3	CrGL	169	2.01	0.53	37.30	27.25	4.46
13	LR2021	3	12	3	C	160	0.78	0.15	27.10	21.75	1.26
14	SR2020	1	13	1	CrF	163	0.96	0.33	34.97	17.25	2.19
15	SR2020	2	14	1	CrGF	163	1.31	0.39	33.10	18.00	2.53
16	SR2020	1	1	1	CrF	168	2.02	0.17	33.70	19.50	0.70
17	SR2020	1	2	1	CrGF	172	0.91	0.17	37.60	20.38	2.25
18	SR2020	1	3	1	CrGL	172	1.07	0.27	31.10	21.00	1.25
19	SR2020	1	4	1	C	174	0.72	0.22	31.30	15.75	0.47
20	SR2020	2	5	2	CrF	167	1.38	0.21	34.20	13.75	0.65
21	SR2020	2	6	2	CrGF	168	1.84	0.24	41.90	19.75	2.17

Fig. 20.3 Data analysis matrix displaying data in different seasons in the same sheet

computer file using predefined codes. You must prepare a data analysis matrix with other data collection methods to facilitate data entry. For example, if we consider data collected from farmers using a questionnaire, you start by listing all the variables you want to handle in your analysis. These could be demographic characteristics of the farmers, farming practices, etc. Each variable you identify will become a column. Name each column clearly to represent the variable it contains. For each farmer surveyed, enter their response into the appropriate columns. Ensure consistency in data entry and use a standardized format (numerical values for variables like age and categorical values).

It is important to decide how to handle any missing responses. You can either leave the cells blank or use "not available" to indicate missing data. Once the data has been entered, it is crucial to review it for quality. This means checking for accuracy, consistency, and outliers or errors that may affect the analysis. To ensure that the data is well-formatted, categorical variables should be converted into numerical codes (coded). Additionally, it is helpful to include relevant details about the survey to provide context for the analysis (metadata). These details can accompany the data matrix and should be considered when creating the data analysis matrix.

Table 20.1 Coding scheme for hypothetical data from a farming household[a]

Module	Question	Response	Code
Household livelihood (1)	Gender of household head	Female	1
		Male	2
	What is the age of the household head?	<20	1
		21–30	2
		31–40	3
		41–50	4
		>50	5
	What is the education level of the household head?	No education	1
		Primary level	2
		Secondary level	3
		Tertiary	4
Crop production (2)	What is the most common spatial arrangement used for cereals?	Monocropping	1
		Rotation	2
		Intercropping	3
		Other	4
	Did you use fertilizer last season?	Yes	1
		No	2
	If yes, which fertilizer?	Nitrogenous	1
		Phosphatic	2
		Potash	3
		Other	4

[a]See how these coded data can be entered in an MS Excel sheet (Fig. 20.4)

20.2.2.4 Coding Categorical Variables

Categorical variables must be coded to facilitate analysis. Although it may seem simple to assign codes, it is good to develop a coding scheme. Table 20.1 shows an example of a coding scheme for categorical variables of data from farmers. You can integrate these codes into your data collection form or online survey platform as predefined options, especially when dealing with a limited number of categories. Even if you choose not to adopt an existing coding scheme due to insufficient detail, ensure that your codes align with previous data for comparison. Coding during data collection is done when there is a finite range of well-defined categories for classifying the data. These categories can be incorporated into your data collection form for respondents to choose from or automatically assigned by your online survey software.

When there is uncertainty about potential responses or when dealing with a vast array of possible responses, it becomes necessary to code post-data collection. To make sure that the coding scheme effectively covers the range of answers and works properly, it is recommended to delay its development until data from the initial 50–100 cases is available. This process, known as creating a codebook, helps researchers organize and document their coding procedures systematically, ensuring consistency and reliability throughout the coding process. A codebook is simply a document that outlines the coding scheme to be used for classifying and interpreting data (Mugenda and Mugenda 2019).

To develop your codebook for each variable, follow the following steps:

1. First, review the data and identify general categories.
2. Refine these broad categories into more specific subgroups based on the analysis objectives.
3. Assign codes to each category at the most detailed level necessary.
4. Document the responses assigned to each category and compile this information into a codebook.
5. Additionally, make sure that categories requiring aggregation are assigned sequential codes to simplify the recoding process if needed.

Experienced researchers often skip the step of preparing codes and directly enter data from the questionnaires into the computer. Some will enter into a spreadsheet or directly into analysis software like SPSS. During this phase, it is crucial to focus on extracting open-ended responses from the questionnaire. They should be extracted and assigned to appropriate categories. Moreover, such responses may require different types of analysis, which must be considered during data entry. Categories that feature *multiple responses* need to be given special attention. At the end of the data entry exer-

Fig. 20.4 Excel spreadsheet with data to demonstrate the entering of the coded data

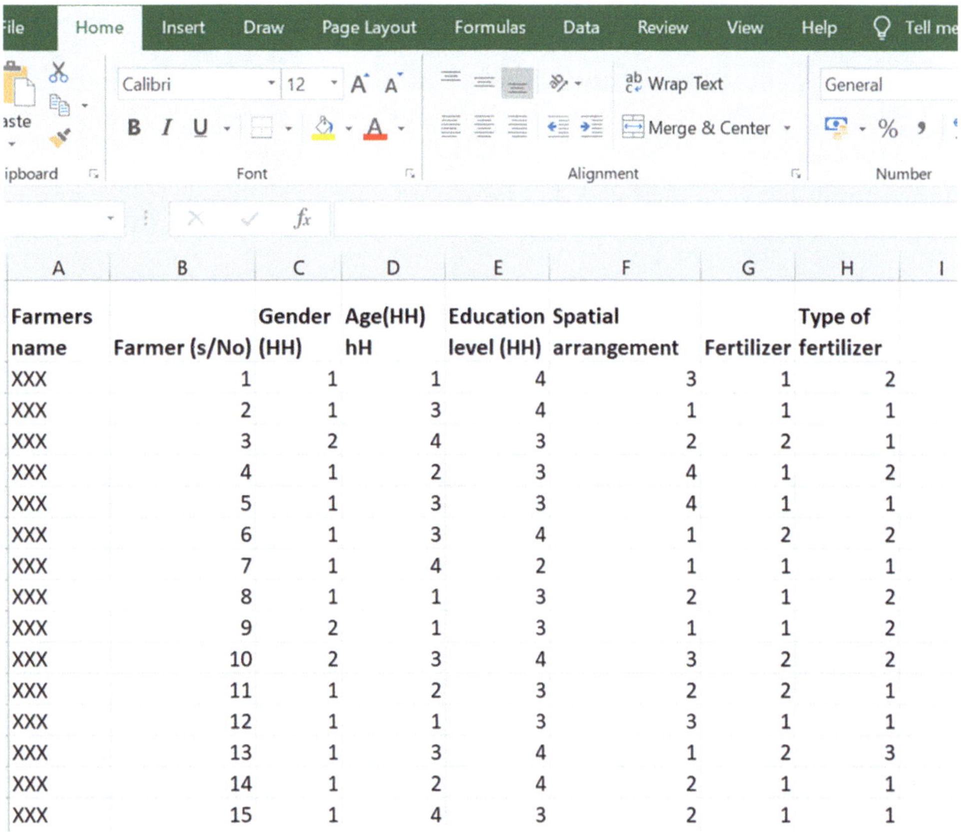

Farmers name	Farmer (s/No)	Gender (HH)	Age(HH) hH	Education level (HH)	Spatial arrangement	Fertilizer	Type of fertilizer
XXX	1	1	1	4	3	1	2
XXX	2	1	3	4	1	1	1
XXX	3	2	4	3	2	2	1
XXX	4	1	2	3	4	1	2
XXX	5	1	3	3	4	1	1
XXX	6	1	3	4	1	2	2
XXX	7	1	4	2	1	1	1
XXX	8	1	1	3	2	1	2
XXX	9	2	1	3	1	1	2
XXX	10	2	3	4	3	2	2
XXX	11	1	2	3	2	2	1
XXX	12	1	1	3	3	1	1
XXX	13	1	3	4	1	2	3
XXX	14	1	2	4	2	1	1
XXX	15	1	4	3	2	1	1

cise, the researcher must ensure that all questionnaire data are accurately captured and organized. Use the tips in Box 20.1 to ensure that you code and enter your data accurately and that your data is retrievable.

Box 20.1: Tips during Coding, Data Entry and Storage

- Ensure you choose an appropriate data entry tool or software based on the data's complexity, nature, and volume.
- Create a data entry matrix within the chosen software, setting up columns to correspond to each variable identified in the coding scheme.
- Ensure accuracy by referring to the codebook and double-checking entries as you progress.
- Accurately input each response for variables with multiple responses.
- Create new variables or expand existing ones for open-ended responses.
- Utilize built-in validation features within the data entry software to identify and correct errors in real time.
- Address any missing or incomplete data entries according to predetermined protocols.
- Always conduct a final review of the entered data to ensure completeness and accuracy.
- Always save the entered data in a secure location, and maintain regular backups to prevent data loss. Some students have suffered trauma as a result of losing their data.

20.2.2.5 Metadata

It is good practice to include metadata in the MS Excel sheet (Fig. 20.5). Metadata tells us about the data in the spreadsheet. It is "data about the data." It contains background information about the data. It includes details such as the data source, the methods used for collection, the time and location of data collection, and any relevant contextual information. Documenting metadata is crucial for maintaining the integrity and reliability of the data, as it allows researchers to trace the origins of the data and understand its context. Additionally, metadata facilitates data sharing and reuse by providing comprehensive information to other researchers who may wish to use the data for their analyses.

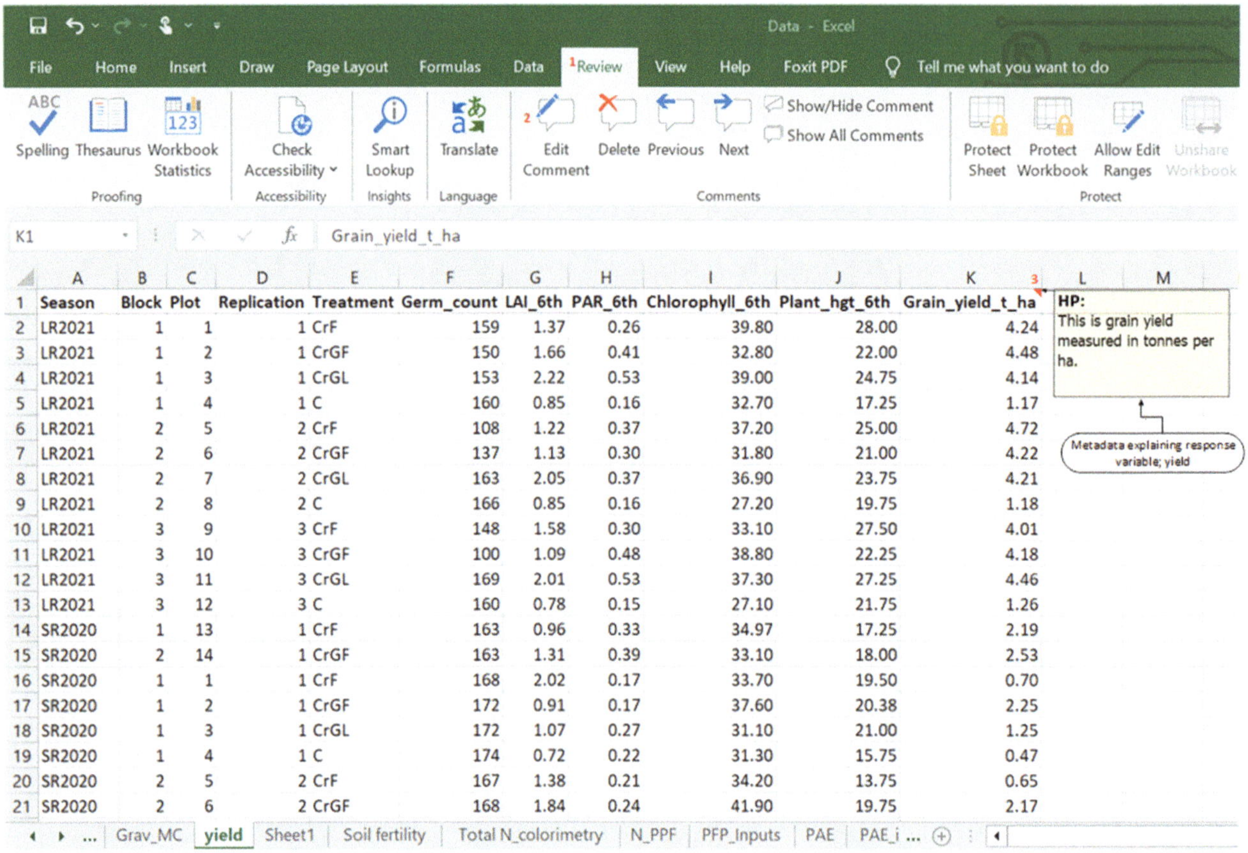

Fig. 20.5 Metadata to describe and contextualize the data in an Excel sheet

Steps to adding metadata

Step 1: Click "Review".

Step 2: Click "Edit comment".

Step 3: Place the cursor on the cell where you want to add metadata. A dialog box will appear. Add the metadata.

20.2.3 Using the Sort and Filter Function as a Data-Checking Tool

MS Excel's spreadsheet sort and filter command is a handy tool for analyzing and checking data. It allows users to quickly sort and/or filter based on specific criteria, making identifying patterns, trends, and outliers within the data set easier. With the command, users can efficiently narrow down large data sets to focus on particular subsets of data that meet their requirements (Fig. 20.6). This functionality is especially useful for data validation, error checking, and exploring relationships between variables. To get the function, follow the steps as follows:

Step 1: Click "Home".

Step 2: Find the Sort and Filter function on the top right-hand side of the screen.

Step 3: Select a command from the options provided.

20.2.4 Data Is a Pattern

In Chap. 8, we explored the fundamental principles of experimentation and examined the various sources of variation inherent in an experiment, i.e., due to treatments and extraneous variables. The data we obtain is a response to these sources of variation. The treatment factors, carefully controlled and manipulated by the experimenter, typically reveal recognizable *patterns* within the data set. The extraneous factors, commonly known as noise or residuals, contribute to the residual variability in the data that cannot be solely attributed to the treatment factors. This is the unexplained variation (error term).

Thus, we can describe the data we obtain as combining two components: "Data = pattern + residual" (Patel et al. 2004). In analysis, our primary objective is to identify and interpret this underlying pattern, which represents the *systematic effects of the treatments* on the response variable.

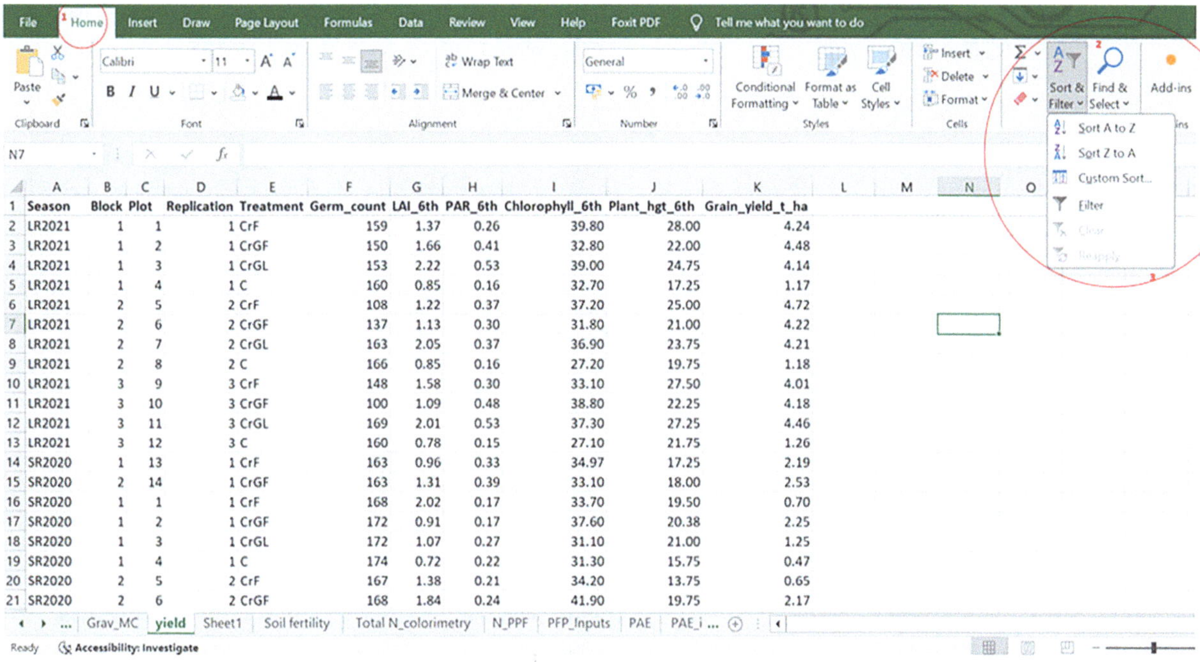

Fig. 20.6 Sort and filter button for checking data

Residuals should be patternless; any visible pattern may indicate a missing factor in the design, treatments, or measurements, affecting the analysis (Patel et al. 2004). The various descriptive and inferential tools used aim to identify this pattern.

Later in Chap. 21, we explain the use of ANOVA as a descriptive tool, in addition to testing hypotheses, and present the concept of data as pattern + residual.

20.3 Descriptive Analysis

Descriptive statistics, as mentioned at the start of this Chapter, are used to organize and summarize data. They allow a researcher to describe what exists in the data. It answers the question: What does the data show? This is presented in the form of data summaries, summary tables, graphs, summary statistics, and data patterns.

The key descriptive statistics include:

- Counts, frequencies, and percentages.
- Measures of central tendency (MoC).
- Measures of variability (MoT).

20.3.1 Counts, Frequencies, and Frequency Distribution Tables

We encounter count data when quantifying various aspects, e.g., pest and disease infestation, number of fruits, number of animals in a certain area, etc. Frequencies expand on counts by presenting the number of occurrences for each category or value as a proportion or percentage of the total, aiding in the comparison of relative frequencies across different categories. In frequencies, we are concerned about counts, e.g., how many males and females?; what was the average number of plants infected by the disease? These are mainly univariate data types.

Frequency distribution tables further organize this information by tabulating the counts or frequencies of each category or value in a structured format, often accompanied by additional summary statistics such as percentages or cumulative frequencies (Manikandan 2011a, b, c; Ishaq 2025). These tables highlight patterns, trends, and outliers that may warrant further investigation. They are useful for nominal and ordinal data.

Example 1 Let's consider a hypothetical data set in Table 20.2 representing the number of dairy goats data collected from 30 farmers. We can organize these data from the lowest to the highest and summarize the data into a frequency distribution showing the proportion of the total observations. The proportion can be presented out of 100 or out of 1. For example, a relative number of farmers having two (2) dairy goats is 3.3% out of the total observations. This also translates to 0.33 out of 1 (Table 20.3).

With a frequency table, we can easily see the pattern. Another method of summarizing count data involves creating a class interval distribution table, which entails grouping

Table 20.2 Number of goats per farmer

Farmer	Number of dairy goats
X1	5
X2	3
X3	6
X4	6
X5	4
X6	8
X7	9
X8	4
X9	5
X10	5
X11	4
X12	3
X13	3
X14	2
X15	6
X16	5
X17	4
X18	3
X19	5
X20	5
X21	6
X22	7
X23	4
X24	3
X25	5
X26	5
X27	6
X28	5
X29	4
X30	4

RF Frequency/summation of N

Table 20.3 Frequency and relative frequency distribution of number of goats per farmer

Number of goats	F (frequency)	rf (relative frequency/100	rf (relative frequency/1
2	1	3.333	0.033
3	5	16.667	0.167
4	7	23.333	0.233
5	9	30.000	0.301
6	5	16.667	0.167
7	1	3.333	0.033
8	1	3.333	0.033
9	1	3.333	0.033
	N=30	100	1.00

the data into intervals (Table 20.4). The first step to constructing such a table is determining the appropriate number of interval classes. Subsequently, the width of each interval is calculated by dividing the range of the data by the number of intervals. This ensures that all data points are adequately

Table 20.4 Class interval distribution table of number of dairy goats

Class interval	f	Rf/100	Rf/1
2–4	6	20	0.20
4–5	16	53	0.53
6–7	6	20	0.20
8–9	2	6	0.6
	N = 30	100	1.00

Table 20.5 Class interval frequency distribution of soybean yields from farmers in Region A

Yield (kg ha^{-1})	Frequency	f/100	f/1
300–500	3	4	0.04
501–700	6	8	0.08
701–900	5	6	0.06
901–1100	12	15	0.15
1101–1300	10	13	0.13
1301–1500	17	21	0.21
1501–1700	13	16	0.16
1701–1900	5	6	0.06
1901–2100	5	6	0.06
2101–2300	4	5	0.05
	N = 80	100	1.00

represented within the intervals, promoting an accurate data set summary. Always ensure that each interval is non-overlapping and encompasses the entire range of the data.

Example 2 We can use another example of data from a study involving 80 farmers growing soybeans in Region A. Demographic and yield data were collected. We shall use the variables yield of soybeans in t ha^{-1} to plot a frequency table (Table 20.5).

20.3.2 Measures of Central Tendency

The three most common measures of central tendency (MCT) are used, i.e., *Mode*, *Median*, and *Mean* (Manikandan 2011b). These are usually discussed in most books on statistics. The choice of measure of central tendency depends on the type of data you have: nominal, ordinal, interval, or ratio data. The suitable measure of central tendency for each data type is shown in Table 20.6. For interval and ratio scales, all three measures, mean, median, and mode, can be used. For data on the nominal scale, only the mode is applicable and meaningful.

The choice for interval/ratio data will also depend on the distribution itself. The mean, median, and mode should be very close if it is a normal distribution. The mean would be the best description of central tendency. However, if you have a few extremely high or low scores, the mean may not accurately represent the center of the data distribution.

Table 20.6 Type and data, and appropriate measure of central tendency

Type of data	Choose this measure of central tendency
Nominal	Mode
Ordinal	Mode or median
Numerical (interval/ ratio)	Mode, median, mean

Mode is the most frequently occurring value in a data set and is particularly useful for nominal variables and less commonly applied to other levels. It serves as an effective measure for identifying the most common category or value in nominal data. For instance, in a survey assessing farmers' preferred crop varieties or livestock breeds, the mode can identify the most commonly selected options.

The Median is the central value in an ordered data set of observations, splitting the data into two equal halves. It is particularly useful for continuous or ordinal data as it summarizes the distribution's central point. Unlike the mean, the median is not influenced by extreme values or outliers, making it a robust measure of central tendency. Therefore, it is a valuable measure in agricultural contexts where data distribution may be skewed with extreme scores. In other words, it is less affected by extreme values or outliers compared to the mean. Because of this, it is often used to analyze and present data related to farm size, household incomes, and other measures where there may be a wide range of values.

Example Given the following milk yields of 11 farms each from Region A and Region B, find their median (Table 20.7).

Solution We must first order the given numbers from the smallest to the largest

Region A farmers: 7, 8, 10, 11, 15, 16, 16, 16, 20, 20, 20, 21.
Region B farmers: 9, 10, 12, 12, 16, 18, 18, 18, 25, 25, 25, 32.
The middle value for Region A farmers is 16. Therefore, the median is 16.
The middle value for Region A Farmers is 18. Therefore, the median is 18.

What about when we even have observations, e.g., if our sample was 12 farmers? You get the median by averaging the two numbers in the middle.

Mean The arithmetic average of a set of scores, making it the most widely used and accepted measure of central tendency. It is calculated by summing up all the values in a data set and then dividing this number by the number of scores. The mean is favored for its mathematical properties, such as its involvement in various statistical analyses and its ability

Table 20.7 Milk sales (liters) of 11 farmers in Region A and Region B

Farmer no	Sample no	Milk sales	
		Region A	Region B
1	X1	16	12
2	X2	10	9
3	X3	8	25
4	X4	20	32
5	X5	16	25
6	X6	21	18
7	X7	20	16
8	X8	7	25
9	X9	11	10
10	X10	20	12
11	X11	15	18

to be influenced by all data points. This makes it a valuable indicator for summarizing and analyzing data, particularly when the data distribution is symmetric and free from extreme outliers. The mean is calculated as follows:

$$\text{Mean, } \bar{x} = \frac{\sum_{i=1}^{n} a_i}{n}$$

$\bar{x}$ = arithmetic mean
n = number of values
a_i = data set values
$\sum$ = summation

The mean for milk yields of Region A farmers is the total yields of the 11 farmers divided by 11 = 164/11 = 14.9 liters, while for Region B farmers, it is 202/11 = 18.4 liters.

20.3.2.1 Utility of the Measures of Central Tendency

Among the three MCTs, the mean is regarded as the most stable (Jankowski and Flannelly 2015). This stability means that when drawing random samples from a population, the means of those samples tend to fluctuate less than the mode or median. Thus, the mean is considered the most reasonable estimate of population characteristics. In reporting or published work, median and mode are rarely used.

The effectiveness of the three MCTs in describing distributions varies based on the shape of the data. For uniform and normal distributions, any of the three measures can be used effectively. However, with skewed distributions, the mean's utility as a descriptive statistic is diminished because every data point influences it and can be distorted by extreme values. The median and the mode are less affected by outliers, which may provide a more accurate reflection of the distribution's central location in such cases. Therefore, the choice of which MCT to use should be guided by the distribution shape (either normal or skewed) and the nature of the data.

20.3.3 Measures of Variability

The measure of variability (MoV) is about data variability indices (Li et al. 2019). It is important because it answers the question: How do scores/observations differ among themselves in magnitude? The measure of central tendency does not show this variability in distribution. Hence, the measure of dispersion is also essential. MoV measures the dispersion of scores around the mean. With just the presentation with MCT, the sample description is incomplete because it does not describe the differences that exist among the scores. The commonly used MoVs are the range, percentiles, box plots, and standard deviation.

The Range The *range* is the most straightforward measure of variation. It is calculated as the difference between the highest and lowest values in a distribution. To accurately convey the range of the data, the range is typically presented along with its upper and lower limits. This provides a clear picture of the extent of variation within a data set.

Range = Maximum value – Minimum value

- The range of Region A's milk yield would be 21 (Maximum)—8 (minimum) = 13.
- The range of the milk yield of Region B farmers would be 32 (Maximum)—9 (minimum) = 23.

The range's applicability is limited because it is determined using only two scores in the distribution. It does not show the dispersion of the scores between the two extremes. Therefore, though it is easily computed, it is only used as a rough measure and is often reported together with other variability measures.

20.3.3.1 Percentiles

Percentile often shows that a given percentage falls below that percentile. Consider a study that evaluates the yield of a new bean variety across multiple farms. After gathering data from various farms, researchers can calculate percentiles to determine how each farm's yield compares to others in the distribution. This is a kind of benchmarking. In a study in Northern China, crop "Normalized Difference Vegetation Index" (*NDVI)* percentiles were used for real-time crop growth monitoring (Li et al. 2019). For instance, Farm A has a yield that falls at the 75th percentile, meaning that 75% of the farms in the study achieved lower yields than Farm A. This percentile ranking offers valuable insights into the performance of Farm A relative to other farms in the study. Similarly, if Farm B falls at the 25th percentile, it implies that 75% of the farms had higher yields than Farm B, indicating that Farm B may have lower productivity compared to others. This information can guide decision-making processes, such as identifying high-performing farms for further study or implementing strategies to improve the productivity of lower-ranking farms.

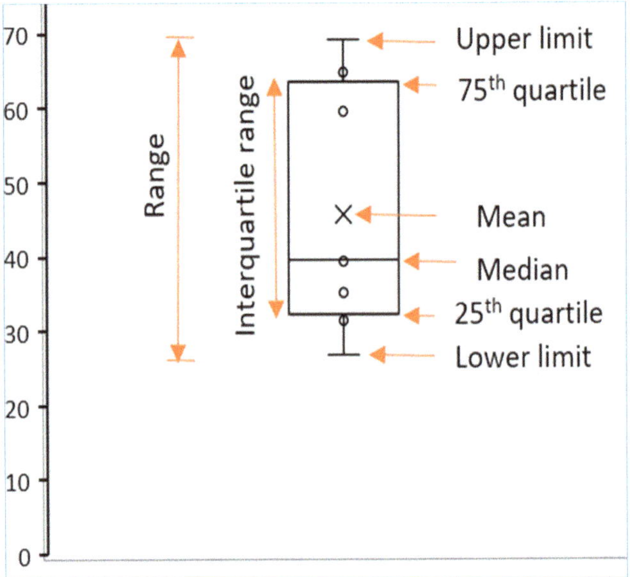

Fig. 20.7 Typical Box plots in vertical structure

20.3.3.2 Box Plots and Quartiles

Box plots provide another useful way of summarizing data and showing the variability of data (Williamson et al. 2008; Abt et al. 2024, 2025) (Fig. 20.7). They display the range, median, upper, and lower quartiles (each containing three-quarters of the data) in diagrammatical form (Fig. 20.7). The box in the plot represents the interquartile range (IQR), which encompasses the middle 50% of the data. The line within the box signifies the median of the data set. The whiskers extending from the edges of the box show the data's minimum and maximum values. Because of the whiskers, box plots are also referred to as "Box and whiskers" plots.

Quartiles The median splits a data set into two halves. Similarly, quartiles partition the ordered data set equally into four equal parts, which is useful when dealing with large amounts of data (Hyndman and Fan 1996; Langford 2006). The lower quartile, denoted as Q1, marks a position whereby 25% of the values fall below it, and 75% are above it. The second quartile corresponds to the median and divides data into two halves (Fig. 20.7).

In summary, the first, second, and third quartiles are represented as Q1, Q2, and Q3, respectively:

Q1 = "median" of the lower half of the ordered data.
Q2 = "median" of the whole data set.
Q3 = "median of the upper half of the ordered data set, excluding the median.

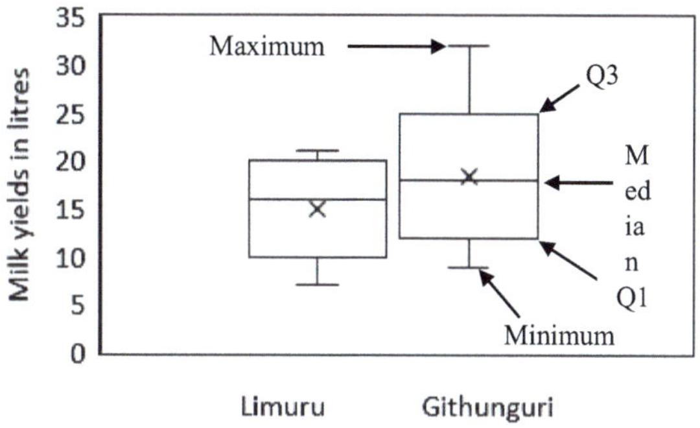

	Limuru	Githunguri
Maximum	21	32
Q1	10	12
Median	16	18
Q3	20	25
Minimum	7	9

Fig. 20.8 Typical box plots in vertical structure plotted using data in Table 20.7 using MS Excel

The interquartile range (IQR) is calculated as Q3–Q1, providing a measure of the spread of the middle 50% of the data.

Here, we draw box plots using the data from Region A and Region B farmers (Fig. 20.8). According to this figure, we observe the following:

- The median values for Region A and Region B farmers are 16 and 18, respectively.
- No outliers in both data sets.
- The "minimum and maximum" values for Region A farmers are 8 and 21.
- The "minimum and maximum" values for Region B farmers are 9 and 32.
- The interquartile range for Region B is higher than for Region A.

20.3.3.3 Use of Box Plots

Researchers commonly use box plots in *data exploration to visually identify potential outliers* (Abt et al. 2024). Once outliers are identified, researchers further investigate and address them as needed to guarantee the validity and reliability of the research findings. These include verifying whether the outliers are genuine data points or if they stem from errors in data collection or recording. This may involve checking the original data sources (raw data) or conducting additional checks to confirm the accuracy of the outliers. They also seek to examine contextual factors, such as the experimental conditions or data collection methods, to identify potential explanations for the outliers. Depending on the nature of the outliers and their impact on the analysis, researchers may handle them differently. Options for handling outliers include removing them from the analysis, transforming the data to reduce their influence, or conducting sensitivity analyses to assess their impact on the results (Osborne and Overbay 2004).

Researchers also use box plots in scientific publications to highlight comparisons and enhance the interpretation of research findings. By visually presenting data distributions and group differences, box plots complement textual descriptions and statistical analyses presented in research papers.

20.3.3.4 Standard Deviation

Standard deviation (SD) quantifies the extent of dispersion or spread in a data set relative to its mean (Manikandan 2011c). The SD is computed as the square root of the variance, which represents the average squared deviation from the mean.

The formula for calculating standard deviation is as follows:

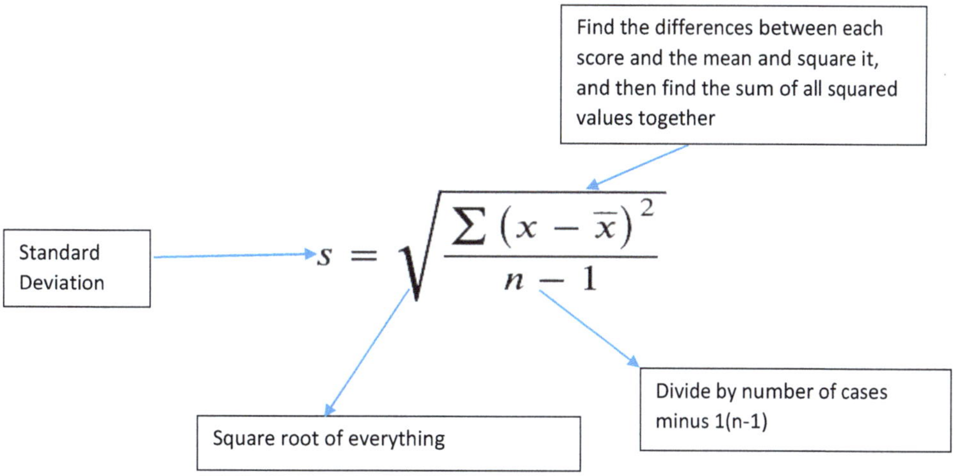

$$s = \sqrt{\frac{\sum (x - \bar{x})^2}{n - 1}}$$

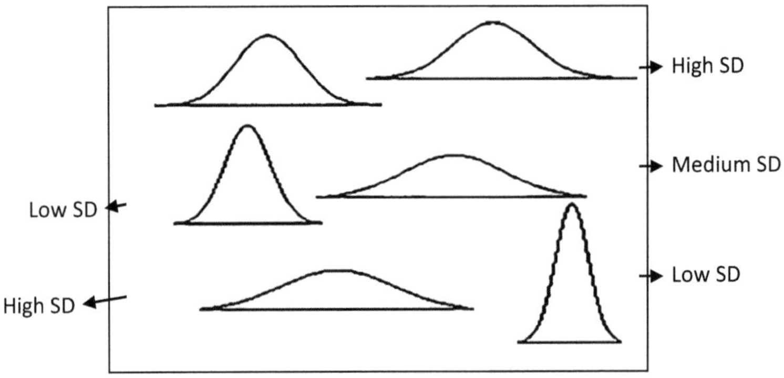

Fig. 20.9 Normal curves drawn from populations with varying standard deviations

A larger SD indicates that the data points are more widely spread from the mean, while a smaller SD suggests that they are closer to the mean (Fig. 20.9).

Reporting Standard Deviation

The standard deviation of sample data is usually reported along with the mean to characterize the data according to other central tendencies and variability. A mean may be expressed as $x = 45 + 5.2$), which refers to the average of the deviations on either side of the mean as 5.2.

We frequently use SD to *compare samples* (Wilson and MacLean 2009). For instance, suppose two samples of wheat yields from different experimental treatments are collected. The mean yield for sample A is 4.5 t ha^{-1} with an SD of 1.5 tons/hectare, while sample B's is 3.2 tons/hectare with an SD of 0.8 t ha^{-1}. In this scenario, the SD allows us to compare the variability in crop yields between the two samples. A smaller SD shows that data points within the sample are closer to the mean, signifying less variability or more consistent yields. Conversely, a larger SD suggests greater variability, with data points dispersed over a broader range of values. By comparing the SDs of the two samples, researchers can assess which treatment or condition leads to *more consistent crop yields* and identify potential factors contributing to variability in agricultural outcomes.

Some researchers are unsure about the SD's unit. The standard deviation (SD) is expressed in the same unit as the original data. For example, if the original data represents measurements in meters (m), then the standard deviation is in meters. If the original data represents kilogram (kg) measurements, then the standard deviation is in kilograms. In the case of other data, such as the age and income level of farmers, the SD is in years and incomes, respectively.

20.3.3.5 Coefficient of Variation

The coefficient of variation (CV) is a statistical measure that assesses the relative variability of data points in relation to the mean (Ospina and Marmolejo-Ramos 2019; Arachchige et al. 2022). To calculate the CV, you divide the standard deviation by the mean and then multiply the result by 100.

The CV is useful for comparing variability between different units or significantly different means. As the CV is dimensionless, it provides a standardized measure of dispersion, making it easier to compare diverse data sets (Ospina and Marmolejo-Ramos 2019). The CV is a practical statistic for comparing distributions that are recorded in different units. For instance, we can compare the distribution of weight expressed in kg and height measured in meters. Therefore, the coefficient of variation is most meaningful when comparing two distributions, unlike the SD. For instance, we cannot compare distributions of the height of maize expressed in meters and the weight of maize expressed in grams using the SD. In the context of animal science, particularly in poultry research, the calculated CV serves as a vital statistic for evaluating growth performance and flock uniformity in broiler and layer chickens. Its inclusion in growth curve analysis allows researchers to assess the degree of variability in body weight and other performance traits relative to the mean, providing a standardized indication of consistency within experimental groups (Maina et al. 2024; Tsementzis et al. 2025). Lower CV values denote higher flock uniformity and greater reliability of experimental outcomes, whereas higher values indicate increased heterogeneity, which may reflect differences in management, genetics, or environmental conditions. Therefore, the use of the CV in animal experiments is essential for accurately interpreting growth data and for ensuring that treatment effects are evaluated within a framework of biological and statistical precision. CV is calculated using the formula: CV = (SD/mean)*100.

20.3.3.6 Scatter Plots

A scatter plot is a pictorial representation that displays individual data points as dots showing the relationship between two quantitative variables (Lesik 2021). In this plot, individual points are depicted as dots, with one variable plotted on

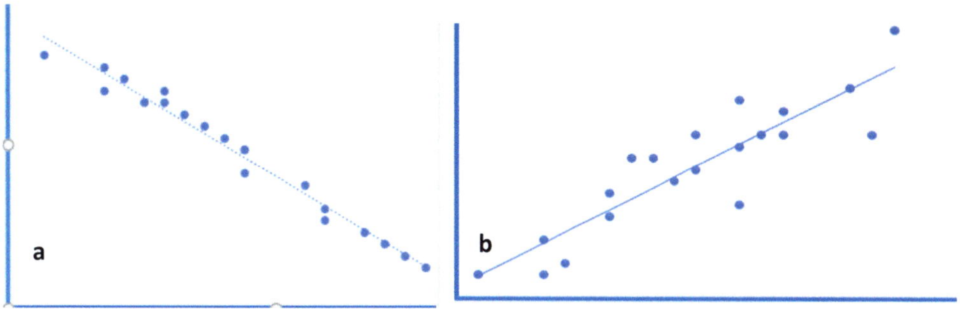

Fig. 20.10 Scatter plot showing data sets (**a**, **b**) with different variabilities; data set (**b**) is more variable than data set (**a**)

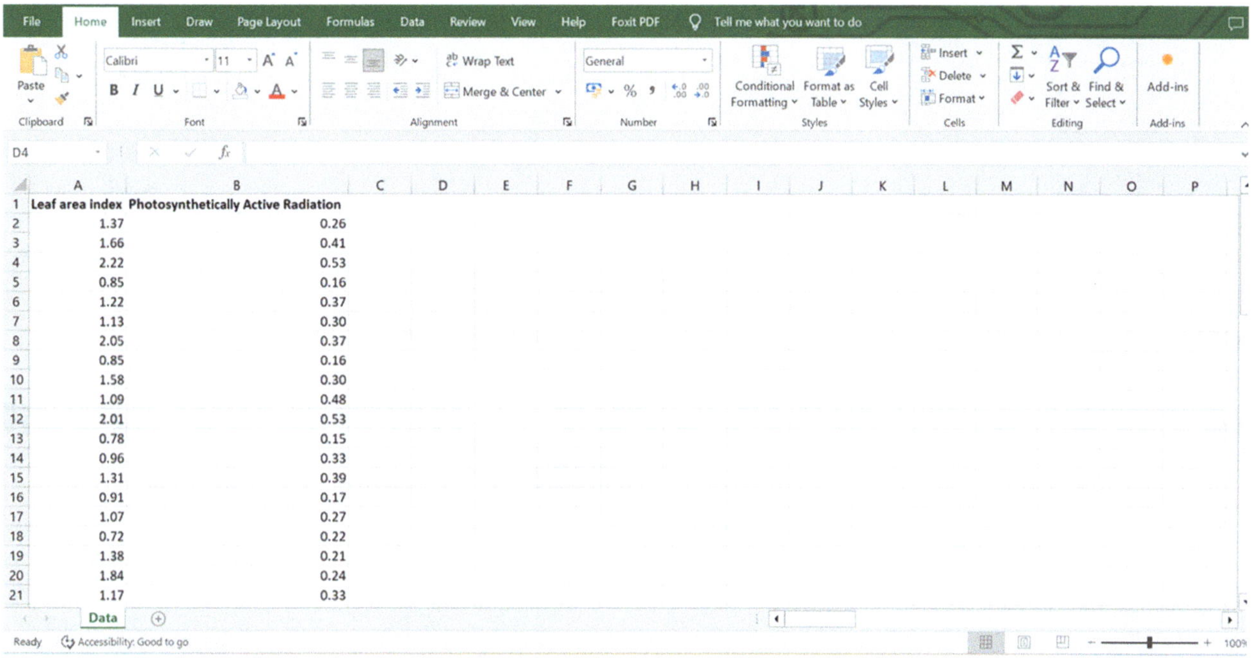

	A	B
1	Leaf area index	Photosynthetically Active Radiation
2	1.37	0.26
3	1.66	0.41
4	2.22	0.53
5	0.85	0.16
6	1.22	0.37
7	1.13	0.30
8	2.05	0.37
9	0.85	0.16
10	1.58	0.30
11	1.09	0.48
12	2.01	0.53
13	0.78	0.15
14	0.96	0.33
15	1.31	0.39
16	0.91	0.17
17	1.07	0.27
18	0.72	0.22
19	1.38	0.21
20	1.84	0.24
21	1.17	0.33

Fig. 20.11a Data in Excel showing two quantitative variables

the x-axis (horizontal axis), and the other on the y-axis (vertical axis). Each dot represents a pair of values, showing how one variable changes in relation to the other (Figs. 20.10a and 20.10b).

Scatter plots are useful for assessing *data variability* and understanding how data points are distributed across different value ranges. A wider spread of points indicates higher variability within the data set, whereas a narrower spread suggests lower variability. Additionally, the pattern formed by the data points can reveal the *"nature and strength" of the relationship* between the variables, which further helps to understand the variability in the data. Scatter plots can also reveal any outliers or anomalies in the data points that deviate from the general pattern, providing additional information about the data set's variability.

Scatter plots are particularly easy to construct, requiring minimal effort. To create a scatter plot, one must plot pairs of data points, forming a dot, with one variable represented along the x-axis and the other along the y-axis. This straightforward process can quickly be done in Excel (see Figs. 20.11a, 20.11b and 20.11c).

Constructing a scatter plot involves five simple steps as follows:

1. To create a graph, first decide which variable you want to represent on the x-axis (horizontal axis) and which variable will be on the y-axis (vertical axis). You should also consider which variable you want to be the independent (predictor) variable and which one will be the dependent (response)variable. Once you have made these decisions, plot the graph with the independent variable on the horizontal axis and the dependent variable on the vertical axis.
2. Plot each data point on a scatter plot by marking the point at its paired x and y values.
3. Adjust point size, color, and marker style to differentiate data and highlight specific features, and enhance the scatter plot's clarity and visual appeal.

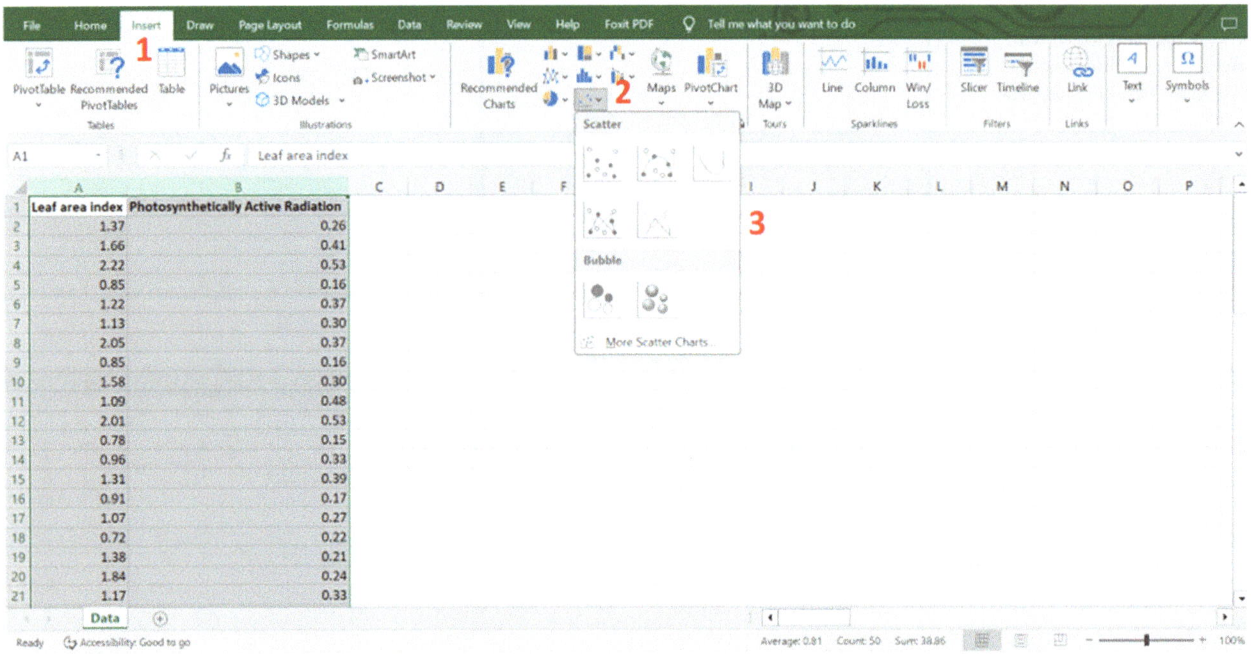

Fig. 20.11b Scatter plot will be drawn. The plot should be further processed, including adding the axes labels

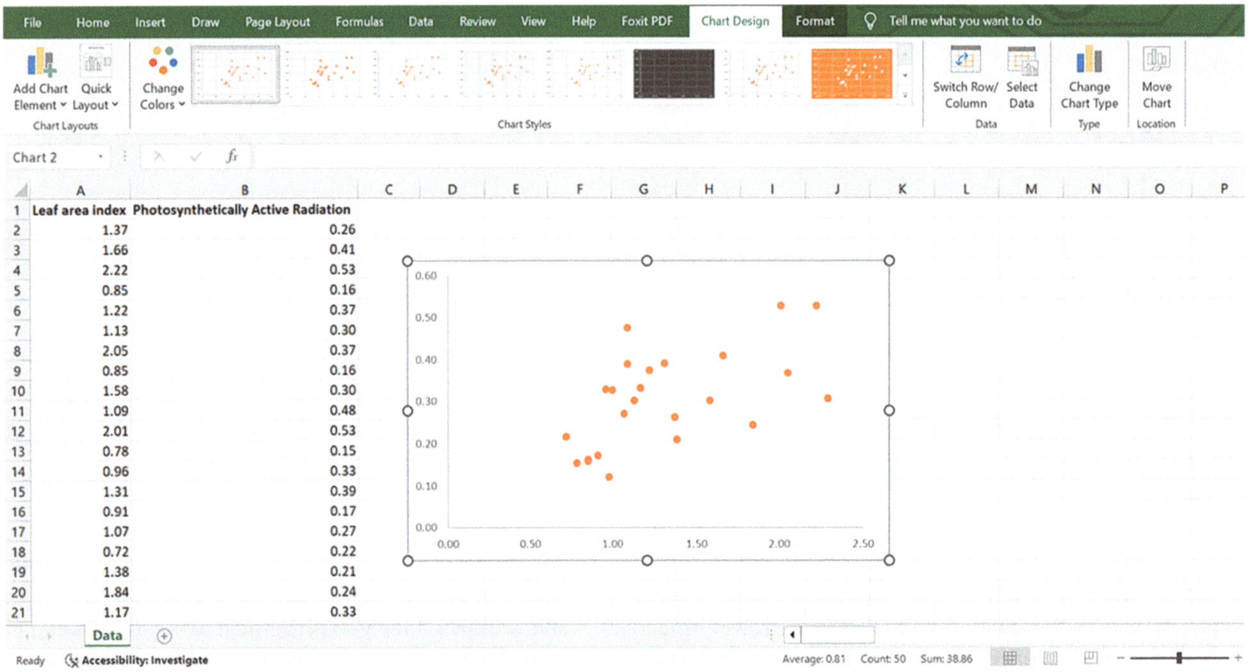

Fig. 20.11c Scatter plot (one variable is plotted on the x-axis (horizontal axis), and another is plotted on the y-axis (vertical axis). Each data point represents the value of both variables simultaneously. This data set shows high variability with a weak correlation

4. Label the x-axis and y-axis with descriptive names that indicate the represented variables.
5. If you have multiple groups or categories, use a legend to clarify their meaning.

How you Interpret the Scatter Plots

First, examine the plot and identify any patterns or trends among the data points. After this, determine the *directionality* of the relationship between the variables. This will help you determine if the data points tend to cluster in a specific direction, indicating a positive or negative correlation. Next, the strength of the *relationship* will be analyzed by looking at the concentration and dispersion of data points on the plot. Finally, identify any *outliers* or data points that significantly deviate from the overall pattern of the scatter plot. These anomalies can provide insights into further investigation regarding potential errors or unique observations.

To interpret scatter plots accurately, you need to contextualize the results. Consider the broader context of the data and variables under examination and scrutinize the underlying factors that may influence the observed relationships. While correlation may imply a relationship between variables, avoid inferring causality solely based on the observed relationship in a scatter plot.

Plotting a scatterplot lets us see whether a relationship exists between leaf area index and photosynthetically active radiation.

Step 1: Range your two continuous variables as follows (as shown in Fig. 20.11a) and select all of the data.
Step 2: Click Insert.
Step 3: Click the Scatter icon's dropdown button (2). A dialog box with various scatter options will appear.
Step 4: Click on the first option (3).

20.4 The Concept of Normal Distribution

The "normal distribution" is a foundational statistical concept that describes how data points are distributed in a data set (Whitley and Ball 2002; Lovric 2025). It features a symmetrical bell-shaped curve, with the majority of the data clustered around the mean and fewer data points appearing at greater distances from the mean in both directions. Many natural and social phenomena exhibit distributions that closely resemble the normal curve. While it serves as a theoretical model, we typically deal with sample data rather than entire populations. As such, while real-world data may not perfectly conform to the "normal curve," the properties of the normal distribution allow statisticians to make informed estimates about populations based on sample data.

In this distribution, the mean, median, and mode are all equal and positioned at the center of the curve (Fig. 20.12). In theory, the normal distribution curve extends infinitely in both directions, with scores potentially ranging to infinite magnitudes above and below the mean. As a result, the curve's tails never truly reach the baseline but instead continue indefinitely.

Statisticians have shown that within a "normal distribution" curve, approximately 34.13% of the area under the curve lies between the mean and one standard deviation above or below it.

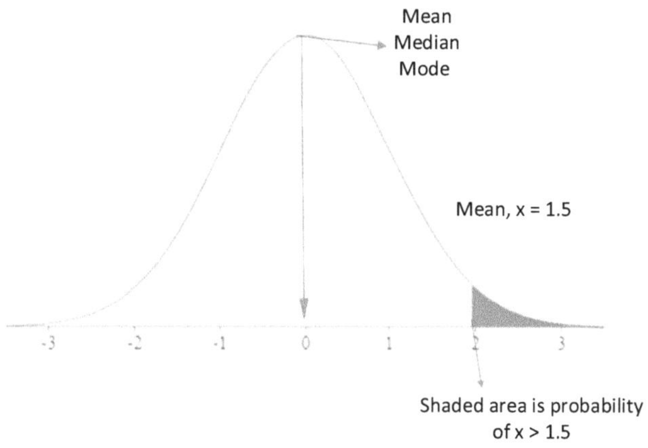

Fig. 20.12 Annotated presentation of the "normal" curve

This suggests that roughly 68.26% of the entire distribution will contain scores within one standard deviation of the mean. Likewise, about 95.44% of the total distribution lies within two standard deviations from the mean, covering a broader spectrum of scores. This pattern continues, with approximately 99.73% of the distribution within three standard deviations from the mean. These proportions illustrate the consistent dispersion of data around the mean in a normal distribution and are essential for understanding data variability within a data set. Since extreme values at either end of the distribution cannot be discounted, the entire 100% of the distribution is not accounted for. This can be a foundation for interpreting standard deviations (SDs).

20.5 Graphing and Preparing a Pivot Table

Pivot table is a powerful MS Excel tool for conducting descriptive or summary statistics like averages and counts. It provides a quick check of the data regarding missing values, completeness, and distribution. In addition to creating summary tables you can also create a pivot chart, allowing you to visualize data patterns and comparisons. You can generate a pivotal table using four steps (Figs. 20.13a, 20.13b, 20.13c, and 20.13d)

Step 1: Select all the values in the sheet. This gives the range of the elements to be included in the pivotal table. The values can be selected by clicking point 1.
Step 2: Click on the Insert menu. The pivot table will appear.
Step 3: Click on the pivot table and pick "From the Table/Range."
Step 4: A dialog box will appear.
Step 5: Click on "New worksheet," a new sheet will be created.

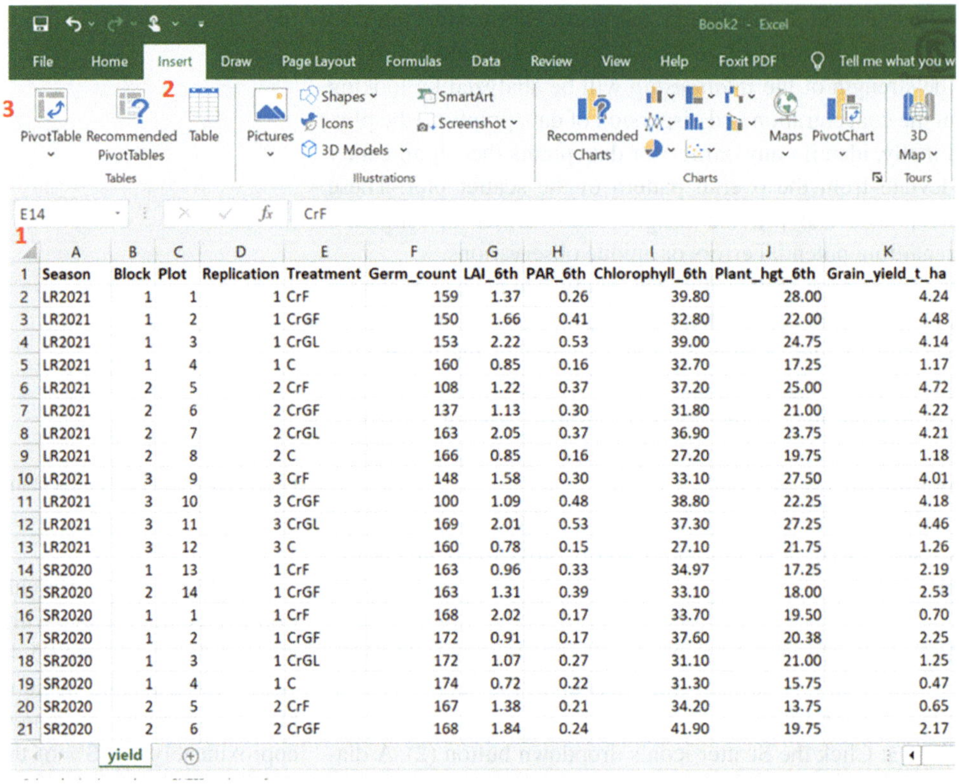

Fig. 20.13a Step 1 in generating pivot table

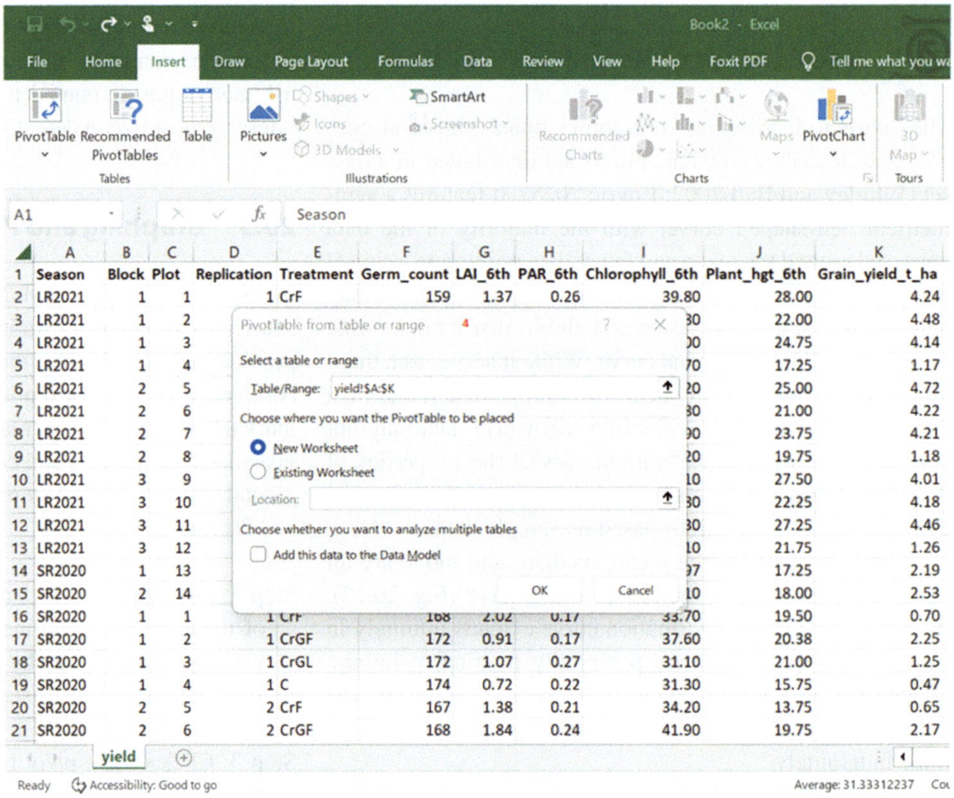

Fig. 20.13b Step 2 in generating pivot table

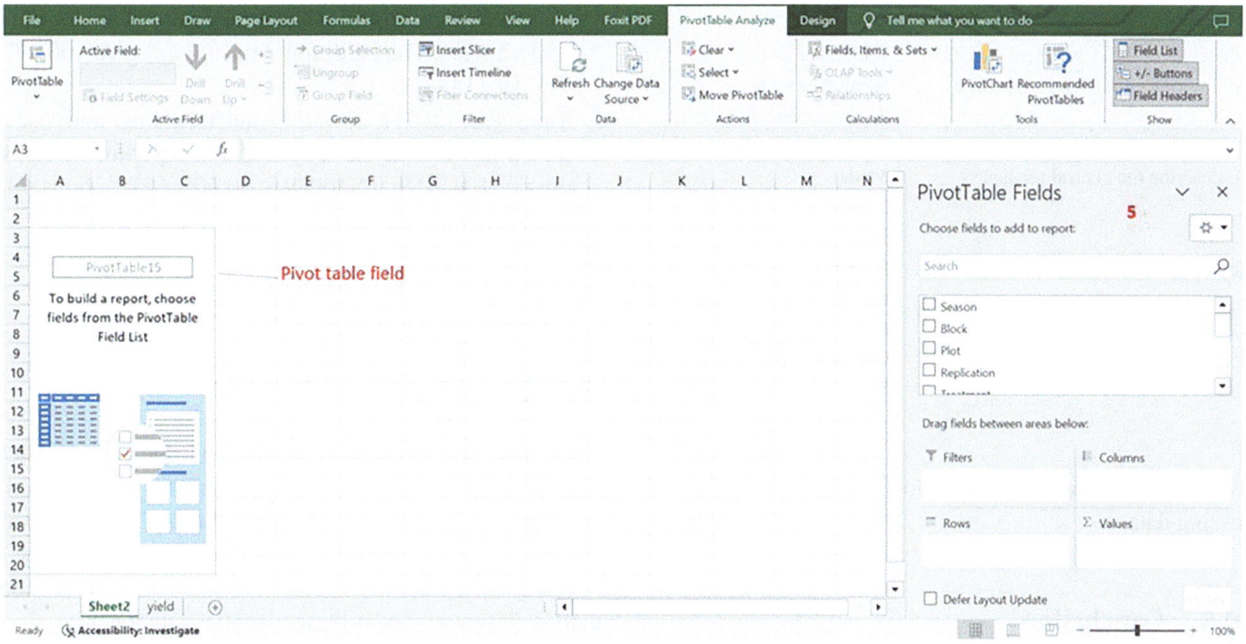

Fig. 20.13c Step 3 in generating pivot table

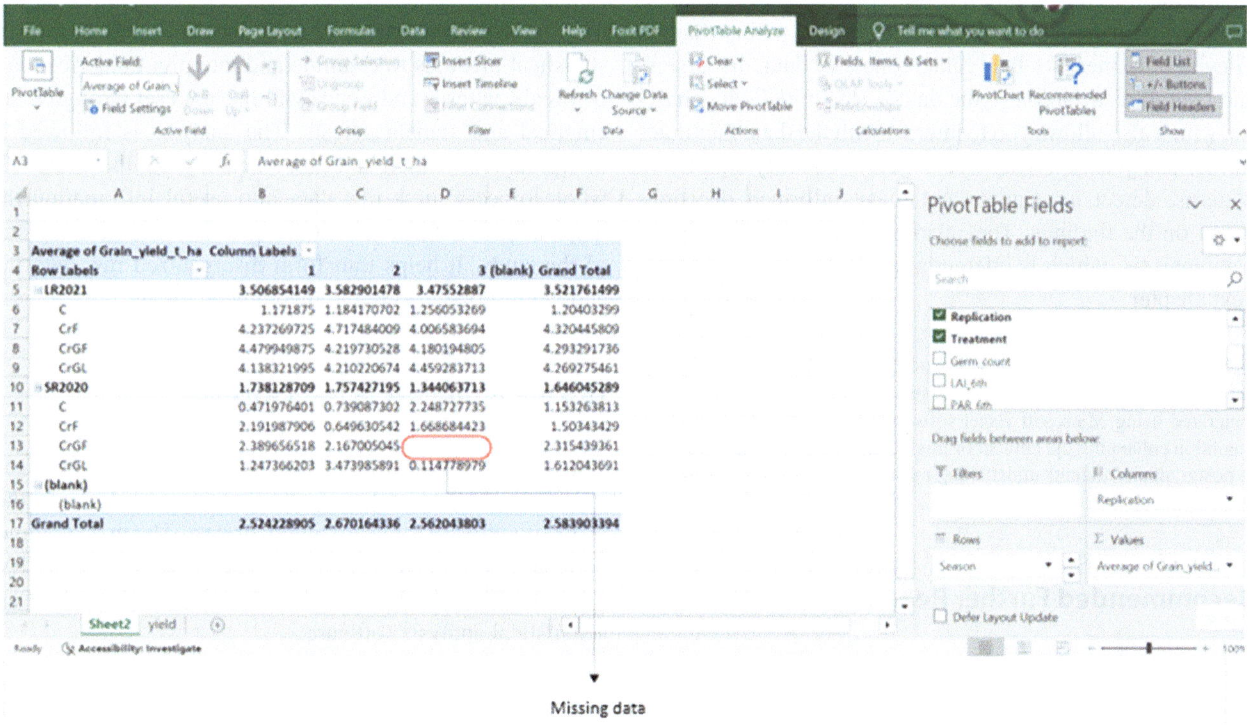

Fig. 20.13d Step 4 in generating pivot table

Field table fields provide a selection field to allow you to choose and add variables to the pivot table. Use columns and rows to organize your table (which parameters do you want in rows and which ones in columns). The "value" function allows the response variable to be inputted. The function provides several arithmetic operations under "functions settings."

Step 6: Create the pivot table by dragging the variables to the drag field. Let us pick our response variable to be grain yield, replication to be in columns, while season and treatment we drag to Rows.

The pivot table above shows the treatments per season, replications (3 replications), and the average grain yield. It

Table 20.8 Summary of descriptive analysis

Objective of the descriptive statistic	Which one to use
To describe counts and frequencies	Frequency table
	Relative frequency table
	Class interval table
To describe the central tendency	Mode
	Median
	Mean
To describe variability	Range
	Percentiles
	Box plots
	Coefficient of variation
	Scatter plots

also quickly shows that replication 3 of treatment CrGF is missing data.

20.6 Conclusion

Descriptive statistics are extremely important for exploring your data as they summarize the main features and reveal patterns and trends within the data set. Table 20.8 summarizes the main descriptive statistics discussed in this chapter. They mainly measure the "centeredness of data, dispersion, and overall distribution of the data". This initial data analysis, which also allows you to interact with the data, facilitates further analysis, allowing researchers to identify significant patterns, detect anomalies, and make informed decisions based on the findings. They also guide the next process of data analysis, which is inferential analysis, discussed in the next chapter.

Acknowledgments The screenshots used to demonstrate how to create data analysis matrices, scatter plots, pivot tables, and box plots were generated using Microsoft Excel software. This software was instrumental in enhancing the content of this book, which is primarily aimed at postgraduate students undertaking master's and doctoral research.

Recommended Further Reading

1. Bate ST, Clark RA (2014) The Design and Statistical Analysis of Animal Experiments. The Design and Statistical Analysis of Animal Experiments 1–310. https://doi.org/10.1017/CBO9781139344319
2. Dean A, Voss D, Draguljić D (2017) Design and Analysis of Experiments. https://doi.org/10.1007/978-3-319-52250-0
3. Hunt N, Tyrrell S (2002) Cleaning Dirty Data in Excel. Teach Stat 24:90–92. https://doi.org/10.1111/1467-9639.00096
4. Toutenburg H (2009) Statistical Analysis of Designed Experiments, Third Edition. https://doi.org/10.1007/978-1-4419-1148-3
5. Yang RC (2011) Towards understanding and use of mixed-model analysis of agricultural experiments. https://doi.org/104141/CJPS10049 90:605–627. https://doi.org/10.4141/CJPS10049

Key Points in This Chapter

Laying the Foundation for Data Analysis

Many students postpone data analysis due to overwhelm after fieldwork and unclear starting points. To begin, relax and input your data into a spreadsheet. Consult existing guidelines, your supervisor, or a biometrician for support. Revising basic statistics is also recommended, as most data analysis resources cover these essential components.

What Is Quantitative Analysis?

Quantitative data analysis is a systematic approach to using statistical methods to examine and interpret numerical data. It reveals patterns and trends within a data set through mathematical and statistical tests. The analysis is followed by interpretation, which attaches meaning to the data. The whole process turns raw data into useful information that helps us provide answers to the questions raised at the start of the study. It helps transform disorganized raw data into actionable conclusions and recommendations.

Use of Computers in Data Analysis

The computer is an invaluable tool in data handling, cleaning, management, and analysis. Students should learn data analysis software for analyzing their data. The most common computer programs include MS Excel, SPSS, GENSTAT, SAS, STATA, and R Studio. Learn at least a spreadsheet and a statistical analysis software.

Steps in Quantitative Data Analysis

Every experiment and study is unique because objectives differ. The following are the six generic steps in data analysis:

1. Preparing data for analysis.
2. Define the analysis objectives.
3. Selecting variables for use in the analysis.

4. Data exploration and descriptive analysis.
5. Confirmatory or inferential data analysis.
6. Interpretation and reporting.

Organizing the Data

Start by organizing data into a data analysis matrix that you populate with experimental or survey data. Survey data has to be coded appropriately. As you enter data, be keen on missing data and multiple responses.

Data Is a Pattern

"Data = pattern + residual." In analysis, our primary objective is to identify and interpret this underlying pattern, which represents the systematic effects of the treatments on the response variable. The various descriptive and inferential tools used aim to identify this pattern by separating the residuals from the pattern.

Descriptive Analysis

Descriptive statistics organize and summarize data. They allow a researcher to describe what exists in the data.

The key descriptive statistics include:

- Counts, frequencies, and percentages.
- Measures of central tendency (MoC).
- Measures of variability (MoT).

Measures of Central Tendency

The three most common measures of central tendency (MCT) used are the *Mode*, *Median*, and *Mean*. The type of measure of central tendency to use depends on the type of data you have: nominal, ordinal and interval, or ratio data. Mean is the most stable and is widely used.

Measures of Variability

The measure of variability concerns data variability indices. These mainly measure the dispersion of scores around the mean. The commonly used MoVs are the Range, Percentiles, Coefficient of variation, Box plots, and Standard deviation. Briefly:

- The range is the difference between the highest and the lowest value in a data set.
- Percentiles are values below which a certain percentage of data falls such as 25th, 50th, and 75th percentiles, which are known as quartiles.
- Coefficient of variation is a standardized measure of dispersion calculated as the ratio of the Standard deviation of the mean, expressed as a percentage.
- Box plots are a graphical representation that displays the distribution of a data set through five main statistics: minimum, first quartile, third quartile, and maximum. It also displays outliers.
- Standard Deviation quantifies the amount of variation or dispersion in a set of data values, showing how much individual points deviate from the mean.

Scatter Plots

A scatter plot is a pictorial representation that displays individual data points as dots showing the relationship between two quantitative variables. The pattern formed by the data points can reveal the nature and strength of the relationship between the variables, which further helps to understand the variability in the data. Scatter plots can also highlight any outliers or unusual data points. Scatter plots can easily be constructed using MS Excel.

The Concept of Normal Distribution

The "normal distribution" is a foundational statistical concept that describes how data points are distributed in a data set. It features a symmetrical shape, resembling a bell, with most data clustered around the mean and fewer data points appearing at greater distances from the mean in both directions. Many natural and social phenomena exhibit distributions that closely resemble the normal curve. Predictive data distribution is helpful for calculating probabilities and making inferences about the population. This is the basis of hypothesis testing.

Graphing and Conducting Pivot Table

Pivot table is a powerful MS Excel tool for data conducting descriptive or summary statistics like averages and counts. It provides a quick check of the data regarding missing values, completeness, and distribution.

Review Questions

1. Explain, in your own words, what quantitative data analysis entails. How does it transform "disorganized raw data into actionable conclusions and recommendations" in the

context of an agricultural experiment (e.g., a fertilizer trial)?
2. Differentiate between "analysis" (revealing patterns) and "interpretation" (attaching meaning).
3. You have just collected data on pest incidence and management practices from a survey of 200 farmers using a questionnaire. What would be your first three actions in "preparing data for analysis" (Step 1)?
4. Why is "descriptive statistics" considered foundational, allowing a researcher to "describe what actually exists in the data," before performing inferential analysis?.
5. For a variable like "weekly milk yield per cow" (in liters), explain what the range, standard deviation, and coefficient of variation would tell you about the variability within your dairy herd data. When would the coefficient of variation be particularly useful compared to standard deviation?
6. Suppose you have collected data on crop yields for three different maize varieties under varying soil conditions. How would creating box plots for each variety help you visually compare their yield distributions and understand their variability and presence of outliers?
7. You suspect a relationship between the amount of rainfall (mm) a farm receives and the incidence of a specific crop disease (number of affected plants). How would plotting a scatter plot help you visualize this relationship? What are you looking for in the scatter plot that would indicate a strong or weak, positive or negative relationship?
8. You have two sets of data on the weight of a local chicken breed at 12 weeks: Group A has a mean weight of 2.5 kg with a standard deviation of 0.2 kg. Group B has a mean weight of 2.5 kg with a standard deviation of 0.8 kg. What do these standard deviations tell you about the "dispersion of scores around the mean" in each group? Which group shows more consistent growth?

Chapter Exercise

1. Imagine you have just completed an experiment designed to evaluate the effectiveness of a new organic fertilizer on maize yields. Your experiment followed a Randomized Complete Block Design (RCBD) with four replications, each containing plots randomly assigned to different fertilizer treatments. Briefly describe the following:
   (a) What would "preparing data for analysis" involve for this data set.
   (b) Give an example of an "analysis objective" for this study.
   (c) What kind of "variables" would you "select for use in analysis" to meet your objective?
   (d) What would "data exploration and descriptive analysis" look like for this data set?
2. A box plot displays five main statistics and outliers. For a data set on the amount of water consumed by individual cattle in a feeding trial, draw a conceptual box plot and label where each of these five statistics would be represented. How would an outlier be visually identified?
3. Create a small hypothetical data set in Excel simulating data from an RCBD experiment on maize yields with three blocks and four fertilizer nitrogen treatments (0 N, 30 kg N ha^{-1}, 60 ha^{-1}, 90 kg N ha^{-1}). Each block has three replications for each treatment.
   (a) Beyond just average yield, how could a Pivot Table help you visually inspect the data for potential issues or initial insights regarding missing values, completeness, or distribution, particularly in an RCBD context?
   (b) Think about how you might use the "Count" function or filtering within the Pivot Table to check for these.

Answers to Review Questions

- Q1: Data analysis = Use of statistical methods to analyze numerical data.
  Raw data is disorganized; analysis is a process that transforms and reduces the raw data into simple, actionable conclusions. Discuss data from a fertilizer trial to fertilizer recommendations. Reference 20.1.2
- Q2. Descriptive analysis examines raw data to identify trends and relationships. It organizes data and uses statistics to figure out what the data contains in the form of patterns. Interpretation is about outcomes. It shows the trends, differences, relationships and attaches meaning. What is the implication of the data? What can be concluded from the data set? Reference 20.1.2
- Q3: Data analysis matrix: Reference: 20.2.2.3 and Figure 20.4
- Q4: Descriptive is exploratory. It allows the researcher to know what is in the data. Summarize complex data, and allow spotting of errors and outliers (Box plots). Avoid rushing to inferential statistics before carrying out descriptive statistics. Reference: 20.3
- Q5: Range, see Reference: 20.3.3
  Standard Deviation: See Reference 20.3.3
  Coefficient of variation: See reference 20.3.3
  CV provides a relative measure of variability. It is ideal for comparing dispersion across different variables or groups that have different scales, e.g., comparing milk yields from two dairy breeds.
- Q6: Box plots: Reference; 20.3.3
- Q7: Scatter plot; Reference 20.3.3.6
- Q8: A shows more consistent results: Reference: 20.3.3.4

References

Abt M, Leuders T, Loibl K, Reinhold F (2024) Developing initial notions of variability when learning about box plots. Math Think Learn. https://doi.org/10.1080/10986065.2024.2421412

Abt M, Loibl K, Leuders T et al (2025) Understanding student errors in comparing data sets with boxplots. Educ Stud Math 1–25. https://doi.org/10.1007/S10649-025-10387-Z/FIGURES/11

Al-Benna S, Al-Ajam Y, Way B, Steinstraesser L (2010) Descriptive and inferential statistical methods used in burns research. Burns 36:343–346. https://doi.org/10.1016/J.BURNS.2009.04.030

Arachchige CNPG, Prendergast LA, Staudte RG (2022) Robust analogs to the coefficient of variation. J Appl Stat 49:268–290. https://doi.org/10.1080/02664763.2020.1808599

Bate ST, Clark RA (2014) The design and statistical analysis of animal experiments 1–310. https://doi.org/10.1017/CBO9781139344319

Bauzon J, Murphy C, Wahi-Gururaj S (2021) Using macros in microsoft excel to facilitate cleaning of research data. J Community Hosp Intern Med Perspect 11:653–657. https://doi.org/10.1080/20009666.2021.1954282

Coe R, Huwe B, Schroth G, Sinclair F (2003) Designing experiments and analyzing data. World Agroforestry Centre (ICRAF)

Creswell JW (2014) Research and design qualitative, quantitative, and mixed methods approaches. SAGE Publication Inc

Dean A, Voss D, Draguljić D (2017) Design and Analysis of Experiments. https://doi.org/10.1007/978-3-319-52250-0

Fisher DJ, Zwahlen M, Egger M, Higgins JPT (2022) Meta-analysis in Stata. Systematic Reviews in Health Research: Meta-Analysis in Context: Third Edition 483–509. https://doi.org/10.1002/9781119099369.CH25

Fribourg HA, Carlisle RJ (1981) Teaching statistical methods with the statistical analysis system (SAS). J Agron Educ 10:19–23. https://doi.org/10.2134/JAE.1981.0019

Giorgi FM, Ceraolo C, Mercatelli D (2022) The R language: an engine for bioinformatics and data science. Life 12:648. https://doi.org/10.3390/LIFE12050648/S1

Gutierrez RG (2010) Stata. Wiley Interdiscip Rev Comput Stat 2:728–733. https://doi.org/10.1002/WICS.116

Hau J, Schapiro SJ (2021) Handbook of Laboratory Animal Science. Handbook of Laboratory Animal Science. https://doi.org/10.1201/9780429439964/HANDBOOK-LABORATORY-ANIMAL-SCIENCE-JANN-HAU-STEVEN-SCHAPIRO

Herzberg PA (1990) How SAS works: a comprehensive introduction to the SAS system (2nd ed.). New York, NY: Springer. https://doi.org/10.1007/978-1-4615-9670-7

Hunt N, Tyrrell S (2002) Cleaning dirty data in excel. Teach Stat 24:90–92. https://doi.org/10.1111/1467-9639.00096

Hyndman RJ, Fan Y (1996) Sample quantiles in statistical packages. Am Stat 50:361–365. https://doi.org/10.1080/00031305.1996.10473566

Ishaq AR (2025) Nature of data, data collection, and management. Epidemiology and Environmental Hygiene in Veterinary Public Health 215–228. https://doi.org/10.1002/9781394208180.CH17

Jankowski KRB, Flannelly KJ (2015) Measures of central tendency in chaplaincy, health care, and related research. J Health Care Chaplain 21:39–49. https://doi.org/10.1080/08854726.2014.989799

Kiboi M, Fliessbach A, Muriuki A, Ngetich F (2022) Data on the response of Zea Mays L. and soil moisture content to tillage and soil amendments in the sub-humid tropics. Data Brief 43:108381. https://doi.org/10.1016/J.DIB.2022.108381

Lai J, Lortie CJ, Muenchen RA et al (2019) Evaluating the popularity of R in ecology. Ecosphere 10:e02567. https://doi.org/10.1002/ECS2.2567

Langford E (2006) Quartiles in elementary statistics. J Stat Educ 14. https://doi.org/10.1080/10691898.2006.11910589

Lesik SA (2021) Applied statistical inference with MINITAB®, 2nd edn. CRC PRESS

Li H, Chen Z, Zhu W (2019) Variability: human nature and its impact on measurement and statistical analysis. J Sport Health Sci 8:527–531. https://doi.org/10.1016/J.JSHS.2019.06.002

Little TM, Hills FJ (1978) Agricultural experimentation: design and analysis:350

Lovric M (ed) (2025) International encyclopedia of statistical science. Springer, Berlin, Heidelberg

Maina AN, Lewis E, Kiarie EG (2023a) Egg production, egg quality, and fatty acids profiles in eggs and tissues in Lohmann LSL lite hens fed algal oils rich in docosahexaenoic acid (DHA)—Department of Animal Biosciences, vol 102. Borealis

Maina AN, Lewis E, Kiarie EG (2023b) Egg production, egg quality, and fatty acids profiles in eggs and tissues in Lohmann LSL lite hens fed algal oils rich in docosahexaenoic acid (DHA). Poult Sci 102:102921. https://doi.org/10.1016/J.PSJ.2023.102921

Maina AN, Schulze H, Kiarie EG (2024) Response of broiler breeder pullets when fed hydrolyzed whole yeast from placement to 22 wk of age. Poult Sci 103:103383. https://doi.org/10.1016/J.PSJ.2023.103383

Maina AN, Schulze H, Kiarie EG (2025a) Supplemental data for: Effects of lifetime feeding of hydrolyzed yeast to broiler breeders on egg production, quality, and hatchling attributes—Department of Animal Biosciences. https://borealisdata.ca/dataset.xhtml?persistentId=doi:10.5683/SP3/ZOBJ1B

Maina AN, Schulze H, Kiarie EG (2025b) Effects of lifetime feeding of hydrolyzed yeast to broiler breeders on egg production, quality, and hatchling attributes. Poult Sci 104826. https://doi.org/10.1016/J.PSJ.2025.104826

Manikandan S (2011a) Frequency distribution J Pharmacol. Pharmacother 2:54–56. https://doi.org/10.4103/0976-500X.77120/ASSET/535FB422-B0B4-4676-B5EA-6770B8675C5B/ASSETS/0976-500X.77120.FP.PNG

Manikandan S (2011b) Measures of central tendency: median and mode. J Pharmacol Pharmacother 2:214. https://doi.org/10.4103/0976-500X.83300

Manikandan S (2011c) Measures of dispersion. J Pharmacol Pharmacother 2:315–316. https://doi.org/10.4103/0976-500X.85931

McCormick K, Salcedo J (2017) SPSS statistics for data analysis and visualization. Wiley

Mead R, Curnow RN, Hasted AM (2017) Statistical methods in agriculture and experimental biology: third edition. Statistical methods in agriculture and experimental biology: third edition 1–472. https://doi.org/10.1201/9780203738559/STATISTICAL-METHODS-AGRICULTURE-EXPERIMENTAL-BIOLOGY-ROGER-MEAD/RIGHTS-AND-PERMISSIONS

Morris T (1999) Experimental Design and Analysis in Animal Sciences. https://doi.org/10.1079/9780851993492.0000

Mugenda A, Mugenda G (2019) Research methods: qualitative and mixed methods approach. 100–102

Nwaogwugwu UC, Udoh UH (2014) The use of SPSS and GenStat computer software in data analysis in agriculture. J Agric Food Environ 1:97–109

Onwuegbuzie AJ, Da Ros D, Ryan JM (1997) The components of statistics anxiety: a phenomenological study. Focus Learn Probl Math 19:11–35

Osborne JW, Overbay A (2004) The power of outliers (and why researchers should ALWAYS check for them). Pract Assess Res Eval 9. https://doi.org/10.7275/QF69-7K43

Ospina R, Marmolejo-Ramos F (2019) Performance of some estimators of relative variability. Front Appl Math Stat 5:421406. https://doi.org/10.3389/FAMS.2019.00043/XML/NLM

Palaniswamy U (2005) Handbook of statistics for teaching and research in plant and crop science. https://doi.org/10.1201/9781482277814

Patel BK, Muir-Leresche K, Coe R, Hainsworth SD (2004) Green book a guide to effective graduate research in African agriculture, environment and rural development

Ralston K, Gorton V, MacInnes J et al (2020) Anxious women or complacent men? Anxiety of statistics in a sample of UK sociology undergraduates. Int J Soc Res Methodol 24:79–91. https://doi.org/10.1080/13645579.2020.1761186

Stoudt S, Vásquez VN, Martinez CC (2021) Principles for data analysis workflows. PLoS Comput Biol 17:e1008770. https://doi.org/10.1371/JOURNAL.PCBI.1008770

Toutenburg H (2009) Statistical analysis of designed experiments. Third Edition. https://doi.org/10.1007/978-1-4419-1148-3

Tsementzis AK, Maina AN, De Cloet CA, et al (2025) Comparative impact of partial replacement of soybean meal with select specialty protein ingredients in broiler chicken starter feeding program on growth, organ, intestinal, plasma, and litter attributes to 49 days of age. Poult Sci 105408. https://doi.org/10.1016/J.PSJ.2025.105408

Vargas M, Combs E, Alvarado G et al (2013) META: a suite of SAS programs to analyze multienvironment breeding trials. Agron J 105:11–19. https://doi.org/10.2134/AGRONJ2012.0016

Whitley E, Ball J (2002) Statistics review 2: samples and populations. Crit Care 6:143–148. https://doi.org/10.1186/CC1473/COMMENTS

Williamson DF, Parker RA, Kendrick JS (2008) The box plot: a simple visual method to interpret data. 110:916–921. https://doi.org/10.7326/0003-4819-110-11-916

Wilson S, MacLean R (2009) Research methods and data analysis for psychology 1:640

Yang RC (2011) Towards understanding and use of mixed-model analysis of agricultural experiments. 90:605–627. https://doi.org/10.4141/CJPS10049

Quantitative Data Analysis II (Inferential Analysis)

21

Abstract

This chapter focuses on inferential data analysis, which builds on the foundational work of descriptive analysis covered in Chap. 20. While descriptive analysis organizes and summarizes data, inferential analysis goes further to make predictions and generalizations about the population based on sample data. In inferential analysis, researchers test hypotheses using samples and make inferences about the population. The main inferential analysis includes means comparison, determining relationships and making predictions. In comparing means, we use a t-test or analysis of variance (ANOVA) depending on the number of means to be compared. T-tests are used for comparing means of two groups, while ANOVA is used for comparing more than two means. T-tests are of three categories: One sample t-test, independent t-test, and correlated t-test. In ANOVA too, there are various types depending on the number of factors (number of independent variables). For instance, we have "one-way ANOVA" with one independent variable, "Two-way ANOVA" with two independent variables, and "Three-way ANOVA" with three independent variables. In testing relationships, the scale of measurement is considered. If dealing with two categorical variables, the Chi-square statistic is used. If the variables are ranked, Spearman rank Correlation is employed, while for two quantitative variables, Pearson product-Moment is used. In making predictions and regression, the main ones include using simple linear and multiple regression models.

Keywords

ANOVA tests · Binary regression analysis · Chi-square test of association · Hypothesis testing · Null hypothesis · One-way ANOVA · Parametric and non-parametric test · Pearson product-moment Correlation · Probit regression · Regression analysis · Significance level · Significance test · Spearman rank correlation · T-tests

Learning Outcomes

After reading this chapter, you should be able to:

1. Demonstrate an understanding of inferential data analysis.
2. Differentiate between "parametric and non-parametric" statistical tests.
3. Describe the steps in "hypothesis testing."
4. Distinguish between "type I" and "type I" errors.
5. Apply statistical tests to test differences between the two groups.
6. Apply statistical tests to determine differences between more than two groups.
7. Explain the meaning of ANOVA and the assumptions in carrying out ANOVA.
8. Describe components of a typical ANOVA Table.
9. Differentiate between data analysis in "One-Way ANOVA" and "Two-way ANOVA."
10. Carry out ANOVA for Completely Randomized design and Randomized complete block design (RCBD).
11. Explain the meaning of means separation.
12. Conduct ANOVA for a factorial and split-plot experiment.
13. Carry out a Chi-square test of association for categorical variables.
14. Demonstrate the use of Spearman Rank correlation.
15. Distinguish between causation and correlation.
16. Apply simple linear regression and multiple regression to evaluate relationships.
17. Appreciate the use of more advanced regression models such as the logistic regression model.

J. Njeri Mugwe, S. Runo, *Research Methodology in Agricultural Sciences*, https://doi.org/10.1007/978-981-95-1892-0_21

21.1 Fundamentals of Inferential Statistics and Hypothesis Testing

21.1.1 Why Inferential Statistics?

In Chap. 19, we discussed the main descriptive analysis. This chapter focuses on inferential data analysis, which involves applying statistical techniques to test hypotheses, estimate population parameters, and assess relationships between variables based on simple observations. By leveraging these techniques, researchers can infer patterns, relationships, and trends within a larger population, which is crucial for making informed decisions and advancing scientific knowledge. Before discussing the various inferential analyses, we will cover an essential concept of "random samples and population."

21.1.2 Concept of Sampling Errors and Confidence Intervals

In empirical research, numerical data introduce complexity and inherent error (Guolo 2008; Buonaccorsi 2010; Brakenhoff et al. 2018). Ideally, researchers would gather data from entire populations to minimize this error, but this is often impractical, leading to reliance on samples. The reliance on samples introduces sampling error, which is the discrepancy between a sample and the population. For instance, in a scenario with ten participants, calculating the mean of milk yields would reveal variations among individual scores, creating a sampling distribution. To assess the error in this distribution compared to the overall mean, researchers calculate the standard error of the mean. This is done by dividing the standard deviation by the square root of the sample size (Kim 2013; Goyal 2018). This measure helps researchers understand the inherent variability within their sample, a crucial aspect of statistical analysis.

To account for this variability, most researchers use a 95% confidence interval (CI), meaning that researchers expect their results to be accurate 95% of the time, meaning inaccurate 5% (Algina et al. 2005; Badenes-Ribera et al. 2018). In some cases, such as some medical studies, where greater precision is required, confidence intervals may be increased to 99.9% for greater certainty. Nevertheless, the 95% confidence interval remains the standard in statistically based research, ensuring a balance between confidence and practicality and allowing the researchers to make reliable inferences while acknowledging the inherent limitations of sampling.

To illustrate this concept, consider a scenario where a drug company investigates whether a drug is lethal. With a 95% confidence rate, the drug would be expected to result in fatality for 1 in every 20 individuals. Conversely, at a 99.9% confidence level, only one out of every 10,000 individuals taking the drug would be anticipated to experience a fatal outcome. Despite the higher safety allowance provided by the 99.9% confidence interval, the 95% confidence interval remains the standard for most statistically driven research endeavors. This is because the 95% CI strikes a balance between precision and practicability, making it widely accepted in the scientific community.

21.1.3 Non-parametric and Parametric Tests

Non-parametric tests, sometimes referred to as "distribution-free tests," are employed when data do not follow the normal distribution or when the assumptions required for parametric tests cannot be met (Garrocho-Rangel et al. 2024). Examples include the "Chi-square test of association," the "Kruskal-Wallis test," the "Mann-Whitney U Test," and the Spearman rank correlation coefficient (Table 21.1) (Garrocho-Rangel et al. 2024). They are particularly valuable when the underlying populations are unknown, though this limits the extrapolation of findings to other populations (See Box 21.1). Despite being less statistically powerful than parametric tests, non-parametric tests offer flexibility and reliability, especially when dealing with ordinal or nominal data commonly derived from surveys and participatory rural appraisal research approaches.

When utilizing non-parametric tests, a researcher should consider three key points: (1) understanding the specific conditions under which each test should be applied, (2) knowing how the tests are computed, and (3) interpreting the results accurately (Schober and Vetter 2020). Therefore, researchers must carefully select non-parametric tests, especially when aiming to detect small effects. In cases where study results are critical, conducting multiple related non-parametric tests

Table 21.1 A comparison of parametric and non-parametric test

Quality	Non-parametric	Parametric
Data	Nominal/ordinal	Ratio/interval
Assumption	No specific distribution assumption Offer more flexibility	Specific assumption: Normal distribution Less flexible
Sample size	Small (<30)	Large; normally >30
Estimates	Less powerful statistically: Less precise	More powerful statically More precise
Examples	Chi-square "Mann Whitney U test" "Kruskal Wallis H test" "Spearman rank correlations" (data are ranked during analysis)	T-tests ANOVA Pearson correlation

Box 21.1: The Mann-Whitney U Test and Kruskal-Wallis H test

The Mann-Whitney U Test assesses whether a statistically significant distinction exists between two independent data groups that do not adhere to a normal distribution (Vickers 2005). In this test, observations from both groups are ranked, and a U statistic is computed to compare the rankings. It is a substitute for the t-test (on the parametric side) when the data fail to fulfill the assumptions of normality or when dealing with small sample sizes.

The Kruskal-Wallis H test evaluates whether significant differences exist among two or more groups based on a continuous or ordinal outcome variable (Harris et al. 2008). It serves as the nonparametric equivalent to the one-way ANOVA, extending the Mann-Whitney U test to allow for comparison across multiple independent groups. Each data point is assigned a rank, and these ranks are utilized to assess whether the groups exhibit similar distributions. In essence, the Kruskal-Wallis test evaluates whether the groups share the same median.

may be necessary to identify the most suitable approach for analysis.

Parametric tests are designed to analyze data based on the assumption that the underlying distribution follows a specific probability distribution, mostly normal. These tests rely on parameters such as the mean and standard deviation to make inferences about the population. Parametric tests often offer more precise estimates than non-parametric tests (Table 21.1).

Determining the normality of the data distribution is crucial to ensure the validity of parametric tests. This can be achieved by using statistical tests such as the Shapiro-Wilk test or visual inspection of histograms and Q-Q plots with SPSS or R software (Thode 2002; Ghasemi and Zahediasl 2012; Avram and Mărușteri 2022). Residual plots are plotted to provide a way of checking the normality and the assumption of equal variance, especially with experimental data. Parametric tests are mainly applicable to interval and ratio data. Examples of parametric tests are shown in Table 21.1.

21.1.4 Choice of Inferential Statistics

The selection of inferential statistics hinges on various factors, including the nature of the data, the research question, the measurement scale, the sample size, and the assumptions underlying the statistical tests. Tests like the "Chi-square" or "Fisher Exact test" are used if the data are categorical. If continuous, tests such as t-tests, ANOVA, and regression analysis are suitable. In some cases, researchers may need to consider the possibility of data transformation to meet the assumptions of parametric tests. Overall, choosing inferential statistics depends on carefully considering these factors to ensure the most appropriate analysis for the research question. Table 21.2 gives the main inferential statistics based on the objective of the analysis and the type of data.

Table 21.2 Key inferential statistics (parametric and non-parametric tests) for different objectives

Objective of analysis	Statistical test	Non-parametric	Example
Compare a sample mean ($\bar{x}$) with a known population fixed mean, (u) (targeted mean, historical mean)	One sample Z test	Parametric	Confirming if there is adequate evidence to back the claim that a sample of 75 farmers randomly selected had a mean sorghum yield of 2 t ha^{-1}
Compare one group sample mean ($\bar{x}$) to a known population value (fixed, targeted value)	One sample t-test	Parametric	Comparing the mean of maize yields in region A (3 t ha^{-1}) with primary data collected in another region B.
Compare two unpaired groups (independent)	Unpaired t-test	Parametric	Comparing means of soil pH of soil samples taken in different fields (lfarmers in region A and farmers in region B)
Compare two paired groups	Paired t-test	Parametric	Comparing means of soil samples taken in the same field before and after an experiment
Compare means of three or more unmatched groups	One-way ANOVA	Parametric	One-factor experiment
Compare means of three or more matched groups	Two-way ANOVA, repeated measures ANOVA	Parametric	Two-factor experiment
Relationship between two numerical (interval/ratio) variables	Pearson product-moment (PPM) correlation	Parametric	Age of farmers and number of trees from farmers' data

(continued)

Table 21.2 (continued)

Objective of analysis	Statistical test	Non-parametric	Example
Compare two independent samples in cases where the outcome is not normally distributed and the sample sizes are small.	Mann-Whitney U test	Non-parametric	Comparing the average milk yields of six dairy cows fed a supplement diet with those of six cows fed a normal diet.
Compares means among multiple independent groups	Kruskal-Wallis test	Non-parametric	Assessing if attitudes (ordinal) among farmers toward technology adoption varied based on education level.
Association between categorical variables	Chi-square Fishers exact test	Non-parametric	Test association between gender and education level of household head
Relationship between two sets of ranked data	Spearman rank correlation	Non-parametric	Rank of food crops by males and females
Relationship between one independent variable and one dependent variable	Simple linear regression	Parametric	Nitrogen and maize yields
Relationship between more than two independent variables and one dependent variable	Multiple linear regression	Parametric	Nitrogen and phosphorus (independent variables) and maize yields (dependent variable)
Relationship between more than two independent variables and one dependent variable (binary variable)	Binary regression model (Probit or logistic)	Parametric	Adoption decision (yes/no) and socio-economic variables of farmers
Relationship between more than two independent variables and one dependent variable (quantitative)	Tobit or OLS	Parametric	Adoption rate (numerical) and socioeconomic variables of farmers

21.2 Hypothesis Testing

21.2.1 Writing and Testing Hypothesis

To remind ourselves, we have many types of hypotheses (Chap. 3): research hypothesis, statistical hypothesis, and null and alternative hypotheses. Hypothesis testing is a method used in inferential statistics to determine whether there is enough evidence to reject the null hypothesis. This process involves analyzing sample data to determine the likelihood of observing the results if the null hypothesis were true. If the probability of observing the results under the null hypothesis is sufficiently low (typically below a predetermined threshold, known as the "significance level," researchers will reject the null hypothesis in favor of the alternative hypothesis (Ruxton et al. 2015). If the probability is not low enough, the researcher will fail to reject the null hypothesis, suggesting that the observed results are not significantly different from what would be expected by chance alone.

Depending on the university, you may be required to have both a null hypothesis and an alternative hypothesis. You should always remember that regardless of the types of hypotheses you write, the testing process always focuses on the null hypothesis (Whitley and Ball 2002; Pugh and Molinaro 2016; McNulty 2022). Additionally, as long as you have stated that you have a hypothesis in your study, you must test it, as this is part of the parameters examined in a thesis. Examiners will often check your understanding of the hypothesis and seek to know how you did the testing.

In a journal article, the scenario might be different, as you may not be required to explicitly state or show that the hypothesis was tested. It is often implied within the study's methodology and results. Depending on the researcher's preferences, the research objective, hypothesis, or research question is typically placed at the end of the introduction section of a journal article (Chap. 25). The placement ensures that the readers understand the focus of the study before reading the methods and findings. Even if the hypothesis is strictly not demanded or discussed by journal publications, it remains crucial to clearly articulate it. The hypothesis serves as a guiding framework for analyzing the data and presentation of the results, ultimately informing the study's conclusion.

21.2.1.1 Testing for Significance

The process of determining whether observed differences or relationships are due to chance is called "significance testing" (Whitley and Ball 2002; McNulty 2022). In significance testing, the goal is to test whether our null hypothesis holds true. Typically, it involves analyzing quantitative data to determine the likelihood that the null hypothesis is correct. It is important to recognize that there is always a possibility of error in significance testing, as it involves probabilities and uncertainties (McNulty 2022).

After you have collected data from a sample, the initial step in significance testing is set a significance level. In other words, a researcher must determine how large a relationship or difference must be before it is considered significant. Just like we discussed in confidence intervals, researchers prefer to be at least 95% confident. A 95% confidence in significance testing indicates that a researcher could expect to see a difference by accident as a result of chance about 5 times out of 100.

In significance testing, we often refer to the "Probability level" when discussing how confident we are. To determine the probability level, we subtract the percentile number (95) from 1 or 1–0.95 = 0.05. The probability level is sometimes called a p-value or probability value or even referred to by the Greek letter alpha (α) (Pugh and Molinaro 2016).

We should note that sampling error is one of the major reasons why replication of studies is important. If a study's findings are accurate 95% of the time, then there is a 5% chance that the results obtained in a study happen because of error. For this reason, scientists often replicate studies or research multiple times to ensure the results are consistent. Often, researchers find contrasting results even when the research process, measures, and participants are identical. Instead of conducting multiple repeated trials, researchers may adopt a metanalysis that may be used to combine several different studies looking at the scale phenomenon to achieve stronger support for the existence of a relationship or difference or the lack of a relationship.

21.2.2 Steps in Testing Hypothesis

Regardless of the type of hypothesis being considered, the process of testing the hypothesis is done in five basic steps (Fig. 21.1).

21.2.2.1 Step 1: Set Analytical Question

You start by determining the analysis question from your data set "what do you want to find out?" The analytical question should be precise and narrow enough to be addressed through statistics; for instance, instead of asking "does fertilizer affect crop yields?" you refine it to "Is there a difference in crop yields between crops treated with fertilizer A and those treated with fertilizer B.

Depending on the data you have, you could have the following questions:

- If comparing two means, you could ask, "Is there a statistically significant difference between the means of maize yields in fertilizer treatment and maize yields in manure treatment?"

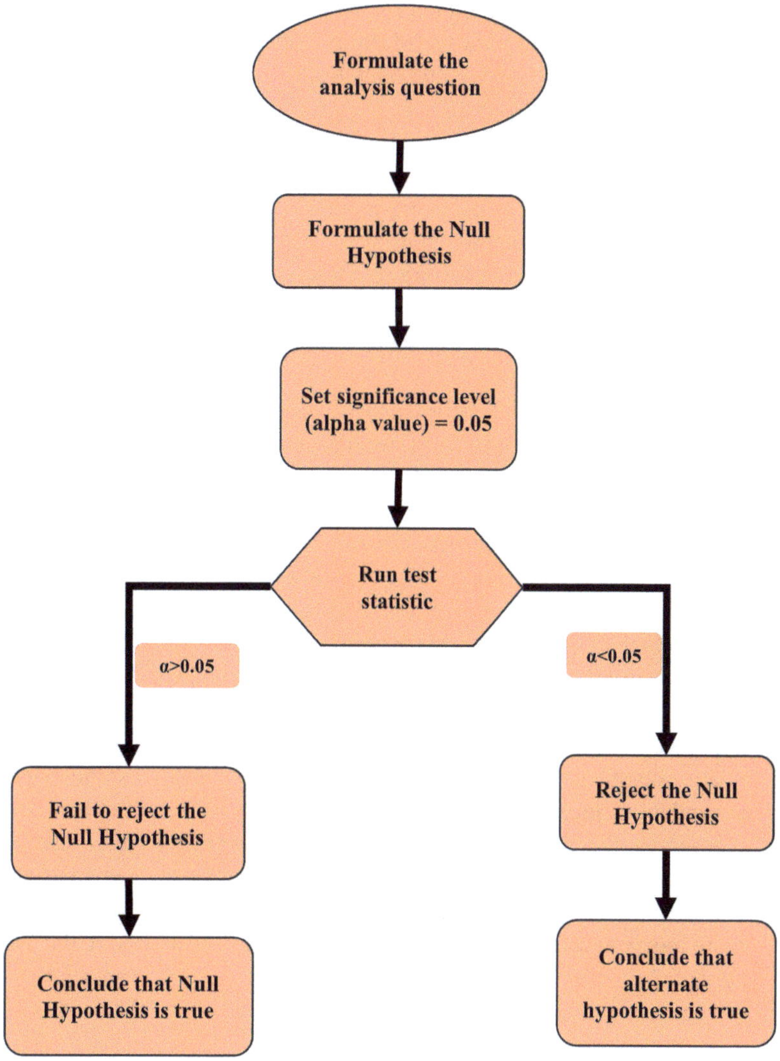

Fig. 21.1 Steps in conducting inferential statistics (hypothesis testing)

- If comparing a pre-test and a post-test, you might ask "Is there a statistically significant difference between the average scores on the pretest and the post-test?"
- If testing for association between two categorical variables, e.g., gender and education, might ask, "Is there a statistically significant relationship between gender and education?"
- If testing for a correlation between two numerical variables like height and crop yields, you might ask, "Is there a statistically significant correlation between height and crop yields?"
- When evaluating the predictive ability of an independent variable on a dependent variable, the key question is: "Is there a significant relationship between the independent variable (X) and the dependent variable (Y)?"
- If assessing the predictive power of multiple independent variables, you might ask "Is there a significant predictive relationship between two or more independent variables (A, B, C, etc.) and dependent variable Y?"

21.2.2.2 Step 2: Test the Null Hypothesis

This involves conducting the appropriate statistical test to determine whether the observed data provide sufficient evidence to reject the null hypothesis (McNulty 2022).

We could have the following null hypothesis:

- If comparing two means, you could have a null hypothesis (Ho): "There is no statistically significant difference between the means of maize yields in fertilizer treatment and maize yields in manure treatment?"
- If comparing a pre-test and a post-test, you might have a null hypothesis (Ho): "There is no statistically significant difference between the average scores on the pretest and the post-test?"
- If testing for association between two categorical variables, e.g., gender and education, we have a null hypothesis (Ho): There is no statistically significant relationship between gender and education"
- If testing for a correlation between two numerical variables like height and crop yields, you could have, a null hypothesis (H_0): "There is no statistically significant correlation between height and crop yields."
- When evaluating the predictive ability of an independent variable on a dependent variable, you can state the null hypothesis (H_0) as: "There is no statistically significant relationship between the independent variable (X) and the dependent variable (Y)."
- If assessing the predictive power of multiple independent variables, you could have a null hypothesis (Ho): "There is no significant predictive relationship between two or more independent variables (A, B, C, etc.) and dependent variable Y."

21.2.2.3 Stage 3: Set the Significance Level

After formulating the null hypothesis, the next step is to establish the significance (α), which is the threshold of rejecting for deciding whether to reject the null hypothesis (Pugh and Molinaro 2016). The significance level represents the probability of rejecting the null hypothesis when it is actually true, known as a type I error. Commonly used levels are 0.05, 0.01, and 0.1, corresponding to 5%, 1%, and 10% chances of committing a type II error, respectively. The choice of significance level should align with the research context and the acceptable risk of making an error.

For instance, if you set alpha = 0.05, you are allowing a 5% chance of incorrectly rejecting the null hypothesis. This level is critical in determining the threshold for statistical significance in your test results. In on-farm studies, for example, a survey, we can go up to 10%.

21.2.2.4 Stage 4: Run the Test Statistic

After setting the significance level, you perform the test statistic. This involves applying the chosen analytical method for your data and computing the test statistic. For example:

- For comparing two means, such as maize yields in fertilizer versus manure treatment, you could use a t-test and calculate the statistic.
- For comparing pre-test and post-test scores, you can calculate the t-statistic of a paired t-test.
- For testing the association between two categorical variables, you could perform Chi-square tests and obtain Chi-square statistic.
- For testing for a correlation between two numerical variables like height and crop yields, you can compute the correlation coefficient associated test statistic.
- For evaluating the predictive ability of one independent variable on a dependent variable, you can use simple linear regression.
- For evaluating the predictive relationship of one or more independents, you could use multiple linear regression analysis with the associated test statistic.

21.2.2.5 Step 5: Draw a Conclusion

After calculating the test statistic, compare it to the critical value associated with the significance level. If the test statistic exceeds the critical value or if the p-value is less than the significance level α, you reject the null hypothesis. If the test statistic does not exceed the critical value (or the p-value is less than α), you fail to reject the null hypothesis. However, when using computer or statistical software to generate the p value, we make direct comparison. We compare the p value and the alpha (0.05) rather than referring to the critical values.

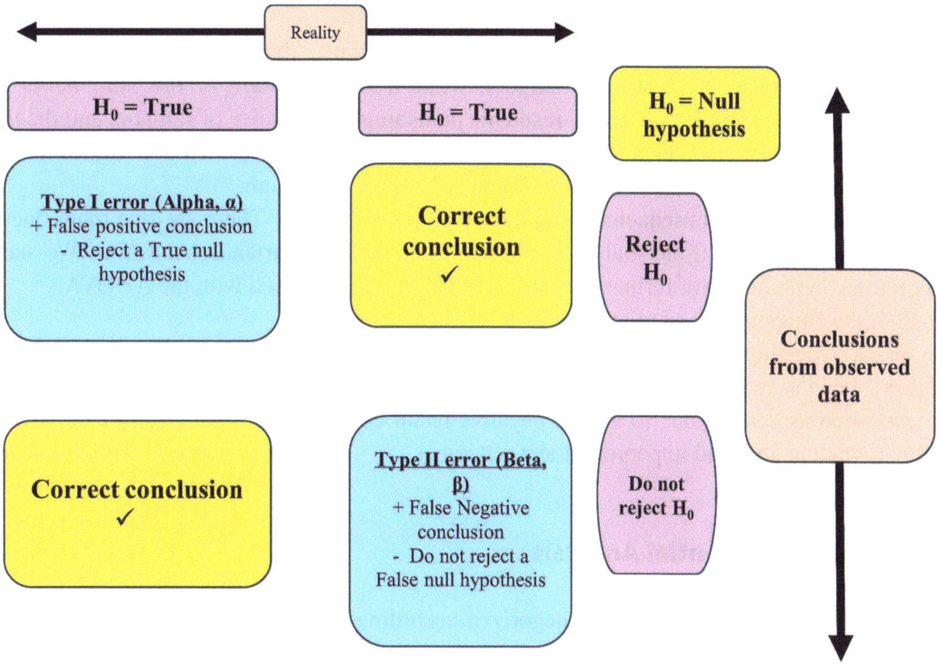

Fig. 21.2 Type I and Type II errors

21.2.3 Statistical Significance Versus Practical Significance

Suppose in your results of hypothesis testing, you interpret that four means of maize yield from four treatments (T1, T2, T3, and T4; fertilizers) are not significantly different. But when you look closely at the means, you realize that the difference between T1 and T2 is 1.3 t ha^{-1}. We know that a difference of 1.3 t ha^{-1} is 1300 kg of maize grains; this is practically significant to a farmer who would fill 14 bags of 90 kgeach. This makes a big difference. You could ask why the difference was not detected. The answer is that hypothesis testing is mathematical and is affected by sample size and variability in the data. Whenever we have a large enough random sample, almost any result would turn out to be statistically significant. The more variable a data set is, the more difficult in detecting statistical significance. In this case, the data could have been variable due to extraneous variables; thus, hypothesis testing failed to detect a significant difference.

Suppose you present these data to a stakeholders' forum, with farmers present, and say that a difference of 14 bags (90 kg *14) yield is insignificant. Interpreting in layman's language, you tell them that this difference does not make a difference. This will sound unbelievable and ridiculous. We have to be careful when using and interpreting statistical significance. Statistical significance and practical significance are two distinct concepts. While statistical significance assesses whether an observed effect or difference is likely due to chance, practical significance or biological significance refers to the actual size and importance of the effect in the real-world context (Lovell 2013; Anderson 2019). Evaluating practical significance requires subject-matter experts to judge whether the effect is substantial enough to be meaningful in practical terms. Subject matter specialists are able to contextualize the statistical results within the practical realities of the subject area.

You can also get results showing a statistical difference when, in fact, the difference is not of practical or agricultural significance. This can happen when the sample size is large (Lantz 2013; Zhu and Spanos 2023). Therefore, we should carefully interpret the results of significance testing to avoid misleading stakeholders and practitioners.

21.2.4 Avoiding Type I and II Errors

Type I error, often referred to as a false positive, occurs when we reject the null hypothesis and incorrectly infer that there are significant differences between groups or variables, despite no actual difference being present (Banerjee et al. 2009; Kim 2015; Forstmeier et al. 2017) (Fig. 21.2). It means that we wrongly assert that a relationship or effect exists when it does not. Type I errors are especially serious because they can lead us to believe that an intervention, treatment, or relationship is effective or significant, even though it is not. This can result in misguided decisions or actions based on flawed conclusions. Therefore, researchers must be careful and use appropriate statistical methods to minimize the likelihood of making these errors.

The majority of researchers believe that a Type I error is a more serious mistake to make than a Type II error. They would rather miss a result (Type II) than say that there is a significant difference when, in reality, it does not exist (Type I error) (Jackson 2016). However, depending on the context, Type I errors could have severe consequences in certain areas like medical testing or criminal justice (Peterson et al. 2015). For example, a false positive medical diagnosis could lead to unnecessary medical procedures or surgery, while a false conviction in a criminal trial could result in punishing an innocent person (Peterson et al. 2015). On the other hand, in other contexts such as scientific research, Type I errors might not be as severe if the consequences of a false-positive result are relatively minor or if the study is exploratory and meant to generate hypotheses for further investigation. On the other hand, if the outcomes are significant, such as medical screening or safety testing, Type II errors could have serious implications. For instance, failing to detect a disease in its early stages (such as cancer) due to a false negative result could delay treatment and lead to poorer patient outcomes.

21.3 Main Inferential Analysis

The main inferential statistics are categorized according to what we want to test. They fall into the following three categories:

1. Testing for differences.
2. Assessing relationships.
3. Assessing regression and predictions.

The discussion in this section will follow these categories.

21.3.1 Testing for Differences

Researchers commonly seek to assess whether the means of two or more groups are different. This answers questions such as: "Does sorghum yield vary significantly across six treatments? Is there a significant difference in milk yield from cattle fed solely on dairy meal versus those fed on other supplements?"

The process of analyzing for differences follows similar steps discussed earlier on hypothesis testing; it starts by asking an analysis question, formulating the null hypothesis, and proceeding with carrying out the test statistic.

Test statistics for assessing differences are typically categorized based on whether they measure differences between *two groups or among more than two groups* and whether the groups are independent or related. Independent groups are entirely distinct, such as farmers' income from avocado sales in different regions or the yield of beans from different treatment applications. On the other hand, related groups consist of the same set of subjects observed at different points in time or under different conditions, like pretest and post-test measurements of the same group of farmers or soil pH levels before and after treatment application.

Decisions regarding the appropriate test statistic hinge on these distinctions, along with two additional factors: *The nature of the data being analyzed*, which influences the choice of statistic, and the *number of groups under comparison*, which determines the assessment of differences among them.

There are two main ones: the T-tests and the analysis of variance (ANOVA). We start with the T-tests and then proceed with the ANOVA.

21.3.1.1 The T-Test for Measuring Difference in Two Groups

The t-test determines whether there are significant differences between the two means of two samples or groups at a specified level of probability (Jankowski et al. 2018; Thukral et al. 2023). We discuss the following three types of T-tests:

1. One-sample t-test.
2. Independent t-test.
3. Paired t-tests or correlated groups.

The One-Sample T-Test

The one-sample t-test formula determines whether the mean of a single variable significantly deviates from a given constant (Thukral et al. 2023). We present two cases: Yields (biophysical) and age of farmers (social).

Example 1

Suppose a researcher wishes to ascertain whether a newly developed soybean variety produces significantly more grain per hectare than the average yield of existing soybeans in the region, which is known as 3.0 t ha^{-1}. To do so, the researcher collects yield data from a sample of fields in which the new soybean was grown and finds out that the average yield is 3.3 t ha^{-1}.

The researcher can use the one-sample T-test to compare the mean 3.3 t ha^{-1} to the known population mean (3.0 t ha^{-1}). Suppose the calculated t-statistic indicates a significant difference at a predetermined level of significance ($p < 0.05$). In that case, the researcher can conclude that the new soybean variety yields significantly more than the regional average.

The analysis question will be: Is there a statistically significant difference in yield between the new soybean variety and the known average yield of 3.0 t ha^{-1}? The null hypothesis (H0) states that there is no significant difference in yield between the new soybean and the expected average yield of 3.0 t ha^{-1}.

If, after computing the one-sample t-test, the researcher obtains a p-value less than the set alpha level of 0.05, the conclusion is that the new soybean variety yields significantly higher than the existing soybean. This information can prove valuable for farmers who are considering whether or not to adopt the new variety for higher yields.

Example 2

Suppose we want to evaluate whether the mean age of farmers in Region A (known average already is different from mean age of farmers in Region B). We employ a one-sample t-test.

The analysis question is: Is there a significant difference between a specified constant (e.g., mean age of farmers in region A (45 years) and the age of farmers in region B determined by a researcher?

The null hypothesis is that there is no statistically significant difference between the age of farmers in Region A (45 years) and the researcher-determined average of farmers in Region B (49 years). Suppose we set the level of significance, alpha at 0.05.

If we get a p-value <0.05, we will reject the null hypothesis. We therefore conclude that the difference between 45 and 49 years is true and that the age of farmers in Region A is significantly different from the mean age of farmers in Region B.

Independent Two Samples T-Test

The independent samples t-test procedure determines whether there is a significant difference between the means of the two independent groups (Thukral et al. 2023). It is a widely used statistic. Researchers collect data from two separate (independent) groups or conditions, such as treatment and control groups, male and female participants, trained and untrained participants, or experimental and control conditions. This test assumes that the two groups are unrelated and the samples are drawn from populations that are normally distributed with equal variances.

To illustrate the use of this test, we present examples of agronomy research and present other examples in Box 21.2: livestock feeding experiments, veterinary research, and farmer surveys. Here we are capturing examples across diverse disciplines in agriculture.

Box 21.2: Examples to Illustrate Use of Independent T-Test Across Disciplines

Variety testing: Imagine a researcher testing two genetically modified strains of cotton: one with a resistance gene (strain X) and another without (strain Y), are planted in separate plots. The plots are then exposed to a common pathogen known to infect cotton. Disease severity or incidence is recorded for each plot after a designated period.

Analysis question in this experiment: Is there a statistically significant difference in disease resistance between the two genetically modified strains of cotton?

The null hypothesis: There is no statistically significant difference in disease resistance between Strain X and Strain Y.

(continued)

Box 21.2 (continued)

An independent t-test is then used to compare the disease resistance levels between Strain X and Strain Y.

Feeding experiments: Researcher conducts a feeding trial where one group of cattle is fed with Diet A, while another group is fed with Diet B. After a specified period, the weight gain of each individual animal is recorded. An independent t-test is then employed to compare the average weight gain between the two groups.

Analysis question: Is there a significant difference in average weight gain between two groups of cattle fed with different diets? Method:

The null hypothesis is that there is no statistically significant difference in average weight gain between cattle fed with Diet A and Diet B.

Testing efficacy of drugs:

In a livestock research study, a researcher evaluated the efficacy of two antibiotics: Antibiotic A and Antibiotic B, in treating a specific bacterial infection in cattle. One group of infected cattle was administered antibiotic A while another group was given antibiotic B. After a defined treatment period, the researcher measured the reduction in bacterial count in the infected tissues.

Analysis Question: Is there a statistically significant difference in the mean reduction of bacterial count between cattle treated with Antibiotic A and those treated with Antibiotic B?

The null hypothesis: There is no statistically significant difference in the mean reduction of bacterial count between cattle treated with Antibiotic A and those treated with Antibiotic B.

In agronomy research, suppose a researcher conducts field experiments to compare the effectiveness of fertilizers A and B on maize yields. One group of plots is treated with Fertilizer A, while another group receives Fertilizer B. After the harvest, the yield in kilograms per hectare from each plot is measured.

Analysis question: Is there a statistically significant difference in the average yield of maize between the two fertilizer treatments?

The null hypothesis: There is no statistically significant difference in the mean yield between Fertilizer A and Fertilizer B.

To test this hypothesis, the researcher performs an independent sample t-test to compare the mean yield between the two fertilizer treatments.

Table 21.3 Repeated measurements of soil pH in an experiment evaluating the long-term effect of applying different soil amendments

	Year			
Treatment*	2010	2011	2012	2013
A	5.2	5.2	5.3	5.4
B	5.4	5.5	5.6	5.6
C	5.3	5.4	5.3	5.5
D	5.1	5.0	5.1	5.1
E	5.4	5.1	5.0	5.1

*For pairwise comparison between two specific years within the same treatment, use *correlated T-test*. To compare soil pH across all years use a *repeated measures ANOVA*

If the resulting p-value associated with the t statistic is less than the predetermined significance level (e.g., $\alpha = 0.05$), we reject the null hypothesis. This indicates a statistically significant difference in the average yield of maize between the two fertilizer treatments. Conversely, if the p-value is greater than α, (e.g., $p = 0.356$) we fail to reject the null hypothesis, suggesting that there is no significant difference in yield between Fertilizer A and Fertilizer B. Any differences observed are thus due to chance.

Correlated Groups T-Test

The correlated groups' t-test, also known as the paired t-test, compares the means of two related groups. This test is commonly employed in research where "participants" or subjects are measured on the same outcome variable at "two different time points" or under two different conditions. Examples include pre-test/post-test designs, where participants are measured before and after receiving an intervention, or within-subjects designs, where participants are exposed to different conditions in a repeated-measures experiment. Examples include;

1. *Measuring changes in soil pH after application of poultry manure*: Compare soil samples collected from the same locations before (pre) and after the application of the soil amendment (post) (Tables 21.3 and 21.4). In this case correlated (paired) t test is used because the samples represent two related measurements taken from the same experimental units. Note that, if you are comparing more than two measurements, you use a repeated measures ANOVA or mixed models framework instead of a paired t-test. These measures allow you to assess the effect of the different soil amendments over multiple points or conditions while considering the correlations between the repeated measures on the same plots or samples.
2. *Evaluating the impact of introducing a new maize variety among farmers*: measure crop yields from farmers before (pre-treatment) and after implementing the new maize variety (post-treatment).

Table 21.4 Repeated measurements of soil pH evaluating changes following application of liming material

	Week			
Treatment*	Two weeks after applying lime	Four weeks after applying lime	Six weeks after applying lime	Eight weeks after applying lime
A				
B				
C				
D				
E				

*For pairwise comparison between two specific weeks; use Correlated T-test. To compare soil pH across all weeks simultaneously; use *Repeated measures ANOVA*

3. *Measuring the effectiveness of a new vaccination protocol on livestock health:* Measure blood samples that may be collected from individual animals before and after administering a new vaccine
4. Assessing the effectiveness of a newly developed insecticide on tomatoes: Measure the number of aphids before (pre-treatment) and after the application of pesticide on the same plants (post-treatment)

21.3.1.2 Analysis of Variance

The Concept of ANOVA

Analysis of variance (ANOVA) is a widely used method in experiments to determine differences between more than two groups of means. Developed by Ronald Fisher in the 1920s, ANOVA separates the total variation in the data into components attributed to the independent variables and to the error or residual variation (Parolini 2015; Visscher and Goddard 2019). The approach helps in assessing whether the mean difference between the groups is statistically significant.

There are *three important assumptions* of ANOVA.

1. The sample was randomly selected from the population, and the subjects or treatments were randomly assigned to the groups (randomness and independence of errors).
2. The variance in each group is the same (homogeneity of variance).
3. The sample came from a population in which the dependent variable is normally distributed (Normality) Fig. 21.3.

Testing for Differences Using ANOVA

Let us use the example of sorghum yield from treatment 1 (T1) with manure and treatment 2 without manure (T2), Treatment with manure plus fertilizer (T3), and Treatment with fertilizer (T4). The analysis question is: Is there a sig-

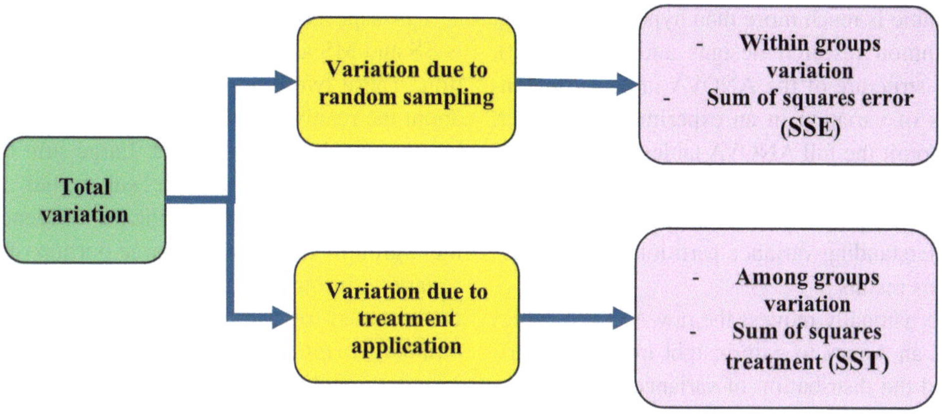

Fig. 21.3 ANOVA separates total variation into two: due to treatments and due to random sampling

nificant difference between sorghum yields in T1, T2, T3, and T4?

The null hypothesis would be that there is no statistically significant difference between sorghum yields in T1, T2, T3, and T4. We use alpha = 0.05.

Ho: μT1 = μT2 = μT3 = μT4

After running the ANOVA test, if we get a p-value of less than 0.05, we will reject the null hypothesis and conclude that there is a statistically significant difference between sorghum yields in T1, T2, T3, and T4. If, on the other hand, we get a p-value of more than 0.05, say 0.356, we conclude that there is no statistically significant difference between sorghum yields in T1, T2, T3, and T4. We conclude that any differences observed are thus due to chance or error as briefly shown in Box 21.3.

Box 21.3: Null Hypothesis and Alternative Hypothesis

Null Hypothesis in ANOVA

- It is stated as: H_0: $\mu_1 = \mu_2 = \mu_3 \ldots\ldots \mu_k$

All population means are equal

No effect due to treatments.

Any differences observed is due to chance or error

Alternative Hypothesis

- It is stated as: HA: $\mu_1 \neq \mu_2 = \mu_3 \ldots\ldots \mu_k$.

At least one population mean is different from the others

There is a treatment effect or response

Treatments (independent variable) affect the dependent variable

Comparison of Groups of Means Using ANOVA

Between-group variance refers to the deviation of the group means from what is called the grand mean (the mean score of all individuals in all the groups) (J. Salkind 2007; Johnstone and Manly 2014). If there is no real difference between the groups, then their means will be similar and close to the overall mean. In this case, the variance between the groups will be small (because the groups do not vary much from each other). As the difference between the groups increases, so too will the between-groups variance.

The question answered by ANOVA is, what is the probability that the variation among a group of sample means has occurred due to randomly selecting the sample from a common population? Are there differences among the groups due to the treatments given or by chance?

There are *three important assumptions* of ANOVA.

1. The sample was randomly selected from the population, and the subjects or treatments were randomly assigned to the groups.
2. The variance in each group is the same (homogeneity of variance).
3. The sample came from a population where the dependent variable is normally distributed.

ANOVA Table

We began this section on quantitative data analysis by highlighting the importance of utilizing computers for statistical analysis because of their efficiency and practicality in handling large datasets and complex calculations. However, we also recognize the significance of traditional statistical methods taught in statistics and biometry courses, such as manual calculation of ANOVA and interpretation of ANOVA tables. These traditional methods enhance understanding as the theory is translated into practice.

We have, therefore, included ANOVA tables in our discussion to bridge theory and practice. ANOVA tables are a valuable tool for summarizing analysis results, providing students with a clear visual representation of key statistical measures such as sources of variation, degrees of freedom, sum of squares, mean squares, F-statistic, and p-values. By including ANOVA tables, we aim to help students better understand and interpret the results of ANOVA analyses, strengthening their statistical skills and knowledge.

The ANOVA table is much more than hypothesis testing. In Chap. 9 (Common research designs and plot layout), we presented the structure of the ANOVA table to demonstrate the sources of variation in an experiment. However, researchers often omit the full ANOVA tables in their thesis or journal papers. This does not mean the details in the ANOVA table are unimportant. Those details are extremely important for understanding variance partitioning and interpreting the analysis results.

Supervisors occasionally request the raw ANOVA tables to be included as an Annex to gain insight into the experimental design and the distribution of variance components. This allows for a more thorough evaluation of the experiments' robustness and reliability of the findings.

To illustrate this, let us examine a typical ANOVA table (Table 21.5) using the example of a randomized complete block design (RCBD) with three replications.

ANOVA table contains the following components;

- *Source of Variation: Identifies the different sources of variability in the experiment, such as treatments, blocks, and residual error.*
- *Degrees of Freedom (DF): Indicates the degrees of freedom associated with each source of variation. Degrees of freedom reflect the number of independent pieces of information available for estimating a parameter or conducting a statistical test (Gomez and Gomez* 1984*).*
- *Sum of Squares (SS): Displays the sum of squares associated with each source of variation. The sum of squares measures the variability explained by each factor or source of variation in the experiment.*
- *Mean Square (MS): The mean square is obtained by dividing the sum of squares by the corresponding degrees of freedom. It represents the average amount of variability explained by each factor or source of variation.*
- *F-Ratio (F): The F-ratio is calculated by dividing the mean square for each factor by the mean square for the residual error. It provides a measure of the ratio of treatment effects to residual variability and is used to test the statistical significance of each factor.*
- *P-value: The p-value associated with each F-ratio indicates the probability of observing an F-value as extreme as the one calculated, assuming that the null hypothesis (no effect of the factor) is true. A low p-value (<0.05) indicates that the factor has a significant effect on the response variable.*

One aspect of interest for supervisors in the ANOVA table is SS and MS associated with each possible source of variation. These two components provide valuable information about the relative importance of different factors in explaining the variability in the data. Large sum of squares attributed to treatments indicate substantial impacts on the response variable, with significant differences among treatment groups. Conversely, a large portion of the total sum of squares attributed to the residual component suggests unaccounted-for variability, representing *noise or unexplained variation* in the data or noise. GENSTAT, for instance, can identify this noise by flagging units with large residuals. Such large residual components can mask treatment effects, potentially yielding insignificant results despite significant treatment effects.

We, therefore, recommend that you always check the ANOVA table details and analyze the SS and MS associated with each source of variation. This is important because if most of the variability is in the "Noise," you are likely not to detect statistical significance when there is a difference. Therefore, you will be committing a "Type II" error, where you incorrectly accept the null hypothesis (no difference) when there is a difference. Considering that Data = Pattern + residuals, large residuals make identifying the pattern difficult. If you find large residuals, you need to recheck the data thoroughly to identify sources of error and variability. Essentially, ANOVA does more than hypothesis testing.

One-Way ANOVA

One-way ANOVA is used to test for differences among two or more independent groups (Ntumi 2021). Typically, however, it is used to test for differences among at least three groups since a t-test can cover the two-group case. One-way ANOVA analyses variance where groups are compared on one variable (factor) that is measured at either nominal or ordinal.

Example 1

Using the milk yield data of a sample of 11 farmers, we take two more samples from Region C, and Region D (Table 21.6). This makes four samples with four sample means.

The analysis question is: Is there a statistically significant difference among the four regions?

We set the null hypothesis that there is no statistically significant difference in milk sales among the four regions. In

Table 21.5 Typical ANOVA table for randomized complete block design

Source of variation	Degrees of freedom (Df)	Sum of squares SS	Mean sum of squares MS	F	P
Blocks	DF (blocks)	SS (blocks)	MS (blocks)	MS (blocks)	P (blocks)
Treatments	DF (treatments)	SS (treatments)	MS (treatments)	MS (treatments)	P (treatments)
Error	DF (residual error)	SS (residuals—Error)	MS (residuals—Error)		
Total	DF (Total)	SS (Total)			

other words, the milk sales from the four regions are statistically similar i.e., Ho: The four means are significantly similar, and Ha: At least one of the four means differs from the others.

Suppose the ANOVA results gave a p-value less than 0.05. This indicates a statistically significant difference among the means of milk sales. The p-value means that the probability of getting the observed differences in means due to random chance alone is very low, suggesting that regions have a true effect on the response variable. As a result, we reject the null hypothesis and conclude that there are significant differences among the groups. The ANOVA table for one way ANOVA will have treatments and residuals (errors) as the source of variation (Table 21.7).

Let us look at another example of one-way ANOVA with data collected from an experiment testing the response of tomato yields to four treatments; TR1, TR2, TR3, and TR4 (Table 21.8).

Objective: To evaluate the effect of four fertilizer treatments on tomato yields.

Design: Randomized complete block replicated three times.

Our analysis question would be: Is there a statistically significant difference among the four fertilizer treatments?

The null hypothesis is that there is a statistically significant difference among the four fertilizer treatments.

Suppose we get the above table of means and p-values more than 0.05. We would interpret it as that:

- The impact of treatments on tomato yield was similar among the four treatments,
- The differences recorded are probably due to random chance or background "noise,"
- The trends in tomato yields observed in TR1, TR2, TR3, and TR4 are unlikely to be replicated in future experiments with the same treatments, and
- The Null hypothesis is accepted (Fail to reject the null hypothesis).

Suppose we get the "opposite" p-value that is less than 0.05. We would interpret it as that:

- The impact of treatment on tomato yield differed among the four treatments,
- The differences recorded are true,
- The apparent trends in treatment yields in TR1, TR2, TR3, and TR4 would probably, be replicated in subsequent experiments involving similar treatment, and
- The Null hypothesis is Rejected.

For illustration, we will use R software to analyze the effect of four treatments (T1, T2, T3, and T4) on maize yield. We will consider Completely Randomized Design (CRD) (Fig. 21.4a) and RCBD (Fig. 21.4b). We hypothesize that the treatment means were not significantly different (Ho).

Sources of variation under CRD are partitioned into two: Treatment and residuals (error) (Forthofer et al. 2006). After running the analysis, we find that the calculated p-value is less than the tabulated p-value (0.05). This means that treatment means are significantly different. Consequently, we reject our hypothesis (null; Ho). Looking at the SS values, we see that treatments have a larger value (21.518) than the residuals (0.377). This means that treatments explained the variations in the yield to a larger extent.

```
Source of variation Df Sum Sq Mean Sq F value    Pr(>F)
Treatment            3 21.518   7.173   152.2 2.15e-07
Residuals            8  0.377   0.047
```

However, for RCBD, the sources of variation are partitioned into three; block, treatments, and residuals (error) (Salinas Ruíz et al. 2024). Similarly, we find that the calculated p-value is less than the tabulated p-value. Hence, the treatment means are significantly different. We thus, reject our null hypothesis. Still, we see that treatment SS (21.518) is larger than SS for block and the error. We see that treatments have larger values than the residuals. This means that

Table 21.6 Milk sales from farmers in Regions A, B, C, and D

Farmer no.	Sample no.	Region A	Region B	Region C	Region D
1	X1	16	12	10	13
2	X2	10	9	23	14
3	X3	8	25	9	9
4	X4	20	32	36	18
5	X5	16	25	34	17
6	X6	21	18	26	26
7	X7	20	16	22	29
8	X8	7	25	14	11
9	X9	11	10	23	14
10	X10	20	12	17	16
11	X11	15	18	36	12
Total		**164**	**202**	**250**	**179**
Mean		**14.9**	**18.4**	**22.7**	**16.3**

Table 21.7 ANOVA Table for one-way ANOVA

Source of variation	Degrees of freedom (Df)	Sum of squares SS	Mean sum of squares MSS	F	P
Treatments	DF (regions)	SS (regions)	MS (regions)	MS (regions)	P (regions)
Error	DF (error)	SS (error)	MS (error)		
Total	DF (Total)	SS (Total)			

treatments explained the variations in the yield to a larger extent than the block and error.

Source of variation	Df	Sum Sq	Mean Sq	F value	Pr(>F)
Block	2	0.023	0.012	0.199	0.825
Treatment	3	21.518	7.173	121.711	9.19e-06
Residuals	6	0.354	0.059		

Once we ascertain significant differences, we conduct a post hoc test also known as means separation. Moreover, you may run the ANOVA text and fail to find significant differences (the p-value calculated is larger than the p-value tabulated). Let us use RCBD to illustrate where treatments have no significant effect on yield. We retain our earlier null hypothesis.

Source of variation	Df	Sum Sq	Mean Sq	F value	Pr(>F)
Block	2	0.0234	0.01170	0.199	0.825
Treatment	3	0.0234	0.00781	0.132	0.937
Residuals	6	0.3536	0.05893		

Table 21.8 Data collected from an experiment testing four treatments

Treatment	Tomato yield (t ha^{-1})
TR 1	38
TR 2	34
TR 3	41
TR 4	45

From the ANOVA table, we see that the p-value calculated is larger (0.825) than the tabulated value (p = 0.05). This infers that the treatments had no significant effect on yield, thus, we fail to reject our null hypothesis. Note the SS values in this table and compare them with those in the two ANOVA tables above. It is also important to note that a post hoc test is not performed in cases where there are no significant differences in means.

Two-Way Analysis of Variance

Whereas one-way analysis of variance (ANOVA) tests measure significant effects of one factor only, two-way analysis of variance (ANOVA) tests (also called two-factor analysis of variance) measure the effects of two factors simultaneously (S.Lewis-Beck et al. 2004; Forthofer et al. 2006). For example, an experiment might be defined by two parameters, such as treatment and time point. One-way ANOVA tests would be able to assess only the treatment effect or the time effect. Two-way ANOVA, on the other hand, would not only be able to assess both time and treatment in the same test but also, whether there is an interaction between the parameters. We use two-way analyses of variance in factorial and split plot experiments (Chap. 8).

Suppose we have the following experiment in a randomized complete block replicated four times (Fig. 21.5).

Objective: To evaluate the effect of fertilizer and variety on tomato yields: The two factors are fertilizer and variety.

The analysis questions, in this case, are more than one: (1) Is there a statistically significant difference in yields due to

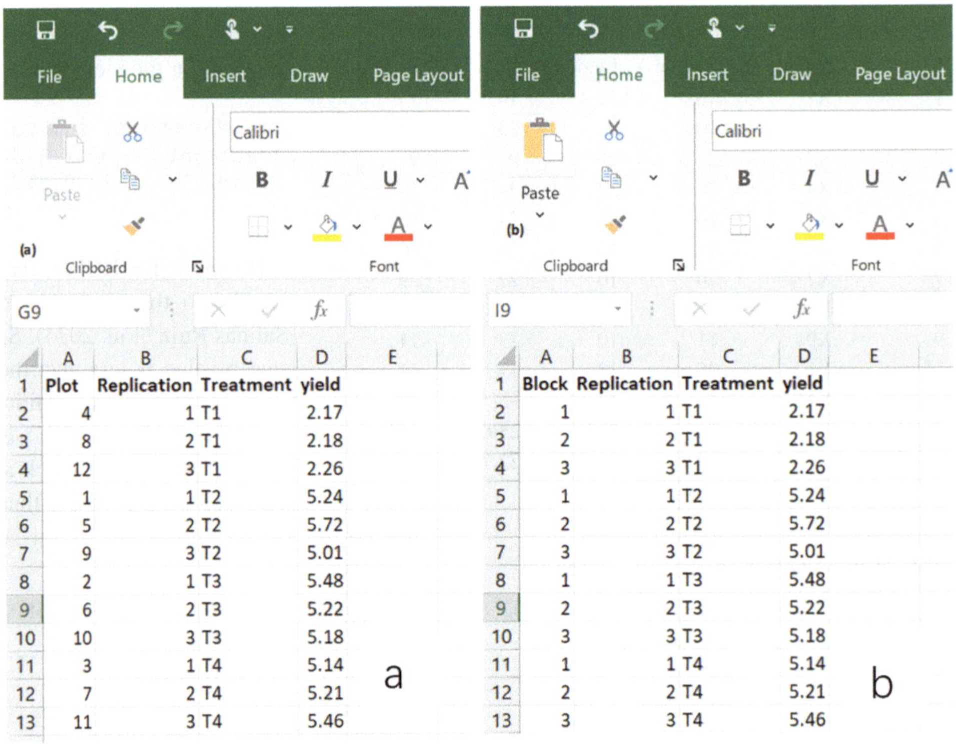

Fig. 21.4 (**a** and **b**) Screenshot of Excel worksheet showing data for CRD and RCBD, respectively

	A	B	C	D
1	Block	Fertilizer_level	Variety	yield
2	1	20	V1	53.36
3	2	20	V1	61.00
4	3	20	V1	67.77
5	4	20	V1	61.88
6	1	20	V2	62.23
7	2	20	V2	70.43
8	3	20	V2	78.63
9	4	20	V2	65.43
10	1	40	V1	98.99
11	2	40	V1	80.62
12	3	40	V1	58.01
13	4	40	V1	74.21
14	1	40	V2	85.22
15	2	40	V2	85.28
16	3	40	V2	84.48
17	4	40	V2	82.93
18	1	60	V1	40.28
19	2	60	V1	53.55
20	3	60	V1	46.49
21	4	60	V1	41.78
22	1	60	V2	57.35
23	2	60	V2	59.34
24	3	60	V2	66.44
25	4	60	V2	56.04

Fig. 21.5 Data for a randomized complete block design experiment (Two-factor experiment)

Fertilizer? (2) is there a statistically significant difference in yields due to variety? and (3) Is there a statistically significant interaction between fertilizer and variety?

We set the null hypothesis that there is no statistically significant difference in yield among the treatments. (1) There is a statistically no significant difference in yields due to fertilizer, (2) there is a statistically no significant difference in yields due to variety, and (3) Is there a statistically no significant interaction between fertilizer and variety.

In running the ANOVA test, a two-way test generates three p-values, one for each parameter independently and one measuring the interaction between the two parameters (see ANOVA results below). This experiment with a factorial arrangement generates the following:

1. Effect of fertilizer levels on tomato yields.
2. Effect of fertilizer levels on tomato yields.
3. Effect of interactions between fertilizer and varieties on tomato yields.

Source of variation	Df	Sum Sq	Mean Sq	F value	Pr(>F)
Block	3	69	22.9	0.280	0.8391
Fertilizer_level	2	3281	1640.3	20.071	5.75e-05 ***
Variety	1	559	559.3	6.844	0.0195 *
Fertilizer_level:Variety	2	67	33.3	0.407	0.6726
Residuals	15	1226	81.7		

Making a conclusion to our three set null hypotheses, we observe that the p-value for fertilizer levels and varieties are less than the tabulated p-value of 0.05. We, therefore, reject hypotheses associated with fertilizer level and variety and conclude that the fertilizer levels and varieties had statistically significant effects on the yields. Yet, the p-value for the interaction effect (fertilizer level * variety) is (0.67) larger than 0.05 hence, we fail to reject the null hypothesis linked to the interaction effect. We, therefore, conclude that the interaction had a statistically no significant effect on the yield. In this case, the effects of fertilizer levels and varieties are discussed individually. Nevertheless, assuming we had a significant interaction effect, the effects of the two factors (fertilizer levels and varieties) are not discussed, even if they are significant. This is because their individual effects are nested in the interaction effect. In this instance, only the interaction effect is discussed.

Three-Way Analysis of Variance

Three-way ANOVA is an extension of the two-way where a third variable is included. We refer to this as a three-factor experiment. Examples include the following:

- Factorial experiment of 2 × 3 × 2 (Two varieties, 3 planting densities, and 2 herbicide applications).
- Factorial experiment of 2 × 3 × 2 (2 tillage systems, 3 manure rates, and 2 varieties).
- Split-split plot with the main plot, sub-plot, and sub-subplot; main plot (three irrigation frequencies), sub-plot (3 planting densities), and sub-sub-plot (2 herbicide applications).
- Split-split plot with 2 × 3 × 2 (2 varieties, 3 planting densities, and 2 herbicide applications).

Let us consider a study that seeks to determine the effects of two maize varieties (V1 and V2), three fertilizer rates (0, 20, and 40 kg/ha), and two tillage methods (CT and RT) on maize yield. The following null hypotheses are set: (1) there is no statistically significant difference in maize yield due to fertilizer rate, (2) there is no statistically significant difference in maize yield due to variety, (3) there is no statistically significant difference in maize yield due to tillage method, and (4) there are no statistically significant interaction effects among variety, fertilizer rate, and tillage method on maize yield (Fig. 21.6).

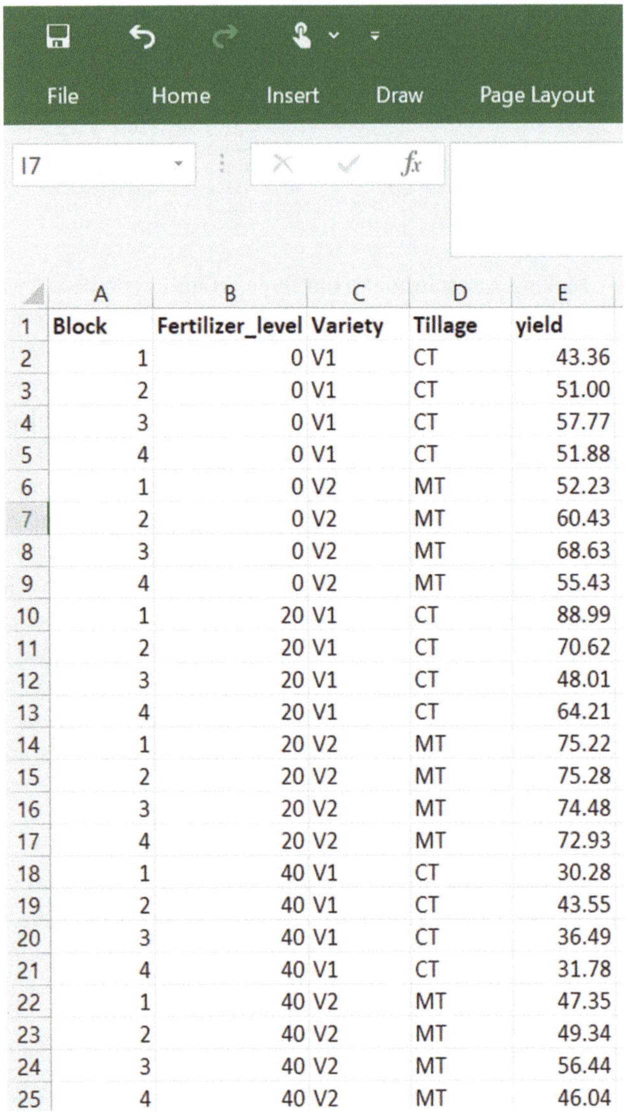

	A	B	C	D	E
1	Block	Fertilizer_level	Variety	Tillage	yield
2	1	0	V1	CT	43.36
3	2	0	V1	CT	51.00
4	3	0	V1	CT	57.77
5	4	0	V1	CT	51.88
6	1	0	V2	MT	52.23
7	2	0	V2	MT	60.43
8	3	0	V2	MT	68.63
9	4	0	V2	MT	55.43
10	1	20	V1	CT	88.99
11	2	20	V1	CT	70.62
12	3	20	V1	CT	48.01
13	4	20	V1	CT	64.21
14	1	20	V2	MT	75.22
15	2	20	V2	MT	75.28
16	3	20	V2	MT	74.48
17	4	20	V2	MT	72.93
18	1	40	V1	CT	30.28
19	2	40	V1	CT	43.55
20	3	40	V1	CT	36.49
21	4	40	V1	CT	31.78
22	1	40	V2	MT	47.35
23	2	40	V2	MT	49.34
24	3	40	V2	MT	56.44
25	4	40	V2	MT	46.04

Fig. 21.6 Data in a randomized complete block design (three-factor experiment)

In running the ANOVA test, a three-way test generates four p-values for each parameter independently and four for the interactions between parameters (see ANOVA results below). We can see that fertilizer rates and varieties had significant effects on yield since they exhibited lower p values than the $\alpha = 0.05$. In this case, we reject the null hypotheses related to fertilizer rates and varieties. But looking at the p-values of tillage and the various interactions, they are larger than $\alpha = 0.05$. In this instance, we fail to reject the null hypotheses associated with the respective parameters and conclude that they have no significant effects on maize yield.

Source of variation

```
                                Df Sum Sq Mean Sq F value  Pr(>F)
Block                            3    161      54   0.668  0.5793
Fertilizer_rate                  2   6425    3212  39.974 8.5e-09 ***
Variety                          1    510     510   6.348  0.0180 *
Tillage                          1     95      95   1.178  0.2873
Fertilizer_rate:Variety          2    504     252   3.134  0.0597 .
Fertilizer_rate:Tillage          1     72      72   0.902  0.3507
Variety:Tillage                  1      1       1   0.006  0.9377
Fertilizer_rate:Variety:Tillage  1      1       1   0.006  0.9377
Residuals                       27   2170      80
```

21.3.2 Mean Separation

As we have seen above examples, the p-value from an ANOVA test indicates whether there is a significant difference among the means of the groups being compared. However, to determine which specific means differ, we perform mean separation (post hoc test) using multiple comparison tests (Midway et al. 2020).

Multiple comparison tests, allow us to compare the means of different groups pairwise and identify statistically significant differences between them (Lee and Lee 2018). These tests are necessary because ANOVA only tells us that there is a difference somewhere among the means, but it does not specify which means differ.

There are several common multiple comparison tests (Midway et al. 2020) used for this purpose including the few listed below. All the mentioned procedures are post hoc tests; however, Fisher's LSD, Tukey's HSD, Duncan's MRT, and Scheffé's method primarily focus on pairwise or group mean comparisons, Dunnett's test is designed for comparing treatment means against a single control, and the Bonferroni correction serves as a general adjustment method applicable to any set of multiple comparisons:

- Fisher's Least Significant Difference (LSD) test (Also, may be referred to as: Protected t-test, Protected Fisher's LSD test, LSD Method, pairwise t-test for pooled variance and/or Fisher's multiple comparison test).
- Tukey's Honest Significant Difference (HSD) test (also known as Tukey's multiple comparison test, Tukey's post hoc test, Tukey's range test, Tukey's studentized range test, Tukey's pairwise comparison method, or the Tukey–Kramer method when sample sizes are unequal.).
- Duncan's Multiple Range Test (also called the Duncan Multiple Comparison Test, Duncan's New Multiple Range Test (DMRT), or simply Duncan's Test).
- Bonferroni correction (also known as the Bonferroni adjustment, Bonferroni method, or Bonferroni multiple comparison correction).
- Scheffe's method (also known as Scheffé's test, Scheffé's multiple comparison procedure, or Scheffé's post hoc test).
- Dunnett's test (also known as the Dunnett multiple comparison test, Dunnett's post hoc test, or Dunnett's procedure).

Here we use HSD to separate the means in the one-way RCBD experiment above.

```
Minimun Significant Difference: 0.6861605

Treatments with the same letter are not significantly different

      yield groups
T2 5.323333      a
T3 5.293333      a
T4 5.270000      a
T1 2.203333      b
```

21.4 Correlations and Relationships

21.4.1 Why Correlations?

In real-world situations, researchers are interested in describing relationships between variables to see whether a change in one affects the other (Guolo 2008; Akoglu 2018). There are three main reasons why researchers test relationships among variables. First, it aids researchers to know how changes in one variable may be associated with changes in another. By identifying these associations, researchers can better understand patterns and dependencies within their data, which can inform theories, models, and further research. Second, predictive modeling allows researchers to develop models that can forecast outcomes or behaviors based on the values of other variables. By examining the strength and direction of relationships, researchers can build models that accurately predict future trends, outcomes, or events, aiding in decision-making and planning. Lastly, testing relationships among variables can also help researchers identify causal relationships between variables. By controlling for other factors, researchers can determine if one variable is causing changes in another variable.

Researchers commonly evaluate the validity of their theoretical assumptions and predictions regarding the relationships between variables through hypothesis testing. By subjecting these hypotheses to empirical scrutiny using correlational analysis, researchers can assess whether the observed data supports or contradicts their proposed relationships. This process helps researchers confirm or refute their hypotheses and contributes to the advancement of knowledge by refining theoretical frameworks and generating new insights into the complex interplay between variables.

21.4.2 Measures of Association Between Two Variables

A variety of bivariate (two variables) correlational statistics are available, the choice of which technique to use depends on:

- Whether the variables of the study are discrete or continuous.
- Whether the scale of measurement used is nominal, ordinal, or interval.

Analysis questions would read as follows:

- Does an association exist between X and Y? Or is X related to Y?
- What is the extent of the association?
- What is the direction of the association?
- What is the nature of the association?

The three most commonly used correlational analyses are (Prematunga 2012):

1. Pearson's Chi-square, Phi Cramer's V for categorical variables.
2. Spearman Rho for ordinal variables (Ranks).
3. Pearson product-Moment for continuous variables.

21.4.2.1 Pearson Chi-Square Test of Association

The Chi-square test of association is a commonly used test in agronomy, studies involving livestock studies, pests and diseases, and also adoption studies when data lack parametric characteristics. Chi-square (χ) is the Greek symbol, Chi. It is employed to determine whether two categorical variables, e.g., gender and education level, are independent or associated. This test, also called bivariate, evaluates the probability that the observed data or more extreme data could change by comparing it to what would be expected if the two variables were independent. This can be framed as the null hypothesis. There is no dependence between the variables. It assesses the likelihood of the data in your table, or data more extreme, occurring by chance alone by comparing it with what you would expect if the two variables were independent of each other.

The Chi-square test analyses a contingency table with rows and columns to see if the observed cell frequencies are significantly different from the expected frequencies. It checks whether the number of individuals or objects in each category significantly deviates from the expected values. The test helps determine if the observed differences between expected and observed are due to chance or represent real differences.

Data used in a Chi-square analysis has to satisfy the following conditions:

1. The data must be randomly drawn from the population.
2. The data should be reported as frequency counts.
3. The variables measured must be independent of each other.
4. Observed frequencies should not be too small; no more than 25% of the cells in the table should have expected values of less than 5.
5. The values of independent and dependent variables must be mutually exclusive.

The Chi-square test offers several advantages, including data distribution robustness, computation simplicity, detailed information, and applicability in studies where parametric assumptions are not satisfied (see Slayi et al. (2023) and Musafiri et al. (2022) as examples). It has the capability of handling that in two groups or multiple groups. However, it also has some limitations, such as the need for a sufficiently large sample size, the complexity of interpretation when there are numerous categories in the independent or dependent variable, and the tendency of Cramer's V to yield low correlation measures even when results are highly significant (Kothari and Garg 2019).

Computation of Chi-Square

The analysis question is: Is there an association between categorical variables? You start by constructing a bivariate cross-tabulation or a contingency table (two-way table). The null and alternative hypotheses are:

Ho: The two categorical variables are independent.
Ha: The two categorical variables are related.

The cross-tabulation table provides a visual representation of the relationship between the two categorical variables by displaying the frequency counts of observations for each combination of categories (Fig. 21.7a). Each cell in the table represents the count of occurrence for a specific combination of the two variables. This table allows you to observe how the data are distributed across the different categories and can highlight any potential apparent or associated between variables.

Assessing the Significance of the Relationship

By analyzing the frequencies in the cross-tabulation, you can assess whether there seems to be an association. However, to formally test the association, you will compute the Chi-square statistic. The test will give a Chi-square statistic, χ, which measures the discrepancy between the observed and the expected. Along with the Chi-square statistics will have a p-value to help you assess if the association is significant (Table 21.9).

1	**Chlorophyll**	**yield**
2	39.80	4.24
3	32.80	4.48
4	39.00	4.14
5	32.70	1.17
6	37.20	4.72
7	31.80	4.22
8	36.90	4.21
9	27.20	1.18
10	33.10	4.01
11	38.80	4.18
12	37.30	4.46
13	27.10	1.26
14	34.97	2.19
15	33.10	2.53
16	33.70	0.70
17	37.60	2.25
18	31.10	1.25
19	31.30	0.47
20	34.20	0.65
21	41.90	2.17
22	38.37	3.47
23	37.70	0.74
24	30.90	1.67

Fig. 21.7a Data set with two quantitative variables subjected to Pearson product-moment correlation

Table 21.9 provides preliminary conclusions. Apart from the gender of the household head, all the other variables show significant relations between them. This means most variables showed a significant association with users and non-users of manure. For instance, the p-value for education is 0.001, meaning the association is statistically significant. Non-users of manure in the educated group were 58%, while users of manure in the uneducated group were 51%. In the educated group, users (56%) than non-users (30%). This indicates that educated farmers are more likely to use a combination of manure and fertilizer than non-educated farmers. Practically, this translates to education encouraging the adoption of using a combination of manure and fertilizer.

The computer also gives Cramer's V and Phi coefficients. This complements the Chi-square by giving you an idea of how strong the association is between those categorical variables (Kothari and Garg 2019). The Chi-square test results should be reported accurately.

The Phi ranges from −1 (perfect negative association) to 1 (perfect association), with 0 indicating no association. However, Phi may pose challenges when comparing the strengths of significant associations between pairs of variables, especially when one or both variables have more than two categories. Therefore, Phi is best suited for comparing pairs of dichotomous variables (Ben-Shachar et al. 2023).

21.4.2.2 Measures of Association in Cohort and Case-Control Studies in Veterinary Research

Whether prospective or retrospective, cohort studies are usually analyzed using a 2 × 2 contingency table to assess asso-

Table 21.9 Socio-economic variables cross-tabulated with use of manure (user or non-user) of a combination of animal manure + inorganic fertilizer

		Used Manure + Fertilizer		
Characteristic		No	Yes	χ^2 P value
Household head gender	Male	58.0 (45.3)	70.0 (54.7)	Ns
	Female	44.0 (48.9)	46.0 (51.1)	
Education	Not educated	72.0 (58.5)	51.0 (41.5)	0.001[a]
	Educated	30.0 (31.6)	65.0 (68.4)	
Any on-farm income	No	17.0 (70.8)	7.0 (29.2)	0.011[a]
	Yes	85.0 (43.8)	109.0 (56.2)	
Benefits of combination	Fertile soil	12.0 (46.2)	14.0 (53.8)	0.046[a]
	High yields	29.0 (24.4)	90.0 (75.6)	
	Early maturity	6.0 (42.9)	8.0 (57.1)	
Source of knowledge	Extension	15.0 (55.6)	12.0 (44.4)	0.040[a]
	Researchers	76.0 (42.7)	102.0 (57.3)	
	Other farmers	8.0 (80.8)	2.0 (20.0)	
Most effective method for teaching	Demonstrations	22.0 (21.6)	80.0 (78.4)	0.001[a]
	Meetings	37.0 (72.5)	14.0 (27.5)	
	Media	40.0 (64.5)	22.0 (35.5)	

Source: Mwebia (2019)
[a] Association significant at $\alpha = 0.05$

Table 21.10 Inputs ranking based on preference by farmers and calculated net benefits in Mukuuni and Murugi sites, central Kenya

Treatment	Ranking based on farmers' choices	Ranking based on net benefits
Tithonia + manure + fertilizer	1	8
Manure + fertilizer	2	5
Tithonia	3	2
Tithonia + fertilizer	4	1
Tithonia + manure	5	3
Calliandra	6	9
Manure	7	4
Calliandra + fertilizer	8	7
Fertilizer	9	6
Control	10	10

ciations between exposure and disease (Albert 2017; Pérez-Guerrero et al. 2024). In these studies, researchers follow groups of animals that are exposed to a certain factor and compare them to groups that are not exposed, tracking the evidence of disease over time.

In case-control studies, researchers start with animals that have the disease (cases) and compare them with animals without the disease (control), looking backward to determine the studies. These studies are analyzed using an odds ratio (OR) to measure the association between exposure and disease. Examples of studies using the odds ratio include Cullen et al. (2016) and Anka et al. (2014). More information about analyzing data from case-control studies can be found in several veterinary epidemiology books.

21.4.2.3 Spearman Rank Correlation

As already shown, Spearman's rank correlation coefficient, or Spearman's rho (ρ), is often a non-parametric measure of the strength and direction of the association between two ranked variables (Vickers 2005). In the computation of the measure of association, data are first ranked, each variable independently, and then used to calculate Spearman's rho (ρ). The Spearman's rho (ρ) ranges from −1 to 1:

- **ρ = +1** indicates a perfect *positive* correlation, meaning as one variable increases, the other variable also increases consistently.
- **ρ = −1** indicates a perfect *negative* correlation, meaning as one variable increases, the other variable consistently decreases.
- **ρ = 0** indicates *no correlation* between the variables.

We illustrate this using a simple example start. Let us use an example where we want to examine the preferences of males and females regarding different crops. In this case, we ask a group of males and females to rank five food crops in order of importance in the community. The crops are maize, cassava, sweet potato, beans, and sorghum. The ranking recorded by males from 1 to 5 are as follows: sorghum, maize, beans, sweet potato, and cassava. In contrast, the females ranked sorghum, sweet potato, maize, beans, and cassava.

The Null hypothesis would be that there is no significant difference between the ranking provided by males and females.

Here we use a published example by Muriuki (2011): We use a study carried out in Meru County, Kenya (Table 21.10), within a project titled "Enhancing farmers' benefits from integrated soil fertility management through improved water use efficiency and better marketing strategies in the highlands of central Kenya," funded through the Flemish Interuniversity Council (VLIR) Own Initiatives Program.

The objective of the study was to compare farmer selection of soil fertility treatments with the economic assessment results.

The Null hypothesis was that ranking based on calculated economic returns is not significantly different from ranking based on farmer preferences and selection.

Spearman correlation was carried out to assess the strength and direction of the association between the ranking of treatments based on farmers' choices and net benefits obtained from experiments.

Results: Correlation analysis between the ranking based on the farmers' choices and calculated net benefit gave a weak positive Spearman's correlation coefficient ($\rho = 0.394$, $p = 0.260$) that was insignificant at the 5% level. The positive coefficient indicates that there is concurrence of the technologies based on the farmers' perceived benefits (the basis of their choices) and ranking based on calculated economic benefits.

The positive correlation between rankings based on farmers' choice of inputs and the calculated net benefits implies that economic benefits significantly influence farmers' choices of technologies to a small extent. The lack of significant correlation suggests that economic benefits alone may not be the primary factor influencing farmer preferences. Other factors, such as, ease of application, perceived health benefits, availability of the inputs may be playing a critical role in shaping their choices.

21.4.2.4 The Pearson Product-Moment

The Pearson Product-Moment (PPM) Correlation is one of the measures of correlation (linear dependence) between two variables X and Y, giving a value between +1 and −1 inclusive (Puth et al. 2014).

The following conditions need to be satisfied:

- The relationship between variables should be linear (evidenced by a scatter graph).

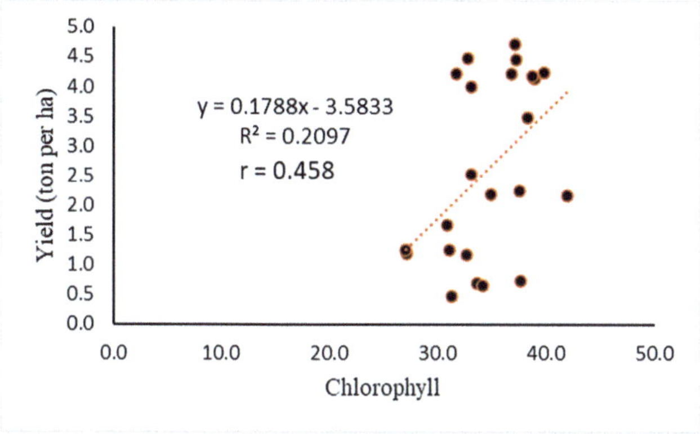

Fig. 21.7b Results of Pearson product-moment correlation analysis with $r = 0.458$

- Data points within samples should be independent of each other.
- The data are on an interval or ratio scale
- Both data sets should exhibit a normal distribution.
- Individuals are selected at random from the population

Let us look at the data in Fig. 21.7a and study whether there is a relationship between the yield and chlorophyll content.

```
> correlation <- cor(rcbd$Chlorophyll, rcbd$yield, method = 'pearson')
> correlation
[1] 0.4579374
```

The analysis results show that there is a moderate positive relationship ($r = 0.458$) between yield and chlorophyll content. A similar relationship can be obtained by plotting a scatter plot (Fig. 21.7b).

Interpretation of Correlation Coefficient

The computation yields a correlation coefficient (r) that ranges from −1 to 1. The coefficient reveals both the direction and strength of the relationship. Specifically, a positive correlation ($r > 0$) indicates a positive relationship, meaning that as Variable X1 increases, Variable X2 also increases. Conversely, a negative correlation ($r < 0$) indicates a negative relationship, where an increase in variable X1 results in a decrease in variable X2 and vice versa. On the other hand, if the correlation coefficient is zero ($r = 0$), it means there is no significant relationship.

In practice, it is rare to observe either perfect or zero correlation between variables due to random effects. Instead, correlations typically fall within the range of 0.00 to +1.00. These correlation coefficients are expressed as decimals, often rounded to two decimal places, such as $r = 0.85$ or $r = 0.46$.

Is the Relationship Strong or Weak?

There are no agreed-upon standards for defining a strong versus weak relationship. However, the following is recommended:

- Correlations ranging from 0.00 to 0.30 indicate little or no relationship, meaning that changes in one variable are not strongly associated with the change in the other variable.
- Correlations ranging from 0.31 to 0.7 suggest a moderate or fair degree of relationship; noticeable relationship but not strong.
- Correlations ranging from 0.61 to 1.00 indicate good to excellent; the change in one variable is strongly related to the change in the other variable.

These values should not be considered rig thresholds, as they can be influenced by sample size, measurement error, and the type of variables being studied (Puth et al. 2014). In addition, different fields of study may have different interpretations of what constitutes a strong correlation. For example, in social studies, lower correlations may be considered meaningful due to the complexity of social phenomena. On the other hand, in biophysical data, higher correlations are usually preferred, especially when the experiment is highly controlled. It is important to always interpret correlations with caution, considering the specific research domain and its context.

Causation and Correlation

A statistical association between two variables does not necessarily imply a causal relationship, meaning it does not mean that X causes Y or Y causes X. It only means that a change in one variable is associated with changes in another variable, but it does not indicate the direction of causality or the presence of underlying mechanisms. Correlational relationships are often misinterpreted, especially in mass media (newspapers and television) (Jackson 2016). Oversimplified explanations presented in these channels may lead to erroneous belief that a certain correlation implies causation. We need to be cautious because other variables, known as confounding variables, influence both the variables under study, leading to a spurious correlation. Additionally, correlation does not account for the possibility of reverse causation, where the direction of influence may be the opposite of what is assumed (Box 21.4).

Box 21.4: Association Is Not Causation

Imagine that there is a statistical association between higher temperatures and increased crop growth measured in a particular region. This association could lead to the assumption that *higher temperatures cause enhanced crop growth*. However, several factors could confound this relationship. One confounding factor could be soil moisture. It is possible that areas with higher temperatures also receive more rainfall or have better irrigation systems, resulting in higher soil moisture levels. This increased soil moisture could be the actual driver of enhanced crop growth, rather than temperature alone. Additionally, factors like soil type, nutrient levels, pest infestations, and management practices could all influence crop growth independent of temperature.

Reverse causation could also play a role. For instance, high crop yields from previous seasons may prompt farmers to invest more resources, such as irrigation or fertilization, into their fields in subsequent seasons, leading to the perception of a positive association between temperature and crop growth.

1	LAI	Chlorophyll	Height	Grain_yield_t_ha
2	1.37	39.80	28.00	4.24
3	1.66	32.80	22.00	4.48
4	2.22	39.00	24.75	4.14
5	0.85	32.70	17.25	1.17
6	1.22	37.20	25.00	4.72
7	1.13	31.80	21.00	4.22
8	2.05	36.90	23.75	4.21
9	0.85	27.20	19.75	1.18
10	1.58	33.10	27.50	4.01
11	1.09	38.80	22.25	4.18
12	2.01	37.30	27.25	4.46
13	0.78	27.10	21.75	1.26
14	0.96	34.97	17.25	2.19
15	1.31	33.10	18.00	2.53
16	2.02	33.70	19.50	0.70
17	0.91	37.60	20.38	2.25
18	1.07	31.10	21.00	1.25
19	0.72	31.30	15.75	0.47
20	1.38	34.20	13.75	0.65
21	1.84	41.90	19.75	2.17
22	1.17	38.37	26.25	3.47
23	2.29	37.70	10.13	0.74
24	1.09	30.90	24.50	1.67
25	0.97	33.50	19.50	1.80

Fig. 21.8 Data in Excel used to demonstrate multiple linear regression

Research in humans illustrates this misinterpretation, for example, research on smoking and cancer and many other research studies on humans (Jackson 2016). Concluding that smoking causes cancer may be misleading. A correlation between tobacco smoking and cancer could be argued that individuals predisposed to cancer are more attracted to smoking (Jackson 2016). This is a correct argument by the tobacco industry. Real causal relationships can only be established in experimental research, where researchers deliberately manipulate one or more variables to observe their effect on another variable and control the confounding factors.

In most cases, we also need additional support for causality, including biological credibility of the association, logical time sequence, consistency across studies, and alignment with relevant theories. In social sciences, understanding causality involves considering complex social dynamics and employing diverse methodologies and theoretical frameworks to investigate causal relationships in social phenomena (see Chap. 11).

The Correlation Matrix

Studies using correlation analysis often examine several variables at one time and may include a matrix of intercorrelations for example as shown in this study (Anderson et al. 2019). These intercorrelations are used to analyze the relationships between pairs of variables within the dataset (Fig. 21.8). Researchers use the intercorrelation matrix to identify patterns, dependencies, and associations among variables, providing valuable insights into the underlying structure of the data and guiding further analyses and interpretations.

21.5 Prediction and Regression Analysis

21.5.1 Definition and Purpose

The correlational analysis is a statistical method used to assess the strength and direction of the relationship between two or more variables (Prematunga 2012; Akoglu 2018). It helps determine whether and how strongly variables are related, but it does not imply causality.

Regression analysis involves techniques for modeling and analyzing the relationship between a dependent variable and one or more independent variables (Wisniewski and Brannan 2024). These techniques allow researchers to understand how changes in the dependent variable are associated with changes in the independent variable, to make predictions, and to identify the strength and nature of the relationship. Regression can also be used to control for confounding variables and assess the impact of multiple factors simultaneously and trends in the data.

Additionally, regression techniques are essential for identifying and quantifying the influence of various predictors in the outcome variable; They can be used to estimate the relative importance of an independent variable, helping provide factors that have the most significant impact. Regression also provides insights into potential interactions between variables revealing how the effect of one predictor variable might change in the presence of another.

To conduct a linear regression analysis, begin by defining the variables of interest and identifying the dependent variable (the Outcome to be predicted, e.g., crop yields) and independent variable predictors (e.g., soil N). Next explore the data to understand its distribution, detect any outliers, and make sure there is a linear relationship between the variables. After this, move to model specification, where you define the form of the regression model.

Next fit the model to estimate the coefficients that minimize the difference between the observed and predicted values of the dependent variable. Determine an equation for the regression line and then use that equation to predict values of Y, that takes the following forms:

$$Y = \beta o + \beta_1 X_1 + \beta_2 X_2 + \ldots\ldots\ldots\ldots\ldots\ldots\ldots\ldots \beta n X n + e$$

- Dependent variable (Y): The single variable being explained/predicted by the regression model.
- Independent variable (X): The explanatory variables (2) are used to predict the dependent variable.
- Coefficients (β) value computed by the regression tool, reflecting an explanatory of the dependent variable relationship.
- Error Term (e): The portion of the dependent variable that is not explained by the model
- Linear regression: The regression line is straight. There are others that are not straight e.g., logarithmic relationships.

21.5.2 Simple Linear Regression Model

A simple linear regression model is characterized by several key features (Wisniewski and Brannan 2024). It involves a single independent variable, denoted as X, which is used to predict the behaviour of a dependent variable, denoted as Y. The fundamental assumption is that variations in Y are linearly associated with, and can be explained by, changes in X, allowing the model to estimate the effect of X on Y with a high degree of precision.

21.5.3 Multiple Linear Regression

Multiple linear regression is a direct extension of the simple linear model and is the most widely used multivariate technique (Wisniewski and Brannan 2024). While simple linear regression employs single variables, multiple regression involves one dependent variable and several independent variables. e.g., maize yields (dependent variable) affected by soil nitrogen, soil phosphorus, intercropping, and maize variety (dependent variables).

To perform multiple linear regression, you first define the dependent variable and the independent variable. After which, you specify the model's performance by checking the R-squared value, which explains the proportion of the variance explained by the independent variables.

Let us consider an example where we want to establish the effect of leaf area index (LAI), chlorophyll content, and height on yield. The data are provided (Fig. 21.8).

Multiple regression model takes the form of

$$\text{Yield} = \alpha + \beta_1 \,\text{x}\,\text{LAI} + \beta_2 \,\text{x}\,\text{Chlorophyll} + \beta_3 \,\text{x}\,\text{Height} + \text{e}.$$

The analysis output;

```
Coefficients:
            Estimate Std. Error t value Pr(>|t|)
(Intercept) -6.40835    1.93162  -3.318  0.00343 **
LAI          0.26600    0.47716   0.557  0.58339
Chlorophyll  0.10912    0.06145   1.776  0.09101 .
Height       0.23026    0.04516   5.099 5.47e-05 ***
---
Signif. codes:  0 '***' 0.001 '**' 0.01 '*' 0.05 '.' 0.1 ' ' 1

Residual standard error: 0.9421 on 20 degrees of freedom
Multiple R-squared:  0.6591,    Adjusted R-squared:  0.608
F-statistic: 12.89 on 3 and 20 DF,  p-value: 6.506e-05
```

21.5.3.1 Interpretation

First, examine the F-statistic and its associated p-value, which are found at the bottom of the model summary, to check if the model is best suited to explain the change in the dependent variable. In our example, it is 6.506×10^{-05}, which is highly significant. The significance shows that at least one of the predictor variables is significantly associated with the response variable (yield). The estimated values are R^2 coefficients. While all the predictor variables had positive coefficients, only height significantly explained change in yield.

The fitted model now takes the form of

$$\text{Yield} = -6.408 + 0.266 \times \text{LAI} + 0.109 \times \text{Chlorophyll} + 0.230 \times \text{Height} + \text{e}.$$

21.5.4 Binary Regression Techniques

Binary regression is a regression analysis in which Y (dependent variable) is binary. In this case, Y is not measured on an interval or ratio scale (like a Likert scale), but instead represents two distinct categories (Harris 2021; Srimaneekarn et al. 2022). These categories are typically coded as "1" and

"0" for ease of processing in statistical software. The binary outcomes can represent any two-group scenario, such as "adopted a technology" versus "did not adopt." Or "diseased" versus "not diseased." Traditional linear regression models, whether simple or multiple, are not suitable in such cases because they assume a continuous dependent variable, which is not met when dealing with binary outcomes (Nathanson and Higgins 2008; Pinheiro-Guedes et al. 2024).

The common models are logistic and probit models. They are both econometric models that estimate the probability of a certain event occurring based on one or more predictor variables. Logistic regression uses the logit link function to model the log odds of the outcome, while probit regression uses the cumulative normal distribution to estimate the probability. Though they are mathematically different, the results they produce are often quite similar (Esmaily et al. 2013; Jose et al. 2020).

These models are helpful in biophysical and social studies for predicting the likelihood of events based on various factors. For instance, these models are widely used in studies on the adoption of agricultural technologies to help identify the factors that drive farmers to adopt new practices or tools (Mugwe et al. 2009; Khonje et al. 2015), informing strategies to enhance technology uptake and agricultural productivity. Suppose we are studying the adoption of a new agricultural technology. We might use a logistic regression model, with the dependent variable being whether a farmer adopts the technology (1 = yes, 0 = no). Predictor variables could include socioeconomic factors. The logistic model will estimate the probability of adoption based on these predictors. Section 22.7 in Chap. 22 demonstrates how to conduct binary logistic regression using SPSS.

In studies on the presence or absence of diseases in crops or livestock, the outcome is coded as 1 for the presence of the disease and 0 for its absence. By using a logistic or probit model, the researchers can assess the probability of disease occurrence based on environmental conditions, geographic location, and management practices. This type of data analysis can be conducted using SPSS or STATA software. This software allows researchers to input data, specify the model, and obtain estimates of the probability of the diseases occurring based on predictor variables.

Recommended Further Reading

1. Zhu W, Spanos A (2023) Revisiting the Large n (Sample Size) Problem: How to Avert Spurious Significance Results. Stats 2023, Vol 6, Pages 1323–1338 6:1323–1338. https://doi.org/10.3390/STATS6040081
2. Prematunga RK (2012) Correlational analysis. Australian Critical Care 25:195–199. https://doi.org/10.1016/J.AUCC.2012.02.003
3. Forthofer RN, Lee ES, Hernandez M (2006) Biostatistics: A Guide to Design, Analysis and Discovery, Second Edition. Biostatistics: A Guide to Design, Analysis and Discovery, Second Edition 1–502. https://doi.org/10.1016/C2009-0-03861-6
4. Gomez KA, Gomez AA (1984) Statistical procedures for agricultural research. Statistical procedures for agricultural research with emphasis on rice 7–8
5. Johnstone P, Manly BFJ (2014) Analysis of Variance Through Examples. Wiley StatsRef: Statistics Reference Online. https://doi.org/10.1002/9781118445112.STAT07533

Acknowledgements

The screenshots used to demonstrate how to arrange data in analyzing data from various experimental designs were generated using Microsoft Excel software. This software was instrumental in enhancing the content of this book, which is primarily aimed at postgraduate students undertaking masters and doctoral research.

Key Points in This Chapter

Why Inferential Statistics?

The goal is to make inferences that extend beyond the immediate sample dataset. These inferences aid in assessing the probability that any observed differences between groups are reliable and not merely due to chance.

Concept of Sampling Errors and Confidence Intervals

Sampling error arises when using samples or when there is a discrepancy between a sample and the population. In most research, a 95% confidence interval (CI) is standard, indicating researchers expect results to deviate inaccurately 5% of the time. This is related to the "test of significance," where we use alpha = 0.05.

Non-parametric and Parametric tests

Non-parametric tests, also known as distribution-free tests, are used when data do not adhere to a normal distribution or when the assumptions necessary for parametric tests cannot be met. Examples include the Chi-square test of association, the Kruskal-Wallis test, and the Mann-Whitney U Test. *Parametric tests* are statistical methods used to analyze data in which researchers assume that the underlying distribution follows a specific probability distribution, typically the normal distribution. Examples include the ANOVA Test.

Hypothesis testing

Hypothesis testing is using inferential statistics to determine whether or not we reject or accept the null hypothesis. Determining if chance causes a difference or relationship is called "significance testing." In hypothesis testing, there are two types of errors: a Type I error, also known as a false positive, and type 2, when we say that there is a significant difference when, in reality, it does not exist or when we reject the null hypothesis and mistakenly conclude that there are significant differences between groups or variables, even when there are none.

Main Inferential Analysis

The main inferential statistics are categorized according to what we want to test. They fall into the following three categories:

1. Testing for differences.
2. Relationships.
3. Regression and predictions.

There are two types of testing for differences: t-tests and ANOVA. The t-test determines whether significant differences exist between two means derived from two samples or groups at a specified probability level. We discuss three types of T-tests:

1. One-sample t-test.
2. Independent t-test.
3. Paired t-tests or correlated groups.

ANOVA is used to test the significance of more than two groups. It has three important assumptions:

1. The sample was randomly selected from the population, and the subjects or treatments were randomly assigned to the groups (randomness and independence of errors).
2. The variance in each group is the same (homogeneity of variance).
3. The sample came from a population in which the dependent variable is normally distributed (Normality).

ANOVA partitions total variation into variation due to treatments and variation due to random sampling:

- One-way ANOVA tests for significant differences among two or more independent groups.
- Two-way ANOVA, is an extension of One-way ANOVA and tests for significant differences when the independent variables are two (two factors).
- Three-way ANOVA is an extension of the two-way where a third variable is included. We refer to this as a three-factor experiment.

Mean separation allow us to compare the means of different groups pairwise and identify statistically significant differences between them. These tests are necessary because ANOVA only tells us that there is a difference somewhere among the means, but it does not specify which means differ.

Measures of Association Between Two Variables

A variety of bivariate (two variables) correlational statistics are available, the choice of which technique to use depends on:

- Whether the variables of the study are discrete or continuous?
- Whether the scale of measurement used is nominal, ordinal, or interval

The Chi-square test of association is used to test whether two categorical variables e.g., gender and household head, are independent or associated.

Spearman's rank correlation coefficient or Spearman's rho (ρ), often denoted by the Greek letter ρ (rho), is a non-parametric measure of statistical independence between two variables.

Pearson Product-Moment (PPM) Correlation is one measure of correlation (linear dependence) between two variables X and Y, giving a value between +1 and −1 inclusive.

Prediction and Regression Analysis

Correlational analysis is useful for describing the relative strength of a relationship between two variables. But when a researcher wants to establish this relationship as a basis for prediction, regression analysis is used.

Regression analysis includes any techniques for modelling and analyzing several variables when the focus is on the relationship between a dependent variable and one or more independent variables. They are of two types:

- *Simple Linear Regression model:* In simple regression, there is one dependent variable and one independent variable.
- *Multiple linear regression* is a multivariate technique in involving many independent variables.

Binary Regression Techniques

Binary regression models, like logistic and probit, are widely used in several studies. They help researchers analyze data with binary outcomes, such as disease presence in crops or adoption of agricultural technologies.

Review Questions

1. How does inferential data analysis go beyond "organizing and summarizing data?"
2. What are the three main objectives of inferential analysis?
3. Differentiate clearly between "Type I" and "Type II" errors in the context of hypothesis testing. Provide a practical example of each error in an agricultural study (e.g., testing a new pest control method). What are the consequences of each type of error for farmers or agricultural policy?
4. What is the difference between "parametric" and "non-parametric" statistical tests?
5. For the following studies, choose the appropriate type of t-test to use. Select from:
   A. One-sample t-test.
   B. Independent samples t-test.
   C. Paired (correlated) t-test.
   - A soil scientist wants to determine if the average pH of soils in a farm differs from the standard recommended pH of 6.5 for optimal crop production.
   - A researcher measures the milk yield of cows before and after being switched to a new high-protein diet, using the same cows.
   - A veterinary study compares the average weight gain of two different breeds of goats over a 60-day feeding trial.
   - A plant pathologist tests whether the average disease severity index in sampled leaves differs significantly from a benchmark index of 3.0.
6. What is the purpose of "means separation" tests (e.g., Tukey's HSD, LSD) following a significant ANOVA result?
7. State the correct statistical test to use in each scenario listed here:
   (a) Testing association between gender and farming method
   (b) Relationship between crop height and yield (both quantitative)
   (c) Relationship between soil quality rank and yield rank
8. You are interested in predicting maize yield, consider and address the following:
   (a) If you use only the amount of fertilizer applied as a predictor, which regression model would you use?
   (b) If you also include other factors like rainfall amount, soil pH, and planting density as predictors, which regression model would be more appropriate?
   (c) What additional insights does the latter provide?

Chapter Exercise

1. A student wants to determine if a new organic fertilizer significantly increases the average growth rate of kale plants (measured in cm/week) compared to a traditional inorganic fertilizer. The student has collected data from 25 plots using the organic fertilizer and 25 plots using the inorganic fertilizer. Address the following:
   (a) Which type of t-test (one-sample, independent, or paired) would be most appropriate?
   (b) State your H0 and H1.
2. A researcher is investigating the impact of three water management regimes (Drip Irrigation, Sprinkler Irrigation, Rainfed-non irrigated) on the final yield of tomatoes (in kg/plant) in a controlled experiment with RCBD replicated four times. Assess:
   (a) Which ANOVA test (one-way, two-way, or three-way) would be most appropriate if you are *only* interested in comparing the yields across the three irrigation techniques?
   (b) State your H0 and H1.
3. Suppose an agricultural economist is studying factors influencing smallholder farmer income. They have collected data on: farmer income (continuous), farm size (continuous), access to extension services (Yes/No), education level (Ordinal: Primary, Secondary, Tertiary), and number of livestock. Critically evaluate:
   (a) Which measure of association would be most appropriate to determine if there is a relationship between Farm Size and Farmer Income? Explain why. What kind of value would you expect from this test if larger farms tend to have higher incomes?
   (b) Which measure of association would you use to test if there is an association between Access to extension services and Education Level? Explain your choice.
   (c) If the economist wants to predict Farmer Income based on Farm size and Number of livestock, which type of inferential analysis would be most suitable?
   (d) If the economist wants to model the probability of a farmer adopting a new sustainable farming practice

(Binary outcome: Yes/No) based on their education level and farm size, which specific type of regression technique would be appropriate?

4. You conducted an experiment to compare yield of five different maize varieties in a Completely Randomized Design (CRD). Explain how you would set up and conduct an ANOVA for this design.

Answer to Review Questions

- Q1: Inferential: Draw conclusions and make predictions about the larger population. Descriptive is to the sample, inferential is to the population. Descriptive will show whether there is a difference; Inferential will determine if the difference or relationship is statistically significant. *Reference 21.1.1*
- Q2: Main objectives: Generalize to a population, Test hypotheses, and make predictions. *Reference: 21.1*
- Q3: Type I and Type II; *Reference; 21.2.4*
- Q4: Parametric and non-parametric tests *Reference 21.3.3*
- Q5: T-tests*; Reference 21.3.3.1*
- Q6: Means separation, *Reference 21.3.2*
- Q7 Tests of association *Reference 21.3.1*
- Q8: Simple and multiple regression (21.4.1)

References

Akoglu H (2018) User's guide to correlation coefficients. Turk J Emerg Med 18:91–93. https://doi.org/10.1016/J.TJEM.2018.08.001

Albert A (2017) Biostatistics: facing the interpretation of 2 × 2 tables. J Belg Soc Radiol 101. https://doi.org/10.5334/JBR-BTR.1399

Algina J, Keselman HJ, Penfield RD (2005) An alternative to Cohen's standardized mean difference effect size: a robust parameter and confidence interval in the two independent groups case. Psychol Methods 10:317–328. https://doi.org/10.1037/1082-989X.10.3.317

Anderson AA (2019) Assessing statistical results: magnitude, precision, and model uncertainty. Am Stat 73:118–121. https://doi.org/10.1080/00031305.2018.1537889

Anderson C, Xiao L, Checkley W (2019) Using data from multiple studies to develop a child growth correlation matrix. Stat Med 38:3540–3554. https://doi.org/10.1002/SIM.7696

Anka MS, Hassan L, Khairani-Bejo S et al (2014) A case-control study of risk factors for bovine brucellosis seropositivity in peninsular Malaysia. PLoS One 9:e108673. https://doi.org/10.1371/JOURNAL.PONE.0108673

Avram C, Mărușteri M (2022) Normality assessment, few paradigms and use cases. Rev Rom Med Lab 30:251–260. https://doi.org/10.2478/RRLM-2022-0030

Badenes-Ribera L, Frias-Navarro D, Iotti NO et al (2018) Perceived statistical knowledge level and self-reported statistical practice among academic psychologists. Front Psychol 9:349696. https://doi.org/10.3389/FPSYG.2018.00996/BIBTEX

Banerjee A, Chitnis U, Jadhav S et al (2009) Hypothesis testing, type I and type II errors. Ind Psychiatry J 18:127. https://doi.org/10.4103/0972-6748.62274

Ben-Shachar MS, Patil I, Thériault R et al (2023) Phi, Fei, Fo, Fum: effect sizes for categorical data that use the chi-squared statistic. Mathematics 11:1982. https://doi.org/10.3390/MATH11091982

Brakenhoff TB, Mitroiu M, Keogh RH et al (2018) Measurement error is often neglected in medical literature: a systematic review. J Clin Epidemiol 98:89–97. https://doi.org/10.1016/j.jclinepi.2018.02.023

Buonaccorsi JP (2010) Measurement error: models, methods, and applications. CRC Press, Boca Raton, pp 1–438. https://doi.org/10.1201/9781420066586

Cullen JN, Sargeant JM, Makielski KM, O'Connor AM (2016) The case-control design in veterinary sciences: a survey. Prev Vet Med 134:179–187. https://doi.org/10.1016/J.PREVETMED.2016.10.008

Esmaily H, Shakeri MT, Avazzadeh Z et al (2013) Comparison between applicability of the logit and probit models to diagnose influencing risk factors of cardiovascular disease in Mashhad. Health Scope 2(22):67–72. https://doi.org/10.17795/JHEALTHSCOPE-8622

Forstmeier W, Wagenmakers EJ, Parker TH (2017) Detecting and avoiding likely false-positive findings – a practical guide. Biol Rev 92:1941–1968. https://doi.org/10.1111/BRV.12315

Forthofer RN, Lee ES, Hernandez M (2006) Biostatistics: a guide to design, analysis and discovery, 2nd edn. Elsevier Academic, Amsterdam, pp 1–502. https://doi.org/10.1016/C2009-0-03861-6

Garrocho-Rangel A, Aranda-Romo S, Martínez-Martínez R et al (2024) Fundamentals of nonparametric statistical tests for dental clinical research. Dent J 12:314. https://doi.org/10.3390/DJ12100314

Ghasemi A, Zahediasl S (2012) Normality tests for statistical analysis: a guide for non-statisticians. Int J Endocrinol Metab 10:486–489

Gomez KA, Gomez AA (1984) Statistical procedures for agricultural research. In: Statistical procedures for agricultural research with emphasis on rice. Wiley, New York, pp 7–8

Goyal MR (2018) Scientific and technical terms in bioengineering and biotechnology. Apple Academic Press, Oakville

Guolo A (2008) Robust techniques for measurement error correction: a review. Stat Methods Med Res 17:555–580. https://doi.org/10.1177/0962280207081318

Harris JK (2021) Primer on binary logistic regression. Fam Med Community Health 9:1290. https://doi.org/10.1136/FMCH-2021-001290

Harris JE, Boushey C, Bruemmer B, Archer SL (2008) Publishing nutrition research: a review of nonparametric methods, part 3. J Am Diet Assoc 108:1488–1496. https://doi.org/10.1016/j.jada.2008.06.426

Jackson SL (2016) Research methods and statistics: a critical thinking approach – product details – Cengage Instructor Center, 5th edn. Cengage

Jankowski KRB, Flannelly KJ, Flannelly LT (2018) The t-test: an influential inferential tool in chaplaincy and other healthcare research. J Health Care Chaplain 24:30–39. https://doi.org/10.1080/08854726.2017.1335050

Johnstone P, Manly BFJ (2014) Analysis of variance through examples. Wiley StatsRef: Statistics Reference Online. https://doi.org/10.1002/9781118445112.STAT07533

Jose A, Philip M, Prasanna LT, Manjula M (2020) Comparison of Probit and logistic regression models in the analysis of dichotomous outcomes. Curr Res Biostat 10:1–19. https://doi.org/10.3844/AMJBSP.2020.1.19

Khonje M, Manda J, Alene AD, Kassie M (2015) Analysis of adoption and impacts of improved maize varieties in Eastern Zambia. World Dev 66:695–706. https://doi.org/10.1016/J.WORLDDEV.2014.09.008

Kim H-Y (2013) Statistical notes for clinical researchers: understanding standard deviations and standard errors. Restor Dent Endod 38:263–265. https://doi.org/10.5395/RDE.2013.38.4.263

Kim H-Y (2015) Statistical notes for clinical researchers: type I and type II errors in statistical decision. Restor Dent Endod 40:249–252. https://doi.org/10.5395/RDE.2015.40.3.249

Kothari CR, Garg G (2019) Research methodology: methods and techniques

Lantz B (2013) The large sample size fallacy. Scand J Caring Sci 27:487–492. https://doi.org/10.1111/J.1471-6712.2012.01052.X

Lee S, Lee DK (2018) What is the proper way to apply the multiple comparison test? Korean J Anesthesiol 71:353–360. https://doi.org/10.4097/KJA.D.18.00242

Lewis-Beck MS, Bryman A, Futing LT (2004) Two-way Anova. In: The SAGE encyclopedia of social science research methods. Sage, Thousand Oaks, pp 1150–1151. https://doi.org/10.4135/9781412950589

Lovell DP (2013) Biological importance and statistical significance. J Agric Food Chem 61:8340–8348. https://doi.org/10.1021/JF401124Y/ASSET/IMAGES/LARGE/JF-2013-01124Y_0001.JPEG

McNulty R (2022) A logical analysis of null hypothesis significance testing using popular terminology. BMC Med Res Methodol 22:1–9. https://doi.org/10.1186/S12874-022-01696-5/TABLES/4

Midway S, Robertson M, Flinn S, Kaller M (2020) Comparing multiple comparisons: practical guidance for choosing the best multiple comparisons test. PeerJ 8:e10387. https://doi.org/10.7717/PEERJ.10387/SUPP-1

Mugwe J, Monicah AE, Ae M-M et al (2009) Adoption potential of selected organic resources for improving soil fertility in the central highlands of Kenya. Agrofor Syst 76:467–485. https://doi.org/10.1007/s10457-009-9217-y

Muriuki JP (2011) An evaluation of organic and inorganic inputs for soil nutrient replenishment in Mukuuni and Murugi Central Kenya

Musafiri CM, Kiboi M, Macharia J et al (2022) Adoption of climate-smart agricultural practices among smallholder farmers in Western Kenya: do socioeconomic, institutional, and biophysical factors matter? Heliyon 8. https://doi.org/10.1016/j.heliyon.2021.e08677

Mwebia FW (2019) Influence of farmer groups' characteristics on up scaling uptake of soil fertility Technologies in Eastern Kenya. Kenyatta University

Nathanson BH, Higgins TL (2008) An introduction to statistical methods used in binary outcome modeling. Semin Cardiothorac Vasc Anesth 12:153–166. https://doi.org/10.1177/1089253208323415

Ntumi S (2021) Reporting and interpreting one-way analysis of variance (ANOVA) using a data-driven example: a practical guide for social science researchers. J Res Educ Sci 12:38–47. https://doi.org/10.14505/JRES.V12.14.04

Parolini G (2015) The emergence of modern statistics in agricultural science: analysis of variance, experimental design and the reshaping of research at Rothamsted Experimental Station, 1919–1933. J Hist Biol 48:301–335. https://doi.org/10.1007/S10739-014-9394-Z/METRICS

Pérez-Guerrero EE, Guillén-Medina MR, Márquez-Sandoval F et al (2024) Methodological and statistical considerations for cross-sectional, case–control, and cohort studies. J Clin Med 13:4005. https://doi.org/10.3390/JCM13144005

Peterson A, Cruse D, Naci L et al (2015) Risk, diagnostic error, and the clinical science of consciousness. Neuroimage Clin 7:588–597. https://doi.org/10.1016/J.NICL.2015.02.008

Pinheiro-Guedes L, Martinho C, O Martins MR (2024) Logistic regression: limitations in the estimation of measures of association with binary health outcomes. Acta Medica Port 37:697–705. https://doi.org/10.20344/AMP.21435

Prematunga RK (2012) Correlational analysis. Aust Crit Care 25:195–199. https://doi.org/10.1016/J.AUCC.2012.02.003

Pugh SL, Molinaro A (2016) The nuts and bolts of hypothesis testing. Neurooncol Pract 3:139–144. https://doi.org/10.1093/NOP/NPV052

Puth MT, Neuhäuser M, Ruxton GD (2014) Effective use of Pearson's product–moment correlation coefficient. Anim Behav 93:183–189. https://doi.org/10.1016/J.ANBEHAV.2014.05.003

Ruxton GD, Wilkinson DM, Neuhäuser M (2015) Advice on testing the null hypothesis that a sample is drawn from a normal distribution. Anim Behav 107:249–252. https://doi.org/10.1016/J.ANBEHAV.2015.07.006

Salinas Ruíz J, Montesinos López OA, Crossa J (2024) Randomized complete block design. In: Introduction to experimental designs with PROC GLIMMIX of SAS. Springer, Cham, pp 75–98. https://doi.org/10.1007/978-3-031-65575-3_4

Salkind NJ (2007) Grand mean. In: Encyclopedia of Measurement and Statistics. Sage, Thousand Oaks, p 412. https://doi.org/10.4135/9781412952644

Schober P, Vetter TR (2020) Nonparametric statistical methods in medical research. Anesth Analg 131:1862–1863. https://doi.org/10.1213/ANE.0000000000005101

Slayi M, Zhou L, Jaja IF (2023) Smallholder farmers' adoption and perception of communally established cattle feedlots for climate change resilience in the Eastern Cape, South Africa. Front Sustain Food Syst 7:1239766. https://doi.org/10.3389/FSUFS.2023.1239766/BIBTEX

Srimaneekarn N, Hayter A, Liu W, Tantipoj C (2022) Binary response analysis using logistic regression in dentistry. Int J Dent 2022:5358602. https://doi.org/10.1155/2022/5358602

Thode HC (2002) Testing for normality. https://doi.org/10.1201/9780203910894

Thukral S, Kovac S, Paturu M (2023) T test. In: Translational interventional radiology. Academic, London, pp 139–143. https://doi.org/10.1016/B978-0-12-823026-8.00104-8

Vickers AJ (2005) Parametric versus non-parametric statistics in the analysis of randomized trials with non-normally distributed data. BMC Med Res Methodol 5:1–12. https://doi.org/10.1186/1471-2288-5-35/COMMENTS

Visscher PM, Goddard ME (2019) From R.A. Fisher's 1918 paper to GWAS a century later. Genetics 211:1125–1130. https://doi.org/10.1534/GENETICS.118.301594

Whitley E, Ball J (2002) Statistics review 3: hypothesis testing and P values. Crit Care 6:222–225. https://doi.org/10.1186/CC1493/FIGURES/1

Wisniewski SJ, Brannan GD (2024) Correlation (coefficient, partial, and spearman rank) and regression analysis. StatPearls

Zhu W, Spanos A (2023) Revisiting the large n (sample size) problem: how to avert spurious significance results. Stats 6:1323–1338. https://doi.org/10.3390/STATS6040081

Survey Data Analysis Using Statistical Packages for Social Sciences (SPSS)

22

Abstract

Data from the agricultural surveys include various types ranging from numeric to non-numeric data. These mainly include crop yield data, livestock production data, farmer demographics, and farming practices. Statistical Package for the Social Sciences (SPSS) is well-suited for analyzing both numeric and non-numeric data from such surveys. Despite the availability of many resources on data analysis using SPSS, the complexity and multifaceted nature of agricultural survey data require precise statistical methods to achieve meaningful conclusions. The chapter aims to bridge this gap by providing a step-by-step guide on analyzing survey data, supported by practical-based field data. It starts by providing an overview of SPSS, showing its structure and key functionalities. We start by demonstrating how to compute the various descriptive statistics, including frequencies, central tendency measures, and variability. Following the descriptives, the chapter covers inferential statistics, demonstrating how to calculate chi-square statistics (test of association) between two categorical variables. It also covers the Pearson correlation to assess the relationship between two numeric variables. The chapter also demonstrates how to compute simple linear and binary logistic regression. By the end of this chapter, students are expected to have a solid understanding of how to use SPSS to analyze survey data in agricultural research effectively.

Keywords

Chi-square statistic · Cross tabulation · Data view window, Frequency table, Logistic regression · Pearson product-moment correlation · Two-way table · Toolbar · Variable view window

Learning Outcomes

After reading this chapter, you should be able to:

1. Appreciate the different types of data from a survey.
2. Demonstrate a basic understanding of the SPSS structure.
3. Conduct basic descriptive analysis using SPSS.
4. Carry out a univariate data analysis and a cross-tabulation.
5. Compute the mean, standard deviation, minimum and maximum values in a data set.
6. Plot box plots with associated measures of variability.
7. Conduct a Pearson Chi-square cross-tabulation using two categorical variables.
8. Conduct a Pearson Product-Moment using two quantitative variables.
9. Appreciate the calculation of simple linear regression and binary logistic regression.
10. Demonstrate some understanding of how to conduct inferential analysis of selected statistics using SPSS.

22.1 Overview of SPSS

22.1.1 Why Survey Data Analysis Using SPSS

The survey is one of the most commonly used designs in agriculture, especially by students in the social disciplines of agricultural research (Chap. 12). Data from the survey includes various types, such as crop yields, soil health parameters, pest management, farmer demographics, and practices. Although many books cover data analysis using SPSS (Landau and Everitt 2003, 2021; Carver and Nash 2011; Kirkpatrick and Feeney 2016), agricultural survey data can be complex and multifaceted, as outlined in Sect. 12.1.3. The complexity arises due to the diverse nature,

J. Njeri Mugwe, S. Runo, *Research Methodology in Agricultural Sciences*, https://doi.org/10.1007/978-981-95-1892-0_22

which involves multiple variables and interactions between different factors, requiring unique statistical methods to sense of the data collected.

Students carrying out surveys for the first time and trying to learn SPSS from general resources may encounter significant challenges. These general resources often lack specific guidelines in agricultural data analysis, leading to difficulties in applying SPSS effectively. This chapter aims to provide step-by-step guidance, using practical examples and tailored advice to help students navigate these challenges and use SPSS more effectively in their agricultural research projects.

22.1.2 Nature of SPSS

Statistical Package for the Social Sciences (SPSS) is a data analysis software widely used in social science studies. It is a versatile software that can analyze data from different sources and open various file formats for structured data. These formats include, but are not limited to, spreadsheets from MS Excel (.xls.xlxs), plain text files (.txt or.csv), Stata, and SAS files.

The SPSS software is also a powerful user-friendly software that can analyze, manipulate, and present data (McCormick and Salcedo 2017) and it is widely used to analyze survey data (Sweet and Grace-Martin 2003). Its extensive array of features performs a wide range of statistical tests, manages and transforms data, and generates detailed reports and visualizations (McCormick and Salcedo 2017). It is simple to use as it contains a straightforward interface with dialogue boxes, which makes it accessible even for those with limited statistical or computing packages (McNamara 2019). Additionally, SPSS includes a help menu and tutorials to help first-time users get started.

22.1.3 What Data Do We Have?

Let us consider the household survey questionnaire presented in Chap. 18, which was used to assess factors influencing the adoption of integrated soil fertility management. The data generated was numeric and non-numeric, i.e., quantitative and qualitative. Examples of such data include the following:
Numeric data:

- Age of household head in years.
- Farm size in hectares or acres.
- Number of trees.
- Number of active members in a household.
- Average land dedicated to integrated soil fertility management (ISFM).
- Changes in crop yield.

Non-numeric data:

- Education level of household head.
- Gender of household head.
- Type of crops grown.
- Type of ISFM practices.
- Perceived benefits of practices.

The analysis of each of these data depends on the data type and what we want to find out, i.e., the analysis objective. Numeric data is subjected to quantitative assessments and statistical analysis, such as testing for differences and correlations (Morris et al. 2019; Pérez-Guerrero et al. 2024). For example, we might analyse how farm size correlates with crop yields. In contrast, non-numeric data will mainly be analyzed using frequencies to understand distribution patterns and associations to explore relationships between categorical variables. For example, we might examine how the household level is associated with the type of ISFM practiced or the perceived benefits of ISFM.

22.1.3.1 Steps in Conducting the Analysis

We have structured the discussion of this chapter into three sub-sections: 1. A brief overview of SPSS software to show the key features and functionalities for data analysis; 2. How to carry out descriptive statistics and generate main ones; and 3. How to carry out inferential statistics, covering differences in means and simple correlations.

22.1.4 Structure of SPSS

22.1.4.1 Data View and Variable View Window

Upon opening SPSS, you are presented with the main interface consisting of two windows: A *Data View window* and a *Variable View window* (Fig. 22.1a) (IBM 2021). The data view window is used for setting up your data for accurate analysis. It shows cases or observations and columns representing variables. The variable window is where you define and manage the properties of each data set. It is used to describe the variables associated with the data in the data view interface window (Fig. 22.1b).

22.2 The Toolbar

The toolbar provides access to various functions and tools for managing, transforming, and analyzing data. We have labelled these main components from 1 to 9 (Fig. 22.2a). We also present a brief description of this toolbar menu. Figures from 22.2b, 22.2c, 22.2d, 22.2e, 22.2f, 22.2g, 22.2h, 22.2i, and 22.2j show how the toolbar's functions open and the various commands users can use in the analysis.

Fig. 22.1a Data and variable view interface in SPSS

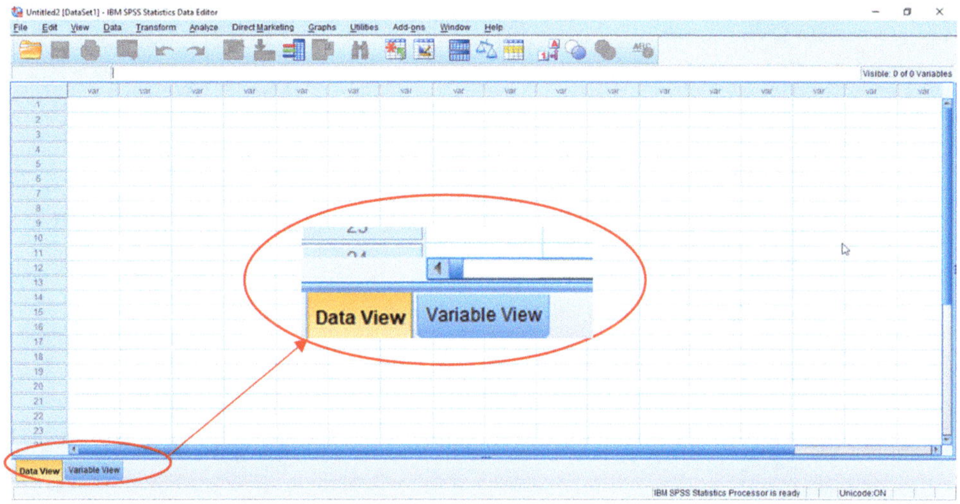

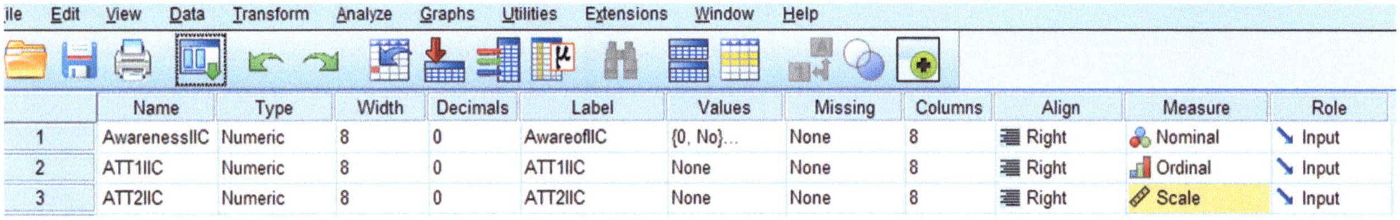

Fig. 22.1b Variable view interface in SPSS with some data already entered

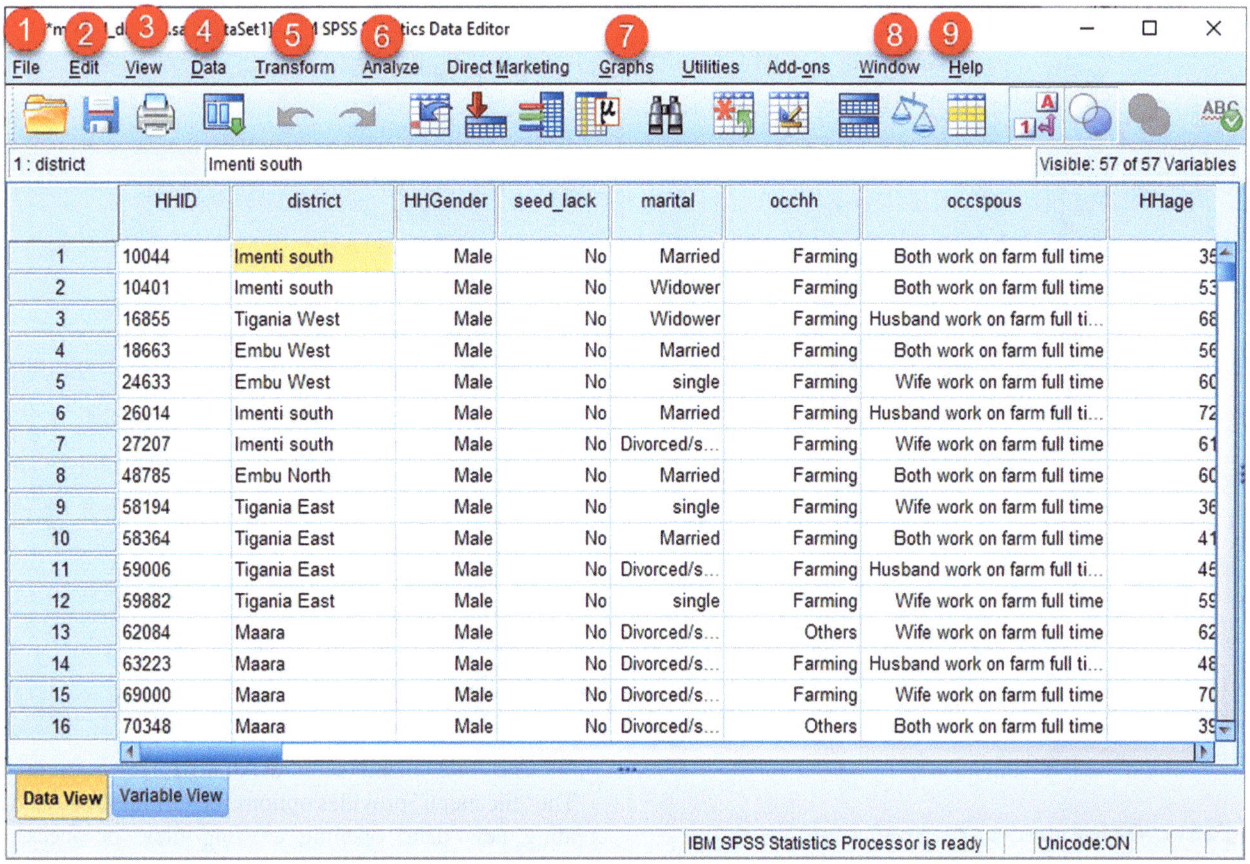

Fig. 22.2a Toolbar showing the various menus in SPSS

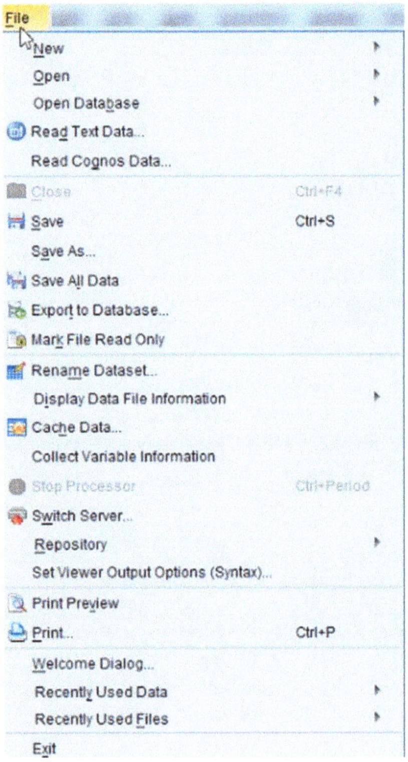

Fig. 22.2b The File menu

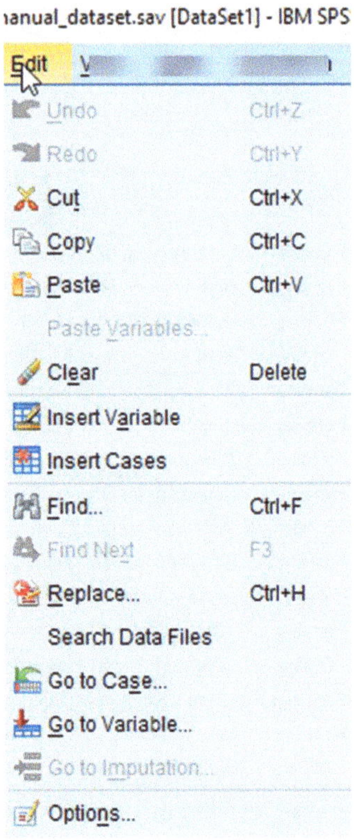

Fig. 22.2c The Edit menu

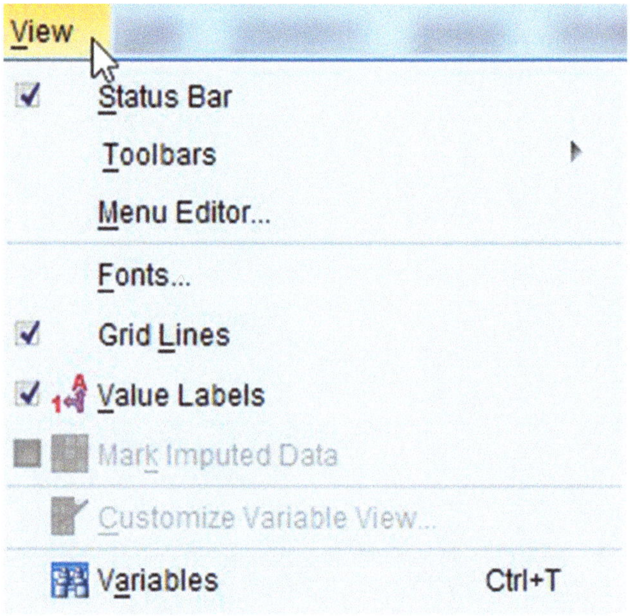

Fig. 22.2d The View menu

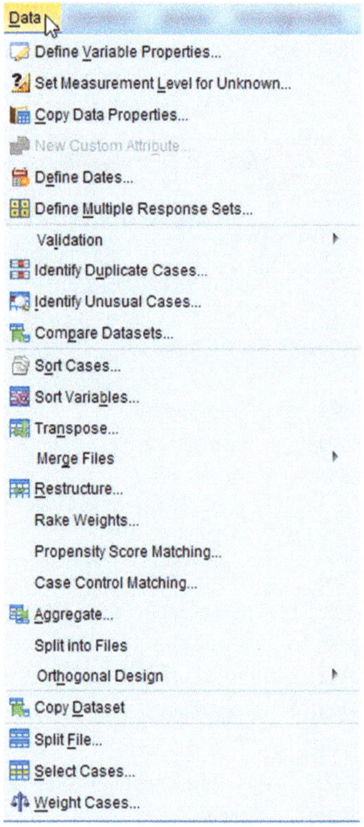

Fig. 22.2e The Data menu

1 The "File" menu (shortcut Alt + F)

The "file menu" provides options for various tasks such as creating new data, opening existing files, or accessing recently opened files. Additionally, it enables the importation of data from other file formats like.dta. Saving the data and

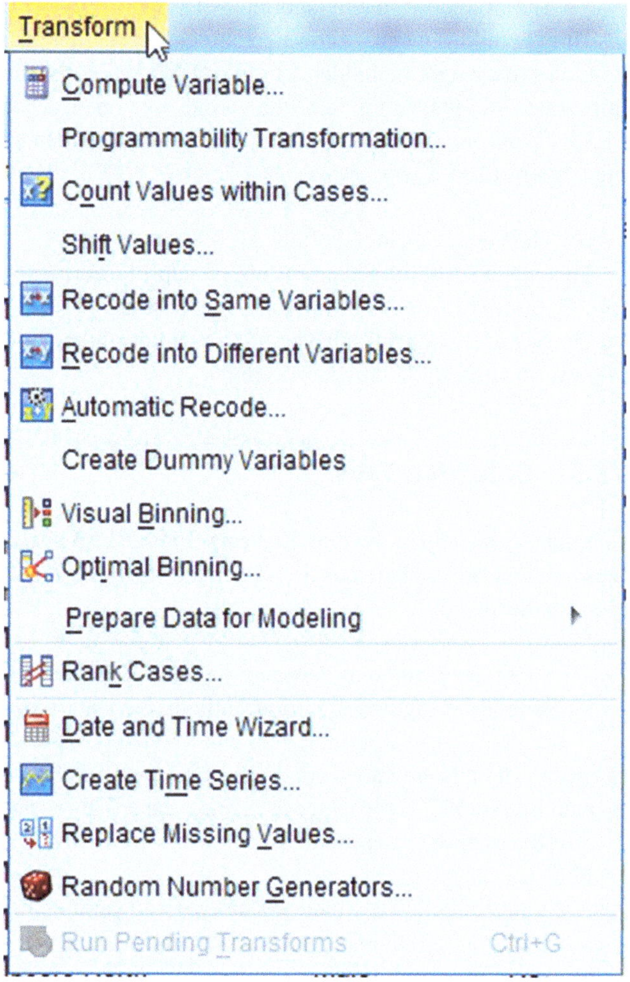

Fig. 22.2f The Transform menu

exiting the program are also essential functions facilitated by this menu.

2 The “Edit” menu (shortcut Alt + E)

The “edit menu” resembles that of a word processor, so if you have prior experience with word processors, it will feel familiar. The Undo and Redo functions allow you to reverse or repeat executed commands, aiding in correcting any errors you may encounter. Additionally, the Find and Go to Case options assist in locating specific cases within the dataset.

3 The “View” menu (shortcut Alt + V)

The “View menu” manages the visual presentation of the spreadsheet, including aspects such as displayed toolbars, fonts, grid lines presence or absence, and whether value labels are visible for variables. The final option permits viewing or inputting details for various variables.

4 The “Data” menu (shortcut Alt + D)

This “Data Menu” allows you to organize your data file.

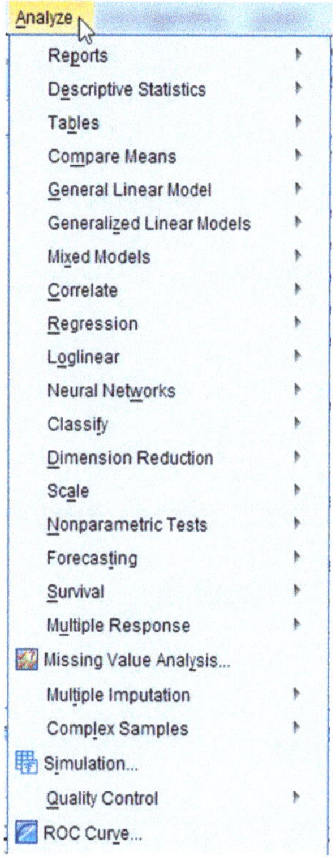

Fig. 22.2g The Analyze menu

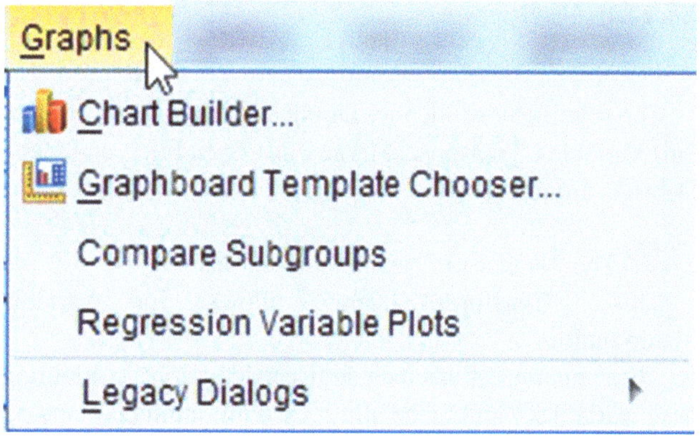

Fig. 22.2h The Graphs menu

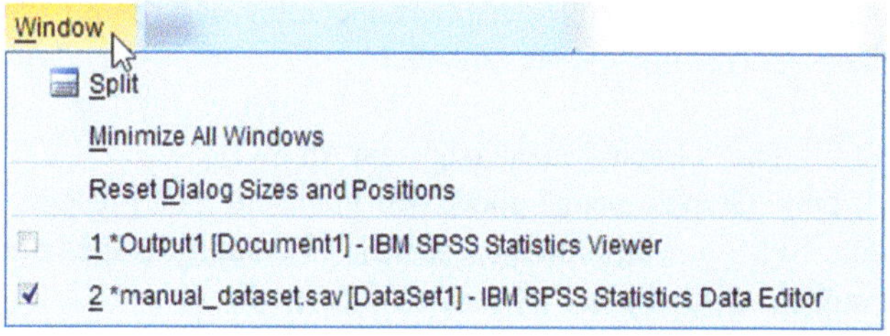

Fig. 22.2i The Window menu

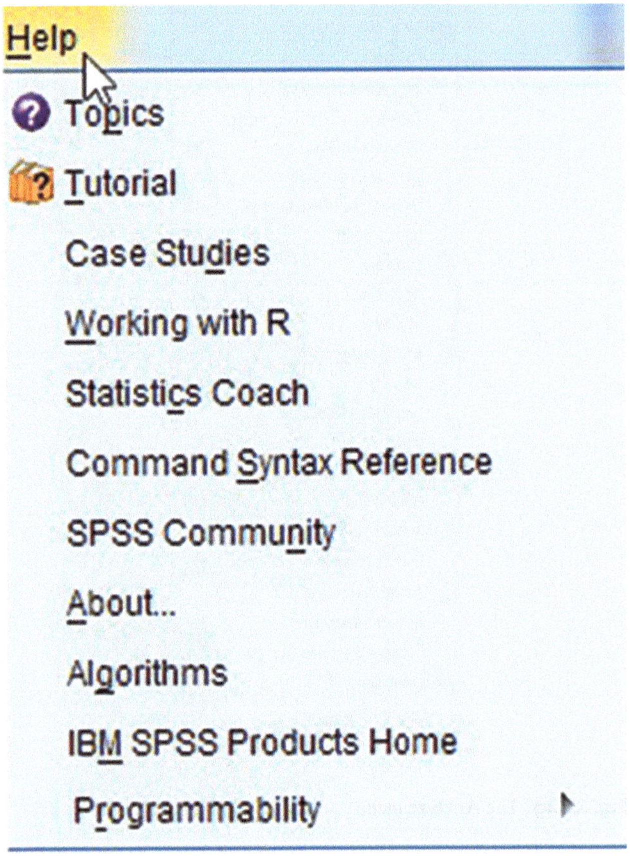

Fig. 22.2j The Help menu

The role includes Identify Duplicate Cases, Sort Cases or Sort Variables, Transpose, Merge Files, Split Files, and Copy Datasets among others

5 The “Transform” menu (shortcut Alt + T)

This “Transform menu” allows for variable manipulation.

These manipulations include: Recode options, computing new variables, and generating random numbers, among others.

6 The “Analyze” menu (shortcut Alt + A)

“Analyze” is the most used tab in SPSS. The statistical options and tests are located here including descriptive statistics, correlations, regression etc.

7 The “Graphs” menu (shortcut Alt + G)

The “Graphs menu” allows for graphical data presentation. SPSS has different approaches for creating graphs but graphs is a good place to start.

The Chart Builder allows for graph creation by dragging and dropping variables on a pre-set virtual canvas creating a greater flexibility in graphing.

8 The “Window” menu (shortcut Alt + W)

This “window menu” allows for quick access to other windows that might be hidden from view. The menu is also important in managing all the windows opened on SPSS. Therefore, the windows menu is important in managing clutter of SPSS windows.

9 The “Help” menu (shortcut Alt + H)

This menu is useful because it offers helpful information for SPSS software and the available statistic tests.

The Statistics Coach is also helpful in understanding how to perform specific statistical procedures.

22.3 Data View Tabs

The data view tab has several elements. Figure 22.3 shows these elements (numbered 1–5). The functions of each part are also discussed.

① The tabs for switching between the Data View and the Variable View. Figure 22.3 is currently on the Data View.

② Rows of cells are called *cases*. Each respondent/participant in a study is represented as a single case. This is also referred to as an observation.

③ Columns of cells are called *variables*. Each variable has a unique name (“hhid”) which is shown in the column header.

④ The cell contents are referred to as *values*.

⑤ *Status bar* gives handy information on the data, e.g., the Unicode mode is in effect.

22.4 Components of the Variable View Window

The Variable View does not show the data; rather, it shows information associated with it, which is also called *metadata* (Fig. 22.4).

To summarize this section, SPSS contains two main components: *The Data View*, where data values are displayed, and *the Variable View*, where information associated with the data is displayed, such as variable names and measurement levels (Numeric, Nominal). It allows users to define and edit the properties of each variable. For viewing the data already entered, you need so switch to the data view window.

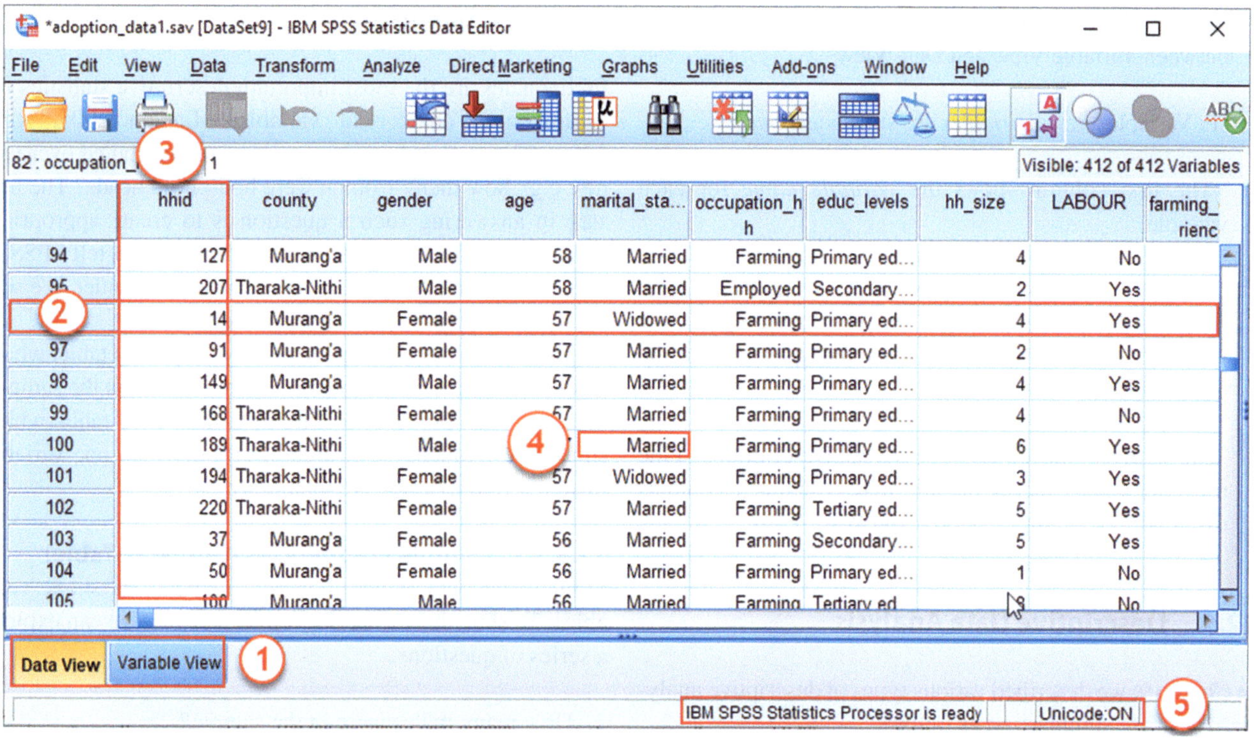

Fig. 22.3 Components of the Data View tab showing data

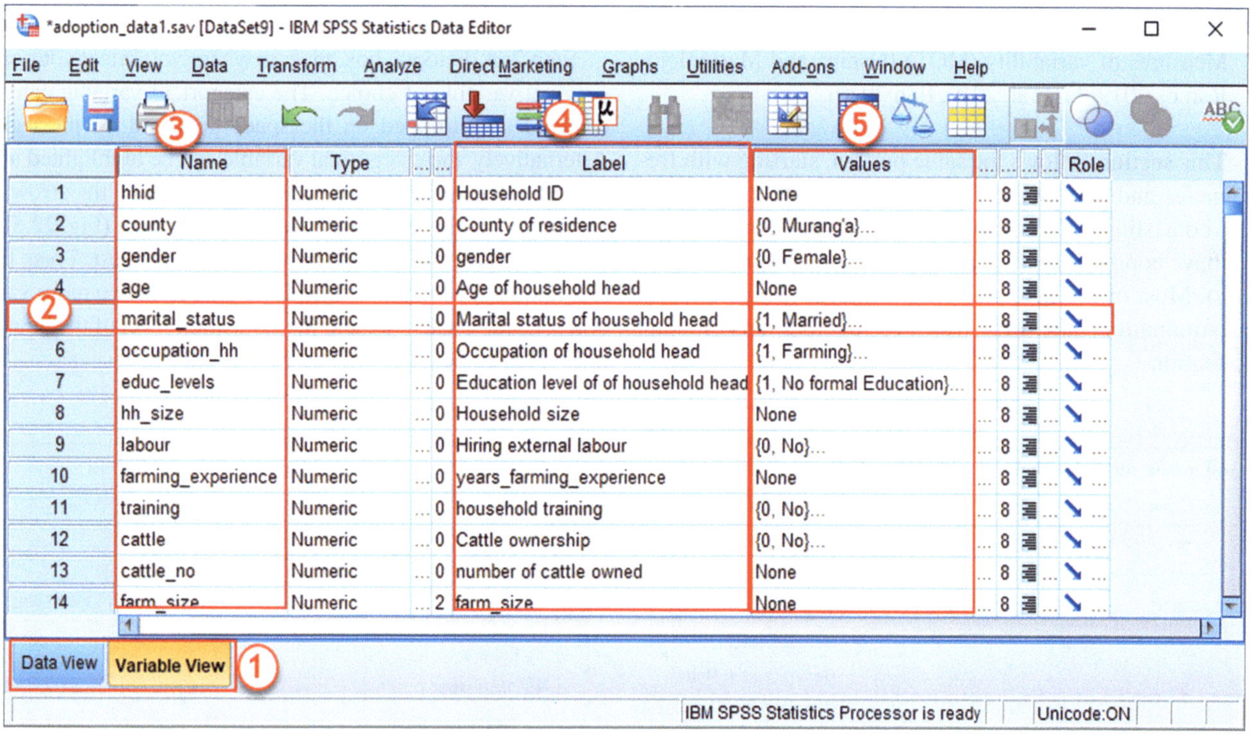

Fig. 22.4 Components of the Variable View tab

① In the left bottom corner, we find *tabs* for switching between Variable View and Data View.

② In Variable View, *variables* are shown as rows

③ The first column shows the *variable name* for each variable.

④ The fifth column contains a *variable label* (which is optional). This label describes the exact meaning of each variable, e.g., it can be read as the question in the questionnaire.

⑤ The sixth column shows *value labels*: Descriptions of the meaning values that a variable may contain. It contains the code. For example, No = 0 and Yes = 1.

22.5 Descriptive Data Analysis

In Chap. 19, we described various types of descriptive analysis. These are summarized as follows;

- Counts, frequencies, and percentages (Salkind 2007; Manikandan 2011a; 2014).
- Measures of central tendency (MOC) (Manikandan 2011b; Jankowski and Flannelly 2015).
- Measures of variability (MOT) (Ospina and Marmolejo-Ramos 2019; Li et al. 2019; Ilan 2020).

This section follows the same pattern, starting with frequencies and then MOC and MOV. The data used is varied, consisting of hypothetical and real data from projects we have coordinated (Listed in the acknowledgments section). Most of the data are borrowed from (Mugwe 2007) questionnaire, part of Chap. 18 of the survey data collection.

22.5.1 Frequencies

Frequencies are the most important descriptive statistics for summarizing categorical variables (Larson 2006; Ross 2009). They tell us how frequencies are distributed over values, e.g., how many women were household heads? The first step in answering such a question is to create appropriate tables and charts. The process will, for instance, tell SPSS to count the number of females in the variable gender. We will explore two ways to look at frequency tables.

First, we will do a simple one-way frequency table, which answers the question of the number of females in the sample. Then, we will proceed to a multiple-frequency table, which shows the frequency distribution of more than one variable (also referred to as cross-tabulation).

22.5.1.1 Simple Frequencies (One-Way Table)

Simple frequency tables show the frequency of single variables. To show how to perform this analysis, we can explore a series of questions:

1. How many males were in the sample?
2. How many households had cattle?
3. How many household heads had secondary education?

Step 1: Click on analyze > Descriptive statistics > Frequencies steps 1, 2, 3 in Fig. 22.5a.

Step 2: A dialogue box appears where you select your categorical variable of choice. The categorical variable can be dragged and dropped on the space provided on the right. Alternatively, the categorical variable can be highlighted and moved to the variable side on the right by using the arrow in the middle. In this case, we have selected gender (Fig. 22.5b).

Step 3: After selecting the variable of interest, Press the OK button to generate results. Here we select two more variables: Cattle ownership and the Education level of the household (Fig. 22.5c).

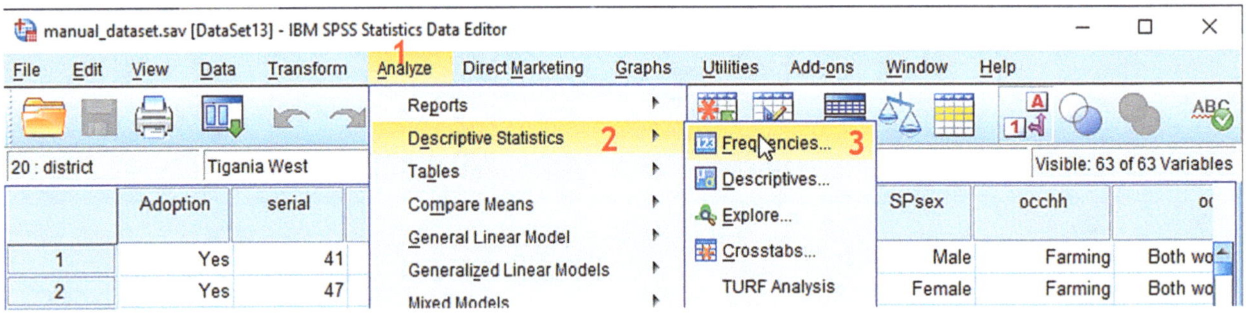

Fig. 22.5a Calculating frequencies

Step 4: The output shows the analysis results. The highlighted results answer the questions we posed earlier. It is important to note that we included the whole variable when presenting the results for categorical variables. For example, even though we can say *x* number of females were household heads, it is more insightful to include the number of male household heads (Fig. 22.5d).

Presenting frequency table from SPSS

We use the set of questions in the questionnaire in Chap. 18 to plot a one-way table. In the module on characterizing farmers, we collected data on gender variables. We had question 2 as this;

2. Gender of household head Male ☐ Female ☐

From the SPSS results, we present data category-wise where they stand alone. The frequency distribution table from SPSS provides the frequency (number of occurrences), percentage (proportion of each value relative to the total observations), and cumulative frequency (Fig. 22.5d). However, in our presentation, we include the frequency and the proportion (Table 22.1). This approach allows comparison of the different categories of the variable. We call this *univariate analysis*, which involves one variable at a time.

Note that we can carry out univariate analysis for all other non-numeric variables. This includes the Education level of the household head, types of ISFM, etc.

22.5.1.2 Multiple Frequency Tables

We use multiple frequency tables to describe how two or more categorical variables are distributed (Cooksey 2020a, b). However, the frequency table used in this approach differs from the one-way tables we discussed earlier. In this case, we use a special type of frequency table called a *cross-tabulation*. It is also called a *crosstab*, *two-way table*, or *contingency table*. It allows the examination of the relationship between two or more categorical variables.

In a cross-tabulation, one variable's category defines the rows of the frequency table, while the other variable's categories determine the columns (Cooksey 2020a). The cells within the table display the frequency of occurrences for each combination of each category. This provides a clear view of how categories interact and overlap, making it easier to analyze relationships between variables.

In this part, let us ask the following two questions;

1. How many males and females are in different districts?
2. How many household heads were in different levels of education?

Step 1: Click on analyze > Descriptive statistics > Crosstabs (Fig. 22.6a).

Step 2: Assign the variables of interest to Rows and Columns; then select the district variable for the Rows and the Gender variable for the Column (Fig. 22.6b).

Step 3: At this point, if you click ok, you will get the results, but the results will show the distribution of gender in

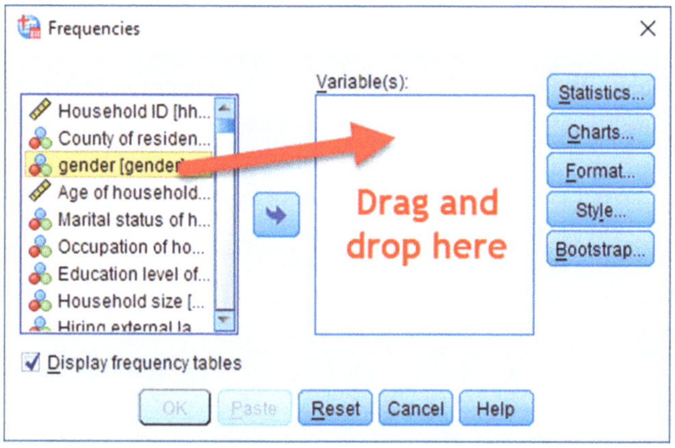

Fig. 22.5b Gender variable selected

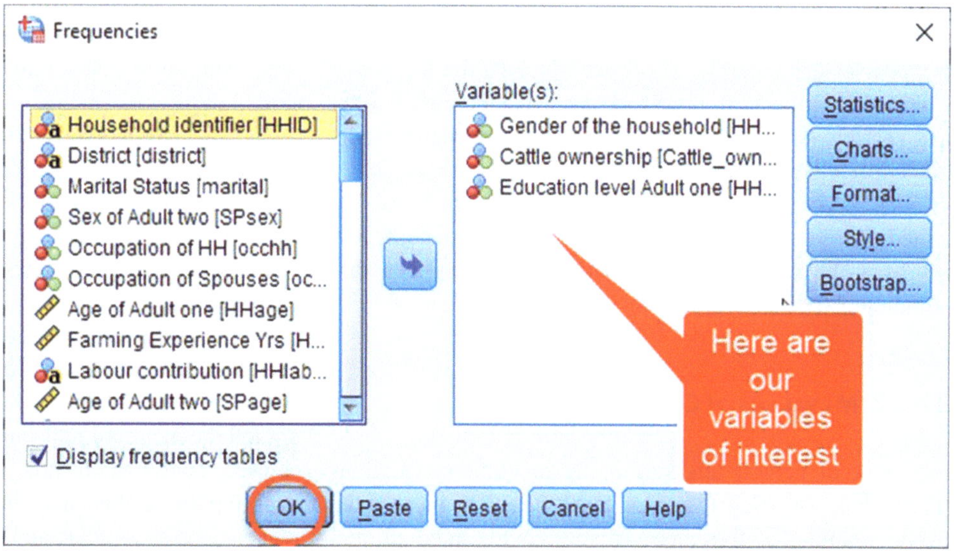

Fig. 22.5c Gender of household head, cattle ownership and Education level selected as our variables of interest

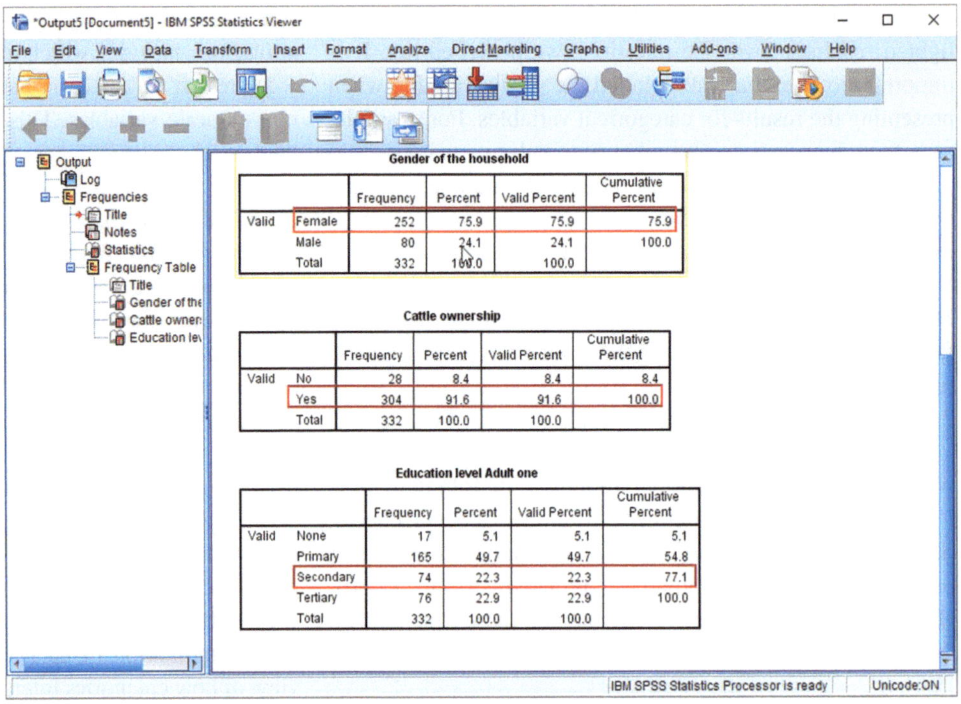

Fig. 22.5d Output of analysis of frequencies for Gender, Cattle ownership, and Education level

Table 22.1 Results showing one-way table showing the frequency of gender

Gender	Frequency	Percent
Female	13	11
Male	93	87
Total	106	100

the districts. In such cases, we would want to see the number of males and females distributed over the districts and the percentages associated with the actual counts. Click Cells > Check the row and the Column box > Continue > Ok on the crosstab window to get the percentages. See Fig. 22.6c.

Step 4: After clicking OK on the Crosstab window, SPSS generates an output as shown in Fig. 22.6d. Apart from the frequencies generated by SPSS, we illustrate the Rows and Columns. For instance, if we had placed Gender on the Row and District on the Column, we would have the same results but a different table setting.

Step 5: The generated output can be copied to Word and edited. Please note that we must not include all the generated results in the final table; we consider only the number that communicates our message (Table 22.2)

22.5.2 Measures of Central Tendency

Measures of central tendency apply to numeric variables (continuous variables) (Manikandan 2011b). The mean is the most important of the three statistics (mode, median, and mean). We will perform the analysis to answer the following questions to demonstrate how to calculate the mean.

1. What is the mean age of the household head?
2. What is the mean age of males?
3. What is the mean age of females?
4. What is the mean number of cattle in different districts?

Step 1: Click on analyze > Descriptive statistics > Descriptives (Fig. 22.6e).

Step 2: On the Descriptives window, select the numeric variable of interest. In our case, age of the household head (Fig. 22.6f).

Step 3: Click on the *options*; when the options window appears, check the following boxes: mean, standard deviation, minimum, and maximum. Click continue > Ok (Fig. 22.6g)

Step 4: The previous steps (1–3) will give you the final descriptive statistics (Fig. 22.6h)

Step 5: This step involves presenting the above table in an editable table (Table 22.3).

We will use the split case approach to answer the question of the mean age of males and females in the sample. The split case option in SPSS allows the program to present the results in subgroups. For example, the above results Fig. 22.6h shows the mean age of the households. In this case, one cannot tell the mean for male and female headed household categories. Therefore, in this case, we will walk you through how to perform this analysis, to generate mean for the separate categories for male and female.

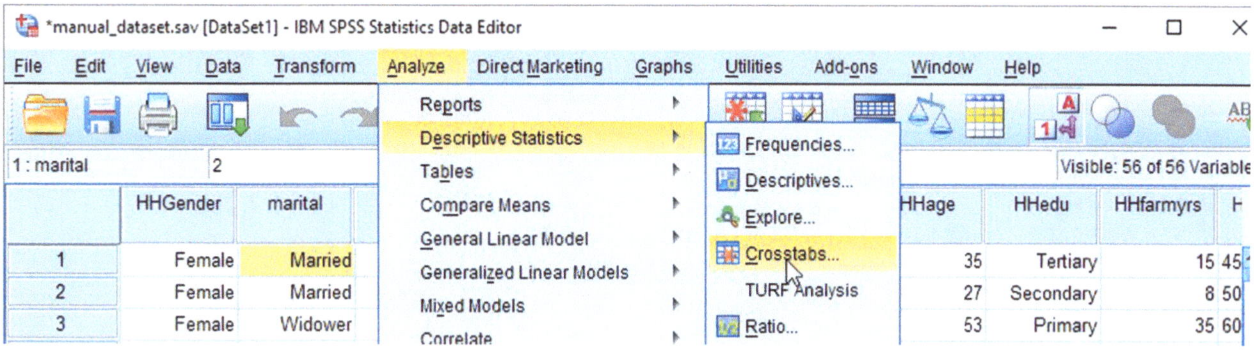

Fig. 22.6a Creating a cross-tabulation

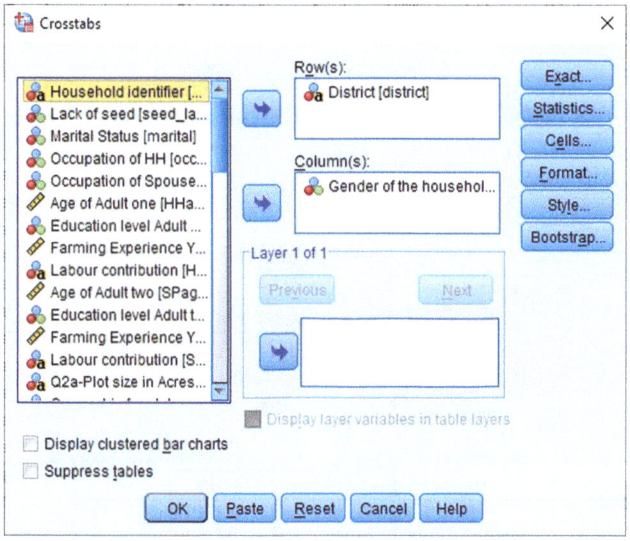

Fig. 22.6b Selecting rows and columns in cross-tabulation

Step 1: Click on Data > Split File (Fig. 22.6i)

Step 2: On the split file window, select the categorical variable in which the means should be presented. In this case, we choose gender. Then, click Ok. This does not bring numbers as output, but it shows that it has executed the "setting" (Fig. 22.6j).

Step 3: Click on analyze > Descriptive statistics > Descriptives (Fig. 22.6k).

Step 4: On the Descriptives window, select the numeric variable of interest. In our case, age of the household head (Fig. 22.6l).

Step 5: Click on the *options*; when the options window appears, check the following boxes: mean, Std. deviation, minimum, and maximum. Click continue > Ok (Fig. 22.6m).

Step 6: The output generated from the procedure is as shown in Fig. 22.6n.

The output in Fig. 22.6n can be presented in one table as shown in Table 22.3.

22.5.3 Measures of Variability

22.5.3.1 Range, Standard Deviation and Variance

In this section, we will derive the three main measures of dispersion: *Range, standard deviation, and variance*. The range is the value equal to the largest value (Max) minus the smallest value (Min). We will consider the following continuous variables;

1. Age of the households
2. Number of cattle owned by households

Step 1: Click on analyze > Descriptive statistics > Frequencies (see the steps in the display) (Fig. 22.7a)

Step 2: On the frequency window, select the age of the household head by dragging and dropping or by using the middle arrow. Deselect the display frequency tables (Fig. 22.7b).

Step 3: After selecting the age variable, click on statistics. In the statistics window that appears, check the following boxes under Dispersion: Standard deviation, Variance, and Range (Fig. 22.7c). Then, click continue.

Step 4: Net Click Ok on the frequency window to obtain output (Fig. 22.7d).

Step 5: The following output will be generated, showing standard deviation, range, and variance (Fig. 22.7e)

22.5.3.2 Box Plots

For box plots, we use variables in pairs; one variable is categorical (often on the x-axis), and the other is numeric (often on the y-axis). Please note that you can use one numeric variable, too. The variables in the pair considered for this demonstration are:

- Age and gender of the households.
- Number of cattle and districts.

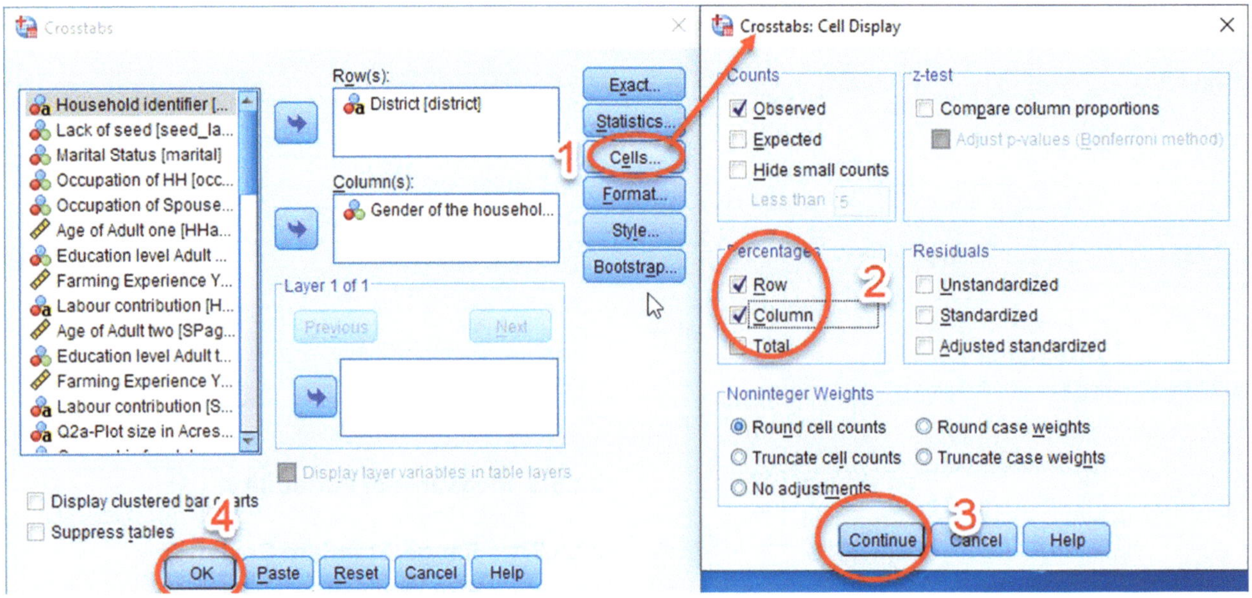

Fig. 22.6c Cross-tabulation steps

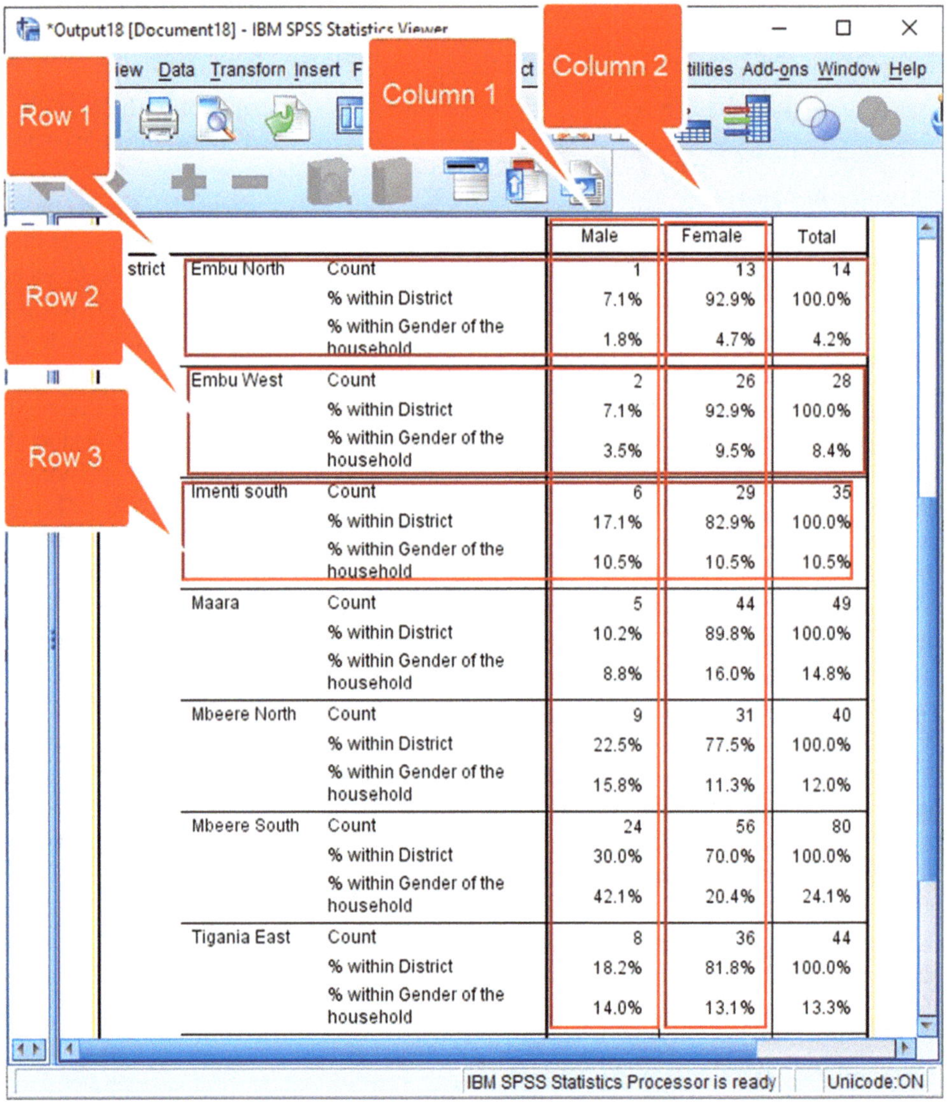

Fig. 22.6d Output of cross-tabulation

Single variables considered are:

- Age of the household head.
- Number of cattle.

Step 1: Click on Graphs > Chart builder (Fig. 22.7f)

Step 2: Select the boxplot on the chart builder window, then a simple box plot. On the variables, drag the gender variable on the x-axis and the age variable on the y-axis, then select OK (Fig. 22.7g).

Step 3: A boxplot is generated (Fig. 22.7h).

The same output as above with additional illustration (Fig. 22.7i).

Table 22.2 Results showing cross-tabulation of gender distribution across districts

Districts	Male	Female	Total
Embu North	1 (7.1)	13 (92.9)	14 (100)
Embu West	2 (7.1)	26 (92.9)	28 (100)
Imenti south	6 (17.1)	29 (82.9)	28 (100)
Maara	5 (10.2)	44 (89.8)	35 (100)
Mbeere North	9 (22.5)	31 (77.5)	40 (100)
Mbeere South	24 (30)	56 (70)	80 (100)
Tigania East	8 (18)	36 (81)	44 (100)
Tigania West	2 (4.8)	40 (95.2)	42 (100)
Total	57 (17.2)	275 (82.8)	332 (100)

NB Numbers in parenthesis are percentages

22.6 Inferential Statistics

22.6.1 Means Comparison

22.6.1.1 Multiple Means

We start with comparisons (means) while dealing with many groups using *Analysis of Variance (ANOVA)*. To perform an ANOVA, we need three or more groups to compare and a single quantitative variable (numeric variable). The data should be random, and the outcome data should be distributed normally. Also, we assume that there is equal variance among the groups.

In our case we compare three groups based on the occupation of the household head (e.g., Farming, Business, and Others) and the years of farming experience. In essence, we want to test whether the mean number of years differs significantly across these three occupational groups.

Step 1: Click analyze > Compare means > One-Way ANOVA (Fig. 22.8a)

Step 2: Assign the years of experience to the dependent variable list and the occupation variable to the factor (Fig. 22.8b).

Step 3: Click on the options; on the options dialogue box, check the following boxes: Descriptive, Homogeneity of variance tests, Brown-Forsythe Tests on equality of group variances based on executing an ANOVA on a transformation of the response variable), Welch T-Test (to compare the means of two groups when the variances of the two groups are not assumed to be equal) and Means plot. Click continue,

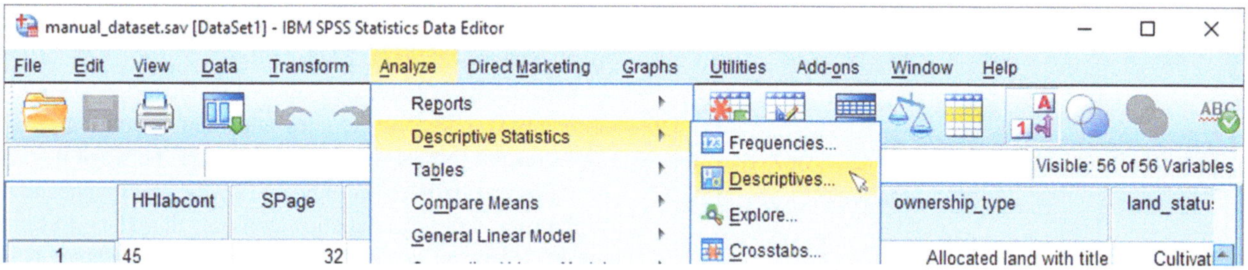

Fig. 22.6e Starting off descriptive statistics Command

Fig. 22.6f Steps for obtaining descriptive statistics in SPSS include selecting variables, choosing desired statistical options, and confirming to generate the output

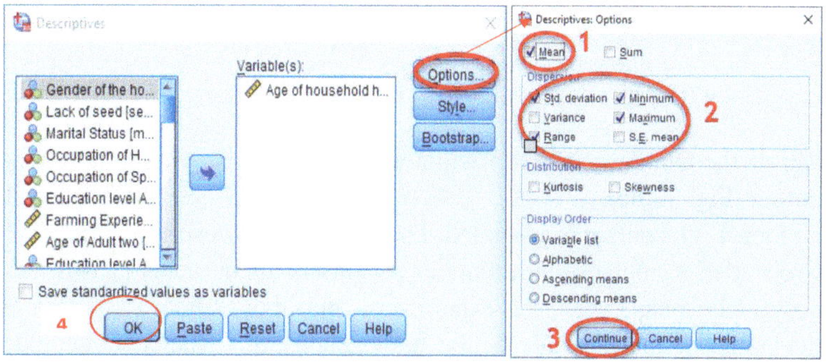

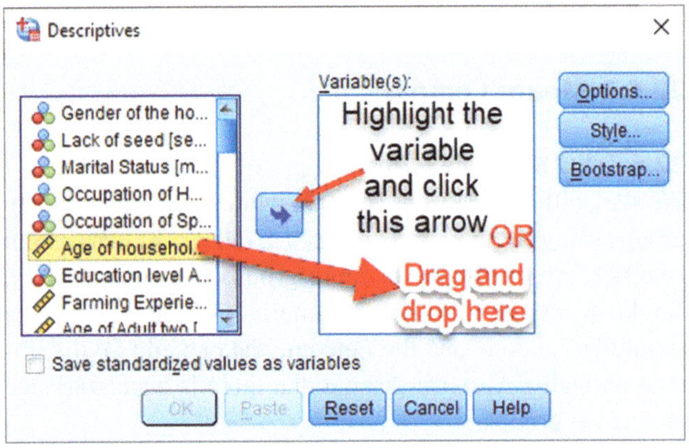

Fig. 22.6g Marking the mean, SD, Minimum, and Maximum

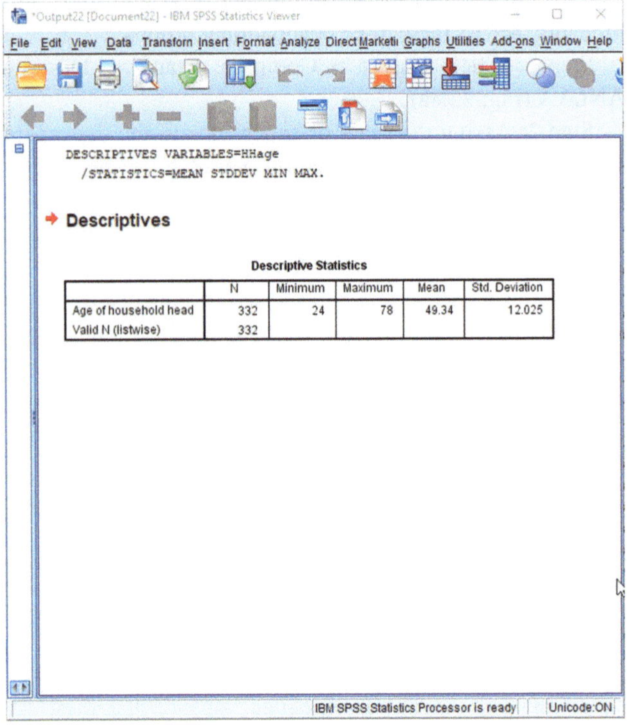

Fig. 22.6h Final output of the descriptives

Table 22.3 Results of the mean age of males and females

Gender	N	Minimum	Maximum	Mean	Std. deviation
Male	57	25	72	46.26	10.46
Female	275	24	78	49.98	12.245

and on the One-Way ANOVA window, click OK to generate output (Fig. 22.8c).

Step 4: The output is generated. The output shows descriptive statistics, including the number of households per category, the mean years of experience, and the standard deviation. Second, results on the homogeneity of variance tests are presented. Third, the ANOVA table is presented. The ANOVA table shows different statistics, including the sum of squares, the degrees of freedom, the mean square, the F statistics, and the significance (Fig. 22.8d).

The p-value is 0.026, which is less than the standard significance level of 0.05. We therefore reject the null hypothesis and conclude that the three means are statistically different. Practically, this suggests that the mean number of years differs significantly across the three occupational groups: farming, business, and others. Overall, this implies that the average number of years individuals have spent in farming, business, or other occupations is not the same in the wider population represented by this study.

22.6.1.2 Two Groups: T-tests

The test is used to establish whether there is a significant difference between the means of the two groups. In SPSS, we use the independent t-test to compare means that are independent of each other (not a case of pre- and post-analysis). In this example, we will explore whether there is a significant difference in the mean age of males and females across gender.

Step 1: Click on Analyze > Compare means > Independent sample T-test (Fig. 22.9a)

Step 2: On the window that pops up, move the age of the household to the *test variable* and the gender of the household variable to the *grouping variable*. Define Groups by clicking the Define Groups tab and insert 1 for Group One and 2 for Group Two. Note the number for grouping the categorical variable should be coded under the data view. For instance, if 0 was coded for female and 1 for male, then under the defined group, group 1 should be 0 and group 2 should be 1. After inputting the details correctly, click on OK to generate results (Fig. 22.9b).

Step 3: The output gives summary statistics showing the number of males and females and their average age. The T-test results show a significant difference between the ages of males and females, as indicated by the significance level. When presenting the statistics, consider that since Levene's test is significant, we conclude that an equal variance is not assumed. Therefore, we will use the degrees of freedom (df) from which the variance is not assumed equal, i.e., 90.385.

$$t(90.385) = -2.37.$$

The p-value of 0.033 is less than the significance level (α) of 0.05 (95% confidence level). That means we reject the null hypothesis and conclude that there is a significant difference in the mean age of males and females across genders. Practically, the mean age of females (49.98) is statistically different from the mean age of Males 46.26. It means that, within this population, female farmers tend to be, on average, older than male individuals, and this difference (approximately 3.72 years) as shown in Fig. 22.9c.

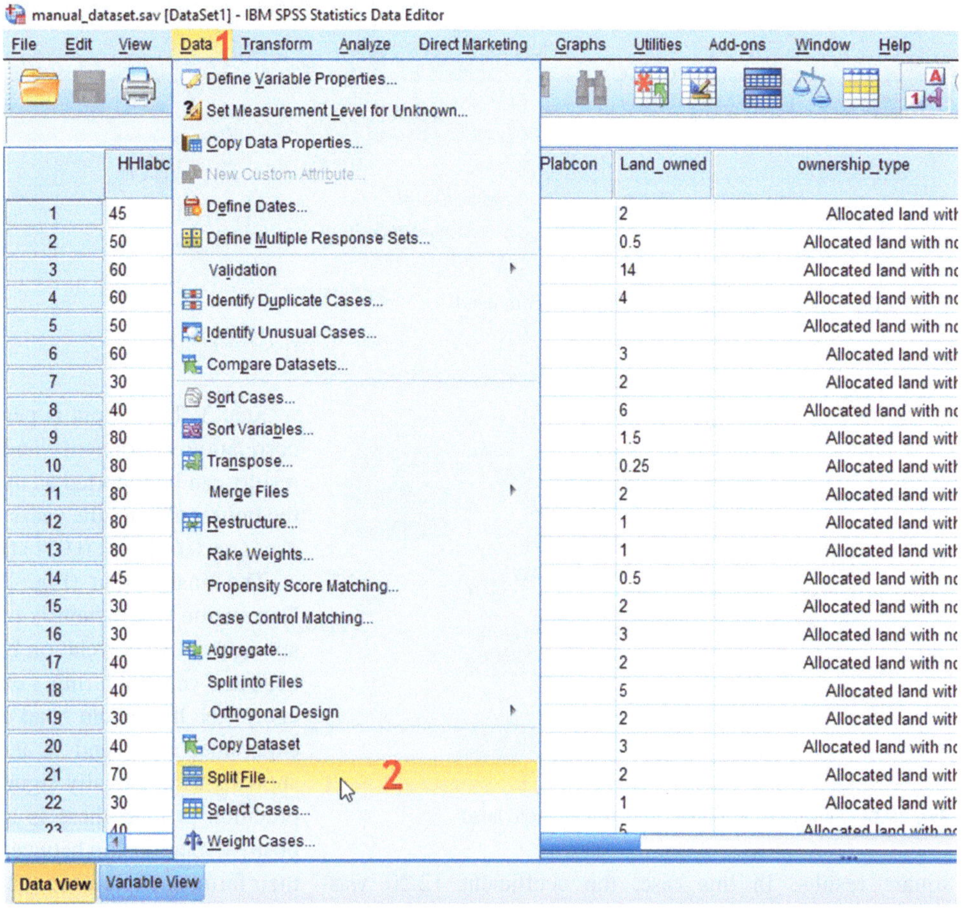

Fig. 22.6i Splitting file command

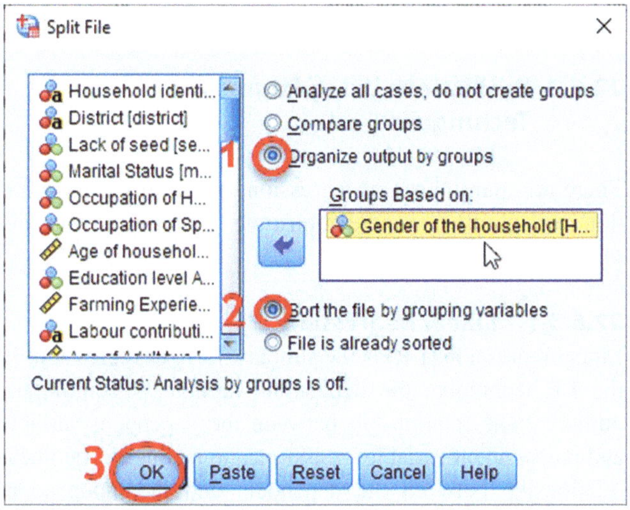

Fig. 22.6j Selecting categorical variable Gender of household head

22.6.2 Relationships: Correlations

22.6.2.1 Pearson Chi-square for Categorical Data

We start with Pearson Chi-square (Categorical), which is used to establish if there is a significant relationship (an association/correlation) between two categorical variables. In this case, we will explore whether there is any relationship between gender and the lack of seed.

Step 1: Click on Analyze > Descriptive Statistics > Crosstabs (display Fig. 22.10a)

Step 2: On the crosstabs window, take the lack of seed variable to the row and gender to the column as shown in Fig. 22.10b. Remember, because the two variables have two categories each, they will form a 2×2 contingency table.

Step 3: From the crosstab window click on statistics. The window shown in Fig. 22.10c appears. Here check the Chi-square then click continue to go back to the crosstab window.

Step 4: In the crosstab window, click on cells. Check the observed and Expected undercounts in the cell window and then click continue to return to the crosstabs window. In the crosstabs window, click OK to generate results (Fig. 22.10d).

Step 5: The Chi-square output shows the sample size and confirms that no missing values were found. Second, it shows a contingency table with the frequency/counts and expected count. Note that, as a rule of thumb, the expected count should not be less than five. If the expected value is less than five, then a Fisher's exact test is calculated in SPSS automatically (Kim 2017). The third table presents the Chi-

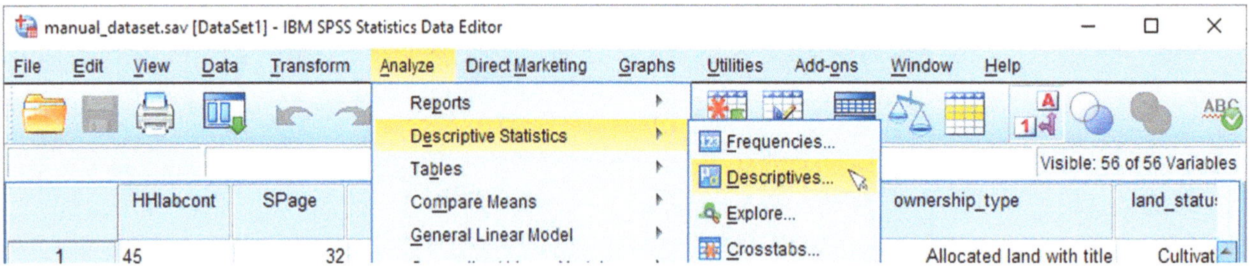

Fig. 22.6k Clicking the Descriptive statistics command

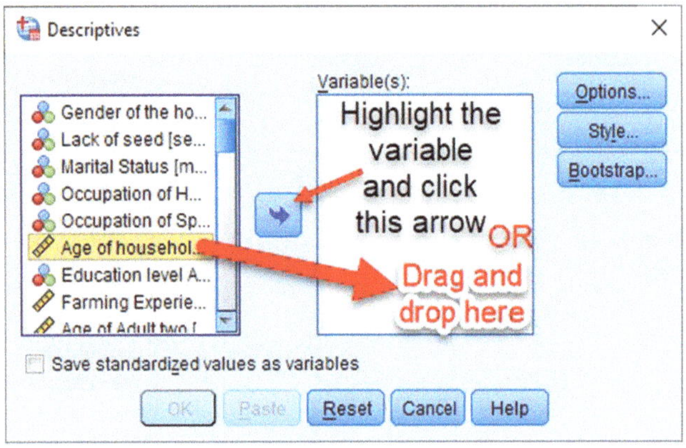

Fig. 22.6l Selecting "Variable" Age of household head

square results. In this case, the coefficient 12.20 was significant at $p < 0.001$ at 1 degree of freedom (Fig. 22.10e). The association was thus significant.

The final output in Fig. 22.10e shows that the p-value was 0.0001, meaning that the association between the lack of seed and gender was significant. Our standard significance level (α) is set at 0.05 (or 5%). Therefore, we reject the null hypothesis (H0). This means that the observed association between "lack of seed" and "gender" is statistically significant.

In practical terms this implies that men and women farmers likely face different challenges or have different levels of access regarding seed availability in the study area.

22.6.2.2 Pearson Product Moment with Ratio Data

The Pearson Product Moment Correlation is used to explore the correlation between two numeric variables (Puth et al. 2014), such as the household's age and years of farming experience.

Step 1: Click Analyze > Correlate > Bivariate (Fig. 22.11a)

Step 2: On the Bivariate Correlations window that appears, move the variables of interest under the variables. Second, ascertain that Pearson is checked under the correlation coefficient, the two-tailed test of significance is selected, and the flag significance correlations are checked. Then, press OK to generate output (Fig. 22.11b).

Step 3: The output is generated as a table presenting the correlation coefficients and their significance levels. The results can be reported as the correlation between the age of the household and the years of farming experience is significant ($r = 0.647$, $p < 0.001$) (Fig. 22.11c).

The final output (Fig. 22.11c) shows that the Pearson Correlation Coefficient is positive ($r = 0.647$); indicating a strong positive correlation between the age of the household head and years of farming experience. This means that as the age of the household head increases, their years of farming experience also tend to increase. The *p*-value is 0.0001, showing a statistically significant correlation. Overall interpretation is that, there is a statistically significant, strong positive relationship between the age of household heads and their farming experience. This means older farmers are likely to have more years of experience in farming than younger farmers.

22.6.3 Predictions Using Regression Techniques

There are many types of regression, but in this demonstration, we will discuss and demonstrate two types: *Linear and logistic regression.*

22.6.3.1 Linear Regression

Linear regression (LR) is the simplest form of regression. In the LR technique, the dependent variable is continuous/numeric. The relationship between the dependent variable and independent variables is assumed to be linear. The linear relationship between the dependent and the independent variable means that if the independent variable increases in some unit, the dependent variable will also increase and vice versa. When only one independent variable and one dependent variable are under consideration, the linear regression is called *simple linear regression.* In comparison, when more than one independent variable and one dependent variable are under consideration, the linear regression is called *multiple linear regression.*

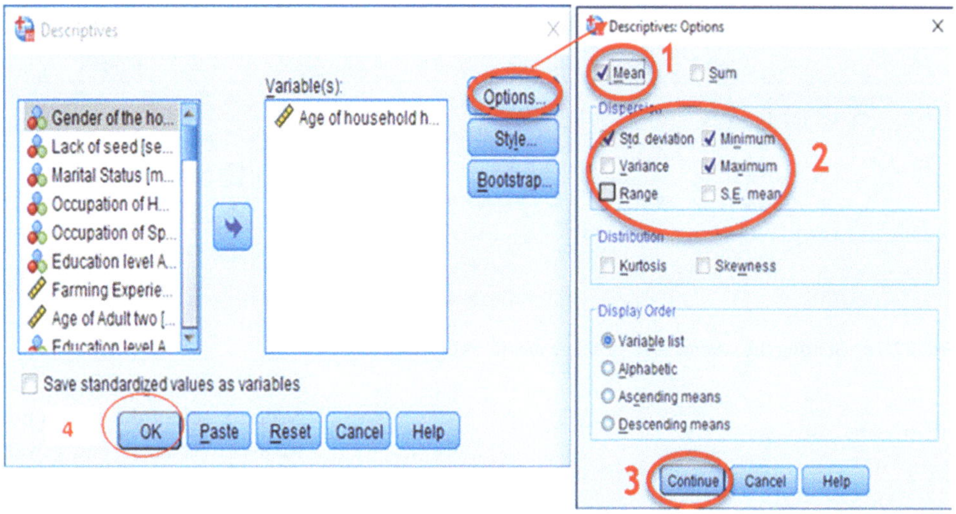

Fig. 22.6m Selecting mean, SD, minimum, and maximums

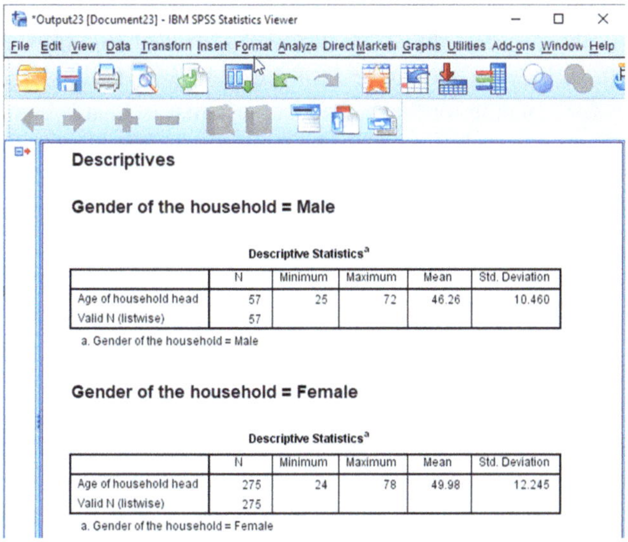

Fig. 22.6n Output calculating means

In the case of *simple linear regression*, we will consider the ages of the household and their years of farming experience.

Step 1: Click on Analyze > Regression > Linear (Fig. 22.12a)

Step 2: On the Linear Regression window, assign a variable to the dependent and independent as shown below and click OK to generate results (Fig. 22.12b).

Step 3: The generated results have three main components: Model summary, ANOVA, and coefficients. Each component will be interpreted (Fig. 22.12c).

Interpretation

Model Summary

The model summary shows $R = 0.647$ correlation coefficient between age and experience. R-squared is the percentage of influence explained by the independent variable (age) on the dependent variable (years of farming experience). Therefore, it is interpreted that the age of the household head can explain 41.9% of the years of farming experience (Table 22.4).

ANOVA table

The rule of thumb is that for a linear regression equation to be valid, the ANOVA should be significant. In our case the ANOVA is significant (0.0001) (Table 22.5)

Coefficients table

We will use the B under the Unstandardized Coefficients to interpret the influence of age on years of farming experience. The results show that a unit increase in age leads to a 0.712 unit increase in the years of farming experience. The beta is interpreted in terms of standard deviation such that a unit increase in the standard deviation will lead to a 0.647 change in the farming experience. The regression equation can be written as: $y = -12.226 + 0.712 * \text{Age of household head}$. Y is farming experience in years. This means that there is a strong positive relationship between age and farming experience (Table 22.6).

22.6.3.2 Logistic Regression

In logistic regression, the dependent variable is binary/dummy in nature (having two categories). The independent variables can be continuous or dummy. In binary Logistic regression, the outcome has two categories (e.g., adopter and non-adopter), while in multinomial logistic regression, you can have more than two categories in your dependent variable. In this demonstration, we are going to focus on the binary logistic regression.

Suppose we are interested in the factors influencing whether a farmer adopts an agricultural technology. The outcome (response) variable is binary (0/1): adopter or non-adopter. The predictor variables of interest are the farmer's age, access to credit, group membership, etc.

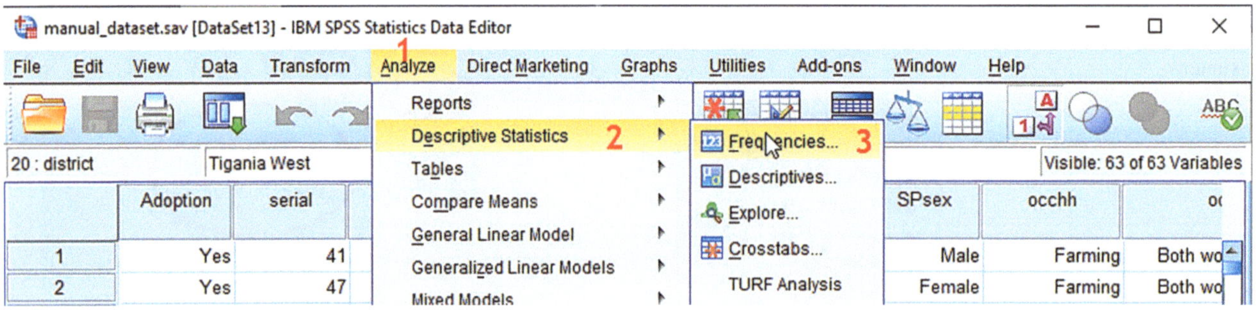

Fig. 22.7a Starting the command to generate measures of variability

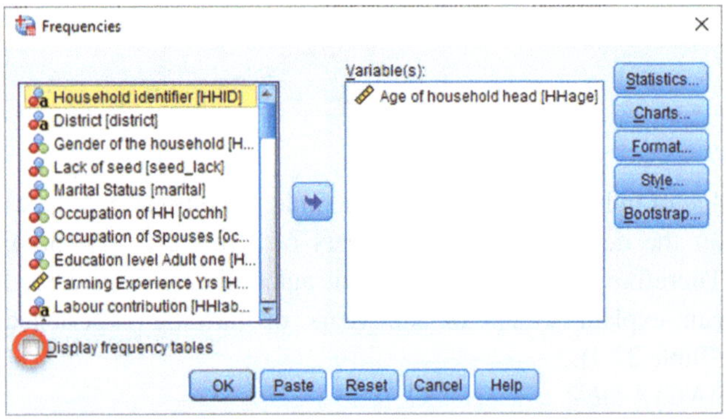

Fig. 22.7b Selecting "variable" age of household head

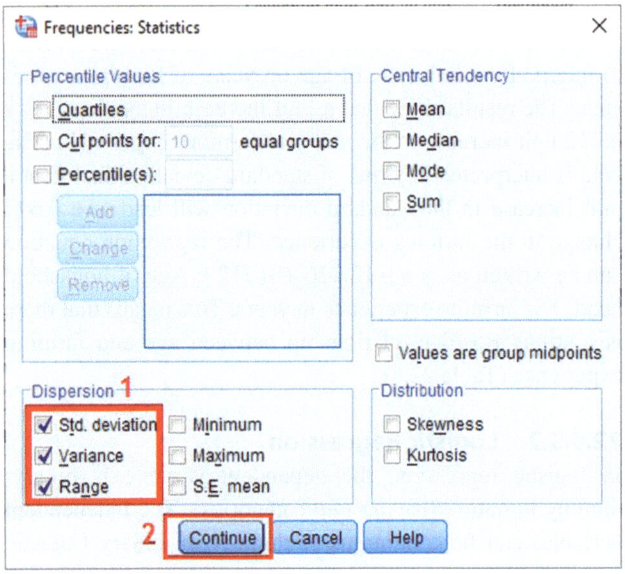

Fig. 22.7c Selecting and clicking continue

Step 1: Click on Analyze > Regression > Binary Logistic (Fig. 22.13a).

Step 2: On the logistic regression window, select the adoption variable and move it to the dependent marked as 1. Second, select the independent variable, which SPSS refers to as covariates. The covariates include cattle ownership, access to credit, training, agricultural group membership, lack of quality seed, age of household, and land tillage method. Then, click OK to generate output (Fig. 22.13b).

Step 3: The generated output has Block 0 and Block 1 results. For the sake of this discussion, we will consider the Block 1 results. More specifically, we will consider variables in the equation for this section because that is what is commonly reported. Figure 22.13c shows the independent variables the B, standard error (S.E.), Wald, degree of freedom (df), Significance level (Sig.), Exp (B), and Lower and Upper 95% CI for EXP (B)

We have extracted the variables and captured them in table format (Table 22.7). We will use the cattle ownership variable for demonstration. The p-value for cattle ownership is less than 0.05 ($p = 0.0001$), indicating that the relationship between cattle ownership and the dependent variable is statistically significant. The B (Coefficient = 1.661) is called the log-odds of the dependent variable of the dependent variable (Field 2018). The B coefficient is positive, indicating that cattle ownership increases the likelihood of the outcome; in other words, cattle ownership increased the logit of adoption by 1.661, $p < 0.0001$. The same can be interpreted using the odds ratio: households that owned cattle were 5.262 times more likely to adopt the agricultural technologies than households that did not have cattle at $p < 0.001$.

Practically, we can say that cattle ownership is a strong and statistically significant predictor of the likelihood of the target outcome (e.g., technology adoption). Farmers with cattle are over five times more likely to adopt the technology than those without.

Binary logistic regression is widely used in many publications studying adoption of agricultural technologies due to its ability to model a binary outcome (Mugwe et al. 2009; Marie et al. 2020; Sarr et al. 2023; Ai et al. 2023; Kule et al. 2025).

Fig. 22.7d: Clicking ok after selection of the variable

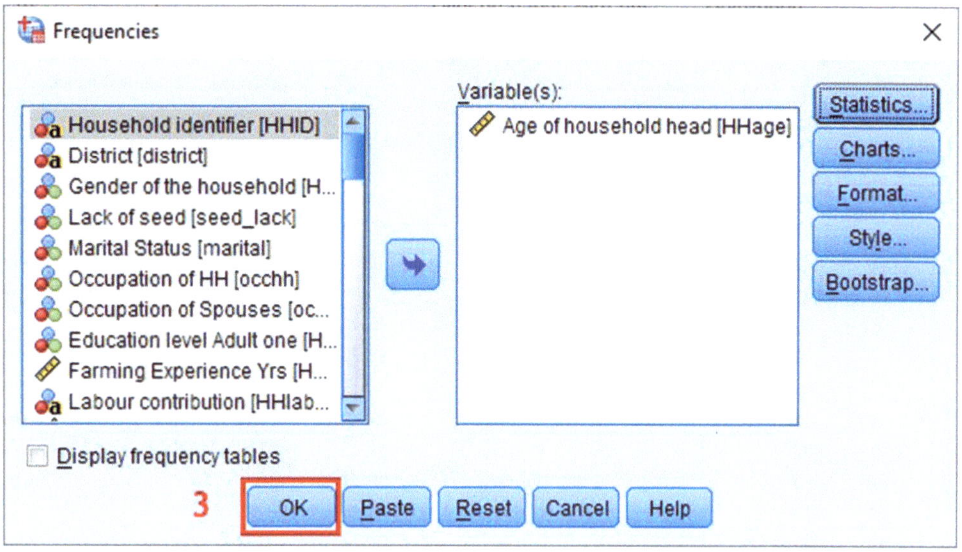

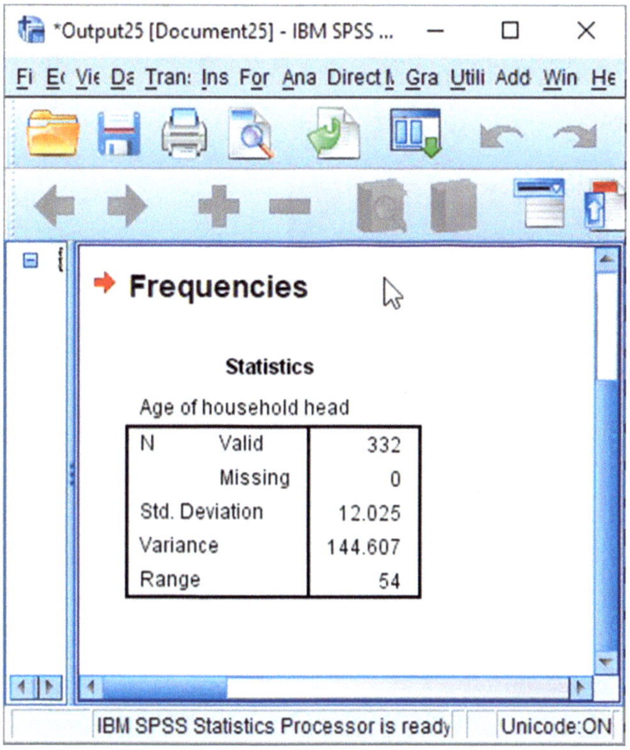

Fig. 22.7e Final Output showing standard deviation, variance, and range of variable "age of household head"

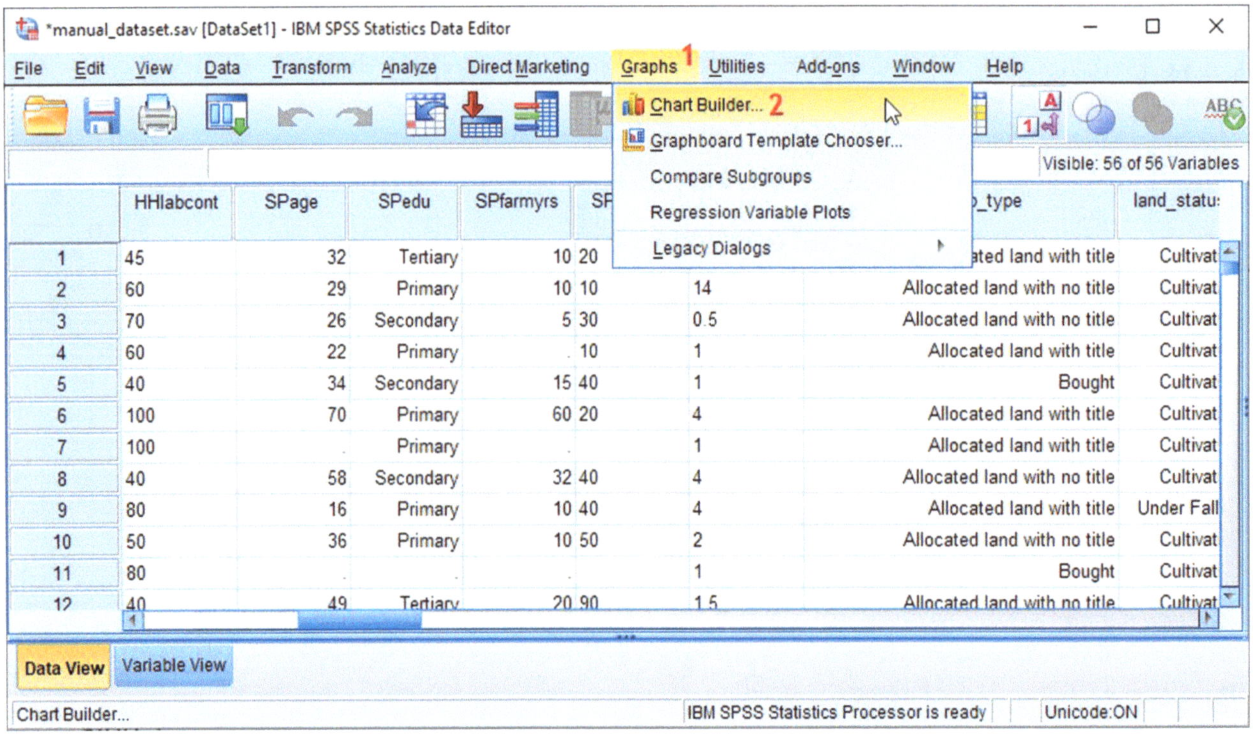

Fig. 22.7f Starting off to plot box plots through the Chart builder command

Fig. 22.7g Selecting the variables and command box plot

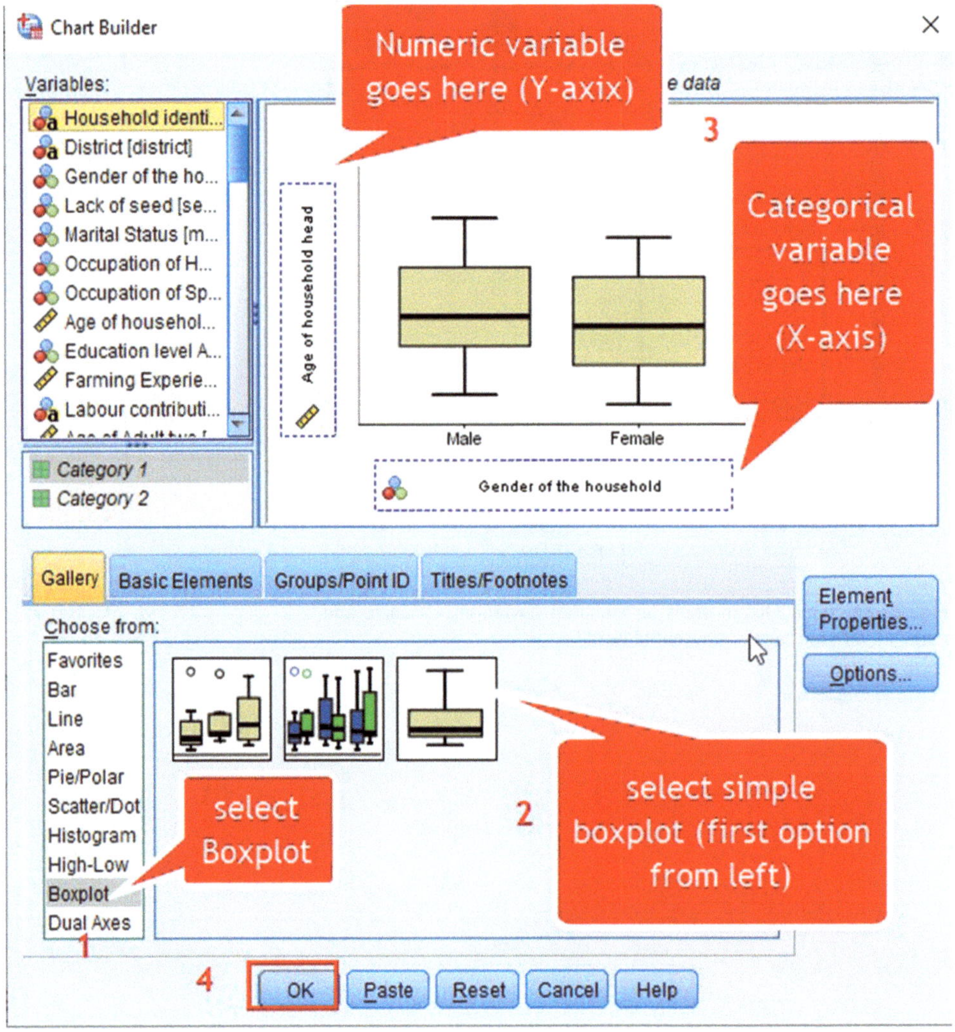

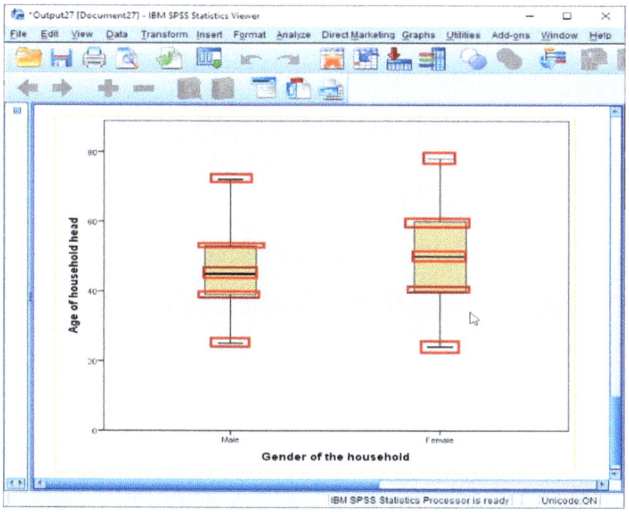

Fig. 22.7h Final output

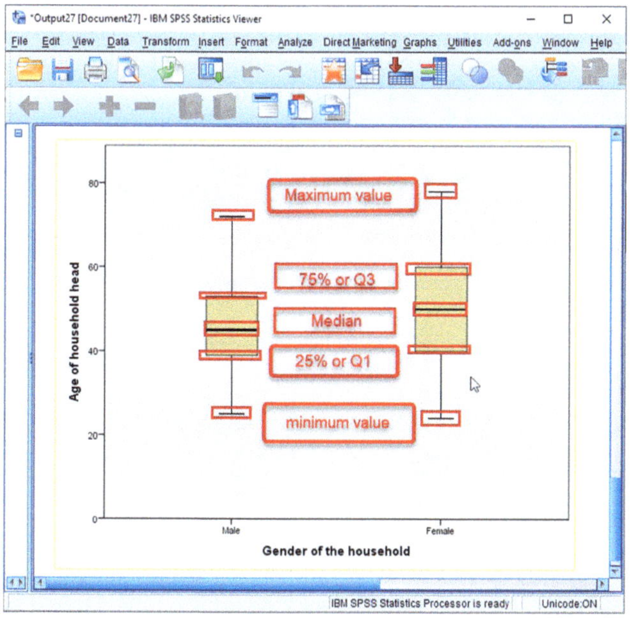

Fig. 22.7i Box plots showing the interquartile ranges

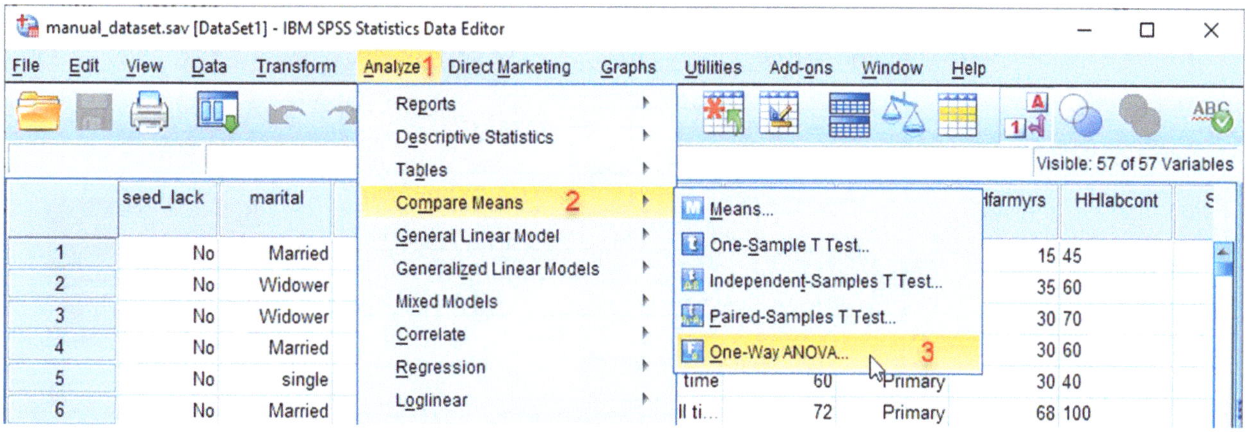

Fig. 22.8a Starting off comparing means in one-way ANOVA command

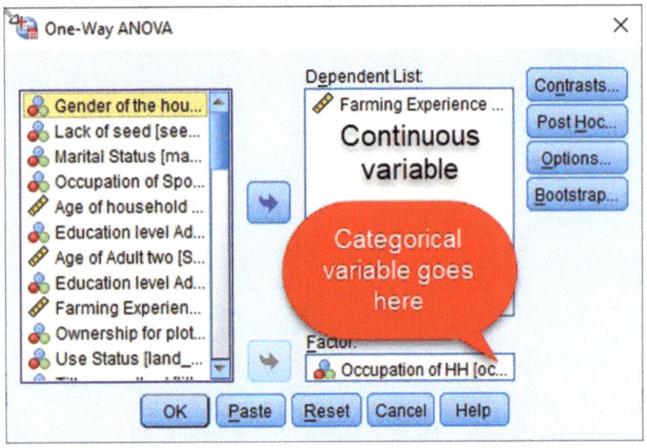

Fig. 22.8b Selecting the categorical variable

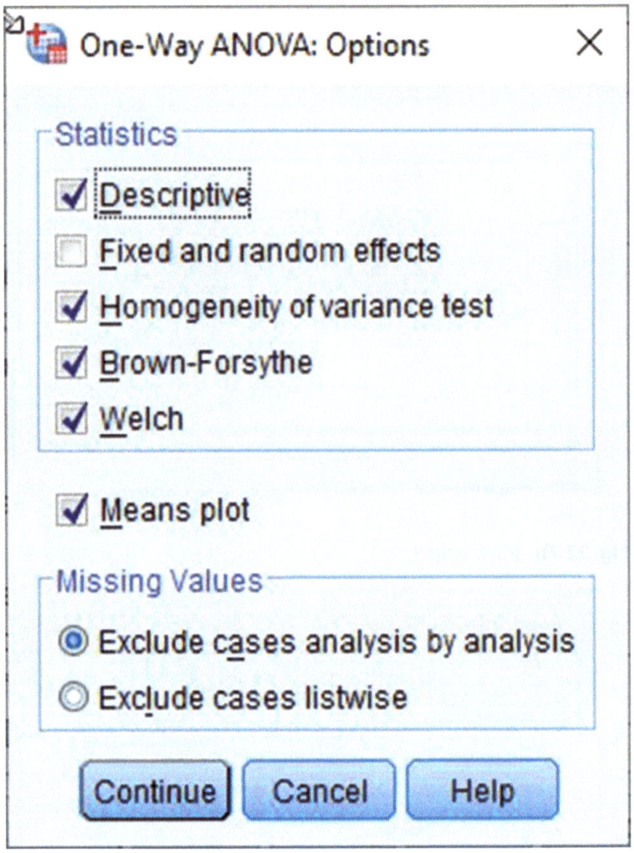

Fig. 22.8c Clicking on the options of the dialogue box

Fig. 22.8d Final output of ANOVA comparing means of three groups (Farming, Business and Others)

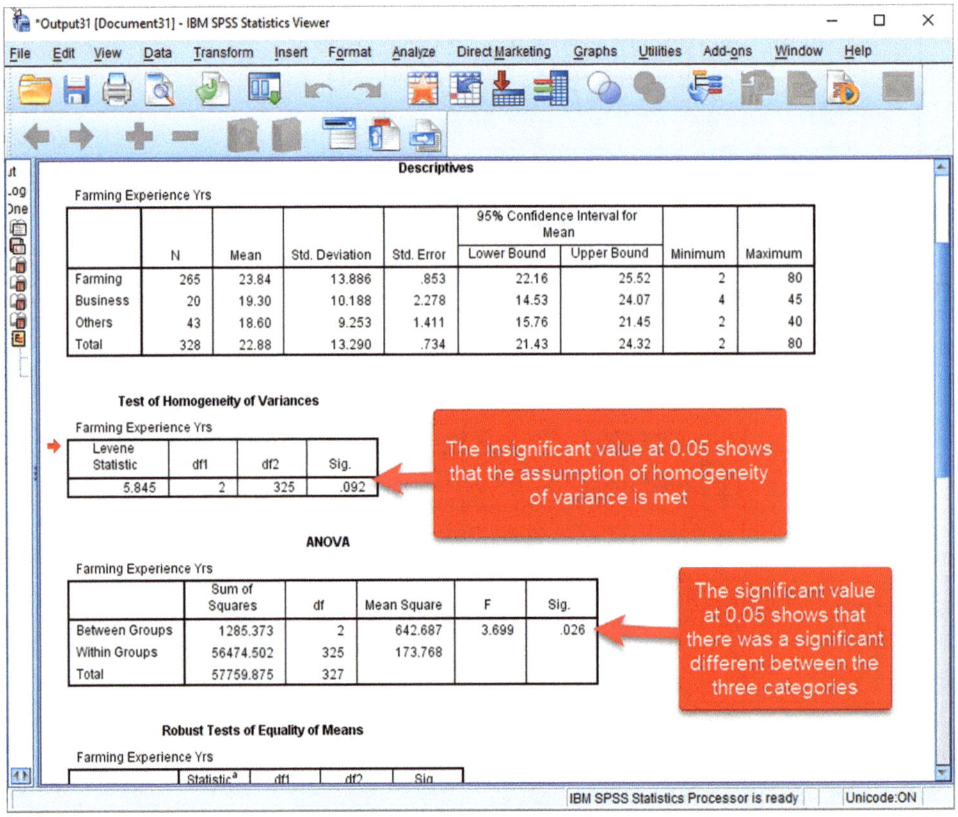

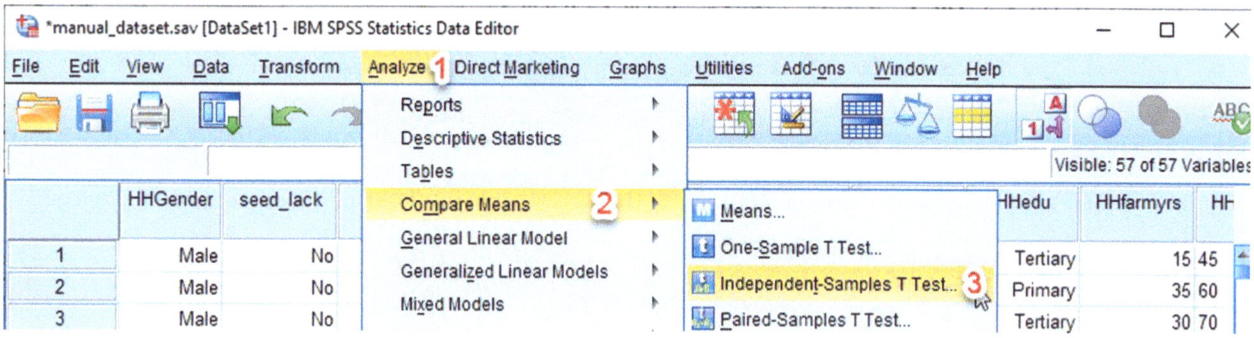

Fig. 22.9a Starting off to compute t-statistics

Fig. 22.9b Arranging the test variable and the grouping variable

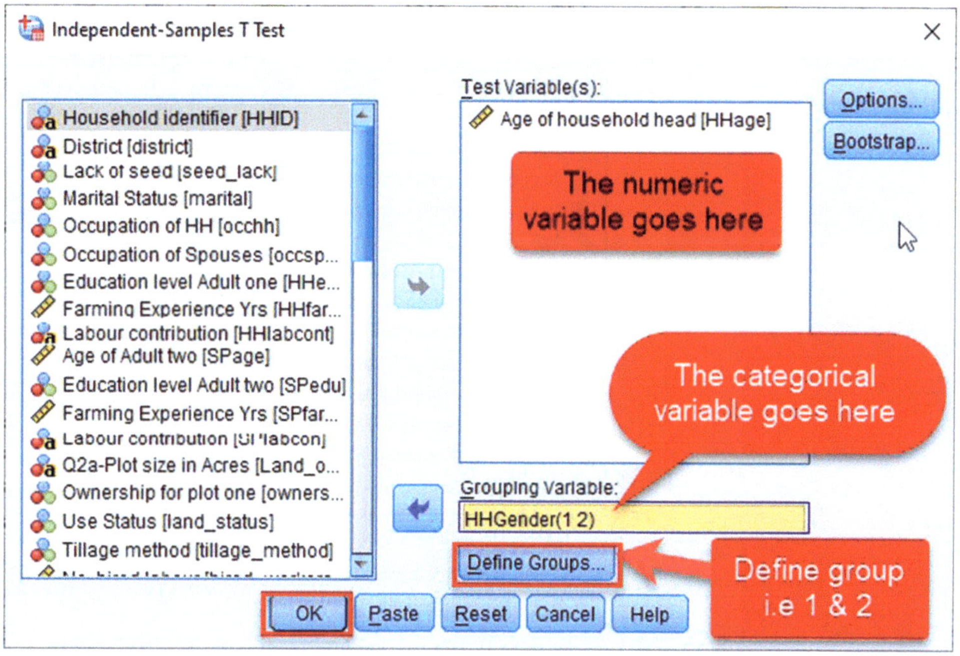

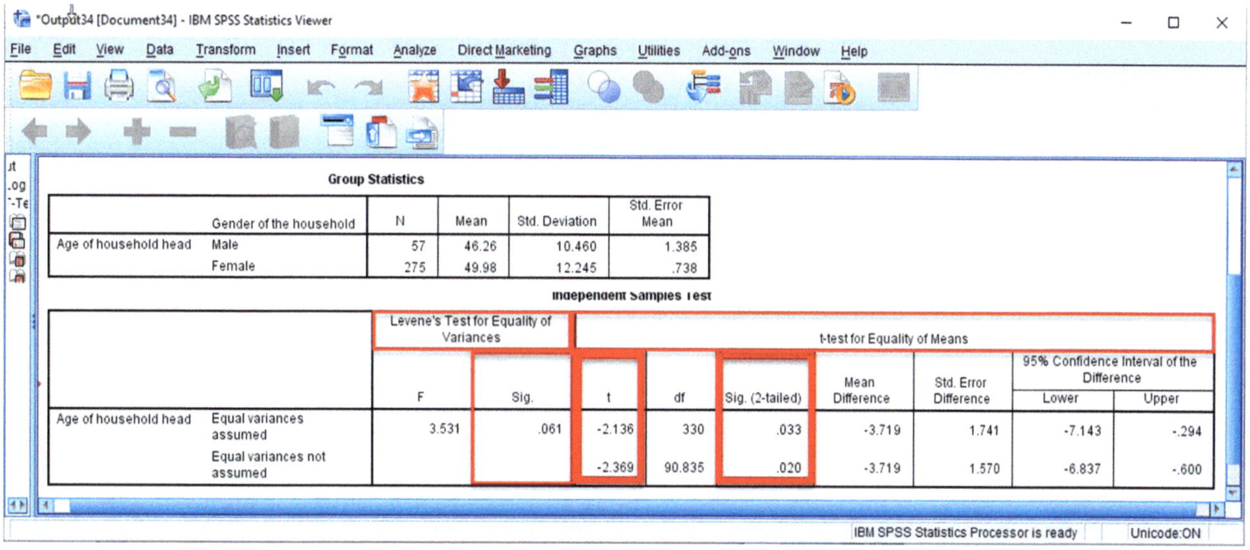

Fig. 22.9c Final output of T-test comparing "age of males" and "age of females" using independent T-test

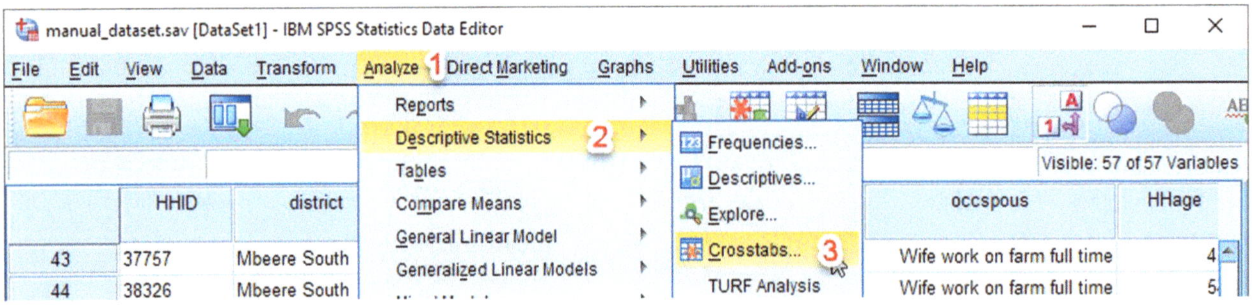

Fig. 22.10a Starting to calculate Chi-square statistic

Fig. 22.10b Selecting the two categorical variables

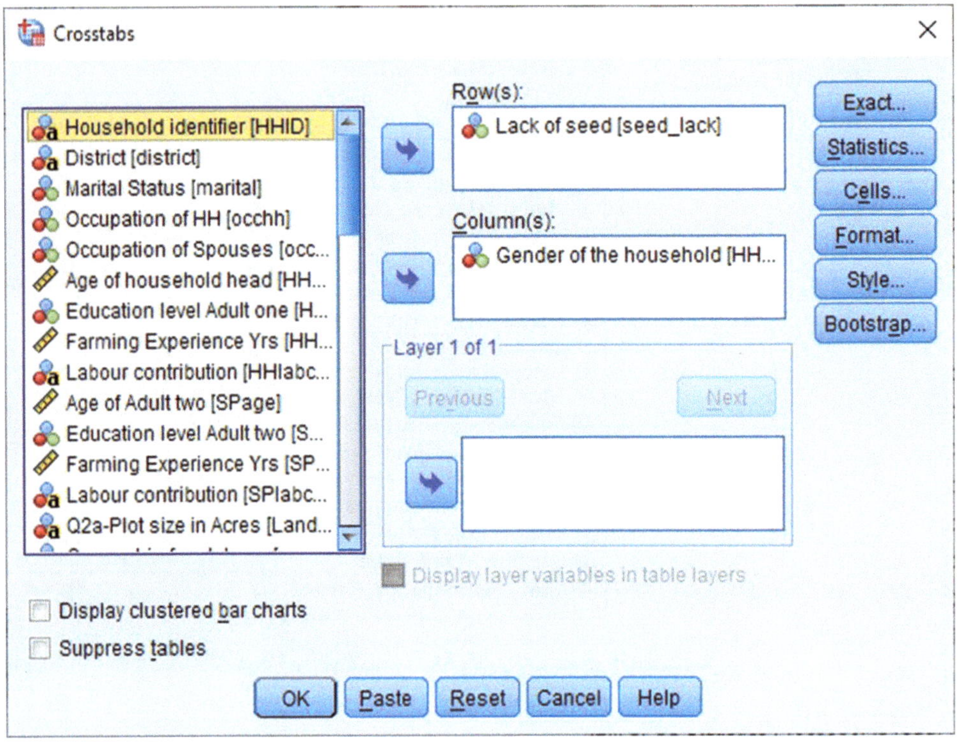

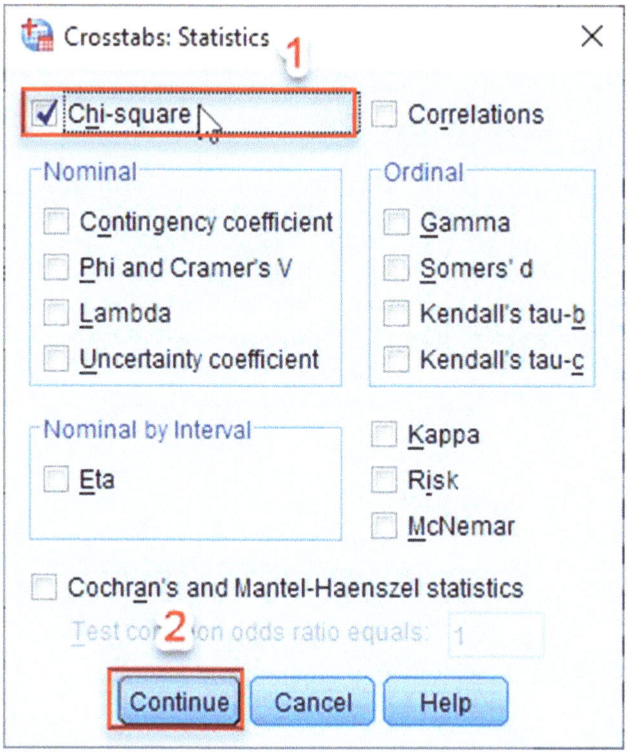

Fig. 22.10c Clicking the command Chi-square statistic

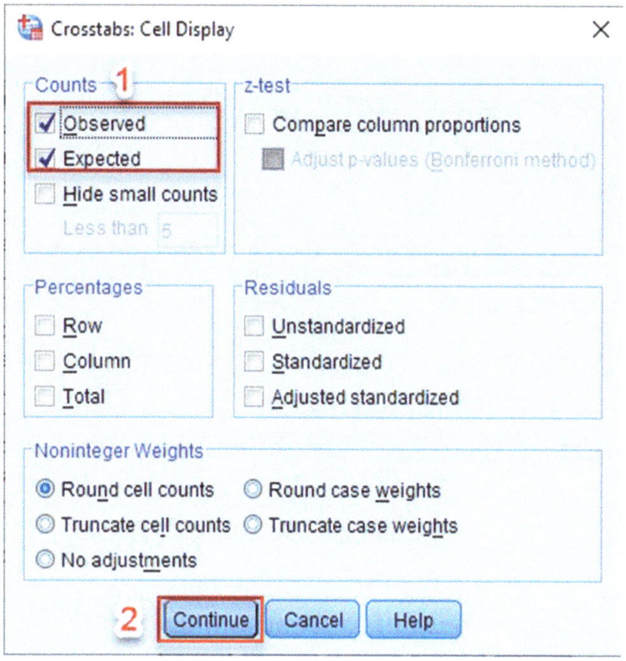

Fig. 22.10d Checking observed and expected

Fig. 22.10e Final output of the Chi-square test between gender and lack of seed

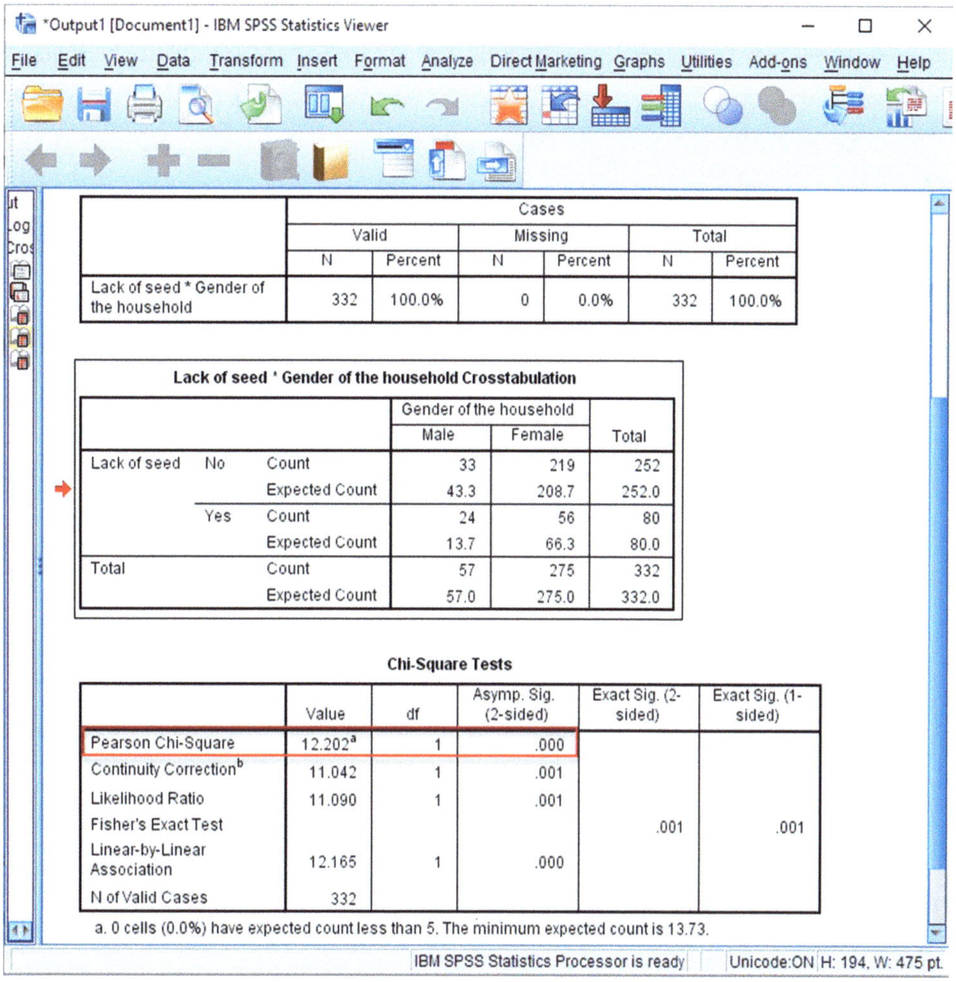

	Cases					
	Valid		Missing		Total	
	N	Percent	N	Percent	N	Percent
Lack of seed * Gender of the household	332	100.0%	0	0.0%	332	100.0%

Lack of seed * Gender of the household Crosstabulation

			Gender of the household		Total
			Male	Female	
Lack of seed	No	Count	33	219	252
		Expected Count	43.3	208.7	252.0
	Yes	Count	24	56	80
		Expected Count	13.7	66.3	80.0
Total		Count	57	275	332
		Expected Count	57.0	275.0	332.0

Chi-Square Tests

	Value	df	Asymp. Sig. (2-sided)	Exact Sig. (2-sided)	Exact Sig. (1-sided)
Pearson Chi-Square	12.202[a]	1	.000		
Continuity Correction[b]	11.042	1	.001		
Likelihood Ratio	11.090	1	.001		
Fisher's Exact Test				.001	.001
Linear-by-Linear Association	12.165	1	.000		
N of Valid Cases	332				

a. 0 cells (0.0%) have expected count less than 5. The minimum expected count is 13.73.

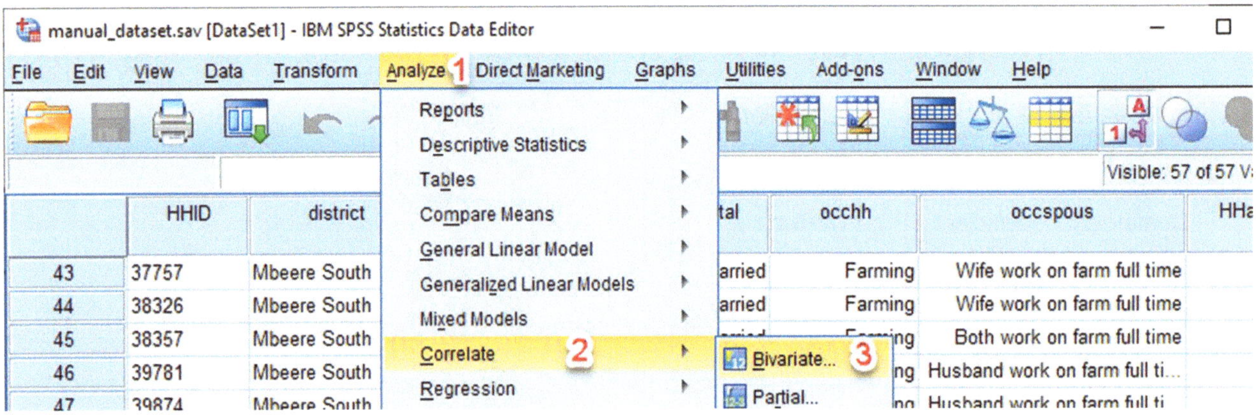

Fig. 22.11a Starting to calculate Pearson Product Moment correlation

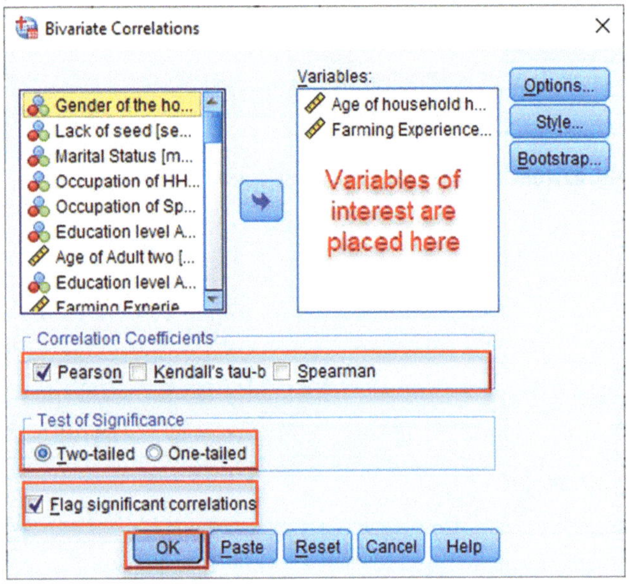

Fig. 22.11b Moving variable of interest (Age of household head and farming experience)

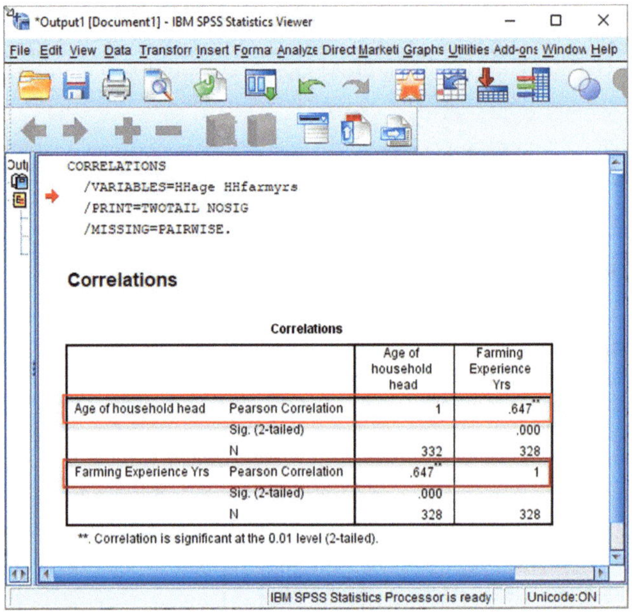

Fig. 22.11c Final output of Pearson Product Moment correlation test between age of household head and farming experience

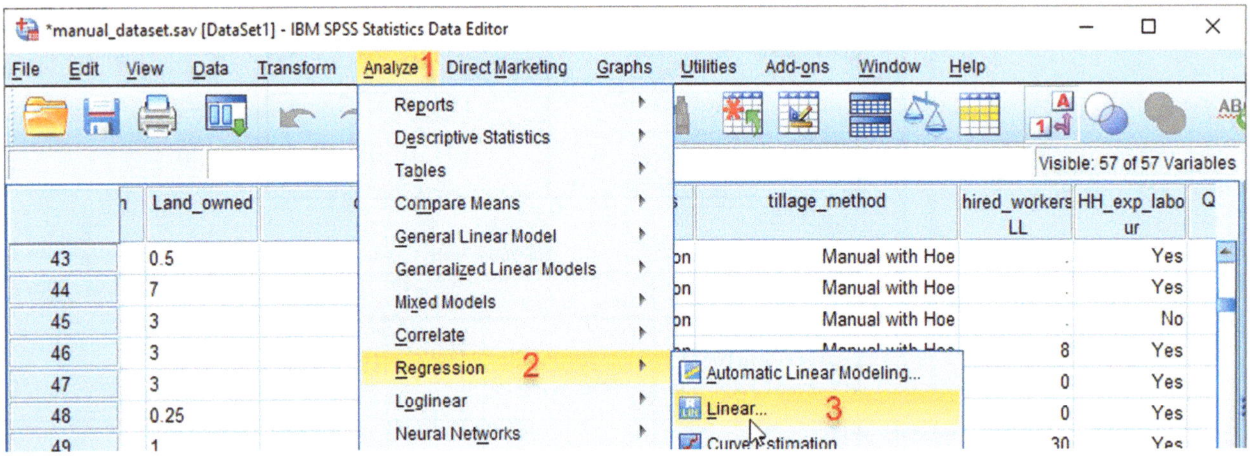

Fig. 22.12a Starting to calculate regression

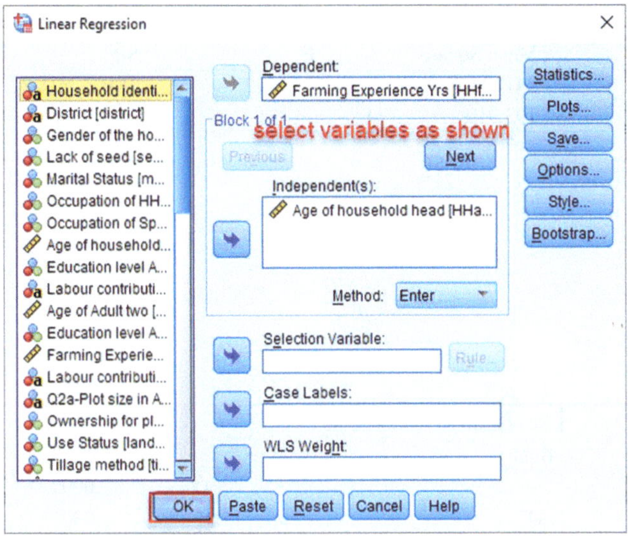

Fig. 22.12b Selecting independent and dependent variables

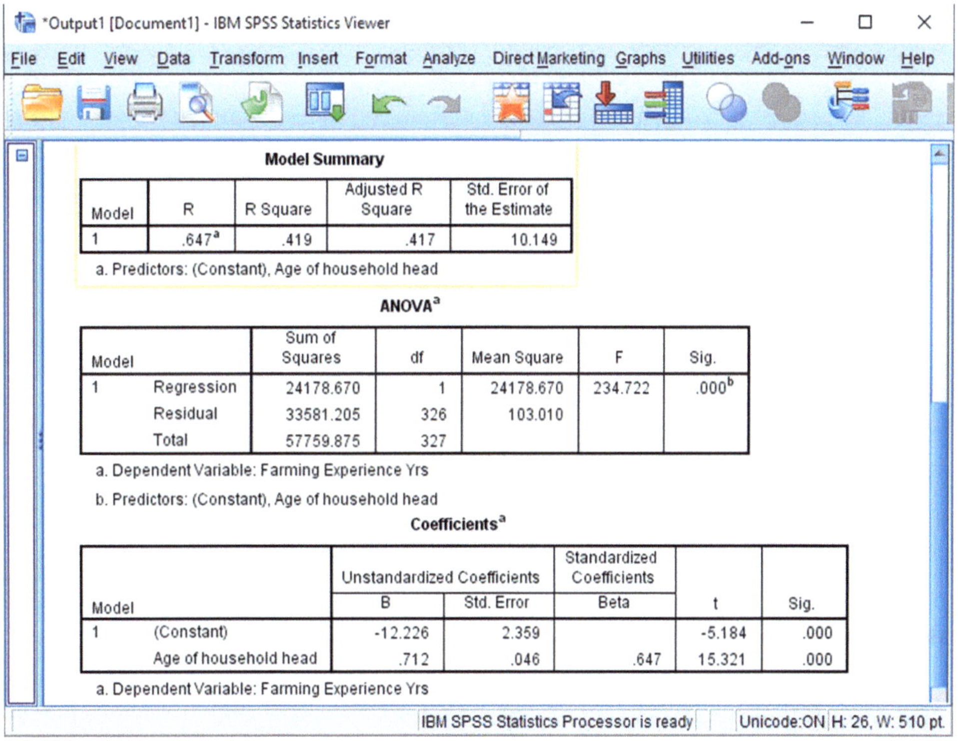

Fig. 22.12c Final Output of simple linear regression between the age of the household head (independent variable) and farming experience (dependent variable)

Table 22.4 Model summary

Model	*R*	*R* square	Adjusted *R* square	Std. error of the estimate
1	.647[a]	.419	.417	10.149

[a]Predictors: (Constant), Age of household head

Table 22.5 ANOVA[a]

Model		Sum of Squares	df	Mean Square	F	Sig.
1	Regression	24178.670	1	24178.670	234.722	.000[b]
	Residual	33581.205	326	103.010		
	Total	57759.875	327			

[a]Dependent variable: Farming experience years
[b]Predictors: (Constant), Age of household head

Table 22.6 Coefficients[a]

Model		Unstandardized Coefficients		Standardized Coefficients	t	Sig.
		B	Std. Error	Beta		
1	(Constant)	−12.226	2.359		−5.184	.000
	Age of household head	.712	.046	.647	15.321	.000

[a]Dependent variable: Farming experience years

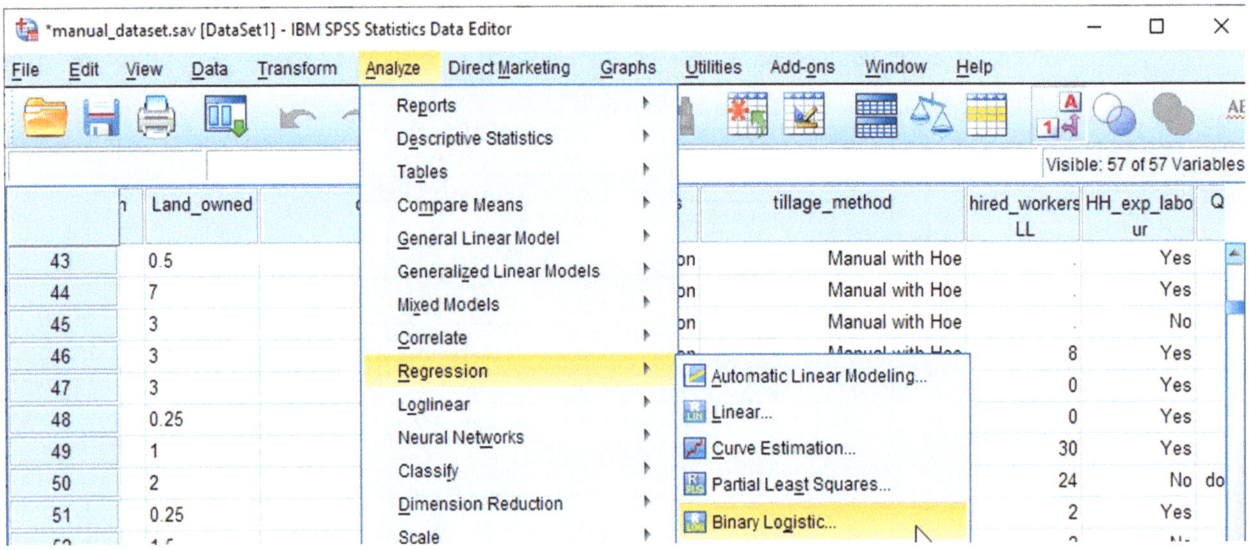

Fig. 22.13a Selecting binary logistic command

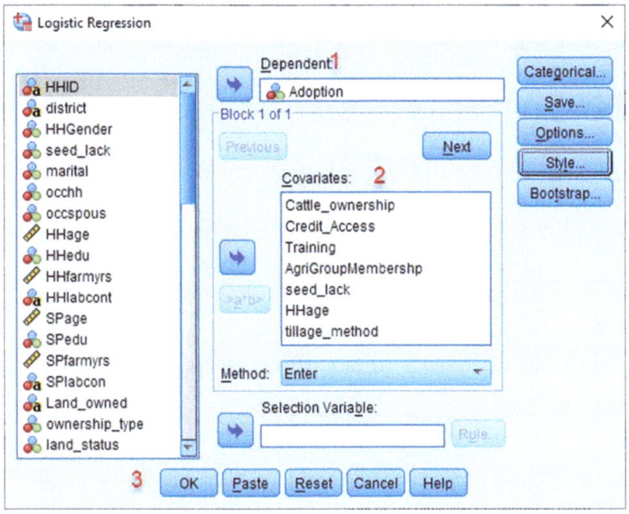

Fig. 22.13b Selecting "Variable" Adoption

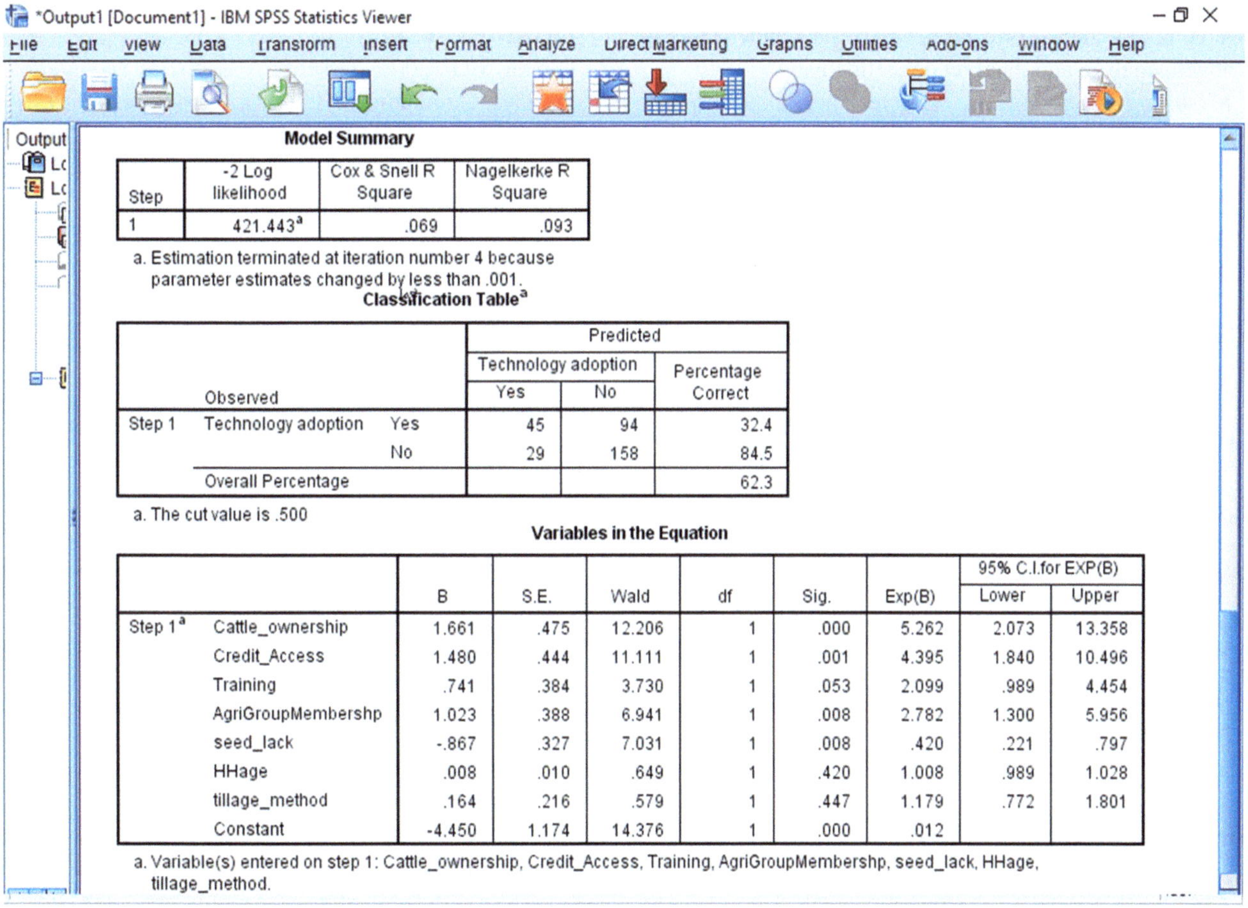

Model Summary

Step	-2 Log likelihood	Cox & Snell R Square	Nagelkerke R Square
1	421.443[a]	.069	.093

a. Estimation terminated at iteration number 4 because parameter estimates changed by less than .001.

Classification Table[a]

Observed			Predicted: Technology adoption Yes	No	Percentage Correct
Step 1	Technology adoption	Yes	45	94	32.4
		No	29	158	84.5
	Overall Percentage				62.3

a. The cut value is .500

Variables in the Equation

		B	S.E.	Wald	df	Sig.	Exp(B)	95% C.I.for EXP(B) Lower	Upper
Step 1[a]	Cattle_ownership	1.661	.475	12.206	1	.000	5.262	2.073	13.358
	Credit_Access	1.480	.444	11.111	1	.001	4.395	1.840	10.496
	Training	.741	.384	3.730	1	.053	2.099	.989	4.454
	AgriGroupMembershp	1.023	.388	6.941	1	.008	2.782	1.300	5.956
	seed_lack	-.867	.327	7.031	1	.008	.420	.221	.797
	HHage	.008	.010	.649	1	.420	1.008	.989	1.028
	tillage_method	.164	.216	.579	1	.447	1.179	.772	1.801
	Constant	-4.450	1.174	14.376	1	.000	.012		

a. Variable(s) entered on step 1: Cattle_ownership, Credit_Access, Training, AgriGroupMembershp, seed_lack, HHage, tillage_method.

Fig. 22.13c Final Output of binary logistic regression to determine the influence of several predictor variables on technology adoption

Table 22.7 Variables in the equation

		B	S.E.	Wald	df	Sig.	Exp (B)	95% CI for EXP (B)	
								Lower	Upper
Step 1[a]	Cattle_ownership	1.661	.475	12.206	1	.000	5.262	2.073	13.358
	Credit_Access	1.480	.444	11.111	1	.001	4.395	1.840	10.496
	Training	.741	.384	3.730	1	.053	2.099	.989	4.454
	AgriGroupMembership	1.023	.388	6.941	1	.008	2.782	1.300	5.956
	seed-_lack	−.867	.327	7.031	1	.008	.420	.221	.797
	HHage	.008	.010	.649	1	.420	1.008	.989	1.028
	tillage_method	.164	.216	.579	1	.447	1.179	.772	1.801
	Constant	−4.450	1.174	14.376	1	.000	.012		

[a]Variable(s) entered in step 1: Cattle_ownership, Credit_Access, Training, AgriGroupMembershp, seed_lack, HHage, tillage_method

22.7 Conclusions

This chapter explored the analysis of survey data using SPSS. By applying SPSS, you can gain valuable insights into survey data, from summarizing descriptive statistics to performing complex inferential tests. The tools and techniques covered in this chapter provide a foundational understanding of how to analyze and interpret data effectively.

Mastering SPSS, like any other technical skill, requires continuous practice. Engaging with various data sets and exploring different statistical methods will significantly enhance your proficiency. Additionally, the correct interpretation of statistical results is crucial for drawing accurate conclusions. Thus, it is highly recommended that additional statistics resources and reference books be consulted. Doing so will refine your analytical skills and abilities and ensure your interpretations are accurate.

Acknowledgements Throughout this chapter, several screenshots generated using SPSS have been used to demonstrate key procedures and steps in data analysis. These include creating frequency tables and cross-tabulations and performing essential analyses such as measures of central tendency, measures of variability, ANOVA, T-tests, Chi-square, Spearman rank correlation and binary logistic regression. The use of SPSS has significantly enhanced the value of this book by providing students an experience of practical application of these analytical techniques.

For more information
https://www.ibm.com/products/spss-statistics/resources
https://www.spss-tutorials.com/

Further Reading Materials

1. Cooksey RW (2020) Illustrating Statistical Procedures: Finding Meaning in Quantitative Data, Third Edition. Illustrating Statistical Procedures: Finding Meaning in Quantitative Data, Third Edition 1–737. https://doi.org/10.1007/978-981-15-2537-7
2. Kirkpatrick LA, Feeney BC (2016) A simple guide to IBM SPSS statistics for version 23.0, 14th edn. Cengage Learning
3. Landau S, Everitt BS (2021) A Brief Introduction to SPSS. A Handbook of Statistical Analyses Using SPSS 12–37. https://doi.org/10.1201/9780203009765-3/BRIEF-INTRODUCTION-SPSS-SABINE-LANDAU-BRIAN-EVERITT
4. Larson MG (2006) Descriptive statistics and graphical displays. Circulation 114:76–81. https://doi.org/10.1161/CIRCULATIONAHA.105.584474/ASSET/D72614F9-7C44-404A-971B-0B7713D7E63B/ASSETS/GRAPHIC/15FF4.JPEG
5. McCormick K, Salcedo J (2017) SPSS Statistics for Data Analysis and Visualization. Wiley

Key Points in This Chapter

Need for Survey Data Analysis Using SPSS

Data from survey includes various types including crop yields, soil health, pest management, farmer demographics, and practices. Due to their diverse nature, these data can be complex and multifaceted, involving multiple variables and interactions between different factors.

Nature of SPSS?

SPSS (Statistical Package for the Social Sciences) is widely used to analyze survey data. It is a powerful and user-friendly software, and analyzes, manipulates, and presents data. It contains a straightforward interface with dialogue boxes and provides a help menu and tutorials to help first-time users get started. It can also handle data from different sources and open various file formats used for structured data.

The Survey Data

The data generated from the survey comprises numeric and non-numeric, i.e., quantitative and qualitative. Examples of numeric data include the household head's age in years, Farm size in hectares or acres. Examples of non-numeric data are the Education level of the household head, Gender of the household head and Type of crops grown.

Structure of SPSS

SPSS has two main windows: a Data View window and a Variable View window. The data view window is used for setting up your data, while the variable window is where you define and manage the properties of each data set.

Another key functionality in SPSS is the toolbar, which has various functions and tools for managing, transforming, and analyzing data. The toolbar contains File Menu, Edit Menu, View Menu, Data Menu, and others designed to streamline the data analysis process.

Descriptive Data Analysis

The main descriptive statistics tackled include Counts, frequencies, and percentages, Measures of central tendency (MOC) and Measures of variability (MOT). Analyzing frequencies requires you to first do a simple one-way frequency table before proceeding to a multiple-frequency table, which shows the frequency distribution of more than one variable (also called cross-tabulation).

Measures of variability computed in this demonstration include range, standard deviation, variance, and box plots. The generated output provides a comprehensive understanding of the data's dispersion and spread.

Inferential Statistics

We use multiple means to analyze variance (ANOVA) and a single quantitative variable (numeric variable). The data should be random, and the outcome data should be distributed normally. Also, we assume equal variance among the groups. The output generates descriptive Tests of homogeneity of variance and an ANOVA table.

The t-test is used to establish whether there is a significant difference between the means of the two groups. In SPSS, we use the independent t-test to compare means independent of each other (not a case of pre- and post-analysis).

Predictions and Regression Techniques

Simple linear regression (LR) is the simplest form of regression that examines the relationship between an independent variable and one dependent variable. When the analysis involves more than one independent variable predicting a single dependent variable, it is referred to as multiple linear regression**.**

Logistic Regression

In logistic regression, the dependent variable is binary/dummy in nature (having two categories). The independent variables can be continuous or dummy.

Review Questions

1. What are the main types of data collected from agricultural surveys? Give examples of each.
2. Describe the main purpose and difference between the "Data View" window and the "Variable View" window in SPSS.
3. Describe the key functions of SPSS's Toolbar and their roles in data analysis.
4. What insights can you gain from plotting a box plot, and which measures of variability does it visually display?
5. Differentiate between the Pearson Chi-square test and the Pearson Product Moment Correlation. Provide appropriate use cases for each.
6. In which scenario would simple linear regression be more appropriate than binary logistic regression**?**

Chapter Exercise

1. How would you compute and interpret the mean, standard deviation, minimum, and maximum values in SPSS?
2. Imagine you have completed a survey among 250 farmers and you want to investigate if there is an association between a farmer's gender (Categorical: Male/Female) and their adoption of certified maize seeds (Categorical: Adopted/Not Adopted).
   (a) State the null hypothesis (H0) and alternative hypothesis (H1) for this test.
   (b) Describe the purpose of a cross-tabulation table (also known as a two-way table) in this context. What kind of information would you expect to see in such a table to begin your analysis?

(c) If the SPSS output for the Pearson Chi-square test gave you a p-value of 0.003, how would you interpret this result at a 5% significance level ($\alpha = 0.05$)?
(d) What practical conclusion could you draw regarding gender and certified seed adoption in this hypothetical scenario?

3. A researcher wants to use farm size (acres) to predict total annual maize yield (tonnes) from a hypothetical survey of farmers
   (a) Write down the general equation for a simple linear regression model. Identify which term represents the dependent variable, independent variable, slope, and intercept.
   (b) If the hypothetical SPSS regression output provided the following key results:
      (i) Constant (Intercept): 0.5
      (ii) Farm Size (Slope): 0.8
      (iii) R-squared: 0.56
      (iv) Overall F-test p-value: <0.001
   (c) Write the specific regression equation.
   (d) Interpret the meaning of the slope coefficient (0.8) in practical terms for the maize yield and farm size relationship.
   (e) What does the R-squared value of 0.56 tell you about the model's ability to explain variation in maize yield?
   (f) Is the overall regression model statistically significant?
4. An agricultural extension officer wants to understand factors influencing whether a farmer adopts a new coffee pruning technique (Binary: 1=Adopted, 0=Not Adopted). They hypothesize that farmer's age (quantitative, in years) might be a predictor.
   (a) Why is binary logistic regression more appropriate than simple linear regression for a dependent variable like "adoption of a new pruning technique (Yes/No)"?
   (b) If the hypothetical SPSS output provided an Odds Ratio (Exp(B)) for "farmer's age" of 1.05 with a p-value of 0.02, interpret this result.
   (c) Based on this finding, what practical insight could you offer to the extension officer about targeting farmers for future training on this new pruning technique?

Answers to Review questions

- Q1. References: 22.1.3
- Q2. References: 22.1.4
- Q3. References: 22.2
- Q4. References: 22.2.3
- Q5. References: 22.3.2
- Q6. References: 23.3.3

References

Ai J, Hu L, Xia S et al (2023) Analysis of factors influencing the adoption behavior of agricultural productive services based on logistic—ISM model: a case study of rice farmers in Jiangxi Province, China. Agriculture 13:162. https://doi.org/10.3390/AGRICULTURE13010162

Carver RH, Nash JG (2011) Doing data analysis with SPSS: version 18.0, 5th edn. Cengage Learning

Cooksey RW (2020a) Descriptive statistics for summarising data. In: Illustrating statistical procedures: finding meaning in quantitative data, pp 61–139. https://doi.org/10.1007/978-981-15-2537-7_5

Cooksey RW (2020b) Illustrating statistical procedures: finding meaning in quantitative data, 3rd edn, pp 1–737. https://doi.org/10.1007/978-981-15-2537-7

IBM (2021) IBM SPSS Statistics 28 Brief Guide

Ilan Y (2020) Order through disorder: the characteristic variability of systems. Front Cell Dev Biol 8:495391. https://doi.org/10.3389/FCELL.2020.00186/XML/NLM

Jankowski KRB, Flannelly KJ (2015) Measures of central tendency in chaplaincy, health care, and related research. J Health Care Chaplain 21:39–49. https://doi.org/10.1080/08854726.2014.989799

Kim H-Y (2017) Statistical notes for clinical researchers: Chi-squared test and Fisher's exact test. Restor Dent Endod 42:152–155. https://doi.org/10.5395/RDE.2017.42.2.152

Kirkpatrick LA, Feeney BC (2016) A simple guide to IBM SPSS statistics for version 23.0, 14th edn. Cengage Learning

Kule EK, Obia A, Agole D et al (2025) Farmer perceptions and their implications for adoption of sustainable agricultural intensification practices. Discov Sustain 6:1–15. https://doi.org/10.1007/S43621-025-00929-Z/TABLES/5

Landau S, Everitt BS (2003) A handbook of statistical analyses using SPSS. Chapman and Hall/CRC, New York. https://doi.org/10.1201/9780203009765

Landau S, Everitt BS (2021) A brief introduction to SPSS. In: A handbook of statistical analyses using SPSS, pp 12–37. https://doi.org/10.1201/9780203009765-3/BRIEF-INTRODUCTION-SPSS-SABINE-LANDAU-BRIAN-EVERITT

Larson MG (2006) Descriptive statistics and graphical displays. Circulation 114:76–81. https://doi.org/10.1161/CIRCULATIONAHA.105.584474/ASSET/D72614F9-7C44-404A-971B-0B7713D7E63B/ASSETS/GRAPHIC/15FF4.JPEG

Li H, Chen Z, Zhu W (2019) Variability: human nature and its impact on measurement and statistical analysis. J Sport Health Sci 8:527–531. https://doi.org/10.1016/J.JSHS.2019.06.002

Manikandan S (2011a) Frequency distribution. J Pharmacol Pharmacother 2:54–56. https://doi.org/10.4103/0976-500X.77120/ASSET/535FB422-B0B4-4676-B5EA-6770B8675C5B/ASSETS/0976-500X.77120.FP.PNG

Manikandan S (2011b) Measures of central tendency: Median and mode. J Pharmacol Pharmacother 2:214. https://doi.org/10.4103/0976-500X.83300

Marie M, Yirga F, Haile M, Tquabo F (2020) Farmers' choices and factors affecting adoption of climate change adaptation strategies: evidence from northwestern Ethiopia. Heliyon 6:e03867. https://doi.org/10.1016/j.heliyon.2020.e03867

McCormick K, Salcedo J (2017) SPSS statistics for data analysis and visualization. Wiley

McNamara A (2019) Key attributes of a modern statistical computing tool. Am Stat 73:375–384. https://doi.org/10.1080/00031305.2018.1482784

Morris TP, White IR, Crowther MJ (2019) Using simulation studies to evaluate statistical methods. Stat Med 38:2074–2102. https://doi.org/10.1002/SIM.8086

Mugwe JN (2007) An evaluation of integrated soil fertility management practices in Meru South District, Kenya. Kenyatta University

Mugwe J, Monicah AE, Ae M-M et al (2009) Adoption potential of selected organic resources for improving soil fertility in the central highlands of Kenya. Agroforest Syst 76:467–485. https://doi.org/10.1007/s10457-009-9217-y

Ospina R, Marmolejo-Ramos F (2019) Performance of some estimators of relative variability. Front Appl Math Stat 5:421406. https://doi.org/10.3389/FAMS.2019.00043/XML/NLM

Pérez-Guerrero EE, Guillén-Medina MR, Márquez-Sandoval F et al (2024) Methodological and statistical considerations for cross-sectional, case–control, and cohort studies. J Clin Med 13:4005. https://doi.org/10.3390/JCM13144005

Puth MT, Neuhäuser M, Ruxton GD (2014) Effective use of Pearson's product–moment correlation coefficient. Anim Behav 93:183–189. https://doi.org/10.1016/J.ANBEHAV.2014.05.003

Ross SM (2009) Introduction to probability and statistics for engineers and scientists, pp 1–664. https://doi.org/10.1016/B978-0-12-370483-2.X0001-X

Salkind NJ (2007) Grand Mean. Encyclopedia of Measurement and Statistics:412. https://doi.org/10.4135/9781412952644

Sarr M, Majili Z, Khalili N et al (2023) Adoption of processing technologies and innovative food preservation techniques: findings from smallholders in the Lindi Region in Tanzania. Front Sustain Food Syst 7:1169578. https://doi.org/10.3389/FSUFS.2023.1169578/BIBTEX

Sweet AS, Grace-Martin K (2003) Data analysis with SPSS: a first course in applied statistics, 2nd edn. Allyn and Bacon

Wiley StatsRef: Statistics Reference Online (2014) Wiley StatsRef: statistics reference online. https://doi.org/10.1002/9781118445112

Qualitative Data Analysis

23

Abstract

This chapter explores the process of qualitative data analysis. We start by highlighting the goals and distinctive features of qualitative data analysis. We break down our discussion into multiple sections. First, we explain the various techniques of qualitative research, including narrative analysis, grounded theory, and thematic analysis, each of which provides a different perspective on the data. Subsequently, we explore the complexities of coding, which involves methodically labelling data segments to detect recurring themes and patterns. The crucial significance of data preparation, exploration, and organization is next discussed, focusing on the careful organizing and structuring required to support in-depth analysis. The next step is to navigate data synthesis, which involves creating narratives, creating links between codes and themes, or creating theoretical frameworks to explain the data. We conclude by looking at computer-assisted data analysis techniques. The last section emphasizes the need for data presentation and visualization.

Keywords

Coding · Axial coding · Conceptual mapping · Content analysis · Constant comparison · Content analysis · Data transcription · Discourse analysis · Inductive analysis · Open coding · Narrative analysis · Selective coding · Thematic analysis · Nvivo

Learning Outcomes

Reading of this chapter should enable you to:

1. Explain the meaning of qualitative data analysis (QDA)
2. Differentiate between qualitative and quantitative data analysis
3. Describe the general steps in qualitative data analysis
4. Carry out data organization, preparation, and transcription of qualitative data
5. Use the various types of codes and approaches to qualitative data to analyze data
6. Explain the main approaches to qualitative data analytic techniques
7. Distinguish between thematic analysis and discourse analysis
8. Differentiate between axial coding, open coding, and selective coding techniques
9. Explain how to analyze data collected using various strategies, such as focus group discussions and visual data and artifacts
10. Describe the different computer-assisted qualitative data analytic tools and their basic steps, such as NVivo and ATLAS.ti: MAXQDA, etc.

23.1 Overview of Qualitative Data Analysis?

23.1.1 What Is Qualitative Data Analysis?

Qualitative data analysis is a multi-process that involves **gathering**, **structuring**, and **interpreting** qualitative data to understand their underlying meanings. Unlike quantitative analysis, which focuses on numerical data and statistical methods (discussed in Chaps. 20, 21, and 22), qualitative analysis emphasizes in-depth, subjective, context-specific, rich meaning by the participants. The rich information is presented in the form of words. The analysis explores dimensions of human experiences, perceptions, and behaviors and reduces data to key points of the phenomenon.

J. Njeri Mugwe, S. Runo, *Research Methodology in Agricultural Sciences*, https://doi.org/10.1007/978-981-95-1892-0_23

Researchers use **diverse coding**, **categorization**, and **thematic analysis** techniques to organize, reduce, and interpret qualitative data.

Some relevant chapters that can be read together with this chapter to enhance understanding include Chaps. 6 on ethics, Chaps. 13 and 14 on qualitative research approaches, and Chap. 19 on qualitative data collection procedures.

What Data are Available?

Qualitative data consist of **unstructured textual data** sourced from various sources such as interview transcripts, field records, focus group discussions, observational notes, and diary entries from farmers and other stakeholders (refer to Chap. 19). Additionally, qualitative data may encompass visual elements like images or audio and video recordings, offering insights into various agricultural practices and experiences (Box 23.1). As we can see, the data are varied and multidimensional, demonstrating the complexity of human experiences and viewpoints.

Box 23.1 Common qualitative data from a study involving farmers

1. Interview transcripts: recordings or transcriptions of interviews with farmers.
2. Field records: documentation maintained by farmers, such as, farmers observations and field notes; on farmers experiences about weather and markets etc.
3. Focus group discussions: group conversations facilitated by researchers among farmers
4. Observational notes: detailed observations made by researchers during field visits.
5. Diary entries: written records maintained by farmers documenting their daily activities.
6. Visual data: photographs and pictures or videos taken by farmers or researchers depicting various aspects of farming activities, crops, and animals.
7. Oral histories: narratives or stories shared by farmers orally.
8. Drawings by farmers, manuals, and email messages
9. Farmers records: Minutes of group meetings, duty roosters, etc.
10. Portfolio of evidence, such as, plans for agriculture for different seasons
11. Video recordings, such as, goat ear pegging or planting in conservation agriculture field

Philosophies in Qualitative Data Analysis

In Chaps. 7 and 13, we found that qualitative research is often associated with an interpretive philosophy, because researchers aim to understand participants' subjective and socially constructed **meanings about the phenomenon** being studied. Interpretivism emphasizes that meanings and realities are shaped by individuals' interpretations of their environment. Consequently, qualitative data reflect this social interaction, and tend to be diverse, flexible, and complex compared to quantitative data. The data collected are also voluminous (Tenny et al. 2022). Therefore, analyzing qualitative data requires sensitivity to these attributes to ensure their significance. It involves using techniques such as coding, thematic analysis, and constant comparison. These techniques help to explore **relationships and meanings** found in the data.

23.1.2 Purpose and Nature of Qualitative Data Analysis

Researchers often aim to **identify patterns and themes** in each data piece as well as the overall dataset. This helps to reveal understanding and larger patterns of knowledge. Thorne (2008) describes the analytical process as shifting "from pieces to patterns" through organizing, coding, reflection, and thematic derivation.

There is no single or correct way to analyze qualitative data (Patton 2002). It is influenced by the **kind of qualitative study** (ethnography, case study), the **number of data sets**, and the **people** from whom the data were collected. This is referred to as "fitness of purpose". The section that follows discusses the various steps that can be used to analyze qualitative data.

Distinctive Features of Qualitative Data Analysis

- No statistics: Do not use statistics like the quantitative, but rather it is subjective and the researcher and participants bring their perspectives and interpretation.
- Focusses on extracting meaning and understanding from the non-numerical data, audio, images, video, etc., from interviews and field notes.
- Data collection and analysis go hand-in-hand.
- Whereas quantitative uses quantitative software like GENSAT, SAS, and SPSS, qualitative data analysis uses qualitative software like N-Vivo, ATLAS.ti: MAXQDA, etc.
- Flexible data analysis, allowing for iterative data. The process may go through several cycles of coding and recoding (Berthet et al. 2023).

- Thematic analysis is one of the primary techniques in data analysis. Themes represent emerging commonalities and meaningful concepts (Creswell and Tashakkori 2007).
- Produces rich and detailed narratives, descriptions, and interpretations.
- Triangulate from multiple sources-enhanced validity.
- Uses narratives and storytelling.
- Data are large and usually challenging to handle.
- Inductive reasoning is used; uses data to generate an idea.
- *Use of direct quotes*: Participants' direct quotes are used, and this adds authority to the findings.
- Uses coding and categorizations.

23.1.3 General Steps in Qualitative Data Analysis

Qualitative data analysis typically begins early in the data-gathering stage and continues throughout the study, allowing for concurrent theory development. Unlike quantitative data analysis, qualitative data analysis often involves a more iterative and interactive process. This **iterative process** of collecting and analyzing data enables researchers to refine their understanding of the research study and explore emerging themes and patterns. Initial interviews with participants can inform further investigations on emerging trends. Additionally, the memoing process forms part of preliminary qualitative data analysis that is useful in guiding further inquiry. Researchers **engage closely with the data throughout the research process** to uncover patterns, themes, and categories. Here, we describe five general steps to achieve this as follows:

1. *Prepare and organize the data for analysis*: This is similar to quantitative data analysis discussed in Chap. 20, where we start by preparing a data analysis matrix. Qualitative data involve transcribing interviews, organizing field notes, or compiling other qualitative data sources.
2. *Read through and get familiar with data*: This is more about exploring data to gain a preliminary understanding. Identifying initial patterns or themes helps researchers develop a sense of the data and begin to formulate initial impressions or ideas. During this stage, **writing memos about observations, thoughts**, and insights helps inform preliminary analyses and guides further data collection.
3. *Code and describe data*: Once familiar with the data, proceed to code and describe it and its themes. You may use inductive and deductive coding approaches to identify emergent themes and organize the data accordingly.
4. *Categorize data into categories*. Finally, create categories or broader themes by synthesizing the coded segments to develop a comprehensive understanding of the data. This will involve organizing the coded data into meaningful clusters or groups, which may be further refined or revised through ongoing analysis and interpretation.
5. *Synthesize, present, and report findings*.

23.1.4 Data Organization and Preparation

Data organization should be carefully conceived at the beginning of the research to take care of the enormous amount of data. The purpose is to ensure that it is easy to retrieve data during the analysis stage. Data organization entails deciding on the type of storage for the data that suits specific data or the quantity of data that are being considered. Data backup and protection are also important at this point to avoid mishaps or exposure of confidential information shared by respondents. Shared drives or saving in the cloud could be considered for research involving a group of people. Voice recordings should be stored in files that specify details such as the date and time of the interview, as well as the interviewee, while fieldwork notes should also specify the date, time, and name of the interviewee.

Data organization ensures that all the data will be easily accessible in the subsequent analysis process. Organization involves creating a structure and accessible data repository. This may include physical files or digital folders to store the transcripts, motes, and other resources. As you create the data repository, assign meaningful and consistent names or labels to each transcript or label for easy retrieval during analysis.

Some common methods of organizing data include;

- Chronological orders: organized according to the different periods in which data were gathered
- Site or location: from which data were collected, common with studies where multiple sites or locations were observed
- Type of event or issue: address different issues in different areas that are grouped together
- Person or group: Organized by the individual or group, data with similar backgrounds are grouped together
- Type of data: interview transcripts might be assembled separately from field notes

The first task in data preparation is to ensure that data is in a form that can be analyzed, much like quantitative data. After collecting all the data sources, including interview recordings, field notes, and other data, the first task is to **transcribe the data**. This involves converting video and audio recordings, focus group discussion data, and any other data into written or typed format and **readable text**. The researcher usually opts for **verbatim transcription**, which captures participants' spoken words. These include **vocal**

cues, nonverbal expressions, and **background noise** that may be relevant to the analysis. Ensuring the accuracy and quality of the transcriptions is essential. You should never allow biases to influence the findings. It is important to have a key/ legend of the symbols that are applied during the transcription so that any person accessing this information can understand it and for future reference (Tracy 2019).

Fieldwork notes are sometimes taken quickly, abbreviations are used, and some words could be misspelled. These errors and missing details can be added during transcription. However, it is advisable to edit the fieldwork notes during the fieldwork, in the evenings, when the interview is still fresh to avoid forgetting some key details. See an example of a transcription of an interview (Box 23.2). During this editing process, inconsistencies in the data identified and corrected.

Transcription is tedious work, requiring a lot of time and patience. On average, it can take 4–5 h to transcribe a one-hour interview involving two speakers, with the exact time varying with the quality of the information, number of speakers, and the type of interview; FGDs can take up to 8 h to transcribe (Tracy 2019). The transcription process offers a unique opportunity for the researcher to interact with the data, which helps in the initial interpretation. The transcription process also refreshes the memory of the fieldwork and ignites reflections by the research concerning the data collected. It is important that the researcher notes down and critically reflects on the issues coming out of the data through **analytic notes**. These analytic notes could be information or comments on the Word document or computer-assisted qualitative data analysis software (CAQDAS) and are a form of conversation between the researcher and the data. In instances where the transcription is outsourced, it is important to listen to the interviews as you go through the transcripts to check for errors and consistency in annotations.

A key question students ask is: Is it necessary to transcribe the verbatim interview? This depends on the study; detailed transcription may be necessary for conversation and discourse analysis studies. Such studies can detail the tone, interruptions, and pauses during the conversation. In other cases, researchers summarize the interview and only note the key quotations from the respondents (Tracy 2019).

23.2 Coding in Qualitative Analysis

Any researcher who wants to become skilled at conducting qualitative analysis needs to have a solid understanding of and ease with **coding**. The coding quality plays a major role in the overall quality of the research (Strauss 1987). According to Creswell (2015), coding is necessary because reading through dense data and making sense of text data takes a lot of time. In this case, coding is a method akin to indexing or mapping data that gives researchers a summary

Box 23.2 Transcription of an interview between Farm Radio and a Dairy Farmer

Storing dairy fodder for the dry season increases farmers' income

March 22, 2019 | Interview from Resource Pack 111

Interviewer:

The division of roles is interesting. In most farms I have visited where both husband and wife are farmers, I find the men drifting towards dairy while women tend to prefer poultry.

Respondent:

(LAUGHING) When we started out, we looked at what interested each of us, and we found that I had a liking for cows and that my husband liked birds. There is no need to do something you have no passion for. Because I love cows, I take good care of them. Even when there are challenges, I keep at it until things change for the better.

Interviewer:

How much milk do you sell?

Respondent:

At the moment, I deliver 300 kg of milk to Fresha (Local dairy cooperative) each day from 11 milking cows. This allows me to pay my workers, maintain the animals, and make a profit. I want to continue expanding the number of animals, but maintaining them on this one-acre farm is not easy.

Interviewer:

This gives an average production of 27 kgs of milk per cow per day. This is quite high for a zero grazing production system and the size of the farm you have. (Editor's note: In a zero-grazing system, cows are provided with all their feed and water. They are confined in a cowshed and do not graze in pastures). How do you manage to achieve and maintain this?

Respondent:

As you can see, I have Napier grass growing on the lower section of the farm and another two acres leased from neighbors to provide fresh fodder for the animals. But it's not enough. These animals need much more.

> I buy baled grass from farmers in Nakuru between August and November when the prices are low, and the grass is of good quality. I have established good working relations with some of the farmers there, and I am able to get a bale of Boma Rhodes grass to my farm for 150 Kenyan shillings (US $1.50). Sometimes, I lease land and grow the grass. But this isn't easy because of the distance between my dairy farm in Githunguri and the leased lands in Nakuru, so I prefer buying grass that is already baled.
>
> (Farm Radio 2019)

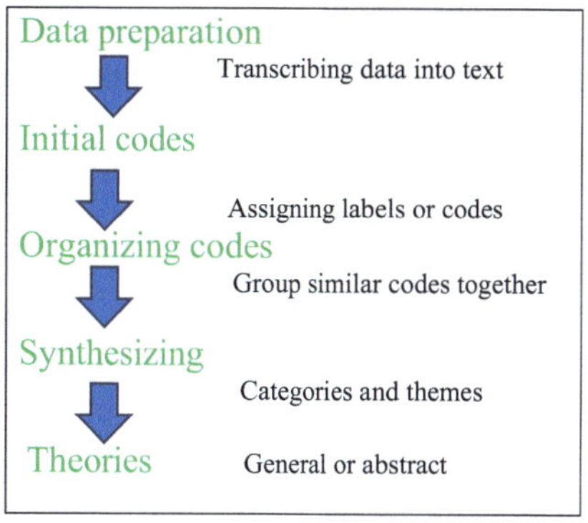

Fig. 23.1 Summary of the coding processes

of diverse data and helps them answer their research questions.

23.2.1 What Is a Code?

In qualitative research, a code is typically a word or brief phrase that designates a summative, notable, essence-capturing, and/or emotive attribute for a subset of textual or visual data (Saldana 2008). Labels or tags called codes denote **themes, concepts, or patterns** in qualitative data and are applied to specific data segments. Transcripts of interviews, field notes from participant observation, journals, papers, books, artifacts, photos, videos, webpages, emails, and more can all be included in the data.

Deriving a code takes several cycle coding procedures where in the first cycle, one word, a phrase, a page of text, or a stream of moving images can all be considered part of the data to be coded. When coding in the second cycle procedures, the coded sections may consist of the same units, lengthier text passages, or even a reorganization of the already generated codes. A code expresses and catches a datum's essential substance and essence in the same way a title reflects and captures the main idea of a book, movie, or poem.

An example of a coded entry, when reviewing farmer notes regarding their experiences implementing new irrigation systems during a drought, might be: " Ever since installing drip irrigation, I have been able to save water and devote less time physically watering my plants."

The code "water conservation/efficiency" sums up the excerpt's key idea. It's acceptable if, when reading that participant's response, you thought of another code (Holton 2007). Keep in mind that codes can sometimes summarize or condense data instead of only decreasing it.

These foundational codes are used as building blocks to find more extensive patterns or trends within the dataset (Saldana 2013). Researchers can find connections between codes, classify related codes into themes or categories, and eventually create patterns that shed light on underlying events or experiences through iterative coding and analysis.

Thematic analysis here helps researchers to find **recurrent themes or patterns** in the collection after the first codes have been given and arranged. This involves going over coded data methodically to find similarities, variations, and connections between codes. Researchers can find broad themes in the data that reflect common experiences, attitudes, or behaviors by grouping comparable codes and analyzing their frequencies and contexts. By doing this, qualitative data coding makes it easier to find and analyze patterns that shed light on the goals and issues of the study, which in turn helps researchers create complex and nuanced understandings of the phenomena they are studying. See a summary of the coding process in Fig. 23.1.

23.2.2 Types of Codes

Various coding techniques exist. They assist researchers in sifting through the data in search of themes and patterns. Every coding technique has a distinct function, such as explaining the contents of the data, understanding its significance, or relating it to accepted theories. The coding techniques include the following:

1. **Open/Initial Coding Process**

The open or initial coding process is necessary for removing complexity from the gathered information. This procedure essentially identifies and labels initial concepts or themes through a systematic review of raw material, such as transcripts or field notes.

Strategies Employed in Open/Initial Coding Process

In Vivo

In vivo, coding involves labelling important concepts with the participants' own words. Rather than the researcher assigning abstract labels, the research subjects assign labels themselves in their own words. This method preserves the data's authenticity and faithfulness to the participants' expressions. For instance, if the farmer's response was, "I felt like I was drowning in farm work," we might choose "drowning in farm work" as the code instead of summarizing it with a different phrase. The analysis is more accurate and perceptive when we use this method to capture the complexity and nuance of the farmers' experiences.

In vivo coding generally involves allowing the participants to direct the process of labeling and comprehending the data. By utilizing their phrases as codes, we can better appreciate their viewpoints and comprehend their experiences. This allows us to respect the perspectives of the people we are researching and ensure that our analysis is based on their actual experiences. In vivo coding enhances the authenticity and depth of our analysis, deepening our comprehension of the data and eventually producing more insightful conclusions.

Descriptive

Descriptive coding is a method of information organization whereby labels are assigned to the data to characterize its contents. It is similar to labelling or labelling certain sections of the data to make it easier to grasp. For instance, in an investigation into how farmers feel about their occupation if a farmer responds, "The responsibilities of tending to my cattle and crops are making me feel overwhelmed". We might name the piece of data as "workload stress". Researchers can identify the primary themes or subjects that emerge by breaking the data down into digestible chunks with the aid of descriptive coding.

To put it simply, descriptive coding is the process of labelling various data points to make it easier to identify important information. It is a method of arranging the data to improve our understanding of it. Descriptive codes help researchers find the most important concepts or subjects inside the data and begin to interpret it all. It's a crucial stage in the research process because it makes it easier for us to identify themes and patterns in the data, which we can then go into more detail about in our analysis.

Subcoding

Subcoding enables researchers to explore the data more thoroughly and spot deeper differences and variations that might not be obvious at first glance. By refining the analysis, researchers can capture the particulars of participants' experiences or viewpoints and increase the study's precision. Subcoding allows researchers to delve further into the study by revealing layers of meaning and addressing the intricacies present in the data.

2. **Grounded Theory Coding Process**

In qualitative research, researchers can examine data without initial predetermined beliefs using the grounded theory coding process. Rather, they allow the facts to direct their research, systematically classifying and organizing data to produce fresh insights or hypotheses. To create a cohesive theoretical framework, this approach entails finding themes and patterns in the data, investigating the connections between these patterns, and iteratively improving the analysis. Instead of depending on preexisting theories or frameworks, grounded theory coding enables researchers to gain insights straight from the data, resulting in the formation of theories rooted in the particulars of the study's findings. Three coding methods for text data analysis are described by Strauss and Corbin (1998) as open, axial, and selective.

Open Coding

Open coding involves researchers segmenting the data into smaller pieces and giving each piece a code or name. The primary thoughts or ideas found in the data are represented by these codes. Researchers can systematically arrange the data and begin to identify the major themes or categories that emerge by employing open coding.

Open coding aids in laying the groundwork for subsequent study by scholars. It enables them to thoroughly examine the data and pinpoint the original notions or thoughts that are there. Open coding is frequently an iterative process in which academics switch back and forth between their codes and the data, improving and changing them as they comprehend the data more fully. Grounded theories that originate from the data itself might be developed as a result of researchers using open coding to find fresh viewpoints and ideas that might not have been obvious at first.

Axial Coding

Axial coding entails examining the connections between the codes found during open coding to create more organized categories or topics. This helps in comprehending how they relate to one another and create logical patterns or relationships. Through this technique, researchers might discover overarching themes or notions that arise from the data and investigate the connections that underlie these themes. Axial coding is an iterative process; thus, researchers can modify and improve their categories or themes as they learn more about the data. Researchers can create grounded theories that are deeply based on the data itself to gain a more thorough and nuanced knowledge of the phenomena they are studying.

Selective Coding

Selective coding is employed to enhance and incorporate important themes or categories that were discovered in the initial phases of analysis, including axial and open coding. It entails concentrating on the most important and core categories that have been revealed by the data and expanding upon them to create a coherent theoretical framework.

23.2.3 Overall Approaches to Coding

Linneberg and Korsgaard (2019) explain the qualitative study coding process. They stress the significance of approaching coding systematically and utilizing both deductive and inductive techniques. Similarly, Goodrick & Rogers (2015. Pg 562) in their qualitative data analysis handbook state that "*the coding process can be done in various ways, from structured processes with codes assigned a priori and applied to the data, to emergent coding frameworks that identify codes directly from the data.*".

(a) **Inductive technique**

The inductive strategy of coding represents an approach to qualitative data analysis where researchers refrain from imposing predefined categories or theories onto the data. Instead, they allow themes, patterns, and concepts to emerge organically from the data itself. This method embraces an open-minded and exploratory stance towards the data, aiming to capture the richness and complexity of participants' experiences or perspectives without biasing the analysis.

The inductive approach allows researchers to engage and look at the data line by line to identify recurring themes or patterns that emerge across different sources of information. Through this process, they generate codes to represent the key concepts or ideas present in the data, allowing for a structured analysis while remaining true to the inductive nature of the approach. The inductive approach can lead to unexpected findings and novel insights that may challenge or expand existing knowledge.

(b) **Deductive Techniques**

In the deductive coding process, researchers apply the predetermined codes to the data, looking for instances or patterns that correspond to the established theoretical categories. This method provides a structured approach to data analysis, allowing researchers to systematically organize and interpret the data within the framework of existing theories. However, researchers must remain vigilant to ensure that their interpretations are not overly constrained by preexisting theories, as these may limit the exploration of alternative perspectives or unexpected findings within the data. While deductive coding offers a systematic framework for analysis, researchers may need to balance the use of predefined codes with the flexibility to adapt their coding scheme in response to the nuances and complexities of the data (See Fig. 23.2).

23.3 Approaches to Data Analysis

In this section, we explore several approaches to qualitative data analysis that provide different perspectives that researchers can use to understand their data. Every approach has a unique set of methods, presumptions, and resources that influence how researchers interact with and interpret qualitative data. These methods, ranging from **narrative** and **grounded theory** to **thematic analysis**, offer flexible frameworks for identifying **patterns, themes, and meanings** in qualitative data.

23.3.1 Content Analysis

Content analysis is a common **qualitative data analysis** approach used for words, phrases, or images in different pieces of writing or communication. It is a widely used approach that is defined as the methodical, objective, and quantitative investigation of messages (Neuendorf and Kumar 2015). For example, we could look at a bunch of newspaper articles, original data, or speeches from farmers (see an example in Box 23.3). With content analysis, we might count the number of times a particular crop variety is mentioned on social media. Or we could look for patterns in how people interpret things. Because content analysis may be applied in many different contexts, it is critical to approach the study with a clear query and objective; otherwise, one risks becoming lost in the haze.

Content analysis can code, summarize, and possibly even tabulate large volumes of text to determine the frequency of

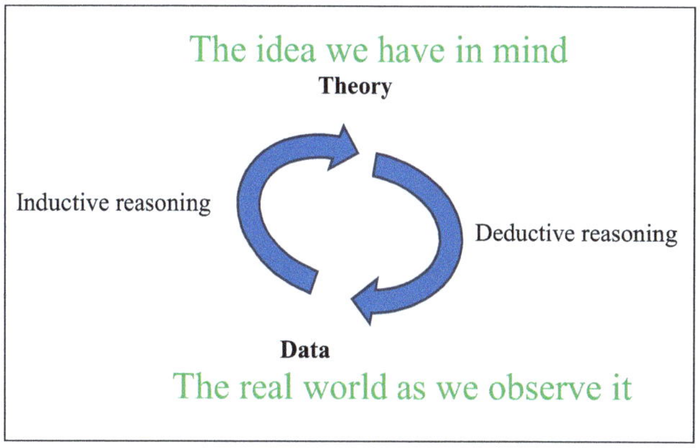

Fig. 23.2 Coding Techniques

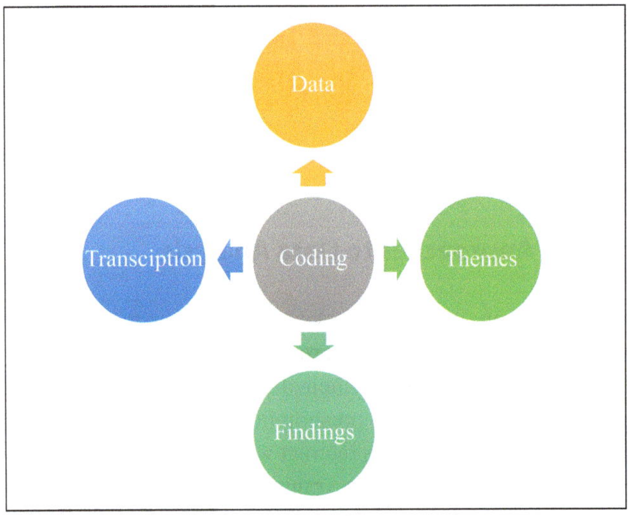

Fig. 23.3 Iterative process of narrative analysis

specific concepts or variables. For this reason, it offers a tiny dash of quantitative reasoning mixed with a qualitative approach. Although content analysis is quite helpful, it has its drawbacks. One major issue is that it requires a lot of time due to the need to read and reread several materials. Furthermore, because it considers both meanings and numbers, it may overlook certain crucial information. Another thing is that it focuses on a specific time and ignores events that occurred before or later, instead concentrating on a single point in time. While it's not necessarily terrible, it is something to consider.

Box 23.3 Farm education and the effect of a farm visit on children's conception of agriculture. European Journal of Educational Research, 4(1), 1–13

Due to structural changes in agriculture over the past few decades, personal links to the sector have significantly declined in Finland. The study sought to clarify urban students' understanding and conception of agriculture among those who have grown up in the twenty-first century in this study. Pre- and post-assessments were part of the research strategy, which involved an intervention in the form of farm visits. Drawings were used as data collection tools to provide a more comprehensive picture of how students perceived, comprehended, and interpreted agriculture. Visual content analysis was used to evaluate the drawings. Inductive content analysis was applied to interviews to further elucidate the findings. Farms' qualities as real learning environments were identified and investigated. The study found that whereas 38% of students had unreasonable ideas about agriculture before the farm tour, these beliefs dramatically declined following the visit. The impact of media sources was thought to be one of the explanations for these illogical beliefs. They were using farms as instructional spaces that allowed teachers to dispel students' misconceptions. Becoming an informed and conscientious customer and pursuing a career in agriculture greatly benefit from having a realistic perception of agriculture.

Smeds et al. (2015).

23.3.2 Narrative Analysis

Narrative analysis, as its name implies, is the process of listening to individuals tell stories and interpreting what they say (Creswell and Poth 2016). Analyzing people's stories and the methods in which they are recounted can provide us with insights into how individuals cope with and make sense of reality since stories have the practical function of assisting us in making sense of the outside world. For instance, one can explore the way something is said. For example, listening to a farmer's account of moving from conventional to organic farming practices demonstrates more than technical improvements. It also shows how farmers' perceptions of sustainability are evolving. By closely examining these stories, we can gain insight into the challenges and reasons behind farmers' decisions. We also note that narrative analysis is iterative, where researchers move repeatedly between stories to refine understanding (Fig. 23.3).

Narrative analysis studies stories individuals tell and, more significantly, the methods by which they tell them. Of course, there are drawbacks to the narrative approach. Because gathering narratives is time-consuming, sample sizes are typically relatively small (Chap. 13). This makes narrative analysis challenging to replicate in later studies because the wide range of social and lifestyle differences among participants can limit the generalizability of the findings. Additionally the researcher's beliefs can also significantly impact the findings, so it's critical to take extreme caution while applying this method to prevent biases in the analysis. Notwithstanding its drawbacks, narrative analysis remains a very valuable research method.

23.3.3 Discourse Analysis

Gill (2000) looks at discourse analysis as the use of various techniques for **text analysis** that have emerged from various academic contexts and theoretical traditions. Talking or writing more than one sentence to one another is called discourse. It goes beyond words alone. While "language" refers to all forms of words and symbols, discourse examines the collective meaning of those words in a given context.

"Context" refers to the conversation's context, such as the relevant society or historical period. Recognizing this makes it easier to understand what the words mean. Discourse is frequently thought of as language utilized in specific social contexts. This point of view holds that language can be used to achieve objectives or alter society.

Discourse analysis, for example, could be used to look at the words and ways that farmers communicate in meetings or conversations on agriculture. Transcripts of meetings where decisions about adjustments to agricultural techniques or crop choices are made could be gathered by researchers. Researchers could look at how language is used in these sessions to establish **consensus, negotiate power dynamics, or question ideas** through discourse analysis. They could examine the distribution of speaking turns among various participants, the usage of specialized language, or the frequency of interruptions. Researchers can learn more about how choices are made, authority is established, and consensus is attained within agricultural settings by looking at the discursive practices of agriculturalists.

Discourse analysis is thus the study of language in the context of its social environment; in another way, examining language with the culture and society in which it is used, such as in a speech, discussion, etc. Considerations regarding the communication partners are crucial. For example, informal discussions regarding crop management tactics may take place between farm owners and farmworkers on a farm that value inclusive decision-making. In a similar vein, agricultural officials might concentrate more on pest management methods after a significant pest infestation in the area. Discourse analysis is useful in understanding how debates regarding farming challenges and practices are shaped by various elements such as **power dynamics, agricultural history, and local customs**.

Discourse analysis has many applications since several social elements influence human speech patterns. However, when examining data, it is critical to keep a specific research topic in mind. If not, one could find themselves becoming bogged down in the details. Examining talks can be time-consuming because researchers have to keep going through the material until they find something new or nothing at all.

Box 23.4 Prevention and control of African swine fever in the smallholder pig value chain in Northern Uganda: Thematic analysis of stakeholders' perceptions. Frontiers in Veterinary Science, 8, 707819

African swine fever (ASF) is a serious threat to Uganda's pig industry because there is currently no vaccine, making biosecurity essential for both its prevention and control. A qualitative interview study was carried out in northern Uganda to comprehend the present application of biosecurity practices and the perspectives of pig value chain actors towards prevention and control. Six towns were chosen for focus group talks based on their prior ASF outbreaks, willingness to control ASF, and representation of different value chain actors. The findings showed that certain villages had adopted biosecurity measures such as sharing illness information, limiting trade, and practicing essential cleanliness.

Two authors (TA and DMO) performed a thematic analysis on the master notes, separately coding sections of the data pertinent to notions of control and prevention. This was the first conceptualization of the transcribed data. Repetitive readings led to the inductive emergence of primary codes, which were then modified through several iterations of analysis to combine the two sets of codes. After that, TA and DMO worked together to apply axial coding to group these primary codes into themes. To guarantee traceability to the original data, and to demonstrate the connection between themes and focus group discussions (FGDs), TA further aggregated relevant themes into six overarching categories. The categories were then summarized in a matrix that also displayed the frequency of frequencies across FGD. Finally a logical framework was used to depict the resulting categories, their relationships, additive implications for ASF (African Swine Fever) prevention and control, and recommendations for enhancing biosecurity implementation.

Aliro et al. (2022)

23.3.4 Thematic Analysis

Thematic analysis examines the patterns of meaning within a set of data, such as transcripts from focus groups or interviews (Creswell and Tashakkori 2007). However, what does that mean? It simply means grouping a set of data, which are frequently rather vast, based on commonalities, or themes. These themes assist us in understanding the material and drawing conclusions from it (see an example of a study that used thematic analysis: in Box 23.4).

For instance, suppose a group of researchers wish to know farmers' experiences using a new irrigation system. During the interviews, they record their answers and before doing a thematic analysis of these interviews, the transcripts are examined to comprehend the overall data. After that, coding takes place, at which time recurrent themes like "effectiveness of irrigation," "challenges in execution," "cost implications," and "environmental impacts" are found and grouped into subgroups. To obtain insight into farmers' experiences with the new irrigation system, the data are then examined to find patterns and correlations among these themes.

Thematic analysis is a valuable tool for learning about people's perspectives, experiences, and ideas. Thematic analysis might therefore be an excellent choice when the research's goals and objectives include understanding people's experiences or points of view. The research topic may evolve during thematic analysis because it is somewhat of an exploratory process. Although this is sometimes expected in exploratory research, it can also be considered as a drawback because it necessitates a new examination of the data whenever a study topic is modified.

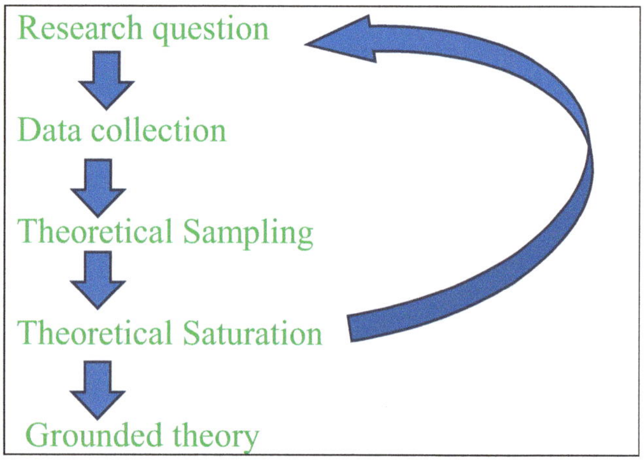

Fig. 23.4 Grounded theory cyclical process

23.3.5 Grounded Theory (GT)

Grounded theory (GT) is a powerful qualitative analysis method where the intention is to create a new theory (or theories) using the data at hand, through a series of tests and revisions (Creswell and Poth 2016). Strictly speaking, GT is more of a research design (See Chap. 14) than an analysis method, but we have included it here as it is often referred to as a method. (See an example of its application in Fig. 23.4).

What is most important with grounded theory is that researchers go into the analysis with an open mind and let the data speak for itself; rather than dragging existing hypotheses or theories into the analysis. In other words, the analysis must develop from the ground up (hence the name).

For instance: imagine a researcher who wants to know how farmers create cooperative networks during community meetings. Researchers tend to discover from their studies that farmers typically bond with one another around shared interests, such as agricultural methods or crop preferences. They also see that certain farmers build relationships through unplanned exchanges, such as helping out with communal chores or joining in on group conversations. The researcher consistently employs grounded theory concepts to gather and examine data. They formulate a theory that suggests mutual agricultural interests and random meetings influence farmer relationships within the community. By conducting continuous interviews and observations with farmers, they enhance their hypothesis and better understand how friendships flourish in this agricultural community.

Researchers use grounded theory to ask a broad question about a **particular population**, like our example of agricultural communities, and then interview a **small representative sample**. The data are then analyzed to look for **patterns or themes** that may appear. If the observed trends continue, researchers look for confirmation by looking at a smaller sample, such as five farmers from a different location; if not, they look for similar characteristics to iteratively develop and evolve the theory. Grounded theory's drawback is its circular nature; although it's generally regarded as unwise to disregard previous literature, it's best to begin with little prior knowledge to minimize bias. Despite this, it's a useful technique, particularly for the first study of novel or understudied subjects.

23.3.6 Phenomenological Analysis

The purpose of phenomenological analysis is to assist in comprehending the individual experiences of an issue (a person or group of people) regarding a significant event, experience, or circumstance (Creswell and Poth 2016). The "phenomenon" is this occurrence or experience. These phenomena could include more exceptional events like the successful development of a rare heritage tomato variety and more commonplace ones like planting a crop or managing a rare pest infestations. It may also feature unique experiences, such as a farmer's story of surviving a natural disaster or their first-hand report of introducing sustainable agricultural methods in a remote location. Therefore, if research entails examining people's first-hand accounts of events they have personally experienced, phenomenological analysis is an excellent option.

It is important to remember that phenomenological analysis is subject-centered. In other words, it's focused on the experiences. This means that, while it is likely to use a coding system to identify commonalities, it is important not to lose the depth of experience or meaning by trying to reduce everything to codes. Also, keep in mind that since the sample size will generally be very small with phenomenological analysis, we will not often be able to draw broad conclusions about the generalizability of the findings. But that's okay as long as it aligns with the research aims and objectives.

Phenomenological analysis involves important steps. First, we need to deeply understand what participants went through by studying the data, often collected through observations or asking questions. Then, we identify key points or "units of meaning" in the data, highlighting important words or phrases that show what the participants experienced. Next, we group these points into themes that show common patterns in what the participants experienced. We then think and interpret to find the main ideas that define what we are studying.

Finally, researchers report their findings in written form, drawing on the participants' experiences and providing a detailed description of the phenomena. They may illustrate key points with quotes. This procedure gives researchers more insight into the subjects of their research and improves their understanding of people's experiences.

23.4 Types of Data and their Analysis

To find fresh information and develop comprehension, it is essential to understand the complex nature of qualitative data types and their analysis methods. Each sort of qualitative data e.g. from observational notes outlining soil health indicators to interview transcripts documenting farmers' experiences, needs a unique analytical framework. In this sub-section we look at how we analyse data from various methods.

23.4.1 Focus Group Discussions Data

Focus group data can be analyzed through various approaches, as discussed above. Focus groups have been applied widely in agricultural research and are important in data triangulation. (Refer: Silverman 2024). There are several approaches to the analysis of data from FGDs that were discussed earlier in this chapter. These include:

(a) Content analysis: This approach analyzes the frequency and distribution of codes and themes within the data. It is useful in identifying patterns, trends, or outliers in the data. Comparative keyword analysis (CKA) can be used to identify frequently used words or phrases to understand what participants are emphasizing or prioritizing.
(b) Thematic analysis is useful in identifying overarching themes and patterns within the data. Themes should capture the essence of participants' responses.
(c) Comparative analysis is used to compare responses between different focus groups or subgroups within focus groups to identify similarities and differences.
(d) Discourse analysis is applied in analyzing the data to pick the tones and sentiments of participants' responses. This can provide insights into their attitudes and emotions. It can also be used to pick out the power plays among various stakeholders in a particular group setup.

23.4.2 Visual Data and Artifacts

Documents, web-based data, and artifacts collected during fieldwork should be analyzed to extract the deeper meanings they portray and it includes the details of the authors of the documents. Why the documents are written? Who are the target audience? These data can be analyzed through content and thematic analysis. Content analysis has a limitation of focusing on the frequency of preestablished categories and lacking in depth in terms of going beyond what is visible from the image to encompass how the image was taken and how the image is used. An in-depth account concerning visual data can be obtained through ethnographical research (Silverman 2024).

In an agricultural study, such data include farm records, group reports, minutes of various kinds of meetings, government reports, and communication by regulatory bodies among others. Web-based data include photographs, videos and communication in farmers' blogs, TV programmes, email communication, and online farming groups among others. When analyzing documents, it is important to decipher the truths, half-truths, or falsehoods in the documents in comparison with other data obtained from fieldwork.

23.4.3 Observation Data

Observational data are gathered by seeing and documenting actions, events, or other phenomena as they naturally arise; that is, without intervention or modification (Creswell et al. 2004). Analyzing observation data in agricultural research involves a systematic and rigorous process to derive meaningful insights. Initially, researchers must establish clear objectives and hypotheses to guide their observations, ensuring that data collection aligns with research goals. Once observations are conducted, researchers meticulously record their findings, noting significant events, patterns, and trends within the observed context. Subsequently, data analysis entails organizing and categorizing observational data, often through thematic coding or pattern recognition techniques. This involves identifying recurring themes, behaviors, or characteristics within the observations and assigning descriptive labels or codes to facilitate interpretation.

To extract deeper meanings and insights from the observational data, researchers can also use qualitative analysis techniques like content or narrative analysis. Researchers remain reflexive throughout the analytic process, evaluating their interpretations critically and taking into account any biases or prejudices that might have an impact on their results. The data analysis ultimately leads to the synthesis of findings, when researchers make links between observed occurrences, formulate theories or explanations, and advance knowledge of agricultural systems and practices.

23.4.4 Concept Maps

A concept map is a diagrammatic tool used to illustrate a set of conceptual meanings that are incorporated into a hypothetical context (Watson 1989). Analyzing concept maps entails closely examining the illustrations individuals create to convey their thoughts or concepts. These maps are examined by researchers to learn more about how people structure their ideas and link disparate ideas. They might search for recurring themes or patterns in several maps and pinpoint important ideas. Researchers may also examine the connections between concepts, examining which ones are more frequently linked together, or which ones appear to be essential to the map's general organization. To make sure they fully convey the meaning behind the concept maps, researchers critically evaluate their interpretations and any biases they may have during this process. In the end, concept map analysis advances our knowledge of how people view and conceive difficult subjects by offering insightful information for future study or useful applications.

23.4.5 Semi-Structured Interviews Data

Semi-structured interview analysis is paying close attention to what participants say during loosely organized conversations. One must closely monitor the themes that surface from these interviews, taking note of the subjects that participants discuss most frequently as well as the relationships between various subjects. They may also search for certain narratives or examples that highlight the experiences or viewpoints of the participants. To ensure they fully convey the meaning behind the participants' statements, researchers consider their interpretations and any biases they may have throughout this process. In the end, semi-structured interview analysis offers insightful information for future research or application while also assisting researchers in developing a deeper grasp of people's attitudes, experiences, and ideas around a certain subject.

23.5 Principles of Qualitative Data Analysis

Bryman and Burgess (1994) offer insightful explanations of the underlying principles of qualitative data analysis. These scholars established a basic paradigm that facilitates the performance of rigorous and perceptive qualitative data analysis. These principles include:

1. **People have varied experiences and perceptions of reality**

This principle recognizes the subjectivity of human experiences and perceptions, acknowledging the various ways in which people interpret and make sense of the world. This variation in experiences and understandings is welcomed as a useful component in qualitative research. The diversity of participants' viewpoints, worldviews, and ways of interpreting reality are acknowledged by researchers as contributing factors to the complexity and depth of the data. For instance, farmers' perspectives on climate change vary based on their farming practices and place of residence, which has an impact on how they plan for and respond to it. Researchers take an inclusive stance, respecting the range of viewpoints and experiences that participants have to offer. They understand that every person contributes their background, culture, and context to the research process, which affects how they view and understand the phenomena being studied. By using techniques like open-ended interviews, participant observation, or document analysis, researchers aim to capture and respect these varied points of view by letting individuals express their opinions in their own words.

2. **You cannot comprehend a social phenomenon outside its specific context**

This idea emphasizes how crucial it is to look at the larger social, cultural, historical, and contextual elements that affect and form the phenomenon being studied. Context is a key factor in determining people's experiences, viewpoints, and behaviors in qualitative research. As a result, to fully comprehend a social event, researchers need to take into account the unique context in which a phenomenon takes place.

Researchers investigate how participants' perceptions, attitudes, and behaviors are shaped by the interaction between the influence of various contextual elements on the phenomenon. In addition, researchers can collect rich contextual data that shed light on the larger social environment in which the phenomena are placed by using techniques like participant observation, interviews, or document analysis. Researchers can do analyses that are sensitive to the complexity of the social world by taking into account the context in which a social phenomenon evolves. This allows for more thorough and nuanced interpretations of the data.

3. **Qualitative research can be used to build data-driven theories or to characterize phenomena**

This idea emphasizes how qualitative research may be used to both create theoretical frameworks that are derived from the data itself and to give in-depth descriptions of phenomena. The phenomenon under research can be richly detailed and understood through the use of certain patterns and themes.

Moreover, grounded theory, a theoretical framework or theory developed by researchers based on data, can be produced through qualitative research. By carefully analyzing the data, grounded theory develops hypotheses that are born out of the data itself, as opposed to being imposed by preexisting theories or frameworks. Researchers can gain new insights and a deeper understanding of the phenomenon they are studying by letting ideas arise from the data. This process also advances the body of knowledge in the topic. For instance, qualitative research in agriculture can be used to describe various farming practices and their consequences or to construct ideas based on facts.

4. **Human Behavior is Slowly and Nonlinearly Understood**

This principle emphasizes that understanding human behavior is a process that takes time and is not always clear-cut or linear. This is due to dynamic and complicated character of human behavior. In principle then, understanding human behavior in qualitative research entails exploring the complex interactions between social environments, individual beliefs, values, and emotions. Being patient and prepared to spend a lot of time studying the facts is necessary for this process because comprehending human behavior is a continuous and iterative process. Therefore, researchers must use a flexible and iterative strategy that enables the investigation of human behavior from many viewpoints and angles. When we immerse in the data and carry out repeated cycles of data gathering and analysis, insights into human behavior may eventually become apparent. Techniques like discourse analysis, narrative analysis, or thematic analysis are useful in uncovering underlying themes, patterns, and meanings that influence human behavior.

5. **Exceptional cases can provide new information or insights into an issue for future research**

This principle acknowledges that oddities or exceptional instances within the data might offer insightful information that tests or broadens the body of knowledge already known in the subject. Exceptional cases, as used in qualitative research, are occurrences or examples that depart from the norm or anticipated trends. These cases frequently give distinctive viewpoints or first-hand experiences that provide insight into the phenomenon being studied. These rare situations deserve special attention from researchers because they can include important information that opens up new study directions or inspires new ideas.

It is important to look closely at the data for patterns, themes, or insights that are exceptional situations (anomalies) that may not be immediately obvious. Exploring the reasons behind these anomalies and considering how they could contribute to or enhance current ideas can lead to deeper understanding in the field. Through critical engagement with such exceptional situations , researchers might reveal novel insights, and generate fresh ideas for additional investigation. This idea emphasizes how crucial it is to keep an open mind and be open to surprising discoveries in the data because they can provide insightful new information and further the field's understanding.

23.6 Computer-Assisted Qualitative Data Analytic Tools (CAQDAS)

Software programs greatly aid the administrative tasks associated with the repetitive act of reading, comparing, and reflecting on text parts. The analysis of qualitative data can be supported by several software packages, including NVivo, MaxQDA, and Atlas.ti. Although you will still need to sort the data manually, these can be a helpful aid in quickly arranging the information. As no programme can understand the data, these programmes should not be seen as shortcuts to thorough and methodical study. Consequently, to further the investigation, the researcher still needs analytical abilities.

1. **ATLAS.ti: www.atlasti.com**

ATLAS.ti is a research tool designed for word and narrative analysis. It facilitates the organization and comprehension of data from research notes or interviews. You can use ATLAS.ti to segment interviews or notes into manageable chunks and assign distinct categories to each one. Making links between various concepts or terms is another benefit of this tool. Along with versions for iPads and Android smartphones, it can be downloaded for Windows and Mac PCs.

To get a Trial Copy, you need to register, visit https://my.atlasti.com. Verify the email address you use. Click Trial Desktop to request a trial license.

General Procedures for Using ATLAS.ti

Naturally, this sequence of steps is not required, but it does provide an overview of a typical script:

- Make a project, an idea container, to hold all of your information, research findings, codes, memoranda, and organizational frameworks under one roof.
- Next, populate your ATLAS.ti project with documents, text, graphics, audio, and video files, or geodocuments.
- Assign keywords (codes), read and highlight text passages, point out places in a picture, or pick out time segments from an audio or video clip that arouse your interest further. You can also create notes and memos that reflect your thoughts on the information.
- Compare data segments using the codes you've been given; consider using more data files in the project.
- Make use of the various tools that ATLAS.ti offers to query the data according to your research questions.

- By creating networks from the codes and other entities you have built, you can further conceptualize your data. An emerging theory is framed by these networks as well as your codes and memoranda.
- Lastly, put together a written report using the memos you wrote during the project's different stages and the networks you established.

2. **NVivo**

NVivo (www.qsrinternational.com) is a software program made specifically for analyzing qualitative data. It helps analysts and researchers code, organize, and explore many kinds of qualitative data, including text, audio, video, and image data. It gives consumers a platform for methodically going through massive amounts of data to find patterns, themes, and connections within their datasets. To make data analysis and interpretation easier, NVivo has capabilities including coding, annotation, visualization tools, and query functionalities. Regardless of the type of research being conducted—academic, market, or otherwise—NVivo is a powerful instrument for deriving conclusions from qualitative data and helping researchers make sense of complicated phenomena.

General Procedures for Using NVivo

Your data must first be imported into the NVivo program. This can apply to photos, movies, audio files, and text documents. You can begin coding your data as soon as it has been imported. Within your data, you can find themes, subjects, or concepts and give them labels or codes by coding them. To represent these codes, you can establish nodes in NVivo, which will facilitate data organization and analysis.

Once your data have been coded, you can begin examining relationships and trends. For this, NVivo provides a number of tools, including matrices and queries. Whereas matrices let you see the links between various codes or themes, queries let you look for particular patterns or relationships within your coded data. You can uncover important facts and obtain a deeper understanding of your data by executing queries and building matrices.

Lastly, NVivo can be used to create reports or visualizations that help you properly present your research. Charts, graphs, or summaries of your analysis could be used to assist you convey your findings understandably and persuasively.

MaxQDA (www.maxqda.com) is a software program that aids in the understanding of data. It helps when dealing with research materials such as written papers, surveys, and interviews. One can label key sections of information or arrange them into groups with MaxQDA. It also facilitates the identification of links or trends in the data, which makes it easier to interpret. MaxQDA is a useful tool for scholars, learners, and anyone requiring extensive data analysis to gain deeper insights into a subject.

General Procedures for Using MaxQDA

We must first import the data into the program. Transcripts of interviews, observational notes, or text-based data could be included here. We can begin coding the data as soon as it is in MaxQDA. Coding is assigning labels or tags to various data segments that stand in for themes, concepts, or ideas. To capture people's attitudes, one may categorize utterances as "positive" or "negative" if researching people's perspectives on a particular subject.

Once the data have been coded, we can explore it further. First, we can run queries to identify particular patterns or create visualizations of the coded data using the tools provided by MaxQDA. As we go along, one may also record their views or interpretations using features like memos. With the help of MaxQDA's tools and examining the coded data, we may learn more about the research questions and make inferences based on the findings. MaxQDA generally facilitates the organization, analysis, and comprehension of qualitative data, enabling researchers to find important insights.

3. **Dedoose**

Dedoose is a software application that facilitates information analysis and comprehension. It is helpful for evaluating several kinds of data, including audio, video, and text. Data can be categorized or thematically organized with Dedoose to help see patterns and relationships. Additionally, it enables group collaboration, allowing more than one person to work on projects with others.

General Procedures for Using Dedoose

Data are first imported into the program. Items such as field notes, survey responses, and transcripts of interviews may fall under this category. Coding begins soon after importation in Dedoose. Coding is the process of reviewing your data and assigning tags or categories to various sections. To classify people's attitudes, one may label utterances as "positive," "negative," or "neutral" if researching people's perspectives on a particular subject.

Once all the data are coded, we use Dedoose's tools to evaluate it by making charts or conducting searches to identify particular trends. Dedoose allows one to record their thoughts or insights using features like annotations. By going over your coded data and utilizing Dedoose's tools, you can learn more about your study topic and make insightful conclusions. All things considered, Dedoose makes qualitative data easier to access for professionals, students, and researchers by streamlining the process of organizing, analyzing, and interpreting it.

Other qualitative data analysis tools such as QDA Miner (www.provalisresearch.com) also utilize the same procedures as the above four; they only differ mostly in terms of features, user interfaces, and intended users.

In summary, NVivo and MaxQDA are comprehensive solutions for analyzing qualitative data and are mainly used by researchers with complicated analytical needs. The researchers can benefit from the broad data management functionality, powerful coding capabilities, and visualization tools provided by these tools. QDA Miner offers strong analysis features and is more user-friendly. It is a good option for researchers and students who need a balance between functionality and ease of use. Dedoose, on the other hand, is notable for its collaborative features, which let several users work on the same project at once, which makes it perfect for group research projects and team-based analysis. Every computer-assisted qualitative data analysis tool has its advantages and accommodates varying user tastes, research needs, and degrees of proficiency.

23.7 Data Synthesis and Presentation

The process of combining, evaluating, and interpreting qualitative data from many sources to produce fresh perspectives or create a thorough grasp of a research issue is known as **qualitative data synthesis** (Rycroft-Malone and Burton 2015). According to Drisko (2020), a wide range of techniques are used in qualitative research synthesis to integrate the information from several studies on a subject to provide new theories, applications, and knowledge.

Qualitative data synthesis combines **textual, visual, or audio data** from diverse qualitative studies, interviews, observations, or documents, in contrast to quantitative data synthesis, which usually uses statistical approaches to aggregate numerical data. Finding recurring themes, patterns, or connections across disparate data sources and examining opposing viewpoints or contradicting conclusions may all be part of this synthesis. To provide a comprehensive and in-depth picture of the phenomenon being studied, qualitative data synthesis attempts to go beyond specific studies or data sources. This process frequently results in the creation of new theories or frameworks based on the data.

23.7.1 Principles of Qualitative Data Synthesis

(a) **Grounded theory approach**: This is a fundamental tenet of qualitative data synthesis that directs researchers in the methodical integration and interpretation of qualitative data. Using this method, themes and patterns naturally arise from the data as researchers become fully immersed in the material without imposing predetermined categories or assumptions. This implies that investigators approach the data with an open mind, delving deeply into it to find important patterns and understandings that might not have been expected when the study began. Through adaptability and receptiveness to the data, researchers can create a thorough, empirically based understanding of the phenomena they are studying.

(b) **Reflexivity**: emphasizes the importance of researchers being conscious of their own prejudices, presumptions, and viewpoints at all stages of the analytical process. When synthesizing qualitative data, researchers know how their personal histories, experiences, and values may affect how the data are interpreted. Through a reflective approach, researchers consistently examine how they contribute to the synthesis process, guaranteeing openness and responsibility in their analytical choices. This idea enables more distinctive and contextually sensitive analysis by pushing researchers to critically consider how their subjectivities may affect how the data are interpreted.

(c) **Transparency**: Transparency in research synthesis emphasizes clarity and openness. Throughout the synthesis process, researchers uphold transparency by clearly documenting their methodological decisions, analytical techniques, and decision-making processes. This entails openly disclosing the procedures followed to gather, examine, and evaluate the data, as well as recording any deviations or difficulties found along the route. Researchers help people comprehend and assess the reliability and rigor of their aggregated findings by being transparent. Additionally, transparency makes it easier for the research to be replicated and verified, enabling other academics to build on the findings and further the field's knowledge.

(d) **Iterative Process**: This principle highlights the importance of ongoing improvement and modification throughout the analytical process. To expand their knowledge and find fresh ideas, researchers use an iterative procedure that involves going back and reviewing the data, coding schemes, and analytical frameworks several times. Through this repeated process, researchers can resolve any ambiguities or contradictions in the data, uncover new themes or patterns, and gradually improve their interpretations. Researchers who embrace iteration recognize that understanding is a dynamic process that requires constant evaluation and modification. Using this cyclical procedure, scholars can cultivate a thorough and refined comprehension of the phenomenon being examined, based on the actual data.

In summary, the principles of data synthesis stress the crucial importance of grounding the synthesis process on the material itself so that themes and patterns develop naturally. Transparency is crucial in ensuring that researchers record their methodological decisions and analytical processes to foster responsibility and credibility. The iterative process of synthesizing qualitative data emphasizes the necessity of ongoing improvement and modification as knowledge grows over time. By encouraging researchers to critically examine their prejudices and presumptions, reflexivity fosters openness and rigor in the analytical process.

23.7.2 Methods of Qualitative Data Synthesis

Qualitative data are drawn from multiple sources and require combining and analyzing to make sense of it. Methods of qualitative data synthesis aid in interpreting the wide range of data acquired through observations, interviews, and document review. These techniques help researchers find common themes, patterns, or connections among many qualitative data sources. Some popular methods of qualitative data synthesis include thematic analysis, meta-ethnography, framework synthesis, and narrative synthesis.

(a) **Thematic Analysis**

Thematic synthesis is creating descriptive and analytical themes through methodical data coding. The inductive approach is crucial given the objective of producing higher-order themes and the main message. According to Thomas and Harden (2008), thematic analysis is a three-step procedure that starts with line-by-line text coding, in which the results of the materials gathered are inputted into the database word-for-word, with each line of text being coded as per its content and significance. The next stage is the creation of descriptive themes, which entails transferring the ideas from one study to another. The codes will then be grouped according to their similarities and differences to establish a hierarchical framework. Finally, generating analytical themes that extend beyond the substance of the original materials requires first identifying the main messages through descriptive themes. More abstract messages and themes that go beyond the content of the source materials can be developed, thanks to these analytical topics. After examining each theme's and message's implications independently, they will be discussed collectively.

(b) **Meta-Ethnography**

Synthesis is an activity in which discrete elements are combined to create a "whole" (Noblit and Hare 1988). Creating the whole is fundamentally marked by some degree of invention, resulting in an outcome that is more significant than the sum of its parts. For meta-ethnography, three distinct synthesis techniques are employed. One is the 'translation' of ideas from one study into another also called 'reciprocal translational analysis', leading to the **evolution of general ideas or metaphors**. The second synthesis technique is the process of refutational synthesis that entails finding and **clarifying inconsistencies** among distinct investigations. In contrast, lines-of-argument synthesis involves **building a picture of the whole** (such as culture, organization, etc.) from studies of its pieces.

Meta-ethnography is an alternative to conventional aggregative methods of synthesis. It is an **inductive and interpretive method**, similar to the qualitative methodologies used in the studies it seeks to synthesize. This kind of synthesis results in the translation of studies into one another, which motivates the researcher to comprehend and apply concepts, ideas, and metaphors from other studies (See Box 23.5).

Box 23.5 Experiences and perceptions of nutritional health and wellbeing amongst food insecure women in Europe: A qualitative meta-ethnography. Social Science & Medicine, 311, 115313

Food insecurity has increased since the global financial crisis of 2008, disproportionately impacting pregnant women, single parents, and mothers of small children. The qualitative research on women's experiences with food insecurity and its effects on their nutritional health and wellbeing was reviewed in this systematic review and meta-ethnography. From January 1, 2008, to July 10, 2021, we searched six electronic databases and grey literature, adhering to PRISMA and eMERGe principles and performing a quality assessment using the Critical Appraisal Skills Programme qualitative checklist. Twenty-three articles from 22 distinct studies with 647 women were included out of 11,589 entries. Using Noblit & Hare's seven phases of meta-ethnography to synthesize the data, we were able to identify two main themes, each with seven subthemes. In meta-ethnography, studies are arranged side by side to interpret major themes in a variety of contexts and to methodically compare meanings to identify a variety of themes. This interpretive method goes beyond simply summarizing results to synthesize insight. We were able to obtain deeper insights into food insecurity and nutritional health than we could have from individual investigations by combining the results of several studies (Bell et al. 2022).

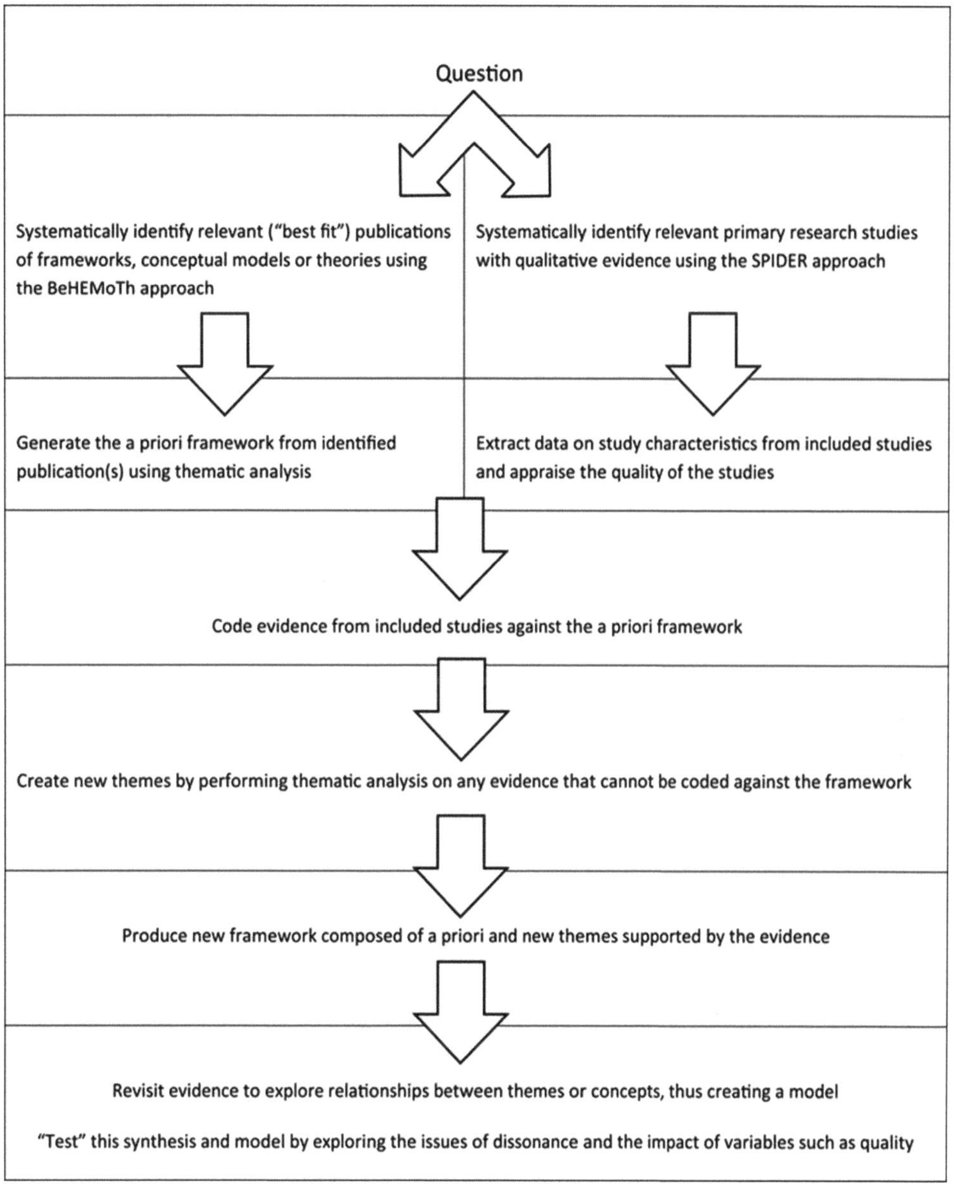

Fig. 23.5 "Best fit" framework synthesis. (Carroll et al. 2013)

(c) **Framework Synthesis**

Framework synthesis is a form of qualitative evidence synthesis that arranges and examines data from various sources using an established or newly developed **conceptual framework**. Framework synthesis can help produce recommendations or implications, combining different viewpoints and providing answers to challenging or policy-relevant topics. Framework synthesis uses a methodical approach to find existing frameworks, models, or theories to build the framework for the synthesis which sets it apart from other methods for synthesizing qualitative evidence (Carroll et al. 2013).

Stages of Framework Synthesis

1. **Familiarization stage:** In this stage, researchers begin by learning more about the subject of their study. This is achieved through the examination of a wide range of information sources. They talk to experts in the field and read books, articles, or reports about the subject. Scholars continue to revisit these sources, gaining new insights with each visit. This aids in their comprehension of what other people have stated about the subject and highlight the key points. It's similar to assembling a puzzle; after gathering all the components, kids begin to see how they fit together to represent the subject they are learning (Fig. 23.5).
2. **Framework selection stage:** Here researchers choose the framework they will use to comprehend and explain the subject they are investigating. This framework aids in structuring their thoughts on the problem, which could be a theory, model, or collection of concepts. They may opt for a framework that has already been established by

other scholars, such as a model or theory that describes how things function in a certain field. As an alternative, individuals could develop their framework using ideas or theories from other sources. This preliminary framework directs the data processing and interpretation, serving as a sort of study roadmap. It gives the findings a framework for organization and aids in the researchers' interpretation of the data they collect. Comparable to picking the finest tool for the task, researchers choose the framework that best suits the particular problem they are investigating, enabling them to approach the research methodically and systematically.

3. **Indexing stage:** Here researchers look for studies that address their research questions. To locate papers, reports, or other materials that contain information regarding the subject they are studying, they meticulously search databases, journals, and other sources. After compiling several papers, they start the process of screening them to identify which ones are most pertinent to their study questions. This entails examining the study titles, abstracts, and occasionally whole texts to determine whether they satisfy specific requirements. Researchers use their original conceptual framework to obtain data from the studies after screening them. This implies that they carefully review each study, noting important details and conclusions that are pertinent to their research concerns. Data regarding study procedures, participant profiles, important discoveries, and any other information that advances their understanding of the subject matter may be extracted. Researchers start classifying the studies according to their primary characteristics and how relevant they are to the review questions as they extract data. This aids in the organization of the data and the discovery of themes or patterns that appear across the investigations.
4. **Charting stage:** In this stage, researchers explore each study's features in further detail that they found in the Indexing step. They classify and carefully examine each study's key features by putting them into groups. These categories could contain, among other things, the study design, participant demographics, intervention specifics, outcome measurements, and important findings. Researchers might start to see recurring themes and patterns in all of the studies by classifying these attributes. Researchers begin to identify patterns straight from the data as they map out the salient features of each study. Finding recurrent themes or patterns that surface from the study's features is part of this process. For instance, researchers may see that the results of multiple studies are consistent or reflect comparable intervention strategies. Themes that aid in organizing and making sense of the data are developed based on these recurrent patterns and concepts.
5. **Mapping and interpretation stage:** Researchers use the themes they have identified from the data and compare them to the initial research questions. This entails assessing the themes' suitability for answering the research questions and their ability to offer insightful analysis of the subject being studied. To determine whether the themes appropriately represent the major problems and ideas found in the data, researchers look at the themes' coherence and applicability. Researchers then go on to present their findings from the review in a variety of formats after mapping the themes against the research questions. Visual aids like tables, figures, forest plots, and narratives may be examples of this. The results of several experiments are graphically summarized in forest plots, which facilitate cross-study comparison of results. Key information, such as study characteristics, main findings, and thematic analysis results, can be succinctly presented using tables and figures. Researchers can provide context and insight for the themes observed by using narratives, which provide a descriptive summary of the findings.

(d) **Narrative Synthesis**

This method tells a cogent tale by condensing and synthesizing qualitative findings (Sutton and Austin 2015). Narrative synthesis stresses the narrative structure, as opposed to other synthesis techniques that concentrate on discovering recurring themes or patterns in studies. This enables researchers to combine the results from several studies into a coherent and captivating story. Using this method, researchers first methodically go over and extract pertinent material, such as quotes, major conclusions, and contextual details, from pertinent qualitative studies. After that, they arrange the data into a logical story that highlights significant themes, revelations, and points of view while narrating the research issue.

Narrative synthesis is appealing because researchers can convey the depth and richness of the data narratively, which gives a flexible and natural way to communicate qualitative findings. Researchers can captivate readers and convey the main ideas and consequences of their work. Furthermore, the integration of several qualitative data sources, such as research with various demographics, methodologies, or settings, is made possible leading to a variety of viewpoints and insights. This holistic approach helps researchers to grasp the research issue as a whole.

Techniques of Qualitative Data Synthesis

Qualitative data synthesis techniques comprise a range of approaches for systematically analyzing and interpreting qualitative data from numerous sources. Goodrick & Rogers (2015. pg 590) offer numerous interpretative strategies for qualitative analysis but we shall focus on constant comparison, conceptual mapping, members checking, and peer debriefing.

- **Constant comparison**: researchers look for patterns, similarities, and differences by continuously comparing data from one study to another. As a result of the continuous comparison, themes and insights gradually surface as researchers gradually improve their comprehension of the data. Throughout the synthesis process, this method actively engages the data, allowing researchers to get a deeper grasp of the research issue.
- **Conceptual mapping**: Another useful tool for synthesizing qualitative data is conceptual mapping, which entails creating concept diagrams or mind maps to show the connections between themes and concepts. Conceptual mapping enables researchers to arrange and examine relationships between various concepts and themes found in the data visually. This method aids in the synthesis process by assisting researchers in identifying overall patterns and linkages.
- **Member checking or respondent validation**: Member checking is a method that asks participants or stakeholders for input on the synthesized findings to increase the validity and reliability of the qualitative data synthesis. Researchers ask participants or stakeholders to offer comments, revisions, or new ideas after presenting their findings to them. The process of member verification contributes to the validity and credibility of research findings by ensuring that the synthesis findings appropriately reflect the viewpoints and experiences of the participants.
- **Peer debriefing**: This involves reviewing and debating the combined results with peers or coworkers. Scholars showcase their discoveries to colleagues and participate in analytical dialogues to investigate diverse perspectives, detect any prejudices, and assess the level of precision in the synthesis procedure. Peer debriefing gives researchers the chance to get helpful criticism and recommendations for raising the credibility and reliability of the combined results. By participating in peer debriefing, researchers can improve the robustness and credibility of their qualitative data synthesis and ensure their conclusions are reliable and rigorous.

Key Points in This Chapter

Getting Started in Qualitative Data Analysis?

Qualitative data analysis gathers, structures, and interprets qualitative data to understand its underlying meanings. Qualitative research focuses on understanding participants' subjective and socially constructed meanings.

Purpose and Nature of Qualitative Data Analysis

Researchers aim to identify patterns and themes and develop larger patterns of knowledge. The method to use in data analysis is based on the fitness of purpose.

Distinctive features of qualitative data analysis include the non-use of statistics, flexibility, and production of rich and detailed narratives, descriptions, and interpretations. Triangulation is emphasized for enhancing validity.

Data Organization and Preparation

Data preparation and organization should be from the beginning of the research to ensure easy data retrieval. Some common methods of organizing data include:

- Chronological orders: organized according to the different periods in which they were gathered.
- Site or location: Data is grouped by the specific sites or locations of data collection: particularly useful in studies involving multiple sites or communities.
- Type of event or issue: Information related to similar events or issues are grouped together.
- Person or group: Organized by the individual or group; data with similar backgrounds are grouped together.
- Type of data: interview transcripts might be assembled separately from field notes.

The first task in data preparation is to ensure that data are in a form that can be analyzed, much like quantitative data followed by transcription. Transcription is tedious work, requiring a lot of time and patience but extremely useful.

Coding for Qualitative Analysis

Coding summarizes diverse data and is used to denote themes, concepts, or patterns in qualitative data and is applied to specific data segments. The coding techniques include the following:

1. **Open/Initial coding process:** Removes complexity, identifies and labels initial concepts or themes through a systematic review. They include: a) **In vivo coding** where participants label important concepts in their own words, b) **Descriptive coding,** which identifies key concepts or subjects within the data, and c) **Subcoding** to identify hidden differences and variations and refine analysis to capture participant experiences and viewpoints.
2. **Grounded theory coding process:** Examines data without predetermined beliefs, systematically classifying and organizing it to produce fresh insights. This identifies themes, investigates connections, and iteratively improves analysis. Three coding methods are used: open, axial, and selective:
   Open coding: Data are segmented into smaller pieces and given a code representing the main ideas.
   Axial coding: The iterative process involves examining the connections between codes found during open coding to create organized categories or topics.
   Selective coding: Incorporates important themes or categories that were discovered in the initial phases of analysis.

Approaches to Data Analysis

Six main strategies are key:

1. **Content analysis** involves analyzing words used in phrases or images in different pieces of writing or communication.
2. **Narrative analysis** involves listening to and interpreting people's stories to gain insights into their coping mechanisms and perceptions of reality.
3. **Discourse analysis** examines the collective meaning of words and symbols in a given context. It goes beyond words and goes beyond language itself, examining the collective meaning of words in a specific social context.
4. **Thematic analysis:** examines the patterns of meaning within a data set.
5. **Grounded theory (GT):** aims to create a new theory (or theories) using the data at hand through a series of tests and revisions.
6. **Phenomenological Analysis** assists in comprehending the individual experiences. The "phenomenon" is this occurrence or experience.

Types of Data and Their Analysis

Understanding the complexity of qualitative data types and their analysis methods is crucial for gaining fresh information and developing comprehension:

- **Semi-structured interview**—content analysis, thematic analysis, comparative keyword analysis, and discourse analysis.
- **Focus group discussion data**—content analysis, thematic analysis, comparative keyword analysis, and discourse analysis.
- **Visual data and artifacts**—Content and thematic analysis.
- **Observational data**—Content or narrative analysis.
- **Concept maps**—diagrammatic tool used to illustrate a set of conceptual meanings.

Principles of Qualitative Data Analysis

Scholars established principles that facilitate rigorous data analysis. These principles include:

1. People have varied experiences and perceptions of reality.
2. You cannot comprehend a social phenomenon outside its specific context.
3. Qualitative research can be used to build data-driven theories or to characterize phenomena.
4. Human behavior is slowly and nonlinearly understood.
5. Exceptional cases can provide new information or insights into an issue for future research.

Computer-Assisted Qualitative Data Analytic Tools (CAQDAS)

Software programs aid in reading, comparing, and reflecting on text, supporting qualitative data analysis. However, they don't replace thorough and methodical study, and researchers still need analytical abilities. They include:

1. ATLAS.ti
2. NVivo
3. MaxQDA
4. Dedoose

Data Synthesis and Presentation

Data synthesis and presentation combine, evaluate, and interpret qualitative data from various sources to create fresh perspectives on a research issue. This method is rapidly spreading across academic fields, focusing on recurring themes, patterns, and connections across diverse data sources, often leading to the creation of new theories or frameworks. The Principles of Qualitative Data Synthesis lay emphasis on a grounded theory approach, reflexivity, transparency, and an iterative process.

Methods of Qualitative Data Synthesis

- Thematic analysis involves identifying key messages and abstract themes from data, often using an inductive approach to develop higher-order themes.
- Meta-ethnography is an inductive, interpretive method that synthesizes qualitative studies using reciprocal translational analysis, refutational synthesis, and lines-of-argument synthesis to generate deeper conceptual insights.
- Framework synthesis uses an existing conceptual framework to analyze qualitative data systematically, producing policy-relevant recommendations. Its steps include familiarization, framework selection, indexing, charting, and mapping/interpretation.
- Narrative synthesis organizes qualitative findings into a coherent story by extracting and arranging key content from multiple studies to highlight major themes.

Techniques of Qualitative Data Synthesis

Qualitative data synthesis techniques involve systematic analysis of data from various sources. The most important ones are as follows:

1. Continuous comparison: identifies patterns across studies.
2. Conceptual mapping: visualizes relationships between concepts and themes.
3. Member checking: solicits input from participants to ensure findings reflect their viewpoints.
4. Peer debriefing: reviews results with peers to explore diverse perspectives.

Review Questions

1. Why is qualitative data analysis considered interpretive and iterative rather than linear?
2. Outline the general sequence of steps involved in qualitative data analysis
3. How does coding (open, axial, and selective) help in structuring and understanding qualitative data?
4. Compare and contrast content analysis, thematic analysis, narrative analysis, grounded theory, and discourse analysis. What are the strengths and limitations of each?
5. How can visual data or artifacts be analyzed qualitatively? Provide an example.
6. How does conceptual mapping assist in understanding relationships between themes and codes?
7. Compare manual coding with computer-assisted coding. What are the advantages and disadvantages of each?

Chapter Exercise

1. Imagine you are a researcher studying the adoption of drought-resistant maize varieties by smallholder farmers. Briefly describe a scenario where you would use qualitative data analysis and another scenario where you would use quantitative data analysis. Highlight the key differences in the type of data collected and the analytical goals for each. Tip (You can also refer to Chaps. 20 and 21).
2. Suppose you are conducting a study on the challenges faced by women farmers in accessing agricultural extension services.
   (a) If your goal is to understand the lived stories and experiences of individual women farmers as they navigate these challenges, which technique would you choose and why?
   (b) If you want to identify recurring themes and patterns in the types of challenges faced across multiple interviews with women farmers, which technique would you employ and why?
3. A graduate student has completed an interview with a smallholder farmer about their transition to organic farming. Here is a brief excerpt from the transcripts: "Initially, I was quite skeptical about organic fertilizers, thinking they would not be as effective as chemical ones. This is because I have always used organics. But after a few seasons, I started seeing the benefits, especially with soil health improving and fewer pest problems. My neighbors were also noticing the healthier crops."
   (a) Applying the concept of open coding, suggest at least three initial codes you might assign to this excerpt. Explain your reasoning for each code in an agricultural context.
   (b) You have identified several initial codes through open coding from interviews with farmers about climate change adaptation strategies. Now, consider the following:
   (c) Explain how you would move from open coding to axial coding with respect to these codes. Provide an example of a potential relationship between codes

(e.g., 'lack of irrigation' and 'reliance on rain-fed agriculture').

4. A researcher is studying online forum discussions among farmers regarding the impact of new agricultural policies.
   (a) If they are primarily interested in identifying the main topics and recurring ideas expressed about the policies' effects (e.g., 'subsidies,' 'market access,' 'input costs'), which analytic technique (thematic or discourse analysis) would be more suitable? Justify your answer.
5. Suppose you have collected data through focus group discussions with livestock keepers about barriers to accessing veterinary services.
   (a) Briefly describe how you would approach analyzing this type of data, considering the group dynamics and potential for diverse perspectives among the keepers.
   (b) Explain the crucial significance of data preparation, exploration, and organization before you begin your in-depth analysis.
   (c) What are some specific tasks involved in this stage when dealing with interview or focus group transcripts from farmers/pastoralists (e.g., related to transcription, anonymization)?

Answers to Review Questions

- Q1. Reference: 23.1.1 and 23.1.2
- Q2. Reference 23.1.3
- Q3. Reference 23.2
- Q4. Reference: 23.2
- Q5. Reference: 23.4.3
- Q6. Reference: 23.4.4
- Q7. Reference: 23.6

References

Aliro T, Chenais E, Odongo W, Okello DM, Masembe C, Ståhl K (2022) Prevention and control of African swine fever in the smallholder pig value chain in northern Uganda: thematic analysis of stakeholders' perceptions. Front Vet Sci 8:707819

Bell Z, Scott S, Visram S, Rankin J, Bambra C, Heslehurst N (2022) https://journals.plos.org/plosone/article?id=10.1371/journal.pone.0292178

Bryman A, Burgess RG (1994) Analyzing qualitative data, vol 11. Routledge, London. https://api.taylorfrancis.com/content/books/mono/download?identifierName=doi&identifierValue=10.4324/9780203413081&type=googlepdf

Carroll C, Booth A, Leaviss J, Rick J (2013) "Best fit" framework synthesis: refining the method. BMC Med Res Methodol 13(1):37. https://doi.org/10.1186/1471-2288-13-37

Creswell JW (2015) Revisiting mixed methods and advancing scientific practices. https://academic.oup.com/edited-volume/34493/chapter/292662219

Creswell JW, Poth CN (2016) Qualitative inquiry and research design: choosing among five approaches. Sage Publications

Creswell JW, Tashakkori A (2007) Editorial: differing perspectives on mixed methods research. J Mixed Methods Res 1(4):303–308. https://doi.org/10.1177/1558689807306132

Creswell JW, Fetters MD, Ivankova NV (2004) Designing a mixed methods study in primary care. Ann Family Med 2(1):7–12

Drisko JW (2020) Qualitative research synthesis: an appreciative and critical introduction. Qual Soc Work 19(4):736–753. https://doi.org/10.1177/1473325019848808

Farm Radio International (2019) Storing dairy fodder for the dry season increases farmers' income. Retrieved 2 November 2021, from http://scripts.farmradio.fm/; website: http://scripts.farmradio.fm/radio-resource-packs/111-farm-radio-resource-pack/storing-dairy-fodder-dry-season-increases-farmers-income/

Gill R (2000) Discourse analysis. In: Qualitative researching with text, image and sound, vol 1, pp 172–190

Goodrick D, Rogers PJ (2015) Qualitative data analysis. In: Newcomer KE, Hatry HP, Wholey JS (eds) Handbook of practical program evaluation, 1st edn. Wiley, pp 561–595. https://doi.org/10.1002/9781119171386.ch22

Holton JA (2007) The coding process and its challenges. In: The Sage handbook of grounded theory, vol 3, pp 265–289

Linneberg MS, Korsgaard S (2019) Coding qualitative data: a synthesis guiding the novice. Qual Res J 19(3):259–270

Neuendorf KA, Kumar A (2015) Content analysis. In: The international encyclopedia of political communication, pp 1–10

Noblit GW, Hare RD (1988) Meta-ethnography: Synthesizing qualitative studies, vol 11. Sage. https://books.google.com/books?hl=en&lr=&id=fQQb4FP4NSgC&oi=fnd&pg=PA5&dq=Noblit+GW,+Hare+RD:+Meta-Ethnography:+Synthesizing+Qualitative+Studies.+1988,+London:+Sage&ots=MX0Cppa0Pa&sig=2ATzbYJvsF3ehBPLwtnibTy50q8

Patton MQ (2002) Qualitative research & evaluation methods. Sage

Rycroft-Malone J, Burton CR (2015) The synthesis of qualitative data. In: Complex interventions in health. Routledge, pp 80–87. https://api.taylorfrancis.com/content/chapters/edit/download?identifierName=doi&identifierValue=10.4324/9780203794982-10&type=chapterpdf

Saldana J (2008) Analyzing longitudinal qualitative observational data. In: Handbook of longitudinal research: design, measurement, and analysis, pp 297–311

Saldana J (2013) The coding manual for qualitative researcher, 2nd edn. SAGE Publications

Silverman D (2024) Interpreting qualitative data. https://www.torrossa.com/gs/resourceProxy?an=5730608&publisher=FZ7200

Smeds P, Jeronen E, Kurppa S (2015) Farm education and the effect of a farm visit on children's conception of agriculture. Eur J Educ Res 4(1):1–13

Strauss A, Corbin J (1998) Basics of qualitative research techniques. https://citeseerx.ist.psu.edu/document?repid=rep1&type=pdf&doi=18c7cfe7a46c7771b60dc384b1b4e350f65b13e2

Strauss AL (1987) Qualitative analysis for social scientists. Cambridge University Press. https://books.google.com/books?hl=en&lr=&id=y16ww5ZsJ0AC&oi=fnd&pg=PR11&dq=(Strauss,+1987)coding&ots=gY3xFPk5dX&sig=aFiQ1-oOo24UKtxaJYZQrrfMVPw

Sutton J, Austin Z (2015) Qualitative research: data collection, analysis, and management. Can J Hospital Pharm 68(3):226–231. https://doi.org/10.4212/cjhp.v68i3.1456

Tenny S, Brannan JM, Brannan GD (2022) Qualitative study. StatPearls Publishing. Retrieved from https://www.ncbi.nlm.nih.gov/books/NBK470395/

Thomas J, Harden A (2008) Methods for the thematic synthesis of qualitative research in systematic reviews. BMC Med Res Methodol 8(1):45. https://doi.org/10.1186/1471-2288-8-45

Thorne S (2008) Interpretive description. Left Coast Press, Walnut Creek, CA

Tracy SJ (2019) Qualitative research methods: collecting evidence, crafting analysis, communicating impact. Wiley. https://books.google.com/books?hl=en&lr=&id=ipOgDwAAQBAJ&oi=fnd&pg=PR1&dq=+Tracy,+S.+J.+(2019).+Qualitative+research+methods:+Collecting+evidence,+crafting+analysis,+communicating+impact.+John+Wiley+%26+Sons&ots=Www-nXdyKn&sig=P3s-9BwbXHNC7C4zkLs2UQ3Q6d8

Watson GR (1989) What is... concept mapping? Med Teach 11(3–4):265–269. https://doi.org/10.3109/01421598909146411

Websites

https://methods.sagepub.com/book/research-methods-in-education/i196.xml

www.atlasti.com

www.qsrinternational.com

www.maxqda.com

www.provalisresearch.com

Data Presentation and Visualization

24

Abstract

Data presentation and visualization are crucial aspects of effectively communicating research findings to various stakeholders. Three main methods of presenting and visualizing data are discussed in this chapter. They include graphical, tabular, and narrative presentations. Graphics and presentations are of various types, ranging from simple frequency graphs to complex staked and compound graphs. The type to use depends on factors such as the nature of the data, the intended audience, and the information to be communicated.

Keywords

Graphical · Histogram · Flow charts · Line graph · Pie-chart · R-studio · Stacked bar gap · Trends

Learning Outcomes

After reading this chapter, you should be able to:

1. Appreciate the importance of data presentation and visualization in communicating research findings effectively.
2. Identify different data visualization techniques and their appropriate applications in various research contexts.
3. Use a suitable visualization method based on the nature of the data and research objectives.
4. Recognize the importance of data visualization using R software.
5. Gain knowledge of best practices for designing and formatting visualizations to enhance readability and interpretability.

24.1 Importance of Data Presentation and Visualization Techniques

Effective communication of research findings is key to informing and engaging various stakeholders (Boaz et al. 2018). Data presentation and visualization are two crucial aspects of this process (Li et al. 2020). Although the terms are often used interchangeably, they have different meanings. Data presentation involves summarizing research results through written or verbal explanations and visual aids such as tables and charts. On the other hand, data visualization aims to transform complex datasets into graphical formats that are easier to interpret and understand (Midway 2020; Jensen et al. 2023). As the saying goes, "a picture is worth ten thousand numbers" (Davenport 1986). Both techniques (data presentation and visualization techniques) complement each other, enhancing clarity and the impact of research outcomes.

Here, we present four main reasons why data presentation and visualization are significant in research:

1. *Enhance accessibility of findings to stakeholders:* Visual representations simplify complex concepts and data, making research findings accessible to stakeholders who may not have a technical background. Whether in academic papers, presentations, or reports, well-designed visualizations can captivate audiences and convey information more memorably than text alone. In conferences, researchers transform and present their scientific data in various forms that are appealing to the audience. The aim is to simplify the complex scientific data into captivating visual presentation. For example, captivating visual presentations to illustrate trends in crop yields over time,

J. Njeri Mugwe, S. Runo, *Research Methodology in Agricultural Sciences*, https://doi.org/10.1007/978-981-95-1892-0_24

spatial distribution of agricultural pests, or the impact of climate change on farming communities can enhance communication to farmers. These visualizations are likely to inspire action and drive positive change in agricultural practices and policies.

2. *Help uncover patterns and trends:* Through visual exploration of data, researchers can uncover patterns, trends, and correlations that may not be immediately evident from tabular data alone. For example, line graphs can reveal relationships between variables, helping researchers understand the potential influence of farmer characteristics on adoption rates.
3. *Accountability:* Clear and accessible data visualization promotes accountability. When data are presented in easy-to-understand visualizations, stakeholders are empowered to hold decision-makers accountable. Furthermore, transparent data visualization practices are instrumental in maintaining donor confidence and securing continued support for agricultural research initiatives.

24.2 Data Presentation and Visualization Techniques

Data presentation and visualization techniques vary according to the data's nature and the analysis's objectives (Hehman and Xie 2021; Huang et al. 2024). Likewise, data visualization software offers a range of options. This chapter explores the primary presentation and visualization methods commonly used in agricultural research. Additionally, we demonstrate the application of R software for data visualization, highlighting its versatility and effectiveness in handling various agricultural data types, especially in molecular and biotechnology studies. Each technique serves specific purposes in visualizing different data types and facilitating interpretation and decision-making in agricultural research.

There are three main methods of presenting and visualizing data. These include:

- Graphical: graphs and charts.
- Tabular: tables.
- Narrative: text.

Graphical presentation utilizes charts, graphs, and plots to visually represent data, facilitating the interpretation of trends and relationships. The tabular presentation involves organizing data into structured tables for easy comparison and reference. Narrative presentation complements tabular and graphical methods by providing written explanations and interpretations of the data. Researchers often use a combination of these methods to effectively communicate research findings to diverse audiences, ensuring clarity and relevance in conveying the significance of the data.

24.2.1 Graphical Presentation

Graphs are good for making broad comparisons or showing strong relationships. In summary, they have the following characteristics:

- Facilitate comparison of various characteristics.
- Are attractive and give a bird's eye-view of the data.
- Easily understood even by layperson.
- Impressions created by them are long-lasting.
- Results of statistics can easily be visualized using graphs.

24.2.1.1 Histogram

A histogram is a graphical representation that organizes data points into user-specified ranges for continuous data. Many researchers opt for histograms to display the highest and lowest values. It condenses a data series into an easily interpreted visual diagram by grouping many data points into logical ranges. Before creating a histogram, data often need to be grouped into class intervals. In a histogram, each bar's area corresponds to the frequency of occurrence, and the absence of gaps between bars underscores the continuous nature of the data. When using equal-width class intervals, the height of each bar still represents the frequency of occurrences, facilitating easy differentiation between the highest and lowest values.

We illustrate a histogram using hypothetical data of soybean yields obtained from 80 farmers (Fig. 24.1).

From the same hypothetical data of soybean farmers, we construct a frequency table to show the ages of the 80 farmers and a histogram using the same data (see table in Fig. 24.2). A frequency table displays the frequency of occurrences or counts of different categories or values within a dataset. It allows for a clear understanding of the frequency of each category or value in the dataset. By organizing data into a frequency table, researchers can easily identify the most common categories or values, as well as any outliers or unusual patterns in the data.

24.2.1.2 Bar Graph

Bar graphs display simple results. A bar graph consists of rectangular bars used to represent both numerical and categorical data (Fig. 24.3a–c). In a bar graph, also referred to as a bar chart, each bar's height or length corresponds to the frequency of occurrence. They are not generally useful for large amounts of structured information. The bars often have no inherent order since the horizontal axis represents a discrete categorization. To highlight certain comparisons, it is

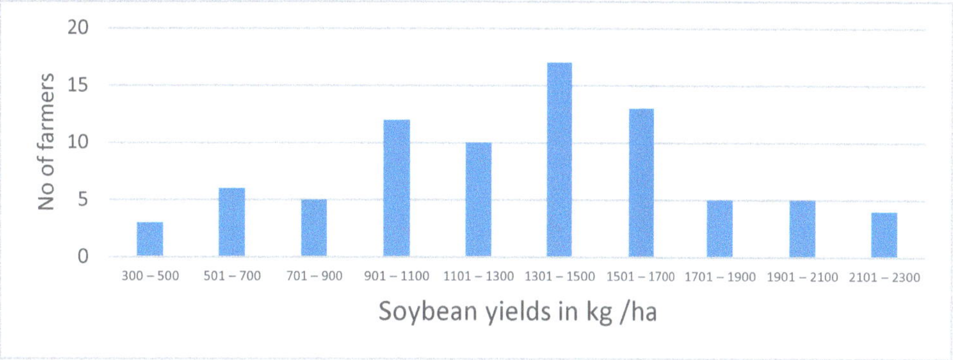

Fig. 24.1 Histogram of 80 smallholder farmers growing soybean in region A

Age in years	No of farmers
26	2
32	3
34	7
39	5
41	9
45	11
47	14
48	9
52	6
54	5
61	6
65	2
72	1
Total	80

a

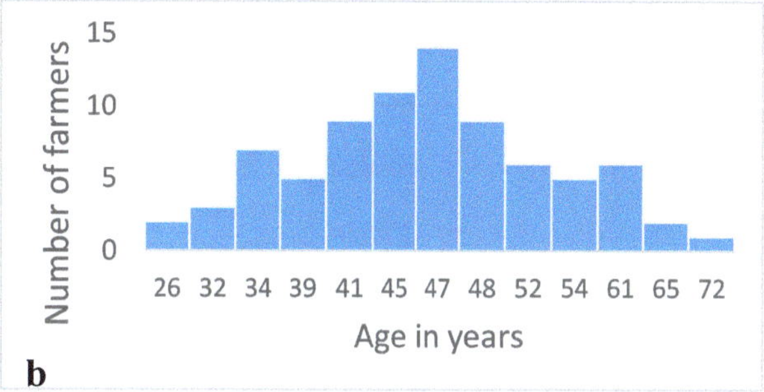

Fig. 24.2 (**a** and **b**) Age in years of 80 farmers in a frequency table and a histogram

frequently useful to "cluster" or group the bars according to the categories they represent. The objective of the bar graph should determine the method of grouping.

When a bar graph is presented using pictures, it is called a pictogram. A bar graph and pictogram are suitable visualization methods for representing categorical and discrete data. Typically, bar graphs offer a more precise depiction and are preferable for research reports, while pictograms capture the audience's attention with a general impression. For instance, pictograms are commonly used in extension materials for farmers, such as pamphlets and brochures.

24.2.1.3 Stacked Bar Graphs

Stacked bar charts, or compound bar diagrams, display multiple variables or categories within each bar. In these diagrams, each bar is divided into segments or sections, with each segment representing a different subgroup or category of the data. The length or height of each segment within the bar corresponds to the value or proportion of that subgroup relative to the total. These bar charts are useful for illustrating the composition or breakdown of a dataset across different dimensions or subgroups.

One variant of a stacked bar graph is a percentage bar graph which represents the percentage or proportion of a particular category within the total. We use data in Table 24.1 to draw a stacked bar graph and a compound bar graph (Fig. 24.4a, b). We can see that the number of males and females adopting new soybean varieties varies across the regions. In regions A and D, more males than females adopted the new soybean variety. The difference in adoption between males and females is largest in region between.

Besides proportions, we can use stacked bar graphs and compound bar graphs to present yield data from different treatments. In this case, we use an example of grain yield from different soybean varieties under different cropping systems (Fig. 24.5a, b).

24.2.1.4 Line Graphs

Line graphs are most effective in presenting trends in data that we want to showcase changes over time or across continuous variables (Kosslyn 2006). For example, they can display the growth of crop yields over multiple planting seasons or illustrate the fluctuations in temperature and rainfall throughout the year. By visually representing these trends, line graphs

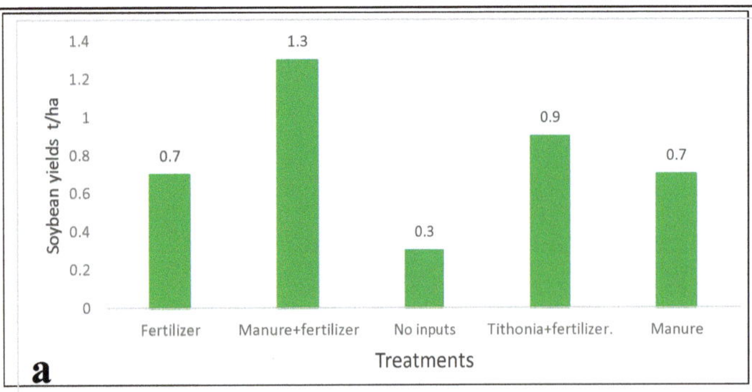

Bar graph (vertical) showing data of an experiment of soybean with different inputs (Five treatments), showing soybean yields response to application of inputs.

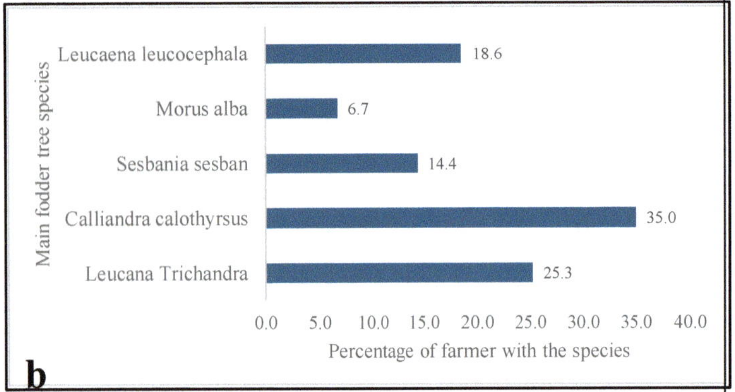

Bar graph (horizontal) showing presence of main fodder tree species in region drawn from the data from the table 24.2 below. Presence of main fodder tree species among farmers in region A.

Main Fodder species	Frequency	%
Leucana Trichandra	91	25.3
Calliandra calothyrsus	126	35.0
Sesbania sesban	52	14.4
Morus alba	24	6.7
Leucaena leucocephala	67	18.6
		100.0

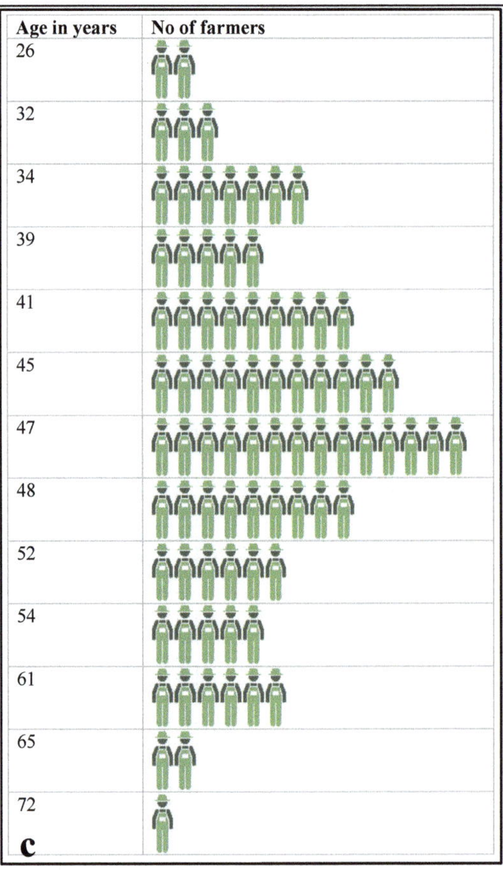

Pictogram of 80 smallholder farmers growing soybean in region A disaggregated by age.

Fig. 24.3 (**a–c**) Bar graphs (vertical and horizontal) and pictograms, respectively. (**a**) Bar graph (vertical) showing data of an experiment of soybean with different inputs (Five treatments), showing soybean yields' response to application of inputs. (**b**) Bar graph (horizontal) showing presence of main fodder tree species in region drawn from the data from the see table illustrated below the graph. Presence of main fodder tree species among farmers in region A. (**c**) Pictogram of 80 smallholder farmers growing soybean in region A disaggregated by age

enable researchers to easily identify patterns and correlations, such as the impact of weather conditions on crop production or the effectiveness of different agricultural practices over time. Other data that can be presented in a line graph include:

- Change in prices of commodities over time.
- Pest infestation over time.
- Soil moisture changes over time or with depth.
- Change in adoption rates over time.

One of the most frequent applications of line graphs is in presenting and visualizing rainfall data, patterns, and trends. Line graphs provide a clear representation of how rainfall changes over time by plotting rainfall measurements at regular intervals along a horizontal axis (such as days, months, or years) and corresponding rainfall amounts on a vertical axis. This visualization allows researchers and meteorologists to identify seasonal trends, fluctuations, and anomalies in precipitation, aiding in understanding weather patterns and their impacts on various ecosystems and agricultural practices.

Specifically, they can show the following:

- Long-term trends such as periods of drought or increased precipitation. Researchers can use this information to assess the overall climate conditions in the area and predict future weather patterns.
- Short-term fluctuations in rainfall, such as seasonal variations or extreme weather events like floods and droughts. By visually representing these patterns, agricultural practitioners can make informed decisions about crop planting schedules, irrigation management, and water conservation strategies.

Using rainfall data collected from the central highlands of Kenya, we plotted a line graph to illustrate the seasonal changes in precipitation. The graph depicted in Fig. 24.6 showcases the distribution of rainfall in millimetres over two distinct seasons. This visualization enables us to observe rainfall intensity and frequency fluctuations throughout the year, offering valuable insights into the region's climatic patterns. These data presentation can help researchers and farmers understand rainfall patterns and optimize crop production by tailoring agricultural practices and water management strategies.

We can also plot rainfall data across a month to examine daily variations in precipitation levels. By plotting daily rainfall measurements over the course of a month on a line graph, we can visualize fluctuations in rainfall intensity and frequency. This type of visualization allows us to identify specific days with particularly high or low rainfall, as well as patterns of rainfall distribution throughout the month. Figure 24.7 shows that during the 9th to 11th day in November the rainfall intensity was high while it was low at the end of the 21st to 30th November, 2014.

We can represent rainfall data in the form of bar graphs where each day of the month is plotted as a bar (Fig. 24.8). When we plot the amount of rainfall for each day of the month, we can easily recognize periods of heavy rainfall, dry spells, or consistent precipitation. In Fig. 24.8, we identify that most of the rainfall in this hypothetical case was received at the beginning of March and remained generally dry towards the end of the month.

Table 24.1 Number of males and females adopting new soybean varieties in five regions

Region	Number of males adopting	Number of females adopting
A	45	55
B	76	24
C	37	63
D	28	72
E	52	48

24.2.1.5 Polygraph

Polygraphs are multiple line graphs that present two or more variables (Figs. 24.9a and 24.8b). They allow viewers to easily observe trends, identify patterns, and make comparisons between different time periods, thus facilitating the identification of correlations or relationships between variables and enabling informed decision-making based on these insights.

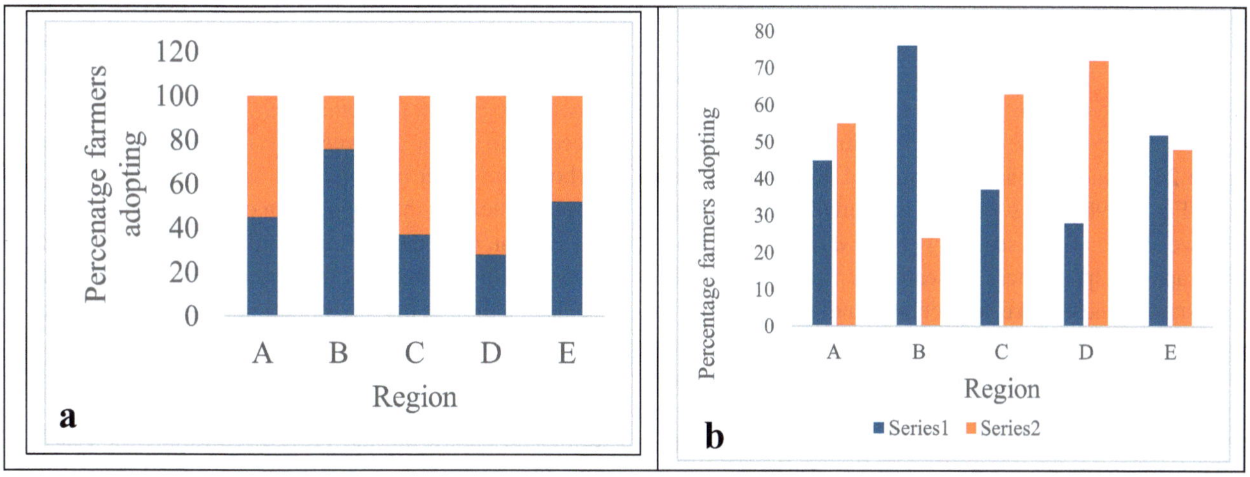

Fig. 24.4 (**a** and **b**) Proportion data (Percentage of farmers) adopting new soybean in two regions plotted in stacked bargraph and compound bargraph

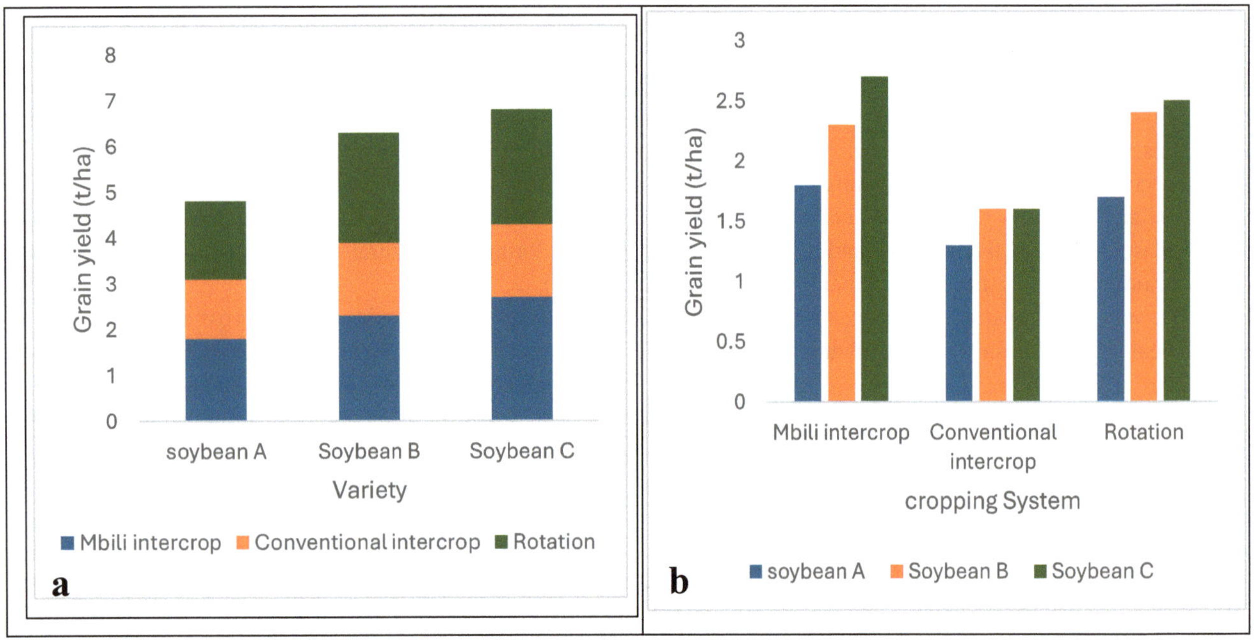

Fig. 24.5 (**a** and **b**) Quantitative data (yield) of soybean varieties under different cropping systems presented in two different types of stacked bargraphs and compound bargraph

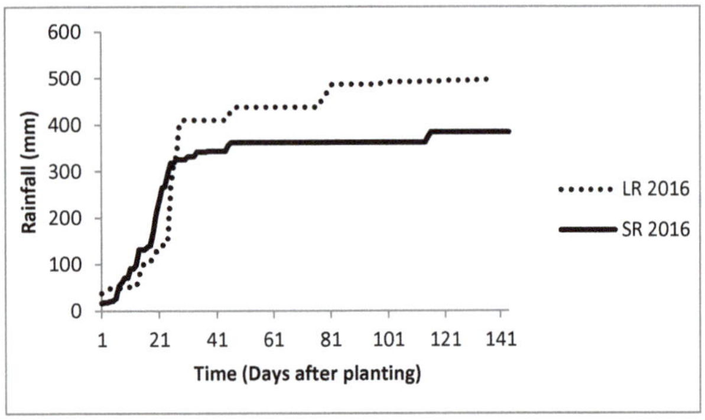

Fig. 24.6 Line graph showing ccumulative rainfall during LR 2016 and SR 2016 seasons at Kigogo

In agriculture, they are commonly used to present data such as crop yields over time, changes in weather patterns, or the relationship between fertilizer application and plant growth. Figure 24.9a shows changes in mineral N across a season, while Fig. 24.9b shows changes in rainfall across a season (Mugwe 2007). By visualizing these variables over time, polygraphs help farmers, researchers, and policymakers understand trends, make predictions, and implement strategies to optimize agricultural practices and maximize productivity. Figure 24.9a shows that mineral N generally increases from 2WAP (2 weeks after planting) in all treatments upto 8WAP, after which it declines steadily as the season ends at 16WAP. Figure 24.9b, on the other hand, shows that rainfall was high at the beginning of the season for about one and half months, after which a dry spell was experienced.

24.2.1.6 Pie Chart

A pie chart displays data in a circular graph. The entire chart represents 100% of a whole, while the pies represent portions of a whole. It is useful when displaying proportions in area, for example, farm sizes and vegetation types.

Examples: Farm sizes in a sample of farmers. Proportion of farmers with different farm sizes (Fig. 24.10).

24.2.1.7 Box Plots

Box plots are powerful tools for visualization. We encountered box plots in the discussion on measures of variability in Chap. 20. Box plots provide a concise summary of the distribution of a dataset, including measures such as the median, quartiles, and any potential outliers. They also allow for easy comparison of multiple groups or categories within the same plot, making them particularly useful for identifying differences or similarities between datasets. In Fig. 24.11, organic carbon was higher in farms with inputs applied than without inputs, indicating that application of inputs enhances organic soil carbon. Both data sets had outliers. Here we see that box plots effectively highlight the spread and central tendency of the data, aiding in the interpretation of variability and potential skewness.

24.2.1.8 Maps and Photos

Maps and photos are indeed powerful tools for visualizing data. One of the main uses of maps is to precisely show location of research sites and field experiments, aiding researchers in logistical planning and data collection (Coetzee et al.

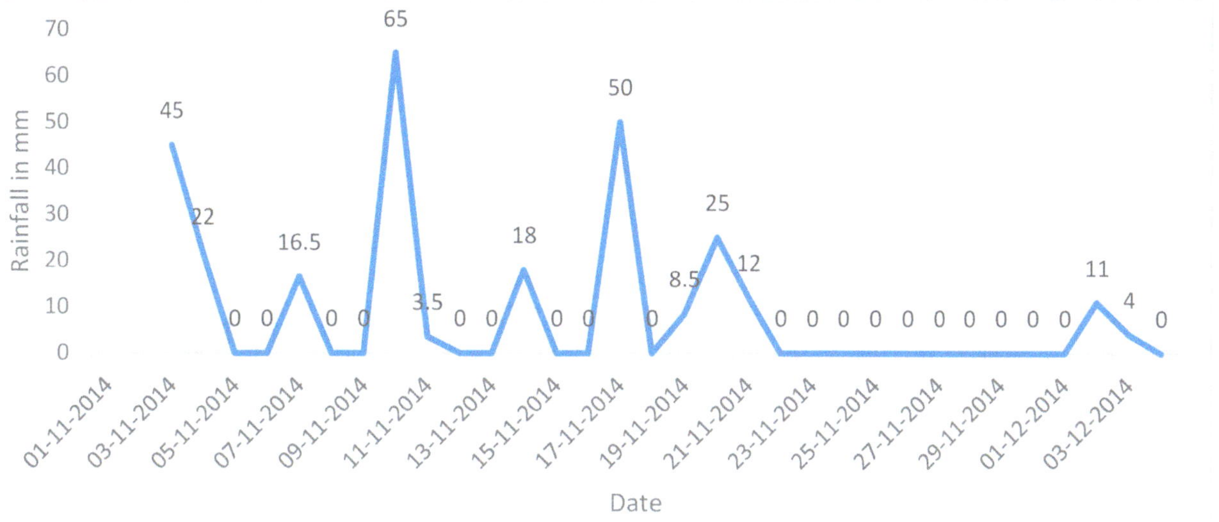

Fig. 24.7 Rainfall distribution for first November 2014 to third December 2014

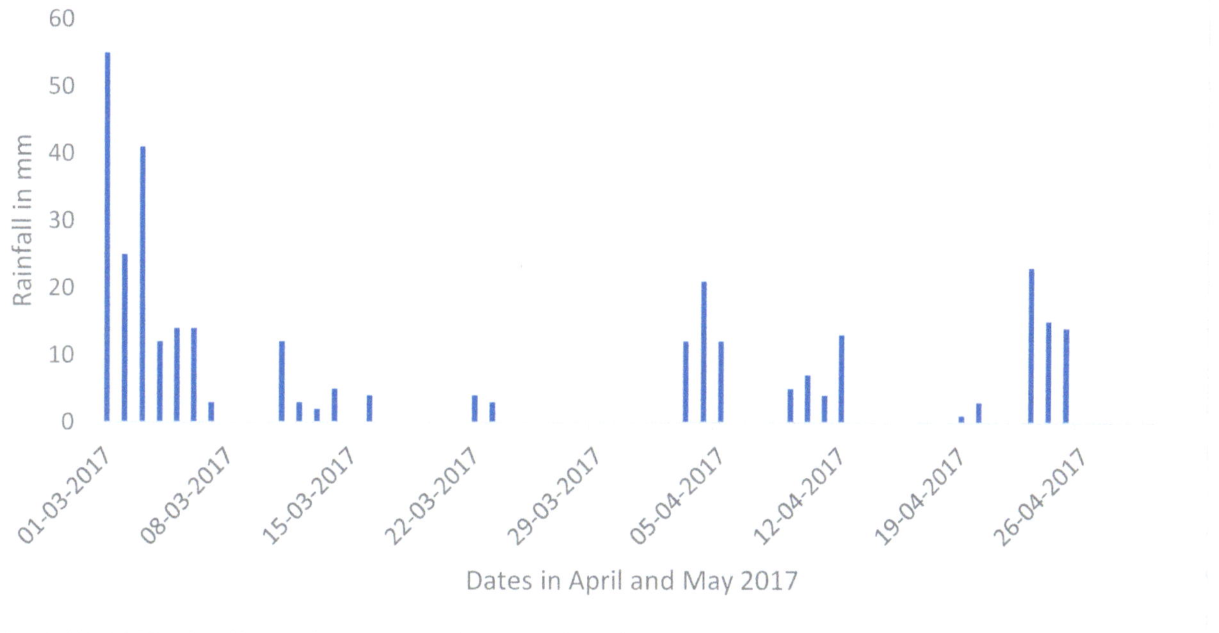

Fig. 24.8 Rainfall data plotted in form of bar graphs with each day of the month is plotted as a bar

2021). For example, Fig. 24.11a shows one of our study sites in Kenya, Ciakariga, along with surrounding areas that may benefit from our research work.

Additionally, maps are used to present spatial representations of various factors such as soil types, weather patterns, and land use. By overlaying spatial data onto maps, farmers and researchers can identify crucial spatial patterns, trends, and correlations essential for informed decisions about crop management. In the central highlands of Kenya, GIS maps were employed to depict the spatial distribution of farmer groups (Fig. 24.12a, b), facilitating the analysis of their geographic distribution and aiding in planning training and interventions by extension officers.

Photographs, on the other hand, depict real-life scenarios and occurrences that may not be adequately conveyed through numbers alone. The photographs aid researchers to present their activities and results with evidence. Figure 24.13 is a photo used in one of the project reports to show the growing and marketing of soybeans. The photos in the reports enhance understanding and engagement among our

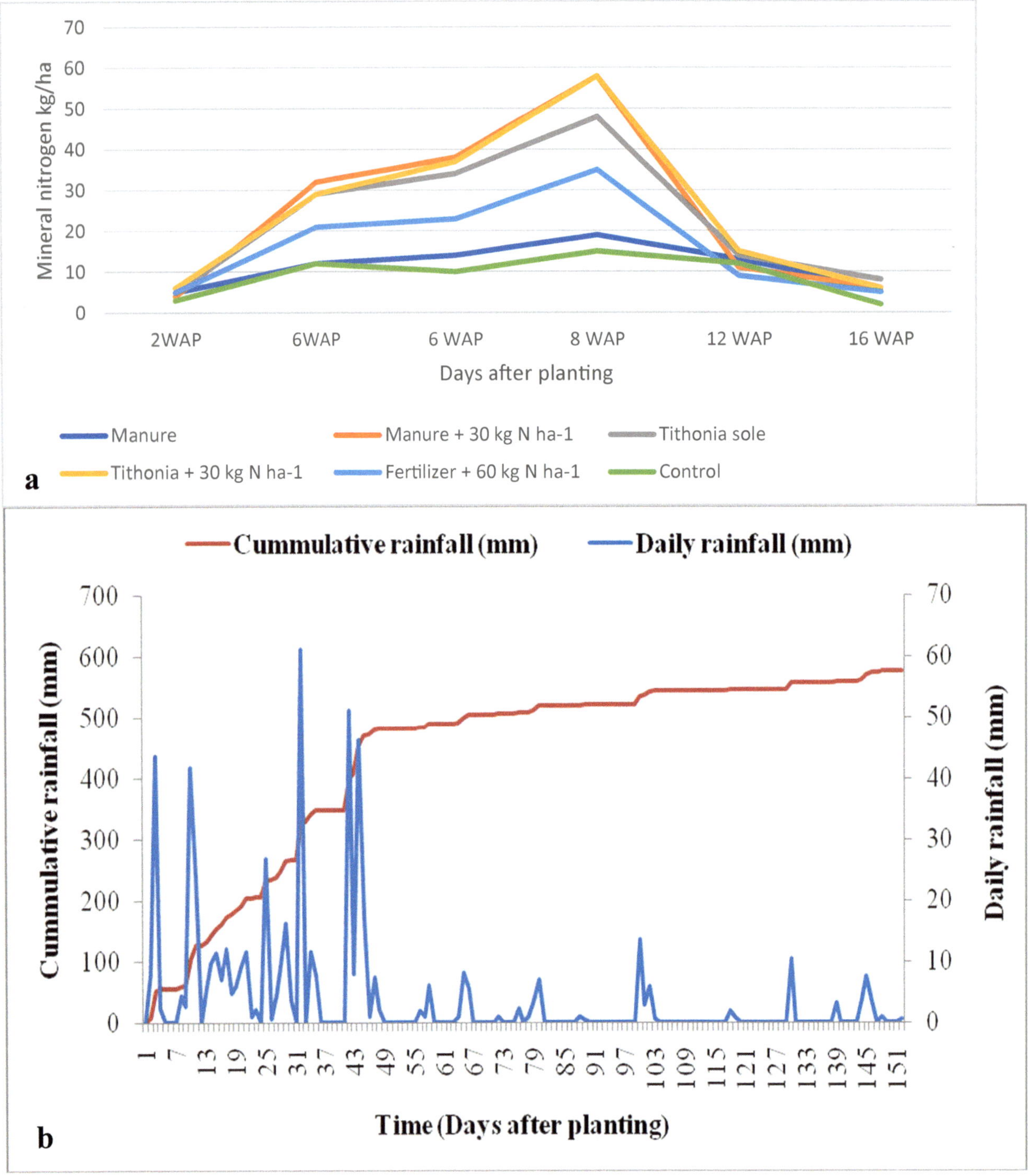

Fig. 24.9 (**a** and **b**) Polygraph showing changes in daily rainfall and cumulative rainfall across the season

stakeholders, allowing them to visualize the agricultural processes firsthand. Furthermore, the inclusion of such photos in project reports enhances donor confidence and trust, creating opportunities for further funding and support of research activities.

24.2.1.9 Flow Charts and Diagrams

A flowchart is a type of chart that represents a process or a workflow. It shows the steps as boxes of various kinds, such as rectangles, diamonds, or ovals, each signifying a specific action or decision point within the process. Arrows connect

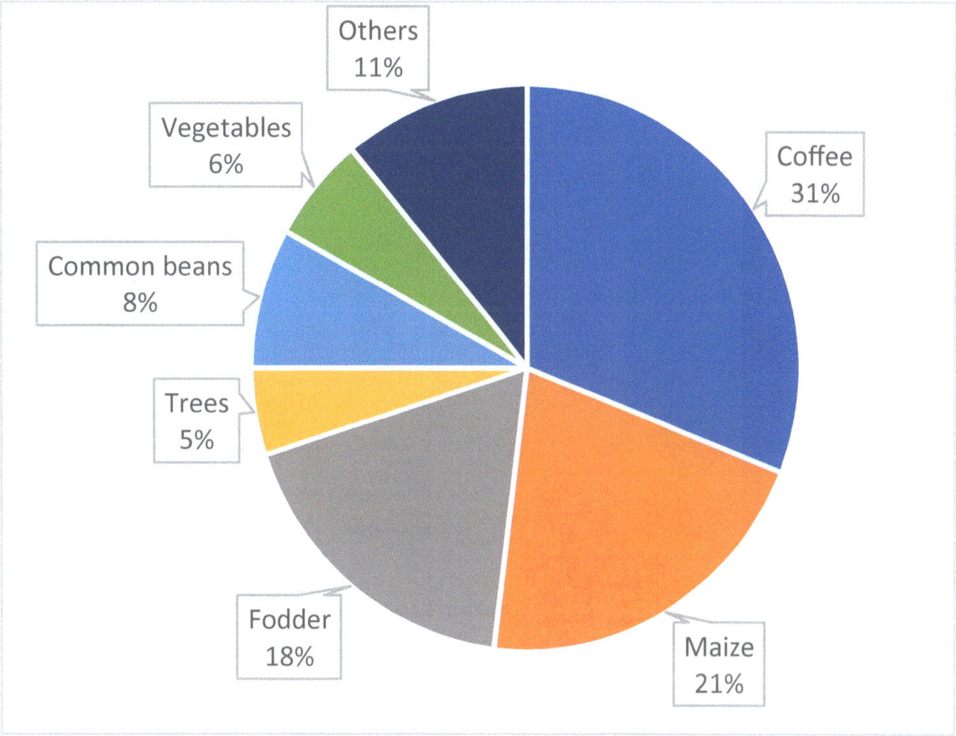

Fig. 24.10 Proportion of land dedicated to different crops

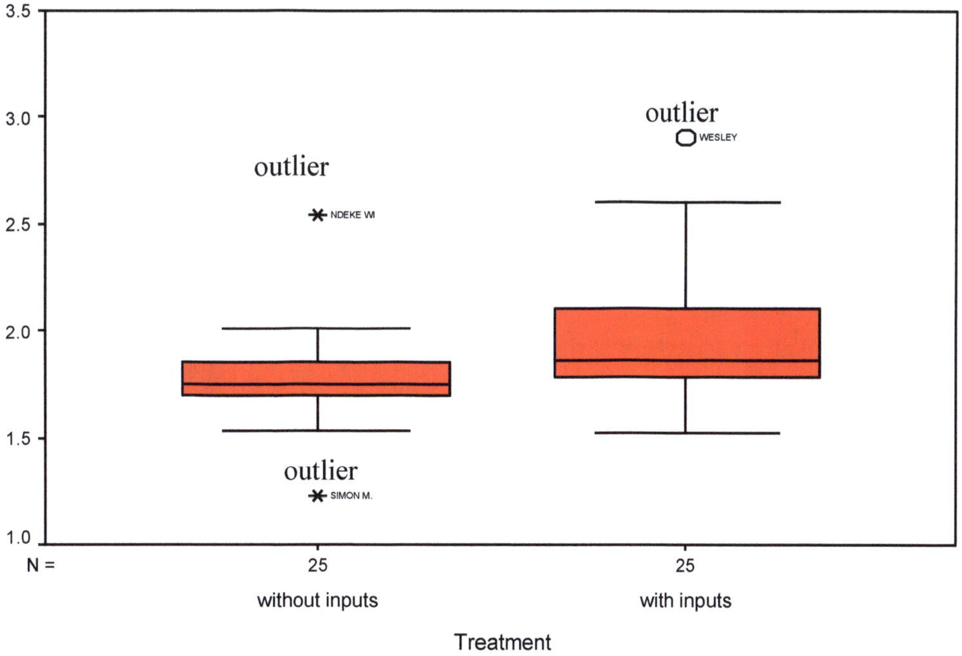

Fig. 24.11 Boxplots showing variation of organic carbon in farms that had inputs and without inputs in (Mugwe 2007: Thesis), Kenyatta University

the boxes, indicating the sequence or flow of the process, and show how one step leads to another. In this book, several flow charts have been used to illustrate processes. Researchers commonly use flowcharts to visualize supply and value chain processes. The value chains encompassing the entire process from production to consumption provide a comprehensive and visual representation of these complex systems.

Here we present a market chain to demonstrate a flow diagram (Fig. 24.14).

24.2.2 Tables

In presentations and data visualization, tables offer a structured way of organizing and presenting detailed datasets concisely. Moreover, tables promote accountability and enhance data accuracy and transparency by clearly indicating the data source and metrics. Tables are thus useful tools for effectively communicating data in reports, publications, and presentations. They help audiences to gain increased

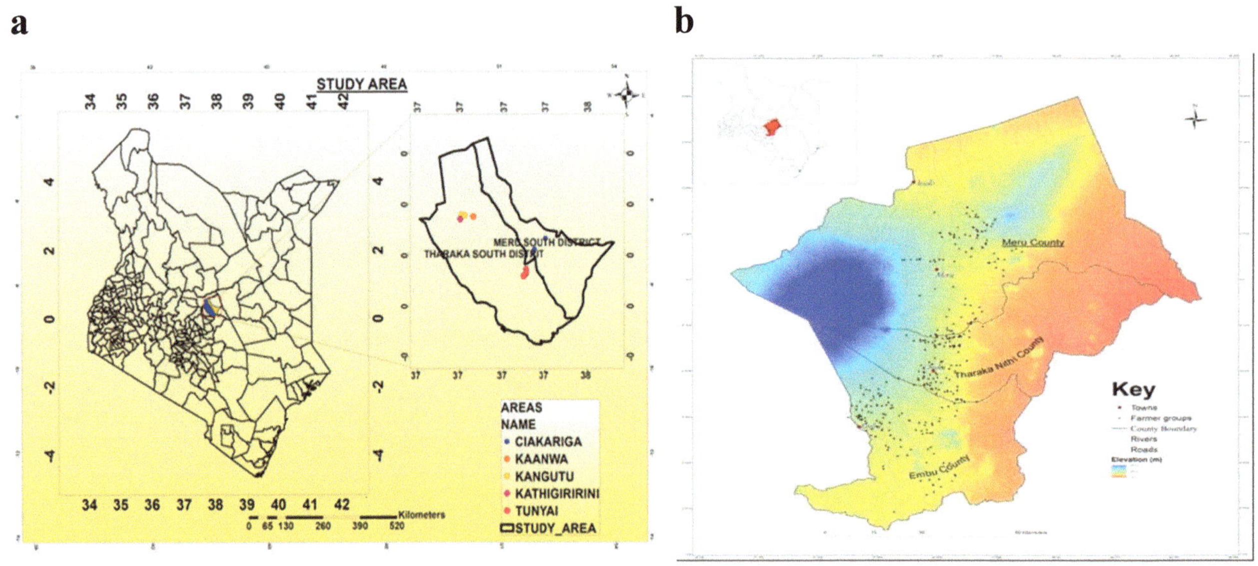

Fig. 24.12 (**a**) Map showing location of a research site in Kenya adapted from Mugwe et al. (2017). (**b**) Farmer groups distribution in the central highlands of Kenya. (Mugwe et al. 2017)

Fig. 24.13 Field photo showing an experimental plot of soybean and marketing of soy products in a roadshow organized by soybean and climbing beans project. (Mugwe et al. 2017)

understanding and make informed decisions. Any data can be presented in the form of a table. It is the most flexible form of presenting data.

We have simple tables and complex tables. Simple tables typically contain straightforward data organized into rows and columns. They summarize data, present key metrics, or display basic comparisons. They are easy to interpret and require minimal effort, making them suitable for various applications. The tables we have used in sections are simple tables, i.e., Tables 24.1 and 24.3.

Here, we present rainfall data in a complex table (Table 24.2).

Do It Yourself

You can easily create your own data presentation and visualization using various computer programs such as MS Excel, MS PowerPoint, MS Word, and many others, including Harvard Graphics for advanced graphics. For example, once data have been entered into an Excel file, you can choose the type of illustration needed, such as a pie chart, line graph, or histogram, and the program will produce the illustration accurately. These programs provide versatile options to create professional-quality visualizations that enhance data interpretation for reports, presentations, or publications.

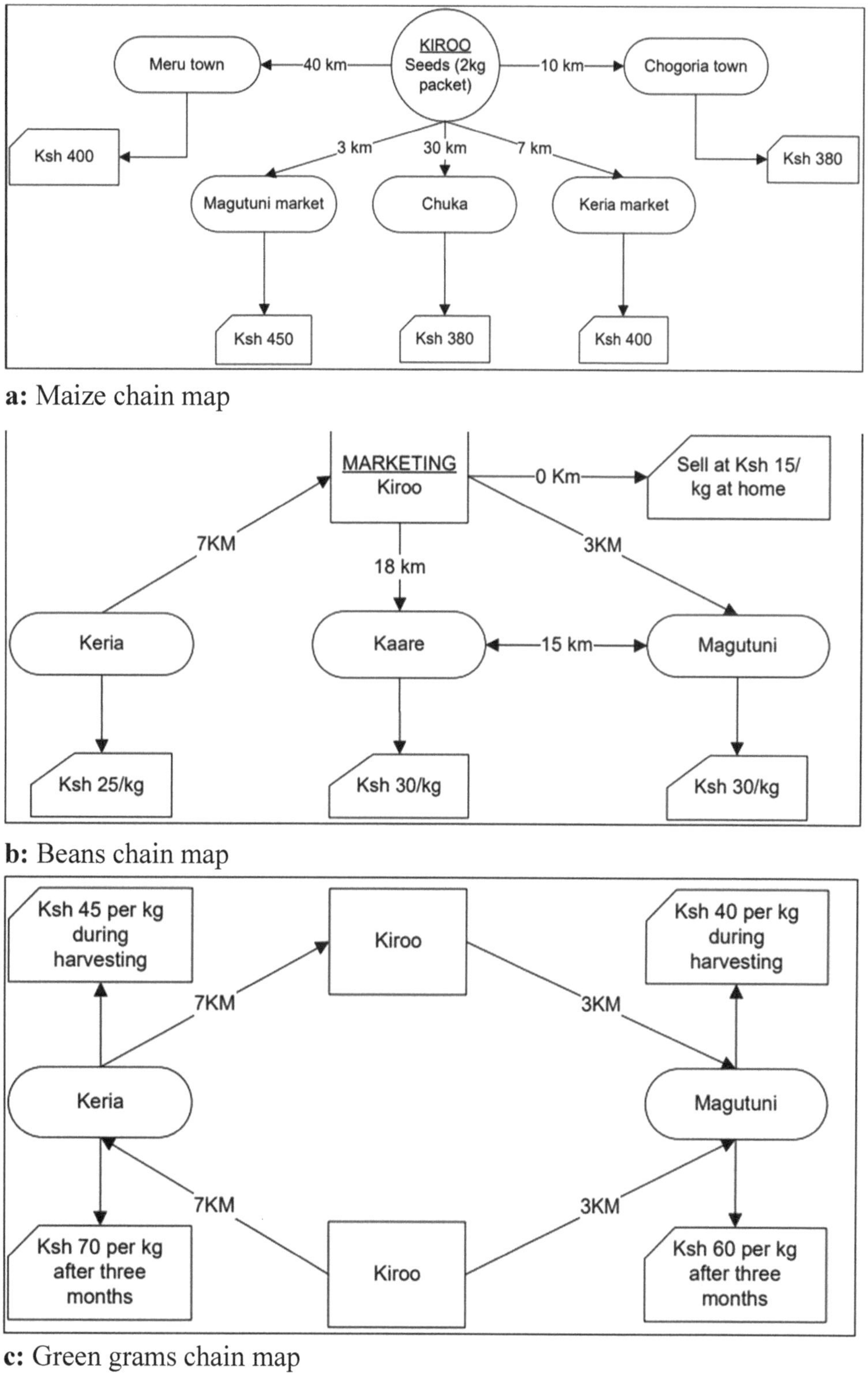

Fig. 24.14 (**a–c**) Chain maps of som ecrops grown in the central highlands of Kenya. The flow diagrams were drawn by farmers in groups. (**a**) Maize chain map. (**b**) Beans chain map. (**c**) Green grams chain map

When creating illustrations, the primary consideration is determining what information should be depicted, it's essential to ask yourself, "What is the most effective way to visually represent the data or information?" By asking this question, you ensure that the chosen visualization method aligns with the objectives of your presentation and effectively communicates the key messages to your audience. Table 24.3 shows a quick summary for reference in choosing the type of table or graph to use.

Table 24.2 Data presentation using a table

Long rains 2015				Short rains 2015			
Month	Rainfall (mm)	No. of rainy days	No of dry days	Month	Rainfall (mm)	No of rainy days	No of dry days
March—2015	76.3	5	26	Nov–2015	266.9	24	6
April—2015	44	8	22	Dec–2015	222.4	13	18
May—2015	14.8	7	24	Jan–2016	30.6	10	21
June—2015	9.7	6	24	Feb–2016	25.4	12	17
July—2015	1.3	4	27	Mar–2016	29.5	6	25
Total	**146.1**	**30**	**123**	**Total**	**574.8**	**65**	**87**

Source: A complex table rainfall pattern at Katumani research station during long rains 2015, short rains 2015, and long rains 2016 (PhD thesis by Dr. Mwende Ngie (Mwende 2019))

Table 24.3 Quick summary for reference in choosing the type of table or graph to use

What do you want to illustrate	Consider using
If you want to show frequency distribution for each class of variables, e.g., crop yields	Histogram
If you want to show frequencies for categorical data	Bar graph
If you want to show the relationship between variables and trend between data sets	Line graphs
If you want to present the proportions	Pie chart
If you want to show the variability of a data set	Box plots
If you want to explain a process	Flow chart
I want to show the actual appearance of something	Photograph
If you have a complex data set with many variables and want to see comparisons	Table
If you want to present structured numerical information	Table
If you want to present frequencies of several variables for comparison	Contingency table

24.2.3 Narrative Presentation

Narrative (text) presentation in scientific publications complements graphs and tables. The text provides interpretation, context, and information derived from the data. While graphs and tables offer visual representations of data, narrative text adds depth and clarity by explaining the significance of the findings, highlighting trends or patterns, and explaining observed phenomena. It guides the reader through the information presented in the visuals. Indeed, scientific reports should always have self-contained captions referenced in the text (Youdeowei et al. 2012). These captions summarize the content presented in the graphs and tables, enabling readers to quickly understand without needing to refer back to the main text. The captions are also reference points for citing figures and tables. Consequently, integrating complete narrative text, graphs, and tables ensures that the report effectively conveys the intended message to the audience.

Example 1 (Survey data): Majority of household heads in Kiambu County male-headed (65%) with the majority (92%) being fulltime farmers. The respondents who had attained primary school education were 45% while 8% had no formal education

Example 2 (Experimental data): During 2014 long and short rains, the application of sole lime recorded the highest value for soil pH water of 5.83 and 5.91, respectively. This was an increase of 15.4% and 15.9% over the control treatment

24.3 Data Visualization in R

R is a software that is widely recognized for its powerful visualization graphics. It enables users to create a wide range of charts ranging from simple histograms and scatter plots to complex heatmaps and interactive visualizations. It also integrates well with other data analysis tools and its support for reproducible research workflows makes it a preferred choice for professionals and researchers seeking to generate publication-quality graphics and presentations. In this section, we discuss how you can use R-software to visualize data.

For basic application on how to use R, the reader is referred to the resources provided by Data Carpentry (https://datacarpentry.org/R-genomics/01-intro-to-R.html). We will demonstrate R usage in visualization using one of R packages called tidyverse in a system called ggplot2. More information is available on the link (https://ggplot2.tidyverse.org/).

Plotting with ggplot2 ggplot2 is a plotting package that makes it simple to create complex plots from data in a data frame. It provides a more programmatic interface for specifying what variables to plot, how they are displayed, and general visual properties (Fig. 24.15). Therefore, we only need minimal changes if the underlying data change or if we decide to change from a bar plot to a scatterplot. This helps create publication-quality plots with minimal adjustments and tweaking.

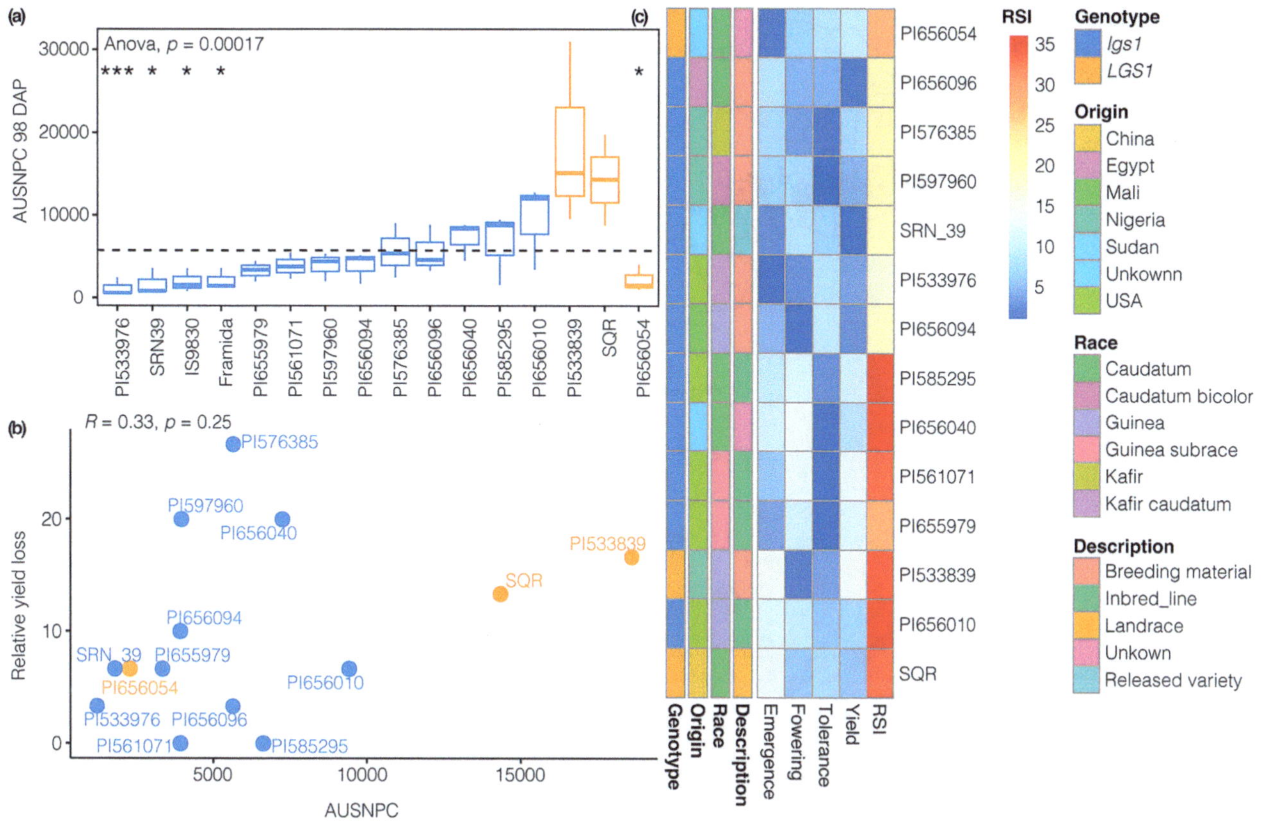

Fig. 24.15 Panel (**a**) is a box plot showing AUSNPC 98 DAP values for different genotypes, with significant differences indicated by asterisks and an ANOVA p-value of 0.00017. Panel (**b**) is a scatter plot depicting the relationship between AUSNPC and relative yield loss, with a correlation coefficient (R) of 0.33 and a p-value of 0.25. Panel (**c**) is a heatmap displaying various attributes such as RSI, genotype, origin, race, and description for each genotype. The color-coded legend provides details on the categories for origin, race, and description. An example of data visualization outputs from R. This figure is obtained from the publication: (Mutinda et al. 2023) "Strigolactone biosynthesis lgs1 mutant alleles mined from the sorghum accession panel are a promising resource of resistance to witchweed (Striga) parasitism" Plants People Planet, First published: 28 September 2023, https://doi.org/10.1002/ppp3.10442. (Source: Mutinda et al. (2023), https://nph.onlinelibrary.wiley.com/doi/full/10.1002/ppp3.10442, https://creativecommons.org/licenses/by/4.0/)

Data Visualization of Population Structure

R can also be used to perform specialized analysis such as maps and population genetic structure. This can be calculated with the APE package of R software using Single Nucleotide Polymorphism (SNP) data. The resulting matrix can be plotted as an unrooted NJ tree using the phytools package.

In this study described in Mutinda et al. (2023), the authors describe the population structure of sorghum genotypes collected from diverse locations. Maps (Fig. 24.16) were created using georeferrenced collection points available at Plant Genetic Resources Conservation Unit, and Griffin, Georgia, USA (https://www.ars-grin.gov/). Box 24.1 illustrates a visualization of Genome-Wide Association Studies (GWAS) Data Using R in a study by Kavuluko et al. (2021) using the Manhattan plot.

Box 24.1: Visualization of Genome-Wide Association Studies (GWAS) Data Using R

The goal of genome-wide association studies, or GWAS, is to screen the entire genome of large numbers of individuals to look for associations between millions of genetic variants within those individuals and their disease outcomes or sometimes for associations between the variants and non-disease traits such as height. There are three requirements: Genotypy data, trait data, and a tool to correlate the two.

Case Study: Striga resistance in sorghum (Kavuluko et al. 2021). In this paper, the authors wanted to determine genetic loci associated with resistance to a para-

(continued)

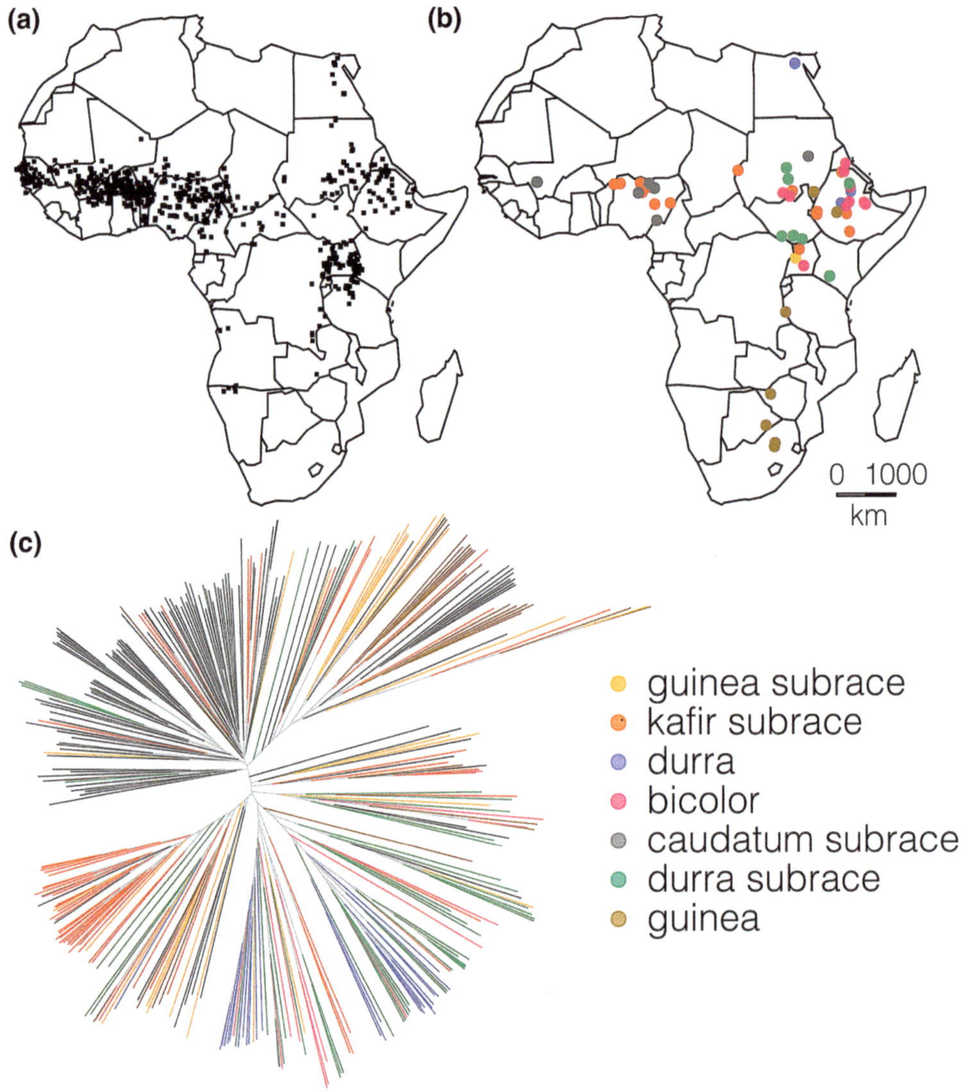

Fig. 24.16 Population genetic structure of sorghum genotypes from diverse regions

Box 24.1 (continued)

sitic plant that diminishes cereal production in Africa. For trait data, they used a number of parasites infecting the host, the length of the parasite infecting the host, and the total biomass of the parasite infecting the host. They then used a diverse collection of sorghum collected from various parts of Africa. The sorghum genotypes were genotyped by sequencing, and the data were available as SNPs.

R was used to produce these figures—called Manhattan plots (Fig. 24.17). These are typically used to describe genetic regions associated with a trait.

https://www.ncbi.nlm.nih.gov/pmc/articles/PMC8379865/
https://creativecommons.org/licenses/by/4.0/

Box 24.1 (continued)

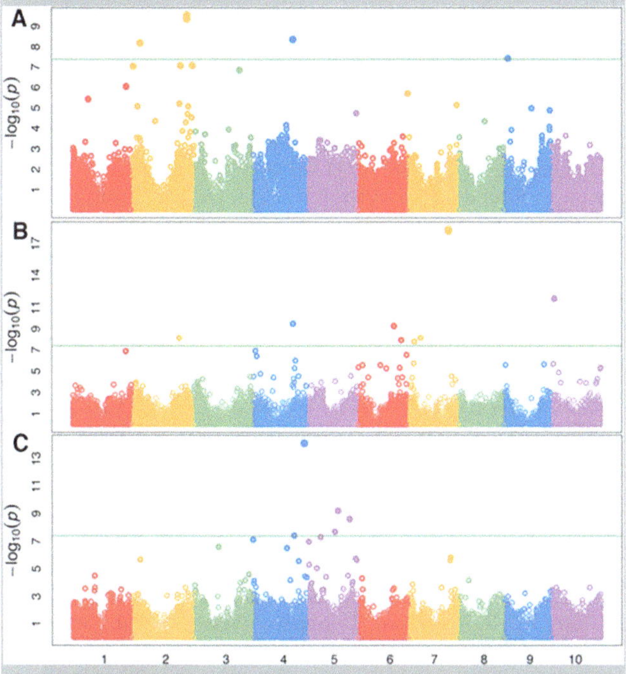

Fig. 24.17 Manhatan plots

24.4 Acknowledgments

Most of the data and graphics presented in this chapter were derived from the following projects. They have greatly enhanced the value of this book by providing real-life examples and supporting the practical application of key concepts.

1. "Enhancing farmers' benefits from integrated soil fertility management through improved water use efficiency and better marketing strategies in the highlands of central Kenya". Funded by The Flemish Interuniversity Council Own Initiative (VLIR—UOS)
2. "Scaling up integrated soil fertility management for improved livelihoods," funded by The Association for Strengthening Agricultural Research in Eastern and Central Africa (ASARECA) and implemented in 2012 to 2015
3. "Enhancing Productivity and Market Development of Soybeans and Climbing Beans in Central Highlands of Kenya" (SHP 024) supported by the Alliance for Green Revolution in Africa (AGRA) (2010–2013).
4. The second one, from 2014 to 2017, was titled "Scaling up Soybean and Climbing Beans Using a Value Chain Approach in Maize-based Systems of Central Highlands of Kenya" (SHP 014) supported by the Alliance for Green Revolution in Africa (AGRA) (2014–2017).

Most graphics have been created using Microsoft excel and R-studio. Data from Mutinda et al. (2023) were supported through funds from:

1. Mawazo fellows fund.
2. Baseline funding from King Abdullah University of Science and Technology and the Bill and Melinda Gates Foundation (grant number OPP1136424).

Data from Kavuluko et al. (2021) were supported by funds from:

1. The National Research Fund (NRF) of Kenya (contract number KU/DVCR/NRF/VOL1/27).
2. Kenyatta University through the Vice Chancellors Research grant number KU/DVCR/VRG/VOL.11/216.
3. Regional Universities Forum for Capacity Building in Agriculture (RUFORUM) through a capacity building competitive grant (RU/2016/DRRG/01/010).
4. NRF and an In-Country/In-Region scholarship from the German Academic Exchange Service—Deutscher Akademischer Austauschdienst (DAAD).

Main Highlights

1. A picture is worth a thousand words. Graphics are indispensable in research.
2. Select appropriate graphics depending on the goal or purpose.
3. R is an important software with enormous versatility.

Review Questions

1. Why is data presentation and visualization important in communicating research findings to stakeholders?
2. What factors should be considered when choosing a method for presenting data?
3. Explain importance of a flow chart to illustrate the process of data collection in a field experiment.

Chapter Exercise

1. Imagine you have conducted a survey on the adoption rates of a new drought-resistant maize variety among 200 farmers. You want to present the percentage of farmers who have adopted it, the percentage who haven't, and the reasons for non-adoption. Which primary "main method" (graphical, tabular, narrative) would you use for each piece of information, and why?

2. A researcher has collected data on the average monthly rainfall over the past 30 years and want to show the long-term patterns and any increasing/decreasing trends.
   (a) What type of graph would be most suitable for visualizing these data?
   (b) Justify your choice based on "the nature of the data" and the "information to be communicated."
3. You are presenting research findings on the various sources of income for rural households in a specific village. You want to show the proportion of total income derived from each source (e.g., crop sales, livestock, off-farm labor, remittances). What type of graphical presentation would best illustrate these proportions, and why?
4. Suppose your research team is presenting complex statistical results from an agricultural experiment to a group of local farmers and extension officers. You need to show differences in crop yield across multiple fertilizer treatments. Would you lean more towards a highly technical, detailed table or a clear, simple graph? Explain your reasoning, considering the "intended audience."

Answers to Review Questions

- Q1: Reference: 24.1.
- Q2: Section on "Do it yourself and Table 24.3.
- Q3: Reference: Section 9: Simplify complex processes, shows pathways.

References

Boaz A, Hanney S, Borst R, O'Shea A, Kok M (2018) How to engage stakeholders in research: design principles to support improvement. Health Res Policy Syst 16(1):60. BioMed Central, London. https://doi.org/10.1186/s12961-018-0337-6

Coetzee S, Carow S, Snyman L (2021) How are maps used in research? An exploratory review of PhD dissertations. Advances in Cartography and GIScience of the International Cartographic Association 3:3. https://doi.org/10.5194/ica-adv-3-3-2021

Davenport JH (1986) Symbolic computation: a formula is worth 10000 numbers. In: Bensoussan A, Lions JL (eds) Analysis and optimization of systems. Lecture notes in control and information sciences (vol 83). Springer, Berlin, Heidelberg. https://doi.org/10.1007/BFb0007581

Hehman E, Xie SY (2021) Doing better data visualization. Adv Meth Pract Psychol Sci 4(3):1–18. https://doi.org/10.1177/25152459211045334

Huang H, Ravi S, Warrington T, Cui H, Wang C, McCreary M, Lauffer B, Lu C (2024) A comparison of data visualization tools: a case study in health-related research. Information Visualization. https://doi.org/10.1177/1473871624127020

Jensen M, Patel R, Lee S, Riche NH (2023) Designing data visualizations for diverse audiences: a review of evidence-based practices. IEEE Trans Vis Comput Graph 29(1):566–579. IEEE, New York. https://doi.org/10.1109/TVCG.2022.3142158

Kavuluko J, Kibe M, Sugut I, Kibet W, Masanga J, Mutinda S, Wamalwa M, Magomere T, Odeny D, Runo S (2021) GWAS provides biological insights into mechanisms of the parasitic plant (Striga) resistance in sorghum. BMC Plant Biol 21:392. https://doi.org/10.1186/s12870-021-03155-7

Kosslyn SM (2006) Graph design for the eye and mind. Oxford University Press, New York. https://doi.org/10.1093/acprof:oso/9780195311846.001.0001

Li X, Chen H, Zhao Y, Zhang X (2020) Visualization and visual analytics in research data communication. Inf Vis 19(3):181–195. SAGE Publications, London. https://doi.org/10.1177/1473871620909483

Midway S (2020) Principles of effective data visualization. BioScience 70(8):763–771. Oxford University Press, Oxford. https://doi.org/10.1093/biosci/biaa064

Mwende N (2019) Soil moisture conservation, cropping systems and soil fertility effects on soil and maize performance in Machakos County, Kenya. PhD thesis, Kenyatta University, Nairobi. Available from Kenyatta University repository: http://ir-library.ku.ac.ke/handle/123456789/19846

Mugwe JN (2007) An evaluation of integrated soil fertility management in the central highland sof Kenya. School of Environmental studies. PhD thesis, Kenyatta University

Mugwe JM, Macharia J, Ngetich FK, Mucheru-Muna MM, Mugendi DN (2017) Annual report for the AGRA-funded project on soybean and climbing beans. Allica e for A Green Revolution final project report

Mutinda S, Jamil M, Wang JY, Berqdar L, Ateka E, Bellis ES, Al-Babili S, Runo S (2023) Strigolactone biosynthesis lgs1 mutant alleles mined from the sorghum accession panel are a promising resource of resistance to witchweed (Striga) parasitism. Plants People Planet. https://doi.org/10.1002/ppp3.10442

Palaniswamy U, Palaniswamy KM (2005) Handbook of statistics for teaching and research in plant and crop science. Haworth Press (now part of CRC Press/Taylor & Francis)

Youdeowei A, Stapleton P, Obubo R (eds) (2012) Scientific writing for agricultural research scientists—a training resource manual. CTA, Wageningen

Part VI

Dissemination of Findings and Scientific Publications

Dissemination is the last and critical stage in the research process. It involves sharing findings for use by end-users. This part of the book has four chapters and focuses on disseminating and publishing research findings.

The first chapter discusses strategies to maximize the dissemination of research findings to various stakeholders through multiple pathways (Chap. 25).

This is followed by Chap. 26, which presents the structure of a scientific paper, using the "Introduction, methodology, results and discussion (IMRAD)" framework for writing journal articles.

Chapter 27 is about publishing and journal selection, guiding readers through identifying suitable academic journals and submitting manuscripts to a journal.

Finally, Chap. 28 presents scientific writingand presentation skills, including seminar, oral and poster presentations.

These chapters provide comprehensive support to researchers who need to disseminate research findings to various stakeholders through various means.

25 Disseminating Research Findings

Abstract

Dissemination of your research findings is the last and critical step in the research process. It involves reporting outcomes and recommendations to the end users. Researchers have a responsibility to share findings with end users for adoption and impact. Most research institutions and universities have dedicated sections for research and dissemination activities. Universities adopt a concept of "laboratory to market" that focuses on disseminating and commercializing products and innovation. Dissemination aims to create a measurable impact, making utilization (adoption) its ultimate goal. We can subdivide the research process for developing solutions for farmers into three phases: research, dissemination, and impact phases. Researchers explain adoption dynamics using many theories. Key among them is the "innovation-diffusion theory" which uses four key elements. We have come up with seven elements that need to be considered in the dissemination of your research findings. These include consideration of the audience or receiver, Message to be communicated, Channel to be used, Packaging of information, Dissemination partners, Dissemination strategy, and Measuring the effectiveness of dissemination. We have also presented various ways of disseminating information using oral, print, practical-oriented, and media platforms. We focused on aligning the dissemination pathways with the needs of our target audience. One case study illustrates how soybean technologies were shared with farmers using a variety of methods, including joint field days, demonstrations, roadshows, business forums, strategic partnerships, bulk SMS, branded materials, and social media.

Keywords

Adoption of agricultural technologies · Dissemination channels · Dissemination triangle · Dissemination elements · Dissemination pathways · Dissemination strategy · Impact · Innovation diffusion theory · Information Packaging

Learning Outcomes

By the end of this chapter, you are expected to be able to:

1. Appreciate the importance of disseminating your research findings
2. Explain the concept of disseminating for impact
3. Explain key elements of dissemination
4. Identify key considerations in the preparation of the message
5. Apply the innovation diffusion theory in relation to dissemination
6. Explain the role of adoption in the dissemination of agricultural technologies
7. Identify various ways of disseminating research findings to beneficiaries
8. Apply the concept of matching messages with clientele
9. Demonstrate the most appropriate channels of dissemination of knowledge
10. Explain dissemination pathways used in a "case study" of a project in Kenya
11. Demonstrate the role of ICT in the dissemination of agricultural information

25.1 Dissemination and Adoption of Technologies

25.1.1 Why Disseminate?

Dissemination of your research findings is the final step in the research process. It involves sharing information about your research outcomes and recommendations with your target audience so that they can use it. Dissemination has been defined as a conscious effort to spread new knowledge to target audiences and the public (Wilson et al. 2010; Turon et al. 2023). The process involves interactive communication

J. Njeri Mugwe, S. Runo, *Research Methodology in Agricultural Sciences*, https://doi.org/10.1007/978-981-95-1892-0_25

of knowledge and aims to enhance changes among members of the intended audience (Wilson et al. 2010). The main audience in agricultural research includes various stakeholders: farmers, practitioners, policymakers, and other key players within the agricultural value chain (Chap. 1). Through effective dissemination, valuable knowledge and solutions derived from research can be applied to real-world contexts (adopted) and consequently drive technological change and improvement in agricultural practices.

This is in line with the reason we use to justify research projects. Researchers always indicate how the information generated through the research activity will be used and who will benefit. This is mostly captured in the significance of the study section. Some of the most common statements in Thesis research proposals include the following:

- Stakeholders will benefit from the information collected; for example, "farmers will benefit through getting recommendations to improve their farming practices and yields." "Extension workers and other stakeholders will receive information on recommendations for increasing yields"
- Policy makers will use the information to formulate suitable policies for improving the agriculture sector
- New information generated will add to the pool of scientific knowledge on a certain aspect
- Information is envisaged to contribute to increasing food security and reducing poverty

The above statements show the critical role of reporting and disseminating the new research information to the beneficiaries.

Dissemination, therefore, is the hallmark of making a positive impact on the agricultural community and society. First, it is a core responsibility of every researcher to share knowledge with end-users if it is to benefit society. Second, information received by the end users may drive technological change ultimately benefitting the poor through increasing productivity and reducing poverty (Alene and Coulibaly 2009; 2021). For example, the dissemination of information by researchers during the era of the "Green Revolution" in the 1960s and 70 s resulted in a huge impact, doubling and tripling yields of the main food grains in many countries (Pingali 2012). Similarly, the dissemination of drought-tolerant maize varieties by the International Maize and Wheat Improvement Centre (CIMMYT) researchers to farmers in drought-prone areas resulted in improved yields and increased resilience against the effects of climate change (Stevenson et al. 2013).

Third, dissemination helps add information to the body of knowledge. Researchers share information with their peers using different forums, such as scientific publications and conferences. These forums unite researchers, enabling them to share expertise and engage in cross-disciplinary dialog. Sharing enhances the building of information and helps formulate topics and meet the needs for further research. Fourth, dissemination benefits the researchers by making them known and building credibility. For example, a researcher who publishes reputable journals increases visibility and promotes career opportunities. Researchers are ranked by the number of citations they receive for their publications (Chap. 27).

Fifth sharing research findings builds trust and strengthens relationships with donors, researchers, and partners, which is crucial for securing future support and advancing ongoing projects. Openly communicating results demonstrates transparency and accountability, highlighting a commitment to impactful outcomes. This reassures donors that their contributions are meaningful.

Finally, dissemination is a way of ensuring ethical responsibility. You have conducted your research, guided by your supervisor, and used time and resources to do so. It is immoral to deny yourself, the supervisor, and the public the results from your research. The objectives of disseminating research findings are summarized in Box 25.1.

Box 25.1 Summary of Objectives for Disseminating Research Findings:

1. Connects researcher with end users: significance of the study is fulfilled.
2. Utility and empowerment with knowledge: Help improve existing practices among the practitioners, especially farmers.
3. Add to body of knowledge: information is deposited in literature.
4. Build credibility: enhance visibility as a professional and researcher.
5. Strengthen relationships: Partners and donors' relations are improved, creating solid multidisciplinary and multi-institutional partnerships.
6. Ethical responsibility: way of accounting for resources used.

Globally, great emphasis is put on disseminating results to end users in donor-funded research projects conducted at research institutions and universities. For example, most donors demand that grant proposals indicate a clear communication and dissemination plan targeting specific beneficiaries. A case in point is EU-funded projects where the project participants are required to provide a comprehensive dissemination plan with clear objectives using a prescribed template. The donors believe that future economic growth and jobs will increasingly have to come from the innovation of products, services, and business models developed from the research projects.

25.1.1.1 Dissemination in Universities and Research Institutions

Most universities and research institutions have dedicated sections for research activities. The research section facilitates the research activities and also plays a key role in disseminating the findings (Fig. 25.1). Terms like technology transfer, innovation, extension, and outreach are used to highlight the role in dissemination. The institutions actively encourage researchers to create innovations, products, and technologies that are demand-driven and aligned with the stakeholders' needs.

While research institutions often have well-structured sections for disseminating their findings, many universities, particularly those in developing countries, often rely on traditional approaches such as publishing journal articles and participating in scientific meetings. However, there is a growing emphasis on diversifying dissemination methods to reach a broader audience, including using mass media platforms. This shift is intended to dispel the perception that universities are "ivory towers" where valuable scientific discoveries remain confined to shelves. By adopting more proactive and inclusive dissemination strategies, Universities can break the barriers that isolate them from society, making research outputs more accessible and impactful. The emphasis is moving research output from the "laboratory to the marketplace" where it can drive real-world solutions and innovations.

In the push to move from the laboratory to the market, there is an increasing emphasis on *commercializing* university research (Etzkowitz 2003; Caulfield and Ogbogu 2015). Commercializing means turning research findings into practical products and services. The push leverages academic expertise and intellectual property for practical applications and economic growth. Universities do this through technology transfer, spin-offs, and partnerships with industry stakeholders. These generate additional funding for research activities and support economic growth.

As a student, your priority is to write reports that fulfill the requirements for the award of your degree. In most cases, for master's and PhD degrees, a thesis plus a journal article is mandatory. The key beneficiaries of thesis and journal articles are the academia. Once you successfully write a thesis and a journal article, you will have fulfilled a part of your responsibility of sharing the research findings. However, there will be a need to go beyond academia and share your findings with other potential beneficiaries. This depends on the nature of the outcomes. Some stakeholders, like farmers, need real recommendations that they can apply in the field.

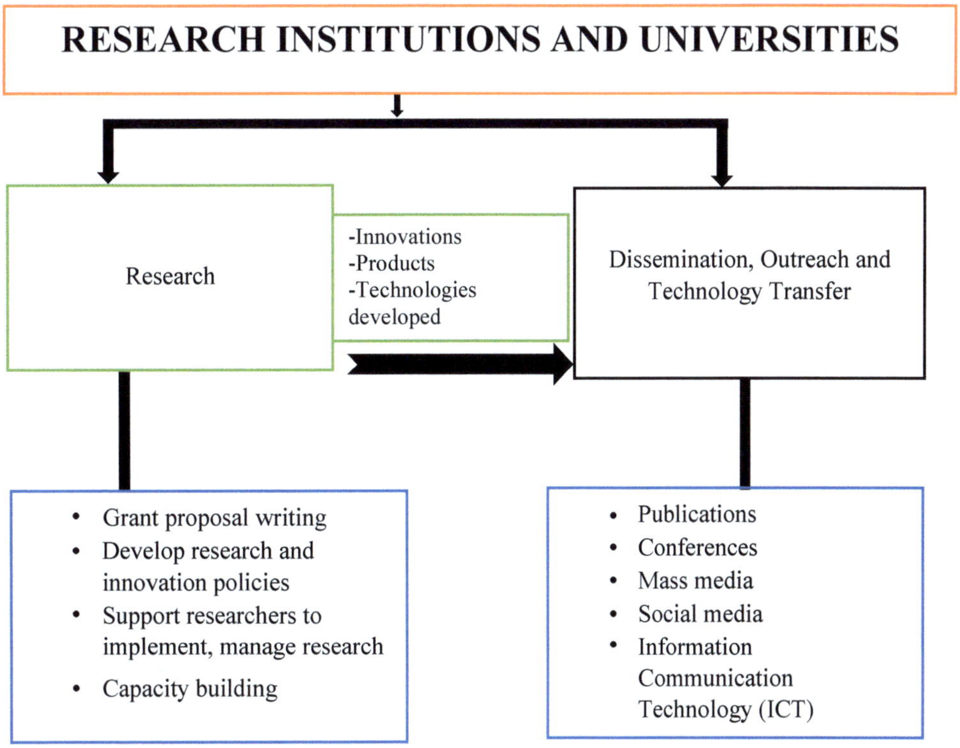

Fig. 25.1 Role of Research Institutions and Universities in research, innovation, and technology transfer

25.1.2 Concept of Disseminating to Create Impact

Dissemination involves a process from the knowledge owner to the receiver. The researcher is the knowledge owner. In our case, we consider the receivers to be mainly farmers and agricultural organizations. To be effective, data from the owner must be interpreted or translated from its complexity into a language, format, and context that the receiver (target audience) can understand. This is because dissemination is "not an end in itself." Rather, the aim is to create impact. The end users must gain an understanding or insight to create an impact, followed by the application. Effective dissemination should, therefore, focus on utilization as the goal (Osterling and Austin 2008; Green et al. 2009).

Looking at the research process, we can draw a framework showing the relationship between data collected, dissemination, and end users/beneficiaries. Figure 25.2 shows such a framework and shows the link from the problem identification stage, at the research conceptualization stage, to offering solutions for farmers. Three phases are presented: the *research phase*, the *dissemination phase*, and the *impact phase*. At the research phase, the main output is raw data obtained at the data collection stage. These data are analyzed and interpreted and linked to create information. The recipients receive information.

In the *dissemination phase*, the first step is to determine the audience, i.e., the recipient of the information. Once the audience is identified, the next stage is to select the channel. After choosing the channel, it is crucial to prepare the message effectively. The information from the research phase is technical and highly scientific; therefore, it is difficult for a farmer to understand if it is unprocessed. The information must, therefore, be processed, organized, and prepared into a message. Finally, the information is delivered to the receiver (farmer).

After receiving the message, we have the impact phase at the tail end. This phase focuses on how the farmer responds to the message received. Upon receiving the message, the farmer has several options: they can adopt and use the message, reject it, or modify it to fit their circumstances. The true impact of the dissemination process is achieved only when the innovation is successfully adopted by the farmer.

We can illustrate the dissemination for impact concept using a dissemination triangle to show the relationship between data and application (Fig. 25.3). The most important concept in this triangle is *"the distilling of raw data"* (pure and simple facts) into "*knowledge*" (condensed), fol-

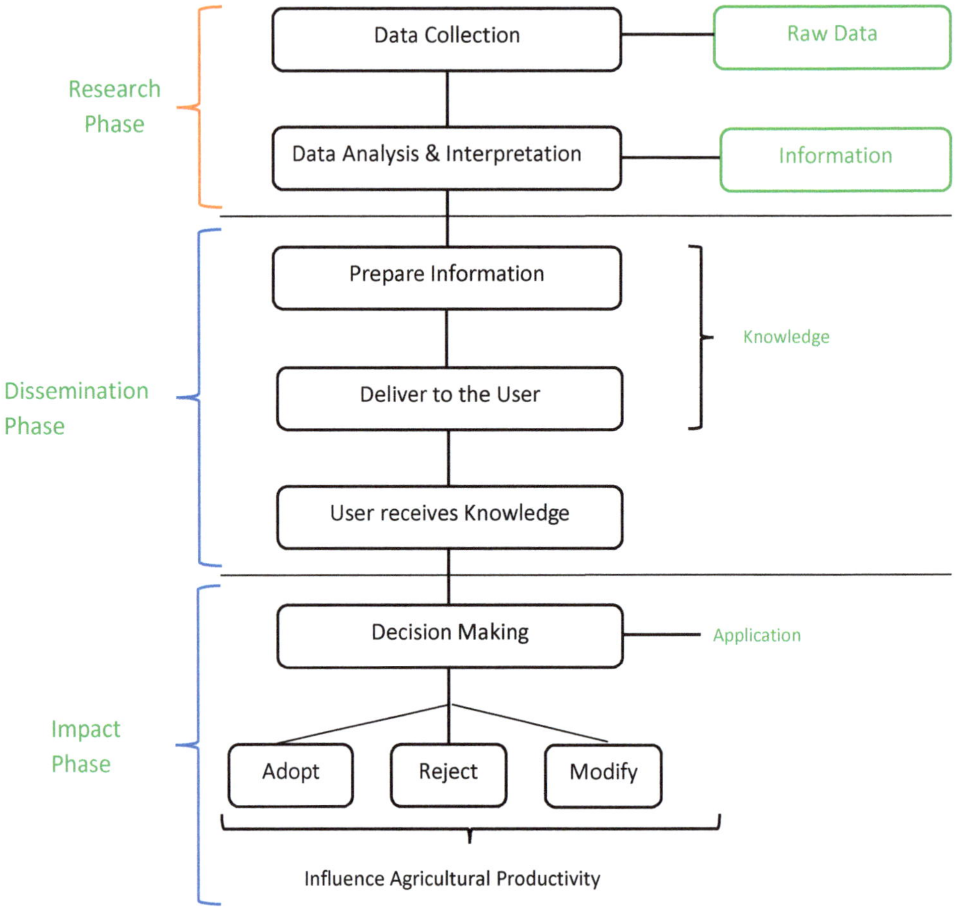

Fig. 25.2 Process of data collection to dissemination and possible adoption of technology (The *research, dissemination,* and *impact phase*)

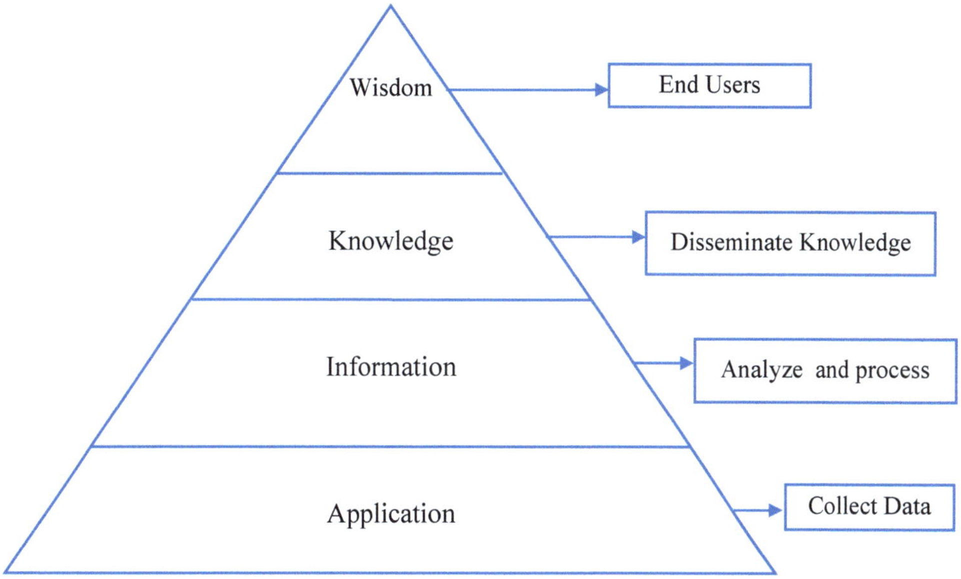

Fig. 25.3 Dissemination triangle showing the relationship between data collected and application by end users

lowed by application by the end users. The simple facts and measurements, e.g., fertilizer treatments increased maize yield by 90%, are processed into knowledge. *Knowledge is organized information* that is factual and gives meaning to the data; it is contextual and has significance, for example, the amount of fertilizer to apply to maize crop to obtain 3 t ha^{-1}. Such knowledge can, therefore, be strategically applied by the clientele. Application is what will bring impact. In this case, we would be looking for a farmer who decided to adopt the knowledge given and find out the increase in maize yield obtained. Within the "knowledge pyramid," Rowley (2007) refers to the application stage as *wisdom*. This is the stage where the recipient gains an understanding and integrates the knowledge into the decision-making process. This final stage is the ultimate goal of research; connecting findings to practical, real world outcomes.

25.1.3 Adoption of Technologies

From the preceding sections, we have seen that the ultimate goal of dissemination is to facilitate the adoption of new agricultural technologies and practices. Adoption is the complete implementation of a new idea or technology. The process of technology adoption is complex and influenced by various factors. It involves implementing different techniques at the farm level, which include farm capital, inputs, education, training, advice, and information, all of which contribute to "farmers' knowledge base". It thus extends beyond the farm level and affects practices within the broader agri-food sector, further impacting farming practices.

Several theories and models are used in the literature to explain adoption behavior. They mainly explain three aspects: (i) User trajectory of new technologies, (ii) Factors that influence users' decision to adopt the technology, and (iii) Factors that influence the level of technology use once adopted (Ruzzante et al. 2021). The common theories and models include "the innovation-diffusion" theory, the "economic constraints" paradigm, the "adopter-perception" paradigm, the "Theory of Planned Behaviour", the "Theory of Reasoned Action", the Technology Adoption Model, the Unified Theory of Acceptance, the Motivational Model, and Social Network Theory (Eweoya et al. 2021; Landmann et al. 2021; Mohr and Kühl 2021; Ahmed et al. 2024). Each paradigm highlights specific factors influencing farmers' pace and pattern of technology adoption. For example, the innovation diffusion theory is widely used to understand how new technologies are adopted and integrated into farming systems. In the next section, we shed light on how researchers can leverage it to enhance their dissemination efforts and effectively promote technology adoption among the farmers. We will also discuss strategies for overcoming barriers to adoption and improving farmers' motivations.

25.1.3.1 Innovation Diffusion Theory

The "*innovation diffusion theory*," developed by Everett Rogers in 1962 (Klingelhöfer 2019), is one of the oldest social science theories. It is relevant in disseminating agricultural technologies and innovations because it explains how information spreads through a population to the adoption of ideas, products, and behaviors. The theory was developed using data from agricultural sciences. Everett Rogers used farmers to develop adopter categories to measure farm-

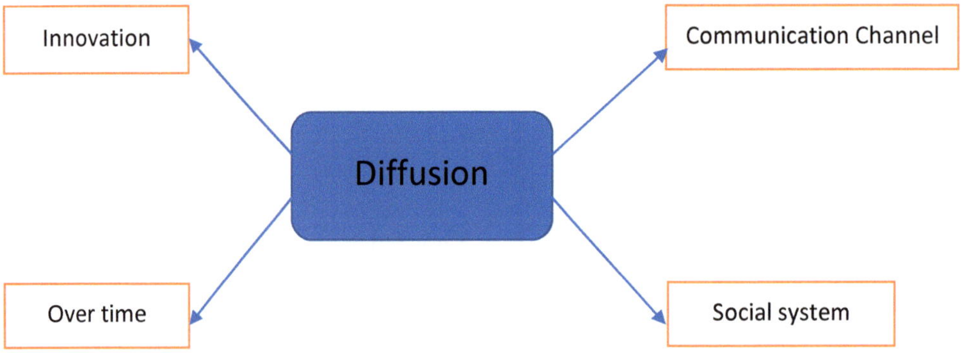

Fig. 25.4 The innovation diffusion theory showing the four main elements that interact to explain the process of dissemination through diffusion and adoption

ers' innovativeness (Klingelhöfer 2019). This theory proposes that the adoption of new technologies follows a predictable order, starting with innovators and early adopters, then moving to the early and late majority, and finally reaching laggards.

Researchers have extensively used this theory in various agricultural disciplines to explain the adoption of technologies and innovations (Fig. 25.4). It has been used to explain the decision-making process, simply the decision to adopt or not to adopt, and the level or intensity of adoption in integrated soil fertility management, organic farming, new crop varieties, plant protection practices, and new farming practices (Wainaina et al. 2016; Kumar et al. 2020).

Four elements are important in influencing adoption (Moseley 2004; Dearing 2009). These are:

(i) The innovative technology or practice
(ii) Communication channels selected
(iii) Time
(iv) The social system

Idea or innovation An idea, practice, or product that is perceived as new by a group of individuals. After conducting research, the aim of distilling the information is to provide practices and products to our clientele. Important components of the innovation include compatibility, complexity relative advantage, testability, and observability (Klingelhöfer 2019). These five components influence adoption/non-adoption and are therefore important in informing barriers to adoption.

Communication channel Serves as the medium through which messages are conveyed between individuals to create an exchange of information for mutual understanding. This refers to the process of sharing information and social interaction among participants. In a study in Kenya, farmer-to-farmer diffusion was reported as an effective channel to communicate (Muchai et al. 2014). Similarly, in Malawi, 78% of primary extension providers utilized the farmer-to-farmer extension method, showing its efficacy in spreading agricultural knowledge (Simpson et al. 2015).

Time Time involved in the innovation-decision process, the time taken to adopt an innovation by the adopter and the adoption rate across the social system (Klingelhöfer 2019).

The social systems of potential adopters are a set of interrelated social units (e.g., individuals, farmers' informal groups, and organizations) that are engaged in problem-solving to achieve a common goal (Klingelhöfer 2019). They determine the boundary for a diffusion process and can be affected by norms and the degree to which individuals can influence one another. The above elements influence the diffusion and adoption of agricultural technologies and innovations among our clientele. We should, therefore, seek to understand these elements and factor them in as we design our dissemination strategy, as this is likely to enhance effectiveness (Fig. 25.4).

Box 25.2 Example of Farmer Dissemination of an Innovation

Idea/practice: development of a water conservation technique for the drylands
Dissemination: transfer of information to potential users
Adoption: decision to adopt, innovative, and teach others
Implementation: actual use of the innovation
Sustainability: are they still using the innovation after 2 years, after 5 years, after 7 years and are the results still visible? Are there any modifications? Modifications enhance sustainability.
Source: Mugwe et al. (2009)

25.1.3.2 Adoption Potential of Agricultural Innovations

Adoption potential is another important term in adoption studies. It refers to the likelihood that a new agricultural technology will be accepted and taken up by the farmers.

Assessment of adoption potential helps researchers design technologies that are not only scientifically sound but also socially acceptable, contextually relevant, and aligned with farmers' needs and capacities (Franzel et al. 2001). This assessment ensures that innovations are practical, user-friendly, and capable of addressing real challenges faced on the ground. Often, this involves further validation and refinement of the technology through participatory, on-farm trials, which provide insights into farmer preferences, performance under local conditions, and barriers to adoption (Franzel et al. 2001; Mugwe et al. 2009).

Key questions in assessment of adoption potential include:

- What are the *practical implications* of the technology? Involves assessing how the technology can be integrated into the current farming practices and the benefits it offers to farmers.
- How *feasible* is its implementation in real-world settings? Involves evaluation of practical aspects, such as availability of necessary resources, compatibility with existing farming systems, and the required level of log knowledge and skills.
- What are the *potential barriers to adoption*, and how can they be addressed? Here we identify obstacles that hinder adoption and develop strategies to overcome them, for instance, high labor requirements.
- What are the *preferences of farmers?* Involves assessment to Understand farmers' needs, preferences, and priorities and ensure that the technology aligns with their goals and expectations.
- Under what *boundaries* is the technology feasible? Determine the environmental, economic, and social conditions under which the technology can be implemented and sustained.
- Studies on adoption potential are excellent opportunities for students allowing them to apply research skills in real world contexts.

25.1.4 Dissemination Research and Adoption Dynamics

Dissemination research is a systematic study of processes and factors that lead to widespread adoption. This study area is a prominent topic among students due to its significance in providing better dissemination strategies. Students can develop more effective strategies for communicating research findings by understanding how innovation spreads within agricultural communities. Common topics in dissemination research include those that: (i) identify factors influencing adoption, (ii) assess barriers to adoption, (iii) evaluate stakeholders involved in adoption, and (iv) assess impact.

Table 25.1 Key topics and how they would help develop effective dissemination strategies

Topic	Develop effective dissemination
Factors influencing adoption (Financial, Social, Demographic, and Psychological) (Kamau et al. 2023)	Tailor dissemination strategies to specific needs
Barriers to adoption (Financial, lack of awareness, infrastructure, lack of resources) (Fadeyi et al. 2022)	Develop strategies to address the specific challenges identified
Role of stakeholders (Governments, Research institutions, Private sector, and nongovernmental organizations) (Ochuku et al. 2023)	Develop recommendations to strengthen the stakeholders and their processes.
Impact of adoption of technologies (Economic, social, and environmental) (Autio et al. 2021)	The positive impact will motivate other farmers Negative impact: Create pathways for correction
Develop innovation diffusion models. (Theories to explain and predict how technologies spread) (Mwangi et al. 2022)	Help predict how a technology will spread and identify early adopters within the community for targeting
Monitoring and evaluation (Methods for assessing effectiveness of dissemination strategies) (Fadeyi et al. 2022)	Develop robust monitoring and evaluation mechanisms

The findings from these studies are used to provide recommendations for enhancing impact. For instance, if we find that one of the barriers to effective dissemination is a lack of awareness, strategies for enhancing adoption will be recommended. This may include focusing more on media campaigns and incorporating workshops and community meetings that provide detailed information. If, on the other hand, the technologies being disseminated are perceived as complex by the farmers, dissemination strategies can incorporate hands-on training programs, demonstrations, and peer learning sessions. Table 25.1 gives some key topics and how they would help to develop effective dissemination strategies.

Adoption studies are particularly popular among students and researchers in agricultural economics due to their significance in addressing the low and uneven adoption rates observed by several authors. Most of these studies focus on identifying the variables motivating adoption and challenges associated with the adoption of effective agricultural solutions. For example, a recent study conducted a meta-analysis using 367 regression models and identified key variables influencing adoption, including household size, education level, land tenure, land size, access to credit, availability of extension services, and organizational membership (Ruzzante et al. 2021). The study highlighted the importance of adapting agricultural technologies to fit the unique agricultural and cultural contexts of the developing world. It emphasizes the need for ongoing research into the factors that influence adoption behavior, as well as the development of effective dissemination strategies that cater to the diverse needs of farmers.

25.2 Enhancing the Reach and Adoption of Innovations

25.2.1 Core Elements of Effective Dissemination

Taking into consideration that we need to communicate to multiple audiences for impact, we have come up with seven elements that need to be considered in the dissemination of your research findings; they include the following:

1. Audience or receiver.
2. Message to be communicated.
3. Channel to be used.
4. Packaging of information.
5. Dissemination partners.
6. Dissemination strategy.
7. Measuring the effectiveness of dissemination.

1. *Recipient or the end user*: This is the person who the message is intended for. The message needs to be relevant. We ask three key questions:
   - What information does the recipient desire or require?
   - What information does the recipient have access to?
   - Does the recipient have any prior knowledge of the subject?

To answer question one, we can think about potential beneficiaries in agriculture and focus on understanding recipient's needs and expectations. These include various stakeholders ranging from academia, the general public, policymakers, and the farming community. For instance, what specific recommendations or data will be available to them. These could include new farming techniques, policy recommendations, or innovative research findings.

Questions two and three ensure that new information adds value. Consider what the recipient already knows and tailor your message to fill gaps or updates.

2. *Decide the message to be communicated*: We ask three questions related to the content of the message:
   - What should the subject matter be?
   - What should be the content?
   - What should the message's format be?
   - How should the message be packaged?

The message and packaging depend on the specific clientele who differ in the type of information needed and how they will use it. The kind of information that each clientele requires varies in *content, depth, and language*. Consequently, at this point of determining the message, one needs to consider the end-user and ensure that the information's content, depth language, and packaging align with the beneficiaries' understanding and needs. To enhance effectiveness, the packaging of the information should be user-friendly. Consequently, researchers should take great care in developing *knowledge-sharing products*.

Dissemination involves communicating relevant information that can be understood by a person and shared for later retrieval and use; moreover, feedback on such information is encouraged and acted upon (Green et al. 2009; Wilson et al. 2010).

3. *Select the pathway and how to package the information*: Regards how the message is transmitted:
   - What are the receivers' expectations?
   - What is the most effective way of disseminating information?
   - Characteristic of the message: does it need visual presentation, demonstration, or in print form?
   - What channels do the receivers have access to?

Many dissemination pathways exist, but none is the "best" way or one-fit-all to disseminate research outputs. The choice of the pathways and how to use them will depend on the *nature of the outputs*, the *characteristics of the potential users*, and the *capacities of the collaborators and institution* (Wilson et al. 2010). End-user uptake is influenced by the relevance or appropriateness of one or more sets of users. The dissemination pathway positively relates to communication effectiveness (Rware et al. 2021; Arslan et al. 2022; Isgren et al. 2023).

They include mainly pathways used to communicate to peers (other scientists and academicians), such as thesis, journal articles, working papers, books and book chapters, research reports, brochures, etc. The oral pathways are varied and may be used for both literate and illiterate persons. Main oral pathways include conferences, symposiums, training workshops, field days, etc. We also have Information technology communication- (ICT) based dissemination pathways.

4. *Decide how to Package the Information*

Deciding how to package the information involves strategic decisions that ensure that research outputs are effectively communicated to target users. The process includes selecting suitable formats, tailoring content, and choosing suitable delivery channels. Information packaging will need to consider the following factors (Njenga et al. 2021):
   - Language.
   - Message display.
   - Translation.
   - Technicality.

- Information sufficiency in quantity.
- Information sufficiency in quality.

In addition to information pathways and information packaging, a third variable that is important is communication factors. These include:

- Accessibility to extension officers.
- Repetition of information.
- Practical orientation.
- Ability to respond to aired programs.
- No. of extension visits.
- Literacy of farmers.
- Attitude of farmers.
- Farmer group membership.

5. Consider Partnering with Dissemination Partners

Dissemination partners play a big role in agriculture because researchers need to communicate with rural areas where most farmers live. The partners on the ground help to disseminate the findings using their existing structures. Moreover, dissemination partners can provide valuable feedback from farmers to help better meet needs and circumstances. They can also assist in organizing various training sessions and field demonstrations making the learning process more interactive and practical. For instance, within a soybean and climbing beans project, we used strategic partners, including churches, community leaders, extension workers, entrepreneurs, and agro-dealers (Fig. 25.5).

6. *Have a dissemination strategy*: the plan should reflect the target audience's needs. A detailed presentation of the dissemination strategy is beyond the scope of this book. We refer you to the following references (Kondo et al. 2020). In brief, a dissemination strategy answers the following questions:

Fig. 25.5 Use of extension officers in the soybean climbing beans project to disseminate technologies for cooking soybean

- What are the existing capabilities, resources, and existing relationships?
- What dissemination techniques will be used? (graphical, written, print, broadcast, or verbal)
- What electronic dissemination exists within the organizations?
- What formats will be used for the techniques and methods of dissemination?
- What information is likely to be of interest to the audiences?
- What quality control measures exist to ensure that information is accurate, relevant, and representative of the study area?
- How will you know that dissemination activities have been successful?

7. *Measuring* the Effectiveness of Dissemination

Evaluating the effectiveness of dissemination measures the overall success of the exercise. This involves evaluating how well the new ideas are being shared and adopted by the farming community. It involves tracking adoption rates and understanding the overall impact on the productivity and sustainability of the growing systems. This process is crucial as it gauges the success of the technology transfer activities.

25.2.2 Communication and Dissemination Pathways in Agriculture

Many pathways or channels for disseminating agricultural research findings exist (Box 25.3). They can be categorized into four:

Box 25.3 Main Dissemination and Communication Pathways Used in Agriculture

- Writing a thesis
- Publishing a policy brief
- Publishing a briefing paper
- Presenting in scientific meetings (Conference, symposium)
- Publishing a journal article
- Writing a working paper
- Writing a book/book chapter
- Writing annual reports
- Preparing flyers, brochures, pamphlets
- Writing results in a research report
- Posting information on organization's website
- Presenting information to farmers and other stakeholders

(continued)

Box 25.3 (continued)

- Organizing a field day
- Organizing a training workshop (Field or indoors)
- Discussing information on mass media platforms (TV, local radio)
- Publishing information in the local newspaper
- Organizing a Webinar
- Participating in agricultural shows
- Using other farmers as trainers, e.g., training them to be trainers of other farmers
- Organizing a training seminar
- Working with agro-input dealers
- Newsletters
- Preparing and mounting market information boards in public places
- ICT-based dissemination pathways (mobile internet, webinars, videos, website, podcasts, etc.)
- Organizing local events (community and stakeholders meetings)
- Preparing fact sheets
- Broadcasting on TV or radio
- Using strategic partners (opinion leaders, religious leaders)
- Organizing community meetings
- Establishing demonstration sites
- Social media (Facebook, blogs, X)
- Roadshows
- Village training workshops
- Farmer groups
- Farmer field schools
- Web-based portal
- Voice-based internet service; Research networks such as ResearchGate

Sources: Nyasimi et al. (2017), Adolwa et al. (2018), Waaswa et al. (2021) and Njenga et al. (2021)

1. Written-related outlets.
2. Oral-related outlets.
3. Practical-oriented outlets.
4. Information Communication Technology (ICT) and Social media-related.

Written-related outlets use documentation with little or no interaction with the recipient. They tend to be prepared by students and professionals in agricultural research though important to a little extent for other clientele as well. They include thesis/dissertations, journal articles, working papers, conference proceedings, brochures for farmers, market information boards, etc. (Box 25.3). Once these documents are released to the public, little assessment is carried out to find out if they are being read and who is reading. However, citations by authors are a reflection of the articles being read and appreciated.

Fig. 25.6 Farmers learning during a field day organized in a demonstration site of a project funded by The Flemish Interuniversity Council Own Initiative (VLIR – UOS) Belgian Government on "Climate smart options allowing agricultural intensification for smallholders in the dry zones of the Central Highlands of Kenya"

Oral-related outlets This involves oral communication. They include conferences, local farmers' meetings, and training seminars whose main communication medium is oral. They are usually referred to as face-to-face communication. They have merit because of their interactive nature. The recipients of the information have an opportunity to comment and ask questions. On the other hand, the owner of the message has an opportunity to assess the level of awareness regarding the message being passed. Overall, it is one of the most effective mediums because of the ability to converse and exchange both verbal and nonverbal information.

Practical-related outlets This involves practicals where recipients see and participate practically. These may include setting demonstration sites in a public place or in farmers' fields, farmer field schools, agricultural shows, etc. It is an effective way of disseminating messages, especially to illiterate farmers. Even with literate farmers, practical-based outlets would still be useful as seeing and doing enhance retention and future use of the new practice. Observation of the practical application of new skills enhances confidence in adopting new techniques (Fig. 25.6).

ICT and social media Since the advent of Information Communication Technology (ICT) and the rise of social media, communication channels have significantly expanded. Platforms such as blogs, videos, webinars, radio, TV, university and organization websites, etc. offer diverse avenues of dissemination of information to a wide audience. However,

Table 25.2 Different clientele and forms of knowledge-sharing products

Clientele	Form of communication/knowledge sharing product
Academia and research institutions (Universities, Scientists)	Research reports, thesis, dissertation, working papers, journal articles, books and book chapters, review papers, conference papers, discussion papers, working and occasional papers
General public	Magazines, newspaper articles, mass media
Farmers	Leaflets, brochures, posters
Extension officers and other development workers	Extension manuals, training manuals, brochures
Policymakers	Policy briefs, briefing meetings, seminars
Employer (institution)	Annual reports, quarterly reports, research reports
Project donor	Baseline and project evaluation reports, annual reports, quarterly and annual reports, research reports

while ICT and social media offer extensive reach and accessibility, they may not always be effective in creating meaningful impact or promoting deep engagement with the audience.

25.2.3 Matching Messages to Beneficiaries

As we have mentioned, disseminating knowledge to beneficiaries requires information (Messages) and a channel to get the message across. Hence, we emphasize that our messages must be clear and matched with the clientele. Table 25.2 shows the different clientele benefitting from agricultural research and, most likely, forms of communication or knowledge-sharing products.

We note that research findings can be communicated using many products and avenues. The following sections provide details of research products and forms of communication that may be of interest to a postgraduate student in agricultural sciences.

25.2.3.1 Research Report

A research report is written to communicate findings mainly to donors or institutions. The scientists spearheading or involved in the research are expected to provide reports on the various research activities. Postgraduate students supported by donor-funded projects will be expected to contribute to the research report written by the principal investigator to the donor. For both donors and institutions, the format and guidelines are usually provided. Whatever the format, the research reports should provide background information, materials, and methods and mainly focus on the findings and implications of the research.

The donor community is mostly interested in the implications of the research and information that could be of direct use. Donors are usually also interested in the number of target groups reached at each point in time. The role of the postgraduate student will be to support the project deliverables and provide the required information to feed into the report.

25.2.3.2 Thesis and Dissertation

Thesis and dissertation are academic requirements for students undertaking postgraduate training under the guidance of a supervisor. Some universities and departments also have research projects instead of a thesis. The difference between a research project and a thesis in this context is the rigorousness and examination processes. Usually the *project is less rigorous* than a thesis and subjected to less external examination. All these academic documents require a student to carry out original research guided by a supervisor. As a student, it is important to select a topic and follow the guidelines provided by your supervisor (Chap. 3). Supervision is an engagement that takes place throughout the research period until you finally submit your final bound copy to the institution. Currently in most universities, the thesis and dissertations are put in library repositories for wide circulation. Therefore, it is important for a student to do quality research and write a good thesis that will not embarrass oneself and the institution. Because of the emphasis of the thesis laid in this book, there is a more detailed discussion of the thesis in Chap. 27.

25.2.3.3 Journal Article

Journal articles and conference papers are the primary forms of sharing research findings to other scientists. A journal paper is a scholarly publication that undergoes a rigorous review process before being published in a journal. Preparing a journal article follows standard procedures that aim to communicate findings objectively, clearly, and concisely. Work must be well justified and supported by a valid methodology. A more detailed explanation of the journal paper is given in Chap. 25.

25.2.3.4 Conference Paper

A conference paper is presented in a scientific meeting of scientists and academicians. It mainly disseminates findings to peers, allowing for an exchange of ideas, feedback, and critique from peers. The papers are usually published in conference proceedings, so dissemination can continue beyond those who attended the conference. However, it is ineffective in disseminating practical knowledge to broader audiences, such as practitioners and policymakers. Discussion about conferences is in Chap. 28.

25.2.3.5 Discussion Paper

The discussion paper is a scientific report presenting an interpretation of results and arguments of a specific subject matter. It aims to stimulate discussion, inform peers about new findings, raise awareness, and encourage further exploration of the subject. It could also aim at proposing solutions to an existing problem. You can choose a subject matter from your research work and document it as a discussion paper. The basis of selection will be the availability of evidence and the ability to provide critical insights, interpretations, and arguments.

25.2.3.6 Occasional Papers and Working Papers

As a postgraduate student, you may occasionally be required to contribute to an occasional paper or a working paper. This occurs mainly if you were funded by an institution or a donor-supported project to carry out your research work. These documents aim to explore and discuss a topic of interest to the donor or organization, especially providing new information from work carried out.

25.2.3.7 Books and Book Chapters

Books and book chapters are another rich area which you can use to share your new information. A book offers a comprehensive way of presenting research outcomes to a broader audience, from experts in the field to general readers interested in the topic.

However, writing a book from your thesis project may not be feasible. The thesis is a focused piece of research and is narrow in scope compared to books. Expanding this content into a stand-alone book may require significant additional research. Finding a publisher will also be difficult. Additionally, it takes great effort and is a major investment of time and effort (Day and Gastel 2012). A book chapter would be a better consideration. A "book chapter" is a focused research piece included in a book's larger edited volume. It explores a specific aspect of the book's broader topic, and each chapter serves as a building block that contributes to the volume's overall theme and objectives. As a researcher, you can dive deeply into a niche or subtopic within your field of expertise, showcasing your specialized knowledge and insights. The editors of the book volume often invite authors first to submit abstracts for review. After peer review, successful authors are invited to contribute to the book. Book chapters are also peer-reviewed, and hence, they are usually credible. Writing a book chapter is almost like writing a review paper (Day and Gastel 2012).

25.2.3.8 Extension Materials for Farmers

Farmers are one of the most important beneficiaries of our research. This is more so in Africa where most of the farmers are smallholders, experience problems of low yields, and lack easily accessible agricultural information. Moreover, farmers in Africa suffer from a lack of adequate technologies for boosting yields. For farmers to benefit from our research findings, they require fully processed, simplified, and ready-to-use information. The most common materials for packaging information for disseminating knowledge to farmers include handbooks, booklets, brochures, field guides, pamphlets, and posters. Some examples of extension materials for farmers can be found in the N_2 Africa project (www.N2africa.org) African Soil Health Consortium (www.ASHC.org). These dissemination materials mostly contain information on how to do things rather than theories of why they happen.

25.2.3.9 Training Manuals

You may need to develop a training manual depending on the research findings. Training manuals are useful for teaching specific topics to different clientele. The clientele for a training manual in agriculture is mostly farmers, farmer groups, or extension workers. The main aim of the training manual should be to provide in-depth information and instructions on how to implement a certain technique or method of farming to improve the performance of the agricultural enterprise. Formats and length of training manuals vary depending on clientele and purpose, but they are always structured to provide a specific skill. There are several examples of agricultural training manuals with different topics and reach (local, regional, and international).

25.2.3.10 Policy Briefs

Policy briefs are generally short documents containing focused information targeted at the policymakers. Policymakers make key decisions for society. They formulate policies. The main aim of a policy brief is to raise awareness of the status, opportunities, and challenges facing agriculture. An effective brief must present the problem to be addressed at the policy level and provide the solution emanating from the research you have carried out. Policymakers are not always technical people; therefore, you should aim at writing simplified direct statements in a summarized way, suggesting policy options and actions. Visuals such as tables, graphs, and charts can present data to enhance understanding and interpretation. This way, the policymakers are likely to read and thus impact policy formulation. The objective should be to persuade them to know and act. One-on-one meetings may be an effective communication method, but the policy brief can still be posted if time does not allow.

25.3 Reaching Farmers with Research Innovations

Farmers are a critical stakeholder in research, playing a key role in generating and utilizing research findings. A farmer's interest in receiving information from the research is driven by the need to address individual challenges in agriculture. Challenges vary across the value chain. These are encountered at both the production and marketing stages. At production, the main problems are soil management, input application, control of pests and diseases, and so on. On the other hand, marketing-related problems include low prices, poor access to markets, postharvest losses, etc. We use a case study from central Kenya where the research findings were disseminated to farmers through various ways.

25.3.1 Case Study from Kenya on Dissemination

This section presents a case study of a research and development project implemented in Kenya between 2013 and 2017. This project aimed to improve smallholder farmers' agricultural productivity and incomes by adopting yield-enhancing integrated soil fertility management practices (ISFM). The purpose was to scale up soybean and climbing beans in maize-based systems of central Kenya highlands. One of the main objectives was to disseminate and scale out ISFM practices among the smallholder farmers.
To facilitate dissemination, the project had a documentation and publicity committee.

The committee implemented the following key activities:

1. Developed *"a communication plan"* for the project period, outlining strategies and timelines for sharing information effectively.
2. Identified and coordinated the development of *knowledge-sharing products* tailored for different stakeholders, ensuring relevant information reached its intended audience.
3. Developed *ICT technologies for sharing information*, including a bulk SMS platform for direct communication with farmers and stakeholders and a Twitter platform hosted on the website
4. Prepared promotional materials such as *branded harvesting sacks and T-shirts* to increase project visibility and awareness.
5. Operationalized the communication plan to ensure seamless execution of dissemination activities throughout the project duration.

Key Dissemination Pathways Used

(a) Joint farmers' field days
At the start of the project, researchers established demonstrations consisting of a central "Mother Demonstration" plot and several on-farm demonstration sites. These served as platforms for testing and showcasing various agricultural practices and innovations directly to farmers. To enhance awareness and promote adoption, field days were organized at both the mother and on-farm demo sites. These events aimed to showcase and expose farmers to a range of soil fertility management options for their informed choice. In addition, the field days provided valuable opportunities for interaction between farmers and key stakeholders along the value chain.

(b) Demonstrations
Researchers used demonstrations as the primary learning platform for farmers. They were, therefore, established in accessible places such as schools, churches, and individual farms. Most of the demonstrations were facilitated by desk officers from the extension wing of the Ministry of Agriculture, ensuring technical accuracy and relevance. Feedback from farmers consistently highlighted demonstrations as one of the most preferred and effective training and dissemination pathways due to their practical, hands-on approach, and relevance to real farming conditions.

(c) Roadshows
Roadshows were strategically organized during market days. The objective of the roadshow was to sensitize farmers on the importance of soybeans and collect data to populate the SoCo Project e-Extension database. In attendance were members of project team and soybean agro-processors. The roadshow caravan used a multi-approach system and visited various locations including markets, schools, and local administrators in the three counties (Figs. 25.7 and 25.8). The strategy was inspired by a ripple effect model adopted by the project.

(d) Business forums with growers
Business forums with entrepreneurs and farmers were organized to promote collaboration. Specific agreements were reached during the forums to streamline the soybean value chain. During one of the forums, it was agreed that the entrepreneur would buy all the soybeans produced by the farmers under the condition that the grains would be clean. They also agreed on the prices, and both parties committed to formalizing the agreement by signing contracts. The business forums overall enhanced market opportunities for the farmers.

(e) Strategic partners
Strategic partner forums were organized with *key* local influencers. These were chiefs, church leaders, pastors, and other community stakeholders. The forums helped spread awareness of the technologies and enhanced collaboration between researchers and farmers. The partner's participation created a more receptive environment for adoption and feedback.

(f) Website
A website was developed to provide basic information about the project and enhance information dissemination. It also provided easy access links to report downloads.

(g) Social media
Social media was mainly used to reach and involve the youth in project activities. It included Facebook and X

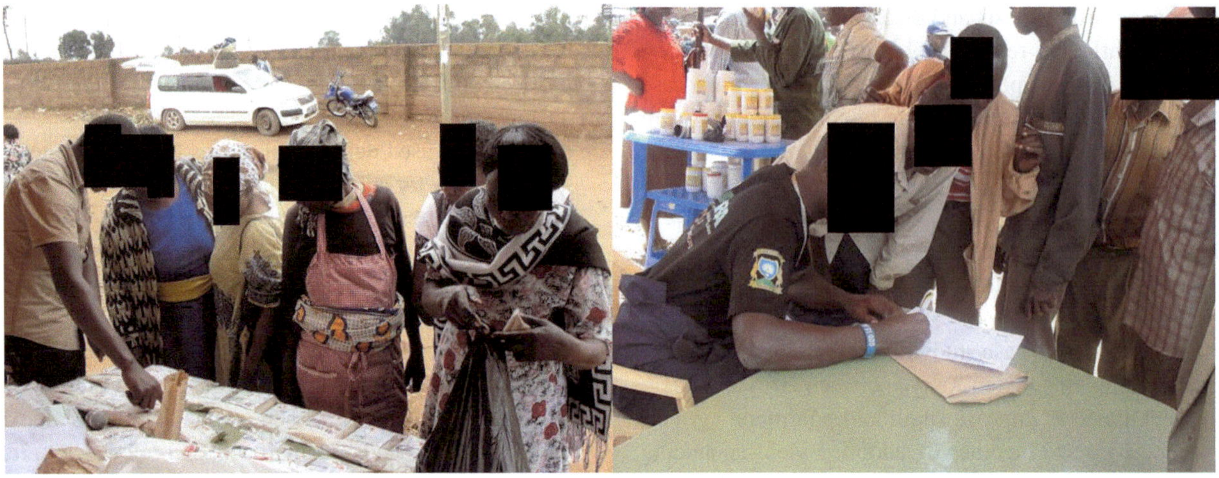

Fig. 25.7 Farmers attending roadshows and showing interest in one of the soybean products sold by entrepreneurs who collaborated with project partners

(Formerly Twitter) networks (Fig. 25.9). These have the advantage of being easy and affordable to build and manage social media platforms. However, it may not effectively reach the project's core beneficiary, the smallholder in rural central Kenya, as internet connectivity is unreliable in these areas.

(h) Bulk SMS

The project built a network of over 11,000 phone numbers for use on the platform within the project period. The bulk SMS was used to invite farmers to field days and training, reflecting the impressive attendance at these events.

(i) Branded materials: The project also prepared branded storage bags and T-shirts, which played a crucial role in raising awareness and significantly contributed to dissemination efforts (Fig. 25.10). Overall, at the end of the project, more farmers than initially targeted during project design were reached.

Fig. 25.8 Roadshow event comprising members of the project team accompanied by an entrepreneur

Fig. 25.9 Some information on posters to promote the use of social media in the project

Fig. 25.10 Students wearing branded T-shirts visit one of the participating farmers

Acknowledgments Most of the photos used in this chapter were derived mainly from two projects funded by the Alliance for a Green Revolution in Africa (AGRA) and implemented between 2010 and 2017. We also acknowledge and appreciate the stakeholders, including farmers, who participated in this project.

Further Reading Materials

1. Wilson PM, Petticrew M, Calnan MW, Nazareth I (2010) Disseminating research findings: what should researchers do? A systematic scoping review of conceptual frameworks. Implement Sci 5:1–16. https://doi.org/10.1186/1748-5908-5-91
2. Day RA, Gastel B (2012) How to Write and Publish a Scientific Paper, 4th edn. Cambridge University Press, Cornwall
3. Klingelhöfer S (2019) Rogers (1962): Diffusion of Innovations. 489–493. https://doi.org/10.1007/978-3-658-21742-6_115
4. Lövei GL (2021) Writing and Publishing Scientific Papers: A Primer for the Non-English Speaker. Writing and Publishing Scientific Papers 214. https://doi.org/10.11647/OBP.0235

Key Points of This Chapter

Why Disseminate?

Dissemination of your research findings is the last step in the research process and involves sharing information about your research outcomes and recommendations with your target audience so they can use it.

Strategic dissemination is the hallmark of creating a positive impact within the agricultural community and broader society. Through the effective sharing of information, technologies are more likely to be adopted, driving change and improvement.

Universities and Research institutions have departments that coordinate the generation of innovations and their dissemination to the relevant clientele.

Concept of Disseminating to Create Impact

Dissemination is not an end in itself but a means to create impact. Effective dissemination focuses on the adoption and impact of innovations to influence productivity.
A decision-making triangle shows how data are linked to application through a process of distilling raw data to knowledge and finally to end user to apply.

Adoption of Innovations

Adoption is the complete implementation of an innovation and is explained by several theories and models. The "*innovation diffusion theory*" is one of the oldest social science theories that is extensively used. It has four key elements that explain adoption. These are:

(v) The innovative technology or practice
(vi) Communication channels selected
(vii) Time
(viii) The social system

Adoption Potential of Agricultural Innovations

The adoption potential of a practice refers to the likelihood that its intended users will adopt a new technology. It helps design research that is relevant.

Dissemination and Adoption Research

Dissemination and adoption research is a systematic study of processes and factors that lead to widespread adoption. The outputs play a key role in developing effective dissemination strategies.

Elements of Effective Dissemination

Taking into consideration that we need to communicate to multiple audiences for impact, we have come up with seven elements that need to be considered in the dissemination of your research findings; they include the following:

1. Audience or receiver.
2. Message to be communicated.
3. Channel to be used.
4. Packaging of information.
5. Dissemination partners.
6. Dissemination strategy.
7. Measuring the effectiveness of dissemination.

Matching Messages to Beneficiaries

Disseminating knowledge to beneficiaries requires information (Messages) and a channel to get the message across. Hence, we should emphasize that our messages should be clear and match the clientele.
The main clientele targeted for disseminating research findings includes academics, the general public, farmers, extension officers, policymakers, institutions, and donors. Various pathways are used including journal papers, field days, policy briefs, workshops, etc.

Dissemination Pathways in Agriculture

Many pathways or channels for disseminating agricultural research findings exist, and they can be categorized into four main ones:
1. Written-related outlets.
2. Oral-related outlets.
3. Practical-oriented outlets.
4. Information Communication Technology (ICT) and Social media-related.

Disseminating Research Findings to Farmers: A Case Study from Kenya
To enhance the dissemination of research findings, the project used several pathways, including publicity committee communication plan, ICT technologies and branded materials, joint farmers' field days, demonstrations, roadshows, business forums, collaboration with strategic partners, and the use of a website and social media.

Review Questions

1. How can researchers ensure their findings are accessible and useful to farmers?
2. How can the concept of "laboratory to market" be implemented by universities?
3. Apply the innovation diffusion theory to explain how a new seed variety could be adopted by farmers.
4. Researchers have gathered extensive raw data on the water requirements of different drought-resistant bean varieties under varying soil conditions.
   (a) Using the concept of the "decision-making triangle," explain the journey of this "raw data" from its initial state to becoming "knowledge" that farmers can "apply" to improve their bean cultivation.
   (b) What transformations occur at each stage of the triangle?
5. A university research team has successfully developed a new bio-pesticide that significantly reduces potato blight. They published their findings in a high-impact international journal.
   (a) Explain why simply publishing in a journal is *not* sufficient for creating a "positive impact within the agricultural community and broader society."
   (b) What additional steps are implied by the term "strategic dissemination"?
6. The main clientele targeted for disseminating research findings includes academics, the general public, farmers, extension officers, policymakers, institutions, and donors. For each of the following "Dissemination pathways," identify which primary *clientele* is best matched for and briefly explain why:
   (a) Publishing a scientific journal paper on a new gene for disease resistance.
   (b) Organizing a "Farmers' Field Day" demonstrating new planting methods.
   (c) Issuing a "Policy Brief" on the economic impact of horticultural exports.
   (d) Hosting a "Workshop" for extension officers on integrated pest management.

Chapter Exercise

1. Suppose a researcher has developed a new, simple, and affordable method for enriching soil fertility for small-scale farmers in a certain region. The target audience includes farmers with varying literacy levels, some active in farmer groups, and others more isolated. Apply the concepts of Audience, Message, Channel, and Packaging from the seven elements of dissemination and:
   (a) Describe at least two distinct segments within your target audience of small-scale farmers.
   (b) For each audience segment, how would you adapt the *message* about the new method to be most relevant and understandable?
   (c) Suggest at least two *dissemination channels* (from the oral, print, practical-oriented, media platforms mentioned) that would be most appropriate for each segment, justifying your choices.
   (d) Identify at least three different types of partners you would seek out (e.g., government, NGOs, private sector, farmer organizations) to help disseminate this technology. Explain the specific role each partner could play.
   (e) After implementing your dissemination strategy for the method for one year, how would you go about Measuring the effectiveness of dissemination? Suggest at least three practical indicators or methods you would use to assess if your efforts are leading toward utilization (adoption).
2. In the case study described in Sect. 25.3, soybean technologies were disseminated using methods like "joint field days, demonstrations, roadshows, business forums, strategic partnerships, bulk SMS, branded materials, and social media."
   (a) For three of these specific methods (e.g., "Roadshows," "Business forums," "Bulk SMS"), describe the likely target audience (clientele) they were best aligned with, and explain why that matching is effective.

Answers to Review Questions

- Q1. Reference: 16.1.2
- Q2. Reference: 16.1.2
- Q3. Reference: 16.1.2
- Q3. Reference: 25.1.2
- Q4. Reference: 25.1.2
- Q5. Reference: 25.2.2
- Q6. Reference: 25.2.1, 25.2.2 and 25.2.3

References

Adolwa IS, Schwarze S, Buerkert A (2018) Best-bet channels for integrated soil fertility management communication and dissemination along the agricultural product value-chain: a comparison of northern Ghana and western Kenya. J Agric Educ Ext 24:435–456. https://doi.org/10.1080/1389224X.2018.1499541

Ahmed H, Ekman L, Lind N (2024) Planned behavior, social networks, and perceived risks: understanding farmers' behavior toward precision dairy technologies. J Dairy Sci 107:2968–2982. https://doi.org/10.3168/jds.2023-23861

Alene AD, Coulibaly O (2009) The impact of agricultural research on productivity and poverty in sub-Saharan Africa. Food Policy 34:198–209. https://doi.org/10.1016/J.FOODPOL.2008.10.014

Arslan C, Wollni M, Oduol J, Hughes K (2022) Who communicates the information matters for technology adoption. World Dev 158:106015. https://doi.org/10.1016/J.WORLDDEV.2022.106015

Autio A, Johansson T, Motaroki L et al (2021) Constraints for adopting climate-smart agricultural practices among smallholder farmers in Southeast Kenya. Agric Syst 194:103284. https://doi.org/10.1016/J.AGSY.2021.103284

Caulfield T, Ogbogu U (2015) The commercialization of university-based research: balancing risks and benefits. BMC Med Ethics 16:1–7. https://doi.org/10.1186/S12910-015-0064-2/PEER-REVIEW

Day RA, Gastel B (2012) How to write and publish a scientific paper, 4th edn. Cambridge University Press, Cornwall

Dearing JW (2009) Applying diffusion of innovation theory to intervention development. Res Soc Work Pract 19:503–518. https://doi.org/10.1177/1049731509335569

Etzkowitz H (2003) Research groups as 'quasi-firms': the invention of the entrepreneurial university. Res Policy 32:109–121. https://doi.org/10.1016/S0048-7333(02)00009-4

Eweoya I, Okuboyejo SR, Odetunmibi OA, Odusote BO (2021) An empirical investigation of acceptance, adoption and the use of E-agriculture in Nigeria. Heliyon 7:e07588. https://doi.org/10.1016/j.heliyon.2021.e07588

Fadeyi OA, Ariyawardana A, Aziz AA (2022) Factors influencing technology adoption among smallholder farmers: a systematic review in Africa. J Agric Rural Dev Trop Subtrop (JARTS) 123:13–30. https://doi.org/10.17170/KOBRA-202201195569

Franzel S, Coe R, Cooper P et al (2001) Assessing the adoption potential of agroforestry practices in sub-Saharan Africa. Agric Syst 69:37–62. https://doi.org/10.1016/S0308-521X(01)00017-8

Green LW, Ottoson JM, García C, Hiatt RA (2009) Diffusion theory and knowledge dissemination, utilization, and integration in public health. Annu Rev Public Health 30:151–174. https://doi.org/10.1146/ANNUREV.PUBLHEALTH.031308.100049/CITE/REFWORKS

Isgren E, Clough Y, Murage A, Andersson E (2023) Are agricultural extension systems ready to scale up ecological intensification in East Africa? A literature review with particular attention to the Push-Pull Technology (PPT). Food Security 15(5):1399–1420. https://doi.org/10.1007/S12571-023-01387-Z

Kamau CN, Majiwa EB, Otieno GO, Kabuage LW (2023) Intention to adopt improved indigenous chicken breeds among smallholder farmers in Machakos county, Kenya. Do socio-psychological factors matter? Heliyon 9:e22381. https://doi.org/10.1016/J.HELIYON.2023.E22381

Klingelhöfer S (2019) Rogers (1962): diffusion of innovations. 489–493. https://doi.org/10.1007/978-3-658-21742-6_115

Kondo K, Cacho O, Fleming E et al (2020) Dissemination strategies and the adoption of improved agricultural technologies: the case of improved cassava varieties in Ghana. Technol Soc 63:101408. https://doi.org/10.1016/J.TECHSOC.2020.101408

Kumar A, Takeshima H, Thapa G et al (2020) Adoption and diffusion of improved technologies and production practices in agriculture: insights from a donor-led intervention in Nepal. Land Use Policy 95:104621. https://doi.org/10.1016/J.LANDUSEPOL.2020.104621

Landmann D, Lagerkvist CJ, Otter V (2021) Determinants of small-scale farmers' intention to use smartphones for generating agricultural knowledge in developing countries: evidence from rural India. Eur J Dev Res 33:1435–1454. https://doi.org/10.1057/S41287-020-00284-X/TABLES/4

Mohr S, Kühl R (2021) Acceptance of artificial intelligence in German agriculture: an application of the technology acceptance model and the theory of planned behavior. Precis Agric 22:1816–1844. https://doi.org/10.1007/S11119-021-09814-X/TABLES/13

Moseley SF (2004) Everett rogers' diffusion of innovations theory: its utility and value in public health. J Health Commun 9:149–151. https://doi.org/10.1080/10810730490271601/ASSET//CMS/ASSET/CDE1A7BF-8EBC-42F9-8A36-BB8BB0FFAB87/10810730490271601.FP.PNG

Muchai SWK, Muna MWM, Mugwe JN et al (2014) Client focused extension approach for disseminating soil fertility management in Central Kenya. Int J Agric Ext 02:129–136

Mugwe J, Monicah AE, Ae M-M et al (2009) Adoption potential of selected organic resources for improving soil fertility in the central highlands of Kenya. Agrofor Syst 76:467–485. https://doi.org/10.1007/s10457-009-9217-y

Mwangi JM, Chitere PO, Kariuki JG (2022) Social inequalities influencing awareness of good Agricultual practices among smallholder farmers in Murang'a County in Kenya. J Agribus Rural Dev 65:189–195–189–195. https://doi.org/10.17306/10.17306/J.JARD.2022.01568

Njenga MW, Mugwe JN, Mogaka H et al (2021) Communication factors influencing adoption of soil and water conservation technologies in the dry zones of Tharaka-Nithi County, Kenya. Heliyon 7:e08236. https://doi.org/10.1016/J.HELIYON.2021.E08236

Nyasimi M, Kimeli P, Sayula G et al (2017) Adoption and dissemination pathways for climate-smart agriculture technologies and practices for climate-resilient livelihoods in Lushoto, Northeast Tanzania. Climate 5:63. https://doi.org/10.3390/CLI5030063

Ochuku B, Osborn TL, Nerima D et al (2023) Testing pathways to scale: study protocol for a three-arm randomized controlled trial of a centralized and a decentralized ("Train the Trainers") dissemination of a mental health program for Kenyan adolescents. Trials 24:1–14. https://doi.org/10.1186/S13063-023-07539-Y/TABLES/2

Osterling KL, Austin MJ (2008) The dissemination and utilization of research for promoting evidence-based practice. J Evid Based Soc Work 5:295–319. https://doi.org/10.1300/J394V05N01_11

Pingali PL (2012) Green revolution: impacts, limits, and the path ahead. Proc Natl Acad Sci 109:12302–12308. https://doi.org/10.1073/PNAS.0912953109

Rowley J (2007) The wisdom hierarchy: representations of the DIKW hierarchy. J Inf Sci 33:163–180. https://doi.org/10.1177/0165551506070706

Ruzzante S, Labarta R, Bilton A (2021) Adoption of agricultural technology in the developing world: a meta-analysis of the empirical literature. World Dev 146:105599. https://doi.org/10.1016/J.WORLDDEV.2021.105599

Rware H, Kansiime MK, Mugambi I et al (2021) Is radio an effective method for delivering actionable information for responding to emerging pest threats? A case study of fall armyworm campaign in Zambia. CABI Agric Biosci 2:1–11. https://doi.org/10.1186/S43170-021-00053-8/TABLES/5

Simpson BM, Franzel S, Degrande A, et al (2015) Farmer-to-farmer extension: issues in planning and implementation

Stevenson JR, Villoria N, Byerlee D et al (2013) Green revolution research saved an estimated 18 to 27 million hectares from being brought into agricultural production. Proc Natl Acad Sci USA 110:8363–8368. https://doi.org/10.1073/PNAS.1208065110/SUPPL_FILE/PNAS.201208065SI.PDF

Turon H, Wolfenden L, Finch M et al (2023) Dissemination of public health research to prevent non-communicable diseases: a scoping review. BMC Public Health 23:1–15. https://doi.org/10.1186/S12889-023-15622-X/TABLES/2

Waaswa A, Nkurumwa AO, Kibe AM, Kipkemoi NJ (2021) Communicating climate change adaptation strategies: climate-smart agriculture information dissemination pathways among smallholder potato farmers in Gilgil Sub-County. Kenya Heliyon 7:e07873. https://doi.org/10.1016/J.HELIYON.2021.E07873/ASSET/8EC60ED7-F4AD-4C70-A6A6-CF153F497127/MAIN.ASSETS/GR6.JPG

Wainaina P, Tongruksawattana S, Qaim M (2016) Tradeoffs and complementarities in the adoption of improved seeds, fertilizer, and natural resource management technologies in Kenya. Agric Econ 47:351–362. https://doi.org/10.1111/AGEC.12235

Wilson PM, Petticrew M, Calnan MW, Nazareth I (2010) Disseminating research findings: what should researchers do? A systematic scoping review of conceptual frameworks. Implement Sci 5:1–16. https://doi.org/10.1186/1748-5908-5-91

Structure of a Scientific Paper

26

Abstract

Scientific papers are vital for sharing findings with the scientific community. Scientists have adopted a standard logical structure called IMRAD; "Introduction, Materials and Methods, Results, and Discussion." It is a predictable and adaptable structure that aids authors in organizing ideas logically and in a standard manner. Following this framework, scientific papers can be structured into three key sections: The Beginning matter, the mid-matter, and the End matter. The beginning matter includes the title and other related sections, such as authors and keywords. The mid-matter contains the main content of the paper (IMRAD), while the end matter includes acknowledgements, references, and other sections. The introduction informs the reader of the relevance of the research study and is written in a "funnel arrangement". The materials and methods section is written in the past tense and represents what you did. The key elements in the methodology section include a description of the study area, research design, and methods of data collection and data analysis. The results section presents key findings with respect to the central research question, while the discussion presents explanations and interpretations. Statistics are an integral part of a scientific paper and should be reported clearly and accurately. A manuscript based on a qualitative research approach follows the IMRAD format, but differs by emphasizing detailed presentations, especially in the results section. In contrast, a mixed methods manuscript also adheres to the IMRAD structure, but must justify the use of a mixed methods design. Additionally, it should include characteristics of the design, and visual illustrations should be incorporated to enhance clarity and comprehension.

Keywords

Abstract · Co-authors, Corresponding author · Back matter: Beginning matter · Data availability statement · Discussion section · Front Matter · IMRAD · Introduction section · Journal paper · Manuscript history · Manuscript submission · Materials and Methods · Mixed methods paper · Peer review · Research ethics declaration · Results section · Statistics reporting · Title · Qualitative research manuscript · Quantitative research manuscript

Learning Outcomes

After reading this chapter, you will be able to:

1. Apply the scientific style of writing using the "IMRAD' Framework.
2. Appreciate the role of the "IMRAD' structure for journal articles.
3. Explain components of the "Beggining matter" of a journal article.
4. Demonstrate components of the "Mid-matter' of a journal article.
5. Illustrate components of the "Back-matter" of a journal article.
6. Generate types of titles that can be used in a journal article.
7. Illustrate how authors should be listed.
8. Develop key components of the results.
9. Generate a clear discussion of a scientific paper.
10. Illustrate how to present statistics in a scientific paper.
11. Describe key considerations in writing a qualitative approach-based journal paper.
12. Explain factors to consider in reporting mixed-method research.

26.1 Foundations of Scientific Writing

Scientific papers are key for sharing research findings with the scientific community. This diverse community comprises fellow researchers, scholars, and professionals in various fields. Therefore, the presentation and packaging of research papers must adhere to the conventions of scientific style.

J. Njeri Mugwe, S. Runo, *Research Methodology in Agricultural Sciences*, https://doi.org/10.1007/978-981-95-1892-0_26

Scientific style, as defined by (Day and Gastel 2012; Lövei 2021), is characterized by *clarity*, *precision*, *objectivity*, and *rigor* in language and presentation. It emphasizes using formal language and standardized formats to effectively convey research findings. Furthermore, scientific writing prioritizes *logical organization* and adherence to *empirical evidence* (Matthews and Matthews 2014).

The logical organization is a priority, and scientists have adopted a standardized format known as IMRAD, which stands for Introduction, Methods, Results, and Discussion (Sollaci and Pereira 2004). This framework has been widely adopted across many disciplines, offering a clear and logical organization for presenting research findings (Okokpujie et al. 2018).

Empirical evidence is crucial in scientific writing, providing robust support for hypotheses and conclusions. The IMRAD framework guides authors to present this evidence clearly and without bias. Authors must explain their methods thoroughly, making it easy for others to understand and replicate their findings. Objectivity is key, allowing readers to judge the evidence for themselves. The use of precise language further enhances the clarity of the presented details.

The IMRAD structure began to be used in 1940 and was adopted as the standard in 1970 (Sollaci and Pereira 2004; Youdeowei et al. 2012). Since then, it has continued to be popular in many disciplines. It has gained popularity because of several benefits. It benefits the *readers, authors, editors,* and *reviewers* in the following ways:

1. Predictable structure: The clear structure helps readers know where to find information in a paper, making scientific writing consistent and easier to understand.
2. Help authors organize ideas: It organizes complex findings, making it simpler for researchers to share their work and for readers to grasp it.
3. Allows quick review: Helps editors and reviewers quickly assess manuscripts, as there are clear criteria for assessment.
4. Adaptable structure: Due to its adaptable structure, it has proven effective in addressing the challenges posed by the exponential growth of scientific research. It is not a cookie-cutter approach; it has room for creativity and innovation (Wu 2011).
5. Used for a long time: IMRAD has been widely used for a long time across different fields and countries, so people are familiar with it.
6. Enhances reproducibility: It facilitates critical comparison between studies and supports reproducibility.

Overall, IMRAD's simplicity, consistency, and adaptability make it a go-to structure in the scientific community (authors, editors, reviewers, and readers). We can think of IMRAD as a farm that is organized and productive. The readers are like getting a good meal from the farm, clear, organized, and ready to be eaten and digested (Sollaci and Pereira 2004). Authors are the farmers, carefully planting and nurturing their seeds into bountiful harvests. Editors are the farm managers overseeing the whole process, ensuring a healthy crop. Reviewers are seasoned agronomists checking for pests (errors) and weeds (flaws) to guarantee a bumper harvest of knowledge.

26.2 The IMRAD Framework

The IMRAD's framework structures your report to say, "Here's *the problem, here's how I studied it, here's what I found, and here's what it means.*" In addition to the IMRaD sections, the paper has other minor components. Here, we subdivide the structure into three main sections: "Front matter," "Mid-matter," and "Back matter". In the Front matter, we have the Title, Authors, authors' affiliations, Abstract, and Keywords, while at the end of the paper, we have acknowledgments, authors' contribution statement, ethical considerations, declaration of conflict of interest, a statement on data availability, references, and appendices (Fig. 26.1).

26.2.1 Shape of IMRAD

The IMRAD structure follows an "hourglass" shape because articles start broad and general, get narrower and more specific in the middle, and return to a general discussion at the end (Sollaci and Pereira 2004; Wu 2011). The introduction first presents a broad problem that a wide range of audiences can understand. An example is "*climate change is one of the main challenges causing declining crop yields in the semi-arid areas.*" After this first broad statement, a selected literature review should be included to show how you connect the broad to theoretical and disciplinary level questions and allow you to narrow it to specific operational questions. The literature review in this section creates the logical path between the big question (broad and general) and the little question (specific). For any broad question, there are a number of ways it could be investigated. The other role of literature is to guide you to filter and arrive at the specific question that can be investigated through research (Fig. 26.2).

IMRAD is a pattern, like organizing seeds in neat rows, to systematically structure and present research findings. It does not represent all the components of a scientific paper (Wu 2011). While it lays the groundwork for organizing the four key sections, additional elements, such as references, abstracts, and acknowledgments, contribute to the complete-

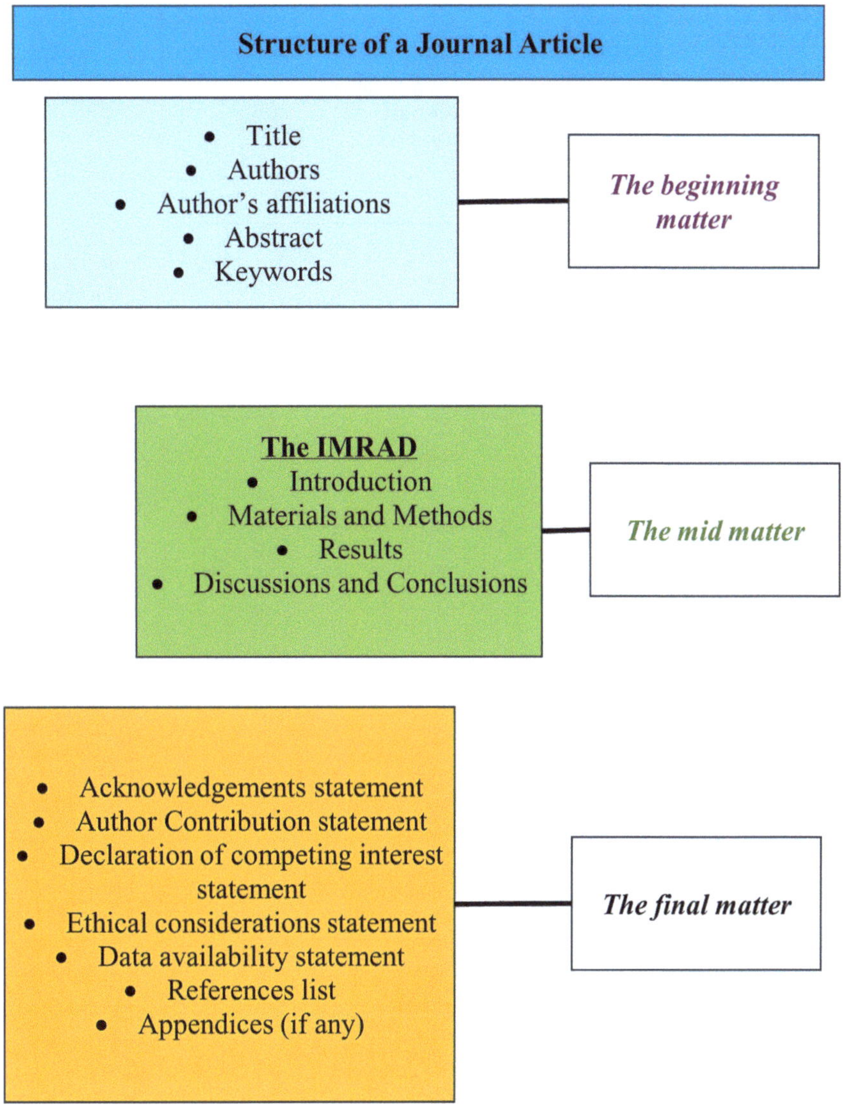

Fig. 26.1 Components of a journal article showing the "beginning, mid, and final matter"

ness of the research paper. Much like a seed that needs water, sunlight, and nutrients to thrive, a research paper requires these supplementary components to flourish and provide a comprehensive understanding of the study.

However, researchers often adapt and modify the IMRAD to suit the specific requirements of the specific study. For example, a qualitative paper will maintain the basic structure of IMRAD, but the author might modify and include other components, such as a detailed exploration of the study area and a conceptual framework to provide context and theoretical underpinning (discussed later in this chapter). This flexibility allows the researcher to tailor the presentation of the findings, ensuring clarity and coherence when communicating the research outcomes. This attribute renders IMRAD a preferred framework, solidifying its enduring presence in scholarly writing (Wu 2011).

The next section discusses the three main parts of a scientific paper, starting with the Beginning matter, the End matter, and the Mid matter (The main part of the paper).

26.3 The Beginning Matter

26.3.1 The Title

The title is a crucial element. It is the first thing readers encounter and often determine whether they will read further. On average, for every 10 readers who look at the title, one reads the abstract, which increases the probability of readers reading the full paper (Nair and Nair 2014). A good title should, therefore, be *informative, appealing,* and *attention-grabbing*.

It should accurately reflect the paper's content and include relevant keywords to facilitate searchability in databases and search engines (Day and Gastel 2012). As an author, you want your article to be read, cited, and recognized within your field. On the other hand, you are writing for a busy reader. By capturing the essence of the study in a concise yet compelling manner, a title not only entices potential readers but also enhances the discoverability of the research in an

Fig. 26.2 Annotated presentation of the IMRaD structure of a scientific paper, in a "hourglass" model

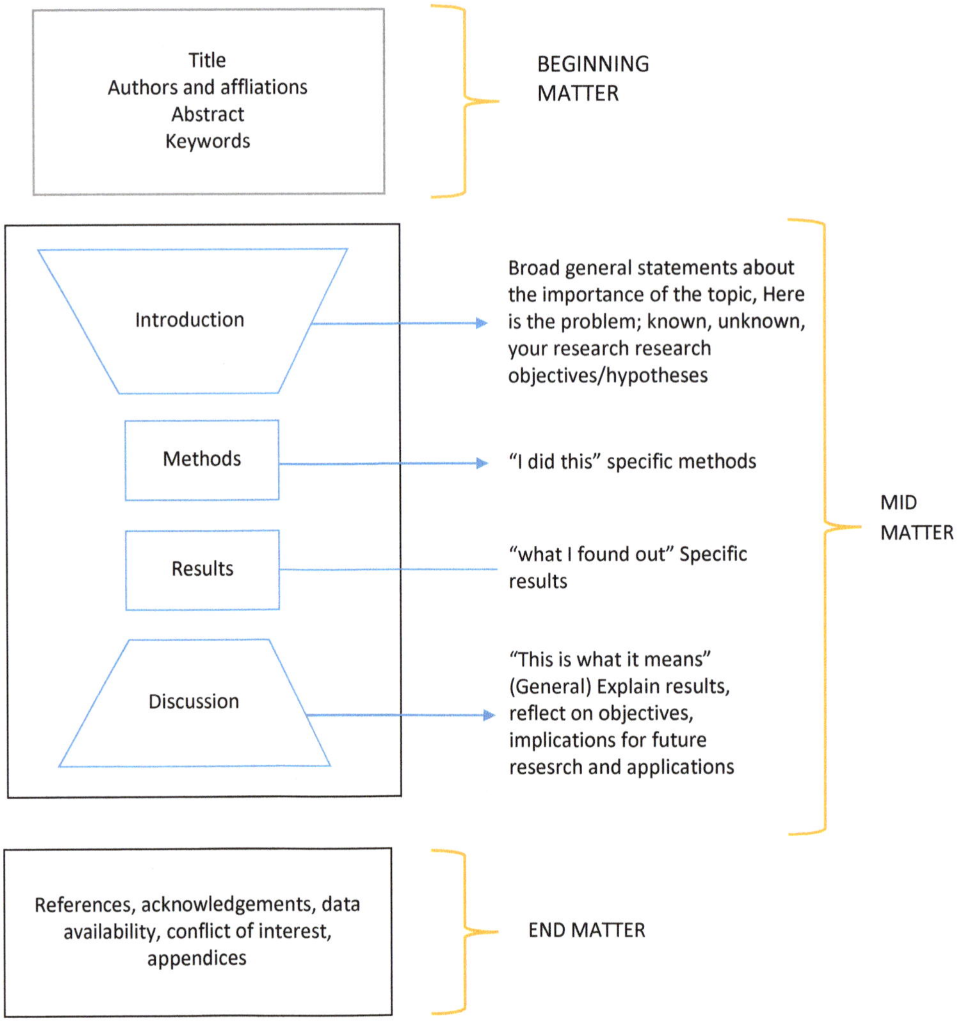

ever-expanding digital landscape. Prioritizing keywords and phrases ensures prominence and accessibility, guiding busy readers directly to the core of your research.

Additionally, the title should be clear and free of jargon or overly technical language to ensure accessibility to a wide audience. A good title for a journal paper possesses several key characteristics:

- *Reflective of content*: The title should accurately represent the content and findings of the paper.
- *Informative but concise*: accurately conveys the main topic or focus of the research but is succinct.
- *Engaging:* It should grab attention and be compelling, encouraging readers to explore the paper further.
- *Clear with no jargon:* should clearly and effectively communicate the main message of the paper, avoiding ambiguity and excessive technical terminology.
- *Avoid "waste words":* should not use like "an investigation of…, A study of…."
- *Contain no abbreviations:* should never contain abbreviations.
- *Memorable*: leaves a lasting impression on readers; use creative language.

Unlike a thesis, where your institution can restrict the type of title used, you can choose from three types of titles in writing a journal paper. The first type is the "Indicative title", which focuses on the subject matter of a paper.

Examples:

- Effectiveness of rock phosphate in increasing soybean yields in the central highlands of Kenya.
- Socio-economic factors influencing the adoption of orange flesh sweet potatoes in Kakamega County, Kenya.
- Adoption potential of selected organic resources for improving soil fertility in the central highlands of Kenya (Mugwe et al. 2009).

The second type of title is "Informative Title." This one focuses on results achieved and helps show the reader what to expect as they read the document. The title communicates concisely the scope and main topic of the paper.

Examples

- Staggered maize-legume intercrop arrangement robustly increases crop yields and economic returns in the highlands of central Kenya (Mucheru-Muna et al. 2010).
- Intercropping optimizes soil temperature and increases crop water productivity and radiation use efficiency of rainfed potato (Nyawade et al. 2019).
- Revisiting intercropping indices with respect to potato-legume intercropping systems (Gitari et al. 2020).
- Enzymatically treated yeast bolstered growth performance of broiler chicks from young broiler breeders linked to improved indices of intestinal function, integrity, and immunity (Maina et al. 2022).

The third type is a *question type* presented as a question. It poses the research question addressed by the study. Such question-type titles are common in research articles based on qualitative or exploratory studies. They have the advantage of engaging the reader and creating interest.

Examples

- Can a combination of UAV-Derived vegetation indices with biophysical variables improve yield variability assessment in smallholder farms? (Adewopo et al. 2020).
- Does rock phosphate increase soybean yields in Humic Nitisols of the central highlands of Kenya?

26.3.2 Authors

Authorship is another important consideration. Who is the author? According to Yan and Snow (2020a), an author is a person who makes a primary contribution to and holds responsibility for the data, concepts, and interpretation of results of the published work. This means that each author should have participated sufficiently in the work to take full responsibility for the content.

According to Vancouver guidelines on authorship, the following qualifies an author:

- Substantial contribution to the paper. Make a major contribution to the concept and design, analysis, and interpretation of the work.
- Participate in drafting the article and revising it critically for intellectual content.
- Participate in final approval of the version to be published.
- Accept responsibility as well as credit for the work.
- Be able to present, discuss, and defend the work, analysis, and discussion.

The *first author* or senior author of a manuscript plays a major role in manuscript preparation, submission, and revision. The main roles are:

- Always responsible for putting the paper together and takes the primary responsibility.
- Decides the key messages of the paper in consultation with the co-authors.
- Supervises data and ensures results are interpreted correctly.
- Submits manuscript to a journal and deals with correspondence.
- In charge of original documents including material documents and data.

On the other hand, *co-authors* hold a major responsibility in terms of making intellectual contributions. Specifically, co-authors do the following:

- Make early decisions about the key message of the paper.
- Keep paper on track in terms of key messages and content by reviewing and giving input.
- Make intellectual contributions to data analysis, interpretation, and content.
- Review each draft.
- Take public responsibility for the content and results.

Any individual who has contributed to the study in some way but does not meet the criteria for authorship should be mentioned under acknowledgement. Such tasks include acquisition of funds, laboratory technicians, research assistant, statistical or technical advice, donor, photography, assistance with data entry and analysis, and participants (respondents). Acknowledgements should specify their role or contribution.

Avoid including the head of a department as an author because they were involved in the administration processes of submission. This can be seen as just gifting them as a favor with no due contribution. Such inclusion will result in your paper being regarded as having "ghost authors."

Corresponding Author

In multiple authorship, one of the authors is designated the corresponding author (Yan and Snow 2020b). This person manages all communications regarding the article and signs agreements for all the authors. The journal editor communicates with the corresponding author. Depending on your agreement, your supervisor can be the corresponding author. However, it is beneficial for the student to be the corresponding author to learn.

Listing of Authors

The listing of authors or ordering in a journal paper is crucial and should reflect each individual's contributions to the study (Yan and Snow 2020a). Typically, authors are listed in order of importance based on their contribution to the research. In the case of a manuscript derived from a graduate thesis, the student who conducted the research takes primary responsibility and is usually listed as the first author. Supervisors and other individuals who have made significant contributions to the study, such as providing guidance, resources, or collaboration, are included as co-authors. A common practice is to have the name of the person who oversaw the work, e.g. the principal investigator of a research project, placed at the end.

Determining the order of authors in a research paper can sometimes result in disagreements or confrontations, particularly when multiple individuals have made significant contributions to the study. Therefore, all authors must discuss and agree on the sequence of authors before submitting the paper for publication. This approach ensures fairness, transparency, and proper recognition of each author's contribution to the research. Ultimately, the order of authors should be decided unanimously and reflect the level of input of each individual to the project.

Number of Authors

The number of authors listed on a journal paper can vary greatly depending on the complexity and scope of the research. Longer studies with extensive collaborations may have over 20 authors, while shorter papers typically have fewer authors who made substantial contributions. However, it's crucial to ensure that each author's involvement and contribution are clearly defined to maintain the research's integrity. Including too many authors without clear roles can raise questions about the validity of the findings. Day and Gastel (2012) give an example of a paper with 27 authors and 12 paragraphs and questions how those authors could all fit in one small lab.

26.3.3 Abstract

An abstract is a summary of a scientific paper. It is a condensed version of the full paper, usually 150–300 words, and usually one paragraph placed at the beginning of the paper. The purpose of the abstract is to inform the reader about the content of the article. This helps the reader decide whether to read the article further. The abstract should be accurate, complete, and concise and follow the structure of the paper.

The abstract should be "definitive" rather than "descriptive' (Nair and Nair 2014). A definitive abstract provides facts and content rather than merely describing something. This means that instead of stating that something is described, the abstract should directly present the content itself. It should offer a succinct summary covering its purpose, methodology, main findings, and implications while avoiding convoluted and lengthy sentences. By staying clear and to the point, the abstract communicates the core of the study without burdening the reader with excessive information. This enables readers to swiftly grasp the importance of the research and decide whether to explore the complete paper further for additional details.

An abstract contains six (6) main components as follows (Table 26.1):

1. *Component 1:* Broad topic and the research problem: Brief background highlighting the research question (1 or 2 sentences).
2. *Component 2:* Main aim and purpose of the study: Objective of the paper (1 or 2 sentences).
3. *Component 3:* Research methods used: Summary of methods used to collect and analyze data (brief and include design).

Table 26.1 Six components of an abstract from the paper titled: Adoption Potential of selected organic resources for improving soil fertility in the Central Highlands of Kenya (Mugwe et al. 2009)

Component 1: Topic and general problem of the study	Soil fertility decline is the major cause of declining crop yields in the central highlands of Kenya and elsewhere within the African continent. This paper reports a study conducted to assess the adoption potential of two leguminous trees, two herbaceous legumes, cattle manure, and *Tithonia diversifolia* either solely applied or combined with inorganic fertilizer, for replenishing soil fertility in the central highlands of Kenya
Component 2: Why is the aim of the study?	The study examined biophysical performance, profitability, feasibility and acceptability, and farmers' experiences in managing and testing the inputs
Component 3: What research methods were used?	The study was based on a series of studies incorporating both sociological and experimental approaches for two and a half years
Component 4: What were the key results?	Results of on-farm trials showed that manure+fertilizer and tithonia+fertilizer treatments increased yields by more than 100% above the control. These treatments were the most profitable, having the highest net benefits and benefit-cost ratios. They were also the most commonly preferred by farmers who used them on larger plots compared to the other inputs
Component 5: What were the conclusions?	In conclusion, cattle manure and tithonia were found to be the organic materials with the highest adoption potential for soil fertility improvement in this area. *Calliandra calothyrsus* and *Leucaena trichandra*, on the other hand, have potential for use as animal fodder. The herbaceous legumes had the least adoption potential due to poor performance recorded on the farms, which possibly led to low farmer preference. However, issues of sustainable seed production could have played a role
Component 6: What were the practical recommendations	This study recommends some policy issues for enhancing adoption and research issues focusing on exploring strategies for increasing biomass production and use efficiency on farms

4. *Component 4:* Key results: key results in tandem with the objective of the study.
5. *Component 5:* Conclusions: Main conclusions filling the gap of the study.
6. *Component 6:* Practical recommendations: Generalized practical application.

Box 26.1 and Table 26.1 show an example of an abstract: Paper in Agroforestry Systems journal (Mugwe et al. 2009).

A good abstract should:

- Be short and to the point, usually ranging from 150 to 300 words, depending on the journal's requirements.
- Be complete and follow the IMRAD format and cover all the components of the study: introduction, objective, methods, results, and conclusions.
- Contain no references except in exceptional cases, e.g. extended abstracts.
- Written in past tense.
- Single paragraph except in rare cases.
- Do not contain verbose, lengthy sentences.

26.3.4 Keywords

Keywords are a list of important words used in the text and the abstract but not already in the title. They help readers search for literature. When indexing the published article in electronic databases keywords are included with the title and the abstract. There should be 3–10 keywords arranged in alphabetical order.

Selecting optimal keywords for your manuscript involves striking a balance, as the aim is to enhance the discoverability of your paper. You should avoid using an excessive number of single words, which can lead to overly broad search results. Conversely, including too many keywords or lengthy phrases may overly restrict the search, potentially reducing the visibility of your paper. Aim for a concise yet comprehensive selection of keywords that accurately represent the content and scope of your research. See the example of a paper Mugwe et al. (2009) with six (8) keywords (Box 26.2) comprising words and short statements. These keywords are short and specific, and other papers in this field can be easily located.

Box 26.2: Keywords in the Paper in Agroforestry Systems Journal; Title: Adoption Potential of Selected Using Organic Resources for Improving Soil Fertility in the Central Highlands of Kenya (Mugwe et al. 2009)

Keywords Biophysical performance · *Calliandra calothyrsus* · *Crotalaria ochroleuca* · Economic returns · Feasibility and acceptability On-farm trials · *Leucaena trichandra* · *Mucuna pruriens*

Avoid general words like maize or agriculture because your paper will get lost among a plethora of similar words. Again, if you are too specific, it will be difficult to find the paper. Choosing a combination of general and slightly specific words is best. For example, in a paper on soil fertility

Box 26.1: Scanned Sample Abstract

Received: 25 April 2008 / Accepted: 31 January 2009 / Published online: 15 March 2009
© Springer Science+Business Media B.V. 2009

Abstract Soil fertility decline is the major cause of declining crop yields in the central highlands of Kenya and elsewhere within the African continent. This paper reports a study conducted to assess adoption potential of two leguminous trees, two herbaceous legumes, cattle manure, and *Tithonia diversifolia* either solely applied or combined with inorganic fertilizer, for replenishing soil fertility in the central highlands of Kenya. The study examined biophysical performance, profitability, feasibility and acceptability, and farmers experiences in managing and testing the inputs. The study was based on a series of studies incorporating both sociological and experimental approaches for two and a half years. Results of on farm trials showed that manure + fertilizer and tithonia + fertilizer treatments increased yields by more than 100% above the control. These treatments were the most profitable having highest net benefits and benefit cost ratios. They were also the most commonly preferred by farmers who used them on larger plots compared to the other inputs. In conclusion, cattle manure and tithonia were found to be the organic materials with the highest adoption potential for soil fertility improvement in this area. *Calliandra calothyrsus* and *Leucaena trichandra*, on the other hand, have potential for use as animal fodder. The herbaceous legumes had the least adoption potential due to poor performance recorded on looking at the "effect of different levels of phosphorus on maize productivity", you can choose keywords like available P, maize grain yield, stover yield, levels of P; instead of soil fertility and maize. You can also test your selected keywords to assess their effectiveness in retrieving your article. Ask yourself if your article is difficult to find using these keywords. Does the search yield results closely align with your research topic? Testing your keywords allows you to refine and optimize them for maximum impact.

26.3.5 Manuscript History

Most manuscripts nowadays contain manuscript history (Fig. 26.3). This history typically includes initial submissions, revisions, resubmissions, and publication history. This helps provide context for the manuscript journey from the initial draft to the final publication. This information is useful when you are selecting your target journal and want to check on how long it is likely to take to publish an article (see Chap. 27 on journal selection).

26.4 The End Matter

References
Crafting a proper reference list is often left until the final stages of preparing a manuscript, as it requires meticulous attention to detail. This process is especially time-consuming because no universal format for citing literature exists (Nair and Nair 2014; Madler and Renwick 2025). With numerous reference, citation styles corresponding to different journals, each publication has its unique requirements. While software programs can assist in formatting reference lists, they cannot cover every aspect, as the format must comply precisely with the journal's guidelines. Therefore, authors must carefully review and ensure that all references are accurately cited, both within the text and in the reference list, to ensure adherence to the specific style and enhance the overall coherence and professionalism of the manuscript.

Acknowledgements
The acknowledgements section is a brief but significant space for expressing gratitude to organizations and individuals who contributed to the research. It is important to acknowledge donors who provided funding for the work and stakeholders and farmers who supplied valuable data or support. Beyond financial assistance and the sources of data, this section allows authors to recognize the various forms of assistance, guidance, or resources received during the study. For instance, authors may thank the laboratory technicians, statisticians, field workers, etc. You do not usually thank the supervisors here if they are part of the authorship. However, you may include them in acknowledgement if you prepare a manuscript that they have not contributed to qualify to be authors, but you feel they helped you somehow in building the study.

Conflict of Interest Statement
The declaration of interest at the end of a paper is a simple statement intended to disclose any potential conflicts of interest that could influence the research or its interpretation. Typically, it employs a statement such as "The authors declare no conflicts of interest" to indicate that there are no financial, personal, or professional affiliations that could bias the study's findings or conclusions.

Research Ethics Declaration
The research ethics declaration formally affirms adherence to ethical standards throughout the research process. It provides a concise overview of the ethical considerations addressed during the study, including the approval or oversight obtained from institutional review boards or ethics committees. This declaration is critical for research involving human subjects, animals, or sensitive data, ensuring that ethical guidelines are followed rigorously.

Reputable journals demand that researchers include a **research ethics declaration**. This declaration demonstrates the research's alignment with established ethical principles, guidelines, and regulations. Failure to include this declaration may prompt editors to request its addition, underscoring its significance in scholarly publishing. In our many years of publishing experience, we have not included an ethical clearance certificate for surveys involving farmers. But recently (2023), we were asked to provide an ethical clearance and signed consent form. This indicates that recent trends will register a growing expectation for researchers to provide evidence of ethical approval or clearance, even for seemingly gentle studies like farm surveys.

Data Availability Statement
The data availability statement demonstrates accountability, integrity, and transparency. It informs editors, reviewers, and readers about the data's accessibility, enabling validation, replication, and further research. It also promotes open science. Data availability statements may be presented in one or more of the following formats:

1. This article does not involve the generation or analysis of datasets; hence, data sharing is not applicable.
2. If data are included, indicate that all data generated or analyzed in this study are included within the published article and its supplementary information files.
3. The datasets produced or analyzed during the current study can be found in the XXX repository, accessible via the YYY (provide link).
4. Upon reasonable request, the datasets utilized in this study are available from the corresponding author.

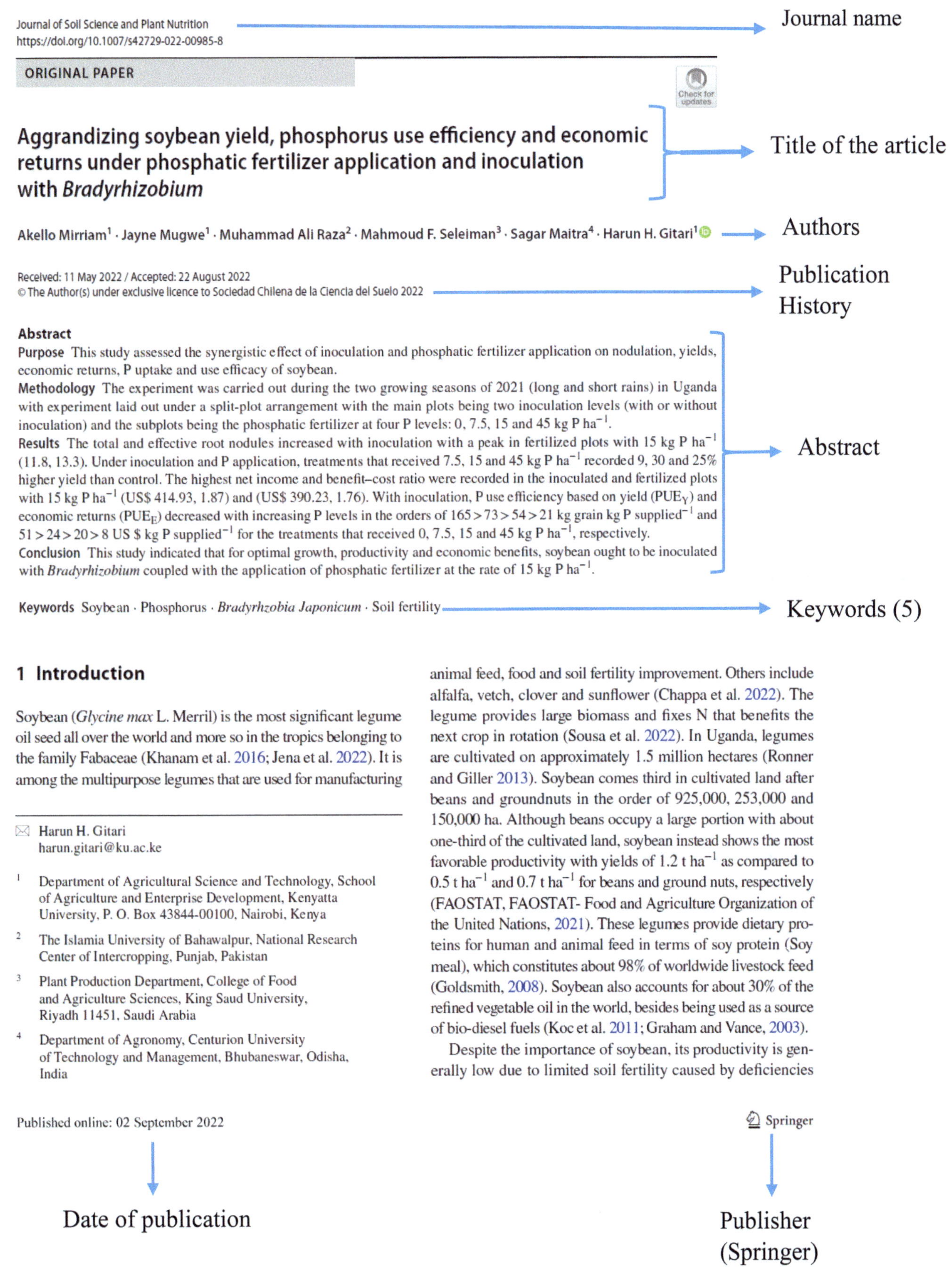

Journal of Soil Science and Plant Nutrition
https://doi.org/10.1007/s42729-022-00985-8

ORIGINAL PAPER

Aggrandizing soybean yield, phosphorus use efficiency and economic returns under phosphatic fertilizer application and inoculation with *Bradyrhizobium*

Akello Mirriam[1] · Jayne Mugwe[1] · Muhammad Ali Raza[2] · Mahmoud F. Seleiman[3] · Sagar Maitra[4] · Harun H. Gitari[1]

Received: 11 May 2022 / Accepted: 22 August 2022
© The Author(s) under exclusive licence to Sociedad Chilena de la Ciencia del Suelo 2022

Abstract

Purpose This study assessed the synergistic effect of inoculation and phosphatic fertilizer application on nodulation, yields, economic returns, P uptake and use efficacy of soybean.

Methodology The experiment was carried out during the two growing seasons of 2021 (long and short rains) in Uganda with experiment laid out under a split-plot arrangement with the main plots being two inoculation levels (with or without inoculation) and the subplots being the phosphatic fertilizer at four P levels: 0, 7.5, 15 and 45 kg P ha^{-1}.

Results The total and effective root nodules increased with inoculation with a peak in fertilized plots with 15 kg P ha^{-1} (11.8, 13.3). Under inoculation and P application, treatments that received 7.5, 15 and 45 kg P ha^{-1} recorded 9, 30 and 25% higher yield than control. The highest net income and benefit–cost ratio were recorded in the inoculated and fertilized plots with 15 kg P ha^{-1} (US$ 414.93, 1.87) and (US$ 390.23, 1.76). With inoculation, P use efficiency based on yield (PUE_Y) and economic returns (PUE_E) decreased with increasing P levels in the orders of 165 > 73 > 54 > 21 kg grain kg P supplied^{-1} and 51 > 24 > 20 > 8 US $ kg P supplied^{-1} for the treatments that received 0, 7.5, 15 and 45 kg P ha^{-1}, respectively.

Conclusion This study indicated that for optimal growth, productivity and economic benefits, soybean ought to be inoculated with *Bradyrhizobium* coupled with the application of phosphatic fertilizer at the rate of 15 kg P ha^{-1}.

Keywords Soybean · Phosphorus · *Bradyrhzobia Japonicum* · Soil fertility

1 Introduction

Soybean (*Glycine max* L. Merril) is the most significant legume oil seed all over the world and more so in the tropics belonging to the family Fabaceae (Khanam et al. 2016; Jena et al. 2022). It is among the multipurpose legumes that are used for manufacturing animal feed, food and soil fertility improvement. Others include alfalfa, vetch, clover and sunflower (Chappa et al. 2022). The legume provides large biomass and fixes N that benefits the next crop in rotation (Sousa et al. 2022). In Uganda, legumes are cultivated on approximately 1.5 million hectares (Ronner and Giller 2013). Soybean comes third in cultivated land after beans and groundnuts in the order of 925,000, 253,000 and 150,000 ha. Although beans occupy a large portion with about one-third of the cultivated land, soybean instead shows the most favorable productivity with yields of 1.2 t ha^{-1} as compared to 0.5 t ha^{-1} and 0.7 t ha^{-1} for beans and ground nuts, respectively (FAOSTAT, FAOSTAT- Food and Agriculture Organization of the United Nations, 2021). These legumes provide dietary proteins for human and animal feed in terms of soy protein (Soy meal), which constitutes about 98% of worldwide livestock feed (Goldsmith, 2008). Soybean also accounts for about 30% of the refined vegetable oil in the world, besides being used as a source of bio-diesel fuels (Koc et al. 2011; Graham and Vance, 2003).

Despite the importance of soybean, its productivity is generally low due to limited soil fertility caused by deficiencies

✉ Harun H. Gitari
harun.gitari@ku.ac.ke

[1] Department of Agricultural Science and Technology, School of Agriculture and Enterprise Development, Kenyatta University, P. O. Box 43844-00100, Nairobi, Kenya

[2] The Islamia University of Bahawalpur, National Research Center of Intercropping, Punjab, Pakistan

[3] Plant Production Department, College of Food and Agriculture Sciences, King Saud University, Riyadh 11451, Saudi Arabia

[4] Department of Agronomy, Centurion University of Technology and Management, Bhubaneswar, Odisha, India

Published online: 02 September 2022

Springer

Fig. 26.3 Paper in journal of Soil Science and Plant Nutrition showing key components of the first-matter of a journal article; title, authorship, publication history, abstract, keywords, authors' address and correspondences, date of publication, and publisher (Mirriam et al. 2022). (Permission for re-use granted-licence number 6035360492481)

5. Although the datasets generated or analyzed during the current study are not publicly accessible due to ZZZ (Give reason why data are not public), they can be obtained from the corresponding author upon reasonable request.

Appendices

Appendices in journal articles are not common. However, author can include additional information that supports the main text in appendix that may be too detailed to include in the main body of the paper. Appendices often present supplementary data, detailed methodologies, lengthy tables, complex mathematical derivations, or extensive literature reviews. While not typically featured prominently, appendices can enhance the comprehensiveness and clarity of the research presented in the article. They also allow authors to maintain the focus and flow of the main text while still providing interested readers with access to relevant supplementary material. Consider using them if you have relevant materials that you feel including them in the main text will obscure the message in the paper.

26.5 The Mid Matter of a Journal Article

The "mid-matter" of a journal paper comprises the main body of the research, where the methodology, results, discussion, and analysis are presented. It details the study's objectives, research design, data collection, findings interpretation, and implications. The section that follows presents the following sections: introduction, materials and methods, results, discussion, and conclusions.

26.5.1 The Introduction Section of a Paper

The purpose of the introduction section is to inform the reader of the relevance of your work. A good introduction, therefore, should grab readers' attention and stimulate interest. It contains the following:

- *Background information*: defines the nature and extent of the problem.
- *Brief of previous work*: a brief review of the literature, but only that which is clearly pertinent and relevant to the problem.
- *Related previous work to the current problem*: summarize results and conclusions of previous work and link it to justification.
- *Justification:* Gives reasons for carrying out the study, i.e. justification. An explanation of why the work being described as needed.
- *Statement of the problem*: Is the gap being filled or the question being answered? (Why was the study carried out?)
- *Objectives/study question or hypothesis:* placed at the end of the introduction.

In the introduction, you want to answer three pertinent questions: 1. "What is the problem, and why is it important?", 2. What is the rationale for carrying out this study (implications, consequences, and potential impacts of the problem), and 3. What is the purpose of writing the paper?

Writing Introduction Section

In writing the introduction of a research paper, it is crucial to lead the reader from a *general understanding of the topic* to the specific focus of the study. The structure of the introduction fits a funnel shape, flowing from broad to specific (Fig. 26.4). This is accomplished by commencing with a comprehensive summary of the subject matter, which offers necessary context and background information (known matters). The introduction then gradually narrows down to emphasize the research questions, knowledge gap, or problem statement the study aims to address. This funnel-shaped transition ensures clarity and coherence by moving from the *broad to the specific.*

The last two paragraphs, or the introduction's final paragraph of the introduction, should clearly articulate the problem statement, knowledge gap, or questions the study aims to solve. This section serves as a bridge between the general introduction and the specific objectives or hypotheses of the study. Here we state the precise objectives or hypotheses being investigated and provide a clear roadmap for the reader. In the case of a qualitative study, however, you can indicate the *central question of the phenomenon.*

Depending on what is used, we may have endings with the following words:

1. The objective of this study was to...
2. This study tested the hypothesis...
3. This study sought to answer the following question. If you have more than one objective, present them in a logical order
4. The central question in this study was to....
5. The key research question...

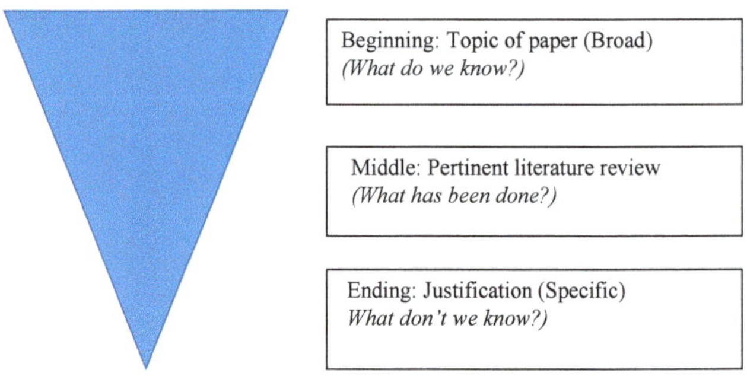

Fig. 26.4 Funnel arrangement of writing the introduction section of a journal paper

Funnel Arrangement of Introduction

Here, we demonstrate funnel arrangement using a sample paper: 1. "Adoption potential of selected organic resources in the central highlands of Kenya" (Mugwe et al. 2009).

Paper 1 adopted from Mugwe et al. (2009).

The introduction has four paragraphs and starts broadly with known information and ends with specific information (unknown). It highlights land intensification and poor soil fertility as serious biophysical constraints threatening food security (Box 26.3).

Box 26.3: The Introduction Starting Broadly

Introduction

Background

The intensification of agriculture in Africa has not been accompanied by sufficient inputs of nutrients through biological N_2 fixation, organic materials and mineral fertilizers to match the outputs of nutrients through harvested products and losses. As a result, poor soil fertility has emerged as one of the greatest biophysical constraint to increasing agricultural productivity hence threatening food security in the African continent. Annual soil loss is high, estimated at 22 kg N ha^{-1}, 5 kg P ha^{-1} and 15 kg N ha^{-1} (Smaling et al. 1997). To reverse the nutrient depletion in African soils, there has been renewed interest among researchers to promote the use of organic resources, with several authors reporting the materials ability to substitute or supplement mineral fertilizer (Vanlauwe et al. 2002; Bationo et al. 2004).

The midsection of the paper, consisting of two paragraphs, supports the constraint with statistics and reduces the information to explain what had been done by other authors regarding addressing the constraint. It gives a solution of using organic resources, but argues that this has not been successful due to low adoption among the farmers. Authoritative citations are given to support the arguments.

The last sections of the introduction (two paragraphs) narrow down the justification for carrying out the study, where the *knowledge gap* is highlighted (Box 26.4). The knowledge gap is the statement of the problem. The authors state that most studies done on organic resources have concentrated on biophysical aspects, and few have determined adoption potential among smallholder farmers. A further justification is that the assessment of adoption potential is

Box 26.4: Last two paragraph highlighting gap as scarce information regarding the multifaceted nature of adoption potential and stating the objectives of the study and clearly showing the questions to be answered by the study (Mugwe et al. 2009)

Most other studies have looked at biophysical performance (e.g., SSSEA 2003) and recommendations based on technologies that have the ability to achieve high crop yield response while ignoring evaluations by farmers and the economic implications of the technologies to the farmers. Few studies, so far, have examined these practices under farmers' management especially to determine their adoption potential. Information on adoption potential is important for improving the effectiveness of the research-development continuum.

Assessment of adoption potential by farmers is multifaceted, requiring an understanding of both biophysical performance under farmers' circumstances, profitability from the farmers' perspectives

Agroforest Syst (2009) 76:467–485

and its acceptability by farmers (Franzel et al. 2002). Such an assessment requires an interdisciplinary approach incorporating both sociological and experimental approaches. This paper adopts such an approach and examines adoption potential of three agroforestry trees, two herbaceous legumes, cattle manure and inorganic fertilizer under farmers' conditions. The study sought to answer the following questions:

- Biophysical performance: what are the effects of applying the soil fertility replenishment inputs on maize yields under farmers conditions?
- Profitability: what are the profitability levels of the soil fertility replenishment inputs introduced?
- Feasibility and acceptability: what inputs do farmers prefer and what are the reasons behind their preferences? How are the farmers managing the inputs on their farms and what are their perceptions about the benefits accrued? What are the constraints that farmers experience as they use the inputs and what opportunities exist for addressing the constraints?

multifaceted, thus requiring sociological and experimental approaches. The last part of the last paragraph raises critical gaps in terms of questions and states the objective of the study, which was to assess the adoption potential of selected organic resources.

26.5.2 Materials and Methods Section

The materials and methods section is usually the easiest to write and written in past tense because it reports what you did. *Description of what was done:* It describes "What was done," how," "where," and "when." The effective methodology section gives sufficient detail to allow any other scientist to repeat the study.

Key Elements in the Methodology Section

The study area: This section describes the study in terms of location, population, soil, climate, etc. It is good to include a map.

The research design: The research design section is extremely important and should be detailed (See details in Chap. 4 on proposal writing).

Methods of data collection: In a study involving experimental design, this section explains the experimental design used, the observations made, and how they were made (Sollaci and Pereira 2004; Wu 2011). The sampling procedure; sampling frame, and sample size should be well explained in a study involving surveys. This section also reports how data were handled and analyzed. This should include descriptive, inferential analysis, and *statistical methods.*

Data Handling and Analysis

The statistical analysis section is usually contained within the Materials and Methods section (Sollaci and Pereira 2004). The section must be written in a way that allows a reader with a reasonable appreciation of statistics to understand how the statistical analysis was conducted. This allows the reader to assess its suitability for the study being described. It is usually simplest when straightforward analytical procedures have been used.

The procedures should be described clearly and completely. Simply saying that "The data were analyzed by GenStat" is insufficient. Always state what kind of statistical analysis was done, e.g. ANOVA or Regression. See the checklist for the methods section (Box 26.5).

Box 26.5: Checklist for the Methods Section

1. Are the methods appropriate, described clearly to allow repetition?
2. Statistical design and analyses—are they appropriate and correct?
3. Experiment: is the design indicated? Are treatments well explained? Are replicates correct, and are significant statements justified?
4. Are the methods described clearly to allow repetition?
5. Survey and qualitative study: clarity of sampling procedure, appropriateness of the sample size and design. Are methods for sampling explained? Is sample size justified?
6. Are data analyses appropriate, correct, and described fully?
7. Are statistical tests correctly applied?
8. Is statistical design clear and appropriate?
9. Is data interpretation correct?

26.5.3 The Results Section

This results section is the core of the paper. Mainly contains three things:

- Text, which is used to draw attention to the main findings.
- Data summaries presented in tables, graphs, charts, etc.
- Relevant statistics.

Key Elements and Presentation Style of Results

General format.

In writing the results, pay attention to include only results that answer the question of what happened (Youdeowei et al. 2012). *You present key findings with respect to the central research question.* Explanations and interpretations are in the discussion. Organize the results by objectives. In mixed-method research, first present the quantitative data followed by the results of qualitative data.

Text

1. *Main results of the study and other trends:* Present the study results but do not attempt to interpret their meaning. You should not present the raw data you collected

but the summarized data, which should be presented in text, tables, and/or figures. Start by presenting in the text the main highlights of the results of your study, then refer the reader to a table or figure where they can see the data for themselves, e.g. "the number of mature cattle negatively influenced adoption with one unit increase of cattle reducing adoption by a factor of 0.6 (Table xxx)".

2. *Concise reporting style:* When presenting the results, start with the most general ideas and relationships, e.g. "the application of nitrogen had a significant effect on maize yields ($p = 0.01$; Table xxx)". After this statement, expand this relationship e.g. maize yields increased by 120% in treatments with the application of organic matter. Do not use sentences like "the ANOVA showed significant differences, or the regression showed." It is better to interpret the results directly and maintain conciseness by using direct statements to report the findings, e.g. "maize yields were higher in tithonia and manure treatments than in fertilizer treatment".
3. *Avoid reporting casual observations:* Do not give results that are not accounted for in Materials and Methods (casual observations do not belong to results, such as raw data). The checklist (Box 26.6) can guide you as you write your results.

Box 26.6: Checklist for Results Section

1. Do the results give answers to your objective?
2. Are there answers for everything you mentioned in the methodology to have been done?
3. Did you attempt to interpret results (answer should be no)?
4. Did you include any raw data or intermediate calculations (answer should be no)?
5. Did you present same data more than once?
6. Are the data presented related to materials and methods?
7. Do not repeat all values in the table, figure as text; Give key highlights.
8. Have you used to appropriate presentation; table/figure?
9. Have you used past tense to describe your results (it is about what you found out)?
10. Are all tables and figures mentioned in the text?
11. Are captions self-contained? What, when, where; Giving variables.
12. Are results interpreted intelligently?
13. Is there a logical flow of data presentation?
14. Are error bars, standard deviations, or confidence intervals included where appropriate?

4. *Account for all data:* Ensure that all variables reported to have been measured in the methodology are accounted for in the results.

Data

5. *Use one presentation method:* Do not include the same data in both a table and a figure. It is best to present the data in a table unless visual information can be gained by using a figure. For example, a figure is useful for reporting a regression analysis (line graph), or comparing several treatment levels (bar graph with error bars). (see detailed information in Chap. 23 on data visualization).
6. *Tables:* Tables are flexible and are commonly used to present any data. Data with multiple variables aiming to show numerical and exact values ideally fit a table presentation. Comparisons and summary statistics are well suited for tables. It is recommended to arrange the dependent variables (the ones you measured) and the independent variables (treatments or categories) as columns. This format makes it easier for readers to interpret relationships, trends, and summaries.
7. *Figures:* Most common figures include line graphs, bar graphs, and pie charts (see detailed presentation in Chap. 23). We give highlights here: Line graphs best show continuous variables, especially trends, showing how data changes over time and how variables are related. Bar charts are good when comparing discrete items, such as categorical data. Pie charts, on the other hand, are used to show proportions; for example, it is good for displaying a number of farmers with different land sizes.
8. *Tables and Figures captions:* Table and figure captions refer to titles or legends in the tables and figures. They should be complete, referred to as "self-contained." This means they should clearly state what is in the table (what, when, and where). Usually, a table's legend appears above it, while the legend for a figure appears below the figure.
9. *Avoid repetition of data presentation:* In reporting the results, emphasize the main point and do not repeat all the information contained in figures or tables. Allow the tables and figures to handle the details (see Checklist in Box 26.9).
10. *Reference to tables and figures:* You must refer to each figure or table you include in your paper in the text and always introduce them in the text before presenting them.
11. *Tenses:* Use past tense to present what has already happened. We use statements like, Results showed…
12. *Organization:* Use logical order throughout the paper. This enhances readability.

Table 26.2 Example of table presentation including *p*-value and LSD

Treatment	Stover yield (t/ha)		Grain yield (t/ha)	
	2012 LR	2012 SR	2012 LR	2012 SR
Sole maize	11.73	8.14	4.95	4.58
Maize-soybean (1 M:1S)	12.64	7.74	5.49	5.16
Maize-soybean (2 M:2S)	13.12	7.62	6.11	5.62
Maize-soybean (2 M:4S)	8.89	4.91	4.05	3.48
Maize-soybean (2 M:6S)	5.26	4.45	2.74	3.16
P-value	0.0001***	<0.0001***	<0.0001***	0.0467*
$LSD_{(0.05)}$	2.59	1.31	0.87	1.79

Source: Matusso (2015) *** p=< 0.001 ; * p=<0.05

Statistics

13. *Make proper use of statistics:* This is because statistics is a tool that assists you in achieving the objectives of your study. The most important thing is what a particular statistic means. Statistics should be included in the tables and graphs ensuring that each statistic presented is relevant and contributes to the overall narrative of your research.
14. *Correct interpretation of statistics:* Always interpret statistics correctly and intelligently. It is advisable, therefore, to revise your statistical books before interpreting and drawing conclusions about your data.
15. *Statistics in an experimental study:* The main statistics presented in an experiment are the summary tables with inferential statistics mainly from ANOVA. Always include p values, plus either SED or LSD (see example-Table 26.2).
16. *Descriptive and inferential statistics in a social/survey study:* Both descriptive and inferential statistics may be presented in a survey. Descriptive statistics are important in summarizing the data from respondents, a kind of characterization, while inferential statistics make conclusions, test hypotheses, and infer about populations (Chaps. 19 and 20).

26.5.4 Discussion Section

26.5.4.1 What to Include in the Discussion Section

The discussion section is the most challenging to write, but it is the most useful in terms of passing new messages and answering questions posed in the study. It requires the greatest amount of thought and logical thinking and takes a lot of effort in thesis and paper writing, and many scientific papers are rejected because of poor discussion.

The discussion section is related to the results section. In the results section, you gave highlights and explained the trends. In the discussion, you review the results and assign meaning based on the objectives and purpose of the study.

The following are important in the discussion section:

- Objectives: Relate the results to the questions that were set out in the introduction. Here consider the objectives of the study: Discuss the data/results in the context of the justification and objectives.
- Implications: What is the significance of the results? What did the results mean (implies, indicates).
- Explanations: Why did they come out the way they did (Attributed to, could be explained).
- Interpretation: Interpretation of the results in light of other work in the field. Here you demonstrate your ability to analyze, evaluate, interpret, and reason effectively. Discuss agreement or contrast to previously published work.
- Principles: Look at principles, relationships, and generalizations: these are most important in terms of wider application by the international audience.
- Recommendations: suggests future research that is planned or is needed to follow up the results.
- It brings the results together by interpreting and analyzing the data in the context of the study objectives and existing literature. It places the findings within the broader field, highlighting the implications and suggesting directions for future research.
- Data: No new data or results: Make sure you do not present any new data in the discussion and do not present data that were not included in the results (Box 26.7).

Box 26.7: Checklist for a Discussion Section

- What are the major patterns in the observations?
- What are the relationships, trends, and generalizations among the results?
- What are the exceptions to these trends, patterns, or generalizations?
- What are the likely causes (mechanisms) underlying these patterns resulting in predictions?
- What is the significance of the results?
- How do the results compare with the theoretical expectations? (Citations needed).
- Is there agreement or disagreement with previous work?
- What is the relationship of the present results to the original questions and hypotheses? Interpret results in terms of background laid out in the introduction and literature review.
- What is the implication of the present results for other unanswered questions in this field?
- What are the things we now know or understand that we didn't know or understand before the present work? Include the evidence or line of reasoning supporting each interpretation.
- What is the significance of the present results?

26.5.4.2 Organization of the Discussion Section

Organization of the discussion section requires clarity and chronological presentation. In most discussion sections, the authors start reviewing the main results and connecting this to the hypothesis and purpose of the paper. Discussion should start with a statement of the study's major findings; it should be a direct declaration of the study results. Example of words that may be used: i) Our results confirm that: ii) Our findings suggest…iii) The results confirm the…. Processes behind the explanations should follow this and also compare with other results.

Ending the discussion requires a proper conclusion that brings all studies together. This should highlight the possible application of the study or possible future research. Sometimes we look at the organization of a discussion as the opposite of the introduction. In the introduction, with a funnel shape, we start broadly and end specifically. In the discussion, we bring in specific studies and finish with a broad conclusion for application to a wider scale.

As stated in the introduction section, reflection on objectives and hypotheses guides the flow of the discussion.

26.5.4.3 Conclusions and Recommendations

The section on conclusions and recommendations presents inferences, deductions, implications, and general statements. We aim to give generalizations that can be applied to populations in the study area and other similar areas. They should appropriately answer specific questions posed at the beginning of the study and in the order they were presented. They should be brief short and concisely written (Box 26.8).

Box 26.8: Checklist for Conclusions and Recommendations

1. What problem was posed in the study? (refer back to the problem).
2. What is the strongest and most important statement you can make from your observations?
3. What are the new observations, interpretations, and insights resulting from the present work?
4. What are the broader implications of your results?
5. What is remedial action to solve the problem?
6. What further research is needed to fill in gaps in our understanding?
7. What are the directions for future investigations on this or related topics?

Finally, as you finish the manuscript, reflect on the whole paper to ensure coherence from beginning to end. Review each section guided, ensuring that each part connects seamlessly to the next. You can use Table 26.3 for a quick check.

26.6 How Do You Present Statistics in a Quantitative Scientific Publication?

Statistics are integral to scientific publications, enabling researchers to analyze data rigorously and draw meaningful conclusions (Lövei 2021). They should be reported clearly and accurately. Consider the following in reporting statistics in a scientific publication:

- *Matching materials and methods with results:* Match the reporting of your statistics in the results section with what you presented in the materials and methods section. In the materials and methods section, you reported how data were analyzed.
- *Use effective illustrations:* Present data using tables and figures when appropriate. Ensure that tables and figures are easy to read and interpret. Clearly label your statistical tables and figures to indicate the presented variables or data. Include descriptive captions that summarize the main findings or interpretations.

Table 26.3 The manuscript from beginning to end

Section	What to include
Title	Nature of the study
Authors	Individuals who wrote the paper
Keywords	Important words that can be used for searching
Abstract	Short summary <300 words
Introduction	Justification for the paper, research problem, knowledge gap, objectives, hypothesis
Materials and methods	How the study was carried out
Results	What the study found out
Discussion	Meaning of the results; interpretation, implication
Conclusions	Conclusions and recommendations to the clientele
Acknowledgements	Individuals and institutions who helped you
Authors contribution statement	Role of the authors in writing of the paper
Ethical considerations	Ethics considered to protect anonymity and confidentiality in studies involving humans and animals (avoidance of harm)
Declaration of competing interest	A statement indicating any conflict of interest
References	All references cited in the text. Use a consistent style as guided by the journal
Appendices	Supplementary materials

- *Report descriptive statistics with caution:* Exercise caution when reporting descriptive statistics. Although it is vital to include relevant statistics to summarize key aspects of the data, presenting every numerical value can make the text cluttered and distract the reader from the main findings. It is better to focus on presenting the most important descriptive statistics that support your research objectives and conclusions. Journals emphasize inferential statistics because they are essential for drawing conclusions and making inferences about the population from which the sample was drawn.
- *Confidently report inferential statistics*: Report inferential statistics (e.g. t-value, F-value, P-value, degrees of freedom, and effect sizes) focusing on the research objectives and main comparisons. Ensure you provide enough detail for readers to understand how the analyses were conducted and interpreted. Additionally, report effect sizes (magnitude of the phenomenon) to help readers determine the practical significance of the results, beyond the statistical significance (See practical aspect in Chap. 20). This approach ensures that readers can assess the statistical significance and magnitude of the observed effects, enhancing the credibility and applicability of the research findings.
- *Report confidence intervals:* Provide confidence intervals when reporting estimates of parameters (such as means or proportions). Confidence intervals provide a range of plausible values for the population parameter and convey the uncertainty associated with the estimate.
- *Be selective in reporting statistics:* Present statistics selectively to avoid overwhelming readers with too much numerical detail. Focus on the key findings and use statistics to support your main conclusions rather than including every single statistical test conducted. Consider including supplementary materials, such as, appendices or online supplements for extensive statistical analyses or additional details. This allows readers to access additional information without cluttering the main text.
- *Review journal guidelines:* Before submitting a manuscript, it is important to carefully review the author's instructions to ensure that it adheres to the specific formatting and reporting guidelines provided by the target journal, including any requirements for presenting statistics.

26.6.1 Data Interpretation

Interpretation takes the results of analysis, makes inferences pertinent to the research relations studied, and draws conclusions about these relations. When scientists interpret data, they try to explain the patterns and trends uncovered through the analysis. You will use your experience and skills while relating data to existing scientific ideas.

1. Connect the study's results in both theory and application by combining the conceptual/theoretical framework, the literature review, and the study's potential significance for application.
2. Examine, summarize, interpret, and justify the results; then, draw inferences. Enables the reader to get the total picture of the findings in summarized form, and helps orient the reader to the discussion that follows.

26.7 Size of the Journal Manuscript

Students frequently seek guidance on the optimal length of a journal paper. Generally, original empirical full papers typically range from 7000 to 12,000 words. Additionally, students often inquire about the distribution of words among different sections. Assuming an average length of 10,000 words, this distribution varies but commonly comprises:

- Abstract: Approximately 250–300 words (2.5–3% of the total word count).
- Introduction: Approximately 1000–1500 words (10–15% of the total word count).
- Materials and Methods: Approximately 1500–2000 words (15–20% of the total word count).
- Results: Approximately 2500–3000 words (25–30% of the total word count).

- Discussion: Approximately 2500–3000 words (25–30% of the total word count).

Additionally, the number of tables and figures should be about 4–6. References typically range from 30 to 50, depending on the scope and depth of the paper. This breakdown aims to present research comprehensively while maintaining reader engagement and clarity.

Avoid having excessive references to enhance the readability of the paper. Too many references can detract from the readability and flow of the paper, making it cumbersome for readers to navigate. Moreover, too many references may indicate a lack of focus or originality in the research. This may suggest that the author relied heavily on existing literature without offering substantial contributions of their own.

26.8 Writing a Manuscript from Qualitative and Mixed Methods

The foregoing discussion predominantly focused on a manuscript from a quantitative perspective. While there are numerous similarities between quantitative and qualitative and mixed-methods research methodologies, key differences exist. When writing a quantitative paper, emphasis is placed on presenting numerical findings, statistical analyses, and objective interpretations of data. In contrast, a qualitative paper explores individuals' underlying meanings, motivations, and experiences and presents rich, descriptive data. A mixed methods paper, on the other hand, integrates both quantitative and qualitative approaches. This requires authors to navigate between different paradigms and methods of inquiry to provide a comprehensive understanding of the research phenomenon. The next section gives some guidelines to consider as you write journal manuscripts originating from these other designs.

26.8.1 Manuscript from Qualitative Research

The IMRAD format described in this chapter is a generic structure for writing journal articles that forms the basis for writing any scientific manuscript. The qualitative paper varies as it attempts to present details. While the qualitative paper will include sections like introduction, methods, results, and discussion, the main variation is often observed in the results section (Korstjens and Moser 2018). In qualitative research, the results section takes on a distinct character compared to quantitative studies. Here, the emphasis shifts toward exploring complex phenomena, understanding contexts, and capturing the perspectives and experiences of participants. Rather than presenting numerical data and statistical analyses, the results section of a qualitative paper delves into rich, descriptive findings, often presented through quotes, themes, or narratives.

The introduction of a qualitative research paper should provide a brief overview of the manuscript, including the research question, research purpose, and the overarching aim (Korstjens and Moser 2018). The researcher should also include a statement justifying the research question and the reason for using qualitative research approaches. In addition, just like a quantitative paper, it should include a relevant literature review from the specific topic under study, informing the readers what other authors have done on the topic and clearly showing the knowledge gap.

In the methods section, a researcher should aim to present a comprehensive overview of their research design and approach. First and foremost, the researcher should highlight the research design, articulating the rationale behind its selection and clearly demonstrating its appropriateness for the study. If more than one approach/method is used, all should be mentioned and their complementarity explained. It is good to include the philosophical underpinnings that align with the chosen qualitative approaches.

The methods section should also include various critical components to thoroughly understand the research process (Wu 2011; Lövei 2021). First, it is essential to address ethics approval, outlining the body responsible for granting clearance and detailing how anonymity, confidentiality, and avoidance of harm to participants were upheld throughout the study. Second, the choice of participants and the recruitment mode should be described, including the sampling strategy employed and any gatekeepers involved. The rationale behind participant selection and recruitment methods should be articulated, elucidating how the sample size was determined. Finally, the methods of data collection should be presented in detail, with careful attention paid to the processes involved and the specific methods utilized.

The section on data analysis must be articulated clearly to enable readers to evaluate the trustworthiness and rigor of the analysis (Nair and Nair 2014). This section should elucidate the analytical approaches employed, establishing a connection with the previously mentioned data collection methods. Analytical techniques such as content analysis, thematic analysis, or other qualitative methods should be detailed, outlining the steps taken to derive meaningful insights from the data. Additionally, researchers should describe any coding software utilized in the analysis process, along with the procedures for coding, theme identification, and category development. Ensuring a transparent and coherent account of the data analysis process enhances confidence in research findings among readers.

26.8.1.1 Results

As always, the results section is the "gist of any research paper" and should be planned and written carefully, linking it to the overall aim of the study. The results should be connected to the research question, following the logic of the theory and methods using consistent concepts.

In your plan, use suitable subheadings guided by the themes and categories to enhance clarity for the reader. In each subsection, present the theme and support it with descriptive narratives, direct quotes, and participant excerpts.

26.8.1.2 Discussion

In the discussion section, as usual, start by repeating the research question and stating the main results to help the reader understand easily. First, interpret your findings and relate them to the research question while discussing the themes and how they contribute to the existing literature. Next, state how your findings relate to earlier research, including studies from both quantitative and qualitative realms. Also, mention how the findings can be generalized to other situations albeit with limitations. A detailed checklist for writing a qualitative paper is presented in Box 26.9.

Box 26.9: Checklist for Authors Writing Papers from Qualitative Research

Introduction

- Research question is well-articulated.
- The significance of the research question is justified, linking it to existing knowledge, including empirical research, theoretical framework, and policy considerations.
- The process of obtaining ethical and/or research/institutional governance approval is detailed and referenced.
- Any specialized terminology used in the manuscript is defined.

Method

- The rationale for selecting the specific method is explained.
- Criteria for choosing participants are outlined and justified.
- Recruitment methods are clearly explained.
- Information is given about individuals who chose not to participate, and reasons are given.
- The research sample and research setting are described.
- The process of obtaining informed consent from the participants is explained.
- Methods for ensuring participants' anonymity and confidentiality are explained.
- The method of data recording (e.g., audio or video recording) and the process of transcribing data are described.
- Research methods are detailed, and examples are included (e.g., observation guides).
- Sampling and the decision to conclude data collection are explained and justified.
- Data analysis and verification are detailed, including who performed them.
- Techniques for identifying/extrapolating themes and concepts from the data are discussed.
- Ways of checking the trustworthiness of data and process of triangulation described.

Results

- Adequate data are provided to enable readers to evaluate whether the data support the interpretation.
- Outlying or deviant cases that do not align with the main interpretation are presented.
- The applicability of the research findings to other contexts is discussed.
- Results are contextualized within the framework of previous research and relevant social theories.

Discussion

- The discussion may be integrated into the results of qualitative papers.
- The relationship between the current research and existing literature is explored, highlighting how the study contributes to the field.
- Strengths and limitations of the research are addressed.
- Consideration is given to how researchers may have influenced the data, including potential biases.

Conclusions

The conclusion summarizes the main results and emphasizes the study contribution to the subject area.

26.8.2 Writing a Manuscript from a Mixed Methods Design

Here, we present specific guidelines for writing each section in the IMRAD format.

26.8.2.1 Introduction

Start with an engaging introduction that provides the background and context of the study. Clearly state the objective or research question and its significance in agriculture. Explain why mixed methods were chosen and how they enhance the comprehensiveness of your study. Include a targeted literature review highlighting findings of mixed methods, quantitative and qualitative studies related to the topic under study. In your write-up, show your understanding of the existing knowledge and how your study contributes to knowledge by filling the knowledge gap. Just like in quantitative and qualitative research papers, ensure that the statement of the problem is justified.

26.8.2.2 Methodology

A paragraph, placed at the beginning of the methods section, introducing the design, should be included to enhance understanding by the readers and reviewers who may not be completely familiar with different types of mixed methods designs (Creswell and Plano Clark 2011). Further, Matthews and Matthews (2014) recommend including five key elements in a mixed methods research study report. This includes i) Intent of the study, e.g. this mixed study addressed…..ii) Design specification, e.g. we used a concurrent triangulation design…iii) Description of how data were collected, e.g. through in-depth interviews, iv) Provide a rationale for why quantitative and qualitative data were collected, e.g. the quantitative study addressed, while the qualitative study examined, and v) A description of how validity and reliability (or trustworthiness) were accomplished in the research design.

In addition to these five elements, Creswell and Plano Clark (2011) recommended including design characteristics, such as level of interaction, timing, priority, and mixing decisions. The researcher should clearly outline the quantitative and qualitative data collection methods, sampling procedures, and data analysis techniques for each component, and put these in separate sections. Moreover, the researcher should provide details of the mixed methods research used, including concurrent, sequential, and transformative methods, showing how they complemented each other (see Chap. 15). It is recommended to include methodological references and a visual diagram or presentation depicting the progression of the research activities. This will help in understanding the complexity and sequence of research activities (Creswell and Plano Clark 2011; Mertens 2014).

Including visual illustrations is crucial for enhancing the clarity and comprehensibility of the research approach. Visual aids, such as diagrams, flowcharts, or tables, effectively convey the sequential or concurrent nature of the design, depicting the integration of quantitative and qualitative components. These visuals help readers grasp the overall structure of the study, the specific phases involved, and the relationship between different data collection and analysis methods. They also enhance transparency by illustrating the timing, priority, and connection between quantitative and qualitative data collection (see studies by Rezaei-Moghaddam and Tohidyan Far (2019) and Acquah et al. (2021)). We provide an example of methodology with a visual illustration using a youth agribusiness study.

This hypothetical study employed an exploratory sequential mixed methods design to investigate barriers to youth engagement in agribusiness in Western Kenya. It began with qualitative research to explore perceptions, motivations, and challenges affecting youth decisions about the agricultural sector. Insights from this phase informed the development of a structured quantitative survey for broader validation of findings.

In the first phase, qualitative data were collected from 30 youth aged 18 to 35, selected from agribusiness training programs, youth cooperatives, and non-engaged rural youth across three counties. Data collection included six focus group discussions (FGDs) of 8–10 participants and eight key informant interviews (KIIs) with stakeholders like youth mentors and agricultural extension officers. Thematic analysis of the transcribed interviews and FGDs using NVivo 12 identified key themes that shaped the quantitative phase.

The second phase involved creating a structured survey based on qualitative findings, focusing on youth attitudes, perceived barriers (economic, social, and institutional), and motivational drivers like peer influence and market access. The survey underwent expert validation and was pilot-tested with 25 rural youth to ensure reliability.

In the final phase, the survey was administered to a stratified random sample of 400 rural youth, using mobile data collection tools. Quantitative data were analyzed with SPSS software, including descriptive statistics and logistic regression to identify predictors of youth engagement. Integration of findings occurred during interpretation, where qualitative insights contextualized quantitative results.

Box 26.10: Checklist for Writing a Mixed Methods Paper

1. Have you provided evidence in the title that it is a mixed methods study and used words like mixed methods, quantitative, qualitative, integrated, triangulation?
2. Have you explained the purpose and function of the research study?
3. Have you explained how you mixed the strands and sequenced the quantitative and qualitative phases?
4. Have you included illustrations to enhance understanding of the integration?
5. Have you given the stages at which the integration of approaches occurred?
6. Have you explained how you analyzed the data separately and the point of integration?
7. Have you given the theoretical perspective in the research?
8. Have you included a visual illustration of the methods and procedures?

(continued)

Box 26.10 (continued)

9. Have you mentioned the organization that approved your research and processed the ethical clearance and how consent from participants was organized?
10. Have you explained how validity and trustworthiness were assessed/audited for the qualitative strand?
11. Have you indicated the typology? How did you define the mixed methods design?
12. Have you given a reason for using both quantitative and qualitative approaches?
13. Have you given the criteria for choosing the mixed methods study?
14. Does the data collection relate to the strategy?

Recommendation for Further Reading

Name of the publication	Key points
Day and Gastel (2012)	Comprehensive guidance on the entire scientific writing publication process in 42 chapters. The book covers topics such as the structure of the scientific papers, the peer review process, and effective communication
Lövei (2021)	Consists of 30 chapters packaged in III parts. Part I covers the basics in 4 chapters, while part II comprises 14 chapters on writing the paper. This is the largest part of the book and mainly covers how to write the various sections of the paper. The last part is about publishing the paper, which has 12 chapters. These chapters take the reader from submission to how to handle the review process up to the final handling of the publishing house. The final three chapters cover how to write other forms of publications (conference proceedings papers, review articles, and book chapters)
Nair and Nair (2014)	The book presents the essentials of good scientific writing, emphasizing manuscript preparation and submission. It gives a discussion of dos and don'ts in scientific writing focusing on agriculture and natural resource management and target early career professionals. The book has 10 chapters presented in 4 parts
Matthews and Matthews (2014)	A comprehensive guide to help scientists improve their writing skills. It provides detailed instructions on how to write and publish scientific papers. From initial planning to submission, it offers practical guidance and emphasizes clarity and conciseness in scientific writing. Though it is targeted at biological and medical students, it is relevant to agricultural scientists
Youdeowei et al. (2012)	Prepared as a training manual to equip agricultural scientists with the necessary skills for scientific communication. It includes topics like structuring a scientific publication and emphasizes writing clearly and concisely. It also includes topics on preparing manuscripts, the submission process, and effective presentation for agricultural scientists
Korstjens and Moser (2018)	Part of a 4 series paper on qualitative approaches. The part guides reporting in a qualitative journal paper. Quality criteria for publishing a qualitative paper are given

Further Reading Materials

1. Lövei GL (2021) Writing and Publishing Scientific Papers: A Primer for the Non-English Speaker. Writing and Publishing Scientific Papers 214. https://doi.org/10.11647/OBP.0235
2. Matthews JR, Matthews RW (2014) Successful Scientific Writing, 4th edn. Cambridge University Press.
3. Nair PKR, Nair VD (2014) Scientific writing and communication in agriculture and natural resources. Scientific Writing and Communication in Agriculture and Natural Resources 1–142. https://doi.org/10.1007/978-3-319-03101-9/COVER
4. Yan Z, Snow C (2020b) Publishing Journal Articles: A Scientific Guide for New Authors Worldwide. Publishing Journal Articles. https://doi.org/10.1017/9781108277426
5. How to write and publish a scientific paper by Gastel and Day (2022).

Key Points in This Chapter

The Scientific Style of a Research Paper

Scientific papers are key for sharing research findings with the scientific community. The scientific style is characterized by clarity, precision, objectivity, and rigor in language and presentation. It also prioritizes logical organization and adherence to empirical evidence.

The IMRAD Framework

The IMRAD's framework: "Introduction, Materials and Methods, Results and Discussion", provides a clear and logical structure for writing scientific papers. It benefits readers, authors, editors, and reviewers by offering a consistent and predictable format that enhances clarity, organization, and ease of navigation throughout the manuscript.

In addition to the IMRAD sections, the paper has other minor components, such as authors and acknowledgements. The IMRAD structure follows an "hourglass" shape because articles start broad and general, get narrower and more specific in the middle, and return to a general discussion at the end

The "Beginning Matter, Middle Matter and End-Matter"

A scientific manuscript consists of three main sections: "Front matter," "Mid-matter," and "End matter." The "Front matter" includes the title, authors, abstract, keywords, and manuscript history. The "Mid-matter" contains the main body detailing methodology, results, discussion, and analysis. The "End matter" includes references, acknowledgements, conflict of interest statement, ethics declaration, data availability statement, and appendices.

The IMRAD Structure

- The introduction section outlines the problem or question that the study aims to address. It provides background information and explains the significance of the research.
- The Materials and Methods section details the research procedures, including the materials utilized and the steps taken to conduct the study.
- The results section: presents the research findings, including data and observations without analysis or interpretation.
- The discussion section: interprets the results, explaining their significance, and how they relate to the broader context. It also addresses the implications and any limitations of the study.

How Do You Present Statistics in a Scientific Publication?

In scientific publications, clear and accurate statistical reporting is essential. Results should be aligned with the methods. Tables and figures should be well-labeled and highlight key descriptive and inferential statistics, such as test statistics, p-values, and effect sizes. Confidence intervals should be included for precision. Selected statistics need to be used to avoid clutter; details can be included as supplementary materials.

Manuscript from a Qualitative Approach

The IMRAD format is a generic structure for writing scientific manuscripts, including qualitative papers. While qualitative studies still include sections like introduction, methods, results, and discussion, the results section notably differs from quantitative papers. Instead of numerical data, it focuses on rich, descriptive findings presented through quotes, themes, or narratives.

Writing a Manuscript from a Mixed Methods Design

When writing a manuscript with a mixed methods design, start with an introduction that outlines the study's background, objectives, significance, and the advantages of using mixed methods. In the methodology section, describe the design, including intent, design type, data collection methods, and the rationale for combining quantitative and qualitative data. Address validity and reliability and include separate sections for each data collection method, sampling, and analysis techniques. A visual diagram or flowchart is recommended to help illustrate the research process.

Review Questions

1. Why is the standardization provided by IMRAD important for the scientific community?
2. Give key characteristics of a scientific paper.
3. Explain the fundamental difference between the Results section and the Discussion section of a scientific paper.
4. What specific types of "detailed presentations" might you expect to see in a qualitative results section (e.g. beyond just numbers)?
5. When writing a mixed methods manuscript, what two key elements, beyond adhering to the IMRAD structure, must you explicitly include or justify, according to the abstract?

Chapter Exercise

1. Select a published journal article in your field and break it down into three main sections: (1) Front Matter, (2) Mid Matter (IMRAD), and (3) Back Matter. Label each section clearly and summarize its content in 1–2 sentences.
2. Propose three different types of titles from a hypothetical paper derived from a study conducted in a specific region, "Socio-economic factors influencing adoption of improved maize varieties": Descriptive, Declarative/Result-Oriented), Question-Based.
3. Create a fictitious author list for a paper you might co-author, including:
   - Full names of the authors.
   - Affiliations.
   - Identification of the corresponding author.
   - A brief note explaining how the order of authorship was determined.
4. Write an abstract (150–200 words) for a hypothetical study you would conduct, and list five relevant keywords.

Answers to Review Questions

- Q1: Clear and organized sequence; benefit: Standardization benefits authors, editors, reviewers, and readers by ensuring consistency and predictability. Reference: 26.1.
- Q2: Clarity, precision, objectivity, precision, logical organization, empirical evidence, consistency. Reference 26.1. and 26.2 (Tip: IMRAD)
- Q3: Reference 26.5.3 and 26.5.4.
- Q4: Reference 26.8.1. Read at least two qualitative journal papers. Examples: Sadati et al (2021) and Shortall (2022)
- Q5: Reference 26.8.2: Read at least two mixed methods journal papers. Examples: 1. Kiptot and Franzel, (2025) (Note the quantitative results in numeric format and the qualitative results in words. 2. Mbunga et al., (2023) used a convergent mixed methods design.

References

Acquah ISK, Naude MJ, Soni S (2021) How the dimensions of culture influence supply chain collaboration: an explanatory sequential mixed-methods investigation. Revista de Gestao 28:241–262. https://doi.org/10.1108/REGE-11-2020-0105/FULL/PDF

Adewopo J, Peter H, Mohammed I et al (2020) Can a combination of UAV-derived vegetation indices with biophysical variables improve yield variability assessment in smallholder farms? Agronomy 10:1934. https://doi.org/10.3390/AGRONOMY10121934

Creswell JW, Plano Clark VL (2011) Designing and conducting mixed methods research. 51–99

Day RA, Gastel B (2012) How to write and publish a scientific paper, 4th edn. Cambridge University Press, Cornwall

Gitari HI, Nyawade SO, Kamau S et al (2020) Revisiting intercropping indices with respect to potato-legume intercropping systems. Field Crop Res 258:107957. https://doi.org/10.1016/J.FCR.2020.107957

Korstjens I, Moser A (2018) Series: practical guidance to qualitative research. Part 4: trustworthiness and publishing. Eu J General Prac 24:120–124. https://doi.org/10.1080/13814788.2017.1375092

Lövei GL (2021) Writing and publishing scientific papers: a primer for the non-English speaker, vol 214. Writing and Publishing Scientific Papers. https://doi.org/10.11647/OBP.0235

Madler AM, Renwick S (2025) Referencing and citation styles. Reference Module in Social Sciences, pp 108–120. https://doi.org/10.1016/B978-0-323-95689-5.00213-3

Maina AN, Thanabalan A, Gasarabwe J et al (2022) Enzymatically treated yeast bolstered growth performance of broiler chicks from young broiler breeders linked to improved indices of intestinal function, integrity, and immunity. Poult Sci 101:102175. https://doi.org/10.1016/J.PSJ.2022.102175

Matthews JR, Matthews RW (2014) Successful scientific writing, 4th edn. Cambridge University Press

Matusso JMM (2015) Effects of maize (zea mays l.)—soybean (glycine max (l.) Merrill) intercropping patterns on yields and soil properties in two contrasting sites of Embu and Meru counties, vol 9, Kenya

Mertens D (2014) Research and evaluation in psychology and education. Libro Electronico, p 4

Mirriam A, Mugwe J, Raza MA et al (2022) Aggrandizing soybean yield, phosphorus use efficiency and economic returns under phosphatic fertilizer application and inoculation with Bradyrhizobium. J Soil Sci Plant Nutr 22:5086–5098. https://doi.org/10.1007/S42729-022-00985-8/METRICS

Mucheru-Muna M, Pypers P, Mugendi D et al (2010) A staggered maize–legume intercrop arrangement robustly increases crop yields and economic returns in the highlands of Central Kenya. Field Crop Res 115:132–139. https://doi.org/10.1016/J.FCR.2009.10.013

Mugwe J, Monicah AE, Ae M-M et al (2009) Adoption potential of selected organic resources for improving soil fertility in the central highlands of Kenya. Agrofor Syst 76:467–485. https://doi.org/10.1007/s10457-009-9217-y

Nair PKR, Nair VD (2014) Scientific writing and communication in agriculture and natural resources, pp 1–142. https://doi.org/10.1007/978-3-319-03101-9/COVER

Nyawade SO, Karanja NN, Gachene CKK et al (2019) Intercropping optimizes soil temperature and increases crop water productivity and radiation use efficiency of rainfed potato. Am J Potato Res 96:457–471. https://doi.org/10.1007/S12230-019-09737-4/FIGURES/3

Okokpujie IP, Fayomi OSI, Leramo RO (2018) The role of research in economic development. IOP Conf Ser Mater Sci Eng 413:012060. https://doi.org/10.1088/1757-899X/413/1/012060

Rezaei-Moghaddam K, Tohidyan Far S (2019) The impact assessment of technologies diffusion: a mixed methods analysis. Chem Biol Technol Agric 6:1–14. https://doi.org/10.1186/S40538-019-0162-3/FIGURES/4

Sollaci LB, Pereira MG (2004) The introduction, methods, results, and discussion (IMRAD) structure: a fifty-year survey. J Med Libr Assoc 92:364

Wu J (2011) Improving the writing of research papers: IMRAD and beyond. Landsc Ecol 26:1345–1349. https://doi.org/10.1007/S10980-011-9674-3/FIGURES/1

Yan Z, Snow C (2020a) Who are the authors of journal articles? Publishing Journal Articles:36–47. https://doi.org/10.1017/9781108277426.007

Yan Z, Snow C (2020b) Publishing journal articles: a scientific guide for new authors worldwide. Publishing Journal Articles. https://doi.org/10.1017/9781108277426

Youdeowei A, Stapleton P, Obubo R (2012) Scientific writing for agricultural research scientists—a training resource manual:190

27 Journal Selection, Submission, and Review of Manuscript

Abstract

Publishing a journal article is a graduation requirement in most universities. It is beneficial to both the institution and the individual. Publishing contributes to the ranking and credibility of the university, while it also aids in the personal development and recognition of the researcher. Many types of journals exist, including international and regional journals, free-to-publish and paid-for journals, general discipline and specialist journals, and credible and predatory journals. Researchers should avoid publishing in predatory journals by all means. Predatory journals are unworthy and lack credibility because they use unethical publishing practices. Researchers start the publishing by drafting a manuscript or by identifying a target journal, writing, and finally submitting to the journal. Investing time to select a suitable journal is worthwhile. Several factors may be considered; consultation with your supervisor and colleagues, the credibility of the journal, scope, and focus, its impact factor or rating, frequency of publication, editorial board, timeliness, publishing model, and scope of readership. Journal finders are one of the latest technologies that allow researchers to identify the most relevant journals for their work. Once submitted, the manuscript undergoes several stages from editor assessment, peer review, addressing comments and author revisions, and finally, production. This process takes 3–12 months, with an average of 6 months depending on the journal. *Journal* editors play an important role as gatekeepers. They uphold journal standards quality and integrity by enforcing ethical publishing practices. As an author, you must respond satisfactorily to the reviewer's comments because failing to do so risks the manuscript being rejected. Key barriers to publishing exist, including a skills gap and a low research infrastructure. Invest persistently in learning and practising writing and publishing.

Keywords

Article processing charges · Barriers to publishing · Bibliometric indicators · Citation index · Cost of publishing · Cover letter · Credible and predatory journals · Free-to-publish journals · Galley proof · Impact factor · Indexed journals · International journals · Journal editor · Manuscript · Open access journals · Paid-for-journals · Peer review · Publishing journey · Regional journals · Rejection · Subscription-based journals · Target journal

Learning Outcomes

After reading this chapter, you are expected to be able to:

1. Outline the importance of publishing a journal paper.
2. Get familiar with key terms used in journal paper publishing.
3. Identify the main types of journals.
4. Differentiate between a subscription and an open-access journal.
5. Distinguish credible and predatory journals.
6. Illustrate factors to consider when selecting a journal.
7. Appreciate the role of "Journal bibliometrics"
8. Explain the steps from submission of the manuscript to publication of a journal paper.
9. Recognize the role of the journal editor.
10. Appreciate what journal reviewers look for.
11. Demonstrate how to respond to editors' feedback.
12. Appreciate barriers to publishing by students.

J. Njeri Mugwe, S. Runo, *Research Methodology in Agricultural Sciences*, https://doi.org/10.1007/978-981-95-1892-0_27

27.1 Basics of Scholarly Publishing

27.1.1 Dilemmas in Publishing

The scholarly publishing journey is not easy. It requires a thorough understanding of academic writing conventions (Discussed in Chap. 28) and familiarity with journal submission guidelines. One must also have the ability to communicate research findings effectively. As instructors teaching topics on manuscript publishing, we frequently encounter numerous student concerns, especially from those publishing for the first time. These concerns range from how to write a manuscript to which is a good journal? and several other concerns (Box 27.1). Students often express worries about meeting the expectations of journal editors and reviewers. They are also worried about meeting the quality and relevance of their research in the target journal. Thus, while the prospect of publishing their work is exciting, students often grapple with uncertainties and anxieties as they embark on this part of their academic journey.

Box 27.1 Common concerns by postgraduate students

1. What does the process of publishing involve?
2. How do I select the right journal for my research?
3. Which is the most appropriate journal?
4. What are the key elements of a well-written manuscript?
5. How do I ensure my manuscript adheres to journal guidelines?
6. Do I pay to get my manuscript published?
7. What about when I cannot afford to pay the publishing fee?
8. What is the process for submitting a manuscript to a journal?
9. How long does the peer review process typically take?
10. What happens after I submit my manuscript to the journal?
11. How should I respond to reviewers' comments and feedback?
12. What are the common reasons for manuscript rejection, and how can I avoid them?
13. Do I need to be aware of any ethical considerations when publishing?
14. How can I increase the visibility and impact of my published work?

This chapter equips students with the knowledge and skills to successfully navigate the *publishing journey*. Specifically, we discuss the journal selection and publication process. Publishing a journal article typically involves several key steps: (i) identifying a suitable journal, (ii) preparing the manuscript according to the journal's guidelines, (iii) undergoing peer review, (iv) revising the manuscript based on reviewers' feedback, and (v) achieving acceptance for publication. Each step requires careful attention to detail and adherence to academic norms to ensure the manuscript's quality and suitability for publication.

27.1.2 Importance of Publishing in a Journal

Universities have three core mandates; to *train, conduct research*, and engage in *community outreach* (Etzkowitz 2003; Youdeowei et al. 2012). Through research, universities make significant contributions to socioeconomic development by informing policies, advancing technologies, and addressing pressing societal challenges. Consequently, many institutions now require graduate students to publish their work before graduating. Besides sharing the findings, publishing benefits the university by enhancing its scholarly reputation, increasing global visibility, and attracting research funding and collaborations.

Further, publications have the potential to generate intellectual property and royalties for the university through patents, licenses, or commercialization opportunities (Goldstein and Rehbogen 2013; Roessner et al. 2013; Sanberg et al. 2014). The generated revenue can be reinvested into research facilities, infrastructure, scholarships, and other initiatives that benefit students and faculty. Moreover, publications attract research funding and grants to the university, as funding agencies prioritize institutions that produce high-impact research. This funding supports ongoing research projects, student scholarships, and research infrastructure development, further boosting the university's academic and financial standing.

Besides benefiting the University, publishing has numerous benefits for individual researchers (Nair and Nair 2014; Yan and Snow 2020; Lövei 2021). The ability to publish in a journal enhances writing and critical thinking skills. It also helps provide satisfaction and confidence to the author that they are able to make a contribution to the expansion of human knowledge. Additionally, publishing raises credibility and enhances personal development. It also increases the likelihood of winning research grants. In summary, publishing is highly advantageous for personal growth and satisfaction. As an author, you reap the most benefits from the published work.

27.1.3 Key Terminologies in Publishing in a Journal

Here are some key terminologies commonly used in the process of publishing in a journal:

- *Manuscript:* The written document submitted to a journal for publication, including the research paper, figures, tables, and any supplementary materials.
- *Peer Review*: The process of evaluating a submitted manuscript by experts in the field (peer reviewers) to assess its quality, validity, and suitability for publication.
- *Journal article*: The published document in a journal. once accepted and published the manuscript becomes an article.
- *Editor*: The individual responsible for overseeing the peer review process, making editorial decisions, and ensuring the quality and integrity of the journal's content.
- *Manuscript revision:* The process of making changes or improvements to a manuscript based on feedback received from peer reviewers or the editor.
- *Acceptance:* The decision by the journal editor to approve a manuscript for publication after it has completed the peer review process and any necessary revisions.
- *Rejection:* The decision by the journal editor not to publish a submitted manuscript, typically due to concerns about its quality, relevance, or originality.
- *Impact factor:* A measure of the average number of citations received by articles published in a particular journal, often used to indicate the journal's prestige and influence within the academic community.
- *Open access:* A publishing model that provides free, unrestricted access to research articles online, allowing anyone to read, download, or distribute the content without charge.
- *Copyright:* Legal protection granted to the author or publisher of a creative work, including research articles, to control its reproduction, distribution, and adaptation.
- *Galley Proof*: The preliminary version of a manuscript that is typeset and formatted by the journal's production team for publication. Authors review the galley proof to check for any errors or discrepancies introduced during typesetting and make final corrections before publication.
- *Indexing*: is the process of cataloguing and organizing journal articles in databases or directories to facilitate researchers' search and retrieval. Indexed journals are included in databases such as PubMed, Scopus, and Web of Science, which increases their visibility and accessibility within the academic community.
- *Editorial board*: experts in the field, usually comprising editor-in-chief and associate editors, knowledgeable about peer review and the publication process.
- *Citation index*: a tool that indexes citations, allowing researchers to assess the impact and influence of research articles, journals, and authors. Examples include Google Scholar indices.
- *Corresponding author:* Primary contact person for communicating with the journal editorial office.
- *Production process*: Process of preparing an accepted manuscript for publication; involves copyediting, formatting, typesetting, proofreading, and preparing the final version.
- *Article Processing Charge (APC):* Fees paid by authors to journals to cover the costs of publishing an article.
- *Bibliometrics,* also known as "publication metrics," are quantitative measures used to evaluate various aspects of scholarly publishing.

27.1.4 Key Steps in Publishing

The publishing process starts with the selection of the target journal. The authors then prepare the manuscript using the guidelines (nearly all journals have a 'guide to authors' document that is downloadable) provided by the target journal. Manuscript preparation involves *writing, editing, and formatting* to ensure that the manuscript meets all requirements specified by the journal, including font size, font type, line spacing, and citation style (Sayer 2019). Authors also align the manuscript's structure, content, number of words, or language to comply with the journal's guidelines. Additionally, authors carefully review the manuscript for clarity, accuracy, and completeness, addressing any errors or inconsistencies before submission. Finally, authors may seek feedback from colleagues or mentors to ensure that the manuscript meets the standards expected by the academic community.

Once satisfied with the final manuscript, the authors submit it to the chosen journal accompanied by a cover letter highlighting the research's significance and suitability for publication. The manuscript then undergoes the peer review process, where experts evaluate its quality. The peer review process is a synergistic relationship involving the *authors*, the *journal editors,* and the *reviewers*, which aims to enhance the quality of the work. Peer reviewers provide valuable feedback and recommendations to the editor, who ultimately decides whether to accept or reject the manuscript for publication (Sovacool et al. 2022; Drozdz and Ladomery 2024).

If the manuscript is accepted, authors may be required to revise it based on the feedback received during the peer review process. This often involves addressing comments, clarifying points, and providing additional data or analysis as necessary. Once the revisions are complete and the manuscript meets the journal's standards, it moves forward for proofreading and production.

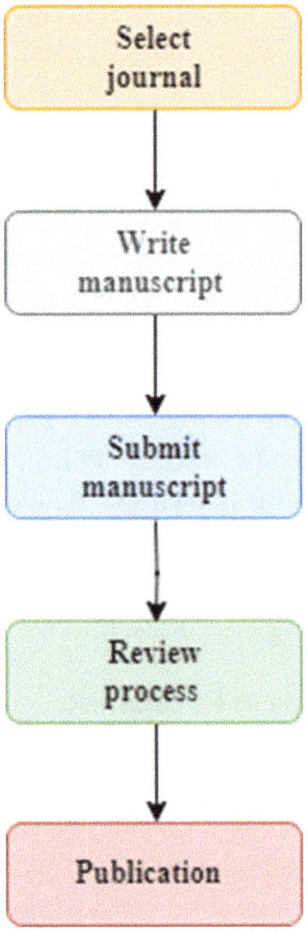

Fig. 27.1 Brief steps in the publication process of a journal paper

During the proofreading and production stage, authors meticulously review the final manuscript for any errors or inconsistencies and make necessary corrections. The manuscript then undergoes copyediting and typesetting by the journal's production team to prepare it for publication. Finally, the manuscript is published in the journal, either in print, online, or both, and becomes accessible to the readership (Fig. 27.1).

Note flexibility in publishing Some writers prefer to start their publishing process by selecting a specific journal in which they would like to publish their paper. They tailor their manuscript to meet the requirements and style of that particular publication, taking into account the journal's target audience, scope, and focus. On the other hand, some writers may opt for a less rigid approach, writing their manuscript first and then modifying it to fit the style of a journal they choose later. The decision to write with a specific journal in mind or adapt the manuscript later depends on individual preferences, publication goals, and the flexibility of their research findings and writing process.

27.2 Understanding Journal Types and Predatory Pitfalls

27.2.1 Type of Journals

Various types of journals exist, each with its characteristics and audience. In our exploration, we examine the following types of journals:

1. *International and regional journals*: *International journals* are globally recognized and often publish good-quality articles. The articles are also highly visible and have a broad readership. Examples include; *Biology and Fertility of Soils* and *Agronomy for Sustainable Development* (Springer), *Field Crops Research*, and *Soil and Tillage* Journals (Elsevier). They cover a wide range of topics across disciplines. The main characteristics include (i) a consistent publication schedule, (ii) a global editorial board, (iii) a wide readership, (iv) a worldwide citation index, and (iv) a high impact factor. Most of these journals are indexed in prestigious databases and abstracting services.
   Regional journals, on the other hand, address local issues and provide a platform for researcher to disseminate their research work within their communities. Examples in the East African region include the *African Crop Science Journal*, the *African Journal for Rural Development*, and University-based journals. Kenyatta University has the Chemi Chemi journal in the School of Pure and Applied Sciences. Regional journals are more localized in scope and target more specific audiences within the region; often local researchers, academicians, and policymakers. Regional journals tend to be limited in readership and lack indexing in prestigious databases (Asubiaro and Onaolapo 2023). They also tend to be irregularly published due to a lack of adequate manuscripts because most researchers prefer to have their papers published by internationally recognized journals that will give them a wide readership.
2. *Free-to-publish and paid-for journals*: Free-to-publish journals usually follow the traditional subscription-based model where the researcher publishes for free. This model is called the traditional model because it has been the dominant scholarly publishing model for many years. In this model, the journal bears the cost of publishing, with the researcher surrendering the copyright to the journal. The journal will also bear the cost of editing, publishing, and online hosting. However, the articles are restricted to

readers and not easily available without payment. This means your article will be locked up and only available to those who have subscribed. Possibly you have noticed many journal articles online that you cannot access. They ask you to pay to access them. Individuals and institutions pay subscription fees to allow them access to these journals. The subscription fee is used to cover the cost of publishing and also make a profit for the publisher. This model of scholarly publishing has been criticized for limiting access to research and contributing to rising publication fees (Burgman 2019; Björk 2021).

In contrast to journals offering free publishing, in paid-for journal authors pay Article Publishing Charges (APC) to get their articles published (Clark et al. 2024). These charges can range from a few hundred to thousands of dollars. Articles published in paid-for journals are usually open-access. Open access means readers can access them for free. The authors typically retain the copyright to their work, and articles are published under licenses that allow unrestricted use, distribution, and reproduction. One key advantage of open-access journals is that they have the potential to reach a wider audience, which can result in higher citation rates for the authors (Clark et al. 2024). Additionally, open-access journals offer various benefits, which are detailed in Box 27.2.

Box 27.2 A comparison of open access and paid-for journals

Advantages of open access journals

1. Avoidance of subscription fees by the readers: articles are freely available.
2. Increased accessibility and visibility.
3. Wide readership and impact.
4. Higher citation rates associated with their access.
5. Promote collaboration and interdisciplinary.
6. Compliance with funding agencies that require researchers to make their work accessible to the publisher (open science principles).

Advantages of subscription journals

1. Traditionally, have a long-established prestige and reputation.
2. Provide editorial and publishing support hence helpful to researchers who may not be able to afford to pay, e.g., students and early career researchers.
3. High level of selectivity of the papers to be published.

(continued)

Box 27.2 (continued)

4. Often, discipline-specific helps researcher with targeted access to content directly relevant to their areas of interest.
5. Generally maintain a rigorous peer-review process.

The publishing landscape has continued to evolve with new publishing models, such as hybrid journals, where researchers can choose between open-access and a paid-for model within the same journal. A recent trend is that most journals are migrating to pure open access model. Despite the many benefits of open access publishing, it could have contributed to increase of predatory journals (Ward 2016).

Open access has become a subject of much discussion because of its many benefits (Burgman 2019; Camkin et al. 2022). The advocates of open access policy argue that it addresses concerns surrounding accessibility, dissemination, and fairness (Björk 2017; Alston 2019). This policy seeks to enhance accessibility by enabling anyone to access articles without financial or technical impediments. The policy uses the Creative Commons licensing framework, where research articles can be repurposed, redistributed, or modified while allowing authors to retain ownership. By expanding readership, open access can amplify visibility and facilitate opportunities for collaboration.

3. *General discipline journals and specialist journals*: General discipline journals, such as *Plant and Soil* (Springer), *Experimental Agriculture* (Cambridge), and *Agricultural Systems* (Elsevier), cover a wide array of topics within a particular field. They allow researchers to publish articles spanning various subdisciplines and specialities. These journals aim to reach a broad audience within the discipline, prioritizing articles that have broad relevance and potential impact across different areas of study. Due to their inclusive nature, general discipline journals often publish issues more frequently and receive more submissions than specialist journals.

   In contrast, specialist journals focus on specific major specialities, subspecialties, or selected research areas within a discipline, catering to a more specialized audience interested in particular aspects of the field. Examples include the "weed science journal" by Weed Science Society of America and and Precision Agriculture Journals. They are typically published quarterly or semi-annual and are often affiliated with professional groups or societies, with scientists frequently serving as editors. Specialist journals prioritize in-depth research within their specialized area of focus, resulting in higher rejection rates. Despite publishing less frequently, specialist journals maintain a high-quality standard within their niche, providing detailed insights and analysis to readers interested in specific topics or subfields.

4. *Credible and predatory journals*: Credible journals are trustworthy publications that follow proper publication ethics. Predatory journals are unworthy and lack credibility and quality because they use unethical publishing practices, e.g., lack of peer review. The term predatory was first coined by Jeffrey Beall (Somoza-Fernández et al. 2016), who released a list of fake or Pseudo-scientific journals. They are also called quasi or questionable journals (Clark and Smith 2015; Cobey et al. 2016). They do not uphold basic standards of quality and control.

Credible journals uphold rigorous standards of quality, transparency, and ethical conduct. They are trusted sources of scholarly information. They undergo a meticulous peer review process, ensuring that the articles meet established criteria for academic excellence. They are transparent, offering clear guidelines for authors, reviewers, and editors. They are open about their peer review practices, publication policies, and potential conflicts of interest. Moreover, ethical conduct is rigorously upheld with adherence to established codes of publication ethics, such as those outlined in the Committee on Publication Ethics (COPE) (COPE 2019). Authors, readers, and the academic community rely on credible journals to uphold these standards.

Predatory journals do not uphold quality standards and engage in unethical practices and deceptive tactics (Box 27.3). They exploit the academic publishing systems for financial gain. Unlike credible journals that prioritize scholarly integrity and adherence to rigorous peer review processes, predatory journals often lack robust quality control mechanisms and prioritize financial gain. The stringent peer review is the most important that distinguishes predatory journals from credible journals (Strinzel et al. 2019). Predatory journals usually accept manuscripts without proper evaluation, leading to the publication of low-quality or even fraudulent research. Shamseer et al. (2017) listed 13 salient features of predatory journals.

Box 27.3 Key features to look for in identification of predatory journals

1. *Bealls'list*: Contained in Bealls list of Predatory publishers and journals Beall's List: https://beallslist.weebly.com/,
2. *Quality of papers:* Low quality in content, formatting, and general layout. Have grammatical errors and poor science. Published work has glaring mistakes.
3. *Scholarly Open access*: https://archive.md/9MAAD
4. *Peer review:* Poor or nonexistent. Check the instructions for authors and find out number of days they cite. Majority will mention fast publication from 2 days; this is not adequate time for peer review
5. *Editorial Team:* Lack professionalism in the editorial team. Usually editors come from the same region and are not professionals, nor experts in their field. The journals keep sending invitations to uninterested individuals to join the editorial board
6. *Indexing*: Not indexed in scholarly databases
7. *Charges:* Charge for publication (APC) without the services
8. *Impact factor*: Nonstandard impact factors (Thompson Reuters-real IF)
9. *Citation:* Rarely cited, and when cited, it is by other predatory journals
10. *Copyright*: They retain the copyright
11. *Journal names:* Extremely broad in an effort to embrace all
12. *Websites*: Poorly designed, some with glaring mistakes
13. *Acceptance rates*: High acceptance rates, even within 2 days
14. *Send emails:* Beware of emails that request papers for publication. Many students have fallen victim to this practice. Predatory journals commonly send frequent emails to a large number of individuals soliciting manuscripts for publication.

Websites to help identify predatory journals

- THINK, CHECK, SUBMIT (https://thinkcheck-submit.org/journals/)
- Bealls list: https://beallslist.net/
- Your University Library
- Open access and scholarly communication by the University of Witwatersrand. Predatory Publishers—Open Access, A2K & Scholarly Communication—LibGuides at University of the Witwatersrand (wits.ac.za)

27.2.2 Dangers of Publishing in Predatory Journals

Publishing in predatory journals poses several dangers for researchers, as well as for the scientific community and policymakers (Clark and Smith 2015; Shamseer et al. 2017). For researchers and scientific communities, the primary danger lies in compromising scientific integrity and credibility. Because of minimal or no peer review, articles may disseminate *inaccurate or misleading research findings*. This can have detrimental effects on researchers' reputations and academic careers, as their work may be perceived as lacking credibility or rigor. Furthermore, paying for publications that do not meaningfully contribute to the field of study wastes resources. Additionally, having articles in predatory journals may hinder researchers' ability to secure funding, collaborations, or career advancement opportunities. These publications are not typically recognized as credible contributions to the scientific literature. We conclude by saying that "Publishing in a predatory journal is a great danger for a researcher."

The decision-making process for policymakers and practitioners in agriculture heavily relies on accurate research. With reliable information, effective policies can be created and informed decisions can be made. However, using flawed research from predatory journals can lead to ineffective policies, wasted resources, and missed opportunities for positive change. This affects not only farmers, extension agents, and agricultural consultants, but also the agricultural industry's overall productivity and financial outcomes. Misinformation from predatory journals can result in poor agricultural practices, lower productivity, and financial losses. So, we find that the dangers of predatory publishing extend beyond the researchers themselves.

27.2.2.1 Why Do Predatory Journals Continue to Flourish?

Despite the drawbacks and condemnation of predatory journals, they continue to flourish. They rapidly increased their publication volumes from 53,000 in 2010 to an estimated 420,000 articles in 2014, published by around 8000 active journals (Shen and Björk 2015). According to this author, more than 5000 German scientists have published in these journals, meaning that even prominent professors are culprits. Some publish knowingly, where some writers consciously decide to submit their work to predatory journals, but a great majority are conned. Higher numbers of authors publishing in predatory journals are found in academic institutions, mainly by students, due to pressure to publish (Yeo-Teh and Tang 2021). Sometimes, students cite confusion in the identification of predatory journals as the reason for publishing in predatory journals. A notable example is when a predatory journal manages to secure indexing in reputable databases, further complicating the process of discerning legitimate scholarly outlets (Strinzel et al. 2019). Additionally, the same papers are being cited in reputable journals, causing contamination and more confusion (Somoza-Fernández et al. 2016).

You may ask who owns predatory journals and where they are located. The answer is that the owners are diverse. They may be individuals like you and me, small companies, or even large publishing entities that seek to capitalize on the growing demand for publishing opportunities. In terms of numbers, it is estimated that Asia plus India (61%) have most journals in predatory journals (Shen and Björk 2015). This is followed by Africa with 16% and by North America and Europe with 9% and 8.8%, respectively. The distribution of publishers is as follows: Asia plus India is 40%, North America is 17.5%, while Europe and Africa are 8.8% and 5.5%, respectively (Shen and Björk 2015).

Multidisciplinary Journals

Multidisciplinary journals, also referred to as megajournals, are ideal for research that integrates various methodologies, theories, or data, such as the intersection of agricultural science with economics, environmental studies, or public health (Solomon 2014). Examples of these journals include *PLOS ONE*, *Scientific Reports* (from the Springer Nature Portfolio), *Heliyon* (published by Cell Press), and various titles under the *MDPI* platform. Prestigious journals like *Nature* and *Science* also fall into this multidisciplinary category, although they typically focus on publishing only groundbreaking research that appeals to a wide scientific audience. Publishing in megajournals can help more people see your work, encourage collaboration across fields, and make your research more useful in real-life situations (Solomon 2014). However, to succeed in these journals, your study must clearly show its relevance to multiple fields and maintain strong research methods and clear writing.

27.3 Selecting the Right Journal

Selecting the right journal is a critical step in the publishing process, and it is essential to dedicate time to this decision. Currently, there are over 30,000 academic journals across various disciplines, and this number is growing by 5–7% each year (Ware and Mabe 2015). The publishing landscape has changed dramatically with the rise of digital journals, open-access models, predatory journals, and an increasing emphasis on interdisciplinarity. Given this vast array of options and the constant evolution of the field, researchers often struggle to identify the most suitable journal for their manuscripts (Sandesh and Wahrekar 2017). Therefore, a thoughtful journal selection process is necessary to ensure that researchers choose the right venue for maximizing their work's visibility and impact within the scholarly community.

27.3.1 Factors to Consider in Selecting the Right Journal

When choosing a journal for publication, several factors come into play (Bahadoran et al. 2021a, b). These include its impact factor or rating, frequency of publication, timeliness, cost, scope of readership (whether international, national, or regional), target audience, and publication language. Researchers may also consider personal preferences, such as submitting to a journal affiliated with their professional organization or one with which they have a membership. These factors collectively influence the decision-making process and help researchers identify the most suitable outlet for disseminating their work effectively. Here, we discuss seven key factors to consider:

1. Consultation with your supervisor
2. Credibility of journal
3. *Scope* and focus
4. Journal metrics
5. Publication process
6. Editorial Board
7. Cost of publishing (Publishing model)

1. Consult your supervisor and colleagues

As a student, consulting your supervisor is one of the best ways to begin selecting a suitable target journal (Sandesh and Wahrekar 2017; Bahadoran et al. 2021a). Your supervisor is experienced in academia and scholarly publishing and can provide valuable insights and recommendations based on their knowledge and familiarity with reputable journals. They can help you identify well-respected journals in your area of research. Additionally, they can offer guidance on factors to consider when selecting a journal, such as the scope and focus of the journal and any specific submission requirements or formatting guidelines that need to be followed.

You could also consult with colleagues. However, we advise caution with colleagues. Colleagues may offer their opinions and experiences, but their perspectives may be subjective or based on limited information. If you consult them, ensure you critically evaluate and cross-reference their suggestions with other sources, such as your supervisor's recommendations. You can also countercheck journal reputations and impact factors before accepting their advice. Many students report being advised by colleagues and end up publishing in journals that are not reputable.

2. Credibility of journal

The integrity of science is paramount when considering the credibility of a journal you are not familiar with. To evaluate a journal's scientific integrity, carefully scrutinize various aspects to ensure its quality and reliability. This involves examining recent publications and assessing research objectives, methodology, data analysis, results, and discussion, which are key indicators of scientific integrity (Cals and Kotz 2014). Check if tables and figures are clearly and appropriately labeled. Also, check if the references are up-to-date. They are crucial markers of scientific quality.

Check if a journal is indexed in renowned databases (Cals and Kotz 2014; Bahadoran et al. 2021a). Indexing signifies that it meets certain standards and has been deemed worthy of inclusion in these authoritative databases. This not only increases its visibility and accessibility, but also improves its credibility within the academic community. Authors should, therefore, prioritize journals that are indexed in well-known databases such as PubMed, Scopus, Web of Science, and Google Scholar. Moreover, database citations provide valuable metrics demonstrating published research's impact and influence. The advantage of publishing in indexed journals is that authors can keep track of the citation count of their articles. Overall, by taking advantage of indexing and database citations, authors can maximize the visibility and impact of their research, thereby enhancing scientific knowledge and scholarly discourse.

A quality journal is ethics compliant; therefore, it will post information about how it deals with research ethics issues on its journal websites. For instance, they will include information such as informed consent for human and animal subjects research and how to deal with it. They will also provide information on issues such as plagiarism, conflicts of interest, and internal review board approval. In addition, credible journals endorse guidelines and best practices for publishers, e.g., COPE.

3. Scope and focus

Consider a scope and focus that align with your research's subject matter and focus (Cals and Kotz 2014). This entails seeking out journals that publish articles relevant to various subdisciplines. Selecting a journal with a scope that closely matches your research topic increases the likelihood of your manuscript being a good fit for the journal's audience and editorial priorities. Selecting an out-of-scope journal is one of the factors contributing to the rejection of the paper. Rejection of a paper that is out of scope is done at the editor's desk. If this happens, the author receives feedback within 2 weeks after submission, informing them of the journal's decision.

Furthermore, publishing in a journal that is within your area of specialization ensures that your research reaches a targeted audience of scholars, practitioners, and policymakers with a vested interest in advancing agricultural knowledge and innovation (Wijewickrema and Petras 2017; Bahadoran et al. 2021a, b). Thus, considering the scope and focus of a journal is essential for maximizing the impact and relevance of your research within the agricultural research community.

4. Journal metrics

One of the most common terms in the academic publishing world is "impact factor'. A journal with a high impact factor is more prestigious than one with a low impact factor (Wijewickrema and Petras 2017). Your supervisor will most likely ask you to publish in a journal with a high impact factor. Publishing in such a journal will enhance your visibility and credibility. This is because articles in high-impact journals are likely to be cited, leading to greater recognition. Thus, the impact factor becomes an extremely important factor to consider in journal selection.

Journal metrics or *bibliometric indicators* measure the impact factor. These are quantitative measures that are statistically calculated using some algorithms. They measure journal influence and reach within the academic community. The main metrics include the Journal Citation Reports (JCR) Impact Factor, Eigenfactor, CiteScore, and Scimago journal rank. All these are good criteria for selecting a journal. Box 27.4 shows the main citation databases (Bibliometric databases). The JCR, commonly called impact factor (JIF), is the oldest and most widely used. It was first proposed in 1955 by Eugene Garfield and adopted by Thomson Reuters in 1970 (Garfield 2006; Larivière and Sugimoto 2019). It is a good indicator of a journal's quality and influence, and it has become synonymous with identifying "high-impact" journals for publication over time. The Eigenfactor (introduced in 2008), CiteScore (launched in 2016), and Scimago journal rank (Bergstrom et al. 2008; James et al. 2018; Myers and Kahn 2021) are recent, but they have emerged as good for assessing journal impact.

Box 27.4 The world of journal bibliometrics

Common bibliometrics databases consist of Web of Science by Clarivate Analytics (paid for), Scopus by Elsevier (Paid for), Google Scholar (free), Open Citations (free), Microsoft Academics (Free), and Dimensions (free) (Martín-Martín et al. 2018; Myers and Kahn 2021). They process citation counts as way of calculating bibliometric metrics. The main bibliometric databases are Web of Science and Scopus (Martín-Martín et al. 2018).

Publication metrics

- Journal impact factor
- SCIMAGO journal rank by SCOPUS. https://www.scimagojr.com/journalrank.php
- Citescore by Scopus; https://www.elsevier.com/products/scopus/metrics#Journal.
- Journal citation reports (JCR); by Clarivate Analytics (Global and largest)
- Science Citation Index Journal Rank (SJR) with impact factor
- Google Scholar Journal Metrics; https://scholar.google.com/citations?view_op=top_venues&hl=en; most widey used because it is free and is all-inclusive
- Source-Normalized Impact per Paper (SNIP)
- Eigenfactor: New and not widely used; http://www.eigenfactor.org/
- Hi-index

The two mostly used publication metrics are "The Journal impact factor" and hi-index":

Journal impact factor

- Is most widely used and created by Scinetometrician Eugene Garfield. Refers to the average number of times published papers are cited up to 2 years after publication.
- Criticized: Not okay to compare journals in different fields (Englander 2014).
- Comprised of science citation index expanded, Social Sciences Citation Index, Conference proceedings citation index, and the Arts and Humanities Citation Index (Larivière and Sugimoto 2019).
- Owned by Clarivate Analytics.
- Limitation: Does not show the quality, scientific accuracy, or impact of any particular article.

Hi-index

- The h-index (Hirsh index) measures the productivity and impact of individual researchers. An author has an h index if h of their publications has been cited h times. For example, a researcher with H 10 has published 10 papers that each have been cited at least 10 times.
- Limitations: The rating is skewed towards researchers' older papers; regardless of authorship position, all researchers are gauged equally. Researchers with a limited number of highly cited publications, though they may have published extensively, receive a lower rating. The rating is not fair when used to compare researchers across fields (Myers and Kahn 2021).

N/B: no publication metrics are perfect; each one is different and limited in scope. We need judicious interpretation.

Resource for further reading
https://www.mdpi.com/2304-6775/9/1/12

Researchers are keen on the impact factor of the journal they publish in because of rewards. Most employers use journal metrics to promote their employees. Employees who have successfully published their work in high-impact journals may be more likely to receive recognition, advancement opportunities, and other career benefits within their organizations. For instance, in China, scientists receive cash prizes, housing perks, and other benefits for publishing in high-impact journals (Englander 2014).

A word of caution. Every scientist would want to publish in high-impact journals. This, however, is not easy as rejection rates and competition are high. For instance, acceptance rates of journals with high impact factors of more than 3.0 range from 5% to 10% (Ware and Mabe 2015; Björk 2021). This has led some scientists to compromise and opt for publishing in low-impact journals, while others prioritize quantity over quality to increase their publication records. However, researchers should strike a balance between quality and quantity. While publishing numerous papers may demonstrate productivity, prioritizing quality over quantity is paramount. One quality publication in reputable journals contributes more to an individual's credibility than many in low-quality journal papers. This is because a good-quality paper will receive more citations than a poor paper. The more citations, the better the paper and the higher the HI Index (Google Scholar) (Carpenter et al. 2014). Thus, a researcher's objective should be to publish good-quality papers in high-impact journals that people will want to cite.

No single metric can accurately measure a researcher's overall impact. Every metric has strengths and limitations. Though they offer valuable insights into a researcher's productivity and influence, they should be combined with other qualitative and quantitative assessments. These may include publication quality, University policy, speed, peer recognition, reliability of reviewing, contributions to the research community, and societal impact (Carpenter et al. 2014; Myers and Kahn 2021). Researchers, funders, and publishers need to understand that it is about looking at the whole picture to understand the true impact of research.

5. Publication Process

The publication process is another crucial aspect to consider when selecting a journal. First, assess the journal's publication frequency and timeliness. Journals that publish regularly and promptly are likely to publish your article fast. Second, examine other factors, such as the journal's policy on article processing charges (APCs) and any associated costs involved in publishing (Burgman 2019; Björk 2021). Some journals may have high APCs, which could be a barrier for authors, especially those from institutions with limited funding. Third, assess if the journal adheres to scholarly publishing practices, including ethical standards, and transparency in editorial processes. These practices help maintain the integrity and credibility of the journal and ensure that published research meets established standards of academic excellence.

6. Editorial Board

Examining the composition of a journal's editorial board can provide valuable insights into its quality and credibility. Check if the editorial board members are recognized authorities in the relevant field. Check their affiliations with reputable institutions and academic qualifications. Additionally, the availability of contact information for editorial staff is crucial for transparency and communication. Missing or incomplete information on the journal's website, particularly regarding editorial board members, raises a red flag. Another indicator of editorial quality is the presence of editorials authored by the Editor-in-Chief or board members. Reputable journals often feature frequent and substantive editorials that offer insights into published articles or discuss updates in journal policies. For example, Springer or Elsevier websites offer information on how to select a journal and other information to enhance the overall scholarly experience for authors and readers alike.

7. Cost of Publishing

One of the major concerns of students when considering publishing papers in journals is the cost of publication fees (Lam and Langer-Gould 2021; Mekonnen et al. 2022). These costs vary depending on several factors. They vary according to the reputation and the impact of the journal and the publisher's services. As earlier mentioned, publication fees typically range from a few hundred to several thousand per article. Such fees pose significant financial barriers for students and affect their ability to publish in credible, high-impact journals. Many will be concerned about where to access such high fees, especially so in developing nations where institutional support grants are rare.

We have also witnessed fundamental shifts in traditional journals that have long operated as subscription-based journals. They have shifted to a new open-access model, and authors must pay to publish after their article has been accepted. A case in point is the Agricultural Water Management Journal. This affects many authors who previously relied on surrendering copyright and publishing their papers for free. One way of overcoming this is by budgeting for publications in our project proposals.

We encourage students not to let cost be a limitation to publishing. If you cannot afford publication fees, consider selecting a reputable Diamond Open Access journal, available through resources like the Directory of Open Access Journals (DOAJ), which lists numerous journals that do not charge authors for publishing. For journals that do require fees, it is still possible to obtain waivers or discounts, especially if you are a student, an unfunded researcher, or come from a low-income country. Many publishers offer these waivers proactively, so always check the journal's website for such opportunities. Additionally, consult your university librarian to identify arrangements your institution may already have with publishers for fee waivers or discounts.

27.3.2 Tools and Resources for Finding Journals

The journal finder is one of the latest technologies in the publishing landscape. It is an Electronic Journal Selector Web-based tool that utilizes advanced algorithmic and semantic technology to assist authors in journal selection (Entrup et al. 2023). These tools assist researchers to assess and compare journals and aid in identifying the most relevant ones for their work. By using a journal finder, users can efficiently generate a list of target journals in related areas by entering the title and abstract. The title and abstract are entered into appropriate search boxes. This functionality significantly streamlines finding a suitable journal for manuscript submission.

Various journal finders are available, including general ones and specific ones offered by publishers. While general journal finders offer a broad search across multiple publishers and disciplines, specific ones offered by publishers may provide more targeted recommendations based on their respective focus areas and editorial policies. Main ones include:

- Find my journal (from major publishers and databases): Springer, Elsevier, Taylor and Francis, Wiley, Pubmed, SAGE, Scopus, Web of Science.
- Springer Journal Suggester.
- Elsevier Journal Finder (Elsevier Journals).
- Enagos Open Access Journal Finder (DOAJ).
- Wiley's Journal Finder.
- Taylor and Francis's journal suggester.

Most of these journal finders use titles, abstracts, and keywords (Entrup et al. 2023). The output gives important information that one can use to decide on journal selection. For instance, using the Springer journal suggester, you get output on publishing model and impact factor (Box 27.5). Key information regarding these two journals "*Journal of Soil Science and Plant Nutrition (JSSPN)*" and "*Agricultural Research*" (AR) includes:

Box 27.5 Screenshot obtained using springer journal suggetser

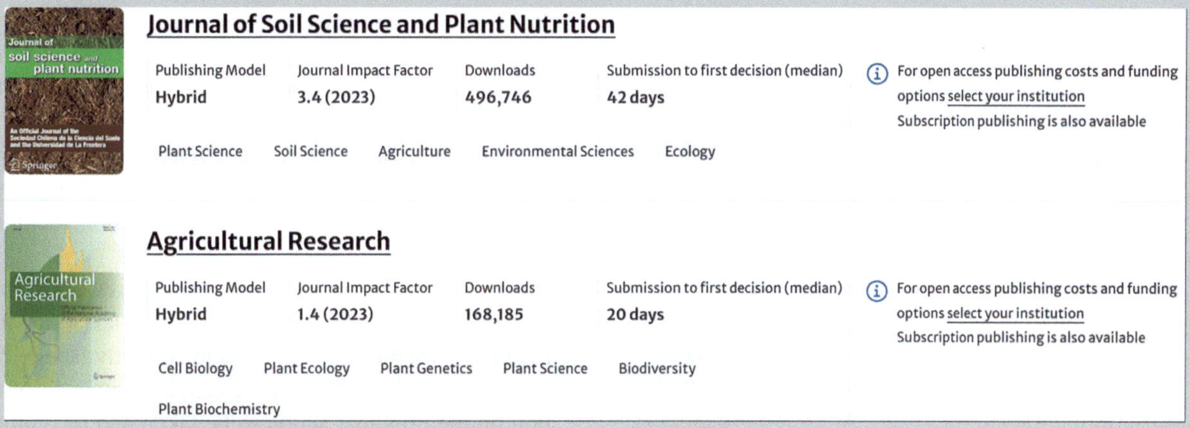

- Both operate a hybrid publishing model, meaning that if you are able to pay, you can choose the open-access model, but if you cannot afford to pay, you can choose the subscription-based model
- Journal impact factor of JSSPN is higher in 2023 (3.4) was higher than for AR (1.4)
- JSSPN has more downloads than AR, suggesting higher readership and visibility
- AR gives quicker feedback than JSSPN
- Both journals are related in scope but AR lean more towards molecular studies

Such information allows you to make your choice of a journal. You may combine this with a quick checklist given in Box 27.6.

Box 27.6 Quick checklist for selecting a journal

1. Check if the journal's scope aligns with the subject matter of your paper. You can do this by checking the journal's scope and aims. Visit the journal's website and look at a recent issue to understand specific topics. Also, consult the submission guidelines (instructions for authors).
2. Check if the journal is peer-reviewed. You can do this by checking previous articles and publishing history. Look for glaring mistakes and check the formatting of tables and graphs.
3. Check who the editorial board members are. You can do this by checking whether the members are researchers or professionals, where they work, their presence in Google Scholar, and their Author ID.
4. Check who publishes in the journal. You can do this by checking if the supervisor or any faculty staff has published in the journal and checking where renowned scientists in your field are published.
5. Check if the journal is by a major publisher or association. You can do this by checking if it is published by Springer, Elsevier, Taylor and Francis, SAGE, John Wiley and Sons, Taylor and Francis, Cambridge University. The world's five major publishers are Springer, Elsevier, SAGE, Taylor and Francis, and Wiley.
6. Check the publishing cycle. You can do this by checking past journal articles, mission, and vision.
7. Check how often it is cited. You can do this by checking if it is available in major databases, such as Web of Science, PubMed, Science Citation Index, SCOPUS, etc.
8. Check if the journal charges publication fees or APC (Article Processing Charges). Ensure that the journal practices fair and responsible publishing by charging authors under the open access model, where authors have an option of choice between paying APC or publishing through a traditional subscription model.

27.3.3 Creating Your Own Database of Target Journals

Once you are in a research career, you must continue publishing. We have emphasized the importance of publishing for career growth (Carpenter et al. 2014). It becomes easier when you have a database that you can easily consider whenever you need to publish a paper. A database is possible because, as a researcher, you must maintain focus in your subject area (Box 27.6). Focus is a secret of personal and professional growth. It keeps you from drifting through your career without clear goals. To quote Zig Ziglar, avoid being a "Wandering generality but be a Meaningful specific' or Robin Sharma "Focus is more important than intelligence". Hence, you will be publishing almost in the same journals. Build your database over time as you grow. This will enhance efficiency in target journal selection.

27.4 Manuscript Submission and Editorial Workflow

27.4.1 Making the Submission

Most journals provide online submission systems that streamline the process for authors. The system requires you to register. Use this platform to upload your manuscript and enter essential metadata, including author information, manuscript title, abstract, keywords, and suggested reviewers (Hartley and Cabanac 2017; Ceci and Williams 2018). Additionally, authors may be required to submit a cover letter detailing the significance of their research. The cover letter is a communication tool between authors and the journal editorial team.

Before submitting the manuscript, do a final check on the following:

1. *Referencing style*: Have you used the referencing style specified by the journal? Always ensure you use the recommended referencing style, check the guidelines but also check a sample paper.
2. *Completeness and consistency of references*: Are the references complete, consistent, and accurate? (authors' names, article title, publication information including dates and page numbers).
3. *Clarity of figures and tables*: Are figures and tables understandable without reference to the text? Are they required? Are they completely and correctly labeled?
4. *Recommendations*: Does your article contain information on the principal action to be taken and recommendations for further research by other researchers?
5. *Authorship*: Have you asked colleagues to read and comment on the text?—friends and peers will give you suggestions for improvement.
6. *Acknowledgments*: Have you acknowledged individuals and organizations?
7. *Take-home message*: What is the take-home message? (What is new?)
8. *Methodology*: Does the methodology correctly and clearly explain how you carried out the research? Can it be replicated by another researcher elsewhere? Does it contain extra methods?

9. *Clear and Concise style*: Is the prose clearly written, is the standard of English good enough to make the article clear and easy to read?
10. *Results*: Are the results correctly and clearly presented? Can they be understood easily without misinterpretation?
11. *Discussion*: Is the discussion clear and does it include sufficient acknowledgement of different perspectives and interpretations?
12. *Reference List*: Does your reference list correctly match the citations given in the text?

27.4.1.1 Cover Letter

A cover letter is needed to submit a manuscript (Bahadoran et al. 2021b). It includes several key elements. First, it should mention which journal the manuscript is being submitted to and provide a brief overview, including its title, authors, and the type of article (e.g., original research, review, case study). Additionally, the cover letter should explain the significance and novelty of the research, highlighting its potential contributions to the field.

Authors should also clearly state that the manuscript has not been previously published and is not under consideration for publication elsewhere. If any related manuscripts have been submitted or published, this information should be disclosed in the cover letter. Other requirements include conflicts of interest statements, funding sources, or ethical considerations related to the research. It is also common to suggest potential reviewers for the manuscript, although this is typically optional. It is courteous to express gratitude to the editor and reviewers for considering the manuscript for publication and provide contact information for correspondence regarding the submission.

27.4.2 Steps from Submission to Publication

The time from submission to publication typically takes ranges from 3 to 12 months, averaging around 6 months. Once a manuscript is submitted, it goes through a series of well-defined steps before publication (Fig. 27.2). The steps include:

1. *Submission*: Submit a manuscript to the editor through the specified submission portal or email address.
2. *Editorial assessment:* The editor makes an initial editorial assessment within 1–2 weeks, checking compliance with journal guidelines, aims, and scope. If compliant, the manuscript proceeds to peer review. If not, it may be returned to the authors for revisions or rejected outright. If returned because it does not fit the journal's scope, the best thing to do is to try another journal.
3. *Peer review:* Manuscripts that pass the editor's initial screening are sent for peer review, which takes about 4–8 weeks (it can vary widely). Two or three reviewers are selected to assess the manuscript's quality. They evaluate significance, originality, and methodological rigor. Finally, they provide detailed feedback and recommendations to the editor.
4. *Editors Decision*: After receiving feedback from reviewers, the editor evaluates the review and decides in about 2–4 weeks. Three options: (i) Acceptance: The manuscript is accepted for publication pending minor or major revisions, (ii) Revision: Authors are requested to revise the manuscript according to reviewers' comments before making a final decision, and (iii) Rejection: The manuscript does not meet the journal's standards and is unsuitable for publication.
5. *Authors revisions*: If the editor's decision is minor or major revisions, the author receives feedback and suggestions on revising and resubmitting. The time frame allocated for the revision is often 1 month. The authors revise the manuscript based on the reviewers' feedback and submit the revised version along with a response letter addressing each reviewer's comment.
6. *Editor's final decision*: Editors make final decisions after the review process is complete, which takes about 1–2 weeks. This is: (i) Acceptance: The manuscript is accepted for publication, (ii) Rejection: The manuscript is rejected if the revisions are inadequate or if it does not meet journal standards.
7. *Production*: The manuscript goes for copy editing and production—The finalized manuscript is published online first.

27.4.3 Role of Journal Editors

Journal editors are the gatekeepers of their respective journals. They have several roles. First, editors are responsible for editorial decision-making, evaluating submitted manuscripts and deciding whether to accept, reject, or request revisions based on the manuscript's originality, significance, methodological soundness, and alignment with the journal's scope.

Second, the editors oversee the peer review process, selecting appropriate reviewers, managing the review timeline, and synthesizing reviewers' feedback to inform editorial decisions. They ensure that the peer review process is fair, rigorous, and conducted per the journal's policies and ethical guidelines (Lim and Bowman 2024). In most journals, two anonymous reviewers are selected. The editors aim to select qualified reviewers for the job. If chosen wisely, the editor will be able to arrive at a worthwhile decision regarding the publication of the manuscript.

Third, editors also uphold the journal's standards of quality and integrity by enforcing ethical publishing practices

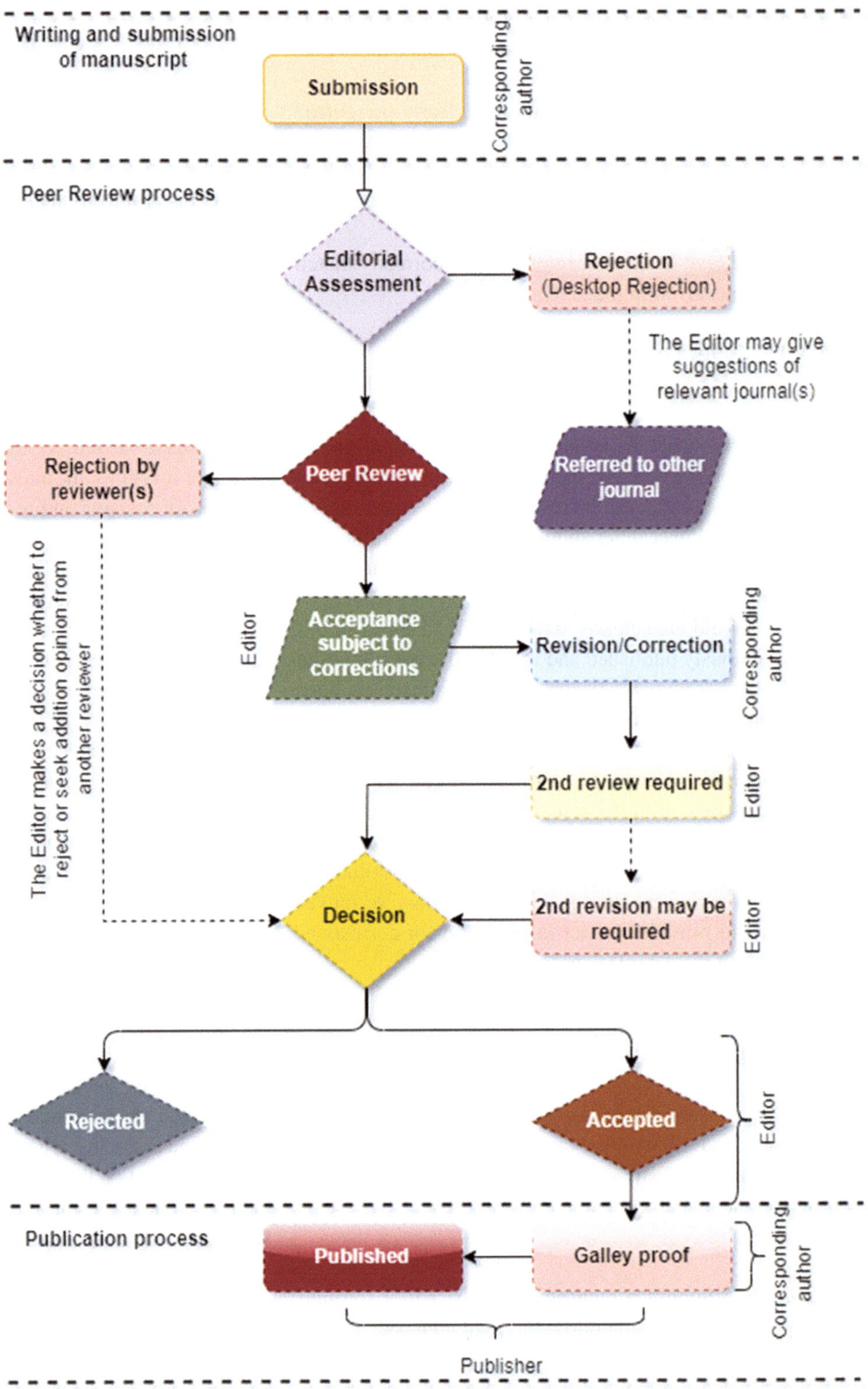

Fig. 27.2 Schematic presentation of the publishing process

(Starfield and Paltridge 2019; Edwards 2022). They address issues of plagiarism or research misconduct and ensure compliance with industry standards and best practices are also important. If they detect any malpractice, they contact the corresponding author for clarification.

Finally, editors establish and communicate editorial policies, guidelines, and criteria for manuscript submission, review, and publication. They define the journal's scope, objectives, and target audience, guiding authors on the types of articles suitable for publication. Editors provide strategic direction for the journal, identifying emerging trends and topics, shaping the editorial agenda, and positioning the journal within the broader academic landscape.

27.4.3.1 Confirmation of Manuscript Receipt

Once received and accepted for review, the editors assign every manuscript a unique code or identifier, and an acknowledgement is promptly sent to the authors via email. This acknowledgement confirms the journal's receipt of the paper and assures the authors that their submission is being processed for review (Box 27.7). All listed authors receive confirmation about the manuscript's progress. They get a link to track the review process and monitor the status of their papers, including updates and decisions from the editorial team and reviewers.

Box 27.7 Sample receipt acknowledgment

Manuscript Number: xxxxxxx

Title of the Manuscript, e.g., Surface runoff and soil erosion from nitisols and ferralsols as influenced by different soil organic carbon levels under simulated rainfall conditions.

Dear Corresponding author,

We have received the above-referenced manuscript you submitted to Name of journal

To track the status of your manuscript, please log in as an author at YYYYYYY, and navigate to the "Revisions Being Processed" folder.

Thank you for submitting your revision to this journal.

Kind regards,

Name of the journal

27.4.4 The Peer Review Process

The reviewers are allocated a specific timeframe for conducting their reviews. Once completed, the reviewers provide feedback to the editor, who then communicates this feedback to the authors. The subsequent process, from receiving the initial comments to reaching a final decision, can be lengthy and often requires all parties' patience. The duration varies depending on the nature of the feedback and the authors' responses to it. From our experience, it ranges from 2 to 6 months. Sometimes it can go beyond 6 months. This iterative process ensures thorough evaluation and consideration of the manuscript before a final decision is made.

27.4.4.1 What Do Reviewers Look for in Journal Papers?

When reviewers receive a manuscript, they carefully evaluate several key aspects to ensure its suitability for publication (Khetan and Sengupta 2017). First, they consider the general significance and interest of the paper to the journal, assessing its relevance and potential impact within the field. They check for novelty in the study. Additionally, reviewers scrutinize the scientific soundness of the research, examining whether the objectives are clearly defined, hypotheses are adequately tested, and methodology is robust. They also assess the manuscript's originality, organization, and clarity. They look for innovative ideas, logical structure, and coherence in the presentation of ideas.

In addition to the general outlook, the reviewers also check keenly the other sections, including the title, abstract, introduction, methods, discussion, and all other sections of the paper (Khetan and Sengupta 2017).

Title and abstract: Reviewers evaluate whether the title accurately reflects the paper's content and effectively captures its nature. They also assess the abstract's informativeness, ensuring that it succinctly outlines the study's objectives, scope, methodology, main findings, and implications. They look for clarity and completeness in each section of the abstract and coherence in conveying the overall message of the research.

Introduction: Reviewers assess whether the background information accurately describes the problem under investigation. They also look for a comprehensive literature review that places the study in a theoretical or historical context, highlighting relevant studies by other researchers with similar inquiries. Reviewers evaluate whether the earlier work is clearly explained and its connection with the current study is established. Importantly, they check the clarity of the problem statement and the articulation of the study's scope. The study's objectives are also scrutinized, along with its overall significance to the field of study.

Methods: Reviewers evaluate the appropriateness and clarity of the methodologies. They check whether the methodologies are detailed enough to allow for replication. Additionally, they assess if the statistical design and analyses are appropriate and correct. This includes paying close attention to issues such as replicates, significant statements, and data analysis procedures. For surveys and qualitative studies, reviewers focus on the clarity of the sampling procedure, appropriateness of the sample size and design, and the soundness of the data analysis approach.

Results section: Reviewers assess the clarity, form, and arrangement of tables and figures, ensuring they effectively present the data. They look for important data highlighted and interpreted in the text without unnecessary repetition. Reviewers also evaluate the accuracy of the analysis and interpretation of the results, addressing any conflicting evidence within the study itself and scrutinizing the data for errors in calculation and interpretation.

Discussion section: Reviewers carefully assess if the discussion conveys the findings' significance, implications, and potential applications. They check if the conclusions are grounded in data. This section is crucial in summarizing the research outcomes, connecting them to existing literature, and guiding future research, ultimately enhancing scholarly discourse and progress in the field. Reviewers often reread the introduction to check if it is clearly linked with the discussion and conclusions made.

27.4.5 Reviewers' Feedback to the Editor and Decision-Making

After reviewing the manuscript, the reviewers provide feedback to the editor (Edwards 2022), typically in the form of a detailed report outlining their comments, critiques, and recommendations. This feedback may address the study's strengths and weaknesses, identify areas for improvement, raise questions or concerns about the methodology or interpretation of results, and suggest revisions or additional experiments if necessary.

The editor carefully considers the feedback from the reviewers in conjunction with their assessment of the manuscript. Based on this evaluation, the editor decides how to treat the manuscript, which typically falls into one of the following categories:

1. Accept: The manuscript is accepted for publication without any further revisions.
2. Minor revisions: The manuscript requires minor revisions to address specific concerns raised by the reviewers.
3. Major revisions: The manuscript requires significant revisions, such as revising the methodology, conducting additional analyses, or rewriting sections of the manuscript.
4. Reject: The manuscript is rejected for publication due to fundamental flaws or lack of novelty, significance, or quality.

Once the editor has made a decision, they communicate this decision to the authors along with the reviewers' feedback. If revisions are requested, the authors are allowed to revise the manuscript and submit a revised version for further review. The revised manuscript may undergo additional rounds of review until it is deemed suitable for publication.

Note that the review process often requires multiple rounds of revisions and review, involving editors, reviewers, and authors, until the manuscript meets the journal's standards and the reviewers' expectations (see Fig. 27.2).

27.4.5.1 Authors' Response to Editor's Feedback

The comments by reviewers, usually conveyed via email, typically include an introduction from the editor and feedback from two or three reviewers. For most people, these verdicts usually evoke emotions, optimism, and apprehension (Shaw 2012). Usually, the reviewers start with general comments, which in most cases are encouraging, but this is followed by questions regarding the work's original intent. The comments are often positive and negative. The positives, at times, seem sarcastic because of the many negative comments that sound harsh. These comments touch on all sections of the paper, focusing on the limitations and shortcomings of the manuscript. Researchers should take this with an open mind and know that this process is a journey of growth and improvement. It is a constructive dialogue aimed at improving quality of the paper.

It is good practice to immediately share the feedback with your colleagues. When ready to start responding, carefully digest the feedback and revisit the content multiple times during revision to ensure that you understand the reviewer's expectations. Addressing these reviews can be a lengthy and challenging process, often likened to the difficulty and time consumption of writing the initial draft of the article (Sundstrom 2023).

This is worse for students publishing for the first time, especially if there's a pressure to publish quickly to meet graduation requirements. Some students have described this as "a nightmare experience". However, to overcome these challenges, commit to staying resilient and focused. Seek guidance and support from mentors and peers who can help you navigate the process with confidence. They will provide the necessary support for you to succeed. If you exceed the time given to revise the manuscript, you may ask for an extension from the editor. Even seasoned researchers fail to meet revision deadlines.

It is extremely crucial to respond satisfactorily to editors' feedback. Many editors have provided guidelines on how to respond to comments. Nahata and Sorkin (2019) identified six common flaws authors should avoid while responding to feedback. These include: (i) overlooking or inadequately addressing comments, (ii) lacking specificity regarding changes made in the manuscript, (iii) struggling to reconcile conflicting suggestions, (iv) offering insufficient evidence to support decisions not to make changes, (v) displaying frustration or impatience with the process or reviewers, and (vi) failing to clearly articulate the revised manuscript's significance and contribution to the field.

On the other hand, Noble (2017) has outlined ten simple rules for responding to reviewers' comments. We have picked five key ones, including beginning with an overview and including the full set of reviews, maintaining politeness and respect towards all reviewers, taking responsibility when necessary, starting each response with a direct answer to the point raised, and clearly indicating changes made compared to the previous version.

Before responding to the comments, carefully review them and organize your thoughts. Then, revisit the comments with a fresh perspective to understand the modifications requested by the reviewers or editors for your manuscript. We advise that reviewers' comments should be copied and pasted into a response table with each comment numbered, followed by the authors' response. Box 27.8 shows a response to the editor in one of our manuscripts (Kwizerimana et al. 2023). Including a kind address and explaining your response to the editor is good practice. As you can see from Box 27.8, we also included the manuscript's unique identifier.

Box 27.8 Sample of response to journal editor by the authors (Kwizerimana et al. 2023)

Ref.: Ms. No. SSHO-D-22-01396
Impact of collective marketing participation on farmers' income: Evidence from smallholder avocado farmers of Murang'a County, Kenya
Social Sciences & Humanities Open

Dear Editor,
Thank you for the feedback in regard to our work
We thank the reviewers for keenly reading our paper and giving useful feedback and comments.

As you guided, we have revised the work and included a table showing how we have addressed the comments raised. Two tables are included: Table 1 addresses general reviewers' comments, and Table 2 addresses major and minor comments

Table 1: General Reviewer's Responses to Questions

	Comment	Response
1	It is not clear whether the data are from a one-time survey (with recall data) or repeated surveys. It is also not clear whether the observations for 2018, 2019, 2020 are panel data or repeated cross-section data	The data was collected using a one-time survey (Cross-sectional survey) in August.2020 Data for 2018, 2019 and 2020 are recall data from the farmers. Avocado is an important

As you can see from Fig. 27.2, the reviewers will again review the paper. Remember, reviewers are not paid, and you should make their lives easier by ensuring that the reply to every comment is self-contained. This means that the reviewers will not have to search through the manuscript to locate the changes you have made. Instead of saying, "We have made the change" see page 7, line 15 of the revised manuscript", provide a direct reference as," we changed (indicate original text) to (indicate edited text) see page 7, line 15". This way, the reviewer will easily confirm the changes made.

Supposing there is a comment you do not agree with and feel may be the reviewer did not clearly understand. You need to explain why you do not agree and provide evidence, such as other published work, tables, or additional data. This evidence should be uploaded together with the response documents to the editor.

After addressing the comments, you need to upload documents following the editor's instructions. Some journals may require authors to use the track-changes mode in word processing software, while others may prefer changes to be highlighted or added in a different color font. Most often, journals may request two copies of the revised manuscript: one showing the changes and one in a clean format, ready for copyediting. Follow instructions as guided by the journal editor to the letter.

27.4.6 Common Reasons for Rejection

As mentioned earlier, rejection rates are generally high. Here are the common reasons for rejection in the publication process (Menon et al. 2022) and ensuring a manuscript avoids these pitfalls can significantly increase its chances of acceptance:

- Inadequate writing and organization, including weak background information, outdated literature reviews, and insufficient explanation of variables.

- A poorly defined research problem.
- Issues with study design, such as incorrect methods, small sample sizes, and other flaws.
- Misalignment with the journal's aims and scope leads to rejection at the editor level (Bordage 2001; Sullivan 2015).
- Lack of originality, novelty, and significance in the research.
- Problems with the sample, including small or poorly chosen samples.
- Weak interpretation of results and discussion and lack of credible evidence.
- Use of outdated data without appropriate justification.

27.4.6.1 What Happens After Acceptance

Once a manuscript is accepted, the editorial process and publishing will commence. The editorial team will conduct a final review, followed by copy-typing. During copy-typing, grammatical and typographical errors are checked and corrected to enhance clarity and readability. Once copy-typing is complete, the authors will receive a galley proof along with an "author query form" from the editorial team. The authors should carefully verify the galley proof and the correctness of figures and tables. It is expected that the authors respond promptly within 48 hours.

27.5 Challenges and Barriers to Publishing a Journal Article

27.5.1 The Problem of Unpublished Research

It is common to find students completing their master's and doctoral studies and failing to publish. Exceptions, however, are situations that require students to publish before they graduate. Completing a master's or doctorate and failing to publish is a huge loss to the university and the individual, as the research work is not idely shared. Even if the thesis is stored in the university repository, its access and readership compared with the journal paper is limited. Consequently, unpublished work is less likely to be cited, reducing its impact.

27.5.2 Publishing is Hard Work

Publishing a journal article in a peer-reviewed journal is hard work (Björk and Solomon 2013). It is a rigorous and challenging process (Day and Gastel 2012). It requires good writing skills, planning, and focus. Many students who have published their work in prime journals will attest that preparing a manuscript and finally publishing is tedious, laborious work that needs commitment. As Stephen King puts it, "Writing well is hard," and emphasizes perseverance and not giving up (King 2020). This becomes even more difficult for postgraduate students and early-career scientists, who might be doing their first serious writing for the first time. The advice is that you develop a growth mindset and approach writing as a skill to be developed over time through learning and persistence.

27.5.3 Disparity in Research Output

On a global scale, there is a notable disparity in the publishing outputs between developed and developing nations. In particular, the number of publications from the African continent remains relatively low (Englander 2014). This scarcity of publications in developing countries results from several significant challenges and constraints that hinder researchers' ability to produce journal articles. These challenges extend from the initial conceptualization of research to its eventual completion. As mentioned in Chap. 5, "a good thesis hinges on a good proposal;" hence, high-quality research is contingent upon well-defined research problems and effective implementation. In the following section, we will highlight these key challenges in the publishing process.

27.5.4 The Underlying Challenges to Publishing

Building on findings from Kumwenda et al. (2017), other scholarly literature (Majid et al. 2022; Hakami 2023), and our own academic experience, we identify five major categories of barriers that hinder successful publishing among graduate students and early-career researchers. These include:

1. *Knowledge and skill gaps*: Students are still under training, and thus, many have not yet developed publication skills. For instance, many students lack knowledge on the publication process, including peer review, editorial decisions, manuscript formatting, and submission protocols, which can result in delays or outright rejections. They also have weak writing skills coupled with poor arguments in their writing. At times, they fail to follow journal guidelines. Methodological problems, such as poorly defined research questions, small sample sizes, and flawed study designs, also undermine the credibility of their work. Additionally, choosing the wrong journal and responding poorly to reviewer comments are also significant errors among these budding researchers.
2. *Systemic barriers:* Reviewers working for free and are usually overburdened. They, therefore, deal with giving feedback causing delays. Further, despite reviewers

working for free, some open-access journals charge high article processing charges (APCs). Many students cannot afford these charges (Forrester 2015). Together with high rejection rates and prolonged turnaround times in prestigious journals, these factors hinder career progression and stifle academic momentum (Frank 2013; Haug 2013).

3. *Time and resource constraints*: Graduate students often have numerous responsibilities. Some will combine coursework, research, and personal issues. Additionally, many students lack access to essential support structures such as writing centers, peer review groups, and regular academic mentorship. This isolation often results in underdeveloped manuscripts and missed opportunities for revision and improvement. Building a strong research profile is further constrained by limited opportunities for collaboration, networking, and professional visibility, especially in resource-scarce institutions or regions.
4. *Psychological and emotional hurdles*: Publishing affects researchers emotionally. Many scholars experience the "Imposter Phenomenon," a psychological pattern characterized by persistent self-doubt and a fear of being exposed as incompetent. Many pull back when they feel they are not meeting the expectations in quality. They develop writer's block, which can delay or stall the writing process entirely (Chintamani 2014). These emotional stressors, when compounded over time, can lead to reduced motivation, burnout, and disengagement from the academic writing process.
5. *Balancing competing publishing pressures:* Researchers face pressure due to the "publish or perish" culture (McGrail et al. 2006; Lee 2014), and as a result, many struggle to balance the need for strategic visibility with the freedom to pursue meaningful, high-impact research.

These five key barriers are unpacked in Box 27.9.

Box 27.9 Quick summary of barriers to publishing by graduate students

1. Inadequate research skills and experience
2. Low funding in research
3. Scarcity of mentors to guide students adequately
4. Inadequate scientific writing skills
5. Limited opportunities for collaboration and networking
6. Low motivation to pursue quality research and to publish
7. Inadequate infrastructure (e.g., field and lab equipment)
8. Inadequate library and literature access
9. Limited understanding of the publication process (e.g., peer review, journal selection)
10. Fear of rejection and criticism
11. Imposter phenomenon and low self-confidence
12. Perfectionism leading to procrastination or writer's block
13. High article processing charges (APCs) for open-access journals
14. Overload from personal responsibilities
15. Poor research design or weak formulation of research questions
16. High rejection rates and long publication timelines
17. Lack of institutional support for editing and publishing
18. Difficulty accessing peer review or writing feedback
19. Difficulty in maintaining a focused academic identity amidst pressure to publish widely
20. Language barriers for non-native English speakers
21. Falling into predatory journal traps
22. Lack of institutional incentives to publish
23. Gaps in research ethics training
24. Academic isolation or lack of support network

The highlighted barriers are complex, and addressing them requires systemic changes in institutions that have a stake in publishing. However, we chose to concentrate on solutions that lie within an individual researcher.

As a student, we encourage you to develop your writing skills. Numerous reference books and resources are available in libraries and online platforms. They can easily be accessed through libraries and online repositories. We have provided a list of key reference books at the end of this chapter that can enhance your writing skills. However, practice is the key to mastering any skill, including writing. Engage in regular writing, attend writing workshops, and seek feedback from peers and mentors. These are not just suggestions but commitments that can significantly enhance your writing skills. So, let us commit to regular practice and improvement.

Always remember that the journal you chose is as important as the quality of the manuscript itself. Take time to choose one that aligns with your work's scope and target audience. Also invest time to understand the key criteria to evaluate manuscripts, including originality, quality of argument, clarity, and whether conclusions are supported by evidence. As you navigate the process from submission to publication, remember that rejections are a natural part of the journey. Rather than taking rejection, view it as an integral and constructive part of the learning and refinement process. Always remain open-minded, maintain positivity, and be persistent in your efforts to publish.

27.5.4.1 Initiative to Support Publications by Breaking the Barriers

Organizations such as *Research for Life* have aided in addressing the challenges in publishing and improving the publication output in low-income countries. They have achieved this through various initiatives that aim to bridge the global research gap and promote inclusivity and diversity in the scientific community. One of their key contributions is providing institutions in low-income countries with access to a vast repository of scientific literature, including journals, articles, and research databases. They support databases, such as AGORA. By removing financial barriers to accessing subscription-based journals, researchers in these countries can keep up-to-date with the latest research findings.

In addition to their research initiatives, these organizations provide training programs and capacity-building initiatives to support researchers in low-income countries. They organize workshops, webinars, and online resources to enhance research skills, writing proficiency, and understanding of publication ethics. Researchers who participate can improve their capacity to produce high-quality publications.

Research for Life and similar organizations also advocate for equal access to research resources. They encourage collaboration and networking among researchers from low-income countries and their global counterparts. By nurturing and knowledge exchange, they create opportunities for researchers to collaborate on international research projects and coauthor publications. They also aid researchers in securing funding opportunities for research projects and publication fees. Such funding initiatives alleviate financial constraints, enabling researchers in low-income countries to conduct research and share their findings through publications.

Recommended Further Reading

1. Youdeowei A, Stapleton P, Obubo R (2012) Scientific writing for agricultural research scientists - A training resource manual. p. 190.
2. How to write and publish a scientific paper by Gastel and Day (2022)
3. Nair PKR, Nair VD (2014) Scientific writing and communication in agriculture and natural resources. Scientific Writing and Communication in Agriculture and Natural Resources 1–142. https://doi.org/10.1007/978-3-319-03101-9/COVERPotential predatory and legitimate biomedical journals. Can you tell the difference? Shamseer et al. (2017)
4. Somoza-Fernández M, Rodríguez-Gairín J-M, Urbano C (2016) Presence of alleged predatory journals in bibliographic databases: analysis of Beall's list. Nature 489:179. https://doi.org/10.1038/489179A
5. Björk BC (2017) Open access to scientific articles: a review of benefits and challenges. Intern Emerg Med 12:247–253. https://doi.org/10.1007/S11739-017-1603-2

Key Points in This Chapter

Importance of Publishing in a Journal

Publishing research work benefits the research community, the institution, and the individual. It also enhances the University's ranking, attracts talent, generates revenue, and attracts funding. It offers personal satisfaction and boosts career growth, enhancing grant support and creating new advancement opportunities.

Types of Journals

We examined the following types of journals:

1. *International and regional journals*: International journals are globally recognized and often publish good-quality articles. Regional journals address local issues and provide a platform for researcher to disseminate their research work within their communities.
2. *Free-to-publish and paid-for journals*: Free-to-publish journals usually follow the traditional subscription-based model, where the researcher publishes for free. In paid-for journals, authors pay Article Publishing Charges (APC) to get their articles published.
3. *General discipline journals and specialist journals*: General discipline journals cover a wide array of topics within a particular field, while speciality journals focus on specific major specialities, subspecialties, or selected research areas within a discipline.
4. *Credible and predatory journals*: Credible journals are trustworthy publications that follow proper publication ethics, while predatory journals are unworthy and lack credibility and quality.
5. *Multidisiciplinary journals*: Multidisciplinary journals are suited for research that combines diverse methodologies, theories, or data. They include *PLOS ONE*, *Scientific Reports* (from the Springer Nature Portfolio), *Heliyon* (published by Cell Press), and various titles under the *MDPI* platform. Others are prestigious journals like *Nature* and *Science*

Selecting the Right Journal

Choosing the right journal is crucial in the publishing process. Key factors include supervisor preferences, journal impact factor, publication frequency, timeliness, costs, readership scope (international, national, or regional), target audience, and language. Personal preferences, such as affiliations with professional organizations, can also play a role.

Submitting the Manuscript

Most journals provide online submission systems that streamline the process for authors. The system requires registration.

A cover letter is needed to submit a manuscript. It includes a brief overview of the manuscript, including its title, authors, and the type of article, its significance and novelty of the research, and stating its potential contributions to the field.

Once a manuscript is submitted, several generic steps are followed. The time from submission to publication can take 3 to 12 months, with an average of 6 months. The steps include Submission, Editorial assessment, Peer review, Editor's Decision, Author revisions, editor's final decision, and production.

Role of Journal Editors

Journal editors oversee the review process, ensuring quality and ethical standards in publishing. They guide the journal's direction, identify trending topics, and enhance its academic presence.

Upon receiving a manuscript, editors assign a unique identifier and promptly acknowledge the submission via email. Authors can track their manuscript's status and updates through the platform.

The Peer Review Process and Authors' Revision

The reviewers are allocated a specific timeframe for conducting their reviews. Once completed, the reviewers provide feedback to the editor, who then communicates this feedback to the authors.

When reviewers evaluate a manuscript for publication, they assess its significance, relevance, novelty, and scientific integrity. They examine all sections of the paper to ensure cohesion.

The editor then considers the reviewers' feedback to decide on the manuscript. Researchers need to receive this feedback with an open mind, recognizing it as part of a journey toward growth and improvement.

Authors must satisfactorily respond to editors' feedback. Before responding, carefully review the comments and organize your thoughts.

Challenges and Barriers to Publishing a Journal Article

- Gaps in knowledge and skills in academic writing and publishing
- Barriers like high article processing charges (APCs) and inconsistent reviews
- Time and resource constraints from diverse responsibilities of graduate students
- Psychological hurdles, including self-doubt and fear of inadequacy, leading to writer's block and delays.
- The pressures to publish frequently in a "publish or perish" culture can stifle intellectual curiosity and innovation.

Publishing in academic journals can be challenging, but researchers can increase their chances of success by improving their writing skills, practicing regularly, and seeking constructive feedback.

Review Questions

1. How is publishing by its faculty and students important for the University?
2. As an individual researcher, how does publishing aid in your "personal development and recognition"?
3. Briefly explain the key distinguishing feature between Free-to-publish vs. Paid-for (Open Access) Journals.
4. Explain why "predatory journals are unworthy and lack credibility.". You receive an email inviting you to publish your research in a journal you've never heard of. It promises very fast peer review and publication and has a very broad-sounding name. List at least three red flags that would make you suspect it might be a predatory journal, based on the chapter's discussion or your understanding.
5. List at least five factors you would consider when selecting your target journal for this paper. For each factor, explain *why* it is important for this specific research topic.
6. What are "Journal finders," and how could they assist you in identifying the most relevant journals for this paper?
7. How do editors "uphold journal standards, quality, and integrity by enforcing ethical publishing practices"?

Chapter Exercise

1. Pick two journals from credible publishers that publish in your field and fill in this and provide the following;
   (a) Journals name
   (b) Scope fit
   (c) Indexing
   (d) Open access, subscription, or hybrid
   (e) Publication fees
   (f) Target audience
   (g) Peer review process
2. Use a journal finder (e.g., Elsevier Journal Finder, Springer Journal Suggester, or Journal/Author Name Estimator [JANE]). Input your title and abstract (Real or hypothetical). List the top 3 suggested journals**.** For each, assess whether it meets your publishing needs (impact, scope, access, and reputation)
3. Suppose you receive the following peer review comment: *"The results are interesting but not clearly linked to the objectives stated in the introduction. Also, avoid vague terms like 'improved significantly' without statistical backing."* Write a polite and professional response to the reviewer, indicating what changes you will make.
4. Describe the typical stages a manuscript undergoes from editor assessment through to production
5. What are some proactive steps you can take *now* to "invest persistently in learning and practising writing and publishing skills" to overcome these barriers?

Answers to Review Questions

- Q1. Reference: 27.1.2
- Q2. Reference: 27.1.2
- Q3. Reference: 27.2.1
- Q4. Reference: 27.2.2
- Q5. Reference: 27.3.2
- Q6. Reference: 27.3.2
- Q7. Reference: 27.4.3

References

Alston JM (2019) Open access principles and practices benefit conservation. Conserv Lett 12. https://doi.org/10.1111/CONL.12672

Asubiaro TV, Onaolapo S (2023) A comparative study of the coverage of African journals in web of science, Scopus, and CrossRef. J Assoc Inf Sci Technol 74:745–758. https://doi.org/10.1002/ASI.24758

Bahadoran Z, Mirmiran P, Kashfi K et al (2021a) Scientific publishing in biomedicine: how to choose a journal? Int J Endocrinol Metab 19:1. https://doi.org/10.5812/IJEM.108417

Bahadoran Z, Mirmiran P, Kashfi K et al (2021b) Scientific publishing in biomedicine: how to write a cover letter? Int J Endocrinol Metab 19:3. https://doi.org/10.5812/IJEM.115242

Bergstrom CT, West JD, Wiseman MA (2008) The Eigenfactor™ metrics. J Neurosci 28:11433–11434. https://doi.org/10.1523/JNEUROSCI.0003-08.2008

Björk BC (2017) Open access to scientific articles: a review of benefits and challenges. Intern Emerg Med 12:247–253. https://doi.org/10.1007/S11739-017-1603-2

Björk BC (2021) Why is access to the scholarly journal literature so expensive? Portal 21:177–192. https://doi.org/10.1353/PLA.2021.0010

Björk BC, Solomon D (2013) The publishing delay in scholarly peer-reviewed journals. J Informetr 7:914–923. https://doi.org/10.1016/J.JOI.2013.09.001

Bordage G (2001) Reasons reviewers reject and accept manuscripts: the strengths and weaknesses in medical education reports. Acad Med 76:889–896. https://doi.org/10.1097/00001888-200109000-00010

Burgman M (2019) Open access and academic imperialism. Conserv Biol 33:5–6. https://doi.org/10.1111/COBI.13248

Cals JWL, Kotz D (2014) Effective writing and publishing scientific papers, part X: choice of journal. J Clin Epidemiol 67:3. https://doi.org/10.1016/j.jclinepi.2013.09.014

Camkin J, Neto S, Bhattarai B et al (2022) Open science for accelerating the sustainable development goals: status and prospects in Asia and the Pacific. Front Polit Sci 4:878761. https://doi.org/10.3389/FPOS.2022.878761/BIBTEX

Carpenter CR, Cone DC, Sarli CC (2014) Using publication metrics to highlight academic productivity and research impact. Acad Emerg Med 21:1160–1172. https://doi.org/10.1111/ACEM.12482

Ceci SJ, Williams WM (2018) Online submission and review systems. In: Guide to publishing in psychology journals. Cambridge University Press, pp 204–216. https://doi.org/10.1017/9781108304443.019

Chintamani (2014) "Challenges in writing"-the writer's block? Indian J Surg 76:3–4. https://doi.org/10.1007/S12262-014-1058-X/METRICS

Clark J, Smith R (2015) Firm action needed on predatory journals. Br Med J 350. https://doi.org/10.1136/bmj.h210

Clark AD, Myers TC, Steury TD et al (2024) Does it pay to pay? A comparison of the benefits of open-access publishing across various sub-fields in biology. PeerJ 12:e16824. https://doi.org/10.7717/PEERJ.16824/FIG-5

Cobey KD, Galipeau J, Shamseer L, Moher D (2016) Report on a pilot project to introduce a publications officer. CMAJ 188:E279–E280. https://doi.org/10.1503/cmaj.151340

COPE (2019) COPE Ethical guidelines for peer reviewers. https://doi.org/10.24318/COPE.2019.1.9

Day RA, Gastel B (2012) How to write and publish a scientific paper, 4th edn. Cambridge University Press, Cornwall

Drozdz JA, Ladomery MR (2024) The peer review process: past, present, and future. Br J Biomed Sci 81:12054. https://doi.org/10.3389/BJBS.2024.12054/BIBTEX

Edwards J (2022) The journal editor as academic custodian. In: The inner world of gatekeeping in scholarly publication. Palgrave Macmillan, Cham, pp 227–244. https://doi.org/10.1007/978-3-031-06519-4_13

Englander K (2014) Writing and publishing science research papers in english. Springer. https://doi.org/10.1007/978-94-007-7714-9

Entrup E, Eppelin A, Ewerth R et al (2023) Comparing different search methods for the open access journal recommendation tool B!SON. Int J Digit Libr 25:505–516. https://doi.org/10.1007/S00799-023-00372-3/TABLES/4

Etzkowitz H (2003) Research groups as 'quasi-firms': the invention of the entrepreneurial university. Res Policy 32:109–121. https://doi.org/10.1016/S0048-7333(02)00009-4

Forrester A (2015) Barriers to open access publishing: views from the library literature. Publications 3:190–210. https://doi.org/10.3390/PUBLICATIONS3030190

Frank M (2013) Open but not free — publishing in the 21st century. N Engl J Med 368:787–789. https://doi.org/10.1056/NEJMP1211259
Garfield E (2006) The history and meaning of the journal impact factor. JAMA 295:90–93. https://doi.org/10.1001/JAMA.295.1.90
Goldstein H, Rehbogen A (2013) University engagement and knowledge commercialization: an analysis of faculty attitudes. In: Knowledge commercialization and valorization in regional economic development. Edward Elgar Publishing, Inc, pp 61–84. https://doi.org/10.4337/9781781004074.00011
Hakami MSA (2023) Barriers to conducting and publishing scientific research among nursing faculty members in Saudi Arabia. J Multidiscip Healthc 16:2733–2743. https://doi.org/10.2147/JMDH.S429478
Hartley J, Cabanac G (2017) The delights, discomforts, and downright furies of the manuscript submission process. Learn Publ 30:167–172. https://doi.org/10.1002/LEAP.1092
Haug C (2013) The downside of open-access publishing. N Engl J Med 368:791–793. https://doi.org/10.1056/NEJMP1214750/SUPPL_FILE/NEJMP1214750_DISCLOSURES.PDF
James C, Colledge L, Meester W et al (2018) CiteScore metrics: creating journal metrics from the Scopus citation index. Learn Publ 32:367–374. https://doi.org/10.1002/leap.1246
Khetan V, Sengupta S (2017) What a reviewer wants? Indian J Ophthalmol 65:656–657. https://doi.org/10.4103/IJO.IJO_699_17
King S (2020) On writing: a memoir of the craft. Scribner
Kumwenda S, Niang EHA, Orondo PW et al (2017) Challenges facing young African scientists in their research careers: a qualitative exploratory study. Malawi Med J 29:1–4. https://doi.org/10.4314/MMJ.V29I1.1
Kwizerimana S, Mugwe J, Nigat B (2023) Impact of collective marketing participation on farmers' income: evidence from smallholder avocado farmers of Murang'a County. Kenya Soc Sci Humanit Open 8:100614. https://doi.org/10.1016/J.SSAHO.2023.100614
Lam K, Langer-Gould A (2021) Open access fees: a barrier to scholarly activity among neurology trainees. Perm J 25. https://doi.org/10.7812/TPP/21.041
Larivière V, Sugimoto CR (2019) The journal impact factor: a brief history, critique, and discussion of adverse effects. In: Springer handbooks of science and technology indicators. Springer, pp 3–24. https://doi.org/10.1007/978-3-030-02511-3_1
Lee I (2014) Publish or perish: the myth and reality of academic publishing. Lang Teach 47:250–261. https://doi.org/10.1017/S0261444811000504
Lim WM, Bowman C (2024) Giving and responding to feedback: guidelines for authors and reviewers. Act Adapt Aging 48:1–20. https://doi.org/10.1080/01924788.2024.2304948
Lövei GL (2021) Writing and publishing scientific papers: a primer for the non-English speaker. Writing and Publishing Scientific Papers 214. Open Book Publishers. https://doi.org/10.11647/OBP.0235
Majid H, Jafri L, Ahmed S et al (2022) Publication dynamics: what can be done to eliminate barriers to publishing full manuscripts by the postgraduate trainees of a low-middle income country? BMC Res Notes 15:1–6. https://doi.org/10.1186/S13104-022-06138-5/TABLES/2
Martín-Martín A, Orduna-Malea E, Thelwall M, Delgado López-Cózar E (2018) Google Scholar, Web of Science, and Scopus: a systematic comparison of citations in 252 subject categories. J Informetr 12:1160–1177. https://doi.org/10.1016/J.JOI.2018.09.002
McGrail MR, Rickard CM, Jones R (2006) Publish or perish: a systematic review of interventions to increase academic publication rates. High Educ Res Dev 25:19–35. https://doi.org/10.1080/07294360500453053
Mekonnen A, Downs C, Effiom EO et al (2022) Can I afford to publish? A dilemma for African scholars. Ecol Lett 25:711–715. https://doi.org/10.1111/ELE.13949
Menon V, Varadharajan N, Praharaj SK, Ameen S (2022) Why do manuscripts get rejected? A content analysis of rejection reports from the Indian Journal of Psychological Medicine. Indian J Psychol Med 44:59–65. https://doi.org/10.1177/0253717620965845
Myers BA, Kahn KL (2021) Practical publication metrics for academics. Clin Transl Sci 14:1705–1712. https://doi.org/10.1111/CTS.13067
Nahata MC, Sorkin EM (2019) Responding to manuscript reviewer and editor comments. Ann Pharmacother 53:959–961. https://doi.org/10.1177/1060028019849941
Nair PKR, Nair VD (2014) Scientific writing and communication in agriculture and natural resources. In: Scientific writing and communication in agriculture and natural resources. Springer, pp 1–142. https://doi.org/10.1007/978-3-319-03101-9/COVER
Noble WS (2017) Ten simple rules for writing a response to reviewers. PLoS Comput Biol 13:e1005730. https://doi.org/10.1371/JOURNAL.PCBI.1005730
Roessner D, Bond J, Okubo S, Planting M (2013) The economic impact of licensed commercialized inventions originating in university research. Res Policy 42:23–34. https://doi.org/10.1016/J.RESPOL.2012.04.015
Sanberg PR, Gharib M, Harker PT et al (2014) Changing the academic culture: valuing patents and commercialization toward tenure and career advancement. Proc Natl Acad Sci USA 111:6542–6547. https://doi.org/10.1073/PNAS.1404094111/SUPPL_FILE/PNAS.201404094SI.PDF
Sandesh N, Wahrekar S (2017) Choosing the scientific journal for publishing research work: perceptions of medical and dental researchers. Clujul Med 90:196. https://doi.org/10.15386/CJMED-704
Sayer EJ (2019) The essentials of effective scientific writing—a revised alternative guide for authors. Funct Ecol 33:1576–1579. https://doi.org/10.1111/1365-2435.13391
Shamseer L, Moher D, Maduekwe O et al (2017) Potential predatory and legitimate biomedical journals: can you tell the difference? A cross-sectional comparison. BMC Med 15:1–14. https://doi.org/10.1186/S12916-017-0785-9/TABLES/10
Shaw JD (2012) Responding to reviewers. Acad Manage J 55:1261–1263. https://doi.org/10.5465/AMJ.2012.4006
Shen C, Björk BC (2015) "Predatory" open access: a longitudinal study of article volumes and market characteristics. BMC Med 13:1–15. https://doi.org/10.1186/S12916-015-0469-2/TABLES/3
Solomon DJ (2014) A survey of authors publishing in four megajournals. PeerJ 2014:e365. https://doi.org/10.7717/PEERJ.365/SUPP-2
Somoza-Fernández M, Rodríguez-Gairín J-M, Urbano C (2016) Presence of alleged predatory journals in bibliographic databases: analysis of Beall's list. Nature 489:179. https://doi.org/10.1038/489179A
Sovacool BK, Axsen J, Delina LL et al (2022) Towards codes of practice for navigating the academic peer review process. Energy Res Soc Sci 89:102675. https://doi.org/10.1016/J.ERSS.2022.102675
Starfield S, Paltridge B (2019) Journal editors: gatekeepers or custodians? In: Novice Writers and Scholarly Publication: Authors, Mentors, Gatekeepers. Palgrave Macmillan, Cham, pp 253–270. https://doi.org/10.1007/978-3-319-95333-5_14
Strinzel M, Severin A, Milzow K, Egger M (2019) Blacklists and whitelists to tackle predatory publishing: a cross-sectional comparison and thematic analysis. MBio 10. https://doi.org/10.1128/MBIO.00411-19/ASSET/9BBB4A91-7260-483C-8C24-BFE94DD9BB14/ASSETS/GRAPHIC/MBIO.00411-19-F0007.JPEG
Sullivan GM (2015) What to do when your paper is rejected. J Grad Med Educ 7:1–3. https://doi.org/10.4300/JGME-D-14-00686.1
Sundstrom A (2023) Responding to reviewers: guidelines and advice. Politics Gender 19:630–635. https://doi.org/10.1017/S1743923X23000168

Ward SM (2016) The rise of predatory publishing: how to avoid being scammed. Weed Sci 64:772–778. https://doi.org/10.1614/WS-D-16-00080.1

Ware M, Mabe M (2015) The STM report: an overview of scientific and scholarly journal publishing. Copyright, Fair Use, Scholarly Communication, etc

Wijewickrema M, Petras V (2017) Journal selection criteria in an open access environment: a comparison between the medicine and social sciences. Learn Publ 30:289–300. https://doi.org/10.1002/LEAP.1113

Yan Z, Snow C (2020) Publishing journal articles: a scientific guide for new authors worldwide. Cambridge University Press. https://doi.org/10.1017/9781108277426

Yeo-Teh NSL, Tang BL (2021) Wilfully submitting to and publishing in predatory journals-a covert form of research misconduct? Biochem Med (Zagreb) 31. https://doi.org/10.11613/BM.2021.030201/FULLARTICLE

Youdeowei A, Stapleton P, Obubo R (2012) Scientific writing for agricultural research scientists - A training resource manual. CTA, Wageningen, p 190

28 Scientific Writing Skills and Presentations

Abstract

Scientific reporting and making presentations in academic forums form the backbone of your academic journey. The reports and presentations are about writing science; therefore, they must be written in a scientific language and pattern. Key scientific writing skills include clarity, conciseness, coherence, logical flow, objectivity, and consistency. A thesis and journal articles are critical scientific reports that graduate students must produce to earn their degree. While these two documents share many similarities, such as the need for rigorous research methodology, they also exhibit distinct differences in purpose, format, and audience. When writing your thesis, ensure you first read a sample thesis to benchmark with others in your area and follow guidelines provided by your institution. A master's thesis differs in scope and aims from a doctoral thesis. While a master's thesis is limited in scope and aims to build research skills, a doctoral thesis has a broader scope and aims to produce original and significant results. Drafting a journal article from a thesis requires first isolating the key message. The key message guides the paper's content, starting by isolating the specific paper's key data, tables, and figures. The examination of your thesis consists of two components: the written document and the oral examination. It is important to prepare adequately for your success. Examiners typically adhere to similar guidelines and generally want you to succeed, which is encouraging. However, it is crucial to maintain scientific rigor throughout your research. During your academic work, you will participate in scientific meetings where presenting your findings, both internally and externally, is essential. Preparing effective visual aids for oral presentations is vital. Additionally, conferences often feature poster sessions; a well-designed poster captures attention and conveys your research's key message. A sample PowerPoint presentation is included in this chapter.

Keywords

Active voice · Clarity and conciseness · Coherent writing · Conferences · Key message · Logical structure · Masters and doctoral theses · Oral presentation · Paragraphs · Paraphrasing · Passive voice · Poster presentation · Power-point slides · Scientific presentations · Seminar presentation · Structured process · Thesis examination

Learning Outcomes

After reading this chapter, the learner is expected to be able to:

1. Explain the importance of scientific reporting.
2. Describe key methods and tools to employ for effective scientific writing.
3. Demonstrate differences between a thesis and a journal paper.
4. Model the structure of a thesis.
5. Demonstrate key considerations in thesis writing.
6. Illustrate the key steps in preparing a journal publication from a thesis.
7. Determine what examiners "look for" in a thesis.
8. Make a powerful oral presentation at internal and external scientific meetings.
9. Develop a good poster presentation.

28.1 Overview

28.1.1 Importance of Scientific Presentations

Scientific reporting and making presentations in academic fora are not just activities but the backbone of your academic journey. As a graduate student, they are the primary channels through which you can disseminate your findings to a broader audience. These are typically delivered through written doc-

J. Njeri Mugwe, S. Runo, *Research Methodology in Agricultural Sciences*, https://doi.org/10.1007/978-981-95-1892-0_28

uments or oral presentations at departmental seminars, conferences, or workshops. These activities allow you to share research outcomes and also help you engage with peers, receive feedback, and cultivate collaborations within the academic community. Because the reports and presentations are about writing science, they must be written scientifically, which necessitates applying *specific scientific writing skills*.

Scientific writing skills are not just a requirement but a powerful tool that empowers you to exchange scientific knowledge. Key skills include the following:

- Writing proficiency
- Structured articulation of research findings
- Adherence to formatting requirements
- Citation conventions adherence
- Effective communication of complex concepts
- Transparency in reporting

Mastery of scientific writing helps you communicate your research findings accurately. This chapter explores three fundamental aspects of scientific writing and presentations: (i) Effective scientific writing skills, (ii) Thesis writing and examination, and (iii) Oral and poster presentations.

28.1.2 What Is Scientific Writing?

Scientific writing (SW) is writing about science to inform others about the findings. It is a two-way process because the objective is communication (Matthews and Matthews 2014; Nair and Nair 2014; Sayer 2019). Scientific writing uses a unique style to make communication as easy as possible. It has three key characteristics;

- *Scientific writing is a structured process:* The research reports and other publications are the final product of a structured process (the scientific process) (Chap. 1). In the same vein, the reporting adopts a structured manner that is organized, logical, and accurate. You should follow a structured way of reporting usually guided by the client you are writing for. For instance, the thesis format will be provided by the university, whereas journals have varied styles that authors are supposed to adopt (IMRAD-Chap. 25).
- *Scientific writing is original:* The final report should be original because the completed research was motivated by a well-defined research problem and supported by a valid methodology. The report contributes to the world of science as it attempts to fill the gap that motivated the research.
- *Scientific writing follows a specific style:* The style is a technical form of writing characterized by precision, clarity, and objectivity to convey research findings effectively within the scientific community. It is also guided by ethical principles.

Freelance Online Writing Is Not Scientific Writing Freelance online writing, especially by young individuals who write to complete assignments for clients and students abroad, is not scientific writing. Though at times it may be presented as science, it deviates significantly from scientific writing. First it prioritizes financial gain, leading to lower standards and a lack of adherence to scientific writing. Second, most writers lack formal training in writing or research methodologies, resulting in inconsistent quality and potential breaches of academic integrity.

28.1.3 General Ideas About Scientific Writing

Poor writing is easy, but good writing is difficult. Good writing is an art (Fennell 1975). This is good news because it can be taught and learned. Good writing is a product of training, hard work, and determination (Grogan 2021). Whenever you write, always remember that authors should write for their readers, which makes the process challenging. A quote from William Zinsser "*Writing is hard work. A clear sentence is no accident. Very few sentences come out right the first time or even the third time. Remember this as a consolation in moments of despair. If you find that writing is hard, it's because it is hard. It's one of the hardest things that people do*" (Yaffe 2023).

Effective scientific writing poses a significant challenge for many postgraduate students. This is partly because of inadequate training. Most academic programs emphasize science and technical training, often neglecting the development of writing skills (Ascheron 2023). Consequently, when students are tasked with writing a thesis or publication, their writing abilities may be underdeveloped. This lack of proficiency can lead to delays in completing their degrees and graduating, as writing is a time-consuming process. It requires adapting content and language for specific audiences while following rules of grammar, spelling, punctuation, and style. To enhance writing skills, students should read widely and practice regularly; a good reader typically becomes a good writer.

28.2 Key Principles of Scientific Writing

The key principles include:

1. Clarity
2. Conciseness
3. Keep it short: Short words, sentences, and paragraphs

4. Coherent writing
5. Logical structure
6. Objectivity
7. Correct word use
8. Use active verbs
9. Paraphrasing
10. Write in correct tense
11. Moderate use of normalization
12. Eliminate double negatives or multiple negatives
13. Avoid filler words
14. Consistency in format
15. Proofreading and editing your work
16. Avoid gender bias

28.2.1 Clarity

Clarity is extremely important because it enhances the reader's understanding. Scientific writing adopts a clear style to facilitate easy reading and understanding. Day and Gastel (2012) stated that clarity is the most important aspect of scientific writing. To achieve clarity, write simple and precise words to communicate the exact meaning of your research. Also, be factual about what you communicate and avoid complicated sentences. Avoid ambiguity, jargon, and uncommon abbreviations that are not explained. Also, ensure you report accurately and avoid ambiguous language, such as approximately and about.

Examples of the Use of Concrete, Simple, and Precise Words

- Prior to—Before
- Due to the fact that—Because
- In a considerable number of cases—Often
- The vast majority—Most
- During the time that—When
- In close proximity to—Near

Words to avoid: Judgmental words, e.g., surprising, unfortunately, interesting.

Avoid when possible	Better
Fact	Evidence
Prove	Support
Plays an important role	Is important because
Due to the fact that	Because
A decreased number of	Fewer
Time period	Time
Longer time period	Longer
Brown in color	Brown
Round in shape	Round
Has been shown to be	Is
It is possible that	May
Demonstrate	Show
Exhibit	Show
Utilize	Use
In other words	Thus/hence/therefore

Examples of Unclear Sentences

- "The crop yields increased significantly after treatment application." It is unclear because it does not specify the scale of increase, the type of treatment, or the crop in question. We can improve it by stating, "*The maize crop yield increased by 27% after applying 50kg N ha*$^{-1}$ *of Nitrogen fertilizer.*"
- "The experimental design followed a randomized complete design (RCBD) with four replications per treatment across three sites, each with variable soil pH, organic matter and cation exchange capacity." It is unclear because of complexity and unnecessary details. An improved version: "*We used a randomized complete design with four replications at three sites with varying conditions.*"

28.2.2 Concise Writing

A concise writing style is also called a "lean writing style." It involves stripping your sentences to just the words that tell. William Strunk gave a famous quote "Vigourous writing is concise". A sentence should contain no unnecessary words, and a paragraph should have no unnecessary sentences." Using the analogy of tree pruning to maximize fruit production, we can say that "just as a well-pruned tree maximizes fruit production by removing unnecessary branches that divert nutrients from the main trunk", concise writing trims away extra words and sentences to focus on key information. But remember, concise writing is not about making every sentence short or skimming over details. Instead, it is about ensuring that every word serves a purpose and contributes meaningfully to the text's overall message (Clapper 2020). We achieve concise writing by avoiding repetition and redundancies Writing guides and manuals offer many other examples and valuable tips on avoiding unnecessary words (Clapper 2020).

28.2.3 Keep It Short or Use Short Words, Sentences, and Paragraphs

Words In the selection of words, you can keep it short by avoiding redundancy. This means removing words that do not add meaning or value to a sentence. There are many examples of redundancy in common misuse. For example

"Please do not hesitate to" = Please
"Up to the present time" = until now
"Prior to the start of" = before
As a matter of fact = in fact

There is also a preference for using shorter words, especially when two words have the same meaning, e.g. "show" instead of "demonstrate."

Sentences Sentences should be short to enhance readability. The average sentence length should be 20 words, balancing long and short sentences to create variation. Mixing long and short sentences adds variety and improves the rhythm of the writing (Youdeowei et al. 2012). Avoid too much information in a sentence, as it makes the message difficult to understand. If a sentence seems too long, break it into shorter, more digestible sentences. This allows the reader to process the information more easily.

Consider a sentence: *"Soybean yields significantly increased in treatments where compost was applied at a rate of 5 t ha^{-1} with irrigation provided at a rate of two times per week during the growing season, demonstrating the combined effects of organic matter addition and optimal moisture management on crop productivity."* This sentence is unclear. We can make this clear by breaking the sentence and having the following:

> *Soybean yields significantly increased in treatments where compost was applied at a rate of 5 t ha^{-1}. Irrigation was provided two times a week during the growing season. This demonstrates the combined effects of organic matter addition and optimal moisture management on crop productivity.*

Another unclear sentence arises when the subject and the verb are separated, making it hard to identify the main subject and its connection to the conditions described.

Examples of Unclear and Clear Sentences

"Lime, present in various forms (powdered, pelletized, and liquid), is used for soil pH adjustment and nutrient supplementation in agriculture" (Unclear).

"Lime is used in agriculture to adjust soil pH and supplement nutrients. It is available in various forms, such as powdered, pelletized, and liquid" (clear).

In this case, the subject is lime, and the authors want to say that it is used in agriculture to adjust soil pH.

Paragraphs Short paragraphs make ideas easier to read and understand. They keep readers engaged and prevent them from feeling overwhelmed with information. By focusing on only one primary idea or argument, short paragraphs maintain flow and coherence, allowing readers to grasp key points more efficiently. It also helps organize information effectively contributing to a more impactful and compelling writing style. While there is no concrete rule on paragraph length, paragraphs should be long enough to develop one complete idea.

28.2.4 Coherence in Writing

Coherence means "to stick together." Therefore, coherence in writing means that all the ideas in a paragraph flow smoothly from one sentence to the next. With coherence, the reader easily understands the ideas that you wish to express.

Coherence is crucial in writing because a researcher is making an argument. This applies regardless of whether we are synthesizing previous research, supporting a new method, or presenting our conclusions. The effectiveness of our arguments depends on how well our sentences connect and build off each other. Readers tend to focus on the ideas presented at the end of sentences, so it is crucial to begin subsequent sentences by expanding upon the preceding idea. This aligns with the end focus, where known information is usually introduced at the beginning of sentences, while new information is presented at the end (Brown et al. 2012; Höhle et al. 2014). Experienced writers skillfully utilize this approach by starting sentences with information already known or previously introduced and concluding them with new information. This structure caters to readers' expectations, as they expect to encounter novel and significant details toward the end of each sentence.

To improve coherence in writing, we use tools of coherence. These are tools that connect sentences. Examples are:
Pronouns (it is, this is …)
Transition terms (examples)
Enumeration (first, second, third)

We can also improve coherence through the use of connector words

Some common connectors:
For example,
For instance,
One example of (this) is
First, Second, Third, etc.
Another example of [xxx] is (that)
Finally,
Similarly,
In conclusion,
To summarize,
On the one hand,
However,
…, but…

Paragraphs are the most important elements in writing that guide the argument. We should use them as the building blocks to convey cohesive ideas and arguments. However, crafting a good paragraph can be challenging due to the need for logical organization and coherence (Plaxco 2010). A well-written paragraph should exhibit unity. This means that every sentence must contribute to the central theme. It should also guide the reader through a coherent progression of thought, maintain focus, and avoid repetition or digression. If we can master the art of writing effective paragraphs, we can go a long way in effectively engaging our readers through our argument.

Maintaining Coherence Within the Paragraph Coherence in writing results from two key elements: paragraph unity and sentence cohesion (Ahmed 2010; Rahman 2013; Chen and Cui 2022). To ensure paragraph unity, a writer must accomplish two objectives. First, the paragraph must revolve around a single idea articulated in a *topic sentence*. This tells the reader about your argument (Plaxco 2010). Second, each subsequent sentence in the paragraph should provide specific details that expand on the topic and maintain its focus (Yakhontova 2021). The content of the paragraph should consistently support the initial argument, discussing only the main idea throughout.

In a cohesive paragraph, every sentence arises logically from the sentence before it and transitions logically into the following sentence. Check these three sentences: (i) Organic matter is vital for sustaining soil fertility and ecosystem health, (ii) Through decomposition, organic matter releases essential nutrients for nourishing plant life, including nitrogen, phosphorus, and potassium, and (iii) These nutrients are readily available for plant uptake, promoting healthy growth and development.

Continuity Between and Among Paragraphs Paragraphs discuss ideas, and there should also be continuity between and among paragraphs:

- Paragraphs containing only one or two sentences are rarely good paragraphs because they cannot develop ideas adequately.
- Read "Hook and eye technique" for continuity and coherence (Creswell 2014).

28.2.5 Logical Structure

Scientific writing adopts a structure with a logical flow of sections and ideas. This is achieved through clear subheadings to ensure that processes follow each other in a logical organization. The use of paragraphs also helps create a logical flow. As you write from one section to the other, ensure that your writing adopts the progression of ideas.

Scientific writing is about making arguments. To maintain the flow of the arguments, build them logically, starting from what is known to the unknown. The IMRAD (Introduction, methods, results, and discussion) structure of writing scientific publications helps scientists write systematically and clearly (Wu 2011). As you write, begin with sufficient background to provide context and then offer concise but enough details to communicate the significance of your research and the findings. Similarly, your results should be presented in the same order as in the materials and methods in the thesis.

Use *parallel* construction to enhance understanding. Parallel construction means that your ideas, which are similar in content and function, should be presented in a consistent grammatical structure (Watling 2015). It helps clarify relationships and improve readability. This structure should be in both sentence and paragraph construction. For example, we can write, "The study aims to assess soil fertility, to evaluate nutrient use efficiency and to assess crop yield improvements". Each of the objectives follows the same grammatical pattern, which helps the reader easily follow the list.

28.2.6 Objectivity

Scientific writing is not biased and remains neutral. It is based on the principles of scientific research methods, and therefore, objectivity and precision are key (Wedderkopp and Rutz 2024). It avoids personal beliefs, emotions, and subjective interpretation. All assumptions, statements, and ideas are supported by appropriate evidence that demonstrates how conclusions have been drawn. That is why citations are commonly used in almost all sections of a journal article.

As you report your scientific work, ensure that all assumptions are supported by evidence. You should also distinguish fact from possibility. For example, if it is an opinion, support it with such words as suggest, probably, seem, and appear.

28.2.7 Correct Word Use

Words have precise meanings, and to use them correctly adds clarity and precision to prose (Luby and Southern 2022a, b). Look at the following pairs of words that are often used in scientific texts. Learn how to use them correctly: Fewer, less; infer, imply; as, because; disinterested, uninterested; discrete, discreet; insignificant, nonsignificant.

Example to show the difference between less and fewer by using the two words in the same sentence. *Less is used to refer to a collective noun, while fewer is used to refer to multiple items, for example "Less grain weight per spike" when discussing grain weight, and "Fewer grains per spike" when counting individual grains."*

- Differ from/Differ with; "Differ from" indicates dissimilarity between two things, while "differ with" suggests individual disagreement.
- "Discreet and discrete": "Discreet" means showing prudent behavior, while "discrete" refers to something separate or distinct.
- "Objective and rationale": Objective refers to the aim or goal, while "rationale" indicates the reason or justification.
- " Persons and people": "Persons" refers to individuals considered separately, while "people" refers to a larger or anonymous group.
- Principal and principle: "Principal" can mean either the amount of money on which interest is earned or a chief official in a school or court proceeding, while "principle" denotes a basic truth or belief. In a research proposal, the lead investigator is referred to as the "Principal Investigator," not the "Principle investigator."

28.2.8 Use Active Voice

In general, the active voice is preferred to the passive voice. In the active voice, the subject of the sentence does the action; in the passive the action is done to the subject.

Examples:

- Passive: The harvesting of the experiment will be done on June 10th

Active: We will harvest the experiment on June 10th
Passive: It was found that maize yields…
Active: We found that maize yields…
Active: The authors found…

In scientific writing, the active voice is preferred because it is

- Stronger,
- Shortens sentences,
- Conveys information concisely and with greater impact than the passive voice.

The passive voice, on the other hand, is

- vague,
- weak,
- awkward.

This makes the active voice, once again quoting Dr. Strunk 'more *vigourous*'. However, in academic writing, the passive voice is sometimes more appropriate and may be required by your institution. For instance, in the methodology section, passive voice is likely to be dominant. This means you have to take care to make it interesting, but whenever possible, use the active voice. Generally, you will find a mix of active and passive voices in scientific papers.

28.2.9 Paraphrasing

Paraphrasing is taking statements from another source and restating them using your own vocabulary. We should mainly use paraphrasing to cite other people's work. This way, you avoid plagiarism and also show that you understand the other person's work. Authors mainly use "authors prominent" or "information prominent" styles (Thomas 2021b).

Direct quotations should only be used when paraphrasing might change the meaning, but quoting large texts should be avoided (Thomas 2021b).

28.2.10 Write in Correct Tense

We use a variety of tenses depending on when the action happened and the section of the report or paper. Past tense is used in the methodology section to describe what was done. Similarly, past tense may be used when citing primary and secondary literature. For example, the Author (Date) reported that treatments with a combined application of P fertilizer plus manure performed better than sole P fertilizer.

The present tense is mainly used for statements that are facts, such as statistics and figures from the Food and Agriculture Organisation (FAO) and general truths and conditions that are continuously true; sometimes, it is even used when the statement concerns published findings.

In the introduction, it is common to use mixed tense. Present tense is used to refer to theory, while past tense is used when referring to previous past findings. For example, *Tithonia diversifolia* is a useful source of green manure for smallholder farmers. Future tense could be used for a hypothesis. In the discussion section, we mainly use two types of tenses. Past tense is used for the present study (we found that…), while present tense relates our work to results of other studies (our results agree…).

The abstract is mainly written in the past tense because we are reporting the current study. Typical sentences could be: The effects of different levels of *Tithonia diversifolia* mulch on kales yield were tested. Soil pH increased in treatments with Tithonia applied and reduced in treatments with sole application of fertilizer.

28.2.11 Moderate Use of Normalization

Normalization is the use of a verb or adjective as a noun. This should be avoided because it obscures clarity and precision. It causes confusion, buries the action, and weakness the sentence, making it harder to understand. Examples of normalized verbs include;

A *decision* was made to harvest the crop before its maturity (weak)
We *decided* to harvest the crop before it matured (strong)
The application of mulch *resulted in an increase* in maize yields (weak)
Application of mulch *increased* maize yields (Strong)
Invasion of the experiment by wild animals led to the destruction of the net plots (Weak)
Wild animals invaded the experiment and destroyed the net plots (strong)
An increase in soil nitrogen took place (weak)
Nitrogen uptake increased (Strong)

Normalized	Verb
Resulted in an increase	Increased
Made an assumption	Assumed
Led to the destruction	Destroyed
Made an analysis	Analyzed
Made a decision	Decided
Was injurious	Injured
Was measured	Measured

Sometimes it is necessary to normalize the verb, and this should be used in moderation.

28.2.12 Eliminate Double or Multiple Negatives

In mathematics, when two negatives are combined, they result in a positive outcome. Similarly, when two "negative" words are used in a sentence, they may cancel each other out, creating a positive meaning. For example, the phrase "not uncommon" means something common. However, it is crucial to maintain clarity and simplicity in scientific writing. Therefore, avoiding using multiple negatives in the same sentence or clause is recommended. Although two negatives do not necessarily negate each other in language as they do in mathematics, they can often confuse instead of providing clarification in scientific discourse.

Consider the following examples:

The phrase "No decline in the absence of not using pesticides" is indeed ambiguous and vague. It can be interpreted in multiple ways, which may lead to confusion. It could mean that something does not decrease when pesticides are not used.

Consider this statement, and you are reading quickly or without close attention: "The maize grain yields were not unaffected by the absence of no fertilizer application." The double negative construction makes it challenging to understand the intended meaning. Does it mean the yields were affected or not by the absence of fertilizer application? The other problem is that the sentence uses complicated words to say a simple thing.

28.2.13 Avoid Filler Words

Filler words are unnecessary and vague additions to sentences. Examples include "a study of," "an investigation of," (often found in titles) and expressions like "it is important to point out that," etc. They add no value to the sentence and should be avoided to make your writing clear and concise.

28.2.14 Consistency

Maintaining consistency in scientific writing is crucial for precision, clarity, and professionalism. Strive to achieve consistency throughout the document; in choice of words, font type and size, numbering, punctuation, abbreviation, spacing, and citation and referencing. For instance, if you use "drought tolerant maize," do not switch to "drought-resistant maize." Similary, use uniform font type and size throughout the document. Journals, conferences, and thesis guidelines specify spacing and font types. Times New Roman is extremely common in academic writing.

28.2.15 Proofreading and Editing

Students commonly submit work with glaring mistakes, indicating that they did not proofread their work with seriousness (Harwood et al. 2009, 2010; Richards and Harwood 2024). You should proofread and edit your reports before making the submission, either to the supervisor or for examination. Proofreading allows you to identify and correct grammatical errors, typographical errors, spelling mistakes, and many other issues. Fixing these issues improves the readability and clarity of your writing. As you proofread and edit, check on the following:

- Sentence structure and clarity: Read a sentence and identify if it says what you meant it to say
- Accuracy of values
- Word choice
- Grammar
- Spelling
- Check for words like from (form), there (their)

28.2.16 Avoiding Gender Bias

Using plural nouns and pronouns helps avoid singling out any particular gender and promotes inclusivity and gender neutrality.

See the following examples

- Instead of "Every scientist should be aware of his responsibilities," use "All scientists should be aware of their responsibilities"
- Instead of "Each farmer received his seed maize" use: "All farmers received their seed maize"
- Instead of "Every student must complete laboratory analyses by the deadline," use "All students must complete their laboratory analysis by the deadline"
- Instead of: "A manager should be prepared to handle his team's challenges." use "Managers should be prepared to handle their teams' challenges"

28.3 Writing a Thesis

Writing a journal scientific paper was discussed in Chap. 26. While journal article and thesis share many similarities, such as the need for rigorous research and adherence to professional standards, they also exhibit distinct differences in terms of purpose, format, and audience (Table 28.1). The thesis typically represents the culmination of a student's academic journey, demonstrating their mastery of a particular subject area and contributing original knowledge to their field (Weaver et al. 2016; Björklund et al. 2024). As such, they tend to be more extensive in scope and detail, often spanning multiple chapters and including comprehensive literature reviews, methodologies, results, and discussions. A thesis often exceeds 20,000 words. In contrast, journal articles are typically more concise and focused, aiming to communicate specific research findings to a targeted audience of experts in the field. They undergo a rigorous peer-review process and must adhere to the formatting and stylistic guidelines of the respective journal.

The primary audience for a thesis is the university, as it is intended for evaluation by a thesis committee and examiners (Björklund et al. 2024). However, beyond merely passing the examination, the quality of the thesis is paramount for establishing credibility as a researcher. The thesis may also be graded in some universities, emphasizing the importance of ensuring its quality. Additionally, upon completion, the thesis is typically preserved in the university's repository, where it becomes part of the institution's scholarly record. A well-written and well-researched thesis has the potential to be cited by other researchers in the field, further contributing to its impact and recognition within the academic community. Therefore, investing time and effort into producing a high-quality thesis fulfils the requirements for academic evaluation, enhances one's reputation as a scholar, and increases the likelihood of making meaningful contributions to advancing knowledge in the field (Björklund et al. 2024).

Table 28.1 Differences between thesis and journal paper

Category	Thesis	Journal Paper
Main audience	University: Earn a degree	Scientific community: Personal development
Content of introduction	Contains many subsections (background, statement of the problem, objectives, significance of the study, etc.)	Combines background, statement of the problem, justification, and literature review in one section
Objectives of the study	Many objectives presented	One or two objectives
Literature review	Most have a detailed chapter on literature review	Usually infused in the introduction
Key messages	Detailed and extensive presentation with many messages	Brief and focused with one or two key messages
Materials and methods	Methodology presents all the materials and methods for all objectives in the thesis.	Methodology is specific only for the objective of the paper
Format	Format is according to university guidelines	Format is usually IMRAD and follow respective journal guidelines
Length	Longer and more than 20,000 words	Shorter and normally ranging from 7000 to 12,000 words
Results	Results contain both descriptive and inferential statistics; detailed results	Results are summarized and mainly concentrate on inferential statistics; gives main results
Quality check	Quality check is done by supervisors and examiners	Quality check is done by peer reviewers and the process is rigorous
Rigorousness	Less rigorous in terms of the quality of writing	More rigorous requiring high skills in writing

28.3.1 Structure and Format of a Thesis

As stated earlier, your university provides guidelines, but generally, the following is the generic structure, outline, and format (IMRAD).

- Preliminary pages
  - Declaration page
  - Dedication
  - Acknowledgement
  - Abstract

  - Table of contents
  - List of tables
  - List of Figures
  - List of abbreviations and acronyms
- Chapter 1: Introduction (Covered under proposal—Chapter 5)
- Chapter 2: Literature Review (Covered under proposal—Chapter 5)
- Chapter 3: Materials and methods (Covered under proposal—Chapter 5)
- Chapter 4: Results (Covered under Scientific paper—Chapter 26)
- Chapter 5: Discussions (Covered under Scientific paper ...)
- Chapter 6: Conclusions and Recommendations (Covered under Scientific paper ...)
  - References
  - Appendices

The thesis is a report where you bring the work you have done together. The writing should be clear, concise, both comprehensive and comprehensible (Patel et al. 2004). It should follow the principle of scientific writing discussed in this chapter. The format is similar to that of the proposal (Chap. 5), with notable differences in the use of tenses and additional chapters on results and discussion, conclusions, and recommendations. The content of the results and discussion is covered in Chap. 24 of the scientific paper.

It is hoped that a student will be able to write a good thesis with guidance from the supervisors, paying attention to differences given in Table 28.1, and borrowing heavily from the discussion on how to write a scientific paper. The following references for further reading are also recommended (Kember and Corbett 2018; Parija and Kate 2018).

Before starting to write, do the following three things:

1. *Read and follow the guidelines for thesis writing of your university*

You must write a thesis that conforms to the format, structure, and style of your University and department. Even if you had read earlier, possibly at the proposal writing stage, consult again the guidelines for writing a thesis. Obtain this from the postgraduate handbook, your department, or the University website. Read the guidelines and take note of the format and structure. In addition, note the writing style in terms of font type, font size, and spacing.

No standard format or structure is universally applicable across all universities. Even within the same institution, guidelines may differ between schools and departments. Therefore, thoroughly familiarize yourself with the specific guidelines and formatting requirements provided by your university, school, or department. You can also seek clarification from faculty advisors or graduate program coordinators to help clarify any uncertainties.

2. *Benchmark with theses written by former students*

Benchmarking is a valuable practice when preparing to write your thesis (Parija and Kate 2018). Obtain and read samples of past theses. This can often be accessed through the university repository or library. As you review these samples, pay close attention to the organization of chapters, the depth of the literature review, the clarity of research methodology, and the presentation of results and analysis. Observing how previous students have tackled similar research questions or topics can provide valuable guidance and inspiration for your thesis writing process. Additionally, analyzing the formatting and citation styles used in these samples can help ensure consistency and adherence to academic standards in your writing. By benchmarking with past theses, you can learn from the successes and challenges of others, ultimately enhancing the quality and effectiveness of your own thesis.

3. *Think of the reader*

Think about the different readers of your thesis. Your thesis's number one reader is the supervisors. The supervisors expect you to understand your topic, follow academic standards, and present findings accurately (Gedamu and Gezahegn 2021; Nangimah and Walldén 2023). The second group of readers is comprised of the examiners who will evaluate your thesis for academic rigor, originality, and contribution to the field. The final is the other scientific community. They will read the thesis after it is finished and approved. Their interest is mainly in the significance of your research. You should, therefore, write your thesis with the readers in mind and aim to satisfy their expectations.

28.4 Drafting a Journal Article from a Thesis

Students often have questions as they write journal papers based on their thesis. Common questions include: (1). Can I use my thesis to write a journal manuscript? How many papers can I write from my thesis? (2) If it is one, how do I compress the thesis into 7000–10,000 words? These are pertinent questions, and hopefully, they are addressed in this section.

Many students struggle to transform their thesis work into a journal paper. Some think they can simply copy and paste their thesis. This is misguided. Journal papers have different guidelines and expectations, making this approach ineffective. Condensing a thesis into a journal paper can be challenging, primarily due to the 7000–10,000 word limit. Therefore, concentrate on the most crucial elements as you write and eliminate the rest. The key is to convey a clear key message to meet the audience and journal's requirements.

28.4.1 Essentials of Writing a Manuscript from a Thesis

Because the thesis will have more information than you require, the following are guidelines for planning a manuscript. Isolate the following:

- Start with the most direct objective and develop a *central message*
- Clearly write out the objective of the paper
- Define the materials and methods for the selected objective
- Define the principle findings and isolate the key data (tables and figures)
- Describe the conclusions and implications
- Organize related ideas together
- Read and assemble key literature related to the objective

After isolating the above, follow the following steps:

Step 1: Determine the Purpose of the Paper Before you start writing, ask the following questions:

- What is the aim of this paper?
- What do I want to say?
- What is the new knowledge solving a problem?
- Why should a journal publish my paper? The answer to this gives the "key Message"

The key message is the main idea that determines whether a paper gets published. It is the most important information you want to communicate. It grabs the attention of readers. In thesis exams, students often struggle to express this effectively. Not having a clear key message in journal papers can lead to rejection (Hobson 2014; Ali 2021). Authors need to communicate this message clearly and concisely, avoiding including too much information. Some journals even require a section with 3–5 key points. This emphasis ensures readers understand the significance of the research. Simplifying the core findings into a clear key message enhances the impact of the paper.

You may find that each significant result or objective can provide a key message. Each of these key messages can potentially be transformed into a journal paper. For a master's thesis, it is reasonable to aim for up to two journal articles, while for a PhD, it's recommended to aim for at least three or four papers with substantial results. To begin, examine all the results and select key messages for various papers. Dividing your research into separate papers increases your work's visibility and enhances its coherence. Each paper can concentrate on a specific aspect or finding, allowing for more in-depth analysis and discussion. Additionally, breaking down your research into multiple papers enables you to target different audiences and journals, increasing the likelihood of publication and impact.

However, as you break your research into multiple papers, refrain from *"salami slicing'* (Szeto and Jull 2022; Wright 2024). This is breaking down findings into smaller pieces to inflate the number of publications artificially. While it may appear beneficial to have an abundance of publications, salami slicing can harm the authenticity and influence of your research. Journals and reviewers may perceive this practice unfavorable and doubt the originality and significance of each paper. Rather, concentrate on producing significant papers. Moreover, prioritizing "quality" over quantity will help you uphold the authenticity of your research and increase its impact on the academic community.

Step 2: Select the Journal Select the journal that you want to publish (see Chap. 26 on how to select a journal). First of all, study the instructions for authors. Follow through with reading the sample paper of the journal you have selected, and finally, plan the structure of your paper. Take note of how the paper is structured and how the results are presented, among other things.

Step 3: Develop an Outline Whether you will use the draft from the thesis or start a new paper, clear writing starts with clear thinking. Developing an outline helps in the writing process by aiding in organizing ideas. It provides a snapshot of each section of the write-up. It shows the flow and aids in the presentation of information logically by showing relationships among ideas.

Step 4: Select Tables and Figures Pick the figures and tables from your essentialist list, making captions self-contained. Your data are the driving force of the paper. Decide on the best form of presentation (see Chap. 24 on data presentation and visualization). Sometimes, you may need to change the mode of presentation in the thesis and even reanalyze the data.

Step 5 Write the Methods Section Proceed to write the methods section of your study, aligned with the central theme, by including relevant tables and figures. This step involves the easiest part of the research, which is describing what you did in your study. Ensure that your methods are written clearly and completely, providing enough details for the readers to understand how the research was conducted.

Step 6 Write the Results Section When writing the results section, make sure that any illustrations, such as tables and figures, are clear and effectively support your findings. Each illustration should have self-contained captions that explain their significance. Maintaining consistency by using only one form of presentation throughout the results section is crucial. Additionally, pay careful attention to statistical analyses, ensuring the accuracy and proper interpretation of the data.

Step 7: Drafting the First Manuscript Use the target journal checklist and instructions for authors. Ensure you have a draft of all the sections of the paper. Take on the role of an editor of your paper, carefully reviewing and revising each section for clarity. Work on it using circular arguments, iteratively, revisiting and refining the ideas, ensuring that each section gradually leads into the next and that your overall argument is easy to follow.

Step 8: Presentable Second Draft Refine the first draft with the aim of presenting it to your co-authors. This second draft should have all the components of the paper. Share with the co-authors and wait for comments. Once you receive feedback, address all comments from co-authors. In case there are comments you are unable to address, discuss them directly with them.

Step 9 Third Draft:
Depending on your writing skills, you are likely to have a good third draft, especially if comments are minor. Work toward finalizing the draft. Share with supervisors and seek approval to submit. You may also share with friends. At this stage, you may engage a scientific editor to check on grammar and editorials.

Step 10: Final Manuscript and Submission Finalize the manuscript using the checklist for the journal. Finalize the title and abstract, ensuring that all sections are flowing and clearly connected. Proofread your paper, clearly reflecting whether you have communicated what you wanted to communicate. Finally, submit through the recommended means.

28.4.2 Ethics in Publishing

To avoid any scientific misconduct, researchers should follow the recommended guidelines for scientific writing. Chap. 6 gives details of ethics in publishing. In summary, authors should keep in mind the following six aspects: refraining from plagiarism, preserving confidentiality, avoiding inaccuracies and bias, disclosing conflicts of interest, acknowledging funding sources, and abstaining from data manipulation, such as falsification and fabrication.

28.5 Differences Between Masters and Doctoral Thesis

While Masters and Doctoral theses present academic work, they differ in scope, depth, and expectations. A master's thesis focuses on demonstrating students' ability to conduct research and write coherently (Parija and Kate 2018). The research question may be a specific issue and narrow in scope. The aim is to show competence in the research process and contribute meaningful knowledge in a limited way. The main aim of the master's program is to build skills in conducting research. Though this is the case, they are still expected to publish work in a journal, in most universities. Conversely, the doctoral thesis is larger in scope and more depth, and readers have bigger expectations. A doctoral thesis must make a significant contribution to the field and show a high level of scholarship. To meet these demands, many universities are increasingly adopting the presentation of a doctoral thesis as a cumulative thesis. This approach consists of a series of publications in reputed refereed journals supplemented with a concise summary of the overall research work. Regardless of the format, the doctoral thesis should show originality and a significant contribution to the scientific community.

28.6 Thesis Examination

The thesis examination has two components: i) examination of the written work, and ii) Oral examination (Tinkler and Jackson 2000; Riches-Suman 2024) (Table 28.2).

Examination of the Written Work
The first part of the thesis examination involves examining the written documents. Two or three examiners conduct the examination, critically assessing several aspects of the thesis. Feedback from this examination process is a detailed report highlighting strengths and identifying areas of improvement. The examiners also give recommendations on corrections or revisions to the thesis.

Oral Examination
The second part of the thesis examination is the oral examination, referred to as "Viva Voci." During this examination, the candidate presents their research to a panel of examiners and answers questions about their work (Lantsoght 2023; Tan 2023). The oral defense tests candidates' understanding of the topic and ability to respond to critical questions. It also tests the ability to articulate their research process and defend their methodology.

Table 28.2 Examination of the written work and oral assessment of a thesis

Examination part	Process	Key activities	Outcome
Written document	Detailed review of written document	A critical review of research quality, methodology, findings, and conclusions	-report with strengths, weaknesses, revisions, and recommendations -judgement on passing or failing (grading)
Oral examination	Presentation, defense, and answering of questions	Presenting thesis work in 15–20 min Defending methodology and conclusions Answering questions	Evaluation of defense performance Potential for future recommendations for revision

Passing both parts of the examination is a requirement for meeting the academic standards for graduation. Therefore, you must prepare adequately to pass. The next section discusses what examiners assess as they read the work and prepare questions for oral defense. The points will help you prepare a good thesis document and deliver an effective oral defense.

28.6.1 What Do Examiners Look for?

Many students have little understanding of what the examiner will assess as they read and examine the thesis. All they know is that work will be evaluated by examiners who are not their supervisors and make a judgement; the details and specifics of the process remain hidden. Students will often learn about the examination process from their supervisor, colleagues, and institutional guidelines, but the details can remain unclear.

Examiners want you to succeed and are reluctant to fail (Mullins and Kiley 2002; Golding et al. 2014). They approach the thesis examination with an aim to assess quality and potential for success. Examiners, at the end of the day, will give both formative and summative verdicts. This dual approach is meant to improve the current work and for future research and publications. Thus, you should approach writing your thesis with seriousness but with a positive note that examiners want you to pass.

Institutions offer criteria for how the thesis will be examined. These criteria and instructions show how the actual process should be done, providing some notes and, at times, a scoring sheet. These guidelines may be vague, inadequate, and inconsistently applied (Kumar and Stracke 2011). For instance, what constitutes a significant contribution often varies across insitutions. Nevertheless, we can draw some common ground on what examiners look for as they examine a thesis (Mullins and Kiley 2002; Golding et al. 2014). They include:

- Evidence of original investigation and achievement of the objectives
- Quality and quantity are usually based on their experience and mastery of the subject
- Meaningful engagement with existing literature; look for classic and recent literature
- Capacity to conduct independent research
- Creative view of the topic
- Coherence and accurate argument
- Adequate interpretation, analysis, and critical appraisal of the findings. Application of proper statistical techniques in quantitative studies. Evidnece based results supported by a rigorous methodology
- Evidence of an independent researcher
- Coherence of theoretical and methodological perspectives
- Use of consistent practices in the whole thesis and in giving the recommendations
- Advancement in the field and publishability of the work

28.6.2 Preparing for the Oral Defense

The oral defense is your last hurdle toward obtaining the master's or doctoral degree. You must prepare adequately and build your confidence. Key steps include reading your thesis thoroughly to understand it inside out and practicing your presentation through mock presentation sessions. You should also familiarize yourself with the form of the defense and common common questions that may arise. Understanding the process and being well-prepared will help you present your research effectively and handle questions confidently. You are encouraged to read "How to Survive Your Viva" by Murray (2015).

28.7 Presentations in Departmental Seminars and Conferences

As part of research project thesis requirements, students are often encouraged or required to participate in scientific meetings, such as seminars, symposiums, conferences, and workshops. These events provide valuable avenues for students to share their research work, learn from experts in their field, and receive feedback from peers and established researchers. The events could be organized internally within the university, such as departmental seminars or externally. Internal university events offer students a supportive environment to gain presentation experience and receive feedback from faculty and fellow students. On the other hand, external events expose students to a wider range of perspectives, methodologies, and research topics beyond their immediate academic environment.

28.7.1 Presenting Research Work in a Departmental Seminar

A departmental seminar is organized by faculty staff to allow students to present the progress of their work to the faculty staff. As a student, you are expected to present your scientific work confidently, demonstrating a clear understanding of your research. Confidently presenting one's scientific work showcases the student's mastery of the subject matter and communicates their commitment to rigorous scholarship and intellectual inquiry.

Adhering to certain guidelines and expectations during the presentation is important. You must clearly articulate the research objectives, methodology, preliminary findings, and potential implications with clarity and precision. Since the seminar serves as a form of examination, commit to presenting within the allocated time frame, typically 15–20 min. Given the time constraints, it is often advisable to focus on a specific portion of your research, as advised by your supervisor. After your presentation, faculty staff and fellow students will have the opportunity to pose questions. Remain calm and composed during this questioning period. Embrace each question with openness and respond to the best of your ability. If you encounter a question that you cannot address satisfactorily, kindly acknowledge it and express your willingness to explore the topic further.

Tips for effective departmental seminar presentations (in-person or virtual):

- Ensure you have a stable internet connection
- Be in a quiet, distraction-free environment to minimize background noise
- Have clear and concise slides to enhance understanding
- Avoid cluttered slides with excessive text or unclear visuals
- Rehearse your presentation to ensure a smooth delivery and avoid stumbling over your words or losing your train of thought
- Avoid being over-defensive
- Be mindful of time constraints and avoid going over your allotted presentation time

28.7.2 Participating and Presenting at a Conference

Conferences are important avenues that allow students to interact with researchers and other stakeholders beyond the institution where they are studying. They provide opportunities to interact with leading figures in the field, establish professional connections, and gain insights into their study area's latest advancements and challenges.

Conferences are usually planned and scheduled by academic or professional organizations, research institutions, universities, government agencies, or industry associations. These entities conceptualize the conference theme, invite keynote speakers, coordinate presentation sessions, manage logistics, such as venue bookings and accommodations, and promote the event to potential participants. They send conference announcements to potential participants, who are expected to respond and send in their abstracts. The organizers review the submissions, select suitable presentations, and compile a program agenda as a book of abstracts. The key steps in preparation for a conference include:

1. *Deciding to participate*

   To participate in a conference, you must initiate and take a first step by actively seeking out opportunities. Conference announcements are often made through academic journals, professional association websites, mailing lists, and social media platforms. Be on the lookout for announcements and select one that aligns with your study and research interests. Once you identify a conference that aligns with your research focus and academic aspirations, take the necessary steps to register for the event and submit your abstract or proposal for presentation.

   Caution: be careful when deciding to participate in a conference. Some conferences are not credible. Always do due diligence.
2. *Sending an abstract*

   Ensure that you send your abstract within the given timelines. If accepted for presentation, you will receive guidelines for preparing your presentation, including details on formatting, time limits, and audiovisual requirements. Adhering to these guidelines ensures a smooth and successful presentation experience. Additionally, make note of any registration deadlines, as well as any requirements for travel arrangements or accommodation bookings if the conference is held in person.

Preparing the Conference Paper

Whether for oral presentation or poster display, preparing a conference paper requires careful consideration of the audience and the significance of your research findings. Conference organizers usually ask for full papers before the conference. They will give guidelines but usually follow the IMRAD (Introduction, Methods, Results, And Discussion) to ensure a structured and coherent presentation of your work (Wu 2011).

Preparing the Slides

You are encouraged to prepare your own slides. If you are not familiar with slide preparation, take the time to learn. Many tutorials, guides, and templates are available online to help you understand the basics of slide design, including layout, formatting, and visual elements. Additionally, exploring examples of well-designed presentations can provide inspiration and insight into effective slide-creation techniques. Slides enhance the delivery of your research findings and engage your audience effectively. However, as Day and Gastel (2012) observe, we get poor slides in presentations.

The slides should be clear, concise, and visually appealing for an in-person or online presentation. Each slide should highlight key sections of your paper, such as the introduction, methodology, results, and discussion, using bullet points, graphs, charts, or images to illustrate important points. Keep text to a minimum, focusing on key phrases or bullet points to reinforce your spoken presentation rather than reading directly from the slides. Use a consistent design theme throughout your slides to maintain visual coherence and professionalism. Pay attention to font size type and color contrast. Use at least 24–28 font size. A Sans Serif typeface is recommended (Day and Gastel 2012; Thomas 2021b). Take consideration of online presentations where participants may view slides on smaller screens. Contrast color well and avoid combinations of red and green and red and blue that create a jarring effect.

The layout should enhance readability. Consider the pacing of your slides to align with the flow of your oral presentation and ensure that each slide transitions smoothly to the next. Avoid crowding the slides by limiting the number of words per line and the number of lines per slide. Consider using not more than seven words per line and not more than six lines per slide. Additionally, do not crowd the tables and figures. Simplifying tables and figures initially presented in a research report is advisable. We present an example of PowerPoint slides used in a conference and tips on how to prepare slides (Box 28.1).

Students usually ask, "How many slides?" The guideline is to avoid too many slides and aim to allocate about one to two minutes per slide. Too many slides might force you to rush through the presentation without delivering the message to the audience. In most cases, presentations at seminars are graded.

Box 28.1: Tips for Preparing the Slides

- Decide on the title, select relevant pictures, and determine the total number of slides needed.
- Budget your slides according to the time allocated, aiming for one slide per minute to maintain a smooth pace.
- Number your slides to facilitate easy reference, and including your name on each slide helps the audience identify you.
- Use Sans Serif typeface fonts, such as, Arial, Tahoma, Verdana, Universal, Helvetica (Thomas 2021a, b).
- Avoid animations and irrelevant photographs.
- Avoid overcrowding slides with excessive text or irrelevant content, as this can make them difficult to read and comprehend.
- Focus on presenting your data logically, following a standard format such as IMRAD,
- Use clear contrast between the background color and text for legibility.
- Opt for lowercase letters, as they are clearer than uppercase letters,
- Aim for a balanced mix of text and graphics.
- Ensure your final slide includes a clear "take-home message" summarizing the key points of your work.

Presenting the Oral Paper

Most conferences allocate 10–15 min for oral presentation. You must, therefore, carefully plan to deliver the presentation concisely. Adhering to the allotted time is crucial to ensure all presenters can share their research findings and engage with the audience effectively. During the presentations, concentrate on the most important points of your research, focusing on key findings and implications that align closely with the conference theme or session topic. Begin with a clear and engaging introduction to grab the audience's attention and provide context for your presentation. Then, clearly outline the objectives of your study and the methodology employed. Present the main results and findings, highlighting their significance and relevance to the broader research field. Finally, conclude with a concise summary of your work's key takeaways and potential implications.

Handling Questions

After completing your presentation, it is common to have a question and answer (Q&A) session. This is an excellent opportunity to interact with the audience, clarify any points, and deepen understanding of your research findings. When approaching the Q&A session, be open, confident, and professional. You should listen carefully to each question and ensure that you understand it before responding. If a question is not clear, you can ask for clarification. Always remain respectful and composed, even if someone asks a challenging or critical question. Avoid getting defensive or dismissive. If you cannot answer a question immediately, you can acknowledge it and express your willingness to follow up later. Engaging with the audience during the Q&A session enhances the impact and effectiveness of your conference presentation.

28.7.3 Online and Hybrid Scientific Meetings

Since the era of COVID-19, many scientific meetings have shifted online or in a hybrid format. While the basics of preparing the slides and making the presentation remain the same, several concerns should be considered for online presentations. In addition to the tips given for departmental seminars, here are additional tips (Day and Gastel 2012) (Boxes 28.1 and 28.2):

Familiarize yourself with the software
Position the camera at eye level and maintain good posture
Use a professional background
Dress professionally
Consider recording your presentation
Have a backup plan and send the slides to the organizers

Box 28.2: Tips During the Oral Presentation

- Ensure thorough preparation and rehearsal before delivering your presentation, as readiness breeds confidence.
- Avoid underestimating the audience's intelligence.
- Engage the audience clearly and audibly.
- Speak slowly, especially at the start of the presentation, to catch the attention of the audience. Aim at 125–150 per minute (Thomas 2021b).
- Avoid reading through the slides; let the powerful points on the slides lead you.
- Briefly explain unfamiliar terms in the title to aid audience comprehension.
- Display confidence and expertise in your subject matter, remaining calm and truthful during Q&A sessions.
- Keep messages concise and prioritize novel information.
- Conclude by expressing gratitude to the audience for their attention.

Box 28.3: Example of a Presentation of a Research Paper in a Conference Using PPT

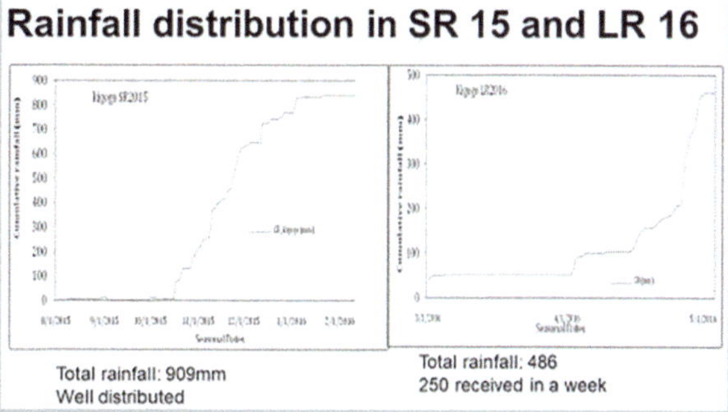

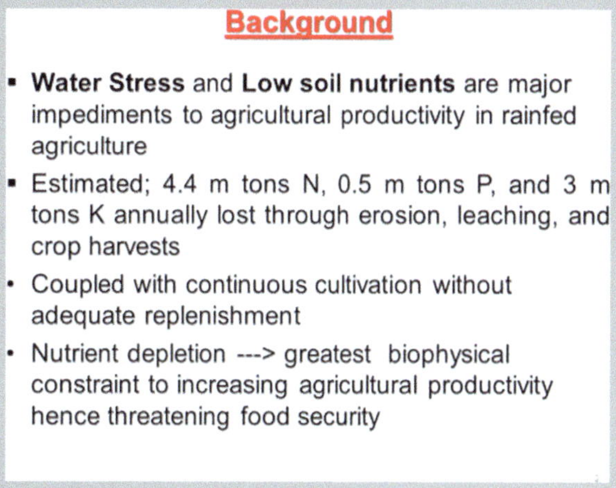

Box 28.3 (continued)

Background cont.....

- Depressed rains; Lack of moisture conservation tillage technologies
- Tied ridging; decrease runoff, increase water infiltration and consequently greater water storage
- Little research; on practices that combine soil fertility replenishment options and soil moisture conservation
- To determine the effect of ISFM technologies and tied ridging on maize and soybean yields

MATERIALS AND METHODS

➢Field trial was set at Kigogo primary school site (00° 23″S, 37° 38″E) in Meru South Sub-County, Kenya

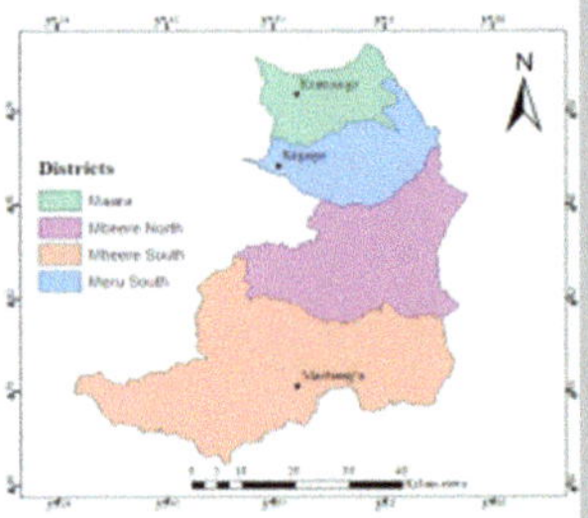

Description of the study area

- Eastern slopes of Mount Kenya; altitude-1,500 m asl; annual mean temps (20°C); Annual rainfall (1200 to 1400mm)
- Rainfall is bi-modal;(SR)-October, November and December and long rains (LR)- March, April and May
- Soils are predominantly humic Nitisols
- Smallholder mixed farming activities; Crops, trees and livestock
- Predominantly maize growing
- Small holdings (0.1 to 2 ha.); average-1.2 ha. per household

(continued)

Box 28.3 (continued)

Soil characteristics of the experimental site

Soil properties	Value
pH_{water}	5.5
Organic C (%)	1.9
Total N (%)	0.2
Available P (ppm)	35.8
Exchangeable K^+ (C mol kg^{-1})	0.6
Exchangeable Ca^{2+} (C mol kg^{-1})	3.7
Exchangeable Mg^{2+} (C mol kg^{-1})	1.4
Exchangeable Na^+ (C mol kg^{-1})	0.4
Bulk density (g/cm^3)	0.95
Sand %	7.5
Silt %	9.5
Clay %	83
Texture class	clay

Design and Treatments

Exptal Design: RCBD with 8 treatments replicated 4 times

No	SWC	ISFM	Treatment	Rate of fertilizer application
1	Tied ridging	Manure+ fertilizer	TR+MF	Half rate
2	Tied ridging	Tithonia+ fertilizer	TR+TF	Half rate
3	Tied ridging	Inorganic fertilizer	TR+IF	Full rate
4	Tied ridging	Control	TR+Con	No
5	No-Tied ridging	Manure+ fertilizer	C+MF	Half rate
6	No-Tied ridging	Tithonia+ fertilizer	C+TF	Half rate
7	No-Tied ridging	Inorganic fertilizer	C+IF	Full rate
8	No-Tied ridging	Control	C+Con	No

Methodology cont........

- Plot size was 4.5 m × 6 m
- Test crops were maize (variety: HD515) and soybean (variety: Gazelle)
- **Spacing-Maize**; 0.75 m X 0.50 m
- **Spacing-Soybean**: 0.50 m by 0.10 m
- Mirror rotation
- Experiment was carried out for four seasons
- **2014 SR, 2015 LR, 2015 SR and 2016 LR**

Management of the experiment

- ➢ Cattle manure and *Tithonia diversifolia*- applied 2 weeks
- ➢ Tied-ridges; seasonally; height ridge-0.20m, height-tie: 0.15 m; distance between ties - 0.5 m

(continued)

Box 28.3 (continued)

Data Collection

- Daily Rainfall - using a rain gauge
- Agronomic data
- Maize stover and grain were harvested at maturity from a net plot and weighed using a balance
- Soybean biomass and grain were measured at mature state from a net plot and weighed using a balance

Results and Discussion

Rainfall distribution in SR 14 and LR 15

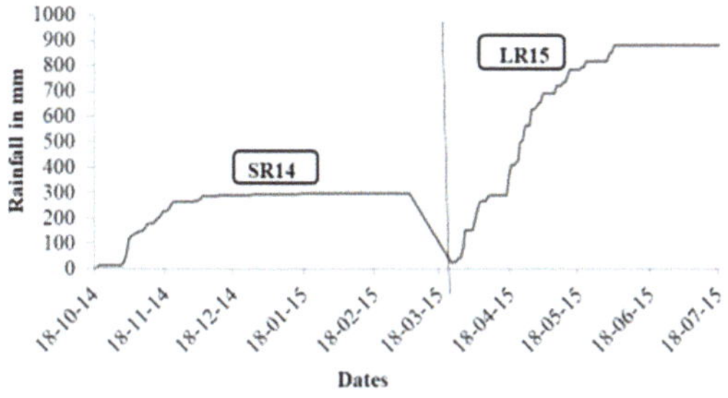

298 mm in SR 14 and 882 mm in LR 2015
Almost 99% received in a month (SR 14)

(continued)

Box 28.3 (continued)

Rainfall distribution in SR 15 and LR 16

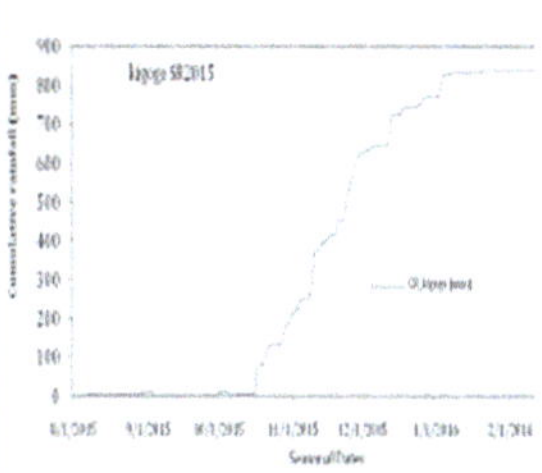

Total rainfall: 909mm
Well distributed

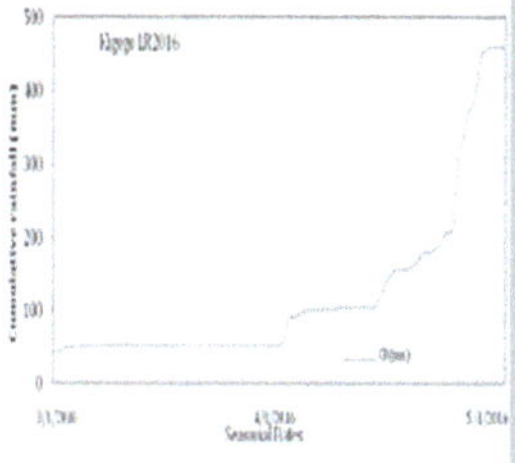

Total rainfall: 486
250 received in a week

Maize grain yields in different treatments at Kigogo site during SR 2014, LR 2015, SR 2015 and LR 2016

Treatment	SR14	LR15	SR15	LR16
	Grain t/ha			
+Tied ridging+No inputs	2.4^{bc}	2.6^{c}	2.8 d	2.3 c
+Tied ridging+fertilizer	3.1^{abc}	4.7^{ab}	5.0 bc	3.6 b
+Tied ridging+Man+Fert	3.3^{ab}	4.9^{a}	4.2 c	5.4 a
+Tied ridging+Tith+Fert	3.7^{a}	5.2^{a}	6.6 a	4.0 b
_Tied ridging+No inputs	2.1^{c}	3.2^{c}	2.5 d	2.2 c
_Tied ridging + Fert	3.2^{abc}	5.2^{a}	5.1 abc	5.1 a
_Tied ridging+Man+fert	3.4^{ab}	5.3^{a}	5.2 ab	5.5 a
_Tiedridging+Tith+Fert	3.7^{a}	5.4^{a}	5.8 a	5.4 a
p-value	0.042*	0.017*	0.042*	0.015*

Soybean grain yields in different treatments at Kigogo site during SR 2014, LR 2015, SR 2015 and LR 2016

Treatment	SR14	LR15	SR15	LR16
	Grain t/ha			
+Tied ridging+No inputs	0.7^{bc}	1.0^{a}	2.2 bc	1.6 b
+Tied ridging+fertilizer	0.8^{abc}	1.0^{a}	3.0 bc	1.8 ab
+Tied ridging+Man+Fert	0.9^{abc}	1.5^{a}	3.3 a	2.6 a
+Tied ridging+Tith+Fert	1.1^{a}	1.1^{a}	3.2 ab	1.5 b
-Tied ridging+No inputs	0.6^{c}	1.2^{a}	2.4 b	2.2 ab
-Tied ridging + Fert	0.8^{abc}	1.5^{a}	3.0 ab	2.5 a
-Tied ridging+Man+fert	1.0^{ab}	1.8^{a}	3.1 a	2.5 a
-Tied ridging+Tith+Fert	1.1^{a}	1.9^{a}	3.3 a	2.5 a
p-value	0.039	0.158	0.03	0.02

(continued)

Box 28.3 (continued)

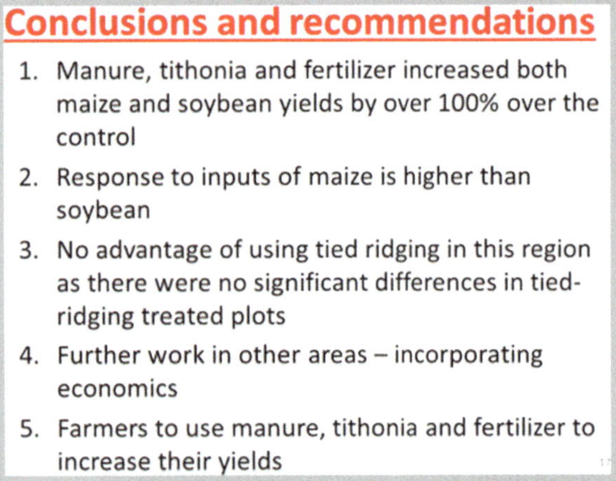

28.8 Poster Presentation of Research Work

Besides oral presentations, poster presentations provide another avenue for presenting research work at conferences (Gastel and Day, 2022; Soon et al. 2022). Poster sessions are often organized to accommodate the large number of potential presenters in a conference, allowing researchers to showcase their work in a visually engaging format. Unlike oral presentations, which typically involve presenting to a larger audience in a structured time slot, poster presentations offer a more informal setting where attendees can interact directly with presenters. This format encourages one-on-one discussions, facilitates networking, and allows for a more detailed exploration of the research findings. Additionally, poster presentations allow presenters to receive feedback, exchange ideas, and engage in discussions with colleagues and fellow researchers in their field. Overall, poster presentations play a valuable role in the conference experience, providing a platform for researchers to share their work, receive input from peers, and contribute to advancing knowledge in their respective disciplines.

Sometimes, conferences organize special sessions for students to present their posters, offering invaluable training opportunities to refine their presentation skills and highlight their research projects. These sessions serve as platforms for students to showcase their work in a supportive environment conducive to learning and growth. As part of these sessions, awards are often incorporated, with posters evaluated by judges or peers based on clarity, creativity, and academic rigor. Winning an award can give students recognition for their achievements and further incentivize excellence in research and presentation. Participating in these sessions also allows students to receive constructive feedback from peers and experts, fostering personal and academic development. Overall, these special sessions offer students a chance to shine and contribute to the overall richness and diversity of the conference program, enhancing the educational experience for all attendees.

28.8.1 Preparing a Poster

A good poster presentation has several key features to effectively engage the audience and convey the main message (Taggart and Arslanian 2000). The following are key considerations:

- It should attract attention through visually appealing design elements such as vibrant colors, compelling visuals, and clear titles.
- Key messages should be prominently highlighted to make them stand out and memorable to viewers.
- Maintain simplicity and readability by avoiding excessive text and unnecessary secondary elements, particularly detailed methodology sections, unless the paper specifically focuses on methods.
- Use graphics and visuals to illustrate complex issues and enhance understanding.
- Keep the message simple and concise, preferably presented in bullet points.

A standard poster uses the key elements of IMRAD structure (Taggart and Arslanian 2000): Title, Background, Methodology, Results (enhanced with appropriate graphics), Conclusions, and Recommendations. However, additional editorial work is necessary to optimize the poster's effectiveness after preparing the initial complete draft. This involves shortening the text by at least 50% to promote precision and conciseness. The take-home message should be condensed

Fig. 28.1 Components of a poster

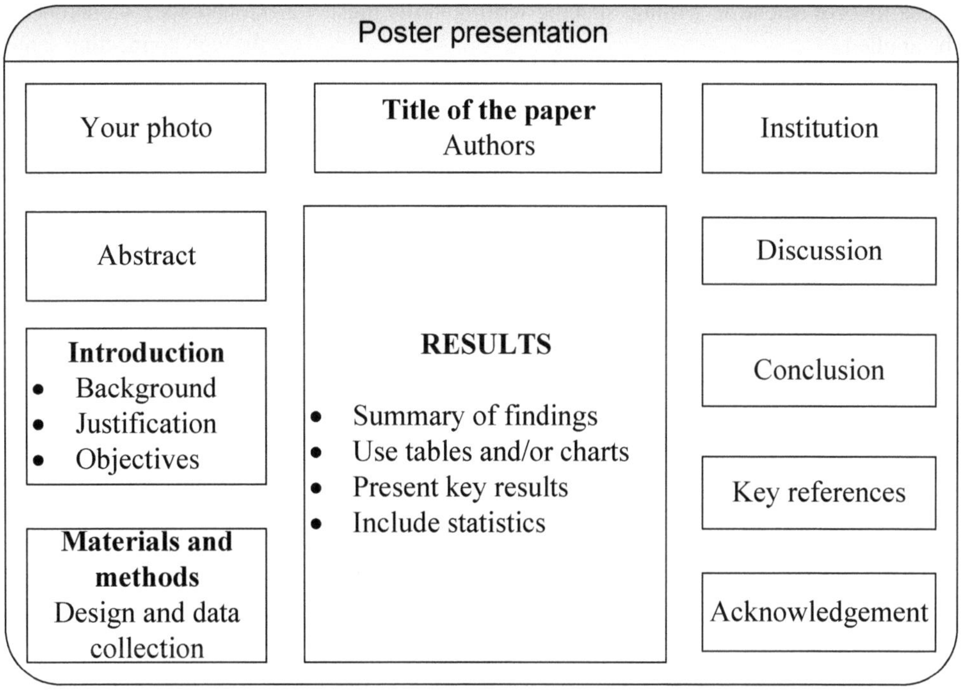

to its most essential points to maximize impact. Limiting graphics and tables to only what is necessary helps prevent overcrowding and maintains focus on the main content. Utilizing a common font type for all text enhances visual consistency and readability, while adjusting the font size to the largest possible without sacrificing coherence increases legibility, ensuring that the poster effectively communicates its message to the audience.

28.8.2 Formatting the Poster

When arranging and placing poster components, it is important to adhere to basic principles to effectively fulfill the intended purpose of posters. The title should be prominently positioned at the top of the poster, capturing the viewer's attention and clearly indicating the research topic. Following the title, the researchers' names and affiliations should be placed either above the poster, immediately after the title, or at the bottom, as the last item after acknowledgments and donor logos, if applicable (Fig. 28.1). This arrangement ensures clarity, readability, and visual appeal, effectively communicating your research findings to the audience.

Recommended Further Reading

1. Kember D, Corbett M (2018) Structuring the thesis: Matching method, paradigm, theories and findings. Structuring the Thesis: Matching Method, Paradigm, Theories and Findings 1–427. https://doi.org/10.1007/978-981-13-0511-5/COVER
2. Parija SC, Kate V (2018) Thesis Writing for Master's and Ph.D. Program. Thesis Writing for Master's and PhD Program 1–317. https://doi.org/10.1007/978-981-13-0890-1/COVER
3. Golding C, Sharmini S, Lazarovitch A (2014) What examiners do: what thesis students should know. Assess Eval High Educ 39:563–576. https://doi.org/10.1080/02602938.2013.859230
4. Youdeowei A, Stapleton P, Obubo R (2012) Scientific writing for agricultural research scientists - A training resource manual. 190.
5. Gastel, B., & Day, R. A. (2022). *How to write and publish a scientific paper* (9th ed.). Greenwood.
6. Nair PKR, Nair VD (2014) Scientific writing and communication in agriculture and natural resources. Scientific Writing and Communication in Agriculture and Natural Resources 1–142. https://doi.org/10.1007/978-3-319-03101-9/COVER
7. Taggart HM, Arslanian C (2000) Creating an effective poster presentation. Orthop Nurs 19:47–52. https://doi.org/10.1097/00006416-200019030-00007

Key Points in This Chapter

Importance of Scientific Presentations

Scientific reporting and presentations are essential to your academic journey, often delivered through written documents or oral presentations at seminars, conferences, or workshops. They aim to share research findings, engage with peers, receive feedback, and foster collaborations. Since these involve scientific writing, specific writing skills must be applied.

What Is Scientific Writing?

Scientific writing is writing about science to inform others about the findings. It is a two-way process because the objective is communication. It has three key characteristics:

- *Process:* Reports are the final product of a structured process, including delineating a topic, formulating a problem, and collecting and analyzing data.
- *Original:* addressing a well-delimited research problem supported by a valid methodology. The report contributes new knowledge and fills a significant gap.
- *Follows a specific style:* technical writing that is precise, clear, concise, and objective. It follows structured formats to communicate effectively within the scientific community.

Key Principles of Scientific Writing

The key principles include:

1. Clarity
2. Conciseness.
3. Keep it short; Short words, sentences, and paragraphs
4. Coherent writing
5. Logical structure
6. Objectivity
7. Correct word use
8. Use active verbs
9. Paraphrasing
10. Write in correct tense
11. Moderate use of normalization
12. Eliminate double negatives or multiple negatives
13. Avoid filler words
14. Consistency in format
15. Proofreading and editing your work
16. Avoiding gender bias

The following are the most important principles:

1. Clarity and Conciseness: Clarity and conciseness are fundamental for effective writing. Clear writing employs straightforward, precise statements that convey facts. Concise writing uses the fewest words possible, opting for shorter words and brief sentences. This approach gets straight to the point while avoiding unnecessary elaboration.
2. Coherence: Coherence ensures ideas' flow logically through connecting sentences and paragraphs to maintain smooth transitions, with each idea building on the last. Use pronouns, transition words, and enumeration to enhance coherence. Additionally, each paragraph should focus on a single idea, with supporting sentences maintaining unity. Logical links between sentences and continuity between paragraphs are essential for thorough idea development.
3. Logical Flow and Objectivity: Scientific writing should have a clear structure with organized sections. Utilizing subheadings and logically arranged paragraphs aids in maintaining a logical flow, guiding the reader from known to unknown information. The IMRAD format (Introduction, Methods, Results, and Discussion) is effective for structuring scientific papers. Consistent presentation of ideas, such as maintaining a parallel grammatical structure in lists, enhances clarity. Objectivity is key; writing must remain unbiased and evidence-based, with personal opinions clearly labeled.

Your Paragraph Should Include the Following

1. A clear topic sentence that introduces the main idea.
2. Supporting details that elaborate on and support the topic.
3. A concluding sentence that wraps up or reinforces the main idea.
4. Logical coherence and unity—all sentences should be related and flow smoothly.
5. Use of descriptive language and sensory details where appropriate.

Writing a Thesis

A thesis and journal articles are essential scientific reports for students earning their degrees. A thesis showcases a student's mastery of a subject and contributes original knowledge, while journal articles convey specific research findings to experts and undergo peer review. Both enhance a scholar's reputation and advance knowledge in the field. Before writing a thesis, it's important to read a benchmark thesis in your discipline.

Essentials of Writing a Manuscript from a Thesis

Because the thesis will have more information than you require, the following are guidelines for planning a manuscript. Isolate the following:

- Starting with the most direct objective and develop a *central message*
- Clearly write out the objective of the paper
- Define the materials and methods for the selected objective
- Define the principle findings and isolate the key data (tables and figures)
- Describe the conclusions and implications
- Organize related ideas together
- Read and assemble key literature related to the objective

Thesis Examination

The thesis examination consists of two components: the written work and an oral examination. The written work is assessed by two or three examiners who provide a detailed report on strengths and areas for improvement. Their evaluations aim to enhance the thesis quality and inform future research. Examiners focus on originality, objective achievement, overall quality, engagement with literature, coherence, and accuracy.

Presentations in Departmental Seminars and Conferences

As part of thesis requirements, students engage in scientific meetings such as seminars, symposiums, and conferences. These events allow students to share their research, learn from experts, and receive feedback from peers.

At departmental seminars, always demonstrate confidence and a clear understanding of your work. Conferences offer opportunities to connect with leading figures in the field and stay updated on recent advancements and challenges.

Presentations should follow the IMRAD structure and be delivered using clear, visually appealing slides or posters. Focus on simplicity, font size, and color contrast. Stick to the allotted time and emphasize key messages during your presentation.

Afterward, approach the question-and-answer session confidently and avoid being defensive. Since the Covid-19 pandemic, many meetings have moved online. Although the setup remains similar to in-person presentations, remember to maintain good posture and use a professional background.

Poster Presentation

Poster presentation is another way of presenting research work in a conference. Some conferences organize special session for students, where awards for top posters are included. Receiving an award can give student recognition and enhance educational experience. A good poster should attract attention, highlight key messages, and maintain simplicity and readability.

Review Questions

1. Explain, in your own words, why accurately and effectively communicating your scientific findings is so critical for your academic progression.
2. Why is "reading a sample thesis" a helpful practice before writing yours? What specific insights can you gain from doing so?
3. What aspects of your *written thesis document* do you think examiners will "look for" to assess its quality and rigour? List at least three.
4. For the oral examination, what is the primary goal of the examiners, and how does knowing this help you prepare effectively?
5. Explain why identifying the "key message" is the crucial initial step in preparing a journal manuscript from a thesis.
6. List at least three other key considerations for making a powerful oral presentation in a scientific meeting (e.g. related to delivery, content, audience engagement).
7. What are two specific design principles or content strategies you would employ to ensure your poster "captures attention and conveys your research's key message"?
8. Create a side-by-side table comparing a *thesis* and a *journal article* across at least *8 dimensions* (e.g. scope, audience, length, structure, depth of literature review, etc.).

Chapter Exercise

1. Evaluate a journal article from a credible journal versus a master's thesis using the following criteria:
   (a) Clarity
   (b) Conciseness
   (c) Logical flow
   (d) Objectivity
   (e) Scientific tone
2. Write one well-developed paragraph (6–8 sentences) on one of the following topics:

(a) The importance of selecting appropriate control treatments in agronomic field trials
(b) Challenges encountered while applying Randomized Complete Block Design (RCBD) in on-farm trials
(c) Balancing statistical power and field constraints when designing my PhD experiment
(d) Using Theory of Change to guide the design of my impact evaluation on climate-smart agriculture
(e) Difficulties faced in maintaining homogeneity of treatments in a greenhouse experiment

(f) How peer reviewer feedback improved the clarity of my results discussion section
(g) Applying quasi-experimental designs in evaluating agricultural development interventions
(h) Designing and implementing a cross-sectional survey for assessing household food security
(i) The importance of randomization in ensuring the internal validity of agricultural experiments

Answers to Review Questions

- Reference: 28.1.1
- Reference: 28.3.1
- Reference: 28.6 and 28.6.1
- Reference: 28.6.1
- Reference: 28.4.1
- Reference: 28.7.2 and 28.7.3
- Reference: The whole of 28.8
- Reference: 28.3 (Tip; You could also consider Chapter 26)

References

Ahmed AH (2010) Students' problems with cohesion and coherence in EFL essay writing in Egypt: different perspectives. Lit Inf Comput Educ J:211–221. https://doi.org/10.20533/LICEJ.2040.2589.2010.0030

Ali M (2021) The science and philosophy of manuscript rejection. Indian J Ophthalmol 69:1934–1935. https://doi.org/10.4103/IJO.IJO_1097_21

Ascheron C (2023) Scientific publishing and presentation: a practical guide with advice on doctoral studies and career planning:1–157. https://doi.org/10.1007/978-3-662-66404-9/COVER

Björklund M, Massoumi R, Ohlsson B (2024) From master's thesis to research publication: a mixed-methods study of medical student publishing and experiences with the publishing process. BMC Med Educ 24:1–8. https://doi.org/10.1186/S12909-024-05060-7/FIGURES/2

Brown M, Savova V, Gibson E (2012) Syntax encodes information structure: evidence from on-line reading comprehension. J Mem Lang 66:194–209. https://doi.org/10.1016/J.JML.2011.08.006

Chen M, Cui Y (2022) The effects of AWE and peer feedback on cohesion and coherence in continuation writing. J Second Lang Writ 57:100915. https://doi.org/10.1016/J.JSLW.2022.100915

Clapper TC (2020) Vigorous writing is concise, research is rigorous. Simul Gaming 51:739–743. https://doi.org/10.1177/1046878120954567

Creswell JW (2014) Research and design qualitative, quantitative, and mixed methods approaches. SAGE Publication Inc

Day RA, Gastel B (2012) How to write and publish a scientific paper, 4th edn. Cambridge University Press, Cornwall

Fennell FL (1975) Writing as art. Coll Compos Commun 26:177. https://doi.org/10.2307/357112

Gastel B, Day RA (2022) How to write and publish a scientific paper, 9th edn, Greenwood

Gedamu AD, Gezahegn TH (2021) EFL supervisors' written feedback focus and language functions: a mixed methods study. Asia-Pac J Second Foreign Lang Educ 6:1–18. https://doi.org/10.1186/S40862-021-00125-2/TABLES/6

Golding C, Sharmini S, Lazarovitch A (2014) What examiners do: what thesis students should know. Assess Eval High Educ 39:563–576. https://doi.org/10.1080/02602938.2013.859230

Grogan KE (2021) Writing science: what makes scientific writing hard and how to make it easier. Bull Ecol Soc Am 102:e01800. https://doi.org/10.1002/BES2.1800

Harwood N, Austin L, Macaulay R (2009) Proofreading in a UK university: proofreaders' beliefs, practices, and experiences. J Second Lang Writ 18:166–190. https://doi.org/10.1016/J.JSLW.2009.05.002

Harwood N, Austin L, Macaulay R (2010) Ethics and integrity in proofreading: findings from an interview-based study. Engl Specif Purp 29:54–67. https://doi.org/10.1016/J.ESP.2009.08.004

Hobson AJ (2014) Why manuscripts submitted to an international peer reviewed journal in education are rejected. Int J Mentor Coach Educ 3. https://doi.org/10.1108/IJMCE-04-2014-0014

Höhle B, Hörnig R et al (2014) Effects of focus and definiteness on children's word order: evidence from German five-year-olds' reproductions of double object constructions*. J Child Lang 41:780–810. https://doi.org/10.1017/S0305000913000196

Kember D, Corbett M (2018) Structuring the thesis: matching method, paradigm, theories and findings. Structuring the Thesis: Matching Method, Paradigm, Theories and Findings, pp 1–427. https://doi.org/10.1007/978-981-13-0511-5/COVER

Kumar V, Stracke E (2011) Examiners' reports on theses: feedback or assessment. J Engl Acad Purp 10(4):211–222. https://doi.org/10.1016/j.jeap.2011.06.001

Lantsoght EOL (2023) Doctoral defence formats. Stud High Educ 48:343–355. https://doi.org/10.1080/03075079.2022.2137123

Luby S, Southern DL (2022a) General research and writing practices. The Pathway to Publishing: A Guide to Quantitative Writing in the Health Sciences, pp 21–39. https://doi.org/10.1007/978-3-030-98175-4_2

Luby S, Southern DL (2022b) Achieving clarity and conciseness. The Pathway to Publishing: A Guide to Quantitative Writing in the Health Sciences, pp 73–86. https://doi.org/10.1007/978-3-030-98175-4_6

Matthews JR, Matthews RW (2014) Successful Scientific Writing, 4th edn. Cambridge University Press

Mullins G, Kiley M (2002) "It's a PhD, not a Nobel Prize": how experienced examiners assess research theses. Stud High Educ 27:369–386. https://doi.org/10.1080/0307507022000011507

Murray R (2015) How to survive your viva: defending a thesis in an oral examination. Open University Press

Nair PKR, Nair VD (2014) Scientific writing and communication in agriculture and natural resources. Scientific Writing and Communication in Agriculture and Natural Resources, pp 1–142. https://doi.org/10.1007/978-3-319-03101-9/COVER

Nangimah M, Walldén R (2023) How supervisors provide and students react to EAL thesis supervision: voices from Sweden and Indonesia. Front Educ (Lausanne) 8:1118436. https://doi.org/10.3389/FEDUC.2023.1118436/BIBTEX

Parija SC, Kate V (2018) Thesis writing for master's and Ph.D. Program. Thesis Writing for Master's and PhD Program:1–317. https://doi.org/10.1007/978-981-13-0890-1/COVER

Patel BK, Muir-Leresche K, Coe R, Hainsworth SD (2004) Green book—a guide to effective graduate research in African agriculture, environment and rural development

Plaxco KW (2010) The art of writing science. Protein Sci 19:2261–2266. https://doi.org/10.1002/PRO.514

Rahman ZAAA (2013) The use of cohesive devices in descriptive writing by Omani student-teachers. SAGE Open 3. https://doi.org/10.1177/2158244013506715/ASSET/7BBF6536-

7179-4C5C-A4AE-E70F71134B23/ASSETS/IMAGES/LARGE/10.1177_2158244013506715-FIG3.JPG

Richards F, Harwood N (2024) Proofreading student writing: a research-based stakeholder tool focused on ethical practice. Int J Educ Integr 20:1–16. https://doi.org/10.1007/S40979-024-00165-4/FIGURES/2

Riches-Suman K (2024) Supporting doctoral candidates through completion and final examination. Encyclopedia 4:836–846. https://doi.org/10.3390/ENCYCLOPEDIA4020053

Sayer EJ (2019) The essentials of effective scientific writing—a revised alternative guide for authors. Funct Ecol 33:1576–1579. https://doi.org/10.1111/1365-2435.13391

Soon CSL, Tudor Car L, Ng CJ et al (2022) What is the utility of posters? Qualitative study of participants at a regional primary healthcare conference in Asia. Med Sci Educ 32:1405–1412. https://doi.org/10.1007/S40670-022-01657-Z/METRICS

Szeto G, Jull G (2022) Defining the scope of your manuscript—Beware of "salami slicing". Musculoskelet Sci Pract 58:102540. https://doi.org/10.1016/j.msksp.2022.102540

Taggart HM, Arslanian C (2000) Creating an effective poster presentation. Orthop Nurs 19:47–52. https://doi.org/10.1097/00006416-200019030-00007

Tan WC (2023) Purpose-driven oral examination: insights from doctoral viva examiners. Discov Educ 2:1–10. https://doi.org/10.1007/S44217-023-00083-6/FIGURES/1

Thomas CG (2021a) Research: the search for knowledge. Research Methodology and Scientific Writing, pp 1–32. https://doi.org/10.1007/978-3-030-64865-7_1

Thomas GC (2021b) Research methodology and scientific writing:1–620. https://doi.org/10.1007/978-3-030-64865-7/COVER

Tinkler P, Jackson C (2000) Examining the doctorate: institutional policy and the PhD examination process in Britain. Stud High Educ 25:167–180. https://doi.org/10.1080/713696136

Watling C (2015) The power of parallel structure. Perspect Med Educ 4:329–330. https://doi.org/10.1007/S40037-015-0227-3

Weaver KF, Morales V, Nelson M, et al (2016) The benefits of peer review and a multisemester capstone writing series on inquiry and analysis skills in an undergraduate thesis. CBE Life Sci Educ 15:ar51.1–ar51.9. https://doi.org/10.1187/CBE.16-01-0072/ASSET/IMAGES/LARGE/AR51FIG2.JPEG

Wedderkopp N, Rutz E (2024) Scientific integrity and transparency in academic writing: the foundation of credible science. Children 11:1191. https://doi.org/10.3390/CHILDREN11101191

Wright DE (2024) Five problems plaguing publishing in the life sciences—and one common cause. FEBS Lett 598:2227–2239. https://doi.org/10.1002/1873-3468.15018

Wu J (2011) Improving the writing of research papers: IMRAD and beyond. Landsc Ecol 26:1345–1349. https://doi.org/10.1007/S10980-011-9674-3/FIGURES/1

Yaffe P (2023) On writing well…and not so well. Ubiquity 2023:1–6. https://doi.org/10.1145/3597454

Yakhontova T (2021) Paragraphing in English articles: tips for non-Anglophone authors. J Korean Med Sci 36:1–5. https://doi.org/10.3346/JKMS.2021.36.E62

Youdeowei A, Stapleton P, Obubo R (2012) Scientific writing for agricultural research scientists—a training resource manual. CTA, p 190

MIX
Papier aus verantwortungsvollen Quellen
Paper from responsible sources
FSC® C105338

If you have any concerns about our products,
you can contact us on
ProductSafety@springernature.com

In case Publisher is established outside the EU,
the EU authorized representative is:
Springer Nature Customer Service Center GmbH
Europaplatz 3, 69115 Heidelberg, Germany

Printed by Libri Plureos GmbH
in Hamburg, Germany